Introductory and Intermediate Algebra

Introductory and Intermediate Algebra

K. Elayn Martin-Gay

University of New Orleans

Upper Saddle River, New Jersey 07458

Library of Congress Cataloging-in-Publication Data

MARTIN-GAY, K. ELAYN, (date)
Introductory and intermediate algebra/K. Elayn Martin-Gay.
p. cm.
Includes index.
ISBN 0-13-341504-X
1. Algebra I. Title.
QA152.2.M3684 1996 95-21970
512.9—dc20 CIP

Acquisitions editor: *Melissa S. Acuña*
Director of production and manufacturing: *David W. Riccardi*
Marketing manager: *Karie Jabe*
Production supervision: *Kathleen M. Lafferty/Roaring Mountain Editorial Services*
Proofreaders: *Bruce D. Colegrove, Roberta H. Lewis*
Indexer: *Nancy M. Fulton*
Interior designer: *Judith A. Matz-Coniglio*
Cover designer: *Heather Scott*
Photo Editor: *Lorinda Morris-Nance*
Cover photo: *Al Tielemans, Duomo Photograph*
Creative director: *Paula Maylahn*
Art director: *Amy Rosen*
Manufacturing buyer: *Alan Fischer*
Supplements editor: *Audra Walsh*
Editorial assistant: *April Thrower*

Photograph credits: Stacy Pick/Stock Boston, p. 2; Chervenky/The Image Works, p. 115; M. Richards/Photo Edit, p. 174; Luis Castandeda/The Image Bank, p. 238; Lawrence Migdale/The Photo Researchers, p. 298; James D. Wilson/Woodfin Camp & Associates, p. 354; Art Stein/ Photo Researchers, Inc., p. 412; Steve Krongard/The Image Bank, p. 464; Jeff Hunter/The Image Bank, p. 551; Alan Oddie/Photo Edit, p. 608; Timothy Egan/Woodfin Camp & Associates, p. 653.

Simon & Schuster/A Viacom Company
Upper Saddle River, New Jersey 07458

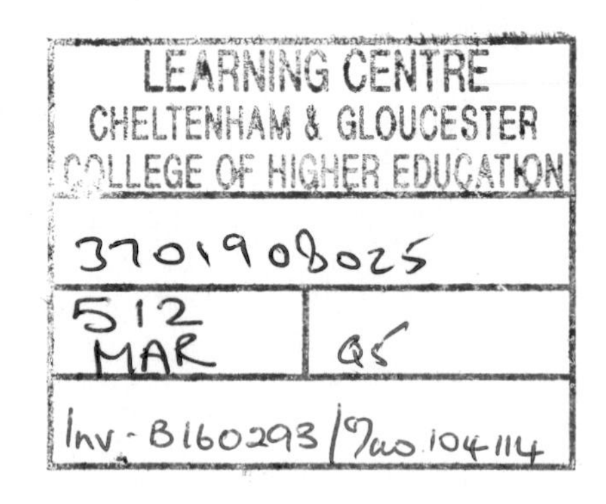

Printed in the United States of America

10 9 8 7 6 5 4 3 2 1

ISBN 0-13-341504-X

Prentice-Hall International (UK) Limited, *London*
Prentice-Hall of Australia Pty. Limited, *Sydney*
Prentice-Hall Canada Inc., *Toronto*
Prentice-Hall Hispanoamericana, S.A., *Mexico*
Prentice-Hall of India Private Limited, *New Delhi*
Prentice-Hall of Japan, Inc., *Tokyo*
Simon & Schuster Asia Pte. Ltd., *Singapore*
Editora Prentice-Hall do Brasil, Ltda., *Rio de Janeiro*

To my two brave friends,
Mary Catherine Dooley
and
Donna Phillips Thieme

Contents

CHAPTER 3 Inequalities, Absolute Value, and Problem Solving *115*

CHAPTER 4 Graphing *174*

CHAPTER 5 Exponents and Polynomials *238*

CHAPTER 6 Factoring Polynomials 298

CHAPTER 7 Rational Expressions 354

CHAPTER 8 Roots and Radicals 412

CHAPTER 9 Quadratic Equations and Equality 464

CHAPTER 10 Systems of Equations and Inequalities 551

CHAPTER 11 Exponential and Logarithmic Functions 608

CHAPTER 12 Sequences, Series, and the Binomial Theorem *653*

Preface

Why This Book Was Written

This book is intended for a two-semester course in introductory and intermediate algebra. Specific care has been taken to prepare students to go on to their next course in algebra. I have tried to achieve this by writing a user-friendly text keyed to objectives containing many of worked-out examples. Functions are introduced in this text, and applications and geometric concepts are emphasized throughout the book.

How This Book Was Written

Throughout the writing and developing of this book, I had the help of many people. Seven instructors, who teach courses similar to this one, were involved in the actual writing of the text, contributing their ideas for helpful examples, interesting applications, and useful exercises.

Once the first draft was complete, Prentice Hall held a focus group with four reviewers, the author, and editors from Prentice Hall. We spent many hours going over the manuscript with a fine-toothed comb, refining the project's focus and enhancing its pedagogical value.

Finally, a full-time development editor worked with me to make the writing style as clear as possible while still retaining the mathematical integrity of the content.

Key Content Features

In addition to the traditional topics taught in introductory and intermediate algebra courses, this text contains a strong emphasis of geometric concepts, reading and interpreting graphs, and problem solving integrated throughout. The geometric concepts covered are those that are most important to a student's understanding of algebra, and I have included many applications and exercises devoted to this topic. Also, geometric figures and a review of angles, lines, and special triangles are covered in the appendices. I have also integrated reading and interpreting line and bar graphs throughout much of this text. Not only does this naturally lead to the rectangular coordinate system, but it gives students practice at interpreting real data. Problem solving is, of course, emphasized by devoting single sections to this concept (such as Sections 2.4,

2.5, and 3.3 on formulas and solving problems that lead to linear equations) as well as by including problem-solving exercises throughout this text.

Key Pedagogical Features

Exercise Sets. Each exercise set is divided into two parts. Both parts contain graded problems. The first part is carefully keyed to worked examples in the text. Once a student has gained confidence in a skill, the second part contains exercises not keyed to examples. There are ample exercises throughout this book, including end-of-chapter reviews, tests, and cumulative reviews. In addition, each exercise set contains one or more of the following features.

Mental Mathematics. These problems are found at the beginning of an exercise set. They are mental warmups that reinforce concepts found in the accompanying section and increase students' confidence before they tackle an exercise set. By relying on their own mental skills, students learn not only confidence in themselves, but also number sense and estimation ability.

Skill Review. At the end of each section after Chapter 1, these problems are keyed to earlier sections and review concepts learned earlier in the text.

Writing in Mathematics. These writing exercises can be used to check a student's comprehension of an algebraic concept. They are located at the end of many exercise sets, where appropriate. Guidelines recommended by the National Council of Teachers of Mathematics and other professional groups recommend incorporating writing in mathematics courses to reinforce concepts.

Applications. This book contains a wealth of practical applications found throughout the book in worked-out examples and exercise sets.

A Look Ahead. These are examples and problems similar to those found in college algebra books. "A Look Ahead" is presented as a natural extension of the material and contains an example followed by advanced exercises. I strongly suggest that any student who plans to take another algebra course work these problems.

Graphing Calculator Boxes. Graphing calculator boxes are placed appropriately throughout the text to instruct students on proper use of the graphing calculator. These boxes, entirely optional, contain examples and exercises to reinforce the material introduced.

Helpful Hint Boxes. These boxes contain practical advice on problem solving. Helpful hints appear in the context of material in the chapter and give students extra help in understanding and working problems. They are set off in a box for easy referral.

Chapter Glossary and Summary. Found at the end of each chapter, the chapter glossary contains a list of definitions of new terms introduced in the chapter, and the summary contains a list of important rules, properties, or steps introduced in the chapter.

Chapter Review and Test. The end of each chapter contains a review of topics introduced in the chapter. These review problems are keyed to sections. The chapter test is not keyed to sections.

Cumulative Review. Each chapter after the first contains a cumulative review. Each problem contained in the cumulative review is actually an earlier worked example in the text that is referenced in the back of the book along with the answer. Students who need to see a complete worked-out solution with explanation can do so by turning to the appropriate example in the text.

Supplements

The following supplements are available to qualified adopters of *Introductory and Intrermediate Algebra*:

For the Instructor

Instructor Solutions Manual provides even-numbered solutions.

TestPro (IBM, Mac) generates test questions and drill worksheets from algorithms keyed to the learning objectives in the book and allows you to edit and add your own questions. Available free upon adoption in 3.5″ and 5.25″ formats.

Test Item File contains a hard copy of test questions on TestPro.

For the Student

Student Solutions Manual contains odd-numbered solutions and solutions to all chapter tests and cumulative tests.

Math Master Tutor software (IBM, Mac) provides text-specific, tutorial exercises graduated in difficulty that are generated new each time, fully worked-out examples, and a timed quiz.

Videotapes with class lectures by the author are closely keyed to the book itself.

Acknowledgments

Writing this book has been a humbling experience, an effort requiring the help of many more people than I originally imagined. I will attempt to thank them here.

First, I would like to thank my husband, Clayton. Without his constant encouragement, this project would not have become a reality. I would also like to thank my children, Eric and Bryan, for continuing to eat my burnt bacon. Writing a book while raising two small children is an experience that requires an infinite amount of patience and a good sense of humor.

I would like to thank my extended family for their invaluable help. Their contributions are too numerous to list. They are Peter, Karen, Michael, Christopher, Matthew, and Jessica Callac; Stuart, Earline, Melissa, and Mandy Martin; Mark and Sabrina Martin; Leo and Barbara Miller; and Jewett Gay.

I would like to thank the following reviewers for their suggestions:

Carol Achs, *Mesa Community College*
Gabrielle Andries, *University of Wisconsin–Milwaukee*
Jan Archibald, *Ventura College*
Carol Atnip, *University of Louisville*
Sandra Beken, *Horry-Georgetown Technical College*
Nancy J. Bray, *San Diego Mesa College*
Helen Burrier, *Kirkwood Community College*
Celeste Carter, *Richland College*
Dee Ann Christianson, *The University of the Pacific*
John Coburn, *St. Louis Community College*

Iris DeLoach-Johnson, *Miami University*
Omar L. DeWitt, *University of New Mexico*
Catherine Folio, *Brookdale Community College*
Robert W. Gesell, *Cleary College*
Dauhrice Gibson, *Gulf Coast Community College*
Marian Glasby, *Anne Arundel Community College*
Margaret (Peg) Greene, *Florida Community College at Jacksonville*
Frank Gunnip, *Oakland Community College*
Doug Jones, *Tallahassee University*
Mike Mears, *Manatee Community College*
James W. Newsom, *Tidewater Community College*
Randy Pittman, *J. Sargeant Reynolds Community College*
Mary Kay Schippers, *Fort Hays State University*
Mary Lee Seitz, *Erie Community College–City Campus*
Ken Seydel, *Skyline College*
Edith Silver, *Mercer County Community College*
Ventura Simmons, *Medgar Evers College*
Bonnie Simon, *Naugatuck Valley Community Technical College*
Debbie Singleton, *Lexington Community College*
Ronald Smith, *Edison Community College*
Richard Spangler, *Tacoma Community College*
Lauren Syda, *Yuba Community College*
Diane Trimble, *Collin County Community College*
Patrick C. Ward, *Illinois Central College*
John C. Wenger, *City College of Chicago–Harold Washington College*
Jerry Wilkerson, *Missouri Western State College*

Laurel Fischer and Karen Schwitters wrote the answers and the solutions manual, contributing to the accuracy as well. Finally, I would like to thank production editor Kathleen Lafferty and acquisitions editor Melissa Acuña for their invaluable contributions.

K. Elayn Martin-Gay

About the Author

Elayn Martin-Gay has taught mathematics at the University of New Orleans for 16 years and has received numerous teaching awards, including the local University Alumni Association's Award for Excellence in Teaching.

Over the years, Elayn has developed videotaped lecture series to help her students understand algebra material better. This highly successful video material is the basis for the three-book series, *Prealgebra, Beginning Algebra,* and *Intermediate Algebra.*

Introductory and Intermediate Algebra

CHAPTER 1

Review of Real Numbers

Sidewalks are constructed from separate concrete blocks rather than one continuous concrete slab, because concrete expands in the heat of the sun. Engineers must account for this expansion when planning the dimensions of the blocks.

INTRODUCTION

In arithmetic, everyday situations are described using numbers. Algebra differs from arithmetic in that letters are used to represent unknown numbers. An important part of learning algebra is learning the symbols and words—the language—of algebra. Much of this language is familiar to you already as the language of arithmetic. We begin our study of algebra with a review of arithmetic: its symbols, words, and patterns. This review is essential in forming the tools needed to learn the language of algebra.

1.1 Symbols

OBJECTIVES

Tape BA 1

1 Identify the symbols used for natural and whole numbers, and picture them on a number line.

2 Define the meaning of the symbols $=$, $\neq$, $<$, $>$, $\leq$, and $\geq$.

3 Translate sentences into mathematical statements.

4 Define the meaning of the symbols used for addition, subtraction, multiplication, and division.

1 We begin with a review of natural numbers and whole numbers and how we use symbols to compare these numbers.

The **natural numbers** are 1, 2, 3, 4, 5, 6, 7, 8, 9, 10, 11, 12, and so on.
The **whole numbers** are the natural numbers together with zero.

The whole numbers 0, 1, 2, 3, 4, 5, 6, 7, 8, 9, 10, 11, 12, and so on can be pictured with a **number line.** We will use the number line often to help us visualize objects and relationships. Visualizing mathematical concepts is an important skill and tool, and later we will develop and explore other visualizing tools.

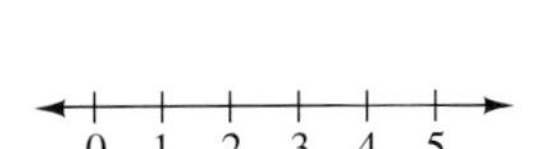

To draw a number line, first draw a line. Choose a point on the line and label it 0. To the right of 0, label any other point 1. Being careful to use the same distance as from 0 to 1, mark off equally spaced distances. Label these points 2, 3, 4, 5, and so on. Since the whole numbers continue indefinitely, it is not possible to show every whole number on the number line. The arrow at the right end of the line indicates that the pattern continues indefinitely.

2 Picturing whole numbers on a number line helps us to see the order of the numbers. Symbols can be used to concisely describe what we see.

The **equal symbol,** $=$, states that one value "is equal to" another.

The **not equal symbol,** $\neq$, states that one value "is not equal to" another. For example,

$2 = 2$ states that "two is equal to two"

$2 \neq 6$ states that "two is not equal to six"

We can use these symbols to form a **mathematical statement.** The statement might be true or it might be false. The above two statements are both true.

If two numbers are not equal, then one number is larger than the other. The **greater than symbol,** $>$, states that one value "is greater than" another. For example,

$2 > 0$ states that "two is greater than zero"

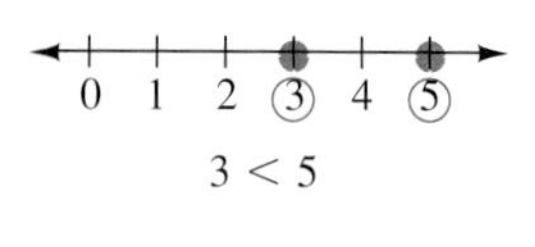

$3 < 5$

0 1 2 3 4 5

$2 > 0$ or $0 < 2$

The **less than symbol,** $<$, states that one value "is less than" another. For example,

$3 < 5$ states that "three is less than five"

On the number line, we see that a number **to the right of** another number is **larger.** Similarly, a number **to the left of** another number is smaller. For example, 3 is to the left of 5 on the number line, which means that 3 is less than 5, or $3 < 5$. Similarly, 2 is to the right of 0 on the number line, which means 2 is greater than 0, or $2 > 0$. Since 0 is to the left of 2, we can also say that 0 is less than 2, or $0 < 2$.

HELPFUL HINT

Notice that $2 > 0$ has exactly the same meaning as $0 < 2$. Switching the order of the numbers and reversing the "direction of the inequality symbol" does not change the meaning. For example,

$5 > 3$ has the same meaning as $3 < 5$

Also notice that, when the statement is true, the inequality arrow "points" to the smaller number.

EXAMPLE 1 Insert $<$, $>$, or $=$ in the space between the paired numbers to make each statement true.

a. 2 3 **b.** 7 4 **c.** 72 27

Solution:

a. $2 < 3$ since 2 is to the left of 3 on the number line.

b. $7 > 4$ since 7 is to the right of 4 on the number line.

c. $72 > 27$ since 72 is to the right of 27 on the number line. ■

Two other symbols are used to compare numbers. The **less than or equal to** symbol, $\leq$, states that one value "is less than or equal to" another value. The **greater than or equal to symbol,** $\geq$, states that one value "is greater than or equal to" another value. For example,

$7 \leq 10$ states that "seven is less than or equal to ten"

This statement is true since $7 < 10$. If either $7 < 10$ or $7 = 10$ is true, then $7 \leq 10$ is true.

$3 \geq 3$ states that "three is greater than or equal to three"

This statement is true since $3 = 3$. If either $3 > 3$ or $3 = 3$ is true, then $3 \geq 3$ is true.

The statement $6 \geq 10$ is false since neither $6 > 10$ nor $6 = 10$

The symbols $<$, $>$, $\leq$, and $\geq$ are called **inequality symbols.**

EXAMPLE 2 Tell whether each statement is true or false.

a. $8 \geq 8$ **b.** $8 \leq 8$ **c.** $23 \leq 0$ **d.** $23 \geq 0$

Solution: **a.** True, since $8 = 8$. **b.** True, since $8 = 8$.

c. False, since neither $23 < 0$ nor $23 = 0$. **d.** True, since $23 > 0$. ■

3 Now let's use the symbols discussed above to translate sentences into mathematical statements.

EXAMPLE 3 Translate each sentence into a mathematical statement.

a. 9 is less than or equal to 11.

b. 8 is greater than 1.

c. 3 is not equal to 4.

Solution: **a.**

9	is less than or equal to	11
9	$\leq$	11

b.

8	is greater than	1
8	$>$	1

c.

3	is not equal to	4
3	$\neq$	4

■

4 Symbols are also used to represent the sum, difference, product, and quotient of numbers, corresponding to the operations of addition, subtraction, multiplication, and division. Before continuing further, you should feel completely comfortable performing these operations on whole numbers. The following table summarizes the symbols and meanings of these basic operations.

Operation Symbols

	Symbols	Meanings
Addition	$6 + 2$	The sum of 6 and 2 or 6 plus 2
Subtraction	$6 - 2$	The difference of 6 and 2 or 6 minus 2
Multiplication	6×2, $6 \cdot 2$, $6(2)$, $(6)2$, $(6)(2)$	The product of 6 and 2 or 6 times 2
Division	$\frac{6}{2}$, $6/2$, $6 \div 2$	The quotient of 6 and 2 or 6 divided by 2

EXAMPLE 4 Translate each sentence into a mathematical statement.

a. The product of 2 and 3 is 6.

b. The difference of 8 and 4 is less than or equal to 4.

c. The quotient of 10 and 2 is not equal to 3.

Solution: **a.**

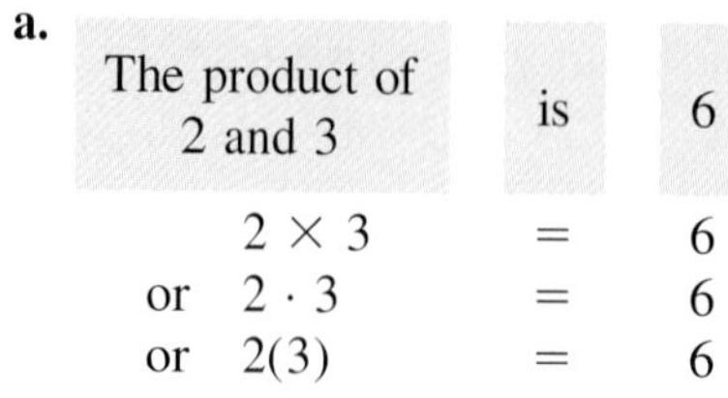

$2 \times 3 = 6$

or $2 \cdot 3 = 6$

or $2(3) = 6$

b.

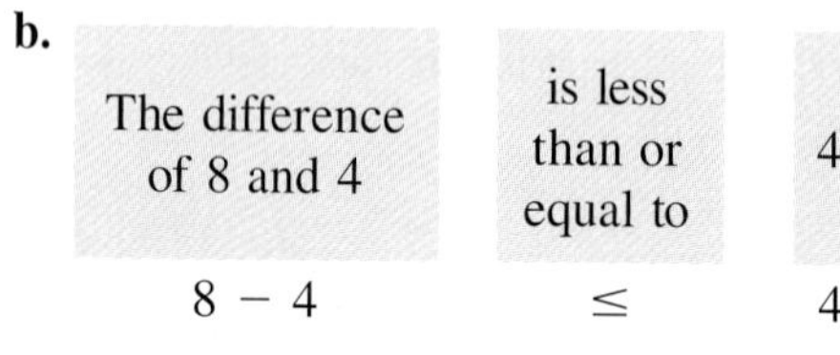

$8 - 4 \leq 4$

c.

The quotient of 10 and 2 | is not equal to | 3

$10 \div 2 \neq 3$

or $\frac{10}{2} \neq 3$ ■

The following is a summary of symbols used to compare numbers.

Symbols Used to Compare Numbers

$=$	is equal to	$\neq$	is not equal to
$<$	is less than	$\leq$	is less than or equal to
$>$	is greater than	$\geq$	is greater than or equal to

EXERCISE SET 1.1

Insert $<$, $>$, or $=$ in the space between the paired numbers to make each statement true. See Example 1.

1. 4 10 **2.** 8 5 **3.** 7 3 **4.** 9 15

5. 18 18 **6.** 213 113 **7.** 0 7 **8.** 20 0

Are the following statements true or false? See Example 2.

9. $11 \leq 11$ **10.** $4 \geq 7$ **11.** $10 > 11$ **12.** $17 > 16$

13. $3 + 8 \geq 3(8)$ **14.** $8 \cdot 8 \leq 8 \cdot 7$ **15.** $7 > 0$ **16.** $4 < 7$

Write each sentence as a mathematical statement. See Examples 3 and 4.

17. 8 is less than 12.
18. 5 is less than 15.
19. 5 is greater than or equal to 4.
20. 10 is greater than or equal to 7.
21. The sum of 2 and 3 is less than 6.
22. The difference of 16 and 4 is greater than 10.
23. The quotient of 10 and 2 is 5.
24. The quotient of 18 and 6 is 3.
25. 4 is less than or equal to the product of 3 and 5.
26. 5 is less than or equal to the product of 4 and 9.

Insert $<$, $>$, or $=$ in the appropriate space to make each statement true.

27. 5   7 **28.** 8   2 **29.** 7   5 **30.** 2   8
31. $3 \cdot 8$ 25 **32.** $4 \cdot 8$ 30 **33.** $15 - 9$ 6 **34.** $19 - 8$ 11
35. $15 - 9$ 5 **36.** $19 - 8$ 12

Are the following statements true or false?

37. $5 - 0 = 5$ **38.** $8(1) = 9$ **39.** $5(0) = 5$ **40.** $12 \neq 12$
41. $0 \leq 7$ **42.** $7 \geq 4$

Rewrite the following inequalities so that the inequality symbol points in the opposite direction and the resulting statement is equivalent to the given one. (The letters in Exercises 47–50 represent numbers.)

43. $25 \geq 25$ **44.** $13 \leq 13$ **45.** $0 < 6$ **46.** $5 > 3$
47. $a \leq b$ **48.** $c \geq d$ **49.** $x > y$ **50.** $m < n$

Write each sentence as a mathematical statement. (The letters in Exercises 67–70 represent numbers.)

51. 4 is greater than 2.5.
52. 5 is greater than 3.2.
53. 8 is less than or equal to 12.
54. 5 is less than or equal to 15.
55. 5 is less than or equal to 6.
56. 8 is less than or equal to 10.
57. The sum of 5 and 6 is greater than 10.
58. The difference of 15 and 7 is less than 9.
59. The product of 3 and 5 is greater than 12.
60. The product of 4 and 7 is less than 30.
61. The quotient of 12 and 6 is greater than 1.
62. The quotient of 25 and 5 is less than 6.
63. 3 is greater than 2.
64. 4 is equal to 3 plus 1.
65. 4 is equal to 4 plus 0.
66. 25 times 1 equals 25.
67. a is less than or equal to 5.
68. b is greater than or equal to 3.
69. c is less than d.
70. d is greater than the product of 3 and c.

71. If you weigh 150 pounds on Earth, your weight on the planet Mercury is approximately 57 pounds. Write an inequality statement comparing the numbers 150 and 57.

72. If you weigh 150 pounds on Earth, your weight on the planet Jupiter is approximately 381 pounds. Write an inequality statement comparing the numbers 150 and 381.

73. The freezing point of water is 32° Fahrenheit. The boiling point of water is 212° Fahrenheit. Write an inequality statement comparing the numbers 32 and 212.

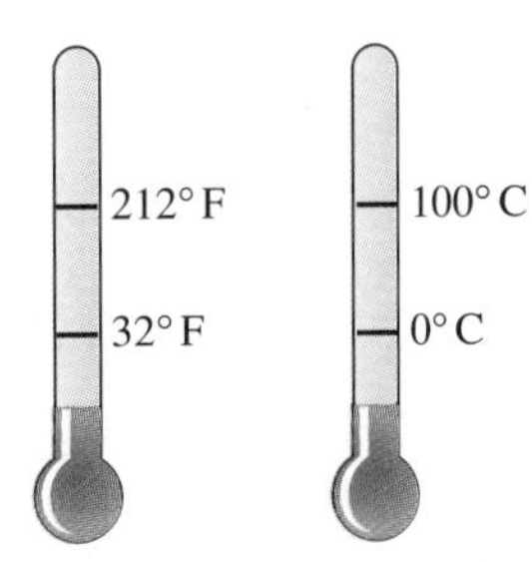

74. The freezing point of water is 0° Celsius. The boiling point of water is 100° Celsius. Write an inequality statement comparing the numbrs 0 and 100.

75. An angle measuring 30° and an angle measuring 45° are shown. Use the inequality symbol $\leq$ or $\geq$ to write a statement comparing the numbers 30 and 45.

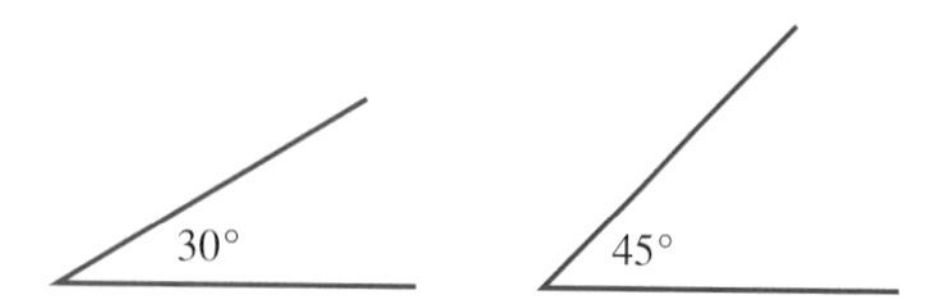

76. The sum of the measures of the angles of a triangle is 180°. The sum of the measures of the angles of a parallelogram is 360°. Use the inequality symbol $\leq$ or $\geq$ to write a statement comparing the numbers 360 and 180.

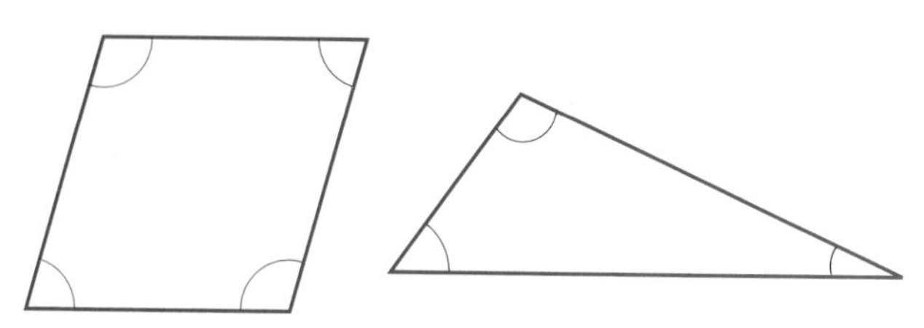

The following graph is a bar graph. This particular graph shows the first three quiz scores of Bill Seggerson in his anatomy class. Each bar represents a different quiz as noted, and the height of each bar represents Bill's score for that particular quiz.

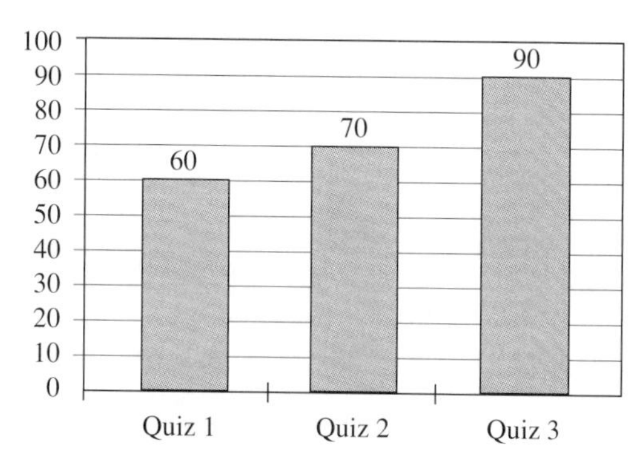

77. What is Bill's highest quiz score shown?

78. What is Bill's lowest quiz score shown?

79. Do you notice any trends shown by this bar graph?

1.2 Sets of Numbers

OBJECTIVES

Tape BA 1

1. Identify natural numbers, whole numbers, integers, rational numbers, irrational numbers, and real numbers.
2. Find the absolute value of real numbers.

1 We now define and explore other **sets** of numbers commonly used in algebra. A set is a collection of objects called **members** or **elements.** A pair of brace symbols { } encloses the list of elements and is translated as "the set of " or "the set containing."

> **Natural Numbers**
>
> The set of **natural numbers** is $\{1, 2, 3, 4, 5, 6, \ldots\}$.

The three dots (an ellipsis) at the end of the list of elements of a set means that the list continues in the same manner indefinitely.

The symbol $\in$ is used to denote that an element is in a particular set. The symbol $\in$ is read as "is an element of." For example, the true statement

3 is an element of $\{1, 2, 3, 4, 5, 6, \ldots\}$

can be written in symbols as

$$3 \in \{1, 2, 3, 4, 5, 6, \ldots\}$$

The symbol $\notin$ is read as "is not an element of." In symbols, we write the true statement

0 is not an element of $\{1, 2, 3, 4, 5, 6, \ldots\}$

as

$$0 \notin \{1, 2, 3, 4, 5, 6, \ldots\}$$

Whole Numbers

The set of **whole numbers** is $\{0, 1, 2, 3, 4, \ldots\}$.

Whole numbers are not sufficient to describe many situations in the real world. For example, quantities smaller than zero must sometimes be represented, such as temperatures less than 0°.

We can picture numbers smaller than zero on the number line as follows:

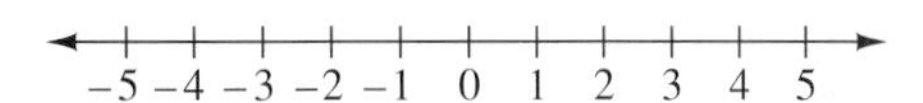

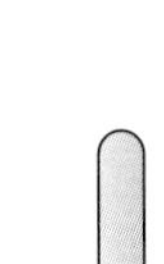

Numbers less than 0 are to the left of 0 and are labeled -1, -2, -3, and so on. A $-$ sign such as the one in -1 tells us that the number is to the left of 0 on the number line. In words, -1 is read "negative one." A $+$ sign or no sign tells us that a number lies to the right of 0 on the number line. For example, 3 and $+3$ both mean positive three. The thermometer to the left shows 20 degrees below 0, or $-20°$.

The numbers we have pictured on the number line above are called the set of **integers.** Integers to the left of 0 are called **negative integers;** integers to the right of 0 are called **positive integers.** The integer 0 is neither positive nor negative.

Integers

The set of **integers** is $\{\ldots, -3, -2, -1, 0, 1, 2, 3, \ldots\}$.

Notice the three dots to the left and to the right of the integers listed. This indicates that the positive integers and the negative integers continue indefinitely.

A problem with integers in real-life settings arises when quantities are smaller than some integer but greater than the next smallest integer. On the number line, these quantities may be visualized by points between integers. Some of these quantities between integers can be represented as a quotient of integers. For example,

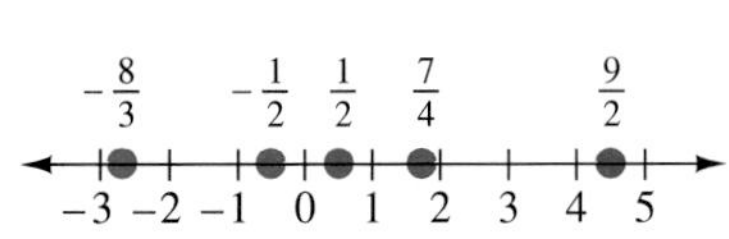

The point on the number line halfway between 0 and 1 can be represented by $\frac{1}{2}$, a quotient of integers.

The point on the number line halfway between 0 and -1 can be represented by $-\frac{1}{2}$. Other quotients of integers and their graphs are shown.

All sets mentioned so far have had their elements listed. When the elements of a set are listed, the set is written in **roster** form. A set can also be written in **set builder notation,** which describes the elements but does not list them. For example, the set $P = \{x \mid x \text{ is a continent}\}$ is written in set builder notation. This set is read as

$$P = \{\underbrace{x}_{\uparrow} \quad \underset{\uparrow}{\mid} \quad \underbrace{x \text{ is a continent}}_{\uparrow}\}$$

"the set of all x such that x is a continent"

The same set P in roster form is

$P = \{$Africa, Antarctica, Asia, Australia, North America, South America, Europe$\}$

The set $\{x \mid x \text{ is a natural number less than } 3\}$ written in roster form is $\{1, 2\}$.

The set of numbers, each of which can be represented as a quotient of integers, is called the set of **rational numbers.** Notice that every integer is also a rational number since each integer can be expressed as a quotient of integers. For example, the integer 5 is also a rational number since $5 = \frac{5}{1}$.

A rational number can be written as a decimal number that either terminates or repeats. For example,

$$\frac{3}{4} = 0.75 \quad \text{(decimal number ends or terminates)}$$

$$\frac{2}{3} = 0.66666\ldots \quad \text{(decimal number repeats in a pattern)}$$

We can describe the set of rational numbers using set builder notation.

Rational Numbers

The set of **rational numbers** is the set of all numbers that can be expressed as a quotient $\frac{a}{b}$, where a and b are integers and $b \neq 0$. In other words, $\left\{\frac{a}{b} \;\middle|\; a \text{ and } b \text{ are integers and } b \neq 0\right\}$.

The number line also contains points that cannot be represented by rational numbers. These numbers are called **irrational numbers** because they cannot be expressed as quotients of integers. For example, $\sqrt{2}$ and π are irrational numbers. An irrational number written as a decimal number will neither terminate nor repeat.

Irrational Numbers

The set of **irrational numbers** is the set of all numbers that correspond to a point on the number line but that are not rational numbers. That is, an irrational number is a number that cannot be expressed as a quotient of integers.

The set of numbers, each of which corresponds to a point on the number line, is called the set of **real numbers.** One and only one point on the number line corresponds to each real number.

> **Real Numbers**
>
> The set of **real numbers** is the set of all numbers each of which corresponds to a point on the number line.

On the following number line, we see that real numbers can be positive, negative, or 0. Numbers to the left of 0 are called **negative numbers;** numbers to the right of 0 are called **positive numbers.** Positive and negative numbers are also called **signed numbers.**

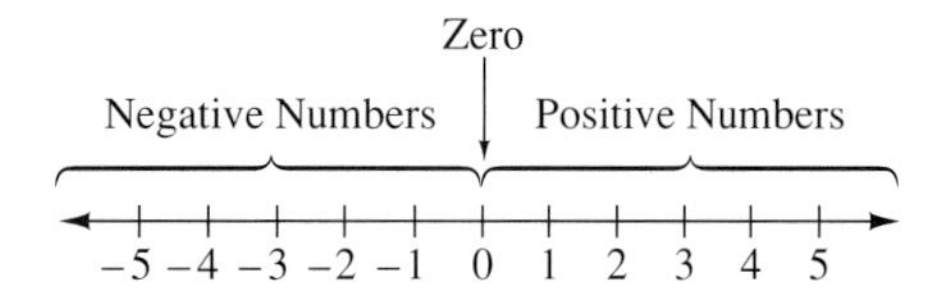

Several different sets of numbers have been discussed in this section. The following diagram shows the relationships between these sets of real numbers.

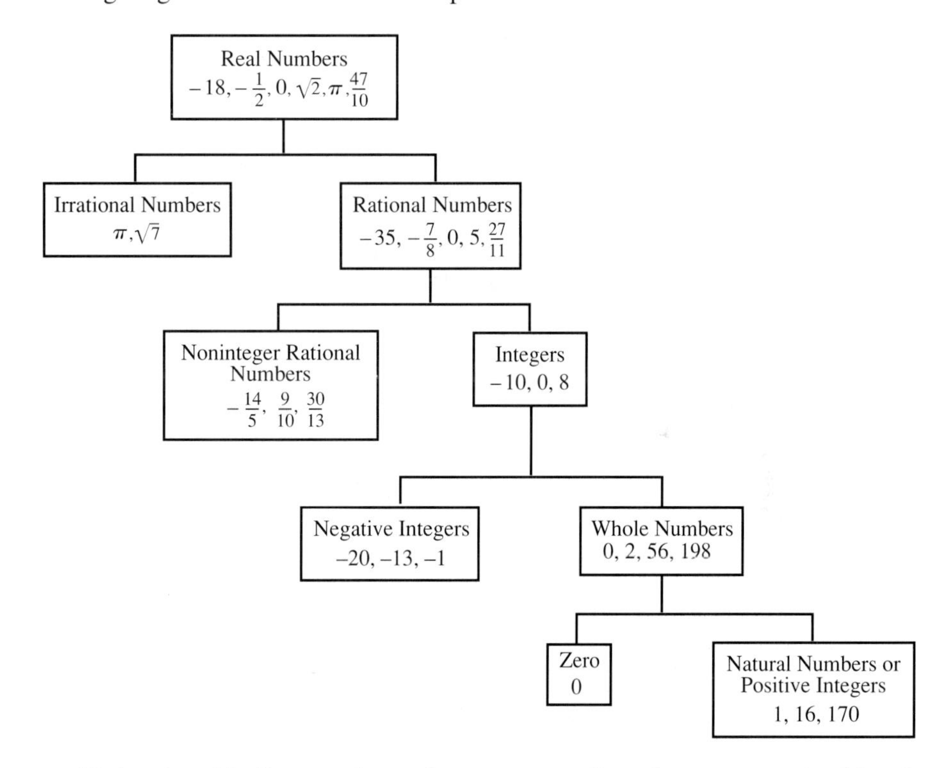

Notice that this diagram shows that some sets of numbers are contained in other sets of numbers. In other words, some sets of numbers are **subsets** of other sets of numbers.

Set X is a subset of set Y, written as $X \subseteq Y$, if all the elements of X are also elements of Y. For example, {2, 3} is a subset of {2, 3, 5, 6}, and we can write this in symbols as

$$\{2, 3\} \subseteq \{2, 3, 5, 6\}$$

↑ is a subset of

Also, if N represents the set of natural numbers and I represents the set of integers, then

$$N \subseteq I$$

EXAMPLE 1 Given the set $\left\{-2, 0, \frac{1}{4}, 112, -3, 11, \sqrt{2}\right\}$, list the numbers in this set that belong to the set of:

a. Natural numbers **b.** Whole numbers **c.** Integers
d. Rational numbers **e.** Irrational numbers **f.** Real numbers

Solution:
a. The natural numbers are 11 and 112.
b. The whole numbers are 0, 11, and 112.
c. The integers are -3, -2, 0, 11, and 112.
d. Recall that integers are rational numbers also. The rational numbers are -3, -2, 0, $\frac{1}{4}$, 11, and 112.
e. The irrational number is $\sqrt{2}$.
f. The real numbers are all numbers in the given set. ■

We can now extend the meaning and use of inequality symbols such as $<$ and $>$ to apply to all real numbers.

Order Property for Real Numbers

Given any two real numbers a and b, $a < b$ if a is to the left of b on the number line. Similarly, $a > b$ if a is to the right of b on the number line.

EXAMPLE 2 Insert $<$, $>$, or $=$ in the space between the paired numbers to make each statement true.

a. $-1 \quad 0$ **b.** $7 \quad \frac{14}{2}$ **c.** $-5 \quad -6$

Solution:
a. $-1 < 0$ since -1 is to the left of 0 on the number line.
b. $7 = \frac{14}{2}$ since $\frac{14}{2}$ simplifies to 7.
c. $-5 > -6$ since -5 is to the right of -6 on the number line. ■

2 The number line not only gives us a picture of the real numbers, but it also helps us visualize the distance between numbers. The distance between a real number a and 0 is given a special name called the **absolute value** of a. "The absolute value of a" is written in symbols as $|a|$.

Absolute Value

The absolute value of a real number a, denoted by $|a|$, is the distance between a and 0 on the number line.

For example, $|3| = 3$ and $|-3| = 3$ since both 3 and -3 are a distance of 3 units from 0 on the number line.

$|-3| = 3 \quad |3| = 3$

3 units 3 units

−3 −2 −1 0 1 2 3

> **HELPFUL HINT**
>
> Since $|a|$ is a distance, $|a|$ will always be either positive or 0, never negative. That is, **for any real number a, $|a| \geq 0$.**

EXAMPLE 3 Find the absolute value of each number.

a. $|4|$ **b.** $|-5|$ **c.** $|0|$

Solution:
a. $|4| = 4$ since 4 is 4 units from 0 on the number line.
b. $|-5| = 5$ since -5 is 5 units from 0 on the number line.
c. $|0| = 0$ since 0 is 0 units from 0 on the number line. ■

EXAMPLE 4 Insert $<$, $>$, or $=$ in the appropriate space to make the statement true.

a. $|0| \quad 2$ **b.** $|-5| \quad 5$ **c.** $|-3| \quad |-2|$ **d.** $|5| \quad |6|$
e. $|-7| \quad |6|$

Solution:
a. $|0| < 2$ since $|0| = 0$ and $0 < 2$.
b. $|-5| = 5$.
c. $|-3| > |-2|$ since $3 > 2$.
d. $|5| < |6|$ since $5 < 6$.
e. $|-7| > |6|$ since $7 > 6$. ■

EXERCISE SET 1.2

Tell which set or sets each number belongs to. Choose among the sets of natural numbers, whole numbers, integers, rational numbers, irrational numbers, and real numbers. See Example 1.

1. 0 **2.** $\frac{1}{4}$ **3.** -2 **4.** $-\frac{1}{2}$

5. 6 **6.** 5 **7.** $\frac{2}{3}$ **8.** $\sqrt{3}$

Insert $<$, $>$, or $=$ in the appropriate space to make a true statement. See Example 2.

9. 10 20 **10.** 100 10 **11.** 0 -2 **12.** 0 2

13. 0.01 0.01 **14.** 0.05 0 **15.** -1.5 2.8 **16.** $\frac{14}{2}$ 7

Insert $<$, $>$, or $=$ in the appropriate space to make a true statement. See Examples 3 and 4.

17. $|20| \quad 20$
18. $|-52| \quad |52|$
19. $-7 \quad -12$
20. $|-2| \quad |-3|$
21. $-500 \quad |-50|$
22. $|0| \quad |-8|$

Tell which set or sets each number belongs to. Choose among the sets of natural numbers, whole numbers, integers, rational numbers, irrational numbers, and real numbers.

23. -9
24. $|-8|$
25. π
26. $\frac{3}{8}$
27. 2
28. $|2|$

Tell whether each statement is true or false.

29. $5 < 6$
30. $7 > 8$
31. $-5 < -6$
32. $-7 > -8$
33. $|-5| < |-6|$
34. $|-7| > |-8|$
35. $|-5| \geq |5|$
36. $|-3| < |0|$
37. $-3 > 2$
38. $-5 < 5$
39. $|8| = |-8|$
40. $|9| = |-9|$
41. $|0| > |-4|$
42. $|0| \leq |0|$

Insert $<$, $>$, or $=$ in the appropriate space to make a true statement.

43. $-10 \quad -100$
44. $-200 \quad -20$
45. $32 \quad 5.2$
46. $7 \quad -7$
47. $\frac{18}{3} \quad \frac{24}{3}$
48. $\frac{8}{2} \quad \frac{12}{3}$
49. $-51 \quad -50$
50. $|-20| \quad -200$
51. $|-5| \quad -4$
52. $0 \quad |0|$
53. $|-1| \quad |1|$
54. $\left|\frac{2}{5}\right| \quad \left|-\frac{2}{5}\right|$

Tell whether each statement is true or false.

55. Every rational number is also an integer.
56. Every natural number is positive.
57. 0 is a real number.
58. Every whole number is an integer.
59. Every negative number is also a rational number.
60. Every rational number is also a real number.
61. Every real number is also a rational number.
62. $\frac{1}{2}$ is an integer.

The apparent magnitude of a star is the measure of its brightness as seen by someone on Earth. The smaller the apparent magnitude, the brighter the star. The apparent magnitude of some stars are listed below.

Star	Apparent Magnitude	Star	Apparent Magnitude
Alpha Centauri	0	Spica	0.98
Sirius	−1.46	Rigel	0.12
Vega	0.03	Regulus	1.35
Antares	0.96	Canopus	−0.72
Sun	−26.7	Hadar	0.61

63. The apparent magnitude of the Sun is -26.7. The apparent magnitude of the star Alpha Centauri is 0. Write an inequality statement comparing the numbers 0 and -26.7.

64. The apparent magnitude of Antares is 0.96. The apparent magnitude of Spica is 0.98. Write an inequality statement comparing the numbers 0.96 and 0.98.

65. Which is brighter, the Sun or Alpha Centauri?

66. Which is dimmer, Antares or Spica?

67. Which star listed is the brightest?

68. Which star listed is the dimmest?

The following bar graph shows the minimum daily temperature for a week in Sioux Falls, South Dakota.

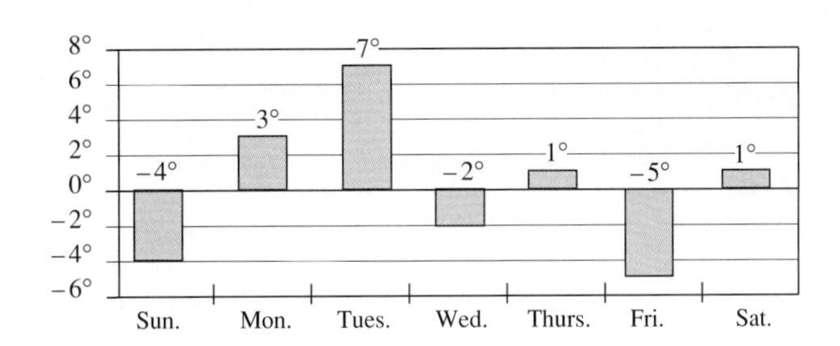

69. On what day of the week was the temperature the highest?

70. On what day of the week was the temperature the lowest?

71. What is the highest temperature shown on the graph?

72. What is the lowest temperature shown on the graph?

Writing in Mathematics

73. In your own words, explain what absolute value is.

74. Give an example of a real-life situation that can be described with integers but not with whole numbers.

1.3 Fractions

OBJECTIVES

Tape BA 1

1. Write fractions in simplest form.
2. Multiply and divide fractions.
3. Add and subtract fractions.

1 A quotient of two integers such as $\frac{2}{9}$ is called a **fraction.** In the fraction $\frac{2}{9}$, the top number, 2, is called the **numerator** and the bottom number, 9, is called the **denominator.**

A fraction may be used to refer to part of a whole. For example, $\frac{2}{9}$ of the circle below is shaded. The denominator 9 tells us how many equal parts the circle is divided into, and the numerator 2 tells us how many equal parts are shaded.

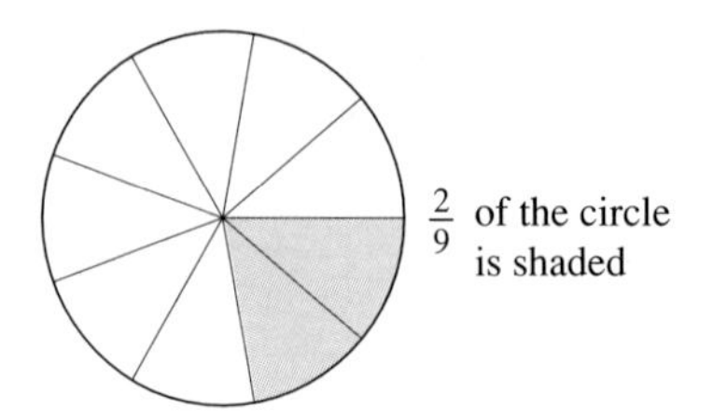

To simplify fractions, we can factor the numerator and the denominator. In the statement $3 \cdot 5 = 15$, 3 and 5 are called **factors** and 15 is the **product.**

$$\underset{\text{factor}}{\underset{\uparrow}{3}} \cdot \underset{\text{factor}}{\underset{\uparrow}{5}} = \underset{\text{product}}{\underset{\uparrow}{15}}$$

To **factor** 15 means to write it as a product. The number 15 can be factored as $3 \cdot 5$ or as $1 \cdot 15$.

A fraction is said to be **simplified** or in **lowest terms** when the numerator and the denominator have no factors in common other than 1. For example, the fraction $\frac{5}{11}$ is in lowest terms since 5 and 11 have no common factors other than 1.

To help us simplify fractions, we will write the numerator and the denominator as a product of **prime numbers.** A prime number is a whole number, other than 1, whose only factors are 1 and itself. The first few prime numbers are

2, 3, 5, 7, 11, 13, 17, 19, 23, 29, and so on

EXAMPLE 1 Write each of the following numbers as a product of primes.

a. 40 **b.** 63

Solution: **a.** Write 40 as the product of any two whole numbers.

$$40 = 4 \cdot 10$$

Next, factor each of these numbers. Continue this process until all of the factors are prime numbers.

$$\begin{aligned} 40 &= 4 \cdot 10 \\ &\swarrow\searrow \quad \swarrow\searrow \\ &= 2 \cdot 2 \cdot 2 \cdot 5 \end{aligned}$$

All the factors are now prime numbers. Then 40 written as a product of primes is

$$40 = 2 \cdot 2 \cdot 2 \cdot 5$$

b.
$$\begin{aligned} 63 &= 9 \cdot 7 \\ &\swarrow\searrow \\ &= 3 \cdot 3 \cdot 7 \end{aligned}$$ ■

To use prime factors to write a fraction in lowest terms, follow these steps:

Writing a Fraction in Lowest Terms

Step 1 Write the numerator and the denominator as a product of primes.
Step 2 Divide the numerator and the denominator by their common factors.

EXAMPLE 2 Write each fraction in lowest terms.

a. $\frac{42}{49}$ **b.** $\frac{11}{27}$ **c.** $\frac{88}{20}$

Solution: **a.** Write the numerator and the denominator as products of primes; then divide both by the common factor 7.

$$\frac{42}{49} = \frac{2 \cdot 3 \cdot 7}{7 \cdot 7} = \frac{6}{7}$$

b. $\frac{11}{27} = \frac{11}{3 \cdot 3 \cdot 3}$

There are no common factors other than 1, so $\frac{11}{27}$ is already in lowest terms.

c. $\frac{88}{20} = \frac{2 \cdot 2 \cdot 2 \cdot 11}{2 \cdot 2 \cdot 5} = \frac{22}{5}$ ■

2 To multiply two fractions, multiply numerator times numerator to obtain the numerator of the product and denominator times denominator to obtain the denominator of the product.

Multiplying Fractions

$$\frac{a}{b} \cdot \frac{c}{d} = \frac{a \cdot c}{b \cdot d}, \quad \text{if } b \neq 0 \text{ and } d \neq 0$$

EXAMPLE 3 Find the product of $\frac{2}{15}$ and $\frac{5}{13}$. Write the answer in lowest terms.

Solution:

$$\frac{2}{15} \cdot \frac{5}{13} = \frac{2 \cdot 5}{15 \cdot 13} \quad \begin{array}{l}\text{Multiply numerators.} \\ \text{Multiply denominators.}\end{array}$$

Next, simplify the product by dividing the numerator and the denominator by any common factors.

$$= \frac{2 \cdot 5}{3 \cdot 5 \cdot 13}$$

$$= \frac{2}{39}$$ ■

Before dividing fractions, we first define **reciprocals.** Two fractions are reciprocals of each other if their product is 1. For example $\frac{2}{3}$ and $\frac{3}{2}$ are reciprocals since

$\frac{2}{3} \cdot \frac{3}{2} = 1$. Also, the reciprocal of 5 is $\frac{1}{5}$ since $5 \cdot \frac{1}{5} = \frac{5}{1} \cdot \frac{1}{5} = 1$.

To divide fractions, multiply the first fraction by the reciprocal of the second fraction.

Dividing Fractions

$$\frac{a}{b} \div \frac{c}{d} = \frac{a}{b} \cdot \frac{d}{c}, \qquad \text{if } b \neq 0, d \neq 0, \text{ and } c \neq 0$$

EXAMPLE 4 Find each quotient. Write all answers in lowest terms.

a. $\frac{4}{5} \div \frac{5}{16}$ **b.** $\frac{7}{10} \div 14$ **c.** $\frac{3}{8} \div \frac{3}{10}$

Solution: **a.** $\frac{4}{5} \div \frac{5}{16} = \frac{4}{5} \cdot \frac{16}{5} = \frac{4 \cdot 16}{5 \cdot 5} = \frac{64}{25}$

b. $\frac{7}{10} \div 14 = \frac{7}{10} \div \frac{14}{1} = \frac{7}{10} \cdot \frac{1}{14} = \frac{7 \cdot 1}{2 \cdot 5 \cdot 2 \cdot 7} = \frac{1}{20}$

c. $\frac{3}{8} \div \frac{3}{10} = \frac{3}{8} \cdot \frac{10}{3} = \frac{3 \cdot 2 \cdot 5}{2 \cdot 2 \cdot 2 \cdot 3} = \frac{5}{4}$ ■

3 To add or subtract fractions with the same denominator, combine numerators and place the sum or difference over the common denominator.

Adding and Subtracting Fractions with the Same Denominator

$$\frac{a}{b} + \frac{c}{b} = \frac{a+c}{b}, \qquad \text{if } b \neq 0$$

$$\frac{a}{b} - \frac{c}{b} = \frac{a-c}{b}, \qquad \text{if } b \neq 0$$

EXAMPLE 5 Add or subtract as indicated. Write each answer in lowest terms.

a. $\frac{2}{7} + \frac{4}{7}$ **b.** $\frac{3}{10} + \frac{2}{10}$ **c.** $\frac{9}{7} - \frac{2}{7}$ **d.** $\frac{5}{3} - \frac{1}{3}$

Solution: **a.** $\frac{2}{7} + \frac{4}{7} = \frac{2+4}{7} = \frac{6}{7}$

b. $\frac{3}{10} + \frac{2}{10} = \frac{3+2}{10} = \frac{5}{10} = \frac{5}{2 \cdot 5} = \frac{1}{2}$

c. $\frac{9}{7} - \frac{2}{7} = \frac{9-2}{7} = \frac{7}{7} = 1$

d. $\frac{5}{3} - \frac{1}{3} = \frac{5-1}{3} = \frac{4}{3}$ ■

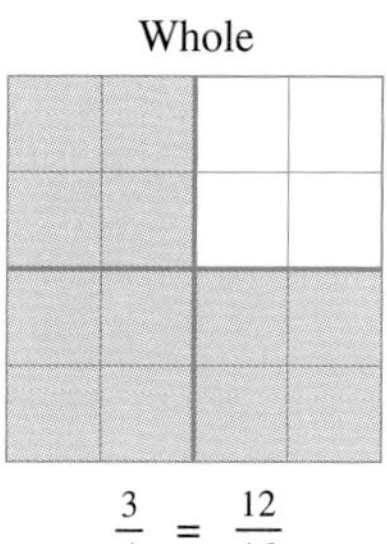

$\frac{3}{4} = \frac{12}{16}$

To add or subtract fractions without the same denominator, first write the fractions as **equivalent fractions** with a common denominator. Equivalent fractions are fractions that represent the same quantity. For example, $\frac{3}{4}$ and $\frac{12}{16}$ are equivalent fractions since they represent the same portion of a whole, as the diagram shows. Count the larger squares and the shaded portion is $\frac{3}{4}$. Count the smaller squares and the shaded portion is $\frac{12}{16}$. Thus, $\frac{3}{4} = \frac{12}{16}$.

We can write equivalent fractions by multiplying a given fraction by 1, as shown in the next example. Multiplying a number by 1 does not change the value of the number.

EXAMPLE 6 Write $\frac{2}{5}$ as an equivalent fraction with a denominator of 20.

Solution: Since $5 \cdot 4 = 20$, multiply the fraction by $\frac{4}{4}$. Multiplying by $\frac{4}{4} = 1$ does not change the value of the fraction.

$$\frac{2}{5} = \frac{2}{5} \cdot \frac{4}{4} = \frac{2 \cdot 4}{5 \cdot 4} = \frac{8}{20} \quad ■$$

EXAMPLE 7 Add or subtract as indicated. Write each answer in lowest terms.

a. $\frac{2}{5} + \frac{1}{4}$ **b.** $\frac{1}{2} + \frac{17}{22} - \frac{2}{11}$ **c.** $3\frac{1}{6} - 1\frac{11}{12}$

Solution: **a.** Fractions must have a common denominator before they can be added or subtracted. Since 20 is the smallest number that both 5 and 4 divide into evenly, 20 is the **least common denominator.** Write both fractions as equivalent fractions with denominators of 20. Since

$$\frac{2}{5} \cdot \frac{4}{4} = \frac{2 \cdot 4}{5 \cdot 4} = \frac{8}{20} \quad \text{and} \quad \frac{1}{4} \cdot \frac{5}{5} = \frac{1 \cdot 5}{4 \cdot 5} = \frac{5}{20}$$

then

$$\frac{2}{5} + \frac{1}{4} = \frac{8}{20} + \frac{5}{20} = \frac{13}{20}$$

b. The least common denominator for denominators 2, 22, and 11 is 22. First write each fraction as an equivalent fraction with a denominator of 22.

$$\frac{1}{2} = \frac{1}{2} \cdot \frac{11}{11} = \frac{11}{22}, \quad \frac{17}{22} = \frac{17}{22}, \quad \text{and} \quad \frac{2}{11} = \frac{2}{11} \cdot \frac{2}{2} = \frac{4}{22}$$

Then

$$\frac{1}{2} + \frac{17}{22} - \frac{2}{11} = \frac{11}{22} + \frac{17}{22} - \frac{4}{22} = \frac{24}{22} = \frac{12}{11}$$

c. To find $3\frac{1}{6} - 1\frac{11}{12}$, first rewrite each mixed number as follows:

$$3\frac{1}{6} = 3 + \frac{1}{6} = \frac{18}{6} + \frac{1}{6} = \frac{19}{6}$$

$$1\frac{11}{12} = 1 + \frac{11}{12} = \frac{12}{12} + \frac{11}{12} = \frac{23}{12}$$

Then

$$3\frac{1}{6} - 1\frac{11}{12} = \frac{19}{6} - \frac{23}{12} = \frac{38}{12} - \frac{23}{12} = \frac{15}{12} = \frac{5}{4}$$ ■

EXERCISE SET 1.3

Represent the shaded part of each geometric figure by a fraction.

1. 

2. 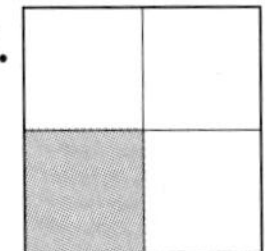

3. 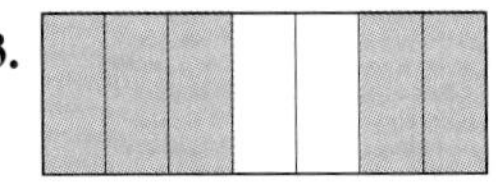

4. 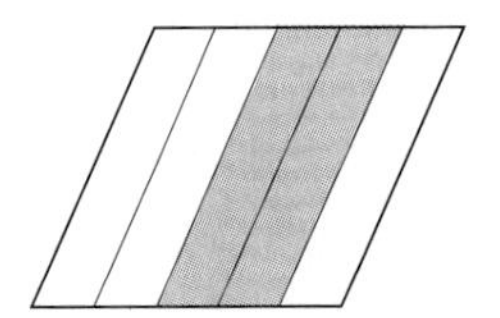

Write each of the following numbers as a product of primes. See Example 1.

5. 20 **6.** 56 **7.** 75

8. 32 **9.** 45 **10.** 24

Write the following fractions in lowest terms. See Example 2.

11. $\frac{2}{4}$ **12.** $\frac{3}{6}$ **13.** $\frac{10}{15}$ **14.** $\frac{15}{20}$

15. $\frac{3}{7}$ **16.** $\frac{5}{9}$ **17.** $\frac{18}{30}$ **18.** $\frac{42}{45}$

Multiply or divide as indicated. Write the answer in lowest terms. See Examples 3 and 4.

19. $\frac{1}{2} \cdot \frac{3}{4}$ **20.** $\frac{10}{6} \cdot \frac{3}{5}$ **21.** $\frac{2}{3} \cdot \frac{3}{4}$ **22.** $\frac{7}{8} \cdot \frac{3}{21}$

23. $\frac{1}{2} \div \frac{7}{12}$ **24.** $\frac{7}{12} \div \frac{1}{2}$ **25.** $\frac{3}{4} \div \frac{1}{20}$ **26.** $\frac{3}{5} \div \frac{9}{10}$

27. $\frac{7}{10} \cdot \frac{5}{21}$ **28.** $\frac{3}{35} \cdot \frac{10}{63}$ **29.** $2\frac{7}{9} \cdot \frac{1}{3}$ **30.** $\frac{1}{4} \cdot 5\frac{5}{6}$

Add or subtract as indicated. Write the answer in lowest terms. See Example 5.

31. $\frac{4}{5} - \frac{1}{5}$ **32.** $\frac{6}{7} - \frac{1}{7}$ **33.** $\frac{4}{5} + \frac{1}{5}$ **34.** $\frac{6}{7} + \frac{1}{7}$

35. $\frac{17}{21} - \frac{10}{21}$ **36.** $\frac{18}{35} - \frac{11}{35}$ **37.** $\frac{23}{105} + \frac{4}{105}$ **38.** $\frac{13}{132} + \frac{35}{132}$

Write each of the following fractions as an equivalent fraction with the given denominator. See Example 6.

39. $\frac{7}{10}$ with a denominator of 30

40. $\frac{2}{3}$ with a denominator of 9

41. $\frac{2}{9}$ with a denominator of 18

42. $\frac{8}{7}$ with a denominator of 56

43. $\frac{4}{5}$ with a denominator of 20

44. $\frac{4}{5}$ with a denominator of 25

Add or subtract as indicated. Write the answer in lowest terms. See Example 7.

45. $\frac{2}{3} + \frac{3}{7}$

46. $\frac{3}{4} + \frac{1}{6}$

47. $\frac{13}{15} - \frac{1}{5}$

48. $\frac{2}{9} - \frac{1}{6}$

49. $\frac{5}{22} - \frac{5}{33}$

50. $\frac{7}{10} - \frac{8}{15}$

51. $\frac{12}{5} - 1$

52. $2 - \frac{3}{8}$

Perform the following operations. Write answers in lowest terms.

53. $\frac{10}{21} + \frac{5}{21}$

54. $\frac{11}{35} + \frac{3}{35}$

55. $\frac{10}{3} - \frac{5}{21}$

56. $\frac{11}{7} - \frac{3}{35}$

57. $\frac{2}{3} \cdot \frac{3}{5}$

58. $\frac{2}{3} \div \frac{3}{4}$

59. $\frac{3}{4} \div \frac{7}{12}$

60. $\frac{3}{5} + \frac{2}{3}$

61. $\frac{5}{12} + \frac{4}{12}$

62. $\frac{2}{7} + \frac{4}{7}$

63. $5 + \frac{2}{3}$

64. $7 + \frac{1}{10}$

65. $\frac{7}{8} \div 3\frac{1}{4}$

66. $3 \div \frac{3}{4}$

67. $\frac{7}{18} \div \frac{14}{36}$

68. $4\frac{3}{7} \div \frac{31}{7}$

69. $\frac{23}{105} - \frac{2}{105}$

70. $\frac{57}{132} - \frac{13}{132}$

71. $1\frac{1}{2} + 3\frac{2}{3}$

72. $2\frac{3}{5} + 4\frac{7}{10}$

73. $\frac{2}{3} - \frac{5}{9} + \frac{5}{6}$

74. $\frac{8}{11} - \frac{1}{4} + \frac{1}{2}$

Determine the unknown part of each circle.

75.

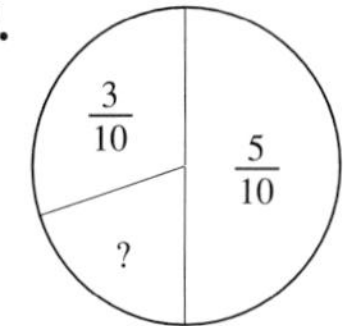

76.

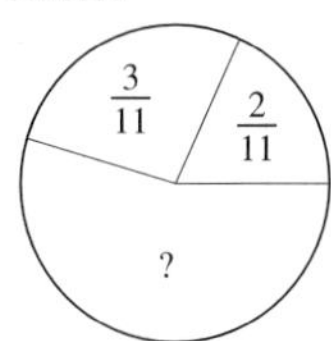

77.

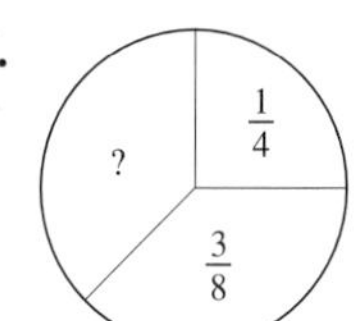

78.

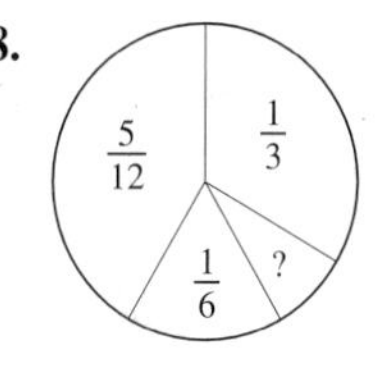

79. One of the New Orleans Saints guards was told to lose 20 pounds during the summer. So far, the guard has lost $11\frac{1}{2}$ pounds. How many more pounds must he lose?

80. In an English class at Cartez College, Stuart Carelton was told to read a 340-page book. So far, he has read $\frac{2}{5}$ of it. How many pages does he *have left to read?*

Each graph below is called a circle graph or pie chart. The circle or pie represents a whole or, in this case, 100%. Each circle is divided into sectors (shaped like pieces of a pie) that represent various parts of the whole 100%.

81. This circle graph represents the annual expenses of owning and operating a car for Betty Solanky, a college student. What percent of Betty's expenses for owning and operating a car are for car payments?

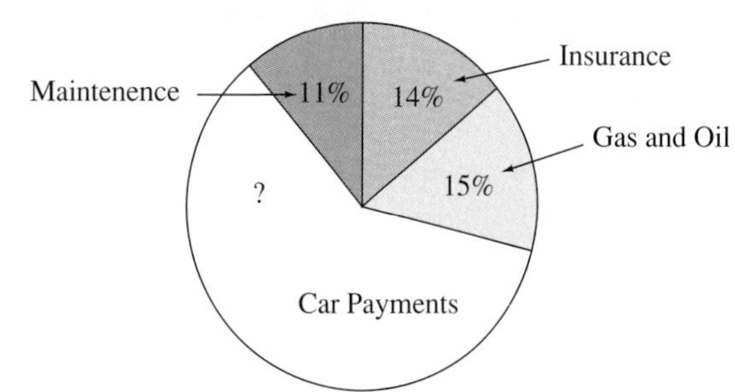

82. This circle graph summarizes a recent survey of milk sales. What percent of milk consumption is skim milk?

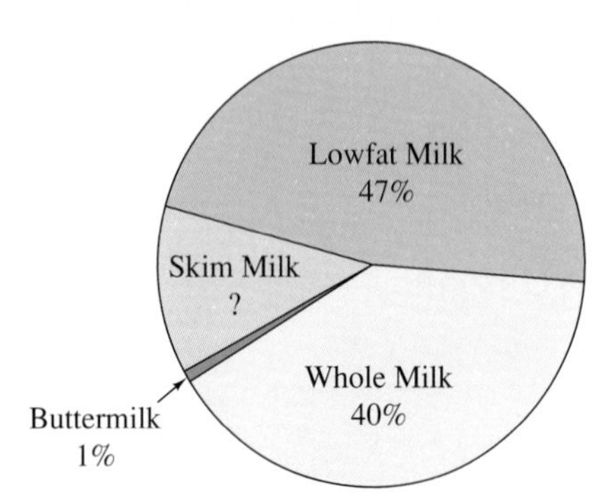

Source: U.S. Department of Agriculture, Economic Research Service, *Food Consumption, Prices, and Expenditures,* annual.

83. Find the perimeter.

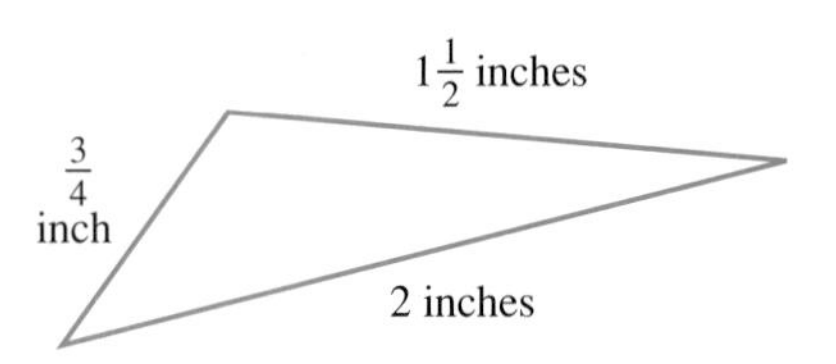

84. Find the area.

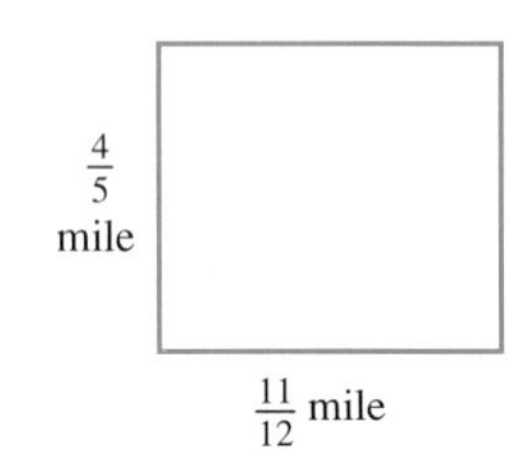

1.4 Exponents, Roots, and Order of Operations

OBJECTIVES

Tape BA 2

1 Use and simplify exponential expressions.

2 Find roots of numbers.

3 Define and use the order of operations.

1 A notation frequently used in algebra is the **exponent.** An exponent is a shorthand notation for repeated multiplication of the same factor. For instance, in $2 \cdot 2 \cdot 2 = 8$ the factor 2 appears three times. Using exponents, $2 \cdot 2 \cdot 2$ can be written as 2^3. The 2 in 2^3 is called the **base;** it is the repeated factor. The 3 in 2^3 is called the **exponent** and is the number of times the base is used as a factor. The expression 2^3 is called an **exponential expression.**

$$2^3 = 2 \cdot 2 \cdot 2 = 8$$

exponent (points to 3)

base (2 is a factor 3 times)

EXAMPLE 1 Evaluate the following:

a. 3^2 [read as "3 squared" or "3 to the second power"]

b. 5^3 [read as "5 cubed" or "5 to the third power"]

c. 2^4 [read as "2 to the fourth power"]

d. 7^1 **e.** $\left(\frac{3}{7}\right)^2$

Solution: **a.** $3^2 = 3 \cdot 3 = 9$, "3 squared equals 9"

b. $5^3 = 5 \cdot 5 \cdot 5 = 125$ **c.** $2^4 = 2 \cdot 2 \cdot 2 \cdot 2 = 16$

d. $7^1 = 7$ **e.** $\left(\frac{3}{7}\right)^2 = \left(\frac{3}{7}\right)\left(\frac{3}{7}\right) = \frac{9}{49}$ ■

HELPFUL HINT

$2^3 \neq 2 \cdot 3$ since 2^3 is a shorthand notation for repeated **multiplication.**

$$2^3 = 2 \cdot 2 \cdot 2 = 8, \text{ whereas } 2 \cdot 3 = 6$$

2 The opposite of squaring a number is taking the **square root** of a number. For example, since the square of 4, or 4^2, is 16, we say that a square root of 16 is 4. The notation $\sqrt{a}$ is used to denote the **positive** or **principal square root** of a nonnegative number a. We then have in symbols that

$$\sqrt{16} = 4$$

EXAMPLE 2 Find the following square roots.

a. $\sqrt{9}$ **b.** $\sqrt{25}$ **c.** $\sqrt{\frac{1}{4}}$

Solution: **a.** $\sqrt{9} = 3$ since 3 is positive and $3^2 = 9$.

b. $\sqrt{25} = 5$ since $5^2 = 25$.

c. $\sqrt{\frac{1}{4}} = \frac{1}{2}$ since $\left(\frac{1}{2}\right)^2 = \frac{1}{4}$. ■

We can find roots other than square roots. Since 2 cubed, written as 2^3, is 8, we say that the cube root of 8 is 2. This is written as

$$\sqrt[3]{8} = 2$$

Also,

$$\sqrt[4]{81} = 3 \qquad \text{since } 3^4 = 81$$

EXAMPLE 3 Find the following roots.

a. $\sqrt[3]{27}$ **b.** $\sqrt[5]{1}$ **c.** $\sqrt[4]{16}$

Solution: **a.** $\sqrt[3]{27} = 3$ since $3^3 = 27$.

b. $\sqrt[5]{1} = 1$ since $1^5 = 1$.

c. $\sqrt[4]{16} = 2$ since $2^4 = 16$. ■

3 Using symbols for mathematical operations is a great convenience. The more operation symbols presented in an expression, the more careful we must be when simplifying. For example, in the expression $2 + 3 \cdot 7$, do we add first or multiply first? To eliminate confusion, **grouping symbols** may be used. Examples of grouping symbols are parentheses ( ), brackets [ ], braces { }, square root $\sqrt{}$, and the fraction bar. If we wish $2 + 3 \cdot 7$ to be simplified by adding first, enclose $2 + 3$ with parentheses.

$$(2 + 3) \cdot 7 = 5 \cdot 7 = 35$$

If we wish to multiply first, $3 \cdot 7$ may be enclosed with parentheses.

$$2 + (3 \cdot 7) = 2 + 21 = 23$$

To eliminate confusion when no grouping symbols are present, use the following agreed-upon order of operations.

Order of Operations

Simplify expressions using the following order. If grouping symbols such as parentheses are present, simplify expressions within those first, starting with the innermost set. If fraction bars are present, simplify the numerator and the denominator separately.

1. Simplify exponential expressions.
2. Perform multiplications or divisions in order from left to right.
3. Perform additions or subtractions in order from left to right.

Now simplify $2 + 3 \cdot 7$. There are no grouping symbols and no exponents, so we multiply and then add.

$$2 + 3 \cdot 7 = 2 + 21 \quad \text{Multiply.}$$
$$= 23 \quad \text{Add.}$$

EXAMPLE 4 Simplify each expression.

a. $6 \div 3 + 5^2$ **b.** $\dfrac{2(12 + 3)}{|-15|}$ **c.** $3 \cdot 10 - 7 \div 7$ **d.** $3 \cdot 4^2$ **e.** $\dfrac{3}{2} \cdot \dfrac{1}{2} - \dfrac{1}{2}$

Solution: **a.** Evaluate 5^2 first.

$$6 \div 3 + 5^2 = 6 \div 3 + 25$$

Next divide, then add.

$$6 \div 3 + 25 = 2 + 25 \quad \text{Divide.}$$
$$= 27 \quad \text{Add.}$$

b. First simplify the numerator and the denominator separately.

$$\frac{2(12 + 3)}{|-15|} = \frac{2(15)}{15} \quad \text{Simplify numerator and denominator separately.}$$
$$= \frac{30}{15}$$
$$= 2 \quad \text{Simplify.}$$

c. Multiply and divide from left to right. Then subtract.

$$3 \cdot 10 - 7 \div 7 = 30 - 1$$
$$= 29 \qquad \text{Subtract.}$$

d. In this example, only the 4 is raised to the second power. The factor of 3 is not part of the base because no grouping symbols include it as part of the base.

$$3 \cdot 4^2 = 3 \cdot 16 \qquad \text{Evaluate the exponential expression.}$$
$$= 48 \qquad \text{Multiply.}$$

e. The order of operations applies to fractions in exactly the same way as it applies to whole numbers.

$$\frac{3}{2} \cdot \frac{1}{2} - \frac{1}{2} = \frac{3}{4} - \frac{1}{2} \qquad \text{Multiply.}$$
$$= \frac{3}{4} - \frac{2}{4} \qquad \text{The least common denominator is 4.}$$
$$= \frac{1}{4} \qquad \text{Subtract.}$$

■

HELPFUL HINT

Be careful when evaluating an exponential expression. In $3 \cdot 4^2$, the exponent 2 applies only to the base of 4. In $(3 \cdot 4)^2$, we multiply first because of parentheses, so the exponent of 2 applies to the product $3 \cdot 4$.

$$3 \cdot 4^2 = 3 \cdot 16 = 48 \qquad (3 \cdot 4)^2 = (12)^2 = 144$$

Expressions that include many grouping symbols can be confusing. When simplifying these expressions, keep in mind that grouping symbols separate the expression into distinct parts. Each is then simplified separately.

EXAMPLE 5 Simplify $\dfrac{|-2|^3}{7^1 - \sqrt{4}}$.

Solution: The fraction bar serves as a grouping symbol and separates the numerator and denominator. Simplify the numerator and the denominator separately; then divide.

$$\frac{|-2|^3}{7^1 - \sqrt{4}} = \frac{2^3}{7 - 2} \qquad \text{Write } |-2| \text{ as 2 and } \sqrt{4} \text{ as 2.}$$
$$= \frac{8}{5} \qquad \text{Simplify the numerator and the denominator.}$$

■

EXAMPLE 6 Simplify $3[4(5 + 2) - 10]$.

Solution: Notice that both parentheses and brackets are used as grouping symbols. Start with the innermost set of grouping symbols.

$3[4(5 + 2) - 10] = 3[4(7) - 10]$ Evaluate the expression in parentheses.

$= 3[28 - 10]$ Multiply.

$= 3[18]$ Subtract inside the brackets.

$= 54$ Multiply. ■

EXERCISE SET 1.4

Evaluate. See Example 1.

1. 3^5 **2.** 5^3 **3.** 3^3 **4.** 4^4

5. 1^5 **6.** 1^8 **7.** 5^1 **8.** 8^1

9. $\left(\frac{1}{5}\right)^3$ **10.** $\left(\frac{6}{11}\right)^2$ **11.** $\left(\frac{2}{3}\right)^4$ **12.** $\left(\frac{1}{2}\right)^5$

Find the following roots. See Examples 2 and 3.

13. $\sqrt{49}$ **14.** $\sqrt{81}$ **15.** $\sqrt{\frac{1}{9}}$ **16.** $\sqrt{\frac{1}{25}}$

17. $\sqrt[3]{64}$ **18.** $\sqrt[5]{32}$ **19.** $\sqrt[4]{81}$ **20.** $\sqrt[3]{1}$

Simplify each expression. See Example 4.

21. $5 + 6 \cdot 2$ **22.** $8 + 5 \cdot 3$ **23.** $4 \cdot 8 - 6 \cdot 2$

24. $12 \cdot 5 - 3 \cdot 6$ **25.** $2(8 - 3)$ **26.** $5(6 - 2)$

27. $2 + (5 - 2) + 4^2$ **28.** $5 \cdot 3^2$ **29.** $2 \cdot 5^2$

30. $\frac{1}{4} \cdot \frac{2}{3} - \frac{1}{6}$ **31.** $\frac{3}{4} \cdot \frac{1}{2} + \frac{2}{3}$

Evaluate each expression. See Examples 5 and 6.

32. $\frac{19 - 3 \cdot 5}{6 - 4}$ **33.** $\frac{|6 - 2| + 3}{8 + 2 \cdot 5}$ **34.** $\frac{15 - |3 - 1|}{12 - 3 \cdot 2}$

35. $\frac{3(2 + 5)}{6 + 2}$ **36.** $\frac{4(8 - 3)}{6 + 3}$ **37.** $\frac{4 \cdot 3 + 2}{4 + 3 \cdot 2}$

38. $5[3(2 + 1) + 4]$ **39.** $4[5(2 + 4) - 8]$ **40.** $6[5(2 + 6) - 9]$

41. $|-18| - |-7|$ **42.** $|-4| + |10|$ **43.** $2 \cdot 5^2$ **44.** $3 \cdot 2^3$

45. $45 - 6^2 \div 3^2 + \sqrt{1}$ **46.** $(12 \div 3)^2 - 10 \div 5$ **47.** $\frac{|-6|^2}{|-8| - 2^2}$ **48.** $\frac{|-4|^3}{3^2 + \sqrt{49}}$

Simplify each expression.

49. $6 + 3(7 - 2)$ **50.** $48 - 3(6 + 5)$ **51.** $3 + (7 - 3) + 3^2$

52. $\frac{6 - \sqrt{4}}{9 - 2}$ **53.** $\frac{8 - 5}{24 - 20}$ **54.** $\frac{7 + 4 \cdot 2}{4 - 1}$

55. $2[5 + 2(8 - 3)]$ **56.** $3[4 + 3(6 - 4)]$ **57.** $(5 \cdot 3)^2$

58. $(2 \cdot 5)^2$

59. $\sqrt[3]{8} + 24 \div 3 \cdot 2 - 6$

60. $\dfrac{6 + 2 \cdot 3}{(6 + 2) \cdot 3}$

61. $\dfrac{3 + 2 \cdot 3^2}{3 + 2^2}$

62. $\dfrac{4 + 3 \cdot 2^2}{4 + 3^2}$

63. $9 + 16 \div 8 \cdot 2 - 4$

64. $\dfrac{1}{4} \cdot \dfrac{4}{3} - \dfrac{1}{6}$

65. $\dfrac{3}{4} \cdot \dfrac{1}{2} + \dfrac{1}{3}$

66. $\dfrac{3 + 3(5 + 3)}{3^2 + 1}$

67. $\dfrac{3 + 6(8 - 5)}{4^2 + 2}$

68. $6[2(3 + 1) + 2]$

69. $6(12 \div 4 + 2) - 3(10 \div 2 + 3)$

70. $5(18 \div 6 + 3) - 2(9 \div 3 + 6)$

71. $\dfrac{6 + |8 - 2| + 3^2}{18 - 3}$

72. $\dfrac{16 + |13 - 5| + 4^2}{17 - 5}$

73. $\dfrac{3 \cdot 6 + 3^2}{12 \cdot 5 - 5}$

74. $\dfrac{5.1 + 2^2}{9 - 2}$

75. $2.6[1.3(4 + 3) - 8]$

76. $\sqrt{\dfrac{1}{4}} + \sqrt{\dfrac{9}{4}}$

77. $\sqrt{\dfrac{25}{9}} - \sqrt{\dfrac{1}{9}}$

78. $\sqrt{100} - \sqrt{36}$

79. $\sqrt[3]{27} + \sqrt{64}$

80. $(15 - 11)(20 - 4)$

81. $(7 + 3)(35 - 2)$

82. $3 \cdot \sqrt{49} + 14 \div 2$

83. $2 \cdot \sqrt{121} - 100 \div 20$

84. $5.2[1.8(2 + 7) - 6]$

85. $\dfrac{6 + 2(1.3)^2}{0.55 + 3(0.65)}$

86. $\dfrac{4 + 2(2.3)^2}{1.8 + 1.6(0.75)}$

87. Find the unknown length of the side of the given triangle by simplifying $\sqrt{3^2 + 4^2}$. (Hint: The square root symbol $\sqrt{\ }$ serves as a grouping symbol. Simplify $3^2 + 4^2$ first.)

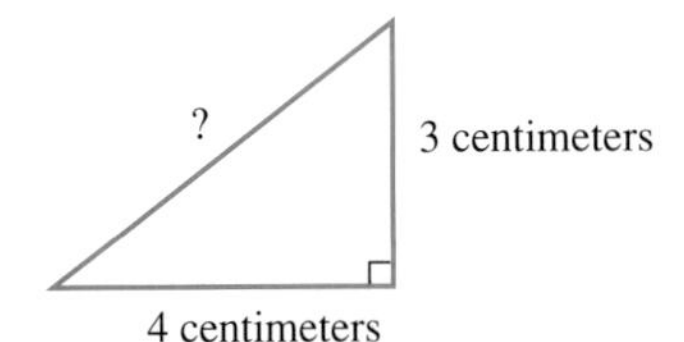

88. Find the unknown length of the side of the given rectangle by simplifying $\sqrt{10^2 - 6^2}$. (Hint: The square root symbol $\sqrt{\ }$ serves as a grouping symbol. Simplify $10^2 - 6^2$ first.)

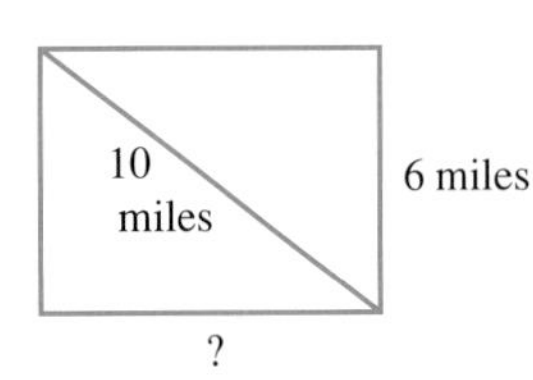

Writing in Mathematics

89. Explain why $-(-2)$ and $-|-2|$ represent different numbers.

1.5 Introduction to Variables and Equations

OBJECTIVES

Tape BA 2

1. Simplify algebraic expressions, given replacement values for variables.
2. Translate word phrases into algebraic expressions.
3. Define equation and solution or root of an equation.
4. Translate word statements into equations.

1 In an earlier exercise set we used letters to represent numbers. A letter that is used to represent a number is called a **variable.** An **algebraic expression** is a collection of numbers, variables, operation symbols, and grouping symbols. For example,

$$2x, \quad -3, \quad 2x - 10, \quad 5(p^2 + 1), \quad \text{and} \quad \frac{3y^2 - 6y + 1}{5}$$

are algebraic expressions. The expression $2x$ means $2 \cdot x$. Also, $5(p^2 + 1)$ means $5 \cdot (p^2 + 1)$ and $3y^2$ means $3 \cdot y^2$. If we give a specific value to a variable, we can **evaluate an algebraic expression.** To evaluate an algebraic expression means to find its numerical value. Make sure the order of operations is followed when evaluating an expression.

EXAMPLE 1 Find the value of each expression if $x = 3$ and $y = 2$.

a. $2x - y$ **b.** $\frac{3x}{2y}$ **c.** $\frac{x}{y} + \frac{y}{2}$ **d.** $x^2 - y^2$

Solution: **a.** Replace x with 3 and y with 2.

$$\begin{aligned} 2x - y &= 2(3) - 2 && \text{Let } x = 3 \text{ and } y = 2. \\ &= 6 - 2 && \text{Multiply.} \\ &= 4 && \text{Subtract.} \end{aligned}$$

b. $\frac{3x}{2y} = \frac{3 \cdot 3}{2 \cdot 2} = \frac{9}{4}$ Let $x = 3$ and $y = 2$.

c. Replace x with 3 and y with 2. Then simplify.

$$\frac{x}{y} + \frac{y}{2} = \frac{3}{2} + \frac{2}{2} = \frac{5}{2}$$

d. Replace x with 3 and y with 2.

$$x^2 - y^2 = 3^2 - 2^2 = 9 - 4 = 5 \quad ■$$

2 Now that we can represent an unknown number by a variable, let's practice translating phrases containing unknown numbers into algebraic expressions.

EXAMPLE 2 Write an algebraic expression that represents each of the following phrases. Let x represent the unknown number.

a. The sum of a number and 3

b. The product of 3 and a number

c. Twice a number

d. 10 decreased by a number

e. 7 more than 5 times a number

Solution: **a.** $x + 3$ since "sum" means to add.

b. $3 \cdot x$ or $3x$ are both ways to denote the product of x and 3.

c. $2 \cdot x$ or $2x$

d. $10 - x$ because "decreased by" means to subtract.

e. $\underbrace{5x}_{\text{5 times a number}} + 7$ ■

3 An **equation** is a mathematical statement that two expressions are equal. The equals symbol = is used to equate the two expressions. For example, $3 + 2 = 5$, $7x = 35$, $\frac{2(x-1)}{3} = 0$, and $I = PRT$ are all equations.

> **HELPFUL HINT**
>
> An equation contains the *is equal to* symbol, = . An algebraic expression does not.

When an equation contains a variable, deciding which values of the variable make an equation a true statement is called **solving** an equation for the variable. A **solution** or **root** of an equation is a value for the variable that makes the equation true. For example, 3 is a solution of the equation $x + 4 = 7$, because if x is replaced by 3 the statement is true.

$$x + 4 = 7$$
$$\downarrow$$
$$3 + 4 = 7 \qquad \text{Replace } x \text{ with 3.}$$
$$7 = 7 \qquad \text{True.}$$

Similarly, 1 is not a solution of the equation $x + 4 = 7$, because $1 + 4 = 7$ is **not** a true statement. The **solution set** of an equation is the set of solutions of the equation. The solution set of the equation $x + 4 = 7$ is $\{3\}$.

EXAMPLE 3 Decide whether 10 is a solution of $3x - 10 = 2x$.

Solution: Replace x with 10 and see if a true statement results.

$$3x - 10 = 2x \qquad \text{Original equation.}$$
$$3(10) - 10 = 2(10) \qquad \text{Replace } x \text{ with 10.}$$
$$30 - 10 = 20 \qquad \text{Simplify each side.}$$
$$20 = 20 \qquad \text{True.}$$

Since we arrived at a true statement after replacing x with 10 and simplifying both sides of the equation, 10 is a solution of the equation. ■

4 We now practice translating sentences into equations.

EXAMPLE 4 Write each sentence as an equation. Let x represent the unknown number.

a. The quotient of 15 and a number is 4.

b. Three subtracted from 12 is a number.

c. Four times a number added to 17 is 21.

Solution: **a.**

The quotient of 15 and a number	is	4
$\frac{15}{x}$	$=$	4

b.

Three subtracted **from** 12	is	a number
$12 - 3$	$=$	x

Care must be taken when the operation is subtraction. The expression $3 - 12$ would be incorrect. Notice that $3 - 12 \neq 12 - 3$.

c.

4 times a number	added to	17	is	21
$4x$	$+$	17	$=$	21

■

EXERCISE SET 1.5

Evaluate each expression if $x = 1$, $y = 3$, and $z = 5$. See Example 1.

1. $3x - 2$
2. $6y - 8$
3. $|2x + 3y|$
4. $5z - 2y$
5. $xy + z$
6. $yz - x$
7. $x^2 + y^2$
8. $|y^2 + z^2|$
9. $|10y - 3z|$

Write each of the following phrases as an algebraic expression. Let x represent the unknown number. See Example 2.

10. Fifteen more than a number
11. One-half times a number
12. Five subtracted from a number
13. The quotient of a number and 9
14. Three times a number increased by 22
15. The product of 8 and a number

Decide whether the given number is a solution to the given equation. See Example 3.

16. $3x - 6 = 9$; 5
17. $2x + 7 = 3x$; 6
18. $2x + 6 = 5x - 1$; 0
19. $4x + 2 = x + 8$; 2
20. $x^2 + 2x + 1 = 0$; 1
21. $x^2 + 2x - 35 = 0$; 5

Write each of the following sentences as an equation. Use x to represent the unknown number. See Example 4.

22. The sum of 5 and a number is 20.
23. Twice a number is 17.
24. Thirteen minus three times a number is 13.
25. Seven subtracted from a number is 0.
26. The quotient of 12 and a number is $\frac{1}{2}$.
27. The sum of 8 and twice a number is 42.

Evaluate each expression if $x = 2$, $y = 6$, and $z = 3$.

28. $3x + 8y$
29. $|4z - 2y|$
30. $\frac{4x}{3y}$
31. $\frac{6z}{5x}$
32. $\frac{y}{x} + \frac{y}{x}$
33. $\frac{9}{z} + \frac{4z}{y}$
34. $3x^2 + y$
35. $y + 4z^2$
36. $x + yz$
37. $z + xy$

Find the value of each expression if $x = 12$, $y = 8$, and $z = 4$.

38. $\frac{x}{z} + 3y$
39. $\frac{y}{\sqrt{z}} + 8x$
40. $3z + z^2$
41. $7y + y^2$
42. $x^2 - 3y + x$
43. $y^2 - 3x + y$

44. $\dfrac{2x + \sqrt{z}}{3y - z}$

45. $\dfrac{3x - \sqrt[3]{y}}{4z + x}$

46. $\dfrac{x^2 + z}{y^2 + 2z}$

47. $\dfrac{y^2 + x}{x^2 + 3y}$

Decide whether the given number is a solution to the given equation.

48. $2x - 5 = 5; 8$

49. $3x - 10 = 8; 6$

50. $x + 6 = x + 6; 2$

51. $x + 6 = x + 6; 10$

52. $x = 5x + 15; 0$

53. $4 = 1 - x; 1$

Write each of the following as an algebraic expression or equation.

54. A number divided by 13

55. Seven times the sum of a number and 19

56. Ten subtracted from twice a number is 18.

57. A number subtracted from four times that number is 75.6.

58. Twenty less the product of 30 and a number

59. Twelve decreased by a number

60. A number times 0.02 equals 1.76.

61. The product of $\frac{3}{4}$ and the sum of a number and 1 equals 9.

62. A number subtracted from 19 is three times that number.

63. Three times a number added to $1\frac{11}{12}$ equals the number increased by 2.

Solve the following.

64. The perimeter of a figure is the distance around the figure. The equation $P = 2l + 2w$ can be used to find the perimeter, P, of a rectangle, where l is its length and w is its width. Find the perimeter of the following rectangle by substituting 8 for l and 6 for w.

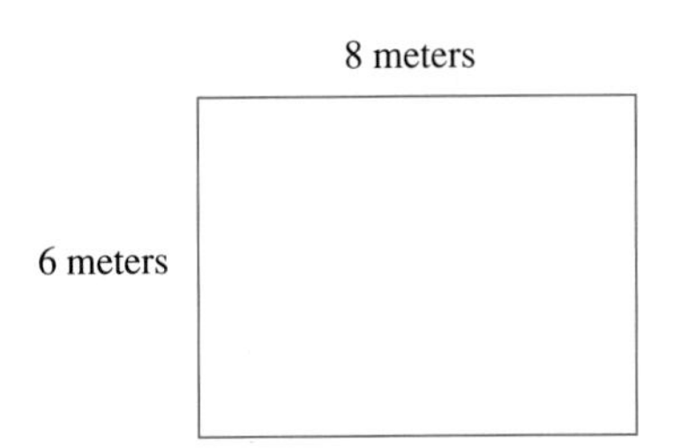

65. The equation $P = a + b + c$ can be used to find the perimeter, P, of a triangle, where a, b, and c are the lengths of its sides. Find the perimeter of the following triangle.

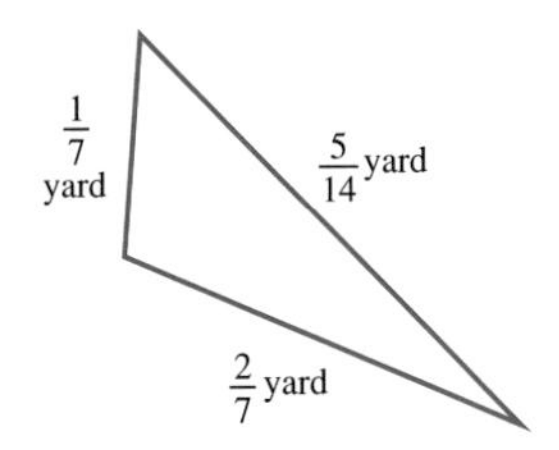

66. The **area** of a figure is the total enclosed surface of the figure. Area is measured in square units. The equation $A = lw$ is used to find the area of a rectangle, where l is its length and w is its width. Find the area of the following rectangular-shaped lot.

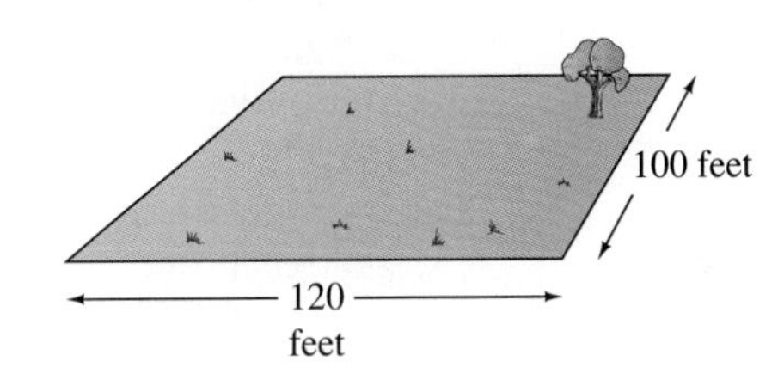

67. A trapezoid is a four-sided figure with exactly one pair of parallel sides. The equation $A = \dfrac{(B + b)h}{2}$ can be used to find its area, where B and b are the lengths of the two parallel sides and h is the height between these sides. Find the area if $B = 15$ inches, $b = 7$ inches, and $h = 5$ inches.

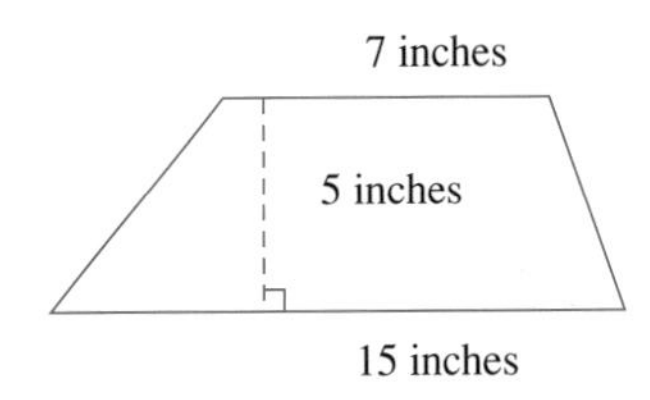

68. The equation $R = \frac{I}{PT}$ can be used to find the rate of interest being charged if a loan of P dollars for T years required I dollars in interest to be paid. Find the interest rate if a \$650 loan for 3 years to buy a used IBM personal computer requires \$126.75 in interest to be paid.

69. The equation $r = \frac{d}{t}$ is used to find the average speed r in miles per hour if a distance of d miles is traveled in t hours. Find the rate to the nearest whole number if the distance between Dallas, Texas, and Kaw City, Oklahoma, is 432 miles, and it takes Barbara Goss 8.5 hours to drive the distance.

70. Peter Callac earns a base salary plus a commission on all sales he makes at St. Joe Brick Company. The equation $I = B + RS$ calculates his gross income, where B is the base income, R is the commission rate, and S is the amount sold. Find Peter's income if $B = \$300$, $R = 8\%$, $S = \$500$. ($8\% = 0.08$.)

1.6
Adding and Subtracting Real Numbers

OBJECTIVES

Tape BA 3

1. Add real numbers with the same sign.
2. Add real numbers with unlike signs.
3. Find the opposite of a number.
4. Subtract real numbers.
5. Evaluate algebraic expressions, given replacement values for variables.

1 Real numbers can be added, subtracted, multiplied, divided, and raised to powers, just as whole numbers can. We will use the number line to help picture the addition of real numbers.

On a number line, a positive number can be represented anywhere by an arrow of appropriate length pointing to the right and a negative number by an arrow pointing to the left. To add $3 + 2$ on a number line, start at 0 on the number line and draw an arrow representing 3. This arrow will be three units long pointing to the right, since 3 is positive. From the tip of this arrow, draw another arrow representing 2. The tip of the second arrow ends at their sum, 5.

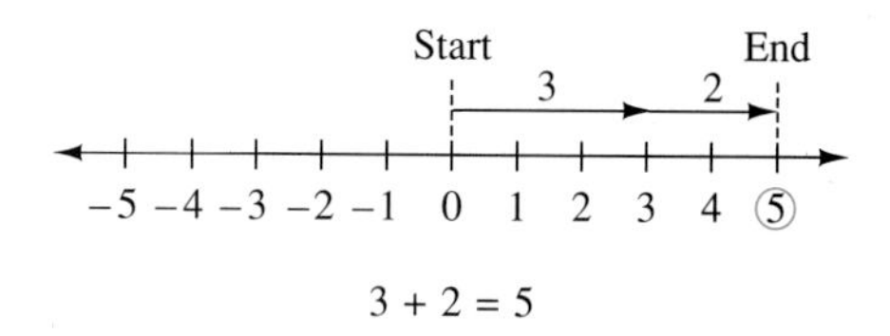

$3 + 2 = 5$

To add $-1 + (-2)$, start at 0 and draw an arrow representing -1. This arrow is one unit long pointing to the left. At the tip of this arrow, draw an arrow representing -2. The tip of the second arrow ends at their sum, -3.

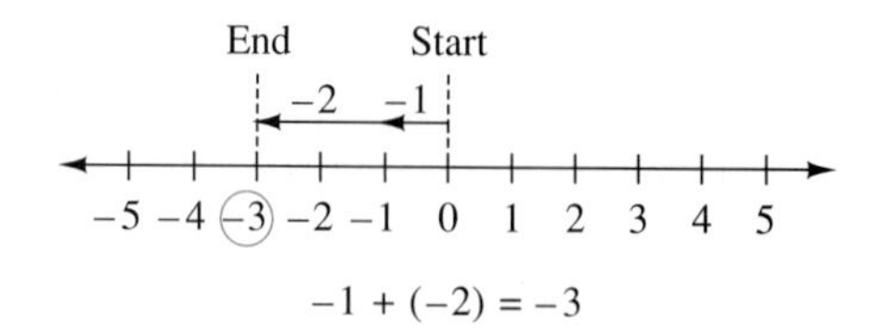

$-1 + (-2) = -3$

Using a number line each time we add two numbers can be time consuming. Instead, we can notice patterns in the previous examples and write rules for adding signed numbers. When adding two numbers with the same sign, notice that the sign of the sum is the same as the sign of the addends.

> **To Add Two Numbers with the Same Sign**
>
> *Step 1* Find the sum of their absolute values.
>
> *Step 2* Use their common sign as the sign of the sum.

Add $(-7) + (-6)$.

Step 1 Find the sum of their absolute values.

$$|-7| = 7, \quad |-6| = 6, \quad \text{and} \quad 7 + 6 = 13$$

Step 2 Use their common sign as the sign of the sum. This means that

$$(-7) + (-6) = -13$$

↑ common sign

Thinking of signed numbers as money earned or lost might help make this rule more meaningful. If \$1 is earned and later another \$3 is earned, the total amount earned is \$4. Earnings can be thought of as positive numbers: $1 + 3 = 4$.

On the other hand, if \$1 is lost and later another \$3 is lost, a total of \$4 is lost. Losses can be thought of as negative numbers: $(-1) + (-3) = -4$.

EXAMPLE 1 Find each sum.

a. $-3 + (-7)$ **b.** $5 + (+12)$ **c.** $(-1) + (-20)$ **d.** $-2 + (-10)$

Solution: **a.** $-3 + (-7) = -10$ **b.** $5 + (+12) = 17$ **c.** $(-1) + (-20) = -21$ **d.** $-2 + (-10) = -12$ ■

2 Adding numbers whose signs are not the same can be pictured on the number line also. To find the sum of $-4 + 6$, begin at 0 and draw an arrow representing -4. This arrow is 4 units long and points to the left. At the tip of this arrow, draw an arrow representing 6. The tip of the second arrow ends at their sum, 2.

Start
End
−4
6
−5 −4 −3 −2 −1 0 1 ② 3 4

$-4 + 6 = 2$

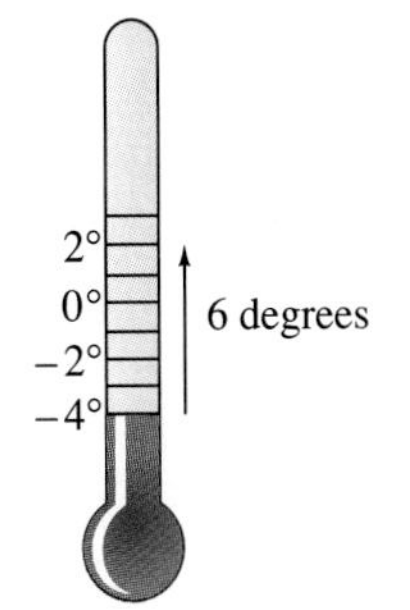

Using temperature as an example, if the thermometer registers 4 degrees below zero and then rises 6 degrees, the new temperature is 2 degrees above zero. Thus, it is reasonable that $-4 + 6 = 2$.

Find the sum of $3 + (-4)$ on the number line. From the number line, the answer is -1.

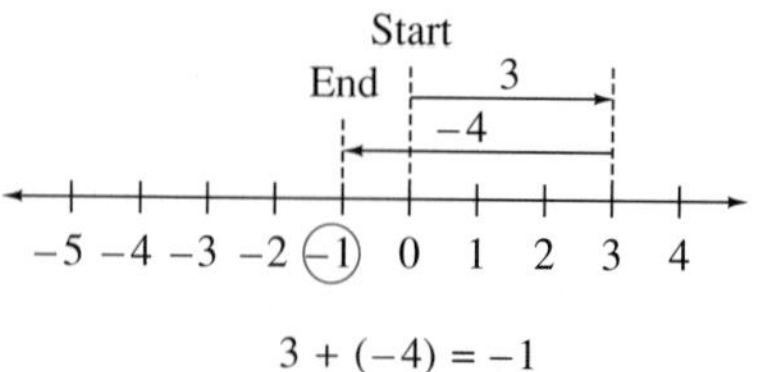

To Add Two Numbers with Unlike Signs

Step 1 Find the difference of the larger absolute value and the smaller absolute value.

Step 2 Use the sign of the addend whose absolute value is larger as the sign of the answer.

To add $4 + (-9)$, two numbers with unlike signs, first recall that $|4| = 4$ and $|-9| = 9$. Their difference is $9 - 4 = 5$. The sign of the sum $4 + (-9)$ will be negative since -9 has the larger absolute value. Therefore, $4 + (-9) = -5$.

EXAMPLE 2 Find each sum.

a. $(+3) + (-7)$ **b.** $(-2) + (10)$ **c.** $2 + (-5)$

Solution: **a.** Since $|-7|$ is larger than $|3|$, the sum will have the same sign as -7. That is, the sum is negative. Then $|-7| = 7$, $|3| = 3$, and $7 - 3 = 4$. The answer is -4.

$$(+3) + (-7) = -4$$

b. Since $|10|$ is larger than $|-2|$, the sign of the sum is the same as the sign of 10, positive. Then $|10| = 10$, $|-2| = 2$, and $10 - 2 = 8$. The answer is $+8$ or 8.

$$10 + (-2) = 8$$

c. $2 + (-5) = -3$ ■

EXAMPLE 3 Simplify the following:

a. $3 + (-7) + (-8)$ **b.** $[7 + (-10)] + [-2 + (-4)]$

Solution: **a.** Perform the additions from left to right.

$$3 + (-7) + (-8) = -4 + (-8) \quad \text{Adding unlike signs.}$$
$$= -12 \quad \text{Adding like signs.}$$

b. Simplify inside brackets first.

$$[7 + (-10)] + [-2 + (-4)] = [-3] + [-6]$$
$$= -9 \quad \text{Add.}$$ ■

3 The graph of 4 and -4 is shown on the number line below.

4 units 4 units

−5 −4 −3 −2 −1 0 1 2 3 4 5

Notice that 4 and -4 lie on opposite sides of 0, and each is 4 units away from 0.

This relationship between -4 and $+4$ is an important one. Such numbers are known as **opposites,** or **additive inverses,** of each other.

> **Opposites or Additive Inverses**
>
> Two numbers that are the same distance from 0 but on opposite sides of 0 are called opposites or additive inverses of each other.

The opposite of 10 is -10.

The opposite of -3 is 3.

The opposite of $\frac{1}{2}$ is $-\frac{1}{2}$.

Notice that the sum of a number and its opposite is 0.

$$10 + (-10) = 0$$

$$-3 + 3 = 0$$

$$\frac{1}{2} + \left(-\frac{1}{2}\right) = 0$$

In general, **if *a* is a number, the opposite or additive inverse of *a* is $-a$.** This means that the opposite of -3 is $-(-3)$. But we said above that the opposite of -3 is 3. This can only be true if $-(-3) = 3$.

> If a is a number, then $-(-a) = a$.

For example, $-(-10) = 10$, $-\left(-\frac{1}{2}\right) = \frac{1}{2}$, and $-(-2x) = 2x$.

EXAMPLE 4 Find the opposite or additive inverse of each number.

a. 5 **b.** 0 **c.** -6

Solution: **a.** The opposite of 5 is -5. Notice that 5 and -5 are on opposite sides of 0 when plotted on a number line and are equal distances away.

b. The opposite of 0 is 0 since $0 + 0 = 0$.

c. The opposite of -6 is 6. ■

EXAMPLE 5 Simplify the following:

a. $-(-6)$ **b.** $-|-6|$

Solution: **a.** $-(-6) = 6$ **b.** $-|-6| = -6$ ■

4 Now that addition of signed numbers has been discussed, we explore subtraction. We know that $9 - 7 = 2$. Notice that $9 + (-7) = 2$, also. This means that

$$9 - 7 = 9 + (-7)$$

In general, any subtraction problem can be written as an equivalent addition problem. To do so, we use an important relationship discussed earlier, that of opposites.

Subtracting Two Real Numbers

If a and b are real numbers, then $a - b = a + (-b)$.

In other words, to find the difference of two numbers, add the first number to the opposite of the second number.

EXAMPLE 6 Find each difference.

a. $-13 - (+4)$ **b.** $5 - (-6)$ **c.** $3 - 6$ **d.** $-1 - (-7)$

Solution:

a. $-13 - (+4) = -13 + (-4)$ Add the opposite of +4, which is −4.

(add: − → +; opposite: +4 → −4)

$= -17$

b. $5 - (-6) = 5 + (6)$ Add the opposite of −6, which is 6.

(add: − → +; opposite: −6 → 6)

$= 11$

c. $3 - 6 = 3 + (-6)$ Add 3 to the opposite of 6, which is −6.

$= -3$

d. $-1 - (-7) = -1 + (7) = 6$ ■

EXAMPLE 7 Subtract 8 from −4.

Solution: Be careful when interpreting. The order of numbers in subtraction is important. Here, 8 is to be subtracted **from** −4.

$$-4 - 8 = -4 + (-8) = -12$$ ■

Expressions containing both sums and differences make good use of grouping symbols. In simplifying these expressions, remember to rewrite differences as sums and follow the standard order of operations.

EXAMPLE 8 Simplify each expression.

a. $-3 + [(-2 - 5) - 2]$ **b.** $2^3 - |10| + [-6 - (-5)]$

Solution: **a.** Start with the innermost sets of parentheses. Rewrite $-2 - 5$ as a sum.

$-3 + [(-2 - 5) - 2] = -3 + [(-2 + (-5)) - 2]$

$= -3 + [(-7) - 2]$ Add $-2 + (-5)$.

$= -3 + [-7 + (-2)]$ Write $-7 - 2$ as a sum.

$= -3 + [-9]$ Add.

$= -12$ Add.

b. Start simplifying the expression inside the brackets by writing $-6 - (-5)$ as a sum.

$$2^3 - |10| + [-6 - (-5)] = 2^3 - |10| + [-6 + 5]$$

$= 2^3 - 10 + [-1]$ Add. Write $|10|$ as 10.

$= 8 - 10 + (-1)$ Evaluate 2^3.

$= 8 + (-10) + (-1)$ Write $8 - 10$ as a sum.

$= -2 + (-1)$ Add.

$= -3$ Add. ■

5 Next we practice evaluating algebraic expressions.

EXAMPLE 9 If $x = 2$ and $y = -5$, find the value of the following expressions.

a. $\dfrac{x - y}{12 + x}$ **b.** $x^2 - y$

Solution: **a.** Replace x with 2 and y with -5. Be sure to put parentheses around -5 to separate signs. Then simplify the resulting expression.

$$\frac{x - y}{12 + x} = \frac{2 - (-5)}{12 + 2} = \frac{2 + 5}{14} = \frac{7}{14} = \frac{1}{2}$$

b. Replace the x with 2 and y with -5 and simplify.

$$x^2 - y = 2^2 - (-5) = 4 - (-5) = 4 + 5 = 9$$ ■

Negative numbers are used in everyday life. Stock market returns show gains and losses as positive and negative numbers. Temperatures in cold climates often dip into the negative range, commonly referred to as "below zero" temperatures. Bank statements report deposits and withdrawals as positive and negative numbers and elevations below sea level are represented by negative numbers.

EXAMPLE 10 The lowest point in North America is in Death Valley, at an elevation of 282 feet below sea level. Nearby, Mount Whitney reaches 14,494 feet, the highest point in the United States outside of Alaska. How much of a variation in elevation is there between these two extremes?

Solution: To find the variation in elevation between the two heights, find the difference of the high point and the low point.

high point	minus	low point	
14,494	$-$	(-282)	$= 14{,}494 + 282$
			$= 14{,}776$ feet

Thus, the variation in elevation is 14,776 feet.

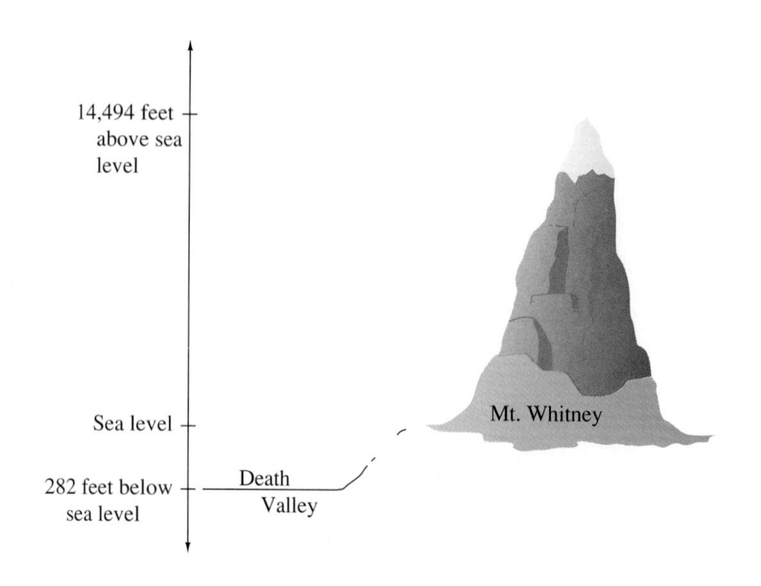

■

EXERCISE SET 1.6

Find the following sums. See Examples 1 through 3.

1. $6 + 3$
2. $9 + (-12)$
3. $-6 + (-8)$
4. $-6 + (-4)$
5. $-8 + (-7)$
6. $6 + (-4)$
7. $-52 + 36$
8. $-94 + 27$
9. $6 + (-4) + 9$
10. $-18 + |-53|$
11. $2\frac{3}{4} + \left(-\frac{1}{8}\right)$
12. $4\frac{4}{5} + \left(-\frac{3}{10}\right)$

Find the additive inverse or the opposite. See Example 4.

13. 6
14. 4
15. -2
16. -8
17. 0
18. $-\frac{1}{4}$
19. $|-6|$
20. $|-11|$

Simplify the following. See Example 5.

21. $-|-2|$
22. $-(-3)$
23. $-|0|$
24. $\left|-\frac{2}{3}\right|$
25. $-\left|-\frac{2}{3}\right|$
26. $-(-7)$

Find each difference. See Example 6.

27. $-6 - 4$
28. $-12 - 8$
29. $4 - 9$
30. $8 - 11$
31. $16 - (-3)$
32. $12 - (-5)$
33. $18 - 33$
34. $14 - 27$
35. $-16 - 18$

Find the following. See Example 7.

36. Subtract -5 from 8.
37. Subtract 3 from -2.
38. Subtract -1 from -6.

Simplify each expression. (Remember the order of operations.) See Example 8.

39. $-6 - (2 - 11)$
40. $-9 - (3 - 8)$
41. $3^3 - 8 \cdot 9$
42. $2^3 - 6 \cdot 3$
43. $2 - 3(8 - 6)$
44. $4 - 6(7 - 3)$
45. $|-3| + 2^2 + [-4 - (-6)]$
46. $|-2| + 6^2 + (-3 - 8)$

If $x = -6$, $y = -3$, and $t = 2$, evaluate each expression. See Example 9.

47. $x - y$
48. $3t - x$
49. $x + t - 12$
50. $y - 3t - 8$
51. $\dfrac{x - (-12)}{y + 6}$
52. $\dfrac{x - (-9)}{t + 1}$
53. $x - t^2$
54. $y - t^3$
55. $\dfrac{4t - y}{3t}$
56. $\dfrac{5t - x}{5t}$

Solve the following. See Example 10.

57. Within 24 hours in 1916, the temperature in Browning, Montana, fell from 44 degrees to −56 degrees. How large a drop in temperature was this?

58. On January 2, 1943, the temperature was −4 degrees at 7:30 A.M. in Spearfish, South Dakota. Incredibly, it got 49 degrees warmer in the next 2 minutes. To what temperature did it rise by 7:32?

59. In some card games, it is possible to have a negative score. Lavonne Davis currently has a score of 15 points. She then loses 24 points. What is her new score?

60. In a series of plays, a football team gains 2 yards, loses 5 yards, and then loses another 20 yards. What is the total gain or loss of yardage?

Simplify each expression.

61. $-6 - 5$
62. $-8 + 4$
63. $7 - (-4)$
64. $-3 + (-6)$
65. $-6 - (-11)$
66. $-4 - (-16)$
67. $-\dfrac{3}{5} - \dfrac{5}{6}$
68. $-\dfrac{3}{4} + \dfrac{1}{9}$
69. $-6.1 - (-5.3)$
70. $-2.6 - (-6.7)$
71. $6 - 11 + (-8)$
72. $4 - (-6) + 9$
73. $-6 - 14 - (-3)$
74. $-11 + (-8) - 4$
75. $-13 + 27$
76. $16 + (-21)$
77. $15 - (-33)$
78. $-44 - (-27)$
79. $-36 - (-51)$
80. $4\dfrac{2}{3} - \dfrac{5}{12}$
81. $-\dfrac{1}{6} - \left(-\dfrac{3}{4}\right)$
82. $-\dfrac{1}{10} + \left(-\dfrac{7}{8}\right)$
83. $8.3 - 11.2$
84. $9.7 - 16.1$
85. $3 - (6[3 - (-2)] - 8)$
86. $4 - (3[9 - (-8)] - 11)$
87. $3 - 6 \cdot 4^2$
88. $2 - 3 \cdot 5^2$
89. $(3 - 6) \cdot 4^2$
90. $(2 - 3) \cdot 5^2$
91. $3 - (6 \cdot 4)^2$
92. $2 - (3 \cdot 5)^2$
93. Subtract 8 from 7.
94. Subtract 9 from −4.
95. Decrease −8 by 15.
96. Decrease 11 by −14.

If $x = -5$, $y = 4$, and $t = 10$, evaluate each expression.

97. $x - y$
98. $y - x$
99. $|x| + 2t - 8y$
100. $|x + t - 7y|$
101. $\dfrac{9 - x}{y + 6}$
102. $\dfrac{15 - x}{y + 2}$
103. $y^2 + x$
104. $t^2 + x$

The following bar graph from an earlier section shows the minimum daily temperatures for a week in Sioux Falls, South Dakota.

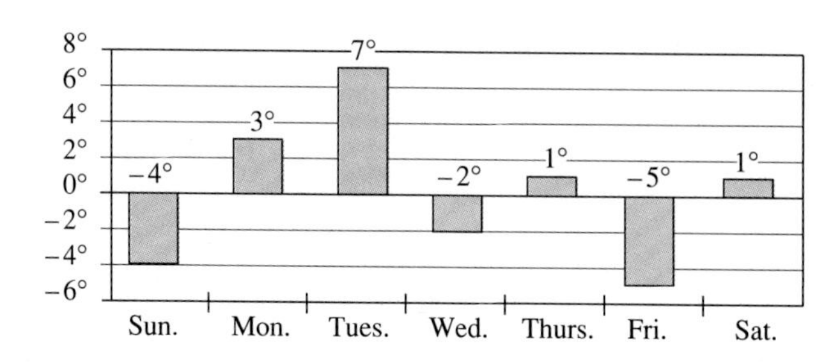

105. Use the bar graph to record the daily increases and decreases in the temperatures.

Day	Daily Increase or Decrease
Monday Tuesday Wednesday Thursday Friday Saturday	

106. Which day of the week had the greatest increase in temperature?

107. Which day of the week had the greatest decrease in temperature?

Decide whether the given number is a solution to the given equation.

108. $x + 9 = 5; -4$

109. $x - 10 = -7; 3$

110. $-x + 6 = -x - 1; -2$

111. $-x + 6 = -x - 1; -10$

112. $-x - 13 = -15; 2$

113. $4 = 1 - x; 5$

114. Augustus Caesar died in A.D. 14 in his 77th year. When was he born?

115. A jet liner hits an air pocket and drops 250 feet. After climbing 120 feet, it drops another 178 feet. What is its overall vertical change?

116. In golf, scores that are under par for the entire round are shown as negative scores; positive scores are shown for scores that are over par. In two rounds in the United States Open, Arnold Palmer had scores of −6 and +4. What was his overall score?

117. The lowest point in Africa is −512 feet at Lake Assal in Djibouti. If we are standing at a point 658 feet above Lake Assal, what is our elevation?

118. Yesterday our stock posted a change of $-1\frac{1}{4}$, but today it showed a gain of $+\frac{7}{8}$. Find the overall change for the two days.

119. In checking the stock market results, Alexis Maritas discovers our stock posted changes of $-1\frac{5}{8}$ and $-2\frac{1}{2}$ over the last two days. What is the combined change?

Writing in Mathematics

120. In your own words, what is an opposite?

121. Explain why 0 is the only number that is its own opposite.

122. Explain why adding a negative number to another negative number produces a negative sum.

123. When a positive and a negative number are added, sometimes the answer is positive and sometimes it is negative. Explain why this happens.

1.7 Multiplying and Dividing Real Numbers

OBJECTIVES

1 Multiply and divide real numbers.

2 Find the value of algebraic expressions.

Tape BA 4

1 In this section, we discover rules for multiplying and dividing signed numbers. The sign rules for multiplying and dividing these numbers will be the same since a quotient can be rewritten as a product.

For example, $\frac{10}{2}$ can be rewritten as $10 \cdot \frac{1}{2}$. In general, we have the following:

> If a and b are real numbers and $b \neq 0$, then
> $$\frac{a}{b} = a \cdot \frac{1}{b}$$

To discover sign rules for multiplication (as well as division), recall that multiplication is repeated addition. Thus $3 \cdot 2$ means that 2 is an addend 3 times. That is,

$$2 + 2 + 2 = 3 \cdot 2$$

which equals 6. Let's use this fact to help us discover a rule for multiplying a positive number and a negative number. To do this, apply the same reasoning above to -2 as an addend 3 times. That is,

$$(-2) + (-2) + (-2) = 3 \cdot (-2)$$

Since $(-2) + (-2) + (-2) = -6$, then $3 \cdot (-2) = -6$. This suggests that the product or quotient of a positive number and a negative number is a negative number.

What about the product of two negative numbers? To find out, consider the following pattern.

$$\left.\begin{aligned} -3 \cdot 2 &= -6 \\ -3 \cdot 1 &= -3 \\ -3 \cdot 0 &= 0 \end{aligned}\right\}$$

Factor decreases by 1 each time

Product increases by 3 each time.

This pattern continues as

$$-3 \cdot -1 = 3$$
$$-3 \cdot -2 = 6$$

This suggests that the product or quotient of two negative numbers gives a positive number.

The rules for multiplying and dividing signed numbers are summarized in the following box.

Multiplying and Dividing Real Numbers

1. The product or quotient of two numbers having the same sign is a positive number.
2. The product or quotient of two numbers having unlike signs is a negative number.

EXAMPLE 1 Find the product.

a. $(-6)(4)$ **b.** $2(-1)$ **c.** $(-5)(-10)$

Solution: **a.** $(-6)(4) = -24$ **b.** $2(-1) = -2$ **c.** $(-5)(-10) = 50$ ■

We know that every whole number multiplied by zero equals zero. This remains true for signed numbers.

Multiplying by Zero

If b is a real number, then $b \cdot 0 = 0$. Also, $0 \cdot b = 0$.

EXAMPLE 2 Simplify each expression.

a. $(7)(0)(-6)$ **b.** $(-2)(-3)(-4)$ **c.** $(-1)(5)(-9)$ **d.** $(-2)^3$

e. $(-4)(-11) - (5)(-2)$

Solution: **a.** By the order of operations, we multiply from left to right. Notice that, because one of the factors is 0, the product is 0.

$$(7)(0)(-6) = 0(-6) = 0$$

b. Multiply two factors at a time, from left to right.

$$(-2)(-3)(-4) = (6)(-4) \quad \text{Multiply } (-2)(-3).$$
$$= -24$$

c. Multiply from left to right.

$$(-1)(5)(-9) = (-5)(-9) \quad \text{Multiply } (-1)(5).$$
$$= 45$$

d. $(-2)^3 = (-2)(-2)(-2)$

$= -8$ Multiply.

e. Follow the rules for order of operation.

$$(-4)(-11) - (5)(-2) = 44 - (-10) \quad \text{Find the products.}$$
$$= 44 + 10 \quad \text{Add 44 to the opposite of } -10.$$
$$= 54 \quad \text{Add.}$$ ■

Multiplying signed decimals or fractions is carried out exactly the same way as multiplying by integers.

EXAMPLE 3 Find each product.

a. $(-1.2)(0.05)$ **b.** $\frac{2}{3} \cdot \left(-\frac{7}{10}\right)$

Solution: **a.** The product of two numbers with opposite signs is negative.

$$(-1.2)(0.05) = -[(1.2)(0.05)]$$
$$= -0.06$$

b. $\frac{2}{3} \cdot \left(-\frac{7}{10}\right) = -\frac{2 \cdot 7}{3 \cdot 10} = -\frac{2 \cdot 7}{3 \cdot 2 \cdot 5} = -\frac{7}{15}$ ■

EXAMPLE 4 Find each quotient.

a. $-18 \div 3$ **b.** $-14 \div -2$ **c.** $20 \div -4$

Solution: **a.** $\frac{-18}{3} = -6$ **b.** $\frac{-14}{-2} = 7$ **c.** $\frac{20}{-4} = -5$ ■

The definition of division shown earlier does not allow for division by 0. How do we interpret $\frac{3}{0}$? If $\frac{3}{0} = a$, then $a \cdot 0$ must equal 3. But recall that, for any real number a, $a \cdot 0 = 0$ and can never equal 3. Thus, we say that division by 0 is not allowed or not defined since $\frac{3}{0}$ does not represent a real number. The denominator of a fraction cannot be 0.

Can the numerator of a fraction be 0? For example, can we simplify a fraction such as $\frac{0}{3}$? Once again, recall that if $\frac{0}{3} = a$, then $a \cdot 3 = 0$. This is true only if a is 0. Thus, $\frac{0}{3} = 0$. In general, we have the following:

Quotients with Zero

1. Division of any nonzero real number by 0 is undefined. In symbols, if $a \neq 0$, $\frac{a}{0}$ is **undefined.**
2. Division of 0 by any real number except 0 is 0. In symbols, if $a \neq 0$, $\frac{0}{a} = 0$.

EXAMPLE 5 Simplify the following if possible.

a. $\frac{1}{0}$ **b.** $\frac{0}{-3}$ **c.** $\frac{0(-8)}{2}$

Solution: **a.** $\frac{1}{0}$ is undefined **b.** $\frac{0}{-3} = 0$ **c.** $\frac{0(-8)}{2} = \frac{0}{2} = 0$ ■

Notice that $\frac{12}{-2} = -6$, $-\frac{12}{2} = -6$, and $\frac{-12}{2} = -6$. This means that $\frac{12}{-2} = -\frac{12}{2} = \frac{-12}{2}$.

In other words, a single negative sign in a fraction can be written in the denominator, in the numerator, or in front of the fraction without changing the value of the fraction. Thus,

$$\frac{1}{-7} = \frac{-1}{7} = -\frac{1}{7}$$

In general, if a and b are real numbers, $b \neq 0$, $\frac{a}{-b} = \frac{-a}{b} = -\frac{a}{b}$.

Examples combining the four basic arithmetic procedures along with the principles of order of operations will help us to review these concepts.

EXAMPLE 6 Simplify each expression.

a. $\frac{(-12)(-3) + 4}{-7 - (-2)}$

b. $\frac{2(-3)^2 - 20}{-5 + 4}$

Solution: **a.** First simplify the numerator and denominator separately.

$$\frac{(-12)(-3) + 4}{-7 - (-2)} = \frac{36 + 4}{-7 + 2}$$

$$= \frac{40}{-5}$$

$$= -8 \quad \text{Divide.}$$

b. Simplify the numerator and denominator separately; then divide.

$$\frac{2(-3)^2 - 20}{-5 + 4} = \frac{2 \cdot 9 - 20}{-5 + 4} = \frac{18 - 20}{-5 + 4} = \frac{-2}{-1} = 2$$ ■

2 Next we find the value of algebraic expressions.

EXAMPLE 7 If $x = -2$ and $y = -4$, find the value of each expression.

a. $5x - y$

b. $x^3 - y^2$

c. $\frac{3x}{2y}$

Solution: **a.** Replace x with -2 and y with -4 and simplify.

$$5x - y = 5(-2) - (-4) = -10 - (-4) = -10 + 4 = -6$$

b. Replace x with -2 and y with -4.

$x^3 - y^2 = (-2)^3 - (-4)^2$ Substitute the given values for the variables.

$= -8 - (16)$ Evaluate exponential expressions.

$= -8 + (-16)$ Write as an equivalent addition problem.

$= -24$ Add.

c. Replace x with -2 and y with -4 and simplify.

$$\frac{3x}{2y} = \frac{3(-2)}{2(-4)} = \frac{-6}{-8} = \frac{3}{4}$$

EXERCISE SET 1.7

Find the following products. See Examples 1 through 3.

1. $(-3)(+4)$
2. $(+8)(-2)$
3. $-6(-7)$
4. $(-3)(-8)$
5. $(-2)(-5)(0)$
6. $(7)(0)(-3)$
7. $2(-9)$
8. $(-5)(3)$
9. $\left(-\frac{3}{4}\right)\left(\frac{8}{9}\right)$
10. $\left(\frac{5}{6}\right)\left(-\frac{3}{10}\right)$
11. $\left(-1\frac{1}{5}\right)\left(-1\frac{2}{3}\right)$
12. $\left(-\frac{5}{6}\right)\left(-\frac{3}{10}\right)$
13. $(-1)(2)(-3)(-5)$
14. $(-2)(-3)(-4)(-2)$
15. $(2)(-1)(-3)(5)(3)$
16. $(3)(-5)(-2)(-1)(-2)$
17. $(-4)^2$
18. $(-3)^3$

Find the following quotients. See Examples 4 and 5.

19. $\frac{18}{-2}$
20. $-\frac{14}{7}$
21. $\frac{-12}{-4}$
22. $-\frac{20}{5}$
23. $\frac{-45}{-9}$
24. $\frac{30}{-2}$
25. $\frac{0}{-3}$
26. $-\frac{4}{0}$
27. $-\frac{3}{0}$
28. $\frac{0}{-4}$

Simplify the following. See Example 6.

29. $\frac{-6^2 + 4}{-2}$
30. $\frac{3^2 + 4}{5}$
31. $\frac{8 + (-4)^2}{4 - 12}$
32. $\frac{6 + (-2)^2}{4 - 9}$
33. $\frac{22 + (3)(-2)}{-5 - 2}$
34. $\frac{-20 + (-4)(3)}{1 - 5}$

If $x = -5$ and $y = -3$, evaluate each expression. See Example 7.

35. $3x + 2y$
36. $4x + 5y$
37. $2x^2 - y^2$
38. $x^2 - 2y^2$
39. $x^3 + 3y$
40. $y^3 + 3x$
41. $\frac{2x - 5}{y - 2}$
42. $\frac{2y - 12}{x - 4}$
43. $\frac{6 - y}{x - 4}$
44. $\frac{4 - 2x}{y + 3}$

Evaluate the following.

45. $(-6)(-2)$

46. $5(-3)$

47. $(-7)(2)$

48. $(-3)(-9)$

49. $\frac{18}{-3}$

50. $\frac{-16}{-4}$

51. $-\frac{6}{0}$

52. $-\frac{16}{2}$

53. $-\frac{15}{-3}$

54. $\frac{48}{-12}$

55. $\frac{0}{-7}$

56. $-\frac{48}{-8}$

57. $(-6)(3)(-2)(-1)$

58. $(-3)(-2)(-1)(-2)$

59. $(-5)^3$

60. $(-2)^5$

61. $(-4)^2$

62. $(-6)^2$

63. -4^2

64. -6^2

65. $\frac{-3 - 5^2}{2(-7)}$

66. $\frac{-2 - 4^2}{3(-6)}$

67. $\frac{6 - 2(-3)}{4 - 3(-2)}$

68. $\frac{8 - 3(-2)}{2 - 5(-4)}$

69. $\frac{-3 - 2(-9)}{-15 - 3(-4)}$

70. $\frac{-4 - 8(-2)}{-9 - 2(-3)}$

71. $-3(2 - 8)$

72. $-4(3 - 9)$

73. $6(3 - 8)$

74. $4(8 - 11)$

75. $-3[(2 - 8) - (-6 - 8)]$

76. $-2[(3 - 5) - (2 - 9)]$

77. $\left(\frac{2}{5}\right)\left(-1\frac{1}{4}\right)$

78. $\left(-4\frac{2}{3}\right)\left(-\frac{8}{21}\right)$

79. $-4\frac{1}{6} + \left(-\frac{1}{30}\right)$

80. $-3\frac{1}{5} + 5$

81. $(1.82)(-4.6)$

82. $(-3.6)(-0.61)$

83. $-22.4 \div (-1.6)$

84. $15.3 \div (-2.4)$

If $x = 2$ and $y = -6$, find the value of each expression.

85. $2x - 3y$

86. $-3x + 2y$

87. $2x^2 + y$

88. $2y^2 + x$

89. $\frac{xy}{-4}$

90. $\frac{-48}{2x - 4}$

91. $\frac{2x + 8}{y + 2}$

92. $\frac{3y - 6}{x - 4}$

93. $\frac{x - y}{3x + 6}$

94. $\frac{8 - y}{2x - 3}$

Write each of the following as an expression and evaluate.

95. Add -2 to the quotient of -15 and 3.

96. Add 1 to the product of -8 and -5.

97. Twice the sum of -5 and -3

98. Subtract 7 from the quotient of 0 and 5.

Decide whether the given number is a solution to the given equation.

99. $-5x = -35$; 7

100. $2x = x - 1$; -4

101. $-\frac{x}{10} = 2$; -20

102. $\frac{45}{x} = -15$; -3

103. $-3x - 5 = -20$; 5

104. $2x + 4 = x + 8$; -4

105. The following graph shows trader's stock consistently decreasing in value by $1.50 per share per day. If this trend continues, when will the stock be worth $20 per share?

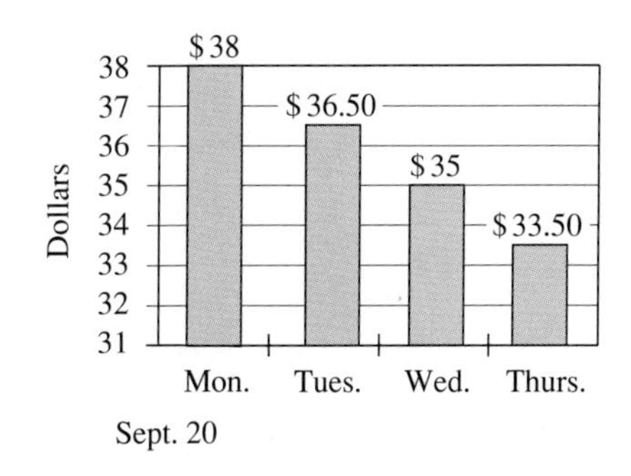

Writing in Mathematics

106. Explain why $\frac{6}{0}$ is undefined.

107. Why must the product of an even number of negative values be positive?

1.8 Properties of Real Numbers

OBJECTIVES

Tape BA 4

1. Identify the commutative property.
2. Identify the associative property.
3. Identify the distributive property.
4. Identify the additive and multiplicative identities.
5. Identify the additive and multiplicative inverse properties.

Specialized terms occur in every area of study. A biologist or ecologist will often be concerned with "symbiotic relationships." "Torque" is a major concern to a physicist. An economist or business analyst might be interested in "marginal revenue." A nurse needs to be familiar with "hemostasis." Mathematics, too, has its own specialized terms and principles.

This section introduces the basic properties of the real number system. Throughout this section, a, b, and c represent real numbers.

When we add, subtract, multiply, or divide two real numbers (except for division by zero), the result is a real number. This is guaranteed by the closure properties.

Closure Properties

If x and y are real numbers, then $x + y$, $x - y$, and xy are real numbers. Also, $\frac{x}{y}$, $y \neq 0$, is a real number.

1 Next we look at the commutative properties. These properties state that the order in which any two real numbers are added or multiplied does not change the value of the sum or product.

Commutative Properties

Addition: $a + b = b + a$

Multiplication: $a \cdot b = b \cdot a$

For example, if we let $a = 3$ and $b = 5$, then the commutative properties guarantee that

$$3 + 5 = 5 + 3 \quad \text{and} \quad 3 \cdot 5 = 5 \cdot 3$$

HELPFUL HINT

Is subtraction also commutative? Try an example. Is $3 - 2 = 2 - 3$? **No!** The left side of this statement equals 1; the right side equals -1. There is no commutative property of subtraction. Similarly, there is no commutative property for division. For example, $\frac{10}{2}$ does not equal $\frac{2}{10}$.

EXAMPLE 1 If $a = -2$, and $b = 5$, show that

a. $a + b = b + a$ **b.** $a \cdot b = b \cdot a$

Solution: **a.** Replace a with -2 and b with 5. Then

$$a + b = b + a$$

becomes

$$-2 + 5 = 5 + (-2)$$

or

$$3 = 3$$

Since both sides represent the same number, $-2 + 5 = 5 + (-2)$ is a true statement.

b. Replace a with -2 and b with 5. Then

$$a \cdot b = b \cdot a$$

becomes

$$-2 \cdot 5 = 5 \cdot (-2)$$

or

$$-10 = -10$$

Since both sides represent the same number, the statement is true. ■

2 When adding or multiplying three numbers, does it matter how we group the numbers? This question is answered by the associative properties. These properties state that when adding or multiplying three numbers, any two adjacent numbers may be grouped together without changing the answer.

Associative Properties

Addition: $(a + b) + c = a + (b + c)$

Multiplication: $(a \cdot b) \cdot c = a \cdot (b \cdot c)$

Illustrate these properties by working the following example.

EXAMPLE 2 If $a = -3$, $b = 2$, and $c = 4$, show that

a. $(a + b) + c = a + (b + c)$ **b.** $(a \cdot b) \cdot c = a \cdot (b \cdot c)$

Solution: Replace a with -3, b with 2, and c with 4.

a. $(a + b) + c = a + (b + c)$

$(-3 + 2) + 4 = -3 + (2 + 4)$ Replace a with -3, b with 2, and c with 4.

$-1 + 4 = -3 + 6$ Simplify inside parentheses.

$3 = 3$ Add.

b. $(a \cdot b) \cdot c = a \cdot (b \cdot c)$

$(-3 \cdot 2) \cdot 4 = -3 \cdot (2 \cdot 4)$ Replace a with -3, b with 2, and c with 4.

$-6 \cdot 4 = -3 \cdot 8$ Simplify inside parentheses.

$-24 = -24$ Multiply. ■

3 The **distributive property of multiplication over addition** is used repeatedly throughout algebra. It is useful because it allows us to write a product as a sum or a sum as a product.

Distributive Property of Multiplication over Addition

$$a(b + c) = ab + ac$$

Since multiplication is commutative, the distributive property can also be written

$$(b + c)a = ba + ca$$

The truth of this property can be illustrated by letting $a = 3$, $b = 2$, and $c = 5$. Then

$$a(b + c) = ab + ac$$

becomes

$$3(2 + 5) = 3 \cdot 2 + 3 \cdot 5$$

$$3 \cdot 7 = 6 + 15$$

$$21 = 21$$

Notice in this example that 3 is "being distributed to" each addend inside the parentheses. That is, 3 is multiplied by each addend.

The distributive property can be extended so that factors can be distributed to more than two addends in parentheses. For example,

$$3(x + y + 2) = 3(x) + 3(y) + 3(2)$$
$$= 3x + 3y + 6$$

EXAMPLE 3 Use the distributive property to write each expression without parentheses.

a. $2(x + y)$ **b.** $-5(-3 + z)$ **c.** $5(x + y - z)$ **d.** $-1(2 - y)$
e. $-(3 + x - w)$

Solution: **a.** $2(x + y) = 2 \cdot x + 2 \cdot y$
$= 2x + 2y$

b. $-5(-3 + z) = -5(-3) + (-5) \cdot z$
$= 15 - 5z$

c. $5(x + y - z) = 5 \cdot x + 5 \cdot y + 5(-z)$
$= 5x + 5y - 5z$

d. $-1(2 - y) = (-1)(2) + (-1)(-y)$
$= -2 + y$

e. $-(3 + x - w) = -1(3 + x - w)$
$= (-1)(3) + (-1)(x) + (-1)(-w)$
$= -3 - x + w$ ■

Notice in the last example that $-(3 + x - w)$ is rewritten as $-\mathbf{1}(3 + x - w)$.

4 Next we look at the **identity properties.** These properties guarantee that two special numbers exist. These numbers are called the **identity element for addition** and the **identity element for multiplication.**

Identities for Addition and Multiplication

0 is the identity element for addition.

$$a + 0 = 0 + a = a$$

1 is the identity element for multiplication.

$$a \cdot 1 = 1 \cdot a = a$$

Notice that 0 is the only number that can be added to any real number with the result that the sum is the same real number. Also, 1 is the only number that can be multiplied by any other real number with the result that the product is the same real number.

5 We were introduced to **additive inverses** or **opposites** in Section 1.6. Two numbers are called additive inverses or opposites if their sum is 0. The additive inverse or opposite of 6 is -6 because $6 + (-6) = 0$. The additive inverse or opposite of -5 is 5 because $-5 + 5 = 0$.

Reciprocals were first introduced in Section 1.3. Another name for reciprocals is **multiplicative inverses.** Two nonzero numbers are called reciprocals or multiplicative inverses if their product is 1. The reciprocal or multiplicative inverse of $\frac{2}{3}$ is $\frac{3}{2}$ because $\frac{2}{3} \cdot \frac{3}{2} = 1$. Likewise, the reciprocal of -5 is $-\frac{1}{5}$ because $-5\left(-\frac{1}{5}\right) = 1$.

> **Additive and Multiplicative Inverses**
>
> The numbers a and $-a$ are additive inverses or opposites of each other because their sum is 0; that is,
>
> $$a + (-a) = 0$$
>
> The numbers b and $\frac{1}{b}$ (for $b \neq 0$) are called reciprocals or multiplicative inverses of each other because their product is 1; that is,
>
> $$b \cdot \frac{1}{b} = 1$$

EXAMPLE 4 Find the additive inverse or opposite of each number.

a. -3 **b.** 5 **c.** 0

Solution: **a.** The additive inverse of -3 is 3 because $-3 + 3 = 0$.

b. The additive inverse of 5 is -5 because $5 + (-5) = 0$.

c. The additive inverse of 0 is 0 because $0 + 0 = 0$. ■

EXAMPLE 5 Find the multiplicative inverse or reciprocal of each number.

a. 7 **b.** $\frac{-1}{9}$

Solution: **a.** The multiplicative inverse of 7 is $\frac{1}{7}$ because $7 \cdot \frac{1}{7} = 1$.

b. The multiplicative inverse of $\frac{-1}{9}$ is $\frac{9}{-1}$, or -9, because $\left(\frac{-1}{9}\right)(-9) = 1$. ■

EXAMPLE 6 Name the property illustrated.

a. $2 \cdot 3 = 3 \cdot 2$ **b.** $3(x + 5) = 3x + 15$ **c.** $2 + (4 + 8) = (2 + 4) + 8$

Solution: **a.** The commutative property of multiplication

b. The distributive property

c. The associative property of addition ■

EXAMPLE 7 Use the indicated property to write each expression in an equivalent form.

a. $2 + 9$; the commutative property of addition

b. $(5 \cdot 8) \cdot 9$; the associative property of multiplication

c. $x + 0$; the additive identity property

Solution: **a.** $2 + 9 = 9 + 2$ **b.** $(5 \cdot 8) \cdot 9 = 5 \cdot (8 \cdot 9)$ **c.** $x + 0 = x$ ■

EXERCISE SET 1.8

Name the properties illustrated by each of the following. See Examples 1, 2, and 6.

1. $3 \cdot 5 = 5 \cdot 3$

2. $4(3 + 8) = 4 \cdot 3 + 4 \cdot 8$

3. $2 + (8 + 5) = (2 + 8) + 5$

4. $4 + 9 = 9 + 4$

5. $9(3 + 7) = 9 \cdot 3 + 9 \cdot 7$

6. $1 \cdot 9 = 9$

Use the distributive property to write each expression without parentheses. See Example 3.

7. $3(6 + x)$

8. $2(x - 5)$

9. $-2(y - z)$

10. $-3(z - y)$

11. $-7(3y - 5)$

12. $-5(2r + 11)$

Find the additive inverse or opposite of each of the following numbers. See Example 4.

13. 16

14. 14

15. -8

16. -3

17. $|-9|$

18. $|11|$

Find the multiplicative inverse or reciprocal of each of the following numbers. See Example 5.

19. $\frac{2}{3}$

20. $\frac{3}{4}$

21. $-\frac{5}{6}$

22. $-\frac{7}{8}$

23. 6

24. 3

25. -2

26. -5

Use the indicated property to write each expression in an equivalent form. See Example 7.

27. $\frac{2}{3} \cdot \frac{3}{2}$; multiplicative inverse property

28. $8 + 16$; commutative property of addition

29. $(-4)(-3)$; commutative property of multiplication

30. $-4 + 4$; additive inverse property

31. $3 + (8 + 9)$; associative property of addition

32. $(4 \cdot 3) \cdot 9$; associative property of multiplication

Name the properties illustrated by each of the following.

33. $(4 \cdot 8) \cdot 9 = 4 \cdot (8 \cdot 9)$

34. $6 \cdot \frac{1}{6} = 1$

35. $0 + 6 = 6$

36. $(4 + 9) + 6 = 4 + (9 + 6)$

37. $-4(3 + 7) = -4 \cdot 3 + (-4) \cdot 7$

38. $11 + 6 = 6 + 11$

39. $-4 \cdot (8 \cdot 3) = (-4 \cdot 8) \cdot 3$

40. $10 + 0 = 10$

Use the distributive property to write each of the following without parentheses.

41. $5(x + 4m + 2)$

42. $8(3y + z - 6)$

43. $-4(1 - 2m + n)$

44. $-4(4 + 2p + 5)$

45. $-(5x + 2)$

46. $-(9r + 5)$

47. $-(r - 3 - 7p)$

48. $-(-q - 2 + 6r)$

Find the additive inverse or opposite of each of the following numbers.

49. 0

50. 7

51. $|2|$

52. $|-5|$

53. $|8|$

54. $|6|$

55. $-(-3)$

56. $-(-4)$

57. $-|-2|$

58. $-|-9|$

Find the multiplicative inverse or reciprocal of each of the following numbers.

59. $\frac{1}{5}$

60. $\frac{1}{8}$

61. $\frac{3}{9}$

62. $\frac{2}{8}$

63. -1

64. 1

65. 0

66. 100

67. $-\left|-\frac{3}{5}\right|$

68. $-\left|-\frac{2}{5}\right|$

69. $3\frac{5}{6}$

70. $2\frac{3}{5}$

Use the indicated property to write each expression in an equivalent form.

71. $y + 0$; additive identity property

72. $1 \cdot x$; multiplicative identity property

73. $x(a + b)$; distributive property

74. $(m + n)y$; distributive property

75. $a(b + c)$; commutative property of multiplication

76. $x + 2$; commutative property of addition

Determine whether each pair of statements is commutative. In other words, is the result the same no matter what order these statements are performed?

77. Put on your left shoe. Put on your right shoe.

78. Put on your socks. Put on your shoes.

79. Bake a cake. Eat the cake.

80. Eat breakfast. Mow the lawn.

Writing in Mathematics

81. Use an example to show that division is not commutative.

82. Explain why 0 does not have a multiplicative inverse.

83. Define the identity for addition.

CHAPTER 1 GLOSSARY

The **absolute value** of a real number a, $|a|$, is the distance between a and 0 on the number line.

An **algebraic expression** is a collection of numbers, variables, operation symbols, and grouping symbols.

An **equation** states that two algebraic expressions are equal.

An **exponent** is a shorthand notation used for repeated multiplication of the same factor.

The **integers** are $\ldots, -3, -2, -1, 0, 1, 2, 3, \ldots$.

Irrational numbers are numbers represented by points on the number line that are not rational numbers. That is, irrational numbers cannot be expressed as a ratio of integers.

The **natural numbers** are $1, 2, 3, 4, 5, 6, \ldots$.

If the sum of two numbers is zero, the two numbers are said to be **opposites** or **additive inverses** of one another.

Rational numbers are numbers that can be expressed as a quotient $\frac{a}{b}$, where a and b are integers and $b \neq 0$.

Real numbers are rational numbers along with irrational numbers.

Two numbers are **reciprocals** or **multiplicative inverses** of each other if their product is 1.

A **solution** or **root** of an equation is a value for the variable that makes the equation true.

A letter that is used to represent a number is called a **variable.**

The **whole numbers** are $0, 1, 2, 3, 4, 5, \ldots$.

CHAPTER 1 SUMMARY

SYMBOLS USED TO COMPARE NUMBERS (1.1)

$=$	is equal to	$\neq$	is not equal to
$<$	is less than	$\leq$	is less than or equal to
$>$	is greater than	$\geq$	is greater than or equal to

MULTIPLYING AND DIVIDING FRACTIONS (1.3)

$$\frac{a}{b} \cdot \frac{c}{d} = \frac{a \cdot c}{b \cdot d}, \qquad b \neq 0, \quad d \neq 0$$

$$\frac{a}{b} \div \frac{c}{d} = \frac{a}{b} \cdot \frac{d}{c} = \frac{a \cdot d}{b \cdot c}, \qquad b \neq 0, \quad d \neq 0, \quad c \neq 0$$

ADDING AND SUBTRACTING FRACTIONS (1.3)

$$\frac{a}{b} + \frac{c}{b} = \frac{a + c}{b}, \qquad b \neq 0$$

$$\frac{a}{b} - \frac{c}{b} = \frac{a - c}{b}, \qquad b \neq 0$$

ORDER OF OPERATIONS (1.4)

Simplify expressions in the order below. If grouping symbols such as parentheses are present, simplify expressions within those first; start with the innermost set. If fraction bars are present, simplify the numerator and the denominator separately.

Step 1 Simplify exponential expressions.
Step 2 Perform multiplications or divisions in order from left to right.
Step 3 Perform additions or subtractions in order from left to right.

ADDING TWO NUMBERS WITH THE SAME SIGN (1.6)

To add two numbers with the same sign, add the absolute values of the two numbers and use their common sign as the sign of the sum.

ADDING TWO NUMBERS WITH UNLIKE SIGNS (1.6)

To add two numbers with unlike signs, find the difference of the larger absolute value and the smaller absolute value. The sign of the answer will be the sign of the number with the larger absolute value.

SUBTRACTING TWO REAL NUMBERS (1.6)

If a and b are real numbers, then $a - b = a + (-b)$.

MULTIPLYING AND DIVIDING REAL NUMBERS (1.7)

The product or quotient of two numbers having the same sign is a positive number, and the product or quotient of two numbers having unlike signs is a negative number.

MULTIPLYING BY ZERO (1.7)

If a is a real number, then $a \cdot 0 = 0$.

QUOTIENTS WITH ZERO (1.7)

$\frac{a}{0}$ is undefined, $a \neq 0$. Also, $\frac{0}{a} = 0$, $a \neq 0$.

COMMUTATIVE PROPERTIES (1.8)

$$a + b = b + a$$

$$a \cdot b = b \cdot a$$

ASSOCIATIVE PROPERTIES (1.8)

$$(a + b) + c = a + (b + c)$$

$$(a \cdot b) \cdot c = a \cdot (b \cdot c)$$

DISTRIBUTIVE PROPERTY (1.8)

$$a(b + c) = ab + ac$$

CHAPTER 1 REVIEW

(1.1) *Insert <, >, or = in the appropriate space to make the following statements true.*

1. 2 6 **2.** 1.2 0.951 **3.** 10.6 10.6 **4.** 0.07 0.7

Translate each statement into symbols.

5. 4 is greater than or equal to 3.

6. 6 is not equal to 5.

7. The sum of 8 and 4 is less than or equal to 12.

8. 32 is greater than three squared.

9. 7 is equal to 3 plus 4.

10. 0.03 is less than 0.3.

11. Lions and hyenas were featured in the Disney film *The Lion King*. For short distances, lions can run at a rate of 50 miles whereas hyenas can run at a rate of 40 miles per hour. Write an inequality statement comparing the numbers 50 and 40.

(1.2) *Given the following sets of numbers, list the numbers in each set that also belong to the set of:*

a. Natural numbers **b.** Whole numbers **c.** Integers

d. Rational numbers **e.** Irrational numbers **f.** Real numbers

12. $\left\{-6, 0, 1, 1\frac{1}{2}, 3, \pi, 9.62\right\}$

13. $\left\{-3, -1.6, 2, 5, \frac{11}{2}, 15.1, \sqrt{5}, 2\pi\right\}$

Insert <, >, or = in the appropriate space to make the statement true.

14. $-4 \quad -5$ **15.** $6 \quad -8$ **16.** $-2 \quad 2$ **17.** $-\frac{3}{2} \quad -\frac{3}{4}$

The following chart shows the gains and losses in dollars of Density Oil and Gas stock for a particular week.

Day	Gain or Loss in Dollars
Monday	+1
Tuesday	−2
Wednesday	+5
Thursday	+1
Friday	−4

18. Which day showed the greatest loss?

19. Which day showed the greatest gain?

(1.3) *Write the number as a product of prime factors.*

20. 36

21. 120

Perform the indicated operations. Write answers in lowest terms.

22. $\frac{8}{15} \cdot \frac{27}{30}$ **23.** $\frac{7}{8} \div \frac{21}{32}$ **24.** $\frac{7}{15} + \frac{5}{6}$ **25.** $\frac{3}{4} - \frac{3}{20}$

26. $2\frac{3}{4} + 6\frac{5}{8}$ **27.** $7\frac{1}{6} - 2\frac{2}{3}$

28. Determine the unknown part of the given circle.

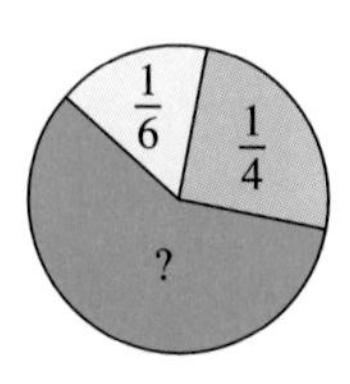

(1.4) *Simplify each expression.*

29. $6 \cdot 3^2 + 2 \cdot 8$ **30.** $24 - 5 \cdot 2^3$ **31.** $3(1 + 2 \cdot 5) + 4$

32. $\sqrt[3]{8} + 3(2 \cdot 6 - 1)$ **33.** $\frac{\sqrt{4} + |6 - 2| + 8^2}{4 + 6 \cdot 4}$ **34.** $5[3(2 - 5) - 5]$

(1.5) *Evaluate each expression if $x = 6$, $y = 2$, and $z = 8$.*

35. $2x + 3y$ **36.** $x(y + 2z)$ **37.** $\frac{x}{y} + \frac{z}{2y}$ **38.** $x^2 - 3y^2$

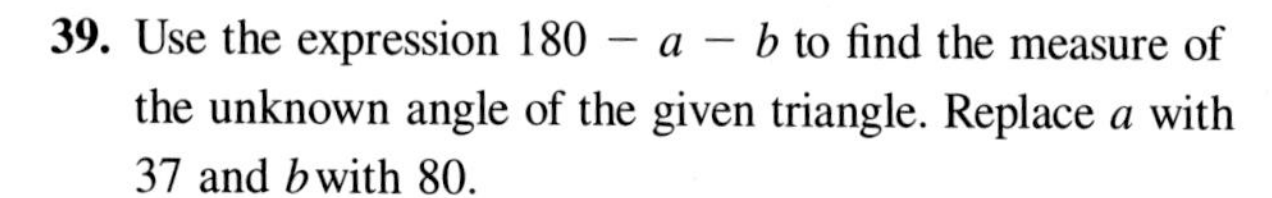
39. Use the expression $180 - a - b$ to find the measure of the unknown angle of the given triangle. Replace a with 37 and b with 80.

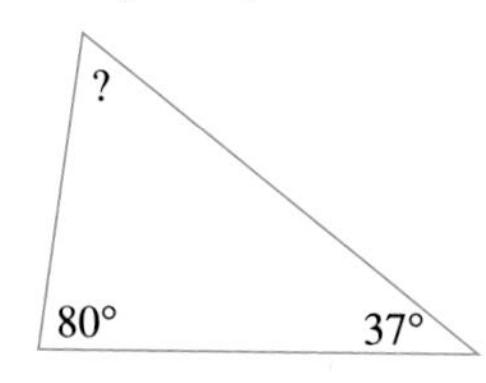

(1.6) *Find the additive inverse or the opposite.*

40. -9 **41.** -3

Find the following sums.

42. $-15 + 4$ **43.** $-6 + (-11)$ **44.** $16 + (-4)$

45. $-8 + |-3|$ **46.** $-4.6 + (-9.3)$ **47.** $-2.8 + 6.7$

Simplify the expression.

48. $6 - 20$ **49.** $-3 - 8$ **50.** $-6 - (-11)$

51. $4 - 15$ **52.** $-21 - 16 + 3(8 - 2)$ **53.** $\frac{11 - (-9) + 6(8 - 2)}{2 + 3 \cdot 4}$

If $x = 3$, $y = -6$, and $z = -9$, evaluate each expression.

54. $2x^2 - y + z$ **55.** $\frac{y - x + 5x}{2x}$

56. At the beginning of the week, the price of Density Oil and Gas stock from exercises 18 and 19 is \$50 per share. Find the price of a share of stock at the end of the week.

(1.7) *Simplify each expression.*

57. $6(-8)$

58. $(-2)(-14)$

59. $\dfrac{-18}{-6}$

60. $\dfrac{42}{-3}$

61. $-3(-6)(-2)$

62. $(-4)(-3)(0)(-6)$

63. $\dfrac{4 \cdot (-3) + (-8)}{2 + (-2)}$

64. $\dfrac{3(-2)^2 - 5}{-14}$

(1.8) *Find the additive inverse or opposite.*

65. -6

66. $-|-7|$

Find the multiplicative inverse or reciprocal.

67. -6

68. $\dfrac{3}{5}$

Name the property illustrated.

69. $-6 + 5 = 5 + (-6)$

70. $6 \cdot 1 = 6$

71. $3(8 - 5) = 3 \cdot 8 + 3 \cdot (-5)$

72. $4 + (-4) = 0$

73. $2 + (3 + 9) = (2 + 3) + 9$

74. $2 \cdot 8 = 8 \cdot 2$

75. $6(8 + 5) = 6 \cdot 8 + 6 \cdot 5$

76. $(3 \cdot 8) \cdot 4 = 3 \cdot (8 \cdot 4)$

77. $4 \cdot \dfrac{1}{4} = 1$

78. $8 + 0 = 8$

79. $4(8 + 3) = 4(3 + 8)$

80. The following circle graph shows the percent distribution of municipal solid waste generated. Find the percent of waste that comes from food.

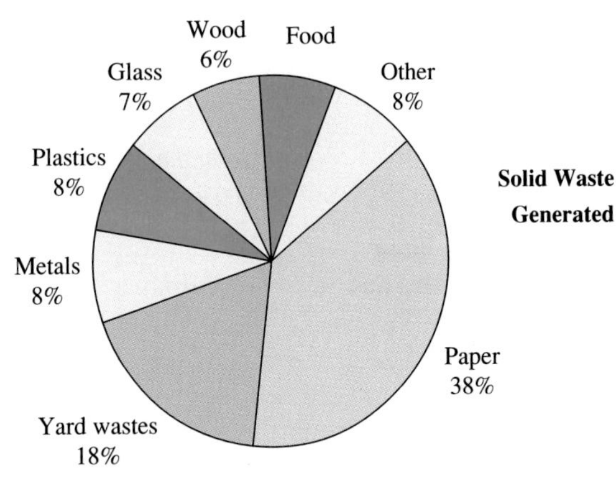

Source: U.S. Bureau of the Census, *Statistical Abstract of the United States: 1993,* 113th ed., Washington, D.C., 1994.

CHAPTER 1 TEST

Translate the statement into symbols.

1. The absolute value of negative seven is greater than five.

2. The sum of nine and five is greater than or equal to four.

Simplify the expression.

3. $-13 + 8$

4. $-13 - (-2)$

5. $6 \cdot 3 - 8 \cdot 4$

6. $(13)(-3)$

7. $(-6)(-2)$

8. $\dfrac{|-16|}{-8}$

9. $\frac{-8}{0}$

10. $\frac{-6 + 2}{5 - 6}$

11. $\frac{1}{2} - \frac{5}{6}$

12. $-1\frac{1}{8} + 5\frac{3}{4}$

13. $-\frac{3}{5} + \frac{15}{8}$

14. $3(-4)^2 - 80$

15. $6[5 + 2(3 - 8) - 3]$

16. $\frac{-12 + 3 \cdot 8}{4}$

17. $\frac{(-2)(0)(-3)}{-6}$

Insert <, >, or = in the appropriate space to make each of the following statements true.

18. $-3 \quad -7$

19. $4 \quad -8$

20. $|-3| \quad 2$

21. $|-2| \quad -1 - (-3)$

22. Given $\left\{-5, -1, \frac{1}{4}, 0, 1, 7, 11.6, \sqrt{7}, 3\pi\right\}$, list the numbers in this set that also belong to the set of:

a. Natural numbers
b. Whole numbers
c. Integers
d. Rational numbers
e. Irrational numbers
f. Real numbers

If $x = 6$, $y = -2$, and $z = -3$, evaluate each expression.

23. $x^2 + y^2$

24. $x + yz$

25. $2 + 3x - y$

26. $\frac{y + z - 1}{x}$

Identify the property illustrated by each expression.

27. $8 + (9 + 3) = (8 + 9) + 3$

28. $6 \cdot 8 = 8 \cdot 6$

29. $-6(2 + 4) = -6 \cdot 2 + (-6) \cdot 4$

30. $\frac{1}{6}(6) = 1$

31. Find the opposite of -9.

32. Find the reciprocal of $-\frac{1}{3}$.

The New Orleans Saints were 22 yards from the goal when the following series of gains and losses occurred.

	Gains and Losses in Yards
First down	5
Second down	−10
Third down	−2
Fourth down	29

33. During which down did the greatest loss of yardage occur?

34. Was a touchdown scored?

35. The temperature at the Winter Olympics was a frigid 14 degrees below zero in the morning, but by noon it had risen 31 degrees. What was the temperature at noon?

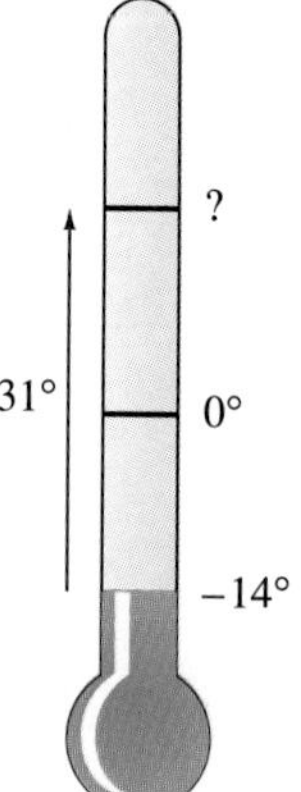

36. Jean Averez decided to sell 280 shares of stock, which decreased in value by $1.50 per share yesterday. How much money did she lose?

CHAPTER 2

Equations and Problem Solving

Many algebraic relationships can be visualized with the use of geometric relationships.

INTRODUCTION

Much of mathematics relates to deciding which statements are true and which are false. When a statement, such as an equation, contains variables, it is usually not possible to decide whether the equation is true or false until the variable has been replaced by a value. For example, the statement $x + 7 = 15$ is an equation stating that the sum $x + 7$ is the same quantity as 15. Is this statement true or false? It is false for some values of x and true for just one value of x, that is, 8. Our purpose in this chapter is to learn ways of deciding which values make an equation or an inequality true. To begin, we spend a bit more time learning about simplifying expressions.

2.1 Simplifying Algebraic Expressions

OBJECTIVES

Tape BA 5

1. Identify terms, like terms, and unlike terms.
2. Combine like terms.
3. Use the distributive property to remove parentheses.
4. Write word phrases as algebraic expressions.

1 Before we practice simplifying expressions, some new language is presented. A **term** is a number or the product of a number and variables raised to powers. For example,

$$-y, \quad 2x^3, \quad -5, \quad 3xz^2, \quad \frac{2}{y}, \quad 0.8z$$

are terms. The **numerical coefficient** of the term $3x$ is 3. Recall that $3x$ means $3 \cdot x$.

Terms	*Numerical Coefficients*
$3x$	3
-5	-5
$\frac{y^3}{5}$	$\frac{1}{5}$ since $\frac{y^3}{5}$ means $\frac{1}{5} \cdot y^3$
$-0.7ab^3c^5$	-0.7
$-y$	-1
z	1

HELPFUL HINT

The term $-y$ means $-1y$ and thus has a numerical coefficient of -1. The term z means $1z$ and thus has a numerical coefficient of 1.

EXAMPLE 1 Find the numerical coefficient.

a. $-3y$ **b.** $22z^4$ **c.** $\frac{2}{y}$ **d.** $-x$

Solution: **a.** The numerical coefficient of $-3y$ is -3.

b. The numerical coefficient of $22z^4$ is 22.

c. The numerical coefficient of $\frac{2}{y}$ is 2.

d. The numerical coefficient of $-x$ is -1 since $-x$ is $-1x$. ■

Terms with the same variables raised to exactly the same powers are **like terms.**

Like Terms	*Unlike Terms*	
$3x, 2x$	$5x, 5x^2$	Same variables, but different powers
$-6x^2y, 2x^2y, 4x^2y$	$7y, 3z, 8x^2$	Different variables
$2ab^2c^3, ac^3b^2$	$6abc^3, 6ab^2$	Different variables and different powers

Each variable and its exponent must match exactly in like terms, but like terms need not have the same numerical coefficients, nor do their factors need to be in the same order. For example, $2x^2y$ and $-yx^2$ are like terms.

EXAMPLE 2 Tell whether the terms are like or unlike.

a. $-x^2, 3x^3$ **b.** $4x^2y, x^2y, -2x^2y$ **c.** $-2yz, -3zy$ **d.** $-x^4, x^4$

Solution: **a.** Unlike terms, since the exponents on x are not the same.

b. Like terms, since each variable and its exponent match.

c. Like terms, since $zy = yz$ by the commutative property.

d. Like terms. ■

2 An expression can be simplified by **combining like terms.** For example, by the distributive property, we rewrite like terms $3x + 2x$ as

$$3x + 2x = (3 + 2)x = 5x$$

Also,

$$-y^2 + 5y^2 = (-1 + 5)y^2 = 4y^2$$

This suggests the following rule for combining like terms.

> To **combine like terms,** add the numerical coefficients and multiply the result by the common variables.

EXAMPLE 3 Simplify the following by combining like terms.

a. $7x - 3x$ **b.** $10y^2 + y^2$ **c.** $8x^2 + 2x - 3x$

Solution: **a.** $7x - 3x = (7 - 3)x = 4x$

b. $10y^2 + y^2 = (10 + 1)y^2 = 11y^2$

c. $8x^2 + 2x - 3x = 8x^2 + (2 - 3)x = 8x^2 - x$ ■

EXAMPLE 4 Simplify each expression by combining like terms.

a. $2x + 3x + 5 + 2$ **b.** $-5a - 3 + a + 2$ **c.** $4y - 3y^2$ **d.** $2.3x + 5x - 6$

Solution: Use the distributive property to combine the numerical coefficients of like terms.

a. $2x + 3x + 5 + 2 = (2 + 3)x + (5 + 2)$
$= 5x + 7$

b. $-5a - 3 + a + 2 = -5a + 1a + (-3 + 2)$
$= (-5 + 1)a + (-3 + 2)$
$= -4a - 1$

c. $4y - 3y^2$ These two terms cannot be combined because they are unlike terms.

d. $2.3x + 5x - 6 = (2.3 + 5)x - 6$
$= 7.3x - 6$ ■

3 Simplifying expressions makes frequent use of the distributive property to remove parentheses.

EXAMPLE 5 Use the distributive property to remove parentheses.

a. $5(x + 2)$ **b.** $-2(y + 0.3z - 1)$ **c.** $-(x + y - 2z + 6)$

Solution: **a.** $5(x + 2) = 5(x) + 5(2)$ Apply the distributive property.
$= 5x + 10$ Multiply.

b. $-2(y + 0.3z - 1) = -2(y) - 2(0.3z) - 2(-1)$ Apply the distributive property.
$= -2y - 0.6z + 2$ Multiply.

c. $-(x + y - 2z + 6) = -1(x + y - 2z + 6)$
$= -1(x) - 1(y) - 1(-2z) - 1(6)$ Distribute -1 over each term.
$= -x - y + 2z - 6$ ■

HELPFUL HINT

If a "−" sign precedes parentheses, the sign of each term inside the parentheses is changed when the distributive property is applied to remove parentheses. For example,

$-(2x + 1) = -2x - 1$ $-(x - 2y) = -x + 2y$

$-(-5x + y - z) = 5x - y + z$ $-(-3x - 4y - 1) = 3x + 4y + 1$

EXAMPLE 6 Simplify the following expressions.

a. $3(2x - 5) + 1$ **b.** $8 - (7x + 2) + 3x$ **c.** $-2(4x + 7) - (3x - 1)$

Solution: **a.** $3(2x - 5) + 1 = 6x - 15 + 1$ Apply the distributive property.

$= 6x - 14$ Combine like terms.

b. $8 - (7x + 2) + 3x = 8 - 7x - 2 + 3x$ Apply the distributive property.

$= -7x + 3x + 8 - 2$

$= -4x + 6$ Combine like terms.

c. $-2(4x + 7) - (3x - 1) = -8x - 14 - 3x + 1$ Apply the distributive property.

$= -11x - 13$ Combine like terms. ■

EXAMPLE 7 Subtract $4x - 2$ from $2x - 3$.

Solution: "Subtract $4x - 2$ **from** $2x - 3$" translates into $(2x - 3) - (4x - 2)$. Next, simplify the algebraic expression.

$(2x - 3) - (4x - 2) = 2x - 3 - 4x + 2$ Apply the distributive property.

$= -2x - 1$ Combine like terms. ■

4 To help prepare us for solving problems, we continue to write algebraic expressions.

EXAMPLE 8 Write the following phrases as algebraic expressions and simplify if possible. Let x represent the unknown number.

a. Twice a number, added to 6

b. The difference of a number and 4, divided by 7

c. Five added to 3 times the sum of a number and 1

Solution: **a.**

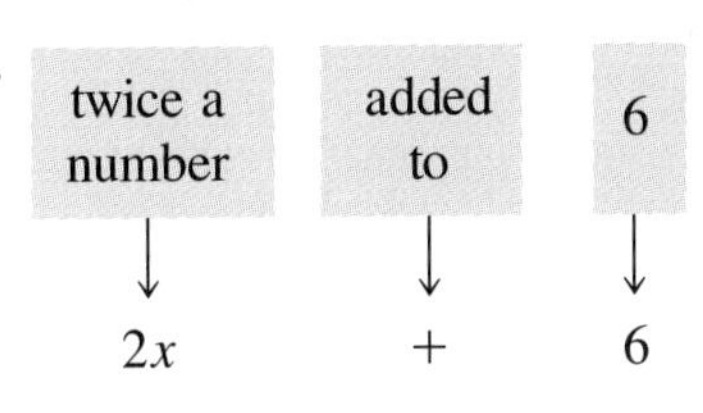

b.

the difference of a number and 4 | divided by | 7

$\frac{x - 4}{7}$

c.

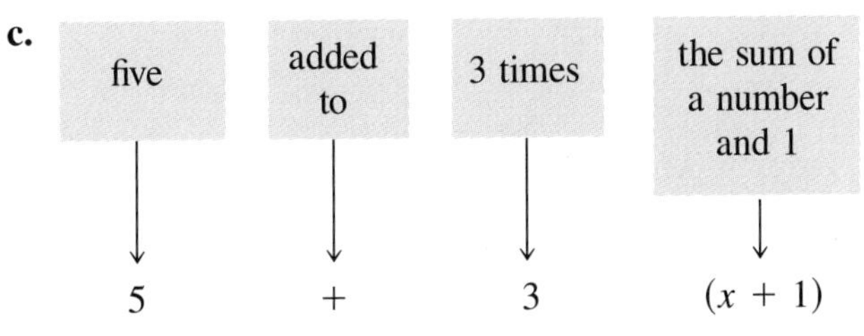

To simplify this expression, we apply the distributive property and then combine like terms.

$$5 + 3(x + 1) = 5 + 3x + 3$$
$$= 8 + 3x$$

■

MENTAL MATH

Give the numerical coefficient of each term. See Example 1.

1. $-7y$
2. $3x$
3. x
4. $-y$
5. $17x^2y$
6. $1.2xyz$

Indicate whether the following lists of terms are like or unlike. See Example 2.

7. $5y, -y$
8. $-2x^2y, 6xy$
9. $2z, 3z^2$
10. $ab^2, -7ab^2$
11. $8wz, 7zw$
12. $7.4p^3q^2, 6.2p^3q^2r$

EXERCISE SET 2.1

Simplify each expression by combining any like terms. See Examples 3 and 4.

1. $7y + 8y$
2. $5x - 2x$
3. $8w - w + 6w$
4. $c - 7c + 2c$
5. $3b - 5 - 10b - 4$
6. $6g + 5 - 3g - 7$
7. $m - 4m + 2m - 6$
8. $a + 3a - 2 - 7a$

Simplify each expression. Use the distributive property to remove any parentheses. See Examples 5 and 6.

9. $5(y - 4)$
10. $7(r - 3)$
11. $7(d - 3) + 10$
12. $9(z + 7) - 15$
13. $-(3x - 2y + 1)$
14. $-(y + 5z - 7)$
15. $5(x + 2) - (3x - 4)$
16. $4(2x - 3) - 2(x + 1)$

Write each of the following as an algebraic expression. Simplify if possible. See Example 7.

17. Add $6x + 7$ to $4x - 10$.
18. Add $3y - 5$ to $y + 16$.
19. Subtract $7x + 1$ from $3x - 8$.
20. Subtract $4x - 7$ from $12 + x$.

Write each of the following phrases as algebraic expressions and simplify if possible. Let x represent the unknown number. See Example 8.

21. Twice a number decreased by four
22. The difference of a number and two, divided by five
23. Three-fourths of a number increased by twelve
24. Eight more than triple the number
25. The sum of -2 and 5 times a number, added to 7 times the number
26. The sum of 3 times a number and 10, **subtracted from** 9 times the number

Simplify each expression.

27. $7x^2 + 8x^2 - 10x^2$
28. $8x + x - 11x$
29. $6x - 5x + x - 3 + 2x$
30. $8h + 13h - 6 + 7h - h$
31. $-5 + 8(x - 6)$
32. $-6 + 5(r - 10)$
33. $5g - 3 - 5 - 5g$
34. $8p + 4 - 8p - 15$
35. $6.2x - 4 + x - 1.2$
36. $7.9y - 0.7 - y + 0.2$
37. $2k - k - 6$
38. $7c - 8 - c$
39. $0.5(m + 2) + 0.4m$
40. $0.2(k + 8) - 0.1k$
41. $-4(3y - 4)$
42. $-3(2x + 5)$
43. $3(2x - 5) - 5(x - 4)$
44. $2(6x - 1) - (x - 7)$
45. $3.4m - 4 - 3.4m - 7$
46. $2.8w - 0.9 - 0.5 - 2.8w$
47. $6x + 0.5 - 4.3x - 0.4x + 3$
48. $0.4y - 6.7 + y - 0.3 - 2.6y$
49. $-2(3x - 4) + 7x - 6$
50. $8y - 2 - 3(y + 4)$
51. $-9x + 4x + 18 - 10x$
52. $5y - 14 + 7y - 20y$
53. $5k - (3k - 10)$
54. $-11c - (4 - 2c)$
55. $(3x + 4) - (6x - 1)$
56. $(8 - 5y) - (4 + 3y)$

Write each of the following as an algebraic expression. Simplify if possible.

57. Subtract $5m - 6$ from $m - 9$.

58. Subtract $m - 3$ from $2m - 6$.

59. Eight times the sum of a number and six

60. Five less than four times the number

61. Double a number minus the sum of the number and ten

62. Half a number minus the product of the number and eight

63. The perimeter of a figure is the total distance around the figure. Given the following rectangle, express the perimeter as an algebraic expression in x.

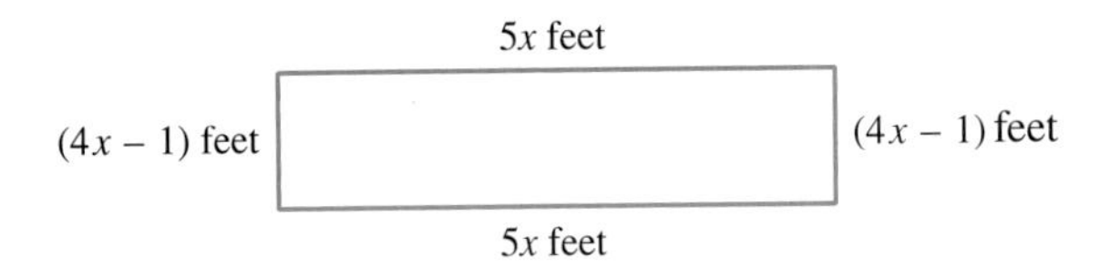

64. Given the following triangle, express its perimeter as an algebraic expression in x.

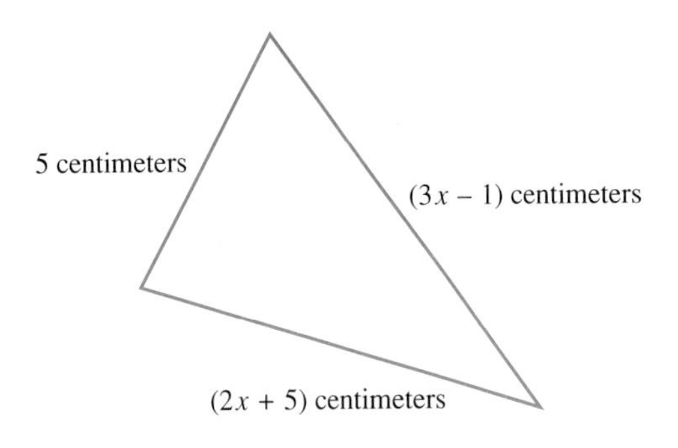

65. Seven multiplied by the quotient of a number and six

66. The product of a number and ten, less twenty

Skill Review

Evaluate the following expressions for the given values. See Section 1.7.

67. If $x = -1$ and $y = 3$, find $y - x^2$.

68. If $a = 2$ and $b = -5$, find $a - b^2$.

69. If $y = -5$ and $z = 0$, find $yz - y^2$.

70. If $g = 0$ and $h = -4$, find $gh - h^2$.

71. If $x = -3$, find $x^3 - x^2 + 4$.

72. If $x = -2$, find $x^3 - x^2 - x$.

Bryan, a 4-year old child, was admitted to Slidell Memorial Hospital with pneumonia and a fever of 104° Fahrenheit. His temperature was taken every hour, and the results are recorded on the bar graph shown. (See Section 1.1.)

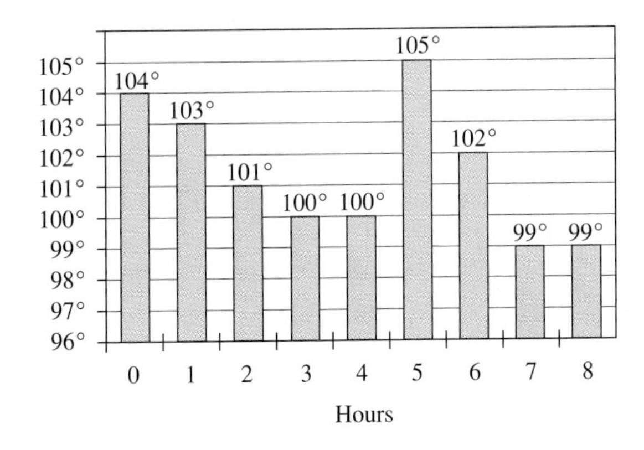

73. How many hours after arrival at the hospital was Bryan's fever the highest?

74. What was Bryan's highest recorded temperature?

75. How many hours after arrival at the hospital was Bryan's temperature 101°?

76. When did Bryan's temperature show the greatest increase?

77. When did Bryan's temperature show the greatest decrease?

A Look Ahead

Simplify each expression. See the following example.

EXAMPLE Simplify $-3xy + 2x^2y - (2xy - 1)$.

Solution: $-3xy + 2x^2y - (2xy - 1) = -3xy + 2x^2y - 2xy + 1$

$= -5xy + 2x^2y + 1$ ■

78. $5b^2c^3 + 8b^3c^2 - 7b^3c^2$

79. $4m^4p^2 + m^4p^2 - 5m^2p^4$

80. $3x - (2x^2 - 6x) + 7x^2$

81. $9y^2 - (6xy^2 - 5y^2) - 8xy^2$

82. $-(2x^2y + 3z) + 3z - 5x^2y$

83. $-(7c^3d - 8c) - 5c - 4c^3d$

Writing in Mathematics

84. Explain why decimal points are lined up vertically when adding or subtracting decimals.

85. In your own words, explain how to combine like terms.

2.2 The Addition and Multiplication Properties of Equality

OBJECTIVES

Tapes BA 5 and BA 6

1. Define linear equation in one variable and equivalent equations.
2. Use the addition property of equality to solve linear equations.
3. Use the multiplication property of equality to solve linear equations.
4. Write sentences as equations and solve.

1 Recall that an **equation** is a statement that two expressions are equal and that a **solution** of an equation is a value for the variable that makes the equation a true statement. The process of finding solutions is called **solving** an equation. In this section, we concentrate on solving **linear equations** in one variable.

Linear Equations in One Variable

$3x = -15$ $\quad 7 - y = 3y$ $\quad 4n - 9n + 6 = 0$ $\quad z = -2$

Linear equations are also called **first-degree equations** since the exponent on the variable is 1.

Linear Equation in One Variable

A linear equation in one variable is an equation that can be written in the form

$$ax + b = c$$

where a, b, and c are real numbers and $a \neq 0$.

To solve a linear equation in x, we will write a series of simpler equations, all equivalent to the original equation, so that the final equation has the form

$$x = \textbf{number} \quad \textbf{or} \quad \textbf{number} = x$$

Equivalent equations have the same solution so that the "number" above will be the solution to the original equation.

2 The first property of equality that will help us write simpler equations is the **addition property of equality.**

Addition Property of Equality

If a, b, and c are real numbers, then

$$a = b \quad \text{and} \quad a + c = b + c$$

are equivalent equations.

This property guarantees that adding the same number to both sides of an equation does not change the solution of the equation. Since subtraction is defined in terms of addition, we may also **subtract the same number from both sides** without changing the solution.

A good way to picture a true equation is as a balanced scale. Since it is balanced, each side of the scale weighs the same amount. If the same weight is added to or subtracted from each side, the scale remains balanced.

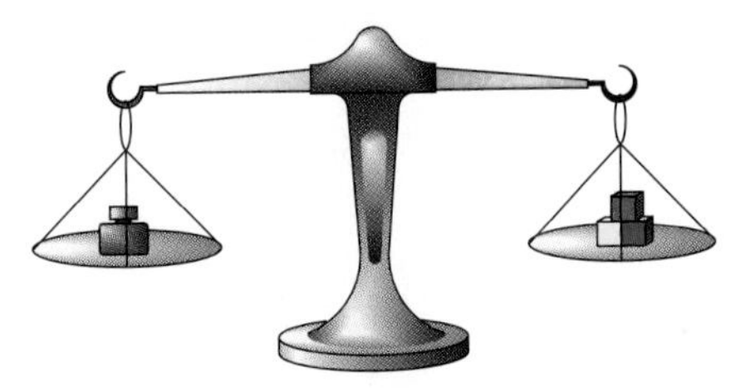

We use the addition property of equality to write equivalent equations until the variable is isolated (by itself on one side of the equation) and the equation looks like "x = number" or "number = x."

EXAMPLE 1 Solve $x - 7 = 10$ for x.

Solution: To solve for x, isolate x on one side of the equation. To do this, add 7 to both sides of the equation.

$$x - 7 = 10$$

$$x - 7 + 7 = 10 + 7 \qquad \text{Add 7 to both sides.}$$

$$x = 17 \qquad \text{Simplify.}$$

To check, replace x with 17 in the original equation.

$$x - 7 = 10$$

$$17 - 7 = 10 \qquad \text{Replace } x \text{ with 17 in the original equation.}$$

$$10 = 10 \qquad \text{True.}$$

Since the statement is true, the solution set is $\{17\}$. ■

If possible, simplify one or both sides of an equation before applying the addition property of equality.

EXAMPLE 2 Solve $2x + 3x - 5 + 7 = 10x + 3 - 6x - 4$ for x.

Solution: First simplify both sides of the equation.

$$2x + 3x - 5 + 7 = 10x + 3 - 6x - 4$$

$$5x + 2 = 4x - 1 \qquad \text{Combine like terms on each side of the equation.}$$

$$5x + 2 - 4x = 4x - 1 - 4x \qquad \text{Subtract } 4x \text{ from both sides.}$$

$$x + 2 = -1 \qquad \text{Combine like terms.}$$

$$x + 2 - 2 = -1 - 2 \qquad \text{Subtract 2 from both sides.}$$

$$x = -3 \qquad \text{Combine like terms.}$$

Check by replacing x with -3 in the original equation.

$$2x + 3x - 5 + 7 = 10x + 3 - 6x - 4$$

$$2(-3) + 3(-3) - 5 + 7 = 10(-3) + 3 - 6(-3) - 4 \qquad \text{Replace } x \text{ with } -3.$$

$$-6 - 9 - 5 + 7 = -30 + 3 + 18 - 4 \qquad \text{Multiply.}$$

$$-13 = -13 \qquad \text{True.}$$

The solution set is $\{-3\}$. ■

If an equation contains parentheses, use the distributive property to remove them.

EXAMPLE 3 Solve $7 = -5(2a - 1) - (-11a + 6)$ for a.

Solution:

$$7 = -5(2a - 1) - (-11a + 6)$$

$$7 = -10a + 5 + 11a - 6 \qquad \text{Apply the distributive property.}$$

$$7 = a - 1 \qquad \text{Combine like terms.}$$

$$7 + 1 = a - 1 + 1 \qquad \text{Add 1 to both sides to isolate } a.$$

$$8 = a \qquad \text{Combine like terms.}$$

Check to see that $\{8\}$ is the solution set. ■

HELPFUL HINT

We may isolate the variable on either side of the equation. For example, $8 = a$ is equivalent to $a = 8$.

When solving equations, we may sometimes encounter an equation such as

$$-x = 5$$

This equation is not solved for x because x is not isolated. One way to solve this equation for x is to recall that

"$-$" can be read as "the opposite of "

We can read the equation $-x = 5$ then as "the opposite of $x = 5$." If the opposite of x is 5, this means that x is the opposite of 5 or -5.

In summary,

$$-x = 5 \quad \text{and} \quad x = -5$$

are equivalent equations and $x = -5$ is solved for x.

3 As useful as the addition property of equality is, it will not help us solve every type of linear equation in one variable. For example, adding or subtracting a value to both sides of the equation does not help us solve

$$\frac{5}{2}x = 15$$

Fortunately, there is a second important property of equality, the **multiplication property of equality.**

> **Multiplication Property of Equality**
>
> If a, b, and c are real numbers and $c \neq 0$, then
>
> $$a = b \quad \text{and} \quad ac = bc$$
>
> are equivalent equations.

This property guarantees that multiplying both sides of an equation by the same nonzero number does not change the solution of the equation. Since division is defined in terms of multiplication, we may also **divide both sides of the equation by the same nonzero number** without changing the solution.

EXAMPLE 4 Solve for x: $\frac{5}{2}x = 15$.

Solution: To isolate x, multiply both sides of the equation by the reciprocal of $\frac{5}{2}$, which is $\frac{2}{5}$.

$$\frac{5}{2}x = 15$$

$$\frac{2}{5} \cdot \frac{5}{2}x = \frac{2}{5} \cdot 15 \qquad \text{Multiply both sides by } \frac{2}{5}.$$

$$\left(\frac{2}{5} \cdot \frac{5}{2}\right)x = \frac{2}{5} \cdot 15 \qquad \text{Apply the associative property.}$$

$$1x = 6 \qquad \text{Simplify.}$$

or

$$x = 6$$

The solution set is $\{6\}$. Check this solution in the original equation. ■

Why did we multiply both sides by the reciprocal of the coefficient of x? Multiplying the coefficient $\frac{5}{2}$ by $\frac{2}{5}$ leaves a coefficient of 1 and thus isolates the variable.

In general, multiplying by the reciprocal of the variable's coefficient is a way to isolate a variable. Multiplying by the reciprocal of a number is, of course, the same as dividing by the number.

EXAMPLE 5 Solve $-3x = 33$ for x.

Solution: Recall that $-3x$ means $-3 \cdot x$. To isolate x, divide both sides by the coefficient of x, that is, -3.

$$-3x = 33$$

$$\frac{-3x}{-3} = \frac{33}{-3} \quad \text{Divide both sides by } -3.$$

$$1x = -11 \quad \text{Simplify.}$$

$$x = -11$$

To check, replace x with -11 in the original equation.

$$-3x = 33$$

$$-3(-11) = 33 \quad \text{Replace } x \text{ with } -11 \text{ in the original equation.}$$

$$33 = 33 \quad \text{True.}$$

The solution set is $\{-11\}$. ■

EXAMPLE 6 Solve $\frac{y}{7} = 20$ for y.

Solution: Recall that $\frac{y}{7} = \frac{1}{7}y$. To isolate y, multiply both sides of the equation by 7, the reciprocal of $\frac{1}{7}$.

$$\frac{y}{7} = 20$$

$$7 \cdot \frac{y}{7} = 7 \cdot 20 \quad \text{Multiply each side by 7.}$$

$$y = 140 \quad \text{Simplify.}$$

To check, replace y with 140 in the original equation.

$$\frac{y}{7} = 20$$

$$\frac{140}{7} = 20 \quad \text{Replace } y \text{ with 140.}$$

$$20 = 20 \quad \text{True.}$$

The solution set is $\{140\}$. ■

EXAMPLE 7 Solve the equation $7(x + 1) - 5 = 6x - 10$.

Solution: First use the distributive property and then simplify.

$7(x + 1) - 5 = 6x - 10$	
$7x + 7 - 5 = 6x - 10$	Apply the distributive property.
$7x + 2 = 6x - 10$	Combine like terms.
$7x + 2 - 6x = 6x - 10 - 6x$	Subtract $6x$ from both sides.
$x + 2 = -10$	Simplify.
$x + 2 - 2 = -10 - 2$	Subtract 2 from both sides.
$x = -12$	Simplify.

The solution set is $\{-12\}$. Check this solution in the original equation. ■

4 We now apply our equation-solving skills to solving problems written in words. In Chapter 1, we provided a chart of common translations, words that are other ways of expressing mathematical symbols. Here, we provide a more extended chart of translations.

Addition	Subtraction	Multiplication	Division	Equality
sum	difference of	product	quotient	equals
plus	minus	times	divide	gives
added to	subtracted from	multiply	into	is/was
more than	less than	twice	ratio	yields
increased by	decreased by	of		amounts to
total	less			represents
				is the same as

Many times, modeling a problem involves a direct translation from a sentence to an equation.

EXAMPLE 8 Twice a number added to seven is the same as three subtracted from the number.

Solution: Translate the sentence into an equation and solve. Let $x =$ the number.

In words:	twice a number	added to	seven	is the same as	three subtracted from the number
	↓	↓	↓	↓	↓
Translate:	$2x$	$+$	7	$=$	$x - 3$

To solve, begin by subtracting x on both sides to isolate the variable term.

$2x + 7 = x - 3$	
$2x + 7 - x = x - 3 - x$	Subtract x from both sides.
$x + 7 = -3$	Simplify.
$x + 7 - 7 = -3 - 7$	Subtract 7 from both sides.
$x = -10$	Simplify.

Check the solution in the **original stated problem.** To do so, replace the unknown number with -10. Twice -10 added to 7 is the same as 3 subtracted from -10.

$$-13 \quad \text{is the same as} \quad -13$$

The unknown number is -10. ■

HELPFUL HINT

When checking solutions, go back to the original stated problem rather than to your equation in case errors have been made in translating to an equation.

MENTAL MATH

Solve each equation mentally. See Example 1.

1. $x + 4 = 6$
2. $x + 7 = 10$
3. $n + 18 = 30$
4. $z + 22 = 40$
5. $b - 11 = 6$
6. $d - 16 = 5$

EXERCISE SET 2.2

Solve each equation. See Example 1.

1. $x + 11 = -2$
2. $y - 5 = -9$
3. $5y + 14 = 4y$
4. $8x - 7 = 9x$
5. $8x = 7x - 8$
6. $x = 2x + 3$

Solve each equation. See Examples 2 and 3.

7. $3x - 6 = 2x + 5$
8. $7y + 2 = 6y + 2$
9. $3t - t - 7 = t - 7$
10. $4c + 8 - c = 8 + 2c$
11. $7x + 2x = 8x - 3$
12. $3n + 2n = 7 + 4n$
13. $-2(x + 1) + 3x = 14$
14. $10 = 8(3y - 4) - 23y + 20$

Solve the following equations. See Example 5.

15. $-5x = 20$
16. $-7x = -49$
17. $3x = 0$
18. $-2x = 0$
19. $-x = -12$
20. $-y = 8$
21. $3x + 2x = 50$
22. $-y + 4y = -10$

Solve the following equations. See Examples 4 and 6.

23. $\frac{2}{3}x = -8$
24. $\frac{3}{4}n = -15$
25. $\frac{1}{6}d = \frac{1}{2}$
26. $\frac{1}{8}v = \frac{1}{4}$
27. $\frac{a}{-2} = 1$
28. $\frac{d}{15} = 2$
29. $\frac{k}{7} = 0$
30. $\frac{f}{-5} = 0$

Solve each equation. See Example 7.

31. $2(x - 4) = x + 3$
32. $3(y + 7) = 2y - 5$
33. $7(6 + w) = 6(2 + w)$
34. $6(5 + c) = 5(c - 4)$
35. $10 - (2x - 4) = 7 - 3x$
36. $15 - (6 - 7k) = 2 + 6k$

Write an equation for each of the following and solve. See Example 8.

37. The sum of twice a number and 7 is equal to the sum of a number and 6. Find the number.

38. Three times the difference of a number and 10 is the same as twice the number added to 14. Find the number.

39. Find a number such that half the number is negative five.

40. Three times a number is equal to twice the number. Find the number.

Solve the following.

41. $x - 2 = -4$

42. $y + 7 = 5$

43. $2y + 10 = y$

44. $4x - 4 = 3x$

45. $y + 0.8 = 9.7$

46. $w + 0.9 = 3.6$

47. $-3w = 18$

48. $5j = -45$

49. $-0.2z = -0.8$

50. $-0.1m = 3.6$

51. $-h = -\frac{3}{4}$

52. $-b = \frac{4}{7}$

53. $5b - 0.7 = 6b$

54. $8n + 1.5 = 9n$

55. $5x - 6 = 6x - 5$

56. $2x + 7 = x - 10$

57. $7t - 12 = 6t$

58. $9m + 14 = 8m$

59. $-5(n - 2) = 8 - 4n$

60. $-4(z - 3) = 2 - 3z$

61. $y - 5y + 0.6 = 0.8 - 5y$

62. $6z + z - 0.9 = 6z + 0.9$

63. $-3(x - 4) = -4x$

64. $-2(x - 1) = -3x$

65. $5x - 7x + 8x = -6$

66. $9y - 12y = -12$

67. $x + 6x - 10x = 0$

Write an equation for each of the following and solve. Check the solution.

68. Three times a number minus 6 is equal to two times a number plus 8. Find the number.

69. The sum of 4 times a number and -2 is equal to the sum of 5 times a number and -2. Find the number.

70. Twice the difference of a number and 8 is equal to three times the sum of a number and 3. Find the number.

71. Five times the sum of a number and -1 is the same as 6 times a number. Find the number.

72. Negative four is half a number. Find the number.

73. The product of a number and six is triple the number. Find the number.

74. The perimeter of a geometric figure is the sum of the lengths of its sides. If the perimeter of the following pentagon (five-sided figure) is 28 centimeters, find the length of each side.

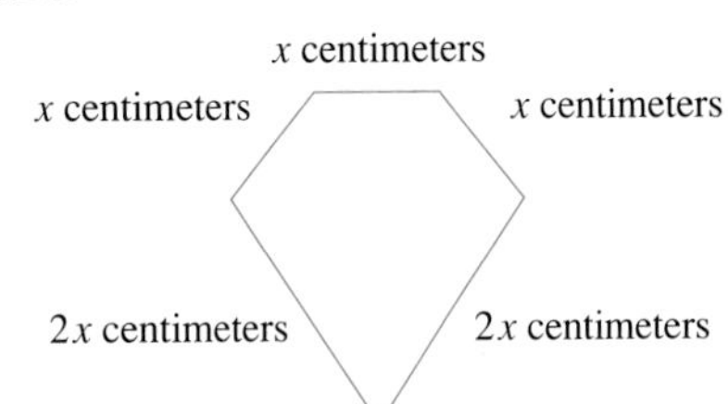

75. The perimeter of the following triangle is 35 meters. Find the length of each side.

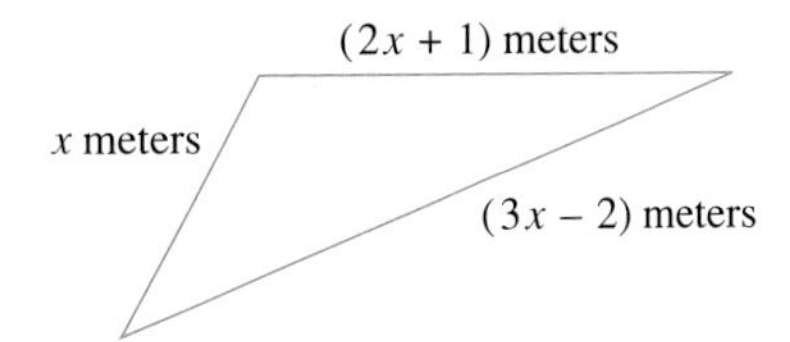

Skill Review

Multiply or divide the following fractions. See Section 1.3.

76. $\frac{3}{5} \cdot \frac{10}{21}$

77. $\frac{7}{8} \cdot \frac{24}{29}$

78. $\frac{12}{15} \div \frac{2}{3}$

79. $\frac{9}{10} \div \frac{4}{3}$

80. $\frac{2}{9} \cdot \frac{12}{14} \cdot \frac{7}{4}$

81. $\frac{6}{15} \cdot \frac{45}{18} \cdot \frac{4}{5}$

Insert <, >, or = in the appropriate space to make the following true. *See Section 1.7.*

82. $(-3)^2 \quad -3^2$

83. $(-2)^4 \quad -2^4$

84. $(-2)^3 \quad -2^3$

85. $(-4)^3 \quad -4^3$

86. $-|-6| \quad 6$

87. $-|-0.7| \quad -0.7$

88. $|-8| - |-5| \quad -|5 - 8|$

89. $-|2| - |-7| \quad |2 - 7|$

Writing in Mathematics

90. Explain why it is important to check the solution to a word problem in the original stated problem rather than in the equation that was written.

2.3 Solving Linear Equations

OBJECTIVES

Tape BA 6

1. Learn the general strategy for solving a linear equation.
2. Recognize identities and equations with no solution.
3. Write sentences as equations and solve.

1 In this section, we present a general strategy for solving linear equations. One new piece of strategy is a suggestion to "clear an equation of fractions" as a first step. Since operating on integers is more convenient than operating on fractions, doing so makes the equation more manageable.

To Solve Linear Equations in One Variable

Step 1 Clear the equation of fractions by multiplying both sides of the equation by the lowest common denominator (LCD) of all denominators in the equation.

Step 2 Use the distributive property to remove any grouping symbols, such as parentheses.

Step 3 Combine like terms on each side of the equation.

Step 4 Use the addition property of equality to write the equation as an equivalent equation with variable terms on one side and numbers on the other side.

Step 5 Use the multiplication property of equality to isolate the variable.

Step 6 Check the solution by substituting it in the original equation.

EXAMPLE 1 Solve for x: $\frac{x}{2} - 1 = \frac{2}{3}x - 3$.

Solution: This equation contains fractions, so we begin by clearing fractions. To do this, multiply both sides of the equation by the LCD of 2 and 3, which is 6.

$$\frac{x}{2} - 1 = \frac{2}{3}x - 3$$

Step 1 $\quad 6\left(\frac{x}{2} - 1\right) = 6\left(\frac{2}{3}x - 3\right)$ Multiply both sides by 6.

Step 2 $\quad 6\left(\frac{x}{2}\right) - 6(1) = 6\left(\frac{2}{3}x\right) - 6(3)$ Apply the distributive property.

$3x - 6 = 4x - 18$ Simplify.

There are no more grouping symbols and no like terms on either side of the equation, so we continue with step 4.

$$3x - 6 = 4x - 18$$

Step 4 $\quad 3x - 6 - 3x = 4x - 18 - 3x$ Subtract $3x$ from both sides.

$-6 = x - 18$ Simplify.

$-6 + 18 = x - 18 + 18$ Add 18 to both sides.

$12 = x$ Simplify.

The equation is solved for x and the solution is 12. To check, replace x with 12 in the original equation.

Step 6 $\quad \frac{x}{2} - 1 = \frac{2}{3}x - 3$ Original equation.

$\frac{12}{2} - 1 = \frac{2}{3} \cdot 12 - 3$ Replace x with 12.

$6 - 1 = 8 - 3$ Simplify.

$5 = 5$ True.

The solution set is $\{12\}$. ■

EXAMPLE 2 Solve for x: $2(x - 3) = 5x - 9$.

Solution: This equation contains no fractions, so we first use the distributive property.

Step 2 $\quad 2(x - 3) = 5x - 9$

$2x - 6 = 5x - 9$

There are no like terms to combine, so we move to step 4. Get variable terms on the same side of the equation by subtracting $5x$ from both sides, then add 6 to both sides.

Step 4 $\quad 2x - 6 - 5x = 5x - 9 - 5x$

$-3x - 6 = -9$ Simplify.

$-3x - 6 + 6 = -9 + 6$ Add 6 to both sides.

$-3x = -3$ Simplify.

Next use the multiplication property of equality to isolate x.

Step 5 $\dfrac{-3x}{-3} = \dfrac{-3}{-3}$ Divide both sides by -3.

$x = 1$

Step 6 Let $x = 1$ in the original equation to see that $\{1\}$ is the solution set. ■

EXAMPLE 3 Solve $4(2x + 3) - 7 = 3x - 5$.

Solution: There are no fractions to clear, so begin with step 2.

$4(2x + 3) - 7 = 3x - 5$

Step 2 $8x + 12 - 7 = 3x - 5$ Apply the distributive property.

Step 3 $8x + 5 = 3x - 5$ Combine like terms.

Step 4 $8x + 5 - 3x = 3x - 5 - 3x$ Subtract $3x$ from both sides.

$5x + 5 = -5$ Simplify.

$5x + 5 - 5 = -5 - 5$ Subtract 5 from both sides.

$5x = -10$ Simplify.

Step 5 $\dfrac{5x}{5} = \dfrac{-10}{5}$ Divide both sides by 5.

$x = -2$ Simplify.

Step 6 *Check:* $4(2x + 3) - 7 = 3x - 5$

$4[2(-2) + 3] - 7 = 3(-2) - 5$ Replace x with -2.

$4(-4 + 3) - 7 = -6 - 5$

$4(-1) - 7 = -11$

$-4 - 7 = -11$

$-11 = -11$ True.

The solution set is $\{-2\}$. ■

EXAMPLE 4 Solve $\dfrac{x + 5}{2} + \dfrac{1}{2} = 2x - \dfrac{x - 3}{8}$.

Solution: Multiply both sides of the equation by 8, the LCD of 2 and 8.

Step 1 $8\left(\dfrac{x + 5}{2} + \dfrac{1}{2}\right) = 8\left(2x - \dfrac{x - 3}{8}\right)$ Multiply both sides by 8.

Step 2 $4(x + 5) + 4 = 16x - (x - 3)$ Apply the distributive property.

$4x + 20 + 4 = 16x - x + 3$ Use the distributive property to remove parentheses.

Step 3 $4x + 24 = 15x + 3$ Combine like terms.

Step 4 $\quad -11x + 24 = 3$ Subtract $15x$ from both sides.

$-11x = -21$ Subtract 24 from both sides.

Step 5 $\quad \dfrac{-11x}{-11} = \dfrac{-21}{-11}$ Divide both sides by -11.

$x = \dfrac{21}{11}$ Simplify.

Step 6 To check, verify that replacing x with $\frac{21}{11}$ makes the original equation true. The solution set is $\left\{\frac{21}{11}\right\}$. ■

2 So far, each linear equation that we have solved has had a single solution.

Not every equation in one variable has a single solution. Some equations have no solution, while others have an infinite number of solutions. For example,

$$x + 5 = x + 7$$

has no solution since, no matter what **real number** we replace x with,

(**real number**) + 5 ≠ (same **real number**) + 7

On the other hand,

$$x + 6 = x + 6$$

has infinitely many solutions since x can be replaced by any real number and the equation is always true. The equation $x + 6 = x + 6$ is called an **identity.** The next few examples illustrate equations like these.

EXAMPLE 5 Solve $-2(x - 5) + 10 = -3(x + 2) + x$.

Solution:

$-2(x - 5) + 10 = -3(x + 2) + x$

$-2x + 10 + 10 = -3x - 6 + x$ Distribute on both sides.

$-2x + 20 = -2x - 6$ Combine like terms.

$-2x + 20 + 2x = -2x - 6 + 2x$ Add $2x$ to both sides.

$20 = -6$ Combine like terms.

Notice that no value for x will make $20 = -6$ a true equation. We conclude that there is **no solution** to this equation. Its solution set is written as either { } or ∅. ■

EXAMPLE 6 Solve $3(x - 4) = 3x - 12$.

Solution:

$3(x - 4) = 3x - 12$

$3x - 12 = 3x - 12$ Apply the distributive property.

The left side of the equation is now identical to the right side. Every real number may be substituted for x and a true statement will result. We arrive at the same conclusion if we continue.

$$3x - 12 = 3x - 12$$

$$3x = 3x \qquad \text{Add 12 to both sides.}$$

$$0 = 0 \qquad \text{Subtract } 3x \text{ from both sides.}$$

Again, one side of the equation is identical to the other side. Thus, $3(x - 4) = 3x - 12$ is an **identity** and every real number is a solution. The solution set may be written as $\{x \mid x \text{ is a real number}\}$. ■

3 As our ability to solve equations becomes more sophisticated, so does our ability to model and solve problems written in words. The next example gives additional practice.

EXAMPLE 7 Twice the sum of a number and 4 is the same as four times the number minus 12. Find the number.

Solution: Translate the sentence into an equation and solve. Let $x =$ the number.

In words:	twice	sum of a number and 4	is the same as	four times the number	minus	12
	↓	↓	↓	↓	↓	↓
Translate:	2	$(x + 4)$	$=$	$4x$	$-$	12

Now solve the equation.

$$2(x + 4) = 4x - 12$$

$$2x + 8 = 4x - 12 \qquad \text{Apply the distributive property.}$$

$$2x + 8 - 4x = 4x - 12 - 4x \qquad \text{Subtract } 4x \text{ from both sides.}$$

$$-2x + 8 = -12 \qquad \text{Simplify.}$$

$$-2x + 8 - 8 = -12 - 8 \qquad \text{Subtract 8 from both sides.}$$

$$-2x = -20 \qquad \text{Simplify.}$$

$$\frac{-2x}{-2} = \frac{-20}{-2} \qquad \text{Divide both sides by } -2.$$

$$x = 10 \qquad \text{Simplify.}$$

Check this solution in the original stated problem. Twice the sum of 10 and 4 is 28, which is the same as 4 times 10 minus 12. The unknown number is 10. ■

Next we practice writing algebraic expressions.

EXAMPLE 8 Write each of the following as an algebraic expression.

a. The sum of two numbers is 8. If one number is 3, find the other number.

b. The sum of two numbers is 8. If one number is x, find the other number in terms of x.

Solution: **a.** The other number is the difference, $8 - 3 = 5$.

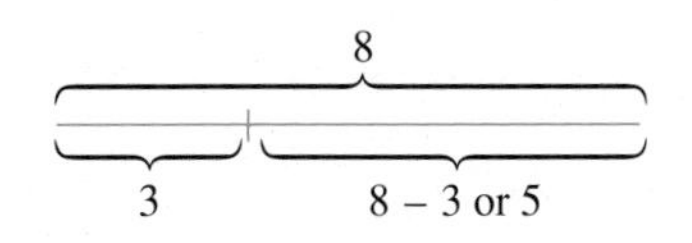

b. The other number is their difference, $8 - x$.

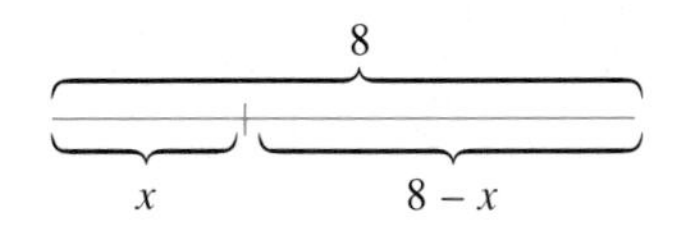

■

EXERCISE SET 2.3

Solve each equation. See Example 1.

1. $\frac{3}{4}x - \frac{1}{2} = 1$

2. $\frac{2}{3}x + \frac{5}{3} = \frac{5}{3}$

3. $x + \frac{5}{4} = \frac{3}{4}x$

4. $\frac{7}{8}x + \frac{1}{4} = \frac{3}{4}x$

5. $\frac{x}{2} - 1 = \frac{x}{5} + 2$

6. $\frac{x}{5} - 2 = \frac{x}{3}$

Solve for the variable. See Examples 2 and 3.

7. $5x + 12 = 2(2x + 7)$

8. $2(x + 3) = x + 5$

9. $3(x - 6) = 5x$

10. $6x = 4(5 + x)$

11. $-2(5y - 1) - y = -4(y - 3)$

12. $-3(2w - 7) - 10 = 9 - 2(5w + 4)$

Solve for the variable. See Example 4.

13. $\frac{x}{2} + \frac{2}{3} = \frac{3}{4}$

14. $\frac{x}{2} + \frac{x}{3} = \frac{5}{2}$

15. $\frac{n - 3}{4} + \frac{n + 5}{7} = \frac{5}{14}$

16. $\frac{2 + h}{9} + \frac{h - 1}{3} = \frac{1}{3}$

Solve each equation. See Examples 5 and 6.

17. $5x - 5 = 2(x + 1) + 3x - 7$

18. $3(2x - 1) + 5 = 6x + 2$

19. $\frac{x}{4} + 1 = \frac{x}{4}$

20. $\frac{x}{3} - 2 = \frac{x}{3}$

21. $3x - 7 = 3(x + 1)$

22. $2(x - 5) = 2x + 10$

Write each of the following sentences as equations. Then solve. See Example 7.

23. Twice the sum of a number and six equals three times the sum of the number and four. Find the number.

24. If the sum of a number and five is tripled, the result is one less than twice the number. Find the number.

Write each as an algebraic expression. See Example 8.

25. Two angles are supplementary if their sum is 180°. If one angle measures $x°$, express the measure of its supplement in terms of x.

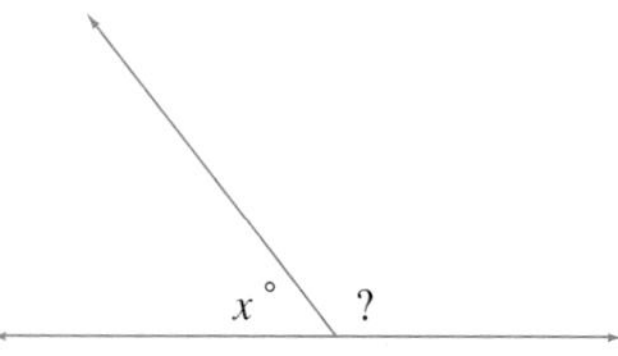

26. Two angles are complementary if their sum is 90°. If one angle measures $x°$, express the measure of its complement in terms of x.

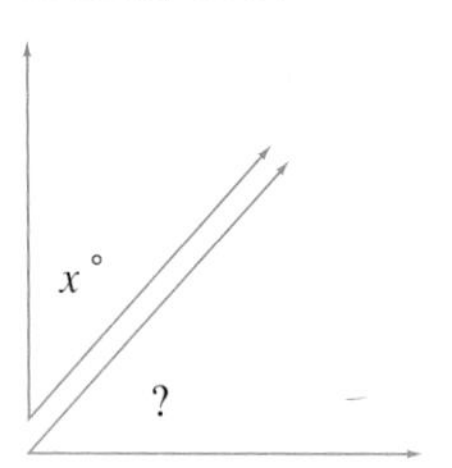

27. The sum of two numbers is 12. If one number is z, express the other number in terms of z.

28. A 10-foot board is cut into two pieces. If one piece is x feet long, express the other length in terms of x.

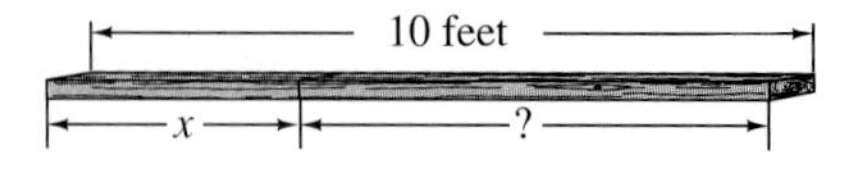

Solve each equation.

29. $-5x + 1 = -19$

30. $-3x - 4 = 11$

31. $4x + 3 = 2x + 11$

32. $6y - 8 = 3y + 7$

33. $-2y - 10 = 5y + 18$

34. $7n + 5 = 10n - 10$

35. $6x - 1 = 5x + 2$

36. $2x - 1 = 6x - 21$

37. $2y + 2 = y$

38. $7y + 4 = -3$

39. $3(5c - 1) - 2 = 13c + 3$

40. $4(3t + 4) - 20 = 3 + 5t$

41. $x + \frac{7}{6} = 2x - \frac{7}{6}$

42. $\frac{5}{2}x - 1 = x + \frac{1}{4}$

43. $2(x - 5) = 7 + 2x$

44. $-3(1 - 3x) = 9x - 3$

45. $\frac{2(z + 3)}{3} = 5 - z$

46. $\frac{3(w + 2)}{4} = 2w + 3$

47. $\frac{4(y - 1)}{5} = -3y$

48. $\frac{5(1 - x)}{6} = -4x$

49. $8 - 2(a - 1) = 7 + a$

50. $5 - 6(2 + b) = b - 14$

51. $2(x + 3) - 5 = 5x - 3(1 + x)$

52. $4(2 + x) + 1 = 7x - 3(x - 2)$

53. $\frac{5x - 7}{3} = x$

54. $\frac{7n + 3}{5} = -n$

55. $\frac{9 + 5v}{2} = 2v - 4$

56. $\frac{6 - c}{2} = 5c - 8$

57. $-3(t - 5) + 2t = 5t - 4$

58. $-(4a - 7) - 5a = 10 + a$

59. $6 - (x + 1) = -2(2 - x)$

60. $-(5x + 4) = 4 - (x + 4)$

61. $\frac{3(x - 5)}{2} = \frac{2(x + 5)}{3}$

62. $\frac{5(x - 1)}{4} = \frac{3(x + 1)}{2}$

63. $\frac{m - 4}{3} - \frac{3m - 1}{5} = 1$

64. $\frac{n + 1}{8} - \frac{2 - n}{3} = \frac{5}{6}$

65. $\frac{5x - 1}{6} - 3x = \frac{1}{3} + \frac{4x + 3}{9}$

66. $\frac{2r - 5}{3} - \frac{r}{5} = 4 - \frac{r + 8}{10}$

67. The difference of three times a number and 1 is the same as twice a number. Find the number.

68. Five times the sum of a number and 3 is equal to 4 times the sum of a number and 8. Find the number.

Write each as an algebraic expression.

69. Terri Santa Coloma must complete a project by the end of September. If x days in September have already passed, express the number of days left to complete the project as an expression in x.

70. All-Star Toyota has 50 Camrys left on their lot to sell by the end of their Labor Day weekend sale. If x Camrys have already been sold, express the remainder of the Camrys on the lot as an expression in x.

71. The limit for the number of red fish one can catch per day in Louisiana is 8. Lou Zawislak has thus far caught x red fish. Express the number of red fish that he can still catch today as an expression in x.

72. Jim Grubbs must pay his property taxes by the end of the year. If x days of the year have already passed, express the number of days left to send in his property taxes as an expression in x.

The three-dimensional bar graph below shows the five most common names of cities, towns, or villages in the United States. Use the information below to complete the graph.

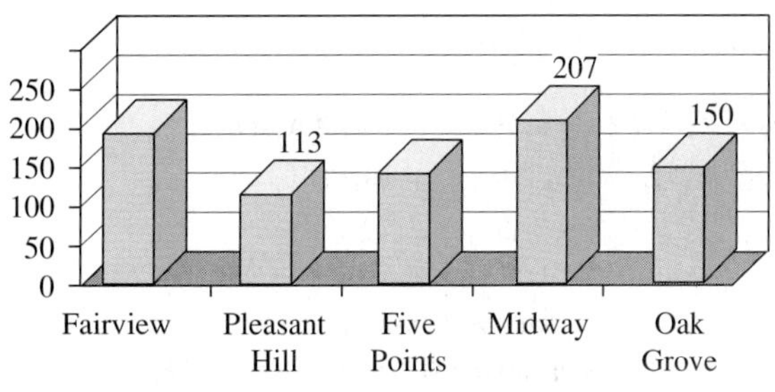

Source: *USA Today,* 1994.

73. Let x represent "the number of towns named Five Points," and use the information given to determine the unknown number. "The number of towns named Five Points" added to 55 is equal to twice "the number of towns named Five Points" minus 90. Check your answer by noticing the height of the bar representing Five Points. Is your answer reasonable?

74. Let x represent "the number of towns named Fairview" and use the information given to determine the unknown number. Three times "the number of towns named Fairview" added to 24 is equal to 168 **subtracted from** 4 times "the number of towns named Fairview." Check your answer by noticing the height of the bar representing Fairview. Is your answer reasonable?

75. What is the most popular name of a city, town, or village in the United States?

76. How many more cities, towns, or villages are named Oak Grove than named Pleasant Hill?

Skill Review

Evaluate. See Sections 1.4 and 1.7.

77. $|2^3 - 3^2| - |5 - 7|$

78. $|5^2 - 2^2| + |9 \div (-3)|$

79. $\frac{\sqrt{25}}{4 + 3 \cdot 7}$

80. $\frac{\sqrt{64}}{24 - 8 \cdot 2}$

See Section 2.1.

81. A plot of land is in the shape of a triangle. If one side is x meters, a second side is $2x - 3$ meters, and a third side is $3x - 5$ meters, express the perimeter of the lot as a simplified expression in x.

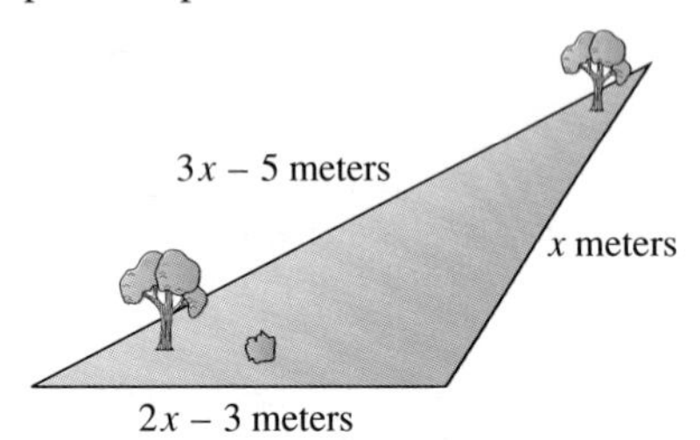

82. A portion of a board has length x feet. The other part has length $7x - 9$ feet. Express the total length of the board as a simplified expression in x.

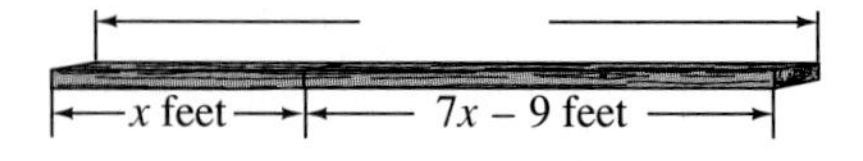

Writing in Mathematics

83. Explain the difference between simplifying an expression and solving an equation.

84. When solving an equation, if the final equivalent equation is $0 = 5$, what can we conclude? If the final equivalent equation is $-2 = -2$, what can we conclude?

85. Make up an equation that has no solution and an equation for which every real number is a solution. Explain the solutions of each.

2.4 Formulas

OBJECTIVES

1. Use formulas to solve problems.
2. Solve a formula for one of its variables.
3. Write phrases as algebraic expressions.

Tape BA 7

1 An equation that describes a known relationship among quantities, such as distance, time, volume, weight, and money, is called a **formula.** These quantities are represented by letters and are thus variables of the formula. Here are some common formulas and their meanings.

$\mathbf{A = lw}$

Area of a rectangle = **length** · **width**

$\mathbf{I = PRT}$

Simple **I**nterest = **P**rincipal · **R**ate · **T**ime

$\mathbf{P = a + b + c}$

Perimeter of a triangle = side **a** + side **b** + side **c**

$\mathbf{d = rt}$

distance = **r**ate · **t**ime

V = lwh

Volume of a rectangular solid = **l**ength · **w**idth · **h**eight

F = (9/5)C + 32

degrees **F**ahrenheit = (9/5) · degrees **C**elsius + 32

Formulas are valuable tools because they allow us to calculate measurements as long as we know certain other measurements. For example, if we know that we traveled a distance of 100 miles at a rate of 40 miles per hour, we can replace the variables d and r in the formula $d = rt$ and find our time, t.

$$d = rt \qquad \text{Formula.}$$

$$100 = 40t \qquad \text{Replace } d \text{ with 100 and } r \text{ with 40.}$$

This is a linear equation in one variable, t. To solve for t, divide both sides of the equation by 40.

$$\frac{100}{40} = \frac{40t}{40} \qquad \text{Divide both sides by 40.}$$

$$\frac{5}{2} = t \qquad \text{Simplify.}$$

The time traveled is $\frac{5}{2}$ hours or $2\frac{1}{2}$ hours.

In this section, we solve problems that can be modeled by known formulas.

EXAMPLE 1 A glacier is a giant mass of rocks and ice that flows downhill like a river. Portage Glacier in Alaska is about 6 miles long and moves 400 feet per year. Icebergs are created when the front end of the glacier flows into Portage Lake. How long does it take for ice at the head (beginning) of the glacier to reach the lake?

Solution: The appropriate formula is the distance formula. Before applying this formula, notice that we have different units of measure given. The glacier is 6 *miles* long and it moves at a rate of 400 *feet* per year. First let's convert 6 miles to feet. Since

$$1 \text{ mile} = 5{,}280 \text{ feet}$$

then

$$6 \text{ miles} = 6(5280 \text{ feet}) = 31{,}680 \text{ feet}$$

Next let's use the formula $d = r \cdot t$ and replace d, distance, with 31,680 and r, rate, with 400.

$$31{,}680 = 400 \cdot t$$

To solve for t, divide both sides by 400.

$$\frac{31{,}680}{400} = \frac{400 \cdot t}{400} \qquad \text{Divide both sides by 400.}$$

$$79.2 = t \qquad \text{Simplify.}$$

6 miles

Check the solution: To check, substitute 79.2 for t and 400 for r in the distance formula and check to see that the distance is 31,680 feet.

State the conclusions: It takes 79.2 years for the ice at the head of Portage Glacier to reach the lake. ■

EXAMPLE 2 A gallon of water sealer covers 480 square feet. How many one-gallon containers of sealer should be bought to protect a rectangular driveway 24 feet wide by 90 feet long?

Solution: We will first find the area of the driveway. The area formula for a rectangle is $A = lw$. Let $l = 90$ feet and $w = 24$ feet.

$$A = lw$$

$$= (90 \text{ feet})(24 \text{ feet}) \qquad \text{Let } l = 90 \text{ feet and } w = 24 \text{ feet.}$$

$$= 2160 \text{ square feet}$$

To find how many gallon containers of sealer are needed, divide the area of the driveway, 2160 square feet, by the number of square feet that each gallon of sealer will cover, or 480 square feet.

$$\text{Number of gallons needed} = \frac{2160 \text{ square feet}}{480 \text{ square feet}}$$

$$= 4.5$$

Since the question asks how many gallon containers should be bought, our proposed answer is 5 one-gallon containers of sealer.

Check: To check, notice that 5 gallons of sealer covers 5(480 square feet) = 2400 square feet while 4 gallons of sealer covers 4(480 square feet) = 1920 square feet. Since the driveway is 2160 square feet, 5 gallons will be enough.

State: Five one-gallon containers should be bought. ■

EXAMPLE 3 The average maximum temperature for January in Algerias, Algeria, is 59° Fahrenheit. Find the equivalent temperature in degrees Celsius.

Solution: The formula $F = \frac{9}{5}C + 32$ relates degrees Fahrenheit to degrees Celsius. Replace the variable F with 59 and solve the equation for C.

$$F = \frac{9}{5}C + 32$$

$$59 = \frac{9}{5}C + 32 \qquad \text{Replace } F \text{ with 59.}$$

$$59 - 32 = \frac{9}{5}C + 32 - 32 \qquad \text{Subtract 32 from both sides.}$$

$$27 = \frac{9}{5}C \qquad \text{Simplify.}$$

$$\frac{5}{9} \cdot 27 = \frac{5}{9} \cdot \frac{9}{5}C \qquad \text{Multiply both sides by } \frac{5}{9}.$$

$$15 = C \qquad \text{Simplify.}$$

Check: To check, replace C with 15 and F with 59 in the conversion formula and see that a true statement results.

State: Then 59° Fahrenheit is equivalent to 15° Celsius. ■

EXAMPLE 4 Four friends are planning now to go to Disneyland in 3 years in celebration of their graduation from college. They have figured the total cost for the trip to be \$3500. If each invests \$800 now in a savings account paying 5% simple interest, will they have enough money in 3 years?

Solution: The appropriate formula is the simple interest formula:

$$I = PRT$$

Replace P, principal, with 3200 (800×4), r, rate, with 0.05 $(5\% = 0.05)$, and t, time, with 3. Then

$$I = 3200(0.05)(3)$$

or

$$I = 480$$

This means that the interest alone will be \$480, or the total amount in the account will be

$$\$3200 + \$480 = \$3680$$

Check: To check, replace I with 480, p with 3200, r with 0.05, and t with 3 in the interest formula and see that a true statement results.

State: They will have at least \$3500 in the account, so they will have enough money.

2 Suppose that we need, for example, to convert many Fahrenheit temperatures to equivalent degrees Celsius. In this case, it is easier to perform this task by first solving the general formula $F = \frac{9}{5}C + 32$ for C. (We will do this in Example 8.) For this reason, an important skill in algebra is to be able to solve a formula for any one of its specified variables. For example, we say the formula $d = rt$ is solved for d in terms of r and t. We can also solve $d = rt$ for t in terms of d and r. To solve for t, divide both sides of the equation by r.

$$d = rt$$

$$\frac{d}{r} = \frac{rt}{r} \quad \text{Divide both sides by } r.$$

$$\frac{d}{r} = t \quad \text{Simplify.}$$ ■

To solve a formula or an equation for a specified variable, we use the same steps as for solving a linear equation. These steps are listed below.

To Solve Equations for a Specified Variable

Step 1 Clear the equation of fractions by multiplying each side of the equation by the lowest common denominator.

Step 2 Use the distributive property to remove grouping symbols such as parentheses.

Step 3 Combine like terms on each side of the equation.

Step 4 Use the addition property of equality to rewrite the equation as an equivalent equation, with terms containing the specified variable on one side and all other terms on the other side.

Step 5 Use the distributive property and the multiplication property of equality to isolate the specified variable.

EXAMPLE 5 Solve $V = lwh$ for l.

Solution: This formula is used to find the volume of a box. To solve for l, divide both sides by wh.

$$V = lwh$$

$$\frac{V}{wh} = \frac{lwh}{wh} \quad \text{Divide both sides by } wh.$$

$$\frac{V}{wh} = l \quad \text{Simplify.}$$

Since we have isolated l on one side of the equation, we have solved for l in terms of V, w, and h. Remember that it does not matter on which side of the equation we isolate the variable. ■

EXAMPLE 6 Solve $y = mx + b$ for x.

Solution: First isolate mx by subtracting b from both sides.

$$y = mx + b$$

$$y - b = mx + b - b \quad \text{Subtract } b \text{ from both sides.}$$

$$y - b = mx \quad \text{Simplify.}$$

Next solve for x by dividing both sides by m.

$$\frac{y - b}{m} = \frac{mx}{m}$$

$$\frac{y - b}{m} = x \quad \text{Simplify.}$$ ■

EXAMPLE 7 Solve $P = 2l + 2w$ for w.

Solution: This formula relates the perimeter of a rectangle to its length and width. To solve for w, begin by subtracting $2l$ from both sides.

$$P = 2l + 2w$$

$$P - 2l = 2l + 2w - 2l \quad \text{Subtract } 2l \text{ from both sides.}$$

$$P - 2l = 2w$$

$$\frac{P - 2l}{2} = \frac{2w}{2} \quad \text{Divide both sides by 2.}$$

$$\frac{P - 2l}{2} = w \quad \text{Simplify.}$$

■

The next example has an equation containing a fraction. We will first clear the equation of fractions and then solve for the specified variable.

EXAMPLE 8 Solve $F = \frac{9}{5}C + 32$ for C.

Solution:

$$F = \frac{9}{5}C + 32$$

$$5(F) = 5\left(\frac{9}{5}C + 32\right) \quad \text{Clear the fraction by multiplying both sides by the LCD.}$$

$$5F = 9C + 160 \quad \text{Distribute the 5.}$$

$$5F - 160 = 9C + 160 - 160 \quad \text{Subtract 160 from both sides.}$$

$$5F - 160 = 9C$$

$$\frac{5F - 160}{9} = \frac{9C}{9} \quad \text{Divide both sides by 9.}$$

$$\frac{5F - 160}{9} = C$$

■

3 Next we sharpen our problem-solving skills by writing algebraic expressions.

EXAMPLE 9 If x is the first of three consecutive integers, express the sum of the three integers in terms of x. Simplify if possible.

Solution: An example of three consecutive integers is 7, 8, 9. (+1, +2) The second consecutive integer is always one unit more than the first, and the third consecutive integer is two units more than the first. If x is the first of three consecutive integers, the three consecutive integers are

$$x, \quad x + 1, \quad x + 2$$

(+1, +2)

Their sum is $x + (x + 1) + (x + 2)$, which simplifies to $3x + 3$. ■

EXERCISE SET 2.4

Substitute the given values into the given formulas and solve for the unknown variable. See Examples 1 through 4.

1. $A = bh$; $A = 45$, $b = 15$
2. $D = rt$; $D = 195$, $t = 3$
3. $I = PRT$; $P = 5000$, $R = 0.08$, $T = 2$
4. $V = lwh$; $l = 14$, $w = 8$, $h = 3$
5. $C = 2\pi r$; $C = 94.2$ (use the approximation 3.14 for π)
6. $A = \pi r^2$; $r = 4$ (use the approximation 3.14 for π)

Solve the following by using known formulas. See Example 1.

7. The distance from the Sun to Earth is approximately 93,000,000 miles. If light travels at a rate of 186,000 miles per second, how long does it take light from the Sun to reach Earth?
8. Light travels at a rate of 186,000 miles per second. If our moon is 238,860 miles from Earth, how long does it take light from the moon to reach Earth? (Round to the nearest tenth of a second.)

Delta Airlines awards "Frequent Flyer" miles equal to the number of miles traveled rounded **up** *to the nearest thousand. Use this for Exercises 9 and 10.*

9. A 5.5-hour nonstop flight from Orlando, Florida, to San Francisco, California, averages 470 mph. Find the "Frequent Flyer" miles earned on this flight.
10. A 45-minute $\left(\frac{3}{4}\text{-hour}\right)$ nonstop flight from New Orleans, Louisiana, to Houston, Texas, averages 500 mph. Find the "Frequent Flyer" miles earned on this flight.

See Example 2.

11. Find how much rope is needed to wrap around Earth at the equator, if the radius of Earth is 4000 miles. (Hint: Use $\pi = 3.14$ and the formula for circumference.)

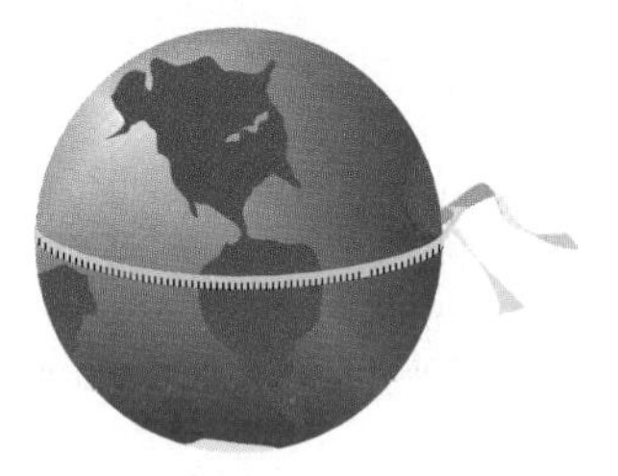

12. If the length of a rectangularly shaped garden is 6 meters and its width is 4.5 meters, find the amount of fencing required.

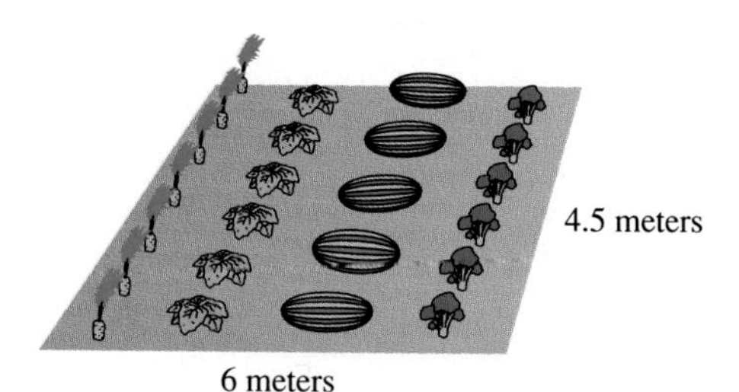

See Example 3.

13. Dry ice is solidified carbon dioxide. At $-78.5°$ Celsius, it changes directly from a solid to a gas. Convert this temperature to degrees Fahreheit.
14. Lightning bolts can reach a temperature of 50,000° Fahrenheit. Convert this to degrees Celsius.
15. Convert Nome, Alaska's 14°F high temperature to Celsius.
16. Convert Paris, France's low temperature of $-5°$C to Fahrenheit.

See Example 4.

17. Bruce and Melanie BlenKarn want to win the Publisher's Clearinghouse Sweepstakes and live off the interest earned by placing it into a savings account paying 8% simple interest annually. Find how much they need to win to have \$40,000 annually to live on. (8% = 0.08.)
18. Julie Tarr needs \$1050 for fall tuition next year. If she places \$985 in a passbook savings account that pays 7% simple interest, will she have enough money in this account after 1 year to pay for fall tuition? (7% = 0.07)

Solve the following formulas for the indicated variable. See Examples 5 through 8.

19. $f = 5gh$ for h

20. $C = 2\pi r$ for r

21. $V = LWH$ for W

22. $T = mnr$ for n

23. $3x + y = 7$ for y

24. $-x + y = 13$ for y

25. $A = p + PRT$ for R

26. $A = p + PRT$ for T

27. $V = \frac{1}{3}Ah$ for A

28. $D = \frac{1}{4}fk$ for k

See Example 9.

29. If x is the first of three consecutive integers, express the sum of the first integer and the third integer as an algebraic expression in terms of x.

30. If x is the first of two consecutive integers, express the sum of 20 and the second consecutive integer as an algebraic expression in terms of x.

Write an algebraic expression.

31. A quadrilateral is a four-sided figure like the one shown next whose angle sum is 360°. If one angle measures $x°$, a second angle measures $3x°$, and a third angle measures $5x°$, express the measure of the fourth angle in terms of x.

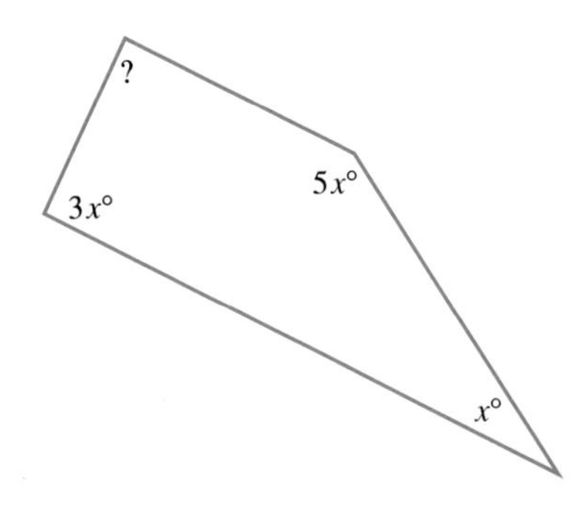

32. The sum of the angles of a triangle is 180°. If one angle of a triangle measures $x°$ and a second angle measures $(2x + 7)°$, express the measure of the third angle in terms of x.

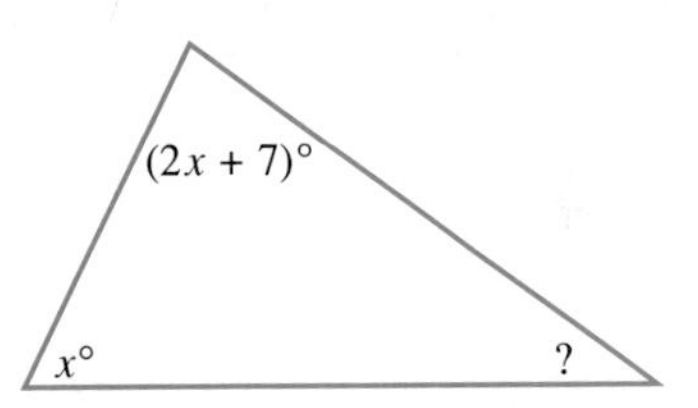

Substitute the given values into the given formulas and solve for the unknown variable.

33. $y = mx + b$; $y = 2$, $m = 3$, and $x = -1$

34. $y - y_1 = m(x - x_1)$; $y_1 = 5$, $m = 0.5$, $x = 4$, $x_1 = 2$

35. $A = 0.5h(b_1 + b_2)$; $A = 40$, $b_1 = 11$, $b_2 = 9$

36. $A = 0.5bh$; $A = 35$, $h = 7$

37. $V = \frac{1}{3}\pi r^2 h$; $V = 565.2$, $r = 6$ (use the approximation 3.14 for π)

38. $V = \frac{4}{3}\pi r^3$; $r = 2$ (use the approximation 3.14 for π)

Solve.

39. The SR-71 is a top-secret spy plane. It is capable of traveling from Rochester, New York to San Francisco, California, a distance of approximately 3000 miles, in $1\frac{1}{2}$ hours. Find the rate of the SR-71.

40. A limousine built in 1968 for the President cost $500,000 and weighed 5.5 tons. This Lincoln Continental Executive could travel at 50 miles per hour with all its tires shot away. At this rate, how long would it take to travel from Charleston, South Carolina, to Washington, D.C., a distance of 375 miles?

41. Charles Douglass can afford enough fencing to enclose a rectangular garden with a perimeter of 140 feet. If he wants one dimension of his garden to be 40 feet, find the other dimension.

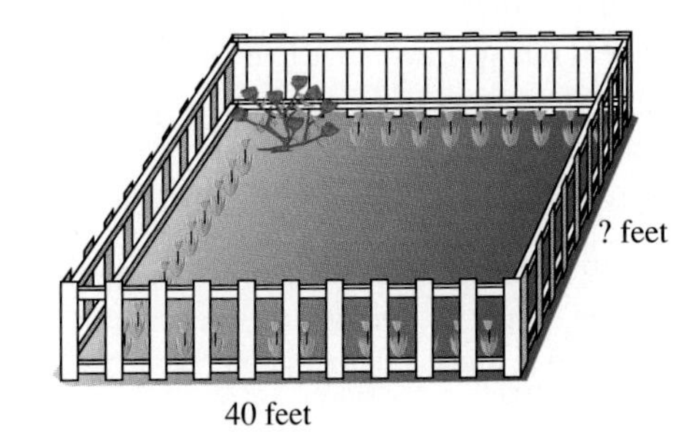

42. If the area of a right triangularly shaped sail is 20 square feet and its base is 5 feet, find the height of the sail.

43. A couple would like to retire eventually and live off the interest earned on a savings account that pays 7% simple interest. How much do they need to place in the savings account in order to have \$35,000 per year to live on? (7% = 0.07.)

44. How much interest will be paid on a \$7000 car loan in 3 years at a bank that charges 11% simple interest? (11% = 0.11.)

Dante II is a spiderlike robot that is used to map the depths of an active Alaskan volcano.

45. The dimensions of *Dante II* are 10 feet long by 8 feet wide by 10 feet high. Find the volume of the smallest box needed to store this robot.

46. *Dante II* traveled 600 feet into an active Alaskan volcano in $3\frac{1}{3}$ hours. Find the traveling rate of *Dante II* in feet per minute. (Hint: First convert $3\frac{1}{3}$ hours to minutes.)

47. Piranhas require 1.5 cubic feet of water per fish to maintain a healthy environment. Find the maximum number of piranhas you could put in a tank measuring 8 feet by 3 feet by 6 feet.

48. Maria's Pizza sells one 16-inch cheese pizza or two 10-inch cheese pizzas for \$9.99. Determine which size gives more pizza.

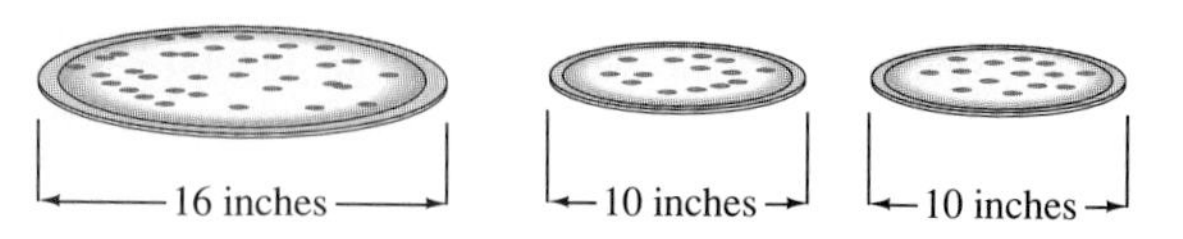

49. Find how long it takes Mark Bassi to drive 135 miles on I-10 if he merges onto I-10 at 10 A.M. and drives nonstop with his cruise control set on 60 mph.

50. Beaumont, Texas, is about 150 miles from Toledo Bend. If Leo Miller leaves Beaumont at 4 A.M. and averages 45 mph, when should he arrive at Toledo Bend?

51. Find the temperature at which the Celsius measurement and Fahrenheit measurement are the same number.

52. Find how many goldfish you can put in a cylindrical tank whose diameter is 8 meters and whose height is 3 meters, if each goldfish needs 2 cubic meters of water.

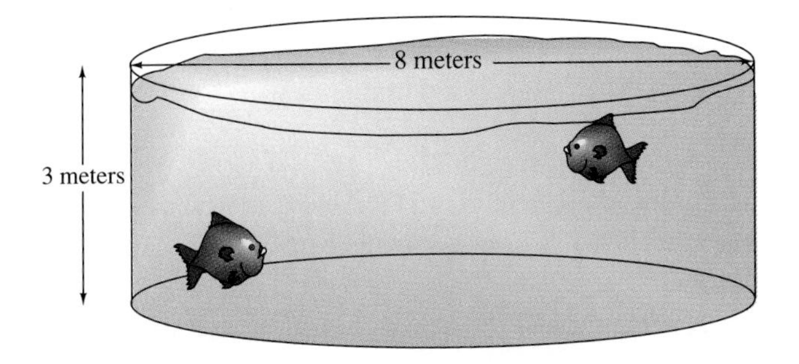

53. Bolts of lightning can travel at 270,000 miles per second. At that rate, how many times can you travel around the world in one second?

54. A glacier is a giant mass of rocks and ice that flows downhill like a river. Exit Glacier, near Seward, Alaska, moves at a rate of 20 inches a day. Find the distance in feet the glacier moves in a year. (Assume 365 days in a year.)

55. On July 16, 1994, the Shoemaker-Levy 9 comet collided with Jupiter. The impact of the largest fragment of the comet, a massive chunk of rock and ice, created a fireball with a radius of 2000 miles. Find the volume of this spherical fireball.

56. The fireball from the largest fragment of the comet (see Exercise 55) immediately collapsed as it was pulled down by gravity. As it fell, it cooled to approximately $-350°$ Fahrenheit. Convert this to degrees Celsius.

57. Stalactites join stalagmites to form columns. A column found at Natural Bridge Caverns near San Antonio, Texas, rises 15 feet and has a **diameter** of only 2 inches. Find the volume of this column. (Hint: Use the formula for the volume of a cylinder.)

58. A lawn in the shape of a trapezoid with a height of 60 feet and bases of 70 feet and 130 feet. How many bags of fertilizer must be purchased to cover the lawn if each bag covers 4000 square feet?

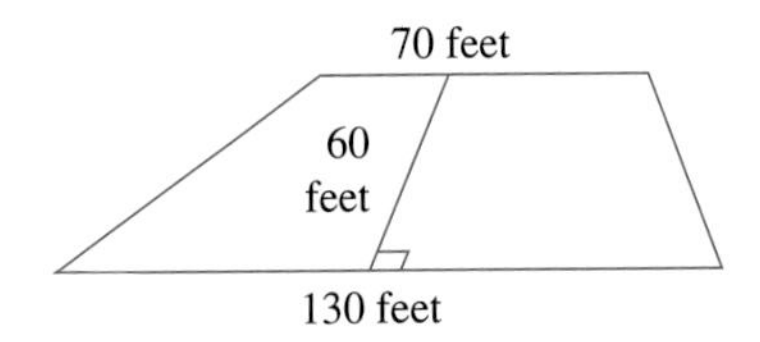

59. Normal room temperature is about 78°F. Write this temperature as degrees Celsius.

60. Flying fish do not fly, but actually glide. They have been known to travel a distance of 1300 feet at a rate of 20 miles per hour. How many seconds did it take to travel this distance? (Hint: First convert miles per hour to feet per second. Recall that 1 mile = 5280 feet.)

61. The X-30 is a new "space plane" being developed that will skim the edge of space at 4000 miles per hour. Neglecting altitude, if the circumference of Earth is approximately 25,000 miles, how long will it take the X-30 to travel around the Earth?

62. In the United States, the longest hang-glider flight was a 303-mile, $8\frac{1}{2}$ hour flight from New Mexico to Kansas. What was the average rate during this flight?

Solve each equation for the specified variable.

63. $D = rt$ for t

64. $W = gh$ for g

65. $I = PRT$ for R

66. $V = LWH$ for L

67. $9x - 4y = 16$ for y

68. $2x + 3y = 17$ for y

69. $P = a + b + c$ for a

70. $A = Prt + P$ for P

71. $T = C(2 + AB)$ for B

72. $A = 5H(b + B)$ for b

73. $C = 2\pi r$ for r

74. $S = 2\pi r^2 + 2\pi rh$ for h

75. $\frac{1}{u} - \frac{1}{v} = \frac{1}{w}$ for w

76. $\frac{1}{r_1} + \frac{1}{r_2} = \frac{1}{R}$ for R

Write each algebraic expression described.

77. In a mayoral election, Erica Martin received 284 more votes than Charles Pecot. If Charles received n votes, how many votes did Erica receive?

78. A 5-foot piece of string is cut into two pieces. If one piece is x feet long, express the other length in terms of x.

79. If x represents the first odd integer, express the next odd integer in terms of x.

80. If x represents the first of two consecutive even integers, express the sum of the two integers in terms of x.

81. If x represents the first even integer, express the next even integer in terms of x.

82. If x represents the first of two consecutive odd integers, express the sum of the two integers in terms of x.

83. Express the sum of three odd consecutive integers as an expression in x. Let x be the first odd integer.

84. If the width of a pad of paper is x inches and its length is 3 more than twice the width, express the perimeter as an expression in x.

85. If the money box in a drink machine contains x nickels, $5x$ dimes, and $30x - 1$ quarters, express their total **value** as an expression in x.

86. If x is the first of four consecutive even integers, write their sum as an expression in x.

Skill Review

Write the following phrases as algebraic expressions. See Section 1.5.

87. Nine divided by the sum of a number and 5

88. Half the product of a number and five

89. Three times the sum of a number and four

90. One-third of the quotient of a number and six

91. Double the sum of ten and four times the number

92. Twice a number divided by three times the number

93. Triple the difference of a number and twelve

94. A number minus the sum of the number and six

Writing in Mathematics

95. Make up one example of a formula that might be used when cooking.

96. When solving an equation, why does it not matter on which side you isolate the variable?

97. Many topics have been introduced in this section's exercises: travel, astronomy, and gardening, just to name a few. Pick a topic in this section or any other topic of interest to you and write a paragraph about how math is used to perform or study this topic.

2.5
An Introduction to Problem Solving

OBJECTIVE

1 Apply the steps for problem solving.

Tape NA

1 In the previous section, we solved problems using known formulas as models. In this section, we solve problems using equation models we create. The problem-solving steps given next may be helpful.

Problem-Solving Steps

1. UNDERSTAND the problem. At this stage, don't work with variables (except for known formulas), but simply become comfortable with the problem. Some ways of accomplishing this are:
   - Read and reread the problem.
   - Construct a drawing.
   - Look up an unknown formula.
   - Guess the answer and check your guess. Pay careful attention as to how to check your guess. This will help later when you model the problem.

2. ASSIGN a variable to an unknown in the problem. Use this variable to represent any other unknown quantities.
3. ILLUSTRATE the problem. A diagram or chart using the assigned variables can often help to visualize the known facts.
4. TRANSLATE the problem into a mathematical model. This is often an equation.
5. COMPLETE the work. This often means to solve the equation.
6. INTERPRET the results. This means to *check* your work in the stated problem and to *state* your conclusion.

EXAMPLE 1 A 10-foot piece of board is to be cut into two pieces so that the longer piece is 4 times the shorter. Find the length of each piece.

Solution: **1.** UNDERSTAND the problem. To do so, read and reread the problem, draw a diagram, and guess the solution. For example, if 3 feet represents the length of the shorter piece, then $4(3) = 12$ feet is the length of the longer piece since it is 4 times the length of the shorter piece. This guess give us a total board length of 3 feet + 12 feet = 15 feet, which is too long. The purpose of guessing a solution is not to guess correctly (although this may happen) but to help us better understand the problem and how to model it. At this point, we have a better understanding of the problem. Although it may be possible for us to guess the solution now, for purposes of practicing problem-solving skills, we will continue with our steps.

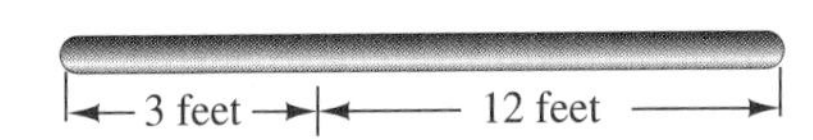

2. ASSIGN a variable. Use this variable to represent any other known quantities. If we let

$$x = \text{length of the shorter piece}$$

then

$$4x = \text{length of longer piece}$$

3. ILLUSTRATE the problem. We do so by drawing a picture of the board and labeling the picture with our assigned variables.

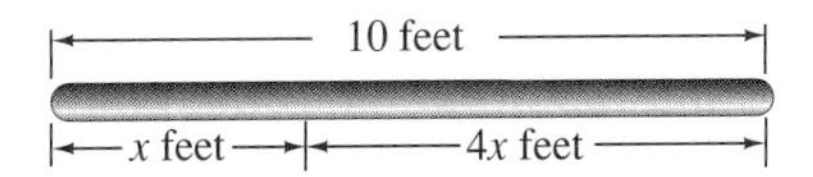

4. TRANSLATE the problem. We first write our equation in words.

In words:	length of shorter piece	added to	length of longer piece	equals	total length of board
Translate:	x	$+$	$4x$	$=$	10

5. COMPLETE the work. Here we solve the equation.

$$x + 4x = 10$$
$$5x = 10$$
$$\frac{5x}{5} = \frac{10}{5}$$
$$x = 2$$

6. INTERPRET the results. First, *check* the solution in the stated problem. If the shorter piece of board is 2 feet, the longer piece is 4 (2 feet) = 8 feet and the sum 2 feet + 8 feet = 10 feet.

   Next *state* the conclusions. The shorter piece of board is 2 feet and the longer piece of board is 8 feet. ■

EXAMPLE 2 In 1994, the U.S. Congress had 12 more Democrat senators than Republican senators. If there are a total of 100 senators, how many senators of each party were there?

Solution:

1. UNDERSTAND the problem. Read and reread the problem. Let's guess that there are 30 Republican senators. Since there are 12 more Democrats than Republicans, there must be 30 + 12 = 42 Democrat senators. The total number of Republicans and Democrats is then 30 + 42 = 72. This is an incorrect guess since the total should be 100, but we now understand the problem. This was our purpose for guessing.
2. ASSIGN a variable. Let

$$x = \text{number of Republican senators}$$

Then

$$x + 12 = \text{number of Democrat senators}$$

3. ILLUSTRATE the problem. No diagram or chart is needed.
4. TRANSLATE the problem. First we write our equation in words.

In words:	number of Republicans	added to	number of Democrats	equals	100
Translate:	x	$+$	$(x + 12)$	$=$	100

5. COMPLETE the work. We solve the equation.

$$x + (x + 12) = 100$$
$$2x + 12 = 100$$
$$2x + 12 - 12 = 100 - 12$$
$$2x = 88$$
$$\frac{2x}{2} = \frac{88}{2}$$
$$x = 44$$

6. INTERPRET the results.

   Check: If there are 44 Republican senators, then there are 44 + 12 = 56 Democrat senators. The total number of senators is then 44 + 56 = 100. The results check.

   State the results: In 1994, there were 44 Republican and 56 Democrat senators. ■

EXAMPLE 3 The length of a rectangular sign is 2 feet less than three times its width. Find the dimensions if the perimeter is 28 feet.

Solution: **1.** UNDERSTAND. Read and reread the problem. Recall that the formula for the perimeter of a rectangle is $P = 2l + 2w$. Draw a rectangle and guess the solution. If the width of the rectangular sign is 5 feet, its length is 2 feet less than 3 times the width or 3(5 feet) − 2 feet = 13 feet. The perimeter of the rectangle is then P = 2(13 feet) + 2(5 feet) = 36 feet, which is too much. We now know that the width is less than 5 feet.

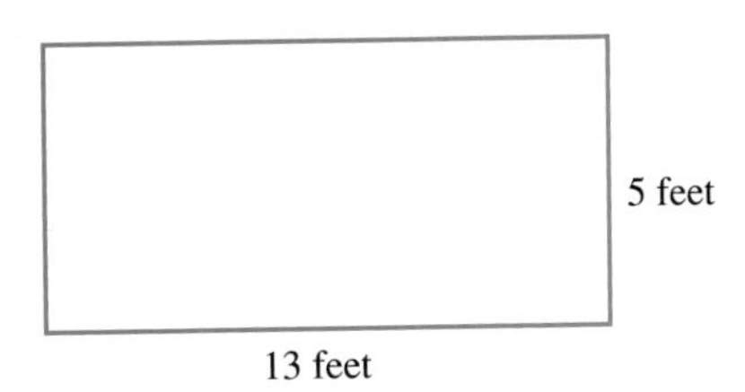

2. ASSIGN. Let

$$x = \text{the width of the rectangular sign}$$

and then

$$3x - 2 = \text{the length of the sign}$$

3. ILLUSTRATE. Draw a rectangle and label it with the assigned variables.

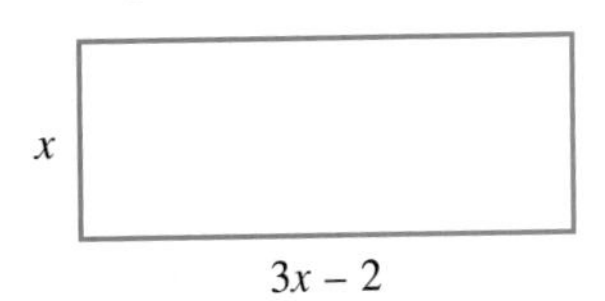

4. TRANSLATE. We use the formula

$$P = 2l + 2w$$

or translate:

$$28 = 2(3x - 2) + 2(x)$$

5. COMPLETE.

$$28 = 6x - 4 \quad + 2x$$

$$28 = 8x - 4$$

$$28 + 4 = 8x - 4 \quad + 4$$

$$32 = 8x$$

$$\frac{32}{8} = \frac{8x}{8}$$

$$4 = x$$

6. INTERPRET. *Check:* If the width of the sign is 4 feet, the length of the sign is $3(4 \text{ feet}) - 2 \text{ feet} = 10$ feet. This gives a perimeter of $P = 2(4 \text{ feet}) + 2(10 \text{ feet}) = 28$ feet, the correct perimeter.
   State: The width of the sign is 4 feet and the length of the sign is 10 feet. ■

EXAMPLE 4 A local cellular phone company charges Elaine Chapoton \$50 per month and \$0.36 per minute of phone use in her usage category. If Elaine budgeted \$100 per month for cellular phone use, determine the number of whole minutes of phone use.

Solution:

1. UNDERSTAND. Read and reread the problem. Next, guess an answer. Let's guess 70 minutes; pay careful attention as to how we check this guess. For 70 minutes of use, Elaine's phone bill will be \$50 plus \$0.36 per minute of use. This is $\$50 + \$0.36(70) = \$75.20$, which is less than \$100. We now understand the problem and know that the number of minutes is greater than 70.
2. ASSIGN. Let x represent the unknown in this problem, or let

$$x = \text{number of minutes}$$

3. ILLUSTRATE. No diagram is needed.
4. TRANSLATE.

In words:	\$50	added to	minute charge	is equal to	\$100
Translate:	50	+	$0.36x$	=	100

5. COMPLETE.

$$50 + 0.36x = 100$$
$$50 + 0.36x - 50 = 100 - 50$$
$$0.36x = 50$$
$$\frac{0.36x}{0.36} = \frac{50}{0.36}$$
$$x = 138 \text{ (rounded down to the nearest whole)}$$

6. INTERPRET. *Check:* If Elaine spends 138 minutes on her cellular phone, her bill is $\$50 + \$0.36(138) = \$99.68$, slightly under her budget of \$100.
   State: Elaine can spend approximately 138 minutes per month on her cellular phone and stay under her budget of \$100. ■

EXERCISE SET 2.5

Solve each word problem. See Examples 1 and 3.

1. An architect designs a rectangular flower garden such that the width is exactly two-thirds of the length. If 260 feet of antique picket fencing are to be used, find the dimensions of the garden.

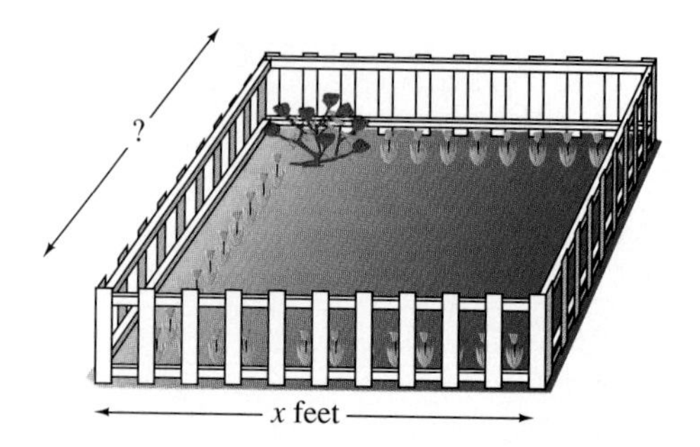

2. If the length of a rectangular parking lot is 10 meters less than twice its width and the perimeter is 400 meters, find the length of the parking lot.

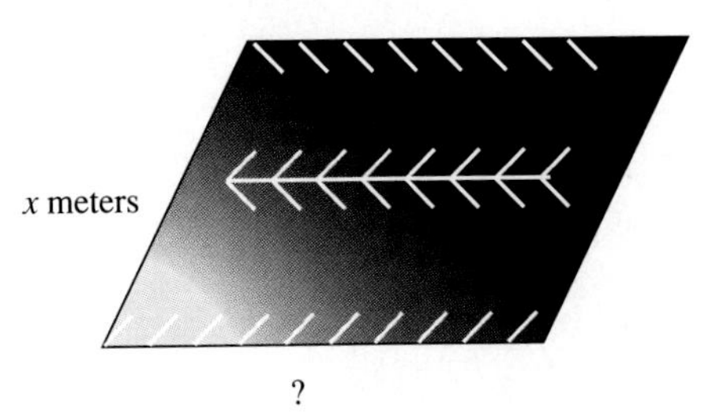

3. The perimeter of a yield sign in the shape of an isosceles triangle is 22 feet. If the shortest side is 2 feet less than the other two sides, find the length of the shortest side. (*Hint:* An isosceles triangle has two sides the same length.

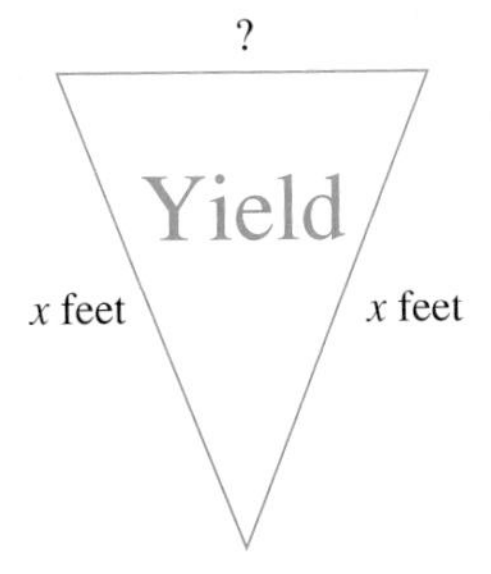

4. The measures of the angles of a triangle are 3 consecutive even integers. Find the measure of each angle.

See Examples 2 and 4.

5. In the 1992 Summer Olympics, the Unified team consisting of athletics from 12 former Soviet republics won 8 more gold medals than the U. S. team. If the total number of gold medals for both is 82, find the number of gold medals that each team won.
6. The governor of New York makes twice as much money as the governor of Nebraska. If the total of their salaries is \$190,000, find the salary of each.
7. A plumber gave an estimate for a renovation of a kitchen. Her hourly pay is \$27 per hour and the plumber's parts will cost \$80. If her total estimate is \$404, how many hours does she expect this job to take?
8. A car rental agency advertised renting a Buick Century for \$24.95 per day and \$0.29 per mile. If you rent this car for 2 days, how many miles can you drive and keep your final bill under \$100?

Solve each problem.

9. The flag of Brazil contains a parallelogram. One angle of the parallelogram is 15° less than twice the measure of the angle next to it. Find the measure of each angle of the parallelogram. (*Hint:* Recall that opposite angles of a parallelogram have the same measure and that the sum of the angles is 360°.)

Brazil

10. The flag of Equatorial Guinea contains an isosceles triangle. (Recall that an isosceles triangle contains two angles with the same measure.) If the measure of the third angle of the triangle is 8° more than either of the other two angles, find the measure of each angle of the triangle.

Equatorial Guinea

11. The perimeter of an equilateral triangle is 7 inches more than the perimeter of a square, and the side of the triangle is 5 inches longer than the side of the square. Find the side of the triangle.
12. A square animal pen and a pen shaped like an equilateral triangle have equal perimeters. Find the length of the sides of each pen if the sides of the triangular pen are fifteen less than twice a side of the square pen. (*Hint:* An equilateral triangle has three sides the same length.

13. On June 20, 1994, John Paxson sank a 3-point shot with 3.9 seconds left to give the Chicago Bulls their third straight National Basketball Association championship. The opposing team was the Phoenix Suns. If the final score of the game was 2 consecutive integers whose sum is 197, find each final score.
14. To make an international telephone call, the code for the country you are calling is needed. The codes for Mali Republic, Cote d'Ivoire, and Niger are 3 consecutive odd integers whose sum is 675. Find the code for each country.

15. Determine whether there are two consecutive odd integers such that 7 times the first exceeds 5 times the second by 54.

16. The sum of three consecutive integers is 13 more than twice the smallest integer. Find the integers.

17. The golden rectangle is a rectangle whose length is approximately 1.6 times its width. Early Greeks thought that a rectangle with these proportions was the most pleasing rectangle to the eye, and examples of the golden rectangle are found in many early works of art. For example, the Parthenon in Athens contains many examples of golden rectangles. Mike Hallahan would like to plant a rectangular garden in the shape of a golden rectangle. If he has 78 feet of fencing available, find the dimensions of the garden.

The Parthenon

18. Dr. Dorothy Smith gave the students in her geometry class at the University of New Orleans the following question. Is it possible to construct a triangle such that the second angle of the triangle has a measure that is twice the measure of the first angle and the measure of the third angle is 3 times the measure of the first? If so, find the measure of each angle. (*Hint:* Recall that the sum of the measures of the angles of a triangle is 180°.)

19. One angle is twice its complement increased by 30°. Find the two complementary angles.

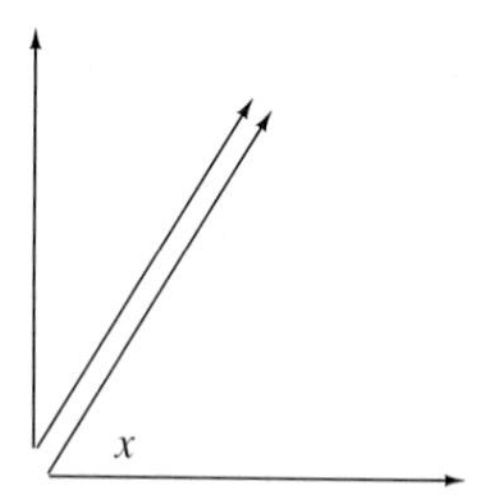

20. Find an angle such that its supplement is equal to twice its complement increased by 50°.

21. On December 7, 1995, a probe launched from the robot explorer called *Galileo* will enter the atmosphere of Jupiter at 100,000 miles per hour. The diameter of the probe is 19 inches less than twice its height. If the sum of the height and the diameter is 83 inches, find each dimension.

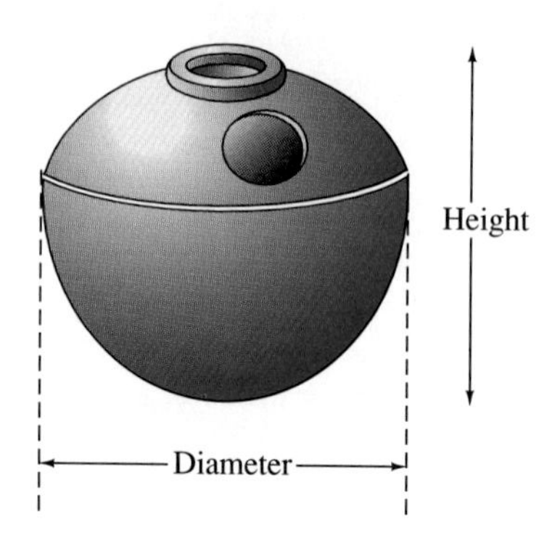

22. Over the past few years, the satellite *Voyager II* has passed by the planets Saturn, Uranus, and Neptune, continually updating information about these planets, including the number of moons for each. Uranus is now believed to have 7 more moons than Neptune, and Saturn is now believed to have 6 more than twice the number of moons of Neptune. If the total number of moons for these three planets is 45, find the number of moons for each planet.

23. Find three consecutive even integers such that the sum of the first integer and three times the third integer is 64 more than the second integer.

24. Find three consecutive odd integers such that four times the first minus the third is the first increased by 10.

25. A flower bed is in the shape of a triangle with one side twice the length of the shortest side, and the third side is 30 feet more than the length of the shortest side. Find the dimensions if the perimeter is 102 feet.

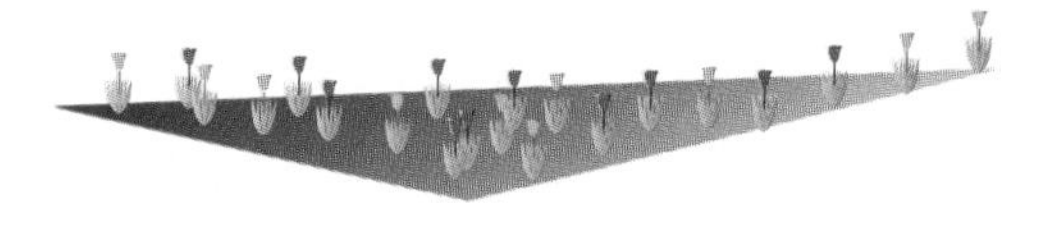

26. Find measures of the angles of a triangle if one angle is twice another angle and the third angle is 105°.

27. Find three consecutive integers such that the sum of the first integer and twice the second integer is 54 more than the third integer.

28. Find four consecutive integers such that their sum is -54.

29. The sum of the angles in a triangle is 180°. Find the measures of the angles of a triangle whose two base angles are equal and whose third angle is 10° less than three times a base angle.

30. Two angles are supplementary if their sum is 180°. One angle measures three times the measure of a smaller angle. If x represents the measure of the smaller angle and these two angles are supplementary, find the measure of each angle.

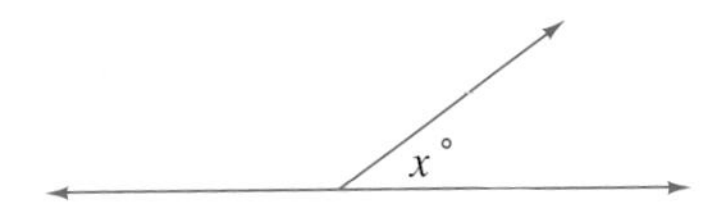

31. A 21-foot beam is to be divided so that the longer piece is 1 foot more than 3 times the shorter piece. If x represents the length of the shorter piece, find the lengths of both pieces.

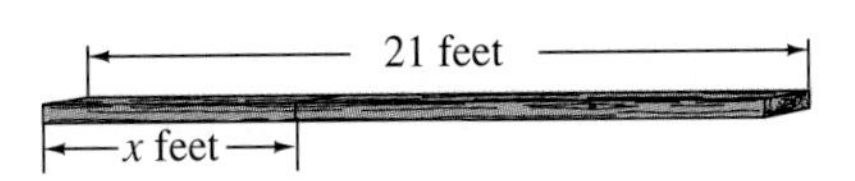

32. A 40-inch board is to be cut into three pieces so that the second piece is twice as long as the first piece and the third piece is 5 times as long as the first piece. If x represents the length of the first piece, find the lengths of all three pieces.

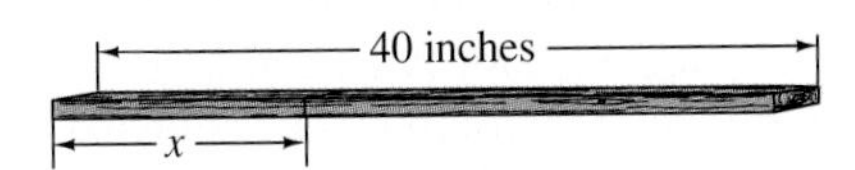

33. A woman's $15,000 estate is to be divided so that her husband receives twice as much as her son. If x represents the amount of money that her son receives, find the amount of money that her husband receives and the amount of money that her son receives.

34. Measure the dimensions of each rectangle and decide which one best approximates the shape of the golden rectangle. (See Exercise 17.)

a.

b.

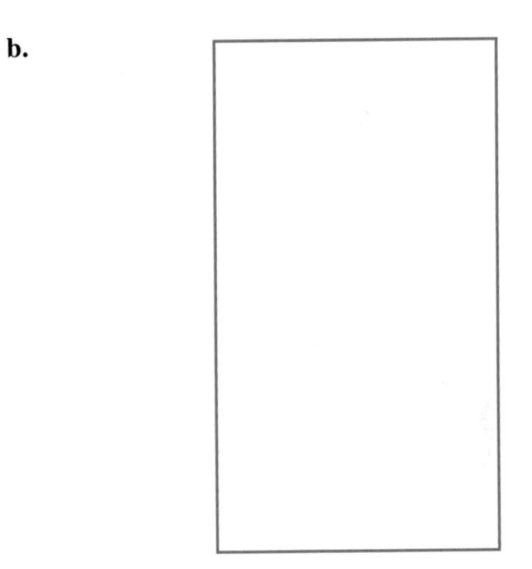

c.

Skill Review

Add or subtract the following. See Section 1.6.

35. $3 + (-7)$

36. $-2 + (-8)$

37. $4 - 10$

38. $-11 + 2$

39. $-5 - (-1)$

40. $-12 - 3$

Translate each sentence into an equation. See Section 1.5.

41. Half of the difference of a number and one is thirty-seven.

42. Five times the opposite of a number is the number plus sixty.

43. If three times the sum of a number and 2 is divided by 5, the quotient is 0.

44. If the sum of a number and 9 is subtracted from 50, the result is 0.

Writing in Mathematics

45. Give an example of how you solved a mathematical application in the last week.

Recall from Exercises 17 and 34 that the golden rectangle is a rectangle whose length is approximately 1.6 times its width.

46. It is thought that about 75% of adults believe that a rectangle in the shape of the golden rectangle is the most pleasing to the eye.

Draw three rectangles, one in the shape of the golden rectangle and poll your class. Do the results agree with the percent above?

47. Examples of golden rectangles can be found today in architecture and manufacturing packaging. Find an example of a golden rectangle in your home. (A few suggestions: Look at the front cover of a book, the floor of a room, or the front of a box of food.)

2.6 Reading Graphs and Ordered Pairs

OBJECTIVES

Tape NA

1. Read bar graphs.
2. Read line graphs.
3. Graph ordered pairs of numbers.

1 In today's world, where the exchange of information is required to be fast and entertaining, graphs are increasingly popular. Thus far, we have practiced reading bar graphs, circle graphs, and pie charts. In this section, we will continue our study of bar graphs and introduce line graphs and ordered pairs of numbers. First let's review reading bar graphs.

EXAMPLE 1 The following bar graph shows selected electricity rates per kilowatt hour.

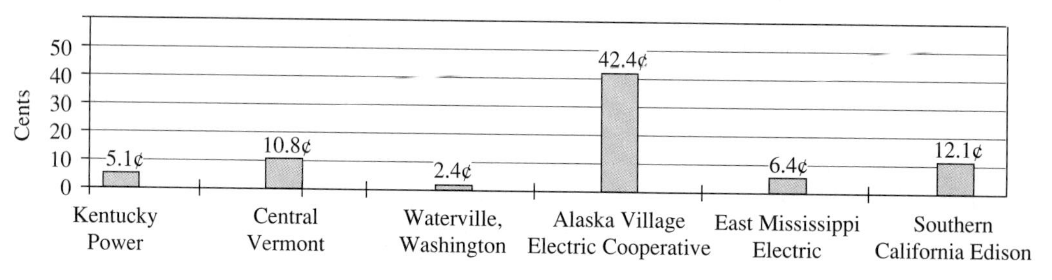

Source: *USA Today*, 1994.

a. Which company charges the highest rate?

b. Which company charges the lowest rate?

c. Find the difference in rates of the companies found in parts a and b.

d. What is the electricity rate charged by Southern California Edison?

Solution: **a.** The highest bar corresponds to the company that charges the highest rate. Alaska Village Electric Cooperative charges the highest rate, which is 42.4¢ per kilowatt hour.

b. The lowest bar corresponds to the company that charges the lowest rate. Waterville, Washington, charges the lowest rate, which is 2.4¢ per kilowatt hour.

c. The difference in rates is 42.4¢ − 2.4¢ = 40¢.

d. To find Southern California Edison's rate, find the bar corresponding to this company and read its height. Southern California Edison charges 12.1¢ per kilowatt hour. ■

A bar graph can consist of vertically arranged bars, as in the graph above, or horizontally arranged bars, as in the graph in Example 2.

EXAMPLE 2 The following bar graph shows Disney's top five animated films before 1993. This graph is determined by U.S. box office revenue.

Disney's Top Five Animated Feature Films

Year First Released

1942 *Bambi* $110

1967 *The Jungle Book* $130

1991 *Beauty and the Beast* $145

1937 *Snow White and the Seven Dwarfs* $176

1992 *Aladdin* $217

100 120 140 160 180 200 220

Dollars (in millions)

Source: *USA Today,* 1994.

a. Which Disney animated film generated the most income for the company?

b. How much more money did the film *Aladdin* generate than *Beauty and the Beast?*

Solution: **a.** Since these bars are arranged horizontally, we look for the bar that extends the farthest to the right. In 1992, the film *Aladdin* generated $217 million, the most income, for Disney.

b. *Beauty and the Beast* generated $145 million. To find how much more money *Aladdin* generated than *Beauty and the Beast,* we subtract: $217 − $145 = $72 million. ■

2 The next graph is called a broken line graph or simply a line graph.

EXAMPLE 3 This particular graph shows the relationship between smoking a cigarette and pulse rate. The horizontal units are minutes, with 0 minute being the moment a smoker lights a cigarette. The vertical units are pulse rates per minute.

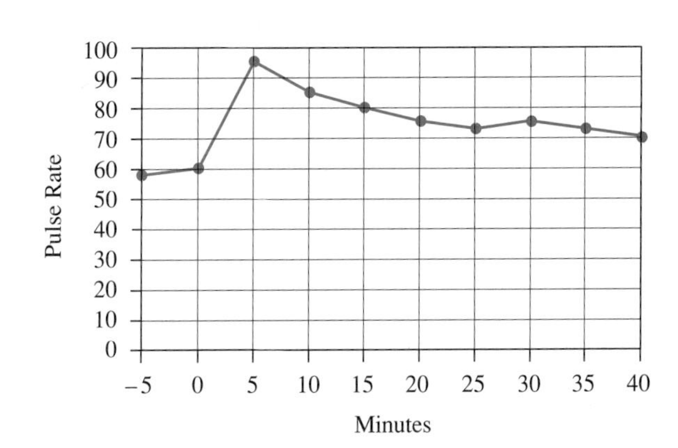

a. What is the pulse rate 15 minutes after lighting a cigarette?
b. When is the pulse rate the lowest?
c. When does the pulse rate show the greatest increase?

Solution: **a.** Locate the number 15 along the minutes axis and move vertically upward until the line is reached. From the line graph, move horizontally to the left until the pulse rate axis is reached. The pulse rate is 80 beats per minute 15 minutes after lighting a cigarette.

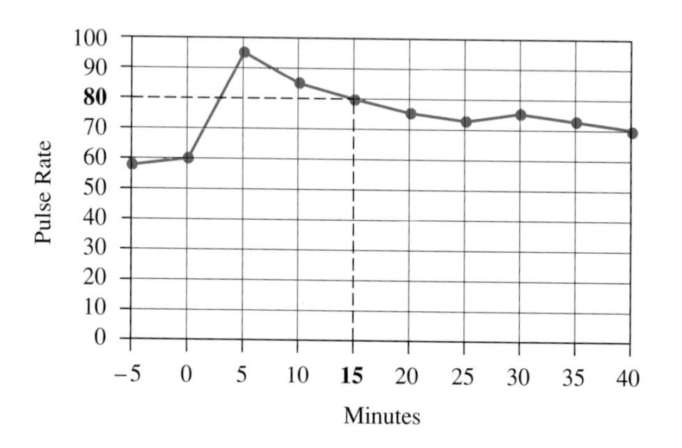

b. Find the lowest point of the graph. This represents the lowest pulse rate. Move vertically downward to the minutes axis. The pulse rate is the lowest at -5, which stands for 5 minutes before lighting a cigarette.

c. The pulse rate shows the greatest increase when the change in the pulse rate from one point to the next is the greatest. This occurs when the line graph is the steepest. The line graph is the steepest between 0 and 5 minutes, so the pulse rate increases the most 5 minutes after lighting a cigarette. ■

3 Notice in this graph that there are two numbers associated with each point of the graph. For example, we discussed earlier that *15* minutes after lighting a cigarette, the pulse rate is *80* beats per minute. If we agree to write the minutes first and the pulse rate second, we can say there is a point on the graph corresponding to the **ordered pair** of numbers (15, 80). A few more ordered pairs are listed alongside their corresponding points.

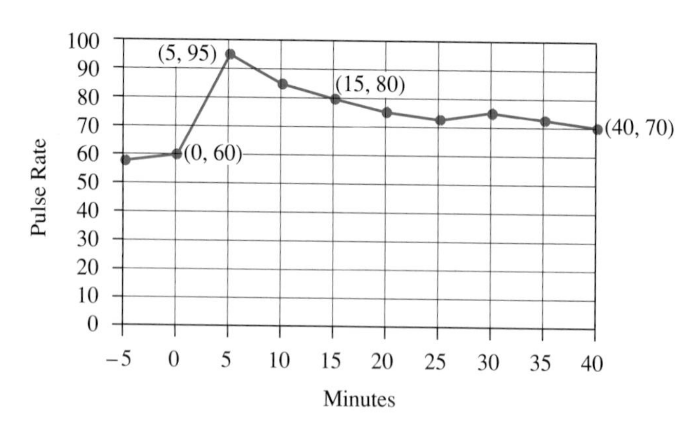

In general, we use this same ordered pair idea to describe the location of a point on a plane (such as a piece of paper). We start with a horizontal and a vertical axis. Each axis resembles a number line, and for the sake of consistency, we construct our axes to intersect at the 0 coordinate of both. This point of intersection is called the **origin.** Notice that these two number lines or axes divide the plane into four regions called quadrants. The quadrants are usually numbered with Roman numerals as shown. The axes are not considered to be in any quadrant.

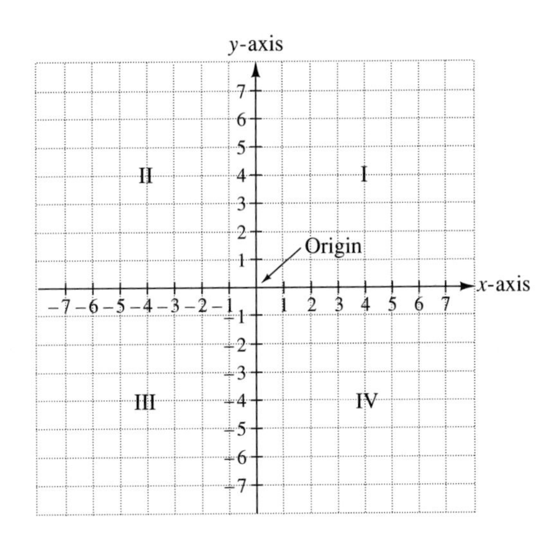

It is helpful to label axes, so we label the horizontal axis the x-axis and the vertical axis the y-axis. We call this system described above the **rectangular coordinate system** or the **Cartesian coordinate system.**

Just as with the pulse rate graph, we can then describe the location of points by ordered pairs of numbers. We list the value on the x-axis first and the value on the y-axis second.

The location of the point shown below can be described by the ordered pair of numbers (3, 2). Here the x-value or x-coordinate is 3 and the y-value or y-coordinate is 2.

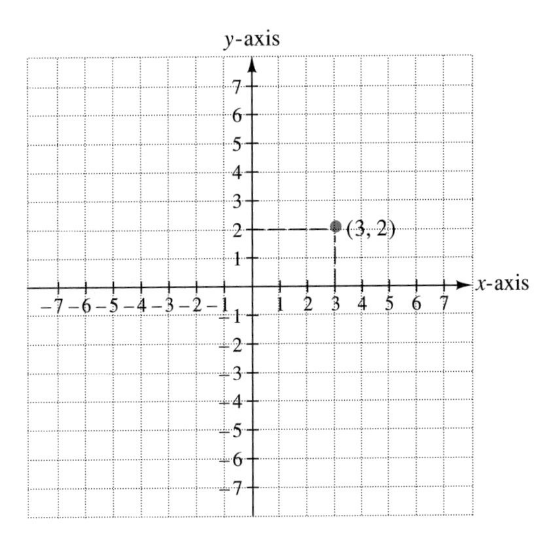

Does the order in which the coordinates are listed matter? Yes! Notice the point corresponding to the ordered pair (2, 3) is in a different location than the point corresponding to (3, 2). These two ordered pairs of numbers describe two different points of the plane.

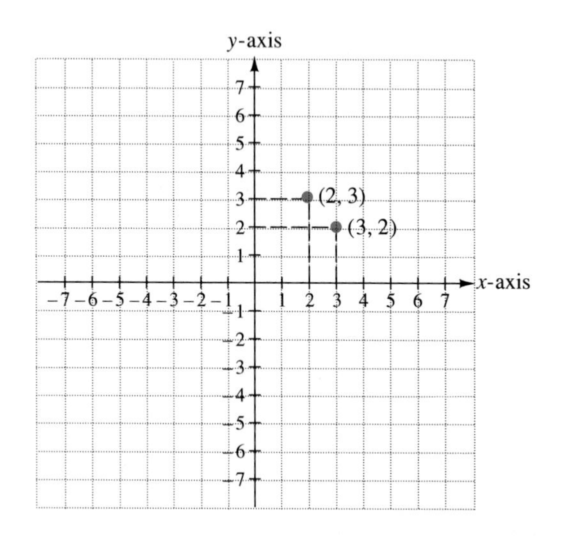

Given an ordered pair, how do we **graph** or **plot** its location? To see, let's graph point A corresponding to the ordered pair $(-2, 5)$. Start at the origin and move 2 units in the negative x direction (left); from there, move 5 units in the positive y direction (up). The ending location is the location of point A.

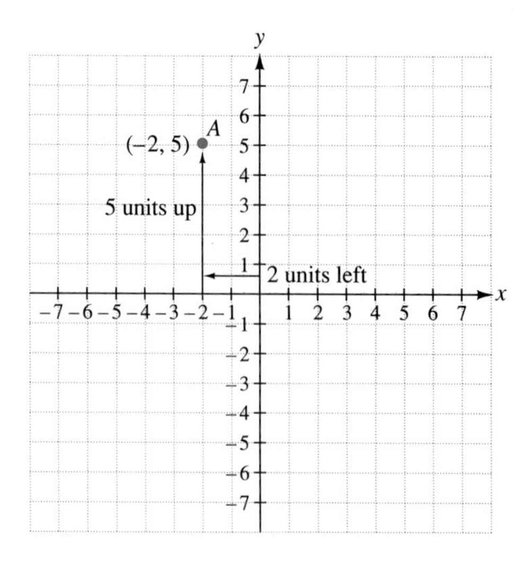

Here are some more points that have been plotted, along with their coordinates.

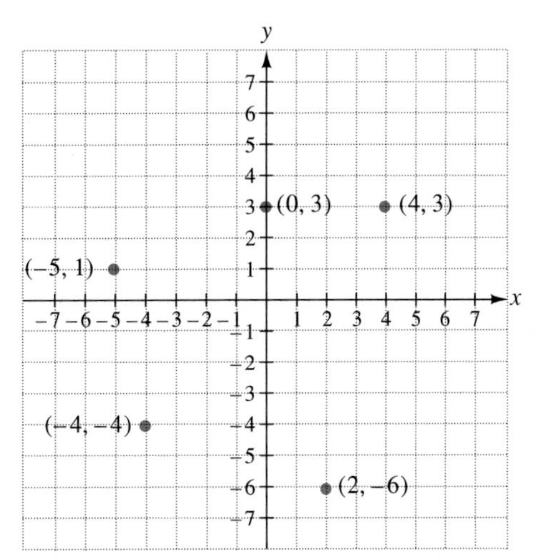

Keep in mind that **each ordered pair corresponds to exactly one point in the real plane and that each point in the plane corresponds to exactly one ordered pair.**

EXAMPLE 4 Plot each ordered pair on a Cartesian coordinate system and name the quadrant in which the point is located.

a. $A(2, -1)$ **b.** $B(0, 5)$ **c.** $C(-3, 5)$ **d.** $D(-2, 0)$ **e.** $\left(-\frac{1}{2}, -4\right)$ **f.** $(0, 0)$

Solution: The five points are graphed as shown:

a. Point A lies in quadrant IV.

b. Point B lies on an axis. It is not in any quadrant.

c. Point C lies in quadrant II.

d. Point D lies on an axis. It is not in any quadrant.

e. Point E is in quadrant III.

f. Point F is the origin. It is not in any quadrant. ■

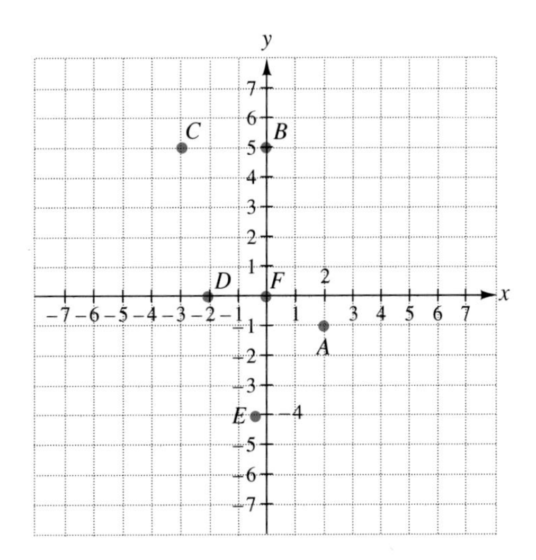

In Chapter 4, we will show how ordered pairs of numbers can be used to record solutions of linear equations in two variables.

EXERCISE SET 2.6

Use the bar graph in Example 1 to answer the following questions.

1. What is the electricity rate charged by Kentucky Power?

2. Find the rate charged by the company shown that is the nearest to your home.

3. Which companies shown charge more than 12¢ per kilowatt hour?

4. Which companies shown charge less than 10¢ per kilowatt hour?

Use the bar graph in Example 2 to answer the following questions.

5. How much money did the film *Bambi* generate?

6. How much more money did the film *Snow White and the Seven Dwarfs* generate than *Bambi?*

7. Before 1990, which Disney film generated the most income?

8. After 1990, which Disney film generated the most income?

Use the line graph in Example 3 to answer the following questions.

9. Approximate the pulse rate 5 minutes before lighting a cigarette.

10. Approximate the pulse rate 10 minutes after lighting a cigarette.

11. Find the difference in pulse rate between 5 minutes before and 10 minutes after lighting a cigarette.

12. What is the highest pulse rate shown on the graph?

Plot the following points. State in which quadrant if any that each point lies. See Example 4.

13. $(1, 5)$

14. $(-5, -2)$

15. $(-6, 0)$

16. $(0, -1)$

17. $(2, -4)$

18. $(-1, 4)$

19. $\left(4\frac{3}{4}, 0\right)$

20. $\left(0, \frac{7}{8}\right)$

The following three-dimensional bar graph shows the team in each sport that has gone the longest without being in a playoff. Use this graph to answer the following questions.

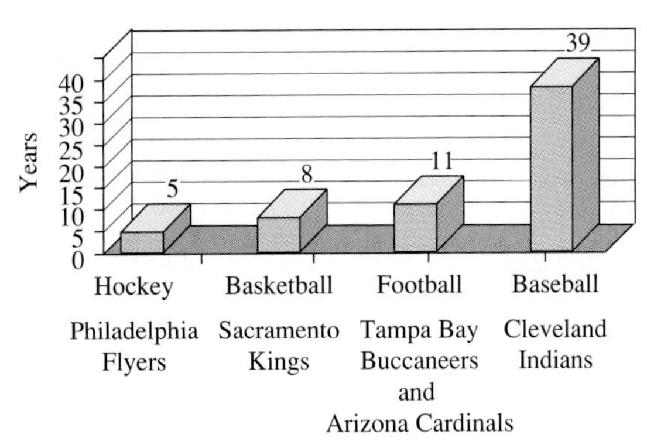

Source: *USA Today,* 1994.

21. Which team for the sports shown has gone the longest without being in a playoff?

22. Which hockey team has gone the longest without being in a playoff?

23. Why are there two football teams listed?

24. What is the most number of years that a basketball team has gone without being in the playoffs?

25. How many more years has a football team gone without being in the playoffs than a basketball team?

26. How many more years has a hockey team gone without being in the playoffs than a football team?

Find the x- and y-coordinates of the following labeled points.

27.

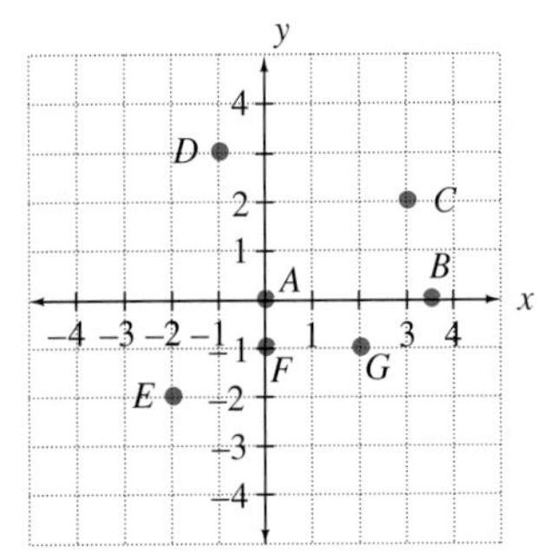

28.

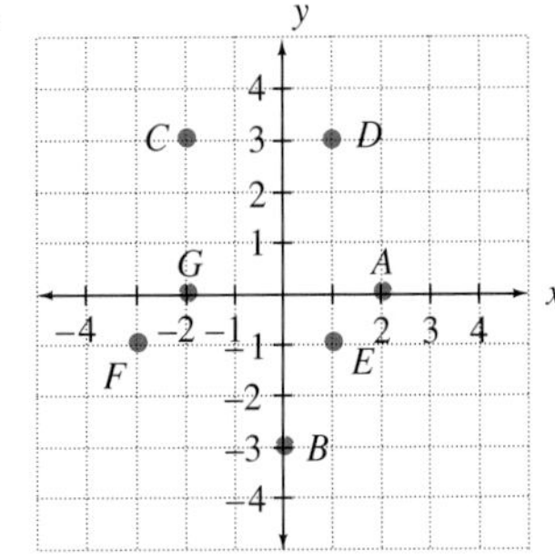

The graph below shows a comparison of chicken and beef consumption in the United States. Use this graph to answer the questions on the following page.

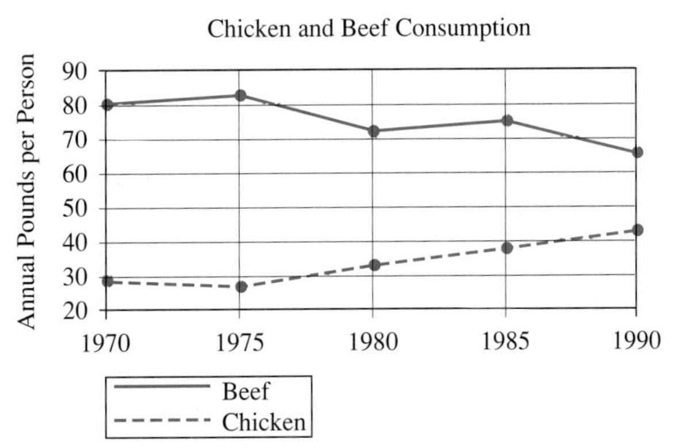

Source: U.S. Bureau of the Census, *Statistical Abstract of the United States: 1993,* 113th ed. Washington, DC, 1994.

29. In what year was the consumption of chicken the lowest.

30. In what year was the consumption of beef the highest?

31. In 1985, how many pounds of chicken did the average person consume?

32. In 1990, how many pounds of beef did the average person consume?

33. In 1970, how many more pounds of beef per person was consumed than chicken?

34. In 1990, how many more pounds of beef per person was consumed than chicken?

35. In what year labeled on the graph was the increase in chicken consumption the greatest?

36. In what year labeled on the graph was the decrease in beef consumption the greatest?

37. In studying this graph, do you notice any trends? To what do you attribute these trends?

Plot each point and state in which quadrant, if any, each point lies.

38. $(3, 2)$

39. $(2, -1)$

40. $(-5, 3)$

41. $(-3, -1)$

42. $(5, -4)$

43. $(2, 3)$

44. $(0, 3)$

45. $(-2, 4)$

46. $(-2, -4)$

47. $(5, 0)$

48. $(0, 0)$

49. Determine the quadrant(s) for which the x-coordinates are greater than 0 and the y-coordinates are less than zero.

50. Determine the quadrant(s) for which both the x- and y-coordinates are negative.

51. Determine the quadrant(s) for which the x-coordinates are positive.

52. Determine the quadrant(s) for which the y-coordinates are positive.

Skill Review

Solve the following equations. See Section 2.3.

53. $3(x - 2) + 5x = 6x - 16$

54. $5 + 7(x + 1) = 12 + 10x$

55. $3x + \frac{2}{5} = \frac{1}{10}$

56. $\frac{1}{6} + 2x = \frac{2}{3}$

Simplify the following fractions. See Section 1.3.

57. $\frac{7}{35}$

58. $\frac{12}{20}$

59. $-\frac{15}{50}$

60. $-\frac{32}{60}$

Writing in Mathematics

61. Name a characteristic common to all ordered pairs whose graphs (points) lie on the x-axis.

62. Name a characteristic common to all ordered paris whose graphs (points) lie on the y-axis.

63. Find the electricity rate charged by your local electric company. How does this rate compare with the rates shown on the bar graph in Example 1?

CHAPTER 2 GLOSSARY

An **equation** states that two algebraic expressions are equal.

A **formula** is a known equation that describes a relation among quantities.

Terms with the same variable raised to the exact same powers are called **like terms.**

A **linear equation** in one variable can be written in the form $ax + b = c$, where a, b, and c are real numbers and $a \neq 0$.

A **solution** or root of an equation is a value for the variable that makes the equation a true statement.

A **term** is a number or the product of a number and variables raised to powers.

CHAPTER 2 SUMMARY

LINEAR EQUATION IN ONE VARIABLE (2.2)

A linear equation in one variable is an equation that can be written in the form

$$ax + b = c$$

where a, b, and c are real numbers and $a \neq 0$.

ADDITION PROPERTY OF EQUALITY (2.2)

If a, b, and c are real numbers, then $a = b$ and $a + c = b + c$ are equivalent equations.

MULTIPLICATION PROPERTY OF EQUALITY (2.2)

If a, b, and c are real numbers and $c \neq 0$, then $a = b$ and $ac = bc$ are equivalent equations.

TO SOLVE LINEAR EQUATIONS (2.3)

Step 1 Clear the equation of fractions by multiplying each side of the equation by the lowest common denominator (LCD) of all denominators in the equation.

Step 2 Use the distributive property to remove any grouping symbols, such as parentheses.

Step 3 Combine any like terms on each side of the equation.

Step 4 Use the addition property of equality to write the equation as an equivalent equation with variable terms on one side and numbers on the other side.

Step 5 Use the multiplication property of equality to isolate the variable.

Step 6 Check the answer by substituting it in the original equation.

PROBLEM-SOLVING STEPS (2.5)

1. UNDERSTAND the problem. At this stage, don't work with variables (except for known formulas), but simply become comfortable with the problem. Some ways of accomplishing this are:
   - Read and reread the problem.
   - Construct a drawing.
   - Look up an unknown formula.
   - Guess the answer and check your guess. Pay careful attention as to how to check your guess. This will help later when you model the problem.
2. ASSIGN a variable to an unknown in the problem. Use this variable to represent any other unknown quantities.
3. ILLUSTRATE the problem. A diagram or chart using the assigned variables can often help to visualize the known facts.
4. TRANSLATE the problem into a mathematical model. This is often an equation.
5. COMPLETE the work. This often means to solve the equation.
6. INTERPRET the results. This means to *check* your work in the stated problem and to *state* your conclusion.

CHAPTER 2 REVIEW

(2.1) *Simplify the following expressions.*

1. $5x - x + 2x$
2. $z - 4x - 7z$
3. $\frac{1}{2}x + 3 + \frac{7}{2}x - 5$
4. $\frac{4}{5}y + 1 + \frac{6}{5}y + 2$
5. $2(n - 4) + n - 10$
6. $3(w + 2) - 12 - w$
7. Subtract $7x - 2$ from $x + 5$
8. Subtract $1.4y - 3$ from $y - 0.7$

Write each of the following as an algebraic expression.

9. Three times a number decreased by 7
10. Twice the sum of a number and 2.8
11. The sum of two numbers is 10. If one number is x, express the other number in terms of x.
12. Mandy is 5 inches taller than Melissa. If x inches represents the height of Mandy, express Melissa's height in terms of x.

(2.2) *Solve the following.*

13. $8x + 4 = 9x$
14. $5y - 3 = 6y$
15. $3x - 5 = 4x + 1$
16. $2x - 6 = x - 6$
17. $4(x + 3) = 3(1 + x)$
18. $6(3 + n) = 5(n - 1)$
19. $\frac{3}{4}x = -9$
20. $\frac{x}{6} = \frac{2}{3}$
21. $-3x + x = 19$
22. $5x + 2x = 20$
23. $5x - 6 - x = 9 + 3x - 1$
24. $8 - y + 4y = 7 + 2y - 3$

Solve each problem.

25. Find a number such that twice the number decreased by 5 is equal to 6 subtracted from three times the number.

26. If twice the sum of a number and 4 is the same as the difference of the number and 8, find the number.

27. Eighteen subtracted from five times a number is the opposite of three. Find the number.

28. Twice a number increased by four is the opposite of ten. Find the number.

29. Double the sum of a number and six is the opposite of the number. Find the number.

(2.3) *Solve the following.*

30. $\frac{2}{7}x - \frac{5}{7} = 1$

31. $\frac{5}{3}x + 4 = \frac{2}{3}x$

32. $-(5x + 1) = -7x + 3$

33. $-4(2x + 1) = -5x + 5$

34. $-6(2x - 5) = -3(9 + 4x)$

35. $3(8y - 1) = 6(5 + 4y)$

36. $\frac{3(2 - z)}{5} = z$

37. $\frac{4(n + 2)}{5} = -n$

38. $5(2n - 3) - 1 = 4(6 + 2n)$

39. $-2(4y - 3) + 4 = 3(5 - y)$

40. $9z - z + 1 = 6(z - 1) + 7$

41. $5t - 3 - t = 3(t + 4) - 15$

42. $-n + 10 = 2(3n - 5)$

43. $-9 - 5a = 3(6a - 1)$

44. $\frac{5(c + 1)}{6} = 2c - 3$

45. $\frac{2(8 - a)}{3} = 4 - 4a$

46. $0.2x - x = 2.4$

47. $y - 0.4y = 0.42$

Solve each problem.

48. Half the difference of a number and three is the opposite of the number. Find the number.

49. If the sum of a number and 5 is divided by 4, the result is twice the number.

50. The quotient of a number and 3 is the same as the difference of the number and two. Find the number.

51. Find three consecutive odd integers such that three times the smallest integer is the sum of the other two integers decreased by 1.

(2.4) *Solve each of the following for the indicated variable.*

52. $y = mx + b$ for m

53. $r = vst - 5$ for s

54. $2y - 5x = 7$ for x

55. $3x - 6y = -2$ for y

56. $C = \pi D$ for π

57. $C = 2\pi r$ for π

Solve each problem.

58. A swimming pool holds 900 cubic meters of water. If its length is 20 meters and its height is 3 meters, find its width.

59. The high temperature in Slidell, Louisiana, one day was 90° Fahrenheit. Convert this temperature to degrees Celsius.

60. How long will it take to run/walk a 10K race (10 kilometers or 10,000 meters) if your average pace is 125 **meters** per minute?

(2.5) *Solve each of the following.*

61. If the length of a rectangular swimming pool is twice the width, and the perimeter is 90 meters, find the length of the pool.

62. One area code in Ohio is 34 more than three times another area code used in Ohio. If the sum of these area codes is 1262, find the two area codes.

63. Find three consecutive even integers whose sum is negative 114.

64. A 12-foot board is to be divided into two pieces so that one piece is twice as long as the other. If x represents the length of the shorter piece, find the length of each piece.

(2.6) *Determine the coordinates of each point on the graph.*

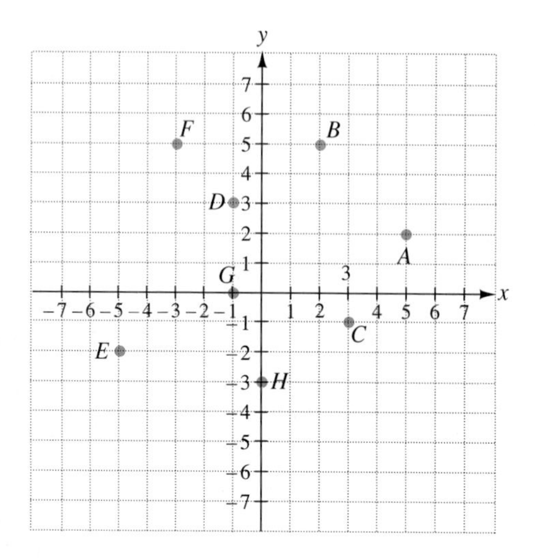

65. Point A **66.** Point B **67.** Point C **68.** Point D

69. Point E **70.** Point F **71.** Point G **72.** Point H

Plot the following points. State in which quadrant, if any, each point lies.

73. $(-1, 6)$ **74.** $(-2, -4)$ **75.** $(-3, 0)$ **76.** $(0, 5)$

77. $(0, 0)$ **78.** $(1, -7)$ **79.** $\left(3\frac{1}{2}, 1\right)$ **80.** $\left(0, \frac{1}{2}\right)$

The following bar graph shows the average miles per gallon for cars during the years shown. Use this graph to answer the following questions.

Fuel Consumption by Automobiles

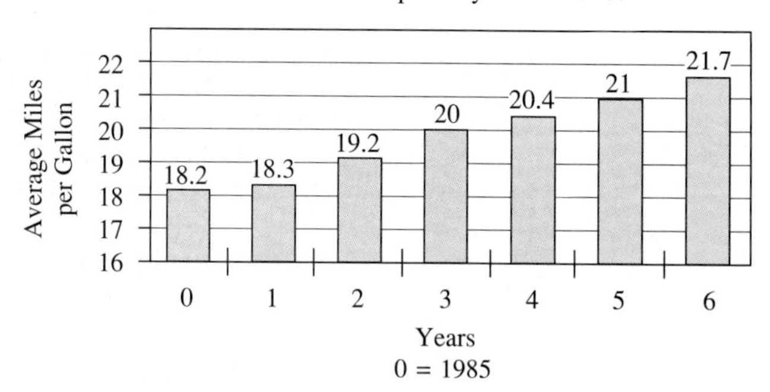

Source: U.S. Bureau of the Census, *Statistical Abstract of the United States: 1993,* 113th ed. Washington, DC, 1994.

81. In 1988, cars averaged how many miles per gallon of fuel?

82. In 1991, cars averaged how many miles per gallon of fuel?

83. Find the increase of average miles per gallon for automobiles for the years 1985 to 1991.

84. Do you notice any trends from this graph?

85. What year showed the greatest increase in miles per gallon?

86. What year showed the smallest increase in miles per gallon?

CHAPTER 2 TEST

Simplify each of the following expressions.

1. $2y - 6 - y - 4$

2. $x + 7 + x - 16$

3. $4(x - 2) - 3(2x - 6)$

4. $-5(y + 1) + 2(3 - 5y)$

Solve each of the following equations.

5. $-\frac{4}{5}x = 4$

6. $4(n - 5) = -(4 - 2n)$

7. $5y - 7 + y = -(y + 3y)$

8. $4z + 1 - z = 1 + z$

9. $\frac{2(x + 6)}{3} = x - 5$

10. $\frac{4(y - 1)}{5} = 2y + 3$

11. $\frac{1}{2} - x + \frac{3}{2} = x - 4$

12. $\frac{2}{3} + n + \frac{1}{3} = n + 2$

13. $\frac{1}{3}(y + 3) = 4y$

14. $\frac{2}{3}(1 - c) = 5c$

15. $-3(x - 4) + x = 5(3 - x)$

16. $-4(a + 1) - 3a = -7(2a - 3)$

Solve each of the following applications.

17. A number increased by two-thirds of the number is 35. Find the number.

18. A gallon of water sealer covers 200 square feet. How many gallons are needed to paint two coats of water sealer on a deck that measures 20 feet by 35 feet?

19. Find two consecutive odd integers such that twice the larger is 15 more than three times the smaller.

20. Javier Sotomayor of Cuba won the gold medal for the men's high jump in the 1992 Olympics. His best jump was 14 inches higher than the jump by John L. Winter of Australia that won the gold metal in 1948. If the total of these two high jumps is 170 inches, find the height of each jump.

Solve each of the following equations for the indicated variable.

21. $V = \pi r^2 h$ for h

22. $W = 6bt$ for t

23. $5g - 2h = p$ for h

24. $3x - 4y = 10$ for y

Plot the following points. State in which quadrant, if any, each point lies.

25. $(-4, -4)$ **26.** $(2, -8)$ **27.** $(0, 3)$ **28.** $(-10, 0)$

Intel is a semiconductor manufacturer that now makes almost one-third of the world's computer chips. (You may have seen the slogan "Intel Inside" in commercials on television.) The line graph below shows Intel's net revenues in billions of dollars. Use this figure to answer the questions below.

29. Estimate Intel's revenue in 1993.

30. Estimate Intel's revenue in 1989.

31. Find the increase in Intel's revenue from 1993 to 1995 (estimated).

32. Which year shows the greatest increase in revenue?

Intel's Revenue

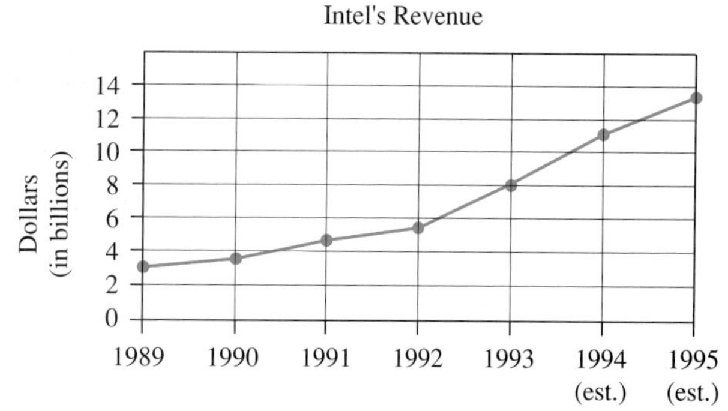

Source: *U.S. News and World Report,* 1994.

CHAPTER 2 CUMULATIVE REVIEW

1. Translate each sentence into a mathematical statement.
  a. 9 is less than or equal to 11.
  b. 8 is greater than 1.
  c. 3 is not equal to 4.

2. Translate each sentence into a mathematical statement.
  a. The product of 2 and 3 is 6.
  b. The difference of 8 and 4 is less than or equal to 4.
  c. The quotient of 10 and 2 is not equal to 3.

3. Insert $<$, $>$, or $=$ in the space between the paired numbers to make each statement true.
  a. $-1 \quad 0$ **b.** $7 \quad \frac{14}{2}$ **c.** $-5 \quad -6$

4. Find the absolute value of each number.
  a. $|4|$ **b.** $|-5|$ **c.** $|0|$

5. Write each fraction in lowest terms.
  a. $\frac{42}{49}$ **b.** $\frac{11}{27}$ **c.** $\frac{88}{20}$

6. Add or subtract as indicated. Write each answer in lowest terms.
  a. $\frac{2}{7}+\frac{4}{7}$ **b.** $\frac{3}{10}+\frac{2}{10}$ **c.** $\frac{9}{7}-\frac{2}{7}$ **d.** $\frac{5}{3}-\frac{1}{3}$

7. Find the following square roots.
  a. $\sqrt{9}$ **b.** $\sqrt{25}$ **c.** $\sqrt{\frac{1}{4}}$

8. Simplify each expression.
  a. $6 \div 3 + 5^2$ **b.** $\frac{2(12+3)}{|-15|}$
  c. $3 \cdot 10 - 7 \div 7$ **d.** $3 \cdot 4^2$ **e.** $\frac{3}{2} \cdot \frac{1}{2} - \frac{1}{2}$

9. Find the value of each expression if $x = 3$ and $y = 2$.
  a. $2x - y$ **b.** $\frac{3x}{2y}$ **c.** $\frac{x}{y}+\frac{y}{2}$ **d.** $x^2 - y^2$

10. Simplify each expression.
  a. $-3 + [(-2 - 5) - 2]$
  b. $2^3 - |10| + [-6 - (-5)]$

11. Simplify the following by combining like terms.
  a. $7x - 3x$ **b.** $10y^2 + y^2$
  c. $8x^2 + 2x - 3x$

12. Subtract $4x - 2$ from $2x - 3$.

13. Solve $2x + 3x - 5 + 7 = 10x + 3 - 6x - 4$ for x.

14. Solve $-3x = 33$ for x.

15. Solve for x: $2(x - 3) = 5x - 9$.

16. Solve $\frac{x+5}{2} + \frac{1}{2} = 2x - \frac{x-3}{8}$.

17. Write each of the following as an algebraic expression.
  a. The sum of two numbers is 8. If one number is 3, find the other number.
  b. The sum of two numbers is 8. If one number is x, find the other number in terms of x.

18. A gallon of water sealer covers 480 square feet. How many one-gallon containers of sealer should be bought to protect a rectangular driveway 24 feet wide by 90 feet long?

19. Solve $P = 2l + 2w$ for w.

20. In 1994, the U. S. Congress had 12 more Democrat senators than Republican senators. If there are a total of 100 senators, how many senators of each party were there?

21. This particular graph shows the relationship between smoking a cigarette and pulse rate. The horizontal units are minutes, with 0 minute being the moment a smoker lights a cigarette. The vertical units are pulse rates per minute.

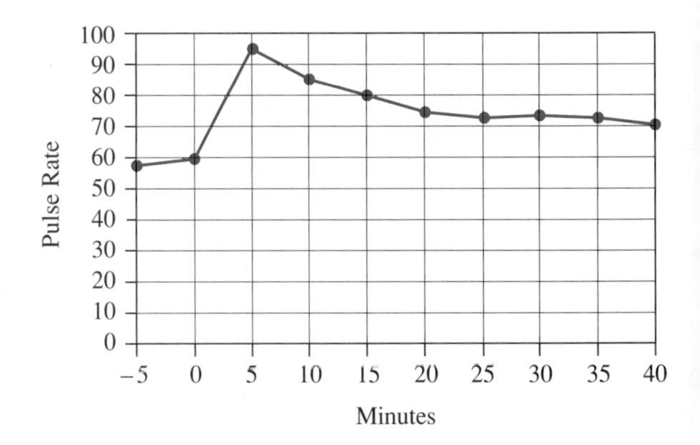

  a. What is the pulse rate 15 minutes after lighting a cigarette?
  b. When is the pulse rate the lowest?
  c. When does the pulse rate show the greatest increase?

22. Plot each ordered pair on a Cartesian coordinate system and name the quadrant in which the point is located.
  a. $A(2, -1)$ **b.** $B(0, 5)$ **c.** $C(-3, 5)$
  d. $D(-2, 0)$ **e.** $\left(-\frac{1}{2}, -4\right)$ **f.** $(0, 0)$

CHAPTER 3

Inequalities, Absolute Value, and Problem Solving

Among the many environmental problems threatening our planet, pollutants dumped into waterways are one of the most sinister. Local authorities must evaluate the extent of pollution from the Chemlife Chemical Company to save a lake downstream.

INTRODUCTION

Mathematics is a tool for solving problems in such diverse fields as transportation, engineering, economics, medicine, business, and biology. We solve problems using mathematics by modeling real-world phenomena with mathematical equations or inequalities. Our ability to solve problems using mathematics, then, depends in part on our ability to solve equations and inequalities. In this chapter, we continue to solve equations and inequalities in one variable and graph their solutions on number lines, and we continue to develop our problem-solving skills.

3.1 Ratio and Proportion

OBJECTIVES

Tape PA 20

1. Rewrite word phrases as ratios using fractional notation.
2. Solve proportions.
3. Use proportions to find unknown sides of similar triangles.
4. Use proportions to solve word problems.

1 A **ratio** is the quotient of two numbers or two quantities.

> **Ratio**
>
> If a and b are two numbers or quantities, $b \neq 0$, the **ratio of a to b** is
>
> $$\frac{a}{b} \quad \text{or} \quad a : b$$

EXAMPLE 1 Write a ratio for each phrase. Use fractional notation.

a. The ratio of 2 parts salt to 5 parts water

b. The ratio of 12 almonds to 16 pecans

Solution: **a.** The ratio of 2 parts salt to 5 parts water is $\frac{2}{5}$.

b. The ratio of 12 almonds to 16 pecans is $\frac{12}{16} = \frac{3}{4}$ in lowest terms. ■

2 If two ratios are equal, we say the ratios are **in proportion** to each other. A **proportion** is a mathematical statement that two ratios are equal.

For example, $\frac{1}{2} = \frac{4}{8}$ is a proportion, as is $\frac{x}{5} = \frac{8}{10}$. When we want to emphasize the equation as a proportion, we

read the proportion $\frac{1}{2} = \frac{4}{8}$ as "one is to two as four is to eight"

In the proportion $\frac{1}{2} = \frac{4}{8}$, the 1 and the 8 are called the **extremes;** the 2 and the 4 are called the **means.**

> **For any proportion, the product of the means equals the product of the extremes.**

To see this, multiply both sides of the proportion $\frac{a}{b} = \frac{c}{d}$ by the LCD bd.

$$\frac{a}{b} = \frac{c}{d}$$

$$bd\left(\frac{a}{b}\right) = bd\left(\frac{c}{d}\right) \qquad \text{Multiply both sides by } bd.$$

$$\underbrace{ad}_{\text{product of extremes}} = \underbrace{bc}_{\text{product of means}} \qquad \text{Simplify.}$$

The products ad and bc are also called **cross products** and can be found by

$$\frac{a}{b} = \frac{c}{d} \qquad \to bc \quad \to ad$$

Equating these cross products is called **cross multiplication.**

> **Cross Multiplication**
>
> For any ratios $\frac{a}{b}$ and $\frac{c}{d}$, if
>
> $$\frac{a}{b} = \frac{c}{d}, \quad \text{then } ad = bc$$

Cross multiplication can be used to solve a proportion for a variable.

EXAMPLE 2 Solve for x: $\frac{45}{x} = \frac{5}{7}$.

Solution: To solve, cross multiply.

$$\frac{45}{x} = \frac{5}{7}$$

$$45 \cdot 7 = 5 \cdot x \qquad \text{Cross multiply.}$$

$$\frac{315}{5} = \frac{5x}{5} \qquad \text{Divide both sides by 5.}$$

$$63 = x \qquad \text{Simplify.}$$

The solution set is $\{63\}$. To check, substitute 63 for x in the original proportion. ■

EXAMPLE 3 Solve for x: $\dfrac{3}{x-2} = \dfrac{5}{x+1}$.

Solution: To solve, cross multiply.

$$\frac{3}{x-2} = \frac{5}{x+1}$$

$$3(x+1) = 5(x-2) \quad \text{Cross multiply.}$$

$$3x + 3 = 5x - 10 \quad \text{Multiply.}$$

$$-2x + 3 = -10 \quad \text{Subtract } 5x \text{ from both sides.}$$

$$-2x = -13 \quad \text{Subtract 3 from both sides.}$$

$$x = \frac{13}{2} \quad \text{Divide both sides by } -2.$$

To check the solution set $\left\{\dfrac{13}{2}\right\}$, substitute $\dfrac{13}{2}$ for x in the original proportion. ■

3 In geometry, proportions can be used to find unknown sides of **similar** triangles. Two triangles are similar if they have the same shape but not necessarily the same size. In similar triangles, the measures of corresponding angles are equal, and corresponding sides are in proportion.

If triangle ABC and triangle XYZ below are similar, then we know that the measure of angle A = the measure of angle X, the measure of angle B = the measure of angle Y, and the measure of angle C = the measure of angle Z, and we also know that corresponding sides are in proportion: $\dfrac{a}{x} = \dfrac{b}{y} = \dfrac{c}{z}$.

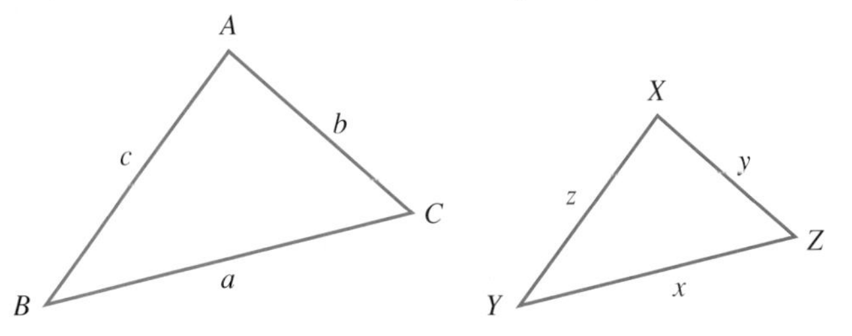

EXAMPLE 4 If triangle ABC is similar to triangle QRS, find the missing length r.

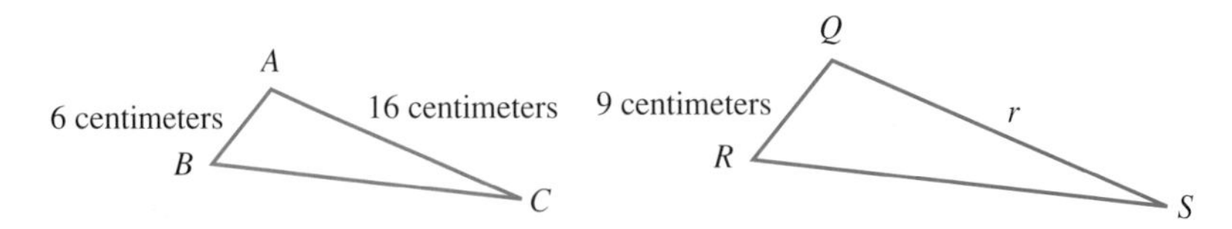

Solution: Since the triangles are similar, corresponding sides are in proportion. Thus,

$$\frac{6}{9} = \frac{16}{r}$$

$$6r = 9(16) \quad \text{Cross multiply.}$$

$$6r = 144$$

$$r = 24 \quad \text{Divide both sides by 6.}$$

Thus, r is 24 centimeters. ■

4 Proportions can be used to model and solve many real-life problems. When using proportions in this way, it is important to judge whether the solution is reasonable. Doing so helps us to decide if the proportion has been formed correctly. In this section, we assume that ratios are constant. We use the same problem-solving steps that were introduced in Section 2.5.

EXAMPLE 5 Three boxes of 3.5″ high density diskettes cost \$37.47. At this rate, how much should 5 boxes cost?

Solution: 1. UNDERSTAND the problem.

To do so, read and reread the problem. Guess the solution and check the guess. Let's guess that 5 boxes cost \$60.00. We know that 5 boxes will cost more than 3 boxes, or \$37.47, and less than 6 boxes, which is double the cost of 3 boxes, or 2(\$37.47) = \$74.94. To check this guess, see if the rates are the same. In other words, see if

$$\frac{\text{price of 3}}{\text{3 boxes}} = \frac{\text{price of 5}}{\text{5 boxes}}$$

or

$$\frac{\$37.47}{3} = \frac{\$60.00}{5}$$

The rates are the same if the proportion is true. To check, we cross multiply.

$$5(37.47) = 3(60.00)$$

or

$$187.35 = 180.00, \quad \textbf{not a true statement}$$

Thus, \$60 is not correct, but we now have a better understanding of the problem.

2. ASSIGN a variable. Let x = price of 5 boxes of diskettes.

3. ILLUSTRATE the problem. No illustration is needed.

4. TRANSLATE the problem.

$$\text{In words:} \quad \frac{\text{price of 3}}{\text{3 boxes}} = \frac{\text{price of 5}}{\text{5 boxes}}$$

$$\text{Translate:} \quad \frac{\$37.47}{3} = \frac{x}{5}$$

5. COMPLETE the work. Here, we solve the proportion.

$$\frac{\$37.47}{3} = \frac{x}{5}$$

$$5(37.47) = 3x \qquad \text{Cross multiply.}$$

$$187.35 = 3x$$

$$62.45 = x \qquad \text{Divide both sides by 3.}$$

6. INTERPRET the results. First *check* the solution. To do so, see that 3 boxes is to \$37.47 as 5 boxes is to \$62.45. Also, notice that our solution is a reasonable one as discussed in step 1.

Next *state* the conclusions. Five boxes of high-density diskettes cost \$62.45. ■

The proportion $\frac{3 \text{ boxes}}{\text{price of } 3} = \frac{5 \text{ boxes}}{\text{price of } 5}$ could also have been used to solve the problem above.

EXAMPLE 6 To estimate the number of people in Jackson, population 50,000, who have no health insurance, 250 people were polled, and 39 of those polled had no insurance. If this rate remains constant, how many people in the city might we expect to be uninsured?

Solution: 1. UNDERSTAND. Read and reread the problem. Guess the solution and check your guess.

2. ASSIGN. Let

x = how many people in the city with no health insurance

3. ILLUSTRATE. No diagram or chart is needed.

4. TRANSLATE.

In words: $\frac{\text{number polled with no insurance}}{\text{total number polled}} = \frac{\text{number in city with no insurance}}{\text{total city population}}$

Translate: $\frac{39}{250} = \frac{x}{50{,}000}$

5. COMPLETE. Solve the proportion for x.

$$\frac{39}{250} = \frac{x}{50{,}000}$$

$$39(50{,}000) = 250x \qquad \text{Cross multiply.}$$

$$1{,}950{,}000 = 250x$$

$$7800 = x \qquad \text{Divide both sides by 250.}$$

6. INTERPRET. *Check* the solution and *state* the conclusion. The city of Jackson has approximately 7800 citizens with no health insurance. ■

EXERCISE SET 3.1

Rewrite the following phrases as ratios in fractional notation. See Example 1.

1. 2 parts flea dip to 15 parts water
2. 4 parts plant food to 18 parts water
3. \$120 to 8 hours of work
4. \$40.00 to 8 hours of work
5. 50 chips to every 1 cup of salsa
6. 3 hot dogs to every 1 hamburger

Solve each proportion. See Examples 2 and 3.

7. $\frac{2}{3} = \frac{x}{6}$
8. $\frac{x}{2} = \frac{16}{6}$
9. $\frac{x}{10} = \frac{5}{9}$
10. $\frac{9}{4x} = \frac{6}{2}$
11. $\frac{9}{5} = \frac{12}{3x + 2}$
12. $\frac{6}{11} = \frac{27}{3x - 2}$
13. $\frac{3}{x + 1} = \frac{5}{2x}$
14. $\frac{7}{x - 3} = \frac{8}{2x}$

15. $\frac{2x - 1}{7} = \frac{x}{2}$

16. $\frac{x - 1}{16} = \frac{x}{9}$

Each pair of triangles is similar. For each pair, find the value of the variable. See Example 4.

17.

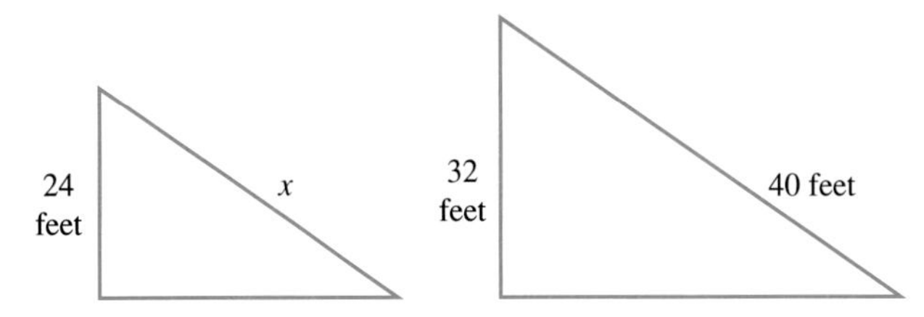

18.

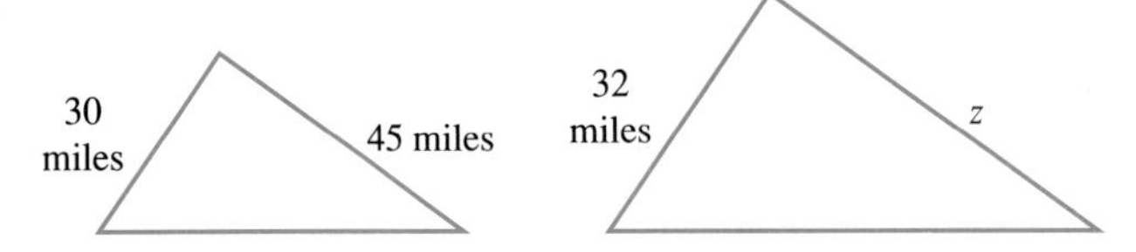

19.

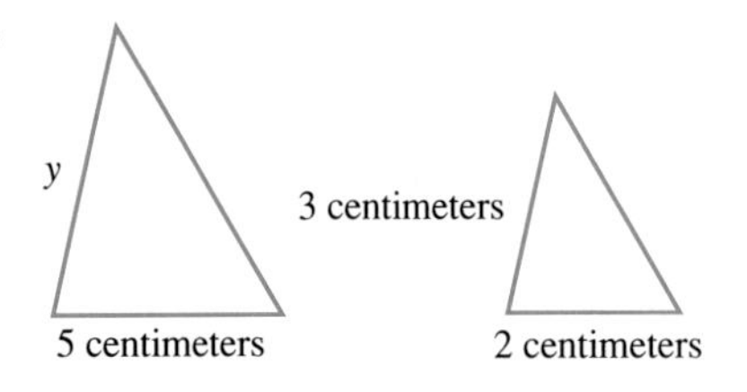

20.

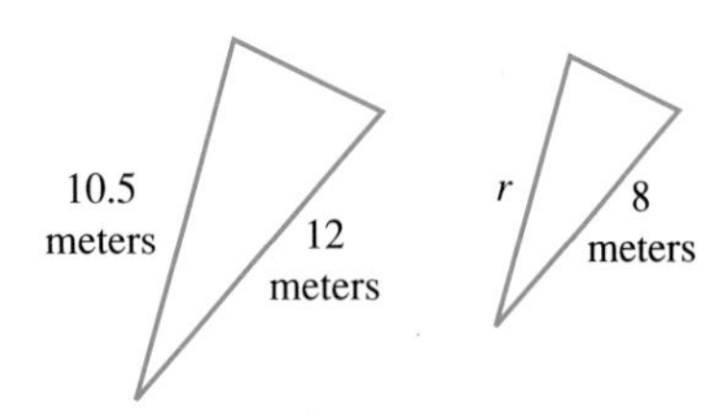

Solve. See Examples 5 and 6.

21. The ratio of the weight of an object on Earth to the weight of the same object on Pluto is 100 to 3. If an elephant weighs 4100 pounds on Earth, find the elephant's weight on Pluto.

22. In a bag of M&M's, 28 out of 80 M&M's were found to be the color brown. How many M&M's would you expect to be brown from a bag containing 208 M&M's? (Round to the nearest whole.)

23. If a 170-pound person weighs approximately 65 pounds on Mars, how much does a 9,000 pound satellite weigh?

24. On an architect's blueprint, 1 inch corresponds to 4 feet. Find the length of a wall represented by a line that is $3\frac{7}{8}$ inches long on the blueprint.

Solve the following proportions.

25. $\frac{4x}{6} = \frac{7}{2}$

26. $\frac{a}{5} = \frac{3}{2}$

27. $\frac{a}{25} = \frac{12}{10}$

28. $\frac{n}{10} = 9$

29. $\frac{x - 3}{x} = \frac{4}{7}$

30. $\frac{y}{y - 16} = \frac{5}{3}$

31. $\frac{5x + 1}{x} = \frac{6}{3}$

32. $\frac{3x - 2}{5} = \frac{4x}{1}$

33. $\frac{x + 1}{2x + 3} = \frac{2}{3}$

34. $\frac{x + 1}{x + 2} = \frac{5}{3}$

Solve the following.

35. There are 110 calories per 28.4 grams of Crispy Rice cereal. Find how many calories are in 42.6 grams of this cereal.

36. There are 1280 calories in a 14-ounce portion of Eagle Brand Milk. Find how many calories are in 2 ounces of Eagle Brand Milk.

37. A box of flea and tick powder instructs the user to mix 4 tablespoons of powder with 1 gallon of water. Find how much powder should be mixed with 5 gallons of water.

38. Miss Babola's new Mazda gets 35 miles per gallon. Find how far she can drive if the tank contains 13.5 gallons of gas.

39. In a week of city driving, Miss Babola noticed that she was able to drive 418.5 miles on a tank of gas (13.5 gallons). Find how many miles per gallon Miss Babola got in city traffic.

40. Ken Hall, a tailback, holds the high school sports record for total yards rushed in a season. In 1953, he rushed for 4045 total yards in 12 games. Find his average rushing yards per game.

41. A recent headline read, "Women earn bigger check in 1 of every 6 couples." If there are 23,000 couples in a nearby metropolitan area, how many women would you expect to earn bigger paychecks?

42. A human factors expert recommends that there be at least 9 square feet of floor space in a college classroom for every student in the class. Find the minimum floor space that 40 students need.

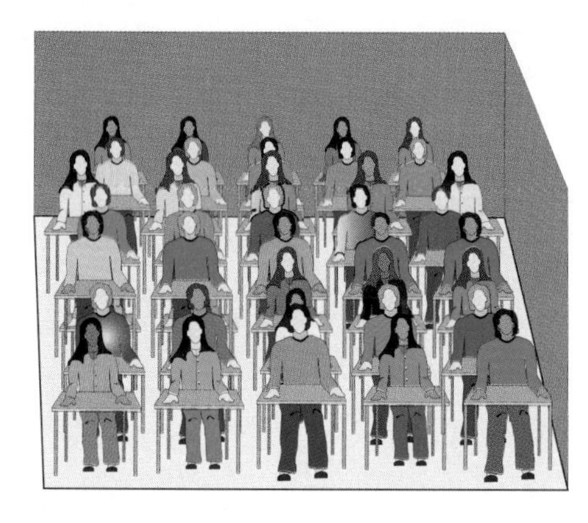

43. Due to space problems at a local university, a 20-foot by 12-foot conference room is converted into a classroom. Find the maximum number of students the room can accomodate. (See Exercise 42.)

44. The manufacturers of cans of salted mixed nuts state that the ratio of peanuts to other nuts is 3 to 2. If 324 peanuts are in a can, find how many other nuts should also be in the can.

45. If Sam Abney can travel 343 miles in 7 hours, find how far he can travel if he maintains the same speed for 5 hours.

46. The instructions on a bottle of plant food read as follows: "Use four tablespoons plant food per 3 gallons of water." Find how many tablespoons of plant food should be mixed into 6 gallons of water.

47. To mix weed killer with water correctly, it is necessary to mix 8 teaspoons of weed killer with 2 gallons of water. Find how many gallons of water are needed to mix with the entire box if it contains 36 teaspoons of weed killer.

48. There are 290 milligrams of sodium per 1-ounce serving of Rice Crispies. Find how many milligrams of sodium are in three 1-ounce servings of the cereal.

49. Mr. Lin's contract states that he will be paid $153 per 8-hour day to teach mathematics. Find how much he earns per hour.

50. Mr. Gonzales, a pool contractor, bases the cost of labor on the volume of the pool to be constructed. The cost of labor on a wading pool of 803 cubic feet is $750.00. If the customer decided to cut the volume by a third, find the cost of labor for the smaller pool.

51. An accountant finds that Country Collections earned $35,063 during its first 6 months. Find how much the business earned **each week** on average.

Each pair of triangles is similar. Find the value of the variable.

52.

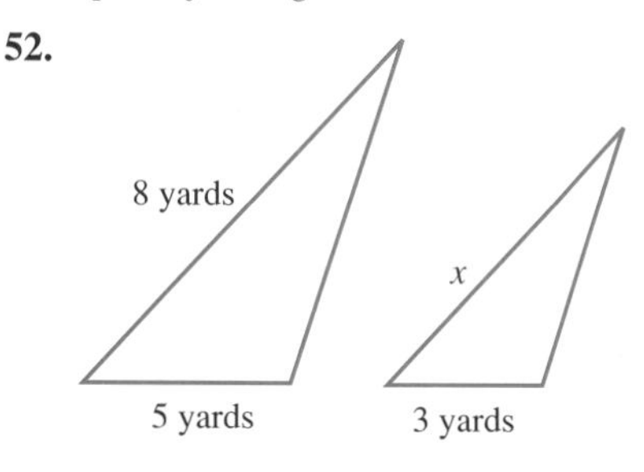

53.

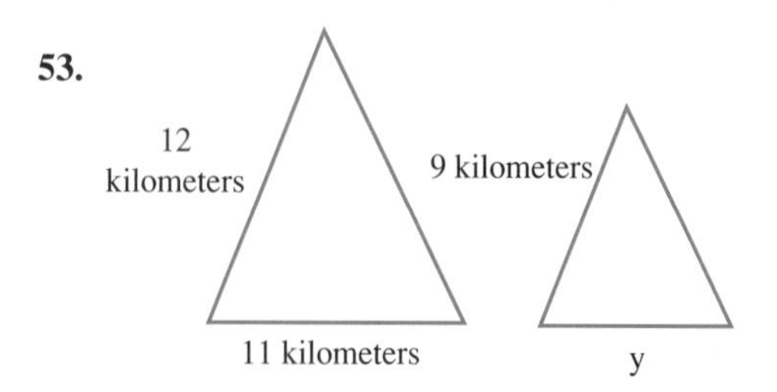

54. Mel Faciane, a 6-foot-tall park ranger at Yosemite National Park, needs to estimate the height of a particular tree. He notices that when his shadow is 10 feet long, the shadow of the tree is 75 feet long. Find the height of the tree.

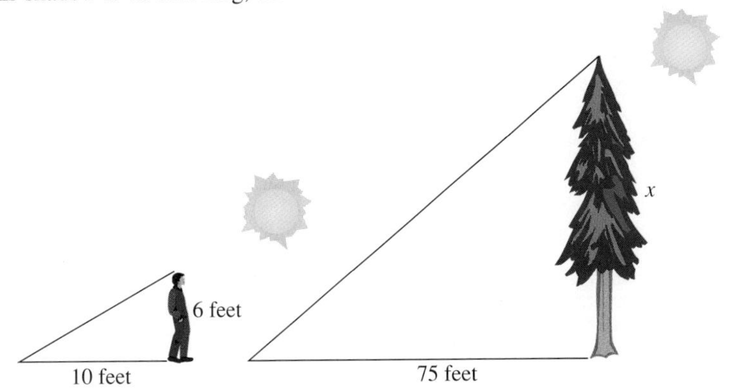

55. Kathleen Williams, a 5-foot-tall engineer in New York City, needs to estimate the height of a building. She notices that the shadow cast by the building is 80 feet when her shadow is 8 feet. Find the height of the building.

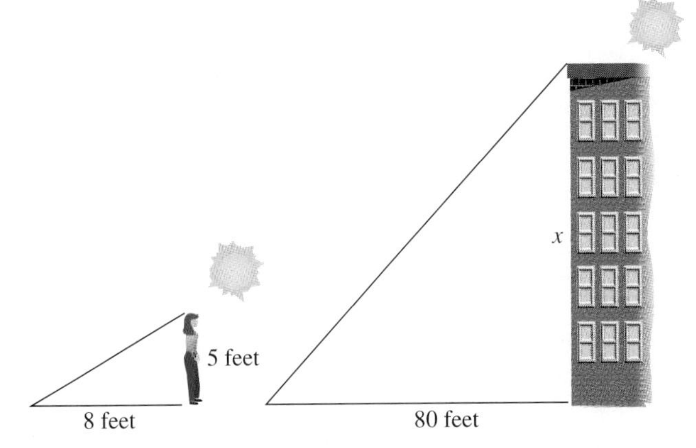

56. University Law School accepts 3 out of every 11 student applicants. If the school received 495 applicants, how many students would you expect the law school to accept?

57. Barbara Hayes is a secretary at a hospital in Slidell. It takes Barbara 30 minutes to type and spell check 4 pages on her word processor. At this rate, how long does it take her to type and spell check 30 pages?

58. Continuing with Exercise 57, how many pages can Barbara type and spell check in 2.5 hours?

The following graph shows the capacity of the world to generate electricity from the wind.

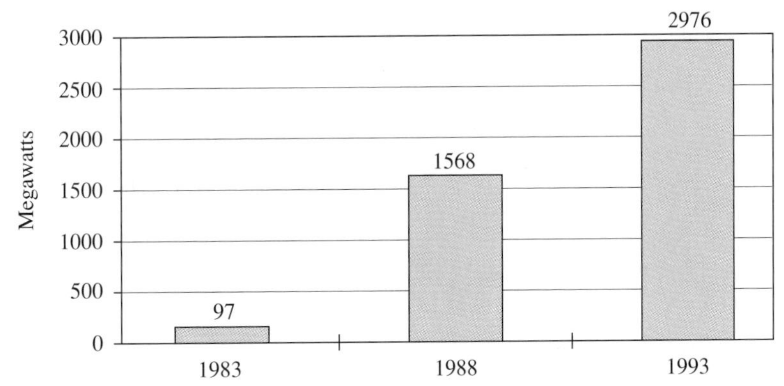

Source: *USA Today*, 1994.

59. Find the increase in megawatt capacity during the 5-year period from 1983 to 1988.

60. Find the increase in megawatt capacity during the 5-year period from 1988 to 1993.

61. If the trend shown on this graph continues, approximate the number of megawatts available from the wind in 1998.

In general, 1000 megawatts will serve the average electricity needs of 560,000 people. Use this fact and the graph on the previous page to answer the following.

62. In 1993, the amount of megawatts that can be generated from wind will serve the electricity needs of how many people?

63. How many megawatts of electricity are needed to serve the city or town in which you live?

Skill Review

Evaluate $\frac{y-b}{x-a}$ *for the given values. See Section 1.5.*

64. $x = 2$, $y = 3$, $a = 5$, and $b = 2$

65. $x = -3$, $y = 5$, $a = 4$, and $b = 5$

Solve. See Section 2.5.

66. Forty is the sum of twice a number and one, multiplied by eight. Find the number.

67. If twice the sum of 4 and a number is added to 10, the sum is 12. Find the number.

68. If the product of 8 and a number is subtracted from 20, the difference is -4. Find the number.

69. Twice the sum of a number and three is forty-two. Find the number.

Evaluate. See Section 1.7.

70. $(-2)^5$

71. -2^5

72. -2^4

73. $(-2)^4$

A Look Ahead

In Exercise Set 2.5, we introduced the golden rectangle as a rectangle whose length is approximately 1.6 times its width. Early Greeks thought that rectangles with these dimensions were the most pleasing to the eye, and many examples of golden rectangles can be found in early Greek art as well as modern-day packaging and architecture. The ratio of the length of a golden rectangle to its width is called the golden ratio. A better approximation of the golden ratio is 1.618 to 1, and its exact value will be found in Section 9.3. Examples of the golden ratio can also be found in measurements of the human body. Find the following ratios of body measurements and see if they approximate the golden ratio.

74. $\frac{A}{B}$

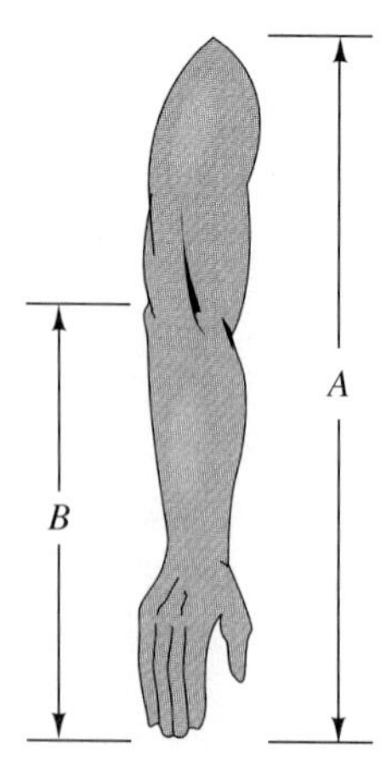

75. $\frac{x}{y}$

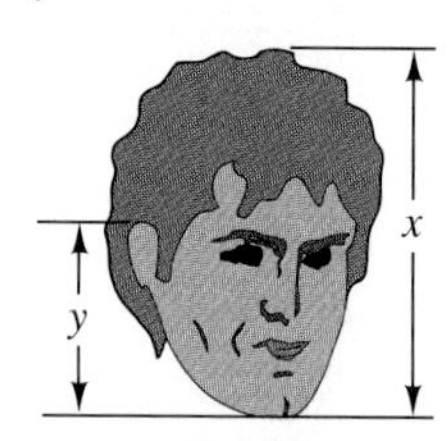

3.2
Percent and Problem Solving

OBJECTIVES

Tape PA 23 and PA 24

1. Write percents as decimals and decimals as percents.
2. Find a percent of a number.
3. Read and interpret graphs containing percents.
4. Solve percent equations.
5. Solve applications involving percents.

1 Many of today's statistics are given in terms of percents; a basketball player's free throw percent, current interest rates, stock market trends, and nutrition labeling are just a few. Our understanding of percents depends in part on our understanding ratios because a percent is a ratio whose denominator is 100. In other words,

the word *percent* means *per hundred* so that 61 *percent* means 61 *per hundred* or $\frac{61}{100}$

We use the symbol % to denote percent, so

$$61\% = \frac{61}{100} \quad or \quad 0.61$$

$$8\% = \frac{8}{100} \quad or \quad 0.08$$

$$100\% = \frac{100}{100} \quad or \quad 1.00$$

This suggests the following procedure:

To Write a Percent as a Decimal

Drop the percent symbol and move the decimal point two places to the left.

EXAMPLE 1 Write each percent as an equivalent decimal.

a. 35% **b.** 89.5% **c.** 150%

Solution: **a.** 35% = 0.35 **b.** 89.5% = 0.895 **c.** 150% = 1.5 ■

To write a decimal as a percent, we reverse the step given above.

To Write a Decimal as a Percent

Move the decimal point two places to the right and attach the percent symbol, %.

EXAMPLE 2 Write each as a percent.

a. 0.73 **b.** 1.39 **c.** $\frac{1}{4}$

Solution: **a.** 0.73 = 73% **b.** 1.39 = 139%

c. First write $\frac{1}{4}$ as a decimal.

$$\frac{1}{4} = 0.25 = 25\% \quad ■$$

2 To find a percent *of* a number, recall that the word *of* means multiply.

EXAMPLE 3 Find 72% of 200.

Solution: To find 72% of 200, we multiply.

$$\begin{aligned} 72\% \text{ of } 200 &= 72\%(200) \\ &= 0.72(200) \\ &= 144 \end{aligned}$$

Thus, 72% of 200 is 144. ■

3 As mentioned earlier, percents are used many times in statistics. Recall that the graph below is called a circle graph or a pie chart. The circle or pie represents a whole, which in this case is 100%. Each circle is divided into sectors (shaped like pieces of a pie) that represent various parts of the whole 100%.

EXAMPLE 4 This circle graph shows how much money homeowners spend annually on maintaining their homes. Use this graph to answer the questions below.

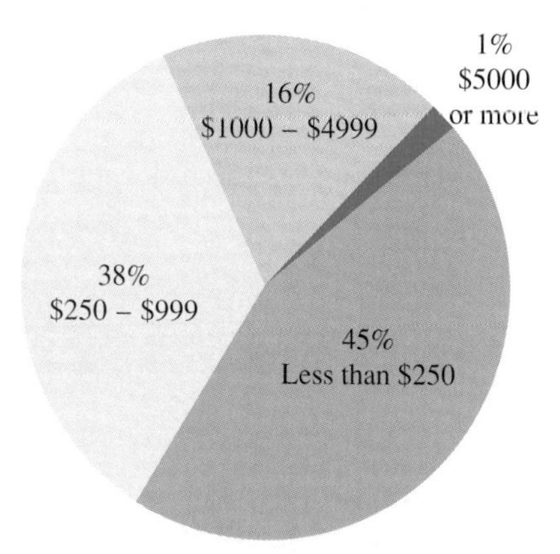

Source: *U.S. Today*; 1994.

a. What percent of homeowners spend less than $250 on yearly home maintenance?
b. What percent of homeowners spend less than $1000 per year on maintenance?

c. How many of the 22,000 homeowners in a town called Fairview might we expect to spend under $250 a year on home maintenance?

d. Find the number of degrees in the 16% sector.

Solution: **a.** From the circle graph, we see that 45% of homeowners spend less than $250 per year on home maintenance.

b. From the circle graph, we know that 45% of homeowners spend less than $250 per year and 38% of homeowners spend $250–$999 per year, so that the sum $45\% + 38\% = 83\%$ of homeowners spend less than $1000 per year.

c. Since 45% of homeowners spend less than $250 per year on maintenance, we find 45% of 22,000.

$$\begin{aligned} 45\% \text{ of } 22{,}000 &= 0.45(22{,}000) \\ &= 9900 \end{aligned}$$

We might then expect that 9900 homeowners in Fairview spend less than $250 per year on home maintenance.

d. To find the number of degrees in the 16% sector, recall that the number of degrees around a circle is 360°. Thus, to find the number of degrees in the 16% sector, we find 16% of 360°.

$$\begin{aligned} 16\% \text{ of } 360^\circ &= 0.16(360)^\circ \\ &= 57.6^\circ \end{aligned}$$

57.6°

The 16% sector contains 57.6°. ■

4 Next we practice writing sentences as percent equations. Since these problems involve direct translations, all our previous problem-solving steps are not needed.

EXAMPLE 5 The number 63 is what percent of 72?

Solution: **1.** UNDERSTAND. Read and reread the problem. Next guess a solution. Suppose we guess that 63 is 80% of 72. We may check our guess by finding 80% of 72, so 80% of $72 = 0.80(72) = 57.6$, which is close, but not 63. At this point, though, we have a better understanding of the problem, we know the correct answer is close to and greater than 80%, and we know how to check our proposed solution later.

2. ASSIGN. Let $x =$ the unknown percent.

4. TRANSLATE. Recall that "is" means equals and "of" means to multiply and translate the sentence directly.

In words:	the number 63	is	what percent	of	72 ?
Translate:	63	=	x	$\cdot$	72

5. COMPLETE.

$$\begin{aligned} 63 &= 72x \\ 0.875 &= x && \text{Divide both sides by 72.} \\ 87.5\% &= x && \text{Write as a percent.} \end{aligned}$$

6. INTERPRET. *Check* by verifying that 87.5% of 72 is 63.
State the results. The number 63 is 87.5% of 72. ■

EXAMPLE 6 The number 120 is 15% of what number?

Solution: 1. UNDERSTAND. Read and reread the problem. Guess a solution and check your guess.

2. ASSIGN. Let x = the unknown number.

4. TRANSLATE. In words: The number 120 is 15% of what number?

Translate: $120 = 15\% \cdot x$

5. COMPLETE.

$120 = 0.15x$

$800 = x$ Divide both sides by 0.15.

6. INTERPRET. *Check* the proposed solution of 800 by finding 15% of 800 and verifying that the result is 120.

State: Thus, 120 is 15% of 800. ■

5 Percent increase and percent decrease is a common way to describe how some measurement has increased or decreased. For example, crime increased by 8%, teachers received a 5.5% increase in salary, or a company decreased its employees by 10%. The next example is a review of percent increase.

EXAMPLE 7 The cost of a large hand-tossed pepperoni pizza at Domino's recently increased from \$5.80 to \$7.03. Find the percent increase.

Solution: 1. UNDERSTAND. Read and reread the problem. Let's guess that the percent increase is 10%. To see if this is the case, we find 10% of \$5.80 to find the increase in price. Then we add this increase to \$5.80 to find the new price. In other words, $10\%(\$5.80) = 0.10(\$5.80) = \$0.58$, the increase in price. The new price is then $\$5.80 + \$0.58 = \$6.38$, not the desired new price of \$7.03. We now know that the increase is greater than 10%, and we know how to check our proposed solution.

2. ASSIGN. Let x = the percent increase.

4. TRANSLATE. First we find the **increase** and then the **percent increase.** The increase in price is found by.

In words:	increase =	new price	−	old price
or translate:	increase =	\$7.03	−	\$5.80
	= \$1.23			

Next, we find the percent increase. The percent increase or percent decrease is always a percent of the original number or, in this case, of the old price.

In words:	increase	is	what percent	of	old price
Translate:	\$1.23	=	x	·	\$5.80

5. COMPLETE:

$1.23 = 5.80x$

$0.212 \approx x$ Divide both sides by 5.80 and round three decimal places.

$21.2\% \approx x$ Write as a percent.

6. INTERPRET. *Check* the proposed solution.
State: The increase in price is approximately 21.2%. ■

EXERCISE SET 3.2

Write each percent as an equivalent decimal. See Example 1.

1. 120%
2. 73%
3. 22.5%
4. 4.2%
5. 0.12%
6. 0.86%

Write each as an equivalent percent. See Example 2.

7. 0.75
8. 0.3
9. 2
10. 5.1
11. $\frac{1}{8}$
12. $\frac{3}{5}$

Use the graph found in Example 4 to answer each question.

13. What percent of homeowners spend \$250–\$999 on yearly home maintenance?
14. What percent of homeowners spend \$5000 or more on yearly home maintenance?
15. What percent of homeowners spend \$250–\$4999 on yearly home maintenance?
16. What percent of homeowners spend \$250 or more on yearly home maintenance?
17. Find the number of degrees in the 38% sector.
18. Find the number of degrees in the 1% sector.
19. How many homeowners in your town might you expect to spend \$250–\$999 on yearly home maintenance?
20. How many homeowners in your town might you expect to spend \$5000 or more on yearly home maintenance?

Solve the following percent equations. See Examples 3, 5, and 6.

21. What number is 16% of 70?
22. What number is 88% of 1000?
23. The number 28.6 is what percent of 52?
24. The number 87.2 is what percent of 436?
25. The number 45 is 25% of what number?
26. The number 126 is 35% of what number?

Solve. See Example 7.

27. Dillard's advertised a 25% off sale. If a London Fog coat originally sold for \$156, find the decrease and the sale price.
28. Time Saver increased the price of a \$0.75 cola by 15%. Find the increase and the new price.
29. Hallahan's Construction Company increased their estimate for building a new house from \$95,500 to \$110,000. Find the percent increase.
30. By buying in quantity, the Cannon family was able to decrease their weekly food bill from \$150 a week to \$130 a week. Find the percent decrease.

Solve the following.

31. Find 23% of 20.
32. Find 140% of 86.
33. The number 40 is 80% of what number?
34. The number 56.25 is 45% of what number?
35. The number 144 is what percent of 480?
36. The number 42 is what percent of 35?

37. According to the World Flying Disc Federation, as of this writing, the women's world record for throwing a plastic disc (like a Frisbee) is held by Anni Kreml of the United States. Her throw was 447.2 feet. The men's record is held by Niclas Bergehamn of Sweden. His throw was 44.8% farther than Anni's. Find the length of his throw.

38. Scoville units are used to measure the hotness of a pepper. An alkaloid, capsaicin, the is the ingredient that makes a pepper hot, and liquid chromatography measures the amount of capsaicin in parts per million. The jalapeño measures around 5000 Scoville units whereas the hottest pepper, the habanero, measures around 3000% hotter. Find the measure of the habanero pepper.

The following graph shows the percent of people in a survey who have used over-the-counter drugs for each of the following ailments in a twelve-month period. Use this graph to answer the questions below.

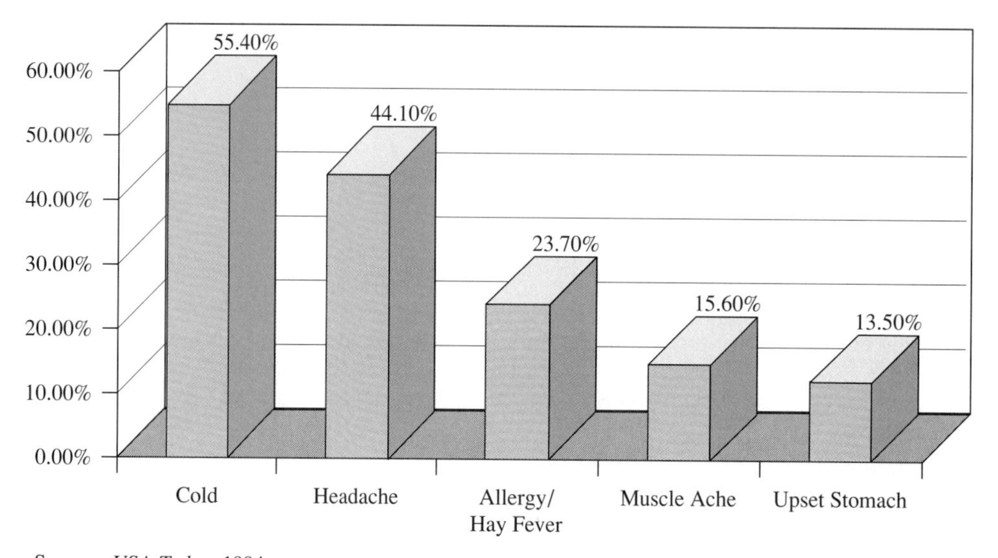

Source: *USA Today*, 1994.

39. What percent of those surveyed used over-the-counter drugs to combat the common cold?

40. What percent of those surveyed used over-the-counter drugs to combat an upset stomach?

41. If 230 people were surveyed, how many of these used over-the-counter drugs for allergies?

42. The city of Chattanooga has a population of approximately 152,000. How many of these people would you expect to have used over-the-counter drugs for relief of a headache?

43. Do the percents shown in the graph have a sum of 100%? Why or why not?

44. Survey your algebra class and find what percent of the class has used over-the-counter drugs for each of the categories listed. Draw a bar graph of the results.

Solve.

45. Iceberg lettuce is grown and shipped to stores for about 40 cents a head whereas consumers purchase it for about 70 cents a head. Find the percent increase.

46. The lettuce consumption per capita in 1968 was about 21.5 pounds whereas in 1992 the consumption rose to 26.1 pounds. Find the percent increase.

Standardized nutrition labeling has been displayed on food items since 1994. Study the label and answer Exercises 47 through 49. The percent column to the right shows the percent of daily values based on a 2000-calorie diet shown at the bottom of the label. For example, a serving of this food contains 4 grams of fat, and the recommended daily value based on a 2000-calorie diet is 65 grams of fat. This means that $\frac{4}{65}$ *or approximately 6% (as shown) of the daily recommended fat is taken in by eating a serving of this food.*

Nutrition Facts
Serving Size 18 crackers (31g)
Servings Per Container About 9

Amount Per Serving	
Calories 130 Calories from Fat 35	
	% Daily Value*
Total Fat 4g	6%
Saturated Fat 0.5g	3%
Polyunsaturated Fat 0g	
Monounsaturated Fat 1.5g	
Cholesterol 0mg	0%
Sodium 230mg	x
Total Carbohydrate 23g	y
Dietary Fiber 2g	8%
Sugars 3g	
Protein 2g	
Vitamin A 0% • Vitamin C 0%	
Calcium 2% • Iron 6%	

*Percent Daily Values are based on a 2,000 calorie diet. Your daily values may be higher or lower depending on your calorie needs.

	Calories	2,000	2,500
Total Fat	Less than	65g	80g
Sat Fat	Less than	20g	25g
Cholesterol	Less than	300mg	300mg
Sodium	Less than	2400mg	2400mg
Total Carbohydrates		300g	375g
Dietary Fiber		25g	30g

47. What percent of daily values of sodium is contained in a serving of this food? In other words, find x.

48. What percent of daily values of total carbohydrate is contained in a serving of this food? In other words, find y.

49. Notice on the nutrition label that a serving of this particular food contains 130 calories and that 35 of these calories are from fat. Find the percent of calories from fat. It is recommended that no more than 30% of calorie intake come from fat. Does this food satisfy this recommendation?

Use the label below to answer Exercises 50 and 51.

NUTRITIONAL INFORMATION PER SERVING
Serving Size: 9.0 oz.

Calories	280	Polyunsaturated Fat	1g
Protein	12g	Saturated Fat	3g
Carbohydrate	45g	Cholesterol	20mg
Fat	6g	Sodium	520mg
Percent of Calories from Fat	?	Potassium	220mg

50. If fat contains approximately 9 calories per gram, find the percent of calories from fat on this label. (Use 6g.)

51. If protein contains approximately 4 calories per gram, find the percent of calories from protein on this label.

52. According to the Better Sleep Council, a recent study showed that 26% of men have dozed off at their place of work. If you currently employ 121 men, how many of these men might you expect to have dozed off at work?

53. According to the Interep Radio Store, a recent study showed that women and girls spend 41% of the money in a household budgeted for clothing. If a family spent $2000 last year on clothing, how much might have been spent on clothing for women and girls?

54. The following table shows where lightning strikes. Use this table to draw a circle graph or pie chart of this information.

Fields and Ballparks	Under Trees	Bodies of Water	Golf Courses	Near Heavy Equipment	Telephone Poles	Other
28%	17%	13%	4%	6%	1%	31%

55. According to the Federal Reserve, over the past 5 years, bank fees for bounced checks have risen from \$12.62 to \$15.65. Find the percent increase. If inflation over the same period has been 16%, do you think that this is a fair increase?

56. During the same period mentioned in Exercise 54, the bank fee for depositing a bad check has risen from \$5.38 to \$6.08. Find the percent increase. Given the inflation rate of 16%, do you think that this is a fair increase?

57. The first Barbie doll was introduced in March 1959 and cost \$3. This same 1959 Barbie doll now costs up to \$5000. Find the percent increase rounded to the nearest whole percent.

58. The ACT Assessment is a college entrance exam taken by about 60% of college-bound students. The national average score of 20.7 in 1993 rose to 20.8 in 1994. Find the percent increase.

The following double bar graph shows selected services and products and the percent of supermarkets offering each. Use this graph to answer the questions below.

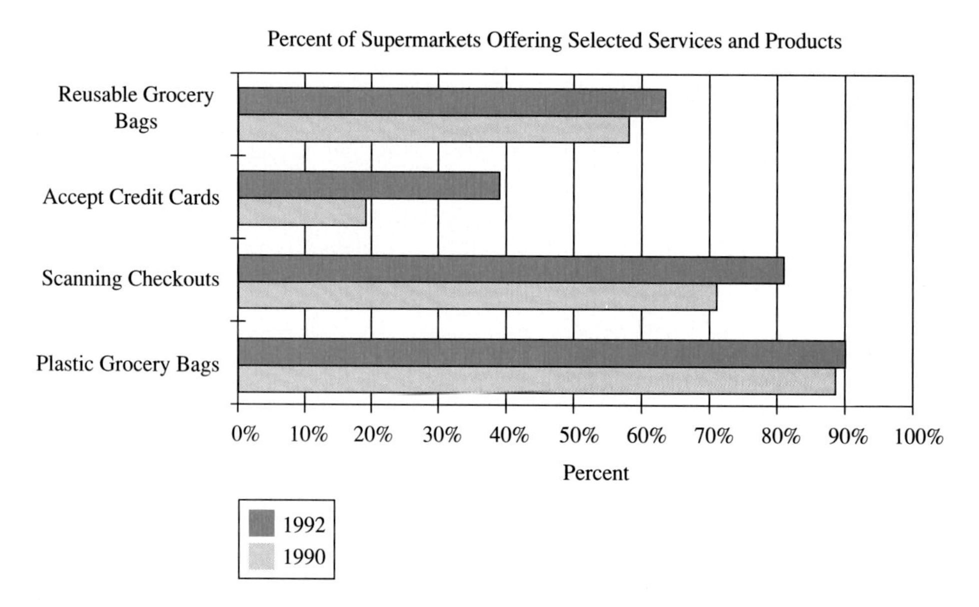

Source: U.S. Bureau of the Census, *Statistical Abstract of the United States: 1993,* 113th ed., Washington, DC, 1994.

59. What percent of supermarkets offered plastic grocery bags in 1990?

60. What percent of supermarkets accepted credit cards in 1992?

61. Suppose that there are 20 supermarkets in your city. How many of these 20 supermarkets might have offered customers plastic grocery bags in 1992?

62. How many more supermarkets (in percent) offered reusable grocery bags in 1992 than in 1990?

63. Do you notice any trends shown in this graph?

Skill Review

Find the value of the following expressions for the given values. See Section 1.7.

64. $2a + b - c$; $a = 5, b = -1$, and $c = 3$

65. $-3a + 2c - b$; $a = -2, b = 6$, and $c = -7$

66. $4ab - 3bc$; $a = -5, b = -8$, and $c = 2$

67. $ab + 6bc$; $a = 0, b = -1$, and $c = 9$

68. $n^2 - m^2$; $n = -3$ and $m = -8$

69. $2n^2 + 3m^2$; $n = -2$ and $m = 7$

Solve. See Section 2.4.

70. Find how much interest \$2000 earns in 3 years in a savings account paying 4% simple interest annually.

71. Find the amount of principal that must be invested in a certificate of deposit (CD) paying 5% annually to earn \$125 in $2\frac{1}{2}$ years.

72. Find the distanced traveled if driving at a speed of 57 miles per hour for 3.5 hours.

73. The distance from Kansas City, Missouri, to Duluth, Minnesota, is approximately 590 miles. How long does it take to travel from Kansas City to Duluth at an average speed of 55 miles per hour?

Writing in Mathematics

74. Find a food that contains more than 30% of its calories per serving from fat. Analyze the nutrition label and verify that the percents shown are correct.

3.3 Further Applications of Linear Equations

OBJECTIVES

1. Solve applications involving distance.
2. Solve applications involving mixtures.
3. Solve applications involving interest.

Tape IA 3

This section is devoted to problem solving in distance, mixtures, and interest. We will use the same problem-solving steps as used in previous sections. They are listed below for review.

Problem-Solving Steps

1. UNDERSTAND the problem. At this stage, don't work with variables (except for known formulas), but simply become comfortable with the problem. Some ways of accomplishing this are:
   - Read and reread the problem.
   - Construct a drawing.
   - Look up an unknown formula.
   - Guess the answer and check your guess. Pay careful attention as to how to check your guess. This will help later when you model the problem.

2. ASSIGN a variable to an unknown in the problem. Use this variable to represent any other unknown quantities.
3. ILLUSTRATE the problem. A diagram or chart using the assigned variables can often help to visualize the known facts.
4. TRANSLATE the problem into a mathematical model. This is often an equation.
5. COMPLETE the work. This often means to solve the equation.
6. INTERPRET the results. This means to *check* your work in the stated problem and to *state* your conclusion.

1

EXAMPLE 1 Marie Antonio, a bicycling enthusiast, rode her ten-speed at an average speed of 18 miles per hour on level roads and then slowed down to an average of 10 miles per hour on the hilly roads of the trip. If she covered a distance of 98 miles, how long did the entire trip take if traveling the level roads took the same time as traveling the hilly roads?

Solution: 1. UNDERSTAND the problem. To do so, read and reread the problem. The formula $d = r \cdot t$ is needed. At this time, let's guess a solution. Suppose that Marie spent 2 hours traveling on the level roads. This means that she also spent 2 hours traveling on the hilly roads since the times spent were the same. What is her total distance? Her distance on the level road is rate · time = 18(2) = 36 miles. Her distance on the hilly roads is rate · time = 10(2) = 20 miles. This gives a total distance of 36 miles + 20 miles = 56 miles, not the correct distance of 98 miles. Remember that the purpose of guessing a solution is not to guess correctly (although this may happen), but to help us better understand the problem and how to model it with an equation.

2. ASSIGN a variable to an unknown in the problem. We are looking for the length of the entire trip, so we begin by letting

$$x = \text{the time spent on level roads}$$

Because the same amount of time is spent on hilly roads, then

$$x = \text{the time spent on hilly roads also}$$

3. ILLUSTRATE the problem. We summarize the information from the problem on the following chart. Fill in the rates given and the variables used to represent the times, and use the formula $d = r \cdot t$ to fill in the distance column.

Trip	Rate	Time	Distance $(= r \cdot t)$
Level	18	x	$18x$
Hilly	10	x	$10x$

4. TRANSLATE the problem into a mathematical model. Since the entire trip covered 98 miles, we have that

In words:	total distance	=	level distance	+	hilly distance
Translate:	98	=	$18x$	+	$10x$

5. COMPLETE the work by solving the equation.

$$98 = 28x$$

$$\frac{98}{28} = \frac{28x}{28}$$

$$3.5 = x$$

6. INTERPRET the results.

Check: Recall that $x = 3.5$ hours is the time of the level portion of the trip and $x = 3.5$ hours is also the time of the hilly portion. If Marie rides for 3.5 hours at 18 mph, her distance is $18(3.5) = 63$ miles. If Marie rides for 3.5 hours at 10 mph, her distance is $10(3.5) = 35$ miles. The total distance 63 miles + 35 miles is the required distance, 98 miles.

State: The time of the entire trip is then 3.5 hours + 3.5 hours or 7 hours. ■

2 Mixture problems involve two or more different quantities being combined to form a new mixture. These applications range from Dow Chemical's need to form a chemical mixture of a required strength to Planter's Peanut Company's need to find the correct mixture of peanuts and cashews, given taste and price constraints.

EXAMPLE 2 A chemist working on his doctorate degree at Massachusetts Institute of Technology needs 12 liters of a 50% acid solution for a lab experiment. The stockroom has only 40% and 70% solutions. How much of each solution should be mixed together to form 12 liters of a 50% solution?

Solution: 1. UNDERSTAND. First, read and reread the problem a few times. Next guess the solution. Suppose that we need 7 liters of the 40% solution. Then we need $12 - 7 = 5$ liters of the 70% solution. To see if this is indeed the solution, let's find the amount of pure acid in 7 liters of the 40% solution, in 5 liters of the 70% solution, and in 12 liters of a 50% solution, the required amount and strength.

number of liters	×	concentration rate	=	amount of pure acid
7	×	0.40	=	2.8 liters
5	×	0.70	=	3.5 liters
12	×	0.50	=	6 liters

Since 2.8 liters + 3.5 liters = 6.3 liters and not 6, our guess is incorrect, but we have gained some invaluable insight into how we model and check this problem.

2. ASSIGN. Let x = number of liters of 40% solution. Then

$12 - x$ = number of liters of 70% solution

3. ILLUSTRATE. The following table summarizes the information given.

	Number of Liters	Concentration Rate	Amount of Pure Acid
First Solution	x	40%	0.40x
Second Solution	$12 - x$	70%	$0.70(12 - x)$
Mixture Needed	12	50%	0.50(12)

4. TRANSLATE. Recall that the amount of acid in each solution is found by multiplying the strength of each solution by the number of liters. The amount of acid in the final mixture is the sum of the amounts of acid in the two solutions.

In words:	acid in first solution	+	acid in second solution	=	acid in mixture
Translate:	$0.40x$	+	$0.70(12 - x)$	=	$0.50(12)$

5. COMPLETE.

$$0.40x + 0.70(12 - x) = 0.50(12)$$
$$0.4x + 8.4 - 0.7x = 6$$
$$-0.3x + 8.4 = 6$$
$$-0.3x = -2.4$$
$$x = 8$$

6. INTERPRET. *Check:* If 8 liters of the 40% solution is needed, then $12 - 8 = 4$ liters of the 70% solution is needed. To check, recall how we checked our guess.

State: If 8 liters of the 40% solution is mixed with 4 liters $(12 - 8)$ of the 70% solution, the result is 12 liters of a 50% solution. ■

3 The next example is an investment problem.

EXAMPLE 3 Rajiv Puri invested part of his $20,000 inheritance in a mutual fund account that pays 7% simple interest yearly and the rest in a certificate of deposit that pays 9% simple interest yearly. At the end of 1 year, Rajiv's investments earned $1550. Find the amount he invested at each rate.

Solution: 1. UNDERSTAND. Read and reread the problem. Next guess a solution. Suppose that Rajiv invested $8000 in the 7% fund and the rest, $12,000, in the fund paying 9%. To check, let's find his interest after 1 year. Recall the formula $I = PRT$ and the interest from the 7% fund is $\$8000(0.07)(1) = \$560$. The interest from the 9% fund is $\$12{,}000(0.09)(1) = \$1080$. The sum of the interests is $\$560 + \$1080 = \$1640$. Our guess does not lead to $1550 and is thus incorrect, but we now have a better understanding of the problem.

2. ASSIGN. Let x = amount of money in the account paying 7%. The rest of the money is $20,000 less x or

$$20{,}000 - x = \text{amount of money in the account paying } 9\%$$

3. ILLUSTRATE. We apply the simple interest formula $I = PRT$ and organize our information in the following chart. Since there are two different rates of interest and two different amounts invested, we apply the formula twice.

	Principal	· Rate	· Time	= Interest
7% Fund	x	0.07	1	$x(0.07)(1)$ or $\mathbf{0.07x}$
9% Fund	$20{,}000 - x$	0.09	1	$(20{,}000 - x)(0.09)(1)$ or $\mathbf{0.09(20{,}000 - x)}$
Total	20,000			1550

4. TRANSLATE. The total interest earned, \$1550, is the sum of the interest earned at 7% and the interest earned at 9%.

In words: interest at 7% + interest at 9% = total interest

Translate: $0.07x + 0.09(20{,}000 - x) = 1550$

5. COMPLETE.

$$0.07x + 0.09(20{,}000 - x) = 1550$$

$$0.07x + 1800 - 0.09x = 1550 \quad \text{Apply the distributive property.}$$

$$1800 - 0.02x = 1550 \quad \text{Combine like terms.}$$

$$-0.02x = -250 \quad \text{Subtract 1800 from both sides.}$$

$$x = 12{,}500 \quad \text{Divide both sides by } -0.02.$$

6. INTERPRET. *Check:* If \$12,500 is invested at 7%, then (20,000 − 12,500) or \$7500 is invested at 9%. The annual interest on \$12,500 at 7% is \$875; the annual interest on \$7500 at 9% is \$675, and \$875 + \$675 = \$1550.

State: The amount invested at 7% is \$12,500. The rest of the money, $20{,}000 - x = 20{,}000 - 12{,}500 =$ \$7500, was invested at 9%. ■

EXERCISE SET 3.3

Solve. See Example 1.

1. A motorcyclist drove 200 miles in 4 hours. He kept a steady speed for the first 3 hours of the trip. Due to rain, he reduced his speed by 20 mph for the last hour of the trip. Find the two speeds.

2. Jose Manualla traveled to San Diego at an average rate of 50 mph. Carlos made the same trip in 1 hour less time at an average rate of 60 mph. Find the distance of the trip.

3. A private plane for Kenco, Inc. traveled to Los Angeles at 200 mph and returned at 300 mph. If the total traveling time was 2 hours, find how far away Los Angeles is.

4. Two trains leave Chicago at the same time and travel in opposite directions. At the end of 4 hours they are 660 miles apart. One train is 15 mph slower than the other. Find their speeds.

Solve. See Example 2.

5. A pharmacist at the Medicine Shoppe needs 6 ounces of a 30% codeine solution, but she only has bottles of 20% and 50% codeine solutions. Find how many ounces from each bottle should be mixed for this prescription.

6. Find how many gallons of a 15% saline solution should be mixed with 10 gallons of a 20% saline solution to produce a 16% saline solution.

7. A lab researcher for a pharmaceutical company has 4 liters of a 10% acid solution. Find how much pure acid should be added to increase the strength to 25%.

8. A doctor needs to dilute 10 centiliters of a 50% glucose solution by adding water to get a 20% glucose solution. Find how much water he should add.

Solve. See Example 3.

9. Lilian Gould invested $24,000 in two accounts: a mutual fund paying 8% annual interest and a CD paying 9% interest. If her annual interest was $2020, find how much she invested in each account.

10. Kim Phong made two investments totaling $50,000. After 1 year, she made an 18% profit on one investment but took an 11% loss on the other investment. Find the amount of each investment if her net gain was $4360.

11. Julio Berman invested money in a passbook savings account paying 6% annual interest and $3200 more in a bond paying 8%. If his total yearly income from both investments was $1656, find how much he invested in the bond.

12. Mr. Goldberg invested his $40,000 bonus in two accounts. He took a 4% loss on one investment and made a 12% profit on the other investment, but ended up breaking even. Find how much he invested at each rate.

Solve.

13. Irene Boesky invested a certain amount of money at 9% annually, twice that amount at 10% annually, and three times that amount at 11% annually. Find the amount invested at each rate if her total yearly income from the investments was $8370.

14. Two cars leave Tuscon at the same time traveling in opposite directions, one driving 55 mph and the other at 65 mph. Find how long Donald and Howard are able to talk on their car phones if the phones have a 300-mile range.

15. Hank Williams put $7000 in an account paying 10%. Find how much he should deposit at 7% annual interest so that the average return on the two investments is 9% annually.

16. Two joggers are 12 miles apart traveling toward each other. Find how long it takes them to meet if one jogger runs at 6 mph and the other runs 2 mph slower.

17. On a 225-mile trip on Germany's autobahn system, Emil Duprey traveled at an average speed of 70 mph, stopped for gas, and then averaged 60 mph for the remainder of the trip. If the entire trip took 3.75 hours and the gas stop took 15 minutes, find how far Emil drove before the gas stop.

18. How many pounds of mint tea, which costs $2.25 per pound, must be mixed with 40 pounds of chamomile tea, which costs $6.00 per pound, to have a blend that costs $3.50 per pound?

19. When Mr. Whipple mixes $6.00-per-pound coffee with $5.80-per-pound coffee, he sells the blend for $5.85 per pound. Find how much of each type of coffee he uses to make 100 pounds of the blend.

20. Zoya Daniell invested part of her $25,000 advance at 8% interest and the rest at 9% interest. If her total yearly interest from both accounts was $2135, find the amount invested at each rate.

21. Michael Schwartz invested part of his $10,000 bonus in a fund that paid an 11% profit and invested the rest in stock that suffered a 4% loss. Find the amount of each investment if his overall net profit was $650.

22. Shirley Hamer invested some money at 9% interest and $250 more than that amount at 10%. If her total yearly interest was $101, how much was invested at each rate?

23. Bruce Cole invested a sum of money at 10% interest and invested twice that amount at 12% interest. If his total yearly income from both investments was $2890, how much was invested at each rate?

24. A jet plane traveling at 500 mph overtakes a propeller plane traveling at 200 mph that had a 2-hour head start. How far from the starting point are the planes?

25. How long will it take a bus traveling at 60 miles per hour to overtake a car traveling at 40 mph if the car had a 1.5-hour head start?

26. The Jones family drove to Disneyland at 50 miles per hour and returned on the same route at 40 mph. Find the distance to Disneyland if the total driving time was 7.2 hours.

27. A bus traveled on a level road for 3 hours at an average speed of 20 miles per hour faster than it traveled on a winding road. The time spent on the winding road was 4 hours. Find the average speed on the level road if the entire trip was 305 miles.

28. How much pure acid should be mixed with 2 gallons of a 40% acid solution to get a 70% acid solution?

29. How many cubic centimeters of a 25% antibiotic solution should be added to 10 cubic centimeters of a 60% antibiotic solution to get a 30% antibiotic solution?

30. Planter's Peanut Company wants to mix 20 pounds of peanuts worth $3 a pound with cashews worth $5 a pound in order to make an experimental mix worth $3.50 a pound. How many pounds of cashews should be added to the peanuts?

31. Community Coffee Company wishes to mix a new flavor of Cajun coffee. How many pounds of coffee worth $7 a pound should be added to 14 pounds of coffee worth $4 a pound to obtain a mixture worth $5 a pound?

32. How can $54,000 be invested, part at 8% and the remainder at 10%, so that the interest earned by the two accounts will be equal?

33. Ms. Mills invested her $20,000 bonus in two accounts. She took a 4% loss on one investment and made a 12% profit on another investment, but ended up breaking even. How much was invested at each rate?

34. If $3000 is invested at 6%, how much should be invested at 9% so that the total income from both investments is $585?

35. Trudy Waterbury, a financial planner, invested a certain amount of money at 9%, twice that amount at 10%, and three times that amount at 11%. Find the amount invested at each rate if her total yearly income from the investments was $2790.

36. Kathleen and Cade Williams leave simultaneously from the same point hiking in opposite directions, Kathleen walking at 4 miles per hour and Cade at 5 mph. How long can they talk on their walkie-talkies if the walkie-talkies have a 20-mile radius?

37. Alan and Dave Schaferkötter leave from the same point driving in opposite directions, Alan driving at 55 miles per hour and Dave at 65 mph. Alan has 1-hour head start. How long will they be able to talk on their car phones if the phones have a 250-mile range?

38. How much of an alloy that is 20% copper should be mixed with 200 ounces of an alloy that is 50% copper to get an alloy that is 30% copper?

39. How much water should be added to 30 gallons of a solution that is 70% antifreeze to get a mixture that is 60% antifreeze?

40. Aaron and Ashleigh Thieme are 12 miles apart hiking toward each other. How long will it take them to meet if Aaron walks at 3 miles per hour and Ashleigh walks 1 mph faster?

41. Charlie and Cindy Truax are 11 miles apart and hiking toward each other. They meet in 2 hours. Find the rate of each hiker if one hiker walks 1.1 miles per hour faster than the other.

42. On a 255-mile trip, Gary Ellis traveled at an average speed of 70 miles per hour, got a speeding ticket, and then traveled at 60 mph for the remainder of the trip. If the entire trip took 4.5 hours and the speeding ticket stop took 30 minutes, how long did Gary speed before getting stopped?

43. Mark Martin can row upstream at 5 miles per hour and downstream at 11 mph. If Mark starts rowing upstream until he gets tired and then rows downstream to his starting point, how far did Mark row if the entire trip took 4 hours?

44. Carla Dorr rented a canoe for 4 hours. She can paddle upstream at 2 mph and downstream at 6 mph. Find how far upstream she should paddle if she wants to return in exactly 4 hours.

45. Jonathan Chace can row a boat at a rate of 3 mph. He plans to row up a river with a current of 1.5 mph, turn around, and row downstream to his starting point in a total of 4 hours. Find how many hours he rows upstream.

46. April Thrower spent $32.25 to have her daughter's birthday party at the movies. Adult tickets cost $5.75 and children tickets cost $3.00. If 8 persons were at the party, how many adult tickets were bought?

47. A butcher mixes ground sirloin worth $2.25 per pound with hamburger worth $1.00 per pound to make 500 pounds of a blend to sell for $1.50 per pound. Find how much ground sirloin and how much hamburger he uses.

48. Eight hundred tickets to a play were sold for a total of $2000. If the adult tickets cost $4 each and student tickets cost $2 each, find how many of each kind of ticket was sold.

49. The Little League had a fund-raiser at which they washed cars for $2 and washed bikes for 75 cents. If $450 was collected from 350 people, find how many bikes were washed.

50. Two hikers on the Appalachian Trail are 21 miles apart and are walking toward each other. They meet in 3 hours. Find the rate of each hiker if one hiker walks 2 mph faster than the other.

51. A 6-gallon radiator contains a 40% antifreeze solution. Find how much needs to be drained and replaced by pure antifreeze to get a 60% solution.

52. A 4-gallon radiator contains a 50% antifreeze solution. Find how much needs to be drained and replaced by water to obtain a 40% antifreeze solution.

To "break even" in a manufacturing business, revenue R (income) **must equal** *the cost C of production. Use this information to answer the following.*

53. The cost C to produce x number of skateboards is $C = 100 + 20x$. The skateboards are sold wholesale for \$24 each, so revenue R is given by $R = 24x$. Find how many skateboards the manufacturer needs to produce and sell to break even.

54. The revenue R from selling x number of computer boards is given by $R = 60x$, and the cost C of producing them is given by $C = 50x + 5000$. Find how many boards must be sold to break even. Find how much money is needed to produce the break-even number of boards.

55. The cost C of producing x number of paperback books is given by $C = 4.50x + 2400$. Income R from these books is given by $R = 7.50x$. Find how many books should be produced and sold to break even.

56. Find the break-even quantity for a company that makes x number of computer monitors at a cost C given by $C = 870 + 70x$ and receives revenue R given by $R = 105x$.

Skill Review

Add or subtract the following. See Section 1.6.

57. $3 + (-7)$ **58.** $(-2) + (-8)$ **59.** $4 - 16$

60. $-11 + 2$ **61.** $-5 - (-1)$ **62.** $-12 - 3$

Solve each equation. See Sections 2.2 and 2.3.

63. $5x + 20 = 8 - x$

64. $6t - 18 = t - 3$

65. $6(x - 3) + 10 = -8$

66. $-4(2 + n) + 9 = 1$

67. $\dfrac{6(3 - z)}{5} = -z$

68. $\dfrac{4(5 - w)}{3} = -w$

Find the coordinates of each point. See Section 2.6.

69.

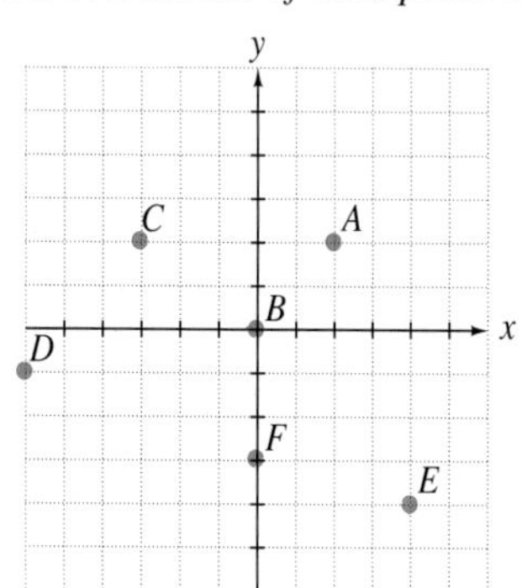

70.

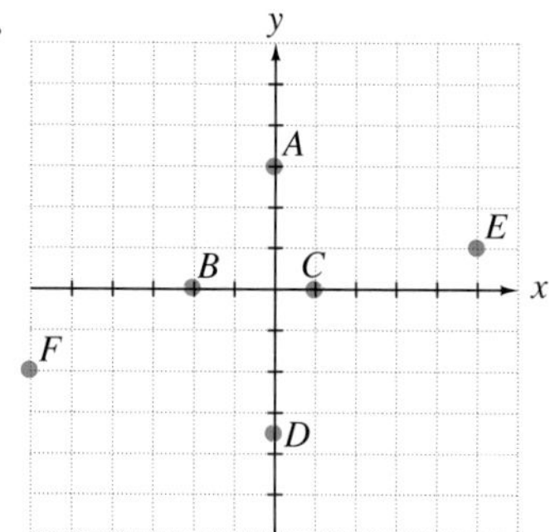

Writing in Mathematics

71. Problems 53 through 56 involve finding the break-even point for manufacturing. Discuss what happens if a company makes and sells fewer products than the break-even point. Discuss what happens if more products than the break-even point are made and sold.

3.4 Absolute Value Equations

OBJECTIVE

1 Solve absolute value equations.

Tape IA 3

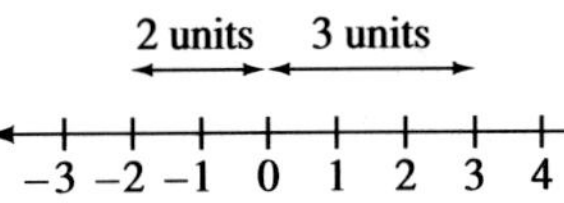

In Chapter 1, we defined the absolute value of a number as its distance on a number line from 0.

$$|-2| = 2 \quad \text{and} \quad |3| = 3$$

1 Let us now consider an equation containing the absolute value of a variable: $|x| = 3$. The solution set of this equation will contain all numbers whose distance from 0 is 3 units. There are two numbers 3 units away from 0 on the number line: 3 and -3:

3 units 3 units

−4 −3 −2 −1 0 1 2 3 4

Thus, the solution set of the equation $|x| = 3$ is $\{3, -3\}$. This suggests the following property:

> **Solving Equations of the Form $|x| = a$**
>
> If a is a positive number, then $|x| = a$ is equivalent to $x = a$ or $x = -a$.

EXAMPLE 1 Solve $|p| = 2$.

Solution: Since 2 is positive, we may use the above rule, and $|p| = 2$ is equivalent to $p = 2$ or $p = -2$.

To check, let $p = 2$ and then $p = -2$ in the original equation.

$|p| = 2$	Original equation.	$|p| = 2$	Original equation.
$|2| = 2$	Let $p = 2$.	$|-2| = 2$	Let $p = -2$.
$2 = 2$	True.	$2 = 2$	True.

The solution set is $\{2, -2\}$. ■

If the expression inside the absolute value bars is more complicated than a single variable such as x, replace x in the preceding rule with the expression inside the absolute value bars and solve the resulting equations.

EXAMPLE 2 Solve $|5w + 3| = 7$.

Solution: Here the expression inside the absolute value bars is $5w + 3$. If we think of the expression $5w + 3$ as x, we have that $|x| = 7$ is equivalent to

$$x = 7 \quad \text{or} \quad x = -7$$

Then substitute $5w + 3$ for x, and we have

$$5w + 3 = 7 \quad \text{or} \quad 5w + 3 = -7$$

Solve these two equations for w.

$$\begin{aligned} 5w + 3 &= 7 \quad \text{or} \quad & 5w + 3 &= -7 \\ 5w &= 4 \quad \text{or} \quad & 5w &= -10 \\ w &= \frac{4}{5} \quad \text{or} \quad & w &= -2 \end{aligned}$$

To check, let $w = -2$ and then $w = \frac{4}{5}$ in the original equation.

Let $w = -2$

$$\begin{aligned} |5(-2) + 3| &= 7 \\ |-10 + 3| &= 7 \\ |-7| &= 7 \\ 7 &= 7 \quad \text{True.} \end{aligned}$$

Let $w = \frac{4}{5}$

$$\begin{aligned} \left|5\left(\frac{4}{5}\right) + 3\right| &= 7 \\ |4 + 3| &= 7 \\ |7| &= 7 \\ 7 &= 7 \quad \text{True.} \end{aligned}$$

Both solutions check and the solution set is $\left\{-2, \frac{4}{5}\right\}$. ■

EXAMPLE 3 Solve $\left|\frac{x}{2} - 1\right| = 11$.

Solution: $\left|\frac{x}{2} - 1\right| = 11$ is equivalent to

$$\frac{x}{2} - 1 = 11 \quad \text{or} \quad \frac{x}{2} - 1 = -11$$

$$2\left(\frac{x}{2} - 1\right) = 2(11) \quad \text{or} \quad 2\left(\frac{x}{2} - 1\right) = 2(-11) \qquad \text{Clear fractions.}$$

$$x - 2 = 22 \quad \text{or} \quad x - 2 = -22 \qquad \text{Apply the distributive property.}$$

$$x = 24 \quad \text{or} \quad x = -20$$

The solution set is $\{24, -20\}$. ■

To apply the absolute value rule, first make sure that the absolute value expression is isolated.

HELPFUL HINT

If the equation has a single absolute value expression containing variables, isolate the absolute value expression first.

EXAMPLE 4 Solve $|2x| + 5 = 7$.

Solution: We want the absolute value expression alone on one side of the equation, so begin by subtracting 5 from both sides. Then proceed as usual.

$$|2x| + 5 = 7$$

$$|2x| = 2 \qquad \text{Subtract 5 from both sides.}$$

$$2x = 2 \quad \text{or} \quad 2x = -2$$

$$x = 1 \quad \text{or} \quad x = -1$$

The solution set is $\{-1, 1\}$. ■

EXAMPLE 5 Solve $|y| = 0$.

Solution: We are looking for all numbers whose distance from 0 is zero units. The only number is 0. The solution set is $\{0\}$. ■

EXAMPLE 6 Solve $|g| = -1$.

Solution: The absolute value of a number is never negative, so this equation has no solution. The solution set is $\{\ \}$ or $\emptyset$. ■

EXAMPLE 7 Solve $\left|\frac{3x + 1}{2}\right| = -2$.

Solution: Again, the absolute value of any expression is never negative, so no solution exists. The solution set is $\{\ \}$ or $\emptyset$. ■

Given two absolute value expressions, we might ask, When are the absolute values of two expressions equal? To see the answer, notice that

$$|2| = |2|, \quad |-2| = |-2|, \quad |-2| = |2|, \quad \text{and} \quad |2| = |-2|$$

same — same — opposites — opposites

Two absolute value expressions are equal when the expressions inside the absolute value bars are equal or are opposites of each other.

EXAMPLE 8 Solve $|3x + 2| = |5x - 8|$.

Solution: This equation is true if the expressions inside the absolute value bars are equal or are opposites of each other.

$$3x + 2 = 5x - 8 \quad \text{or} \quad 3x + 2 = -(5x - 8)$$

Next, solve each equation.

$$3x + 2 = 5x - 8 \quad \text{or} \quad 3x + 2 = -5x + 8$$

$$-2x + 2 = -8 \quad \text{or} \quad 8x + 2 = 8$$

$$-2x = -10 \quad \text{or} \quad 8x = 6$$

$$x = 5 \quad \text{or} \quad x = \frac{3}{4}$$

The solution set is $\left\{\frac{3}{4}, 5\right\}$. ■

EXAMPLE 9 Solve $|x - 3| = |5 - x|$.

Solution:

$$\begin{aligned} x - 3 &= 5 - x &\text{or}\quad x - 3 &= -(5 - x) \\ 2x - 3 &= 5 &\text{or}\quad x - 3 &= -5 + x \\ 2x &= 8 &\text{or}\quad x - 3 - x &= -5 + x - x \\ x &= 4 &\text{or}\quad -3 &= -5 \quad \text{False.} \end{aligned}$$

Recall that when an equation simplifies to a false statement the equation has no solution. Thus, the only solution for x is 4, and the solution set is $\{4\}$. ■

The following box summarizes the methods shown for solving absolute value equations.

Absolute Value Equations

$|x| = a$
1. If a is positive, then solve $x = a$ or $x = -a$.
2. If a is 0, solve $x = 0$.
3. If a is negative, the equation $|x| = a$ has no solution.

$|x| = |y|$ Solve $x = y$ or $x = -y$.

MENTAL MATH

Simplify each expression.

1. $|-7|$
2. $|-8|$
3. $-|5|$
4. $-|10|$
5. $-|-6|$
6. $-|-3|$
7. $|-3| + |-2| + |-7|$
8. $|-1| + |-6| + |-8|$

EXERCISE SET 3.4

Solve each absolute value equation. See Examples 1 through 3.

1. $|x| = 7$
2. $|y| = 15$
3. $|3x| = 12$
4. $|6n| = 12$
5. $|2x - 5| = 9$
6. $|6 + 2n| = 4$
7. $\left|\frac{x}{2} - 3\right| = 1$
8. $\left|\frac{n}{3} + 2\right| = 4$

Solve. See Example 4.

9. $|z| + 4 = 9$
10. $|x| + 1 = 3$
11. $|3x| + 5 = 14$
12. $|2x| - 6 = 4$
13. $|2x| = 0$
14. $|7z| = 0$

Solve. See Examples 5 through 7.

15. $|4n + 1| + 10 = 4$
16. $|3z - 2| + 8 = 1$
17. $|5x - 1| = 0$
18. $|3y + 2| = 0$

Solve. See Examples 8 and 9.

19. $|5x - 7| = |3x + 11|$
20. $|9y + 1| = |6y + 4|$
21. $|z + 8| = |z - 3|$
22. $|2x - 5| = |2x + 5|$

Solve each absolute value equation.

23. $|x| = 4$

24. $|x| = 1$

25. $|y| = 0$

26. $|y| = 8$

27. $|z| = -2$

28. $|y| = -9$

29. $|7 - 3x| = 7$

30. $|4m + 5| = 5$

31. $|6x| - 1 = 11$

32. $|7z| + 1 = 22$

33. $|4p| = -8$

34. $|5m| = -10$

35. $|x - 3| + 3 = 7$

36. $|x + 4| - 4 = 1$

37. $\left|\frac{z}{4} + 5\right| = -7$

38. $\left|\frac{c}{5} - 1\right| = -2$

39. $|9v - 3| = -8$

40. $|1 - 3b| = -7$

41. $|8n + 1| = 0$

42. $|5x - 2| = 0$

43. $|1 + 6c| - 7 = -3$

44. $|2 + 3m| - 9 = -7$

45. $|5x + 1| = 11$

46. $|8 - 6c| = 1$

47. $|4x - 2| = |-10|$

48. $|3x + 5| = |-4|$

49. $|5x + 1| = |4x - 7|$

50. $|3 + 6n| = |4n + 11|$

51. $|6 + 2x| = -|-7|$

52. $|4 - 5y| = -|-3|$

53. $|2x - 6| = |10 - 2x|$

54. $|4n + 5| = |4n + 3|$

55. $\left|\frac{2x - 5}{3}\right| = 7$

56. $\left|\frac{1 + 3n}{4}\right| = 4$

57. $2 + |5n| = 17$

58. $8 + |4m| = 24$

59. $\left|\frac{2x - 1}{3}\right| = |-5|$

60. $\left|\frac{5x + 2}{2}\right| = |-6|$

61. $|2y - 3| = |9 - 4y|$

62. $|5z - 1| = |7 - z|$

63. $\left|\frac{3n + 2}{8}\right| = |-1|$

64. $\left|\frac{2r - 6}{5}\right| = |-2|$

65. $|x + 4| = |7 - x|$

66. $|8 - y| = |y + 2|$

67. $\left|\frac{8c - 7}{3}\right| = -|-5|$

68. $\left|\frac{5d + 1}{6}\right| = -|-9|$

Skill Review

Find the opposite or additive inverse of each. See Section 1.6.

69. 5

70. -12

71. -8

72. $-(-2)$

Find the reciprocal or multiplicative inverse of each. See Section 1.3.

73. -3

74. $\frac{2}{3}$

75. $\frac{5}{9}$

76. $-\frac{1}{7}$

Solve each proportion for x. See Section 3.1.

77. $\frac{x}{3} = \frac{x - 4}{5}$

78. $\frac{x}{4} = \frac{x - 12}{7}$

79. $\frac{2x - 1}{6} = \frac{3x}{10}$

80. $\frac{5x - 5}{6} = \frac{5x}{9}$

Writing in Mathematics

81. In your own words, explain why some absolute value equations have two solutions.

3.5 The Addition and Multiplication Properties of Inequality

OBJECTIVES

Tape IA 4

1. Define linear inequality.
2. Graph intervals on a number line.
3. Use interval notation.
4. Solve linear inequalities.
5. Solve inequality applications.

1 In Chapter 1, we reviewed these inequality symbols and their meanings:

$<$ means "is less than"	$\leq$ means "is less than or equal to"
$>$ means "is greater than"	$\geq$ means "is greater than or equal to"

A linear inequality is similar to a linear equation except that the equality symbol is replaced with an inequality symbol.

Linear Equations	*Linear Inequalities*
$x = 3$	$x < 3$
$5n - 6 = 14$	$5n - 6 \geq 14$
$12 = 7 - 3y$	$12 \leq 7 - 3y$
$\frac{x}{4} - 6 = 1$	$\frac{x}{4} - 6 > 1$

Linear Inequality in One Variable

A linear inequality in one variable is an inequality that can be written in the form

$$ax + b < c$$

where a, b, and c are real numbers and a is not 0.

This definition and all other definitions, properties, and steps in this section also hold true for the inequality symbols, $>$, $\geq$, and $\leq$.

2 A **solution of an inequality** is a value of the variable that makes the inequality a true statement. The **solution set** is the set of all solutions. In the inequality $x < 3$, replacing x with any number smaller than 3 or to the left of 3 on the number line makes the resulting inequality true. This means that any number less than 3 is a solution of the inequality $x < 3$. Since there are infinitely many such numbers, we cannot list all the solutions of the inequality. We can picture them, though, on a number line. To do so, shade the portion of the number line corresponding to numbers less than 3.

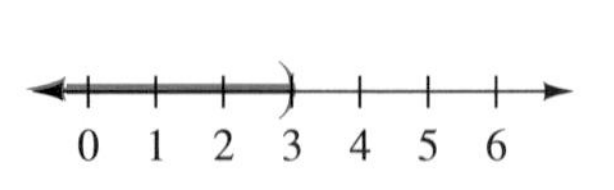

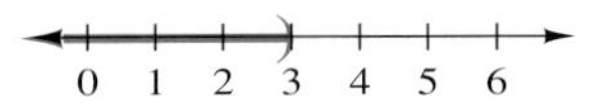

Recall that all the numbers less than 3 lie to the left of 3 on the number line. A parenthesis on the point representing 3 indicates that 3 is not a solution of the inequality: 3 **is not** less than 3. The shaded arrow indicates that the solutions of $x < 3$ continue indefinitely to the left of 3.

Picturing the solutions of an inequality on a number line is called **graphing** the solutions or graphing the inequality, and the picture is called the **graph** of the inequality.

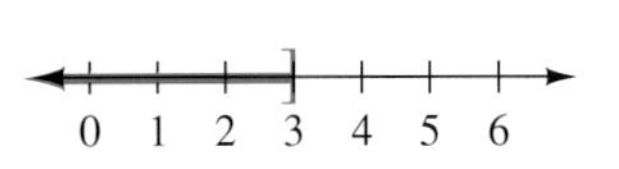

To graph $x \leq 3$, shade the numbers to the left of 3 and place a bracket on the point representing 3. The bracket indicates that 3 is a solution of the inequality $x \leq 3$ and is part of the graph.

EXAMPLE 1 Graph $x \geq -1$.

Solution: We place a bracket at -1 since the inequality symbol is $\geq$ and -1 is greater than or equal to -1. Then shade to the right of -1.

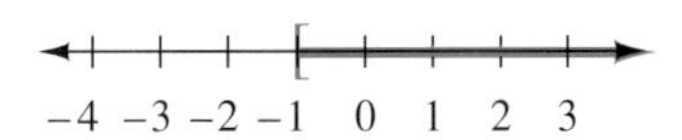

The solution set can be written in set notation as $\{x \mid x \geq -1\}$. Recall that this is read as

$\{x$	$\mid$	$x \geq -1\}$
the set of all x	such that	x is greater than or equal to -1

■

Inequalities containing one inequality symbol are called **simple inequalities,** whereas inequalities containing two inequality symbols are called **compound inequalities.** A compound inequality is really two simple inequalities in one. The compound inequality

$$3 < x < 5 \quad \text{means} \quad 3 < x \textbf{ and } x < 5$$

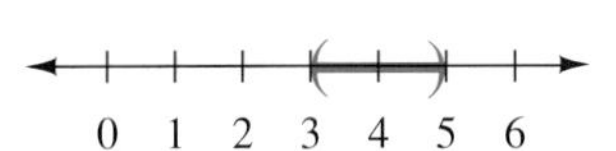

This can be read "x is greater than 3 and less than 5." To graph $3 < x < 5$, place a parenthesis at both 3 and 5 and shade between.

EXAMPLE 2 Graph $2 < x \leq 4$.

Solution: Graph all numbers greater than 2 and less than or equal to 4. Place a parenthesis at 2, a bracket at 4, and shade between.

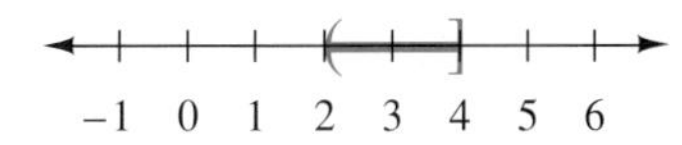

The solution set can be written in set notation as $\{x \mid 2 < x \leq 4\}$. ■

3 Another way to list solutions of a linear inequality is by interval notation.

The graph of $\{x \mid x > 2\}$ looks like

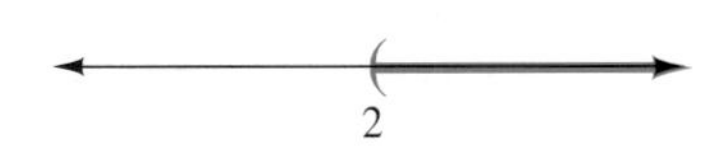

and the solutions can be represented in interval notation as $(2, \infty)$. The symbol ∞ is read "infinity" and indicates that the interval includes **all** numbers greater than 2. The left parenthesis indicates that 2 **is not** included in the interval. Using a left bracket would indicate that 2 **is** included in the interval. The following table shows three forms of describing intervals: in set notation, as a graph, and in interval notation.

Set Notation	Graph	Interval Notation
$\{x \mid x < a\}$	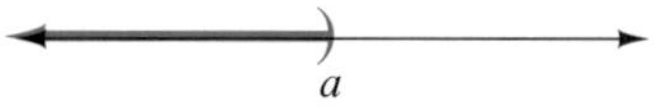 a	$(-\infty, a)$
$\{x \mid x > a\}$	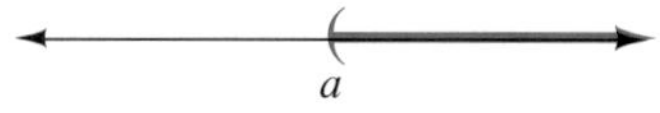 a	(a, ∞)
$\{x \mid x \leq a\}$	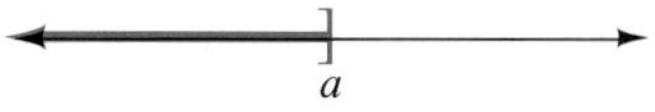 a	$(-\infty, a]$
$\{x \mid x \geq a\}$	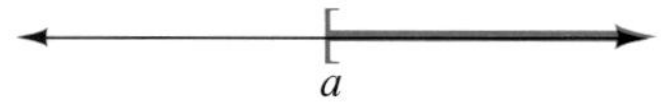 a	$[a, \infty)$
$\{x \mid a < x < b\}$	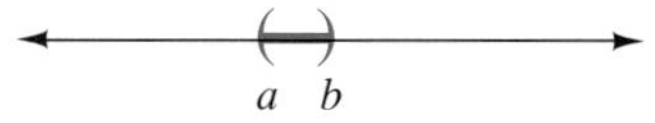 a b	(a, b)
$\{x \mid a \leq x \leq b\}$	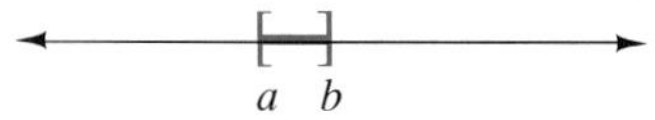 a b	$[a, b]$
$\{x \mid a < x \leq b\}$	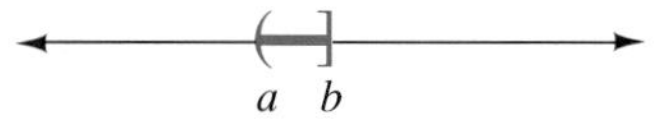 a b	$(a, b]$
$\{x \mid a \leq x < b\}$	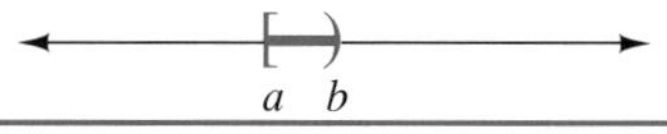 a b	$[a, b)$

HELPFUL HINT

Notice that a parenthesis is always used to enclose ∞ and $-\infty$.

EXAMPLE 3 Graph each set on a number line and then write in interval notation.

a. $\{x \mid x \geq 2\}$ **b.** $\{x \mid x < -1\}$ **c.** $\{x \mid 0.5 < x \leq 3\}$

Solution:

a. 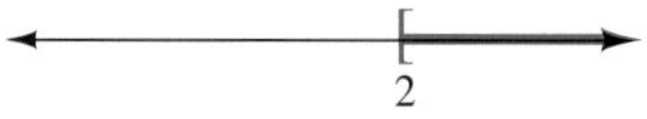 2 $[2, \infty)$

b. -1 $(-\infty, -1)$

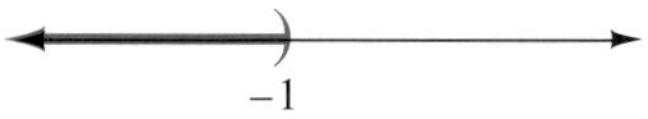

c. 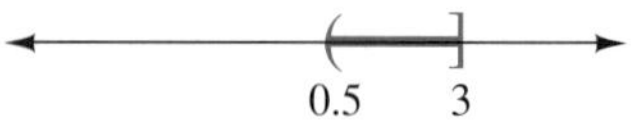 0.5 3 $(0.5, 3]$ ■

4 When solutions to a linear inequality are not immediately obvious, they are found through a process similar to the one used to solve a linear equation. Our goal is to isolate the variable, and we use properties of inequality similar to properties of equality.

> **Addition Property of Inequality**
>
> If a, b, and c are real numbers, then
>
> $$a < b \quad \text{and} \quad a + c < b + c$$
>
> are equivalent inequalities.

This property also holds true for subtracting values since subtraction is defined in terms of addition. In other words, adding or subtracting the same quantity from both sides of an inequality does not change the solutions of the inequality.

EXAMPLE 4 Solve $x + 4 \leq -6$ for x. Graph the solution and write it in interval notation.

Solution: To solve for x, subtract 4 from both sides of the inequality.

$x + 4 \leq -6$	Original inequality.
$x + 4 - 4 \leq -6 - 4$	Subtract 4 from both sides.
$x \leq -10$	Simplify.

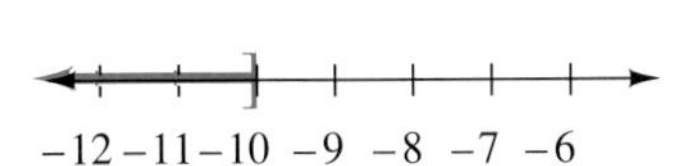

The solution set is $\{x \mid x \leq -10\}$, which in interval notation is $(-\infty, -10]$. ■

An important difference between linear equations and linear inequalities is shown when we multiply or divide both sides of an inequality by a nonzero real number. Let us start with the true statement $6 < 8$ and multiply both sides by 2.

$6 < 8$	True.
$2(6) < 2(8)$	Multiply both sides by 2.
$12 < 16$	True.

The inequality remains true.

But if we start with the same true statement $6 < 8$ and multiply both sides by -2, the resulting inequality is no longer a true statement.

$6 < 8$	True.
$-2(6) < -2(8)$	Multiply both sides by -2.
$-12 < -16$	**False.**

Notice, however, that if we reverse the direction of the inequality symbol, the resulting inequality is true.

$-12 < -16$	**False.**
$-12 > -16$	**True.**

This demonstrates the multiplication property of inequality.

Multiplication Property of Inequality

1. If a, b, and c are real numbers and c is **positive,** then

$$a < b \quad \text{and} \quad ac < bc$$

are equivalent inequalities.

2. If a, b, and c are real numbers and c is **negative,** then

$$a < b \quad \text{and} \quad ac > bc$$

are equivalent inequalities.

Because division is defined in terms of multiplication, **this property also holds true when dividing both sides of an inequality by a nonzero number:** If we multiply or divide both sides of an inequality by a negative number, **the direction of the inequality sign must be reversed for the inequalities to remain equivalent.**

HELPFUL HINT

Whenever both sides of an inequality are multiplied or divided by a negative number, the direction of the inequality symbol **must be** reversed to form an equivalent inequality.

EXAMPLE 5 Solve for x.

a. $2x < -6$

b. $-2x < 6$

Solution: **a.** $2x < -6$

$\frac{2x}{2} < \frac{-6}{2}$ Divide both sides by 2.

$x < -3$ Simplify.

The solution set is $\{x \mid x < -3\}$, which in interval notation is $(-\infty, -3)$. The graph of the solution set is

-3 $(-\infty, -3)$

b. $-2x < 6$

$\frac{-2x}{-2} > \frac{6}{-2}$ Divide both sides by -2 and reverse the inequality symbol.

$x > -3$ Simplify.

The solution set is $\{x \mid x > -3\}$, which is $(-3, \infty)$ in interval notation. The graph of the solution set is

■

To solve linear inequalities in general, we follow steps similar to those for solving linear equations.

To Solve a Linear Inequality in One Variable

Step 1 Clear the equation of fractions by multiplying both sides of the inequality by the lowest common denominator (LCD) of all fractions in the inequality.

Step 2 Use the distributive property to remove grouping symbols such as parentheses.

Step 3 Combine like terms on each side of the inequality.

Step 4 Use the addition property of inequality to write the inequality as an equivalent inequality with variable terms on one side and numbers on the other side.

Step 5 Use the multiplication property of inequality to isolate the variable.

EXAMPLE 6 Solve $-4x + 7 \geq -9$, and graph the solution.

Solution:

$$-4x + 7 \geq -9$$

$$-4x + 7 - 7 \geq -9 - 7 \quad \text{Subtract 7 from both sides.}$$

$$-4x \geq -16 \quad \text{Simplify.}$$

$$\frac{-4x}{-4} \leq \frac{-16}{-4} \quad \text{Divide both sides by } -4 \text{ and reverse the direction of the inequality sign.}$$

$$x \leq 4 \quad \text{Simplify.}$$

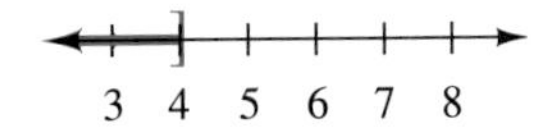

The solution set $\{x \mid x \leq 4\}$ written in interval notation is $(-\infty, 4]$, and its graph is shown. ■

EXAMPLE 7 Solve $2x + 7 \leq x - 11$, and graph the solution.

Solution:

$$2x + 7 \leq x - 11$$

$$2x + 7 - x \leq x - 11 - x \quad \text{Subtract } x \text{ from both sides.}$$

$$x + 7 \leq -11 \quad \text{Simplify.}$$

$$x + 7 - 7 \leq -11 - 7 \quad \text{Subtract 7 from both sides.}$$

$$x \leq -18 \quad \text{Simplify.}$$

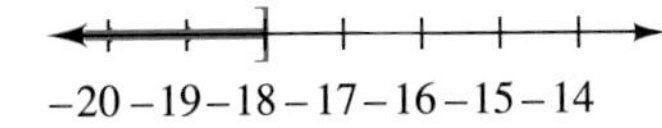

The solution set $\{x \mid x \leq -18\}$ is graphed. In interval notation, the solution is $(-\infty, -18]$. ■

EXAMPLE 8 Solve for x: $-(x - 3) + 2 < 3(2x - 5) + x$.

Solution:

$$-(x - 3) + 2 < 3(2x - 5) + x$$

$$-x + 3 + 2 < 6x - 15 + x \quad \text{Apply the distributive property.}$$

$$5 - x < 7x - 15 \quad \text{Combine like terms.}$$

$$5 - x + x < 7x - 15 + x \quad \text{Add } x \text{ to both sides.}$$

$$5 < 8x - 15$$ Combine like terms.

$$5 + 15 < 8x - 15 + 15$$ Add 15 to both sides.

$$20 < 8x$$ Combine like terms.

$$\frac{20}{8} < \frac{8x}{8}$$ Divide both sides by 8.

$$\frac{5}{2} < x \quad \text{or} \quad x > \frac{5}{2}$$

The solution set written in interval notation is $\left(\frac{5}{2}, \infty\right)$ and its graph is

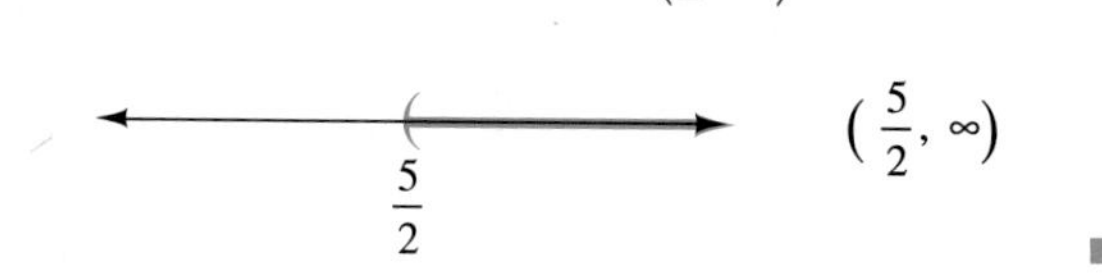

EXAMPLE 9 Solve for x: $\frac{2}{5}(x - 6) \geq x - 1$.

Solution:

$$\frac{2}{5}(x - 6) \geq x - 1$$

$$5\left[\frac{2}{5}(x - 6)\right] \geq 5(x - 1)$$ Multiply both sides by 5 to eliminate fractions.

$$2x - 12 \geq 5x - 5$$ Use the distributive property.

$$-3x - 12 \geq -5$$ Subtract $5x$ from both sides.

$$-3x \geq 7$$ Add 12 to both sides.

$$\frac{-3x}{-3} \leq \frac{7}{-3}$$ Divide both sides by -3 and reverse the inequality symbol.

$$x \leq -\frac{7}{3}$$ Simplify.

The solution set $\left\{x \mid x \leq -\frac{7}{3}\right\}$ is graphed on a number line and is written in interval notation as $\left(-\infty, -\frac{7}{3}\right]$

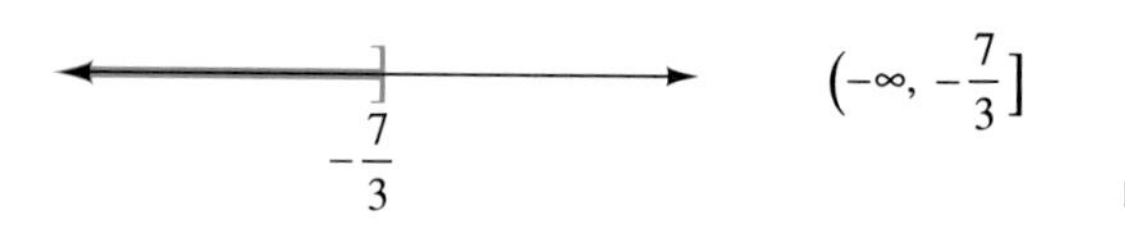

EXAMPLE 10 Solve for x: $2(x + 3) > 2x + 1$.

Solution:

$$2(x + 3) > 2x + 1$$

$$2x + 6 > 2x + 1$$ Distribute on left side.

$$2x + 6 - 2x > 2x + 1 - 2x$$ Subtract $2x$ from both sides.

$$6 > 1$$ Simplify.

$6 > 1$ is a true statement for all values of x, so this inequality and the original inequality are true for all numbers. The solution set is $\{x \mid x \text{ is a real number}\}$, or $(-\infty, \infty)$ in interval notation, and its graph is

5 Problems containing words such as "at least," "at most," "between," "no more than," and "no less than" usually indicate that an inequality, instead of an equation, be solved. In solving applications involving linear inequalities, we use the same procedure as when we solved applications involving linear equations.

EXAMPLE 11 Marie Chase and Jonathan Edwards are having their wedding reception at the Gallery reception hall. They may spend at most \$1000 for the reception. If the reception hall charges a \$100 cleanup fee plus \$14 per person, find the largest number of people that they can invite.

Solution: 1. UNDERSTAND. Read and reread the problem. Next guess a solution. If 50 people attend the reception, the cost is $\$100 + \$14(50) = \$100 + \$700 = \$800$.

2. ASSIGN. Let $x =$ the number of people who attend the reception.

4. TRANSLATE.

In words:	Cleanup fee	+	cost for guests	can't be more than	\$1000
Translate:	100	+	$14x$	$\leq$	1000

5. COMPLETE.

$$100 + 14x \leq 1000$$

$$14x \leq 900 \quad \text{Subtract 100 from both sides.}$$

$$x \leq 64\frac{2}{7} \quad \text{Divide both sides by 14.}$$

6. INTERPRET. *Check:* Since x represents the number of people, we round down to the nearest whole, or 64. Notice that if 64 people attend, the cost is $\$100 + \$14(64) = \$996$. If 65 people attend, the cost is $\$100 + \$14(65) = \$1010$, which is more than the given \$1000.

State: At most, 64 people may attend the reception. ■

EXERCISE SET 3.5

Graph the solution set of each inequality on a number line, and write the solution set in interval notation. See Examples 1 through 3.

1. $\{x \mid x < -3\}$

2. $\{x \mid x \geq -7\}$

3. $\{x \mid x \geq 0.3\}$

4. $\{x \mid x < -0.2\}$

5. $\{x \mid 5 < x\}$

6. $\{x \mid -7 \geq x\}$

7. $\{x \mid -2 < x < 5\}$

8. $\{x \mid -5 \leq x \leq -1\}$

9. $\{x \mid 5 > x > -1\}$

10. $\{x \mid -3 \geq x \geq -7\}$

Solve each inequality for the variable. Graph the solution set and write the solution set in interval notation. See Examples 4 and 5.

11. $2x < -6$

12. $3x > -9$

13. $x - 2 \geq -7$

14. $x + 4 \leq 1$

15. $-8x \leq 16$

16. $-5x < 20$

Solve each inequality for the variable. Graph the solution set and write it in interval notation. See Examples 6 through 9.

17. $15 + 2x \geq 4x - 7$

18. $20 + x < 6x$

19. $\frac{3x}{4} \geq 2$

20. $\frac{5}{6}x \geq -8$

21. $3(x - 5) < 2(2x - 1)$

22. $5(x + 4) \leq 4(2x + 3)$

23. $\frac{1}{2} + \frac{2}{3} \geq \frac{x}{6}$

24. $\frac{3}{4} - \frac{2}{3} > \frac{x}{6}$

Solve each inequality for the variable. Graph the solution set and write it in interval notation. See Example 10.

25. $4(x - 1) \geq 4x - 8$

26. $3x + 1 < 3(x - 2)$

27. $7x < 7(x - 2)$

28. $8(x + 3) \leq 7(x + 5) + x$

Solve. See Example 11.

29. A small plane's maximum take-off weight is 2000 pounds. Six passengers weigh an average of 160 pounds each. Use an inequality to find the maximum weight of luggage and cargo the plane can carry.

30. A clerk uses the freight elevator to move boxes of paper. The elevator's weight limit is 1500 pounds. If each box of paper weighs 66 pounds and the clerk weighs 147 pounds, use an inequality to find the maximum number of boxes she can move on the elevator at one time.

31. To mail an envelope first class, the U.S. Postal Service charges 32 cents for the first ounce and 23 cents per ounce for each additional ounce. Use an inequality to find the maximum weight that can be mailed for $4.23.

32. A parking garage charges $1 for the first half-hour and 60 cents for each additional half-hour or a portion of a half-hour. Use an inequality to find how long you can park if you have only $4.00 in cash.

Solve each inequality for the variable. Graph the solution set and write it in interval notation.

33. $5x + 3 > 2 + 4x$

34. $7x - 1 \geq 6x - 1$

35. $8x - 7 \leq 7x - 5$

36. $12x + 14 < 11x - 2$

37. $5x > 10$

38. $9x < 45$

39. $-4x \leq 32$

40. $-6x \geq 42$

41. $-2x + 7 \geq 9$

42. $8 - 5x \leq 23$

43. $4(2x + 1) > 4$

44. $6(2 - x) \geq 12$

45. $\frac{x + 7}{5} > 1$

46. $\frac{2x - 4}{3} \leq 2$

47. $-6x + 2 \geq 2(5 - x)$

48. $-7x + 4 > 3(4 - x)$

49. $4(3x - 1) \leq 5(2x - 4)$

50. $3(5x - 4) \leq 4(3x - 2)$

51. $\frac{-5x + 11}{2} \leq 7$

52. $\frac{4x - 8}{7} < 0$

53. $8x - 16 \leq 10x + 2$

54. $18x - 24 < 10x + 64$

55. $2(x - 3) > 70$

56. $3(5x + 6) \geq -12$

57. $-5x + 4 \leq -4(x - 1)$

58. $-6x + 2 < -3(x + 4)$

59. $\frac{1}{4}(x - 7) \geq x + 2$

60. $\frac{3}{5}(x + 1) \leq x + 1$

61. $\frac{2}{3}(x + 2) < \frac{1}{5}(2x + 7)$

62. $\frac{1}{6}(3x + 10) > \frac{5}{12}(x - 1)$

63. $4(x - 6) + 2x - 4 \geq 3(x - 7) + 10x$

64. $7(2x + 3) + 4x \leq 7 + 5(3x - 4)$

65. $\frac{5x + 1}{7} - \frac{2x - 6}{4} \geq -4$

66. $\frac{1 - 2x}{3} + \frac{3x + 7}{7} > 1$

67. $\frac{-x + 2}{2} - \frac{1 - 5x}{8} < -1$

68. $\frac{3 - 4x}{6} - \frac{1 - 2x}{12} \leq -2$

Solve.

69. Shureka has scores of 72, 67, 82, and 79 on her algebra tests. Use an inequality to find the minimum score she can make on the final exam to pass the course with an average of 60 or higher, given that the final exam counts as two tests.

70. Northeast Telephone Company offers two billing plans for local calls. Plan 1 charges $25 per month for unlimited calls, and plan 2 charges $13 per month plus 6 cents per call. Use an inequality to find the number of monthly calls for which plan 1 is more economical than plan 2.

71. In a Winter Olympics speed-skating event, Hans scored times of 3.52, 4.04, and 3.87 minutes on his first three trials. Use an inequality to find the maximum time he can score on his last trial so that his average time is under 4.0 minutes.

72. A car rental company offers two subcompact rental plans. Plan A charges $32 per day for unlimited mileage, and plan B charges $24 per day plus 15 cents per mile. Use an inequality to find the number of daily miles for which plan A is more economical than plan B.

73. Ben Holladay bowled 146 and 201 in his first two games. What must he bowl in his third game to have an average of at least 180 and qualify for his bowling team?

74. On an NBA team the two forwards measure 6′8″ and 6′6″, and the two guards measure 6′0″ and 5′9″ tall. How tall a center should they hire if they wish to have a starting team average height of at least 6′5″?

75. The perimeter of a rectangle is to be no greater than 100 centimeters, and the width must be 15 centimeters. Find the maximum length of the rectangle.

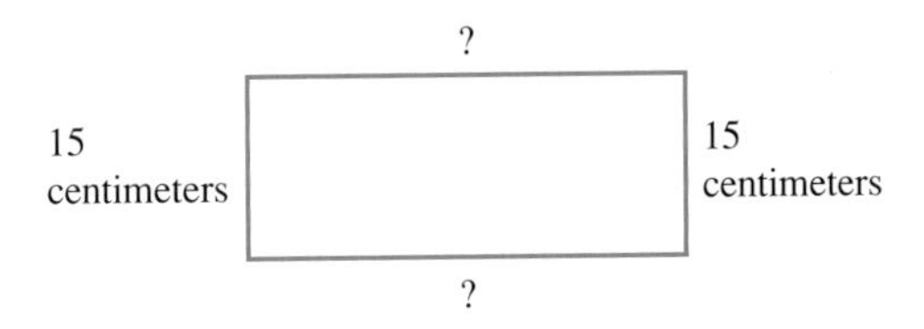

76. One side of a triangle is four times as long as another side, and the third side is 12 inches long. If the perimeter can be no longer than 87 inches, find the maximum lengths of the other two sides.

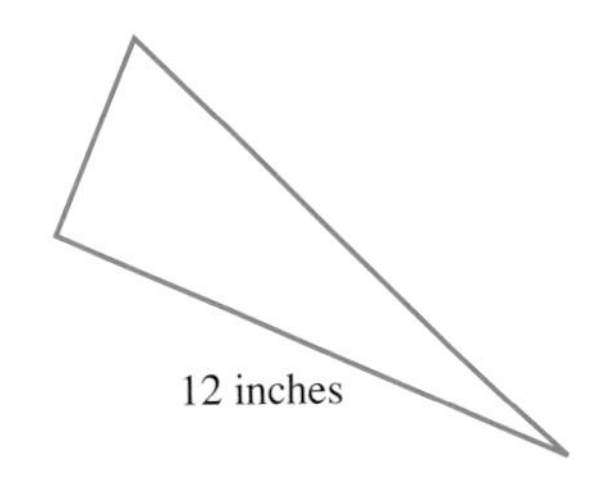

Skill Review

Find the following roots. See Section 1.4.

77. $\sqrt{25}$

78. $\sqrt{49}$

79. $\sqrt[3]{8}$

80. $\sqrt[3]{27}$

81. $\sqrt[4]{16}$

82. $\sqrt[4]{81}$

83. $-\sqrt{4}$

82. $-\sqrt{36}$

Graph each set on a number line and write it in interval notation. See Section 3.5.

85. $\{x \mid 0 \leq x \leq 5\}$

86. $\left\{x \mid -\frac{1}{2} < x < \frac{1}{2}\right\}$

87. $\{x \mid 3 \leq x < 10\}$

88. $\{x \mid -7 < x \leq -1\}$

This broken line graph shows the enrollment of people (members) in a health maintenance organization (HMO). The height of each dot corresponds to the number of members (in millions.) Use this graph to answer the questions below. See Section 2.6.

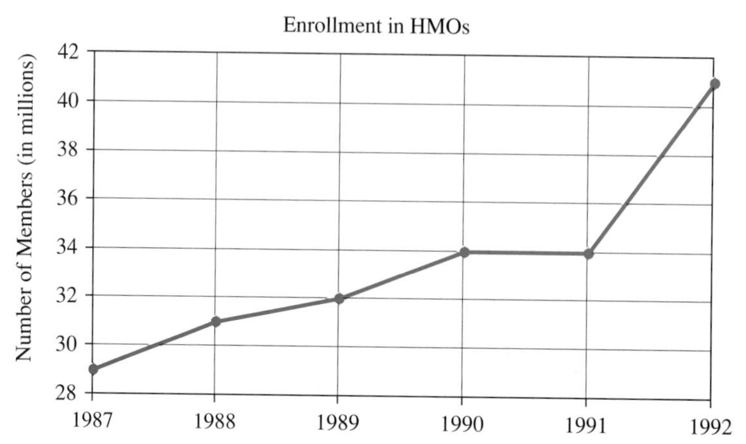

Source: U.S. Bureau of the Census, *Statistical Abstract of the United States: 1994*, 113th ed., Washington, DC, 1994.

89. How many people were enrolled in health maintenance organizations in 1989?

90. How many people were enrolled in health maintenance organizations in 1992?

91. Which year shows the greatest increase in number of members?

Writing in Mathematics

92. Explain how solving a linear inequality is similar to solving a linear equation.

93. Explain how solving a linear inequality is different from solving a linear equation.

3.6 Compound Inequalities

OBJECTIVE **1** Solve compound inequalities.

Tape IA 4

1 Two inequalities joined by the words ***and*** or ***or*** are called **compound inequalities.**

Compound Inequalities

$$x + 3 < 8 \quad \text{and} \quad x > 2$$

$$\frac{2x}{3} \geq 5 \quad \text{or} \quad -x + 10 < 7$$

The solution set of a compound inequality formed by the word ***and*** is the intersection of the solution sets of the two inequalities. The intersection of two sets, denoted by $\cap$, is the set of elements common to both sets. In other words, a solution of a compound inequality formed by the word ***and*** is a number that makes both inequalities true.

The graph of the compound inequality $x \le 5$ and $x \ge 3$ contains all numbers that make the inequality $x \le 5$ a true statement **and** that make the inequality $x \ge 3$ a true statement. The first graph shown next is the graph of $x \le 5$, the second graph is the graph of $x \ge 3$, and the third graph shows the intersection of the two graphs. It is the graph of $x \le 5$ **and** $x \ge 3$.

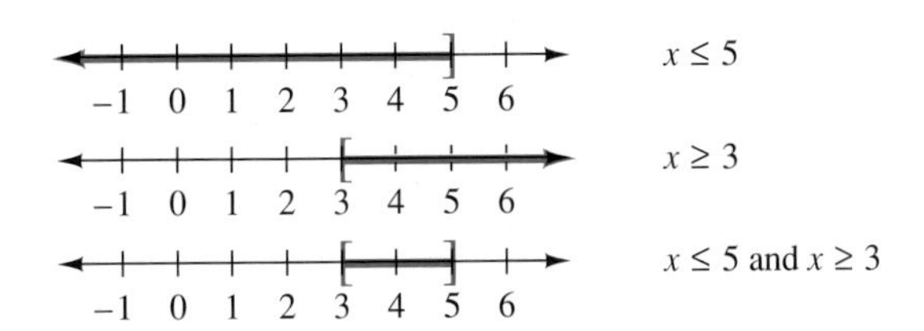

In interval notation, the set $\{x \mid x \le 5 \text{ and } x \ge 3\}$ is written as $[3, 5]$.

EXAMPLE 1 Solve for x: $x - 7 < 2$ and $2x + 1 < 9$.

Solution: First solve each inequality separately.

$$x - 7 < 2 \quad \text{and} \quad 2x + 1 < 9$$

$$x < 9 \quad \text{and} \quad 2x < 8$$

$$x < 9 \quad \text{and} \quad x < 4$$

Graph the two intervals on two number lines and find their intersection.

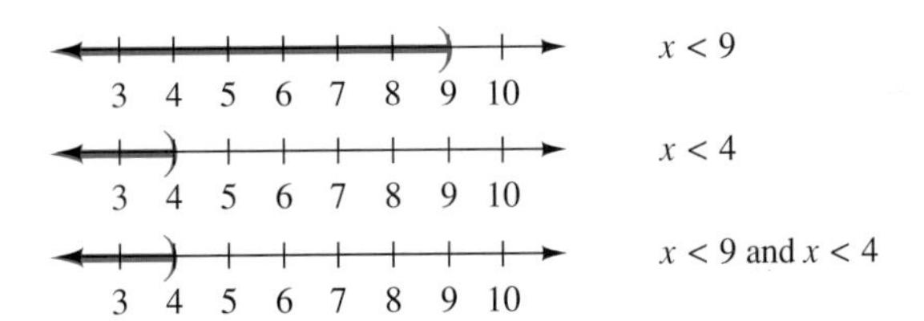

The solution set written as an interval is $(-\infty, 4)$. ■

Compound inequalities containing the word **and** can be written in a more compact form. The compound inequality $2 \le x$ and $x \le 6$ can be written as

$$2 \le x \le 6$$

Recall that the graph of $2 \le x \le 6$ is all numbers between 2 and 6, including 2 and 6.

The set $\{x \mid 2 \le x \le 6\}$ written in interval notation is $[2, 6]$.

To solve a compound inequality like $2 < 4 - x < 7$, we isolate x "on the middle side." Since a compound inequality is really two inequalities in one statement, we must perform the same operation to all three "sides" of the inequality.

EXAMPLE 2 Solve $2 < 4 - x < 7$.

Solution: To isolate x, first subtract 4 from all three sides.

$$2 < 4 - x < 7$$

$$2 - 4 < 4 - x - 4 < 7 - 4 \quad \text{Subtract 4 from all three sides.}$$

$$-2 < -x < 3 \quad \text{Simplify.}$$

$$\frac{-2}{-1} > \frac{-x}{-1} > \frac{3}{-1} \quad \text{Divide all three sides by } -1 \text{ and reverse the inequality symbols.}$$

$$2 > x > -3$$

This is equivalent to $-3 < x < 2$, and its graph is shown below. The solution set in interval notation is $(-3, 2)$.

EXAMPLE 3 Solve for x: $-1 \le \frac{2x}{3} + 5 \le 2$.

Solution: First, clear the inequality of fractions by multiplying all three sides by the LCD of 3.

$$-1 \le \frac{2x}{3} + 5 \le 2$$

$$3(-1) \le 3\left(\frac{2x}{3} + 5\right) \le 3(2) \quad \text{Multiply by the LCD of 3.}$$

$$-3 \le 2x + 15 \le 6 \quad \text{Apply the distributive property and multiply.}$$

$$-3 - 15 \le 2x + 15 - 15 \le 6 - 15 \quad \text{Subtract 15 from all three sides.}$$

$$-18 \le 2x \le -9 \quad \text{Simplify.}$$

$$\frac{-18}{2} \le \frac{2x}{2} \le \frac{-9}{2} \quad \text{Divide all three sides by 2.}$$

$$-9 \le x \le -\frac{9}{2} \quad \text{Simplify.}$$

The graph of the solution is shown below. The solution set in interval notation is $\left[-9, -\frac{9}{2}\right]$.

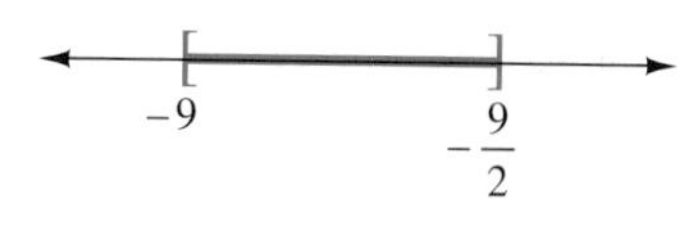

The solution set of a compound inequality formed by the word ***or*** is the **union** of the solution sets of the two inequalities. The union of two sets, denoted by $\cup$, is the set of elements that belong to either of the sets. In other words, a solution of a compound inequality formed by the word ***or*** is a number that makes either inequality true.

The graph of the compound inequality $x \leq 1$ or $x \geq 3$ contains all numbers that make the inequality $x \leq 1$ a true statement **or** that make the inequality $x \geq 3$ a true statement.

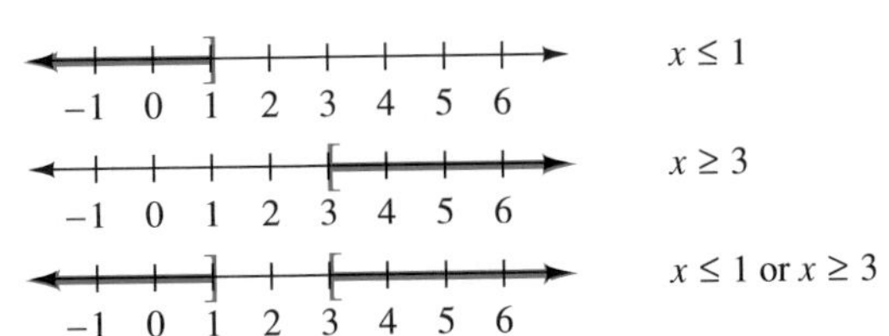

In interval notation, the set $\{x \mid x \leq 1 \text{ or } x \geq 3\}$ is written as $(-\infty, 1] \cup [3, \infty)$.

EXAMPLE 4 Solve $5x - 3 \leq 10$ or $x + 1 \geq 5$.

Solution: Solve each inequality separately.

$$5x - 3 \leq 10 \quad \text{or} \quad x + 1 \geq 5$$
$$5x \leq 13$$
$$x \leq \frac{13}{5} \quad \text{or} \quad x \geq 4$$

Graph each interval on a number line. Then find their union.

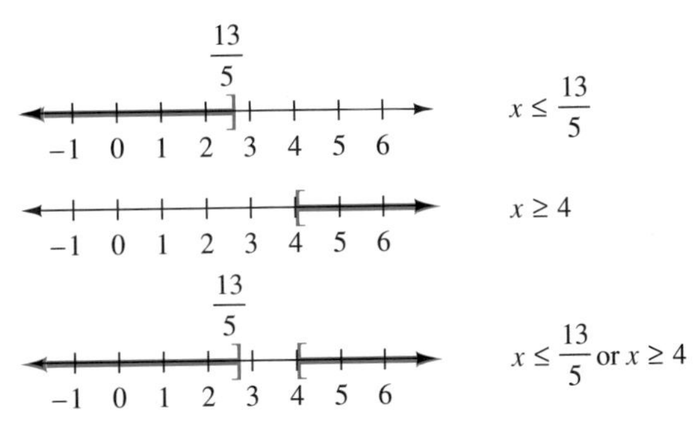

The solution set in interval notation is $\left(-\infty, \frac{13}{5}\right] \cup [4, \infty)$. ■

EXERCISE SET 3.6

Solve each compound inequality. Graph the solution set. See Example 1.

1. $x < 5$ and $x > -2$

2. $x \leq 7$ and $x \leq 1$

3. $x + 1 \geq 7$ and $3x - 1 \geq 5$

4. $-2x < -8$ and $x - 5 < 5$

5. $4x + 2 \leq -10$

6. $x + 4 > 0$ and $4x > 0$

Solve each compound inequality. Graph the solution set. See Examples 2 and 3.

7. $5 < x - 6 < 11$

8. $-2 \leq x + 3 \leq 0$

9. $-2 \leq 3x - 5 \leq 7$

10. $1 < 4 + 2x < 7$

11. $1 \le \frac{2}{3}x + 3 \le 4$

12. $-2 < \frac{1}{2}x - 5 < 1$

13. $-5 \le \frac{x + 1}{4} \le -2$

14. $-4 \le \frac{2x + 5}{3} \le 1$

Solve each compound inequality. Graph the solution set. See Example 4.

15. $x < -1$ or $x > 0$

16. $x \le 1$ or $x \le -3$

17. $-2x \le -4$ or $5x - 20 \ge 5$

18. $x + 4 < 0$ or $6x > -12$

19. $3(x - 1) < 12$ or $x + 7 > 10$

20. $5(x - 1) \ge -5$ or $5 - x \le 11$

Solve each compound inequality. Graph the solution set.

21. $x < 2$ and $x > -1$

22. $x < 5$ and $x < 1$

23. $x < 2$ or $x > -1$

24. $x < 5$ or $x < 1$

25. $x \ge -5$ and $x \ge -1$

26. $x \le 0$ or $x \ge -3$

27. $x \ge -5$ or $x \ge -1$

28. $x \le 0$ and $x \ge -3$

29. $0 \le 2x - 3 \le 9$

30. $3 < 5x + 1 < 11$

31. $\frac{1}{2} < x - \frac{3}{4} < 2$

32. $\frac{2}{3} < x + \frac{1}{2} < 4$

33. $x + 3 \ge 3$ and $x + 3 \le 2$

34. $2x - 1 \ge 3$ and $-x > 2$

35. $3x \ge 5$ or $-x - 6 < 1$

36. $\frac{3}{8}x + 1 \le 0$ or $-2x < -4$

37. $0 < \frac{5 - 2x}{3} < 5$

38. $-2 < \frac{-2x - 1}{3} < 2$

39. $-6 < 3(x - 2) \le 8$

40. $-5 < 2(x + 4) < 8$

41. $-x + 5 > 6$ and $1 + 2x \le -5$

42. $5x \le 0$ and $-x + 5 < 8$

43. $3x + 2 \le 5$ or $7x > 29$

44. $-x < 7$ or $3x + 1 < -20$

45. $5 - x > 7$ and $2x + 3 \ge 13$

46. $-2x < -6$ or $1 - x > -2$

47. $-\frac{1}{2} \le \frac{4x - 1}{6} < \frac{5}{6}$

48. $-\frac{1}{2} \le \frac{3x - 1}{10} < \frac{1}{2}$

49. $\frac{1}{15} < \frac{8 - 3x}{15} < \frac{4}{5}$

50. $-\frac{1}{4} < \frac{6 - x}{12} < -\frac{1}{6}$

51. $0.3 < 0.2x - 0.9 < 1.5$

52. $-0.7 \le 0.4x + 0.8 < 0.5$

The formula for converting Fahrenheit temperatures to Celsius temperatures is $C = \frac{5}{9}(F - 32)$. *Use this formula for Exercises 53 and 54.*

53. The temperatures in Ohio range from −39°C to 45°C. Use a compound inequality to convert these temperatures to Fahrenheit temperatures.

54. In Oslo, the temperature ranges from −10° to 18° Celsius. Use a compound inequality to convert these temperatures to the Fahrenheit scale.

Solve.

55. Mario Lipco has scores of 85, 95, and 92 on his algebra tests. Use a compound inequality to find the range of scores he can make on his final exam to receive an A in the course. The final exam counts as three tests, and an A is received if the final course average is from 90 to 100.

56. Wendy Wood has scores of 80, 90, 82, and 75 on her chemistry tests. Use a compound inequality to find the range of scores she can make on her final exam to receive a B in the course. The final exam counts as two tests, and a B is received if the final course average is from 80 to 89.

57. The formula $C = 3.14d$ can be used to approximate the circumference of a circle given its diameter. Waldo Manufacturing manufactures and sells a certain washer with an outside circumference of 3 centimeters. The company has decided that a washer whose actual circumference is in the interval $2.9 \le C \le 3.1$ centimeters is acceptable. Use a compound inequality to find the corresponding interval for diameters of these washers.

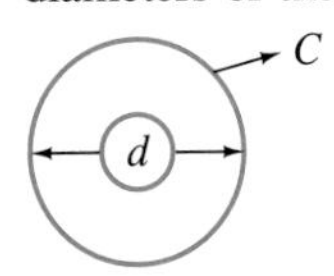

58. Bunnie Supplies manufactures plastic Easter eggs that open. The company has determined that if the circumference of the opening of each part of the egg is in the interval $118 \leq C \leq 122$ millimeters, the eggs will open and close comfortably. Use a compound inequality to find the corresponding interval for diameters of these openings.

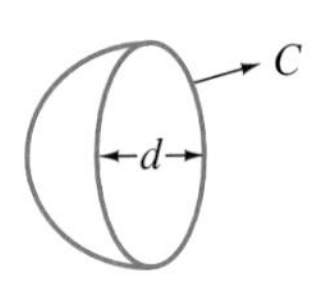

Skill Review

Evaluate. See Section 1.6.

59. $|-3|$

60. $|0|$

61. $|7|$

62. $\left|-\frac{1}{2}\right|$

63. $|7-15|$

64. $|5(-2)-1|$

65. $\left|\frac{10(5)}{-15-10}\right|$

66. $\left|\frac{3(-4)+6}{6}\right|$

Solve. See Section 3.4.

67. $|x-2|=19$

68. $|x+5|=8$

69. $|2x+1|+7=-4$

70. $|6-3x|-1=4$

71. $|x-3|=|x+11|$

72. $|2y-4|=|y-9|$

A Look Ahead

Solve each compound inequality for x. See the following example.

EXAMPLE Solve: $x-6<3x<2x+5$

Solution: Notice that this inequality contains a variable on the left, the right, and in the middle. To solve, rewrite the inequality using the word ***and***.

$$x-6<3x \quad \text{and} \quad 3x<2x+5$$
$$-6<2x \quad \text{and} \quad x<5$$
$$-3<x$$
$$x>-3 \quad \text{and} \quad x<5$$

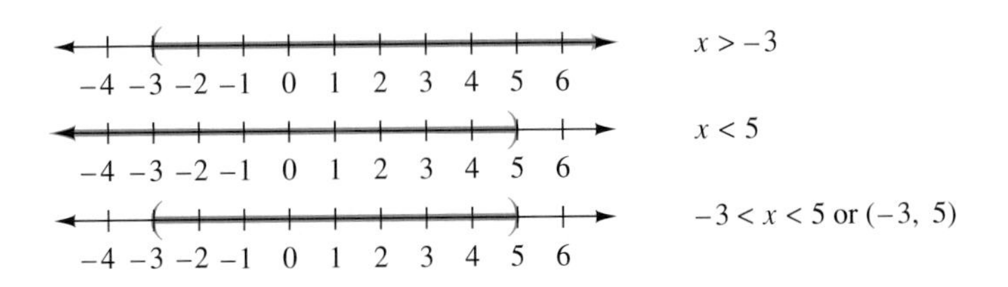

73. $2x-3<3x+1<4x-5$

74. $x+3<2x+1<4x+6$

75. $-3(x-2)\leq 3-2x\leq 10-3x$

76. $7x-1\leq 7+5x\leq 3(1+2x)$

77. $5x-8<2(2+x)<-2(1+2x)$

78. $1+2x<3(2+x)<1+4x$

3.7
Absolute Value Inequalities

OBJECTIVES

1. Solve absolute value inequalities of the form $|x| < a$.
2. Solve absolute value inequalities of the form $|x| > a$.

Tape IA 5

1 The solution set of an absolute value inequality such as $|x| < 2$ will contain all numbers whose distance from 0 is less than 2 units, as shown next.

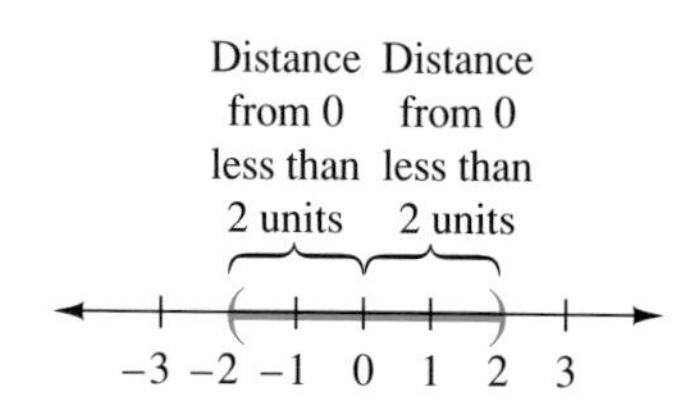

The solution set is $\{x \mid -2 < x < 2\}$ or $(-2, 2)$ in interval notation.

EXAMPLE 1 Solve $|x| \le 3$.

Solution: The solution set of this inequality contains all numbers whose distance from 0 is less than or equal to 3. Thus, 3, -3, and all numbers between 3 and -3 are in the solution set.

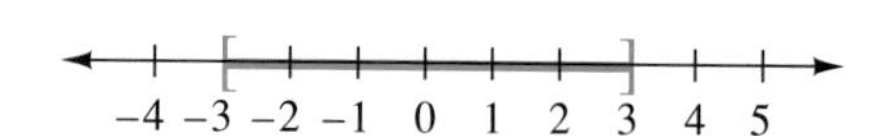

The solution set is $[-3, 3]$. ■

In general, we have the following property.

> **Solving Absolute Value Inequalities of the Form $|x| < a$**
>
> If a is a positive number, then $|x| < a$ is equivalent to $-a < x < a$.

This statement also holds true for the inequality symbol $\le$.

EXAMPLE 2 Solve for m: $|m - 6| < 2$.

Solution: Replace x with $m - 6$ and a with 2 in the preceding rule and we have that

$$|m - 6| < 2 \quad \text{is equivalent to} \quad -2 < m - 6 < 2$$

Solve this compound inequality for m by adding 6 to all three sides.

$$-2 < m - 6 < 2$$

$$-2 + 6 < m - 6 + 6 < 2 + 6 \qquad \text{Add 6 to all three sides.}$$

$$4 < m < 8 \qquad \text{Simplify.}$$

The solution set is $(4, 8)$, and its graph is shown at the left. ■

HELPFUL HINT

Isolate the absolute value expression on one side of the inequality before applying the absolute value inequality rule.

EXAMPLE 3 Solve for x: $|5x + 1| + 1 \leq 10$.

Solution: First isolate the absolute value expression by subtracting 1 from both sides.

$$|5x + 1| + 1 \leq 10$$

$$|5x + 1| \leq 10 - 1 \qquad \text{Subtract 1 from both sides.}$$

$$|5x + 1| \leq 9 \qquad \text{Simplify.}$$

Since 9 is positive, we apply the absolute value inequality rule for $|x| \leq a$.

$$-9 \leq 5x + 1 \leq 9$$

$$-9 - 1 \leq 5x + 1 - 1 < 9 - 1 \qquad \text{Subtract 1 from all three sides.}$$

$$-10 \leq 5x \leq 8 \qquad \text{Simplify.}$$

$$-2 \leq x \leq \frac{8}{5} \qquad \text{Divide all three sides by 5.}$$

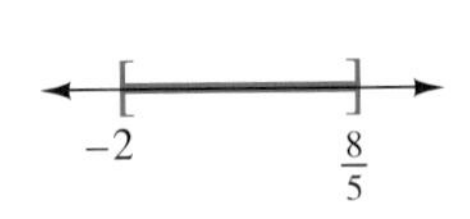

The solution set is $\left[-2, \frac{8}{5}\right]$, and the graph is shown at the left. ■

EXAMPLE 4 Solve for x: $\left|2x - \frac{1}{10}\right| < -13$.

Solution: The absolute value of a number is always nonnegative and will never be less than -13. Thus, this absolute value inequality has no solution. The solution set is { } or Ø. ■

2 Let us now solve an absolute value inequality of the form $|x| > a$, such as $|x| \geq 3$. The solution set contains all numbers whose distance from 0 is 3 or more units. Thus, the graph of the solution set contains 3 and all points to the right of 3 on the number line or -3 and all points to the left of -3 on the number line.

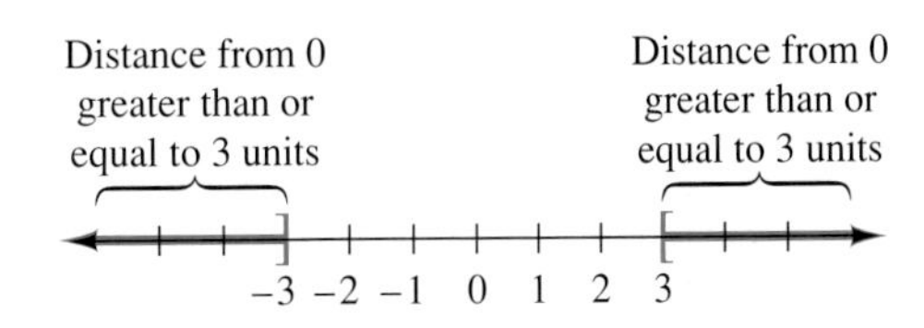

This solution set is written as $\{x | x \leq -3 \text{ or } x \geq 3\}$. In interval notation, the solution set is $(-\infty, -3] \cup [3, \infty)$, since "or" means "union." In general, we have the following:

Solving Absolute Value Inequalities of the Form $|x| > a$

If a is a positive number, then $|x| > a$ is equivalent to $x < -a$ or $x > a$.

This rule also holds true for the inequality symbol $\geq$.

EXAMPLE 5 Solve for y: $|y - 3| > 7$.

Solution: Since 7 is positive, we apply the preceding rule.

$$|y - 3| > 7 \quad \text{is equivalent to} \quad y - 3 < -7 \text{ or } y - 3 > 7$$

Next solve the compound inequality.

$$y - 3 < -7 \qquad \text{or} \qquad y - 3 > 7$$

$$y - 3 + 3 < -7 + 3 \qquad \text{or} \qquad y - 3 + 3 > 7 + 3 \qquad \text{Add 3 to both sides.}$$

$$y < -4 \qquad \text{or} \qquad y > 10 \qquad \text{Simplify.}$$

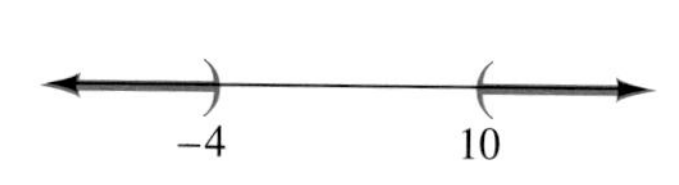

The solution set is $(-\infty, -4) \cup (10, \infty)$, and its graph is shown at the left. ■

EXAMPLE 6 Solve $|2x + 9| + 5 > 3$.

Solution: First isolate the absolute value expression by subtracting 5 from both sides.

$$|2x + 9| + 5 > 3$$

$$|2x + 9| + 5 - 5 > 3 - 5 \qquad \text{Subtract 5 from both sides.}$$

$$|2x + 9| > -2 \qquad \text{Simplify.}$$

The absolute value of any number is always nonnegative and thus is always greater than -2. This inequality and the original inequality are true for all values of x. The solution set is $\{x|x \text{ is a real number}\}$ or $(-\infty, \infty)$, and its graph is at the left. ■

EXAMPLE 7 Solve $\left|\frac{x}{3} - 1\right| - 7 \geq -5$.

Solution: First isolate the absolute value expression by adding 7 to both sides.

$$\left|\frac{x}{3} - 1\right| - 7 \geq -5$$

$$\left|\frac{x}{3} - 1\right| - 7 + 7 \geq -5 + 7 \qquad \text{Add 7 to both sides.}$$

$$\left|\frac{x}{3} - 1\right| \geq 2 \qquad \text{Simplify.}$$

Next write the absolute value inequality as an equivalent compound inequality and solve.

$$\frac{x}{3} - 1 \leq -2 \qquad \text{or} \qquad \frac{x}{3} - 1 \geq 2$$

$$3\left(\frac{x}{3} - 1\right) \leq 3(-2) \quad \text{or} \quad 3\left(\frac{x}{3} - 1\right) \geq 3(2) \qquad \text{Clear the inequalities of fractions.}$$

$$x - 3 \leq -6 \qquad \text{or} \qquad x - 3 \geq 6 \qquad \text{Apply the distributive property.}$$

$$x \leq -3 \qquad \text{or} \qquad x \geq 9 \qquad \text{Add 3 to both sides.}$$

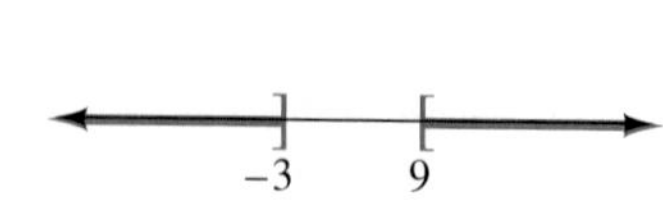

The solution set is $(-\infty, -3] \cup [9, \infty)$, and its graph is shown at the left. ■

EXAMPLE 8 Solve for x: $\left|\frac{2(x+1)}{3}\right| \leq 0.$

Solution: Recall that " $\leq$ " means "less than or equal to." The absolute value of any expression will never be less than 0, but it may be equal to 0. Thus, to solve $\left|\frac{2(x+1)}{3}\right| \leq 0$, we solve $\frac{2(x+1)}{3} = 0.$

$$\frac{2(x+1)}{3} = 0$$

$$3\left[\frac{2(x+1)}{3}\right] = 3(0) \quad \text{Clear the equation of fractions.}$$

$$2x + 2 = 0 \quad \text{Apply the distributive property.}$$

$$2x = -2 \quad \text{Subtract 2 from both sides.}$$

$$x = -1 \quad \text{Divide both sides by 2.}$$

The solution set is $\{-1\}$. ■

The following box summarizes the types of absolute value equations and inequalities.

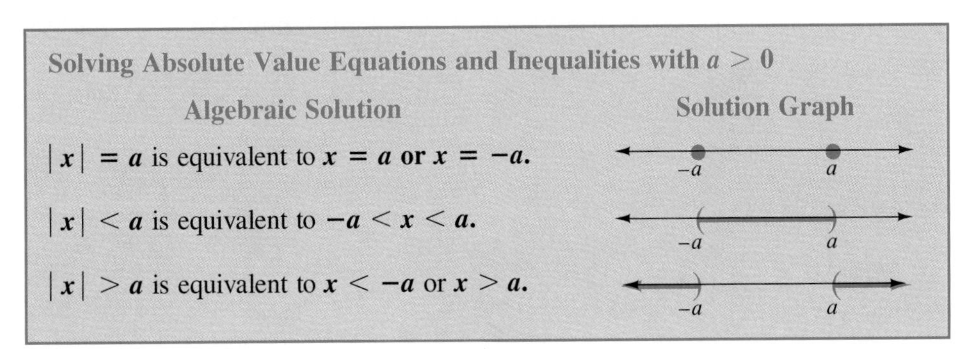

Solving Absolute Value Equations and Inequalities with $a > 0$

Algebraic Solution	Solution Graph
$\|x\| = a$ is equivalent to $x = a$ or $x = -a$.	$-a$, a
$\|x\| < a$ is equivalent to $-a < x < a$.	$-a$, a
$\|x\| > a$ is equivalent to $x < -a$ or $x > a$.	$-a$, a

EXERCISE SET 3.7

Solve each inequality. Then graph the solution set. See Examples 1 and 2.

1. $|x| \leq 4$

2. $|x| < 6$

3. $|x - 3| < 2$

4. $|y| \leq 5$

5. $|x + 3| < 2$

6. $|x + 4| < 6$

7. $|2x + 7| \leq 13$

8. $|5x - 3| \leq 18$

Solve each inequality. Graph the solution set. See Examples 3 and 4.

9. $|x| + 7 \leq 12$

10. $|x| + 6 \leq 7$

11. $|3x - 1| < -5$

12. $|8x - 3| < -2$

13. $|x - 6| - 7 \leq -1$

14. $|z + 2| - 7 < -3$

Solve each inequality. Graph the solution set. See Example 5.

15. $|x| > 3$

16. $|y| \geq 4$

17. $|x + 10| \geq 14$

18. $|x - 9| \geq 2$

Solve each inequality. Graph the solution set. See Examples 6 and 7.

19. $|x| + 2 > 6$

20. $|x| - 1 > 3$

21. $|5x| > -4$

22. $|4x - 11| > -1$

23. $|6x - 8| + 3 > 7$

24. $|10 + 3x| + 1 > 2$

Solve each inequality. Graph the solution set. See Example 8.

25. $|x| \leq 0$

26. $|x| \geq 0$

27. $|8x + 3| > 0$

28. $|5x - 6| < 0$

Solve each inequality. Graph the solution set.

29. $|x| \leq 2$

30. $|z| < 6$

31. $|y| > 1$

32. $|x| \geq 10$

33. $|x - 3| < 8$

34. $|-3 + x| \leq 10$

35. $|6x - 8| > 4$

36. $|1 + 0.3x| \geq 0.1$

37. $5 + |x| \leq 2$

38. $8 + |x| < 1$

39. $|x| > -4$

40. $|x| \leq -7$

41. $|2x - 7| \leq 11$

42. $|5x + 2| < 8$

43. $|x + 5| + 2 \geq 8$

44. $|-1 + x| - 6 > 2$

45. $|x| > 0$

46. $|x| < 0$

47. $9 + |x| > 7$

48. $5 + |x| \geq 4$

49. $6 + |4x - 1| \leq 9$

50. $-3 + |5x - 2| \leq 4$

51. $\left|\frac{2}{3}x + 1\right| > 1$

52. $|5x - 1| \geq 2$

53. $|5x + 3| < -6$

54. $|4 + 9x| \geq -6$

55. $|8x + 3| \geq 0$

56. $|5x - 6| \leq 0$

57. $|1 + 3x| + 4 < 5$

58. $|7x - 3| - 1 \leq 10$

59. $|x| - 3 \geq -3$

60. $|x| + 6 < 6$

61. $|8x| - 10 > -2$

62. $|6x| - 13 \geq -7$

63. $\left|\frac{x + 6}{3}\right| > 2$

64. $\left|\frac{7 + x}{2}\right| \geq 4$

65. $|2(3 + x)| > 6$

66. $|5(x - 3)| \geq 10$

67. $\left|\frac{5(x + 2)}{3}\right| < 7$

68. $\left|\frac{6(3 + x)}{5}\right| \leq 4$

69. $-15 + |2x - 7| \leq -6$

70. $-9 + |3 + 4x| < -4$

71. $\left|2x + \frac{3}{4}\right| - 7 \leq -2$

72. $\left|\frac{3}{5} + 4x\right| - 6 < -1$

Solve each equation or inequality for x.

73. $|2x - 3| < 7$

74. $|2x - 3| > 7$

75. $|2x - 3| = 7$

76. $|5 - 6x| = 29$

77. $|x - 5| \geq 12$

78. $|x + 4| \geq 20$

79. $|9 + 4x| = 0$

80. $|9 + 4x| \geq 0$

81. $|2x + 1| + 4 < 7$

82. $8 + |5x - 3| \geq 11$

83. $|3x - 5| + 4 = 5$

84. $|8x| = -5$

85. $|x + 11| = -1$

86. $|4x - 4| = -3$

87. $\left|\frac{2x - 1}{3}\right| = 6$

88. $\left|\frac{6 - x}{4}\right| = 5$

89. $\left|\frac{3x - 5}{6}\right| > 5$

90. $\left|\frac{4x - 7}{5}\right| < 2$

Skill Review

List the elements in each set. See Section 1.2.

91. $\{x \mid x$ is an odd natural number less than $10\}$

92. $\{x \mid x$ is a negative integer greater than $1\}$

93. $\{x \mid x$ is a natural number that is not a whole number$\}$

94. $\{x \mid x$ is an even integer greater than $5\}$

Simplify the following. See Section 1.7.

95. $(-3)^2 + (-5)^2$

96. $|-3|^2 + |-5|^2$

97. $\frac{2 - (-7)}{3^2}$

98. $\frac{-8 + (-4)}{2^2}$

The following circle graph shows the sources of Walt Disney Company's income. Use this graph to answer the questions below. See Section 3.2.

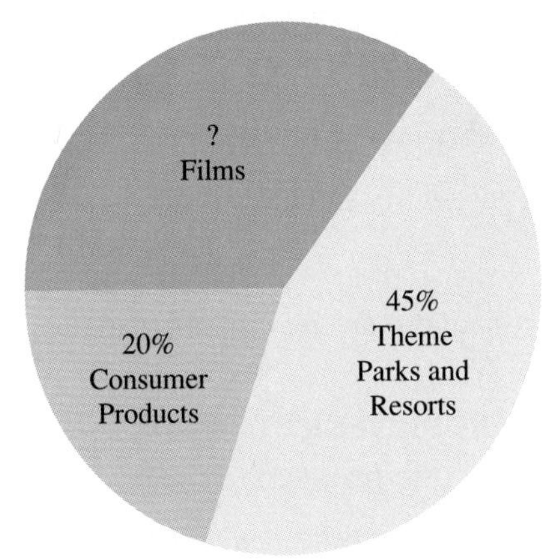

Source: *USA Today*, August 25, 1994.

99. What percent of Disney's income comes from films?

100. Find the number of degrees found in the 20% sector.

101. If Disney's income one year is \$9.75 billion, find the income from theme parks and resorts.

Graph the ordered pairs on a single rectangular coordinate system. Discuss patterns in the ordered pairs and on their graphs. See Section 2.6.

102. $(-2, -4), (-1, -2), (0, 0), (1, 2), (2, 4), (3, 6)$

103. $(-3, -9), (-2, -6), (-1, -3), (0, 0), (1, 3), (2, 6)$

CHAPTER 3 GLOSSARY

Two inequalities joined by the words **and** or **or** are called **compound inequalities.**

In the proportion $\frac{a}{b} = \frac{c}{d}$, the products ad and bc are called **cross products. Cross multiplication** is the process of setting the cross products equal: $ad = bc$.

In the proportion $\frac{1}{2} = \frac{4}{8}$, the 1 and the 8 are called the **extremes;** the 2 and the 4 are called the **means.**

A **proportion** is a mathematical statement that two ratios are equal.

A **ratio** is the quotient of two numbers or quantities.

CHAPTER 3 SUMMARY

CROSS MULTIPLICATION (3.1)

For any ratios $\frac{a}{b}$ and $\frac{c}{d}$, if

$$\frac{a}{b} = \frac{c}{d}, \quad \text{then } ad = bc$$

TO WRITE A DECIMAL AS A PERCENT (3.2)

Move the decimal point two places to the right and attach the percent symbol, %.

TO WRITE A PERCENT AS A DECIMAL (3.2)

Drop the percent symbol and move the decimal point two places to the left.

LINEAR INEQUALITY IN ONE VARIABLE (3.5)

A linear inequality in one variable is an inequality that can be written in the form

$$ax + b < c$$

where a, b, and c are real numbers and a is not 0.

ADDITION PROPERTY OF INEQUALITY (3.5)

If a, b, and c are real numbers, then

$$a < b \quad \text{and} \quad a + c < b + c$$

are equivalent inequalities.

MULTIPLICATION PROPERTY OF INEQUALITY (3.5)

If a, b, and c are real numbers and c is **positive,** then $a < b$ and $ac < bc$ are equivalent inequalities.

If a, b, and c are real numbers and c is **negative,** then $a < b$ and $ac > bc$ are equivalent inequalities.

SOLVING ABSOLUTE VALUE EQUATIONS AND INEQUALITIES WITH $a > 0$ (3.4 and 3.7)

Algebraic Solution | **Solution Graph**

$|x| = a$ is equivalent to $x = a$ or $x = -a$.

$|x| < a$ is equivalent to $-a < x < a$.

$|x| > a$ is equivalent to $x < -a$ or $x > a$.

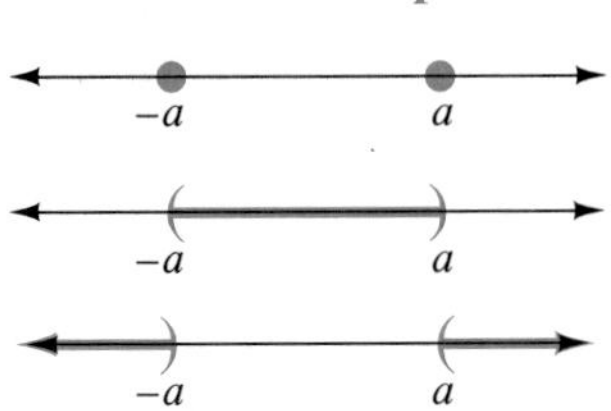

CHAPTER 3 REVIEW

(3.1) *Write each phrase as a ratio in fractional notation.*

1. 20 francs to 1 dollar

2. four parts red to six parts white

Solve each proportion.

3. $\frac{x}{2} = \frac{12}{4}$

4. $\frac{20}{1} = \frac{x}{25}$

5. $\frac{32}{100} = \frac{100}{x}$

6. $\frac{20}{2} = \frac{c}{5}$

7. $\frac{2}{x-1} = \frac{3}{x+3}$

8. $\frac{4}{y-3} = \frac{2}{y-3}$

9. $\frac{y+2}{y} = \frac{5}{3}$

10. $\frac{x-3}{3x+2} = \frac{2}{6}$

Solve.

11. A machine can process 300 parts in 20 minutes. Find how many parts can be processed in 45 minutes.

12. As his consulting fee, Mr. Visconti charges $90.00 per day. Find how much he charges for 3 hours of consulting. Assume an 8-hour work day.

13. One fund raiser can address 100 letters in 35 minutes. Find how many he can address in 55 minutes.

14. The U.S. Census Bureau recently reported that 1 home in 20 has no telephone. If this rate remains constant, find the number of homes that have no telephone in a city with 30,000 homes.

15. If the following pair of triangles is similar, find x.

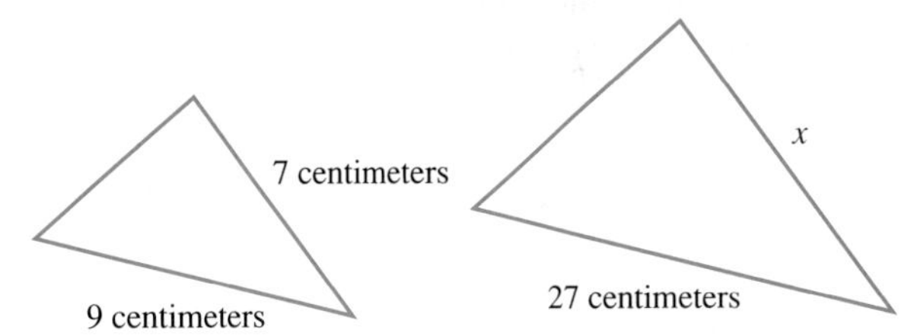

(3.2) *Solve.*

16. Find 12% of 250.

17. Find 110% of 85.

18. The number 9 is what percent of 45?

19. The number 59.5 is what percent of 85?

20. The number 137.5 is 125% of what number?

21. The number 768 is 60% of what number?

22. The state of Mississippi has the highest phoneless rate in the United States, 12.6% of households. If a city in Mississippi has 50,000 households, how many of these would you expect to be phoneless?

The graph below shows how business travelers relax when in their hotel rooms. Use this graph to answer the following questions.

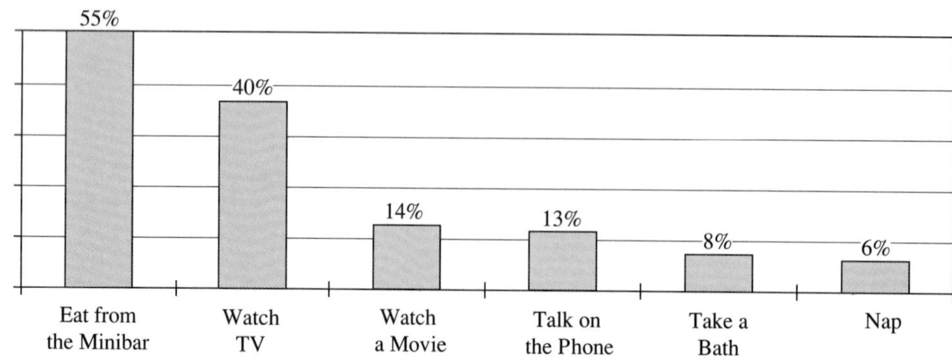

Source: *USA TODAY*, 1994.

23. What percent of business travelers surveyed relax by taking a nap?

24. What is the most popular way to relax according to the survey?

25. If a hotel in New York currently has 300 business travelers, how many might you expect to relax by waching TV?

26. Do the percents in the graph above have a sum of 100%? Why or why not?

27. The number of employees at Arnold's Box Manufacturers just decreased from 210 to 180. Find the percent decrease. Round to the nearest tenth of a percent.

(3.3) *Solve.*

28. Two hikers leave simultaneously from the same point traveling in opposite directions, one walking at 4 mph and the other jogging at 5 mph. Find how long a time they can talk on walkie-talkies that have a 15-mile range.

29. On a 320-mile trip, a car traveled at an average speed of 70 mph, stopped for a speeding ticket, and then traveled at 60 mph for the remainder of the trip. If the entire trip took 5.5 hours and the speeding ticket stop took 30 minutes, find how long the car traveled at 70 mph.

30. LaTonya Harrison invested part of her $25,000 book royalty at 8% annual interest and the rest at 9% annual interest. If her total yearly interest from both accounts is $2135, find the amount invested at each rate.

31. Manuel Cortez invested part of his $10,000 commission check in a mutual fund, which paid an 11% profit annually, and invested the rest in a stock that suffered a 4% annual loss. Find the amount of each investment if his overall net profit was $650.

32. Connie Leigh needs to add pure bleach to 100 cubic centimeters of a 25% bleach solution to increase the strength to 40%. Find how much pure bleach she should add.

33. Find how many gallons of water must evaporate from 200 gallons of a 4% saline solution to strengthen it to a 6% saline solution.

34. A van traveled on a level road for 3 hours at an average speed of 20 mph faster than it traveled on a hilly road. The time spent on the hilly road was 4 hours. Find the average speed on the level road if the entire trip was 305 miles.

35. Alex Weiner can row upstream at 5 mph and downstream at 11 mph. If he starts rowing upstream until he gets tired and then rows downstream to his starting point, find how far he rowed upstream if the entire trip took 4 hours.

36. A coffee merchant blends Colombian roast beans selling for $4.00 per pound with French roast beans selling for $6.50 per pound to make 300 pounds of a blend to sell for $4.75 per pound. Find how many pounds of each bean the merchant uses.

37. Ian Fulton wants to combine a 15% glycerin solution with a 45% glycerin solution to form 150 centiliters of a 35% glycerin solution. Find how much of each solution he should use.

(3.4) *Solve each absolute value equation.*

38. $|x - 7| = 9$

39. $|8 - x| = 3$

40. $|2x + 9| = 9$

41. $|-3x + 4| = 7$

42. $|3x - 2| + 6 = 10$

43. $5 + |6x + 1| = 5$

44. $-5 = |4x - 3|$

45. $|5 - 6x| + 8 = 3$

46. $|7x| - 26 = -5$

47. $-8 = |x - 3| - 10$

48. $\left|\frac{3x - 7}{4}\right| = 2$

49. $\left|\frac{9 - 2x}{5}\right| = -3$

50. $|6x + 1| = |15 + 4x|$

51. $|x - 3| = |7 + 2x|$

(3.5) *Solve each linear inequality.*

52. $3(x - 5) > -(x + 3)$

53. $-2(x + 7) \geq 3(x + 2)$

54. $4x - (5 + 2x) < 3x - 1$

55. $3(x - 8) < 7x + 2(5 - x)$

56. $24 \geq 6x - 2(3x - 5) + 2x$

57. $48 + x \geq 5(2x + 4) - 2x$

58. $\frac{x}{3} + \frac{1}{2} > \frac{2}{3}$

59. $x + \frac{3}{4} < \frac{-x}{2} + \frac{9}{4}$

60. $\frac{x - 5}{2} \leq \frac{3}{8}(2x + 6)$

61. $\frac{3(x - 2)}{5} > \frac{-5(x - 2)}{3}$

Solve.

62. George Boros can pay his housekeeper \$15 per week to do his laundry, or he can have the laundromat do it at a cost of 50 cents per pound for the first 10 pounds and 40 cents for each additional pound. Use an inequality to find the weight at which it is more economical to use the housekeeper than the laundromat.

63. Ceramic firing temperatures usually range from 500° Fahrenheit to 1000° Fahrenheit. Use a compound inequality to convert this range to the Celsius scale. Round to the nearest degree.

64. In the Olympic ice dancing competition, Nana Borsky must score 9.65 to win the silver medal. Seven of the eight judges have reported scores of 9.5, 9.7, 9.9, 9.7, 9.7, 9.6, and 9.5. Use an inequality to find the minimum score that the last judge can give so that Nana wins the silver medal.

65. Carol Guilford would like to pay cash for a car when she graduates from college and estimates that she can afford a car that costs between \$4000 and \$8000. She has saved \$500 so far and plans to earn the rest of the money by working the next two summers. If Carol plans to save the same amount each summer, use a compound inequality to find the range of money she must save each summer to buy the car.

(3.6) *Solve each inequality.*

66. $-3 < -2x < 6$

67. $0 \leq -5x \leq 20$

68. $1 \leq 4x - 7 \leq 3$

69. $-2 \leq 8 + 5x < -1$

70. $-3 < 4(2x - 1) < 12$

71. $-6 < x - (3 - 4x) < -3$

72. $\frac{1}{6} < \frac{4x - 3}{3} \leq \frac{4}{5}$

73. $0 \leq \frac{2(3x + 4)}{5} \leq 3$

74. $x \leq 2$ and $x > -5$

75. $x \leq 2$ or $x > -5$

76. $3x - 5 > 6$ or $-x < -5$

77. $-2x \leq 6$ and $-2x + 3 < -7$

(3.7) *Solve each absolute value inequality. Graph the solution set and write in interval notation.*

78. $|5x - 1| < 9$

79. $|6 + 4x| \geq 10$

80. $|3x| - 8 > 1$

81. $9 + |5x| < 24$

82. $|6x - 5| \leq -1$

83. $|6x - 5| \geq -1$

84. $\left|3x + \frac{2}{5}\right| \geq 4$

85. $\left|\frac{4x - 3}{5}\right| < 1$

86. $\left|\frac{x}{3} + 6\right| - 8 > -5$

87. $\left|\frac{4(x - 1)}{7}\right| + 10 < 2$

CHAPTER 3 TEST

Solve the following.

1. $\dfrac{5}{y+1} = \dfrac{4}{y+2}$

2. $|6x - 5| = 1$

3. $|8 - 2t| = -6$

4. $\left|\dfrac{x+3}{2}\right| = 7$

5. $3(2x - 7) - 4x > -(x + 6)$

6. $8 - \dfrac{x}{2} \le 7$

7. $-3 < 2(x - 3) \le 4$

8. $|3x + 1| > 5$

9. $x \ge 5$ and $x \ge 4$

10. $x \ge 5$ or $x \ge 4$

11. $-x > 1$ and $3x + 3 \ge x - 3$

12. $6x + 1 > 5x + 4$ or $1 - x > -4$

13. Find 85% of 200.

14. The number 35 is what percent of 140?

15. The number 31.5 is 35% of what number?

16. The cost of a hamburger at Ruby's Bar and Grill just increased from \$3.50 to \$3.95. Find the percent increase. Round to the nearest tenth of a percent.

The following graph shows the source of income for charities. Use this graph to answer the following questions.

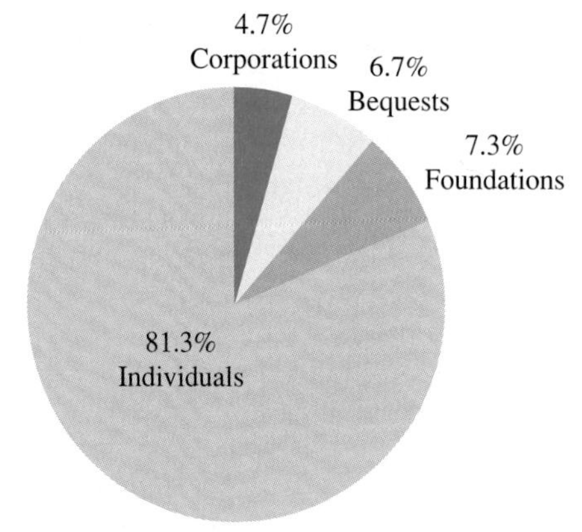

Source: *USA Today*, 1994.

17. What percent of charity income comes from individuals?

18. If the total income for charities is \$126.2 billion, find the amount that comes from corporations.

19. Find the number of degrees in the bequests sector.

20. The company that makes Photoray sunglasses figures that the cost C to make x number of sunglasses weekly is given by $C = 3910 + 2.8x$, and the weekly revenue R is given by $R = 7.4x$. Use an inequality to find the number of sunglasses that must be made and sold to make a profit. (Revenue must exceed cost to make a profit.)

21. Sedric Angell invested an amount of money in Amoxil stock that earned an annual 10% return, and then he invested twice the original amount in IBM stock that earned an annual 12% return. If his total return from both investments was \$2890, find how much he invested in each stock.

22. Two cars leave from the same point at the same time traveling in opposite directions, one driving at 55 mph and the other at 65 mph. Find how long a time they can talk on their car phones if the phones have a 250-mile range.

23. A car wash can wash 5 cars in 30 minutes. Find how many cars can be washed in 2 hours. (*Hint:* Make sure your units are consistent.)

24. Find how many liters of a 12% acid solution must be added to 10 liters of a 25% acid solution to get a 20% acid solution.

CHAPTER 3 CUMULATIVE REVIEW

1. Given the set $\left\{-2, 0, \frac{1}{4}, 112, -3, 11, \sqrt{2}\right\}$, list the numbers in this set that belong to the set of:

a. Natural numbers **b.** Whole numbers
c. Integers **d.** Rational numbers
e. Irrational numbers **f.** Real numbers

2. Write each of the following numbers as a product of primes.

a. 40 **b.** 63

3. Find each quotient. Write all answers in lowest terms.

a. $\frac{4}{5} \div \frac{5}{16}$ **b.** $\frac{7}{10} \div 14$ **c.** $\frac{3}{8} \div \frac{3}{10}$

4. Evaluate the following:

a. 3^2 **b.** 5^3 **c.** 2^4

5. Decide whether 10 is a solution of $3x - 10 = 2x$.

6. Find each sum.

a. $(+3) + (-7)$
b. $(-2) + (10)$
c. $2 + (-5)$

7. Find each product.

a. $(7)(0)(-6)$ **b.** $(-2)(-3)(-4)$
c. $(-1)(5)(-9)$ **d.** $(-2)^3$
e. $(-4)(-11) - (5)(-2)$

8. If $x = -2$ and $y = -4$, find the value of each expression.

a. $5x - y$ **b.** $x^3 - y^2$ **c.** $\frac{3x}{2y}$

9. Use the distributive property to write each expression without parentheses.

a. $2(x + y)$ **b.** $-5(-3 + z)$ **c.** $5(x + y - z)$
d. $-1(2 - y)$ **e.** $-(3 + x - w)$

10. Tell whether the terms are like or unlike.

a. $-x^2, 3x^3$ **b.** $4x^2y, x^2y, -2x^2y$
c. $-2yz, -3zy$ **d.** $-x^4, x^4$

11. Write the following phrases as algebraic expressions and simplify if possible. Let x represent the unknown number.

a. Twice a number, added to 6
b. The difference of a number and 4, divided by 7
c. Five added to 3 times the sum of a number and 1

12. Solve $x - 7 = 10$ for x.

13. Twice a number added to seven is the same as three subtracted from the number. Translate into an equation and solve.

14. Solve $4(2x + 3) - 7 = 3x - 5$.

15. The average maximum temperature for January in Algerias, Algeria, is 59° Fahrenheit. Find the equivalent temperature in degrees Celsius.

16. Solve for x: $\frac{45}{x} = \frac{5}{7}$.

17. Three boxes of 3.5″ high-density diskettes cost $37.47. At this rate, how much should five boxes cost?

18. The number 63 is what percent of 72?

19. Solve $|5w + 3| = 7$.

20. Solve $x + 4 \leq -6$ for x. Graph the solution and write it in interval notation.

21. Marie Chase and Jonathan Edwards are having their wedding reception at the Gallery reception hall. They may spend at most $1000 for the reception. If the reception hall charges a $100 cleanup fee plus $14 per person, find the largest number of people that they can invite.

22. Solve for x: $x - 7 < 2$ and $2x + 1 < 9$.

23. Solve for m: $|m - 6| < 2$.

CHAPTER 4

Graphing

Financial wizards once scoffed at cable television as a concept with limited potential. Today, however, even these doubters see the future possibilities, in part, by analyzing information such as the increase in the amount of money Americans spend on cable television.

INTRODUCTION

The linear equations we explored in Chapters 2 and 3 are statements about a single variable. This chapter examines statements about two variables. We focus particularly on graphs of these equations and inequalities, which in this chapter lead to the notion of relation and to the notion of function, perhaps the single most important and useful notion in all mathematics.

4.1 Graphing Equations

OBJECTIVES

Tape BA 20

1. Plot ordered pairs.
2. Determine whether an ordered pair of numbers is a solution to an equation in two variables.
3. Graph linear equations.
4. Graph nonlinear equations.

1 Recall that the **rectangular coordinate system** consists of two number lines, called the x-axis and the y-axis, that intersect at right angles at their 0 coordinates. The point of intersection of these axes is named the **origin,** and the axes divide the plane into four quadrants as shown. The x-axis and the y-axis are not in any quadrant.

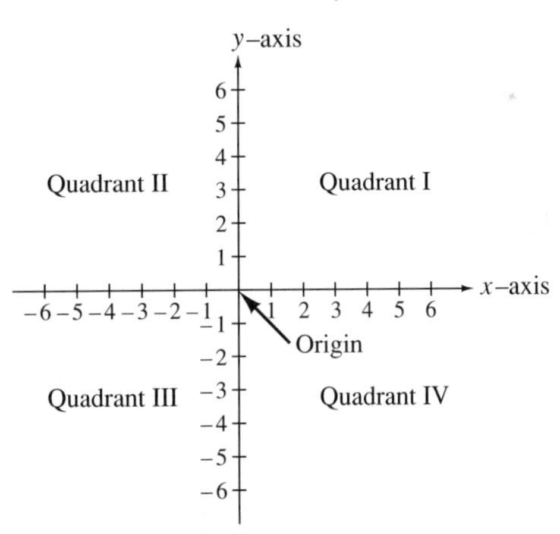

Also recall that each point in the plane can be located or **plotted** by describing its position in terms of distances along each axis from the origin. An **ordered pair,** represented by the notation (x, y), records these distances.

Some points along with their coordinates are shown at the top of page 176. Remember that the x-value of the ordered pair is the x-coordinate and the y-value is the y-coordinate.

Notice that the y-coordinate of any point on the x-axis is 0. For example, the coordinates of point D are $(2, 0)$. Also, the x-coordinate of any point on the y-axis is 0. For example, the coordinates of point B are $(0, -5)$.

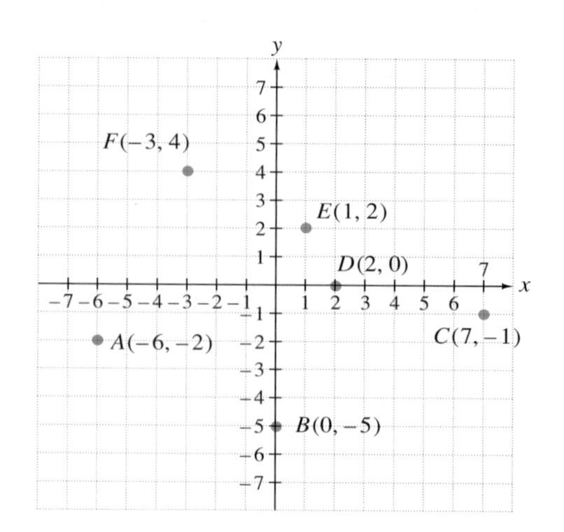

Keep in mind that **each ordered pair corresponds to exactly one point in the real plane and each point in the plane corresponds to exactly one ordered pair.**

2 An equation in one variable such as $x + 1 = 5$ has one solution, which is 4: the number 4 is the value of the variable x that makes the equation true.

An equation in two variables, such as $2x + y = 8$, has solutions consisting of two values, one for x and one for y. For example, $x = 3$ and $y = 2$ is a solution of $2x + y = 8$ because, if x is replaced with 3 and y with 2, we get a true statement.

$$2x + y = 8$$
$$2(3) + 2 = 8$$
$$8 = 8 \qquad \text{True.}$$

We can use ordered pairs of numbers to write solutions of equations in two variables. For example, the solution $x = 3$ and $y = 2$ can be written as (3, 2), an **ordered pair** of numbers. The first number 3 is the x-value and the second number 2 is the y-value.

In general, an ordered pair is a **solution** of an equation in two variables if replacing the variables by the values of the ordered pair results in a true statement.

EXAMPLE 1 Determine whether each ordered pair is a solution to the equation $x - 2y = 6$.

a. (6, 0) **b.** (0, 3) **c.** (2, −2)

Solution: **a.** Let $x = 6$ and $y = 0$ in the equation $x - 2y = 6$.

$$x - 2y = 6$$
$$6 - 2(0) = 6 \qquad \text{Replace } x \text{ with 6 and } y \text{ with 0.}$$
$$6 - 0 = 6 \qquad \text{Simplify.}$$
$$6 = 6 \qquad \text{True.}$$

(6, 0) is a solution, since $6 = 6$ is a true statement.

b. Let $x = 0$ and $y = 3$.

$$x - 2y = 6$$
$$0 - 2(3) = 6 \qquad \text{Replace } x \text{ with 0 and } y \text{ with 3.}$$
$$0 - 6 = 6$$
$$-6 = 6 \qquad \text{False.}$$

(0, 3) is **not** a solution, since $-6 = 6$ is a false statement.

c. Let $x = 2$ and $y = -2$ in the equation.

$$x - 2y = 6$$

$$2 - 2(-2) = 6 \qquad \text{Replace } x \text{ with 2 and } y \text{ with } -2.$$

$$2 + 4 = 6$$

$$6 = 6 \qquad \text{True.}$$

$(2, -2)$ is a solution, since $6 = 6$ is a true statement. ■

3 Notice in the above example that both $(6, 0)$ and $(2, -2)$ are solutions of the equation $x - 2y = 6$. In fact, this equation has an infinite number of solutions. Other solutions include $(0, -3)$, $(4, -1)$, $(-2, -4)$ and $(8, 1)$. If we graph these solutions, notice that a pattern appears.

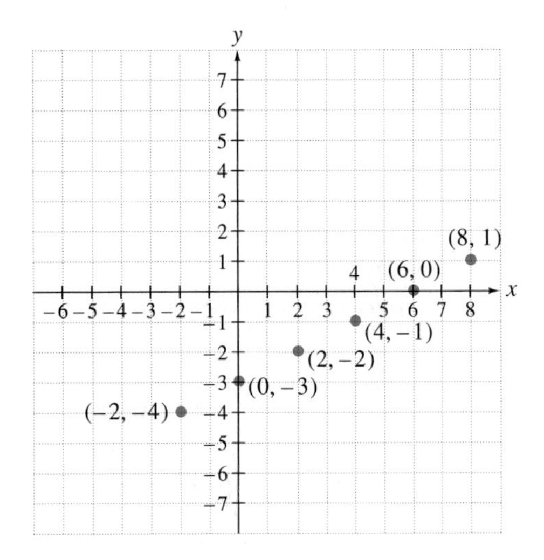

These solutions all appear to lie on the same line. It can be shown that the graph of the solutions of $x - 2y = 6$, or the graph of $x - 2y = 6$, is the single line through these points as shown below. Every ordered pair solution of the equation corresponds to a point on this line, and every point on this line corresponds to an ordered pair solution.

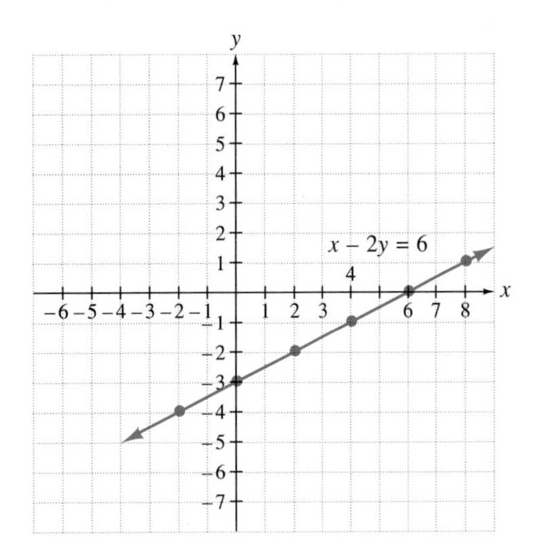

The equation $x - 2y = 6$ is called a linear equation in two variables, and **the graph of every linear equation in two variables is a line.**

Linear Equation in Two Variables

A linear equation in two variables is an equation that can be written in the form

$$Ax + By = C$$

where A, B, and C are real numbers and A and B are not both 0.

The form $Ax + By = C$ is called **standard form.**

Examples of Linear Equations in Two Variables

$$2x + y = 8 \qquad -2x = 7y \qquad y = \frac{1}{3}x + 2 \qquad y = 7$$

From geometry, we know that a straight line is determined by just two points. Graphing a linear equation in two variables, then, requires that we find just two of its infinitely many solutions. Once we do so, we plot the solution points and draw the line connecting the points. Usually, we find a third solution as well, as a check.

EXAMPLE 2 Graph the linear equation $2x + y = 5$.

Solution: Find three ordered pair solutions of $2x + y = 5$. To do this, choose a value for one variable, x or y, and solve for the other variable. For example, let $x = 1$. Then $2x + y = 5$ becomes

$$2(\mathbf{1}) + y = 5$$

$$2 + y = 5 \quad \text{Multiply.}$$

$$y = \mathbf{3} \quad \text{Subtract 2 from both sides.}$$

Since $y = 3$ when $x = 1$, the ordered pair $(1, 3)$ is a solution of $2x + y = 5$. Next, let $x = 0$.

$$2x + y = 5$$

$$2(\mathbf{0}) + y = 5 \quad \text{Replace } x \text{ with 0.}$$

$$0 + y = 5$$

$$y = \mathbf{5}$$

The ordered pair $(0, 5)$ is a second solution.

The two solutions found so far will allow us to draw the straight line that is the graph of all solutions of $2x + y = 5$. We will find a third ordered pair, however, as a check. Let $y = -1$. Then $2x + y = 5$ becomes

$$2x + (\mathbf{-1}) = 5$$

$$2x - 1 = 5$$

$$2x = 6 \quad \text{Add 1 to both sides.}$$

$$x = \mathbf{3} \quad \text{Divide both sides by 2.}$$

The third solution is $(3, -1)$. These three ordered pair solutions can be listed in table form as shown. The graph of $2x + y = 5$ is the line through the three points.

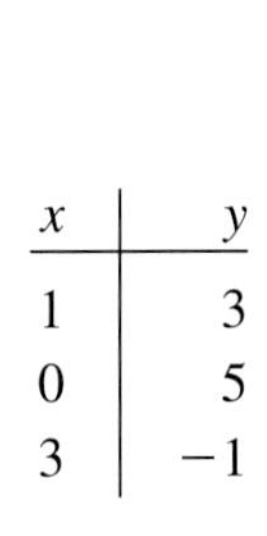

x	y
1	3
0	5
3	−1

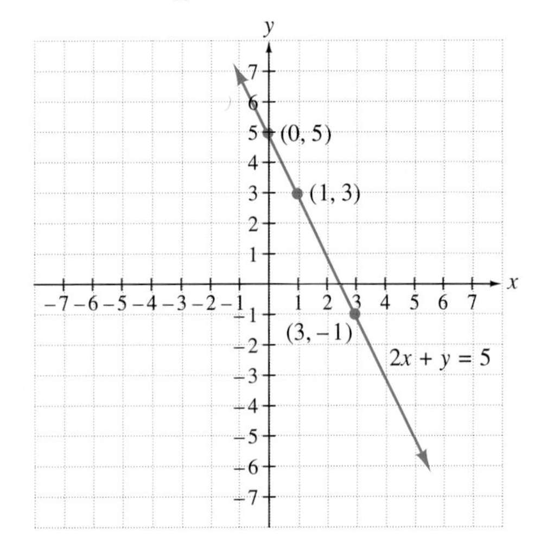

■

EXAMPLE 3 Graph the linear equation $y = 3x$.

Solution: To graph this linear equation we find three ordered pair solutions. Since this equation is solved for y, we will choose three values of x.

If $x = \mathbf{2}$, then $y = 3(2)$, or $\mathbf{6}$.
If $x = \mathbf{0}$, then $y = 3(0)$, or $\mathbf{0}$.
If $x = \mathbf{-3}$, then $y = 3(-3)$, or $\mathbf{-9}$.

x	y
2	6
0	0
−3	−9

Next we graph the ordered pair solutions listed in the table above and draw a line through the plotted points. The line is the graph of $y = 3x$. Every point on the graph represents an ordered pair solution of the equation, and every ordered pair solution is a point on this graph.

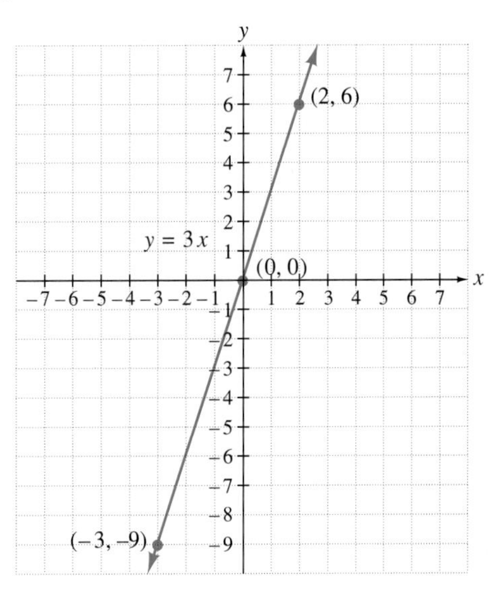

■

EXAMPLE 4 Graph the linear equation $y = \frac{1}{3}x$.

Solution: To graph, we find ordered pair solutions, graph the solutions, and draw a line through the solutions. We will choose x-values and substitute in the equation. To avoid fractions, we choose x-values that are multiples of 3. Recall that the equation is $y = \frac{1}{3}x$.

If $x = \mathbf{6}$, then $y = \frac{1}{3}(6)$, or $\mathbf{2}$.

If $x = \mathbf{0}$, then $y = \frac{1}{3}(0)$, or $\mathbf{0}$.

If $x = \mathbf{-3}$, then $y = \frac{1}{3}(-3)$, or $\mathbf{-1}$.

x	y
6	2
0	0
−3	−1

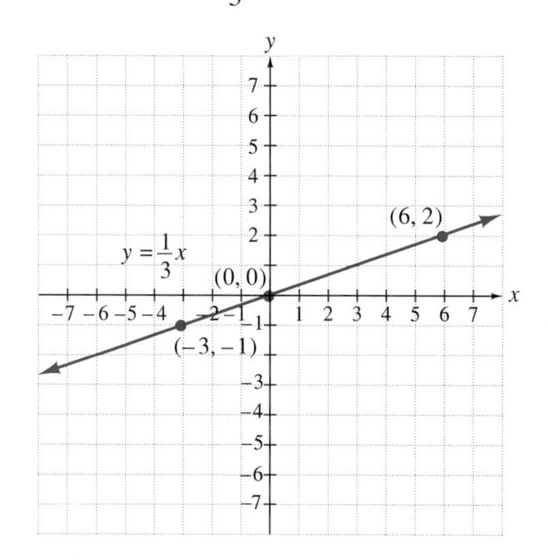

■

EXAMPLE 5 Graph the linear equation $y = 3x + 2$, and compare this graph with the graph of $y = 3x$ in Example 3.

Solution: To graph, we find ordered pair solutions, graph the solutions, and draw a line through the solutions. We will choose x-values and substitute in the equation $y = 3x + 2$. (The graph of $y = 3x$ is shown on the same set of axes to help us compare.)

If $x = \mathbf{-3}$, then $y = 3(-3) + 2$, or $\mathbf{-7}$.
If $x = \mathbf{0}$, then $y = 3(0) + 2$, or $\mathbf{2}$.
If $x = \mathbf{2}$, then $y = 3(2) + 2$, or $\mathbf{8}$.

x	y
−3	−7
0	2
2	8

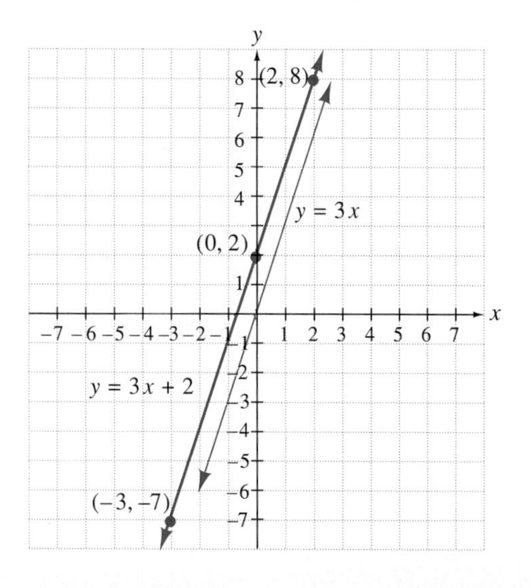

■

Notice above that the graph of $y = 3x$ intersects the y-axis at 0, and the graph of $y = 3x + 2$ intersects the y-axis at 2. Also notice that these graphs tilt or slope the same. **In fact, the graph of $y = 3x + 2$ is simply the graph of $y = 3x$ moved up 2 units.**

4 Not all equations in two variables are linear equations; therefore, not all graphs of equations in two variables are lines.

EXAMPLE 6 Graph the equation $y = |x|$.

Solution: This is not a linear equation and its graph is not a line. Because we do not know the shape of this graph, we find many ordered pair solutions. We will choose x-values and substitute to find corresponding y-values.

If $x = \mathbf{-4}$, then $y = |-4|$, or $\mathbf{4}$.
If $x = \mathbf{-2}$, then $y = |-2|$, or $\mathbf{2}$.
If $x = \mathbf{0}$, then $y = |0|$, or $\mathbf{0}$.
If $x = \mathbf{\frac{1}{2}}$, then $y = \left|\frac{1}{2}\right|$, or $\mathbf{\frac{1}{2}}$.
If $x = \mathbf{1}$, then $y = |1|$, or $\mathbf{1}$.
If $x = \mathbf{3}$, then $y = |3|$, or $\mathbf{3}$.

x	y
-4	4
-2	2
0	0
$\frac{1}{2}$	$\frac{1}{2}$
1	1
3	3

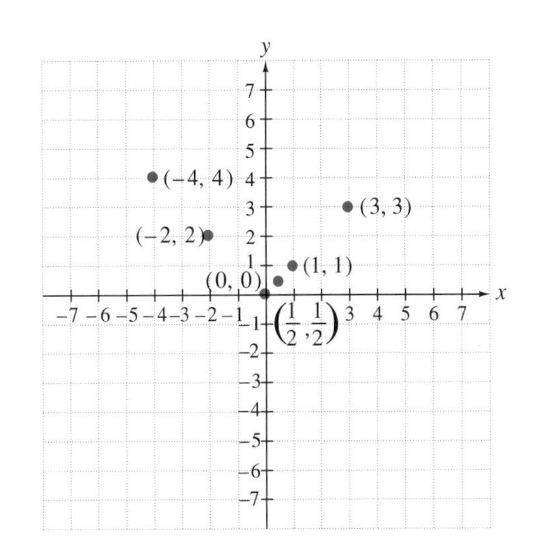

From the plotted ordered pairs, we see that the graph of this absolute value equation is V-shaped. The completed graph is shown below.

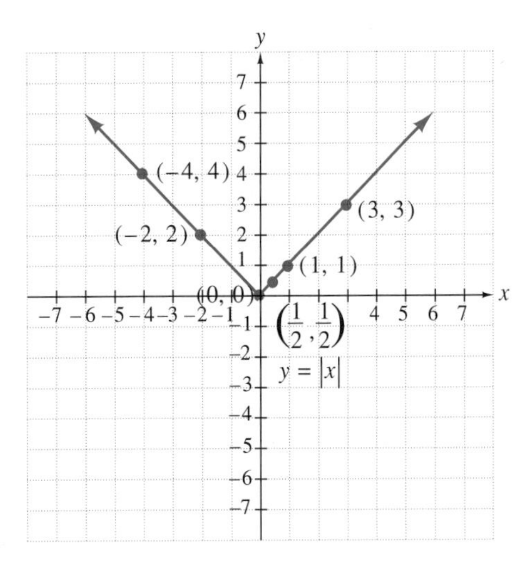

■

EXAMPLE 7 Graph the equation $y = |x| - 3$.

Solution: The graph of $y = |x| - 3$ is simply the graph of $y = |x|$ moved down 3 units. To check, we choose x-values and substitute fo find corresponding y-values.

If $x = \mathbf{-4}$, then $y = |-4| - 3$, or $\mathbf{1}$.
If $x = \mathbf{-2}$, then $y = |-2| - 3$, or $\mathbf{-1}$.
If $x = \mathbf{0}$, then $y = |0| - 3$, or $\mathbf{-3}$.
If $x = \mathbf{2}$, then $y = |2| - 3$, or $\mathbf{-1}$.
If $x = \mathbf{4}$, then $y = |4| - 3$, or $\mathbf{1}$.

x	y
−4	1
−2	−1
0	−3
2	−1
4	1

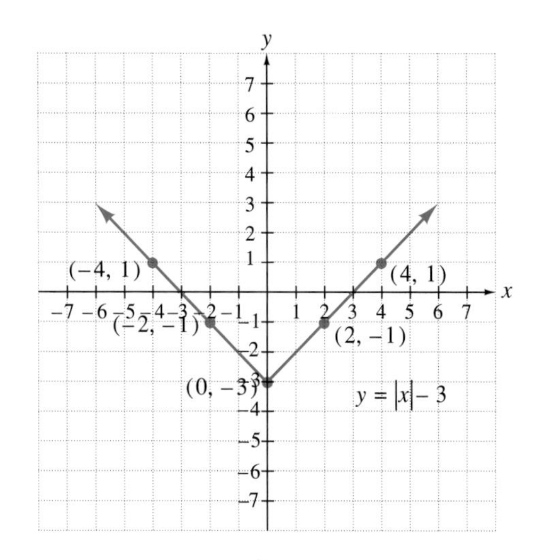

Notice that the graph of $y = |x| - 3$ is the graph of $y = |x|$ moved down 3 units. ■

EXAMPLE 8 Graph $y = x^2$.

Solution: This equation is not linear and its graph is not a line. We begin by finding ordered pair solutions. Because this graph is solved for y, we choose x-values and find corresponding y-values.

If $x = \mathbf{-3}$, then $y = (-3)^2$, or $\mathbf{9}$.
If $x = \mathbf{-2}$, then $y = (-2)^2$, or $\mathbf{4}$.
If $x = \mathbf{-1}$, then $y = (-1)^2$, or $\mathbf{1}$.
If $x = \mathbf{0}$, then $y = 0^2$, or $\mathbf{0}$.
If $x = \mathbf{1}$, then $y = 1^2$, or $\mathbf{1}$.
If $x = \mathbf{2}$, then $y = 2^2$, or $\mathbf{4}$.
If $x = \mathbf{3}$, then $y = 3^2$, or $\mathbf{9}$.

x	y
−3	9
−2	4
−1	1
0	0
1	1
2	4
3	9

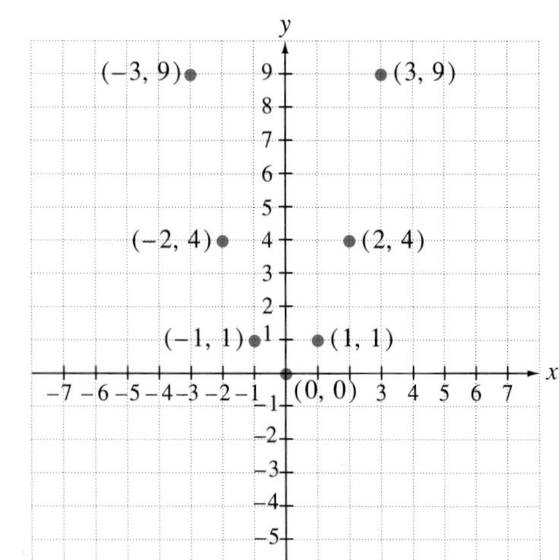

Notice that this graph is not V-shaped. We connect these plotted points with a smooth curve to sketch its graph as shown below.

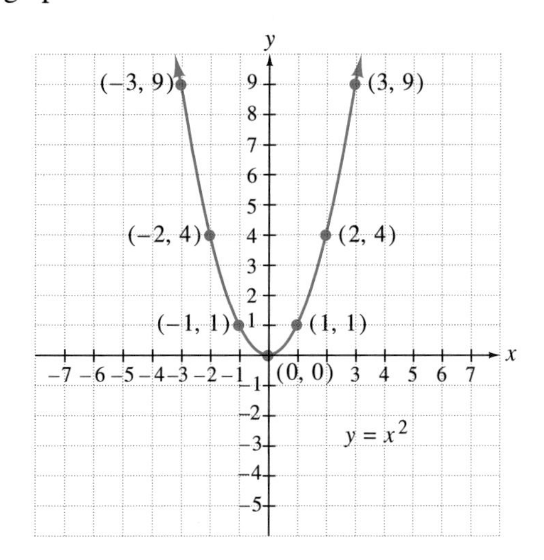

This curve is given a special name, a parabola. We will study more about parabolas in later chapters. ■

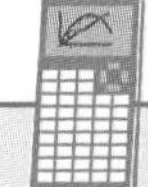

GRAPHING CALCULATOR BOX

In this section, we begin a study of graphing calculators and graphing software packages for computers. These graphers use the same point plotting technique that we introduced in this section. The advantage of this graphing technology is, of course, that graphing calculators and computers can find and plot ordered pair solutions much faster than we can. Note, however, that the features described in these boxes may not be available on all graphing calculators.

The rectangular screen where a portion of the rectangular coordinate system is displayed is called a **window.** We call it a **standard window** for graphing when both the x- and y-axes display coordinates between -10 and 10. This information is often displayed in the window menu on a graphing calculator as

Xmin = −10
Xmax = 10
Xscl = 1 The scale on the x-axis is one unit per tick mark.
Ymin = −10
Ymax = 10
Yscl = 1 The scale on the y-axis is one unit per tick mark.

To use a graphing calculator to graph the equation $y = 2x + 3$, press the [Y =] key and enter the keystrokes [2] [x] [+] [3]. The top row should now read $Y_1 = 2x + 3$. Next press the [GRAPH] key, and the display should look like this:

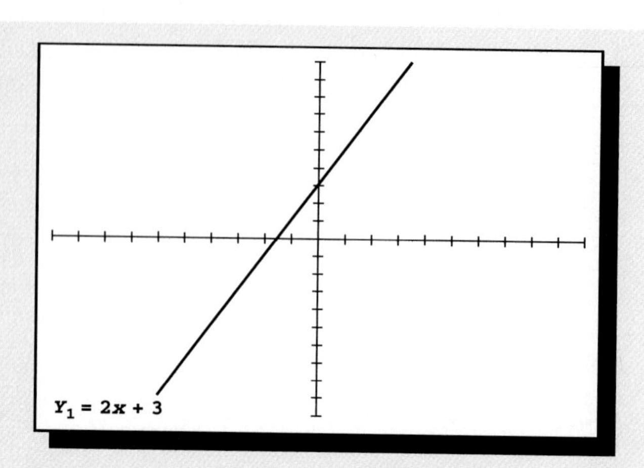

Use a standard window and graph the following equations. (Unless otherwise stated, we will use a standard window when graphing.)

1. $y = -3x + 7$
2. $y = -x + 5$
3. $y = \frac{1}{4}x - 2$
4. $y = \frac{2}{3}x - 1$
5. $y = |x - 3| + 2$
6. $y = |x + 1| - 1$
7. $y = x^2 + 3$
8. $y = (x + 3)^2$

MENTAL MATH

Give two ordered pair solutions for each of the following linear equations.

1. $x + y = 10$
2. $x + y = 6$
3. $x - y = 7$
4. $x - y = 3$
5. $y = 5x$
6. $y = 10x$

EXERCISE SET 4.1

Determine whether each ordered pair is a solution of the given linear equation. See Example 1.

1. $2x + y = 7$; (3, 1), (7, 0), (0, 7)
2. $x - y = 6$; (5, −1), (7, 1), (0, −6)
3. $y = -5x$; (−1, −5), (0, 0), (2, −10)
4. $x = 2y$; (0, 0), (2, 1), (−2, −1)
5. $x = 5$; (4, 5), (5, 4), (5, 0)
6. $y = 2$; (−2, 2), (2, 2), (0, 2)

Complete the ordered pairs so that each is a solution of the given linear equation; then use these ordered pair solutions to graph each equation. See Example 2.

7. $x + 3y = 6$
   a. (0,)
   b. (,0)
   c. (, 1)
8. $2x + y = 4$
   a. (0,)
   b. (, 0)
   c. (, 2)
9. $2x - y = 12$
   a. (0,)
   b. (, −2)
   c. (−3,)
10. $-5x + y = 10$
   a. (, 0)
   b. (, 5)
   c. (2,)

11. $2x + 7y = 5$
a. $(0,\ \ )$
b. $(\ \ , 0)$
c. $(\ \ , 1)$

12. $x - 6y = 3$
a. $(0,\ \ )$
b. $(1,\ \ )$
c. $(\ \ , -1)$

Graph each linear equation. See Examples 3 through 5.

13. $y = 2x$
14. $y = 5x$
15. $y = 2x + 1$
16. $y = 5x - 1$
17. $y = \frac{1}{4}x$
18. $y = \frac{1}{2}x$
19. $y = 3x - 2$
20. $y = 3x + 4$

Graph each nonlinear equation. See Examples 6 through 8.

21. $y = x^2 + 2$
22. $y = x^2 - 2$
23. $y = |x| - 1$
24. $y = |x| + 3$
25. $y = -x^2$
26. $y = -|x|$
27. $y = |x| + 5$
28. $y = x^2 - 5$

Determine whether each ordered pair is a solution of the given linear equation.

29. $x + 2y = 9$; $(5, 2)$, $(0, 9)$
30. $3x + y = 8$; $(2, 3)$, $(0, 8)$
31. $2x - y = 11$; $(3, -4)$, $(9, 8)$
32. $x - 4y = 14$; $(2, -3)$, $(14, 6)$
33. $x = \frac{1}{3}y$; $(0, 0)$, $(3, 9)$
34. $y = -\frac{1}{2}x$; $(0, 0)$, $(4, 2)$
35. $y = 2x^2$; $(1, 2)$, $(3, 18)$
36. $y = 2|x|$; $(-1, 2)$, $(0, 2)$
37. $y = x^3$; $(2, 8)$, $(3, 9)$
38. $y = x^4$; $(-1, 1)$, $(2, 16)$

Determine whether each equation is linear or not. Then graph each equation.

39. $x + y = 3$
40. $y - x = 8$
41. $y = 4x$
42. $y = 6x$
43. $y = 4x - 2$
44. $y = 6x - 5$
45. $y = |x| + 3$
46. $y = |x| + 2$
47. $2x - y = 5$
48. $4x - y = 7$
49. $y = 2x^2$
50. $y = 3x^2$
51. $y = x^2 - 3$
52. $y = x^2 + 3$
53. $y = -2x$
54. $y = -3x$
55. $y = -2x + 3$
56. $y = -3x + 2$

For income tax purposes, Rob Calcutta, owner of Copy Services, uses a method called ***straight-line depreciation*** *to show the loss in value of a copy machine he recently purchased. Rob assumes that he can use the machine for 7 years. The graph below shows the value of the machine over the years. Use this graph to answer the questions below.*

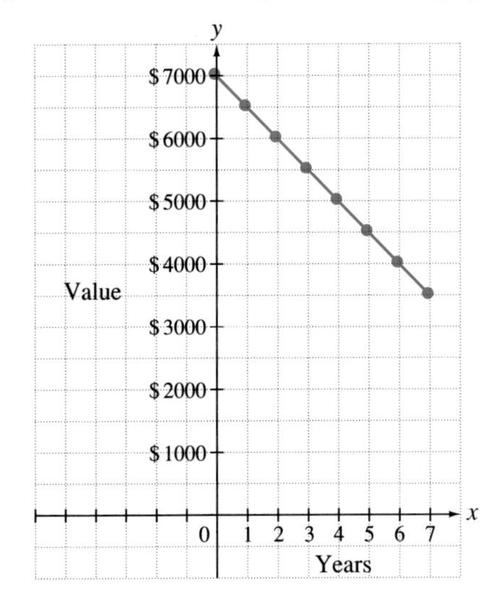

57. What was the purchase price of the copy machine?

58. What is the salvage value of the machine in 7 years?

59. What loss in value occurred during the first year?

60. What loss in value occurred during the second year?

61. Why do you think that this method of depreciating is called straight-line depreciation?

62. Why is the line tilted downward?

63. On the same set of axes, graph $y = 2x$, $y = 2x - 5$, and $y = 2x + 5$. What patterns do you see in these graphs?

64. On the same set of axes, graph $y = 2x$, $y = x$, and $y = -2x$. Describe the differences and similarities in these graphs.

Write each statement as an equation in two variables. Then graph each equation.

65. The y-value is 5 more than the x-value.

66. The y-value is twice the x-value.

Skill Review

67. The coordinates of three vertices of a rectangle are $(-2, 5)$, $(4, 5)$, and $(-2, -1)$. Find the coordinates of the fourth vertex.

68. The coordinates of two vertices of a square are $(-3, -1)$ and $(2, -1)$. Find the coordinates of two pairs of points possible for the third and fourth vertex.

Solve the following equations. See Section 2.3.

69. $3(x - 2) + 5x = 6x - 16$

70. $5 + 7(x + 1) = 12 + 10x$

71. $3x + \frac{2}{5} = \frac{1}{10}$

72. $\frac{1}{6} + 2x = \frac{2}{3}$

Solve the following inequalities. See Section 3.5.

73. $3x \leq -15$

74. $-3x > 18$

75. $2x - 5 > 4x + 3$

76. $9x + 8 \leq 6x - 4$

4.2
Introduction to Functions

OBJECTIVES

Tape IA 8

1 Define relation, domain, and range.

2 Identify functions.

3 Use the vertical line test for functions.

4 Find the domain and range of a function.

---

1 Daily events often have to do with the relationship between two quantities. For example, the amount of money inserted into a parking meter is related to the length of parking time received. Suppose that each quarter buys 15 minutes. If x represents the number of quarters inserted in the meter and y represents the number of minutes bought, then the equation

x Quarters inserted	y Time bought
0 quarters	0 minutes
1 quarters	15 minutes
2 quarters	30 minutes
3 quarters	45 minutes

$$y = 15x$$

models the relationship between quarters inserted and minutes bought. For the model $y = 15x$, the number of quarters, x, is called the **independent variable,** and the number of minutes, y, is called the **dependent variable.** This is because we can **independently** decide how many quarters to insert, but the time bought **depends** on the number of quarters inserted.

The set of possible values for x is the **domain** of the **relation** $y = 15x$, and the set of values of y that correspond to some value of x is the **range** of the relation. The domain of this quarter–time relation is $\{0, 1, 2, 3, \ldots\}$, and the range is $\{0, 15, 30, 45, \ldots\}$.

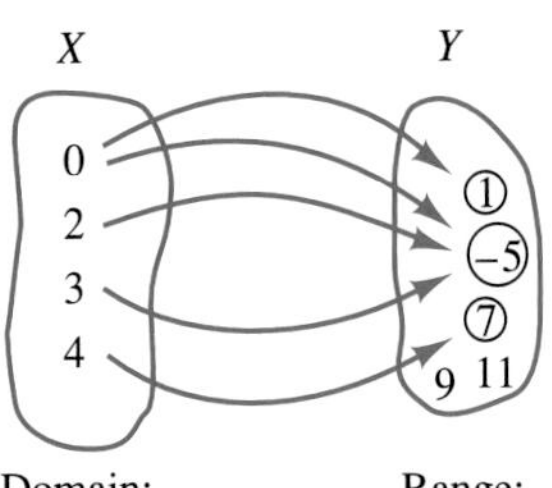

> **Relation, Domain, and Range**
>
> A **relation** is a correspondence between two sets X and Y that assigns to each element of set X an element of set Y. The **domain** of the relation is the set X, and the **range** of the relation is the set of elements of Y that corresponds to some element of X.

An equivalent definition of relation is that a relation is a set of ordered pairs. For example, $\{(0, 1), (0, -5), (2, -5), (3, -5), (4, 7)\}$ is a relation whose domain is $\{0, 2, 3, 4\}$ and whose range is $\{1, -5, 7\}$. Notice that the domain is the set of first coordinates and the range is the set of second coordinates.

EXAMPLE 1 Determine the domain and range of each relation. Write the domain and range in set notation.

a. $\{(2, 3), (2, 4), (0, -1), (3, -1)\}$

b.

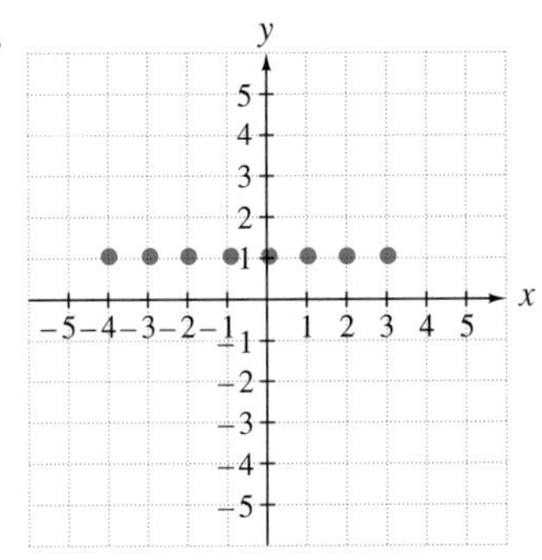

c.

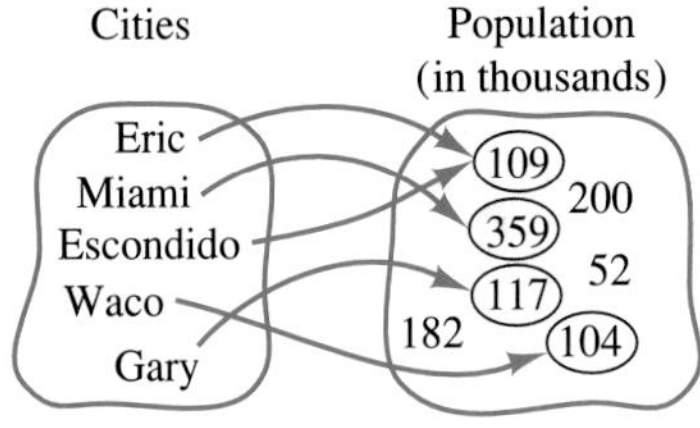

Solution:

a. The domain is the set of all first coordinates of the ordered pairs, $\{2, 0, 3\}$. The range is the set of all second coordinates, $\{3, 4, -1\}$.

b. Ordered pairs are not listed here, but are given in graph form. The relation is $\{(-4, 1), (-3, 1), (-2, 1), (-1, 1), (0, 1), (1, 1), (2, 1), (3, 1)\}$. The domain is $\{-4, -3, -2, -1, 0, 1, 2, 3\}$.
The range is $\{1\}$.

c. The domain is {Eric, Escondido, Gary, Miami, Waco}.
The range is $\{104, 109, 117, 359\}$. ■

2 Some relations are also functions.

> **Function**
>
> A function is a set of ordered pairs that assigns to each x-value exactly one y-value.

EXAMPLE 2 Which of the following relations are also functions?

a. $\{(-1, -1), (2, 3), (7, 3), (8, 6)\}$ **b.** $\{(0, -2), (1, 5), (0, 3), (7, 7)\}$

Solution:

a. Although the ordered pairs $(2, 3)$ and $(7, 3)$ have the same y-value, each x-value is assigned to only one y-value, so this set of ordered pairs is a function.

b. The x-value 0 is assigned to two y-values, -2 and 3, so this set of ordered pairs is not a function. ■

We will call an equation such as $y = 2x + 1$ a **relation** since the equation defines a set of ordered pair solutions.

EXAMPLE 3 Is the relation $y = 2x + 1$ a function?

Solution: The relation $y = 2x + 1$ is a function if each x-value corresponds to just one y-value. For each x-value substituted in the equation $y = 2x + 1$, the multiplication and addition performed gives a single result, so only one y-value will be associated with each x-value. Thus, $y = 2x + 1$ is a function. ■

EXAMPLE 4 Is the equation $x = y^2$ a function?

Solution: In $x = y^2$, if $y = 3$, then $x = 9$. Also, if $y = -3$, then $x = 9$. In other words, the x-value 9 corresponds to two y-values, 3 and -3. Thus $x = y^2$ is not a function. ■

3 As we have seen so far, not all relations are functions. Consider the graphs of $y = 2x + 1$ and $x = y^2$ shown next. On the graph of $y = 2x + 1$, notice that each x-value corresponds to only one y-value. Recall from Example 3 that $y = 2x + 1$ is a function.

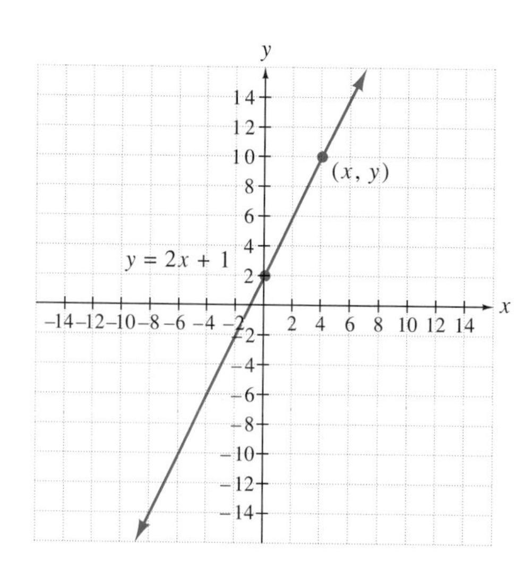

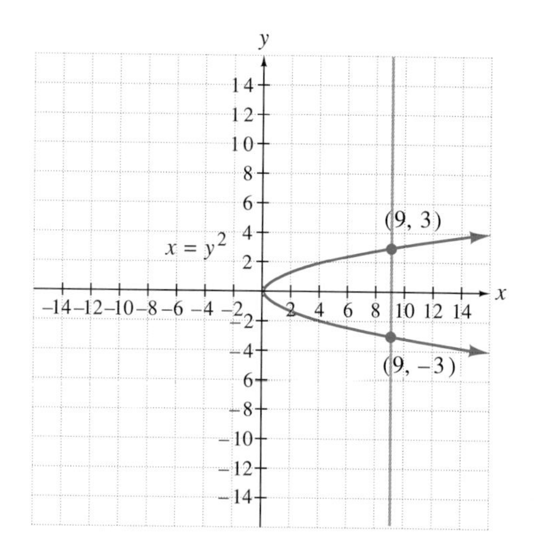

On the graph of $x = y^2$, the x-value 9, for example, corresponds to two y-values, 3 and -3, as shown by the vertical line. Recall from Example 4 that $x = y^2$ is not a function.

Graphs can be used to help determine whether a relation is also a function by the following vertical line test.

Vertical Line Test

If a vertical line can be drawn so that it intersects a graph more than once, the graph is not the graph of a function.

EXAMPLE 5 Which of the following graphs are graphs of functions?

a.

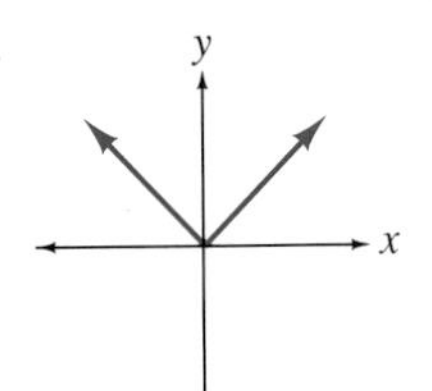

b.

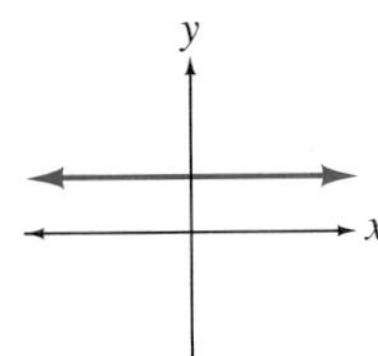

c.

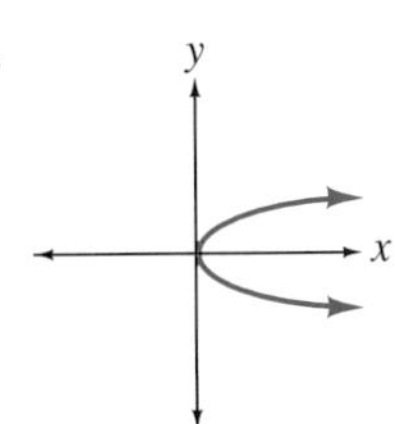

d.

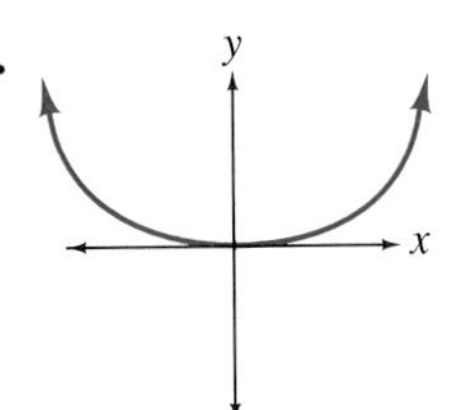

e.

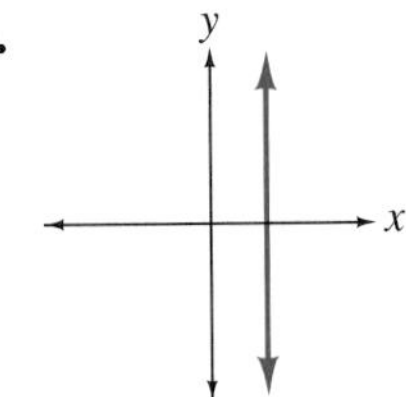

f.

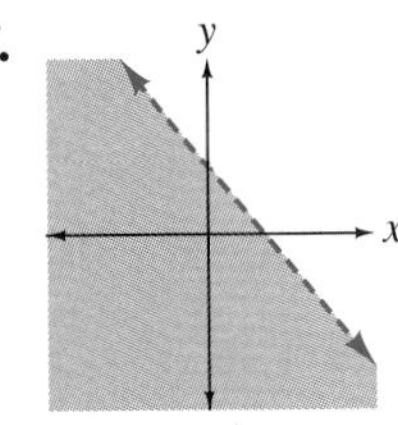

Solution:

a. This graph is the graph of a function since no vertical line will intersect this graph more than once.

b. This graph is also the graph of a function.

c. This graph is not the graph of a function. Note that vertical lines can be drawn that intersect the graph in two points.

d. This graph is the graph of a function.

e. This graph is not the graph of a function. A vertical line can be drawn that intersects this line at every point.

f. This graph is not the graph of a function. ■

Recall that the graph of a linear equation in two variables is a line, and a line that is not vertical will pass the vertical line test. Thus, all linear equations are functions except those whose graph is a vertical line.

4 We can also find the domain and range of a relation from its graph.

EXAMPLE 6 Find the domain and range of each relation. Determine whether the relation is also a function.

a.

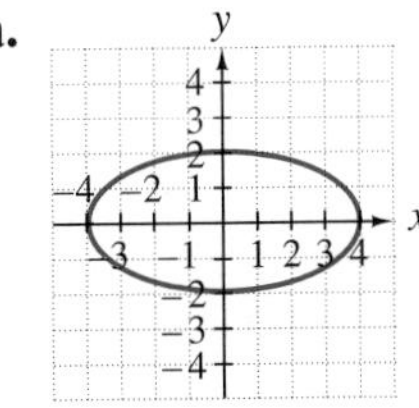

b.

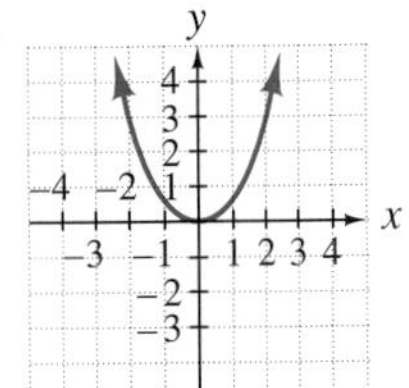

c.

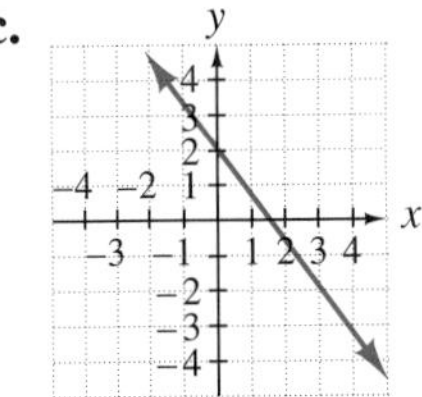

Solution: The domain is the set of values of x and the range is the set of values of y. We read these values from each graph.

a. By the vertical line test, this is not the graph of a function.

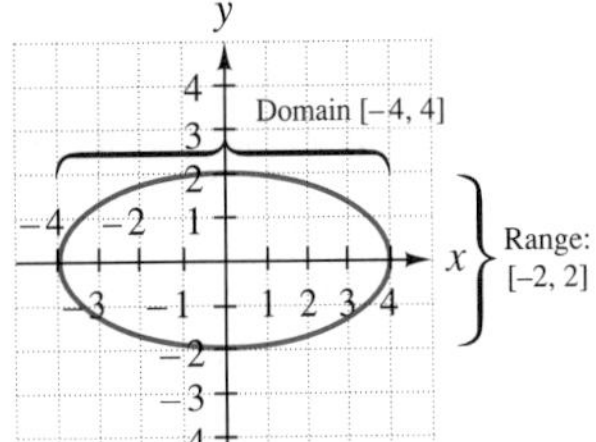

b. By the vertical line test, this is the graph of a function.

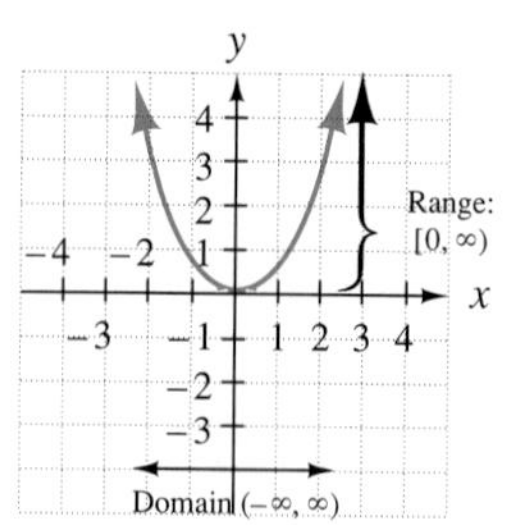

c. By the vertical line text, this is the graph of a function.

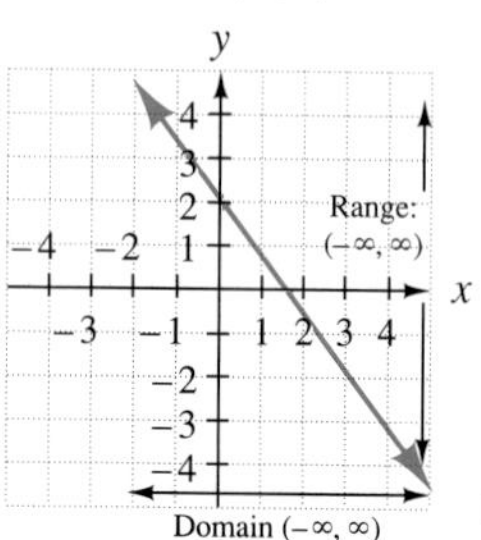

■

Earlier, we decided that $y = 2x + 1$ is a function. Many times letters such as f, g, and h are used to name functions. For example, to denote that y is a function of x in the equation $y = 2x + 1$, we can write $y = f(x)$. Then $y = 2x + 1$ can be written as $f(x) = 2x + 1$. The symbol $\boldsymbol{f(x)}$ means **function of $\boldsymbol{x}$** and is read "f of x." This notation is called **function notation.**

The notation $f(1)$ means to replace x with 1 and find the resulting y or function value. Since

$$f(x) = 2x + 1$$

then

$$f(1) = 2(1) + 1 = 3$$

This means that, when $x = 1$, y or $f(x) = 3$. Now find $f(2)$, $f(0)$, and $f(-1)$.

$$\begin{aligned} f(x) &= 2x + 1 \\ f(2) &= 2(2) + 1 \\ &= 4 + 1 \\ &= 5 \end{aligned} \qquad \begin{aligned} f(x) &= 2x + 1 \\ f(0) &= 2(0) + 1 \\ &= 0 + 1 \\ &= 1 \end{aligned} \qquad \begin{aligned} f(x) &= 2x + 1 \\ f(-1) &= 2(-1) + 1 \\ &= -2 + 1 \\ &= -1 \end{aligned}$$

HELPFUL HINT

Note that $f(x)$ is a special symbol in mathematics used to denote a function. The symbol $f(x)$ is read "f of x." It does **not** mean $f \cdot x$ (f times x).

EXAMPLE 7 Given $g(x) = x^2 - 3$, find the following function values.

a. $g(2)$ **b.** $g(-2)$ **c.** $g(0)$

Solution:

(a) $$\begin{aligned} g(x) &= x^2 - 3 \\ g(2) &= 2^2 - 3 \\ &= 4 - 3 \\ &= 1 \end{aligned}$$

(b) $$\begin{aligned} g(x) &= x^2 - 3 \\ g(-2) &= (-2)^2 - 3 \\ &= 4 - 3 \\ &= 1 \end{aligned}$$

(c) $$\begin{aligned} g(x) &= x^2 - 3 \\ g(0) &= 0^2 - 3 \\ &= 0 - 3 \\ &= -3 \end{aligned}$$

■

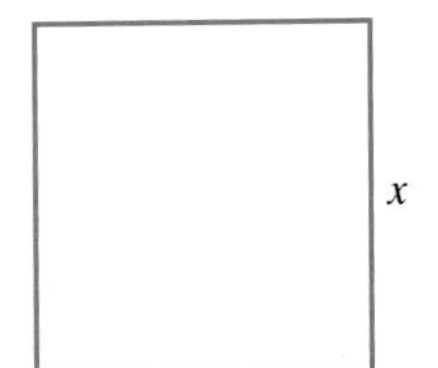

Many formulas that are familiar to you are functions. For example, we know that the formula for the area of a square with side length x is $A = x^2$. The area of the square is actually a function of the length of the side x. If we want to write the formula using function notation, instead of

$$A = x^2$$

we may write

$$A(x) = x^2$$

Then to find the area of the square whose side measures 10 meters, we write

$$\begin{aligned} A(10) &= (10 \text{ meters})^2 \\ &= 100 \text{ square meters} \end{aligned}$$

EXAMPLE 8 The following graph shows the research and development expenditures by the Pharmaceutical Manufacturers Association as a function of time.

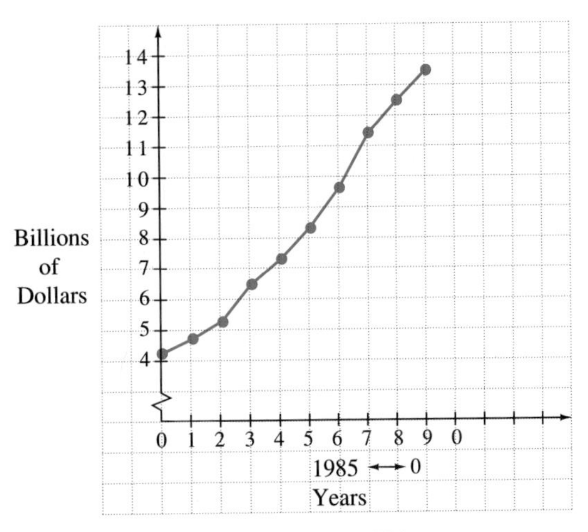

Source: *Science,* vol. 264, May 20, 1994.

a. Approximate the money spent on research and development in 1992.

b. In 1958, research and development expenditures were \$200 million. Find the increase in expenditures from 1958 to 1994.

Solution: **a.** In 1992, approximately \$11.5 billion was spent.

b. In 1994, approximately \$13.8 billion or \$13,800 million was spent. The increase in spending from 1958 to 1994 is \$13,800 − \$200 = \$13,600 million. ■

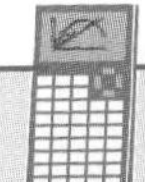

GRAPHING CALCULATOR BOX

It is possible to use a grapher to sketch the graph of more than one equation on the same set of axes. This feature can be used to confirm our findings from Section 4.1. For example, graph the functions $f(x) = x^2$ and $g(x) = x^2 + 4$ on the same set of axes.

To graph on the same set of axes, press the [Y =] key and enter the equations on the first two lines.

$$Y_1 = x^2$$
$$Y_2 = x^2 + 4$$

Then press the GRAPH key as usual. The screen should look like this:

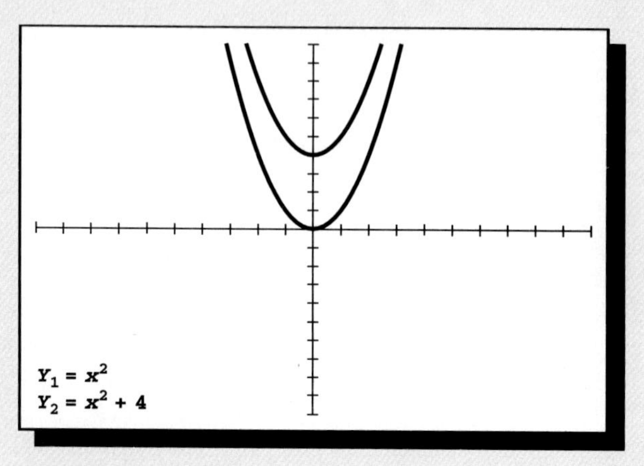

Notice that the graph of y or $g(x) = x^2 + 4$ is the graph of $y = x^2$ moved 4 units upward.

Graph each pair of functions on the same set of axes. Describe the similarities and differences in their graphs.

1. $f(x) = |x|$
   $g(x) = |x| + 1$
2. $f(x) = x^2$
   $h(x) = x^2 - 5$
3. $f(x) = x$
   $H(x) = x - 6$
4. $f(x) = |x|$
   $G(x) = |x| + 3$
5. $f(x) = -x^2$
   $F(x) = -x^2 + 7$
6. $f(x) = x$
   $F(x) = x + 2$

EXERCISE SET 4.2

Find the domain and the range of each relation. See Example 1.

1. $\{(2, 4), (0, 0), (-7, 10), (10, -7)\}$
2. $\{(3, -6), (1, 4), (-2, -2)\}$
3. $\{(0, -2), (1, -2), (5, -2)\}$
4. $\{(5, 0), (5, -3), (5, 4) (5, 3)\}$
5.

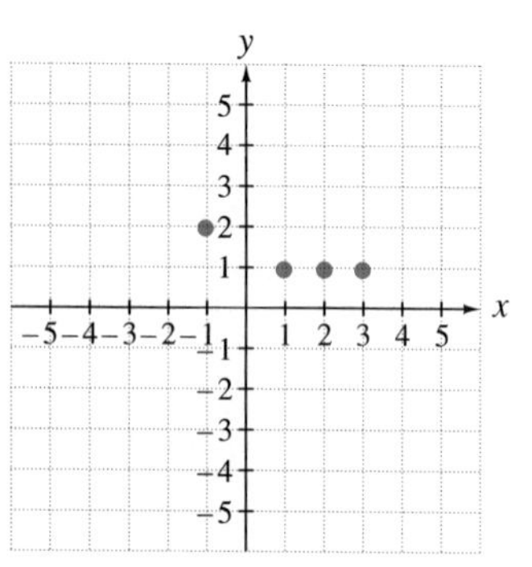

6.

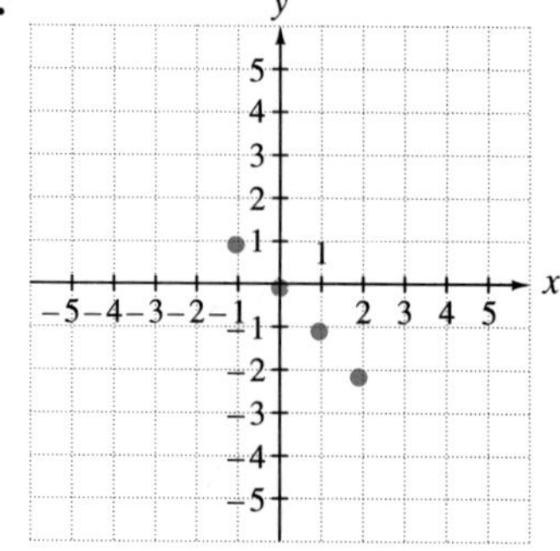

7.

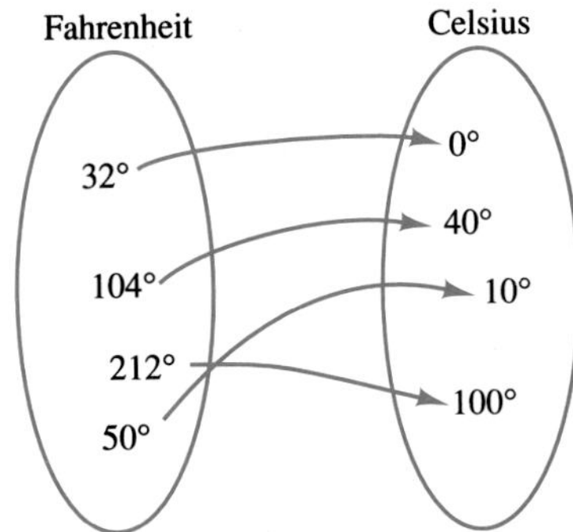

8.

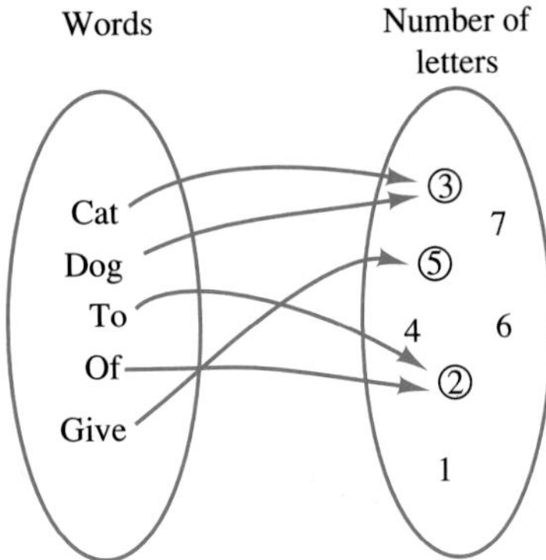

Decide whether each relation is a function. See Example 2.

9. $\{(1, 1), (2, 2), (-3, -3), (0, 0)\}$

10. $\{(1, 2), (3, 2), (4, 2)\}$

11. $\{(-1, 0), (-1, 6), (-1, 8)\}$

12. $\{(11, 6), (-1, -2), (0, 0), (3, -2)\}$

Decide whether each is a function. See Examples 3 and 4.

13. $y = x + 1$

14. $y = x - 1$

15. $x = 2y^2$

16. $y = x^2$

17. $y - x = 7$

18. $2x - 3y = 9$

19. $y = \frac{1}{x}$

20. $y = \frac{1}{x - 3}$

Use the vertical line test to determine whether each graph is the graph of a function. See Example 5.

21.

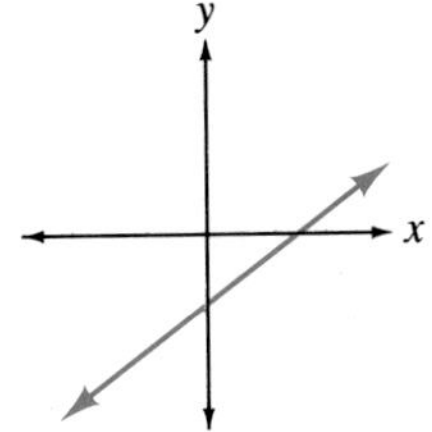

22.

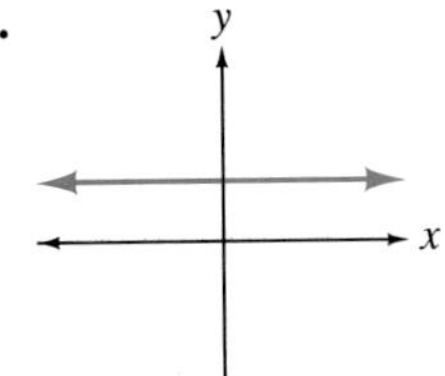

23.

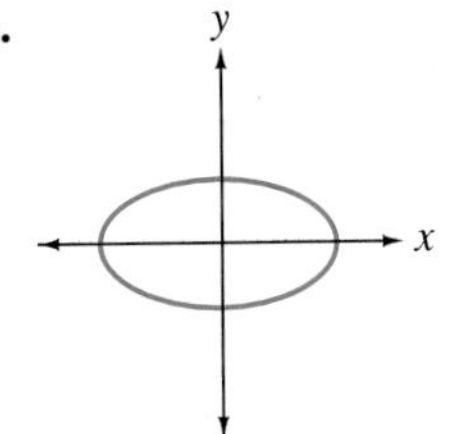

24.

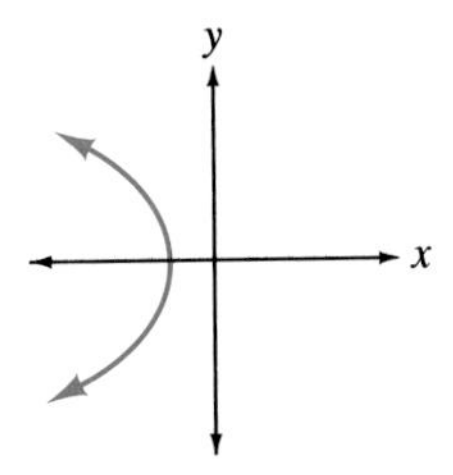

25.

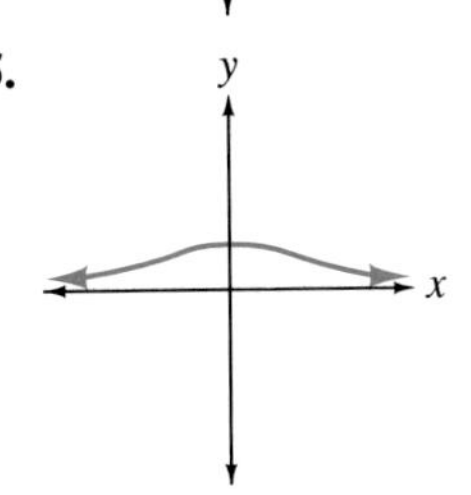

26.

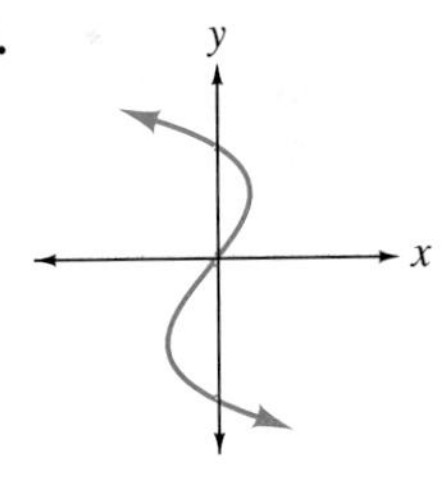

Find the domain and range of each relation. Determine whether the relation is also a function. See Example 6.

27.

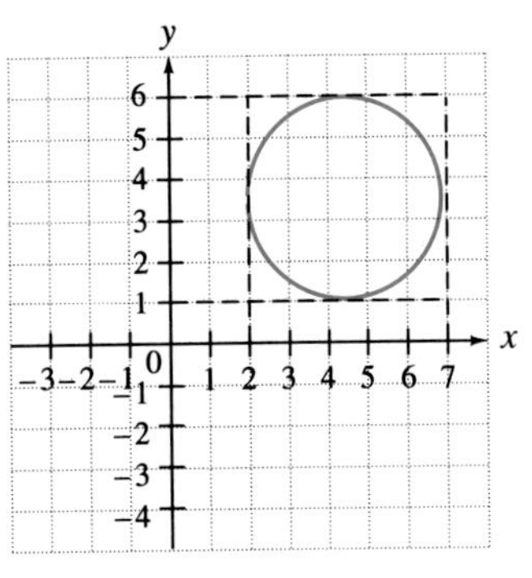

28.

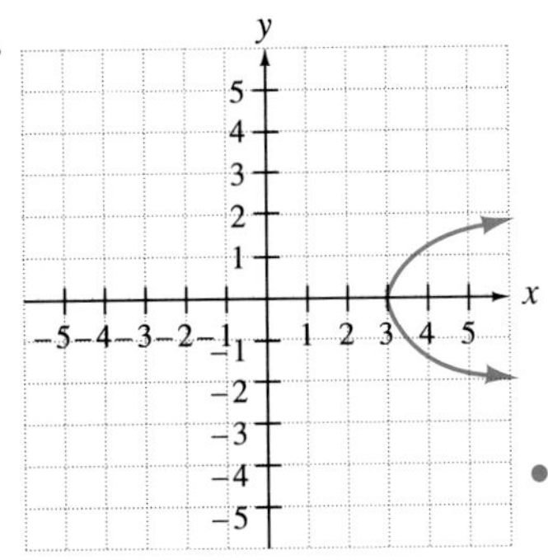

29.

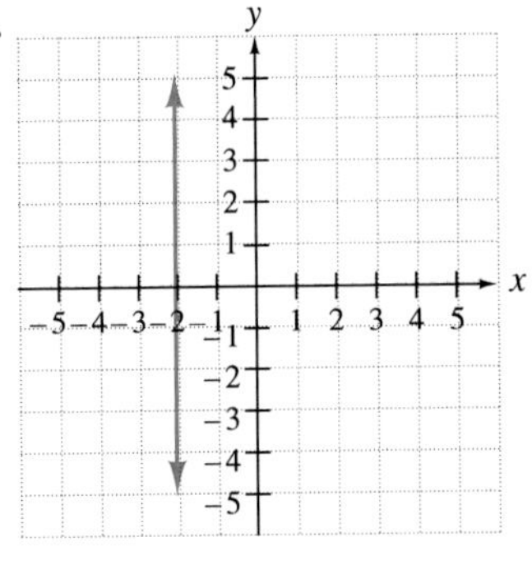

30.

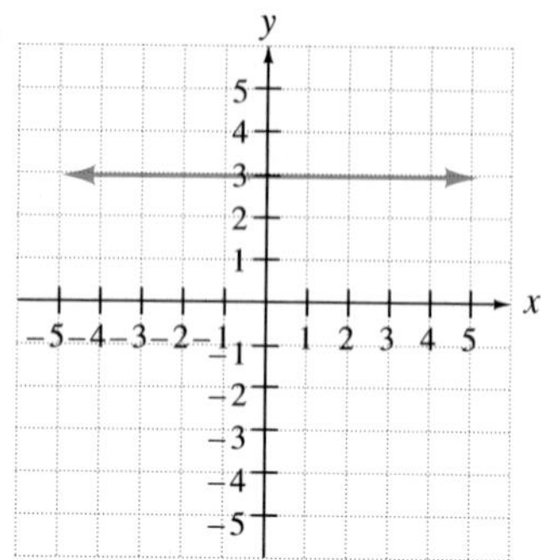

31.

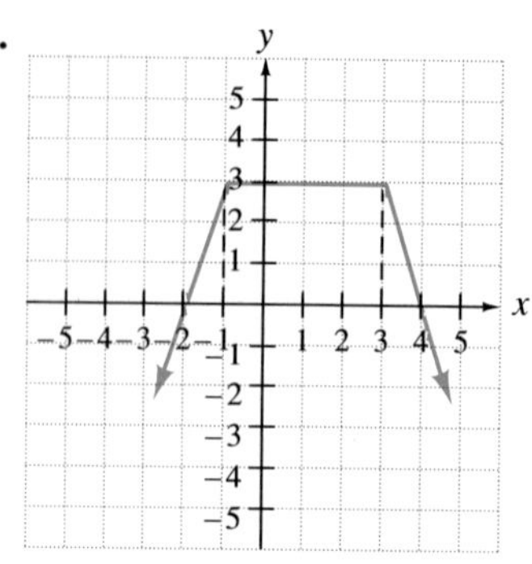

32.

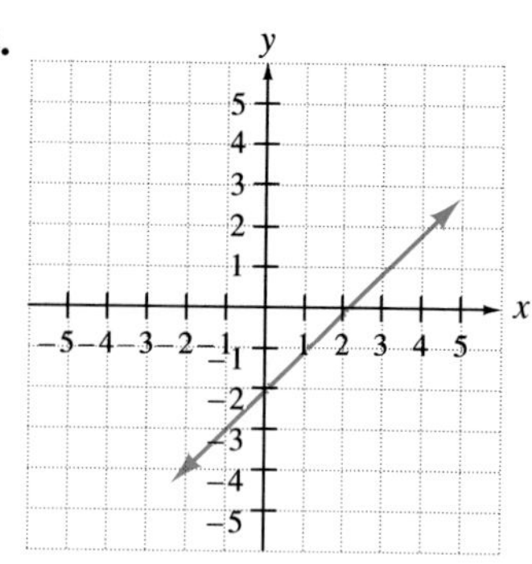

Given the following functions, find the indicated values. See Example 7.

33. Given $f(x) = 2x - 5$; **a.** $f(-2)$ **b.** $f(0)$ **c.** $f(3)$

34. $g(x) = 3 - 7x$; **a.** $g(-7)$ **b.** $g(7)$ **c.** $g\left(\frac{3}{7}\right)$

35. $h(x) = x^2 + 2$; **a.** $h(-3)$ **b.** $h\left(\frac{1}{4}\right)$ **c.** $h(3)$

36. $f(x) = x^2 - 4$; **a.** $f(7)$ **b.** $f(4)$ **c.** $f(-10)$

37. $h(x) = x^3$; **a.** $h(-2)$ **b.** $h(-6)$ **c.** $h(0)$

38. $h(x) = -x^3$; **a.** $h(-2)$ **b.** $h(-6)$ **c.** $h(0)$

39. $f(x) = |x|$; **a.** $f(7)$ **b.** $f(-7)$ **c.** $f(0)$

40. $h(x) = |2 - x|$; **a.** $h(0)$ **b.** $h(-5)$ **c.** $h(6)$

Use the graph in Example 8 to answer the following.

41. Approximate the money spent on research and development in 1988.

42. During the last three years shown on the graph, the increase in spending appears to be linear. If this trend continues, project the amount of money spent on research and expenditures in 1998.

Find the domain and range of each relation. Also determine whether the relation is a function.

43. $\{(-1, 7), (0, 6), (-2, 2), (5, 6)\}$

44. $\{(4, 9), (-4, 9), (2, 3), (10, -5)\}$

45. $\{(-2, 4), (6, 4), (-2, -3), (-7, -8)\}$

46. $\{(6, 6), (5, 6), (5, -2), (7, 6)\}$

47. $\{(1, 1), (1, 2), (1, 3), (1, 4)\}$

48. $\{(1, 1), (2, 1), (3, 1), (4, 1)\}$

49. $\left\{\left(\frac{3}{2}, \frac{1}{2}\right), \left(1\frac{1}{2}, -7\right), \left(0, \frac{4}{5}\right)\right\}$

50. $\{(\pi, 0), (0, \pi), (-2, 4), (4, -2)\}$

51. $\{(-3, -3), (0, 0), (3, 3)\}$

52. $\left\{\left(\frac{1}{2}, \frac{1}{4}\right), \left(0, \frac{7}{8}\right), (0.5, \pi)\right\}$

53.

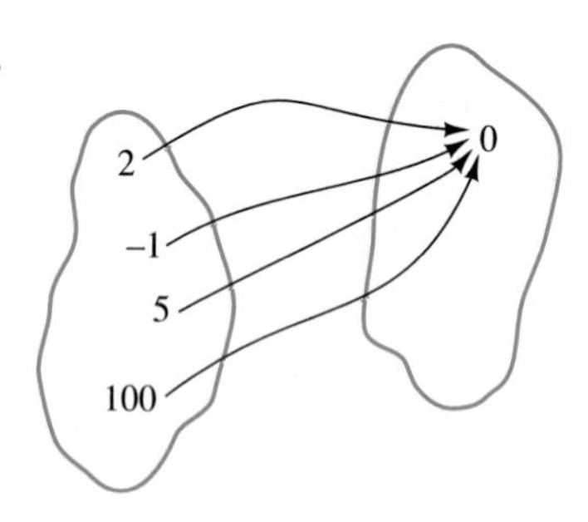

54.

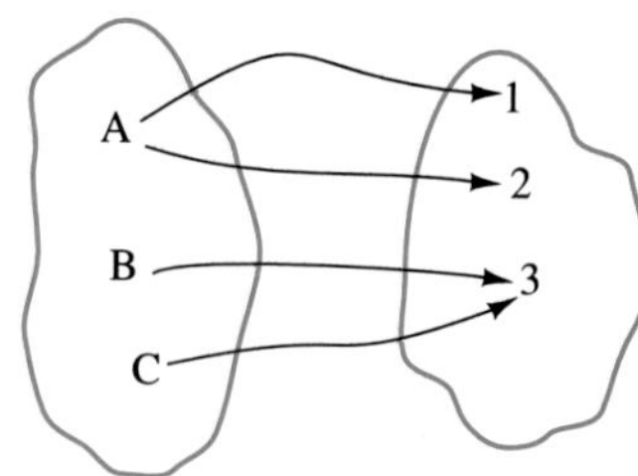

Determine whether the relation is a function.

55. $y = 5x - 12$

56. $y = \frac{1}{2}x + 4$

57. $x = y^2$

58. $x = |y|$

Find the domain and the range of each relation. Use the vertical line test to determine whether each graph is the graph of a function.

59.

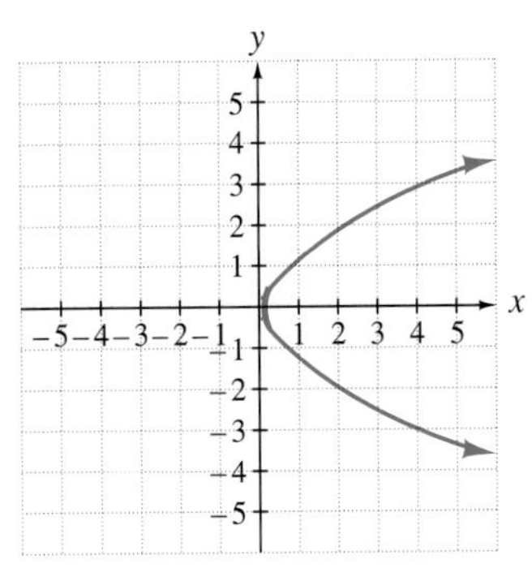

60.

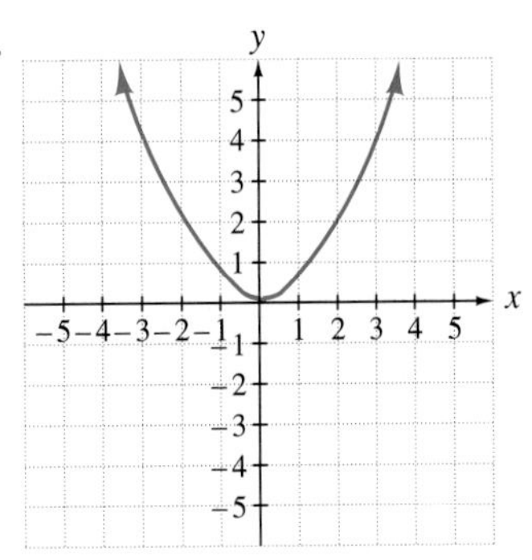

61.

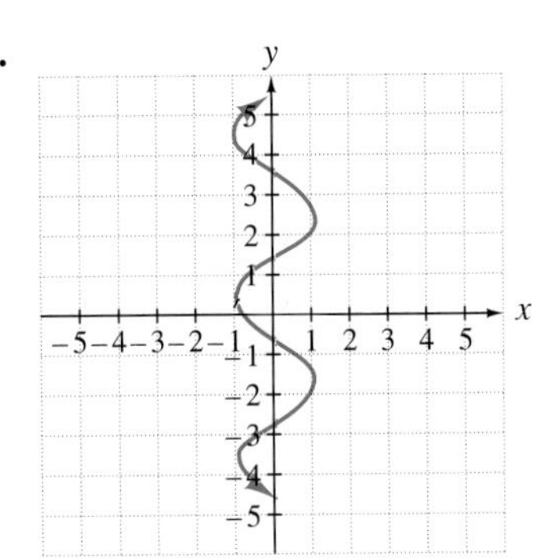

62.

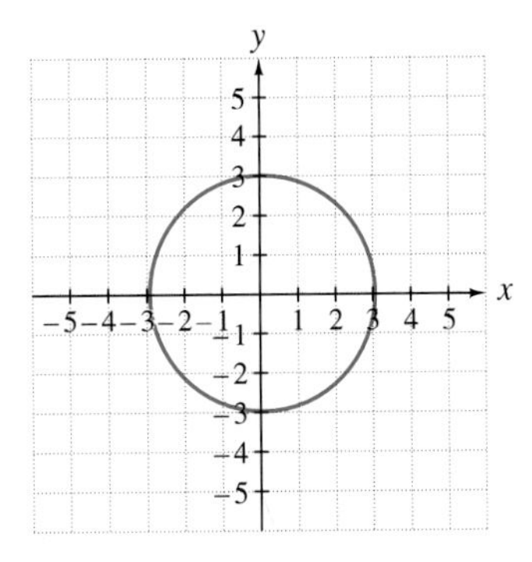

63.

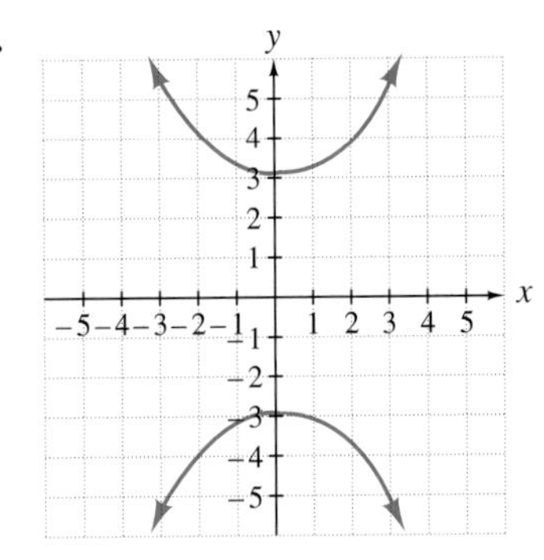

64.

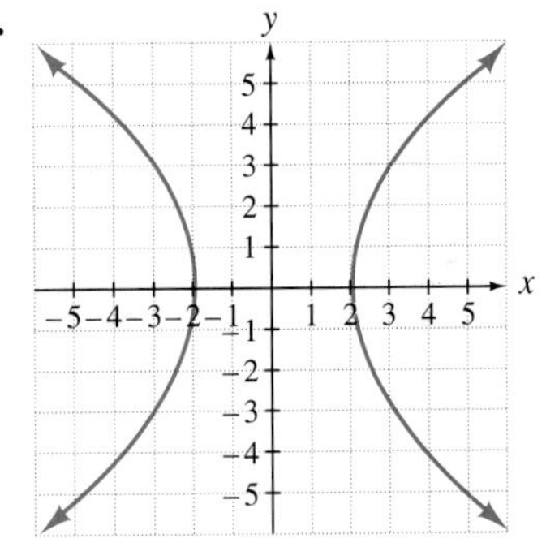

Given the following functions, find the indicated values.

65. $f(x) = 5x$; **a.** $f(0)$ **b.** $f(2)$ **c.** $f(-2)$

66. $g(x) = -3x$; **a.** $g(0)$ **b.** $g(-1)$ **c.** $g(3)$

67. $g(x) = 2x^2 + 3$; **a.** $g(-11)$ **b.** $g(-1)$ **c.** $g\left(\frac{1}{2}\right)$

68. $h(x) = -x^2$; **a.** $h(-5)$ **b.** $h\left(-\frac{1}{3}\right)$ **c.** $h\left(\frac{1}{3}\right)$

69. $f(x) = -x - 2x + 3$; **a.** $f(1)$ **b.** $f(-1)$

70. $g(x) = -x^2 + 4x - 3$; **a.** $g(-6)$ **b.** $g(2)$

71. $f(x) = 6$; **a.** $f(2)$ **b.** $f(0)$ **c.** $f(606)$

72. $h(x) = -12$; **a.** $h(7)$ **b.** $h(542)$ **c.** $h\left(-\frac{3}{4}\right)$

The function $A(r) = \pi r^2$ may be used to find the area of a circle given its radius.

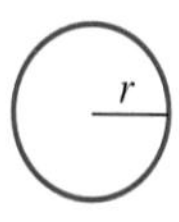

73. Find the area of a circle whose radius is 5 centimeters.

74. Find the area of a circular garden whose radius is 8 feet.

The function $V(x) = x^3$ may be used to find the volume of a cube given the length of x of a side.

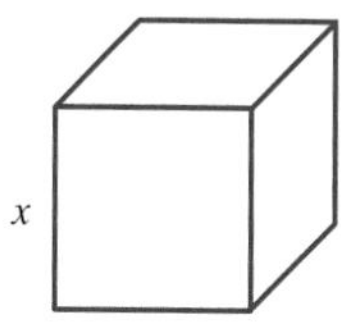

75. Find the volume of a cube whose side is 14 inches.

76. Find the volume of a die whose side is 2 centimeters.

Forensic scientists use the following functions to find the height of a woman given the height of her femur bone f or her tibia bone t in centimeters.

$$H(f) = 2.59f + 47.24$$

$$H(t) = 2.72t + 61.28$$

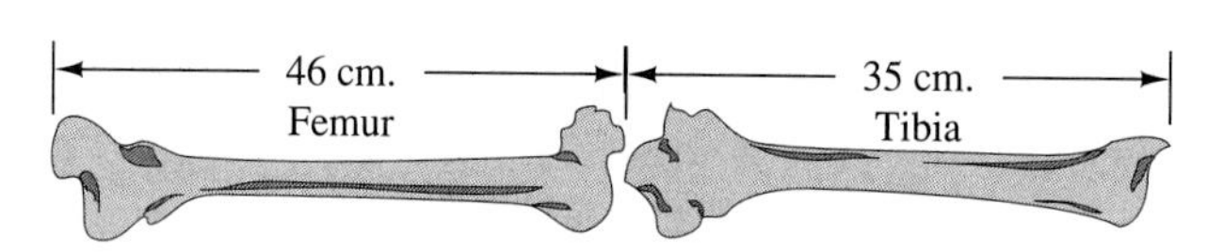

77. Find the height of a woman whose femur measures 46 centimeters.

78. Find the height of a woman whose tibia measures 35 centimeters.

The dosage in milligrams D of Ivermectin, a heartworm preventive, for a dog who weights x pounds is given by $D(x) = \frac{136}{25}x$.

79. Find the proper dosage for a dog that weighs 30 pounds.

80. Find the proper dosage for a dog that weighs 50 pounds.

Skill Review

Solve the following. See Section 3.6.

81. $x < 4$ and $x > -1$

82. $x > -3$ and $x > 0$

83. $-2x > -10$ and $x + 7 < 15$

84. $3x - 1 < 5$ and $2x + 7 > 3$

Solve the following. See Section 2.5.

85. If two angles are supplementary and one angle's measure is three times the measure of the other angle, find the measure of each angle.

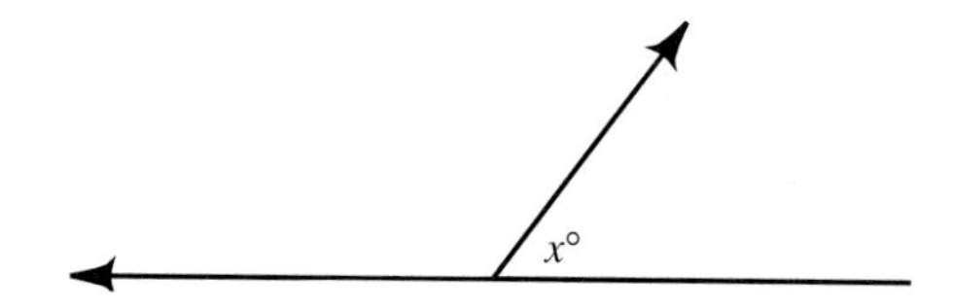

86. Is it possible to find the perimeter of the following geometric figure? If so, find the perimeter.

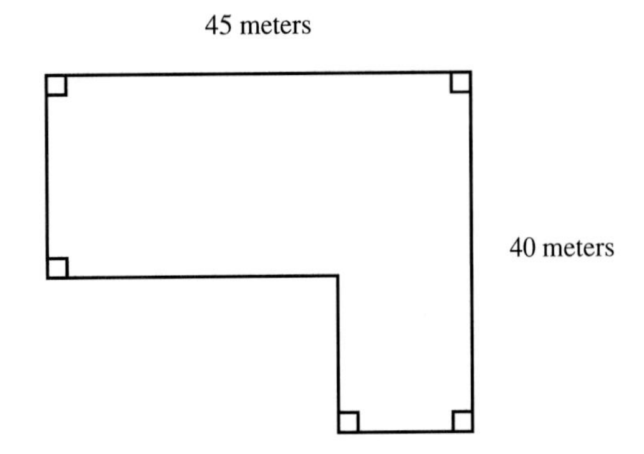

A Look Ahead

Given the following functions, find the indicated values. See the following example.

EXAMPLE If $f(x) = x^2 + 2x + 1$, find the following:

a. $f(\pi)$ **b.** $f(c)$

Solution: **a.** $f(x) = x^2 + 2x + 1$

$f(\pi) = \pi^2 + 2\pi + 1$

b. $f(x) = x^2 + 2x + 1$

$f(c) = (c)^2 + 2(c) + 1$

$= c^2 + 2c + 1$ ■

87. $f(x) = 2x + 7;$ **a.** $f(2)$ **b.** $f(a)$

88. $g(x) = -3x + 12;$ **a.** $g(s)$ **b.** $g(r)$

89. $h(x) = x^2 + 7;$ **a.** $h(3)$ **b.** $h(a)$

90. $f(x) = x^2 - 12;$ **a.** $f(12)$ **b.** $f(a)$

Writing in Mathematics

91. In your own words define **(a)** function; **(b)** domain; **(c)** range.

92. Explain the vertical line test and how it is used.

93. Since $y = x + 7$ is a function, rewrite the equation using function notation.

94. Describe a function whose domain and range are sets of people.

95. Explain how the vertical line test accurately reveals whether a relation is a function.

4.3 Graphing Linear Functions

OBJECTIVES

1. Graph linear functions.
2. Graph linear functions by finding intercepts.
3. Graph vertical and horizontal lines.

Tape BA 20

1 In this section, we identify and graph linear functions. By the vertical line test, we know that all linear equations except those whose graphs are vertical lines are functions. For example, we know from Section 4.1 that the graph of $y = 3x$ is a line that passes through the origin.

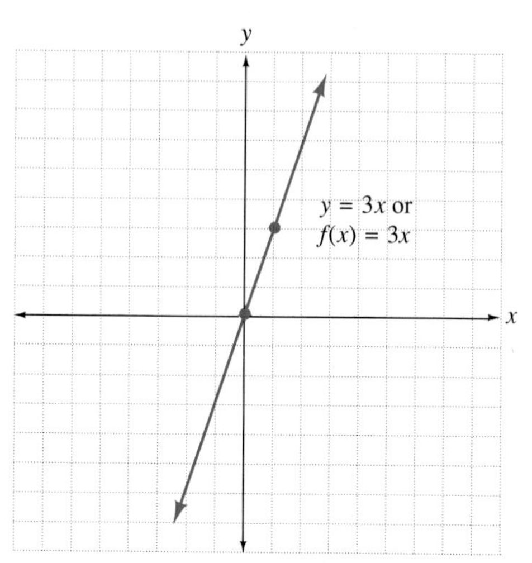

Because this graph passes the vertical line test, we know that $y = 3x$ is a function. If we want to emphasize that this equation describes a function, we may write $y = 3x$ as $f(x) = 3x$.

EXAMPLE 1 Use the graph of $f(x) = 3x$ to graph $f(x) = 3x + 6$.

Solution: From Section 4.1, we know that the graph of $f(x) = 3x + 6$ is the graph of $f(x) = 3x$ moved up 6 units. We will find three ordered pair solutions for practice.

If $x = 1$, then $f(1) = 3(1) + 6$, or $f(1) = 9$.
If $x = 0$, then $f(0) = 3(0) + 6$, or $f(0) = 6$.
If $x = -2$, then $f(-2) = 3(-2) + 6$, or $f(-2) = 0$.

x	y
1	9
0	6
-2	0

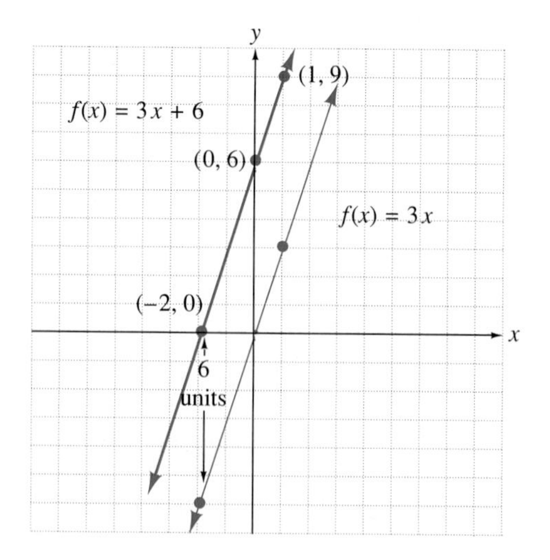

In general, the graph of $y = f(x) + K$ is the graph of $y = f(x)$ shifted $|K|$ units up if K is positive and down if K is negative. ■

2 The graph of $y = 3x + 6$ or $f(x) = 3x + 6$ in the preceding figure crosses the y-axis at the point $(0, 6)$. The y-coordinate of this point, 6, is called the ***y*-intercept.** Likewise, the graph crosses the x-axis at $(-2, 0)$ and the x-coordinate -2 is called the ***x*-intercept.**

To find the y-intercept of a graph, let $x = 0$ since a point on the y-axis has an x-coordinate of 0. To find the x-intercept of a line, let $y = 0$ or $f(x) = 0$ since a point on the x-axis has a y-coordinate of 0.

Finding *x*- and *y*-Intercepts

To find the x-intercept, let $y = 0$ or $f(x) = 0$ and solve for x.
To find the y-intercept, let $x = 0$ and solve for y.

Intercept points are usually easy to find and plot since one coordinate is 0.

EXAMPLE 2 Graph $x - 3y = 6$ by plotting intercept points.

Solution: Let $y = 0$ to find the x-intercept and $x = 0$ to find the y-intercept.

If $y = 0$ then	If $x = 0$ then
$x - 3(0) = 6$	$0 - 3y = 6$
$x - 0 = 6$	$-3y = 6$
$x = 6$	$y = -2$

The x-intercept is 6 and the y-intercept is -2. We find a third ordered pair solution to check our work. If we let $y = -1$, then $x = 3$. Plot the points $(6, 0)$, $(0, -2)$, and $(3, -1)$. The graph of $x - 3y = 6$ is the line drawn through these points, as shown.

x	y
6	0
0	−2
3	−1

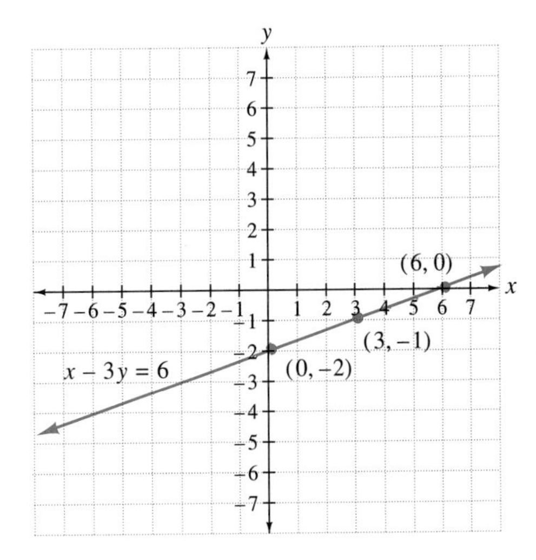

■

If we want to emphasize that $x - 3y = 6$ above describes a function, first solve the equation for y.

$$x - 3y = 6$$

$$-3y = -x + 6 \qquad \text{Subtract } x \text{ from both sides.}$$

$$\frac{-3y}{-3} = \frac{-x}{-3} + \frac{6}{-3} \qquad \text{Divide both sides by } -3.$$

$$y = \frac{1}{3}x - 2 \qquad \text{Simplify.}$$

Next let $y = f(x)$.

$$f(x) = \frac{1}{3}x - 2$$

EXAMPLE 3 Graph $x = -2y$ by plotting intercept points.

Solution: Let $y = 0$ to find the x-intercept and $x = 0$ to find the y-intercept.

If $y = 0$ then	If $x = 0$ then
$x = -2(0)$ or	$0 = -2y$ or
$x = 0$	$0 = y$

Both the x-intercept and y-intercept are 0. In other words, when $x = 0$, then $y = 0$, which gives the ordered pair $(0, 0)$. Also, when $y = 0$, then $x = 0$, which gives the same ordered pair $(0, 0)$. This happens when the graph passes through the origin. Since two points are needed to determine a line, we must find at least one more ordered pair that satisfies $x = -2y$. Let $y = -1$ to find a second ordered pair solution and let $y = 1$ as a checkpoint.

If $y = -1$ then	If $y = 1$ then
$x = -2(-1)$ or	$x = -2(1)$ or
$x = 2$	$x = -2$

The ordered pairs are $(0, 0)$, $(2, -1)$, and $(-2, 1)$. Plot these points to graph $x = -2y$.

x	y
0	0
2	−1
−2	1

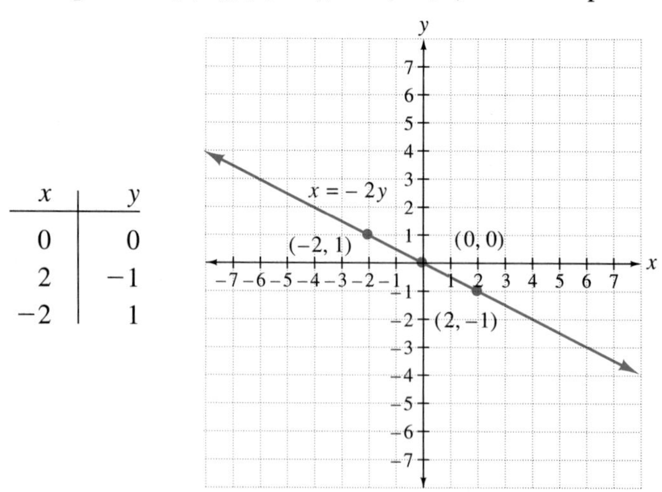

■

EXAMPLE 4 Graph $4x = 3y - 9$.

Solution: Find the x- and y-intercepts, and then choose $x = 2$ to find a third checkpoint.

If $y = 0$ then	If $x = 0$ then	If $x = 2$ then
$4x = 3(0) - 9$ or	$0 = 3y - 9$ or	$4(2) = 3y - 9$ or
$4x = -9$	$9 = 3y$	$8 = 3y - 9$
Solve for x.	Solve for y.	Solve for y.
$x = -\frac{9}{4}$ or $-2\frac{1}{4}$	$3 = y$	$17 = 3y$
		$\frac{17}{3} = y$ or $y = 5\frac{2}{3}$

The ordered pairs are $\left(-2\frac{1}{4}, 0\right)$, $(0, 3)$, $\left(2, 5\frac{2}{3}\right)$. The equation $4x = 3y - 9$ is graphed as follows.

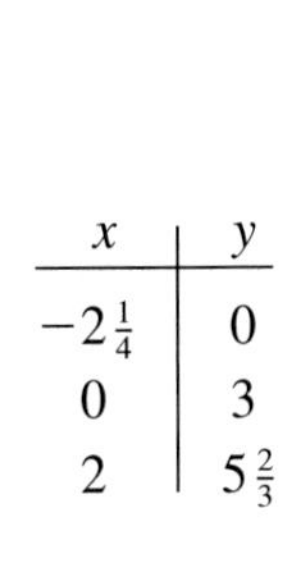

x	y
$-2\frac{1}{4}$	0
0	3
2	$5\frac{2}{3}$

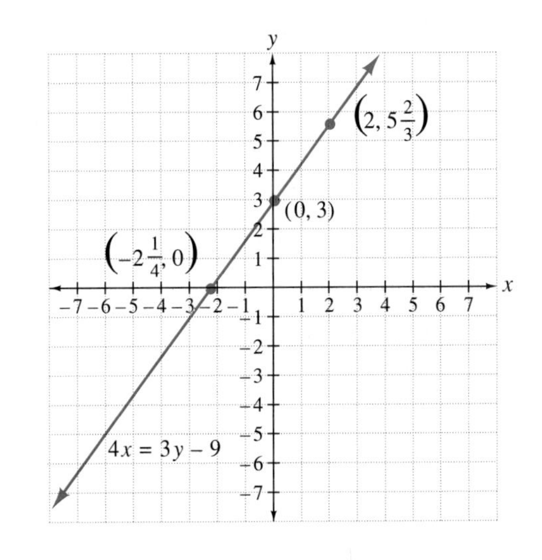

■

3 The equation $x = c$, where c is a real number constant, is a linear equation in two variables because it can be written in the form $x + 0y = c$. The graph of this equation is a vertical line as shown in the next example. Since a vertical line does not pass the vertical line test, the equation $x = c$ does **not** describe a function.

EXAMPLE 5 Graph $x = 2$.

Solution: The equation $x = 2$ can be written as $x + 0y = 2$. For any y-value chosen, notice that x is 2. No other value for x satisfies $x + 0y = 2$. Any ordered pair whose x-coordinate is 2 is a solution to $x + 0y = 2$. We will use the ordered pairs $(2, 3)$, $(2, 0)$ and $(2, -3)$ to graph $x = 2$.

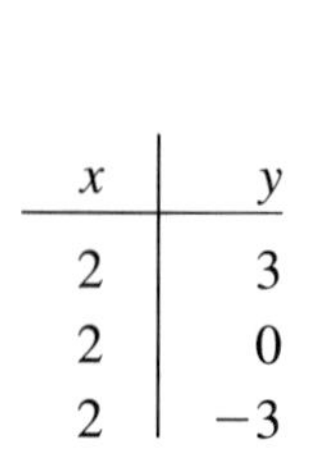

x	y
2	3
2	0
2	−3

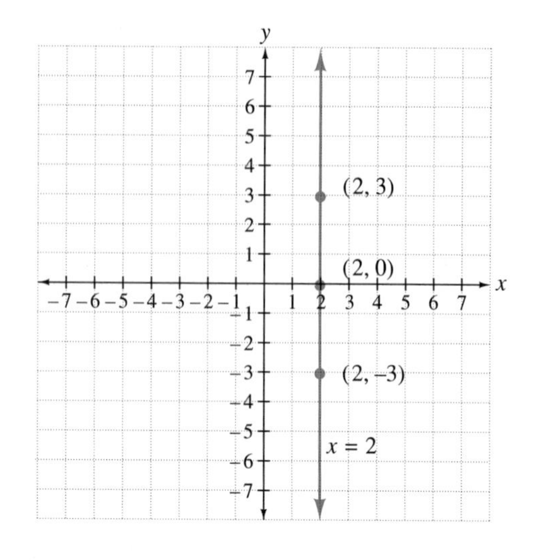

The graph is a vertical line with x-intercept 2. Notice that this graph is not the graph of a function and it has no y-intercept because x is never 0. ■

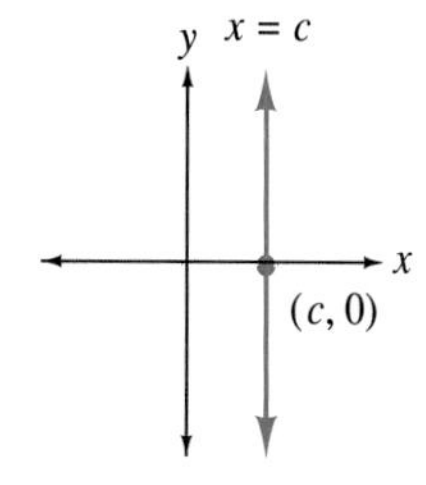

Vertical Lines

The graph of $x = c$, where c is a real number, is a vertical line with x-intercept c.

Does the equation $y = c$ describe a function? Yes, because the graph is a horizontal line, as shown next.

EXAMPLE 6 Graph $y = -3$.

Solution: The equation $y = -3$ can be written as $0x + y = -3$. For any x-value chosen, y is -3. If we choose 4, 1, and -2 as x-values, the ordered pair solutions are $(4, -3)$, $(1, -3)$, and $(-2, -3)$. Use these ordered pairs to graph $y = -3$. The graph is a horizontal line with y-intercept -3 and no x-intercept. Recall that we may write $y = -3$ as $f(x) = -3$.

x	y
4	−3
1	−3
−2	−3

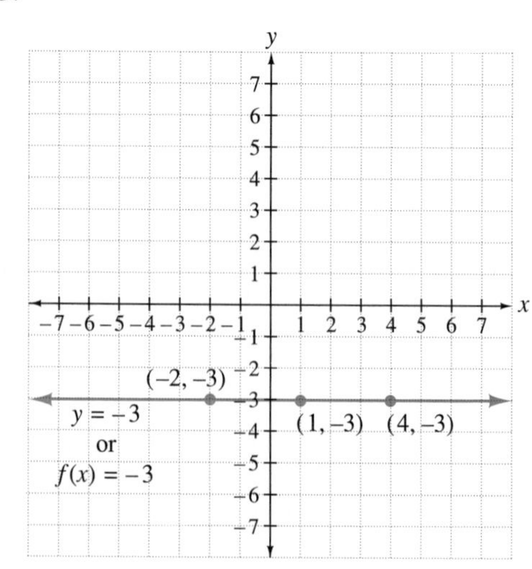

■

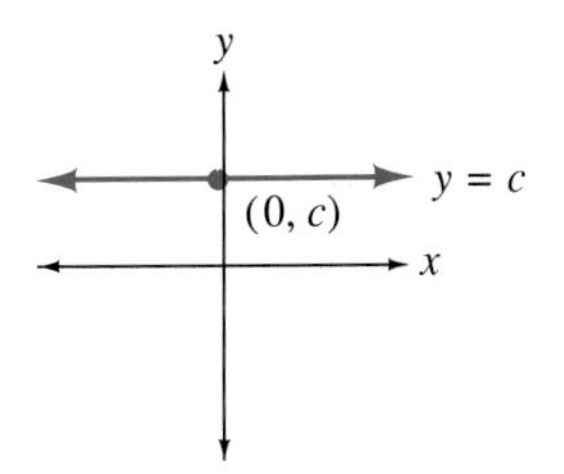

Horizontal Lines

The graph of $y = c$ or $f(x) = c$, where c is a real number, is a horizontal line with y-intercept c.

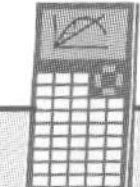

GRAPHING CALCULATOR BOX

You may have noticed by now that to use the Y = key on a grapher to graph an equation, the equation must be solved for y. For example, to graph $2x + 3y = 7$, we solve this equation for y.

$$2x + 3y = 7$$

$$3y = -2x + 7 \qquad \text{Subtract } 2x \text{ from both sides.}$$

$$\frac{3y}{3} = -\frac{2x}{3} + \frac{7}{3} \qquad \text{Divide both sides by 3.}$$

$$y = -\frac{2}{3}x + \frac{7}{3} \qquad \text{Simplify.}$$

To graph $2x + 3y = 7$ or $y = -\frac{2}{3}x + \frac{7}{3}$, press the Y = key and enter

$$Y_1 = -\frac{2}{3}x + \frac{7}{3}$$

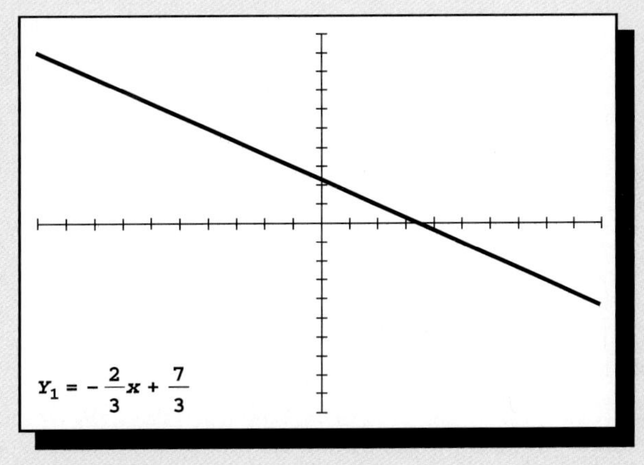

Graph each linear function.

1. $x = 3y$
2. $-2y = x$
3. $3x + 7y = 21$
4. $-4x + 6y = 12$
5. $-2.2x + 6.8y = 15.5$
6. $5.9x - 0.8y = -10.4$

EXERCISE SET 4.3

Graph each linear function. See Example 1.

1. $f(x) = -2x$
2. $f(x) = 2x$
3. $f(x) = -2x + 3$
4. $f(x) = 2x + 6$

5. $f(x) = \frac{1}{2}x$

6. $f(x) = \frac{1}{3}x$

7. $f(x) = \frac{1}{2}x - 4$

8. $f(x) = \frac{1}{3}x - 2$

Graph each linear function by finding x- and y- intercepts. See Examples 2 through 4.

9. $x - y = 3$

10. $x - y = -4$

11. $x = 5y$

12. $2x = y$

13. $-x + 2y = 6$

14. $x - 2y = -8$

15. $2x - 4y = 8$

16. $2x + 3y = 6$

Graph each linear equation. See Examples 5 and 6.

17. $x = -1$

18. $y = 5$

19. $y = 0$

20. $x = 0$

21. $y + 7 = 0$

22. $x - 2 = 0$

Graph each linear equation.

23. $x + 2y = 8$

24. $x - 3y = 3$

25. $f(x) = \frac{3}{4}x + 2$

26. $f(x) = \frac{4}{3}x + 2$

27. $x = -3$

28. $f(x) = 3$

29. $3x + 5y = 7$

30. $3x - 2y = 5$

31. $f(x) = x$

32. $f(x) = -x$

33. $x + 8y = 8$

34. $x - 3y = 9$

35. $5 = 6x - y$

36. $4 = x - 3y$

37. $-x + 10y = 11$

38. $-x + 9 = -y$

39. $y = 1$

40. $x = 1$

41. $f(x) = \frac{1}{2}x$

42. $f(x) = -2x$

43. $x + 3 = 0$

44. $y - 6 = 0$

45. $f(x) = 4x - \frac{1}{3}$

46. $f(x) = -3x + \frac{3}{4}$

47. $2x + 3y = 6$

48. $4x + y = 5$

Solve.

49. The perimeter $P(x)$ of a rectangle whose width is a constant 3 inches and whose length is x inches is given by the function

$$P(x) = 2x + 6$$

a. Draw a graph of this function.
b. Read from the graph the perimeter $P(x)$ of a rectangle whose length x is 4 inches.

50. The distance $D(t)$ traveled in a train moving at a constant speed of 50 miles per hour is given by the function

$$D(t) = 50t$$

where t is the time in hours traveled.

a. Graph this function. Use the horizontal axis for time t.
b. Read from the graph the distance traveled after 6 hours.

51. Broyhill Furniture found that it takes 2 hours to manufacture each table for one of its special dining room sets. Each chair takes 3 hours to manufacture. A total of 1500 hours is available to produce tables and chairs of this style. The linear equation that models this situation is $2x + 3y = 1500$, where x represents the number of tables produced and y the number of chairs produced.

a. Complete the ordered pair solution (0,   ) of this equation. Describe the manufacturing situation this solution corresponds to.

b. Complete the ordered pair (, 0) for this equation. Describe the manufacturing situation this solution corresponds to.

c. If 50 tables are produced, find the greatest number of chairs they can make.

52. While manufacturing two different camera models, Kodak found that the basic model costs \$55 to produce, while the deluxe model costs \$75. The weekly budget for these two models is limited to \$33,000 in production costs. The linear equation that models this situation is $55x + 75y = 33{,}000$, where x represents the number of basic models and y the number of deluxe models.

a. Complete the ordered pair solution (0,) of this equation. Describe the manufacturing situation this solution corresponds to.

b. Complete the ordered pair solution (, 0) of this equation. Describe the manufacturing situation this solution corresponds to.

c. If 350 deluxe models are produced, find the greatest number of basic models that can be made in one week.

Skill Review

Solve the following. See Sections 3.4 and 3.7.

53. $|x - 3| = 6$

54. $|x + 2| < 4$

55. $|2x + 5| > 3$

56. $|5x| = 10$

57. $|3x - 4| \le 2$

58. $|7x - 2| \ge 5$

Simplify.

59. $\dfrac{-6 - 3}{2 - 8}$

60. $\dfrac{4 - 5}{-1 - 0}$

61. $\dfrac{-8 - (-2)}{-3 - (-2)}$

62. $\dfrac{12 - 3}{10 - 9}$

63. $\dfrac{0 - 6}{5 - 0}$

64. $\dfrac{2 - 2}{3 - 5}$

Writing in Mathematics

65. Discuss whether a vertical line ever has a y-intercept.

66. Discuss whether a horizontal line ever has an x-intercept.

67. Explain why it is a good idea to use three points to graph a linear equation.

68. Explain how to find intercepts.

4.4 The Slope of a Line

OBJECTIVES

Tape IA 7

1. Find the slope of a line given two points on the line.
2. Find the slope of a line given the equation of a line.
3. Compare the slopes of parallel and perpendicular lines.
4. Find the slopes of horizontal and vertical lines.

1 Anyone who has experienced the plunging descent of a roller coaster or the rapid climb of a jetliner has some awareness of the steepness of such rides. In mathematics, the steepness or tilt of a line is also known as its **slope.** We measure the slope of a line as a ratio of **vertical change** to **horizontal change.** Slope is usually designated by the letter m.

Suppose that we want to measure the slope of the following line.

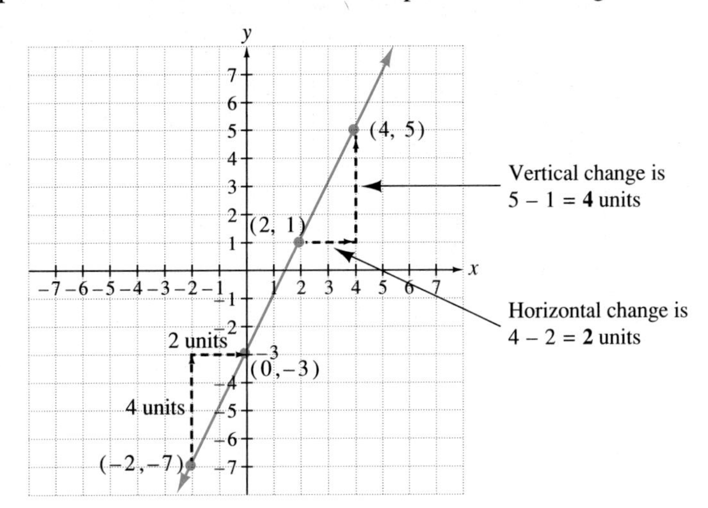

The vertical change between these pairs of points on the line is 4 units per horizontal change of 2 units. Then

$$m = \frac{\text{change in } y \text{ (vertical change)}}{\text{change in } x \text{ (horizontal change)}} = \frac{4}{2} = 2$$

Consider the following line, which passes through the points (x_1, y_1) and (x_2, y_2). (The notation x_1 is read "x-sub-one.") The vertical change or rise between these points is the difference in the y-coordinates: $y_2 - y_1$. The horizontal change or run between the points is the difference of the x-coordinates: $x_2 - x_1$.

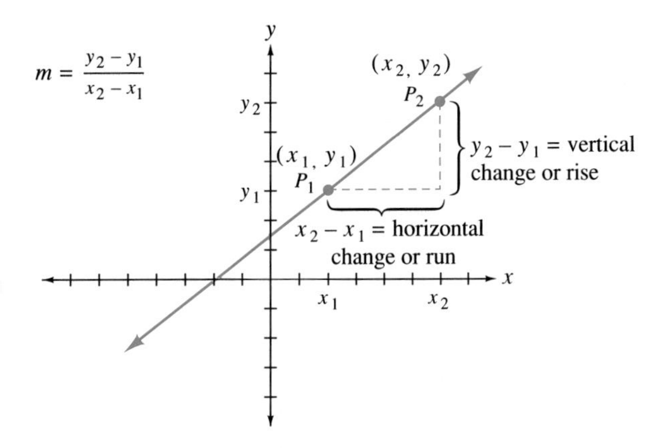

Slope of a Line

Given a line passing through points (x_1, y_1) and (x_2, y_2), the slope (denoted by m) of the line is given by

$$m = \frac{\text{rise}}{\text{run}} = \frac{y_2 - y_1}{x_2 - x_1}, \qquad \text{as long as } x_2 \neq x_1$$

EXAMPLE 1 Find the slope of the line through $(-1, 5)$ and $(2, -3)$. Graph the line.

Solution: If we let $(-1, 5)$ be (x_1, y_1), then $x_1 = -1$ and $y_1 = 5$. Also, let $(2, -3)$ be point (x_2, y_2) so that $x_2 = 2$ and $y_2 = -3$. Then, by the definition of slope,

$$m = \frac{y_2 - y_1}{x_2 - x_1}$$

$$= \frac{-3 - 5}{2 - (-1)}$$

$$= \frac{-8}{3} = -\frac{8}{3}$$

The slope of the line is $-\frac{8}{3}$. Its graph follows.

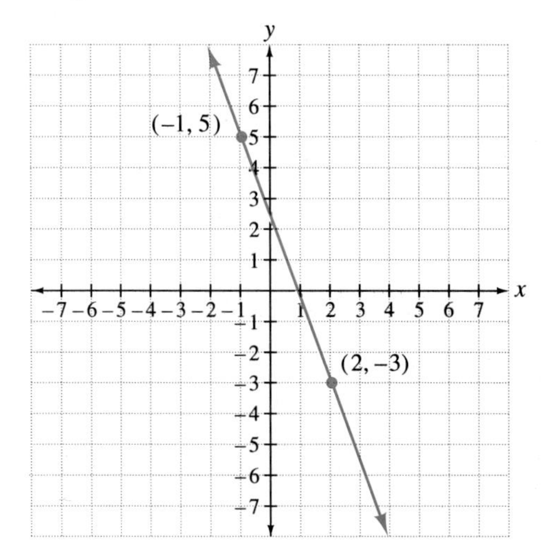

■

In Example 1, we could just as well have identified (x_1, y_1) with $(2, -3)$ and (x_2, y_2) with $(-1, 5)$. It makes no difference which point is called (x_1, y_1) or (x_2, y_2).

HELPFUL HINT

When finding the slope of a line through two given points, it makes no difference which given point is called (x_1, y_1) and which is called (x_2, y_2). Once an x-coordinate is called x_1, however, make sure its corresponding y-coordinate is called y_1.

EXAMPLE 2 Find the slope of the line through $(-1, -2)$ and $(2, 4)$. Graph the line.

Solution: Let $(2, 4)$ be (x_1, y_1) and let $(-1, -2)$ be (x_2, y_2).

$$m = \frac{y_2 - y_1}{x_2 - x_1}$$

$$= \frac{-2 - 4}{-1 - 2}$$

$$= \frac{-6}{-3} = 2$$

The slope is 2.

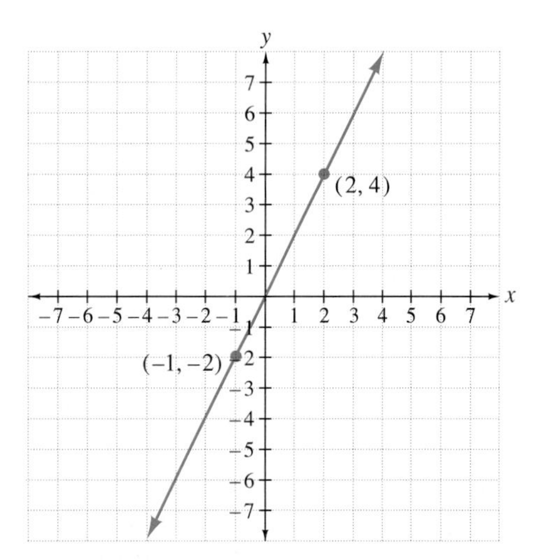

Notice that the slope of the line in Example 1 is negative, whereas the slope of the line in Example 2 is positive. Let your eye follow the line with negative slope from left to right and notice that the line "goes down." Following the line with positive slope from left to right, notice that the line "goes up." This is true in general.

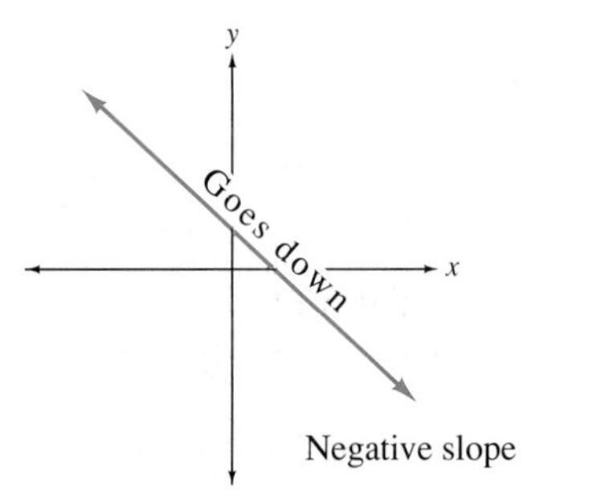

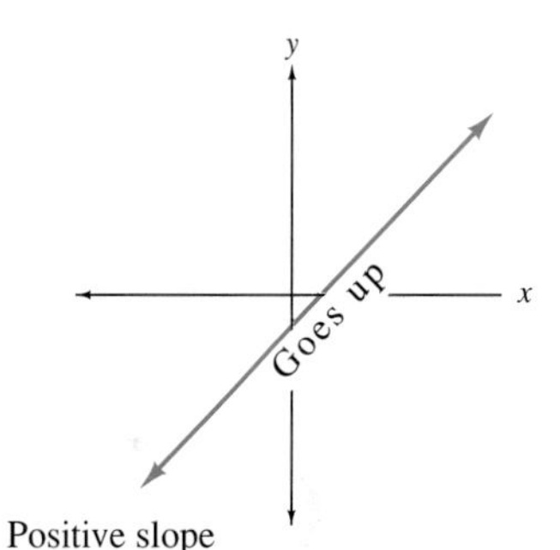

2 As we have seen, the slope of a line is defined by two points on the line. Thus, if we know the equation of a line, we can find its slope.

EXAMPLE 3 Find the slope of the line whose equation is $f(x) = \frac{2}{3}x + 4$.

Solution: Two points are needed on the line defined by $f(x) = \frac{2}{3}x + 4$ or $y = \frac{2}{3}x + 4$ to find its slope. We will use intercepts as our two points.

If $x = 0$, the corresponding y-value is 4, the y-intercept.
If $y = 0$, the corresponding x-value is -6, the x-intercept.

Use the points (0, 4) and $(-6, 0)$ to find the slope. Let (0, 4) be (x_1, y_1) and $(-6, 0)$ be (x_2, y_2). Then

$$m = \frac{y_2 - y_1}{x_2 - x_1} = \frac{0 - 4}{-6 - 0} = \frac{-4}{-6} = \frac{2}{3}$$

Analyzing the results of Example 3, you may notice a striking pattern:

The slope of $y = \frac{2}{3}x + 4$ is $\frac{2}{3}$, the same as the coefficient of x.

Also, the y-intercept is 4, the same as the constant term.

When a linear equation is written in the form $f(x) = mx + b$ or $y = mx + b$, m is the slope of the line and b is its y-intercept. The form $y = mx + b$ is appropriately called the **slope–intercept form,** and $f(x) = mx + b$ describes a linear function.

Slope–Intercept Form

When a linear equation in two variables is written in slope–intercept form,

$$y = mx + b$$

then m is the slope of the line and b is the y-intercept of the line.

EXAMPLE 4 Find the slope and the y-intercept of the line whose equation is $3x - 4y = 4$.

Solution: Write the equation in slope–intercept form by solving for y.

$$3x - 4y = 4$$

$$-4y = -3x + 4 \quad \text{Subtract } 3x \text{ from both sides.}$$

$$\frac{-4y}{-4} = \frac{-3x}{-4} + \frac{4}{-4} \quad \text{Divide both sides by } -4.$$

$$y = \frac{3}{4}x - 1 \quad \text{Simplify.}$$

The coefficient of x, $\frac{3}{4}$, is the slope, and the constant term -1 is the y-intercept. ■

3 Slopes of lines can help us determine whether lines are parallel. Parallel lines have the same steepness, so it follows that they have the same slope.

Parallel Lines

Nonvertical parallel lines have the same slope.

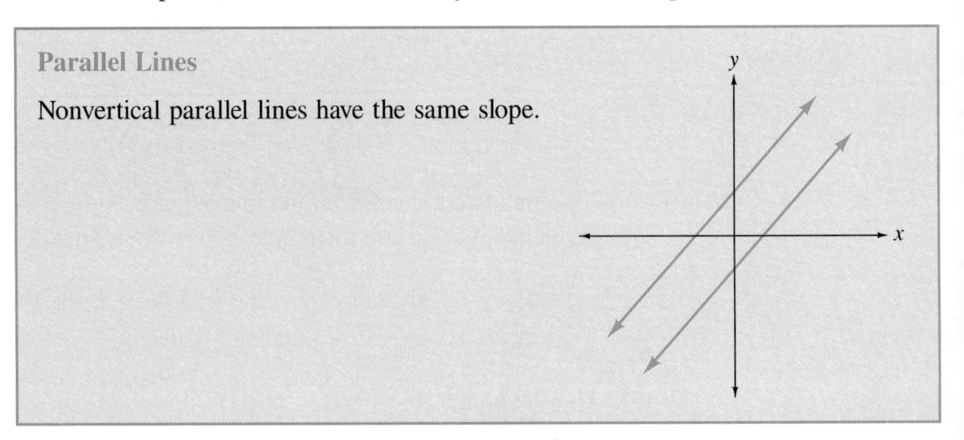

How do the slopes of perpendicular lines compare? Two lines that intersect at right angles are said to be **perpendicular.** The product of the slopes of two perpendicular lines is -1.

Perpendicular Lines

If the product of the slopes of two lines is -1, then the lines are perpendicular.

(Two nonvertical lines are perpendicular if the slope of one is the negative reciprocal of the slope of the other.)

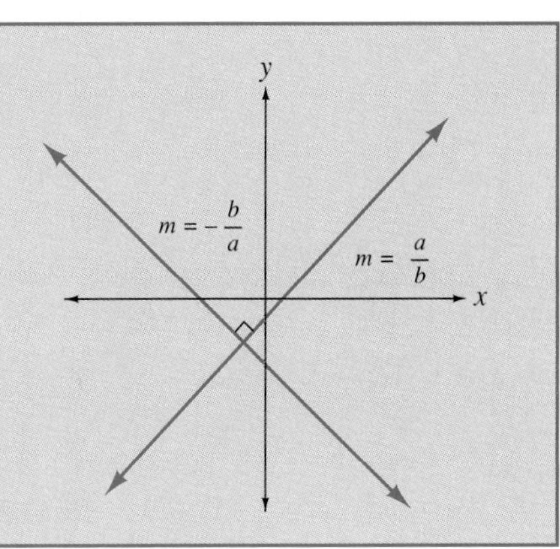

EXAMPLE 5 Are the following pairs of lines parallel, perpendicular, or neither?

a. $3x + 7y = 4$ **b.** $-x + 3y = 2$
$6x + 14y = 7$ $2x + 6y = 5$

Solution: Find the slope of each line by solving each equation for y.

a. $3x + 7y = 4$ $\qquad$ $6x + 14y = 7$

$$7y = -3x + 4 \qquad 14y = -6x + 7$$

$$\frac{7y}{7} = \frac{-3x}{7} + \frac{4}{7} \qquad \frac{14y}{14} = \frac{-6x}{14} + \frac{7}{14}$$

$$y = -\frac{3}{7}x + \frac{4}{7} \qquad y = -\frac{3}{7}x + \frac{1}{2}$$

The slope of both lines is $-\frac{3}{7}$. The lines are parallel.

b. $-x + 3y = 2$ $\qquad$ $2x + 6y = 5$

$$3y = x + 2 \qquad 6y = -2x + 5$$

$$\frac{3y}{3} = \frac{x}{3} + \frac{2}{3} \qquad \frac{6y}{6} = \frac{-2x}{6} + \frac{5}{6}$$

$$y = \frac{1}{3}x + \frac{2}{3} \qquad y = -\frac{1}{3}x + \frac{5}{6}$$

The slope of the line $-x + 3y = 2$ is $\frac{1}{3}$ and the slope of the line $2x + 6y = 5$ is $-\frac{1}{3}$. The slopes are not equal, so the lines are not parallel. The product of the slopes is $\frac{1}{3} \cdot -\frac{1}{3} = -\frac{1}{9}$, not -1, so the lines are not perpendicular. They are neither parallel nor perpendicular. ■

4 Next we find the slopes of vertical and horizontal lines.

EXAMPLE 6 Find the slope of the line $x = 5$.

Solution: Recall that the graph of $x = 5$ is a vertical line with x-intercept 5.

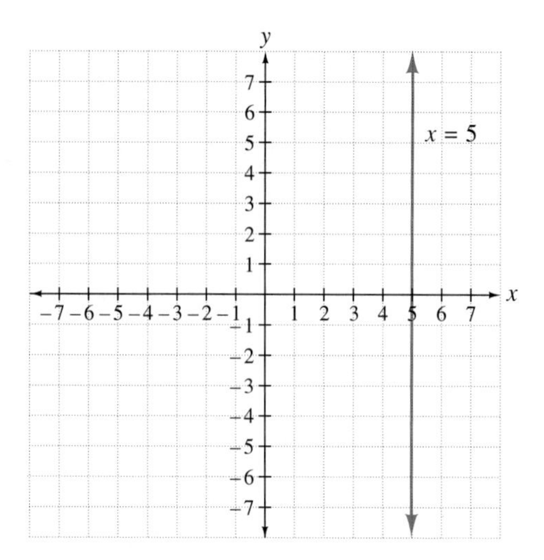

To find the slope, find two ordered pair solutions of $x = 5$. Solutions of $x = 5$ must have an x-value of 5.

Let $(5, 0) = (x_1, y_1)$ and $(5, 4) = (x_2, y_2)$. Then

$$m = \frac{y_2 - y_1}{x_2 - x_1} = \frac{4 - 0}{5 - 5} = \frac{4}{0}$$

Since $\frac{4}{0}$ is undefined, we say the slope of the vertical line $x = 5$ is undefined. Since all vertical lines are parallel, we can say that **vertical lines have undefined slope.** ■

EXAMPLE 7 Find the slope of the line $y = -1$.

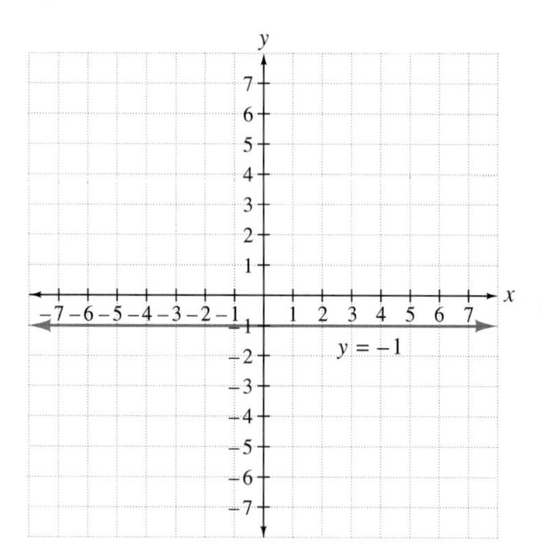

Solution: Recall that $y = -1$ is a horizontal line with y-intercept -1. To find the slope, find two ordered pair solutions of $y = -1$. Solutions of $y = -1$ must have a y-value of -1.

Let $(2, -1) = (x_1, y_1)$ and $(-3, -1) = (x_2, y_2)$. Then

$$m = \frac{y_2 - y_1}{x_2 - x_1} = \frac{-1 - (-1)}{-3 - 2} = \frac{0}{-5} = 0$$

The slope of the line $y = -1$ is 0. Since all horizontal lines are parallel, we can say that **horizontal lines have a slope of 0.** ■

HELPFUL HINT

Slope of 0 and undefined slope are not the same. Vertical lines have undefined slope or no slope, whereas horizontal lines have 0 slope.

Here is a general review of slope.

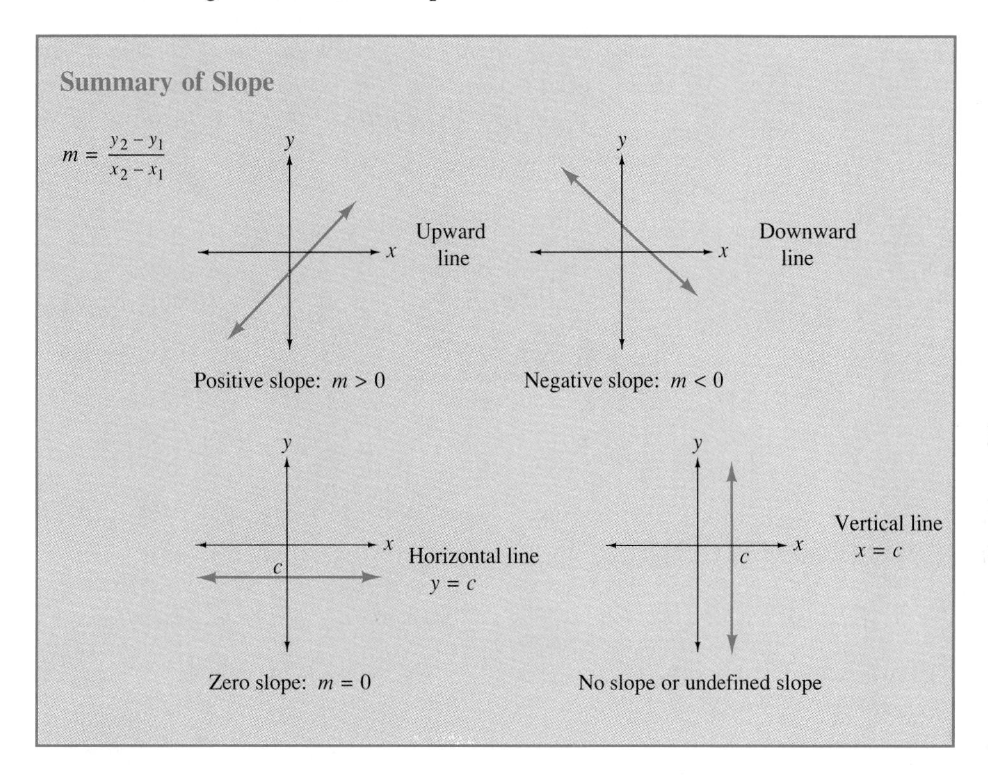

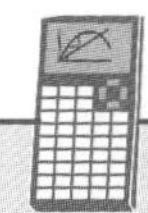

GRAPHING CALCULATOR BOX

A grapher is a very useful tool for discovering patterns. To discover the change in the graph of a linear function caused by a change in slope, try the following. Use a standard window and graph a linear function in the form $f(x)$ or $y = mx + b$. Recall that the graph of such an equation will have slope m and y-intercept b.

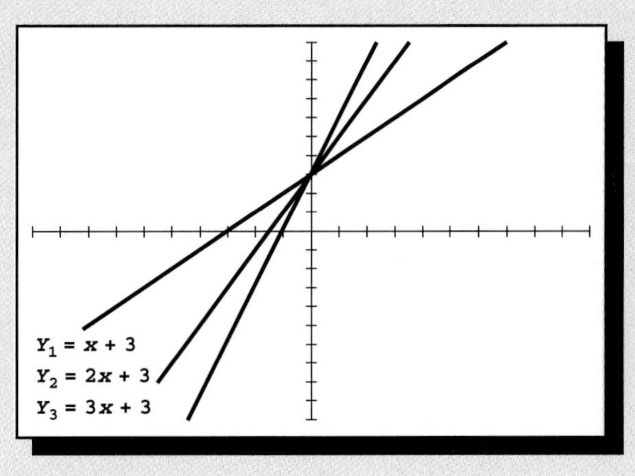

First graph $y = x + 3$. To do so, press the $\boxed{Y =}$ key and enter $Y_1 = x + 3$. Notice that this graph has slope 1 and that the y-intercept is 3. Next, on the same set of axes, graph $y = 2x + 3$ and $y = 3x + 3$ by pressing $\boxed{Y =}$ and entering $Y_2 = 2x + 3$ and $Y_3 = 3x + 3$.

Notice the difference in the graph of each function as the slope changes from 1 to 2 to 3. How would the graph of $y = 5x + 3$ appear? To see the change in the graph caused by a change in negative slope, try graphing $y = -x + 3$, $y = -2x + 3$, and $y = -3x + 3$ on the same set of axes.

Use a grapher to graph the following equations. For each exercise, graph the first equation and use its graph to predict the appearance of the other equations. Then graph the other equations on the same set of axes and check your prediction.

1. $y = x$; $y = 6x$, $y = -6x$

2. $y = -x$; $y = -5x$, $y = -10x$

3. $y = \frac{1}{2}x + 2$; $y = \frac{3}{4}x + 2$, $y = x + 2$

4. $y = x + 1$; $y = \frac{5}{4}x + 1$, $y = \frac{5}{2}x + 1$

5. $y = -7x + 5$; $y = 7x + 5$

6. $y = 3x - 1$; $y = -3x - 1$

MENTAL MATH

Decide whether a line with the given slope is upward, downward, horizontal, or vertical.

1. $m = \frac{7}{6}$

2. $m = -3$

3. $m = 0$

4. m is undefined.

EXERCISE SET 4.4

Find the slope of the line that goes through the given points. See Examples 1 and 2.

1. $(0, 0)$ and $(7, 8)$

2. $(-1, 5)$ and $(0, 0)$

3. $(-1, 5)$ and $(6, -2)$

4. $(-1, 9)$ and $(-3, 4)$

5. $(1, 4)$ and $(5, 3)$

6. $(3, 1)$ and $(2, 6)$

7. $(-4, 3)$ and $(-4, 5)$

8. $(6, -6)$ and $(6, 2)$

Find the slope and the y-intercept of each line. See Examples 3 and 4.

9. $f(x) = 5x - 2$

10. $f(x) = -2x + 6$

11. $2x + y = 7$

12. $-5x + y = 10$

13. $2x - 3y = 10$

14. $-3x - 4y = 6$

15. $f(x) = \frac{1}{2}x$

16. $f(x) = -\frac{1}{4}x$

Determine whether the lines are parallel, perpendicular, or neither. See Example 5.

17. $y = -3x + 6$
$y = 3x + 5$

18. $y = 5x - 6$
$y = 5x + 2$

19. $-4x + 2y = 5$
$2x - y = 7$

20. $2x - y = -10$
$2x + 4y = 2$

21. $-2x + 3y = 1$
$3x + 2y = 12$

22. $x + 4y = 7$
$2x - 5y = 0$

Find the slope of each line. See Examples 6 and 7.

23. $x = 1$

24. $y = -2$

25. $y = -3$

26. $x = 4$

27. $x + 2 = 0$

28. $y - 7 = 0$

Find the slope of the line that goes through the given points.

29. $(-2, 8)$ and $(1, 6)$

30. $(4, -3)$ and $(2, 2)$

31. $(1, 0)$ and $(1, 1)$

32. $(0, 13)$ and $(-4, 13)$

33. $(5, -11)$ and $(1, -11)$

34. $(5, 4)$ and $(0, 5)$

35. $(0, 6)$ and $(-3, 0)$

36. $(5, 2)$ and $(0, 5)$

37. $(-1, 2)$ and $(-3, 4)$

38. $(3, -2)$ and $(-1, -6)$

Find the slope and the y-intercept of each line.

39. $f(x) = -x + 5$

40. $f(x) = x + 2$

41. $-6x + 5y = 30$

42. $4x - 7y = 28$

43. $3x + 9 = y$

44. $2y - 7 = x$

45. $y = 4$

46. $x = 7$

47. $f(x) = 7x$

48. $f(x) = \frac{1}{7}x$

49. $6 + y = 0$

50. $x - 7 = 0$

51. $2 - x = 3$

52. $2y + 4 = -7$

Find each slope.

53. Find the pitch or slope of the roof shown.

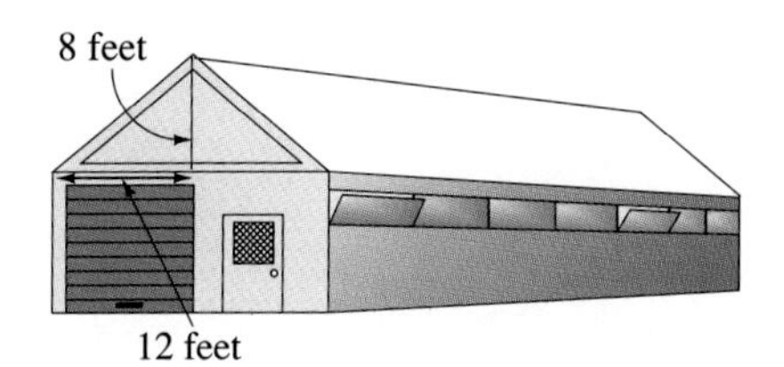

54. Upon takeoff, a Delta Airlines jet climbs to 3 miles as it passes over 25 miles of land below it. Find the slope of its climb.

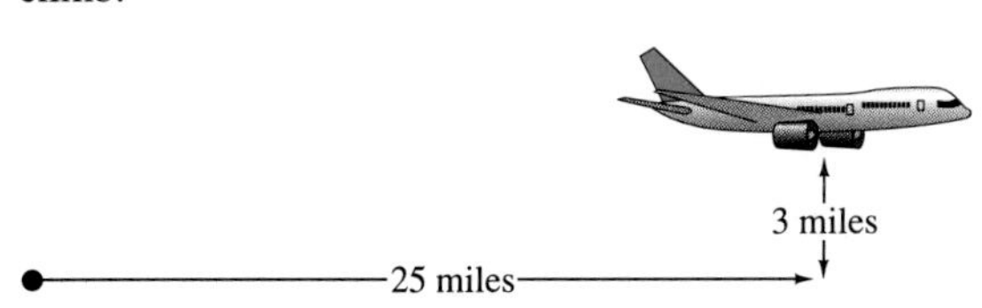

55. Driving down Bald Mountain in Wyoming, Bob Dean finds that he descends 1600 feet in elevation by the time he is 2.5 miles (horizontally) away from the high point on the mountain road. Find the slope of his descent (1 mile = 5280 feet).

56. Find the grade or slope of the road shown.

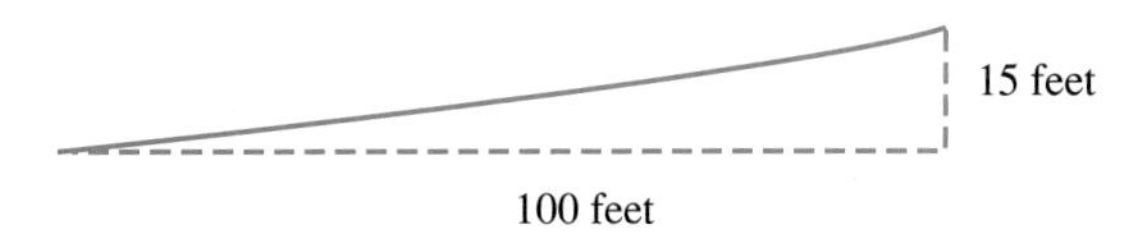

57. Find the slope of the line parallel to the line $y = -\frac{7}{2}x - 6$.

58. Find the slope of the line parallel to the line $y = x$.

59. Find the slope of the line perpendicular to the line $y = -\frac{7}{2}x - 6$.

60. Find the slope of the line perpendicular to the line $y = x$.

61. Find the slope of the line parallel to the line passing through $(-7, -5)$ and $(-2, -6)$.

62. Find the slope of the line parallel to the line passing through the origin and $(-2, 10)$.

63. Find the slope of the line perpendicular to the line passing through the origin and $(1, -3)$.

64. Find the slope of the line perpendicular to the line passing through $(-1, 2)$ and $(5, -3)$.

Skill Review

Evaluate the following. See Section 1.2.

65. $|-6|$

66. $-|7|$

67. $-|-1|$

68. $|-3|$

The following graph shows the average price per share of Xerox stock.

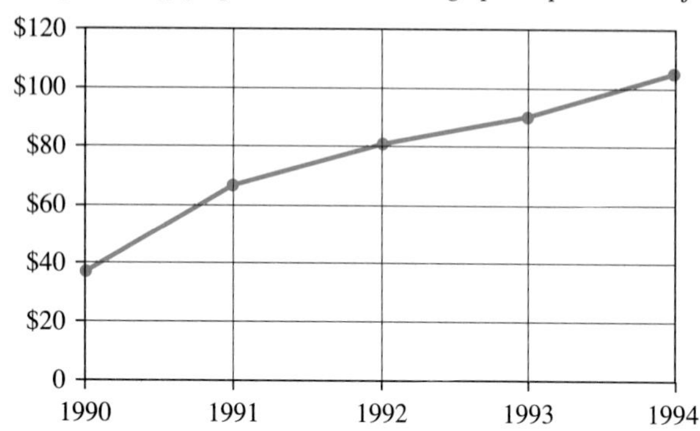

Source: *U.S. News & World Report,* 1994.

69. Approximate the price of stock in 1991.

70. Approximate the price of stock in 1992.

71. Approximate the price of stock in 1993.

72. Find the slope of the line segment between the years 1992 and 1993.

73. Which year shows the greatest increase in stock price?

74. Which segment has the greatest slope?

Simplify and solve for y.

75. $y - 2 = 5(x + 6)$

76. $y - (-1) = 2(x - 0)$

77. $y - 0 = -3[x - (-10)]$

78. $y - 9 = -8[x - (-4)]$

Writing in Mathematics

79. Explain whether two lines, both with positive slopes, can be perpendicular.

80. A horizontal line is perpendicular to a vertcial line. Explain why the product of the slope of a horizontal line and a vertical line is not -1.

81. Explain why the graph of $y = b$ is a horizontal line.

82. Explain why it is reasonable that nonvertical parallel lines have the same slope.

4.5 Equations of Lines

OBJECTIVES

Tape IA 7

1. Use the slope–intercept form to find the equation of a line.
2. Graph a line given its slope and y-intercept.
3. Use the point–slope form to find the equation of a line.
4. Find equations of parallel and perpendicular lines.

1 In the last section, we learned that the slope–intercept form of a linear equation is $y = mx + b$. When an equation is written in this form, the slope of the line is the same as the coefficient, m, of x. Also, the y-intercept of the line is the same as the constant term b. For example, the slope of the line defined by $y = 2x + 3$ is 2 and its y-intercept is 3.

We may also use the slope–intercept form to write the equation of a line given its slope and y-intercept.

EXAMPLE 1 Find the equation of the line with y-intercept -3 and slope of $\frac{1}{4}$.

Solution: We are given the slope and the y-intercept. Let $m = \frac{1}{4}$ and $b = -3$, and write the equation in slope–intercept form, $y = mx + b$.

$$y = mx + b$$

$$y = \frac{1}{4}x + (-3) \qquad \text{Let } m = \frac{1}{4} \text{ and } b = -3.$$

$$y = \frac{1}{4}x - 3 \qquad \text{Simplify.}$$

2 Given the slope and y-intercept of a line, we may graph the line as well as write its equation. Let's graph the line from Example 1. We are given that it has slope $\frac{1}{4}$ and that its y-intercept is -3. First plot the y-intercept point $(0, -3)$. To plot another point on the line, recall that slope is $\frac{\text{rise}}{\text{run}} = \frac{1}{4}$. Another point may then be plotted by starting at $(0, -3)$, rising 1 unit up, and then running 4 units to the right. We are now at the point $(4, -2)$. The graph of $y = \frac{1}{4}x - 3$ is the line through points $(0, -3)$ and $(4, -2)$.

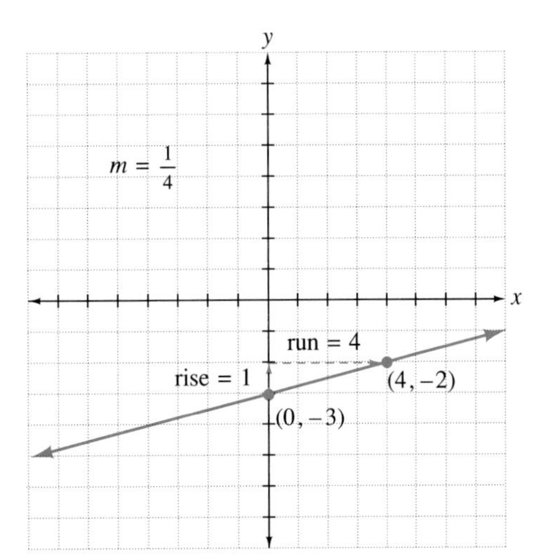

EXAMPLE 2 Graph the line through $(-1, 5)$ with slope -2.

Solution: To graph the line, we need two points. One point is $(-1, 5)$, and we will use the slope -2, which can be written as $\frac{-2}{1}$, to find another point.

$$m = \frac{\text{rise}}{\text{run}} = \frac{-2}{1}$$

To find another point, start at $(-1, 5)$ and move vertically two units down, since the numerator is -2; then move horizontally 1 unit to the right. We stop at the point $(0, 3)$. The line through $(-1, 5)$ and $(0, 3)$ will have the required slope of -2.

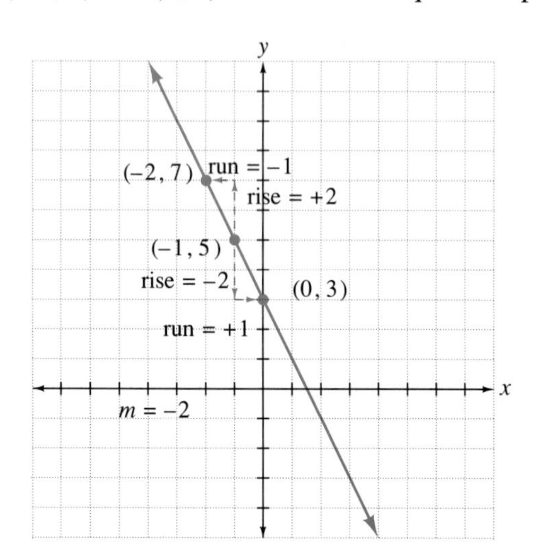

The slope -2 can also be written as $\frac{2}{-1}$, so to find another point we could start at $(-1, 5)$ and move 2 units up and then 1 unit left. We would stop at the point $(-2, 7)$. The line through $(-1, 5)$ and $(-2, 7)$ will have the required slope and will be the same line as shown previously through $(-1, 5)$ and $(0, 3)$.

3 When the slope of a line and a point on the line are known, the equation of the line can also be found. To do this, use the slope formula to write the slope of a line that passes through points (x, y) and (x_1, y_1). We have

$$m = \frac{y - y_1}{x - x_1}$$

Multiply both sides of this equation by $x - x_1$ to obtain

$$y - y_1 = m(x - x_1)$$

This form is called the **point–slope form** of the equation of a line.

> **Point–Slope Form of the Equation of a Line**
>
> The point–slope form of the equation of a line is $y - y_1 = m(x - x_1)$, where m is the slope of the line and (x_1, y_1) is a point on the line.

EXAMPLE 3 Find the equation of the line passing through $(-1, 5)$ with slope -2. Write the equation in standard form: $Ax + By = C$.

Solution: Since the slope and a point on the line are given, use point–slope form $y - y_1 = m(x - x_1)$ to write the equation. Let $m = -2$ and $(x_1, y_1) = (-1, 5)$.

$$y - y_1 = m(x - x_1)$$

$$y - 5 = -2[x - (-1)] \qquad \text{Let } m = -2 \text{ and } (x_1, y_1) = (-1, 5).$$

$y - 5 = -2(x + 1)$	Simplify.
$y - 5 = -2x - 2$	Use the distributive property.
$y = -2x + 3$	Add 5 to both sides.
$2x + y = 3$	Add $2x$ to both sides.

In standard form, the equation is $2x + y = 3$. ■

EXAMPLE 4 Find the equation of the line through (2, 5) and (−3, 4). Write the equation using function notation.

Solution: First use the two given points to find the slope of the line. Let (2, 5) be (x_1, y_1) and (−3, 4) be (x_2, y_2).

$$m = \frac{y_2 - y_1}{x_2 - x_1} = \frac{4 - 5}{-3 - 2} = \frac{-1}{-5} = \frac{1}{5}$$

Next use the slope and either one of the given points to write the equation in point–slope form. We use (2, 5).

$y - y_1 = m(x - x_1)$	Use point–slope form.
$y - 5 = \frac{1}{5}(x - 2)$	Let $x_1 = 2$ and $y_1 = 5$.
$5(y - 5) = 5 \cdot \frac{1}{5}(x - 2)$	Multiply both sides by 5 to clear fractions.
$5y - 25 = x - 2$	Use the distributive property and simplify.
$-x + 5y - 25 = -2$	Subtract x from both sides.
$-x + 5y = 23$	Add 25 to both sides.

The equation in standard form is $-x + 5y = 23$. To write the equation using function notation, we solve for y.

$5y = x + 23$	Add x to both sides.
$y = \frac{1}{5}x + \frac{23}{5}$	Divide both sides by 5.
$f(x) = \frac{1}{5}x + \frac{23}{5}$	Function notation. ■

HELPFUL HINT

Multiply both sides of the equation $-x + 5y = 23$ by -1 and it becomes $x - 5y = -23$. Both $-x + 5y = 23$ and $x - 5y = -23$ are in standard form, and they are equations of the same line.

EXAMPLE 5 Find the equation of the vertical line through (−1, 5).

Solution: The equation of a vertical line can be written in the form $x = c$, so the equation for a vertical line passing through (−1, 5) is $x = -1$. ■

4

EXAMPLE 6 Find the equation of the line containing the point (4, 4) and parallel to the line $2x + 3y = -6$.

Solution: Because the line we want to find is **parallel** to the line $2x + 3y = -6$, the two lines must have equal slopes. Find the slope of $2x + 3y = -6$ by writing it in the form $y = mx + b$.

$$2x + 3y = -6$$

$$3y = -2x - 6 \qquad \text{Subtract } 2x \text{ from both sides.}$$

$$y = \frac{-2x - 6}{3} \qquad \text{Divide by 3.}$$

$$y = -\frac{2}{3}x - 2 \qquad \text{Slope–intercept form.}$$

The slope of this line is $-\frac{2}{3}$. Thus a line parallel to this line will also have a slope of $-\frac{2}{3}$. The equation we are asked to find describes a line containing the point (4, 4) with a slope of $-\frac{2}{3}$. We use the point–slope form.

$$y - y_1 = m(x - x_1)$$

$$y - 4 = -\frac{2}{3}(x - 4) \qquad \text{Let } m = -\frac{2}{3},\ x_1 = 4, \text{ and } y_1 = 4.$$

$$3(y - 4) = -2(x - 4) \qquad \text{Multiply both sides by 3.}$$

$$3y - 12 = -2x + 8 \qquad \text{Apply the distributive property.}$$

$$2x + 3y = 20 \qquad \text{Standard form.}$$ ■

EXAMPLE 7 Write a function that describes the line containing the point (4, 4) and is perpendicular to the line $2x + 3y = -6$.

Solution: Recall that the slope of the line $2x + 3y = -6$ is $-\frac{2}{3}$. A line perpendicular to this line will have a slope that is the negative reciprocal of $-\frac{2}{3}$, or $\frac{3}{2}$. From the point–slope equation, we have

$$y - 4 = \frac{3}{2}(x - 4) \qquad \text{Let } x_1 = 4 \text{ and } y_1 = 4.$$

$$2(y - 4) = 3(x - 4) \qquad \text{Multiply both sides by 2.}$$

$$2y - 8 = 3x - 12$$

$$2y = 3x - 4 \qquad \text{Add 8 to both sides.}$$

$$y = \frac{3}{2}x - 2 \qquad \text{Divide both sides by 2.}$$

$$f(x) = \frac{3}{2}x - 2 \qquad \text{Function notation.}$$ ■

Forms of Linear Equations

$Ax + By = C$	**Standard form** of a linear equation A and B are not both 0.
$y = mx + b$	**Slope–intercept form** of a linear equation The slope is m and the y-intercept is b.
$y - y_1 = m(x - x_1)$	**Point–slope form** of a linear equation The slope is m and (x_1, y_1) is a point on the line.
$y = c$	**Horizontal line** The slope is 0 and the y-intercept is c.
$x = c$	**Vertical line** The slope is undefined and the x-intercept is c.

Parallel and Perpendicular Lines

Nonvertical parallel lines have the same slope.

The product of the slopes of two nonvertical perpendicular lines is -1.

MENTAL MATH

State the slope and the y-intercept for each line for the given equation.

1. $y = -4x + 12$

2. $y = \frac{2}{3}x - \frac{7}{2}$

3. $y = 5x$

4. $y = -x$

5. $y = \frac{1}{2}x + 6$

6. $y = -\frac{2}{3}x + 5$

Decide whether the lines are parallel, perpendicular, or neither.

7. $y = 12x + 6$
$y = 12x - 2$

8. $y = -5x + 8$
$y = -5x - 8$

9. $y = -9x + 3$
$y = \frac{3}{2}x - 7$

10. $y = 2x - 12$
$y = \frac{1}{2}x - 6$

EXERCISE SET 4.5

Use the slope–intercept form of the linear equation to write the equation of each line with given slope and y-intercept. See Example 1.

1. Slope -1; y-intercept 1

2. Slope $\frac{1}{2}$; y-intercept -6

3. Slope 2; y-intercept $\frac{3}{4}$

4. Slope -3; y-intercept $-\frac{1}{5}$

5. Slope $\frac{2}{7}$; y-intercept 0

6. Slope $-\frac{4}{5}$; y-intercept 0

Graph each line passing through the given point with the given slope. See Example 2.

7. Through (1, 3) with slope $\frac{3}{2}$

8. Through (−2, −4) with slope $\frac{2}{5}$

9. Through (0, 0) with slope 5

10. Through (−5, 2) with slope 2

11. Through (0, 7) with slope −1

12. Through (3, 0) with slope −3

Use the point–slope form of the linear equation to find the equation of each line with the given slope and passing through the given point. Then write the equation in standard form. See Example 3.

13. Slope 6; through (2, 2)

14. Slope 4; through (1, 3)

15. Slope −8; through (−1, −5)

16. Slope −2; through (−11, −12)

17. Slope $\frac{1}{2}$; through (5, −6)

18. Slope $\frac{2}{3}$; through (−8, 9)

Find the equation of the line through the given points. Write the equation using function notation. See Example 4.

19. Through (3, 2) and (5, 6)

20. Through (6, 2) and (8, 8)

21. Through (−1, 3) and (−2, −5)

22. Through (−4, 0) and (6, −1)

23. Through (2, 3) and (−1, −1)

24. Through (0, 0) and $\left(\frac{1}{2}, \frac{1}{3}\right)$

Find the equation of each line. See Example 5.

25. Vertical line through (0, 2)

26. Horizontal line through (1, 4)

27. Horizontal line through (−1, 3)

28. Vertical line through (−1, 3)

29. Vertical line through (−7, −2)

30. Horizontal line through (2, 0)

Find the equation of each line. Write the equation using function notation. See Examples 6 and 7.

31. Through (3, 8); parallel to $f(x) = 4x - 2$

32. Through (1, 5); parallel to $f(x) = 3x - 4$

33. Through (2, −5); perpendicular to $3y = x - 6$

34. Through (−4, 8); perpendicular to $2x - 3y = 1$

35. Through (−2, −3); parallel to $3x + 2y = 5$

36. Through (−2, −3); perpendicular to $3x + 2y = 5$

Find the equation of each line described. Write each equation in standard form.

37. With slope $-\frac{1}{2}$, through $\left(0, \frac{5}{3}\right)$

38. With slope $\frac{5}{7}$, through (0, −3)

39. Slope 1, through (−7, 9)

40. Slope 5, through (6, −8)

41. Through (10, 7) and (7, 10)

42. Through (5, −6) and (−6, 5)

43. Through (6, 7), parallel to the x-axis

44. Through (0, −5), parallel to the y-axis

45. Slope $-\frac{4}{7}$, through (−1, −2)

46. Slope $-\frac{3}{5}$, through (4, 4)

47. Slope 2; through (−2, 3)

48. Slope 3; through (−4, 2)

49. Through (1, 6) and (5, 2)

50. Through (2, 9) and (8, 6)

51. With slope $-\frac{1}{2}$; y-intercept 11

52. With slope −4; y-intercept $\frac{2}{9}$

53. Through (−7, −4) and (0, −6)

54. Through (2, −8) and (−4, −3)

55. Slope $-\frac{4}{3}$; through (−5, 0)

56. Slope $-\frac{3}{5}$; through (4, −1)

57. Vertical line; through $(-2, -10)$

58. Horizontal line; through $(1, 0)$

59. Through $(6, -2)$; parallel to the line $2x + 4y = 9$

60. Through $(8, -3)$; parallel to the line $6x + 2y = 5$

61. Slope 0; through $(-9, 12)$

62. Undefined slope; through $(10, -8)$

Solve.

63. A rock is dropped from the top of a 400-foot building. After 1 second, the rock is traveling 32 feet per second. After 3 seconds, the rock is traveling 96 feet per second. Let $R(x)$ be the rate of descent and x be the number of seconds since the rock was dropped.

a. Write a linear function that relates time x to rate $R(x)$. (Hint: Use the ordered pairs $(1, 32)$ and $(3, 96)$ and write an equation.)

b. Use this function to determine the rate of the rock 4 seconds after it was dropped.

64. The Whammo Company has learned that, by pricing a newly released Frisbee at \$6, sales will reach 2000 per day. Raising the price to \$8 will cause the sales to fall to 1500 per day. Assume that the ratio of change in price to change in daily sales is constant, and let x be the price of the Frisbee and S be number of sales.

a. Find the linear function $S(x)$ that models the price–sales relationship for this Frisbee. [Hint: The line must pass through $(6, 2000)$ and $(8, 1500)$.]

b. Predict the daily sales of Frisbees if the price is set at \$7.50.

65. Del Monte Fruit Company recently released a new pineapple sauce. By the end of its first year, profits on this product amounted to \$30,000. The anticipated profit for the end of the fourth year is \$66,000. The ratio of change in time to change in profit is constant. Let x be years and P be profit.

a. Write a linear function $P(x)$ that expresses profit as a function of time.

b. Use this function to predict the company's profit at the end of the seventh year.

c. Predict when the profit should reach \$126,000.

66. The equation relating Celsius temperature and Fahrenheit temperature is linear. Find this linear equation if the freezing point of water is 0°C or 32°F, and the boiling point is 100°C or 212°F. [Hint: This is equivalent to saying that the line passes through points $(0, 32)$ and $(100, 212)$.]

67. The relationship between temperatures on the Kelvin scale and the Celsius scale is modeled by a linear equation. Since 273 K equals 0°C, and 373 K is the same as 100°C, find the linear equation that models the Kelvin–Celsius relationship.

Skill Review

Find the value of each expression if $x = 1$, $y = -4$, and $z = 0$. See Section 1.7.

68. $\dfrac{x^2}{y^2 - zx}$

69. $\dfrac{2x}{zx - y^2}$

70. $\dfrac{|y| + z^2}{2x}$

71. $\dfrac{|x| + 3z}{2y}$

Solve and graph the solution. See Section 3.5.

72. $2x - 7 \leq 21$

73. $-3x + 1 > 0$

74. $5(x - 2) \geq 3(x - 1)$

75. $-2(x + 1) \leq -x + 10$

76. $\dfrac{x}{2} + \dfrac{1}{4} < \dfrac{1}{8}$

77. $\dfrac{x}{5} - \dfrac{3}{10} \geq \dfrac{x}{2} - 1$

4.6 Graphing Linear Inequalities

OBJECTIVES

Tape IA 8

1 Graph linear inequalities.

2 Graph the intersection or union of two linear inequalities.

1 Recall that the graph of a linear equation in two variables is the graph of all ordered pairs that satisfy the equation, and we determined that the graph is a line. Here we graph **linear inequalities** in two variables; that is, we graph all the ordered pairs that satisfy the inequality.

If the equal sign in a linear equation in two variables is replaced with an inequality symbol, the result is a linear inequality in two variables.

Examples of Linear Inequalities

$$3x + 5y \geq 6 \qquad 2x - 4y < -3$$
$$4x > 2 \qquad y \leq 5$$

To graph the linear inequality $x + y \leq 5$, recall that this inequality means

$$x + y = 5 \quad \text{or} \quad x + y < 5$$

The graph of $x + y = 5$ is a line. This line is called a **boundary** because it separates the plane into two **half-planes.** All points "above" the boundary line $x + y = 5$ have coordinates that satisfy the inequality $x + y > 5$, and all points "below" the line have coordinates that satisfy the inequality $x + y < 5$. Thus, the graph of $x + y \leq 5$ is the boundary line together with the half-plane below it.

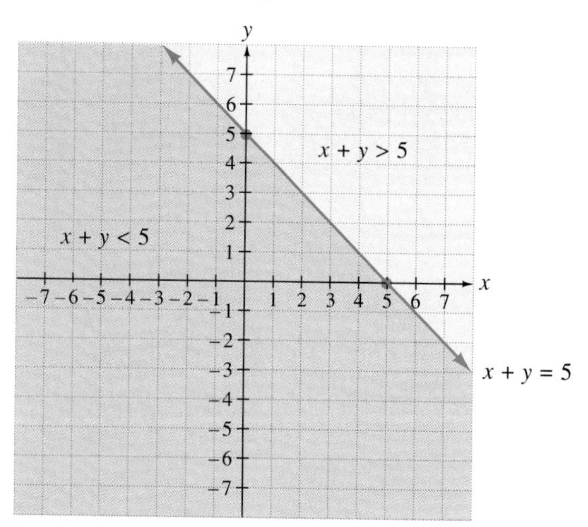

The following steps may be used to graph linear inequalities in two variables.

To Graph a Linear Inequality

Step 1 Graph the boundary line found by replacing the inequality sign with an equal sign. If the inequality sign is $<$ or $>$, graph a dashed line indicating that points on the line are not solutions of the inequality. If the inequality sign is $\leq$ or $\geq$, graph a solid line indicating that points on the line are solutions of the inequality.

Step 2 Choose a **test point not on the boundary line** and substitute the coordinates of this test point into the **original inequality.**

Step 3 If a true statement is obtained in Step 2, shade the half-plane that contains the test point. If a false statement is obtained, shade the half-plane that does not contain the test point.

EXAMPLE 1 Graph $2x - y < 6$.

Solution: First, the boundary line for this inequality is the graph of $2x - y = 6$. Graph a dashed boundary line because the inequality symbol is $<$. Next, choose a test point on either side of the boundary line. The point (0, 0) is not on the boundary line so we use this point. Replacing x with 0 and y with 0 in the **original inequality** $2x - y < 6$ leads to the following:

$$2x - y < 6$$

$$2(0) - 0 < 6 \quad \text{Let } x = 0 \text{ and } y = 0.$$

$$0 < 6 \quad \text{True.}$$

Because (0, 0) satisfies the inequality, so does every point on the same side of the boundary line as (0, 0). Shade the half-plane that contains (0, 0). The half-plane graph of the inequality is shown next.

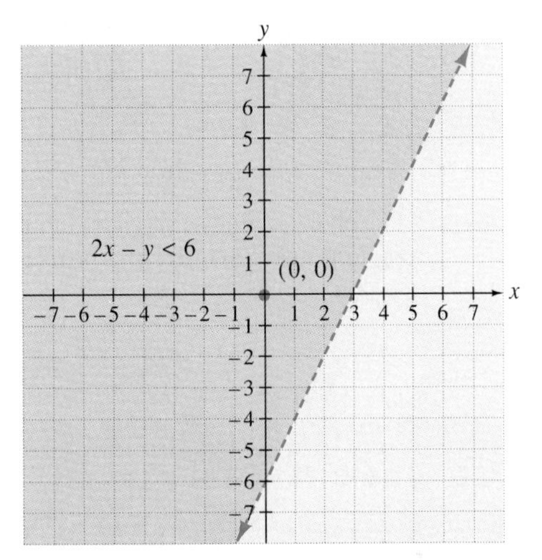

Every point in the shaded half-plane satisfies the original inequality. Notice that the inequality $2x - y < 6$ does not describe a function since its graph does not pass the vertical line test. ■

In general, linear inequalities of the form $Ax + By \leq C$ when A and B are not both 0 do not describe functions.

EXAMPLE 2 Graph $3x \geq y$.

Solution: First graph the boundary line $3x = y$. Graph a solid boundary line because the inequality symbol is $\geq$. Test a point not on the boundary line to determine which half-plane contains points that satisfy the inequality. We choose (0, 1) as our test point.

$$3x \geq y$$

$$3(0) \geq 1 \quad \text{Let } x = 0 \text{ and } y = 1.$$

$$0 \geq 1 \quad \text{False.}$$

This point does not satisfy the inequality, so the correct half-plane is on the opposite side of the boundary line from (0, 1). The graph of $3x \geq y$ is the boundary line together with the shaded region shown next.

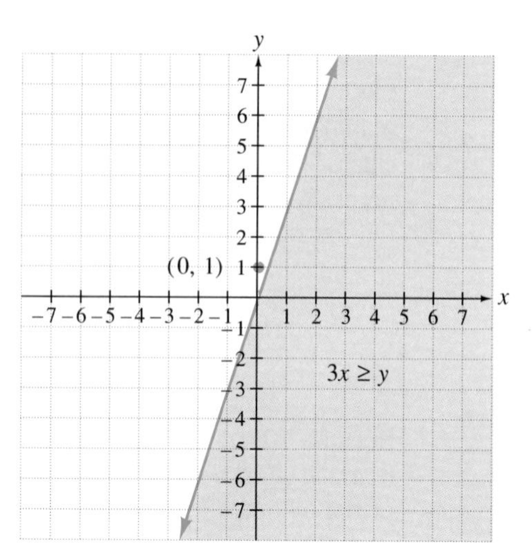

2 The intersection and the union of linear inequalities can also be graphed, as shown in the next two examples.

EXAMPLE 3 Graph the intersection of $x \geq 1$ and $y \geq 2x - 1$.

Solution: Graph each inequality. The intersection of the two graphs is all points common to both regions, as shown by the heaviest shading in the third graph.

$x \geq 1$

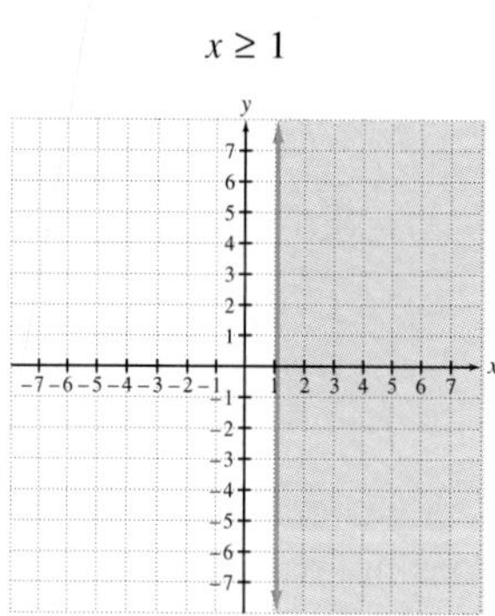

$y \geq 2x - 1$

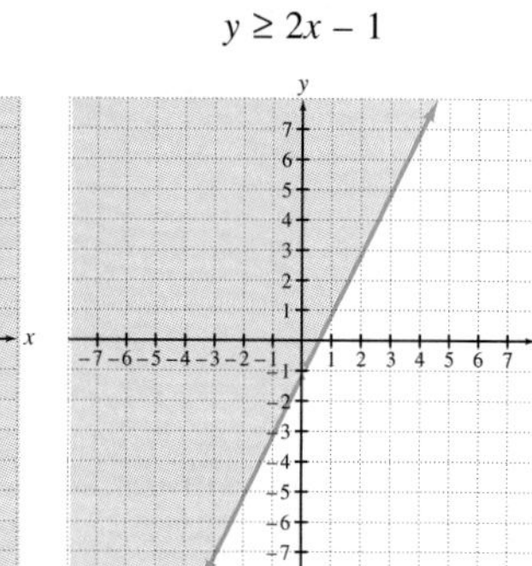

$x \geq 1$ and $y \geq 2x - 1$

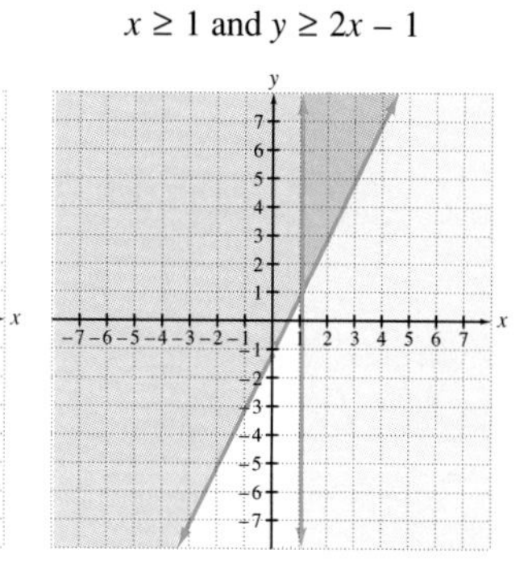

EXAMPLE 4 Graph the union of $x + \frac{1}{2}y \geq -4$ or $y \leq -2$.

Solution: Graph each inequality. The union of the two inequalities is both shaded regions, including the solid boundary lines shown in the third graph.

$x + \frac{1}{2}y \geq -4$

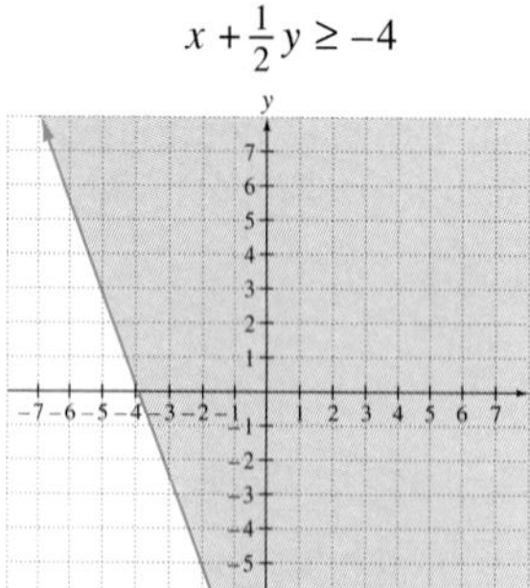

$y \leq -2$

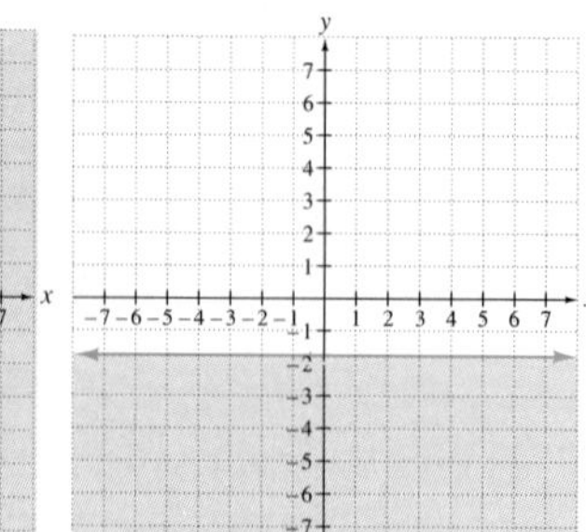

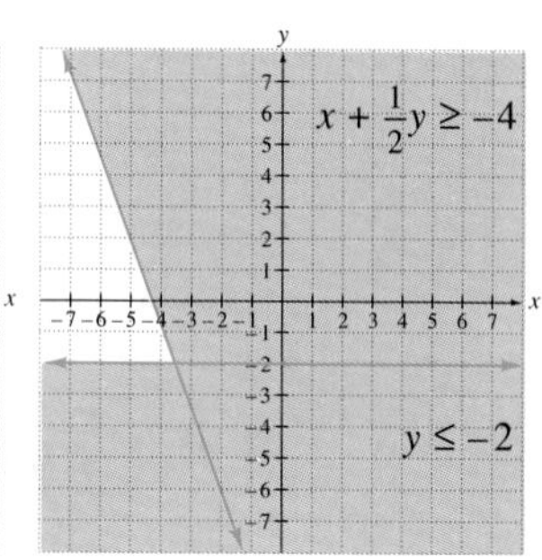

EXERCISE SET 4.6

Graph each inequality. See Examples 1 and 2.

1. $x < 2$

2. $x > -3$

3. $x - y \geq 7$

4. $3x + y \leq 1$

5. $3x + y > 6$

6. $2x + y > 2$

7. $y \leq -2x$

8. $y \leq 3x$

9. $2x + 4y \geq 8$

10. $2x + 6y \leq 12$

11. $5x + 3y > -15$

12. $2x + 5y < -20$

Graph each union or intersection. See Examples 3 and 4.

13. The intersection $x \geq 3$ and $y \leq -2$

14. The union $x \geq 3$ or $y \leq -2$

15. The union $x \leq -2$ or $y \geq 4$

16. The intersection $x \leq -2$ and $y \geq 4$

17. The intersection $x - y < 3$ and $x > 4$

18. The intersection $2x > y$ and $y > x + 2$

19. The union $x + y \leq 3$ or $x - y \geq 5$

20. The union $x - y \leq 3$ or $x + y > -1$

Graph each inequality.

21. $y \geq -2$

22. $y \leq 4$

23. $x - 6y < 12$

24. $x - 4y < 8$

25. $x > 5$

26. $y \geq -2$

27. $-2x + y \leq 4$

28. $-3x + y \leq 9$

29. $x - 3y < 0$

30. $x + 2y > 0$

31. $3x - 2y \leq 12$

32. $2x - 3y \leq 9$

33. The union of $x - y \geq 2$ or $y < 5$

34. The union of $x - y < 3$ or $x > 4$

35. The intersection of $x + y \leq 1$ and $y \leq -1$

36. The intersection of $y \geq x$ and $2x - 4y \geq 6$

37. The union of $2x + y > 4$ or $x \geq 1$

38. The union of $3x + y < 9$ or $y \leq 2$

39. The intersection of $x \geq -2$ and $x \leq 1$

40. The intersection of $x \geq -4$ and $x \leq 3$

41. The union of $x + y \leq 0$ or $3x - 6y \geq 12$

42. The intersection of $x + y \leq 0$ and $3x - 6y \geq 12$

43. The intersection of $2x - y > 3$ and $x \geq 0$

44. The union of $2x - y > 3$ or $x \geq 0$

Write the inequality whose graph is given.

45.

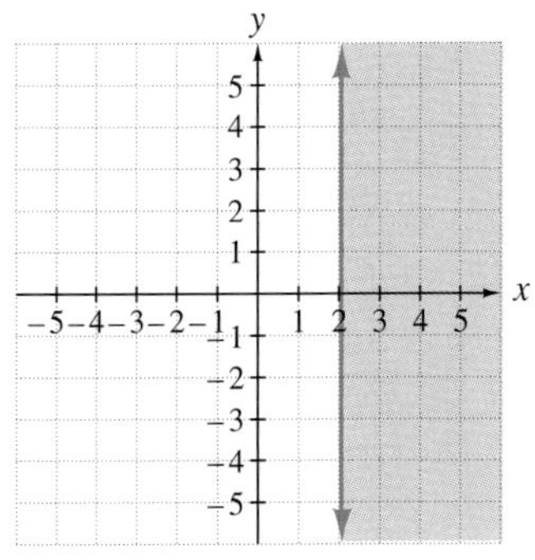

46.

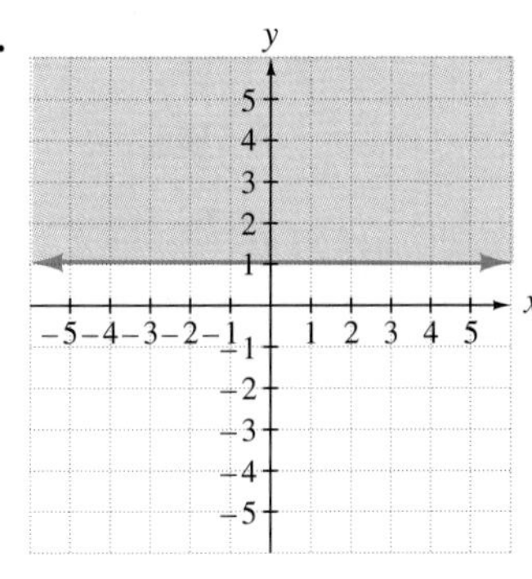

47.

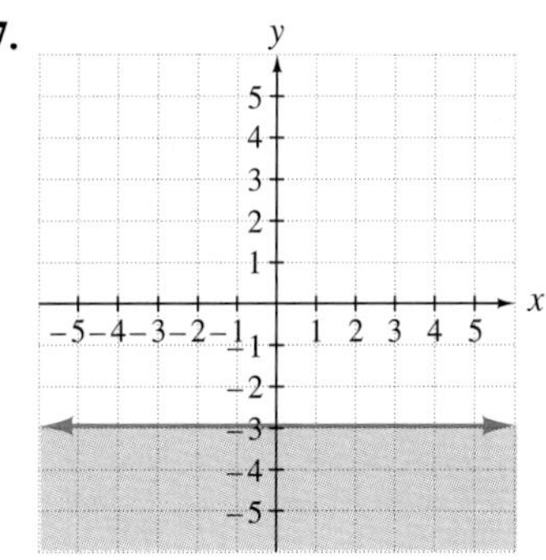

48.

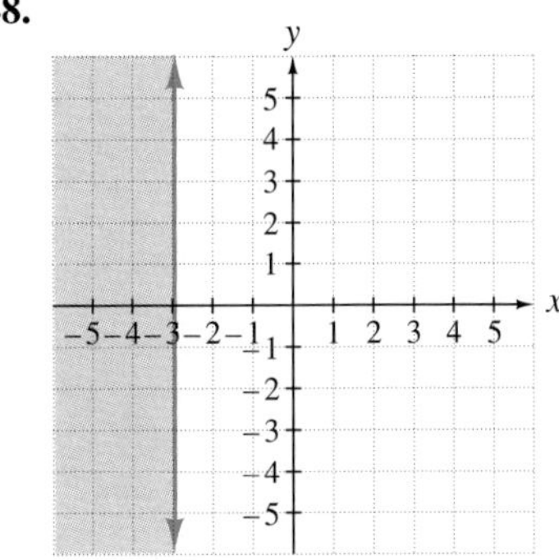

49.

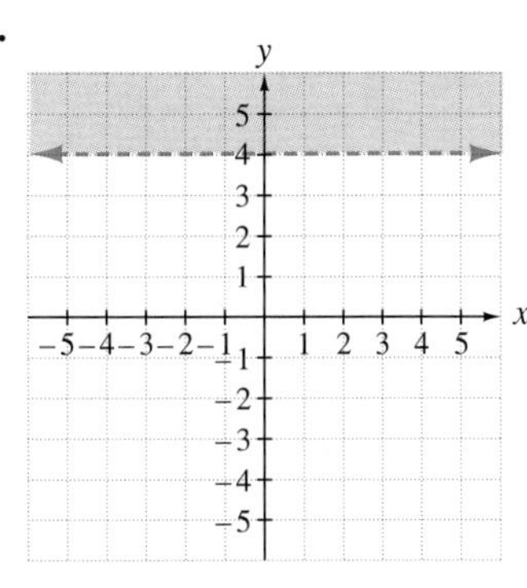

50.

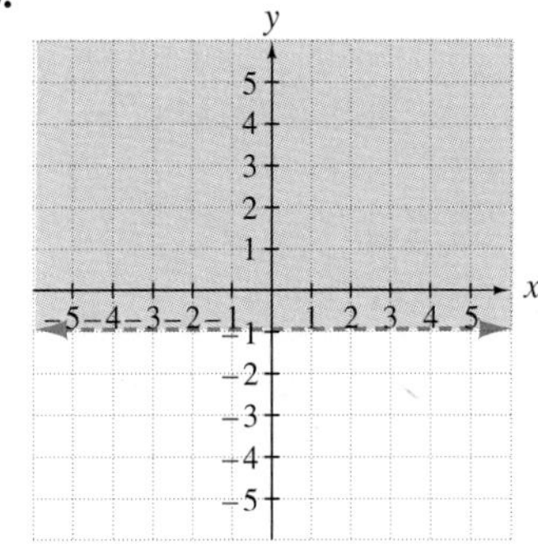

51.

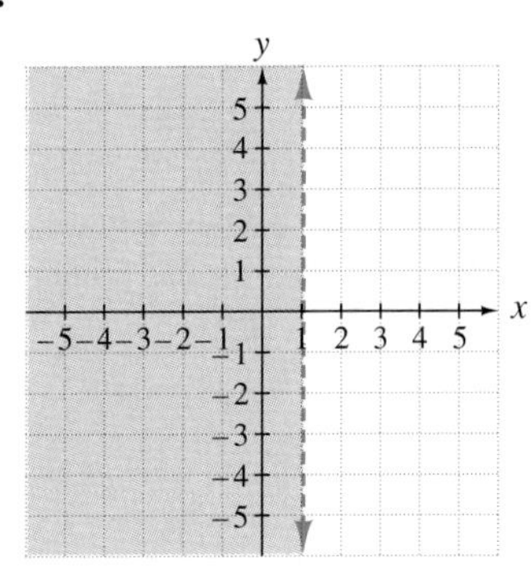

52.

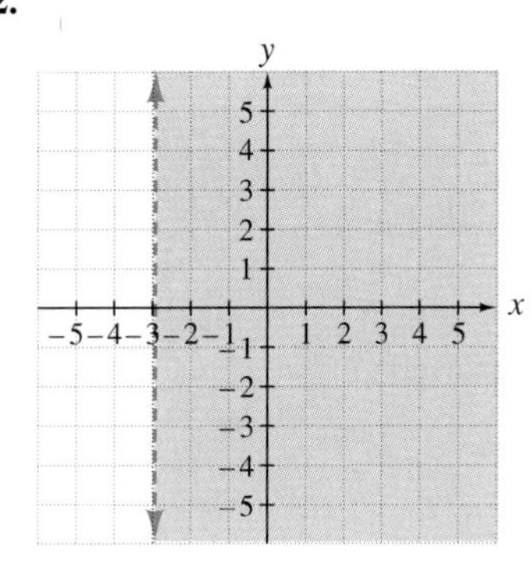

Solve.

53. Rheem Abo-Zahrah decides that she will study at most 20 hours every week and that she must work at least 10 hours every week. Let x represent the hours studying and y represent the hours working. Write two inequalities that model this situation and graph their intersection.

54. The movie and TV critic for the *New York Times* spends between 2 and 6 hours daily reviewing movies and fewer than 5 hours reviewing TV shows. Let x represent the hours watching movies and y represent the time spent watching TV. Write two inequalities that model this situation and graph their intersection.

55. Chris-Craft manufactures boats out of fiberglass and wood. Fiberglass hulls require 2 hours work, whereas wood hulls require 4 hours work. Employees work fewer than 40 hours a week. The following inequalities model these restrictions, where x represents the number of fiberglass hulls produced and y represents the number of wood hulls produced.

$$\begin{cases} x \geq 0 \\ y \geq 0 \\ 2x + 4y \leq 40 \end{cases}$$

Graph the intersection of these inequalities.

Skill Review

Evaluate each expression. See Sections 1.3 and 1.4.

56. 2^3

57. 3^2

58. -5^2

59. $(-5)^2$

60. $(-2)^4$

61. -2^4

62. $\left(\frac{3}{5}\right)^3$

63. $\left(\frac{2}{7}\right)^2$

The following pairs of triangles are similar. Find the unknown lengths. See Section 3.1.

64.

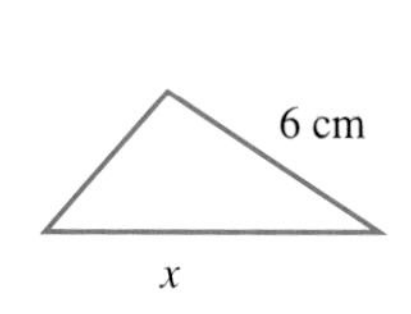

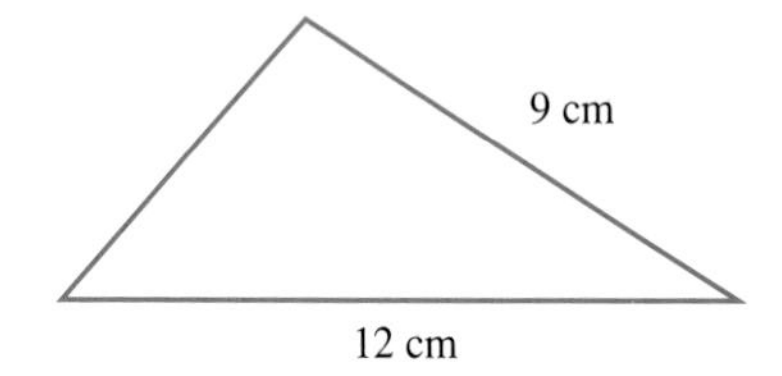

65.

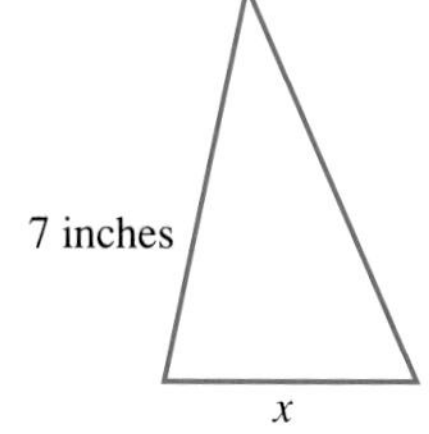

5 inches
3 inches

Find the domain and the range of each relation. Determine whether the relation is also a function. See Section 4.2.

66.

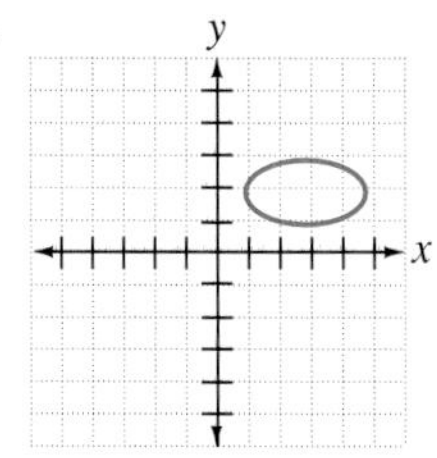

67.

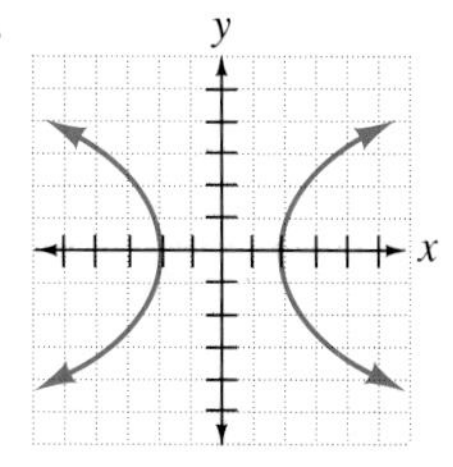

Writing in Mathematics

68. Explain a dashed boundary line in the graph of an inequality.

69. When the union of two inequalities is graphed, explain why all shaded regions are included, not just the overlap of the shaded regions.

70. After the boundary line is sketched, explain why we test a point on either side of this boundary in the original inequality.

CHAPTER 4 GLOSSARY

A **function** is a set of ordered pairs that assigns to each x-value exactly one y-value.

The notation $f(x)$ is called **function notation.**

The notation (x, y), where x and y represent numbers, is called an **ordered pair.**

The **rectangular coordinate system** consists of two number lines perpendicular to each other, intersecting at point 0 on each.

The **slope** of a line measures its steepness.

If a graph crosses the x-axis at $(a, 0)$, the **x-intercept** of the graph is a.

If a graph crosses the y-axis at $(0, b)$, the **y-intercept** of the graph is b.

CHAPTER 4 SUMMARY

(4.1)

Linear Equation in Two Variables

A linear equation in two variables is of the form

$$Ax + By = C$$

when A, B, and C are real numbers and A and B are not both 0.

The graph of a linear equation in two variables is a straight line.

(4.2)

Vertical Line Test

If a vertical line can be drawn so that it intersects a graph more than once, the graph is not the graph of a function.

(4.4)

The slope of a line through the points (x_1, y_1) and (x_2, y_2) is

$$m = \frac{y_2 - y_1}{x_2 - x_1} \quad \text{as long as } x_1 \neq x_2$$

(4.4) and (4.5)

Forms of Linear Equations

$Ax + By = C$	**Standard form** of a linear equation A and B are not both 0.
$y = mx + b$	**Slope–intercept form** of a linear equation The slope is m and the y-intercept is b.
$y - y_1 = m(x - x_1)$	**Point–slope form** of a linear equation The slope is m and (x_1, y_1) is a point on the line.
$y = c$	**Horizontal line** The slope is 0 and the y-intercept is c.
$x = c$	**Vertical line** The slope is undefined and the x-intercept is c.

Parallel and Perpendicular Lines

Nonvertical parallel lines have the same slope.
The product of the slopes of two nonvertical perpendicular lines is -1.

TO GRAPH A LINEAR INEQUALITY (4.6)

Step 1 Graph the boundary line.

Step 2 Shade the half-plane that contains solutions to the inequality.

CHAPTER 4 REVIEW

(4.1) *Determine whether each ordered pair is a solution to the given equation.*

1. $7x - 8y = 56$; (0, 56), (8, 0)

2. $-2x + 5y = 10$; (−5, 0), (1, 1)

3. $x = 13$; (13, 5), (13, 13)

4. $y = 2$; (7, 2), (2, 7)

Complete the ordered pairs so that each is a solution of the given equation.

5. $-2 + y = 6x$; (7,)

6. $y = 3x + 5$; (, −8)

Complete the ordered pairs so that each is a solution of the given equation; then plot the ordered pairs. Use a single coordinate system for each exercise.

7. $9 = -3x + 4y$
- **a.** (, 0)
- **b.** (, 3)
- **c.** (9,)

8. $y = -2x$
- **a.** (7,)
- **b.** (−7,)
- **c.** (0,)

9. $x = 2y$
- **a.** (, 0)
- **b.** (, 5)
- **c.** (, −5)

Determine whether each equation is linear or not. Then graph each equation.

10. $3x - y = 4$

11. $x - 3y = 2$

12. $y = |x| + 4$

13. $y = x^2 + 4$

14. $y = -\frac{1}{2}x + 2$

15. $y = -x + 5$

(4.2) *Find the domain and range of each relation. Also determine whether the relation is a function.*

16. $\left\{\left(-\frac{1}{2}, \frac{3}{4}\right), (6, 0.75), (0, -12), (25, 25)\right\}$

17. $\left\{\left(\frac{3}{4}, -\frac{1}{2}\right), (0.75, 6), (-12, 0), (25, 25)\right\}$

18.

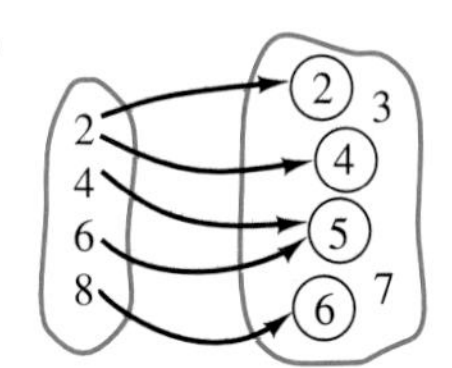

19.

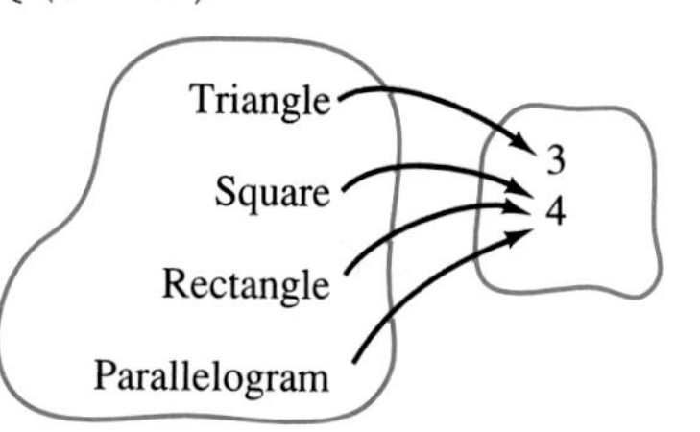

Find the domain and the range of each relation. Use the vertical line test to determine whether each graph is the graph of a function.

20.

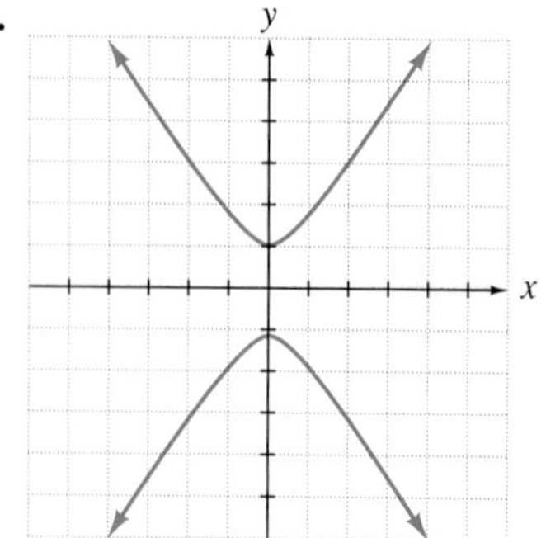

21.

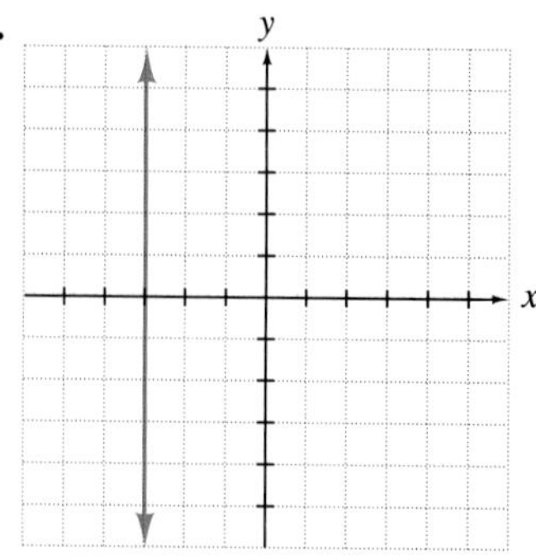

22.

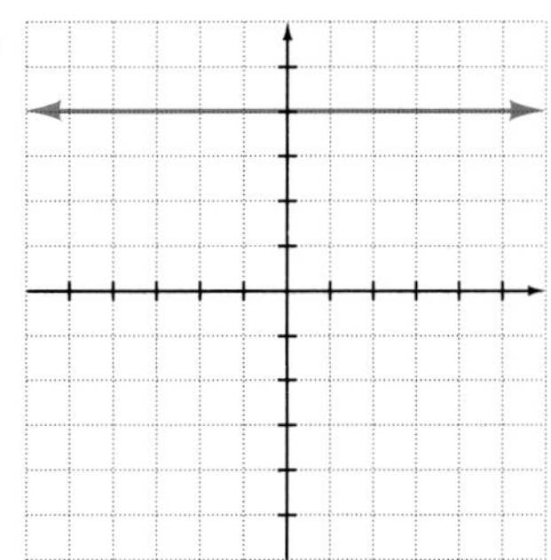

23.

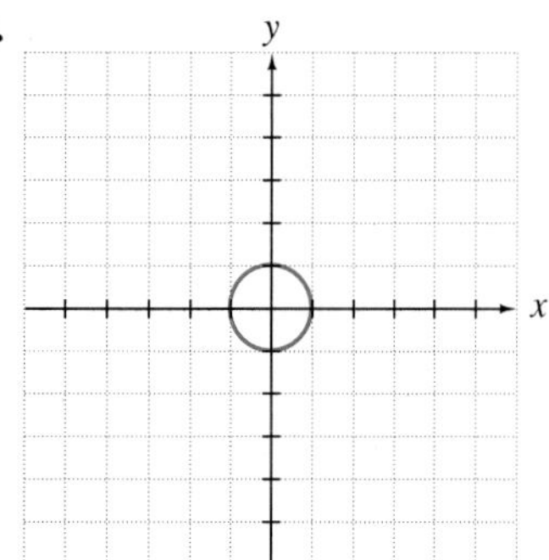

Given the following functions, find the indicated function value.

24. Given $f(x) = -2x + 6$, find **(a)** $f(0)$ **(b)** $f(-2)$ **(c)** $f\left(\frac{1}{2}\right)$

25. Given $h(x) = -5 - 3x$, find **(a)** $h(2)$ **(b)** $h(-3)$ **(c)** $h(0)$

26. Given $g(x) = x^2 + 12x$, find **(a)** $g(3)$ **(b)** $g(-5)$ **(c)** $g(0)$

27. Given $h(x) = 6 - x^2$, find **(a)** $h(-1)$ **(b)** $h(1)$ **(c)** $h(-4)$

The function $J(x) = 2.54x$ may be used to calculate the weight of an object on Jupiter J given its weight on Earth x.

28. If a person weights 150 pounds on Earth, find the equivalent weight on Jupiter.

29. A 2000-pound probe on Earth weighs how many pounds on Jupiter?

(4.3) *Graph each linear equation.*

30. $x - y = 1$

31. $x + y = 6$

32. $f(x) = \frac{1}{3}x - 4$

33. $f(x) = 5x + 8$

34. $x = 3y$

35. $y = -2x$

36. $x = 3$

37. $y = -2$

38. $2x - 3y = 6$

39. $4x - 3y = 12$

(4.4) *Find the slope of the line that goes through the given points.*

40. (2, 5) and (6, 8)

41. (4, 7) and (1, 2)

42. (1, 3) and (−2, −9)

43. (−4, 1) and (3, −6)

Find the slope and the y-intercept of each line.

44. $f(x) = 3x + 7$

45. $f(x) = \frac{1}{2}x - 2$

46. $2x + 3y = 5$

47. $4x - 5y = 9$

48. $y = 7$

49. $x = -3$

Determine whether the lines are parallel, perpendicular, or neither.

50. $x - y = -6$
$x + y = 3$

51. $3x + y = 7$
$-3x - y = 10$

(4.5) *Find an equation of the line satisfying the conditions given.*

52. Horizontal; through (3, −1)

53. Vertical; through (−2, −4)

54. Parallel to the line $x = 6$; through (−4, −3)

55. Slope 0; through (2, 5)

Find the equation of each line satisfying the conditions given. Write each equation using function notation, if possible.

56. Through (−3, 5); slope 3

57. Slope 2; through (5, −2)

58. Slope $-\frac{2}{3}$; y-intercept 4

59. Slope −1; y-intercept −2

60. Through (2, −6); parallel to $6x + 3y = 5$

61. Through (−4, −2); parallel to $3x + 2y = 8$

62. Through (−6, −1); perpendicular to $4x + 3y = 5$

63. Through (−4, 5); perpendicular to $2x - 3y = 6$

64. Through (−6, −1) and (−4, −2)

65. Through (−5, 3) and (−4, −8)

66. Through (−2, 3); perpendicular to $x = 4$

67. Through (−2, −5); parallel to $y = 8$

(4.6) *Graph each linear inequality.*

68. $3x + y > 4$

69. $\frac{1}{2}x - y < 2$

70. $5x - 2y \leq 9$

71. $3y \geq x$

72. $y < 1$

73. $x > -2$

74. Graph the union $y > 2x + 3$ or $x \leq -3$.

75. Graph the intersection $2x < 3y + 8$ and $y \geq -2$.

CHAPTER 4 TEST

Complete the ordered pairs for the following equations.

1. $12y - 7x = 5$; $(1, \quad)$

2. $y = 17$; $(-4, \quad)$

Determine whether the ordered pairs are solutions to the equations.

3. $x - 2y = 3$; $(1, 1)$

4. $2x + 3y = 6$; $(0, -2)$

Find the slopes of the following lines.

5. $-3x + 5 = y$

6. Through $(6, -5)$ and $(-1, 2)$

Graph the following.

7. $f(x) = -2x + 8$

8. $f(x) = \frac{1}{4}x + \frac{5}{4}$

9. $x - y \geq -2$

10. $y \geq -4x$

11. $5x - 7y = 10$

12. $2x - 3y > -6$

13. $6x + y > -1$

14. $y = -1$

15. $x - 3 = 0$

16. $5x - 3y = 15$

17. $y = -x^2$

18. $y = 3|x| + 2$

Find the domain and range of each relation. Which of the following graphs are graphs of functions?

19.

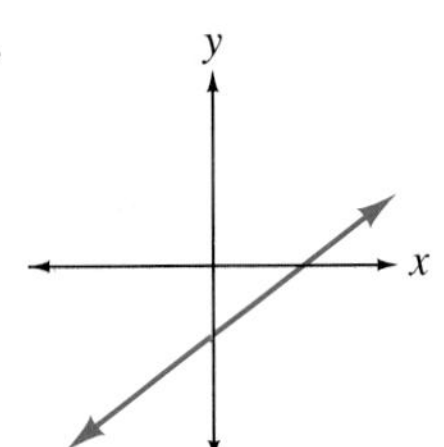

20.

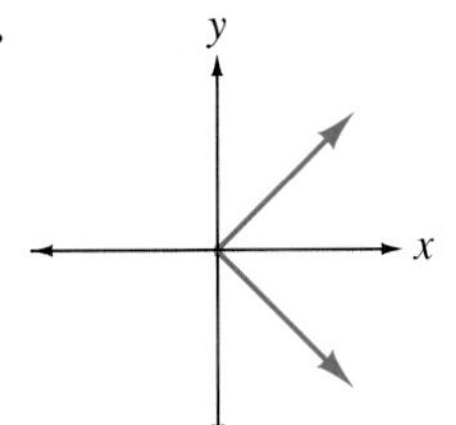

21.

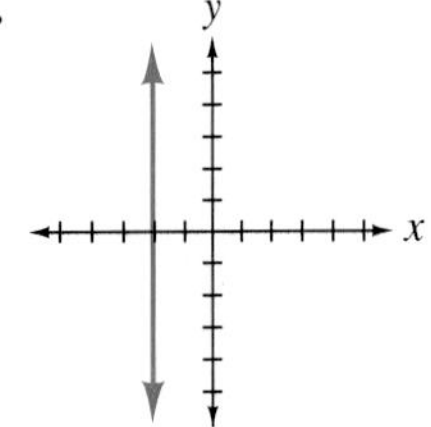

Given the following functions, find the indicated function values.

22. $f(x) = 2x - 4$ **(a)** $f(-2)$ **(b)** $f(0.2)$ **(c)** $f(0)$

23. $h(x) = x^3 - x$ **(a)** $h(-1)$ **(b)** $h(0)$ **(c)** $h(4)$

24. $g(x) = 6$ **(a)** $g(0)$ **(b)** $g(a)$ **(c)** $g(242)$

Find equations of the following lines. Write the equation using function notation.

25. Through the origin and $(6, -7)$

26. Through $(2, -5)$ and $(1, 3)$

27. Through $(-5, -1)$ and parallel to $y = 7$

28. With slope $\frac{1}{8}$; through $(0, 12)$

29. Through $(-1, 2)$; perpendicular to $3x - y = 4$

30. Parallel to $2y + x = 3$; through $(3, -2)$

The function $C(d) = \pi d$ may be used to find the circumference C of a circle given its diameter d.

31. Find the circumference of a circle whose diameter is 5 centimeters.

32. Find the circumference of a circle whose radius is 8 meters.

d C

CHAPTER 4 CUMULATIVE REVIEW

1. Find the product of $\frac{2}{15}$ and $\frac{5}{13}$. Write the answer in lowest terms.

2. Find the following roots.

a. $\sqrt[3]{27}$ **b.** $\sqrt[5]{1}$ **c.** $\sqrt[4]{16}$

3. Write an algebraic expression that represents each of the following phrases. Let x represent the unknown number.

a. The sum of a number and 3

b. The product of 3 and a number

c. Twice a number

d. 10 decreased by a number

e. 7 more than 5 times a number

4. Find each sum.

a. $-3 + (-7)$ **b.** $5 + (+12)$

c. $(-1) + (-20)$ **d.** $-2 + (-10)$

5. Simplify each expression.

a. $\dfrac{(-12)(-3) + 4}{-7 - (-2)}$

b. $\dfrac{2(-3)^2 - 20}{-5 + 4}$

6. Find the numerical coefficient.

a. $-3y$ **b.** $22z^4$ **c.** $\frac{2}{y}$ **d.** $-x$

7. Solve for x: $\frac{5}{2}x = 15$

8. A glacier is a giant mass of rocks and ice that flows downhill like a river. Portage Glacier in Alaska is about 6 miles long and moves 400 feet per year. Icebergs are created when the front end of the glacier flows into Portage Lake. How long does it take for ice at the head (beginning) of the glacier to reach the lake?

9. The length of a rectangular sign is 2 feet less than three times its width. Find the dimensions if the perimeter is 28 feet.

10. The following bar graph shows Disney's top five animated films before 1993. This graph is determined by U.S. box office revenue.

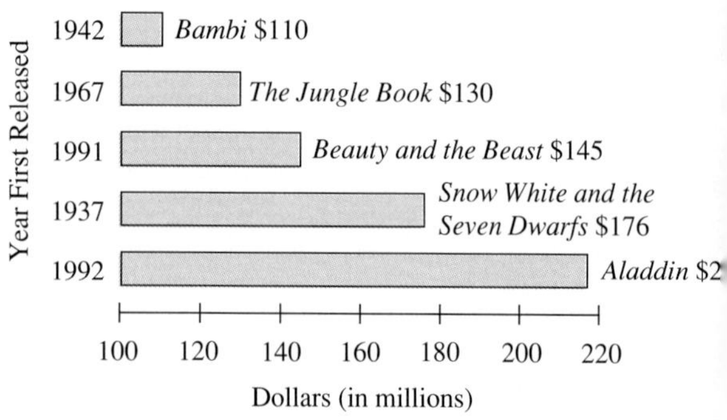

Source: *USA Today*, 1994.

a. Which Disney animated film generated the most income for Disney?
b. How much more money did the film *Aladdin* generate than *Beauty and the Beast?*

11. If triangle ABC is similar to triangle QRS, find the missing length r.

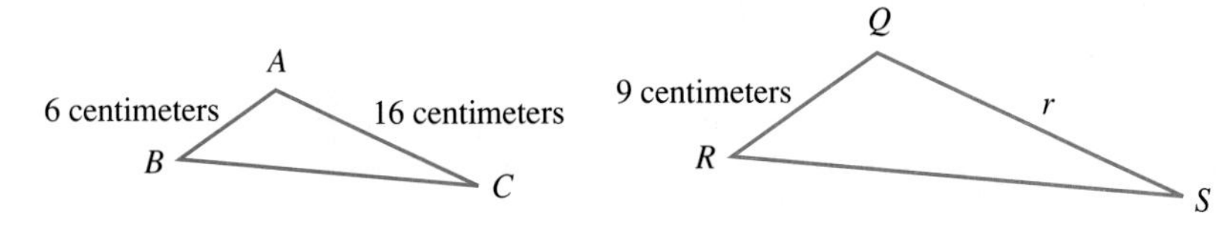

12. Write each percent as an equivalent decimal.
  a. 35% b. 89.5% c. 150%
13. Marie Antonio, a bicycling enthusiast, rode her ten-speed at an average speed of 18 miles per hour on level roads and then slowed down to an average of 10 miles per hour on the hilly roads of the trip. If she covered a distance of 98 miles, how long did the entire trip take if traveling the level roads took the same time as traveling the hilly roads?
14. Solve $|2x| + 5 = 7$.
15. Solve for x: $-(x - 3) + 2 < 3(2x - 5) + x$.
16. Graph the linear equation $2x + y = 5$.
17. Which of the following relations are also functions?
  a. $\{(-1, -1), (2, 3), (7, 3), (8, 6)\}$
  b. $\{(0, -2), (1, 5), (0, 3), (7, 7)\}$
18. Graph $x - 3y = 6$ by plotting intercept points.
19. Find the slope of the line through $(-1, -2)$ and $(2, 4)$. Graph the line.
20. Find the equation of the line through $(2, 5)$ and $(-3, 4)$. Write the equation using function notation.
21. Graph $2x - y < 6$.

CHAPTER 5

Exponents and Polynomials

Many people enjoy chain letters, but certain chain letters violate laws of the U.S. Postal Service. Subtle reasoning involving exponential growth justifies the law.

INTRODUCTION

Linear equations are important for solving problems. They are not sufficient, however, to solve all problems. If, for example, we want to express the area of a square field whose side has length s, a linear expression will not do. Many nonlinear expressions are written with exponents, a shorthand notation for repeated multiplication. The first two sections of this chapter refresh your skills working with exponents. The next sections are devoted to operations on polynomials. Polynomials model many real-world phenomena, and hence your ability to work with them can apply to solving many problems. We conclude the chapter with additional concepts relating to functions, particularly polynomial functions.

5.1 Exponents

OBJECTIVES

Tape BA 9

1. Evaluate a number raised to a power.
2. Use the product rule for exponents.
3. Use the power rule for exponents.
4. Use power rules for products and quotients.
5. Use the quotient rule for exponents, and define a number raised to the 0 power.

1 As we reviewed in Section 1.4, an exponent is shorthand notation for repeated factors. For example, $2 \cdot 2 \cdot 2 \cdot 2 \cdot 2$ can be written as 2^5.

$$\underbrace{5 \cdot 5 \cdot 5 \cdot 5 \cdot 5 \cdot 5}_{\text{6 factors of 5}} = 5^6 \quad \text{and} \quad \underbrace{(-3) \cdot (-3) \cdot (-3) \cdot (-3)}_{\text{4 factors of } -3} = (-3)^4$$

The **base** of an exponential expression is the repeated factor. The **exponent** is the number of times that the base is a factor.

5^6 ← exponent (6), base (5) $(-3)^4$ ← exponent (4), base (−3)

> **Definition of a^n**
>
> If a is a real number and n is a positive integer, then **a raised to the nth power,** written a^n, is the product of n factors of a.
>
> $$a^n = \underbrace{a \cdot a \cdot a \cdot a \cdot a \cdot \cdots \cdot a}_{n \text{ factors of } a}$$

EXAMPLE 1 Evaluate each expression.

a. 2^3 **b.** 3^1 **c.** $(-4)^2$ **d.** -4^2 **e.** $\left(\frac{2}{3}\right)^3$ **f.** $2 \cdot 4^2$

Solution: **a.** $2^3 = 2 \cdot 2 \cdot 2 = 8$

b. To raise 3 to the first power means that 3 will be used as a factor only once. Therefore, $3^1 = 3$. Also, when no exponent is shown, the exponent is assumed to be 1.

c. $(-4)^2 = (-4)(-4) = 16$

d. $-4^2 = -(4 \cdot 4) = -16$

e. $\left(\frac{2}{3}\right)^3 = \frac{2}{3} \cdot \frac{2}{3} \cdot \frac{2}{3} = \frac{8}{27}$

f. $2 \cdot 4^2 = 2 \cdot 4 \cdot 4 = 32$ ■

Notice how similar -4^2 is to $(-4)^2$. The difference between the two is the parentheses. In $(-4)^2$ the parentheses tells us that the base, or repeated factor, is -4. In -4^2 only 4 is the base.

HELPFUL HINT

Be careful when identifying the base of an exponent.

$(-3)^2$	-3^2	$2 \cdot 3^2$
Base is -3	Base is 3	Base is 3
$(-3)^2 = (-3)(-3) = 9$	$-3^2 = -(3 \cdot 3) = -9$	$2 \cdot 3^2 = 2 \cdot 3 \cdot 3 = 18$

2 Exponents used with variables have the same meaning as they do with numbers. If x is a real number and n is a positive integer, then x^n is the product of n factors of x.

$$x^n = \underbrace{x \cdot x \cdot x \cdot x \cdot x \cdot \cdots \cdot x}_{n \text{ factors of } x}$$

Exponential expressions can be multiplied, divided, added, subtracted, and themselves raised to powers. Suppose we multiply $5^4 \cdot 5^3$.

$$\begin{aligned} 5^4 \cdot 5^3 &= \underbrace{(5 \cdot 5 \cdot 5 \cdot 5)}_{4 \text{ factors of } 5} \cdot \underbrace{(5 \cdot 5 \cdot 5)}_{3 \text{ factors of } 5} \\ &= \underbrace{5 \cdot 5 \cdot 5 \cdot 5 \cdot 5 \cdot 5 \cdot 5}_{7 \text{ factors of } 5} \\ &= 5^7 \end{aligned}$$

Or we might multiply x^2 by x^3:

$$\begin{aligned} x^2 \cdot x^3 &= (x \cdot x) \cdot (x \cdot x \cdot x) \\ &= x \cdot x \cdot x \cdot x \cdot x \\ &= x^5 \end{aligned}$$

In both cases, notice that the result is exactly the same if the exponents are added.

$$5^4 \cdot 5^3 = 5^{4+3} = 5^7 \qquad \text{and} \qquad x^2 \cdot x^3 = x^{2+3} = x^5$$

The following property states this result.

Product Rule for Exponents

If m and n are positive integers and a is a real number, then

$$a^m \cdot a^n = a^{m+n}$$

In other words, to multiply two exponential expressions with a **common base,** keep the base and add the exponents.

EXAMPLE 2 Simplify each product.

a. $4^2 \cdot 4^5$ **b.** $x^2 \cdot x^5$ **c.** $y^3 \cdot y$ **d.** $y^3 \cdot y^2 \cdot y^7$ **e.** $(-5)^7 \cdot (-5)^8$

Solution: **a.** $4^2 \cdot 4^5 = 4^{2+5} = 4^7$

b. $x^2 \cdot x^5 = x^{2+5} = x^7$

c. $y^3 \cdot y = y^3 \cdot y^1$ Recall that if no exponent is written, it is assumed to be 1.

$= y^{3+1}$

$= y^4$

d. $y^3 \cdot y^2 \cdot y^7 = y^{3+2+7} = y^{12}$

e. $(-5)^7 \cdot (-5)^8 = (-5)^{7+8} = (-5)^{15}$. Although we do not want to evaluate $(-5)^{15}$, we can simplify this expression. Because $(-5)^{15}$ is the product of an odd number of negative numbers, the product will be negative, so that

odd number ↓

$(-5)^{15}$ can also be written as -5^{15}.

Both expressions have the same value. ■

EXAMPLE 3 Simplify $(2x^2)(-3x^5)$.

Solution: Recall that $2x^2$ means $2 \cdot x^2$ and $-3x^5$ means $-3 \cdot x^5$.

$(2x^2)(-3x^5) = 2 \cdot x^2 \cdot -3 \cdot x^5$ Remove parentheses.

$= 2 \cdot -3 \cdot x^2 \cdot x^5$ Group factors with like bases.

$= -6x^7$ Multiply. ■

3 Exponential expressions can themselves be raised to powers. Let's try to discover a rule that simplifies an expression like $(7^2)^3$. By the definition of a^n,

$$(7^2)^3 = \underbrace{(7^2)(7^2)(7^2)}_{3 \text{ factors of } 7^2}$$

which can be simplified by the product rule by adding exponents. Then

$$(7^2)^3 = (7^2)(7^2)(7^2) = 7^{2+2+2} = 7^6$$

Notice that the result is exactly the same if we multiply the exponents.

$$(7^2)^3 = 7^{2 \cdot 3} = 7^6$$

The following property states this result.

Power Rule for Exponents

If m and n are positive integers and a is a real number, then

$$(a^m)^n = a^{mn}$$

In other words, to raise an exponential expression to a power, keep the base and multiply the exponents.

EXAMPLE 4 Simplify each of the following expressions.

a. $(x^2)^5$ **b.** $(y^8)^2$ **c.** $[(-5)^3]^4$

Solution: **a.** $(x^2)^5 = x^{2\cdot5} = x^{10}$

b. $(y^8)^2 = y^{8\cdot2} = y^{16}$

c. $[(-5)^3]^4 = (-5)^{12}$. Because $(-5)^{12}$ is the product of an even number of negative numbers, the product is a positive number, so that

$$\underset{\uparrow \text{ even number}}{(-5)^{12}} \quad \text{can be written as} \quad 5^{12}$$

Both expressions have the same value. ■

4 When a base contains more than one factor, the definition of a^n still applies. To simplify $(x^2y)^3$, for example,

$$
\begin{aligned}
(x^2y)^3 &= \underbrace{(x^2y)(x^2y)(x^2y)}_{3 \text{ factors of } x^2y} \\
&= (x^2 \cdot x^2 \cdot x^2)(y^1 \cdot y^1 \cdot y^1) && \text{Group common bases.} \\
&= (x^{2+2+2})(y^{1+1+1}) && \text{Add exponents of common bases.} \\
&= x^6y^3 && \text{Simplify.}
\end{aligned}
$$

The result is exactly the same if each factor of the base is raised to the third power.

$$(x^2y)^3 = (x^2)^3 \cdot y^3 = x^6y^3$$

The definition of a^n applies as well when the base is a quotient. To simplify an expression like $\left(\frac{x}{7}\right)^4$,

$$
\begin{aligned}
\left(\frac{x}{7}\right)^4 &= \underbrace{\left(\frac{x}{7}\right)\left(\frac{x}{7}\right)\left(\frac{x}{7}\right)\left(\frac{x}{7}\right)}_{4 \text{ factors of } \frac{x}{7}} \\
&= \frac{x \cdot x \cdot x \cdot x}{7 \cdot 7 \cdot 7 \cdot 7} && \text{Multiply fractions.} \\
&= \frac{x^4}{7^4} && \text{Simplify.}
\end{aligned}
$$

Notice that the result is exactly the same if we raise both the numerator and the denominator of the base to the fourth power.

$$\left(\frac{x}{7}\right)^4 = \frac{x^4}{7^4}$$

In general, we have the following rules.

Power Rules for Products and Quotients

If n is a positive integer and a, b, and c are real numbers, then

$$(ab)^n = a^n b^n \quad \text{and} \quad \left(\frac{a}{c}\right)^n = \frac{a^n}{c^n}$$

as long as c is not 0.

In other words, to raise a product to a power, raise each factor of the product to the power. Also, to raise a quotient to a power, raise both the numerator and the denominator to the power.

EXAMPLE 5 Simplify each expression.

a. $(st)^4$ **b.** $\left(\frac{m}{n}\right)^7$ **c.** $(2a)^3$ **d.** $(-5x^2y^3z)^2$ **e.** $\left(\frac{2x^4}{3y^5}\right)^4$

Solution:

a. $(st)^4 = s^4 \cdot t^4 = s^4t^4$ Use the power rule for a product.

b. $\left(\frac{m}{n}\right)^7 = \frac{m^7}{n^7}, n \neq 0$ Use the power rule for a quotient.

c. $(2a)^3 = 2^3 \cdot a^3 = 8a^3$ Use the power rule for a product.

d. $(-5x^2y^3z)^2 = (-5)^2 \cdot (x^2)^2 \cdot (y^3)^2 \cdot (z^1)^2$ Use the power rule for a product.

$= 25x^4y^6z^2$

e. $\left(\frac{2x^4}{3y^5}\right)^4 = \frac{2^4 \cdot (x^4)^4}{3^4 \cdot (y^5)^4}$ Use the power rules for products and quotients.

$= \frac{16x^{16}}{81y^{20}}, y \neq 0$ Use the power rule for exponents.

■

5 Another pattern for simplifying exponential expressions relates to quotients.

To simplify an expression like $\frac{x^5}{x^3}$, in which the numerator and the denominator have a common base, we can divide the numerator and the denominator by common factors. For the remainder of this section, assume that denominators are not 0.

$$\frac{x^5}{x^3} = \frac{x \cdot x \cdot x \cdot x \cdot x}{x \cdot x \cdot x}$$

$$= \frac{x \cdot x \cdot x \cdot x \cdot x}{x \cdot x \cdot x}$$

$$= x \cdot x = x^2$$

Notice that the result is exactly the same if we subtract exponents of the common base.

$$\frac{x^5}{x^3} = x^{5-3} = x^2$$

The quotient rule for exponents states this result in a general way.

Quotient Rule for Exponents

If m and n are positive integers and a is a real number, then

$$\frac{a^m}{a^n} = a^{m-n}$$

as long as a is not 0.

In other words, to divide one exponential expression by another with a common base, keep the base and subtract exponents.

EXAMPLE 6 Simplify each quotient.

a. $\frac{x^5}{x^2}$ **b.** $\frac{4^7}{4^3}$ **c.** $\frac{(-3)^5}{(-3)^2}$ **d.** $\frac{2x^5y^2}{xy}$

Solution:

a. $\frac{x^5}{x^2} = x^{5-2} = x^3$ Use the quotient rule.

b. $\frac{4^7}{4^3} = 4^{7-3} = 4^4 = 256$ Use the quotient rule.

c. $\frac{(-3)^5}{(-3)^2} = (-3)^3 = -27$

d. Begin by grouping common bases.

$$\frac{2x^5y^2}{xy} = 2 \cdot \frac{x^5}{x^1} \cdot \frac{y^2}{y^1}$$

$$= 2 \cdot (x^{5-1}) \cdot (y^{2-1})$$ Use the quotient rule.

$$= 2x^4y^1 \quad \text{or} \quad 2x^4y$$ ■

Let's look at one more case. To simplify $\frac{x^3}{x^3}$, we use the quotient rule and subtract exponents.

$$\frac{x^3}{x^3} = x^{3-3} = x^0$$

But our definition of a^n does not include the possibility that n might be 0. What is the meaning when 0 is an exponent? To find out, simplify $\frac{x^3}{x^3}$ by dividing the numerator and denominator by common factors.

$$\frac{x^3}{x^3} = \frac{x \cdot x \cdot x}{x \cdot x \cdot x} = 1$$

Since $\frac{x^3}{x^3} = x^0$ and $\frac{x^3}{x^3} = 1$, we conclude that $x^0 = 1$ as long as x is not 0.

Zero Exponent

$a^0 = 1$, as long as a is not 0.

EXAMPLE 7 Simplify the following expressions.

a. 3^0 **b.** $(ab)^0$ **c.** $(-5)^0$ **d.** -5^0

Solution: **a.** $3^0 = 1$

b. Assume that neither a nor b is zero.

$$(ab)^0 = a^0 \cdot b^0 = 1 \cdot 1 = 1$$

c. $(-5)^0 = 1$ **d.** $-5^0 = -1 \cdot 5^0 = -1 \cdot 1 = -1$ ■

HELPFUL HINT

These examples will remind you of the differences between adding and multiplying terms.

Addition	*Multiplication*
$5x^3 + 3x^3 = (5 + 3)x^3 = 8x^3$	$(5x^3)(3x^3) = 5 \cdot 3 \cdot x^3 \cdot x^3 = 15x^{3+3} = 15x^6$
$7x + 4x^2 = 7x + 4x^2$	$(7x)(4x^2) = 7 \cdot 4 \cdot x \cdot x^2 = 28x^{1+2} = 28x^3$

In the next example, exponential expressions are simplified using two or more of the exponent rules presented in this section.

EXAMPLE 8 Simplify the following:

a. $\left(\dfrac{-5x^2}{y^3}\right)^2$ **b.** $\dfrac{(x^3)^4x}{x^7}$ **c.** $\dfrac{(2x)^5}{x^3}$ **d.** $\dfrac{(a^2b)^3}{a^3b^2}$

Solution: **a.** Use the power rules for products and quotients, then use the power rule for exponents.

$$\left(\frac{-5x^2}{y^3}\right)^2 = \frac{(-5)^2(x^2)^2}{(y^3)^2} = \frac{25x^4}{y^6}$$

b. $\dfrac{(x^3)^4x}{x^7} = \dfrac{x^{12} \cdot x}{x^7} = \dfrac{x^{12+1}}{x^7} = \dfrac{x^{13}}{x^7} = x^{13-7} = x^6$

c. Use the power rules for products and quotients, then use the quotient rule.

$$\frac{(2x)^5}{x^3} = \frac{2^5 \cdot x^5}{x^3} = 2^5 \cdot x^{5-3} = 32x^2$$

d. Begin by applying the power rule for a product to the numerator.

$$\frac{(a^2b)^3}{a^3b^2} = \frac{(a^2)^3 \cdot b^3}{a^3 \cdot b^2}$$

$= \dfrac{a^6b^3}{a^3b^2}$ Use the power rule for exponents.

$= a^{6-3}b^{3-2}$ Use the quotient rule.

$= a^3b^1$ or a^3b ■

MENTAL MATH

State the base and the exponent for each of the following expressions.

1. 3^2 **2.** 5^4 **3.** $(-3)^6$
4. -3^7 **5.** -4^2 **6.** $(-4)^3$
7. $2 \cdot 5^3$ **8.** $9y^2$

EXERCISE SET 5.1

Evaluate each expression. See Example 1.

1. 7^2 **2.** -3^2 **3.** $(-5)^1$ **4.** $(-3)^2$
5. -2^4 **6.** -4^3 **7.** $(-2)^4$ **8.** $(-4)^3$

Simplify each expression. See Examples 2 and 3.

9. $x^2 \cdot x^5$ **10.** $y^2 \cdot y$ **11.** $(5y^4)(3y)$ **12.** $(-2z^3)(-2z^2)$
13. $(4z^{10})(-6z^7)(z^3)$ **14.** $(12x^5)(-x^6)(x^4)$

Simplify each expression. See Examples 4 and 5.

15. $(pq)^7$ **16.** $(4s)^3$ **17.** $\left(\frac{m}{n}\right)^9$ **18.** $\left(\frac{xy}{7}\right)^2$
19. $(x^2y^3)^5$ **20.** $(a^4b)^7$ **21.** $\left(\frac{-2xz}{y^5}\right)^2$ **22.** $\left(\frac{y^4}{-3z^3}\right)^3$

Simplify each expression. See Example 6.

23. $\frac{x^3}{x}$ **24.** $\frac{y^{10}}{y^9}$ **25.** $\frac{(-2)^5}{(-2)^3}$ **26.** $\frac{(-5)^{14}}{(-5)^{11}}$
27. $\frac{p^7q^{20}}{pq^{15}}$ **28.** $\frac{x^8y^6}{y^5}$ **29.** $\frac{7x^2y^6}{14x^2y^3}$ **30.** $\frac{9a^4b^7}{3ab^2}$

Simplify the following. See Example 7.

31. $(2x)^0$ **32.** $-4x^0$ **33.** $-2x^0$ **34.** $(4y)^0$
35. $5^0 + y^0$ **36.** $-3^0 + 4^0$

Simplify the following. See Example 8.

37. $\frac{5p^3q^2}{pq}$ **38.** $\frac{4x^{11}y^8}{(2x^2y)^5}$ **39.** $\frac{(3x)^4yz^3}{zx^2}$ **40.** $\frac{4x(a^2b^3)^2}{3(a^2)^2}$
41. $\frac{(6mn)^5}{mn^2}$ **42.** $\frac{(6xy)^2x}{9x^2y^2}$ **43.** $\frac{a^5b^7}{a^4b^3}$ **44.** $\frac{10y^3x^3}{2xy}$

Simplify the following.

45. -5^1 **46.** $(6x)^1$ **47.** $(-4)^3$ **48.** -4^3
49. $(6b)^0$ **50.** $(5ab)^0$ **51.** $2^3 + 2^5$ **52.** $7^2 - 7^0$

53. b^4b^2 **54.** y^4y^1 **55.** $a^2a^3a^4$ **56.** $x^2x^{15}x^9$

57. $(2x^3)(-8x^4)$ **58.** $(3y^4)(-5y)$ **59.** $2x^3 - 8x^4$ **60.** $b^5 + 2b^3$

61. $(4a)^3$ **62.** $(2ab)^4$ **63.** $(-6xyz^3)^2$ **64.** $(-3xy^2a^3b)^3$

65. $\left(\frac{3y^5}{6x^4}\right)^3$ **66.** $\left(\frac{2ab}{6yz}\right)^4$ **67.** $\frac{x^5}{x^4}$ **68.** $\frac{5x^9}{x^3}$

69. $\frac{2x^3y^2z}{xyz}$ **70.** $\frac{(ab^2)^4}{a^3b^9}$ **71.** $\frac{10a^9y^{12}}{35a^4y^7}$ **72.** $\frac{x^{12}y^{13}}{x^5y^7}$

73. $\frac{3x^4y^6}{-2xy^2}$ **74.** $\frac{2x^3y^5}{5x^2y^4}$ **75.** $\left(\frac{a^3}{q^2}\right)^5$ **76.** $\left(\frac{9x}{7b}\right)^2$

77. $\left(\frac{y^3y^5}{xy^2}\right)^3$ **78.** $\left(\frac{a^9b^4}{a^3b^4}\right)^4$ **79.** $\left(\frac{5xy}{5^2x^2}\right)^2$ **80.** $\left(\frac{3x^4b^8}{3x}\right)^5$

81. $\frac{(3x^2)^5}{x^3}$ **82.** $\frac{(4a^2)^4}{a^4b}$ **83.** $\frac{7(xy)^3}{(7xy)^2}$ **84.** $\left(\frac{5}{-q}\right)^2$

85. $\frac{(-5)^2}{-q}$ **86.** $\left(\frac{w}{-v}\right)^5$

Solve.

87. The following rectangle has width $4x^2$ feet and length $5x^3$ feet. Find its area.

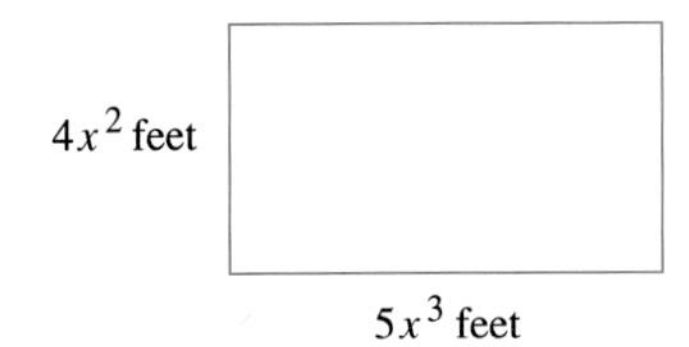

88. Given the following circle with radius $5y$ centimeters, find its area. Do not approximate π.

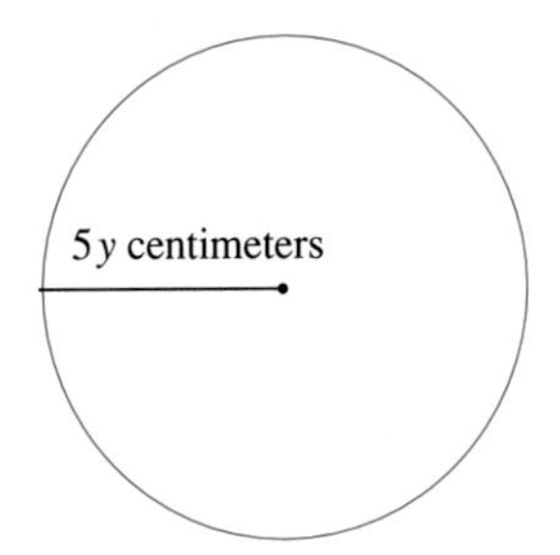

89. Given the following vault in the shape of a cube, if each side is $3y^4$ feet, find its volume.

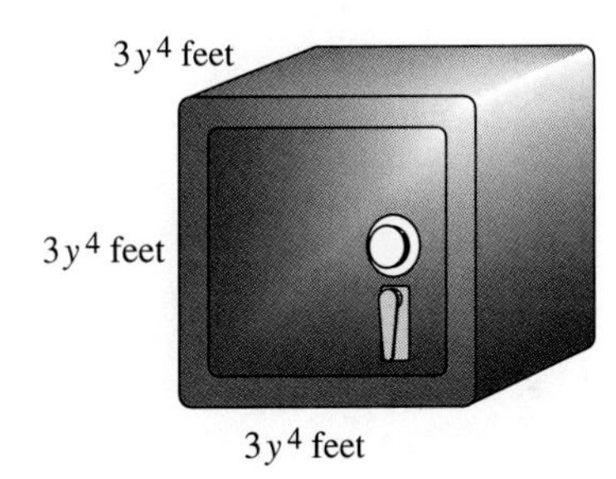

90. The silo shown is in the shape of a cylinder. If its radius is $4x$ meters and its height is $5x^3$ meters, find its volume.

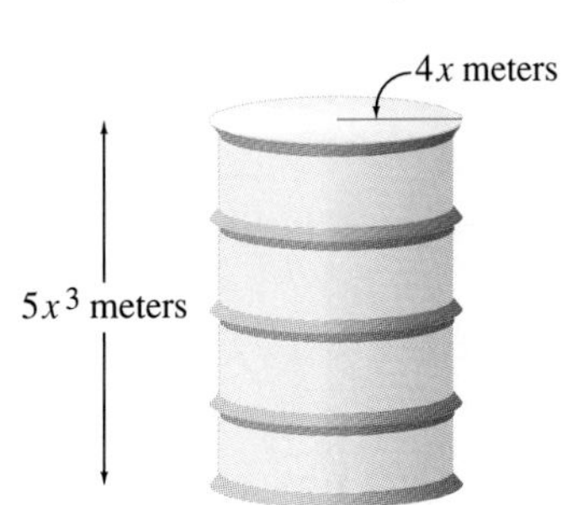

Skill Review

Use the slope–intercept form of a line, $y = mx + b$, to find the slope of each line. See Section 4.4.

91. $y = -2x + 7$ **92.** $y = \frac{3}{2}x - 1$ **93.** $3x - 5y = 14$ **94.** $x + 7y = 2$

Use the vertical line test to determine which of the following are graphs of functions. See Section 4.2.

95.

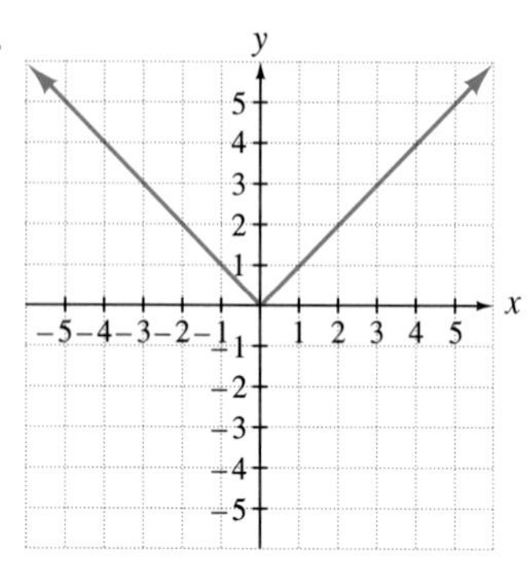

96.

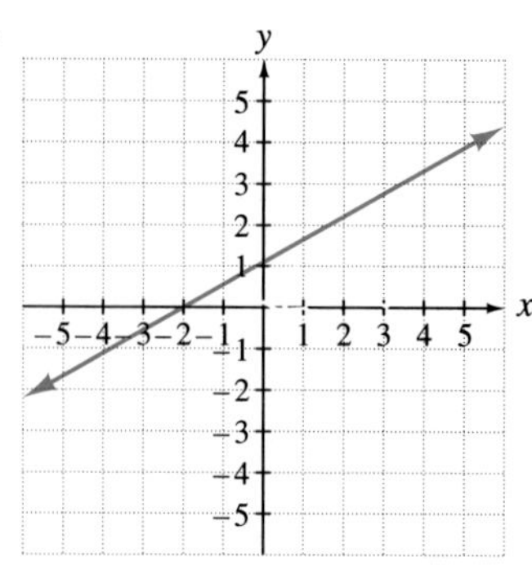

Solve. See Section 3.2.

97. A survey by the Electronic Industries Association concluded that 33% of households have a home computer. If your city contains 59,000 households, how many of these would you expect to have a home computer?

98. In 1989, the number of talk radio stations in the United States was 308. In 1994, the number was 1028. Find the percent increase in the number of talk radio stations.

A Look Ahead

Simplify each expression. Assume that variables represent positive integers. See the following example.

EXAMPLE Simplify $x^a \cdot x^{3a}$.

Solution: Like bases, so add exponents.

$$x^a \cdot x^{3a} = x^{a+3a} = x^{4a}$$ ■

99. $x^{5a}x^{4a}$

100. $b^{9a}b^{4a}$

101. $(a^b)^5$

102. $(2a^{4b})^4$

103. $\frac{x^{9a}}{x^{4a}}$

104. $\frac{y^{15b}}{y^{6b}}$

105. $(x^a y^b z^c)^{5a}$

106. $(9a^2b^3c^4d^5)^{ab}$

Writing in Mathematics

107. Explain why $(-5)^4 = 625$, whereas $-5^4 = -625$.

5.2 Negative Exponents and Scientific Notation

OBJECTIVES

Tape BA 9

1. Evaluate numbers raised to negative integer powers.
2. Use all exponents rules and definitions to simplify exponential expressions.
3. Write numbers in scientific notation.
4. Convert numbers from scientific notation to standard form.

1 Our work with exponential expressions so far has been limited to exponents that are positive integers or 0. Here we expand to give meaning to an expression like x^{-3}.

Suppose that we wish to simplify the expression $\frac{x^2}{x^5}$. If we use the quotient rule for exponents, we subtract exponents:

$$\frac{x^2}{x^5} = x^{2-5} = x^{-3},\ x \neq 0$$

But what does x^{-3} mean? Let's simplify $\frac{x^2}{x^5}$ using the definition of a^n.

$$\frac{x^2}{x^5} = \frac{x \cdot x}{x \cdot x \cdot x \cdot x \cdot x}$$

$$= \frac{x \cdot x}{x \cdot x \cdot x \cdot x \cdot x}$$ Divide numerator and denominator by common factors.

$$= \frac{1}{x^3}$$

If the quotient rule is to hold true for negative exponents, then x^{-3} must equal $\frac{1}{x^3}$.

From this example, we state the definition for negative exponents.

Negative Exponents

If a is a real number other than 0 and n is an integer, then

$$a^{-n} = \frac{1}{a^n}$$

EXAMPLE 1 Simplify each expression. Write answers using positive exponents only.

a. 3^{-2} **b.** $2x^{-3}$ **c.** $\frac{1}{2^{-5}}$

Solution: **a.** $3^{-2} = \frac{1}{3^2} = \frac{1}{9}$ Use the definition of negative exponent.

b. $2x^{-3} = 2 \cdot \frac{1}{x^3} = \frac{2}{x^3}$ Use the definition of negative exponent.

Since there are no parentheses, notice that the exponent -3 applies only to the base of x.

c. $\frac{1}{2^{-5}} = \frac{1}{\frac{1}{2^5}}$ Use the definition of negative exponent.

$$= 1 \div \frac{1}{2^5}$$

$$= 1 \cdot \frac{2^5}{1} = 2^5 \quad \text{or} \quad 32$$ ■

HELPFUL HINT

It may help you to think about negative exponents in this way: When a **factor** is moved from numerator to denominator or denominator to numerator (as in the previous example), the **sign** of its **exponent** changes. For example,

$$x^{-2} = \frac{1}{x^2}, \qquad 2^{-3} = \frac{1}{2^3} \text{ or } \frac{1}{8}$$

$$\frac{1}{y^{-4}} = \frac{1}{\frac{1}{y^4}} = y^4, \qquad \frac{1}{5^{-2}} = 5^2 \text{ or } 25$$

EXAMPLE 2 Simplify each expression. Write answers with positive exponents.

a. $\left(\frac{2}{3}\right)^{-4}$ **b.** $2^{-1} + 4^{-1}$ **c.** $(-2)^{-4}$

Solution: **a.** $\left(\frac{2}{3}\right)^{-4} = \frac{2^{-4}}{3^{-4}} = \frac{3^4}{2^4} = \frac{81}{16}$ **b.** $2^{-1} + 4^{-1} = \frac{1}{2} + \frac{1}{4} = \frac{2}{4} + \frac{1}{4} = \frac{3}{4}$

c. $(-2)^{-4} = \frac{1}{(-2)^4} = \frac{1}{(-2)(-2)(-2)(-2)} = \frac{1}{16}$ ■

EXAMPLE 3 Simplify each expression. Write answers with positive exponents.

a. $\frac{y}{y^{-2}}$ **b.** $\frac{p^{-4}}{q^{-9}}$ **c.** $\frac{x^{-5}}{x^7}$

Solution: **a.** $\frac{y}{y^{-2}} = y^{1-(-2)} = y^3$ **b.** $\frac{p^{-4}}{q^{-9}} = \frac{q^9}{p^4}$ **c.** $\frac{x^{-5}}{x^7} = x^{-5-7} = x^{-12} = \frac{1}{x^{12}}$ ■

2 The following is a summary of the rules and definitions for exponents.

Summary of Exponent Rules

If m and n are integers and a, b, and c are real numbers, then:

Product rule for exponents:	$a^m \cdot a^n = a^{m+n}$
Power rule for exponents:	$(a^m)^n = a^{m \cdot n}$
Power rules for products and quotients:	$(ab)^n = a^n b^n$ and $\left(\frac{a}{c}\right)^n = \frac{a^n}{c^n}, c \neq 0$
Quotient rule for exponents:	$\frac{a^m}{a^n} = a^{m-n}, a \neq 0$
Zero exponent:	$a^0 = 1, a \neq 0$
Negative exponent:	$a^{-n} = \frac{1}{a^n}, a \neq 0$

EXAMPLE 4 Simplify the following expressions. Write each answer using positive exponents only.

a. $(2x^3)(5x)^{-2}$ **b.** $\left(\frac{3a^2}{b}\right)^{-3}$ **c.** $\frac{4^{-1}x^{-3}y}{4^{-3}x^2y^{-6}}$ **d.** $(y^{-3}z^6)^{-6}$ **e.** $\left(\frac{-2x^3y}{xy^{-1}}\right)^3$

Solution: **a.** $(2x^3)(5x)^{-2} = 2x^3 \cdot 5^{-2}x^{-2}$ Use the power rule.

$= \dfrac{2x^{3+(-2)}}{5^2}$ Use the product and quotient rules and the definition of a negative exponent.

$= \dfrac{2x}{25}$

b. $\left(\dfrac{3a^2}{b}\right)^{-3} = \dfrac{3^{-3}a^{-6}}{b^{-3}} = \dfrac{b^3}{3^3a^6} = \dfrac{b^3}{27a^6}$

c. $\dfrac{4^{-1}x^{-3}y}{4^{-3}x^2y^{-6}} = 4^{-1-(-3)}x^{-3-2}y^{1-(-6)} = 4^2x^{-5}y^7 = \dfrac{16y^7}{x^5}$

d. $(y^{-3}z^6)^{-6} = y^{18} \cdot z^{-36} = \dfrac{y^{18}}{z^{36}}$

e. $\left(\dfrac{-2x^3y}{xy^{-1}}\right)^3 = \dfrac{(-2)^3x^9y^3}{x^3y^{-3}} = \dfrac{-8x^9y^3}{x^3y^{-3}} = -8x^{9-3}y^{3-(-3)} = -8x^6y^6$ ■

3 Both very large and very small numbers frequently occur in many fields of science. For example, the distance between the Sun and Pluto is approximately 5,906,000,000 kilometers, but the mass of a proton is approximately 0.000000000000000000000000165 gram. It can be tedious to write these numbers in standard notation like this, so **scientific notation** is used as a convenient shorthand for expressing very large and very small numbers.

Scientific Notation

A positive number is written in scientific notation if it is written as the product of a number a, where $1 \le a < 10$, and an integer power r of 10:

$$a \times 10^r$$

The following numbers are written in scientific notation. The $\times$ sign for multiplication is used as part of the notation.

2.03×10^2 $\quad$ 7.362×10^7

1×10^{-3} $\quad$ 8.1×10^{-5}

To write the distance between the Sun and Pluto in scientific notation, begin by moving the decimal point to the left until we have a number between 1 and 10.

5,906,000,000.

Next count the number of places the decimal point is moved.

5,906,000,000.

9 decimal places

We moved the decimal point **9** places **to the left.** This count is used as the power of 10.

$$5{,}906{,}000{,}000 = 5.906 \times 10^9$$

To express the mass of a proton in scientific notation, move the decimal point until the number is between 1 and 10.

0.000000000000000000000000165

The decimal point was moved 24 places **to the right,** so the exponent on 10 is **−24.**

$$0.000\ 000\ 000\ 000\ 000\ 000\ 000\ 001\ 65 = 1.65 \times 10^{-24}$$

To Write a Number in Scientific Notation

Step 1 Move the decimal point in the original number so that the new number has a value between 1 and 10.

Step 2 Count the number of decimal places the decimal point is moved in step 1. If the decimal point is moved to the left, the count is positive. If the decimal point is moved to the right, the count is negative.

Step 3 Multiply the new number in step 1 by 10 raised to an exponent equal to the count found in step 2.

EXAMPLE 5 Write the following numbers in scientific notation.

a. 367,000,000 **b.** 0.000003 **c.** 20,520,000,000 **d.** 0.00085

Solution: **a.** *Step 1* Move the decimal point until the number is between 1 and 10.

367,000,000.

Step 2 The decimal point is moved 8 places to the left so the count is positive 8.

Step 3 $367{,}000{,}000 = 3.67 \times 10^{8}$.

b. *Step 1* Move the decimal point until the number is between 1 and 10.

0.000003

Step 2 The decimal point is moved 6 places to the right, so the count is −6.

Step 3 $0.000003 = 3.0 \times 10^{-6}$.

c. $20{,}520{,}000{,}000 = 2.052 \times 10^{10}$

d. $0.00085 = 8.5 \times 10^{-4}$ ■

4 A number written in scientific notation can be rewritten in standard form. To write 8.63×10^{3} in standard form, recall that $10^{3} = 1000$.

$$8.63 \times 10^{3} = 8.63(1000) = 8630$$

Notice that the exponent on the 10 is +3 and we moved the decimal point 3 places to the right.

To write 8.63×10^{-3} in standard form, recall that $10^{-3} = \frac{1}{10^{3}} = \frac{1}{1000}$.

$$8.63 \times 10^{-3} = 8.63\left(\frac{1}{1000}\right) = \frac{8.63}{1000} = 0.00863$$

The exponent on the 10 is −3, and we moved the decimal point to the left 3 places.

In general, **to write a scientific notation number in standard form,** move the decimal point the same number of places as the exponent on 10. If the exponent is positive, move the decimal point to the right; if the exponent is negative, move the decimal point to the left.

EXAMPLE 6 Write the following numbers in standard notation, without exponents.

a. 1.02×10^5 **b.** 7.358×10^{-3} **c.** 8.4×10^7 **d.** 3.007×10^{-5}

Solution: **a.** Move the decimal point 5 places to the right.

$$1.02 \times 10^5 = 102{,}000.$$

b. Move the decimal point 3 places to the left.

$$7.358 \times 10^{-3} = 0.007358$$

c. $8.4 \times 10^7 = 84{,}000{,}000$

7 places to the right

d. $3.007 \times 10^{-5} = 0.00003007$

5 places to the left ■

To perform operations on numbers written in scientific notation, we make use of the rules and definitions of exponents.

EXAMPLE 7 Write each number without exponents.

a. $(8 \times 10^{-6})(7 \times 10^3)$ **b.** $\dfrac{12 \times 10^2}{6 \times 10^{-3}}$

Solution: **a.** $(8 \times 10^{-6})(7 \times 10^3) = 8 \cdot 7 \cdot 10^{-6} \cdot 10^3$

$$= 56 \times 10^{-3}$$

$$= 0.056$$

b. $\dfrac{12 \times 10^2}{6 \times 10^{-3}} = \dfrac{12}{6} \times 10^{2-(-3)} = 2 \times 10^5 = 200{,}000$ ■

MENTAL MATH

State each expression using positive exponents.

1. $5x^{-2}$

2. $3x^{-3}$

3. $\dfrac{1}{y^{-6}}$

4. $\dfrac{1}{x^{-3}}$

5. $\dfrac{4}{y^{-3}}$

6. $\dfrac{16}{y^{-7}}$

EXERCISE SET 5.2

Simplify each expression. Write answers with positive exponents. See Examples 1 and 2.

1. 4^{-3}

2. 6^{-2}

3. $7x^{-3}$

4. $(7x)^{-3}$

5. $\left(\dfrac{1}{4}\right)^{-3}$

6. $\left(\dfrac{1}{8}\right)^{-2}$

7. $3^{-1} + 2^{-1}$

8. $4^{-1} + 4^{-2}$

9. $\dfrac{1}{p^{-3}}$

10. $\dfrac{1}{q^{-5}}$

Simplify each expression. Write answers with positive exponents. See Example 3.

11. $\frac{p^{-5}}{q^{-4}}$ **12.** $\frac{r^{-5}}{s^{-2}}$ **13.** $\frac{x^{-2}}{x}$

14. $\frac{y}{y^{-3}}$ **15.** $\frac{z^{-4}}{z^{-7}}$ **16.** $\frac{x^{-4}}{x^{-1}}$

Simplify the following. Write each answer with positive exponents only. See Example 4.

17. $(a^{-5})^{-6}$ **18.** $(4^{-1})^{-2}$ **19.** $\left(\frac{x^{-2}y^4}{x^3y^7}\right)^2$

20. $\left(\frac{a^5b}{a^7b^{-2}}\right)^{-3}$ **21.** $\frac{4^2z^{-3}}{4^3z^{-5}}$ **22.** $\frac{3^{-1}x^4}{3^3x^{-7}}$

Write each number in scientific notation. See Example 5.

23. 78,000

24. 9,300,000,000

25. 0.00000167

26. 0.00000017

27. The distance between Earth and the Sun is 93,000,000 miles.

28. The population of the world is 5,506,000,000.

Write each number without using exponents. See Example 6.

29. 7.86×10^8

30. 1.43×10^7

31. 8.673×10^{-10}

32. 9.056×10^{-4}

33. One coulomb of electricity is 6.25×10^{18}.

34. The mass of a hydrogen atom is 1.7×10^{-24} grams.

Evaluate the following expressions using exponential rules. Write the answers without using exponents. See Example 7.

35. $(4 \times 10^4)(2 \times 10^6)$

36. $(1.8 \times 10^8)(3.4 \times 10^{-4})$

37. $\frac{3 \times 10^7}{2 \times 10^3}$

38. $\frac{1.5 \times 10^{14}}{5 \times 10^6}$

Simplify the following. Write answers with positive exponents.

39. $(-3)^{-2}$ **40.** $(-2)^{-4}$ **41.** $\frac{-1}{p^{-4}}$ **42.** $\frac{-1}{y^{-6}}$

43. $-2^0 - 3^0$ **44.** $5^0 + (-5)^0$ **45.** $\frac{r}{r^{-3}r^{-2}}$ **46.** $\frac{p}{p^{-3}q^{-5}}$

47. $(x^5y^3)^{-3}$ **48.** $(z^5x^5)^{-3}$ **49.** $2^0 + 3^{-1}$ **50.** $4^{-2} - 4^{-3}$

51. $\frac{2^{-3}x^{-4}}{2^2x}$ **52.** $\frac{5^{-1}z^7}{5^{-2}z^9}$ **53.** $\frac{7ab^{-4}}{7^{-1}a^{-3}b^2}$ **54.** $\frac{6^{-5}x^{-1}y^2}{6^{-2}x^{-4}y^4}$

55. $\left(\frac{a^{-5}b}{ab^3}\right)^{-4}$ **56.** $\left(\frac{r^{-2}s^{-3}}{r^{-4}s^{-3}}\right)^{-3}$ **57.** $\frac{(xy^3)^5}{(xy)^{-4}}$ **58.** $\frac{(rs)^{-3}}{(r^2s^3)^2}$

59. $\frac{(-2xy^{-3})^{-3}}{(xy^{-1})^{-1}}$ **60.** $\frac{(-3x^2y^2)^{-2}}{(xyz)^{-2}}$

Write each number in scientific notation.

61. 0.00635 **62.** 0.00194 **63.** 1,160,000 **64.** 700,000

65. The temperature at the interior of Earth is 20,000,000 degrees Celsius.

66. The half-life of a carbon isotope is 5000 years.

Write each number in standard notation.

67. 3.3×10^{-2}

68. 4.8×10^{-6}

69. 2.032×10^{4}

70. 9.07×10^{10}

71. The distance light travels in 1 year is 9.460×10^{12} kilometers.

72. The population of the United States is 2.58×10^{8}.

Evaluate the following expressions using exponential rules. Write the answers without exponents.

73. $(1.2 \times 10^{-3})(3 \times 10^{-2})$

74. $(2.5 \times 10^{6})(2 \times 10^{-6})$

75. $(4 \times 10^{-10})(7 \times 10^{-9})$

76. $(5 \times 10^{6})(4 \times 10^{-8})$

77. $\dfrac{8 \times 10^{-1}}{16 \times 10^{5}}$

78. $\dfrac{25 \times 10^{-4}}{5 \times 10^{-9}}$

79. $\dfrac{1.4 \times 10^{-2}}{7 \times 10^{-8}}$

80. $\dfrac{0.4 \times 10^{5}}{0.2 \times 10^{11}}$

Use scientific notation to find each quotient. Express the quotient in standard notation without exponents.

81. $\dfrac{6000 \times 0.006}{0.009 \times 400}$

82. $\dfrac{0.00016 \times 300}{0.064 \times 100}$

83. $\dfrac{0.00064 \times 2000}{16{,}000}$

84. $\dfrac{0.00072 \times 0.003}{0.00024}$

85. $\dfrac{66{,}000 \times 0.001}{0.002 \times 0.003}$

86. $\dfrac{0.0007 \times 11{,}000}{0.001 \times 0.0001}$

Solve.

87. The fastest computer can add two numbers in about 10^{-8} second. Express in scientific notation how long it would take this computer to do this task 200,000 times.

88. The density D of an object is equivalent to the quotient of its mass M and volume V. Thus, $D = \dfrac{M}{V}$. Express in scientific notation the density of an object whose mass is 500,000 pounds and whose volume is 250 cubic feet.

89. The average amount of water flowing past the mouth of the Amazon River is 4.2×10^{6} cubic feet per second. How much water flows past in an hour? (1 hour equals 3600 seconds.)

90. A beam of light travels 9.460×10^{12} kilometers per year. How far does light travel in 10,000 years?

91. Each side of the cube shown is $\dfrac{2x^{-2}}{y}$ meters. Find its volume.

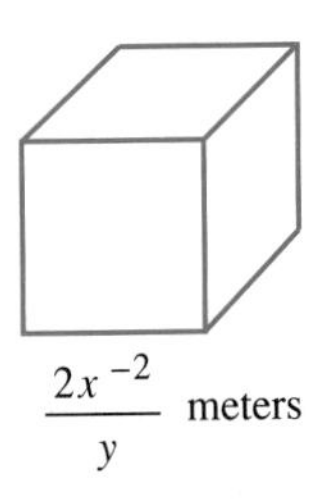

92. The lot shown is in the shape of a parallelogram with base $\dfrac{3x^{-1}}{y^{-3}}$ feet and height $5x^{-7}$ feet. Find its area.

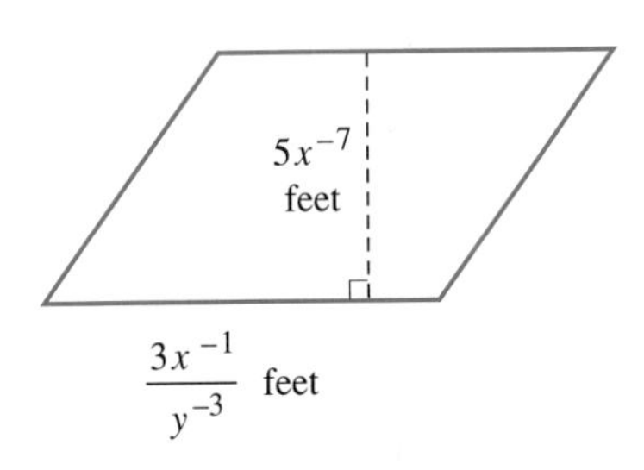

Skill Review

Simplify each expression. Use the distributive property to remove any parentheses. See Section 2.1.

93. $7x + 5 - 9x - 10$

94. $-x - 3x + 14 - x - 1$

95. $-3(y - 6) + 2(y + 4)$

96. $5(2y - 1) - (4y - 10)$

Given the following square and triangle, express each perimeter as an algebraic expression in x. See Section 2.1.

97. 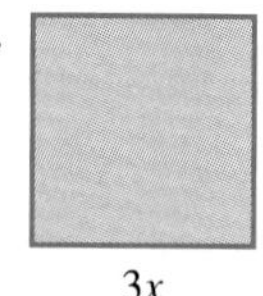

98. 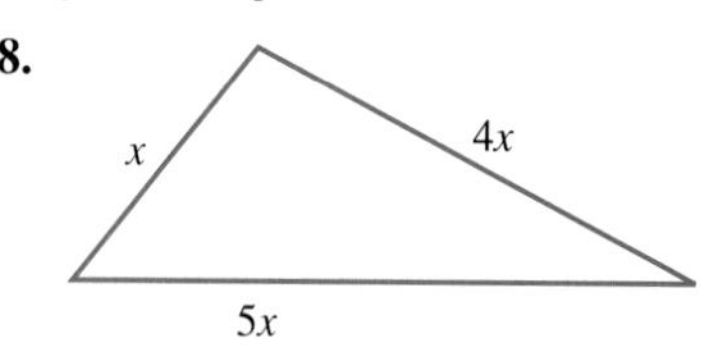

Solve each equation. See Section 2.3.

99. $5x + 2 = 3x + 8$

100. $9x - 14 = -6x + 1$

101. $\frac{x}{2} + \frac{1}{4} = 6$

102. $\frac{2x - 4}{6} + 5 = \frac{1}{3}$

A Look Ahead

Simplify each expression. Assume that variables represent positive integers. See the following example.

EXAMPLE Simplify the following expressions. Assume that the variable in the exponent represents a positive integer.

a. $x^{m+1} \cdot x^m$ **b.** $(z^{2x+1})^x$ **c.** $\frac{y^{6a}}{y^{4a}}$

Solution: **a.** $x^{m+1} \cdot x^m = x^{(m+1)+m} = x^{2m+1}$ **b.** $(z^{2x+1})^x = z^{(2x+1)x} = z^{2x^2+x}$

c. $\frac{y^{6a}}{y^{4a}} = y^{6a-4a} = y^{2a}$ ■

103. $a^{-4m} \cdot a^{5m}$

104. $(x^{-3s})^3$

105. $(3y^{2z})^3$

106. $a^{4m+1} \cdot a^4$

107. $\frac{y^{4a}}{y^{-a}}$

108. $\frac{y^{-6a}}{zy^{6a}}$

109. $(z^{3a+2})^{-2}$

110. $(a^{4x-1})^{-1}$

5.3 Polynomial Functions and Addition and Subtraction of Polynomials

OBJECTIVES

Tape IA 10

1. Define *term, constant, polynomial, monomial, binomial,* and *trinomial.*
2. Identify the degree of a term and of a polynomial.
3. Define polynomial functions.
4. Add polynomials.
5. Subtract polynomials.

1 Recall from Section 2.1 that a **term** of an algebraic expression is a number or the product of a number and one or more variables raised to powers. The **numerical coefficient,** or simply the **coefficient,** is the numerical factor of a term.

Term	*Numerical Coefficient*
$-12x^5$	-12
x^3y	1
$-z$	-1
2	2

If a term contains only a number, it is called a **constant term,** or simply a **constant.**

A **polynomial** is a finite sum of terms in which all variables have exponents raised to nonnegative integer powers and no variables appear in the denominator.

Polynomials	*Not Polynomials*	
$4x^5 + 7xz$	$5x^{-3} + 2x$	(negative integer exponent)
$-5x^3 + 2x + \frac{2}{3}$	$\frac{6}{x^2} - 5x + 1$	(variable in denominator)

A polynomial that contains only one variable is called a **polynomial in one variable.** For example, $3x^2 - 2x + 7$ is a **polynomial in x.** This polynomial in x is written in **descending order** since the terms are listed in descending order of the variable's exponents. (The term 7 can be thought of as $7x^0$.) The following examples are polynomials in one variable written in **descending order.**

$$4x^3 - 7x^2 + 5, \qquad y^2 - 4, \qquad 8a^4 - 7a^2 + 4a$$

A **monomial** is a polynomial consisting of one term. A **binomial** is a polynomial consisting of two terms. A **trinomial** is a polynomial consisting of three terms.

Monomials	*Binomials*	*Trinomials*
ax^2	$x + y$	$x^2 + 4xy + y^2$
$-3x$	$6y^2 - 2$	$-x^4 + 3x^3 + 1$
4	$\frac{5}{7}z^3 - 2z$	$8y^2 - 2y - 10$

By definition, all monomials, binomials, and trinomials are also polynomials.

2 Each term of a polynomial has a **degree.**

Degree of a Term

The **degree of a term** is the sum of the exponents on the variables contained in the term.

EXAMPLE 1 Find the degree of each term.

a. $3x^2$ **b.** -2^3x^5 **c.** y **d.** $12x^2yz^3$ **e.** 5

Solution:

a. The exponent on x is 2, so the degree of the term is 2.

b. The exponent on x is 5, so the degree of the term is 5.

c. The degree of y or y^1 is 1.

d. The degree is the sum of the exponents on the variables, or $2 + 1 + 3 = 6$.

e. The degree of 5, which can be written as $5x^0$, is 0. ■

From the preceding example, we can say that the degree of a constant is 0. Also, the term 0 has no degree.

Each polynomial also has a degree.

Degree of a Polynomial

The **degree of a polynomial** is the largest degree of all its terms.

EXAMPLE 2 Find the degree of each polynomial and indicate whether the polynomial is a monomial, binomial, trinomial, or none of these.

a. $7x^3 - 3x + 2$ **b.** $-xyz$ **c.** $x^2 - 4$ **d.** $2xy + x^2y^2 - 5x^2 - 6$

Solution: **a.** The degree of the trinomial $7x^3 - 3x + 2$ is 3, the largest degree of any of its terms.

b. The degree of the monomial $-xyz$ or $-x^1y^1z^1$ is 3.

c. The degree of the binomial $x^2 - 4$ is 2.

d. The degree of each term of the polynomial $2xy + x^2y^2 - 5x^2 - 6$ is:

Term	*Degree*
$2xy$ or $2x^1y^1$	$1 + 1 = 2$
x^2y^2	$2 + 2 = 4$
$-5x^2$	2
-6	0

The highest degree of any term is 4, so the degree of this polynomial is 4. ■

3 At times, it is convenient to represent polynomials using function notation. For example, we may write $P(x)$ to represent the polynomial $7x^2 - 3x + 1$. In symbols, this is

$$P(x) = 7x^2 - 3x + 1$$

This function is called a **polynomial function** because the expression $7x^2 - 3x + 1$ is a polynomial.

HELPFUL HINT

Recall that the symbol $P(x)$ **does not mean** P times x. It is a special symbol used to denote a function.

EXAMPLE 3 If $P(x) = 7x^2 - 3x + 1$, find the following:

a. $P(1)$ **b.** $P(-2)$

Solution: **a.** Substitute 1 for x in $P(x) = 7x^2 - 3x + 1$ and simplify.

$$P(x) = 7x^2 - 3x + 1$$

$$P(1) = 7(1)^2 - 3(1) + 1 = 5$$

b. $P(x) = 7x^2 - 3x + 1$

$$P(-2) = 7(-2)^2 - 3(-2) + 1 = 35$$ ■

Many applications are modeled by polynomial functions. If the polynomial function is given, the solution of the problem is often found by evaluating the function at a certain value.

EXAMPLE 4 The CN Tower in Toronto, Ontario, is 1821 feet tall and is the world's tallest self-supporting structure. An object is dropped from the top of this building. Neglecting air resistance, the height of the object at time t seconds is given by the polynomial function $P(t) = -16t^2 + 1821$. Find the height of the object when $t = 1$ second and when $t = 10$ seconds.

Solution: To find the height of the object at 1 second we find $P(1)$.

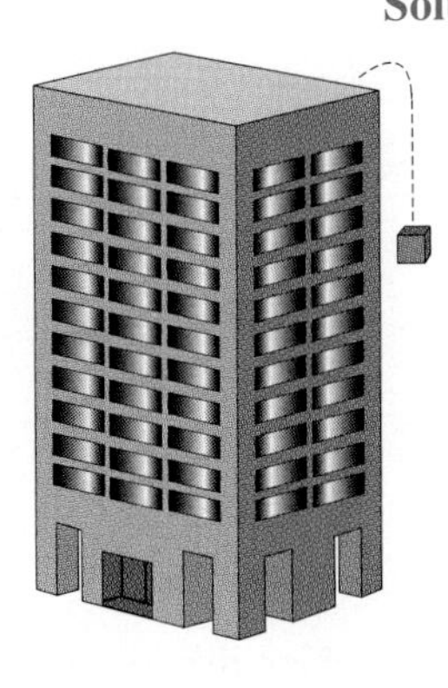

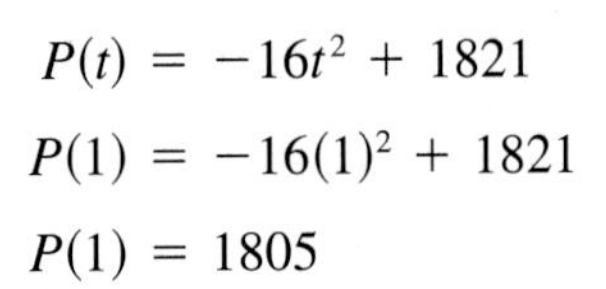

$$P(t) = -16t^2 + 1821$$
$$P(1) = -16(1)^2 + 1821$$
$$P(1) = 1805$$

When $t = 1$ second, the height of the object is 1805 feet.

To find the height of the object at 10 seconds we find $P(10)$.

$$P(t) = -16t^2 + 1821$$
$$P(10) = -16(10)^2 + 1821$$
$$P(10) = -1600 + 1821$$
$$P(10) = 221$$

When $t = 10$ seconds, the height of the object is 221 feet.

Notice that when time t increases, the height of the object decreases. ■

4 To add and subtract polynomials, we **combine like terms.** Recall that terms are considered to be **like terms** if they contain exactly the same variables raised to exactly the same powers.

Like Terms	*Unlike Terms*
$-5x^2, -x^2$	$4x^2, 3x$
$7xy^3z, -2xzy^3$	$12x^2y^3, -2xy^3$

We **combine like terms** by using the distributive property. For example, by the distributive property,

$$5x + 7x = (5 + 7)x = 12x$$

EXAMPLE 5 Combine like terms.

a. $-12x^2 + 7x^2 - 6x$ **b.** $3xy - 2x + 5xy - x$

Solution: By the distributive property,

a. $-12x^2 + 7x^2 - 6x = (-12 + 7)x^2 - 6x = -5x^2 - 6x$

b. Use the associative and commutative properties to group together like terms; then combine.

$$\begin{aligned} 3xy - 2x + 5xy - x &= 3xy + 5xy - 2x - x \\ &= (3 + 5)xy + (-2 - 1)x \\ &= 8xy - 3x \end{aligned}$$ ■

Now we have reviewed the necessary skills to add polynomials.

To Add Polynomials

Combine all like terms.

EXAMPLE 6 Add.

a. $(7x^3y - xy^3 + 11) + (6x^3y - 4)$ **b.** $(3a^3 - b + 2a - 5) + (a + b + 5)$

Solution: **a.** To add, remove the parentheses and group like terms.

$$(7x^3y - xy^3 + 11) + (6x^3y - 4)$$

$= 7x^3y - xy^3 + 11 + 6x^3y - 4$

$= 7x^3y + 6x^3y - xy^3 + 11 - 4$ Group like terms.

$= 13x^3y - xy^3 + 7$ Combine like terms.

b. $(3a^3 - b + 2a - 5) + (a + b + 5)$

$= 3a^3 - b + 2a - 5 + a + b + 5$

$= 3a^3 - b + b + 2a + a - 5 + 5$ Group like terms.

$= 3a^3 + 3a$ Combine like terms.

■

EXAMPLE 7 Add $11x^3 - 12x^2 + x - 3$ and $x^3 - 10x + 5$.

Solution: $(11x^3 - 12x^2 + x - 3) + (x^3 - 10x + 5)$

$= 11x^3 + x^3 - 12x^2 + x - 10x - 3 + 5$ Group like terms.

$= 12x^3 - 12x^2 - 9x + 2$ Combine like terms. ■

Sometimes it is more convenient to add polynomials vertically. To do this, line up like terms underneath one another and add like terms.

EXAMPLE 8 Add $11x^3 - 12x^2 + x - 3$ and $x^3 - 10x + 5$ vertically.

Solution:

$$\begin{array}{rrrr} 11x^3 & -\ 12x^2 & +\ \ x & -\ 3 \\ x^3 & & -\ 10x & +\ 5 \\ \hline 12x^3 & -\ 12x^2 & -\ \ 9x & +\ 2 \end{array}$$

Line up like terms.

Combine like terms.

Notice that this example is the same as Example 7, only here we added vertically.

■

5 The definition of subtraction of real numbers can be extended to apply to polynomials. To subtract a number, we add its opposite.

$$a - b = a + (-b)$$

To subtract a polynomial, we add its opposite. In other words, if P and Q are polynomials, then

$$P - Q = P + (-Q)$$

The polynomial $-Q$ is the **opposite** or **additive inverse** of the polynomial Q. We can find $-Q$ by changing the sign of each term of Q.

To Subtract Polynomials

To subtract two polynomials, change the sign of each term of the polynomial that is being subtracted; then add.

For example,

To subtract, change the signs.

Then add

$$(3x^2 + 4x - 7) - (3x^2 - 2x - 5) = (3x^2 + 4x - 7) + (-3x^2 + 2x + 5)$$
$$= 3x^2 + 4x - 7 - 3x^2 + 2x + 5$$
$$= 6x - 2 \quad \text{Combine like terms.}$$

EXAMPLE 9 Simplify $(12z^5 - 12z^3 + z) - (-3z^4 + z^3 + 12z)$.

Solution: To subtract, change the sign of each term of the second polynomial and add the result to the first polynomial.

$$(12z^5 - 12z^3 + z) - (-3z^4 + z^3 + 12z)$$
$$= 12z^5 - 12z^3 + z + 3z^4 - z^3 - 12z \quad \text{Change signs and add.}$$
$$= 12z^5 + 3z^4 - 12z^3 - z^3 + z - 12z \quad \text{Group like terms.}$$
$$= 12z^5 + 3z^4 - 13z^3 - 11z \quad \text{Combine like terms.}$$

■

EXAMPLE 10 Subtract $4x^3y^2 - 3x^2y^2 + 2y^2$ from $10x^3y^2 - 7x^2y^2$.

Solution: If we subtract 2 from 8, the difference is $8 - 2 = 6$. Notice the order of the numbers, and then write "Subtract $4x^3y^2 - 3x^2y^2 + 2y^2$ from $10x^3y^2 - 7x^2y^2$" as a mathematical expression.

$$(10x^3y^2 - 7x^2y^2) - (4x^3y^2 - 3x^2y^2 + 2y^2)$$
$$= 10x^3y^2 - 7x^2y^2 - 4x^3y^2 + 3x^2y^2 - 2y^2 \quad \text{Remove parentheses.}$$
$$= 6x^3y^2 - 4x^2y^2 - 2y^2 \quad \text{Combine like terms.}$$

■

EXAMPLE 11 Perform the subtraction $(10x^3y^2 - 7x^2y^2) - (4x^3y^2 - 3x^2y^2 + 2y^2)$ vertically.

Solution: Add the opposite of the second polynomial.

$$\begin{array}{r} 10x^3y^2 - 7x^2y^2 \\ -(4x^3y^2 - 3x^2y^2 + 2y^2) \\ \hline \end{array} \quad \text{is equivalent to} \quad \begin{array}{r} 10x^3y^2 - 7x^2y^2 \\ -4x^3y^2 + 3x^2y^2 - 2y^2 \\ \hline 6x^3y^2 - 4x^2y^2 - 2y^2 \end{array}$$

■

EXERCISE SET 5.3

Find the degree of each term. See Example 1.

1. 4
2. 7
3. $5x^2$
4. $-z^3$
5. $-3xy^2$
6. $12x^2z$

Find the degree of each polynomial and indicate whether the polynomial is a monomial, binomial, trinomial, or none of these. See Example 2.

7. $6x + 3$
8. $7x - 8$
9. $3x^2 - 2x + 5$
10. $5x^2 - 3x^2y - 2x^3$
11. $-xyz$
12. -9
13. $x^2y - 4xy^2 + 5x + y$
14. $-2x^2y - 3y^2 + 4x + y^5$

If $P(x) = x^2 + x + 1$ and $Q(x) = 5x^2 - 1$, find the following. See Example 3.

15. $P(7)$
16. $Q(4)$
17. $Q(-10)$
18. $P(-4)$
19. $P(0)$
20. $Q(0)$

Refer to Example 4 for Exercises 21 through 24.

21. Find the height of the object at $t = 2$ seconds.
22. Find the height of the object at $t = 4$ seconds.
23. Find the height of the object at $t = 6$ seconds.
24. Approximate (to the nearest second) how long it takes before the object hits the ground. (Hint: The object hits the ground when height $P(x) = 0$.)

Simplify by combining like terms. See Example 5.

25. $5y + y$
26. $-x + 3x$
27. $4x + 7x - 3$
28. $-8y + 9y + 4y^2$
29. $4xy + 2x - 3xy - 1$
30. $-8xy^2 + 4x - x + 2xy^2$

Add. See Examples 6, 7, and 8.

31. $(9y^2 - 8) + (9y^2 - 9)$
32. $(x^2 + 4x - 7) + (8x^2 + 9x - 7)$
33. $(x^2 + xy - y^2)$ and $(2x^2 - 4xy + 7y^2)$
34. $(4x^3 - 6x^2 + 5x + 7)$ and $(2x^2 + 6x - 3)$
35. $\begin{array}{r} x^2 - 6x + 3 \\ + \quad (2x + 5) \\ \hline \end{array}$
36. $\begin{array}{r} -2x^2 + 3x - 9 \\ + \quad (2x - 3) \\ \hline \end{array}$

Subtract. See Examples 9, 10, and 11.

37. $(9y^2 - 7y + 5) - (8y^2 - 7y + 2)$
38. $(2x^2 + 3x + 12) - (5x - 7)$
39. $(6x^2 - 3x)$ from $(4x^2 + 2x)$
40. $(xy + x - y)$ from $(xy + x - 3)$
41. $\begin{array}{r} 3x^2 - 4x + 8 \\ - \quad (5x^2 - 7) \\ \hline \end{array}$
42. $\begin{array}{r} -3x^2 - 4x + 8 \\ - \quad (5x + 12) \\ \hline \end{array}$

Perform the indicated operations.

43. $(5x - 11) + (-x - 2)$
44. $(3x^2 - 2x) + (5x^2 - 9x)$
45. $(7x^2 + x + 1) - (6x^2 + x - 1)$
46. $(4x - 4) - (-x - 4)$
47. $(7x^3 - 4x + 8) + (5x^3 + 4x + 8x)$
48. $(9xyz + 4x - y) + (-9xyz - 3x + y + 2)$
49. $(9x^3 - 2x^2 + 4x - 7) - (2x^3 - 6x^2 - 4x + 3)$
50. $(3x^2 + 6xy + 3y^2) - (8x^2 - 6xy - y^2)$
51. Add $(y^2 + 4y + 7)$ and $(-19y^2 + 7y + 7)$.
52. Subtract $(x - 4)$ from $(3x^2 - 4x + 5)$.
53. $(3x^3 - b + 2a - 6) + (-4x^3 + b + 6a - 6)$
54. $(5x^2 - 6) + (2x^2 - 4x + 8)$
55. $(4x^2 - 6x + 2) - (-x^2 + 3x + 5)$
56. $(5x^2 + x + 9) - (2x^2 - 9)$
57. $(-3x + 8) + (-3x^2 + 3x - 5)$
58. $(5y^2 - 2y + 4) + (3y + 7)$
59. $(-3 + 4x^2 + 7xy) + (2x^3 - x^2 + xy)$
60. $(-3xy + 4) - (-7xy - 8y)$

61. $\begin{array}{r} 6y^2 - 6y + 4 \\ -\ (-y^2 - 6y + 7) \\ \hline \end{array}$

62. $\begin{array}{r} -4x^3 + 4x^2 - 4x \\ -\ (2x^3 - 2x^2 + 3x) \\ \hline \end{array}$

63. $\begin{array}{r} 3x^2 + 15x + 8 \\ +\ (2x^2 + 7x + 8) \\ \hline \end{array}$

64. $\begin{array}{r} 9x^2 + 9x - 4 \\ +\ (7x^2 - 3x - 4) \\ \hline \end{array}$

65. Find the sum of $(5q^4 - 2q^2 - 3q)$ and $(-6q^4 + 3q^2 + 5)$.

66. Find the sum of $(5y^4 - 7y^2 + x^2 - 3)$ and $(-3y^4 + 2y^2 + 4)$.

67. Subtract $(3x + 7)$ from the sum of $(7x^2 + 4x + 9)$ and $(8x^2 + 7x - 8)$.

68. Subtract $(9x + 8)$ from the sum of $(3x^2 - 2x - x^3 + 2)$ and $(5x^2 - 8x - x^3 + 4)$.

69. Find the sum of $(4x^4 - 7x^2 + 3)$ and $(2 - 3x^4)$.

70. Find the sum of $(8x^4 - 14x^2 + 6)$ and $(-12x^6 - 21x^4 - 9x^2)$.

If $P(x) = 3x + 3$, $Q(x) = 4x^2 - 6x + 3$, *and* $R(x) = 5x^2 - 7$, *find the following.*

71. $P(4)$

72. $Q(-1)$

73. $R(-3)$

74. $P(0)$

75. $Q(2)$

76. $R(1)$

77. $Q(0)$

78. $R(-2)$

Express each of the following as a polynomial. Simplify if possible.

79. Given the following triangle, find its perimeter.

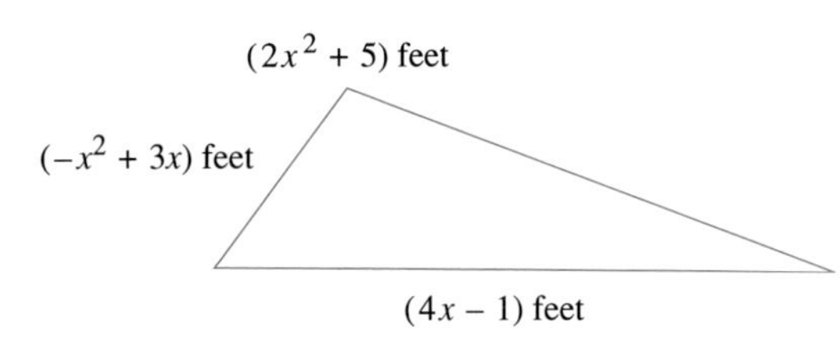

80. Given the following quadrilateral, find its perimeter.

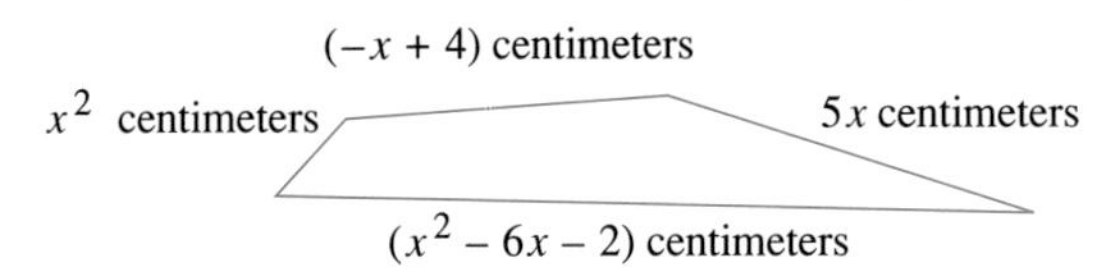

81. A wooden beam is $(4y^2 + 4y + 1)$ meters long. If a piece $(y^2 - 10)$ meters is cut, express the length of the remaining piece of beam as a polynomial in y.

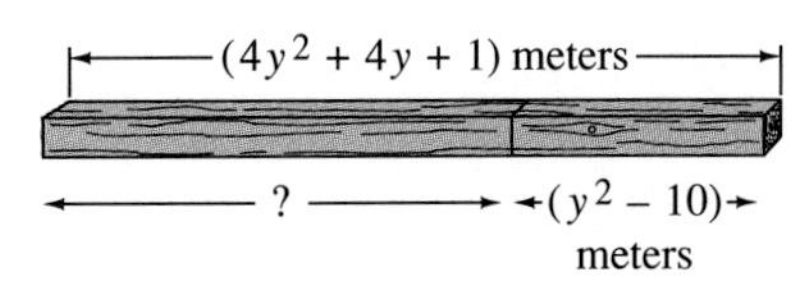

82. A piece of taffy is $(13x - 7)$ inches long. If a piece $(2x + 2)$ inches is removed, express the length of the remaining piece of taffy as a polynomial in x.

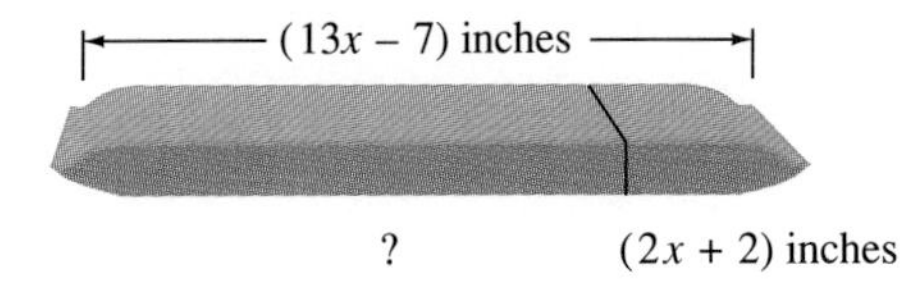

83. A piece of cable $3x^2 + 2x - 5$ centimeters is cut from a cable of length $7x^2 - 4$ centimeters. Express the length of the remaining piece of cable as a polynomial in x.

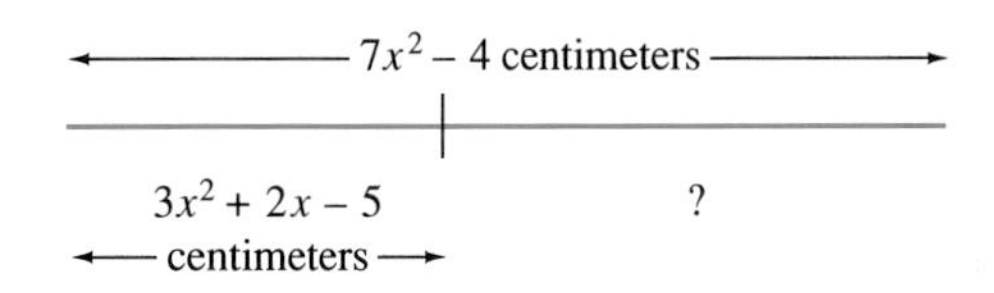

84. Boards are to be placed around the border of a proposed driveway (see figure) before cement can be poured. To determine the number of feet of board strips to order, find the perimeter of the proposed driveway.

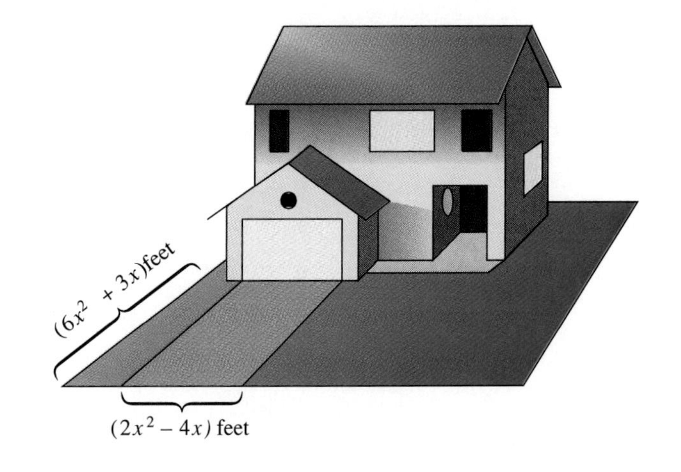

A projectile is fired upward from the ground with an initial velocity of 300 feet per second. Neglecting air resistance, the height of the projectile at any time t can be described by the polynomial function

$$P(t) = -16t^2 + 300t$$

85. Find the height of the projectile at the given times.
a. $t = 1$ second **b.** $t = 2$ seconds
c. $t = 3$ seconds **d.** $t = 4$ seconds

86. Explain why the height increases and then decreases as time passes.

87. Approximate (to the nearest second) how long before the object hits the ground.

Solve.

88. An object is thrown upward with an initial velocity of 50 feet per second from the top of the 350-foot-high City Hall in Milwaukee, Wisconsin. The height of the object at any time t can be described by the polynomial function $P(t) = -16t^2 + 50t + 350$. Find the height of the projectile when $t = 1$ second, $t = 2$ seconds, and $t = 3$ seconds.

89. The total cost (in dollars) for MCD, Inc. Manufacturing Company to produce x blank audiocassette tapes per week is given by the polynomial function $C(x) = 0.8x + 10{,}000$. Find the total cost in producing 20,000 tapes per week.

90. The total revenues (in dollars) for MCD, Inc. Manufacturing Company to sell x blank audiocassette tapes per week is given by the polynomial function $R(x) = 2x$. Find the total revenue in selling 20,000 tapes per week.

Skill Review

Multiply. See Section 5.1.

91. $3x(2x)$ **92.** $-7x(x)$ **93.** $5y^2(-2y)$ **94.** $2z(30z^3)$

95. $(12x^3)(-x^5)$ **96.** $6r^3(7r^{10})$

Solve. See Section 3.1.

97. $\frac{1}{5} = \frac{x}{9}$ **98.** $\frac{x}{3} = \frac{14}{6}$ **99.** $\frac{x + 6}{16} = \frac{3x - 2}{8}$ **100.** $\frac{x + 11}{10} = \frac{3x - 9}{2}$

101. The Defense Department found that 15 out of 100 people serving in the U.S. Air Force are women. If an air force base contains 1600 air force personnel, how many would you expect to be women?

102. The Defense Department reports that 9 out of every 25 women serving in the U.S. military serve in the army. If 200,000 females currently serve in the military, how many women would you expect to be in the army?

A Look Ahead

If P(x) is the polynomial given, find P(a), P(−x), and P(x + h). See the following example.

EXAMPLE If $P(x) = -3x + 5$, find the following:
a. $P(a)$ **b.** $P(-x)$ **c.** $P(x + h)$

Solution: **a.** $P(x) = -3x + 5$

$$P(a) = -3a + 5$$

b. $P(x) = -3x + 5$

$$P(-x) = -3(-x) + 5$$

$$= 3x + 5$$

c. $$P(x) = -3x + 5$$
$$P(x + h) = -3(x + h) + 5$$
$$= -3x - 3h + 5 \quad \blacksquare$$

103. $P(x) = 2x - 3$
104. $P(x) = 8x + 3$
105. $P(x) = 4x$
106. $P(x) = -4x$
107. $P(x) = 4x - 1$
108. $P(x) = 3x - 2$

Writing in Mathematics

109. Describe how to find the degree of a polynomial.

110. Explain why xyz is a monomial whereas $x + y + z$ is a trinomial.

5.4 Multiplication of Polynomials

OBJECTIVES

Tape IA 10

1 Multiply two polynomials.
2 Multiply binomials.
3 Square a binomial.
4 Multiply the sum and difference of two terms.

1 Properties of real numbers and exponents are used continually in the process of multiplying polynomials. To multiply monomials, for example, we apply the commutative and associative properties of real numbers and the product rule for exponents.

EXAMPLE 1 Multiply:
a. $(2x^3)(5x^6)$ **b.** $(7y^4z^4)(-xy^{11}z^5)$

Solution: Group like bases and apply the product rule for exponents.
a. $(2x^3)(5x^6) = 2(5)(x^3)(x^6) = 10x^9$
b. $(7y^4z^4)(-xy^{11}z^5) = 7(-1)x(y^4y^{11})(z^4z^5) = -7xy^{15}z^9$ ■

To multiply a monomial by a polynomial other than a monomial, we use an expanded form of the distributive property:

$$a(b + c + d + \cdots + z) = ab + ac + ad + \cdots + az$$

Notice that the monomial a is multiplied by each term of the polynomial.

EXAMPLE 2 Find the following products.
a. $2x(5x - 4)$ **b.** $-3x^2(4x^2 - 6x + 1)$ **c.** $-xy(7x^2y + 3xy - 11)$

Solution: Apply the distributive property.
a. $2x(5x - 4) = 2x(5x) + 2x(-4)$
$$= 10x^2 - 8x$$

b. $-3x^2(4x^2 - 6x + 1) = -3x^2(4x^2) + (-3x^2)(-6x) + (-3x^2)(1)$
$= -12x^4 + 18x^3 - 3x^2$

c. $-xy(7x^2y + 3xy - 11) = -xy(7x^2y) + (-xy)(3xy) + (-xy)(-11)$
$= -7x^3y^2 - 3x^2y^2 + 11xy$ ■

To multiply any polynomial by a polynomial, we again use the distributive property, multiplying each term of one polynomial by each term of the other polynomial.

EXAMPLE 3 Multiply and simplify the product if possible.

a. $(x + 3)(2x + 5)$ **b.** $(2x^3 - 3)(5x^2 - 6x + 7)$

Solution: **a.** Multiply each term of $(x + 3)$ by $(2x + 5)$.

$(x + 3)(2x + 5) = x(2x + 5) + 3(2x + 5)$	Apply the distributive property.
$= 2x^2 + 5x + 6x + 15$	Apply the distributive property again.
$= 2x^2 + 11x + 15$	Combine like terms.

b. Multiply each term of $(2x^3 - 3)$ by each term of $(5x^2 - 6x + 7)$.

$$(2x^3 - 3)(5x^2 - 6x + 7) = 2x^3(5x^2 - 6x + 7) + (-3)(5x^2 - 6x + 7)$$
$$= 10x^5 - 12x^4 + 14x^3 - 15x^2 + 18x - 21$$ ■

Sometimes polynomials are easier to multiply vertically, in the same way we multiply real numbers. When multiplying vertically, line up like terms in the **partial products** vertically. This makes combining like terms easier.

EXAMPLE 4 Find the product of $(4x^2 + 7)$ and $(x^2 + 2x + 8)$.

Solution:

$x^2 + 2x + 8$	
$4x^2 + 7$	
$7x^2 + 14x + 56$	$7(x^2 + 2x + 8)$.
$4x^4 + 8x^3 + 32x^2$	$4x^2(x^2 + 2x + 8)$.
$4x^4 + 8x^3 + 39x^2 + 14x + 56$	Combine like terms. ■

EXAMPLE 5 Multiply $(x + 3)$ by $(2x + 5)$ vertically.

Solution:

$x + 3$	
$2x + 5$	
$5x + 15$	$5(x + 3)$.
$2x^2 + 6x$	$2x(x + 3)$.
$2x^2 + 11x + 15$	Combine like terms. ■

2 When multiplying a binomial by a binomial, a special order of multiplying terms, called the **FOIL** order, may be used. The letters of FOIL stand for "**F**irst–**O**uter–**I**nner–**L**ast." To illustrate this method, multiply $(2x - 3)$ by $(3x + 1)$.

Multiply the **F**irst terms of each binomial.

$$(2x - 3)(3x + 1) \qquad \textbf{F} \quad 2x(3x) = 6x^2$$

Multiply the **O**uter terms of each binomial.

$$(2x - 3)(3x + 1) \qquad \textbf{O} \quad 2x(1) = 2x$$

Multiply the **I**nner terms of each binomial.

$$(2x - 3)(3x + 1) \qquad \textbf{I} \quad -3(3x) = -9x$$

Multiply the **L**ast terms of each binomial.

$$(2x - 3)(3x + 1) \qquad \textbf{L} \quad -3(1) = -3$$

Combine like terms.

$$6x^2 + 2x - 9x - 3 = 6x^2 - 7x - 3$$

EXAMPLE 6 Multiply $(x - 1)(x + 2)$. Use the FOIL order.

Solution:

$$\begin{aligned} (x - 1)(x + 2) &= \overset{\text{First}}{x \cdot x} + \overset{\text{Outer}}{2 \cdot x} + \overset{\text{Inner}}{(-1)x} + \overset{\text{Last}}{(-1)(2)} \\ &= x^2 + 2x - x - 2 \\ &= x^2 + x - 2 \qquad \text{Combine like terms.} \end{aligned}$$

■

EXAMPLE 7 Multiply $(2x - 7)(3x - 4)$.

Solution:

$$\begin{aligned} (2x - 7)(3x - 4) &= \overset{\text{First}}{2x(3x)} + \overset{\text{Outer}}{2x(-4)} + \overset{\text{Inner}}{(-7)(3x)} + \overset{\text{Last}}{(-7)(-4)} \\ &= 6x^2 - 8x - 21x + 28 \\ &= 6x^2 - 29x + 28 \end{aligned}$$

■

3 The **square of a binomial** is a special case of the product of two binomials. Find $(a + b)^2$ by the FOIL order for multiplying two binomials.

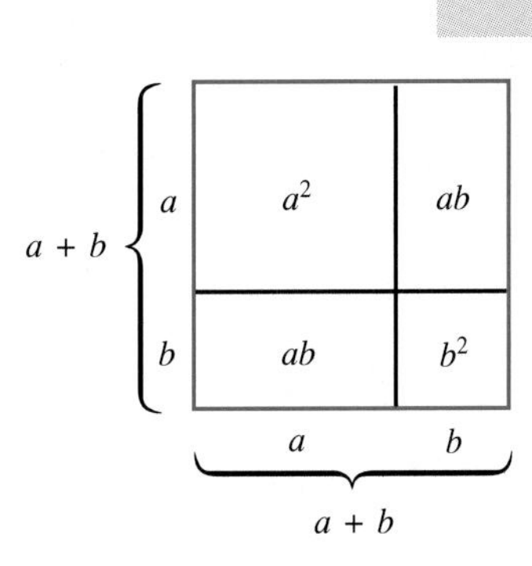

$$\begin{aligned} (a + b)^2 &= (a + b)(a + b) \\ &\quad \;\; \text{F} \quad\; \text{O} \quad\; \text{I} \quad\; \text{L} \\ &= a^2 + ab + ba + b^2 \\ &= a^2 + 2ab + b^2 \end{aligned}$$

This product can be visualized geometrically by analyzing areas. The area of the square shown at left can be written as $(a + b)^2$ or as the sum of the areas of the smaller rectangles.

The sum of the areas of the smaller rectangles is $a^2 + ab + ab + b^2$ or $a^2 + 2ab + b^2$. Thus,

$$(a + b)^2 = a^2 + 2ab + b^2$$

We use this result as the basis of a quick method for squaring a binomial. The method has two forms: the sum of terms and the difference of terms. We call these **special products.**

Square of a Binomial

$$(a + b)^2 = a^2 + 2ab + b^2, \qquad (a - b)^2 = a^2 - 2ab + b^2$$

A binomial squared is the sum of the first term squared, twice the product of both terms, and the second term squared.

EXAMPLE 8 Find the following products.

a. $(x + 5)^2$ **b.** $(x - 9)^2$ **c.** $(3x + 2z)^2$ **d.** $(4m^2 - 3n)^2$

Solution:

a. $(x + 5)^2 = x^2 + 2 \cdot x \cdot 5 + 5^2 = x^2 + 10x + 25$

b. $(x - 9)^2 = x^2 - 2 \cdot x \cdot 9 + 9^2 = x^2 - 18x + 81$

c. $(3x + 2z)^2 = (3x)^2 + 2(3x)(2z) + (2z)^2 = 9x^2 + 12xz + 4z^2$

d. $(4m^2 - 3n)^2 = (4m^2)^2 - 2(4m^2)(3n) + (3n)^2 = 16m^4 - 24m^2n + 9n^2$ ■

HELPFUL HINT

Note that $(a + b)^2 = a^2 + 2ab + b^2$, **not** $a^2 + b^2$. Also, $(a - b)^2 = a^2 - 2ab + b^2$, **not** $a^2 - b^2$.

4 Another special product applies to the sum and difference of the same two terms. Multiply $(a + b)(a - b)$.

$$\begin{aligned}(a + b)(a - b) &= a^2 - ab + ba - b^2 \\ &= a^2 - b^2\end{aligned}$$

Product of the Sum and Difference of Two Terms

$$(a + b)(a - b) = a^2 - b^2$$

The product of the sum and difference of the same two terms is the difference of the first term squared and the second term squared.

EXAMPLE 9 Find the following products.

a. $(x - 3)(x + 3)$ **b.** $(4y + 1)(4y - 1)$ **c.** $(x^2 + 2y)(x^2 - 2y)$

Solution:

a. $(x - 3)(x + 3) = x^2 - 3^2 = x^2 - 9$

b. $(4y + 1)(4y - 1) = (4y)^2 - 1^2 = 16y^2 - 1$

c. $(x^2 + 2y)(x^2 - 2y) = (x^2)^2 - (2y)^2 = x^4 - 4y^2$ ■

EXAMPLE 10 Multiply $[3 + (2a + b)]^2$.

Solution: Think of 3 as the first term and $(2a + b)$ as the second term, and apply the method for squaring a binomial.

$$[3 + (2a + b)]^2 = \underbrace{(3)^2}_{\text{First term squared}} + \underbrace{2(3)(2a + b)}_{\text{Twice the product of both terms}} + \underbrace{(2a + b)^2}_{\text{Last term squared}}$$

$$= 9 + 6(2a + b) + \overbrace{(2a)^2 + 2(2a)(b) + b^2}^{(2a + b)^2}$$

$$= 9 + 12a + 6b + (2a)^2 + 2(2a)(b) + b^2 \quad \text{Square } (2a + b).$$

$$= 9 + 12a + 6b + 4a^2 + 4ab + b^2 \quad ■$$

EXAMPLE 11 Multiply $[(5x - 2y) - 1][(5x - 2y) + 1]$.

Solution: Think of $(5x - 2y)$ as the first term and 1 as the second term, and apply the method for the product of the sum and difference of two terms.

$$[(5x - 2y) - 1][(5x - 2y) + 1] = \underbrace{(5x - 2y)^2}_{\text{First term squared}} - \underbrace{1^2}_{\text{Second term squared}}$$

$$= (5x - 2y)^2 - 1$$

$$= (5x)^2 - 2(5x)(2y) + (2y)^2 - 1 \quad \text{Square } (5x - 2y).$$

$$= 25x^2 - 20xy + 4y^2 - 1 \quad ■$$

Our work in multiplying polynomials is often useful when evaluating polynomial functions.

EXAMPLE 12 If $f(x) = x^2 + 5x - 2$, find $f(a + 1)$.

Solution: To find $f(a + 1)$, replace x with the expression $a + 1$ in the polynomial function $f(x)$.

$$f(x) = x^2 + 5x - 2$$

$$f(a + 1) = (a + 1)^2 + 5(a + 1) - 2$$

$$= a^2 + 2a + 1 + 5a + 5 - 2$$

$$= a^2 + 7a + 4 \quad ■$$

EXERCISE SET 5.4

Multiply. See Example 1.

1. $(-4x^3)(3x^2)$

2. $(-6a)(4a)$

3. $(-xyz)(-9xy^2z^2)$

4. $(-4yt^2)(6yt^3z)$

Multiply. See Example 2.

5. $3x(4x + 7)$

6. $5x(6x - 4)$

7. $-6xy(4x + y)$

8. $-8y(6xy + 4x)$

9. $-4ab(xa^2 + ya^2 - 3)$

10. $-6b^2z(z^2a + baz - 3b)$

Multiply. See Example 3.

11. $(x - 3)(2x + 4)$
12. $(y + 5)(3y - 2)$
13. $(2x + 3)(x^3 - x + 2)$
14. $(a + 2)(3a^2 - a + 5)$
15. $(x^2 + 2x - 1)^2$
16. $(2x^2 - 3x + 2)^2$

Multiply vertically. See Examples 4 and 5.

17. $\begin{array}{r} 3x + 2 \\ \underline{x + 4} \end{array}$
18. $\begin{array}{r} 2s - 3 \\ \underline{s + 2} \end{array}$
19. $\begin{array}{r} 2x^2 - 4x + 2 \\ \underline{x - 5} \end{array}$
20. $\begin{array}{r} -3b^2 + 2b - 4 \\ \underline{b - 3} \end{array}$

21. $\begin{array}{r} a^2 - 4a - 3 \\ \underline{a^2 - 2a - 5} \end{array}$
22. $\begin{array}{r} 3b^2 + 2b - 2 \\ \underline{-3b^2 + b - 4} \end{array}$
23. $\begin{array}{r} 2a^2 + ab + b^2 \\ \underline{3a^2 - ab + b^2} \end{array}$
24. $\begin{array}{r} a^2 - 4ab + 3b^2 \\ \underline{4a^2 + 4ab - 5} \end{array}$

Multiply the binomials. See Examples 6 and 7.

25. $(x - 3)(x + 4)$
26. $(c - 3)(c + 1)$
27. $(2x - 8)(2x - 4)$
28. $(3n - 9)(n + 7)$
29. $(3x - 1)(x + 3)$
30. $(5d - 3)(d + 6)$
31. $\left(3x + \frac{1}{2}\right)\left(3x - \frac{1}{2}\right)$
32. $\left(2x - \frac{1}{3}\right)\left(2x + \frac{1}{3}\right)$

Multiply using special product methods. See Examples 8 and 9.

33. $(x + 4)^2$
34. $(x - 5)^2$
35. $(6y - 1)(6y + 1)$
36. $(x - 9)(x + 9)$
37. $(3x - y)^2$
38. $(4x - z)^2$
39. $(3b - 6y)(3b + 6y)$
40. $(2x - 4y)(2x + 4y)$

Multiply using special product methods. See Examples 10 and 11.

41. $[3 + (4b + 1)]^2$
42. $[5 - (3b - 3)]^2$
43. $[(2s - 3) - 1][(2s - 3) + 1]$
44. $[(2y + 5) + 6][(2y + 5) - 6]$
45. $[(xy + 4) - 6]^2$
46. $[(2a^2 + 4a) + 1]^2$

If $f(x) = x^2 - 3x$, find the following. See Example 12.

47. $f(a)$
48. $f(c)$
49. $f(a + 5)$
50. $f(b - 2)$

Multiply.

51. $(3ab)(-4b)$
52. $(-6x^2y)(2x^3)$
53. $4y(x^2 + y + z)$
54. $3x(6x^2 + 4y - 3)$
55. $(3x + 1)(3x + 5)$
56. $(4x - 5)(5x + 6)$
57. $(2x^3 + 5)(5x^2 + 4x + 1)$
58. $(3y^3 - 1)(3y^3 - 6y + 1)$
59. $(7x - 3)(7x + 3)$
60. $(4x + 1)(4x - 1)$
61. $\begin{array}{r} 3x^2 + 4x - 4 \\ \underline{3x + 6} \end{array}$
62. $\begin{array}{r} 6x^2 + 2x - 1 \\ \underline{3x - 6} \end{array}$

63. $\left(4x + \frac{1}{3}\right)\left(4x - \frac{1}{2}\right)$
64. $\left(4y - \frac{1}{3}\right)\left(3y - \frac{1}{8}\right)$
65. $(6x + 1)^2$
66. $(4x + 7)^2$
67. $(x^2 + 2y)(x^2 - 2y)$
68. $(3x + 2y)(3x - 2y)$
69. $-6a^2b^2(5a^2b^2 - 6a - 6b)$
70. $7x^2y^3(-3ax - 4xy + z)$
71. $(a - 4)(2a - 4)$
72. $(2x - 3)(x + 1)$
73. $(7ab + 3c)(7ab - 3c)$
74. $(3xy - 2b)(3xy + 2b)$
75. $[4 - (3x + 1)][4 + (3x + 1)]$
76. $[7 - (3x + 5)][7 + (3x + 5)]$
77. $(m - 4)^2$
78. $(x + 2)^2$
79. $[3 + (2y - c)]^2$
80. $[4 - (3x + 2y)]^2$

81. $[(5x - 2y) - 4][(5x - 2y) + 4]$

82. $[(4x - 2y) + 3][(4x - 2y) - 3]$

83. $(3x + 1)^2$

84. $(4x + 6)^2$

85. $(y - 4)(y - 3)$

86. $(c - 8)(c + 2)$

87. $[4 - (2x + y)][4 + (2x + y)]$

88. $[7 - (6x + b)][(7 + (6x + b)]$

89. $(x + y)(2x - 1)(x + 1)$

90. $(z + 2)(z - 3)(2z + 1)$

91. $(3x^2 + 2x - 1)^2$

92. $(4x^2 + 4x - 4)^2$

93. $(3x + 1)(4x^2 - 2x + 5)$

94. $(2x - 1)(5x^2 - x - 2)$

Solve.

95. If $R(x) = x + 5$, find the following.

a. $R(x) + R(x)$

b. $R(x) \cdot R(x)$

96. If $C(x) = x^2 - 2$, find the following.

a. $C(x) + C(x)$

b. $C(x) \cdot C(x)$

97. If $P(x) = 3x^2 + 1$, find the following.

a. $P(a)$

b. $P(a - 1)$

98. If $f(x) = -x^2 + x$, find the following.

a. $f(h)$

b. $f(a + h)$

Express each of the following as polynomials.

99. Find the area of the following rectangle.

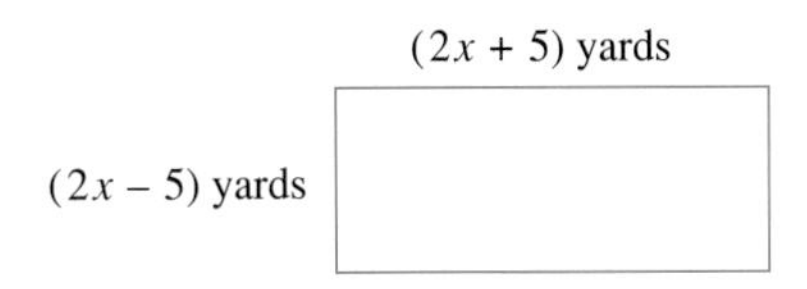

100. Find the area of the square-shaped field.

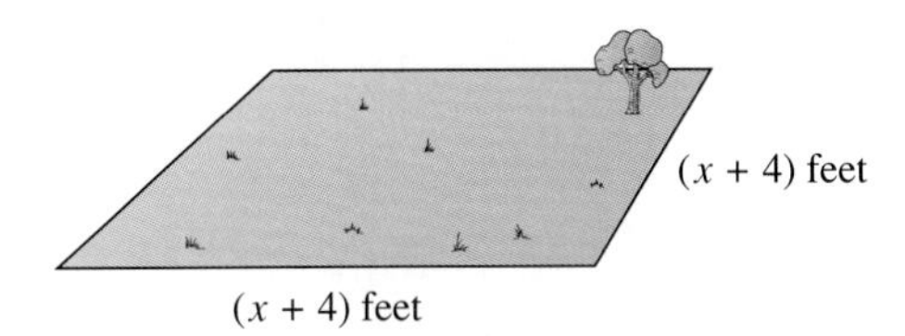

101. Find the area of the following triangle.

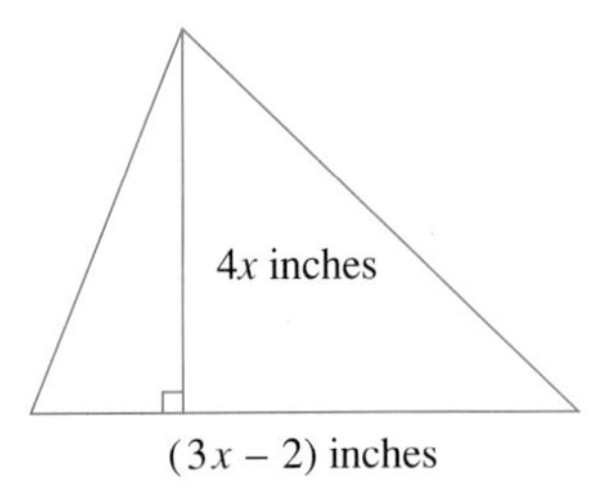

102. Find the volume of the cube-shaped glass block.

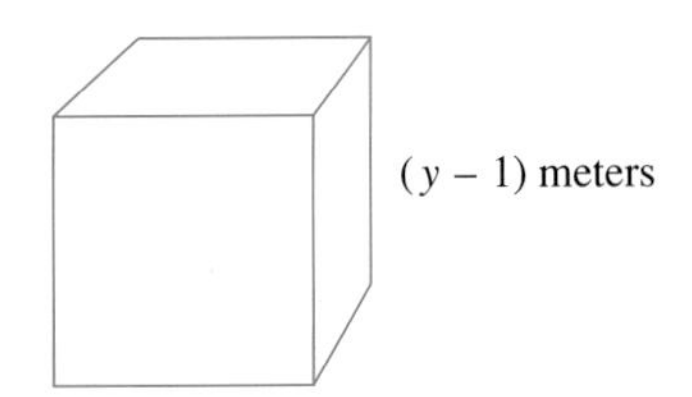

103. Find the volume of the refrigerator.

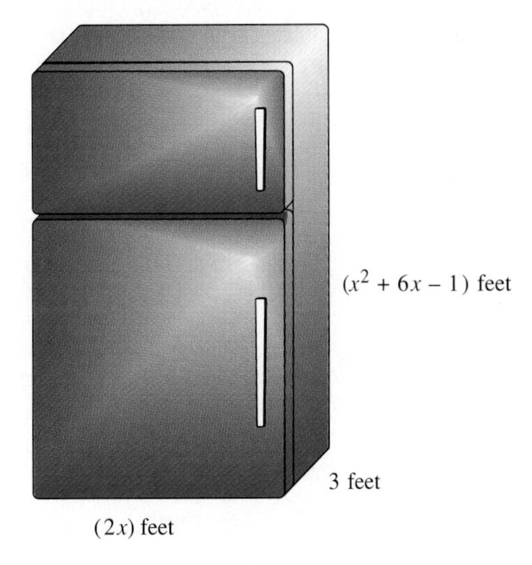

104. Find the area of the circle. Do not approximate π.

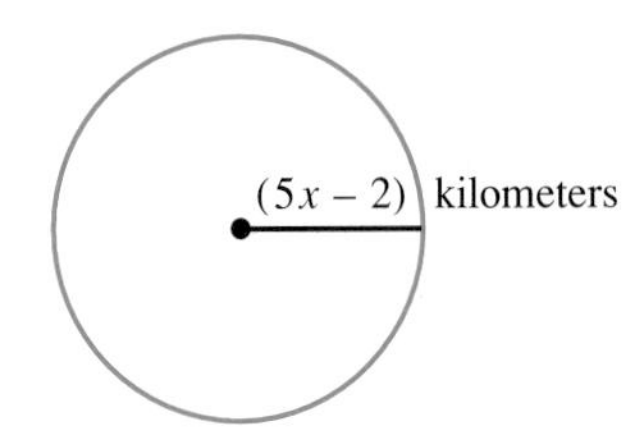

Skill Review

Simplify. See Section 5.1.

105. $\dfrac{3x^3y^2}{12x}$ **106.** $\dfrac{-36xb^3}{9xb^2}$ **107.** $\dfrac{144x^5y^5}{-16x^2y}$ **108.** $\dfrac{48x^3y^2}{-4xy}$

Solve the following. See Section 3.4.

109. $|x - 5| = 9$

110. $|2y + 1| = 1$

111. $\left|\dfrac{x}{2} + 7\right| + 1 = 8$

112. $\left|\dfrac{5x}{3} - 2\right| + 9 = 14$

Solve the following. See Section 3.7.

113. $|x + 3| < 10$

114. $|x - 4| \geq 13$

115. $|3x - 6| \geq 3$

116. $|5x + 8| < 18$

A Look Ahead

Multiply. Assume that variables are not 0. See the following example.

EXAMPLE Multiply.

$(x^{-2} + y^{-1})(x^{-1} - y^3)$

Solution: $(x^{-2} + y^{-1})(x^{-1} - y^3) = x^{-2}x^{-1} - x^{-2}y^3 + x^{-1}y^{-1} - y^{-1}y^3$

$= x^{-3} - x^{-2}y^3 + x^{-1}y^{-1} - y^2$ ■

117. $5x^{-1}(x^{-2} + 3x^{-1} + 2)$ **118.** $x^{-2}(7x^2 + 2x - 6)$ **119.** $(x^{-2} + 2y)(x^{-2} - 2y)$

120. $(4x^{-3} + 6)(x + 1)$ **121.** $(3a + y^{-4})(a + y^{-2})$ **122.** $(6z^{-1} + 2y^{-3})(6z^{-1} - 2y^{-3})$

Multiply. Assume that all variables in the exponents represent integers and that all other variables are not 0. See the following example.

EXAMPLE Multiply.

$6a^n(3a^{2n} - 5)$

Solution: $6a^n(3a^{2n} - 5) = 6a^n(3a^{2n}) + 6a^n(-5)$

$= 18a^{3n} - 30a^n$ ■

123. $6a^4(3a^{n+1} - 6)$ **124.** $4a^{n+1}(3a^{n-1} - 2)$ **125.** $(3x^y + 7)^2$ **126.** $(4y^{2c} - 5)(4y^{2c} + 5)$

Writing in Mathematics

127. Explain how to multiply a polynomial by a polynomial.

128. Explain why $(3x + 2)^2$ does not equal $9x^2 + 4$.

129. Simplify each of the following. Explain the difference between the two problems.

a. $(3x + 5) + (3x + 7)$

b. $(3x + 5)(3x + 7)$

5.5 Division of Polynomials

OBJECTIVES

1. Divide by a monomial.
2. Divide by a polynomial.

Tape IA 11

1 When dividing a **monomial** by a **monomial,** we use the rules for exponents developed at the beginning of this chapter. In this section, assume that a variable in the denominator does not have a value that makes the denominator 0.

EXAMPLE 1 Simplify $\dfrac{2xz^4}{3x^5z^3}$.

Solution: $\dfrac{2xz^4}{3x^5z^3} = \dfrac{2}{3}x^{1-5}z^{4-3} = \dfrac{2}{3}x^{-4}z^1 = \dfrac{2z}{3x^4}$ ■

Dividing a **polynomial** by a **monomial** uses our knowledge of fractions, since a fraction, after all, is a quotient. Recall that two fractions that have a common denominator are added by adding the numerators:

$$\frac{a}{c} + \frac{b}{c} = \frac{a+b}{c}$$

If a, b, and c are monomials, we might read this equation from right to left and gain insight into dividing a polynomial by a monomial.

To Divide a Polynomial by a Monomial

Divide each term in the polynomial by the monomial.

$$\frac{a+b}{c} = \frac{a}{c} + \frac{b}{c}, \qquad c \neq 0$$

EXAMPLE 2 Divide $10x^2 - 5x + 20$ by 5.

Solution: Divide each term of $10x^2 - 5x + 20$ by 5 and simplify.

$$\frac{10x^2 - 5x + 20}{5} = \frac{10x^2}{5} - \frac{5x}{5} + \frac{20}{5} = 2x^2 - x + 4$$

To check, see that (quotient)(divisor) = dividend or

$$(2x^2 - x + 4)(5) = 10x^2 - 5x + 20$$ ■

EXAMPLE 3 Find the quotient: $\dfrac{7a^2b - 2ab^2}{2ab^2}$.

Solution: Divide each term of the polynomial in the numerator by $2ab^2$.

$$\frac{7a^2b - 2ab^2}{2ab^2} = \frac{7a^2b}{2ab^2} - \frac{2ab^2}{2ab^2} = \frac{7a}{2b} - 1$$ ■

EXAMPLE 4 Find the quotient: $\dfrac{3x^5y^2 - 15x^3y - 6x}{6x^2y^3}$.

Solution: Divide each term in the numerator by $6x^2y^3$.

$$\frac{3x^5y^2 - 15x^3y - 6x}{6x^2y^3} = \frac{3x^5y^2}{6x^2y^3} - \frac{15x^3y}{6x^2y^3} - \frac{6x}{6x^2y^3} = \frac{x^3}{2y} - \frac{5x}{2y^2} - \frac{1}{xy^3}$$ ■

2 To divide a polynomial by a polynomial other than a monomial, we use **long division.** Polynomial long division is similar to long division of real numbers. We review long division of real numbers by dividing 7 into 296.

$$\begin{array}{rl} & 42 \\ \text{Divisor:}\quad 7\overline{)296} & \\ -28 & \quad 4(7) = 28. \\ \hline 16 & \quad \text{Subtract and bring down the next digit in the dividend.} \\ -14 & \quad 2(7) = 14. \\ \hline 2 & \quad \text{Subtract. The remainder is 2.} \end{array}$$

The quotient is $42\dfrac{2 \text{ (remainder)}}{7 \text{ (divisor)}}$. To check, notice that

$$42(7) + 2 = 296 \qquad \text{The dividend.}$$

This same division process can be applied to polynomials, as shown next.

EXAMPLE 5 Divide $2x^2 - x - 10$ by $x + 2$.

Solution: $2x^2 - x - 10$ is the dividend while $x + 2$ is the divisor.

Step 1 Divide $2x^2$ by x.

$$\begin{array}{r} 2x \phantom{{}- x - 10} \\ x + 2\overline{)2x^2 - x - 10} \end{array} \qquad \frac{2x^2}{x} = 2x,$$

so $2x$ is the first term of the quotient.

Step 2 Multiply $2x(x + 2)$.

$$\begin{array}{r} 2x \phantom{{}- 10} \\ x + 2\overline{)2x^2 - x - 10} \\ 2x^2 + 4x \phantom{{}- 10} \end{array}$$

$2x(x + 2)$
Like terms are lined up vertically.

Step 3 Subtract $(2x^2 + 4x)$ from $(2x^2 - x - 10)$ by changing the signs of $(2x^2 + 4x)$ and adding.

$$\begin{array}{r} 2x \phantom{{}- 10} \\ x + 2\overline{)\ 2x^2 - x - 10} \\ -2x^2 - 4x \phantom{{}- 10} \\ \hline -5x \phantom{{}- 10} \end{array}$$

Step 4 Bring down the next term, -10, and start the process over.

$$\begin{array}{r} 2x \\ x+2\overline{)\,2x^2 - x - 10} \\ \underline{-2x^2 - 4x} \\ -5x - 10 \end{array}$$

Step 5 Divide $-5x$ by x.

$$\begin{array}{r} 2x - 5\\ x+2\overline{)\,2x^2 - x - 10} \\ \underline{-2x^2 - 4x} \\ -5x - 10 \end{array}$$

$\frac{-5x}{x} = -5$, so -5 is the second term of the quotient.

Step 6 Multiply $-5(x + 2)$.

$$\begin{array}{r} 2x - 5\\ x+2\overline{)\,2x^2 - x - 10} \\ \underline{-2x^2 - 4x} \\ -5x - 10 \\ \underline{-5x - 10} \end{array}$$

$-5(x + 2)$.

Like terms are lined up vertically.

Step 7 Subtract $(-5x - 10)$ from $(-5x - 10)$.

$$\begin{array}{r} 2x - 5\\ x+2\overline{)\,2x^2 - x - 10} \\ \underline{-2x^2 - 4x} \\ -5x - 10 \\ \underline{+5x + 10} \\ 0 \end{array}$$

Then $\frac{2x^2 - x - 10}{x + 2} = 2x - 5$. There is no remainder.

Check this result by multiplying $2x - 5$ by $x + 2$. Their product is $(2x - 5)(x + 2) = 2x^2 - x - 10$, the dividend. ■

EXAMPLE 6 Find the quotient: $\frac{6x^2 - 19x + 12}{3x - 5}$.

Solution:

$$\begin{array}{r} 2x \\ 3x-5\overline{)\,6x^2 - 19x + 12} \\ \underline{6x^2 - 10x} \\ -9x + 12 \end{array}$$

Divide: $\frac{6x^2}{3x} = 2x$.

Multiply: $2x(3x - 5)$.

Subtract by changing the signs of $6x^2 - 10x$ and adding. Bring down $+12$.

$$\begin{array}{r} 2x - 3\\ 3x-5\overline{)\,6x^2 - 19x + 12} \\ \underline{6x^2 - 10x} \\ -9x + 12 \\ \underline{-9x + 15} \\ -3 \end{array}$$

Divide: $\frac{-9x}{3x} = -3$.

Multiply: $-3(3x - 5)$.

Subtract.

When checking, we call the **divisor** polynomial $3x - 5$. The **quotient** polynomial is $2x - 3$. The **remainder** polynomial is -3. See that

$$\textbf{dividend} = \textbf{divisor} \cdot \textbf{quotient} + \textbf{remainder}$$

or

$$\begin{aligned} 6x^2 - 19x + 12 &= (3x - 5)(2x - 3) + (-3) \\ &= 6x^2 - 19x + 15 - 3 \\ &= 6x^2 - 19x + 12 \end{aligned}$$

The division checks, so

$$\frac{6x^2 - 19x + 12}{3x - 5} = 2x - 3 - \frac{3}{3x-5}$$ ■

EXAMPLE 7 Divide $2x^3 + 3x^4 - 8x + 6$ by $x^2 - 1$.

Solution: Before dividing, we will write both the divisor and the dividend in descending order of exponents. Any "missing powers" can be represented by the product of 0 and the variable raised to the missing power. There is no x^2 term in the dividend, so include $0x^2$ to represent the missing term. Also, there is no x term in the divisor, so include $0x$ in the divisor.

$$\begin{array}{r} 3x^2 + 2x + 3 \\ x^2 + 0x - 1\overline{)3x^4 + 2x^3 + 0x^2 - 8x + 6} \\ \underline{3x^4 + 0x^3 - 3x^2} \\ 2x^3 + 3x^2 - 8x \\ \underline{2x^3 + 0x^2 - 2x} \\ 3x^2 - 6x + 6 \\ \underline{3x^2 + 0x - 3} \\ -6x + 9 \end{array}$$

$\frac{3x^4}{x^2} = 3x^2$.
$3x^2(x^2 + 0x - 1)$.
Subtract. Bring down $-8x$.
$2x^3/x^2 = 2x$, a term of the quotient.
$2x(x^2 + 0x - 1)$.
Subtract. Bring down 6.
$3x^2/x^2 = 3$, a term of the quotient.
$3(x^2 + 0x - 1)$.
Subtract.

The division process is finished when the degree of the remainder polynomial is less than the degree of the divisor. Thus,

$$\frac{3x^4 + 2x^3 - 8x + 6}{x^2 - 1} = 3x^2 + 2x + 3 + \frac{-6x + 9}{x^2 - 1}$$

To check, see that

$$3x^4 + 2x^3 - 8x + 6 = (x^2 - 1)(3x^2 + 2x + 3) + (-6x + 9)$$ ■

EXAMPLE 8 Divide $27x^3 + 8$ by $2 + 3x$.

Solution: Write both the divisor and the dividend in descending order of exponents. Replace the missing terms in the dividend with $0x^2$ and $0x$.

$$
\begin{array}{r}
9x^2 - 6x + 4 \\
3x + 2\overline{)27x^3 + 0x^2 + 0x + 8} \\
\underline{27x^3 + 18x^2} \\
-18x^2 + 0x \\
\underline{-18x^2 - 12x} \\
12x + 8 \\
\underline{12x + 8}
\end{array}
\qquad
\begin{array}{l}
\\ \\
9x^2(3x + 2). \\
\text{Subtract. Bring down } 0x. \\
-6x(3x + 2). \\
\text{Subtract. Bring down } 8. \\
4(3x + 2).
\end{array}
$$

Thus, $\dfrac{27x^3 + 8}{3x + 2} = 9x^2 - 6x + 4.$ ∎

EXERCISE SET 5.5

Find each quotient. See Example 1.

1. $\dfrac{50b^{10}}{25b^5}$

2. $\dfrac{x^3y^6}{x^6y^2}$

3. $\dfrac{26x^2y^3z^7}{13x^6bz^5}$

4. $\dfrac{-48ab^{10}}{32a^4b^3}$

5. Divide $-2x^4y^2$ by $6x^4y^4$.

6. Divide $x^{17}y^5$ by $-x^7y^{10}$.

Find each quotient. See Examples 2 through 4.

7. Divide $4a^2 + 8a$ by $2a$.

8. Divide $6x^4 - 3x^3$ by $3x^2$.

9. $\dfrac{12a^5b^2 + 16a^4b}{4a^4b}$

10. $\dfrac{4x^3y + 12x^2y^2 - 4xy^3}{4xy}$

11. $\dfrac{4x^2y^2 + 6xy^2 - 4y^2}{2y^2}$

12. $\dfrac{6x^5 + 74x^4 + 24x^3}{2x^3}$

13. $\dfrac{4x^2 + 8x + 4}{4}$

14. $\dfrac{15x^3 - 5x^2 + 10x}{5x^2}$

Find each quotient. See Examples 5 through 8.

15. $\dfrac{x^2 + 3x + 3}{x + 2}$

16. $\dfrac{y^2 + 7y + 10}{y + 5}$

17. $\dfrac{2x^2 - 6x - 8}{x + 1}$

18. $\dfrac{3x^2 + 19x + 20}{x + 5}$

19. Divide $2x^2 + 3x - 2$ by $2x + 4$

20. Divide $6x^2 - 17x - 3$ by $3x - 9$

21. $\dfrac{4x^3 + 7x^2 + 8x + 20}{2x + 4}$

22. $\dfrac{18x^3 + x^2 - 90x - 5}{9x^2 - 45}$

Find each quotient.

23. Divide $25a^2b^{12}$ by $10a^5b^7$.

24. Divide $12a^2b^3$ by $8a^7b$.

25. $\dfrac{x^6y^6 - x^3y^3}{x^3y^3}$

26. $\dfrac{25xy^2 + 75xyz + 125x^2yz}{-5x^2y}$

27. $\dfrac{a^2 + 4a + 3}{a + 1}$

28. $\dfrac{3x^2 - 14x + 16}{x - 2}$

29. $\dfrac{2x^2 + x - 10}{x - 2}$

30. $\dfrac{x^2 - 7x + 12}{x - 5}$

31. Divide $-16y^3 + 24y^4$ by $-4y^2$.

32. Divide $-20a^2b + 12ab^2$ by $-4ab$.

33. $\dfrac{2x^2 + 13x + 15}{x - 5}$

34. $\dfrac{2x^2 + 13x + 5}{2x + 3}$

35. $\dfrac{20x^2y^3 + 6xy^4 - 12x^3y^5}{2xy^3}$

36. $\dfrac{3x^2y + 6x^2y^2 + 3xy}{3xy}$

37. $\dfrac{6x^2 + 16x + 8}{3x + 2}$

38. $\dfrac{x^2 - 25}{x + 5}$

39. $\dfrac{2y^2 + 7y - 15}{2y - 3}$

40. $\dfrac{3x^2 - 4x + 6}{x - 2}$

41. Divide $4x^2 - 9$ by $2x - 3$.

42. Divide $8x^2 + 6x - 27$ by $4x + 9$.

43. Divide $2x^3 + 6x - 4$ by $x + 4$.

44. Divide $4x^3 - 5x$ by $2x - 1$.

45. Divide $3x^2 - 4$ by $x - 1$.

46. Divide $x^2 - 9$ by $x + 4$.

47. $\dfrac{-13x^3 + 2x^4 + 16x^2 - 9x + 20}{5 - x}$

48. $\dfrac{5x^2 - 5x + 2x^3 + 20}{4 + x}$

49. Divide $3x^5 - x^3 + 4x^2 - 12x - 8$ by $x^2 - 2$.

50. Divide $-8x^3 + 2x^4 + 19x^2 - 33x + 15$ by $x^2 - x + 5$.

51. $\dfrac{3x^3 - 5}{3x^2}$

52. $\dfrac{14x^3 - 2}{7x - 1}$

Express each of the following as a polynomial in x.

53. The perimeter of a square is $(12x^3 + 4x - 16)$ feet. Find the length of its side.

54. The area of the following parallelogram is $(10x^2 + 31x + 15)$ square meters. If its base is $(5x + 3)$ meters, find its height.

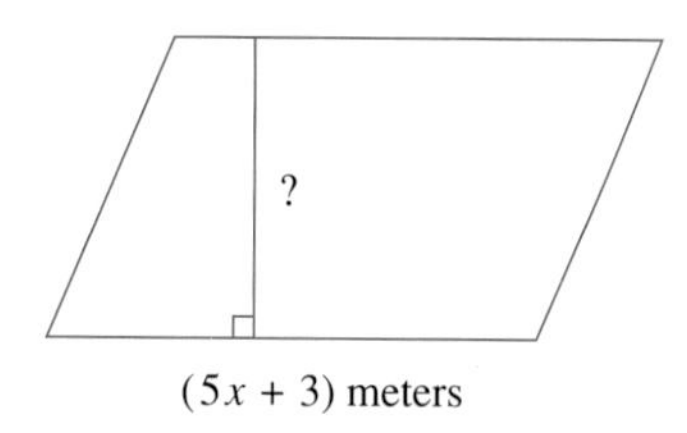

55. The area of the top of a Ping-Pong table is $(49x^2 + 70x - 200)$ square inches. If its length is $(7x + 20)$ inches, find its width.

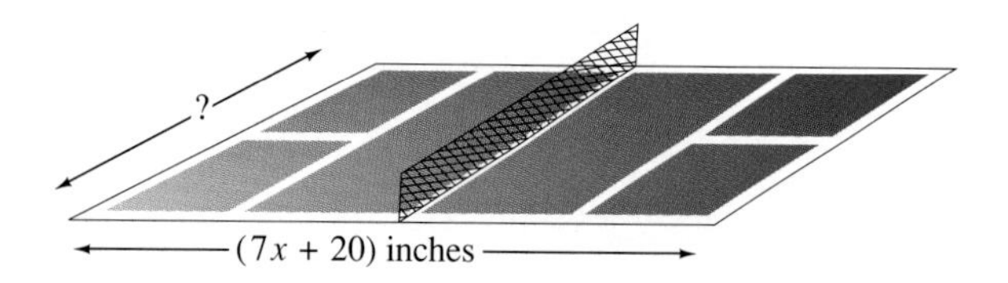

56. If the area of the rectangle is $15x^2 - 29x - 14$ square inches, find its width.

57. The volume of the swimming pool shown is $(36x^5 - 12x^3 + 6x^2)$ cubic feet. If its height is $2x$ feet and its width is $3x$ feet, find its length.

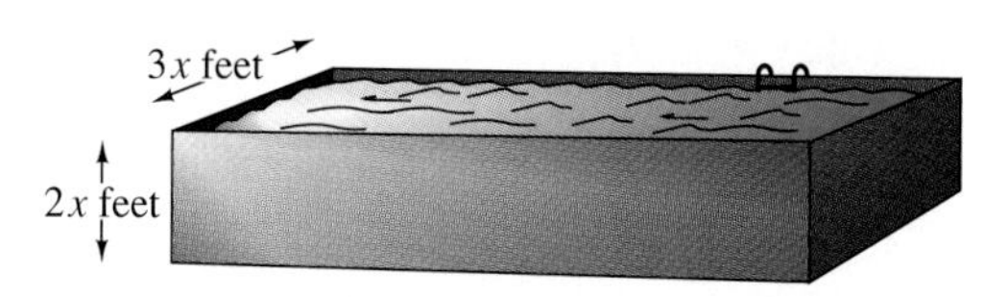

58. A board of length $3x^4 + 6x^2 - 18$ meters is to be cut into three pieces of the same length. Find the length of each piece.

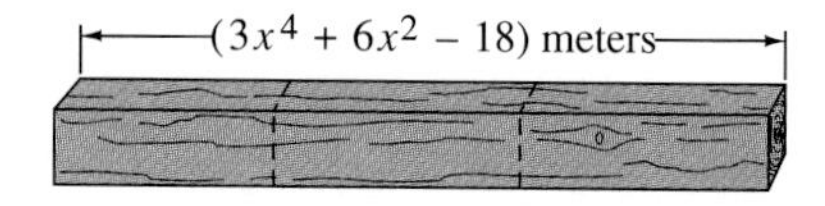

59. If the area of a parallelogram is $2x^2 - 17x + 35$ square centimeters and its base is $2x - 7$ centimeters, find its height.

60. Find $P(1)$ for the polynomial function $P(x) = 3x^3 + 2x^2 - 4x + 3$. Next divide $3x^3 + 2x^2 - 4x + 3$ by $x - 1$. Compare the remainder with $P(1)$.

61. Find $P(-2)$ for the polynomial function $P(x) = x^3 - 4x^2 - 3x + 5$. Next divide $x^3 - 4x^2 - 3x + 5$ by $x + 2$. Compare the remainder with $P(-2)$.

Skill Review

A histogram is a special type of bar graph in which the width of the bar stands for a single number or a range of numbers and the height of each bar represents of number of occurrences. Use the histogram below to answer Exercises 62 through 64. See Section 2.6.

Percentage of People Not Covered by Health Insurance

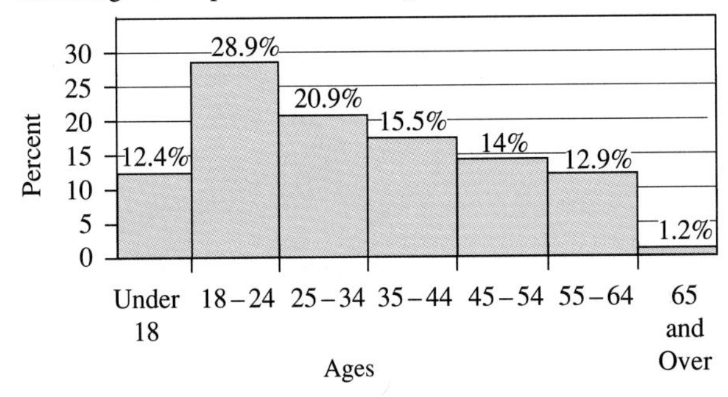

Source: *USA Today,* June 10, 1994.

62. What percent of people aged 45–54 have no health insurance?

63. Find the age group of people who have the highest percent of no health insurance.

64. What percent of people less than 25 years old have no health insurance?

65. Roll a single die 20 times and tally the results. Draw a histogram from your findings. Use the numbers 1 through 6 for your horizontal axis and the number of occurrences for your vertical axis.

66. Compare your histogram from Exercise 65 with a classmate's. Are they the same? Why or why not?

Solve. See Section 2.5.

67. The lowest point of elevation in Vermont is 103 feet higher than the lowest point in Louisiana. If the sum of these elevations is 87 feet, find each elevation.

68. New York has 58 more hazardous waste sites than North Carolina. If the total number of waste sites for these two states is 100, find the number of hazardous waste sites in each state.

See Section 2.6.

69. Plot the ordered pairs $(5, 3)$, $(0, -2)$, $(-1, 4)$, $(7, 0)$, $(-3, -3)$, and $(4, -3)$ on a single set of coordinate axes.

Solve each inequality. See Section 3.5.

70. $2x - 5 > 17$

71. $-3(x - 2) < -4$

72. $\dfrac{3x + 1}{2} \leq \dfrac{-3}{4}$

73. $\dfrac{5x}{20} - 2 \geq \dfrac{x - 2}{10}$

A Look Ahead

Find each quotient. See the following example.

EXAMPLE Divide $x^2 - \frac{7}{2}x + 4$ by $x + 2$.

Solution:

$$
\begin{array}{r}
x - \frac{11}{2} \\
x + 2 \overline{\big)\, x^2 - \frac{7}{2}x + 4} \\
\underline{x^2 + 2x \qquad} \\
-\frac{11}{2}x + 4 \\
\underline{-\frac{11}{2}x - 11} \\
15
\end{array}
$$

The quotient is $x - \dfrac{11}{2} + \dfrac{15}{x + 2}$. ■

74. $\dfrac{x^4 + \frac{2}{3}x^3 + x}{x - 1}$

75. $\dfrac{2x^3 + \frac{9}{2}x^2 - 4x - 10}{x + 2}$

76. $\dfrac{3x^4 - x - x^3 + \frac{1}{2}}{2x - 1}$

77. $\dfrac{2x^4 + \frac{1}{2}x^3 + x^2 + x}{x - 2}$

78. $\dfrac{5x^4 - 2x^2 + 10x^3 - 4x}{5x + 10}$

79. $\dfrac{9x^5 + 6x^4 - 6x^2 - 4x}{3x + 2}$

Writing in Mathematics

80. Explain how to check polynomial long division.

81. Try performing the following division without changing the order of the terms. Describe why this makes the process more complicated. Then perform the division again after putting the terms in the dividend in descending order of exponents.

$$\frac{4x^2 - 12x - 12 + 3x^3}{x - 2}$$

5.6 Synthetic Division and the Remainder Theorem

OBJECTIVES

1. Use synthetic division to divide a polynomial by a binomial.
2. Use the remainder theorem to evaluate polynomials.

Tape IA 11

1 When a polynomial is to be divided by a binomial of the form $x - c$, a shortcut process called **synthetic division** may be used. On the left is an example of long division, and on the right is the same example showing the coefficients of the variables only.

$$\begin{array}{r}
2x^2 + 5x + 2 \\
x - 3\overline{)2x^3 - x^2 - 13x + 1} \\
\underline{2x^3 - 6x^2} \qquad\qquad \\
5x^2 - 13x \qquad \\
\underline{5x^2 - 15x} \qquad \\
2x + 1 \\
\underline{2x - 6} \\
7
\end{array}
\qquad\qquad
\begin{array}{r}
2 \quad 5 \quad 2 \\
1 - 3\overline{)2 - 1 - 13 + 1} \\
\underline{2 - 6} \qquad\qquad \\
5 - 13 \qquad \\
\underline{5 - 15} \qquad \\
2 + 1 \\
\underline{2 - 6} \\
7
\end{array}$$

Notice that as long as we keep coefficients of powers of x in the same column, we can perform division of polynomials by performing algebraic operations on the coefficients only. This shortcut process of dividing with coefficients only in a special format is called synthetic division. To find $(2x^3 - x^2 - 13x + 1) \div (x - 3)$ by synthetic division, follow the example shown.

EXAMPLE 1 Use synthetic division to divide $2x^3 - x^2 - 13x + 1$ by $x - 3$.

Solution: To use synthetic division, the divisor must be in the form $x - c$. Since we are dividing by $x - 3$, c is 3. Write down 3 and the coefficients of the dividend.

$$\begin{array}{r|rrrr} 3 & 2 & -1 & -13 & 1 \\ & \downarrow & & & \\ \hline & 2 & & & \end{array}$$

Next draw a line and bring down the first coefficient of the dividend.

$$\begin{array}{r|rrrr} 3 & 2 & -1 & -13 & 1 \\ & & 6 & & \\ \hline & 2 & & & \end{array}$$

Multiply $3 \cdot 2$ and write down the product 6.

$$\begin{array}{r|rrrr} 3 & 2 & -1 & -13 & 1 \\ & & 6 & & \\ \hline & 2 & 5 & & \end{array}$$

Add $-1 + 6$. Write down the sum 5.

$$\begin{array}{r|rrrr} 3 & 2 & -1 & -13 & 1 \\ & & 6 & 15 & \\ \hline & 2 & 5 & 2 & \end{array}$$

$3 \cdot 5 = 15$.
$-13 + 15 = 2$.

$$\begin{array}{r|rrrr} 3 & 2 & -1 & -13 & 1 \\ & & 6 & 15 & 6 \\ \hline & 2 & 5 & 2 & 7 \end{array}$$

$3 \cdot 2 = 6$.
$1 + 6 = 7$.

The quotient is found in the bottom row. The numbers 2, 5, and 2 are the coefficients of the quotient polynomial, and the number 7 is the remainder. The degree of the quotient polynomial is one less than the degree of the dividend. In our example, the degree of the dividend is 3, so the degree of the quotient polynomial is 2. As we found when we performed the long division, the quotient is

$$2x^2 + 5x + 2, \qquad \text{remainder } 7$$

or

$$2x^2 + 5x + 2 + \frac{7}{x - 3} \qquad \blacksquare$$

EXAMPLE 2 Use synthetic division to divide $x^4 - 2x^3 - 11x^2 + 5x + 34$ by $x + 2$.

Solution: The divisor is $x + 2$, which we write in the form $x - c$ as $x - (-2)$. Thus, c is -2. The dividend coefficients are 1, -2, -11, 5, and 34.

$$\begin{array}{r|rrrrr} -2 & 1 & -2 & -11 & 5 & 34 \\ & & -2 & 8 & 6 & -22 \\ \hline & 1 & -4 & -3 & 11 & 12 \end{array}$$

The dividend is a fourth-degree polynomial, so the quotient polynomial is a third-degree polynomial. The quotient is $x^3 - 4x^2 - 3x + 11$ with a remainder of 12. Thus,

$$\frac{x^4 - 2x^3 - 11x^2 + 5x + 34}{x + 2} = x^3 - 4x^2 - 3x + 11 + \frac{12}{x + 2} \qquad \blacksquare$$

HELPFUL HINT

Before dividing by synthetic division, write the dividend in descending order of variable exponents. Any "missing powers" of the variable should be represented by 0 times the variable raised to the missing power.

EXAMPLE 3 If $P(x) = 2x^3 - 4x^2 + 5$:

a. Find $P(2)$ by substitution.

b. Use synthetic division to find the remainder when $P(x)$ is divided by $x - 2$.

Solution: **a.** $P(x) = 2x^3 - 4x^2 + 5$

$$P(2) = 2(2)^3 - 4(2)^2 + 5$$

$$= 2(8) - 4(4) + 5 = 16 - 16 + 5 = 5$$

Thus, $P(2) = 5$.

b. The coefficients of $P(x)$ are 2, −4, 0, and 5. The number 0 is a coefficient of the missing power of x^1. The divisor is $x - 2$, so c is 2.

$$\begin{array}{r|rrrr} 2 & 2 & -4 & 0 & 5 \\ & & 4 & 0 & 0 \\ \hline & 2 & 0 & 0 & 5 \end{array} \quad \text{remainder}$$

The remainder when $P(x)$ is divided by $x - 2$ is 5. ■

2 Notice in the preceding example that $P(2) = 5$ and that the remainder when $P(x)$ is divided by $x - 2$ is 5. This is no accident. This illustrates the **remainder theorem.**

Remainder Theorem

If a polynomial $P(x)$ is divided by $x - c$, then the remainder is $P(c)$.

EXAMPLE 4 Use the remainder theorem and synthetic division to find $P(4)$ if

$$P(x) = 4x^6 - 25x^5 + 35x^4 + 17x^2$$

Solution: To find $P(4)$ by the remainder theorem, we divide $P(x)$ by $x - 4$. The coefficients of $P(x)$ are 4, −25, 35, 0, 17, 0, and 0. Also, c is 4.

$$c \searrow \begin{array}{r|rrrrrrr} 4 & 4 & -25 & 35 & 0 & 17 & 0 & 0 \\ & & 16 & -36 & -4 & -16 & 4 & 16 \\ \hline & 4 & -9 & -1 & -4 & 1 & 4 & 16 \end{array} \quad \text{remainder}$$

Thus, $P(4) = 16$, the remainder. ■

EXERCISE SET 5.6

Use synthetic division to find each quotient. See Examples 1 through 3.

1. $\dfrac{x^2 + 3x - 40}{x - 5}$

2. $\dfrac{x^2 - 14x + 24}{x - 2}$

3. $\dfrac{x^2 + 5x - 6}{x + 6}$

4. $\dfrac{x^2 + 12x + 32}{x + 4}$

5. $\dfrac{x^3 - 7x^2 - 13x + 5}{x - 2}$

6. $\dfrac{x^3 + 6x^2 + 4x - 7}{x + 5}$

7. $\dfrac{4x^2 - 9}{x - 2}$

8. $\dfrac{3x^2 - 4}{x - 1}$

For the given polynomial $P(x)$ and the given c, find $P(c)$ by (a) direct substitution and (b) the remainder theorem. See Examples 3 and 4.

9. $P(x) = 3x^2 - 4x - 1$; $P(2)$

10. $P(x) = x^2 - x + 3$; $P(5)$

11. $P(x) = 4x^4 + 7x^2 + 9x - 1$; $P(-2)$

12. $P(x) = 8x^5 + 7x + 4$; $P(-3)$

13. $P(x) = x^5 + 3x^4 + 3x - 7$; $P(-1)$

14. $P(x) = 5x^4 - 4x^3 + 2x - 1$; $P(-1)$

Use synthetic division to find each quotient.

15. $\dfrac{x^3 - 3x^2 + 2}{x - 3}$

16. $\dfrac{x^2 + 12}{x + 2}$

17. $\dfrac{6x^2 + 13x + 8}{x + 1}$

18. $\dfrac{x^3 - 5x^2 + 7x - 4}{x - 3}$

19. $\dfrac{2x^4 - 13x^3 + 16x^2 - 9x + 20}{x - 5}$

20. $\dfrac{3x^4 + 5x^3 - x^2 + x - 2}{x + 2}$

21. $\dfrac{3x^2 - 15}{x + 3}$

22. $\dfrac{3x^2 + 7x - 6}{x + 4}$

23. $\dfrac{3x^3 - 6x^2 + 4x + 5}{x - \frac{1}{2}}$

24. $\dfrac{8x^3 - 6x^2 - 5x + 3}{x + \frac{3}{4}}$

25. $\dfrac{3x^3 + 2x^2 - 4x + 1}{x - \frac{1}{3}}$

26. $\dfrac{9y^3 + 9y^2 - y + 2}{y + \frac{2}{3}}$

27. $\dfrac{7x^2 - 4x + 12 + 3x^3}{x + 1}$

28. $\dfrac{x^4 + 4x^3 - x^2 - 16x - 4}{x - 2}$

29. $\dfrac{x^3 - 1}{x - 1}$

30. $\dfrac{y^3 - 8}{y - 2}$

31. $\dfrac{x^2 - 36}{x + 6}$

32. $\dfrac{4x^3 + 12x^2 + x - 12}{x + 3}$

For the given polynomial $P(x)$ and the given c, use the remainder theorem to find $P(c)$.

33. $P(x) = x^3 + 3x^2 - 7x + 4$; 1

34. $P(x) = x^3 + 5x^2 - 4x - 6$; 2

35. $P(x) = 3x^3 - 7x^2 - 2x + 5$; -3

36. $P(x) = 4x^3 + 5x^2 - 6x - 4$; -2

37. $P(x) = 4x^4 + x^2 - 2$; -1

38. $P(x) = x^4 - 3x^2 - 2x + 5$; -2

39. $P(x) = 2x^4 - 3x^2 - 2$; $\frac{1}{3}$

40. $P(x) = 4x^4 - 2x^3 + x^2 - x - 4$; $\frac{1}{2}$

41. $P(x) = x^5 + x^4 - x^3 + 3$; $\frac{1}{2}$

42. $P(x) = x^5 - 2x^3 + 4x^2 - 5x + 6$; $\frac{2}{3}$

We say that 2 is a factor of 8 because 2 divides 8 evenly or with a remainder of 0. In the same manner, the polynomial $x - 2$ is a factor of the polynomial $x^3 - 18x^2 + 24x$ because the remainder is 0 when $x^3 - 18x^2 + 24x$ is divided by $x - 2$. Use this information for Exercises 43 through 45.

43. Use synthetic division to show that $x + 3$ is a factor of $x^3 + 3x^2 + 4x + 12$.

44. Use synthetic division to show that $x - 2$ is a factor of $x^3 - 2x^2 - 3x + 6$.

45. From the remainder theorem, the polynomial $x - c$ is a factor of a polynomial function $P(x)$ if $P(c)$ is what value?

46. If a polynomial is divided by $x - 5$, the quotient is $2x^2 + 5x - 6$ and the remainder is 3. Find the original polynomial.

47. If a polynomial is divided by $x + 3$, the quotient is $x^2 - x + 10$ and the remainder is -2. Find the original polynomial.

48. If the area of a parallelogram is $x^4 - 23x^2 + 9x - 5$ square centimeters and its base is $x + 5$ centimeters, find its height.

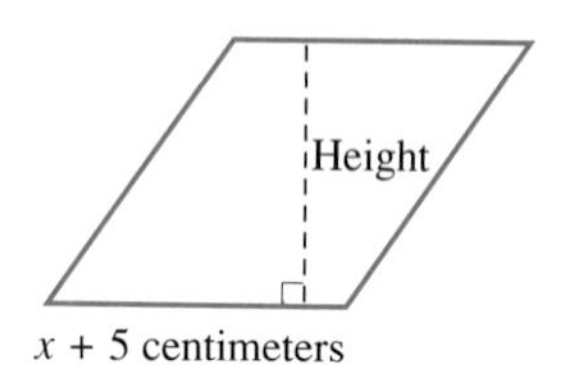

49. If the volume of a box is $x^4 + 6x^3 - 7x^2$ cubic meters, its height is x^2 meters, and its length is $x + 7$ meters, find its width.

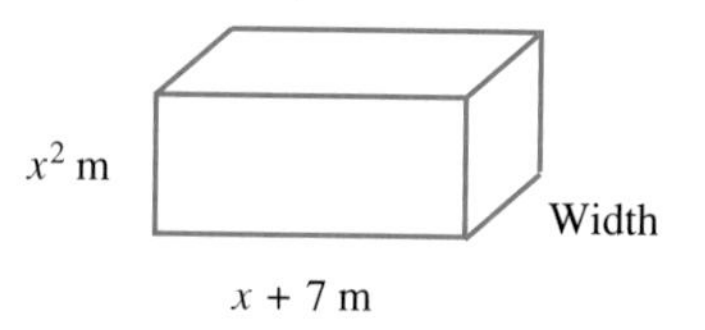

Skill Review

If $f(x) = 3x^2 + 5$, $g(x) = x - 5$, *and* $h(x) = x^2 + 3$, *find the following. See Sections 5.3 and 5.4.*

50. $f(x) + g(x)$

51. $f(x) - h(x)$

52. $2 \cdot g(x) - f(x)$

53. $3 \cdot h(x) + f(x)$

54. $f(x) \cdot g(x)$

55. $h(x) \cdot h(x)$

56. $h(x) \div f(x)$

57. $g(x) \div f(x)$

Insert $<$, $>$, *or* $=$ *to make each statement true. See Section 1.7.*

58. 3^2 _____ $(-3)^2$

59. $(-5)^2$ _____ 5^2

60. -2^3 _____ $(-2)^3$

61. 3^4 _____ $(-3)^4$

Writing in Mathematics

62. Explain an advantage of using the remainder theorem instead of direct substitution.

5.7 Algebra of Functions and Composition of Functions

OBJECTIVES

1. Review function notation.
2. Add, subtract, multiply, and divide functions.
3. Compose functions.

Tape IA 12

1 We begin this section with a review of function notation. Recall that, when ***y*** is a function of ***x***, it means that each value of the independent variable ***x*** corresponds to exactly one value of the dependent variable ***y***. The set of all possible x-values is called the domain of the function, and the set of all corresponding y-values is called the range of the function. To denote that y is a function of x, we write

$$y = f(x)$$

Recall that the symbol $f(x)$ is read "f of x." The letter f is not the only one that can be used, any letter is acceptable. For example, $f(x)$, $g(x)$, $G(x)$, $h(x)$, and $p(x)$ all mean "function of x."

HELPFUL HINT

Recall that the notation $f(x)$ **does not mean** $f \cdot x$. It is a single function notation symbol.

Since the solutions of a two-variable equation such as $y = -x + 2$ are ordered pairs, some two-variable equations "define" a function. The equation $y = -x + 2$ defines a function, because each x-value corresponds to a single y-value. Thus, we are entitled to write

$$y = -x + 2$$

as

$$f(x) = -x + 2$$

Suppose we want to find the **value of the function,** that is, the value of y, when x is 3. This value can be written as $f(3)$.

Since

$$f(x) = -x + 2$$

then

$$f(3) = -3 + 2 = -1$$

Thus, $f(3) = -1$, and the ordered pair $(3, -1)$ is a solution of $f(x) = -x + 2$.

EXAMPLE 1 If $g(x) = 2x + 5$, find:

a. $g(-4)$ **b.** $g(a)$ **c.** $g(x + h)$

Solution: **a.**

$$g(x) = 2x + 5$$
$$g(-4) = 2(-4) + 5$$
$$= -3$$

b.

$$g(x) = 2x + 5$$
$$g(a) = 2(a) + 5$$
$$= 2a + 5$$

c.

$$g(x) = 2x + 5$$
$$g(x + h) = 2(x + h) + 5$$
$$= 2x + 2h + 5$$

■

Recall that equations of the form $f(x) = mx + b$ are called **linear functions.** Their graphs are nonvertical straight lines.

EXAMPLE 2 Graph $H(x) = 3x$.

Solution: If it helps, write $H(x) = 3x$ as $y = 3x$. The function $y = 3x$ or $H(x) = 3x$ is a **linear function** since it can be written in the form $f(x) = mx + b$. This means that its graph is a straight line with slope 3 and y-intercept 0.

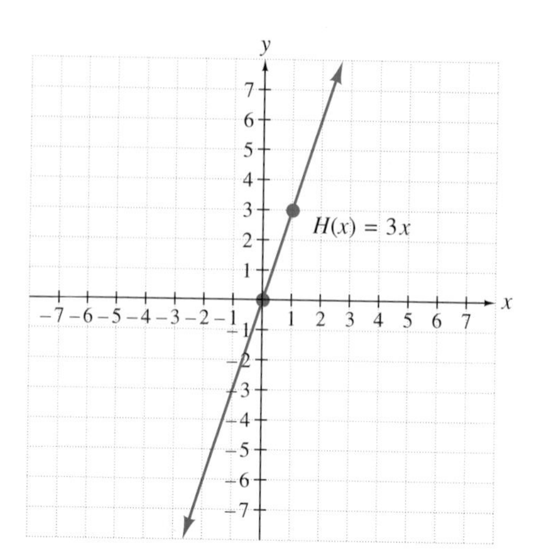

■

2 When we add, subtract, multiply, and divide functions, we form new functions. This **algebra** of functions is defined next.

Algebra of Functions

Let f and g be functions.
Their **sum,** written $f + g$, is defined by

$$(f + g)(x) = f(x) + g(x)$$

Their **difference,** written as $f - g$, is defined by

$$(f - g)(x) = f(x) - g(x)$$

Their **product,** written as $f \cdot g$, is defined by

$$(f \cdot g)(x) = f(x) \cdot g(x)$$

Their **quotient,** written as $\frac{f}{g}$, is defined by

$$\left(\frac{f}{g}\right)(x) = \frac{f(x)}{g(x)}, \qquad g(x) \neq 0$$

The domain of the sum, difference, product, or quotient is the set of all values common to the domains of both functions.

EXAMPLE 3 If $f(x) = x - 1$ and $g(x) = 2x - 3$, find the following and the domain of each.

a. $(f + g)(x)$ **b.** $(f - g)(x)$

c. $(f \cdot g)(x)$ **d.** $\left(\frac{f}{g}\right)(x)$

Solution: The domain of both $f(x)$ and $g(x)$ is the set of all real numbers. Therefore, the domain of $(f + g)$, $(f - g)$, and $(f \cdot g)$ is also the set of all real numbers. The domain of $\frac{f}{g}$ is the set of all real numbers except $\frac{3}{2}$ because the denominator $2x - 3 = 0$ when $x = \frac{3}{2}$.

a. $(f + g)(x) = f(x) + g(x)$
$= (x - 1) + (2x - 3)$
$= 3x - 4$

b. $(f - g)(x) = f(x) - g(x)$
$= (x - 1) - (2x - 3)$
$= x - 1 - 2x + 3$
$= -x + 2$

c. $(f \cdot g)(x) = f(x) \cdot g(x)$
$= (x - 1)(2x - 3)$
$= 2x^2 - 5x + 3$

d. $\left(\dfrac{f}{g}\right)(x) = \dfrac{f(x)}{g(x)} = \dfrac{x - 1}{2x - 3}, \quad x \neq \dfrac{3}{2}$ ■

There is an interesting, but not surprising, relationship between the graphs of functions and the graph of their sum, difference, product, and quotient. For example, the graph of $(f + g)(x)$ can be found by adding the graph of $f(x)$ to the graph of $g(x)$. We add two graphs by adding corresponding y-values.

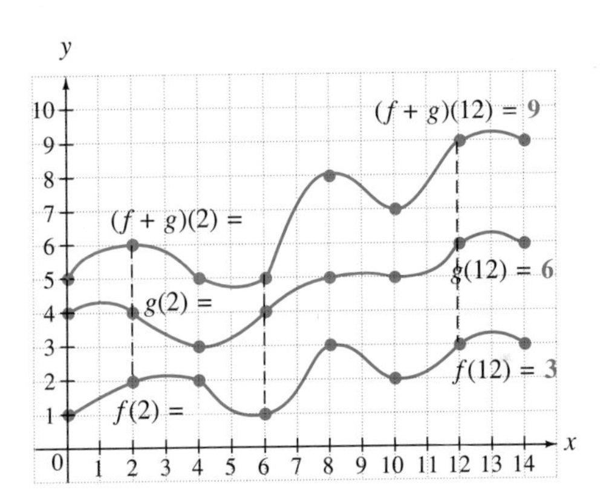

3 We can also form new functions by finding the **composition** of two functions. The notation $\boldsymbol{f[g(x)]}$ means "f composed with g," and it can be written as $\mathbf{(f \circ g)(x)}$. Also $g[f(x)]$, or $(g \circ f)(x)$, means "g composed with f." If $f(x) = 5x - 2$ and $g(x) = 4x$, then

$(f \circ g)(x) = f[g(x)]$

$= f(4x)$	Replace $g(x)$ with $4x$.
$= 5(4x) - 2$	Find $f(4x)$.
$= 20x - 2$	Simplify.

Thus, $(f \circ g)(x) = 20x - 2$. Also,

$$(g \circ f)(x) = g[f(x)] = g(5x - 2) = 4(5x - 2) = 20x - 8$$

Composition of Two Functions

The composition of the functions f and g is defined by

$$(f \circ g)(x) = f[g(x)]$$

In this section, the range of g is in the domain of f.

EXAMPLE 4 If $f(x) = x^2$ and $g(x) = x + 1$, find the following:

a. $(f \circ g)(x)$ **b.** $(g \circ f)(2)$

Solution: **a.** $(f \circ g)(x) = f[g(x)]$

$= f(x + 1)$ Replace $g(x)$ with $x + 1$.

$= (x + 1)^2$ $f(x + 1) = (x + 1)^2$.

$= x^2 + 2x + 1$

b. $(g \circ f)(2) = g[f(2)]$

Next replace $f(2)$ with 4 since $f(2) = 2^2 = 4$.

$$g[f(2)] = g(4) = 4 + 1 = 5$$

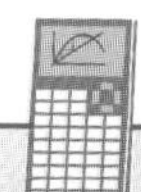

GRAPHING CALCULATOR BOX

We can use a grapher to illustrate the relationship between the graphs of functions and their sum. To see this relationship, graph the function $f(x) = x^2$, $g(x) = x$, and $(f + g)(x) = x^2 + x$. Use the window

Xmin = −4 Ymin = −4
Xmax = 4 Ymax = 4
Xscl = 1 Yscl = 1

The graphs should appear as follows:

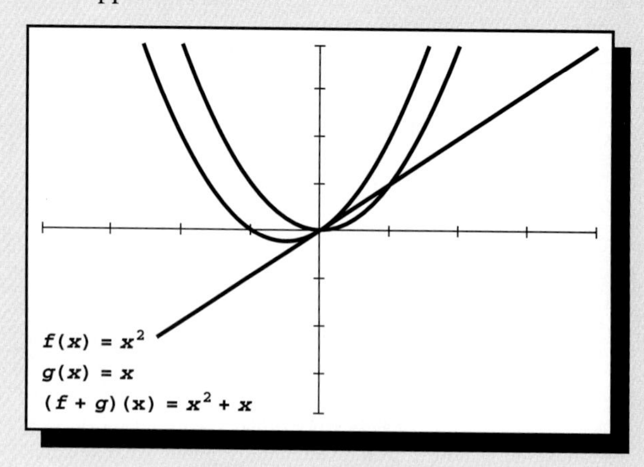

Next press the TRACE key and a cursor will appear on the first selected function in the Y= list. Use the < and > keys to move along this function. Notice that the x and y values are shown at the bottom of the screen. The number shown in the upper-right portion of the screen is the number of the function. To trace along another function, press the ∧ or ∨ key.

After you are comfortable with the cursor, press the ∧ or ∨ keys only and move to the different functions. Notice the change in x- and y-values. The x-value will remain the same while the different y-values are displayed for the different graphs. Notice that the y-value for the graph of $(f + g)(x)$ or $y = x^2 + x$ is the sum of the y-values for the graphs of $f(x)$ or $y = x^2$ and $g(x)$ or $y = x$.

EXERCISE SET 5.7

If $f(x) = 3x$, $g(x) = 2x^2 - 5$, and $h(x) = \sqrt{x}$, find the following. See Example 1.

1. $f(5)$ **2.** $f(-3)$ **3.** $g(-2)$ **4.** $g(2)$

5. $h(16)$ **6.** $h(36)$ **7.** $h\left(\frac{1}{9}\right)$ **8.** $f(\pi)$

Graph each function. See Example 2.

9. $f(x) = 2x$ **10.** $g(x) = -5x$ **11.** $h(x) = x - 3$

12. $p(x) = 4x + 1$ **13.** $g(x) = -1$ **14.** $h(x) = 0$

If $f(x) = x^2 + 1$ and $g(x) = 5x$, find the following and find each domain. See Example 3.

15. $(f + g)(x)$ **16.** $(g \cdot f)(x)$ **17.** $(f - g)(x)$

18. $\left(\frac{f}{g}\right)(x)$ **19.** $\left(\frac{g}{f}\right)(x)$ **20.** $(g - f)(x)$

If $f(x) = x^2 + 1$ and $g(x) = 5x$, find the following. See Example 4.

21. $(f \circ g)(x)$ **22.** $(g \circ f)(x)$ **23.** $(f \circ f)(1)$ **24.** $(g \circ g)(-3)$

If $f(x) = -2x$, $g(x) = x^2 + 2$, and $h(x) = 4x + 3$, find the following.

25. $g(7)$ **26.** $h(-2)$ **27.** $f(0)$ **28.** $g(0)$

29. $(f + g)(x)$ **30.** $(g - f)(x)$ **31.** $\left(\frac{f}{g}\right)(x)$ **32.** $\left(\frac{g}{f}\right)(x)$

33. $(g - h)(x)$ **34.** $(h \cdot g)(x)$ **35.** $h(4)$ **36.** $f(-5)$

37. $(f \circ g)(x)$

38. $(g \circ f)(x)$

39. $(h \circ f)(-3)$

40. $(h \circ g)(4)$

41. $(h + f)(x)$

42. $(h - g)(x)$

43. $(f \circ f)(x)$

44. $(f \circ h)(x)$

45. $f(a + b)$

46. $g(a + b)$

47. $\left(\frac{f}{h}\right)x$

48. $\left(\frac{h}{g}\right)x$

49. $(g \circ h)(x)$

50. $h(a + b)$

Graph each function.

51. $f(x) = -x$

52. $h(x) = x + 7$

53. $g(x) = 3x + 2$

54. $f(x) = -x + 1$

55. $h(x) = -x - 3$

56. $g(x) = 4x - 3$

Sketch the graph of f, g, and f + g on the same axes.

57. $f(x) = x - 1$
$g(x) = x^2$

58. $f(x) = |x|$
$g(x) = 2x$

59. $f(x) = x + 2$
$g(x) = \frac{1}{2}x$

60. $f(x) = -x^2$
$g(x) = x$

The graph below shows a comparison of chicken consumption C(t) and beef consumption B(t) in the United States. Use this graph to answer the questions below.

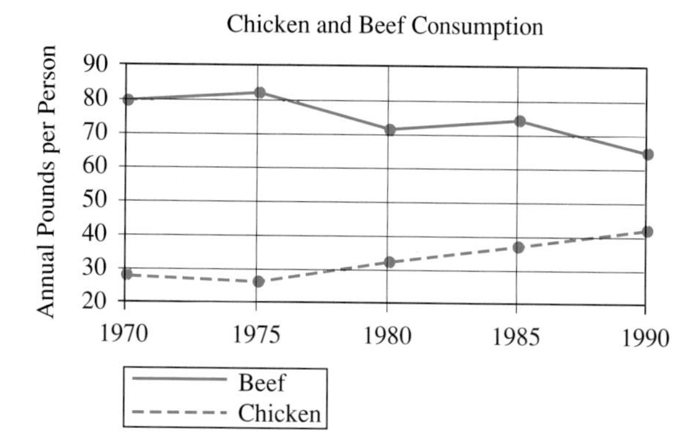

Source: U.S. Bureau of the Census, *Statistical Abstract of the United States,* 113th ed., Washington, DC, 1993.

61. In 1970, how many more pounds of beef per person were consumed than pounds of chicken?

62. In 1990, how many more pounds of beef per person were consumed than pounds of chicken?

63. In studying this graph, do you notice any trends? To what did you attribute these trends?

64. Explain what function $(C + B)(t)$ represents. Has the total chicken and beef consumption changed for the years shown? Why or why not?

Solve.

65. If $f(x) = 2x + 4$ and $g(x) = \frac{1}{2}x - 2$, find $(f \circ g)(x)$ and $(g \circ f)(x)$.

66. If $f(x) = x + 6$ and $g(x) = x - 6$, find $(f \circ g)(x)$ and $(g \circ f)(x)$.

67. Business people are concerned with cost functions, revenue functions, and profit functions. The profit ***P(x)*** obtained from x units of a product **is equal to** the revenue ***R(x)*** from selling the x units **minus** the cost ***C(x)*** of manufacturing the x units. Write an equation expressing this relationship among $C(x)$, $R(x)$, and $P(x)$.

68. Suppose the revenue $R(x)$ for x units of a product can be described by $R(x) = 25x$ and that the cost $C(x)$ can be described by $C(x) = 50 + x^2 + 4x$. Find the profit $P(x)$ for x units.

69. The area of a circle is a function of the radius of the circle. If area is $A(r)$, radius is r, and $A(r) = \pi r^2$, find the area of a circle with radius 10 centimeters. That is, find $A(10)$.

70. Maximum heart rate per minute is a function of age. If maximum heart rate is $M(x)$, age is x, and $M(x) = 220 - x$, find the maximum heart rate for a 20-year-old. That is, find $M(20)$.

71. According to some sports enthusiasts, a female runner in good shape weighs about 2 pounds per inch of height.
a. Write an equation showing the weight of a female runner in good shape as a function of height x (in inches).
b. If Samantha is a runner in good shape and is 5 feet 5 inches, determine her approximate weight.

Skill Review

Multiply the following. See Section 5.4.

72. $4x(3x + 2)$

73. $7x(9x - 1)$

74. $-8xy(y^2 + 2)$

75. $-9xy(-9x + 4)$

Solve the following. See Section 2.3.

76. $3(x - 2) + 5x = 1$

77. $-2(x + 7) + 14 = -3x$

78. $\frac{x}{3} + \frac{1}{2} = 2$

79. $\frac{2x}{5} - \frac{1}{3} = \frac{4}{5}$

Simplify each expression. See Section 1.4.

80. $2 - (8^2 - 6)$

81. $7 - (4^2 - 3)$

See Section 4.5.

82. Find the equation of the line with slope 5 that contains the point $(3, -5)$.

83. Find the equation of the line that contains the points $(-2, 4)$ and $(0, 6)$.

CHAPTER 5 GLOSSARY

A **binomial** is a polynomial with two terms.

The **numerical coefficient,** or simply the **coefficient,** is the numerical factor of a term.

The **composition** of two functions f and g in x, written as $(f \circ g)(x)$, is the function $f[g(x)]$.

The **degree of a term** is the sum of the exponents on the variables contained in the term.

The **degree of a polynomial** is the largest degree of all its terms.

In the expression a^n, a is called the **base** and n is called the **exponent.**

A **monomial** is a polynomial with one term.

A **polynomial** is a finite sum of terms in which all variables have exponents raised to nonnegative integer powers and no variables appear in the denominator.

A polynomial that contains only one variable is called a **polynomial in one variable.**

A positive number is written in **scientific notation** if it is written as the product of a number x, where $1 \le x < 10$, and a power of 10.

A **trinomial** is a polynomial with three terms.

CHAPTER 5 SUMMARY

SUMMARY OF RULES FOR EXPONENTS (5.1 and 5.2)

If a, b, and c are real numbers and m and n are integers, then:

Product rule for exponents $a^m \cdot a^n = a^{m+n}$

Zero exponent $a^0 = 1 \quad (a \neq 0)$

Negative exponent $a^{-n} = \frac{1}{a^n}$, $(a \neq 0)$

Quotient rule $\frac{a^m}{a^n} = a^{m-n}$ $(a \neq 0)$

Power rules $(a^m)^n = a^{mn}$

$(ab)^n = a^n \cdot b^n$

$\left(\frac{a}{c}\right)^n = \frac{a^n}{c^n}$ $(c \neq 0)$

(5.3)

To add polynomials, combine all like terms.
To subtract two polynomials, change the sign of each term of the polynomial that is being subtracted; then add.

(5.4)

To multiply two polynomials, multiply each term of the first polynomial by each term of the second polynomial and then combine products that are like terms.

SQUARE OF A BINOMIAL (5.4)

$(a + b)^2 = a^2 + 2ab + b^2$ and $(a - b)^2 = a^2 - 2ab + b^2$

PRODUCT OF THE SUM AND DIFFERENCE OF TWO TERMS (5.4)

$(a + b)(a - b) = a^2 - b^2$

ALGEBRA OF FUNCTIONS (5.7)

$(f + g)(x) = f(x) + g(x)$, $(f - g)(x) = f(x) - g(x)$

$(f \cdot g)(x) = f(x) \cdot g(x)$, $\left(\frac{f}{g}\right)(x) = \frac{f(x)}{g(x)}$, $g(x) \neq 0$

COMPOSITION OF FUNCTIONS (5.7)

$(f \circ g)(x) = f[g(x)]$
$(g \circ f)(x) = g[f(x)]$

CHAPTER 5 REVIEW

(5.1) *State the base and the exponent for each expression.*

1. 3^2

2. $(-5)^4$

3. -5^4

Evaluate each expression.

4. 8^3

5. $(-6)^2$

6. -6^2

7. $-4^3 - 4^0$

8. $(3b)^0$

9. $\frac{8b}{8b}$

Simplify each expression.

10. $5b^3b^5a^6$

11. $2^3 \cdot x^0$

12. $[(-3)^2]^3$

13. $(2x^3)(-5x^2)$

14. $\left(\frac{mn}{q}\right)^2 \cdot \left(\frac{mn}{q}\right)$

15. $\left(\frac{3ab^2}{6ab}\right)^4$

16. $\frac{x^9}{x^4}$

17. $\frac{2x^7y^8}{8xy^2}$

18. $\frac{12xy^6}{3x^4y^{10}}$

19. $5a^7(2a^4)^3$

20. $(2x)^2(9x)$

21. $\frac{(-4)^2(3^3)}{(4^5)(3^2)}$

22. $\frac{(-7)^2(3^5)}{(-7)^3(3^4)}$

23. $\frac{(2x)^0(-4)^2}{16x}$

24. $\frac{(8xy)(3xy)}{18x^2y^2}$

25. $m^0 + p^0 + 3q^0$

26. $(-5a)^0 + 7^0 + 8^0$

27. $(3xy^2 + 8x + 9)^0$

28. $8x^0 + 9^0$

29. $6(a^2b^3)^3$

30. $\frac{(x^3z)^a}{x^2z^2}$

(5.2) *Simplify each expression.*

31. 7^{-2}

32. -7^{-2}

33. $2x^{-4}$

34. $(2x)^{-4}$

35. $\left(\frac{1}{5}\right)^{-3}$

36. $\left(\frac{-2}{3}\right)^{-2}$

37. $2^0 + 2^{-4}$

38. $6^{-1} - 7^{-1}$

Simplify each expression. Assume that variables in an exponent represent positive integers only. Write each answer using positive exponents.

39. $\frac{1}{(2q)^{-3}}$

40. $\frac{-1}{(qr)^{-3}}$

41. $\frac{r^{-3}}{s^{-4}}$

42. $\frac{rs^{-3}}{r^{-4}}$

43. $\frac{-6}{8x^{-3}r^4}$

44. $\frac{-4s}{16s^{-3}}$

45. $(2x^{-5})^{-3}$

46. $(3y^{-6})^{-1}$

47. $(3a^{-1}b^{-1}c^{-2})^{-2}$

48. $(4x^{-2}y^{-3}z)^{-3}$

49. $\frac{5^{-2}x^8}{5^{-3}x^{11}}$

50. $\frac{7^5y^{-2}}{7^7y^{-10}}$

51. $\left(\frac{bc^{-2}}{bc^{-3}}\right)^4$

52. $\left(\frac{x^{-3}y^{-4}}{x^{-2}y^{-5}}\right)^{-3}$

53. $\frac{x^{-4}y^{-6}}{x^2y^7}$

54. $\frac{a^5b^{-5}}{a^{-5}b^5}$

55. $-2^0 + 2^{-4}$

56. $-3^{-2} - 3^{-3}$

57. $a^{6m}a^{5m}$

58. $\frac{(x^{5+h})^3}{x^5}$

59. $(3xy^{2z})^3$

60. $a^{m+2}a^{m+3}$

Write each number in scientific notation.

61. 0.00027

62. 0.8868

63. 80,800,000

64. −868,000

65. The population of the United States is 258,000,000.

66. The radius of Earth is 4000 miles.

Write each number in standard form.

67. 8.67×10^5

68. 3.86×10^{-3}

69. 8.6×10^{-4}

70. 8.936×10^5

71. The number of photons of light emitted by a 100-watt bulb every second is 1×10^{20}.

72. The real mass of all the galaxies in the constellation Virgo is 3×10^{-25}.

Simplify. Express each answer in standard form.

73. $(8 \times 10^4)(2 \times 10^{-7})$

74. $\frac{8 \times 10^4}{2 \times 10^{-7}}$

(5.3) *Find the degree of each term.*

75. $-3xy^2z$

76. 7

77. $3x$

Find the degree of each polynomial.

78. $x^2y - 3xy^3z + 5x + 7y$

79. $3x + 2$

Simplify.

80. $4x + 8x - 6x^2$

81. $-8xy^3 + 4xy^3 - 3x^3y$

Add or subtract as indicated.

82. $(3x + 7y) + (4x^2 - 3x + 7) + (y - 1)$

83. $(4x^2 - 6xy + 9y^2) - (8x^2 - 6xy - y^2)$

84. $(3x^2 - 4b + 28) + (9x^2 - 30) - (4x^2 - 6b + 20)$

85. Add $(9xy + 4x^2 + 18)$ and $(7xy - 4x^3 - 9x)$.

86. Subtract $(x - 7)$ from the sum of $(3x^2y - 7xy - 4)$ and $(9x^2y + x)$.

87.
$$\begin{array}{r} x^2 - 5x + 7 \\ \underline{-\quad (x + 4)} \end{array}$$

88.
$$\begin{array}{l} x^3 \quad + 2xy^2 - y \\ \underline{+ (x - 4xy^2 \quad - 7)} \end{array}$$

If $P(x) = 9x^2 - 7x + 8$, find the following.

89. $P(6)$

90. $P(-2)$

91. $P(-3)$

92. $P(-x)$

93. An object is dropped from the top of the Bank of China Building in Hong Kong. Neglecting air resistance, the height of the object at time t seconds is given by the polynomial function $P(t) = -16t^2 + 1001$. Find the height of the object when $t = 5$ seconds.

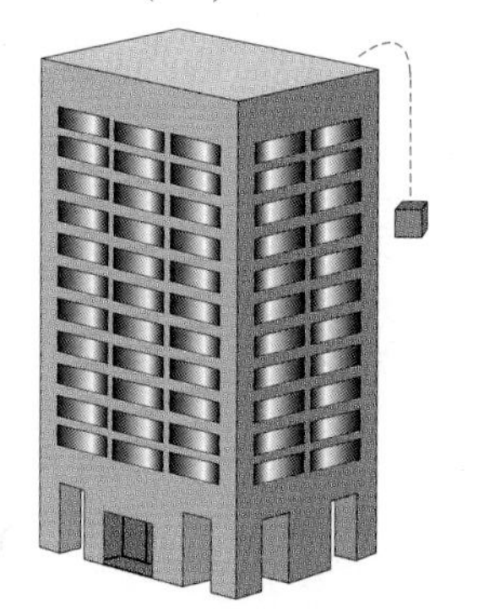

(5.4) *Multiply.*

94. $(-4xy^3)(3xy^2t)$

95. $-6xy^3(4xy - 6x + 1)$

96. $-4ab^2(3ab^3 + 7ab + 1)$

97. $(x - 4)(2x + 9)$

98. $(x^2 + 9x + 1)^2$

99. $(-3xa + 4b)^2$

100. Multiply $9x^2 + 4x + 1$ and $4x - 3$ vertically.

101. Multiply $(5x - 9)$ by $(3x + 9)$ using the FOIL method.

102. Multiply $x - \frac{1}{3}$ by $x + \frac{2}{3}$ using the FOIL method.

Multiply using special products.

103. $(3x - y)^2$

104. $(4x + 9)^2$

105. $(x + 3y)(x - 3y)$

106. $[4 + (3x + y)]^2$

107. $[4 + (3a - b)][4 - (3a - b)]$

108. $[(4y - 2) + x]^2$

109. $[(9y - 3) - y^2]^2$

110. $(x^2 - 9y^3)^2$

111. $(4y^3 + 3x^2)^2$

Multiply. Assume that all variable exponents represent integers.

112. $4a^b(3a^{b+2} - 7)$

113. $(4xy^z - b)^2$

114. $(3x^a - 4)(3x^a + 4)$

115. If $f(x) = 5x^2 + 2x$, find the following.
a. $f(a)$ **b.** $f(a + h)$ **c.** $f(x) \cdot f(x)$

(5.5) *Find each quotient. Write using only positive exponents.*

116. $\dfrac{3x^5yb^9}{9xy^7}$

117. Divide $-9xb^4z^3$ by $-4axb^2$.

118. $\dfrac{4xy + 2x^2 - 9}{4xy}$

119. Divide $12xb^2 + 16xb^4$ by $4xb^3$.

Find each quotient.

120. $\dfrac{3x^4 - 25x^2 - 20}{x - 3}$

121. $\dfrac{-x^2 + 2x^4 + 5x - 12}{x - 3}$

122. $\dfrac{2x^4 - x^3 + 2x^2 - 3x + 1}{x - \frac{1}{2}}$

123. $\dfrac{x^3 + 3x^2 - 2x + 2}{x - \frac{1}{2}}$

124. $\dfrac{3x^4 + 5x^3 + 7x^2 + 3x - 2}{x^2 + x + 2}$

125. $\dfrac{9x^4 - 6x^3 + 3x^2 - 12x - 30}{3x^2 - 2x - 5}$

(5.6) *Use synthetic division to find each quotient.*

126. $\dfrac{3x^3 + 12x - 4}{x - 2}$

127. $\dfrac{3x^3 + 2x^2 - 4x - 1}{x + \frac{3}{2}}$

128. $\dfrac{x^5 - 1}{x + 1}$

129. $\dfrac{x^3 - 81}{x - 3}$

130. $\dfrac{x^3 - x^2 + 3x^4 - 2}{x - 4}$

131. $\dfrac{3x^4 - 2x^2 + 10}{x + 2}$

If $P(x) = 3x^5 - 9x + 7$, find the following using the remainder theorem.

132. $P(4)$

133. $P(-5)$

134. $P\left(\frac{2}{3}\right)$

135. $P\left(-\frac{1}{2}\right)$

136. If the area of the rectangle is $x^4 - x^3 - 6x^2 - 6x + 18$ square miles and its width is $x - 3$ miles, find the length.

$x^4 - x^3 - 6x^2 - 6x + 18$ square miles

$x - 3$ miles

(5.7) *If $f(x) = x^2 - 2$, $g(x) = x + 1$ and $h(x) = x^3 - x^2$, find the following.*

137. $(f + g)(x)$

138. $(h - g)(x)$

139. $\left(\dfrac{h}{g}\right)(x)$

140. $(g \cdot f)(x)$

141. $(f \circ g)(x)$

142. $(g \circ f)(x)$

143. $(h \circ g)(2)$

144. $(f \circ f)(x)$

145. $(f \circ g)(-1)$

146. $(h \circ h)(2)$

Graph.

147. $f(x) = x + 1$

148. $g(x) = -3x$

149. Sketch the graphs of f, g, and $f + g$ on the same set of axes if $f(x) = x - 5$ and $g(x) = x^2$.

CHAPTER 5 TEST

Evaluate the following.

1. $(-2)^3$

2. 6^{-2}

Simplify. Write using only positive exponents.

3. $-3xy^{-2}(4xy^2)z$

4. $(-9x)^{-2}$

5. $\left(\dfrac{-xy^{-5}z}{xy^3}\right)^{-5}$

6. $\dfrac{144(xy^{-3}z)^3}{12(xy)^{-2}}$

Write in scientific notation.

7. 630,000,000

8. 0.01200

9. Write 5.0×10^{-6} without exponents.

10. Use scientific notation to find the quotient.

$$\frac{(0.0024)(0.00012)}{0.00032}$$

Add or subtract.

11. $(4x^3 - 3x - 4) - (9x^3 + 8x + 5)$

12. $(12x^6 - 6xy^2 + 15) + (2x^6 + 3xy^2 + 6y^2)$

13. If $P(x) = -3x^3 - 4x + 2$, find $P(-3)$.

Multiply or divide.

14. $-3xy(4x + y)$

15. $(3x + 4)(4x - 7)$

16. $(9x^2 + 4y + 2)^2$

17. $[3 - (4x - y)]^2$

18. $\dfrac{4x^2y + 9x + z}{3xz}$

19. $\dfrac{x^6 + 3x^5 - 2x^4 + x^2 - 3x + 2}{x - 2}$

20. Use synthetic division to divide $4x^4 - 3x^3 + 2x^2 - x - 1$ by $x - \dfrac{2}{3}$.

21. If $P(x) = 4x^4 + 7x^2 - 2x - 5$, use the remainder theorem to find $P(-2)$.

If $f(x) = x$, $g(x) = x - 7$, and $h(x) = x^2 - 6x + 5$, find the following.

22. $(h - g)(x)$

23. $(h \cdot f)(x)$

24. $(g \circ f)(x)$

25. $(g \circ h)(x)$

Solve.

26. Graph $f(x) = 2x - 4$

27. If $f(x) = 2x^2 - 4$, find the following.

a. $f(a)$ **b.** $f(a + 2)$ **c.** $f(x) \cdot f(x)$

28. Find the area of the shaded square.

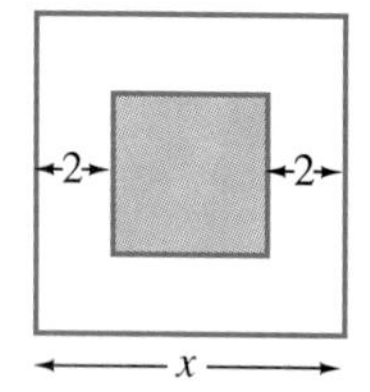

29. The total cost for a manufacturing company to produce x blank audiocassette tapes per week is given by the polynomial function $C(x) = 0.8x + 10{,}000$, and the total revenue for producing these tapes is given by the polynomial function $R(x) = 2x$. Use these functions to write a function for the total profit for producing x blank audiocassettes. (Hint: Recall that profit is equal to revenue minus cost.) Find the profit from selling 20,000 tapes per week.

CHAPTER 5 CUMULATIVE REVIEW

1. Add or subtract as indicated. Write each answer in lowest terms.

a. $\dfrac{2}{5} + \dfrac{1}{4}$ **b.** $\dfrac{1}{2} + \dfrac{17}{22} - \dfrac{2}{11}$ **c.** $3\dfrac{1}{6} - 1\dfrac{11}{12}$

2. Write each sentence as an equation. Let x represent the unknown number.

a. The quotient of 15 and a number is 4.

b. Three subtracted from 12 is a number.

c. Four times a number added to 17 is 21.

3. Subtract 8 from -4.

4. Solve the equation $7(x + 1) - 5 = 6x - 10$.

5. Four friends are planning now to go to Disneyland in 3 years in celebration of their graduation from college. They have figured the total cost for the trip to be \$3500. If each invests \$800 now is a savings account paying 5% simple interest, will they have enough money in 3 years?

6. A local cellular phone company charges Elaine Chapoton \$50 per month and \$0.36 per minute of phone use in her usage category. If Elaine budgeted \$100 per month for cellular phone use, determine the number of whole minutes of phone use.

7. Solve $|p| = 2$.

8. Solve $-4x + 7 \geq -9$, and graph the solution.

9. Solve $2 < 4 - x < 7$.

10. Solve for x: $|5x + 1| + 1 \leq 10$.

11. Graph the linear equation $y = 3x$.

12. Graph $x = 2$.

13. Find the equation of the line passing through $(-1, 5)$ with slope -2. Write the equation in standard form: $Ax + By = C$.

14. Simplify $(2x^2)(-3x^5)$.

15. Simplify each quotient.

a. $\dfrac{x^5}{x^2}$ **b.** $\dfrac{4^7}{4^3}$ **c.** $\dfrac{(-3)^5}{(-3)^2}$ **d.** $\dfrac{2x^5y^2}{xy}$

16. Simplify each expression. Write answers with positive exponents.

a. $\dfrac{y}{y^{-2}}$ **b.** $\dfrac{p^{-4}}{q^{-9}}$ **c.** $\dfrac{x^{-5}}{x^7}$

17. Add $11x^3 - 12x^2 + x - 3$ and $x^3 - 10x + 5$.

18. Multiply $(2x - 7)(3x - 4)$.

19. Find the quotient: $\dfrac{7a^2b - 2ab^2}{2ab^2}$.

20. Divide $2x^3 + 3x^4 - 8x + 6$ by $x^2 - 1$.

21. Use synthetic division to divide $x^4 - 2x^3 - 11x^2 + 5x + 34$ by $x + 2$.

22. If $g(x) = 2x + 5$, find the following.

a. $g(-4)$ **b.** $g(a)$ **c.** $g(x + h)$

CHAPTER 6

Factoring Polynomials

Peter yearns for the luxury of a swimming pool in his backyard, complete with surrounding cement patio. He knows how large a pool he can afford, but now he wonders how large the patio can be.

INTRODUCTION

When two or more polynomials are multiplied, each polynomial is called a **factor** of the product. "Reversing" this process is called **factoring** a polynomial. It is the process of writing a polynomial as a product of simpler factors. In this chapter, we explore factoring polynomials, solving polynomial equations, and graphing polynomial functions.

6.1 The Greatest Common Factor and Factoring by Grouping

OBJECTIVES

Tape BA 12

1. Identify the GCF.
2. Factor out the GCF of a polynomial's terms.
3. Factor polynomials by grouping.

1 **Factoring** is the reverse process of multiplying. When an integer is written as the product of two integers, we call each integer a factor. For example, in the product $3 \cdot 5 = 15$, 3 and 5 are called **factors** of 15, and $3 \cdot 5$ is a **factored form** of 15. This is true of polynomials also. Since

$$(3x - 1)(2x + 5) = 6x^2 + 13x - 5$$

$3x - 1$ and $2x + 5$ are called **factors** of $6x^2 + 13x - 5$, and $(3x - 1)(2x + 5)$ is a **factored form** of $6x^2 + 13x - 5$. The process of writing a polynomial as a product is called **factoring.**

factoring →

$$6x^2 + 13x - 5 = (3x - 1)(2x + 5)$$

← multiplying

Factoring polynomials is closely related to factoring integers, so we begin by reviewing integer factoring and finding the **greatest common factor (GCF)** of a list of integers. The GCF of a list of integers is the largest integer that is a factor of each integer. For example, 4 is the GCF of integers 16, 20, and 80, because 4 is the largest factor common to 16, 20, and 80.

To find the GCF, we write each integer as a product of **prime numbers.** A prime number is a natural number greater than 1 whose only natural number factors are 1 and itself. The set of prime numbers is

$$\{2, 3, 5, 7, 11, 13, 17, \ldots\}$$

EXAMPLE 1 Find the GCF of each list of integers.

a. 40 and 52 **b.** 30, 45, and 75

Solution: **a.** Start by writing each number as a product of primes.

$$40 = 4 \cdot 10 = 2 \cdot 2 \cdot 2 \cdot 5 = 2^3 \cdot 5 \qquad 52 = 4 \cdot 13 = 2 \cdot 2 \cdot 13 = 2^2 \cdot 13$$

The **common factors** are two factors of 2, so the GCF is

$$\text{GCF} = 2^2 = 4$$

b.

$$30 = 3 \cdot 10 = 3 \cdot 2 \cdot 5 = 2 \cdot 3 \cdot 5 \qquad 45 = 9 \cdot 5 = 3 \cdot 3 \cdot 5 = 3^2 \cdot 5 \qquad 75 = 3 \cdot 25 = 3 \cdot 5 \cdot 5 = 3 \cdot 5^2$$

The **common factors** are one factor of 3 and one factor of 5, so the GCF is

$$\text{GCF} = 3 \cdot 5 = 15 \quad ■$$

To find the GCF of a list of variables raised to powers, we use a similar process. The GCF of x^3, x^5, and x^6 is x^3 since each power of x contains at least a factor of x^3.

$$x^3 = x^3$$
$$x^5 = x^3 \cdot x^2$$
$$x^6 = x^3 \cdot x^3$$

In general, the GCF of a list of common variables raised to powers is the common variable raised to an exponent equal to the smallest exponent in the list. To find the GCF of a list of monomials, the following steps can be used.

To Find the GCF of a List of Monomials

Step 1 Find the GCF of the numerical coefficients.
Step 2 Find the GCF of the variable factors.
Step 3 The product of the factors found in steps 1 and 2 is the GCF of the monomials.

EXAMPLE 2 Find the GCF of $20x^3y$, $10x^2y^2$, and $35x^3$.

Solution: The GCF of the numerical coefficients 20, 10, and 35 is 5. The GCF of the variable factors x^3, x^2, and x^3 is x^2. The variable y is not a common factor because it does not appear in the third monomial. The GCF is thus

$$5 \cdot x^2 \quad \text{or} \quad 5x^2 \quad ■$$

2 The goal of this chapter is to present methods of factoring polynomials. In other words, we want to write polynomials as products of simpler polynomials. The first step in factoring polynomials is to use the distributive property and write the polynomial as a product of the GCF of its monomial terms and a simpler polynomial. This is called **factoring out** the GCF.

EXAMPLE 3 Factor.

a. $8x + 4$ **b.** $5y - 2z$ **c.** $6x^2 - 3x^3$

Solution: **a.** The GCF of terms $8x$ and 4 is 4.

$$8x + 4 = 4(2x) + 4(1) \quad \text{Factor out 4 from each term.}$$
$$= 4(2x + 1) \quad \text{Apply the distributive property.}$$

The factored form of $8x + 4$ is $4(2x + 1)$. To check, multiply $4(2x + 1)$ to see that the product is $8x + 4$.

b. There is no common factor of the terms $5y$ and $-2z$ other than 1 (or -1).

c. The greatest common factor of $6x^2$ and $-3x^3$ is $3x^2$. Thus,

$$6x^2 - 3x^3 = 3x^2(2) - 3x^2(x)$$
$$= 3x^2(2 - x) \quad ■$$

HELPFUL HINT

To check that the GCF has been factored out correctly, multiply the factors together and see that their product is the original polynomial.

EXAMPLE 4 Factor $17x^3y^2 - 34x^4y^2$.

Solution: The GCF of the two terms is $17x^3y^2$, which we factor out of each term.

$$17x^3y^2 - 34x^4y^2 = 17x^3y^2(1) - 17x^3y^2(2x)$$
$$= 17x^3y^2(1 - 2x) \quad ■$$

HELPFUL HINT

If the GCF happens to be one of the terms in the polynomial, a factor of 1 will remain for this term when the GCF is factored out. For example, in the polynomial $21x^2 + 7x$ the GCF of $21x^2$ and $7x$ is $7x$, so

$$21x^2 + 7x = 7x(3x) + 7x(1) = 7x(3x + 1)$$

EXAMPLE 5 Factor $-3x^3y + 2x^2y - 5xy$.

Solution: Two possibilities are shown for factoring this polynomial. First, the common factor xy is factored out.

$$-3x^3y + 2x^2y - 5xy = xy(-3x^2 + 2x - 5)$$

Also, the common factor $-xy$ can be factored out as shown.

$$-3x^3y + 2x^2y - 5xy = -xy(3x^2) + (-xy)(-2x) + (-xy)(5)$$
$$= -xy(3x^2 - 2x + 5)$$

Both of these alternatives are correct. ■

EXAMPLE 6 Factor $2(x - 5) + 3a(x - 5)$.

Solution: The GCF is the binomial $(x - 5)$. Thus,

$$2(x - 5) + 3a(x - 5) = (x - 5)(2 + 3a) \quad ■$$

EXAMPLE 7 Factor $7x(x^2 + 5y) - (x^2 + 5y)$.

Solution: The GCF is the expression $(x^2 + 5y)$. Factor this from each term.

$$7x(x^2 + 5y) - (x^2 + 5y) = 7x(x^2 + 5y) - 1(x^2 + 5y) = (x^2 + 5y)(7x - 1)$$

Notice that we write $-(x^2 + 5y)$ as $-1(x^2 + 5y)$ to aid in factoring. ■

3 Sometimes it is possible to factor a polynomial by grouping the terms of the polynomial and looking for common factors in each group. This method of factoring is called **factoring by grouping.**

EXAMPLE 8 Factor $ab - 6a + 2b - 12$.

Solution: First look for the GCF of all four terms. The GCF for all four terms is 1. Next group the first two terms and the last two terms and factor out common factors from each group.

$$ab - 6a + 2b - 12 = (ab - 6a) + (2b - 12)$$

Factor a from the first group and 2 from the second group.

$$= a(b - 6) + 2(b - 6)$$

Now we see a GCF of $(b - 6)$. Factor out $(b - 6)$ to get

$$a(b - 6) + 2(b - 6) = (b - 6)(a + 2)$$

This factorization can be checked by multiplying $(b - 6)$ by $(a + 2)$ to verify that the product is the original polynomial. ■

HELPFUL HINT

Notice that the polynomial $a(b - 6) + 2(b - 6)$ is **not** in factored form. It is a **sum,** not a **product.** The factored form is $(b - 6)(a + 2)$.

EXAMPLE 9 Factor $m^2n^2 + m^2 - 2n^2 - 2$.

Solution: Once again, the GCF of all four terms is 1. Try grouping the first two terms together and the last two terms together.

$$m^2n^2 + m^2 - 2n^2 - 2 = (m^2n^2 + m^2) + (-2n^2 - 2)$$

Factor m^2 from the first group and 2 from the second group.

$$= m^2(n^2 + 1) + 2(-n^2 - 1)$$

There is no common factor in this resulting polynomial, but notice that $(n^2 + 1)$ and $(-n^2 - 1)$ are opposites. Try grouping the terms differently, as follows:

$m^2n^2 + m^2 - 2n^2 - 2 = (m^2n^2 + m^2) - (2n^2 + 2)$ Watch the signs!

$= m^2(n^2 + 1) - 2(n^2 + 1)$ Factor common factors from the groups of terms.

$= (n^2 + 1)(m^2 - 2)$ Factor out a GCF of $(n^2 + 1)$. ■

MENTAL MATH

Find the GCF of each list of monomials.

1. 6, 12
2. 9, 27
3. $15x$, 10
4. $9x$, 12
5. $13x$, $2x$
6. $4y$, $5y$
7. $7x$, $14x$
8. $8z$, $4z$

EXERCISE SET 6.1

Find the GCF of each list of numbers. See Example 1.

1. 24, 30
2. 30, 75
3. 84, 140
4. 90, 225
5. 20, 36, 60
6. 18, 45, 54
7. 30, 48, 72
8. 42, 63, 147

Find the GCF of each list of monomials. See Example 2.

9. a^8, a^5, a^3
10. b^9, b^2, b^5
11. $x^2y^3z^3, y^2z^3, xy^2z^2$
12. $xy^2z^3, x^2y^2z^2, x^2y^3$
13. $6x^3y, 9x^2y^2, 12x^2y$
14. $4xy^2, 16xy^3, 8x^2y^2$
15. $10x^3yz^3, 20x^2z^5, 45xz^3$
16. $12y^2z^4, 9xy^3z^4, 15x^2y^2z^3$

Factor out the GCF in each polynomial. See Examples 3 through 5.

17. $18x - 12$
18. $21x + 14$
19. $4y^2 - 16xy^3$
20. $3z - 21xz^4$
21. $6x^5 - 8x^4 + 2x^3$
22. $9x + 3x^2 - 6x^3$
23. $8a^3b^3 - 4a^2b^2 + 4ab + 16ab^2$
24. $12a^3b - 6ab + 18ab^2 - 18a^2b$

Factor out the GCF in each polynomial. See Examples 6 and 7.

25. $6(x + 3) + 5a(x + 3)$
26. $2(x - 4) + 3y(x - 4)$
27. $2x(z + 7) + (z + 7)$
28. $x(y - 2) + (y - 2)$
29. $3x(x^2 + 5) - 2(x^2 + 5)$
30. $4x(2y + 3) - 5(2y + 3)$

Factor each polynomial by grouping. See Examples 8 and 9.

31. $ab + 3a + 2b + 6$
32. $ab + 2a + 5b + 10$
33. $ac + 4a - 2c - 8$
34. $bc + 8b - 3c - 24$
35. $2xy - 3x - 4y + 6$
36. $12xy - 18x - 10y + 15$
37. $12xy - 8x - 3y + 2$
38. $20xy - 15 - 4y + 3$

Factor out the GCF in each polynomial.

39. $6x^3 + 9$
40. $6x^2 - 8$
41. $x^3 + 3x^2$
42. $x^4 - 4x^3$
43. $8a^3 - 4a$
44. $12b^4 + 3b^2$
45. $-20x^2y + 16xy^3$
46. $-18xy^3 + 27x^4y$
47. $10a^2b^3 + 5ab^2 - 15ab^3$
48. $10ef - 20e^2f^3 + 30e^3f$
49. $9abc^2 + 6a^2bc - 6ab + 3bc$
50. $4a^2b^2c - 6ab^2c - 4ac + 8a$
51. $4x(y - 2) - 3(y - 2)$
52. $8y(z + 8) - 3(z + 8)$

53. $2m(n - 8) - (n - 8)$

54. $3a(b - 4) - (b - 4)$

55. $15x^3y^2 - 18x^2y^2$

56. $12x^4y^2 - 16x^3y^3$

Factor each polynomial by grouping.

57. $6xy + 10x + 9y + 15$

58. $15xy + 20x + 6y + 8$

59. $xy + 3y - 5x - 15$

60. $xy + 4y - 3x - 12$

61. $6ab - 2a - 9b + 3$

62. $16ab - 8a - 6b + 3$

63. $12xy + 18x + 2y + 3$

64. $20xy + 8x + 5y + 2$

65. $2x^2 + 3xy + 4x + 6y$

66. $3x^2 + 12x + 4xy + 16y$

67. $5x^2 + 5xy - 3x - 3y$

68. $4x^2 + 2xy - 10x - 5y$

69. $x^3 + 3x^2 + 4x + 12$

70. $x^3 + 4x^2 + 3x + 12$

71. $x^3 - x^2 - 2x + 2$

72. $x^3 - 2x^2 - 3x + 6$

Factor.

73. The material needed to manufacture a tin can is given by the polynomial

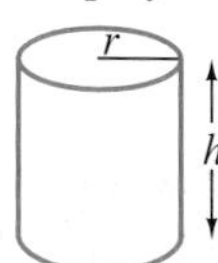

$$2\pi r^2 + 2\pi rh$$

where radius is r and height is h. Factor this expression.

74. The amount E of current in an electrical circuit is given by the formula

$$IR_1 + IR_2 = E$$

Write an equivalent equation by factoring the expression $IR_1 + IR_2$.

75. At the end of T years, the amount of money A in a savings account earning simple interest from an initial investment of P dollars at rate R is given by the formula

$$A = P + PRT$$

Write an equivalent equation by factoring the expression $P + PRT$.

76. An open-topped box has a square base and a height of 10 inches. If each of the bottom edges of the box has length x inches, find the amount of material needed to construct the box. Write the answer in factored form.

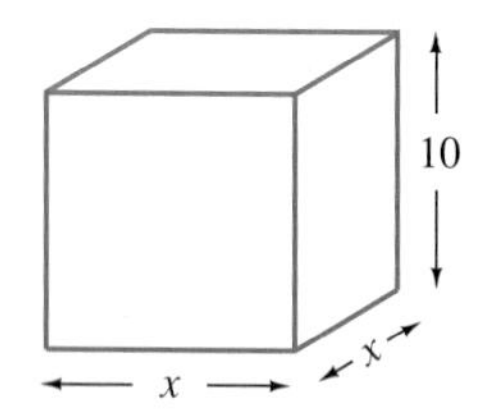

77. An object is thrown upward from the ground with an initial velocity of 64 feet per second. The height $h(t)$ of the object after t seconds is given by the polynomial function

$$h(t) = -16t^2 + 64t$$

Write an equivalent factored expression for the function $h(t)$ by factoring $-16t^2 + 64t$.

78. An object is dropped from the top of a hot air balloon at a height of 224 feet. The height $h(t)$ of the object after t seconds is given by the polynomial function

$$h(t) = -16t^2 + 224$$

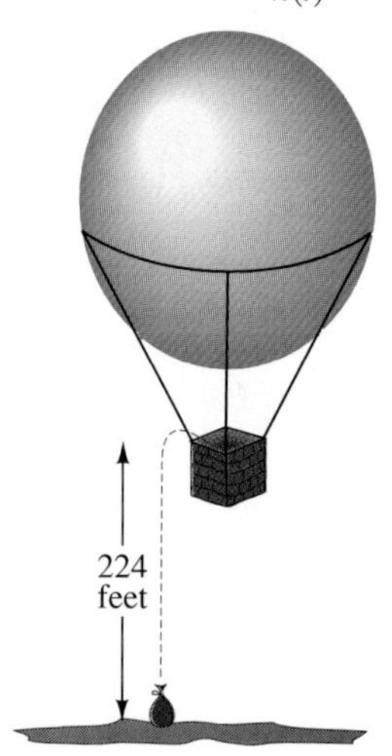

Write an equivalent factored expression for the function $h(t)$ by factoring $-16t^2 + 224$.

Skill Review

Simplify the following. See Section 5.1.

79. $(5x^2)(11x^5)$

80. $(7y)(-2y^3)$

81. $(5x^2)^3$

82. $(-2y^3)^4$

Find each product by using the FOIL order of multiplying binomials. See Section 5.4.

83. $(x + 2)(x - 5)$
84. $(x - 7)(x - 1)$
85. $(x + 3)(x + 2)$
86. $(x - 4)(x + 2)$
87. $(y - 3)(y - 1)$
88. $(s + 8)(s + 10)$

Perform the subtraction. See Section 5.3.

89. Subtract $x^2 - 2x + 1$ from $9x^2 - 6$.
90. Subtract $10xy + 3$ from $-6xy + 8$.

A Look Ahead

Factor. Assume that variables used as exponents represent positive integers. See the following example.

EXAMPLE Factor $x^{5a} - x^{3a} + x^{7a}$.

Solution: The variable x is common to all three terms, and the power $3a$ is the smallest of the exponents. So factor out the common factor x^{3a}.

$$x^{5a} - x^{3a} + x^{7a} = x^{3a}(x^{2a}) - x^{3a}(1) + x^{3a}(x^{4a})$$
$$= x^{3a}(x^{2a} - 1 + x^{4a}) \quad ■$$

91. $x^{3n} - 2x^{2n} + 5x^n$
92. $3y^n + 3y^{2n} + 5y^{8n}$
93. $6x^{8a} - 2x^{5a} - 4x^{3a}$
94. $3x^{5a} - 6x^{3a} + 9x^{2a}$

Factor. See the following example.

EXAMPLE Factor $2x^{-1} + 10x^{-3}$.

Solution: The variable x is common to both terms. The smallest power of x is -3, so the GCF is $2x^{-3}$.

$$2x^{-1} + 10x^{-3} = 2x^{-3}(x^2) + 2x^{-3}(5) = 2x^{-3}(x^2 + 5) \quad ■$$

95. $3x^{-2} + 8x^{-1}$
96. $5y^{-2} - 3y^{-1}$
97. $3x^2y^{-3} + 2xy^{-1}$
98. $2x^2y^{-2} + 7xy^{-3}$
99. $6x^{-2}y^{-1} - 2x^{-2}y^{-4} + 8xy^{-2}$
100. $9x^{-2}y^{-2} - 6x^{-3}y^{-1} - 3x^{-3}y^{-2}$

Writing in Mathematics

101. Define the GCF of a list of integers.

102. When $3x^2 - 9x + 3$ is factored, the result is $3(x^2 - 3x + 1)$. Explain why it is necessary to include the term 1 in this factored form.

103. Consider the following sequence of algebraic steps:

$$x^3 - 6x^2 + 2x - 10 = (x^3 - 6x^2) + (2x - 10)$$
$$= x^2(x - 6) + 2(x - 5)$$

Explain whether the final result is the factored form of the original polynomial.

6.2 Factoring Trinomials of the Form $x^2 + bx + c$

OBJECTIVES

1. Factor trinomials of the form $x^2 + bx + c$.
2. Factor out the greatest common factor before factoring a trinomial of the form $x^2 + bx + c$.

Tape BA 12

1 In this section, we factor trinomials of the form $x^2 + bx + c$, where the numerical coefficient of the squared variable is 1. Recall that factoring a polynomial is the process of writing the polynomial as a product. For example, since $(x + 3)(x + 1) = x^2 + 4x + 3$, we say that the factored form of $x^2 + 4x + 3$ is

$$x^2 + 4x + 3 = (x + 3)(x + 1)$$

Notice that the product of the first terms of the binomials is $x \cdot x = x^2$, the first term of the trinomial. Also, the product of the last two terms of the binomials is $3 \cdot 1 = 3$, the third term of the trinomial. The sum of these same terms is $3 + 1 = 4$, the coefficient of the x term of the trinomial.

$$x^2 = x \cdot x$$
$$x^2 + 4x + 3 = (x + 3)(x + 1)$$
$$3 = 3 \cdot 1$$
$$4 = 3 + 1$$

Many trinomials, such as the preceding one, factor into two binomials. To factor $x^2 + 7x + 10$, assume that it factors into two binomials and begin by writing two pairs of parentheses. The first term of the trinomial is x^2, so we will use x and x as first terms of the binomial factors.

$$(x + \quad)(x + \quad)$$

To determine the last term of each binomial factor, we look for two integers whose product is 10 and whose sum is 7. Since our numbers must have a positive product and a positive sum, we list positive integer factors of 10 only.

Positive Factors of 10	*Sum of Factors*
1, 10	$1 + 10 = 11$
2, 5	$2 + 5 = 7$

The correct pair of numbers is 2 and 5 because their product is 10 and their sum is 7. Now we can fill in the last terms of the binomial factors.

$$x^2 + 7x + 10 = (x + 5)(x + 2)$$

To see if we have factored correctly, multiply.

$$(x + 5)(x + 2) = x^2 + 2x + 5x + 10$$
$$= x^2 + 7x + 10 \qquad \text{Combine like terms.}$$

Since multiplication is commutative, the factored form of $x^2 + 7x + 10$ can also be written as $(x + 2)(x + 5)$.

In general, to factor a trinomial of the form $x^2 + bx + c$, look for two numbers whose product is c and whose sum is b. The factored form of $x^2 + bx + c$ is

$$(x + \textbf{one number})(x + \textbf{other number})$$

EXAMPLE 1 Factor $x^2 + 7x + 12$.

Solution: Begin by writing the first terms of the binomial factors.

$$(x + \quad)(x + \quad)$$

Next look for two numbers whose product is 12 and whose sum is 7. Since our numbers must have a positive product and a positive sum, we look at positive factors of 12 only.

Positive Factors of 12	*Sum of Factors*
1, 12	$1 + 12 = 13$
2, 6	$2 + 6 = 8$
3, 4	$3 + 4 = 7$

The correct pair of numbers is 3 and 4 because their product is 12 and their sum is 7. Use these factors to fill in the last terms of the binomial factors.

$$x^2 + 7x + 12 = (x + 3)(x + 4)$$

To check, multiply $(x + 3)$ by $(x + 4)$. ■

EXAMPLE 2 Factor $x^2 - 8x + 15$.

Solution: Begin by writing the first terms of the binomials.

$$(x + \quad)(x + \quad)$$

Now look for two numbers whose product is 15 and whose sum is -8. Since our numbers must have a positive product and a negative sum, we look at negative factors of 15 only.

Negative Factors of 15	*Sum of Factors*
$-1, -15$	$-1 + (-15) = -16$
$-3, -5$	$-3 + (-5) = -8$

The correct pair of numbers is -3 and -5 because their product is 15 and their sum is -8. Then

$$x^2 - 8x + 15 = (x - 3)(x - 5) \quad ■$$

EXAMPLE 3 Factor $x^2 + 4x - 12$.

Solution:

$$x^2 + 4x - 12 = (x + \quad)(x + \quad)$$

Look for two numbers whose product is -12 and whose sum is 4.

Factors of −12	*Sum of Factors*
$-1, 12$	$-1 + 12 = 11$
$1, -12$	$1 + (-12) = -11$
$-2, 6$	$-2 + 6 = 4$
$2, -6$	$2 + (-6) = -4$
$-3, 4$	$-3 + 4 = 1$
$3, -4$	$3 + (-4) = -1$

The correct pair of numbers is -2 and 6 since their product is -12 and their sum is 4. Hence,

$$x^2 + 4x - 12 = (x - 2)(x + 6) \quad ■$$

EXAMPLE 4 Factor $r^2 - r - 42$.

Solution: Because the variable in this trinomial is r, the first term in each binomial factor is r.

$$r^2 - r - 42 = (r + \quad)(r + \quad)$$

Find two numbers whose product is -42 and whose sum is -1, the numerical coefficient of r. The numbers are 6 and -7. Therefore,

$$r^2 - r - 42 = (r + 6)(r - 7) \quad ■$$

EXAMPLE 5 Factor $a^2 + 2a + 10$.

Solution: Look for two numbers whose product is 10 and whose sum is 2. Neither 1 and 10 nor 2 and 5 give the required sum, 2. We conclude that $a^2 + 2a + 10$ is not factorable with integers. The polynomial $a^2 + 2a + 10$ is called a **prime polynomial.** ■

EXAMPLE 6 Factor $x^2 + 5xy + 6y^2$.

Solution:

$$x^2 + 5xy + 6y^2 = (x + \quad)(x + \quad)$$

Look for two numbers whose product is 6 and whose sum is 5. The numbers are 2 and 3. Since the last term of the trinomial is $6y^2$, we use $2y$ and $3y$. Notice that $2y \cdot 3y = 6y^2$ and $2y + 3y = 5y$, the coefficient of x in the middle term. Therefore,

$$x^2 + 5xy + 6y^2 = (x + 2y)(x + 3y) \quad ■$$

2 The first step in factoring a polynomial is to look for a common factor to factor out.

EXAMPLE 7 Factor $3m^2 - 24m - 60$.

Solution: First factor out the greatest common factor, 3.

$$3m^2 - 24m - 60 = 3(m^2 - 8m - 20)$$

Next factor $m^2 - 8m - 20$ by looking for two factors of -20 whose sum is -8. The factors are -10 and 2.

$$3m^2 - 24m - 60 = 3(m + 2)(m - 10)$$

↑

Remember to write the common factor 3 as part of the answer.

Check by multiplying.

$$3(m + 2)(m - 10) = 3(m^2 - 8m - 20)$$
$$= 3m^2 - 24m - 60 \quad ■$$

HELPFUL HINT

When factoring a polynomial, remember that factored out common factors are part of the final factored form. For example,

$$5x^2 - 15x - 50 = 5(x^2 - 3x - 10)$$
$$= 5(x + 2)(x - 5)$$

Thus, $5x^2 - 15x - 50$ **factored completely** is $5(x + 2)(x - 5)$.

MENTAL MATH

Complete the following.

1. $x^2 + 9x + 20 = (x + 4)(x \quad)$
2. $x^2 + 12x + 35 = (x + 5)(x \quad)$
3. $x^2 - 7x + 12 = (x - 4)(x \quad)$
4. $x^2 - 13x + 22 = (x - 2)(x \quad)$
5. $x^2 + 4x + 4 = (x + 2)(x \quad)$
6. $x^2 + 10x + 24 = (x + 6)(x \quad)$

EXERCISE SET 6.2

Factor each trinomial. See Examples 1 through 5.

1. $x^2 + 7x + 6$
2. $x^2 + 6x + 8$
3. $x^2 + 9x + 20$
4. $x^2 + 13x + 30$
5. $x^2 - 8x + 15$
6. $x^2 - 9x + 14$
7. $x^2 - 10x + 9$
8. $x^2 - 6x + 9$
9. $x^2 - 15x + 5$
10. $x^2 - 13x + 30$
11. $x^2 - 3x - 18$
12. $x^2 - x - 30$
13. $x^2 + 5x + 2$
14. $x^2 - 7x + 5$

Factor each trinomial completely. See Example 6.

15. $x^2 + 8xy + 15y^2$
16. $x^2 + 6xy + 8y^2$
17. $x^2 - 2xy + y^2$
18. $x^2 - 11xy + 30y^2$
19. $x^2 - 3xy - 4y^2$
20. $x^2 - 4xy - 77y^2$

Factor each trinomial completely. See Example 7.

21. $2z^2 + 20z + 32$
22. $3x^2 + 30x + 63$
23. $2x^3 - 18x^2 + 40x$
24. $x^3 - x^2 - 56x$
25. $7x^2 + 14xy - 21y^2$
26. $6r^2 - 3rs - 3s^2$

Factor each trinomial completely.

27. $x^2 + 15x + 36$
28. $x^2 + 19x + 60$
29. $x^2 - x - 2$
30. $x^2 - 5x - 14$
31. $r^2 - 16r + 48$
32. $r^2 - 10r + 21$
33. $x^2 - 4x - 21$
34. $x^2 - 4x - 32$
35. $x^2 + 7xy + 10y^2$
36. $x^2 - 3xy - 4y^2$
37. $r^2 - 3r + 6$
38. $x^2 + 4x - 10$
39. $2t^2 + 24t + 64$
40. $2t^2 + 20t + 50$
41. $x^3 - 2x^2 - 24x$
42. $x^3 - 3x^2 - 28x$
43. $x^2 - 16x + 63$
44. $x^2 - 19x + 88$
45. $x^2 + xy - 2y^2$
46. $x^2 - xy - 6y^2$
47. $3x^2 + 9x - 30$
48. $4x^2 - 4x - 48$
49. $3x^2 - 60x + 108$
50. $2x^2 - 24x + 70$
51. $x^2 - 18x - 144$
52. $x^2 + x - 42$
53. $6x^3 + 54x^2 + 120x$
54. $3x^3 + 3x^2 - 126x$
55. $2t^5 - 14t^4 + 24t^3$
56. $3x^6 + 30x^5 + 72x^4$
57. $5x^3y - 25x^2y^2 - 120xy^3$
58. $3x^2 - 6x - 72$
59. $4x^2 + 4x - 12$
60. $3x^2 - 9x + 45$
61. $x^2 - 10x + 21$
62. $x^2 - 14x + 28$
63. $2x^2y + 30xy + 100y$
64. $3x^2z^2 + 9xz^2 + 6z^2$
65. $-12x^2y^3 - 24xy^3 - 36y^3$
66. $-4x^2t^4 + 4xt^4 + 24t^4$
67. $y^2(x + 1) - 2y(x + 1) - 15(x + 1)$
68. $z^2(x + 1) - 3z(x + 1) - 70(x + 1)$

Skill Review

Simplify. See Sections 5.1 and 5.2.

69. 5^{-2}
70. 7^{-1}
71. $\dfrac{x^{-10}}{x^5}$
72. $\dfrac{y^{-7}}{y^4}$

73. The total force F against the face of a dam that is 100 feet long by 20 feet high is given by the following:

$$F = \frac{(6.24 \times 10)(4 \times 10^4)}{2}$$

Compute the force and express the answer in scientific notation.

74. Suppose that \$1000 is invested at a rate of 9% and compounded monthly. The amount of principal (P) after 1 year is given by

$$P = (1 \times 10^3)(1.09381)$$

Compute the amount of principal.

Multiply the following. See Section 5.4.

75. $(3x - 2)(x + 5)$

76. $(y + 5)^2$

77. $(y - 1)^2$

78. $(2x + 3)(2x - 13)$

If $P(x) = 5x + 1$ and $Q(x) = 4x - 3$, find the following. See Section 5.7.

79. $(P \cdot Q)(x)$

80. $(Q \cdot P)(x)$

81. $(P \cdot P)(x)$

82. $(Q \cdot Q)(x)$

Writing in Mathematics

Complete the following sentences in your own words.

83. If the sign of the constant term of a factorable trinomial such as $x^2 + bx + c$ is negative, then the signs of the last-term factors of the binomial are opposite because . . . .

84. If the sign of the constant term of a factorable trinomial such as $x^2 + bx + c$ is positive, then the signs of the last-term factors of the binomials are the same because . . . .

6.3 Factoring Trinomials of the Form $ax^2 + bx + c$

OBJECTIVES

Tape BA 13

1. Factor trinomials of the form $ax^2 + bx + c$.
2. Factor out a GCF before factoring a trinomial of the form $ax^2 + bx + c$.
3. Factor perfect square trinomials.
4. Factor trinomials of the form $ax^2 + bx + c$ by an alternative method.

1 In this section, we factor trinomials of the form $ax^2 + bx + c$, where the numerical coefficient of the squared variable is any integer, not just 1.

To begin, let's review the relationship between the numerical coefficients of the trinomial and the numerical coefficients of its factored form. For example, since $(2x + 1)(x + 6) = 2x^2 + 13x + 6$, the factored form of $2x^2 + 13x + 6$ is

$$2x^2 + 13x + 6 = (2x + 1)(x + 6)$$

Notice that $2x$ and x are factors of $2x^2$. Also, 6 and 1 are factors of 6, as shown:

$$2x^2 = 2x \cdot x$$

$$2x^2 + 13x + 6 = (2x + 1)(x + 6)$$

$$6 = 1 \cdot 6$$

Also notice that $13x$ is the sum of the following products:

$$2x^2 + 13x + 6 = (2x + 1)(x + 6)$$

$$1x + 12x = 13x$$

Use this information to factor $5x^2 + 7x + 2$. First find factors of $5x^2$. Since all numerical coefficients in this trinomial are positive, we use factors with positive numerical coefficients only.

Factors of $5x^2$ are $5x$ and x. Try these factors as first terms of the binomials. Thus far, we have

$$5x^2 + 7x + 2 = (5x \quad)(x \quad)$$

Next find factors of 2. Factors of 2 are 1 and 2. Try possible combinations of these factors as second terms of the binomials until a middle term of $7x$ is obtained.

$$(5x + 1)(x + 2) = 5x^2 + 11x + 2$$

$$1x + 10x = 11x$$ Incorrect middle term.

Try switching factors 2 and 1.

$$(5x + 2)(x + 1) = 5x^2 + 7x + 2$$

$$2x + 5x = 7x$$ Correct middle term.

The factored form of $5x^2 + 7x + 2$ is

$$5x^2 + 7x + 2 = (5x + 2)(x + 1)$$

EXAMPLE 1 Factor $3x^2 + 11x + 6$.

Solution: Since all numerical coefficients are positive, use factors with positive numerical coefficients. First find factors of $3x^2$.

$$\text{Factors of } 3x^2\text{:} \quad 3x^2 = 3x \cdot x$$

If factorable, the trinomial will be of the form

$$3x^2 + 11x + 6 = (3x \quad)(x \quad)$$

Next factor 6.

$$\text{Factors of 6:} \quad 6 = 1 \cdot 6, \qquad 6 = 2 \cdot 3$$

Try combinations of factors of 6 until a middle term of $11x$ is obtained. First try 1 and 6.

$$(3x + 1)(x + 6) = 3x^2 + 19x + 6$$

$$1x + 18x = 19x$$ Incorrect middle term.

Next try 6 and 1.

$$(3x + 6)(x + 1)$$

Before multiplying, notice that the factor $3x + 6$ has a common factor of 3. The original trinomial $3x^2 + 11x + 6$ has no common factor other than 1, so the factored form of $3x^2 + 11x + 6$ will contain no common factor other than 1, also. This means that $(3x + 6)(x + 1)$ is not the factored form.

Try 2 and 3.

$$(3x + 2)(x + 3) = 3x^2 + 11x + 6$$

$2x$

$+ 9x$

$11x$ Correct middle term.

The factored form of $3x^2 + 11x + 6$ is

$$3x^2 + 11x + 6 = (3x + 2)(x + 3)$$ ■

HELPFUL HINT

If a trinomial has no common factor (other than 1), then none of its binomial factors will contain a common factor (other than 1).

EXAMPLE 2 Factor $8x^2 - 22x + 5$.

Solution: Factors of $8x^2$: $8x^2 = 8x \cdot x$, $8x^2 = 4x \cdot 2x$

Try $8x$ and x.

$$8x^2 - 22x + 5 = (8x \quad)(x \quad)$$

Since the middle term $-22x$ has a negative numerical coefficient, factor 5 into negative factors.

Factors of 5: $5 = -1 \cdot -5$

Try -1 and -5.

$$(8x - 1)(x - 5) = 8x^2 - 41x + 5$$

$-1x$

$+ (-40x)$

$-41x$ Incorrect middle term.

Try -5 and -1.

$$(8x - 5)(x - 1) = 8x^2 - 13x + 5$$

$-5x$

$+ (-8x)$

$-13x$ Incorrect middle term.

Don't give up yet. Try other factors of $8x^2$. Try $4x$ and $2x$ with -1 and -5.

$$(4x - 1)(2x - 5) = 8x^2 - 22x + 5$$

$-2x$

$+ (-20x)$

$-22x$ Correct middle term.

The factored form of $8x^2 - 22x + 5$ is

$$8x^2 - 22x + 5 = (4x - 1)(2x - 5) \quad \blacksquare$$

EXAMPLE 3 Factor $2x^2 + 13x - 7$.

Solution:

$$\text{Factors of } 2x^2\text{:} \quad 2x^2 = 2x \cdot x$$

$$\text{Factors of } -7\text{:} \ -7 = -1 \cdot 7, \qquad -7 = 1 \cdot -7$$

Try possible combinations. The combination that yields the middle term of $13x$ is

$$(2x - 1)(x + 7) = 2x^2 + 13x - 7$$

$-1x$

$+\ 14x$

$13x$ Correct middle term.

The factored form of $2x^2 + 13x - 7$ is

$$2x^2 + 13x - 7 = (2x - 1)(x + 7) \quad \blacksquare$$

EXAMPLE 4 Factor $10x^2 - 13xy - 3y^2$.

Solution:

$$\text{Factors of } 10x^2\text{:} \quad 10x^2 = 2x \cdot 5x, \qquad 10x^2 = 10x \cdot x$$

$$\text{Factors of } -3y^2\text{:} \quad -3y^2 = -3y \cdot y, \qquad -3y^2 = 3y \cdot -y$$

Try possible combinations. The combination that yields the correct middle term is

$$(2x - 3y)(5x + y) = 10x^2 - 13xy - 3y^2$$

$-15xy$

$+\ \ 2xy$

$-13xy$ Correct middle term.

The factored form of $10x^2 - 13xy - 3y^2$ is

$$10x^2 - 13xy - 3y^2 = (2x - 3y)(5x + y) \quad \blacksquare$$

2 Don't forget that the first step in factoring any polynomial is to look for a common factor to factor out.

EXAMPLE 5 Factor $24x^4 + 40x^3 + 6x^2$.

Solution: Notice that all three terms have a common factor of $2x^2$. First factor out $2x^2$.

$$24x^4 + 40x^3 + 6x^2 = 2x^2(12x^2 + 20x + 3)$$

Next factor $12x^2 + 20x + 3$.

$$\text{Factors of } 12x^2\text{:} \quad 12x^2 = 6x \cdot 2x, \qquad 12x^2 = 4x \cdot 3x, \qquad 12x^2 = 12x \cdot x$$

Since all terms in the trinomial have positive numerical coefficients, factor 3 using positive factors only.

$$\text{Factors of 3:} \quad 3 = 1 \cdot 3$$

The correct combination of factors is

$$2x^2(2x + 3)(6x + 1) = 2x^2(12x^2 + 20x + 3)$$

$18x$

$+\ \ 2x$

$20x$ Correct middle term.

The factored form of $24x^4 + 40x^3 + 6x^2$ is

$$24x^4 + 40x^3 + 6x^2 = 2x^2(2x + 3)(6x + 1) \quad ■$$

EXAMPLE 6 Factor $4x^2 - 12x + 9$.

Solution:

$$\text{Factors of } 4x^2\text{:} \quad 4x^2 = 2x \cdot 2x, \qquad 4x^2 = 4x \cdot x$$

Since the middle term $-12x$ has a negative numerical coefficient, factor 9 into negative factors only.

$$\text{Factors of 9:} \quad 9 = -3 \cdot -3, \qquad 9 = -1 \cdot -9$$

The correct combination is

$$(2x - 3)(2x - 3) = 4x^2 - 12x + 9$$

$$\begin{array}{r} -6x \\ + (-6x) \\ \hline -12x \end{array} \quad \text{Correct middle term.}$$

Thus, $4x^2 - 12x + 9 = (2x - 3)(2x - 3)$, which can be written as $(2x - 3)^2$. ■

3 Notice in Example 6 that $4x^2 - 12x + 9 = (2x - 3)^2$. The trinomial $4x^2 - 12x + 9$ is called a **perfect square trinomial** since it is the square of the binomial $2x - 3$.

In the last chapter, we learned a special product for squaring a binomial.

$$(a + b)^2 = a^2 + 2ab + b^2$$

The trinomial $a^2 + 2ab + b^2$ is also a perfect square trinomial, since it is the square of the binomial $a + b$. We can use this pattern to help us factor perfect square trinomials. To use this pattern, we must first be able to recognize a perfect square trinomial. A trinomial is a perfect square when its first term is the square of some quantity a, its last term is the square of some quantity b, and its middle term is twice the product of the quantities a and b.

Perfect Square Trinomials

$$a^2 + 2ab + b^2 = (a + b)^2$$

$$a^2 - 2ab + b^2 = (a - b)^2$$

EXAMPLE 7 Factor $x^2 + 12x + 36$.

Solution: This trinomial is a perfect square trinomial since:

1. The first term is a square: $x^2 = (x)^2$.
2. The last term is a square: $36 = (6)^2$.
3. The middle term is twice the product of x and 6: $12x = 2 \cdot x \cdot 6$.

Thus, $x^2 + 12x + 36 = (x + 6)^2$. ■

EXAMPLE 8 Factor $25x^2 + 25xy + 4y^2$.

Solution: Determine whether or not this trinomial is a perfect square by considering the same three questions.

1. Is the first term a square? Yes, $25x^2 = (5x)^2$.
2. Is the last term a square? Yes, $4y^2 = (2y)^2$.
3. Is the middle term twice the product of $5x$ and $2y$? **No.** $2(5x)(2y) = 20xy$, not $25xy$.

Therefore, try to factor $25x^2 + 25xy + 4y^2$ by other methods. It is factorable, and $25x^2 + 25xy + 4y^2 = (5x + 4y)(5x + y)$. ■

HELPFUL HINT

A perfect square trinomial that is not recognized as such can be factored by other methods.

EXAMPLE 9 Factor $4m^2 - 4m + 1$.

Solution: This is a perfect square trinomial since $4m^2 = (2m)^2$, $1 = (1)^2$, and $4m = 2(2m)(1)$.

$$4m^2 - 4m + 1 = (2m - 1)^2 \quad ■$$

4 There is another method for factoring trinomials of the form $ax^2 + bx + c$. This method is described next.

To factor $2x^2 + 11x + 12$, find two numbers whose product is $2 \cdot 12 = 24$ and whose sum is 11. Since we want a positive product and a positive sum, we consider positive factors of 24 only.

Factors of 24	*Sum of Factors*
1, 24	1 + 24 = 25
2, 12	2 + 12 = 14
3, 8	3 + 8 = 11

The factors are 3 and 8. Use these factors to write the middle term $11x$ as $3x + 8x$. Replace $11x$ with $3x + 8x$ in the original trinomial and factor by grouping.

$$\begin{aligned} 2x^2 + 11x + 12 &= 2x^2 + 3x + 8x + 12 \\ &= x(2x + 3) + 4(2x + 3) \\ &= (2x + 3)(x + 4) \end{aligned}$$

In general, we have the following:

To Factor Trinomials of the Form $ax^2 + bx + c$ by Grouping

Step 1 Find two numbers whose product is $a \cdot c$ and whose sum is b.
Step 2 Write the middle term, bx, using the factors found in step 1.
Step 3 Factor by grouping.

EXAMPLE 10 Factor $8x^2 - 14x + 5$.

Solution: In this trinomial, $a = 8$, $b = -14$, and $c = 5$.

Step 1 Find two numbers whose product is $a \cdot c$ or $8 \cdot 5 = 40$ and whose sum is b or -14. The numbers are -4 and -10.

Step 2 Write $-14x$ as $-4x - 10x$ so that

$$8x^2 - 14x + 5 = 8x^2 - 4x - 10x + 5$$

Step 3 Factor by grouping.

$$8x^2 - 4x - 10x + 5 = 4x(2x - 1) - 5(2x - 1)$$
$$= (2x - 1)(4x - 5)$$

EXAMPLE 11 Factor $3x^2 - x - 10$ by the method just described.

Solution: In $3x^2 - x - 10$, $a = 3$, $b = -1$, and $c = -10$.

Step 1 Find two numbers whose product is $a \cdot c$ or $3(-10) = -30$ and whose sum is b, -1. The numbers are -6 and 5.

Step 2 $3x^2 - x - 10 = 3x^2 - 6x + 5x - 10$

Step 3 $= 3x(x - 2) + 5(x - 2)$

$= (x - 2)(3x + 5)$

MENTAL MATH

State whether or not each trinomial is a perfect trinomial square.

1. $x^2 + 14x + 49$
2. $9x^2 - 12x + 4$
3. $y^2 + 2y + 4$
4. $x^2 - 4x + 2$
5. $9y^2 + 6y + 1$
6. $y^2 - 16y + 64$

EXERCISE SET 6.3

Factor completely. See Examples 1 through 4. (See Example 10 for an alternative method.)

1. $2x^2 + 13x + 15$
2. $3x^2 + 8x + 4$
3. $2x^2 - 9x - 5$
4. $3x^2 + 20x - 63$
5. $2y^2 - y - 6$
6. $8y^2 - 17y + 9$
7. $16a^2 - 24a + 9$
8. $25x^2 + 20x + 4$
9. $36r^2 - 5r - 24$
10. $20r^2 + 27r - 8$
11. $10x^2 + 17x + 3$
12. $21x^2 - 41x + 10$

Factor completely. See Example 5. (See Example 11 for an alternative method.)

13. $21x^2 - 48x - 45$
14. $12x^2 - 14x - 10$
15. $12x^2 - 14x - 6$
16. $20x^2 - 2x + 6$
17. $4x^3 - 9x^2 - 9x$
18. $6x^3 - 31x^2 + 5x$

Factor the following perfect square trinomials. See Examples 6 through 9.

19. $x^2 + 22x + 121$

20. $x^2 + 18x + 81$

21. $x^2 - 16x + 64$

22. $x^2 - 12x + 36$

23. $16y^2 - 40y + 25$

24. $9y^2 + 48y + 64$

25. $x^2y^2 - 10xy + 25$

26. $4x^2y^2 - 28xy + 49$

Factor the following completely.

27. $2x^2 - 7x - 99$

28. $2x^2 + 7x - 72$

29. $4x^2 - 8x - 21$

30. $6x^2 - 11x - 10$

31. $30x^2 - 53x + 21$

32. $21x^3 - 6x - 30$

33. $24x^2 - 58x + 9$

34. $36x^2 + 55x - 14$

35. $9x^2 - 24xy + 16y^2$

36. $25x^2 + 60xy + 36y^2$

37. $x^2 - 14xy + 49y^2$

38. $x^2 + 10xy + 25y^2$

39. $2x^2 + 7x + 5$

40. $2x^2 + 7x + 3$

41. $3x^2 - 5x + 1$

42. $3x^2 - 7x + 6$

43. $-2y^2 + y + 10$

44. $-4x^2 - 23x + 6$

45. $16x^2 + 24xy + 9y^2$

46. $4x^2 - 36xy + 81y^2$

47. $8x^2y + 34xy - 84y$

48. $6x^2y^2 - 2xy^2 - 60y^2$

49. $3x^2 + x - 2$

50. $8y^2 + y - 9$

51. $x^2y^2 + 4xy + 4$

52. $x^2y^2 - 6xy + 9$

53. $49y^2 + 42xy + 9x^2$

54. $16x^2 - 8xy + y^2$

55. $3x^2 - 42x + 63$

56. $5x^2 - 75x + 60$

57. $42a^2 - 43a + 6$

58. $54a^2 + 39ab - 8b^2$

59. $18x^2 - 9x - 14$

60. $8x^2 + 6x - 27$

61. $25p^2 - 70pq + 49q^2$

62. $36p^2 - 18pq + 9q^2$

63. $15x^2 - 16x - 15$

64. $12x^2 + 7x - 12$

65. $-27t + 7t^2 - 4$

66. $4t^2 - 7 - 3t$

67. $-12x^3y^2 + 3x^2y^2 + 15xy^2$

68. $-12r^3x^2 + 38r^2x^2 + 14rx^2$

69. $-30p^3q + 88p^2q^2 + 6pq^3$

70. $3x^3y^2 + 3x^2y^3 - 18xy^4$

71. $4x^2(y - 1)^2 + 10x(y - 1)^2 + 25(y - 1)^2$

72. $3x^2(a + 3)^3 - 28x(a + 3)^3 + 25(a + 3)^3$

73. $3x^2y - 11xy + 8y$

74. $5xy^2 - 9xy + 4x$

75. $2x^2 + 2x - 12$

76. $3x^2 + 6x - 45$

77. $(x - 4)^2 + 3(x - 4) - 18$

78. $(x - 3)^2 - 2(x - 3) - 8$

79. $2x^6 + 3x^3 - 9$

80. $3x^6 - 14x^3 + 8$

81. $72xy^4 - 24xy^2z + 2xz^2$

82. $36xy^2 - 48xyz^2 + 16xz^4$

Solve.

83. The volume $V(h)$ of a box in terms of its height h is given by the function

$$V = 3h^3 - 2h^2 - 8h$$

Factor this expression for $V(h)$.

84. Based on your results from Exercise 83, find the length and width of the box if the height is 5 inches and the dimensions of the box are whole numbers.

Skill Review

Multiply the following. See Section 5.4.

85. $(x - 2)(x^2 + 2x + 4)$

86. $(y + 1)(y^2 - y + 1)$

87. $(x + 3)(x - 3)$

88. $(2x - 9)(2x + 9)$

89. $(5a + b)(5a - b)$

90. $(a - b)(a + b)$

If $P(x) = 3x^2 + 2x - 9$, find the following. See Section 5.3.

91. $P(0)$ **92.** $P(1)$ **93.** $P(-1)$ **94.** $P(-2)$

Graph each inequality. See Section 4.6.

95. $x + y < 2$ **96.** $x \geq 3y$ **97.** $y \leq 2$ **98.** $x > -1$

A Look Ahead

Factor. Assume that variables used as exponents represent positive integers. See the following example.

EXAMPLE Factor $x^{2n} + 7x^n + 12$.

Solution: Factors of x^{2n} are x^n and x^n, so $x^{2n} + 7x^n + 12 = (x^n + \text{one number})(x^n + \text{other number})$. Factors of 12 whose sum is 7 are 3 and 4. Thus

$$x^{2n} + 7x^n + 12 = (x^n + 4)(x^n + 3) \quad ■$$

99. $x^{2n} + 10x^n + 16$

100. $x^{2n} - 7x^n + 12$

101. $x^{2n} - 3x^n - 18$

102. $x^{2n} + 7x^n - 18$

103. $2x^{2n} + 11x^n + 5$

104. $3x^{2n} - 8x^n + 4$

105. $4x^{2n} - 12x^n + 9$

106. $9x^{2n} + 24x^n + 16$

Writing in Mathematics

107. Describe a perfect square trinomial, and then give one example.

6.4 Factoring Binomials

OBJECTIVES

1 Factor the difference of two squares.

2 Factor the sum or difference of two cubes.

Tape IA 14

1 When learning to multiply binomials in Chapter 5, we studied a special product: the product of the sum and difference of two terms, a and b.

$$(a + b)(a - b) = a^2 - b^2$$

For example, the product of $x + 3$ and $x - 3$ is

$$(x + 3)(x - 3) = x^2 - 9$$

The binomial $x^2 - 9$ is called a **difference of squares.** In this section, we use the pattern for the product of a sum and difference to factor the binomial difference of squares.

To use this pattern to help us factor, we must be able to recognize a difference of squares. A binomial is a difference of squares when it is the difference of the square of some quantity a and the square of some quantity b.

Difference of Two Squares

$$a^2 - b^2 = (a + b)(a - b)$$

EXAMPLE 1 Factor $4x^2 - 1$.

Solution: $4x^2 - 1$ is the difference of two squares since $4x^2 = (2x)^2$ and $1 = (1)^2$; therefore,

$$4x^2 - 1 = (2x)^2 - 1^2 = (2x + 1)(2x - 1)$$

Multiply to check. ■

EXAMPLE 2 Factor $25a^2 - 9b^2$.

Solution: $25a^2 - 9b^2 = (5a)^2 - (3b)^2 = (5a + 3b)(5a - 3b)$ ■

EXAMPLE 3 Factor $9x^2 - 36$.

Solution: Remember when factoring to always check first for common factors. If there are common factors, factor out the GCF, and then factor the resulting polynomial.

$$\begin{aligned} 9x^2 - 36 &= 9(x^2 - 4) && \text{Factor out the GCF 9.} \\ &= 9(x^2 - 2^2) \\ &= 9(x + 2)(x - 2) \end{aligned}$$

In this example, if we forget to factor out the GCF first, we still have the difference of two squares.

$$9x^2 - 36 = (3x)^2 - (6)^2 = (3x + 6)(3x - 6)$$

This binomial has not been factored completely since both terms of both binomial factors have a common factor of 3.

$$3x + 6 = 3(x + 2) \quad \text{and} \quad 3x - 6 = 3(x - 2)$$

Then

$$9x^2 - 36 = (3x + 6)(3x - 6) = 3(x + 2)3(x - 2) = 9(x + 2)(x - 2)$$

Factoring is easier if the GCF is factored out first before using other methods. ■

EXAMPLE 4 Factor $x^2 + 4$.

Solution: The binomial $x^2 + 4$ is the **sum** of squares since we can write $x^2 + 4 = x^2 + 2^2$. We might try to factor using $(x + 2)(x + 2)$ or $(x - 2)(x - 2)$. But when we multiply to check, neither factoring is correct.

$$\begin{aligned} (x + 2)(x + 2) &= x^2 + 4x + 4 \\ (x - 2)(x - 2) &= x^2 - 4x + 4 \end{aligned}$$

In both cases, the product is a trinomial, not the required binomial. Thus, $x^2 + 4$ is a prime polynomial. ■

2 Although the sum of two squares usually does not factor, the sum or difference of two cubes can be factored. The pattern for factoring the sum of cubes is illustrated by multiplying the binomial $x + y$ and the trinomial $x^2 - xy + y^2$.

$$\begin{array}{rrrr} & x^2 & - xy & + y^2 \\ & & x & + y \\ \hline & x^2y & - xy^2 & + y^3 \\ x^3 & - x^2y & + xy^2 & \\ \hline x^3 & & & + y^3 \end{array}$$

$$(x + y)(x^2 - xy + y^2) = x^3 + y^3 \qquad \text{Sum of cubes.}$$

The pattern for the difference of two cubes is illustrated by multiplying the binomial $x - y$ by the trinomial $x^2 + xy + y^2$. The result is

$$(x - y)(x^2 + xy + y^2) = x^3 - y^3 \qquad \text{Difference of cubes.}$$

Sum or Difference of Two Cubes

$$a^3 + b^3 = (a + b)(a^2 - ab + b^2)$$
$$a^3 - b^3 = (a - b)(a^2 + ab + b^2)$$

EXAMPLE 5 Factor $x^3 + 8$.

Solution: First write the binomial in the form $a^3 + b^3$.

$$x^3 + 8 = x^3 + 2^3 \qquad \text{Write in the form } a^3 + b^3.$$

If we replace a with x and b with 2 in the formula above, we have

$$\begin{aligned} x^3 + 2^3 &= (x + 2)(x^2 - x \cdot 2 + 2^2) \\ &= (x + 2)(x^2 - 2x + 4) \end{aligned}$$ ■

HELPFUL HINT

When factoring sums or differences of cubes, notice the sign patterns.

same sign

$$x^3 + y^3 = (x + y)(x^2 - xy + y^2)$$

opposite sign — always positive

same sign

$$x^3 - y^3 = (x - y)(x^2 + xy + y^2)$$

opposite sign — always positive

EXAMPLE 6 Factor $y^3 - 27$.

Solution:

$$\begin{aligned} y^3 - 27 &= y^3 - 3^3 \qquad \text{Write in the form } a^3 - b^3. \\ &= (y - 3)(y^2 + y \cdot 3 + 3^2) \\ &= (y - 3)(y^2 + 3y + 9) \end{aligned}$$ ■

EXAMPLE 7 Factor $64x^3 + 1$.

Solution:
$$64x^3 + 1 = (4x)^3 + 1^3$$
$$= (4x + 1)[(4x)^2 - (4x)(1) + 1^2]$$
$$= (4x + 1)(16x^2 - 4x + 1)$$
■

EXAMPLE 8 Factor $54a^3 - 16b^3$.

Solution: Remember to factor out common factors first before using other factoring methods.

$$54a^3 - 16b^3 = 2(27a^3 - 8b^3)$$ Factor out the GCF 2.

$$= 2[(3a)^3 - (2b)^3]$$ Difference of two cubes.

$$= 2(3a - 2b)[(3a)^2 + (3a)(2b) + (2b)^2]$$
$$= 2(3a - 2b)(9a^2 + 6ab + 4b^2)$$
■

MENTAL MATH

State each number as a square.

1. 1 **2.** 25 **3.** 81 **4.** 64
5. 9 **6.** 100

State each number as a cube.

7. 1 **8.** 64 **9.** 8 **10.** 27

EXERCISE SET 6.4

Factor the difference of two squares. See Examples 1 through 3.

1. $25y^2 - 9$ **2.** $49a^2 - 16$ **3.** $121 - 100x^2$
4. $144 - 81x^2$ **5.** $12x^2 - 27$ **6.** $36x^2 - 64$
7. $169a^2 - 49b^2$ **8.** $225a^2 - 81b^2$ **9.** $x^2y^2 - 1$
10. $16 - a^2b^2$

Factor the sum or difference of two cubes. See Examples 5 through 8.

11. $a^3 + 27$ **12.** $b^3 - 8$ **13.** $8a^3 + 1$
14. $64x^3 - 1$ **15.** $5k^3 + 40$ **16.** $6r^3 - 162$
17. $x^3y^3 - 64$ **18.** $8x^3 - y^3$ **19.** $x^3 + 125$
20. $a^3 - 216$ **21.** $24x^4 - 81xy^3$ **22.** $375y^6 - 24y^3$

Factor the binomials completely.

23. $x^2 - 4$ **24.** $x^2 - 36$ **25.** $81 - p^2$
26. $100 - t^2$ **27.** $4r^2 - 1$ **28.** $9t^2 - 1$
29. $9x^2 - 16$ **30.** $36y^2 - 25$ **31.** $16r^2 + 1$
32. $49y^2 + 1$ **33.** $27 - t^3$ **34.** $125 + r^3$
35. $8r^3 - 64$ **36.** $54r^3 + 2$ **37.** $t^3 - 343$

38. $s^3 + 216$

39. $x^2 - 169y^2$

40. $x^2 - 225y^2$

41. $x^2y^2 - z^2$

42. $x^3y^3 - z^3$

43. $x^3y^3 + 1$

44. $x^2y^2 + z^2$

45. $s^3 - 64t^3$

46. $8t^3 + s^3$

47. $18r^2 - 8$

48. $32t^2 - 50$

49. $9xy^2 - 4x$

50. $16xy^2 - 64x$

51. $25y^4 - 100y^2$

52. $xy^3 - 9xyz^2$

53. $x^3y - 4xy^3$

54. $12s^3t^3 + 192s^5t$

55. $8s^6t^3 + 100s^3t^6$

56. $25x^5y + 121x^3y$

57. $27x^2y^3 - xy^2$

58. $8x^3y^3 + x^3y$

59. $x^4 - 16$

60. $b^4 - 81a^4$

61. $a^2 - (2 + b)^2$

62. $(x + 3)^2 - y^2$

63. $(x^2 - 4)^2 - (x - 2)^2$

64. $(x^2 - 9) - (3 - x)$

Solve.

65. What binomial multiplied by $(x - 6)$ gives the difference of two squares?

66. What binomial multiplied by $(5 + y)$ gives the difference of two squares?

67. What binomial multiplied by $(4x^2 - 2xy + y^2)$ gives the sum or difference of two cubes?

68. What binomial multiplied by $(1 + 4y + 16y^2)$ gives the sum or difference of two cubes?

69. The manufacturer of Antonio's Metal Washers needs to determine the cross-sectional area of each washer. If the outer radius of the washer is R and the radius of the hole is r, express the area of the washer as a polynomial. Factor this polynomial completely.

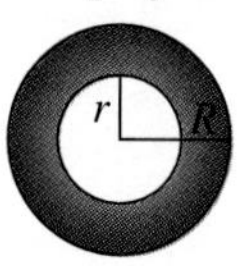

70. The manufacturer of Tootsie Roll Pops plans to change the size of its candy. To compute the new cost, they need a formula for the volume of the candy coating without the Tootsie Roll center. Given the diagram, express the volume as a polynomial. Factor this polynomial completely.

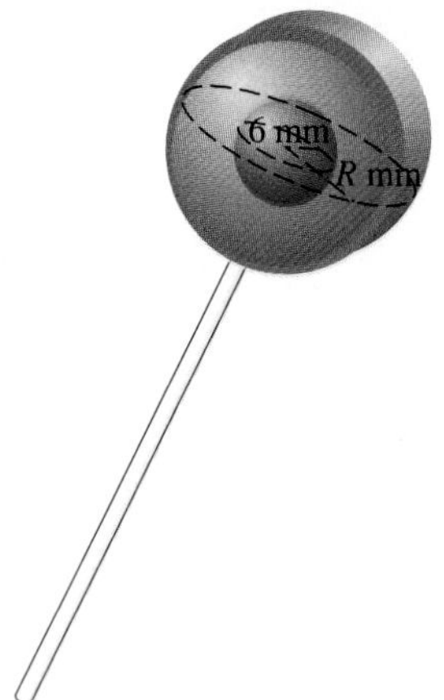

71. Express the area of the shaded region as a polynomial. Factor the polynomial completely.

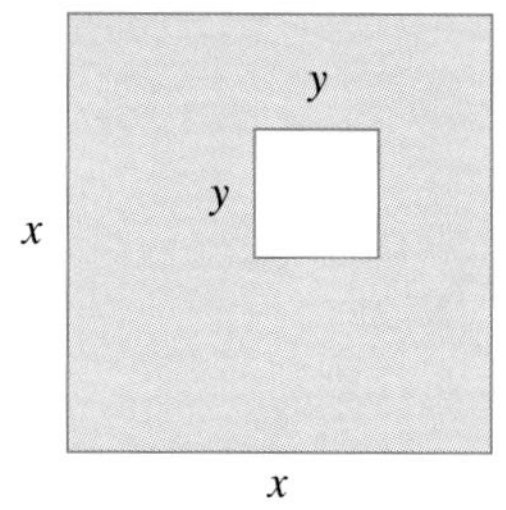

72. Express the area of the shaded region as a polynomial. Factor the polynomial completely.

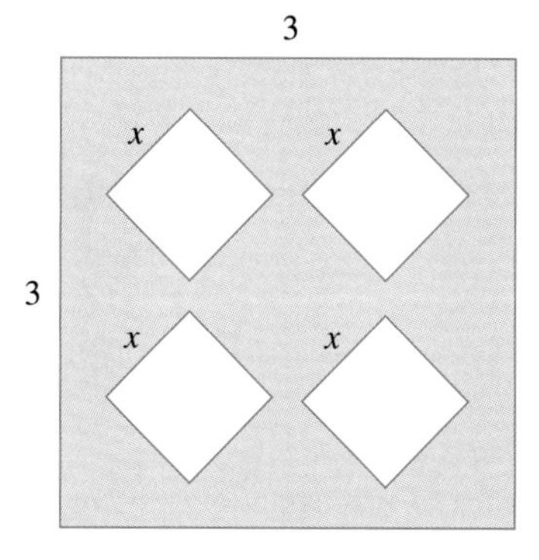

Skill Review

Divide the following. See Section 5.5.

73. $\dfrac{8x^4 + 4x^3 - 2x + 6}{2x}$

74. $\dfrac{3y^4 + 9y^2 - 6y + 1}{3y^2}$

Use long division to divide the following. See Section 5.5.

75. $\dfrac{2x^2 - 3x - 2}{x - 2}$

76. $\dfrac{4x^2 - 21x + 21}{x - 3}$

77. $\dfrac{3x^2 + 13x + 10}{x + 3}$

78. $\dfrac{5x^2 + 14x + 12}{x + 2}$

Solve each compound inequality. Graph the solution set. See Section 3.6.

79. $6 < x - 1 < 7$

80. $-2 < 4x + 2 < 18$

81. $0 \leq \dfrac{x + 2}{6} \leq 5$

82. $-4 \leq \dfrac{3x}{7} \leq 0$

83. $x \geq -5$ or $x \geq -1$

84. $x \leq 6$ or $x \geq 0$

85. $x < -2$ and $x > -10$

86. $x < 14$ and $x < 9$

A Look Ahead

Factor each expression. Assume that variables used as exponents represent positive integers. See the following example.

EXAMPLE Factor $x^{2n} - 100$.

Solution: This binomial is a difference of squares.

$$x^{2n} - 100 = (x^n)^2 - 10^2$$
$$= (x^n + 10)(x^n - 10) \quad ■$$

87. $x^{2n} - 25$

88. $x^{2n} - 36$

89. $x^{2n} - 9$

90. $x^{2n} - 16$

91. $36x^{2n} - 49$

92. $25x^{2n} - 81$

93. $x^{4n} - 16$

94. $x^{4n} - 625$

Writing in Mathematics

95. Knowing that $x - 2$ is a factor of the difference of the fifth powers $x^5 - 32$, explain how to find another factor.

96. Factor $x^6 - 1$ completely, using the following methods from this chapter.

a. Factor the expression treating it as the difference of two squares, $(x^3)^2 - 1^2$.

b. Factor the expression treating it as the difference of two cubes, $(x^2)^3 - 1^3$.

c. Are the answers to parts a and b the same? Why or why not?

6.5 Factoring Polynomials Completely

OBJECTIVE

1 Practice techniques for factoring polynomials.

Tape IA 14

1 The key to proficiency in factoring is to practice until you are comfortable with each technique. A summary of the methods of factoring is given next.

To Factor a Polynomial

Step 1 Are there any common factors? If so, factor them out.

Step 2 How many terms are in the polynomial?

a. If there are **two** terms, decide if one of the following formulas may be applied.

i. Difference of two squares: $a^2 - b^2 = (a - b)(a + b)$.

ii. Difference of two cubes: $a^3 - b^3 = (a - b)(a^2 + ab + b^2)$.

iii. Sum of two cubes: $a^3 + b^3 = (a + b)(a^2 - ab + b^2)$.

b. If there are **three** terms, try one of the following:

i. Perfect square trinomial: $a^2 + 2ab + b^2 = (a + b)^2$.

$a^2 - 2ab + b^2 = (a - b)^2$.

ii. If not a perfect square trinomial, factor using the methods presented in Sections 6.2 and 6.3.

c. If there are **four** or more terms, try factoring by grouping.

Step 3 See if any factors in the factored polynomial can be factored further.

EXAMPLE 1 Factor each polynomial completely.

a. $8a^2b - 4ab$ **b.** $36x^2 - 9$ **c.** $2x^2 - 5x - 7$

Solution: **a.** The terms have a common factor of $4ab$, which we factor out.

$$8a^2b - 4ab = 4ab(2a - 1)$$

b. $36x^2 - 9 = 9(4x^2 - 1)$ Factor out a GCF of 9.

$= 9(2x + 1)(2x - 1)$ Factor the difference of squares.

c. $2x^2 - 5x - 7 = (2x - 7)(x + 1)$ ■

EXAMPLE 2 Factor each polynomial completely.

a. $5p^2 + 5 + qp^2 + q$ **b.** $9x^2 + 24x + 16$ **c.** $y^2 + 25$

Solution: **a.** There is no common factor of all terms of $5p^2 + 5 + qp^2 + q$. The polynomial has four terms, so try factoring by grouping.

$5p^2 + 5 + qp^2 + q = (5p^2 + 5) + (qp^2 + q)$ Group the terms.

$$= 5(p^2 + 1) + q(p^2 + 1)$$

$$= (p^2 + 1)(5 + q)$$

b. The trinomial $9x^2 + 24x + 16$ is a perfect square trinomial and $9x^2 + 24x + 16 = (3x + 4)^2$.

c. There is no common factor of $y^2 + 25$ other than 1. This binomial is the sum of two squares and is prime. ■

EXAMPLE 3 Factor each completely.

a. $27a^3 - b^3$ **b.** $3n^2m^4 - 48m^6$ **c.** $2x^2 - 12x + 18 - 2z^2$
d. $8x^4y^2 + 125xy^2$ **e.** $(x - 5)^2 - 49y^2$

Solution: **a.** This binomial is a difference of two cubes.

$$27a^3 - b^3 = (3a)^3 - b^3$$

$$= (3a - b)[(3a)^2 + (3a)(b) + b^2]$$

$$= (3a - b)(9a^2 + 3ab + b^2)$$

b. $3n^2m^4 - 48m^6 = 3m^4(n^2 - 16m^2)$ Factor out the GCF, $3m^4$.

$= 3m^4(n + 4m)(n - 4m)$ Factor the difference of squares.

c. $2x^2 - 12x + 18 - 2z^2 = 2(x^2 - 6x + 9 - z^2)$ The GCF is 2.

$= 2[(x^2 - 6x + 9) - z^2]$ Group the first three terms together.

$= 2[(x - 3)^2 - z^2]$ Factor the perfect square trinomial.

$= 2[(x - 3) + z][(x - 3) - z]$ Factor the difference of squares.

$= 2(x - 3 + z)(x - 3 - z)$

d. $8x^4y^2 + 125xy^2 = xy^2(8x^3 + 125)$ The GCF is xy^2.

$= xy^2[(2x)^3 + 5^3]$

$= xy^2(2x + 5)[(2x)^2 - (2x)(5) + 5^2]$ Factor the sum of cubes.

$= xy^2(2x + 5)(4x^2 - 10x + 25)$

e. This binomial is a difference of squares.

$$(x - 5)^2 - 49y^2 = (x - 5)^2 - (7y)^2$$
$$= [(x - 5) + 7y][(x - 5) - 7y]$$
$$= (x - 5 + 7y)(x - 5 - 7y)$$ ■

EXERCISE SET 6.5

Factor completely.

1. $x^2 - 9$
2. $6x - 9$
3. $x^2 - 8x - 9$
4. $x^3 + 8$
5. $x^2 + 8$
6. $x^2 - 2xy - 3x + 6y$
7. $x^2 - 8x + 16 - y^2$
8. $12x^2 - 22x - 20$
9. $x^4 - x$
10. $(2x + 1)^2 - 3(2x + 1) + 2$
11. $14x^2y - 2xy$
12. $24ab^2 - 6ab$
13. $8a^2b - 6ab^2$
14. $15x^2y^4 - 6x^3y^2$
15. $x^4 - x^3$
16. $4y^4 - 2y$
17. $x^4 - 4x^2$
18. $x^6 - x^4$
19. $4x^2 - 16$
20. $9x^2 - 81$
21. $x - 9x^3$
22. $y^2 - 25y^4$
23. $3x^2 - 8x - 11$
24. $5x^2 - 2x - 3$
25. $4x^2 + 8x - 12$
26. $6x^2 - 6x - 12$
27. $4x^2 + 23x + 15$
28. $6x^2 + 19x + 10$
29. $6x^2 - 8 + 3x^2y - 4y$
30. $10a^2 - 15 + 2a^2b - 3b$
31. $2xy + 6y + 8x + 24$
32. $6xy + 30x + 3y + 15$
33. $x^3 + 3x^2 - 4x - 12$
34. $x^3 + 2x^2 - 9x - 18$

35. $4x^2 + 36x + 81$
36. $25x^2 + 40x + 16$
37. $9x^2 - 30x + 25$
38. $4x^2 - 28x + 49$
39. $16x^2 + 12x + 9$
40. $18x^2 + 12x + 2$
41. $48x^2 - 24x + 3$
42. $9x^2 - 6x + 4$
43. $x^2 + 16$
44. $x^4 - 16$
45. $2x^4 - 2$
46. $3x^4 + 3$
47. $a^3 - 8b^3$
48. $64a^3 - b^3$
49. $125 - x^3$
50. $27 - y^3$
51. $8x^3 + 27y^3$
52. $125x^3 + 8y^3$
53. $2a^3 + 128b^3$
54. $5x^3 - 40y^3$
55. $6a^2b^4 - 24a^4$
56. $8a^3b^6 - 2a$
57. $8x^2y^5 - 2y^3$
58. $6x^3 - 96x^5$
59. $x^2 + 6x + 9 - y^2$
60. $x^2 - 4x + 4 - y^2$
61. $2x^2 - 20x + 50 - 2y^2$
62. $3y^2 - 24y + 48 - 3x^2$
63. $3a^2 - 6a + 3 - 3b^2$
64. $8a^2 + 8a + 2 - 2b^2$
65. $27x^4 - xy^3$
66. $8xy^5 - x^4y^2$
67. $16x^3y^4 + 54y$
68. $2xy^4 + 250xy$
69. $64x^2y^3 - 8x^2$
70. $27x^5y^4 - 216x^2y$
71. $(3x + 1)^2 + 3(3x + 1) + 2$
72. $(2x + 3)^2 + 6(2x + 3) + 5$
73. $x^4 + 6x^2 - 7$
74. $x^4 - 2x^2 - 8$
75. $x^6 + 4x^3 + 3$
76. $x^6 + 6x^3 - 16$
77. $(x + y)^2 - 9$
78. $z^2 - (x + 2)^2$
79. $(a - 4)^3 + 1$
80. $(a + b)^3 + 8$
81. $(x + 5)^3 + y^3$
82. $(y - 1)^3 + 27x^3$
83. $(2x + 1)^3 + 27$
84. $(3x + 2)^3 + 8$
85. $27 - (x + 3)^3$
86. $64 - (x + 2)^3$

Solve.

87. Three inches of matting is placed around a square picture with each side x inches, as shown. Write an expression, in factored form, for the area of the matting.

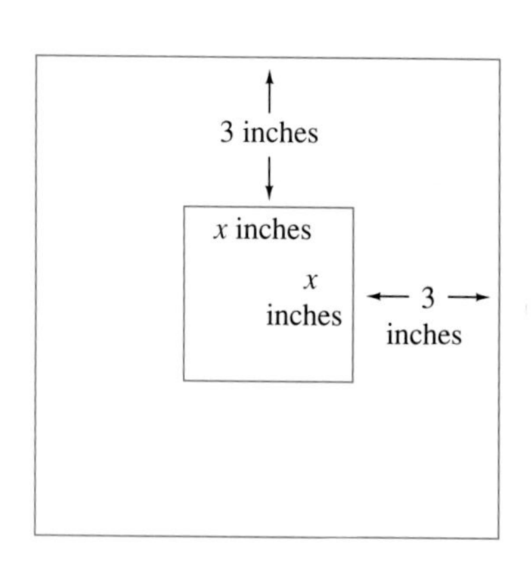

Skill Review

Perform indicated operations. See Section 1.6.

88. $8 - (-3)$
89. $4 + (-5)$
90. $-2 + (-6)$
91. $-7 - 10$

Solve the following equations. See Section 2.2.

92. $x - 5 = 0$
93. $x + 7 = 0$
94. $3x + 1 = 0$
95. $5x - 15 = 0$
96. $-2x = 0$
97. $3x = 0$
98. $-5x + 25 = 0$
99. $-4x - 16 = 0$

Solve the following. See Section 2.4.

100. The following suitcase has a volume of 960 cubic inches. Find x.

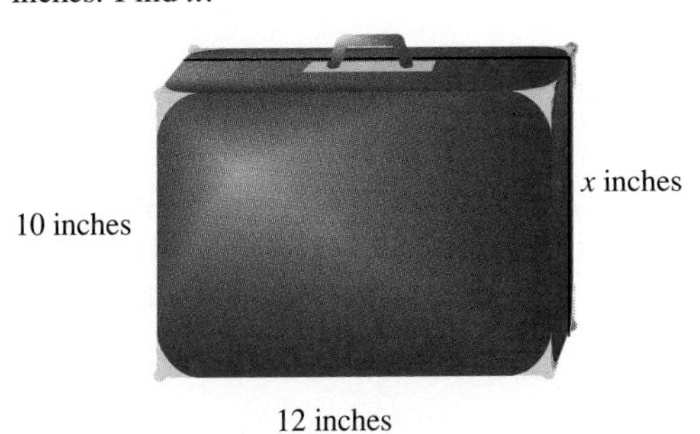

101. The sail shown has an area of 25 square feet. Find its height.

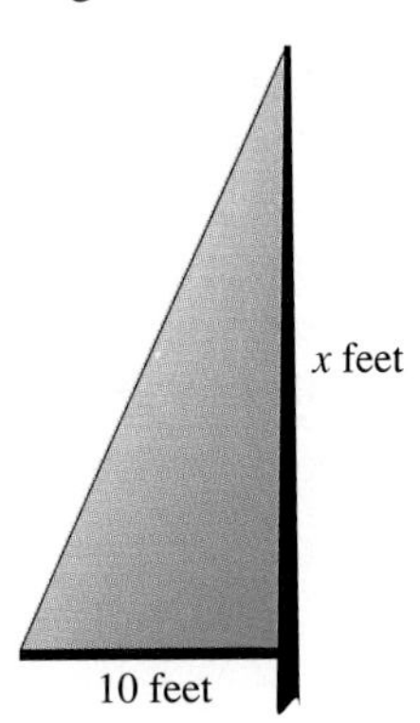

A Look Ahead

See A Look Ahead examples from Section 6.4. Factor completely.

102. $x^{2n} - 9$

103. $x^{2n} - 25$

104. $x^{4n} - 81$

105. $x^{4n} - 16$

106. $x^{3p} - 8$

107. $y^{3n} - 27$

6.6 Solving Equations by Factoring and Problem Solving

OBJECTIVES

Tape IA 15

1. Solve quadratic equations by factoring.
2. Solve higher-degree equations by factoring.
3. Solve problems that can be modeled by quadratic equations.

1 In this section, your efforts to learn factoring start to pay off. We use factoring to solve equations that contain polynomials of degree 2 and higher.

The following equations are examples of

Second Degree or Quadratic Equations

$x^2 = 25$ $\quad y^2 + 7y - 8 = 0$ $\quad 5n^2 = 2n - 12$ $\quad z(z - 3) = 2$

Quadratic Equation in One Variable

A **quadratic equation** in one variable is an equation that can be written in the form

$$ax^2 + bx + c = 0$$

where a, b, and c are real numbers and $a \neq 0$.

The form $ax^2 + bx + c = 0$ is called the **standard form** of a quadratic equation. The following are examples of quadratic equations and the equivalent equations written in standard form.

Quadratic equations:	$x^2 = 25$	$5n^2 = 2n - 12$	$z(z - 3) = 2$
Standard form:	$x^2 - 25 = 0$	$5n^2 - 2n + 12 = 0$	$z^2 - 3z - 2 = 0$

A solution of a quadratic equation in one variable is a value of the variable that makes the equation true. The method presented in this section for solving quadratic equations is called the **factoring method.** This method is based on the **zero-factor property.**

Zero-Factor Property

If a and b are real numbers and $a \cdot b = 0$, then $a = 0$ or $b = 0$.

In other words, if the product of two real numbers is zero, then at least one number must be zero.

EXAMPLE 1 Solve $(x + 2)(x - 6) = 0$.

Solution: By the zero-factor property, $(x + 2)(x - 6) = 0$ only if $x + 2 = 0$ or $x - 6 = 0$.

$$x + 2 = 0 \quad \text{or} \quad x - 6 = 0 \qquad \text{Apply zero-factor property.}$$

$$x = -2 \quad \text{or} \quad x = 6 \qquad \text{Solve each linear equation.}$$

To check, let $x = -2$ and then let $x = 6$ in the original equation.

Let $x = -2$.

Then $(x + 2)(x - 6) = 0$

becomes $(-2 + 2)(-2 - 6) = 0$

$(0)(-8) = 0$

$0 = 0$ True.

Let $x = 6$.

Then $(x + 2)(x - 6) = 0$

becomes $(6 + 2)(6 - 6) = 0.$

$(8)(0) = 0$

$0 = 0$ True.

Both -2 and 6 check, and the solution set is $\{-2, 6\}$. ■

EXAMPLE 2 Solve $2x^2 + 9x - 5 = 0$.

Solution: To use the zero-factor property, one side of the equation must be 0 and the other side must be in factored form.

$$2x^2 + 9x - 5 = 0$$

$$(2x - 1)(x + 5) = 0 \qquad \text{Factor.}$$

$$2x - 1 = 0 \quad \text{or} \quad x + 5 = 0 \qquad \text{Set each factor equal to zero.}$$

$$2x = 1$$

$$x = \frac{1}{2} \quad \text{or} \quad x = -5 \qquad \text{Solve each linear equation.}$$

The solution set is $\left\{-5, \frac{1}{2}\right\}$. To check, let $x = \frac{1}{2}$ in the original equation; then let $x = -5$ in the original equation. ■

The general procedure to solve a quadratic equation by factoring is given next.

To Solve Quadratic Equations by Factoring

Step 1 Write the equation in standard form: $ax^2 + bx + c = 0$.
Step 2 Factor the quadratic expression.
Step 3 Set each factor containing a variable equal to 0.
Step 4 Solve the resulting equations.
Step 5 Check each solution in the original equation.

Since it is not always possible to factor a quadratic polynomial, not all quadratic equations can be solved by factoring. Other methods of solving quadratic equations are presented in Chapter 9.

EXAMPLE 3 Solve $x(2x - 7) = 4$.

Solution: First write the equation in standard form; then factor.

$x(2x - 7) = 4$	
$2x^2 - 7x = 4$	Multiply.
$2x^2 - 7x - 4 = 0$	Write in standard form.
$(2x + 1)(x - 4) = 0$	Factor.
$2x + 1 = 0$ or $x - 4 = 0$	Set each factor equal to zero.
$2x = -1$ or $x = 4$	Solve.
$x = -\frac{1}{2}$	

The solution set is $\left\{-\frac{1}{2}, 4\right\}$. Check both solutions in the original equation. ■

HELPFUL HINT

To apply the zero-factor property, one side of the equation must be 0 and the other side of the equation must be factored. To solve the equation $x(2x - 7) = 4$, for example, you may **not** set each factor equal to 4.

EXAMPLE 4 Solve $3(x^2 + 4) + 5 = -6(x^2 + 2x) + 13$.

Solution: Rewrite the equation so that one side is 0.

$3(x^2 + 4) + 5 = -6(x^2 + 2x) + 13$	
$3x^2 + 12 + 5 = -6x^2 - 12x + 13$	Apply the distributive property.
$9x^2 + 12x + 4 = 0$	Rewrite the equation so that one side is 0.
$(3x + 2)(3x + 2) = 0$	Factor.

$3x + 2 = 0 \quad \text{or} \quad 3x + 2 = 0$ Set each factor equal to 0.

$3x = -2 \quad \text{or} \quad 3x = -2$

$x = -\frac{2}{3} \quad \text{or} \quad x = -\frac{2}{3}$ Solve each equation.

The solution set is $\left\{-\frac{2}{3}\right\}$. Check by substituting $-\frac{2}{3}$ into the original equation. ■

If the equation contains fractions, we clear the equation of fractions as a first step.

EXAMPLE 5 Solve $2x^2 = \frac{17}{3}x + 1$.

Solution:

$2x^2 = \frac{17}{3}x + 1$

$3(2x^2) = 3\left(\frac{17}{3}x + 1\right)$ Clear the equation of fractions.

$6x^2 = 17x + 3$ Apply the distributive property.

$6x^2 - 17x - 3 = 0$ Rewrite the equation in standard form.

$(6x + 1)(x - 3) = 0$ Factor.

$6x + 1 = 0 \quad \text{or} \quad x - 3 = 0$ Set each factor equal to zero.

$6x = -1$

$x = -\frac{1}{6} \quad \text{or} \quad x = 3$ Solve each equation.

The solution set is $\left\{-\frac{1}{6}, 3\right\}$. ■

2 Since the zero-factor property can extend to more than two numbers whose product is 0, we can apply it to solving any equation in which one side of the equation is 0 and the other side is a factored expression.

EXAMPLE 6 Solve $x^3 = 4x$.

Solution:

$x^3 = 4x$

$x^3 - 4x = 0$ Rewrite the equation so that one side is 0.

$x(x^2 - 4) = 0$ Factor out the GCF x.

$x(x + 2)(x - 2) = 0$ Factor the difference of squares.

$x = 0 \quad \text{or} \quad x + 2 = 0 \quad \text{or} \quad x - 2 = 0$ Set each factor equal to 0.

$x = 0 \quad \text{or} \quad x = -2 \quad \text{or} \quad x = 2$ Solve each equation.

The solution set is $\{-2, 0, 2\}$. Check by substituting into the original equation. ■

Notice that the **third**-degree equation of Example 6 yielded **three** solutions.

HELPFUL HINT

In Example 6, it is incorrect to divide both sides of the equation by x. If x is 0, dividing by x is dividing by 0, and this is not allowed.

EXAMPLE 7 Solve $x^3 + 5x^2 = x + 5$.

Solution: First write the equation so that one side is 0.

$$x^3 + 5x^2 - x - 5 = 0$$

$$(x^3 - x) + (5x^2 - 5) = 0 \quad \text{Factor by grouping.}$$

$$x(x^2 - 1) + 5(x^2 - 1) = 0$$

$$(x^2 - 1)(x + 5) = 0$$

$$(x + 1)(x - 1)(x + 5) = 0 \quad \text{Factor the difference of squares.}$$

$$x + 1 = 0 \quad \text{or} \quad x - 1 = 0 \quad \text{or} \quad x + 5 = 0 \quad \text{Set each factor equal to 0.}$$

$$x = -1 \quad \text{or} \quad x = 1 \quad \text{or} \quad x = -5 \quad \text{Solve each equation.}$$

The solution set is $\{-5, -1, 1\}$. Check in the original equation. ■

3 Some problems may be modeled by quadratic equations. To solve these problems, we use the same problem-solving steps that were introduced in Section 2.5. When solving these problems, keep in mind that a solution of an equation that models a problem may not be a solution to the problem. For example, a person's age or the length of a rectangle is always a positive number. Discard solutions that do not make sense as solutions of the problem.

EXAMPLE 8 For a TV commercial, a piece of luggage is dropped from a cliff 256 feet above the ground to show the durability of the luggage. Neglecting air resistance, the height $h(t)$ in feet of the luggage above the ground after t seconds is given by the polynomial function

$$h(t) = -16t^2 + 256$$

Find how long it takes for the luggage to hit the ground.

Solution: **1.** UNDERSTAND. Read and reread the problem. Then draw a picture of the problem.

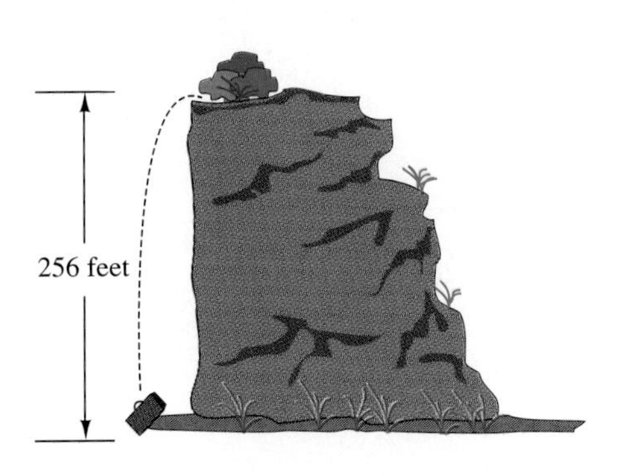

The function $h(t) = -16t^2 + 256$ models the height of the falling luggage. Familiarize yourself with this function by finding a few function values.

When $t = 1$ second, the height of the suitcase is $h(1) = -16(1)^2 + 256 = 240$ feet.

When $t = 2$ seconds, the height of the suitcase is $h(2) = -16(2)^2 + 256 = 192$ feet.

Since we have been given the needed function, we proceed to step 4.

4. TRANSLATE. To find how long it takes the luggage to hit the ground, we want to know for what value of t is the height $h(t) = 0$.

$$0 = -16t^2 + 256$$

5. COMPLETE. Solve the quadratic equation by factoring.

$$0 = -16t^2 + 256$$
$$0 = -16(t^2 - 16)$$
$$0 = -16(t - 4)(t + 4)$$
$$t - 4 = 0 \quad \text{or} \quad t + 4 = 0$$
$$t = 4 \quad \text{or} \quad t = -4$$

6. INTERPRET. Since the time t cannot be negative, the proposed solution is 4 seconds. To *check,* see that the height of the luggage when t is 4 seconds, $h(4)$, is 0.

$$h(4) = -16(4)^2 + 256 = -256 + 256 = 0$$

State: The solution checks and the luggage hits the ground 4 seconds after it is dropped. ■

The next example makes use of the **Pythagorean theorem.** Before we introduce this theorem, recall that a **right triangle** is a triangle that contains a 90° or right angle. The **hypotenuse** of a right triangle is the side opposite the right angle and is the longest side of the triangle. The **legs** of a right triangle are the other sides of the triangle.

Hypotenuse c

Leg b

Leg a

Pythagorean Theorem

In a right triangle, the sum of the squares of the lengths of the two legs is equal to the square of the length of the hypotenuse.

$$(\text{leg})^2 + (\text{leg})^2 = (\text{hypotenuse})^2 \quad \text{or} \quad a^2 + b^2 = c^2$$

EXAMPLE 9 Find the lengths of the sides of a right triangle if the lengths can be expressed by three consecutive even integers.

Solution: **1.** UNDERSTAND. Read and reread the problem, draw a diagram, and guess the solution. For example, if 4 units is the length of one side of the triangle, then the next consecutive even integers are 6 and 8. (Notice that we let 8 represent the length of the hypotenuse since the hypotenuse is always the longest side.) To see if these values represent the lengths of the sides of a right triangle, we see if these values satisfy the

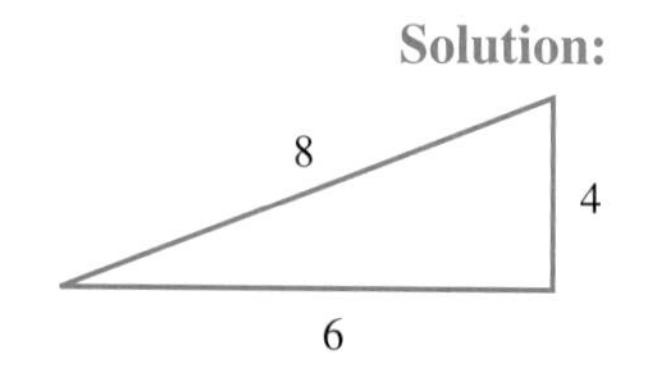

Pythagorean theorem. The value $4^2 + 6^2 = 52$, not 8^2, so although we have guessed incorrectly, we have a better understanding of the problem.

2. ASSIGN. Let $x, x + 2$, and $x + 4$ be three consecutive even integers. Since these integers represent lengths of the sides of a right triangle, we have

$$x = \text{one leg}$$
$$x + 2 = \text{other leg}$$
$$x + 4 = \text{hypotenuse (longest side)}$$

3. ILLUSTRATE.

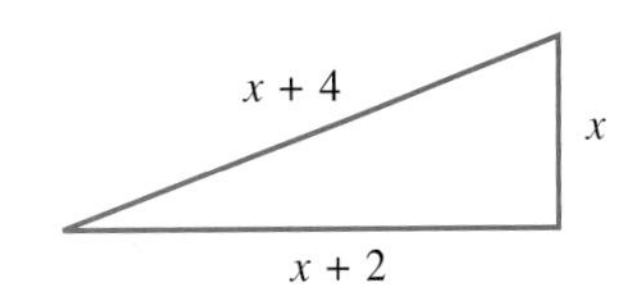

4. TRANSLATE. By the Pythagorean theorem, we have

In words: $\quad (\text{hypotenuse})^2 = (\text{leg})^2 + (\text{leg})^2$

Translate: $\quad (x + 4)^2 = (x)^2 + (x + 2)^2$

5. COMPLETE. Solve the equation.

$$(x + 4)^2 = x^2 + (x + 2)^2$$
$$x^2 + 8x + 16 = x^2 + x^2 + 4x + 4 \quad \text{Multiply.}$$
$$x^2 + 8x + 16 = 2x^2 + 4x + 4$$
$$x^2 - 4x - 12 = 0 \quad \text{Write in standard form.}$$
$$(x - 6)(x + 2) = 0$$
$$x - 6 = 0 \quad \text{or} \quad x + 2 = 0$$
$$x = 6 \quad \text{or} \quad x = -2$$

6. INTERPRET. Discard $x = -2$ since length cannot be negative. If $x = 6$, then $x + 2 = 8$ and $x + 4 = 10$.

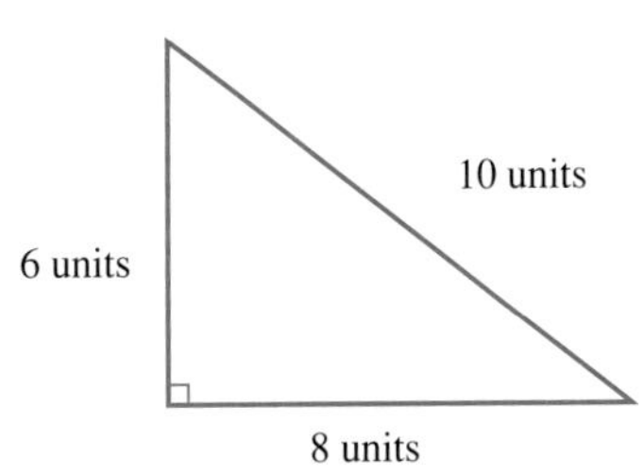

Check: To check, see that $(\text{hypotenuse})^2 = (\text{leg})^2 + (\text{leg})^2$,

$$10^2 = 6^2 + 8^2$$

or

$$100 = 36 + 64 \quad \text{True.}$$

State: The lengths of the sides of the right triangle are 6, 8, and 10 units. ■

GRAPHING CALCULATOR BOX

We can use a grapher to approximate real number solutions of any quadratic equation in standard form, whether the associated polynomial is factorable or not. For example, let's solve the quadratic equation $x^2 - 2x - 4 = 0$. The solutions of this equation will be the x-intercepts of the graph of the function $f(x) = x^2 - 2x - 4$. (Recall that to find x-intercepts, we let $f(x)$ or $y = 0$.) When we use a standard window, the graph of this function looks like this:

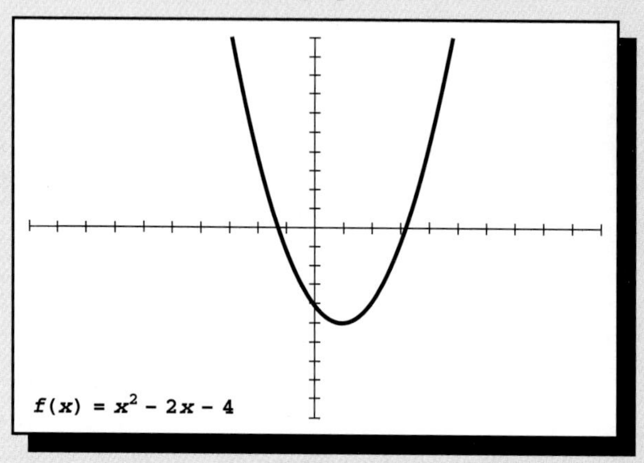

The graph appears to have one x-intercept between -2 and -1 and one between 3 and 4. To find the x-intercept between 3 and 4 to the nearest hundredth, we can use a Zoom feature, which magnifies a portion of the graph around the cursor, or we can redefine our window. If we redefine our window to

$$\begin{array}{ll} \text{Xmin} = 2 & \text{Ymin} = -1 \\ \text{Xmax} = 5 & \text{Ymax} = 1 \\ \text{Xscl} = 1 & \text{Yscl} = 1 \end{array}$$

the resulting screen is

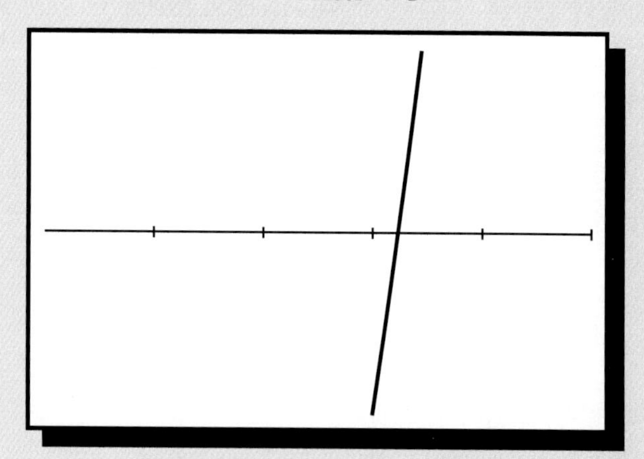

By using the Trace feature, we can now see that one of the intercepts is between 3.21 and 3.25. To approximate to the nearest hundredth, Zoom again or redefine the window to

$$\begin{array}{ll} \text{Xmin} = 3.2 & \text{Ymin} = -0.1 \\ \text{Xmax} = 3.3 & \text{Ymax} = 0.1 \\ \text{Xscl} = 1 & \text{Yscl} = 1 \end{array}$$

If we use the Trace feature again, we see that, to the nearest hundredth, the x-intercept is 3.23. By repeating this process, we can approximate the other x-intercept to be -1.23.

To check, find $f(3.23)$ and $f(-1.23)$. Both of these values should be close to 0. (They will not be exactly 0 since we approximated these solutions.)

$$f(3.23) = -0.027 \quad \text{and} \quad f(-1.23) = -0.0271$$

Solve each of these quadratic equations by graphing a related function and approximating the x-intercepts.

1. $x^2 + 3x - 2 = 0$
2. $5x^2 - 7x + 1 = 0$
3. $2.3x^2 - 4.4x - 5.6 = 0$
4. $0.2x^2 + 6.2x + 2.1 = 0$
5. $0.09x^2 - 0.13x - 0.08 = 0$
6. $x^2 + 0.08x - 0.01 = 0$

MENTAL MATH

Solve each equation for the variable. See Example 1.

1. $(x - 3)(x + 5) = 0$
2. $(y + 5)(y + 3) = 0$
3. $(z - 3)(z + 7) = 0$
4. $(c - 2)(c - 4) = 0$
5. $x(x - 9) = 0$
6. $w(w + 7) = 0$

EXERCISE SET 6.6

Solve each equation. See Example 1.

1. $(x + 3)(3x - 4) = 0$
2. $(5x + 1)(x - 2) = 0$
3. $3(2x - 5)(4x + 3) = 0$
4. $8(3x - 4)(2x - 7) = 0$

Solve each equation. See Example 2.

5. $x^2 + 11x + 24 = 0$
6. $y^2 - 10y + 24 = 0$
7. $12x^2 + 5x - 2 = 0$
8. $3y^2 - y - 14 = 0$

Solve each equation. See Examples 3 and 4.

9. $z^2 + 9 = 10z$
10. $n^2 + n = 72$
11. $x(5x + 2) = 3$
12. $n(2n - 3) = 2$
13. $x^2 - 6x = x(8 + x)$
14. $n(3 + n) = n^2 + 4n$

Solve each equation. See Example 5.

15. $\frac{z^2}{6} - \frac{z}{2} - 3 = 0$
16. $\frac{c^2}{20} - \frac{c}{4} + \frac{1}{5} = 0$
17. $\frac{x^2}{2} + \frac{x}{20} = \frac{1}{10}$
18. $\frac{y^2}{30} = \frac{y}{15} + \frac{1}{2}$
19. $\frac{4t^2}{5} = \frac{t}{5} + \frac{3}{10}$
20. $\frac{5x^2}{6} - \frac{7x}{2} + \frac{2}{3} = 0$

Solve each equation. See Examples 6 and 7.

21. $(x + 2)(x - 7)(3x - 8) = 0$
22. $(4x + 9)(x - 4)(x + 1) = 0$
23. $y^3 = 9y$
24. $n^3 = 16n$
25. $x^3 - x = 2x^2 - 2$
26. $m^3 = m^2 + 12m$

Solve. See Example 8.

27. An object is thrown upward from the top of an 80-foot building with an initial velocity of 64 feet per second. The height $h(t)$ of the object after t seconds is given by the polynomial function

$$h(t) = -16t^2 + 64t + 80$$

When will the object hit the ground?

28. A hang-glider pilot accidentally drops her compass from the top of a 400-foot cliff. The height $h(t)$ of the compass after t seconds is given by the polynomial function

$$h(t) = -16t^2 + 400$$

When will the compass hit the ground?

29. If $N(x) = \dfrac{x(x-3)}{2}$ is the formula for the number of diagonals N of a polygon with x sides, find the number of sides for a polygon with 5 diagonals.

30. If a switchboard handles n telephones, the number C of telephone connections it can make simultaneously is given by the equation $C(n) = \dfrac{n(n-1)}{2}$. Find how many telephones are handled by a switchboard making 120 telephone connections simultaneously.

Solve the following. See Example 9.

31. Find the lengths of the sides of a right triangle if the hypotenuse is 10 centimeters longer than the short leg and 5 centimeters longer than the long leg.

32. Find the lengths of the sides of a right triangle if the length of the hypotenuse is 12 kilometers longer than the short leg and 6 kilometers longer than the long leg.

33. Find the length of the short leg of a right triangle if the long leg is 12 feet more than the short leg and the hypotenuse is 12 feet less than twice the short leg.

34. Find the length of the short leg of a right triangle if the long leg is 10 miles more than the short leg and the hypotenuse is 10 miles less than twice the short leg.

Solve each equation.

35. $(2x + 7)(x - 10) = 0$

36. $(x + 4)(5x - 1) = 0$

37. $3x(x - 5) = 0$

38. $4x(2x + 3) = 0$

39. $x^2 - 2x - 15 = 0$

40. $x^2 + 6x - 7 = 0$

41. $12x^2 + 2x - 2 = 0$

42. $8x^2 + 13x + 5 = 0$

43. $w^2 - 5w = 36$

44. $x^2 + 32 = 12x$

45. $25x^2 - 40x + 16 = 0$

46. $9n^2 + 30n + 25 = 0$

47. $2r^3 + 6r^2 = 20r$

48. $-2t^3 = 108t - 30t^2$

49. $z(5z - 4)(z + 3) = 0$

50. $2r(r + 3)(5r - 4) = 0$

51. $2z(z + 6) = 2z^2 + 12z - 8$

52. $3c^2 - 8c + 2 = c(3c - 8)$

53. $(x - 1)(x + 4) = 24$

54. $(2x - 1)(x + 2) = -3$

55. $\dfrac{x^2}{4} - \dfrac{5}{2}x + 6 = 0$

56. $\dfrac{x^2}{18} + \dfrac{x}{2} + 1 = 0$

57. $y^2 + \dfrac{1}{4} = -y$

58. $\dfrac{x^2}{10} + \dfrac{5}{2} = x$

59. $y^3 + 4y^2 = 9y + 36$

60. $x^3 + 5x^2 = x + 5$

61. $2x^3 = 50x$

62. $m^5 = 36m^3$

63. $x^2 + (x + 1)^2 = 61$

64. $y^2 + (y + 2)^2 = 34$

65. $m^2(3m - 2) = m$

66. $x^2(5x + 3) = 26x$

67. $3x^2 = -x$

68. $y^2 = -5y$

69. $x(x - 3) = x^2 + 5x + 7$

70. $z^2 - 4z + 10 = z(z - 5)$

71. $3(t - 8) + 2t = 7 + t$

72. $7c - 2(3c + 1) = 5(4 - 2c)$

73. $-3(x - 4) + x = 5(3 - x)$

74. $-4(a + 1) - 3a = -7(2a - 3)$

Solve.

75. The sum of the squares of two consecutive negative integers is 221. Find the integers.

76. The sum of the squares of two consecutive even positive integers is 100. Find the integers.

77. The length of the base of a triangle is twice its altitude. If the area of the triangle is 100 square kilometers, find the altitude.

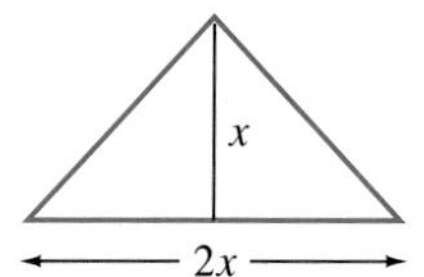

78. The altitude of a triangle is 2 millimeters less than the base. If the area is 60 square millimeters, find the base.

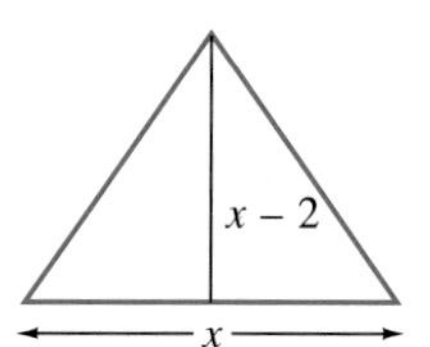

79. If the cost, $C(x)$, for manufacturing x units of a certain product is given by $C(x) = x^2 - 15x + 50$, find the number of units manufactured at a cost of $9500.

80. If $N(x) = \dfrac{x(x-3)}{2}$ is the formula for the number of diagonals N of a polygon with x sides, find the number of sides for a polygon with 35 diagonals.

81. Marie Mulroney has a rectangular board 12 inches by 16 inches around which she wants to put a uniform border of shells. If she has enough shells for a border whose area is 128 square inches, determine the width of the border.

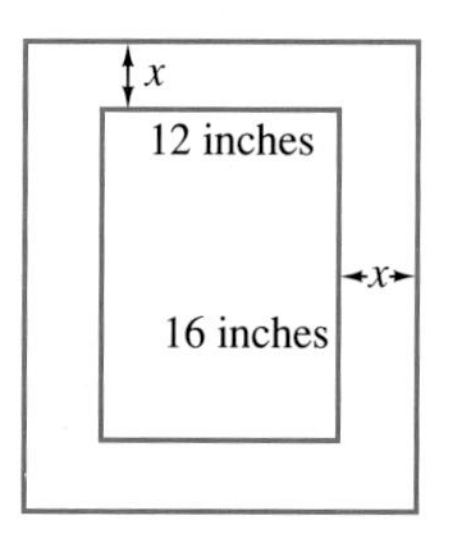

82. A gardener has a rose garden that measures 30 feet by 20 feet. He wants to put a uniform border of pine bark around the outside of the garden. Find how wide the border should be if he has enough pine bark to cover 336 square feet.

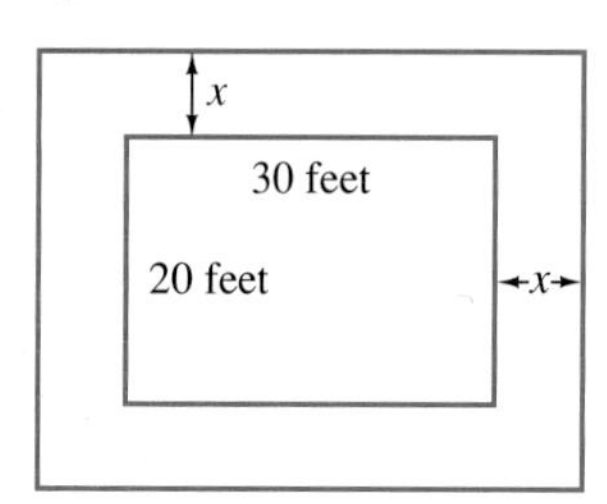

83. After t seconds, the height $h(t)$ of a model rocket launched from the ground into the air is given by the function

$$h(t) = -16t^2 + 80t$$

Find how long it takes the rocket to reach a height of 96 feet.

84. While hovering near the top of Ribbon Falls in Yosemite National Park at 1600 feet, a helicopter pilot accidentally drops his sunglasses. The height $h(t)$ of the sunglasses after t seconds is given by the polynomial function

$$h(t) = -16t^2 + 1600$$

When will the sunglasses hit the ground?

85. One leg of a right triangle is 4 millimeters more than the smaller leg and the hypotenuse is 8 millimeters more than the smaller leg. Find the lengths of the sides of the triangle.

86. One leg of a right triangle is 9 centimeters longer than the other leg and the hypotenuse is 45 centimeters. Find the lengths of the legs of the triangle.

Skill Review

Write each fraction in simplest form. *See Section 1.3.*

87. $\dfrac{20}{35}$

88. $\dfrac{24}{32}$

89. $\dfrac{27}{18}$

90. $\dfrac{15}{27}$

Simplify. See Section 5.1.

91. $\dfrac{10x^2y}{5x^3}$ **92.** $\dfrac{24xy^3}{6y^4}$ **93.** $\dfrac{45a^2b}{9b^3}$ **94.** $\dfrac{36a^3b}{12b^3}$

Determine whether each graph is the graph of a function. See Section 4.2.

95.

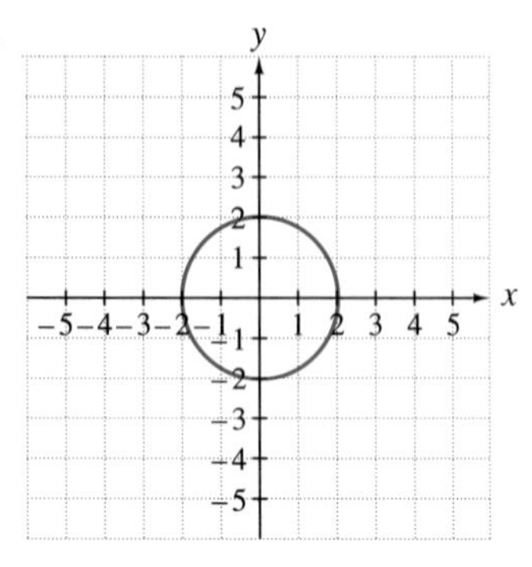

96.

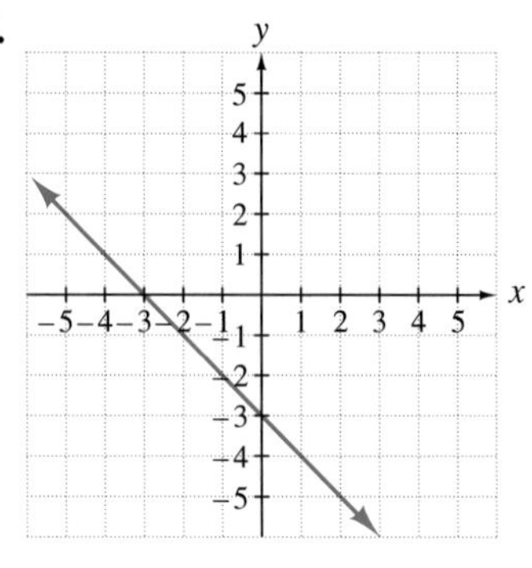

97.

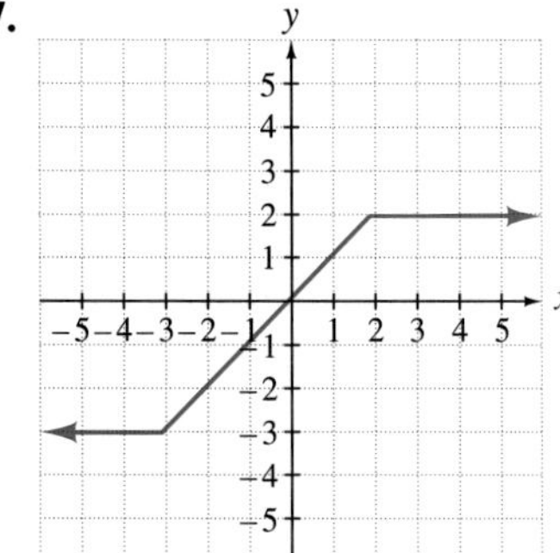

98.

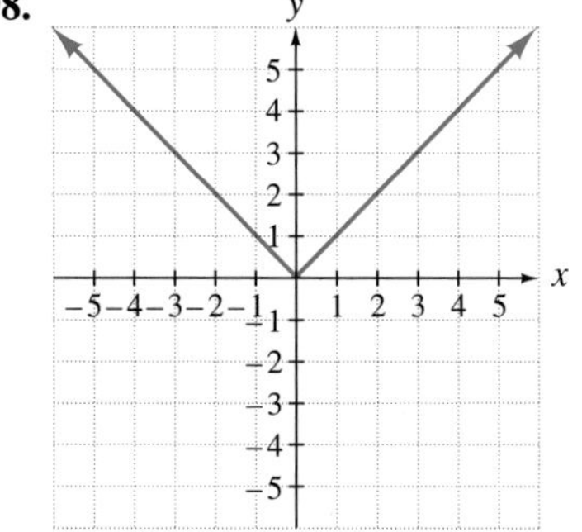

Writing in Mathematics

99. Describe two ways a linear equation differs from a quadratic equation.

100. Explain how solving $2(x - 3)(x - 1) = 0$ differs from solving $2x(x - 3)(x - 1) = 0$.

101. Explain why the zero-factor property works for more than two numbers whose product is 0.

6.7 An Introduction to Graphing Polynomial Functions

OBJECTIVES

Tape IA 15

1. Define a polynomial function.
2. Graph quadratic functions.
3. Find the vertex of a parabola by using the vertex formula.
4. Graph cubic functions.

1 We discussed linear functions defined by equations of the form $f(x) = mx + b$ in Chapter 4. In this section, we discuss two other special cases of **polynomial functions,** that is, **quadratic functions** and **cubic functions.** For example,

$f(x) = 2x - 6$ is a **linear function** since its **degree is one.**
$f(x) = 5x^2 - x + 3$ is a **quadratic function** since its **degree is two.**
$f(x) = 7x^3 + 3x^2 - 1$ is a **cubic function** since its **degree is three.**

All the above functions are also polynomial functions.

The graph of any polynomial function (linear, quadratic, cubic, and so on) can be sketched by plotting a sufficient number of ordered pairs that satisfy the function and connecting them to form a smooth curve. The graph of all polynomial functions will pass the vertical line test since they are graphs of functions. To graph a linear function defined by $f(x) = mx + b$, recall that two ordered pair solutions will suffice since its graph is a line. To graph other polynomial functions, we need to find and plot more ordered pair solutions to ensure a reasonable picture of its graph.

2 Since we know how to graph linear functions (see Chapter 4), we will now discuss and graph quadratic functions.

Quadratic Function

A quadratic function is a function that can be written in the form

$$f(x) = ax^2 + bx + c$$

where a, b, and c are real numbers and $a \neq 0$.

We know that $y = f(x)$, so an equation of the form $f(x) = ax^2 + bx + c$ may be written as $y = ax^2 + bx + c$. Thus, both $f(x) = ax^2 + bx + c$ and $y = ax^2 + bx + c$ define quadratic functions as long as a is not 0.

Graph the quadratic function defined by $f(x) = x^2$ by plotting points. Choose $-3, -2, -1, 0, 1, 2,$ and 3 as x-values and find corresponding $f(x)$ or y-values.

x	$y = f(x)$
-3	9
-2	4
-1	1
0	0
1	1
2	4
3	9

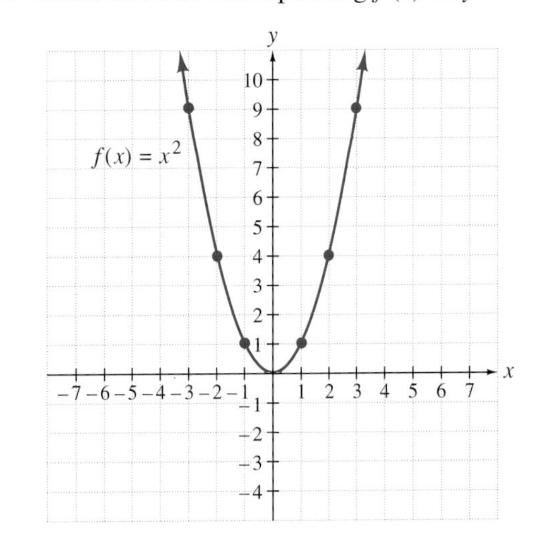

Notice that the graph passes the vertical line test as it should since it is a function. This curve is called a **parabola.** The highest point on a parabola that opens downward or the lowest point on a parabola that opens upward is called the **vertex** of the parabola. The vertex of this parabola is (0, 0), the lowest point on the graph. If we fold the graph along the y-axis, we can see that the two sides of the graph coincide. This means that this curve is symmetric about the y-axis, and the y-axis or the line $x = 0$ is called the **axis of symmetry.** The graph of every quadratic function is a parabola and has an axis of symmetry: the vertical line that passes through the vertex of the parabola.

EXAMPLE 1 Graph the quadratic function $f(x) = -x^2 + 2x - 3$ by plotting points.

Solution: To graph, choose values for x and find corresponding $f(x)$ or y-values.

x	$y = f(x)$
-2	-11
-1	-6
0	-3
1	-2
2	-3
3	-6

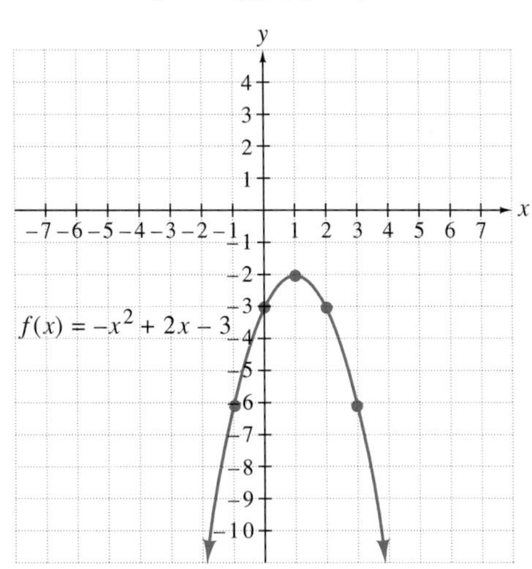

The vertex of this parabola is $(1, -2)$, the highest point on the graph. The vertical line $x = 1$ is the axis of symmetry. ■

Notice that the parabola $f(x) = -x^2 + 2x - 3$ opens downward, whereas $f(x) = x^2$ opens upward. When the equation of a quadratic function is written in the

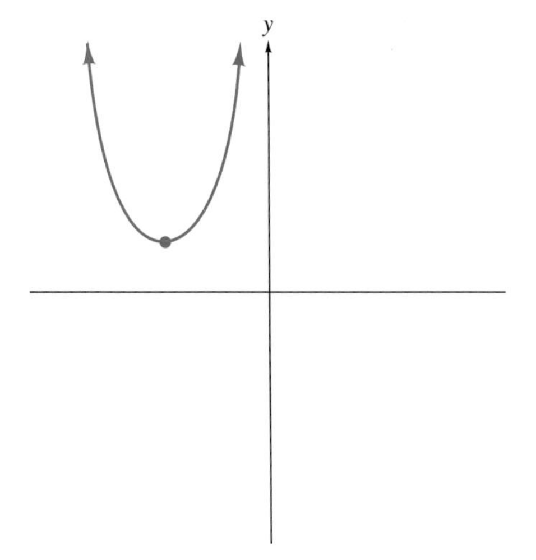

$f(x) = ax^2 + bx + c$, $a > 0$, opens upward

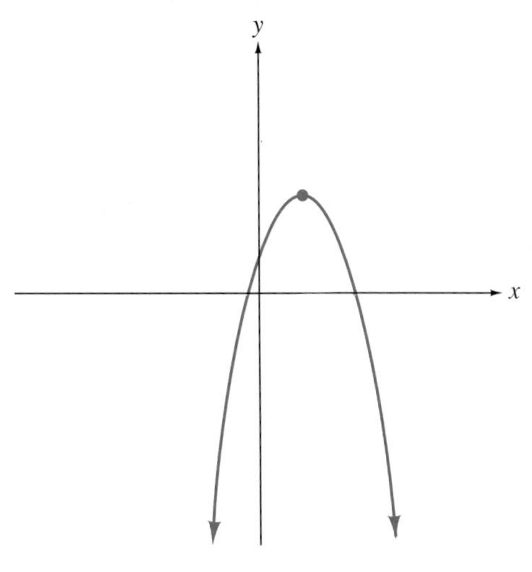

$f(x) = ax^2 + bx + c$, $a < 0$, opens downward

form $f(x) = ax^2 + bx + c$, the coefficient of the squared variable, a, determines whether the parabola opens downward or upward. If $a > 0$, the parabola opens upward, and if $a < 0$, the parabola opens downward.

3 In both $f(x) = x^2$ and $f(x) = -x^2 + 2x - 3$, the vertex happens to be one of the points we chose to plot. Since this is not always the case, and since plotting the vertex allows us to quickly draw the graph, we need a consistent method for finding the vertex. One method is to use the following formula, which we derive in Chapter 9.

Vertex Formula

The graph of $f(x) = ax^2 + bx + c$, $a \neq 0$, is a parabola with vertex

$$\left(\frac{-b}{2a}, f\left(\frac{-b}{2a}\right)\right)$$

We can also find the x- and y-intercepts of a parabola to aid in graphing. Recall that x-intercepts of the graph of any equation may be found by letting $y = 0$ in the equation and solving for x. Also, y-intercepts may be found by letting $x = 0$ in the equation and solving for y or $f(x)$.

EXAMPLE 2 Graph $f(x) = x^2 + 2x - 3$. Find the vertex and any intercepts.

Solution: To find the vertex, use the vertex formula. For the function $f(x) = x^2 + 2x - 3$, $a = 1$ and $b = 2$. Thus,

$$x = \frac{-b}{2a} = \frac{-2}{2(1)} = -1$$

Next find $f(-1)$.

$$\begin{aligned} f(-1) &= (-1)^2 + 2(-1) - 3 \\ &= 1 - 2 - 3 \\ &= -4 \end{aligned}$$

The vertex is $(-1, -4)$, and since $a = 1$ is greater than 0, this parabola opens upward. This parabola will have two x-intercepts because its vertex lies below the x-axis and it opens upward. To find the x-intercepts, let y or $f(x) = 0$ and solve for x.

$$\begin{aligned} f(x) &= x^2 + 2x - 3 & & \\ 0 &= x^2 + 2x - 3 & & \text{Let } f(x) = 0. \\ 0 &= (x + 3)(x - 1) & & \text{Factor.} \\ x + 3 = 0 \quad &\text{or} \quad x - 1 = 0 & & \text{Set each factor equal to 0.} \\ x = -3 \quad &\text{or} \quad x = 1 & & \text{Solve.} \end{aligned}$$

The x-intercepts are -3 and 1.

To find the y-intercept, let $x = 0$.

$$\begin{aligned} f(x) &= x^2 + 2x - 3 \\ f(0) &= 0^2 + 2(0) - 3 \\ f(0) &= -3 \end{aligned}$$

The y-intercept is -3.

Now plot these points and connect them with a smooth curve.

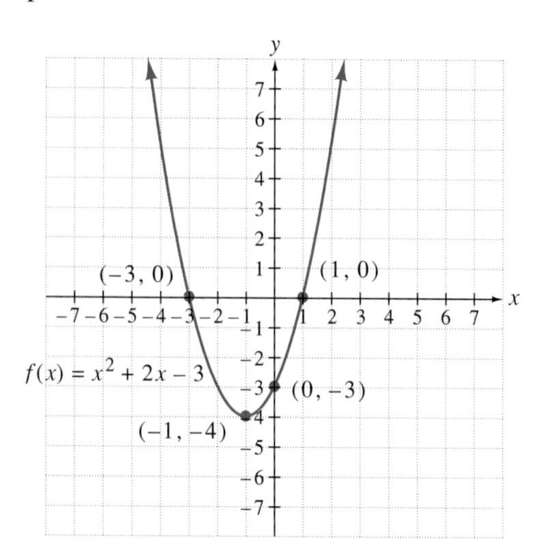

■

HELPFUL HINT

Not all graphs of parabolas have x-intercepts. To see this, first plot the vertex of the parabola and decide whether the parabola opens upward or downward. Then use this information to decide whether the graph of the parabola has x-intercepts.

EXAMPLE 3 Graph $f(x) = 3x^2 - 12x + 13$. Find the vertex and any intercepts.

Solution: To find the vertex, use the vertex formula. For the function $y = 3x^2 - 12x + 13$, $a = 3$ and $b = -12$. Thus,

$$x = \frac{-b}{2a} = \frac{-(-12)}{2(3)} = \frac{12}{6} = 2$$

Next find $f(2)$.

$$\begin{aligned} f(2) &= 3(2)^2 - 12(2) + 13 \\ &= 3(4) - 24 + 13 \\ &= 1 \end{aligned}$$

The vertex is (2, 1). Also, this parabola opens upward, since $a = 3$, which is greater than 0. Notice that this parabola has no x-intercepts: its vertex lies above the x-axis and it opens upward.

To find the y-intercept, let $x = 0$.

$$\begin{aligned} f(0) &= 3(0)^2 - 12(0) + 13 \\ &= 0 - 0 + 13 \\ &= 13 \end{aligned}$$

The y-intercept is 13. Use this information along with symmetry of a parabola to sketch the graph of $f(x) = 3x^2 - 12x + 13$.

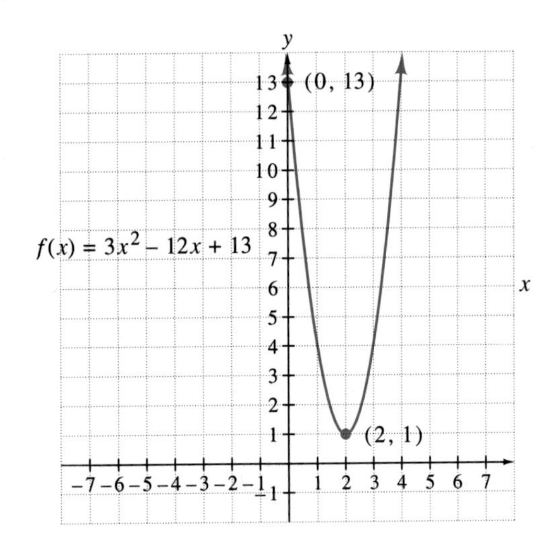

■

In Section 9.5, we study quadratic functions further.

4 To sketch the graph of a cubic function, we again plot points and then connect the points with a smooth curve.

EXAMPLE 4 Graph $f(x) = x^3 - 4x$. Find any intercepts.

Solution: To find x-intercepts, let y or $f(x) = 0$ and solve for x.

$$f(x) = x^3 - 4x$$

$$0 = x^3 - 4x \qquad \text{Let } f(x) = 0.$$

$$0 = x(x^2 - 4)$$

$$0 = x(x + 2)(x - 2) \qquad \text{Factor.}$$

$$x = 0 \quad \text{or} \quad x + 2 = 0 \quad \text{or} \quad x - 2 = 0 \qquad \text{Set each factor equal to 0.}$$

$$x = 0 \quad \text{or} \quad x = -2 \quad \text{or} \quad x = 2 \qquad \text{Solve.}$$

This graph has three x-intercepts. They are 0, -2 and 2.

To find the y-intercept, let $x = 0$.

$$f(0) = 0^3 - 4(0) = 0$$

Next select some x-values and find their corresponding $f(x)$ or y-values.

$$f(x) = x^3 - 4x$$

$$f(-3) = (-3)^3 - 4(-3) = -27 + 12 = -15$$

$$f(-1) = (-1)^3 - 4(-1) = -1 + 4 = 3$$

$$f(1) = 1^3 - 4(1) = 1 - 4 = -3$$

$$f(3) = 3^3 - 4(3) = 27 - 12 = 15$$

x	$f(x)$
-3	-15
-1	3
1	-3
3	15

Plot the intercepts and points and connect them with a smooth curve.

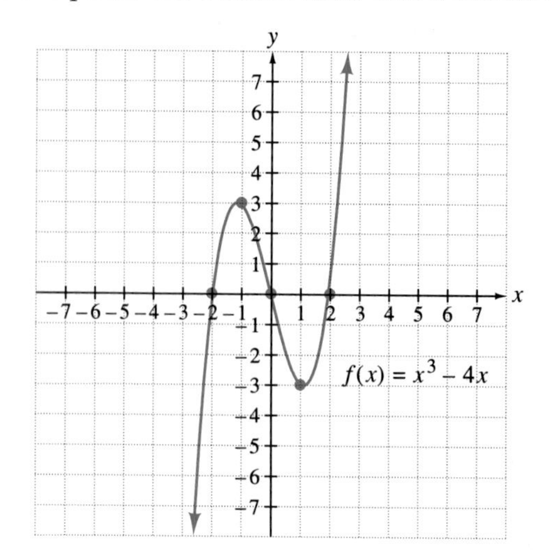

■

HELPFUL HINT

When a graph has an x-intercept of 0, notice that the y-intercept will also be 0.

HELPFUL HINT

If unsure about the graph of a function, plot more points.

EXAMPLE 5 Graph $f(x) = x^3$. Find any intercepts.

Solution: To find x-intercepts, let y or $f(x) = 0$ and solve for x.

$$f(x) = x^3$$
$$0 = x^3$$
$$0 = x$$

The only x-intercept is 0. This means that the y-intercept is 0 also.

Next choose some x-values and find corresponding y-values.

$$f(x) = x^3$$
$$f(-2) = (-2)^3 = -8$$
$$f(-1) = (-1)^3 = -1$$
$$f(1) = (1)^3 = 1$$
$$f(2) = 2^3 = 8$$

x	$f(x)$
-2	-8
-1	-1
1	1
2	8

Plot the points and sketch the graph of $f(x) = x^3$.

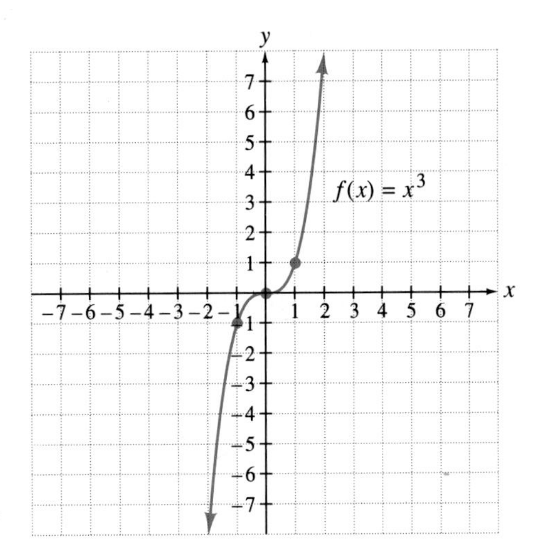

MENTAL MATH

State whether the graph of each quadratic function, a parabola, opens upward or downward.

1. $f(x) = 2x^2 + 7x + 10$

2. $f(x) = -3x^2 - 5x$

3. $f(x) = -x^2 + 5$

4. $f(x) = x^2 + 3x + 7$

EXERCISE SET 6.7

Graph each quadratic function by plotting points. See Example 1.

1. $f(x) = 2x^2$

2. $f(x) = -3x^2$

3. $f(x) = x^2 + 1$

4. $f(x) = x^2 - 2$

5. $f(x) = -x^2$

6. $f(x) = \frac{1}{2}x^2$

Find the vertex of the graph of each function. See Examples 2 and 3.

7. $f(x) = x^2 + 8x + 7$

8. $f(x) = x^2 + 6x + 5$

9. $f(x) = 3x^2 + 6x + 4$

10. $f(x) = -2x^2 + 2x + 1$

11. $f(x) = -x^2 + 10x + 5$

12. $f(x) = -x^2 - 8x + 2$

Graph each quadratic function. Find and label the vertex and intercepts. See Examples 2 and 3.

13. $f(x) = x^2 + 8x + 7$

14. $f(x) = x^2 + 6x + 5$

15. $f(x) = x^2 - 2x - 24$

16. $f(x) = x^2 - 12x + 35$

17. $f(x) = 2x^2 - 6x$

18. $f(x) = -3x^2 + 6x$

Graph each cubic function. Find any intercepts. See Examples 4 and 5.

19. $f(x) = 4x^3 - 9x$

20. $f(x) = 2x^3 - 5x^2 - 3x$

21. $f(x) = x^3 + 3x^2 - x - 3$

22. $f(x) = x^3 + x^2 - 4x - 4$

Graph each function. Find intercepts. If a quadratic function, find the vertex.

23. $f(x) = x^2 + 4x - 5$

24. $f(x) = x^2 + 2x - 3$

25. $f(x) = (x - 2)(x + 2)(x + 1)$

26. $f(x) = x^3 - 4x^2 + 3x$

27. $f(x) = x^2 + 1$

28. $f(x) = x^2 + 4$

29. $f(x) = -5x^2 + 5x$

30. $f(x) = 3x^2 - 12x$

31. $f(x) = x^3 - 9x$

32. $f(x) = x^3 + x^2 - 12x$

33. $f(x) = -x^3 - x^2 + 2x$

34. $f(x) = x^3 + x^2 - 9x - 9$

35. $f(x) = x^2 - 4x + 4$

36. $f(x) = x^2 - 2x + 1$

37. $f(x) = -x^3 + x$

38. $f(x) = x^2 + 6x$

39. $f(x) = 2x^2 - x - 3$

40. $f(x) = (x + 2)(x - 2)$

41. $f(x) = -x^3 + 3x^2 + x - 3$

42. $f(x) = -x^3 + 25x$

43. $f(x) = x^2 - 10x + 26$

44. $f(x) = x^2 + 2x + 4$

45. $f(x) = x(x - 4)(x + 2)$

46. $f(x) = 3x(x - 3)(x + 5)$

Skill Review

Find the domain and the range of each function graphed. See Section 4.2.

47.

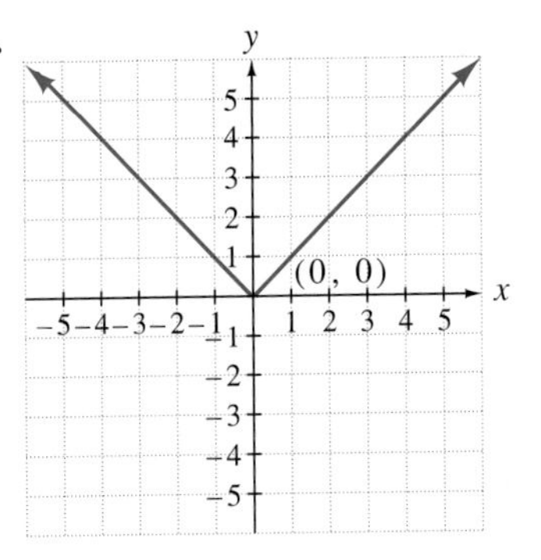

48.

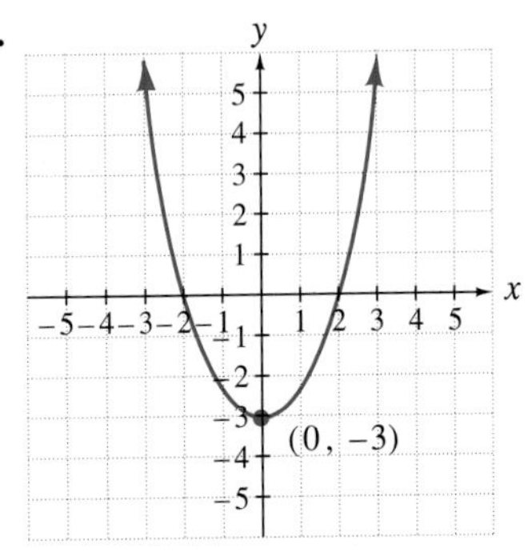

49.

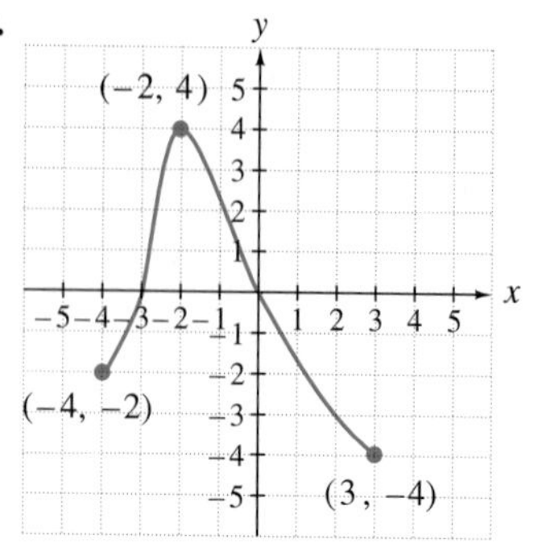

50.

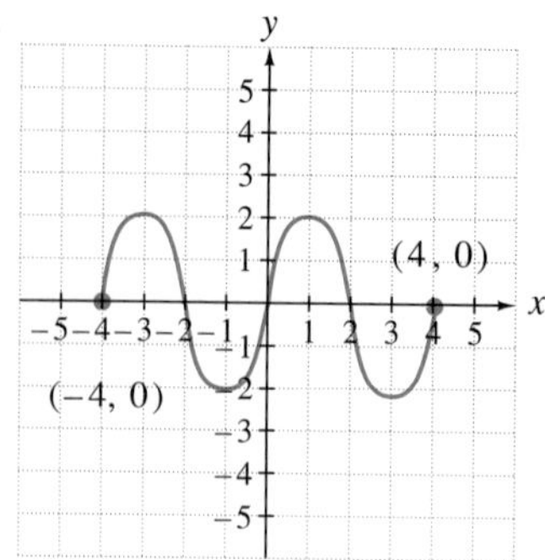

Multiply. See Section 5.1.

51. $(x^2y)(3x)$

52. $(6x^4y^6)(9x^2y^3)$

53. $(x^2y^2)(-16xy)$

54. $(21x^2y)(-2xy)$

Square each binomial. See Section 5.4.

55. $(x + 6)^2$

56. $(3x + 2)^2$

57. $(2x - 1)^2$

58. $(x - 5)^2$

Writing in Mathematics

59. Explain why the graph of a function never has two or more y-intercepts.

CHAPTER 6 GLOSSARY

The vertical line that passes through the vertex of a parabola opening upward or downward is called the **axis of symmetry.**

In the product $a \cdot b = c$, a and b are called **factors** of c, and $a \cdot b$ is a **factored form** of c.

The process of writing a polynomial as a product is called **factoring.**

The **greatest common factor (GCF)** of a list of integers is the largest integer that is a factor of each integer.

The graph of a quadratic function is a **parabola** that opens upward or downward.

A **prime number** is a natural number greater than 1 whose only natural number factors are 1 and itself.

A **quadratic equation** in one variable is an equation that can be written in the form $ax^2 + bx + c = 0$, where a, b, and c are real numbers and $a \neq 0$.

A **quadratic function** can be written in the form $f(x) = ax^2 + bx + c$, where a, b, and c are real numbers and $a \neq 0$.

The highest point on a parabola that opens downward or the lowest point on a parabola that opens upward is called the **vertex** of the parabola.

CHAPTER 6 SUMMARY

TO FACTOR A POLYNOMIAL (6.5)

Step 1 Are there any common factors? If so, factor them out.

Step 2 How many terms are in the polynomial?

a. If there are **two** terms, decide if one of the following formulas may be applied.

i. Difference of two squares: $a^2 - b^2 = (a - b)(a + b)$.

ii. Difference of two cubes: $a^3 - b^3 = (a - b)(a^2 + ab + b^2)$.

iii. Sum of two cubes: $a^3 + b^3 = (a + b)(a^2 - ab + b^2)$.

b. If there are **three** terms, try one of the following:

i. Perfect square trinomial: $a^2 + 2ab + b^2 = (a + b)^2$.
$a^2 - 2ab + b^2 = (a - b)^2$.

ii. If not a perfect square trinomial, factor using the methods presented in Sections 6.2 and 6.3.

c. If there are **four** or more terms, try factoring by grouping.

Step 3 See if any factors in the factored polynomial can be factored further.

TO SOLVE QUADRATIC EQUATIONS BY FACTORING (6.6)

Step 1 Write the equation in standard form: $ax^2 + bx + c = 0$.

Step 2 Factor the quadratic expression.

Step 3 Set each factor containing a variable equal to 0.

Step 4 Solve the resulting equations.

Step 5 Check each solution in the original equation.

PYTHAGOREAN THEOREM (6.6)

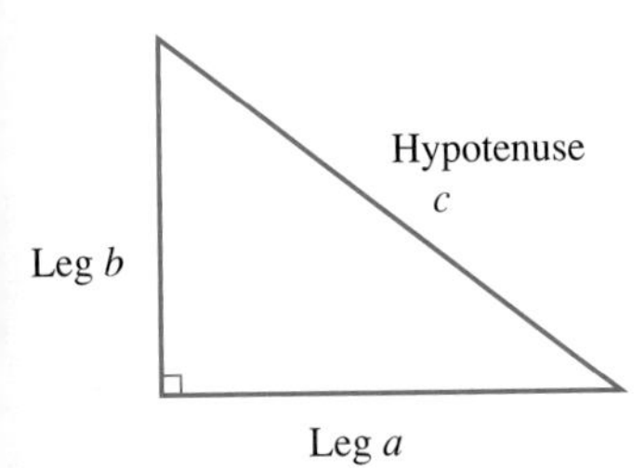

In a right triangle, the sum of the squares of the lengths of the two legs is equal to the square of the length of the hypotenuse.

$$(\text{leg})^2 + (\text{leg})^2 = (\text{hypotenuse})^2 \quad \text{or} \quad a^2 + b^2 = c^2$$

VERTEX FORMULA (6.7)

The graph of $f(x) = ax^2 + bx + c$, $a \neq 0$, is a parabola with vertex

$$\left(\frac{-b}{2a}, f\left(\frac{-b}{2a}\right)\right)$$

CHAPTER 6 REVIEW

(6.1) *Find the GCF of the monomials in the list.*

1. 24, 60, 84

2. 90, 135, 225

3. x^6, x^8, x^2, x^4

4. y^5, y^3, y^7, y^4

5. $6x^2y^3, 16xy^2, 8x^3$

6. $9x^3y^5, 12y^2, 6x^3y^4$

Factor each polynomial.

7. $16x^3 - 24x^2$

8. $36y - 24y^2$

9. $15x^3y^4z - 3xy^2z^2 + 6x^2y^2z$

10. $20x^4yz^3 - 6x^2y^2z^4 + 4x^3yz^2$

11. $6ab^2 + 8ab - 4a^2b^2$

12. $14a^2b^2 - 21ab^2 + 7ab$

13. $6a(a + 3b) - 5(a + 3b)$

14. $4x(x - 2y) - 5(x - 2y)$

15. $xy - 6y + 3x - 18$

16. $ab - 8b + 4a - 32$

17. $pq - 3p - 5q + 15$

18. $xy - 2x - 5y + 10$

19. $x^3 - x^2 - 2x + 2$

20. $x^3 - x^2 - 5x + 5$

Solve.

21. A smaller square is cut from a larger square of plastic. Write the area of the shaded region as a factored polynomial.

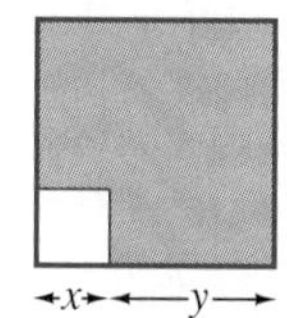

(6.2) *Factor each trinomial.*

22. $x^2 + 6x + 8$

23. $x^2 - 11x + 24$

24. $x^2 + x + 2$

25. $x^2 - 5x - 6$

26. $x^2 + 2x - 8$

27. $x^2 + 4x - 12y^2$

28. $x^2 + 8xy + 15y^2$

29. $3x^2y + 6xy^2 + 3y^3$

30. $72 - 18x - 2x^2$

31. $32 + 12x - 4x^2$

(6.3) *Factor each trinomial.*

32. $2x^2 + 11x - 6$

33. $4x^2 - 7x + 4$

34. $4x^2 + 4x - 3$

35. $6x^2 + 5xy - 4y^2$

36. $6x^2 - 25xy + 4y^2$

37. $18x^2 - 60x + 50$

38. $2x^2 - 23xy - 39y^2$

39. $4x^2 - 28xy + 49y^2$

40. $18x^2 - 9xy - 20y^2$

41. $36x^3y + 24x^2y^2 - 45xy^3$

(6.4) *Factor each binomial.*

42. $4x^2 - 9$

43. $9t^2 - 25s^2$

44. $16x^2 + y^2$

45. $x^3 - 8y^3$

46. $8x^3 + 27$

47. $2x^3 + 8x$

48. $54 - 2x^3y^3$

49. $9x^2 - 4y^2$

50. $16x^4 - 1$

51. $x^4 + 16$

Solve.

52. The volume of the cylindrical shell is $\pi R^2h - \pi r^2h$ cubic units. Write this volume as a factored expression.

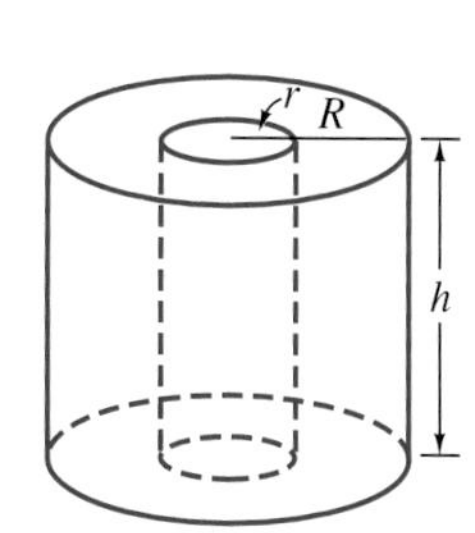

(6.5) *Factor each polynomial.*

53. $6xy^2 - 3xy$

54. $2xy^2 - 8x^2y^2$

55. $25x^2 - 100$

56. $16x^2 - 36$

57. $3x^2 - 10x + 8$

58. $2x^2 + 13x + 18$

59. $x^2y + 3x^2 - 4y - 12$

60. $x^2 - 4x - 45$

61. $4x^2 - 14x + 49$

62. $9x^2 + 30x + 25$

63. $4x^2 - 25$

64. $4x^2 + 36$

65. $8x^3 - y^3$

66. $x^3 + 27y^3$

67. $2x^6y + 54x^3y^4$

68. $4xy^2 - 500xy^5$

69. $4a^2 - 24a + 36 - 4b^2$

70. $3x^2 - 36x + 108 - 3y^2$

71. $2x^3y^8 - 128x^3y^5$

72. $54x^7y^4 + 2xy$

(6.6) *Solve each quadratic or higher-degree equation for the variable.*

73. $(3x - 1)(x + 7) = 0$

74. $3(x + 5)(8x - 3) = 0$

75. $5x(x - 4)(2x - 9) = 0$

76. $6(x + 3)(x - 4)(5x + 1) = 0$

77. $2x^2 = 12x$

78. $4x^3 - 36x = 0$

79. $(1 - x)(3x + 2) = -4x$

80. $2x(x - 12) = -40$

81. $3x^2 + 2x = 12 - 7x$

82. $2x^2 + 3x = 35$

83. $x^3 - 18x = 3x^2$

84. $19x^2 - 42x = -x^3$

85. $12x = 6x^3 + 6x^2$

86. $8x^3 + 10x^2 = 3x$

Solve.

87. The base of a triangle is four times its altitude. If the area of the triangle is 162 square feet, find the base.

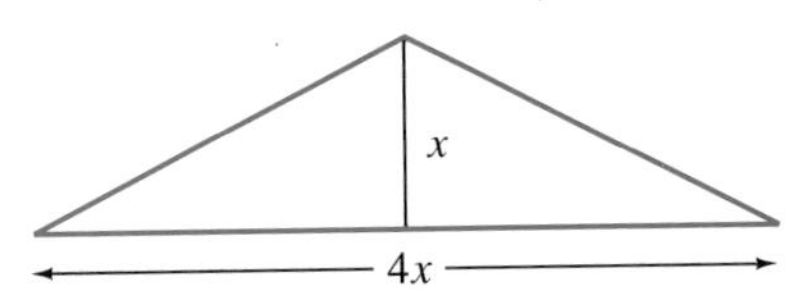

88. The sum of a number and twice its square is 105. Find the number.

89. Find two consecutive positive integers whose product is 380.

90. Find the length of the long leg of a right triangle if the hypotenuse is 8 feet longer than the long leg and the short leg is 8 feet shorter than the long leg.

91. The length of a rectangular piece of carpet is 2 meters less than 5 times its width. Find the dimensions of the carpet if its area is 16 square meters.

(6.7) *Graph each polynomial function defined by the equation. Find all intercepts. If the function is a quadratic function, find the vertex.*

92. $f(x) = x^2 + 6x + 9$

93. $f(x) = x^2 - 5x + 4$

94. $f(x) = (x - 1)(x^2 - 2x - 3)$

95. $f(x) = (x + 3)(x^2 - 4x + 3)$

96. $f(x) = 2x^2 - 4x + 5$

97. $f(x) = x^2 - 2x + 3$

98. $f(x) = x^3 - 16x$

99. $f(x) = x^3 + 5x^2 + 6x$

CHAPTER 6 TEST

Factor each polynomial completely. If the polynomial cannot be factored, write "prime."

1. $16x^3y - 12x^2y^4$
2. $20a^3b^2 - 35ab^3$
3. $xy - 5x + 3y - 15$
4. $x^3 + 2x^2 + 3x + 6$
5. $x^2 - 11x + 24$
6. $x^2 - 13x - 30$
7. $2x^2 + 17x - 9$
8. $4y^2 + 20y + 25$
9. $6x^2 - 15x - 9$
10. $x^5 + 3x^3 + 2x$
11. $4x^2 - 25$
12. $(3x + 1)^2 - 4$
13. $x^3 + 64$
14. $8x^3 - 125$
15. $3x^2y - 27y^3$
16. $8x^2 - 6x - 9$
17. $6x^2 + 24$
18. $4x^6y^2 - 32y^5$
19. $x^2 - 2x + 1 - y^2$
20. $8x^2 - 44x + 20$

Solve the equation for the variable.

21. $3(n - 4)(7n + 8) = 0$
22. $(x - 7)(x + 2) = -20$
23. $3m^3 = 12m$
24. $2x^3 + 5x^2 - 8x - 20 = 0$
25. $\frac{3x^2}{5} - \frac{2x}{5} = 1$

Solve.

26. Find the dimensions of a rectangular rug whose length is 5 feet longer than its width and whose area is 66 square feet.
27. Write the area of the shaded region as a factored polynomial.

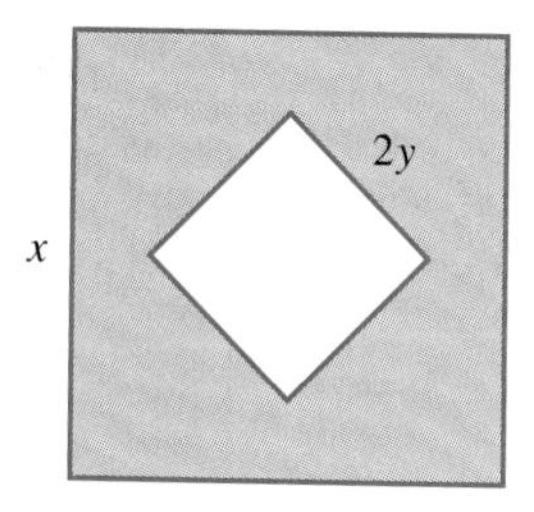

28. A pebble is hurled upward from the top of the Canada Trust Tower, which is 880 feet tall, with an initial velocity of 96 feet per second. Neglecting air resistance, the height $h(t)$ of the pebble after t seconds is given by the polynomial function

$$h(t) = -16t^2 + 96t + 880$$

When will the pebble hit the ground?

Graph.

29. $f(x) = x^2 - 4x - 5$
30. $f(x) = x^3 - 1$

CHAPTER 6 CUMULATIVE REVIEW

1. Write $\frac{2}{5}$ as an equivalent fraction with a denominator of 20.
2. Simplify the following:
   a. $3 + (-7) + (-8)$ **b.** $[7 + (-10)] + [-2 + (4)]$
3. If $a = -3$, $b = 2$, and $c = 4$, show that
   a. $(a + b) + c = a + (b + c)$
   b. $(a \cdot b) \cdot c = a \cdot (b \cdot c)$
4. Simplify each expression by combining like terms.
   a. $2x + 3x + 5 + 2$

b. $-5a - 3 + a + 2$
c. $4y - 3y^2$ **d.** $2.3x + 5x - 6$

5. Solve $\frac{y}{7} = 20$ for y.

6. Write a ratio for each phrase. Use fractional notation.
a. The ratio of 2 parts salt to 5 parts water
b. The ratio of 12 almonds to 16 pecans

7. The cost of a large hand-tossed pepperoni pizza at Domino's recently increased from \$5.80 to \$7.03. Find the percent increase.

8. Solve $\left|\frac{x}{2} - 1\right| = 11$.

9. Determine whether each ordered pair is a solution to the equation $x - 2y = 6$.
a. $(6, 0)$ **b.** $(0, 3)$ **c.** $(2, -2)$

10. The following graph shows the research and development expenditures by the Pharmaceutical Manufacturers Association as a function of time.
a. Approximate the money spent on research and development in 1992.
b. In 1958, research and development expenditures were \$200 million. Find the increase in expenditures from 1958 to 1994.

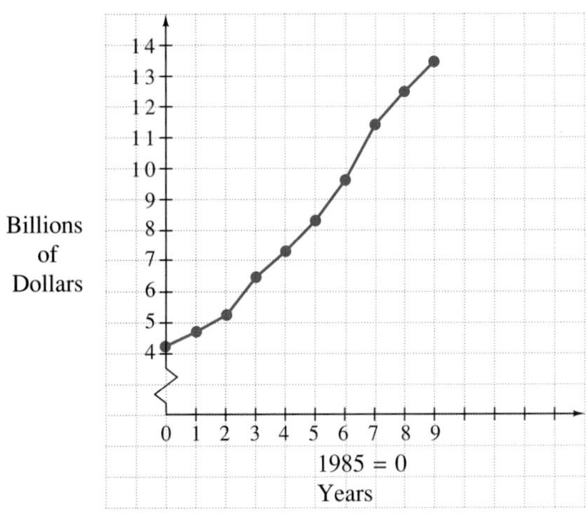

Source: *Science,* vol. 264, May 20, 1994.

11. Find the equation of the line with y-intercept -3 and the slope of $\frac{1}{4}$.

12. Graph $3x \geq y$.

13. Simplify the following expressions. Write each answer using positive exponents only.
a. $(2x^3)(5x)^{-2}$ **b.** $\left(\frac{3a^2}{b}\right)^{-3}$
c. $\frac{4^{-1}x^{-3}y}{4^{-3}x^2y^{-6}}$ **d.** $(y^{-3}z^6)^{-6}$ **e.** $\left(\frac{-2x^3y}{xy^{-1}}\right)^3$

14. If $P(x) = 7x^2 - 3x + 1$ find the following:
a. $P(1)$ **b.** $P(-2)$

15. Find the quotient: $\frac{6x^2 - 19x + 12}{3x - 5}$.

16. If $P(x) = 2x^3 - 4x^2 + 5$:
a. Find $P(2)$ by substitution.
b. Use synthetic division to find the remainder when $P(x)$ is divided by $x - 2$.

17. Find the GCF of $20x^3y$, $10x^2y^2$, and $35x^3$.

18. Factor $r^2 - r - 42$.

19. Factor each polynomial completely.
a. $8a^2b - 4ab$ **b.** $36x^2 - 9$
c. $2x^2 - 5x - 7$

20. Solve $x(2x - 7) = 4$.

21. For a TV commercial, a piece of luggage is dropped from a cliff 256 feet above the ground to show the durability of the luggage. Neglecting air resistance, the height $h(t)$ in feet of the luggage above the ground after t seconds is given by the polynomial function

$$h(t) = -16t^2 + 256$$

Find how long it takes for the luggage to hit the ground.

CHAPTER 7

Rational Expressions

Windmills generate electricity. Progressive communities are experimenting with fields of windmills for communal electric needs, examining exactly how windspeed affects the amount of electricity produced.

INTRODUCTION

When we divided one polynomial by another in Sections 5.5 and 5.6, we found quotients of polynomials. When the remainder part of the quotient wasn't 0, the remainder was a fraction, such as $\frac{x+1}{x^2}$. This fraction is not a polynomial since it cannot be written as the sum of whole number powers. Instead, it is called a **rational expression.** In this chapter, we present techniques for operating on these rational expressions. Since they are fractions, operating on rational expressions depends on your ability to work with number fractions and on the factoring techniques of Chapter 6.

7.1 Simplifying Rational Expressions

OBJECTIVES

Tape IA 16

1. Define a rational expression and a rational function.
2. Find the domain of a rational function.
3. Write a rational expression in lowest terms.
4. Write a rational expression equivalent to a rational expression with a given denominator.

1 Recall that a rational number, or fraction, is the quotient $\frac{p}{q}$ of two integers p and q as long as q is not 0. A **rational expression** is the quotient $\frac{P}{Q}$ of two polynomials P and Q as long as Q is not 0.

Examples of Rational Expressions

$$\frac{3x+7}{2} \qquad \frac{5x^2-3}{x-1} \qquad \frac{7x-2}{2x^2+7x+6}$$

Rational expressions are sometimes used to describe functions. For example, we call the function $f(x) = \frac{x^2+2}{x-3}$ a **rational function** since $\frac{x^2+2}{x-3}$ is a rational expression.

EXAMPLE 1 ICL Production Company uses the rational function $C(x) = \frac{2.6x + 10{,}000}{x}$ to describe the average cost per disc of pressing x compact discs. Find the average cost per disc for pressing:

a. 100 compact discs
b. 1000 compact discs

Solution: **a.** $C(100) = \dfrac{2.6(100) + 10{,}000}{100} = \dfrac{10{,}260}{100} = 102.6$

The average cost per disc for pressing 100 compact discs is \$102.60.

b. $C(1000) = \dfrac{2.6(1000) + 10{,}000}{1000} = \dfrac{12{,}600}{1000} = 12.6$

The average cost per disc for pressing 1000 compact discs is \$12.60.

Notice that as the level of production increases, the cost per disc decreases. ■

2 As with fractions, a rational expression is **undefined** if the denominator is 0. If a variable in a rational expression is replaced with a number that makes the denominator 0, we say that the rational expression is **undefined** for this value of the variable. For example, the rational expression $\dfrac{x^2 + 2}{x - 3}$ is not defined when x is 3, because replacing x with 3 results in a denominator of 0. Similarly, we exclude 3 from the domain of the function $f(x) = \dfrac{x^2 + 2}{x - 3}$.

The domain of f is

$$\{x \mid x \text{ is a real number and } x \neq 3\}$$

"The set of all x such that x is a real number and x is not equal to 3"

We assume that the domain of a function is the set of all real numbers for which the equation is defined.

EXAMPLE 2 Find the domain of each rational function.

a. $f(x) = \dfrac{8x^3 + 7x^2 + 20}{2}$ **b.** $g(x) = \dfrac{5x^2 - 3}{x - 1}$ **c.** $f(x) = \dfrac{7x - 2}{2x^2 + 7x + 6}$

Solution: The domain of each function will contain all real numbers except those values that make the denominator 0.

a. The denominator of $f(x) = \dfrac{8x^3 + 7x^2 + 20}{2}$ is never 0, so the domain of f is $\{x \mid x \text{ is a real number}\}$.

b. To find the values that make the denominator of $g(x)$ equal to 0, we solve.

$$x - 1 = 0 \quad \text{or} \quad x = 1$$

The domain of $g(x)$ must exclude the value of 1 since the rational expression is undefined when x is 1. The domain of g is $\{x \mid x \text{ is a real number and } x \neq 1\}$.

c. We find the domain by setting the denominator equal to 0.

$$2x^2 + 7x + 6 = 0 \qquad \text{Set the denominator equal to 0.}$$

$$(2x + 3)(x + 2) = 0 \qquad \text{Factor.}$$

$$2x + 3 = 0 \quad \text{or} \quad x + 2 = 0 \qquad \text{Set each factor equal to 0.}$$

$$x = -\frac{3}{2} \quad \text{or} \quad x = -2 \qquad \text{Solve.}$$

The domain of f is $\{x \mid x \text{ is a real number and } x \neq -\frac{3}{2} \text{ and } x \neq -2\}$. ■

3 Recall that a fraction is in lowest terms or simplest form if the numerator and denominator have no common factors other than 1 (or -1). For example, $\frac{3}{13}$ is in lowest terms since 3 and 13 have no common factors other than 1 (or -1).

To **simplify** a rational expression, or to write it in lowest terms, we use the fundamental principle of rational expressions.

Fundamental Principle of Rational Expressions

For any rational expression $\frac{P}{Q}$ and any polynomial R, $R \neq 0$,

$$\frac{P}{Q} = \frac{P \cdot R}{Q \cdot R} \quad \text{and} \quad \frac{P}{Q} = \frac{P \div R}{Q \div R}$$

Thus, the fundamental principle says that multiplying or dividing the numerator and denominator of a rational expression by a nonzero polynomial yields an equivalent rational expression.

To simplify a rational expression such as $\frac{(x+2)^2}{x^2-4}$, factor the numerator and the denominator and then use the fundamental principle of rational expressions to divide out common factors.

$$\frac{(x+2)^2}{x^2-4} = \frac{(x+2)(x+2)}{(x+2)(x-2)}$$

$$= \frac{x+2}{x-2}$$

In general, the following steps may be used to write rational expressions in lowest terms or simplest form.

To Write a Rational Expression in Lowest Terms

Step 1 Completely factor the numerator and denominator of the rational expression.

Step 2 Divide both the numerator and denominator by their GCF.

For now, we assume that variables in a rational expression do not represent values that make the denominator 0.

EXAMPLE 3 Write each rational expression in lowest terms.

a. $\frac{24x^6y^5}{8x^7y}$ **b.** $\frac{2x^2}{10x^3 - 2x^2}$

Solution: **a.** The GCF of the numerator and denominator is $8x^6y$.

$$\frac{24x^6y^5}{8x^7y} = \frac{(8x^6y)\,3y^4}{(8x^6y)\,x} = \frac{3y^4}{x}$$

b. Factor out $2x^2$ from the denominator. Then divide numerator and denominator by their GCF, $2x^2$.

$$\frac{2x^2}{10x^3 - 2x^2} = \frac{2x^2}{2x^2(5x - 1)} = \frac{1}{5x - 1}$$ ■

EXAMPLE 4 Write each rational expression in lowest terms.

a. $\frac{2 + x}{x + 2}$ **b.** $\frac{2 - x}{x - 2}$ **c.** $\frac{18 - 2x^2}{x^2 - 2x - 3}$

Solution: **a.** By the commutative property of addition, $2 + x = x + 2$, so

$$\frac{2 + x}{x + 2} = \frac{x + 2}{x + 2} = 1$$

b. The terms in the numerator of $\frac{2 - x}{x - 2}$ differ by sign from the terms of the denominator, so the polynomials are opposites of each other and the expression simplifies to -1. To see this, factor out -1 from the numerator or the denominator. If -1 is factored from the numerator, then

$$\frac{2 - x}{x - 2} = \frac{-1(-2 + x)}{x - 2} = \frac{-1(x - 2)}{x - 2} = -1$$

> **HELPFUL HINT**
>
> When the numerator and the denominator of a rational expression are opposites of each other, the expression simplifies to -1.

c. $$\frac{18 - 2x^2}{x^2 - 2x - 3} = \frac{2(9 - x^2)}{(x + 1)(x - 3)}$$

$$= \frac{2(3 + x)(3 - x)}{(x + 1)(x - 3)} \quad \text{Factor.}$$

Notice the opposites $3 - x$ and $x - 3$. We write $3 - x$ as $-1(x - 3)$ and simplify.

$$\frac{2(3 + x)(3 - x)}{(x + 1)(x - 3)} = \frac{2(3 + x) \cdot -1(x - 3)}{(x + 1)(x - 3)} = -\frac{2(3 + x)}{x + 1}$$ ■

> **HELPFUL HINT**
>
> Recall from Section 1.7 that, for a fraction $\frac{a}{b}$,
>
> $$\frac{a}{-b} = \frac{-a}{b} = -\frac{a}{b}$$
>
> For example,
>
> $$\frac{-(x + 1)}{(x + 2)} = \frac{(x + 1)}{-(x + 2)} = -\frac{x + 1}{x + 2}$$

EXAMPLE 5 Write each rational expression in lowest terms.

a. $\dfrac{x^3 + 8}{2 + x}$

b. $\dfrac{2y^2 + 2}{y^3 - 5y^2 + y - 5}$

Solution: **a.** $\dfrac{x^3 + 8}{2 + x} = \dfrac{(x + 2)(x^2 - 2x + 4)}{x + 2}$ Factor the sum of two cubes.

$= x^2 - 2x + 4$ Divide out common factors.

b. First factor the denominator by grouping.

$$\begin{aligned} y^3 - 5y^2 + y - 5 &= (y^3 - 5y^2) + (y - 5) \\ &= y^2(y - 5) + (y - 5) \\ &= (y - 5)(y^2 + 1) \end{aligned}$$

Then

$$\frac{2y^2 + 2}{y^3 - 5y^2 + y - 5} = \frac{2(y^2 + 1)}{(y - 5)(y^2 + 1)} = \frac{2}{y - 5}$$ ■

4 The fundamental property of fractions also allows us to write a rational expression as an equivalent rational expression with a given denominator. Doing so is necessary to add and subtract rational expressions.

EXAMPLE 6 Write each rational expression as an equivalent rational expression with the given denominator.

a. $\dfrac{3x}{2y}$, denominator $10xy^3$ **b.** $\dfrac{3x + 1}{x - 5}$, denominator $2x^2 - 11x + 5$

Solution: **a.** $\dfrac{3x}{2y} = \dfrac{?}{10xy^3}$

If the denominator $2y$ is multiplied by $5xy^2$, the result is the given denominator $10xy^3$.

$$\underbrace{2y}_{\substack{\uparrow \\ \text{original} \\ \text{denominator}}}(5xy^2) = \underbrace{10xy^3}_{\substack{\uparrow \\ \text{given} \\ \text{denominator}}}$$

Use the fundamental principle of rational expressions and multiply the numerator and the denominator of the original rational expression by $5xy^2$. Then

$$\frac{3x}{2y} = \frac{3x(5xy^2)}{2y(5xy^2)} = \frac{15x^2y^2}{10xy^3}$$

b. The factored form of the given denominator, $2x^2 - 11x + 5$, is $(x - 5)(2x - 1)$.

$$\frac{3x + 1}{x - 5} = \frac{?}{(x - 5)(2x - 1)}$$

Use the fundamental principle of rational expressions and multiply the numerator and denominator of the original rational expression by $2x - 1$.

$$\frac{3x+1}{x-5} = \frac{(3x+1)(2x-1)}{(x-5)(2x-1)} = \frac{6x^2 - x - 1}{(x-5)(2x-1)}$$

To prepare for adding and subtracting rational expressions, we multiply the binomials in the numerator but leave the denominator in factored form. ■

MENTAL MATH

Find any real numbers for which each rational expression is undefined. See Example 2.

1. $\frac{x+5}{x}$
2. $\frac{x^2-5x}{x-3}$
3. $\frac{x^2+4x-2}{x(x-1)}$
4. $\frac{x+2}{(x-5)(x-6)}$

EXERCISE SET 7.1

Find each function value. See Example 1.

1. $f(x) = \frac{x+8}{2x-1}$; $f(2), f(0), f(-1)$
2. $f(y) = \frac{y-2}{-5+y}$; $f(-5), f(0), f(10)$
3. $g(x) = \frac{x^2+8}{x^3-25x}$; $g(3), g(-2), g(1)$
4. $s(t) = \frac{t^3+1}{t^2+1}$; $s(-1), s(1), s(2)$

Find the domain of each rational function. See Example 2.

5. $f(x) = \frac{5x-7}{4}$
6. $g(x) = \frac{4-3x}{2}$
7. $s(t) = \frac{t^2+1}{2t}$
8. $v(t) = -\frac{5t+t^2}{3t}$
9. $C(x) = \frac{x+3}{x^2-4}$
10. $R(x) = \frac{5}{x^2-7x}$

Write each rational expression in lowest terms. See Example 3.

11. $\frac{10x^3}{18x}$
12. $-\frac{48a^7}{16a^{10}}$
13. $\frac{9x^6y^3}{18x^2y^5}$
14. $\frac{10ab^5}{15a^3b^5}$
15. $\frac{8q^2}{16q^3-16q^2}$
16. $\frac{3y}{6y^2-30y}$

Write each rational expression in lowest terms. See Example 4.

17. $\frac{x+5}{5+x}$
18. $\frac{x-5}{5-x}$
19. $\frac{x-1}{1-x^2}$
20. $\frac{10+5x}{x^2+2x}$
21. $\frac{7-x}{x^2-14x+49}$
22. $\frac{x^2-9}{2x^2-5x-3}$

Write each rational expression in lowest terms. See Example 5.

23. $\frac{x+3}{x^3+27}$
24. $\frac{x^3-64}{4-x}$
25. $\frac{2x^3-16}{3x-6}$
26. $\frac{x^2-x+1}{2x^3+2}$
27. $\frac{xy-3y+2x-6}{x^2-6x+9}$
28. $\frac{2x^3-5x^2+2x-5}{3x^2+3}$

Write each rational expression as an equivalent rational expression with the given denominator. See Example 6.

29. $\frac{5}{2y}$, $4y^3z$

30. $\frac{1}{z}$, $5z^5$

31. $\frac{3x}{2x - 1}$, $2x^2 + 9x - 5$

32. $\frac{5}{3x + 2}$, $3x^2 - 13x - 10$

33. $\frac{x - 2}{1}$, $x + 2$

34. $\frac{x - 5}{1}$, $x + 1$

Find the domain of each rational function.

35. $f(x) = \frac{3x}{7 - x}$

36. $f(x) = \frac{-4x}{-2 + x}$

37. $g(x) = \frac{2 - 3x^2}{3x^2 + 9x}$

38. $C(x) = \frac{5 - 4x^2}{4x - 2x^2}$

39. $R(x) = \frac{3 + 2x}{x^3 + x^2 - 2x}$

40. $h(x) = \frac{5 - 3x}{2x^2 - 14x + 20}$

Solve.

41. The total revenue from the sale of a popular music cassette tape is approximated by the rational function

$$R(x) = \frac{200x^2}{x^2 + 5}$$

where x is the number of years since the cassette tape has been released and $R(x)$ is the total revenue in millions of dollars.

a. Find the total revenue generated by the end of the first year.

b. Find the total revenue generated by the end of the second year.

c. Find the total revenue generated by the end of the third year.

d. Find the revenue generated in the second year only.

42. The average cost $C(x)$ in dollars per desk of manufacturing x computer desks is given by the rational function

$$C(x) = \frac{25x + 2000}{x}$$

a. Find the average cost per desk of manufacturing 100 desks.

b. Find the average cost per desk of manufacturing 1000 desks.

Write each rational expression in lowest terms.

43. $\frac{6x^3}{27x^3}$

44. $\frac{18y^7}{30y}$

45. $\frac{6a^2b^3}{3a^2b^2}$

46. $\frac{27m^4p^2}{3m^4p}$

47. $\frac{7 - y}{y - 7}$

48. $\frac{17 + x}{x + 17}$

49. $\frac{4x - 8}{3x - 6}$

50. $\frac{12 - 6x}{30 - 15x}$

51. $\frac{2x - 14}{7 - x}$

52. $\frac{9 - x}{5x - 45}$

53. $\frac{x^2 - 2x - 3}{x^2 - 6x + 9}$

54. $\frac{x^2 + 10x + 25}{x^2 + 8x + 15}$

55. $\frac{2x^2 + 12x + 18}{x^2 - 9}$

56. $\frac{x^2 - 4}{2x^2 + 8x + 8}$

57. $\frac{3x + 6}{x^2 + 2x}$

58. $\frac{3x + 4}{9x^2 + 4}$

59. $\frac{x + 2}{x^2 - 4}$

60. $\frac{x^2 - 9}{x - 3}$

61. $\frac{2x^2 - x - 3}{2x^3 - 3x^2 + 2x - 3}$

62. $\frac{3x^2 - 5x - 2}{6x^3 + 2x^2 + 3x + 1}$

63. $\frac{x^4 - 16}{x^2 + 4}$

64. $\frac{x^2 + y^2}{x^4 - y^4}$

65. $\frac{x^2 + 6x - 40}{10 + x}$

66. $\frac{x^2 - 8x + 16}{4 - x}$

67. $\frac{2x^2 - 7x - 4}{x^2 - 5x + 4}$

68. $\frac{3x^2 - 11x + 10}{x^2 - 7x + 10}$

69. $\frac{x^3 - 125}{5 - x}$

70. $\frac{4x + 4}{2x^3 + 2}$

71. $\frac{8x^3 - 27}{4x - 6}$

72. $\frac{9x^2 - 15x + 25}{27x^3 + 125}$

73. $\frac{x + 5}{x^2 + 5}$

74. $\frac{5x}{5x^2 + 5x}$

Write each rational expression as an equivalent rational expression with the given denominator.

75. $\frac{5}{m}$, $6m^3$

76. $\frac{3}{x}$, $3x$

77. $\frac{7}{m - 2}$, $5m - 10$

78. $\frac{-2}{x + 1}$, $10x + 10$

79. $\frac{y + 4}{y - 4}$, $y^2 - 16$

80. $\frac{5}{x - 1}$, $x^2 - 1$

81. $\dfrac{12x}{x+2}$, $x^2 + 4x + 4$

82. $\dfrac{x+6}{x-4}$, $x^2 - 8x + 16$

83. $\dfrac{1}{x+2}$, $x^3 + 8$

84. $\dfrac{x}{3x-1}$, $27x^3 - 1$

85. $\dfrac{a}{a+2}$, $ab - 3a + 2b - 6$

86. $\dfrac{5}{x-y}$, $2x^2 + 5x - 2xy - 5y$

Skill Review

Perform the indicated operations. See Section 1.3.

87. $\dfrac{6}{35} \cdot \dfrac{28}{9}$

88. $\dfrac{3}{8} \cdot \dfrac{4}{27}$

89. $\dfrac{8}{35} \div \dfrac{4}{5}$

90. $\dfrac{6}{11} \div \dfrac{2}{11}$

91. $\left(\dfrac{1}{2} \cdot \dfrac{1}{4}\right) \div \dfrac{3}{8}$

92. $\dfrac{3}{7} \cdot \left(\dfrac{1}{5} \div \dfrac{3}{10}\right)$

Graph the following linear inequalities. See Section 4.6.

93. $x + y \leq 7$

94. $x - y > 2$

Recall that a ***histogram*** *is a special type of bar graph where the width of the bar stands for a single number or a range of numbers and the height of each bar represents of number of occurrences. For example, the width of each bar in the histogram below represents a range of ages, and the height represents the number of occurrences, or percents. Use this graph to answer Exercises 95 through 97. See Section 5.6.*

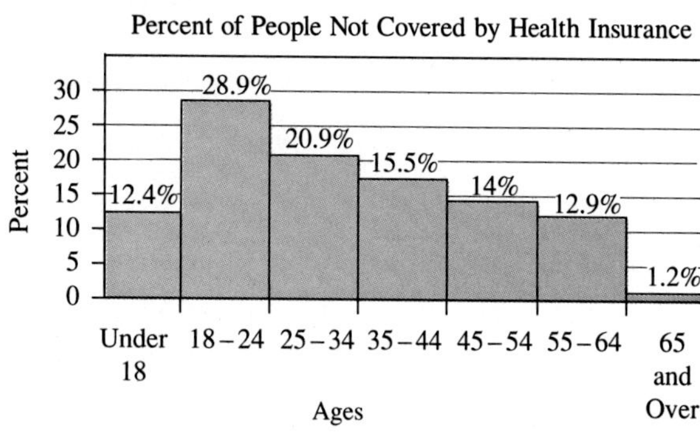

Source: *USA Today*, June 10, 1994.

95. What percent of people aged 45 and over have no health insurance?

96. Find the age group of people who have the lowest percent of no health insurance.

97. What percent of people aged 18 to 34 have no health insurance?

A Look Ahead

Write each rational expression in lowest terms. Assume that no denominator is 0.

EXAMPLE Write $\dfrac{x^{2n} + 2x^n y^n + y^{2n}}{x^n + y^n}$ in lowest terms.

Solution: Factor the numerator.

$$\frac{x^{2n} + 2x^n y^n + y^{2n}}{x^n + y^n} = \frac{(x^n + y^n)(x^n + y^n)}{(x^n + y^n)} = x^n + y^n$$ ■

98. $\dfrac{p^x - 4}{4 - p^x}$

99. $\dfrac{3 + q^n}{q^n + 3}$

100. $\dfrac{x^n + 4}{x^{2n} - 16}$

101. $\dfrac{x^{2k} - 9}{3 + x^k}$

102. $\dfrac{x^{2k} - 4x^k + 16}{x^{3k} + 64}$

103. $\dfrac{4x^k - 12}{x^{2k} + 4}$

Writing in Mathematics

104. The rational expression $\frac{x^2 - 4}{x + 2}$ is equivalent to the expression $x - 2$ for all real number values of x except -2. In other words, $\frac{x^2 - 4}{x + 2} = x - 2$ as long as $x \neq -2$. How do you think the graph of $y = \frac{x^2 - 4}{x + 2}$ differs from the graph of $y = x - 2$?

7.2 Multiplying and Dividing Rational Expressions

OBJECTIVES

1. Multiply rational expressions.
2. Divide by a rational expression.

Tape IA 16

Arithmetic operations on rational expressions are performed in the same way as they are on rational numbers.

1

Multiplying Rational Expressions

Let P, Q, R, and S be polynomials. Then

$$\frac{P}{Q} \cdot \frac{R}{S} = \frac{PR}{QS}$$

as long as $Q \neq 0$ and $S \neq 0$.

To multiply rational expressions, the product of their numerators is the numerator of the product and the product of their denominators is the denominator of the product.

EXAMPLE 1 Multiply.

a. $\frac{2x^3}{9y} \cdot \frac{y^2}{4x^3}$

b. $\frac{1 + 3n}{2n} \cdot \frac{2n - 4}{3n^2 - 2n - 1}$

Solution: **a.** $\frac{2x^3}{9y} \cdot \frac{y^2}{4x^3} = \frac{2x^3y^2}{36x^3y}$

To simplify, divide the numerator and the denominator by the common factor $2x^3y$.

$$\frac{2x^3y^2}{36x^3y} = \frac{y\,(2x^3y)}{18\,(2x^3y)} = \frac{y}{18}$$

b. $$\frac{1+3n}{2n} \cdot \frac{2n-4}{3n^2-2n-1} = \frac{1+3n}{2n} \cdot \frac{2(n-2)}{(3n+1)(n-1)}$$ Factor.

$$= \frac{(1+3n) \cdot 2\,(n-2)}{2\,n\,(3n+1)\,(n-1)}$$ Multiply.

$$= \frac{n-2}{n(n-1)}$$ Divide out common factors. ■

The following steps may be used to multiply rational expressions.

To Multiply Rational Expressions

Step 1 Completely factor the numerators and denominators.

Step 2 Multiply the numerators and multiply the denominators.

Step 3 Write the product in lowest terms by dividing both the numerator and the denominator by their GCF.

EXAMPLE 2 Multiply.

a. $\frac{2x^2+3x-2}{-4x-8} \cdot \frac{16x^2}{4x^2-1}$

b. $(ac-ad+bc-bd) \cdot \frac{a+b}{d-c}$

Solution: **a.** $$\frac{2x^2+3x-2}{-4x-8} \cdot \frac{16x^2}{4x^2-1} = \frac{(2x-1)(x+2)}{-4(x+2)} \cdot \frac{16x^2}{(2x+1)(2x-1)}$$ Factor.

$$= \frac{4 \cdot 4x^2\,(2x-1)(x+2)}{-1 \cdot 4(x+2)\,(2x+1)\,(2x-1)}$$ Multiply.

$$= -\frac{4x^2}{2x+1}$$ Divide out common factors.

b. First factor $ac-ad+bc-bd$ by grouping.

$$ac-ad+bc-bd = (ac-ad)+(bc-bd)$$ Group terms.

$$= a(c-d)+b(c-d)$$

$$= (c-d)(a+b)$$

To multiply, write $(c-d)(a+b)$ as a fraction whose denominator is 1.

$$(ac-ad+bc-bd) \cdot \frac{a+b}{d-c} = \frac{(c-d)(a+b)}{1} \cdot \frac{a+b}{d-c}$$ Factor.

$$= \frac{(c-d)(a+b)(a+b)}{d-c}$$ Multiply.

Write $(c-d)$ as $-1(d-c)$ and simplify.

$$\frac{-1\,(d-c)\,(a+b)(a+b)}{d-c} = -(a+b)^2$$ ■

2 To divide by a rational expression, multiply by its reciprocal. Recall that two numbers are reciprocals of each other if their product is 1. Similarly, if $\frac{P}{Q}$ is a rational expression, then $\frac{Q}{P}$ is its **reciprocal,** since

$$\frac{P}{Q} \cdot \frac{Q}{P} = \frac{P \cdot Q}{Q \cdot P} = 1$$

The following are examples of expressions and their reciprocals.

Expression	*Reciprocal*
$\frac{3}{x}$	$\frac{x}{3}$
$\frac{2 + x^2}{4x - 3}$	$\frac{4x - 3}{2 + x^2}$
x^3	$\frac{1}{x^3}$
0	no reciprocal

Division by rational expressions is defined as follows.

Dividing Rational Expressions

Let P, Q, R, and S be polynomials. Then

$$\frac{P}{Q} \div \frac{R}{S} = \frac{P}{Q} \cdot \frac{S}{R} = \frac{PS}{QR}$$

as long as $Q \neq 0$, $S \neq 0$, and $R \neq 0$.

Notice that division of rational expressions is the same as for rational numbers.

EXAMPLE 3 Divide.

a. $\frac{3x}{5y} \div \frac{9y}{x^5}$

b. $\frac{8m^2}{3m^2 - 12} \div \frac{40}{2 - m}$

Solution: **a.** $\frac{3x}{5y} \div \frac{9y}{x^5} = \frac{3x}{5y} \cdot \frac{x^5}{9y}$ Multiply by the reciprocal of the divisor.

$= \frac{x^6}{15y^2}$ Simplify.

b. $\frac{8m^2}{3m^2 - 12} \div \frac{40}{2 - m} = \frac{8m^2}{3m^2 - 12} \cdot \frac{2 - m}{40}$ Multiply by the reciprocal of the divisor.

$= \frac{8m^2(2 - m)}{3(m + 2)(m - 2) \cdot 40}$ Factor and multiply.

$$= \frac{8 m^2 \cdot -1 (m - 2)}{3(m + 2)(m - 2) \cdot 8 \cdot 5}$$ Write $(2 - m)$ as $-1(m - 2)$.

$$= -\frac{m^2}{15(m + 2)}$$ Simplify. ■

HELPFUL HINT

When dividing rational expressions, do not divide out common factors until the division problem is rewritten as a multiplication problem.

EXAMPLE 4 Divide $\frac{8x^3 + 125}{x^4 + 5x^2 + 4} \div \frac{2x + 5}{2x^2 + 8}$.

Solution:

$$\frac{8x^3 + 125}{x^4 + 5x^2 + 4} \div \frac{2x + 5}{2x^2 + 8} = \frac{8x^3 + 125}{x^4 + 5x^2 + 4} \cdot \frac{2x^2 + 8}{2x + 5}$$

$$= \frac{(2x + 5)(4x^2 - 10x + 25) \cdot 2(x^2 + 4)}{(x^2 + 1)(x^2 + 4) \cdot (2x + 5)}$$

$$= \frac{2(4x^2 - 10x + 25)}{x^2 + 1}$$ ■

EXERCISE SET 7.2

Multiply as indicated. Write all answers in lowest terms. See Examples 1 and 2.

1. $\frac{4}{x} \cdot \frac{x^2}{8}$

2. $\frac{x}{3} \cdot \frac{9}{x^3}$

3. $\frac{2a^2b}{6ac} \cdot \frac{3c^2}{4ab}$

4. $\frac{5ab^4}{6abc} \cdot \frac{2bc^2}{10ab^2}$

5. $\frac{2x}{5} \cdot \frac{5x + 10}{6(x + 2)}$

6. $\frac{3x}{7} \cdot \frac{14 - 7x}{9(2 - x)}$

7. $\frac{2x - 4}{15} \cdot \frac{6}{2 - x}$

8. $\frac{10 - 2x}{7} \cdot \frac{14}{5x - 25}$

9. $\frac{18a - 12a^2}{4a^2 + 4a + 1} \cdot \frac{4a^2 + 8a + 3}{4a^2 - 9}$

10. $\frac{a - 5b}{a^2 + ab} \cdot \frac{b^2 - a^2}{10b - 2a}$

11. $\frac{x^2 - 6x - 16}{2x^2 - 128} \cdot \frac{x^2 + 16x + 64}{3x^2 + 30x + 48}$

12. $\frac{2x^2 + 12x - 32}{x^2 + 16x + 64} \cdot \frac{x^2 + 10x + 16}{x^2 - 3x - 10}$

13. $\frac{4x + 8}{x + 1} \cdot \frac{2 - x}{3x - 15} \cdot \frac{2x^2 - 8x - 10}{x^2 - 4}$

14. $\frac{3x - 15}{2 - x} \cdot \frac{x + 1}{4x + 8} \cdot \frac{x^2 - 4}{2x^2 - 8x - 10}$

Divide as indicated. Write all answers in lowest terms. See Examples 3 and 4.

15. $\frac{4}{x} \div \frac{8}{x^2}$

16. $\frac{x}{3} \div \frac{x^3}{9}$

17. $\frac{4ab}{3c^2} \div \frac{2a^2b}{6ac}$

18. $\frac{6abc}{5ab^4} \div \frac{2bc^2}{10ab^2}$

19. $\frac{2x}{5} \div \frac{6x + 12}{5x + 10}$

20. $\frac{7}{3x} \div \frac{14 - 7x}{18 - 9x}$

21. $\frac{2(x + y)}{5} \div \frac{6(x + y)}{25}$

22. $\frac{4}{x - 2y} \div \frac{9}{3x - 6y}$

23. $\frac{x^2 - 6x + 9}{x^2 - x - 6} \div \frac{x^2 - 9}{4}$

24. $\frac{x^2 - 4}{3x + 6} \div \frac{2x^2 - 8x + 8}{x^2 + 4x + 4}$

25. $\frac{x^2 - 6x - 16}{2x^2 - 128} \div \frac{x^2 + 10x + 16}{x^2 + 16x + 64}$

26. $\frac{a^2 - a - 6}{a^2 - 81} \div \frac{a^2 - 7a - 18}{4a + 36}$

27. $\dfrac{14x^4}{y^5} \div \dfrac{2x^2}{y^7} \div \dfrac{2x^4}{7y^5}$

28. $\dfrac{x^2}{7y^2} \div \left(\dfrac{2x^2}{y^7} \div \dfrac{2x^2}{7y^5}\right)$

Perform the indicated operation. Write all answers in lowest terms.

29. $\dfrac{3xy^3}{4x^3y^2} \cdot \dfrac{-8x^3y^4}{9x^4y^7}$

30. $-\dfrac{2xyz^3}{5x^2z^2} \cdot \dfrac{10xy}{x^3}$

31. $\dfrac{8a}{3a^4b^2} \div \dfrac{4b^5}{6a^2b}$

32. $\dfrac{3y^3}{14x^4} \div \dfrac{8y^3}{7x}$

33. $\dfrac{a^2b}{a^2 - b^2} \cdot \dfrac{a + b}{4a^3b}$

34. $\dfrac{3ab^2}{a^2 - 4} \cdot \dfrac{a - 2}{6a^2b^2}$

35. $\dfrac{x^2 - 9}{4} \div \dfrac{x^2 - 6x + 9}{x^2 - x - 6}$

36. $\dfrac{a - 5b}{a^2 + ab} \div \dfrac{15b - 3a}{b^2 - a^2}$

37. $\dfrac{9x + 9}{4x + 8} \cdot \dfrac{2x + 4}{3x^2 - 3}$

38. $\dfrac{x^2 - 1}{10x + 30} \cdot \dfrac{12x + 36}{3x - 3}$

39. $\dfrac{a + b}{ab} \div \dfrac{a^2 - b^2}{4a^3b}$

40. $\dfrac{6a^2b^2}{a^2 - 4} \div \dfrac{3ab^2}{a - 2}$

41. $\dfrac{2x^2 - 4x - 30}{5x^2 - 40x - 75} \div \dfrac{x^2 - 8x + 15}{x^2 - 6x + 9}$

42. $\dfrac{4a + 36}{a^2 - 7a - 18} \div \dfrac{a^2 - a - 6}{a^2 - 81}$

43. $\dfrac{2x^3 - 16}{6x^2 + 6x - 36} \cdot \dfrac{9x + 18}{3x^2 + 6x + 12}$

44. $\dfrac{x^2 - 3x + 9}{5x^2 - 20x - 105} \cdot \dfrac{x^2 - 49}{x^3 + 27}$

45. $\dfrac{15b - 3a}{b^2 - a^2} \div \dfrac{a - 5b}{ab + b^2}$

46. $\dfrac{4x + 4}{x - 1} \div \dfrac{x^2 - 4x - 5}{x^2 - 1}$

47. $\dfrac{a^3 + a^2b + a + b}{a^3 + a} \cdot \dfrac{6a^2}{2a^2 - 2b^2}$

48. $\dfrac{a^2 - 2a}{ab - 2b + 3a - 6} \cdot \dfrac{8b + 24}{3a + 6}$

49. $\dfrac{5a}{12} \cdot \dfrac{2}{25a^2} \cdot \dfrac{15a}{2}$

50. $\dfrac{4a}{7} \div \dfrac{a^2}{14} \cdot \dfrac{3}{a}$

51. $\dfrac{3x - x^2}{x^3 - 27} \div \dfrac{x}{x^2 + 3x + 9}$

52. $\dfrac{x^2 - 3x}{x^3 - 27} \div \dfrac{2x}{2x^2 + 6x + 18}$

53. $\dfrac{4a}{7} \div \left(\dfrac{a^2}{14} \cdot \dfrac{3}{a}\right)$

54. $\dfrac{a^2}{14} \cdot \dfrac{3}{a} \div \dfrac{4a}{7}$

55. $\dfrac{8b + 24}{3a + 6} \div \dfrac{ab - 2b + 3a - 6}{a^2 - 4a + 4}$

56. $\dfrac{2a^2 - 2b^2}{a^3 + a^2b + a + b} \div \dfrac{6a^2}{a^3 + a}$

57. $\dfrac{4}{x} \div \dfrac{3xy}{x^2} \cdot \dfrac{6x^2}{x^4}$

58. $\dfrac{4}{x} \cdot \dfrac{3xy}{x^2} \div \dfrac{6x^2}{x^4}$

59. $\dfrac{3x^2 - 5x - 2}{y^2 + y - 2} \cdot \dfrac{y^2 + 4y - 5}{12x^2 + 7x + 1} \div \dfrac{5x^2 - 9x - 2}{8x^2 - 2x - 1}$

60. $\dfrac{x^2 + x - 2}{3y^2 - 5y - 2} \cdot \dfrac{12y^2 + y - 1}{x^2 + 4x - 5} \div \dfrac{8y^2 - 6y + 1}{5y^2 - 9y - 2}$

61. $\dfrac{5a^2 - 20}{3a^2 - 12a} \div \dfrac{a^3 + 2a^2}{2a^2 - 8a} \cdot \dfrac{9a^3 + 6a^2}{2a^2 - 4a}$

62. $\dfrac{5a^2 - 20}{3a^2 - 12a} \div \left(\dfrac{a^3 + 2a^2}{2a^2 - 8a} \cdot \dfrac{9a^3 + 6a^2}{2a^2 - 4a}\right)$

63. $\dfrac{5x^4 + 3x^2 - 2}{x - 1} \cdot \dfrac{x + 1}{x^4 - 1}$

64. $\dfrac{3x^4 - 10x^2 - 8}{x - 2} \cdot \dfrac{3x + 6}{15x^2 + 10}$

Solve.

65. Find the area of the rectangle.

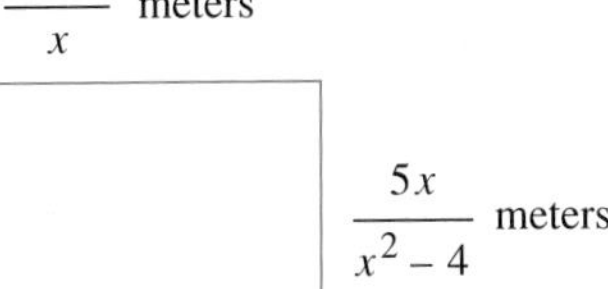

66. Find the area of the triangle.

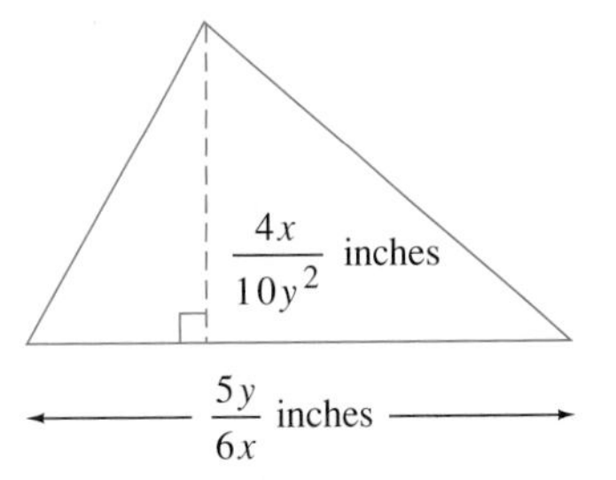

67. A parallelogram has area $\frac{x^2 + x - 2}{x^3}$ square feet and height $\frac{x^2}{x - 1}$ feet. Express the length of its base as a rational expression in x ($A = b \cdot h$).

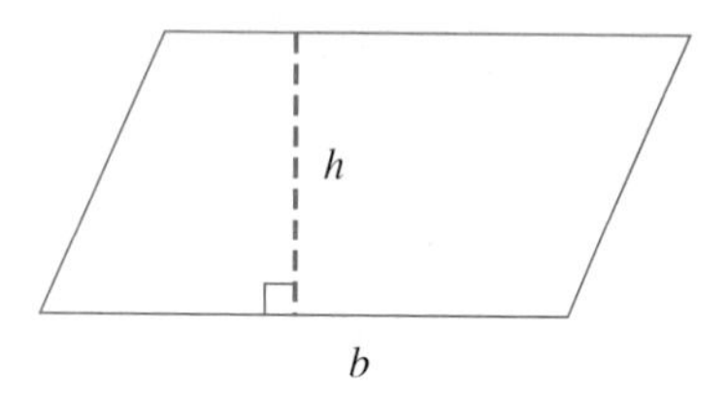

68. A lottery prize of $\frac{15x^3}{y^2}$ dollars is to be divided among $5x$ people. Express the amount of money each person is to receive as a rational expression in x and y.

Skill Review

Perform the indicated operation. See Section 1.3.

69. $\frac{4}{5} + \frac{3}{5}$

70. $\frac{4}{10} - \frac{7}{10}$

71. $\frac{5}{28} - \frac{2}{21}$

72. $\frac{5}{13} + \frac{2}{7}$

73. $\frac{3}{8} + \frac{1}{2} - \frac{3}{16}$

74. $\frac{2}{9} - \frac{1}{6} + \frac{2}{3}$

Use synthetic division to divide the following. See Section 5.6.

75. $(x^3 - 6x^2 + 3x - 4) \div (x - 1)$

76. $(5x^4 - 3x^2 + 2) \div (x + 2)$

Thirty people were recently polled about their average monthly balance in their checking account. The results of this poll are shown in the histogram below. Use this graph to answer Exercises 77 through 82. See Sections 7.1 and 3.2.

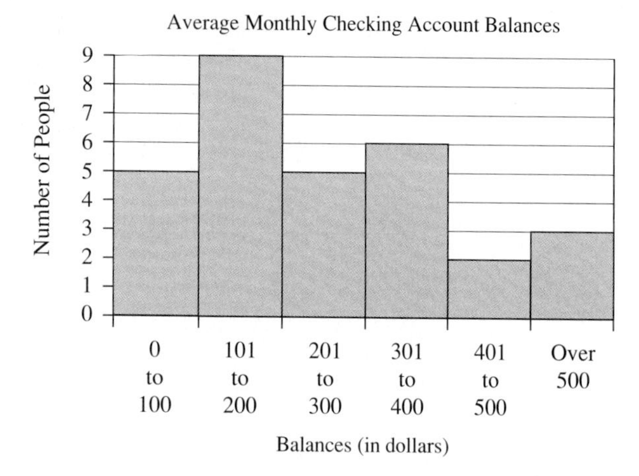

77. How many people polled reported an average checking balance of 201 to 300 dollars?

78. How many people polled reported an average checking balance of 0 to 100 dollars?

79. How many people polled reported an average checking balance of \$200 or less?

80. How many people polled reported an average checking balance of \$301 or more?

81. What percent of people polled reported an average checking balance of 201 to 300 dollars?

82. What percent of people polled reported an average checking balance of 0 to 100 dollars?

A Look Ahead

Perform the indicated operation. Write all answers in lowest terms. See the following example.

EXAMPLE Perform the following operation.

$$\frac{x^{2n} - 3x^n - 18}{x^{2n} - 9} \cdot \frac{3x^n + 9}{x^{2n}}$$

Solution:

$$\frac{x^{2n} - 3x^n - 18}{x^{2n} - 9} \cdot \frac{3x^n + 9}{x^{2n}} = \frac{(x^n + 3)(x^n - 6) \cdot 3(x^n + 3)}{(x^n + 3)(x^n - 3) \cdot x^{2n}}$$

$$= \frac{3(x^n - 6)(x^n + 3)}{x^{2n}(x^n - 3)} \quad ■$$

83. $\dfrac{x^{2n} - 4}{7x} \cdot \dfrac{14x^3}{x^n - 2}$

84. $\dfrac{x^{2n} + 4x^n + 4}{4x - 3} \cdot \dfrac{8x^2 - 6x}{x^n + 2}$

85. $\dfrac{y^{2n} + 9}{10y} \cdot \dfrac{y^n - 3}{y^{4n} - 81}$

86. $\dfrac{y^{4n} - 16}{y^{2n} + 4} \cdot \dfrac{6y}{y^n + 2}$

87. $\dfrac{y^{2n} - y^n - 2}{2y^n - 4} \div \dfrac{y^{2n} - 1}{1 + y^n}$

88. $\dfrac{y^{2n} + 7y^n + 10}{10} \div \dfrac{y^{2n} + 4y^n + 4}{5y^n + 25}$

7.3 Adding and Subtracting Rational Expressions

OBJECTIVES

Tape IA 17

1. Add or subtract rational expressions with common denominators.
2. Identify the least common denominator of two or more rational expressions.
3. Add and subtract rational expressions with unlike denominators.

1 Rational expressions, like rational numbers, may be added or subtracted. We define the sum or difference of rational expressions in the same way that we defined the sum or difference of rational numbers.

> **Adding or Subtracting Rational Expressions with Common Denominators**
>
> If $\dfrac{P}{Q}$ and $\dfrac{R}{Q}$ are rational expressions, then
>
> $$\frac{P}{Q} + \frac{R}{Q} = \frac{P + R}{Q} \quad \text{and} \quad \frac{P}{Q} - \frac{R}{Q} = \frac{P - R}{Q}$$

To add or subtract rational expressions with common denominators, add or subtract the numerators and write the sum or difference over the common denominator.

EXAMPLE 1 Add or subtract.

a. $\dfrac{5}{7} + \dfrac{x}{7}$ **b.** $\dfrac{x}{4} + \dfrac{5x}{4}$

c. $\dfrac{x^2}{x + 7} - \dfrac{49}{x + 7}$ **d.** $\dfrac{x}{3y^2} - \dfrac{x + 1}{3y^2}$

Solution: We have common denominators, so add or subtract the numerators and place the sum or difference over the common denominator.

a. $\dfrac{5}{7} + \dfrac{x}{7} = \dfrac{5 + x}{7}$

b. $\frac{x}{4} + \frac{5x}{4} = \frac{x + 5x}{4} = \frac{6x}{4} = \frac{3x}{2}$

c. $\frac{x^2}{x + 7} - \frac{49}{x + 7} = \frac{x^2 - 49}{x + 7}$

Next write this rational expression in lowest terms.

$$\frac{x^2 - 49}{x + 7} = \frac{(x + 7)(x - 7)}{x + 7} = x - 7$$

d. $\frac{x}{3y^2} - \frac{x + 1}{3y^2} = \frac{x - (x + 1)}{3y^2}$ Subtract numerators.

$= \frac{x - x - 1}{3y^2}$ Apply the distributive property.

$= -\frac{1}{3y^2}$ Simplify. ■

2 To add or subtract rational expressions with unlike denominators, first write the rational expressions as equivalent rational expressions with common denominators.

The **least common denominator (LCD)** is usually the easiest common denominator to work with. The LCD of a list of rational expressions is a polynomial of least degree whose factors include all the denominator factors in the list.

Use the following steps to find the LCD.

To Find the Least Common Denominator (LCD)

Step 1 Factor each denominator completely.

Step 2 The LCD is the product of all unique factors formed in step 1, each raised to a power equal to the greatest number of times that the factor appears in any one factored denominator.

EXAMPLE 2 Find the LCD of the rational expressions in each list.

a. $\frac{2}{15x^5y^2}, \frac{3z}{5xy^3}$

b. $\frac{7}{z + 1}, \frac{z}{z - 1}$

c. $\frac{m - 1}{m^2 - 25}, \frac{2m}{2m^2 - 9m - 5}, \frac{7}{m^2 - 10m + 25}$

Solution: **a.** Factor each denominator.

$$15x^5y^2 = 3 \cdot 5 \cdot x^5 \cdot y^2$$

$$5xy^3 = 5 \cdot x \cdot y^3$$

The unique factors are 3, 5, x, and y.

The greatest number of times that 3 appears in one denominator is 1.

The greatest number of times that 5 appears in one denominator is 1.

The greatest number of times that x appears in one denominator is 5.

The greatest number of times that y appears in one denominator is 3.

The LCD is the product of $3^1 \cdot 5^1 \cdot x^5 \cdot y^3$, or $15x^5y^3$.

b. The denominators $z + 1$ and $z - 1$ do not factor further. Each factor appears once, so the

$$\text{LCD} = (z + 1)(z - 1)$$

c. First factor each denominator.

$$m^2 - 25 = (m + 5)(m - 5)$$
$$2m^2 - 9m - 5 = (2m + 1)(m - 5)$$
$$m^2 - 10m + 25 = (m - 5)(m - 5)$$

The LCD $= (m + 5)(2m + 1)(m - 5)^2$, which is the product of each unique factor raised to a power equal to the greatest number of times it appears in one factored denominator. ■

3 To add or subtract rational expressions with unlike denominators, follow the steps shown.

To Add or Subtract Rational Expressions with Unlike Denominators

Step 1 Find the LCD of the rational expressions.

Step 2 Write each rational expression as an equivalent rational expression whose denominator is the LCD found in step 1.

Step 3 Add or subtract numerators, and write the sum or difference over the common denominator.

Step 4 Write the answer in lowest terms.

EXAMPLE 3 Perform the indicated operation.

a. $\dfrac{2}{x^2y} + \dfrac{5}{3x^3y}$ **b.** $\dfrac{3z}{z + 2} + \dfrac{2z}{z - 2}$ **c.** $\dfrac{5k}{k^2 - 4} - \dfrac{2}{k^2 + k - 2}$

Solution: **a.** The LCD is $3x^3y$. Write each fraction as an equivalent fraction with denominator $3x^3y$.

$$\frac{2}{x^2y} + \frac{5}{3x^3y} = \frac{2 \cdot 3x}{x^2y \cdot 3x} + \frac{5}{3x^3y}$$
$$= \frac{6x}{3x^3y} + \frac{5}{3x^3y}$$
$$= \frac{6x + 5}{3x^3y} \qquad \text{Add the numerators.}$$

b. The LCD is the product of the two denominators: $(z + 2)(z - 2)$.

$$\frac{3z}{z + 2} + \frac{2z}{z - 2} = \frac{3z \cdot (z - 2)}{(z + 2) \cdot (z - 2)} + \frac{2z \cdot (z + 2)}{(z - 2) \cdot (z + 2)} \qquad \text{Write equivalent rational expressions.}$$
$$= \frac{3z(z - 2) + 2z(z + 2)}{(z + 2)(z - 2)} \qquad \text{Add.}$$
$$= \frac{3z^2 - 6z + 2z^2 + 4z}{(z + 2)(z - 2)}$$
$$= \frac{5z^2 - 2z}{(z + 2)(z - 2)} \qquad \text{Simplify the numerator.}$$

c. $\dfrac{5k}{k^2 - 4} - \dfrac{2}{k^2 + k - 2}$

$$= \frac{5k}{(k + 2)(k - 2)} - \frac{2}{(k + 2)(k - 1)}$$ Factor each denominator to find the LCD.

The LCD is $(k + 2)(k - 2)(k - 1)$. Write equivalent rational expressions with the LCD as denominators.

$$\frac{5k}{(k + 2)(k - 2)} - \frac{2}{(k + 2)(k - 1)} = \frac{5k(k - 1)}{(k + 2)(k - 2)(k - 1)} - \frac{2(k - 2)}{(k + 2)(k - 1)(k - 2)}$$

$$= \frac{5k(k - 1) - 2(k - 2)}{(k + 2)(k - 2)(k - 1)}$$ Subtract.

$$= \frac{5k^2 - 5k - 2k + 4}{(k + 2)(k - 2)(k - 1)}$$

$$= \frac{5k^2 - 7k + 4}{(k + 2)(k - 2)(k - 1)}$$ Simplify the numerator.

Since the numerator polynomial is prime, this rational expression is in lowest terms. ■

EXAMPLE 4 Perform the indicated operation.

a. $\dfrac{x}{x - 3} - 5$ **b.** $\dfrac{7}{x - y} + \dfrac{3}{y - x}$

Solution: **a.** $\dfrac{x}{x - 3} - 5 = \dfrac{x}{x - 3} - \dfrac{5}{1}$ Write 5 as $\dfrac{5}{1}$.

The LCD is $x - 3$.

$$= \frac{x}{x - 3} - \frac{5 \cdot (x - 3)}{1 \cdot (x - 3)}$$

$$= \frac{x - 5(x - 3)}{x - 3}$$ Subtract.

$$= \frac{x - 5x + 15}{x - 3}$$

$$= \frac{-4x + 15}{x - 3}$$ Simplify.

b. Notice that the denominators $x - y$ and $y - x$ are opposites of one another. To write equivalent rational expressions with the LCD, write one denominator such as $y - x$ as $-1(x - y)$.

$$\frac{7}{x - y} + \frac{3}{y - x} = \frac{7}{x - y} + \frac{3}{-1(x - y)}$$

$$= \frac{7}{x - y} + \frac{-3}{x - y}$$

$$= \frac{4}{x - y}$$ Add the numerators. ■

EXAMPLE 5 Add.

$$\frac{2x - 1}{2x^2 - 9x - 5} + \frac{x + 3}{6x^2 - x - 2}$$

Solution:

$$\frac{2x - 1}{2x^2 - 9x - 5} + \frac{x + 3}{6x^2 - x - 2} = \frac{2x - 1}{(2x + 1)(x - 5)} + \frac{x + 3}{(2x + 1)(3x - 2)}$$ Factor the denominators.

The LCD is $(2x + 1)(x - 5)(3x - 2)$.

$$= \frac{(2x - 1) \cdot (3x - 2)}{(2x + 1)(x - 5) \cdot (3x - 2)} + \frac{(x + 3) \cdot (x - 5)}{(2x + 1)(3x - 2) \cdot (x - 5)}$$

$$= \frac{6x^2 - 7x + 2}{(2x + 1)(x - 5)(3x - 2)} + \frac{x^2 - 2x - 15}{(2x + 1)(x - 5)(3x - 2)}$$

$$= \frac{6x^2 - 7x + 2 + x^2 - 2x - 15}{(2x + 1)(x - 5)(3x - 2)}$$ Add.

$$= \frac{7x^2 - 9x - 13}{(2x + 1)(x - 5)(3x - 2)}$$ Simplify the numerator.

The numerator polynomial is prime, and the rational expression is in lowest terms. ■

EXAMPLE 6 Perform the indicated operations.

$$\frac{7}{x - 1} + \frac{2(x + 1)}{(x - 1)^2} - \frac{2}{x^2}$$

Solution: The LCD is $x^2(x - 1)^2$.

$$\frac{7}{x - 1} + \frac{2(x + 1)}{(x - 1)^2} - \frac{2}{x^2} = \frac{7 \cdot x^2(x - 1)}{(x - 1) \cdot x^2(x - 1)} + \frac{2(x + 1) \cdot x^2}{(x - 1)^2 \cdot x^2} - \frac{2 \cdot (x - 1)^2}{x^2 \cdot (x - 1)^2}$$

$$= \frac{7x^3 - 7x^2}{x^2(x - 1)^2} + \frac{2x^3 + 2x^2}{x^2(x - 1)^2} - \frac{2x^2 - 4x + 2}{x^2(x - 1)^2}$$

$$= \frac{7x^3 - 7x^2 + 2x^3 + 2x^2 - 2x^2 + 4x - 2}{x^2(x - 1)^2}$$ Watch your signs!

$$= \frac{9x^3 - 7x^2 + 4x - 2}{x^2(x - 1)^2}$$ Add or subtract like terms in the numerator.

Since the numerator and denominator have no common factors, this rational expression is in lowest terms. ■

EXERCISE SET 7.3

Perform the indicated operation. Write each answer in lowest terms. See Example 1.

1. $\frac{2}{x} - \frac{5}{x}$

2. $\frac{4}{x^2} + \frac{2}{x^2}$

3. $\frac{2}{x - 2} + \frac{x}{x - 2}$

4. $\frac{x}{5-x}+\frac{2}{5-x}$

5. $\frac{x^2}{x+2}-\frac{4}{x+2}$

6. $\frac{4}{x-2}-\frac{x^2}{x-2}$

7. $\frac{2x-6}{x^2+x-6}+\frac{3-3x}{x^2+x-6}$

8. $\frac{5x+2}{x^2+2x-8}+\frac{2-4x}{x^2+2x-8}$

Find the LCD of the rational expressions in the list. See Example 2.

9. $\frac{2}{7}, \frac{3}{5x}$

10. $\frac{4}{5y}, \frac{3}{4y^2}$

11. $\frac{3}{x}, \frac{2}{x+1}$

12. $\frac{5}{2x}, \frac{7}{2+x}$

13. $\frac{12}{x+7}, \frac{8}{x-7}$

14. $\frac{1}{2x-1}, \frac{x}{2x+1}$

15. $\frac{5}{3x+6}, \frac{2x}{2x-4}$

16. $\frac{2}{3a+9}, \frac{5}{5a-15}$

17. $\frac{2a}{a^2-b^2}, \frac{1}{a^2-2ab+b^2}$

18. $\frac{2a}{a^2+8a+16}, \frac{7a}{a^2+a-12}$

Perform the indicated operation. Write each answer in lowest terms. See Example 3.

19. $\frac{4}{3x}+\frac{3}{2x}$

20. $\frac{10}{7x}-\frac{5}{2x}$

21. $\frac{5}{2y^2}-\frac{2}{7y}$

22. $\frac{4}{11x^4y}-\frac{1}{4x^2y^3}$

23. $\frac{x-3}{x+4}-\frac{x+2}{x-4}$

24. $\frac{x-1}{x-5}-\frac{x+2}{x+5}$

25. $\frac{1}{x-5}+\frac{x}{x^2-x-20}$

26. $\frac{x+1}{x^2-x-20}-\frac{2}{x+4}$

Perform the indicated operation. Write each answer in lowest terms. See Example 4.

27. $\frac{1}{a-b}+\frac{1}{b-a}$

28. $\frac{1}{a-3}-\frac{1}{3-a}$

29. $x+1+\frac{1}{x-1}$

30. $5-\frac{1}{x-1}$

31. $\frac{5}{x-2}+\frac{x+4}{2-x}$

32. $\frac{3}{5-x}+\frac{x+2}{x-5}$

Perform the indicated operation. Write each answer in lowest terms. See Example 5.

33. $\frac{y+1}{y^2-6y+8}-\frac{3}{y^2-16}$

34. $\frac{x+2}{x^2-36}-\frac{x}{x^2+9x+18}$

35. $\frac{x+4}{3x^2+11x+6}+\frac{x}{2x^2+x-15}$

36. $\frac{x+3}{5x^2+12x+4}+\frac{6}{x^2-x-6}$

37. $\frac{7}{x^2-x-2}+\frac{x}{x^2+4x+3}$

38. $\frac{a}{a^2+10a+25}+\frac{4}{a^2+6a+5}$

Perform the indicated operation. Write each answer in lowest terms. See Example 6.

39. $\frac{2}{x+1}-\frac{3x}{3x+3}+\frac{1}{2x+2}$

40. $\frac{5}{3x-6}-\frac{x}{x-2}+\frac{3+2x}{5x-10}$

41. $\frac{3}{x+3}+\frac{5}{x^2+6x+9}-\frac{x}{x^2-9}$

42. $\frac{x+2}{x^2-2x-3}+\frac{x}{x-3}-\frac{4}{x+1}$

Add or subtract as indicated. Write each answer in lowest terms.

43. $\dfrac{4}{3x^2y^3} + \dfrac{5}{3x^2y^3}$

44. $\dfrac{7}{2xy^4} + \dfrac{1}{2xy^4}$

45. $\dfrac{x-5}{2x} - \dfrac{x+5}{2x}$

46. $\dfrac{x+4}{4x} - \dfrac{x-4}{4x}$

47. $\dfrac{3}{2x+10} + \dfrac{8}{3x+15}$

48. $\dfrac{10}{3x-3} + \dfrac{1}{7x-7}$

49. $\dfrac{-2}{x^2-3x} - \dfrac{1}{x^3-3x^2}$

50. $\dfrac{-3}{2a+8} - \dfrac{8}{a^2+4a}$

51. $\dfrac{ab}{a^2-b^2} + \dfrac{b}{a+b}$

52. $\dfrac{x}{25-x^2} + \dfrac{2}{3x-15}$

53. $\dfrac{5}{x^2-4} - \dfrac{3}{x^2+4x+4}$

54. $\dfrac{3z}{z^2-9} - \dfrac{2}{3-z}$

55. $\dfrac{2}{a^2+2a+1} + \dfrac{3}{a^2-1}$

56. $\dfrac{9x+2}{3x^2-2x-8} + \dfrac{7}{3x^2+x-4}$

Perform the indicated operation. Write each answer in simplest form.

57. $\left(\dfrac{2}{3} - \dfrac{1}{x}\right) \cdot \left(\dfrac{3}{x} + \dfrac{1}{2}\right)$

58. $\left(\dfrac{2}{3} - \dfrac{1}{x}\right) \div \left(\dfrac{3}{x} + \dfrac{1}{2}\right)$

59. $\left(\dfrac{1}{x} + \dfrac{2}{3}\right) - \left(\dfrac{1}{x} - \dfrac{2}{3}\right)$

60. $\left(\dfrac{1}{2} + \dfrac{2}{x}\right) - \left(\dfrac{1}{2} - \dfrac{1}{x}\right)$

61. $\left(\dfrac{2a}{3}\right)^2 \div \left(\dfrac{a^2}{a+1} - \dfrac{1}{a+1}\right)$

62. $\left(\dfrac{x+2}{2x} - \dfrac{x-2}{2x}\right) \cdot \left(\dfrac{5x}{4}\right)^2$

63. $\left(\dfrac{2x}{3}\right)^2 \div \left(\dfrac{x}{3}\right)^2$

64. $\left(\dfrac{2x}{3}\right)^2 \cdot \left(\dfrac{3}{x}\right)^2$

65. $\dfrac{x}{x^2-9} + \dfrac{3}{x^2-6x+9} - \dfrac{1}{x+3}$

66. $\dfrac{3}{x^2-9} - \dfrac{x}{x^2-6x+9} + \dfrac{1}{x+3}$

67. $\left(\dfrac{x}{x+1} - \dfrac{x}{x-1}\right) \div \dfrac{x}{2x+2}$

68. $\dfrac{x}{2x+2} \div \left(\dfrac{x}{x+1} + \dfrac{x}{x-1}\right)$

69. $\dfrac{4}{x} \cdot \left(\dfrac{2}{x+2} - \dfrac{2}{x-2}\right)$

70. $\dfrac{1}{x+1} \cdot \left(\dfrac{5}{x} + \dfrac{2}{x-3}\right)$

Solve.

71. Find the perimeter and the area of the square.

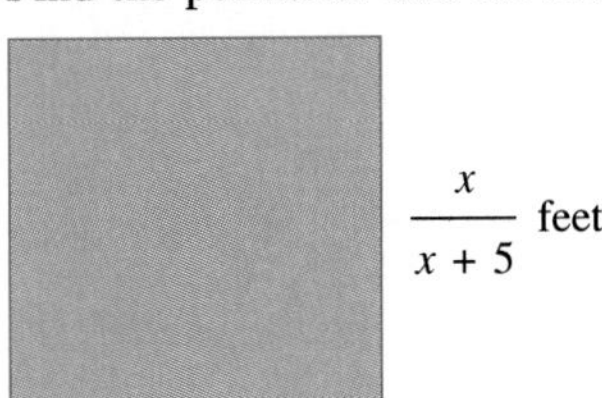

72. Find the perimeter of the quadrilateral.

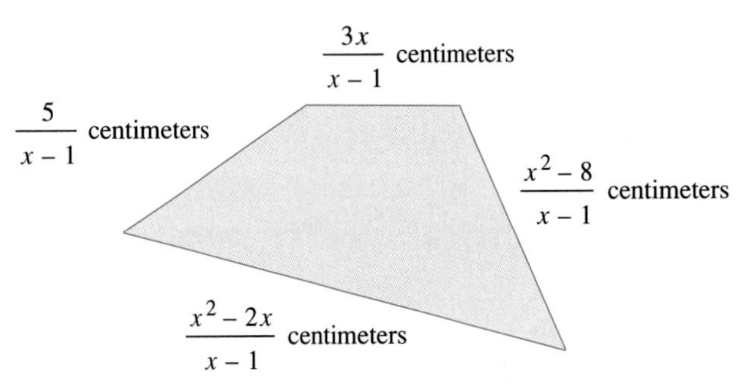

If $f(x) = \dfrac{3x}{x^2-4}$ *and* $g(x) = \dfrac{6}{x^2+2x}$, *find the following:*

73. $(f+g)(x)$

74. $(f \cdot g)(x)$

75. $\left(\dfrac{f}{g}\right)(x)$

76. $(f-g)(x)$

Skill Review

Use the distributive property to multiply the following. See Section 1.8.

77. $12\left(\frac{2}{3} + \frac{1}{6}\right)$

78. $14\left(\frac{1}{7} + \frac{3}{14}\right)$

79. $x^2\left(\frac{4}{x^2} + 1\right)$

80. $5y^2\left(\frac{1}{y^2} - \frac{1}{5}\right)$

Find each root. See Section 1.4.

81. $\sqrt{100}$

82. $\sqrt{25}$

83. $\sqrt[3]{8}$

84. $\sqrt[3]{27}$

85. $\sqrt[4]{81}$

86. $\sqrt[4]{16}$

Use the Pythagorean theorem to find each unknown length of a right triangle. See Section 6.6.

87.

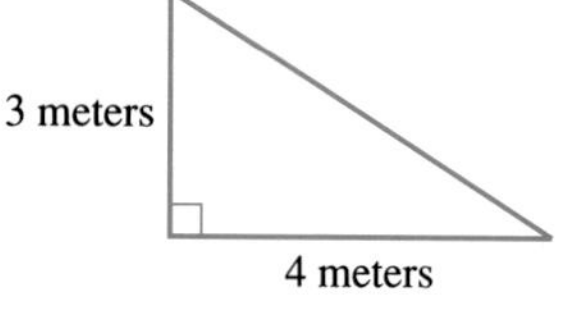

88.

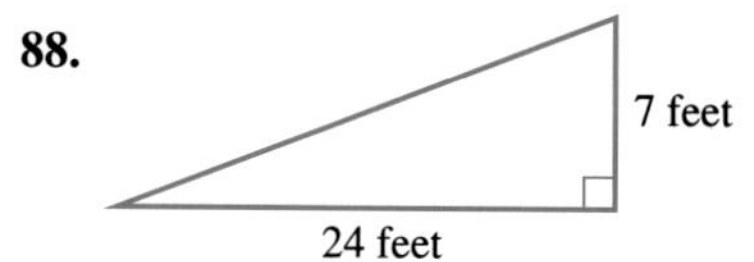

A Look Ahead

Perform the indicated operation. See the following example.

EXAMPLE Add $x^{-1} + 3x^{-2}$.

Solution:
$$x^{-1} + 3x^{-2} = \frac{1}{x} + \frac{3}{x^2}$$
$$= \frac{1 \cdot x}{x \cdot x} + \frac{3}{x^2}$$
$$= \frac{x}{x^2} + \frac{3}{x^2}$$
$$= \frac{x + 3}{x^2}$$ ■

89. $x^{-1} + (2x)^{-1}$

90. $3y^{-1} + (4y)^{-1}$

91. $4x^{-2} - 3x^{-1}$

92. $(4x)^{-2} - (3x)^{-1}$

93. $x^{-3}(2x + 1) - 5x^{-2}$

94. $4x^{-3} + x^{-4}(5x + 7)$

7.4 Simplifying Complex Fractions

OBJECTIVES

Tape IA 18

1. Identify complex fractions.
2. Simplify complex fractions by simplifying the numerator and denominator and then dividing.
3. Simplify complex fractions by multiplying by a common denominator.
4. Simplify expressions with negative exponents.

1 A fraction whose numerator, denominator, or both contain one or more rational expressions is called a **complex fraction.**

Examples of Complex Fractions

$$\frac{\frac{1}{a}}{\frac{b}{2}} \qquad \frac{\frac{x}{2y^2}}{\frac{6x-2}{9y}} \qquad \frac{x+\frac{1}{y}}{y+1}$$

The parts of a complex fraction are

$$\frac{\left.\frac{x}{y+2}\right\}}{\left.7+\frac{1}{y}\right\}}$$

← Numerator of complex fraction.
← Main fraction bar.
← Denominator of complex fraction.

2 Two methods for simplifying complex fractions are introduced. The first method evolves from the definition of a fraction as a quotient.

To Simplify a Complex Fraction: Method I

Step 1 Simplify the numerator and the denominator of the complex fraction so that each is a single fraction.

Step 2 Perform the indicated division by multiplying the numerator of the complex fraction by the reciprocal of the denominator of the complex fraction.

Step 3 Simplify if possible.

EXAMPLE 1 Simplify each complex fraction.

a. $\dfrac{\frac{2x}{27y^2}}{\frac{6x^2}{9}}$ **b.** $\dfrac{\frac{5x}{x+2}}{\frac{10}{x-2}}$ **c.** $\dfrac{x+\frac{1}{y}}{y+\frac{1}{x}}$

Solution: **a.** The numerator of the complex fraction is already a single fraction and so is the denominator. Perform the indicated division by multiplying the numerator $\frac{2x}{27y^2}$ by the reciprocal of the denominator $\frac{6x^2}{9}$. Then simplify.

$$\frac{\frac{2x}{27y^2}}{\frac{6x^2}{9}} = \frac{2x}{27y^2} \div \frac{6x^2}{9}$$

$$= \frac{2x}{27y^2} \cdot \frac{9}{6x^2} \qquad \text{Multiply by the reciprocal of } \frac{6x^2}{9}.$$

$$= \frac{2x \cdot 9}{27y^2 \cdot 6x^2}$$

$$= \frac{1}{9xy^2}$$

b. $\dfrac{\dfrac{5x}{x+2}}{\dfrac{10}{x-2}} = \dfrac{5x}{x+2} \cdot \dfrac{x-2}{10}$ Multiply by the reciprocal of $\dfrac{10}{x-2}$.

$= \dfrac{5x(x-2)}{2 \cdot 5(x+2)}$

$= \dfrac{x(x-2)}{2(x+2)}$ Simplify.

c. First simplify the numerator and the denominator of the complex fraction separately.

$\dfrac{x + \dfrac{1}{y}}{y + \dfrac{1}{x}} = \dfrac{\dfrac{x \cdot y}{1 \cdot y} + \dfrac{1}{y}}{\dfrac{y \cdot x}{1 \cdot x} + \dfrac{1}{x}}$ The LCD is y. The LCD is x.

$= \dfrac{\dfrac{xy+1}{y}}{\dfrac{yx+1}{x}}$ Add. Add.

$= \dfrac{xy+1}{y} \cdot \dfrac{x}{xy+1}$ Multiply by the reciprocal of $\dfrac{yx+1}{x}$.

$= \dfrac{x\,(xy+1)}{y\,(xy+1)}$

$= \dfrac{x}{y}$ ■

3 Next we look at another method for simplifying complex fractions. With this method we multiply the numerator and the denominator of the complex fraction by the LCD of all fractions in the complex fraction.

To Simplify a Complex Fraction: Method II

Step 1 Multiply the numerator and the denominator of the complex fraction by the LCD of the fractions in both the numerator and the denominator.

Step 2 Simplify.

EXAMPLE 2 Simplify each complex fraction.

a. $\dfrac{\dfrac{5x}{x+2}}{\dfrac{10}{x-2}}$ **b.** $\dfrac{x + \dfrac{1}{y}}{y + \dfrac{1}{x}}$

Solution: **a.** The least common denominator of $\frac{5x}{x+2}$ and $\frac{10}{x-2}$ is $(x+2)(x-2)$. Multiply both the numerator $\frac{5x}{x+2}$ and the denominator $\frac{10}{x-2}$ by the LCD.

$$\frac{\frac{5x}{x+2}}{\frac{10}{x-2}} = \frac{\left(\frac{5x}{x+2}\right)\cdot(x+2)(x-2)}{\left(\frac{10}{x-2}\right)\cdot(x+2)(x-2)} \quad \text{Simplify.}$$

$$= \frac{5x\cdot(x-2)}{2\cdot 5\cdot(x+2)} \quad \text{Simplify.}$$

$$= \frac{x(x-2)}{2(x+2)}$$

b. The least common denominator of $\frac{1}{y}$ and $\frac{1}{x}$ is xy.

$$\frac{x+\frac{1}{y}}{y+\frac{1}{x}} = \frac{\left(x+\frac{1}{y}\right)\cdot xy}{\left(y+\frac{1}{x}\right)\cdot xy}$$

$$= \frac{x\cdot xy + \frac{1}{y}\cdot xy}{y\cdot xy + \frac{1}{x}\cdot xy} \quad \text{Apply the distributive property.}$$

$$= \frac{x^2y + x}{xy^2 + y} \quad \text{Simplify.}$$

$$= \frac{x(xy+1)}{y(xy+1)} \quad \text{Factor.}$$

$$= \frac{x}{y} \quad \text{Simplify.} \quad ■$$

4

EXAMPLE 3 Simplify.

$$\frac{x^{-1} + 2xy^{-1}}{x^{-2} - x^{-2}y^{-1}}$$

Solution: This fraction does not appear to be a complex fraction. If we write it using only positive exponents, however, we see that it is a complex fraction.

$$\frac{x^{-1} + 2xy^{-1}}{x^{-2} - x^{-2}y^{-1}} = \frac{\frac{1}{x} + \frac{2x}{y}}{\frac{1}{x^2} - \frac{1}{x^2y}}$$

The LCD of $\frac{1}{x}$, $\frac{2x}{y}$, $\frac{1}{x^2}$, and $\frac{1}{x^2y}$ is x^2y. Multiply both the numerator and the denominator by x^2y.

$$= \frac{\left(\frac{1}{x} + \frac{2x}{y}\right) \cdot x^2y}{\left(\frac{1}{x^2} - \frac{1}{x^2y}\right) \cdot x^2y}$$

$$= \frac{\frac{1}{x} \cdot x^2y + \frac{2x}{y} \cdot x^2y}{\frac{1}{x^2} \cdot x^2y - \left(\frac{1}{x^2y}\right) \cdot x^2y}$$ Apply the distributive property.

$$= \frac{xy + 2x^3}{y - 1}$$ Simplify. ■

EXERCISE SET 7.4

Simplify each complex fraction. See Examples 1 and 2.

1. $\dfrac{\frac{1}{3}}{\frac{2}{5}}$ **2.** $\dfrac{\frac{3}{5}}{\frac{4}{5}}$ **3.** $\dfrac{\frac{4}{x}}{\frac{5}{2x}}$ **4.** $\dfrac{\frac{5}{2x}}{\frac{4}{x}}$

5. $\dfrac{\frac{10}{3x}}{\frac{5}{6x}}$ **6.** $\dfrac{\frac{15}{2x}}{\frac{5}{6x}}$ **7.** $\dfrac{1 + \frac{2}{5}}{2 + \frac{3}{5}}$ **8.** $\dfrac{2 + \frac{1}{7}}{3 - \frac{4}{7}}$

9. $\dfrac{\frac{4}{x - 1}}{\frac{x}{x - 1}}$ **10.** $\dfrac{\frac{x}{x + 2}}{\frac{2}{x + 2}}$ **11.** $\dfrac{1 - \frac{2}{x}}{x - \frac{4}{9x}}$ **12.** $\dfrac{5 - \frac{3}{x}}{x + \frac{2}{3x}}$

13. $\dfrac{\frac{1}{x + 1} - 1}{\frac{1}{x - 1} + 1}$ **14.** $\dfrac{1 + \frac{1}{x - 1}}{1 - \frac{1}{x + 1}}$

Simplify. See Example 3.

15. $\dfrac{x^{-1}}{x^{-2} + y^{-2}}$ **16.** $\dfrac{a^{-3} + b^{-1}}{a^{-2}}$ **17.** $\dfrac{2a^{-1} + 3b^{-2}}{a^{-1} - b^{-1}}$

18. $\dfrac{x^{-1} + y^{-1}}{3x^{-2} + 5y^{-2}}$ **19.** $\dfrac{1}{x - x^{-1}}$ **20.** $\dfrac{x^{-2}}{x + 3x^{-1}}$

Simplify.

21. $\dfrac{\frac{x + 1}{7}}{\frac{x + 2}{7}}$ **22.** $\dfrac{\frac{y}{10}}{\frac{x + 1}{10}}$ **23.** $\dfrac{\frac{1}{2} - \frac{1}{3}}{\frac{3}{4} + \frac{2}{5}}$ **24.** $\dfrac{\frac{5}{6} - \frac{1}{2}}{\frac{1}{3} + \frac{1}{8}}$

25. $\dfrac{\frac{x+1}{3}}{\frac{2x-1}{6}}$

26. $\dfrac{\frac{x+3}{12}}{\frac{4x-5}{15}}$

27. $\dfrac{\frac{x}{3}}{\frac{2}{x+1}}$

28. $\dfrac{\frac{x-1}{5}}{\frac{3}{x}}$

29. $\dfrac{\frac{2}{x}+3}{\frac{4}{x^2}-9}$

30. $\dfrac{2+\frac{1}{x}}{4x-\frac{1}{x}}$

31. $\dfrac{1-\frac{x}{y}}{\frac{x^2}{y^2}-1}$

32. $\dfrac{1-\frac{2}{x}}{x-\frac{4}{x}}$

33. $\dfrac{\frac{-2x}{x-y}}{\frac{y}{x^2}}$

34. $\dfrac{\frac{7y}{x^2+xy}}{\frac{y^2}{x^2}}$

35. $\dfrac{\frac{2}{x}+\frac{1}{x^2}}{\frac{y}{x^2}}$

36. $\dfrac{\frac{5}{x^2}-\frac{2}{x}}{\frac{1}{x}+2}$

37. $\dfrac{\frac{x}{9}-\frac{1}{x}}{1+\frac{3}{x}}$

38. $\dfrac{\frac{x}{4}-\frac{4}{x}}{1-\frac{4}{x}}$

39. $\dfrac{\frac{x-1}{x^2-4}}{1+\frac{1}{x-2}}$

40. $\dfrac{\frac{2}{x+5}+\frac{4}{x+3}}{\frac{3x+13}{x^2+8x+15}}$

41. $\dfrac{\frac{4}{5-x}+\frac{5}{x-5}}{\frac{2}{x}+\frac{3}{x-5}}$

42. $\dfrac{\frac{3}{x-4}-\frac{2}{4-x}}{\frac{2}{x-4}-\frac{2}{x}}$

43. $\dfrac{\frac{x+2}{x}-\frac{2}{x-1}}{\frac{x+1}{x}+\frac{x+1}{x-1}}$

44. $\dfrac{\frac{5}{a+2}-\frac{1}{a-2}}{\frac{3}{2+a}+\frac{6}{2-a}}$

45. $\dfrac{\frac{x-2}{x+2}+\frac{x+2}{x-2}}{\frac{x-2}{x+2}-\frac{x+2}{x-2}}$

46. $\dfrac{\frac{x-1}{x+1}-\frac{x+1}{x-1}}{\frac{x-1}{x+1}+\frac{x+1}{x-1}}$

47. $\dfrac{\frac{2}{y^2}-\frac{5}{xy}-\frac{3}{x^2}}{\frac{2}{y^2}+\frac{7}{xy}+\frac{3}{x^2}}$

48. $\dfrac{\frac{2}{x^2}-\frac{1}{xy}-\frac{1}{y^2}}{\frac{1}{x^2}-\frac{3}{xy}+\frac{2}{y^2}}$

49. $\dfrac{a^{-1}+1}{a^{-1}-1}$

50. $\dfrac{a^{-1}-4}{4+a^{-1}}$

51. $\dfrac{3x^{-1}+(2y)^{-1}}{x^{-2}}$

52. $\dfrac{5x^{-2}-3y^{-1}}{x^{-1}+y^{-1}}$

53. $\dfrac{2a^{-1}+(2a)^{-1}}{a^{-1}+2a^{-2}}$

54. $\dfrac{a^{-1}+2a^{-2}}{2a^{-1}+(2a)^{-1}}$

55. $\dfrac{5x^{-1}+2y^{-1}}{x^{-2}y^{-2}}$

56. $\dfrac{x^{-2}y^{-2}}{5x^{-1}+2y^{-1}}$

57. $\dfrac{5x^{-1}-2y^{-1}}{25x^{-2}-4y^{-2}}$

58. $\dfrac{3x^{-1}+3y^{-1}}{4x^{-2}-9y^{-2}}$

59. $(x^{-1}+y^{-1})^{-1}$

60. $\dfrac{xy}{x^{-1}+y^{-1}}$

61. $\dfrac{x}{1-\frac{1}{1+\frac{1}{x}}}$

62. $\dfrac{x}{1-\frac{1}{1-\frac{1}{x}}}$

In the study of calculus, the difference quotient $\dfrac{f(a+h)-f(a)}{h}$ *is often found and simplified. Find and simplify this quotient for each function* $f(x)$ *below by following steps a through d.*

a. *Find* $f(a+h)$.

b. *Find* $f(a)$.

c. *Use steps a and b to find* $\dfrac{f(a+h)-f(a)}{h}$.

d. *Simplify the result of step c.*

63. $f(x)=\dfrac{1}{x}$

64. $f(x)=\dfrac{5}{x}$

65. $\dfrac{3}{x+1}$

66. $\dfrac{2}{x^2}$

Solve.

67. In electronics, when two resistors R_1 and R_2 are connected in parallel, the total resistance is given by the complex fraction $\dfrac{1}{\dfrac{1}{R_1}+\dfrac{1}{R_2}}$. Simplify this fraction.

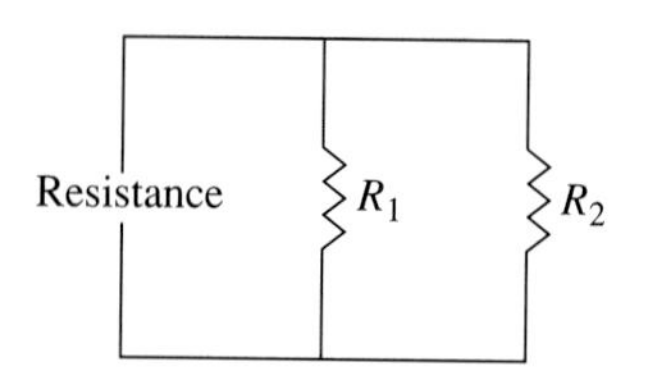

Skill Review

Solve each equation for x. See Sections 2.3 and 6.6.

68. $7x + 2 = x - 3$
69. $4 - 2x = 17 - 5x$
70. $x^2 = 4x - 4$
71. $5x^2 + 10x = 15$
72. $\dfrac{x}{3} - 5 = 13$
73. $\dfrac{2x}{9} + 1 = \dfrac{7}{9}$

Factor the following. See Sections 6.1 and 6.4.

74. $x^3 - 1$
75. $8y^3 + 1$
76. $125z^3 + 8$
77. $a^3 - 27$
78. $xy + 2x + 3y + 6$
79. $x^2 - x + xy - y$
80. $x^3 - 9x$
81. $2x^3 - 32x$

A Look Ahead

Simplify. See the following example.

EXAMPLE Simplify.

$$\frac{2(a+b)^{-1} - 5(a-b)^{-1}}{4(a^2-b^2)^{-1}}$$

Solution:

$$\frac{2(a+b)^{-1} - 5(a-b)^{-1}}{4(a^2-b^2)^{-1}} = \frac{\dfrac{2}{a+b} - \dfrac{5}{a-b}}{\dfrac{4}{a^2-b^2}}$$

$$= \frac{\left(\dfrac{2}{a+b} - \dfrac{5}{a-b}\right)\cdot(a+b)(a-b)}{\left[\dfrac{4}{(a+b)(a-b)}\right]\cdot(a+b)(a-b)}$$

$$= \frac{\dfrac{2}{a+b}\cdot(a+b)(a-b) - \dfrac{5}{a-b}\cdot(a+b)(a-b)}{\dfrac{4(a+b)(a-b)}{(a+b)(a-b)}}$$

$$= \frac{2(a-b) - 5(a+b)}{4}$$

$$= \frac{-3a-7b}{4} \quad \text{or} \quad -\frac{3a+7b}{4} \quad ■$$

82. $\dfrac{1}{1-(1-x)^{-1}}$

83. $\dfrac{1}{1+(1+x)^{-1}}$

84. $\dfrac{(x+2)^{-1}+(x-2)^{-1}}{(x^2-4)^{-1}}$

85. $\dfrac{(y-1)^{-1}-(y+4)^{-1}}{(y^2+3y-4)^{-1}}$

86. $\dfrac{3(a+1)^{-1}+4a^{-2}}{(a^3+a^2)^{-1}}$

87. $\dfrac{9x^{-1}-5(x-y)^{-1}}{4(x-y)^{-1}}$

7.5 Solving Equations Containing Rational Expressions

OBJECTIVE

1 Solve equations containing rational expressions.

Tape IA 18

1 To solve equations containing rational expressions, we first clear the equation of fractions by multiplying both sides of the equation by the LCD of all rational expressions.

EXAMPLE 1 Solve $\dfrac{8x}{5}+\dfrac{3}{2}=\dfrac{3x}{5}$.

Solution: The LCD of $\dfrac{8x}{5}, \dfrac{3}{2}$, and $\dfrac{3x}{5}$ is 10. Multiply both sides of the equation by 10.

$$\frac{8x}{5}+\frac{3}{2}=\frac{3x}{5}$$

$$10\left(\frac{8x}{5}+\frac{3}{2}\right)=10\left(\frac{3x}{5}\right) \quad \text{Multiply by the LCD.}$$

$$10\cdot\frac{8x}{5}+10\cdot\frac{3}{2}=10\cdot\frac{3x}{5} \quad \text{Apply the distributive property.}$$

$$16x+15=6x \quad \text{Simplify.}$$

$$15=-10x$$

$$-\frac{15}{10}=x \quad \text{or} \quad x=-\frac{3}{2} \quad \text{Solve.}$$

Verify this solution by substituting $-\dfrac{3}{2}$ for x in the original equation. The solution set is $\left\{-\dfrac{3}{2}\right\}$. ■

The important difference in the equations in this section is that the denominator of a rational expression may contain a variable. Recall that a rational expression is undefined for values of the variable that make the denominator 0. Thus, special precautions must be taken when an equation contains rational expressions with variables in the denominator. If a proposed solution makes the denominator 0, then it must be rejected as a solution. Such proposed solutions are called **extraneous solutions.**

EXAMPLE 2 Solve $\frac{3}{x} - \frac{x+21}{3x} = \frac{5}{3}$.

Solution: The LCD of denominators x, $3x$, and 3 is $3x$. Multiply both sides by $3x$.

$$\frac{3}{x} - \frac{x+21}{3x} = \frac{5}{3}$$

$$3x\left(\frac{3}{x} - \frac{x+21}{3x}\right) = 3x\left(\frac{5}{3}\right)$$

$$3x\left(\frac{3}{x}\right) - 3x\left(\frac{x+21}{3x}\right) = 3x\left(\frac{5}{3}\right) \quad \text{Apply the distributive property.}$$

$$9 - (x + 21) = 5x$$

$$9 - x - 21 = 5x$$

$$-12 = 6x$$

$$-2 = x \quad \text{Solve.}$$

The proposed solution is -2. Check the solution in the original equation.

$$\frac{3}{-2} - \frac{-2+21}{3(-2)} = \frac{5}{3}$$

$$-\frac{9}{6} + \frac{19}{6} = \frac{5}{3}$$

$$\frac{10}{6} = \frac{5}{3} \quad \text{True.}$$

The solution set is $\{-2\}$. ■

> **To Solve an Equation Containing Rational Expressions**
>
> *Step 1* Multiply both sides of the equation by the LCD of all rational expressions in the equation.
>
> *Step 2* Use the distributive property to remove any grouping symbols such as parentheses.
>
> *Step 3* Determine whether the equation is linear or quadratic and solve accordingly.
>
> *Step 4* Check the solution in the original equation.

EXAMPLE 3 Solve $\frac{x+6}{x-2} = \frac{2(x+2)}{x-2}$.

Solution: First multiply both sides of the equation by the LCD, $x - 2$.

$$\frac{x+6}{x-2} = \frac{2(x+2)}{x-2}$$

$$(x-2)\left(\frac{x+6}{x-2}\right) = (x-2)\left[\frac{2(x+2)}{x-2}\right] \quad \text{Multiply both sides by } x - 2.$$

$$x + 6 = 2(x + 2) \quad \text{Simplify.}$$

$$x + 6 = 2x + 4 \quad \text{Use the distributive property.}$$

$$x = 2 \quad \text{Solve.}$$

Now check the proposed solution 2 in the **original equation.**

$$\frac{2 + 6}{2 - 2} = \frac{2(2 + 2)}{2 - 2}$$

$$\frac{8}{0} = \frac{2(4)}{0}$$

Since the denominators are 0, 2 is an extraneous solution. There is no solution to the original equation. The solution set is ∅ or { }. ■

EXAMPLE 4 Solve $\frac{z}{2z^2 + 3z - 2} - \frac{1}{2z} = \frac{3}{z^2 + 2z}$.

Solution: Factor the denominators to find that the LCD is $2z(z + 2)(2z - 1)$. Multiply both sides by the LCD.

$$\frac{z}{2z^2 + 3z - 2} - \frac{1}{2z} = \frac{3}{z^2 + 2z}$$

$$\frac{z}{(2z - 1)(z + 2)} - \frac{1}{2z} = \frac{3}{z(z + 2)}$$

$$2z(z + 2)(2z - 1)\left[\frac{z}{(2z - 1)(z + 2)} - \frac{1}{2z}\right]$$

$$= 2z(z + 2)(2z - 1)\left[\frac{3}{z(z + 2)}\right]$$

$$2z(z + 2)(2z - 1)\left[\frac{z}{(2z - 1)(z + 2)}\right] - 2z(z + 2)(2z - 1)\left(\frac{1}{2z}\right)$$

$$= 2z(z + 2)(2z - 1)\left[\frac{3}{z(z + 2)}\right] \quad \text{Apply the distributive property.}$$

$$2z(z) - (z + 2)(2z - 1) = 3 \cdot 2(2z - 1) \quad \text{Simplify.}$$

$$2z^2 - (2z^2 + 3z - 2) = 12z - 6$$

$$2z^2 - 2z^2 - 3z + 2 = 12z - 6$$

$$-3z + 2 = 12z - 6$$

$$-15z = -8$$

$$z = \frac{8}{15} \quad \text{Solve.}$$

The proposed solution $\frac{8}{15}$ does not make any denominator 0; the solution set is $\left\{\frac{8}{15}\right\}$. ■

EXAMPLE 5 Solve $\frac{2x}{x-3} + \frac{6-2x}{x^2-9} = \frac{x}{x+3}$.

Solution: Factor the second denominator to find that the LCD is $(x+3)(x-3)$. Multiply both sides of the equation by $(x+3)(x-3)$. By the distributive property, this is the same as multiplying each term by $(x+3)(x-3)$.

$$\frac{2x}{x-3} + \frac{6-2x}{x^2-9} = \frac{x}{x+3}$$

$$(x+3)(x-3)\left(\frac{2x}{x-3}\right) + (x+3)(x-3)\left[\frac{6-2x}{(x+3)(x-3)}\right]$$

$$= (x+3)(x-3)\left(\frac{x}{x+3}\right)$$

$$2x(x+3) + (6-2x) = x(x-3) \quad \text{Simplify.}$$

$$2x^2 + 6x + 6 - 2x = x^2 - 3x \quad \text{Apply the distributive property.}$$

$$x^2 + 7x + 6 = 0$$

Next we solve this quadratic equation by the factoring method.

$$(x+6)(x+1) = 0 \quad \text{Factor.}$$

$$x = -6 \quad \text{or} \quad x = -1 \quad \text{Set each factor equal to 0.}$$

Neither -6 nor -1 makes any denominator 0. The solution set is $\{-6, -1\}$. ■

EXERCISE SET 7.5

Solve each equation. See Example 1.

1. $\frac{x}{2} - \frac{x}{3} = 12$

2. $x = \frac{x}{2} - 4$

3. $\frac{x}{3} = \frac{1}{6} + \frac{x}{4}$

4. $\frac{x}{2} = \frac{21}{10} - \frac{x}{5}$

Solve each equation. See Example 2.

5. $\frac{2}{x} + \frac{1}{2} = \frac{5}{x}$

6. $\frac{5}{3x} + 1 = \frac{7}{6}$

7. $\frac{x+3}{x} = \frac{5}{x}$

8. $\frac{4-3x}{2x} = -\frac{8}{2x}$

Solve each equation. See Example 3.

9. $\frac{x+5}{x+3} = \frac{8}{x+3}$

10. $\frac{5}{x-2} - \frac{2}{x+4} = -\frac{4}{x^2+2x-8}$

11. $\frac{1}{x-1} + \frac{1}{x+1} = \frac{2}{x^2-1}$

Solve each equation. See Example 4.

12. $\frac{1}{x-1} = \frac{2}{x+1}$

13. $\frac{6}{x+3} = \frac{4}{x-3}$

14. $\frac{1}{x-4} - \frac{3x}{x^2-16} = \frac{2}{x+4}$

15. $\frac{3}{2x+3} - \frac{1}{2x-3} = \frac{4}{4x^2-9}$

Solve each equation. See Example 5.

16. $\frac{1}{x-4} = \frac{8}{x^2-16}$

17. $\frac{2}{x^2-4} = \frac{1}{2x-4}$

18. $\frac{1}{x-2} - \frac{2}{x^2-2x} = 1$

19. $\frac{12}{3x^2+12x} = 1 - \frac{1}{x+4}$

Solve each equation.

20. $\frac{5}{x} = \frac{20}{12}$

21. $\frac{2}{x} = \frac{10}{5}$

22. $1 - \frac{4}{a} = 5$

23. $7 + \frac{6}{a} = 5$

24. $\frac{1}{2x} - \frac{1}{x+1} = \frac{1}{3x^2+3x}$

25. $\frac{2}{x-5} + \frac{1}{2x} = \frac{5}{3x^2-15x}$

26. $\frac{1}{x} - \frac{x}{25} = 0$

27. $\frac{x}{4} + \frac{5}{x} = 3$

28. $5 - \frac{2}{2y-5} = \frac{3}{2y-5}$

29. $1 - \frac{5}{y+7} = \frac{4}{y+7}$

30. $\frac{x-1}{x+2} = \frac{2}{3}$

31. $\frac{6x+7}{2x+9} = \frac{5}{3}$

32. $\frac{x+3}{x+2} = \frac{1}{x+2}$

33. $\frac{2x+1}{4-x} = \frac{9}{4-x}$

34. $\frac{1}{a-3} + \frac{2}{a+3} = \frac{1}{a^2-9}$

35. $\frac{12}{9-a^2} + \frac{3}{3+a} = \frac{2}{3-a}$

36. $\frac{64}{x^2-16} + 1 = \frac{2x}{x-4}$

37. $2 + \frac{3}{x} = \frac{2x}{x+3}$

38. $\frac{-15}{4y+1} + 4 = y$

39. $\frac{36}{x^2-9} + 1 = \frac{2x}{x+3}$

40. $\frac{28}{x^2-9} + \frac{2x}{x-3} + \frac{6}{x+3} = 0$

41. $\frac{x^2-20}{x^2-7x+12} = \frac{3}{x-3} + \frac{5}{x-4}$

42. $\frac{x+2}{x^2+7x+10} = \frac{1}{3x+6} - \frac{1}{x+5}$

43. $\frac{3}{2x-5} + \frac{2}{2x+3} = 0$

*Perform the indicated operation and simplify **or** solve the equation for the variable.*

44. $\frac{2}{x^2-4} = \frac{1}{x+2} - \frac{3}{x-2}$

45. $\frac{3}{x^2-25} = \frac{1}{x+5} + \frac{2}{x-5}$

46. $\frac{5}{x^2-3x} + \frac{4}{2x-6}$

47. $\frac{5}{x^2-3x} \div \frac{4}{2x-6}$

48. $\frac{x-1}{x+1} + \frac{x+7}{x-1} = \frac{4}{x^2-1}$

49. $\left(1 - \frac{y}{x}\right) \div \left(1 - \frac{x}{y}\right)$

50. $\frac{a^2-9}{a-6} \cdot \frac{a^2-5a-6}{a^2-a-6}$

51. $\frac{2}{a-6} + \frac{3a}{a^2-5a-6} - \frac{a}{5a+5}$

52. $\frac{2x+3}{3x-2} = \frac{4x+1}{6x+1}$

53. $\frac{5x-3}{2x} = \frac{10x+3}{4x+1}$

54. $\frac{a}{9a^2-1} + \frac{2}{6a-2}$

55. $\frac{3}{4a-8} - \frac{a+2}{a^2-2a}$

56. $\frac{-3}{x^2} - \frac{1}{x} + 2 = 0$

57. $\frac{x}{2x+6} + \frac{5}{x^2-9}$

58. $\frac{x-8}{x^2-x-2} + \frac{2}{x-2}$

59. $\frac{x-8}{x^2-x-2} + \frac{2}{x-2} = \frac{3}{x+1}$

60. $\frac{3}{a} - 5 = \frac{7}{a} - 1$

61. $\frac{7}{3z-9} + \frac{5}{z}$

Solve each equation. Begin by writing each equation with positive exponents only.

62. $x^{-2} - 19x^{-1} + 48 = 0$

63. $x^{-2} - 5x^{-1} - 36 = 0$

64. $p^{-2} + 4p^{-1} - 5 = 0$

65. $6p^{-2} - 5p^{-1} + 1 = 0$

Skill Review

Write each sentence as an equation and solve. See Section 2.5.

66. Four more than 3 times a number is 19.

67. The sum of two consecutive integers is 147.

68. The length of a rectangle is 5 inches more than the width. Its perimeter is 50 inches. Find the length and width.

69. The sum of a number and its reciprocal is $\frac{5}{2}$.

Simplify the following. See Section 1.4.

70. $-|-6| - (-5)$

71. $\sqrt{49} - (10 - 6)^2$

72. $|4 - 8| + (4 - 8)$

73. $(-4)^2 - 5^2$

The following is from a survey of state prisons. Use this histogram to answer Exercises 74 through 78. See Section 7.1.

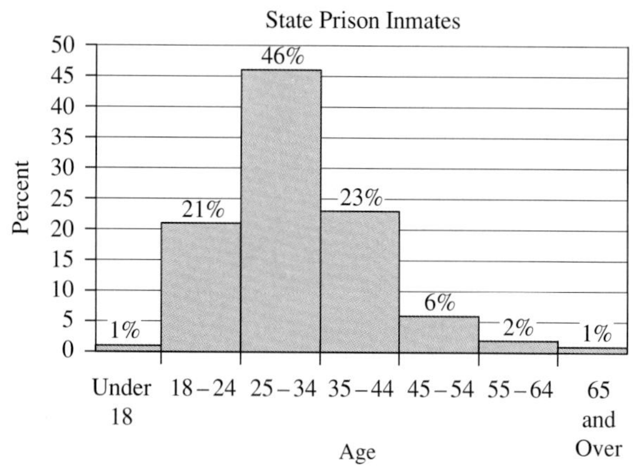

Source: U.S. Bureau of the Census, *Statistical Abstract of the United States: 1994,* 113th ed., Washington, DC, 1994.

74. What percent of state prison inmates are aged 45 to 54?

75. What percent of state prison inmates are 65 years old or older?

76. What age category shows the highest percent of prison inmates?

77. What percent of state prison inmates are 18 to 34 years old?

78. A state prison in Louisiana houses 2000 inmates. Approximately how many 25- to 34-year-old inmates might you expect this prison to hold?

A Look Ahead

Solve each equation by substitution. See the following example.

EXAMPLE Solve $\left(\frac{x}{x+1}\right)^2 - 7\left(\frac{x}{x+1}\right) + 10 = 0.$

Solution: Let $u = \frac{x}{x+1}$ and solve for u. Then substitute back and solve for x.

$$\left(\frac{x}{x+1}\right)^2 - 7\left(\frac{x}{x+1}\right) + 10 = 0$$

$$u^2 - 7u + 10 = 0 \qquad \text{Let } u = \frac{x}{x+1}.$$

$$(u - 5)(u - 2) = 0 \qquad \text{Factor.}$$

$$u = 5 \quad \text{or} \quad u = 2 \qquad \text{Solve.}$$

Since $u = \frac{x}{x+1}$, we have that $5 = \frac{x}{x+1}$ and $2 = \frac{x}{x+1}$. Thus, there are two rational equations to solve.

1.

$$5 = \frac{x}{x+1}$$
$$5 \cdot (x + 1) = x$$
$$5x + 5 = x$$
$$5 = -4x$$
$$x = -\frac{5}{4}$$

2.

$$2 = \frac{x}{x+1}$$
$$2 \cdot (x + 1) = x$$
$$2x + 2 = x$$
$$2 = -x$$
$$x = -2$$

Since neither $-\frac{5}{4}$ nor -2 makes the denominator 0, the solution set is $\left\{-\frac{5}{4}, -2\right\}$. ■

79. $(x - 1)^2 + 3(x - 1) + 2 = 0$

80. $(4 - x)^2 - 5(4 - x) + 6 = 0$

81. $\left(\frac{3}{x - 1}\right)^2 + 2\left(\frac{3}{x - 1}\right) + 1 = 0$

82. $\left(\frac{5}{2 + x}\right)^2 + \left(\frac{5}{2 + x}\right) - 20 = 0$

7.6 Rational Equations and Problem Solving

OBJECTIVES

1. Solve an equation containing rational expressions for a specified variable.
2. Solve problems by writing equations containing rational expressions.

Tape IA 19

1 In Section 2.4, we solved equations for a specified variable. In this section, we continue practicing this skill by solving equations containing rational expressions for a specified variable. The steps given in Section 2.4 for solving these equations are repeated here.

To Solve Equations for a Specified Variable

Step 1 Clear the equation of fractions or rational expressions by multiplying each side of the equation by the least common denominator (LCD) of all denominators in the equation.

Step 2 Use the distributive property to remove grouping symbols such as parentheses.

Step 3 Add or subtract like terms on the same side of the equation.

Step 4 Use the addition property of equality to rewrite the equation as an equivalent equation with terms containing the specified variable on one side and all other terms on the other side.

Step 5 Use the distributive property and the multiplication property of equality to isolate the specified variable.

EXAMPLE 1 Solve $\frac{1}{x} + \frac{1}{y} = \frac{1}{z}$ for x.

Solution: To clear this equation of fractions, multiply both sides of the equation by xyz, the LCD of $\frac{1}{x}$, $\frac{1}{y}$, and $\frac{1}{z}$.

$$\frac{1}{x} + \frac{1}{y} = \frac{1}{z}$$

$$xyz\left(\frac{1}{x} + \frac{1}{y}\right) = xyz\left(\frac{1}{z}\right)$$

$$xyz\left(\frac{1}{x}\right) + xyz\left(\frac{1}{y}\right) = xyz\left(\frac{1}{z}\right)$$

$$yz + xz = xy$$

Next subtract xz from both sides so that all terms containing the specified variable x are on one side of the equation and all other terms are on the other side.

$$yz = xy - xz$$

Use the distributive property to factor x from $xy - xz$ and then the multiplication property of equality to solve for x.

$$yz = x(y - z)$$

$$\frac{yz}{y - z} = x \quad \text{or} \quad x = \frac{yz}{y - z} \qquad \text{Divide both sides by } y - z.$$

■

2 Problem solving sometimes involves modeling a described situation with an equation containing rational expressions. In Examples 2 through 5, we practice solving such problems and use the problem-solving steps first introduced in Section 2.5.

EXAMPLE 2 Find the number that, when subtracted from the numerator and added to the denominator of $\frac{9}{19}$, changes $\frac{9}{19}$ into a fraction equivalent to $\frac{1}{3}$.

Solution: **1.** UNDERSTAND the problem. To do so, read and reread the problem and try guessing the solution. For example, if the unknown number is 3, we have the following:

$$\frac{9 - 3}{19 + 3} = \frac{1}{3}$$

To see if this is a true statement, simplify the fraction on the left.

$$\frac{6}{22} = \frac{1}{3}$$

or

$$\frac{3}{11} = \frac{1}{3}$$

when the fraction on the left is simplified further. Since this is not a true statement, 3 is not the correct number. Remember that the purpose of this step is not to guess the correct solution but to gain a better understanding of the problem posed.

2. ASSIGN a variable. Let n be the number to be subtracted from the numerator and added to the denominator.

3. ILLUSTRATE. No illustration is needed.

4. TRANSLATE the problem.

In words:	when the number is subtracted from the numerator and the denominator	this fraction is equivalent to	$\frac{1}{3}$
Translate:	$\frac{9 - n}{19 + n}$	$=$	$\frac{1}{3}$

5. COMPLETE the work. Here we solve the equation.

$$\frac{9 - n}{19 + n} = \frac{1}{3}$$

To solve for n, begin by multiplying both sides by the LCD, $3(19 + n)$.

$$3(19 + n) \cdot \left(\frac{9 - n}{19 + n}\right) = 3(19 + n)\left(\frac{1}{3}\right) \quad \text{Multiply by the LCD.}$$

$$3(9 - n) = 19 + n \quad \text{Simplify.}$$

$$27 - 3n = 19 + n$$

$$8 = 4n$$

$$2 = n \quad \text{Solve.}$$

6. INTERPRET the results. First *check* the stated problem. If we subtract 2 from the numerator and add 2 to the denominator of $\frac{9}{19}$, we have $\frac{9 - 2}{19 + 2} = \frac{7}{21} = \frac{1}{3}$, and the problem checks. Next, *state* the conclusions. The unknown number is 2. ■

EXAMPLE 3 The intensity $I(x)$ in foot-candles of light x feet from its source is given by the rational function

$$I(x) = \frac{320}{x^2}$$

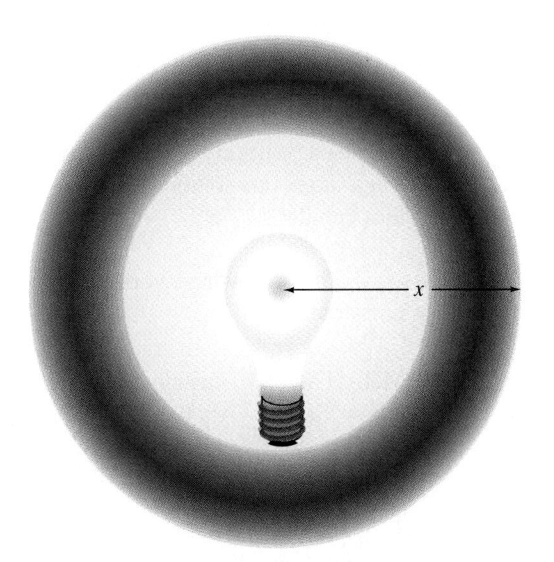

How far away is the source if the intensity of light is 5 foot-candles?

Solution: **1.** UNDERSTAND. Read and reread the problem. Since a function has been given that describes the relationship between $I(x)$ and x, we will replace x with a few values to help us become familiar with the function.

To find the intensity $I(x)$ of light 1 foot from the source, we find $I(1)$.

$$I(1) = \frac{320}{1^2} = \frac{320}{1} = 320 \text{ foot-candles}$$

To find the intensity $I(x)$ of light 4 feet from the source, we find $I(4)$.

$$I(4) = \frac{320}{4^2} = \frac{320}{16} = 20 \text{ foot-candles}$$

Notice that as distance from the light x increases, the intensity $I(x)$ decreases as expected. Steps 2 and 3 are not needed because a formula has been given.

4. TRANSLATE. We are given that the intensity $I(x)$ is 5 foot-candles, and we are asked to find how far away is the light source, x. To do so, let $I(x) = 5$.

$$I(x) = \frac{320}{x^2}$$

$$5 = \frac{320}{x^2} \qquad \text{Let } I(x) = 5.$$

5. COMPLETE. Here we solve the equation for x.

$$5 = \frac{320}{x^2}$$

$$x^2(5) = x^2\left(\frac{320}{x^2}\right) \qquad \text{Multiply both sides by } x^2.$$

$$5x^2 = 320 \qquad \text{Simplify.}$$

$$x^2 = 64 \qquad \text{Divide both sides by 5.}$$

Then, since $8^2 = 64$ and also $(-8)^2 = 64$, we have that

$$x = 8 \quad \text{or} \quad x = -8$$

6. INTERPRET. Since x represents distance and distance cannot be negative, the proposed solution -8 must be rejected. *Check* the solution 8 feet in the given formula. Then *state* the solution: The source of light is 8 feet away when the intensity is 5 foot-candles. ■

The following work example leads to an equation containing rational expressions.

EXAMPLE 4 Melissa Scarlatti can clean the house in 4 hours while her husband Zack can do the same job in 5 hours. They have agreed to clean together so that they can finish in time to watch a movie on TV that starts in 2 hours. How long will it take them to clean the house together? Can they finish before the movie starts?

Solution: **1.** UNDERSTAND. Read and reread the problem. The key idea here is the relationship between the **time** (hours) it takes to complete the job and the **part of the job** completed in 1 unit of time (hour). For example, if the **time** it takes Melissa to complete the job is 4 hours, the **part of the job** she can complete in 1 hour is $\frac{1}{4}$. Similarly, Zack can complete $\frac{1}{5}$ of the job in 1 hour.

2. ASSIGN. Let t represent the **time** in hours it takes Melissa and Zack to clean the house together. Then $\frac{1}{t}$ represents the **part of the job** they complete in 1 hour.

3. ILLUSTRATE. Here we summarize the information discussed above on a chart.

	Hours to Complete	**Part of Job** Completed in 1 Hour
Melissa	4	$\frac{1}{4}$
Zack	5	$\frac{1}{5}$
Together	t	$\frac{1}{t}$

4. TRANSLATE.

In words:	part of job Melissa completed in 1 hour	added to	part of job Zack completed in 1 hour	is equal to	part of job they completed together in 1 hour
Translate:	$\frac{1}{4}$	$+$	$\frac{1}{5}$	$=$	$\frac{1}{t}$

5. COMPLETE.

$$\frac{1}{4} + \frac{1}{5} = \frac{1}{t}$$

$$20t\left(\frac{1}{4} + \frac{1}{5}\right) = 20t\left(\frac{1}{t}\right) \quad \text{Multiply both sides by the LCD } 20t.$$

$$5t + 4t = 20$$

$$9t = 20$$

$$t = \frac{20}{9} \text{ or } 2\frac{2}{9} \quad \text{Solve.}$$

6. INTERPRET. *Check:* The proposed solution is $2\frac{2}{9}$. That is, Melissa and Zack take $2\frac{2}{9}$ hours to clean the house together. This proposed solution is reasonable since $2\frac{2}{9}$ hours is more than half of Melissa's time and less than half of Zack's time. Check this solution in the originally stated problem.

State: Can they finish before the movie starts? If they can clean the house together in $2\frac{2}{9}$ hours, they could not complete the job before the movie starts. ■

EXAMPLE 5 In his boat, Steve Deitmer takes $1\frac{1}{2}$ times as long to go 72 miles upstream as he does to return. If the boat cruises at 30 mph in still water, what is the speed of the current?

Solution: **1.** UNDERSTAND. Read and reread the problem. Next guess a solution. Suppose that the current is 4 mph. The speed of the boat upstream is then $30 - 4$ or 26 mph and the speed of the boat downstream is $30 + 4$ or 34 mph. Next let's find out how long it takes to travel 72 miles upstream and 72 miles downstream. To do so, we use the formula $d = r \cdot t$.

Upstream	*Downstream*
$d = r \cdot t$	$d = r \cdot t$
$72 = 26 \cdot t$	$72 = 34 \cdot t$
$\frac{72}{26} = t$	$\frac{72}{34} = t$
$2\frac{10}{13} = t$	$2\frac{2}{17} = t$

Since the time upstream $\left(2\frac{10}{13}\text{ hours}\right)$ is not $1\frac{1}{2}$ times the time downstream $\left(2\frac{2}{17}\text{ hours}\right)$, our guess is not correct. We do, however, have a better understanding of the problem.

2. ASSIGN. Since we are asked to find the speed of the current, **let x represent the current's speed.** The speed of the boat traveling downstream is made faster by the current and is represented by $30 + x$. The speed of the boat traveling upstream is made slower by the current and is represented by $30 - x$.

3. ILLUSTRATE. This information is summarized in the following chart.

	Distance	**Rate**	**Time** $\left(\frac{d}{r}\right)$
Upstream	72	$30 - x$	$\frac{72}{30 - x}$
Downstream	72	$30 + x$	$\frac{72}{30 + x}$

4. TRANSLATE. Since the time spent traveling upstream is $1\frac{1}{2}$ times the time spent traveling downstream, we have

In words:	time upstream	is	$1\frac{1}{2}$	times	time downstream
Translate:	$\frac{72}{30 - x}$	$=$	$\frac{3}{2}$	$\cdot$	$\frac{72}{30 + x}$

5. COMPLETE.

$$\frac{72}{30 - x} = \frac{3}{2} \cdot \frac{72}{30 + x}$$

Multiply both sides by the LCD $2(30 + x)(30 - x)$.

$$2(30 + x)(30 - x)\left(\frac{72}{30 - x}\right) = 2(30 + x)(30 - x)\left(\frac{3}{2} \cdot \frac{72}{30 + x}\right)$$

$$72 \cdot 2(30 + x) = 3 \cdot 72 \cdot (30 - x) \qquad \text{Simplify.}$$

$$\begin{aligned} 2(30 + x) &= 3(30 - x) && \text{Divide by 72.} \\ 60 + 2x &= 90 - 3x && \text{Simplify.} \\ 5x &= 30 && \\ x &= 6 && \text{Solve.} \end{aligned}$$

6. INTERPRET. *Check* this proposed solution of 6 mph in the originally stated problem. *State*: The current's speed is 6 mph. ■

EXERCISE SET 7.6

Solve each equation for the specified variable. See Example 1.

1. $F = \frac{9}{5}C + 32$; C

2. $V = \frac{1}{3}\pi r^2 h$; h

3. $\frac{1}{R} = \frac{1}{R_1} + \frac{1}{R_2}$; R

4. $\frac{1}{R} = \frac{1}{R_1} + \frac{1}{R_2}$; R_1

5. $S = \frac{n(a + L)}{2}$; n

6. $S = \frac{n(a + L)}{2}$; a

Solve. See Example 2.

7. The sum of a number and 5 times its reciprocal is 6. Find the number(s).

8. The quotient of a number and 9 times its reciprocal is 1. Find the number(s).

9. If a number is added to the numerator of $\frac{12}{41}$ and twice the number is added to the denominator of $\frac{12}{41}$, the resulting fraction is equivalent to $\frac{1}{3}$. Find the number.

10. If a number is subtracted from the numerator of $\frac{13}{8}$ and added to the denominator of $\frac{13}{8}$, the resulting fraction is equivalent to $\frac{2}{5}$. Find the number.

In electronics, the relationship among the combined resistance r of two resistors r_1 and r_2 wired in a parallel circuit is described by the formula $\frac{1}{r} = \frac{1}{r_1} + \frac{1}{r_2}$. Use this formula to solve Exercises 11 through 13. See Example 3.

11. If the combined resistance is 2 ohms and one of the two resistances is 3 ohms, find the other resistance.

12. Find the combined resistance of two resistors of 12 ohms each when they are wired in a parallel circuit.

13. The relationship among resistance and two resistors wired in a parallel circuit may be extended to three resistors, r_1, r_2, and r_3. Write an equation you believe may describe the relationship, and use it to find the combined resistance if r_1 is 5, r_2 is 6, and r_3 is 2.

Solve. See Example 4.

14. Alan Cantrell can word process a research paper in 6 hours. With Steve Isaac's help, the paper can be processed in 4 hours. Find how long it takes Steve to word process the paper alone.

15. An experienced roofer can roof a house in 26 hours. A beginning roofer needs 39 hours for the same job. Find how long it takes for the two to do the job working together.

16. A new printing press can print newspapers twice as fast as the old one. The old one can print the afternoon edition in 4 hours. Find how long it takes to print the afternoon edition if both printers are operating.

17. Three computers can do a sorting task in 20 minutes, 30 minutes, and 60 minutes, respectively. Find how long it takes to do the sorting task if all three computers run together.

Solve. See Example 5.

18. An F-100 plane and a car leave the same town at sunrise and head for a town 450 miles away. The speed of the plane is three times the speed of the car, and the plane arrives 6 hours ahead of the car. Find the speed of the car.

19. Mattie Evans drove 150 miles in the same amount of time that it took a turbopropeller plane to travel 600 miles. The speed of the plane was 150 mph faster than the speed of the car. Find the speed of the plane.

20. The speed of a boat in still water is 24 mph. If the boat travels 54 miles upstream in the same time that it takes to travel 90 miles downstream, find the speed of the current.

21. The speed of Lazy River's current is 5 mph. If a boat travels 20 miles downstream in the same time that it takes to travel 10 miles upstream, find the speed of the boat in still water.

Solve each equation for the specified variable.

22. $A = \dfrac{h(a + b)}{2}; b$

23. $A = \dfrac{h(a + b)}{2}; h$

24. $\dfrac{P_1V_1}{T_1} = \dfrac{P_2V_2}{T_2}; T_2$

25. $H = \dfrac{kA(T_1 - T_2)}{L}; T_2$

26. $f = \dfrac{f_1f_2}{f_1 + f_2}; f_2$

27. $I = \dfrac{E}{R + r}; r$

28. $\lambda = \dfrac{2L}{n}; L$

29. $S = \dfrac{a_1 - a_nr}{1 - r}; a_1$

30. $\dfrac{\theta}{\omega} = \dfrac{2L}{c}; c$

31. $F = \dfrac{-GMm}{r^2}; M$

Solve.

32. The sum of the reciprocals of two consecutive odd integers is $\dfrac{20}{99}$. Find the two integers.

33. The sum of the reciprocals of two consecutive integers is $-\dfrac{15}{56}$. Find the two integers.

34. If Sarah Clark can do a job in 5 hours and Dick Belli and Sarah working together can do the same job in 2 hours, find how long it takes Dick to do the job alone.

35. One hose can fill a goldfish pond in 45 minutes, and two hoses can fill the same pond in 20 minutes. Find how long it takes the second hose alone to fill the pond.

36. The speed of a bicyclist is 10 mph faster than the speed of a walker. If the bicyclist travels 26 miles in the same amount of time that the walker travels 6 miles, find the speed of the bicyclist.

37. Two trains leave at the same time going in opposite directions. One train travels 15 mph faster than the other. In 6 hours the trains are 630 miles apart. Find the speed of each.

38. The numerator of a fraction is 4 less than the denominator. If both the numerator and the denominator are increased by 2, the resulting fraction is equivalent to $\dfrac{2}{3}$. Find the fraction.

39. Fabio Casartelli, from Italy, won the individual road race in cycling during the 1992 Summer Olympics. An amateur cyclist rode the first 20-mile portion of his workout at a constant rate. For the 16-mile cooldown portion of his workout, he reduced his speed by 2 miles per hour. Each portion of the workout took equal time. Find the cyclist's rate during the first portion and his rate during the cool-down portion.

40. The denominator of a fraction is 1 more than the numerator. If both the numerator and the denominator are decreased by 3, the resulting fraction is equivalent to $\dfrac{4}{5}$. Find the fraction.

41. Moo Dairy has three machines to fill half-gallon milk cartons. The machines can fill the daily quota in 5 hours, 6 hours, and 7.5 hours, respectively. Find how long it takes to fill the daily quota if all three machines are running.

42. The inlet pipe of an oil tank can fill the tank in 1 hour 30 minutes. The outlet pipe can empty the tank in 1 hour. Find how long it takes to empty a full tank if both pipes are open.

43. A plane flies 465 miles with the wind and 345 miles against the wind in the same length of time. If the speed of the wind is 20 mph, find the speed of the plane in still air.

44. Two rockets are launched. The first travels at 9000 mph. Fifteen minutes later the second is launched at 10,000 mph. Find the distance at which both rockets are an equal distance from Earth.

45. Two joggers, one averaging 8 mph and one averaging 6 mph, start from a designated initial point. The slower jogger arrives at the end of the run a half-hour after the other jogger. Find the distance of the run.

46. A semitruck travels 300 miles through the flatland in the same amount of time that it travels 180 miles through the Great Smoky Mountains. The rate of the truck is 20 miles per hour slower in the mountains than in the flatland. Find both the flatland rate and mountain rate.

47. Smith Engineering found that an experienced surveyor surveys a roadbed in 4 hours. An apprentice surveyor needs 5 hours to survey the same stretch of road. If the two work together, find how long it takes them to complete the job.

48. An experienced bricklayer constructs a small wall in 3 hours. An apprentice completes the job in 6 hours. Find how long it takes if they work together.

49. A marketing manager travels 1080 miles in a corporate jet and then an additional 240 miles by car. If the car ride takes 1 hour longer, and if the rate of the jet is 6 times the rate of the car, find the time the manager travels by jet and find the time the manager travels by car.

50. Gary Marcus and Tony Alva work for Lombardo's Pipe and Concrete. Mr. Lombardo is preparing an estimate for a customer. He knows that Gary lays a slab of concrete in 6 hours. Tony lays the same size slab in 4 hours. If both work on the job and the cost of labor is $45.00 per hour, decide what the labor estimate should be.

51. In 2 minutes, a conveyor belt moves 300 pounds of recyclable aluminum from the delivery truck to a storage area. A smaller belt moves the same quantity of cans the same distance in 6 minutes. If both belts are used, find how long it takes to move the cans to the storage area.

52. Mr. Dodson can paint his house by himself in four days. His son needs an additional day to complete the job if he works by himself. If they work together, find how long it takes to paint the house.

53. While road testing a new make of car, the editor of a consumer magazine finds that she can go 10 miles into a 3-mile-per-hour wind in the same amount of time that she can go 11 miles with a 3 miles per hour wind behind her. Find the speed of the car in still air.

54. The world record for the largest white bass caught is held by Ronald Sprouse from Virginia. The bass weighed 6 pounds 13 ounces. If Ronald rows 9 miles downstream in the same amount of time that he rows 3 miles upstream and if the current is 6 miles per hour, find how long it takes him to cover the 12 miles.

Skill Review

Solve the equation for x. See Section 3.1.

55. $\frac{x}{5} = \frac{x + 2}{3}$

56. $\frac{x}{4} = \frac{x + 3}{6}$

57. $\frac{x - 3}{2} = \frac{x - 5}{6}$

58. $\frac{x - 6}{4} = \frac{x - 2}{5}$

Factor the following. See Section 6.5.

59. $2x^3 - 9x^2 - 18x$

60. $5yx^2 - 45y$

61. $x^3 + 8$

62. $y^3 - 27$

Use the circle graph to answer Exercises 63 through 68.

How long have you worn your hair in its current style?

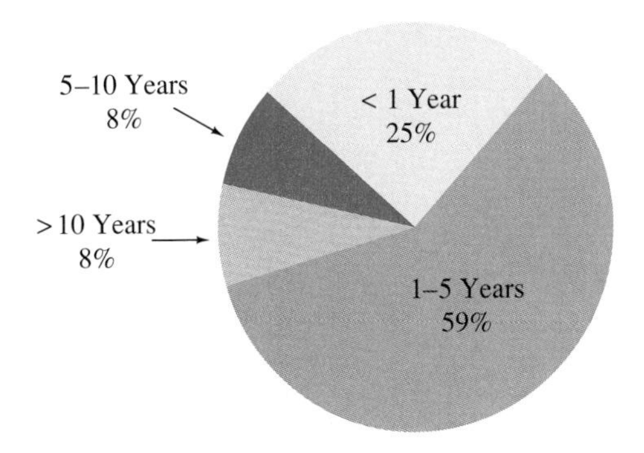

Source: *Self Magazine,* June, 1993.

63. What percent of those polled have worn their hair in its current style for less than 1 year?

64. What percent of those polled have worn their hair in its current style for more than 10 years?

65. What percent of those polled have worn their hair in its current style for 5 years or less?

66. What percent of those polled have worn their hair in its current style for 5 years or more?

67. Poll your algebra class with this same question and compute the percents in each category.

68. Use the results of Exercise 67 and construct a circle graph. (Hint: Recall that the number of degrees around a circle is 360°. Then, for example, the number of degrees in a sector representing 8% of a whole should be $0.08(360°) = 28.8°$.)

7.7 Variation and Problem Solving

OBJECTIVES

Tape IA 19

1. Write an equation expressing direct variation.
2. Write an equation expressing inverse variation.
3. Write an equation expressing joint variation.

In this section, we solve problems that can be modeled by using the concept of direct variation, inverse variation, or joint variation.

1 A very familiar example of direct variation is the relationship of the circumference C of a circle to its radius r. The formula $C = 2\pi r$ expresses that the circumference is always 2π times the radius. In other words, C is always a constant multiple (2π) of r. Because it is, we say that **C varies directly as r** or that **C is directly proportional to r.**

Direct Variation

y* varies directly as *x or ***y* is directly proportional to *x*** if there is a nonzero constant k such that

$$y = kx$$

The number k is called the **constant of variation** or the **constant of proportionality.**

Recall that the relationship described between x and y is a linear one. In other words, the graph of $y = kx$ is a line. The slope of the line is k, and the line passes through the origin.

The graph of our direct variation example is shown below. The horizontal axis represents the radius, r, and the vertical axis is the circumference, C.

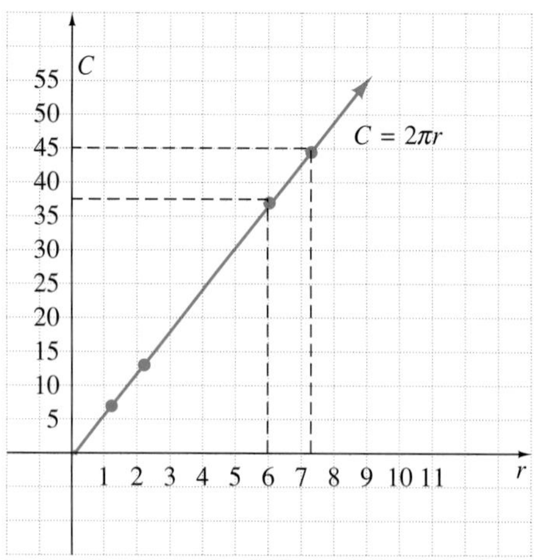

From the graph we can read that when the radius is 6 units, the circumference is approximately 38 units. Also, when the circumference is 45 units, the radius is between 7 and 8 units. Notice that as the radius increases, the circumference increases.

EXAMPLE 1 Suppose that y varies directly as x. If y is 5 when x is 30, find the constant of variation. Also, find y when $x = 90$.

Solution: Since y varies directly as x, we write $y = kx$. If $y = 5$ when $x = 30$, we have that

$$y = kx$$

$$5 = k(30) \quad \text{Replace } y \text{ with 5 and } x \text{ with 30.}$$

$$\frac{1}{6} = k \quad \text{Solve for } k.$$

The constant of variation is $\frac{1}{6}$.

After finding the constant of variation k, the direct variation equation can be written as $y = \frac{1}{6}x$. Next find y when x is 90.

$$y = \frac{1}{6}x$$

$$= \frac{1}{6}(90) \quad \text{Let } x = 90.$$

$$= 15$$

When $x = 90$, y must be 15. ■

Notice that the direct variation equation $y = kx$ is not only a linear equation, but y is also a function of x.

EXAMPLE 2 Hooke's law states that the distance a spring stretches is directly proportional to the weight attached to the spring. If a 40-pound weight attached to the spring stretches the spring 5 inches, find the distance that a 65-pound weight attached to the spring stretches the spring.

Solution: **1.** UNDERSTAND. Read and reread the problem. Notice that we are given that the distance a spring stretches is **directly proportional** to the weight attached.

2. ASSIGN. Let d represent the distance stretched and let w represent the weight attached. The constant of variation is represented by k.

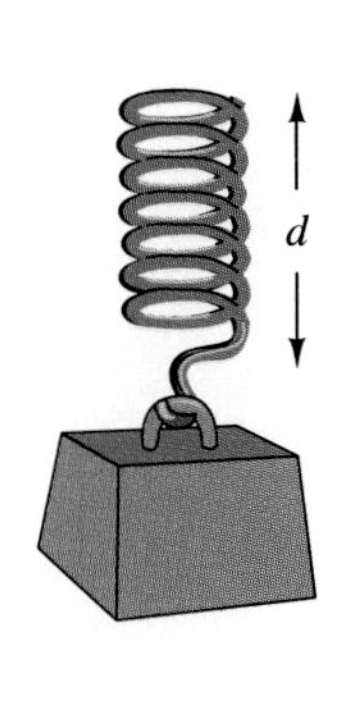

4. TRANSLATE. Because d is directly proportional to w, we write

$$d = kw$$

When 40 pounds is attached, the spring stretches 5 inches. That is, when $w = 40$, $d = 5$.

$$5 = k(40) \quad \text{Replace by the known values.}$$

$$\frac{1}{8} = k \quad \text{Solve for } k.$$

Thus,

$$d = \frac{1}{8}w$$

To find the stretch when a weight of 65 pounds is attached, replace w with 65, find d.

$$d = \frac{1}{8}(65)$$

5. COMPLETE.

$$d = \frac{1}{8}(65)$$

$$= \frac{65}{8} = 8\frac{1}{8} \quad \text{or} \quad 8.125$$

6. INTERPRET. *Check* the proposed solution of 8.125 inches. *State:* The spring stretches 8.125 inches when a 65-pound weight is attached. ■

2 When y is proportional to the **reciprocal** of another variable x, we say that y **varies inversely as** x, or that y **is inversely proportional to** x. An example of the inverse variation relationship is the relationship between the pressure that a gas exerts and the volume of its container. As the volume of a container decreases, the pressure of the gas it contains increases.

Inverse Variation

y **varies inversely as** x or y **is inversely proportional to** x if there is a nonzero constant k such that

$$y = \frac{k}{x}$$

The number k is called the **constant of variation** or the **constant of proportionality.**

EXAMPLE 3 Suppose that u varies inversely as w. If u is 3 when w is 5, find u when w is 30.

Solution: Since u varies inversely as w, we have $u = \frac{k}{w}$. Let $u = 3$, $w = 5$, and solve for k.

$$u = \frac{k}{w}$$

$$3 = \frac{k}{5} \quad \text{Let } u = 3 \text{ and } w = 5.$$

$$15 = k \quad \text{Multiply both sides by 5 or cross multiply.}$$

The constant of variation k is 15. This gives the inverse variation equation:

$$u = \frac{15}{w}$$

Now find u when $w = 30$.

$$u = \frac{15}{30} \quad \text{Let } w = 30.$$

$$= \frac{1}{2}$$

Thus, when $w = 30$, $u = \frac{1}{2}$. ■

EXAMPLE 4 Boyle's law says that, if the temperature stays the same, the pressure P of a gas is inversely proportional to the volume V. If a cylinder in a steam engine has a pressure of 960 kilopascals when the volume is 1.4 cubic meters, find the pressure when the volume increases to 2.5 cubic meters.

Solution: **1.** UNDERSTAND. Read and reread the problem. Notice that we are given that the pressure of a gas is **inversely proportional** to the volume.

2. ASSIGN. Let P represent the pressure and let V represent the volume. The constant of variation is represented by k.

4. TRANSLATE. Because P is inversely proportional to V, we write

$$P = \frac{k}{V}$$

When $P = 960$ kilopascals, the volume $V = 1.4$ cubic meters. Use this information to find k.

$$960 = \frac{k}{1.4} \qquad \text{Let } P = 960 \text{ and } V = 1.4.$$

$$1344 = k \qquad \text{Cross multiply.}$$

Thus, the value of k is 1344. Replace k by 1344 in the variation equation:

$$P = \frac{1344}{V}$$

Next find P when V is 2.5 cubic meters.

$$P = \frac{1344}{2.5} \qquad \text{Let } V = 2.5.$$

5. COMPLETE.

$$P = \frac{1344}{2.5}$$

$$P = 537.6$$

6. INTERPRET. *Check* the proposed solution. *State:* When the volume is 2.5 cubic meters, the pressure is 537.6 kilopascals. ■

3 Sometimes the ratio of a variable to the product of many other variables is constant. For example, the ratio of distance traveled to the product of speed and time traveled is constantly 1.

$$\frac{d}{rt} = 1 \quad \text{or} \quad d = rt$$

Such a relationship is called **joint variation.**

Joint Variation

If the ratio of a variable y to the product of two or more variables is constant, then **y varies jointly as** or **is jointly proportional to** the other variables. If

$$y = kxz$$

then the number k is the **constant of variation** or the **constant of proportionality.**

EXAMPLE 5 The lateral surface area of a cylinder varies jointly as its radius and height. Express surface area S in terms of radius r and height h.

Solution: Because the surface area varies jointly as the radius r and the height h, we equate S to the constant multiple, of r and h:

$$S = krh$$

EXERCISE SET 7.7

Write each statement as an equation. See Examples 1 through 5.

1. A is directly proportional to B.
2. C varies inversely as D.
3. X is inversely proportional to Z.
4. G varies directly with M.
5. N varies directly with the square of P.
6. A varies jointly with D and E.
7. T is inversely proportional to R.
8. G is inversely proportional to H.
9. P varies directly with R.
10. T is directly proportional to S.

Solve. See Examples 1 and 2.

11. A varies directly as B. If A is 60 when B is 12, find A when B is 9.
12. C varies directly as D. If C is 42 when D is 14, find C when D is 6.
13. Charles's law states that, if the pressure P stays the same, the volume V of a gas is directly proportional to its temperature T. If a balloon is filled with 20 cubic meters of a gas at a temperature of 300 K, find the new volume if the temperature rises to 360 K while the pressure stays the same.
14. The amount P of pollution varies directly with the population N of people. Kansas City has a population of 450,000 and produces 260,000 tons of pollutants. Find how many tons of pollution we should expect St. Louis to produce, if we know that its population is 980,000.

Solve. See Examples 3 and 4.

15. H is inversely proportional to J. If H is 4 when J is 5, find H when J is 2.
16. D varies inversely as A. If D is 16 when A is 2, find D when A is 8.
17. If the voltage V in an electric circuit is held constant, the current I is inversely proportional to the resistance R. If the current is 40 amperes when the resistance is 270 ohms, find the current when the resistance is 150 ohms.
18. Pairs of markings a set distance apart are made on highways so that police can detect drivers exceeding the speed limit. Over a fixed distance, the speed R varies inversely with the time T. In one particular pair of markings, R is 45 mph when T is 6 seconds. Find the speed of a car that travels the given distance in 5 seconds.

Write each statement as an equation. See Example 5.

19. x varies jointly as y and z.
20. P varies jointly as R and the square of S.
21. r varies jointly as s and the cube of t.
22. a varies jointly as b and c.

Solve.

23. Q is directly proportional to R. If Q is 4 when R is 20, find Q when R is 35.
24. S is directly proportional to T. If S is 4 when T is 16, find S when T is 40.
25. M varies directly with P. If M is 8 when P is 20, find M when P is 24.
26. F is directly proportional to G. If F is 18 when G is 10, find F when G is 16.

27. B is inversely proportional to C. If B is 12 when C is 3, find B when C is 18.

28. U varies inversely with V. If U is 14 when V is 4, find U when V is 7.

29. W varies inversely with X. If W is 18 when X is 6, find W when X is 40.

30. Z is inversely proportional to Y. If Z is 9 when Y is 6, find Z when Y is 24.

31. The weight of a synthetic ball varies directly with the cube of its radius. A ball with a radius of 2 inches weighs 1.20 pounds. Find the weight of a ball of the same material with a 3-inch radius.

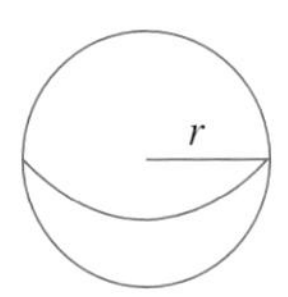

32. At sea, the distance to the horizon is directly proportional to the square root of the elevation of the observer. If a person who is 36 feet above the water can see 7.4 miles, find how far a person 64 feet above the water can see.

33. Because it is more efficient to produce larger numbers of items, the cost of producing Dysan computer disks is inversely proportional to the number produced. If 4000 can be produced at a cost of $1.20 each, find the cost per disk when 6000 are produced.

34. The weight of an object on or above the surface of the earth varies inversely as the square of the distance between the object and the center of Earth. If a person weighs 160 pounds on the surface of Earth, find the individual's weight if he moves 200 miles above Earth. (Assume that the radius of Earth is 4000 miles.)

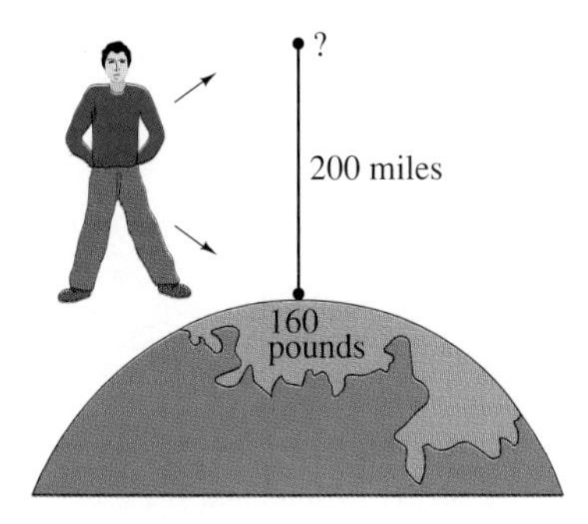

35. The number of cars manufactured on an assembly line at a General Motors plant varies jointly as the number of workers and the time they work. If 200 workers can produce 60 cars in 2 hours, find how many cars 240 workers should be able to make in 3 hours.

36. The volume of a cone varies jointly as the square of its radius and its height. If the volume of a cone is 32π when the radius is 4 inches and the height is 6 inches, find the volume of a cone when the radius is 3 inches and the height is 5 inches.

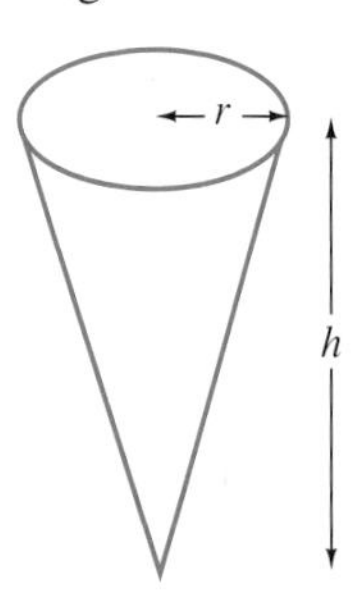

37. When a wind blows perpendicularly against a flat surface, its force is jointly proportional to the surface area and the velocity of the wind. A sail whose surface area is 12 square feet experiences a 20-pound force when the wind speed is 10 miles per hour. Find the force on an 8-square-foot sail if the wind speed is 12 miles per hour.

38. The horsepower that can be safely transmitted to a shaft varies jointly as its angular speed of rotation (in revolutions per minute) and the cube of its diameter. A 2-inch shaft making 120 revolutions per minute safely transmits 40 horsepower. Find how much horsepower can be safely transmitted by a 3-inch shaft making 80 revolutions per minute.

39. A circular column has a *safe load* that is directly proportional to the fourth power of its diameter and inversely proportional to the square of its length. An 8-inch pillar 10 feet long can safely support a 16-ton load. Find the load that a 6-inch pillar made of the same material can support, if it is 8 feet long.

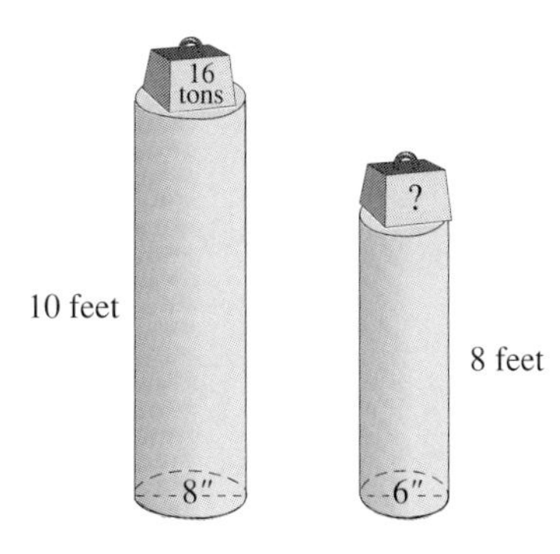

40. The maximum safe load for a rectangular beam varies jointly as its width and the square of its height and inversely as its length. If a beam 6 inches wide, 4 inches high, and 10 feet long supports 12 tons, find how much a similar beam can support if the beam is 8 inches wide, 5 inches high, and 16 feet long.

41. The area of a circle is directly proportional to the square of its radius. If the radius is tripled, determine how the area changes.

42. The horsepower to drive a boat varies directly as the cube of the speed of the boat. If the speed of the boat is to double, determine the corresponding increase in horsepower required.

43. The intensity I of light varies inversely as the square of the distance d from the light source. If the distance from the light source is doubled (see figure), determine what happens to the intensity of light at the new location.

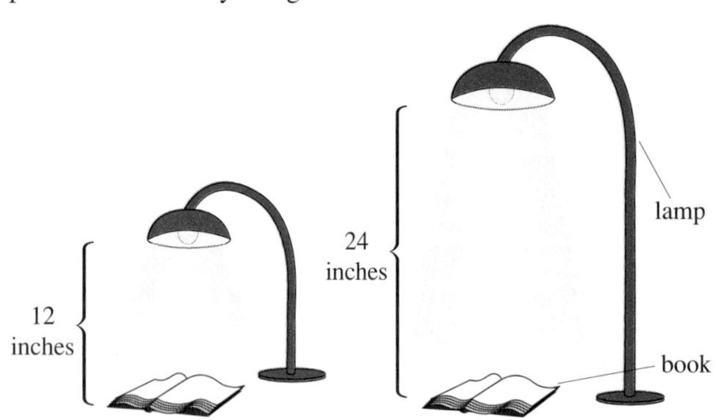

44. The volume of a cylinder varies jointly as the height and the square of the radius. If the height is halved and the radius is doubled, determine what happens to the volume.

Skill Review

Find the circumference and area of each circle.

45.

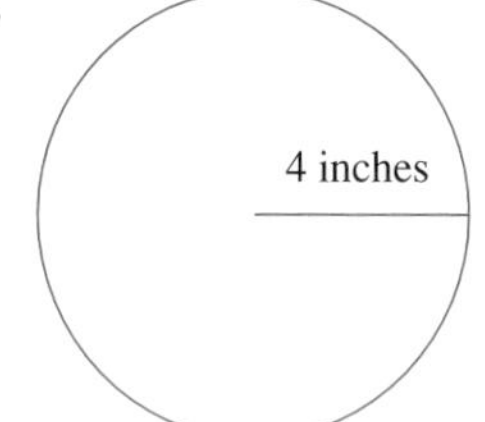

46.

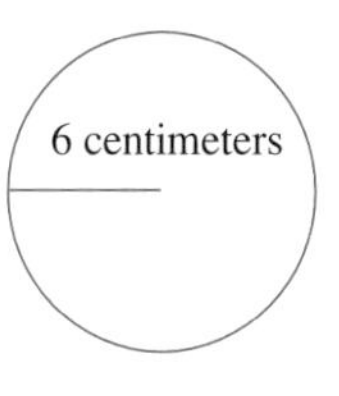

47.

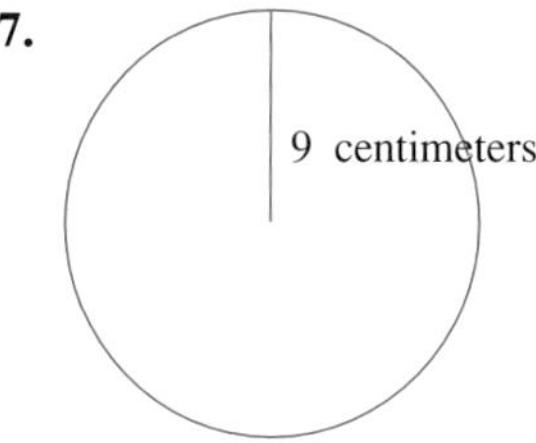

48.

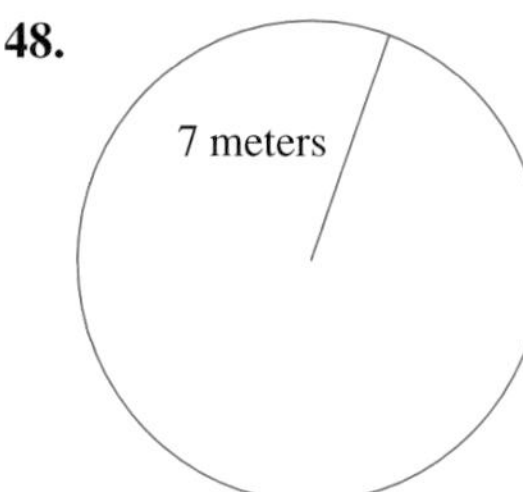

Find the slope of the line containing each pair of points. See Section 4.4.

49. $(-5, -2,), (0, 7)$ **50.** $(3, 6), (-2, 6)$ **51.** $(2, 1), (2, -3)$ **52.** $(4, -1), (5, -2)$

Graph each function. See Sections 4.1 and 4.2.

53. $f(x) = 2x - 3$ **54.** $f(x) = x + 4$ **55.** $g(x) = |x|$ **56.** $h(x) = |x| + 2$

57. $h(x) = x^2$ **58.** $f(x) = x^2 - 1$

Writing in Mathematics

59. It has been said that, in the long run, lifetime income is directly proportional to the number of years of education an individual has. Explain what is meant by this statement.

60. The number of careless errors found in a production run is inversely proportional to the amount of time needed to perform the job. Explain what is meant by this statement.

CHAPTER 7 GLOSSARY

A fraction whose numerator, denominator, or both contains one or more rational expressions is called a **complex fraction.**

A proposed solution that makes the denominator of a rational expression 0 is called an **extraneous solution.**

The **least common denominator (LCD)** of a list of rational expressions is a polynomial of least degree whose factors include all the denominator factors in the list.

A **rational expression** is the quotient $\frac{P}{Q}$ of two polynomials P and Q as long as Q is not 0.

A rational expression is **undefined** if the denominator is 0.

We say that **y varies directly as x** or **y is directly proportional to x** if there is a nonzero constant k such that $y = kx$. The number k is called the **constant of variation** or the **constant of proportionality.**

The variable **y varies inversely as x** or **y is inversely proportional to x** if there is a nonzero constant k such that $y = \frac{k}{x}$. The number k is called the **constant of variation** or the **constant of proportionality.**

If the ratio of a variable y to the product of two or more variables is constant, then **y varies jointly as** or **is jointly proportional to** the other variables. If $y = kxz$, then the number k is the **constant of variation** or the **constant of proportionality.**

CHAPTER 7 SUMMARY

Let $\frac{P}{Q}$ and $\frac{R}{S}$ be rational expressions.

FUNDAMENTAL PRINCIPLE OF RATIONAL EXPRESSIONS (7.1)

For any rational expression $\frac{P}{Q}$ and any polynomial R, $R \neq 0$,

$$\frac{P}{Q} = \frac{P \cdot R}{Q \cdot R} \quad \text{and} \quad \frac{P}{Q} = \frac{P \div R}{Q \div R}$$

MULTIPLYING RATIONAL EXPRESSIONS (7.2)

$$\frac{P}{Q} \cdot \frac{R}{S} = \frac{PR}{QS}$$

DIVIDING RATIONAL EXPRESSIONS (7.2)

$$\frac{P}{Q} \div \frac{R}{S} = \frac{P}{Q} \cdot \frac{S}{R} = \frac{PS}{QR}, \quad \text{as long as } R \text{ is not } 0$$

ADDING OR SUBTRACTING RATIONAL EXPRESSIONS WITH COMMON DENOMINATORS (7.3)

If $\frac{P}{Q}$ and $\frac{R}{Q}$ are rational expressions, then

$$\frac{P}{Q} + \frac{R}{Q} = \frac{P + R}{Q} \quad \text{and} \quad \frac{P}{Q} - \frac{R}{Q} = \frac{P - R}{Q}$$

TO ADD OR SUBTRACT RATIONAL EXPRESSIONS WITH UNLIKE DENOMINATORS (7.3)

Step 1 Find the LCD of the rational expressions.

Step 2 Write each rational expression as an equivalent rational expression whose denominator is the LCD found in step 1.

Step 3 Add or subtract numerators, and write the sum or difference over the common denominator.

Step 4 Write the answer in lowest terms.

TO SOLVE AN EQUATION CONTAINING RATIONAL EXPRESSIONS (7.5)

Step 1 Multiply both sides of the equation by the LCD of all rational expressions in the equation.

Step 2 Use the distributive property to remove any grouping symbols such as parentheses.

Step 3 Determine whether the equation is linear or quadratic and solve accordingly.

Step 4 Check the solution in the original equation.

CHAPTER 7 REVIEW

(7.1) *Find the domain of each rational function.*

1. $f(x) = \frac{3 - 5x}{7}$
2. $g(x) = \frac{2x + 4}{11}$
3. $F(x) = \frac{-3x^2}{x - 5}$
4. $h(x) = \frac{4x}{3x - 12}$
5. $f(x) = \frac{x^3 + 2}{x^2 + 8x}$
6. $G(x) = \frac{20}{3x^2 - 48}$

Write each rational expression in lowest terms.

7. $\frac{15x^4}{45x^2}$
8. $\frac{x + 2}{2 + x}$
9. $\frac{18m^6p^2}{10m^4p}$
10. $\frac{x - 12}{12 - x}$
11. $\frac{5x - 15}{25x - 75}$
12. $\frac{22x + 8}{11x + 4}$
13. $\frac{2x}{2x^2 - 2x}$
14. $\frac{x + 7}{x^2 - 49}$
15. $\frac{2x^2 + 4x - 30}{x^2 + x - 20}$
16. $\frac{xy - 3x + 2y - 6}{x^2 + 4x + 4}$
17. The average cost for manufacturing x bookcases is given by the rational function

$$C(x) = \frac{35x + 4200}{x}$$

   a. Find the average cost per bookcase for manufacturing 50 bookcases.

   b. Find the average cost per bookcase for manufacturing 100 bookcases.

(7.2) *Perform the indicated operation. Write answers in lowest terms.*

18. $\frac{5}{x^3} \cdot \frac{x^2}{15}$
19. $\frac{3x^4yz^3}{15x^2y^2} \cdot \frac{10xy}{z^6}$
20. $\frac{4 - x}{5} \cdot \frac{15}{2x - 8}$
21. $\frac{x^2 - 6x - 9}{2x^2 - 18} \cdot \frac{4x + 12}{5x - 15}$
22. $\frac{a - 4b}{a^2 + ab} \cdot \frac{b^2 - a^2}{8b - 2a}$
23. $\frac{x^2 - x - 12}{2x^2 - 32} \cdot \frac{x^2 + 8x + 16}{3x^2 + 21x + 36}$
24. $\frac{2x^3 + 54}{5x^2 + 5x - 30} \cdot \frac{6x + 12}{3x^2 - 9x + 27}$
25. $\frac{3}{4x} \div \frac{8}{2x^2}$
26. $\frac{4x + 8y}{3} \div \frac{5x + 10y}{9}$

27. $\frac{5ab}{14c^3} \div \frac{10a^4b^2}{6ac^5}$

28. $\frac{2}{5x} \div \frac{4 - 18x}{6 - 27x}$

29. $\frac{x^2 - 25}{3} \div \frac{x^2 - 10x + 25}{x^2 - x - 20}$

30. $\frac{a - 4b}{a^2 + ab} \div \frac{20b - 5a}{b^2 - a^2}$

31. $\frac{7x + 28}{2x + 4} \div \frac{x^2 + 2x - 8}{x^2 - 2x - 8}$

32. $\frac{3x + 3}{x - 1} \div \frac{x^2 - 6x - 7}{x^2 - 1}$

33. $\frac{2x - x^2}{x^3 - 8} \div \frac{x^2}{x^2 + 2x + 4}$

34. $\frac{5a^2 - 20}{a^3 + 2a^2 + a + 2} \div \frac{7a}{a^3 + a}$

35. $\frac{2a}{21} \div \frac{3a^2}{7} \cdot \frac{4}{a}$

36. $\frac{5x - 15}{3 - x} \cdot \frac{x + 2}{10x + 20} \cdot \frac{x^2 - 9}{x^2 - x - 6}$

37. $\frac{4a + 8}{5a^2 - 20} \cdot \frac{3a^2 - 6a}{a + 3} \div \frac{2a^2}{5a + 15}$

(7.3) *Find the LCD of the rational expressions in the list.*

38. $\frac{4}{9}, \frac{5}{2}$

39. $\frac{5}{4x^2y^5}, \frac{3}{10x^2y^4}, \frac{x}{6y^4}$

40. $\frac{5}{2x}, \frac{7}{x - 2}$

41. $\frac{3}{5x}, \frac{2}{x - 5}$

42. $\frac{1}{5x^3}, \frac{4}{x^2 + 3x - 28}, \frac{11}{10x^2 - 30x}$

Perform the indicated operation. Write answers in lowest terms.

43. $\frac{2}{15} + \frac{4}{15}$

44. $\frac{4}{x - 4} + \frac{x}{x - 4}$

45. $\frac{4}{3x^2} + \frac{2}{3x^2}$

46. $\frac{1}{x - 2} - \frac{1}{4 - 2x}$

47. $\frac{2x + 1}{x^2 + x - 6} + \frac{2 - x}{x^2 + x - 6}$

48. $\frac{7}{2x} + \frac{5}{6x}$

49. $\frac{1}{3x^2y^3} - \frac{1}{5x^4y}$

50. $\frac{1}{10 - x} + \frac{x - 1}{x - 10}$

51. $\frac{x - 2}{x + 1} - \frac{x - 3}{x - 1}$

52. $\frac{x}{9 - x^2} - \frac{2}{5x - 15}$

53. $2x + 1 - \frac{1}{x - 3}$

54. $\frac{2}{a^2 - 2a + 1} + \frac{3}{a^2 - 1}$

55. $\frac{x}{9x^2 + 12x + 16} - \frac{3x + 4}{27x^3 - 64}$

Perform the indicated operation. Write answers in lowest terms.

56. $\frac{2}{x - 1} - \frac{3x}{3x - 3} + \frac{1}{2x - 2}$

57. $\frac{3}{2x} \cdot \left(\frac{2}{x + 1} - \frac{2}{x - 3}\right)$

58. $\left(\frac{3x}{4}\right)^2 \cdot \left(\frac{2}{x}\right)^3$

59. $\left(\frac{2}{x} - \frac{1}{5}\right) \cdot \left(\frac{2}{x} + \frac{1}{3}\right)$

60. $\frac{2}{x^2 - 16} - \frac{3x}{x^2 + 8x + 16} + \frac{3}{x + 4}$

61. $\left(\frac{x}{x + 5}\right)^2 - 1$

62. $\frac{x}{x^2 - 6x + 9} + \frac{3}{x - 3} \cdot \frac{2x^2 - 18}{4x}$

(6.4) *Simplify each complex fraction.*

63. $\dfrac{\frac{2}{5}}{\frac{3}{5}}$

64. $\dfrac{1 - \frac{3}{4}}{2 + \frac{1}{4}}$

65. $\dfrac{\frac{1}{x} - \frac{2}{3x}}{\frac{5}{2x} - \frac{1}{3}}$

66. $\dfrac{\frac{x^2}{15}}{\frac{x + 1}{5x}}$

67. $\dfrac{\frac{3}{y^2}}{\frac{6}{y^3}}$

68. $\dfrac{\frac{x + 2}{3}}{\frac{5}{x - 2}}$

69. $\dfrac{2 - \frac{3}{2x}}{x - \frac{2}{5x}}$

70. $\dfrac{1 + \frac{x}{y}}{\frac{x^2}{y^2} - 1}$

71. $\dfrac{\frac{5}{x} + \frac{1}{xy}}{\frac{3}{x^2}}$

72. $\dfrac{\frac{x}{3} - \frac{3}{x}}{1 + \frac{3}{x}}$

73. $\dfrac{\frac{1}{x - 1} + 1}{\frac{1}{x + 1} - 1}$

74. $\dfrac{2}{1 - \frac{2}{x}}$

75. $\dfrac{1}{1+\dfrac{2}{1-\dfrac{1}{x}}}$

76. $\dfrac{\dfrac{x^2+5x-6}{4x+3}}{\dfrac{(x+6)^2}{8x+6}}$

77. $\dfrac{\dfrac{x-3}{x+3}+\dfrac{x+3}{x-3}}{\dfrac{x-3}{x+3}-\dfrac{x+3}{x-3}}$

78. $\dfrac{\dfrac{3}{x-1}-\dfrac{2}{1-x}}{\dfrac{2}{x-1}-\dfrac{2}{x}}$

79. If $f(x)=\dfrac{3}{x}$, find each of the following:

a. $f(a+h)$ **b.** $f(a)$

c. Use parts a and b to find $\dfrac{f(a+h)-f(a)}{h}$.

d. Simplify the results of part c.

(7.5) *Solve each equation for x.*

80. $\dfrac{2}{5}=\dfrac{x}{15}$

81. $\dfrac{3}{x}+\dfrac{1}{3}=\dfrac{5}{x}$

82. $4+\dfrac{8}{x}=8$

83. $\dfrac{2x+3}{5x-9}=\dfrac{3}{2}$

84. $\dfrac{1}{x-2}-\dfrac{3x}{x^2-4}=\dfrac{2}{x+2}$

85. $\dfrac{7}{x}-\dfrac{x}{7}=0$

86. $\dfrac{x-2}{x^2-7x+10}=\dfrac{1}{5x-10}-\dfrac{1}{x-5}$

Solve the equations for x or perform the indicated operation. Simplify.

87. $\dfrac{5}{x^2-7x}+\dfrac{4}{2x-14}$

88. $3-\dfrac{5}{x}-\dfrac{2}{x^2}=0$

89. $\dfrac{4}{3-x}-\dfrac{7}{2x-6}+\dfrac{5}{x}$

(7.6) *Solve the equation for the specified variable.*

90. $A=\dfrac{h(a+b)}{2}$, a

91. $\dfrac{1}{R}=\dfrac{1}{R_1}+\dfrac{1}{R_2}$, R_2

92. $I=\dfrac{E}{R+r}$, R

93. $A=P+Prt$, r

94. $H=\dfrac{kA(T_1-T_2)}{L}$, A

Solve.

95. The sum of a number and twice its reciprocal is 3. Find the number(s).

96. If a number is added to the numerator of $\frac{3}{7}$, and twice that number is added to the denominator of $\frac{3}{7}$, the result is equivalent to $\frac{10}{21}$. Find the number.

97. Mary Willis is three-fourths as old as her friend Mark Snow. If the sum of their ages is 42, find their ages.

98. The denominator of a fraction is 2 more than the numerator. If the numerator is decreased by 3 and the denominator is increased by 5, the resulting fraction is equivalent to $\frac{2}{3}$. Find the fraction.

99. The sum of the reciprocals of two consecutive integers is $\frac{29}{210}$. Find the two integers.

100. The sum of the reciprocals of two consecutive even integers is $-\frac{9}{40}$. Find the two integers.

101. Three boys can paint a fence in 4 hours, 5 hours, and 6 hours, respectively. Find how long it will take all three boys to paint the fence.

102. If Sue Katz can word process mailing labels in 6 hours and Tom Nielson and Sue working together can word process the same number of mailing labels in 4 hours, find how long it takes Tom to word process the mailing labels alone.

103. The inlet pipe of a water tank can fill the tank in 2 hours and 30 minutes. The outlet pipe can empty the tank in 2 hours. Find how long it takes to empty a full tank if both pipes are open.

104. Timmy Garnica drove 210 miles in the same amount of time that it took a Boeing 757 jet to travel 1715 miles. The speed of the jet was 430 mph faster than the speed of the car. Find the speed of the jet.

105. The combined resistance R of two resistors, r_1 and r_2, in parallel is given by the formula $\frac{1}{R} = \frac{1}{r_1} + \frac{1}{r_2}$. If the combined resistance is $\frac{30}{11}$ ohms and one of the two resistors is 5 ohms, find the other resistor.

106. The speed of a Ranger boat in still water is 32 mph. If the boat travels 72 miles upstream in the same time that it takes to travel 120 miles downstream, find the current of the stream.

107. A plane flies 445 miles with the wind and 355 miles against the wind in the same length of time. If the speed of the plane in still air is 400 mph, find the speed of the wind.

108. The speed of a jogger is 3 mph faster than the speed of a walker. If the jogger travels 14 miles in the same amount of time that the walker travels 8 miles, find the speed of the walker.

109. Two Amtrack trains leave Tucson, Arizona, at the same time traveling on parallel tracks. In 6 hours the faster train is 382 miles from Tucson and the trains are 112 miles apart. Find how fast each train is traveling.

(7.7) *Solve each of the following variation problems.*

110. A is directly proportional to B. If $A = 6$ when $B = 14$, find A when $B = 21$.

111. C is inversely proportional to D. If $C = 12$ when $D = 8$, find C when $D = 24$.

112. According to Boyle's law, the pressure exerted by a gas is inversely proportional to the volume, as long as the temperature stays the same. If a gas exerts a pressure of 1250 pounds per square inch when the volume is 2 cubic feet, find the volume when the pressure is 800 pounds per square inch.

113. The surface area of a sphere varies directly as the square of its radius. If the surface area is 36 square inches when the radius is 3 inches, find the surface area when the radius is 4 inches.

CHAPTER 7 TEST

Find the domain of each rational function.

1. $f(x) = \frac{5x^2}{1 - x}$

2. $g(x) = \frac{9x^2 - 9}{x^2 + 4x + 3}$

Write each rational expression in lowest terms.

3. $\frac{5x^7}{3x^4}$

4. $\frac{7x - 21}{24 - 8x}$

5. $\frac{2x + 6}{x^2 - 9}$

6. $\frac{x^2 - 4x}{x^2 + 5x - 36}$

Perform the indicated operation. Write answers in lowest terms.

7. $\frac{x}{x - 2} \cdot \frac{x^2 - 4}{5x}$

8. $\frac{2x^2 - x - 3}{5x + 10} \cdot \frac{3x + 6}{4x - 6}$

9. $\frac{2x^3 + 16}{6x^2 + 12x} \cdot \frac{5}{x^2 - 2x + 4}$

10. $\frac{26ab}{7c} \div \frac{13a^2c^5}{14a^4b^3}$

11. $\frac{3x^2 - 12}{x^2 + 2x - 8} \div \frac{6x + 18}{x + 4}$

12. $\frac{4x - 12}{2x - 9} \div \frac{3 - x}{4x^2 - 81} \cdot \frac{x + 3}{5x + 15}$

Find the LCD of the rational expressions in the list.

13. $\frac{2}{25x}, \frac{3}{15x^3}$

Perform the indicated operation.

14. $\frac{5}{4x^3} + \frac{7}{4x^3}$

15. $\frac{3 + 2x}{10 - x} + \frac{13 + x}{x - 10}$

16. $\frac{3}{x^2 - x - 6} + \frac{2}{x^2 - 5x + 6}$

17. $\dfrac{x}{4x^2 - 2x + 1} - \dfrac{2x + 1}{8x^3 + 1}$

18. $\dfrac{5}{x - 7} - \dfrac{2x}{3x - 21} + \dfrac{x}{2x - 14}$

19. $\dfrac{3x}{5} \cdot \left(\dfrac{5}{x} - \dfrac{5}{2x}\right)$

20. $\dfrac{5x + 3y}{x^2 - y^2} \cdot \left(\dfrac{x}{y} - \dfrac{y}{x}\right)$

21. $\dfrac{3}{x + y} + \dfrac{2x^2 - 2xy + 4x - 4y}{4x + 8} \div \dfrac{y^2 - x^2}{2y}$

Simplify each complex fraction.

22. $\dfrac{\frac{4x}{13}}{\frac{20x}{13}}$

23. $\dfrac{\frac{5}{x} - \frac{7}{3x}}{\frac{9}{8x} - \frac{1}{x}}$

24. $\dfrac{\frac{7}{2x} + \frac{2}{xy}}{\frac{3}{xy^2}}$

25. $\dfrac{\frac{x^2 - 5x + 6}{x + 3}}{\frac{x^2 - 4x + 4}{x^2 - 9}}$

26. The product of one more than a number and twice the reciprocal of the number is $\frac{12}{5}$. Find the number.

27. If Jan Shields can weed the garden in 2 hours and her husband can weed the garden in 1 hour and 30 minutes, find how long it takes them to weed the garden together.

28. Suppose that W is inversely proportional to V. If $W = 20$ when $V = 12$, find W when $V = 15$.

29. Suppose that Q is jointly proportional to R and the square of S. If $Q = 24$ when $R = 3$ and $S = 4$, find Q when $R = 2$ and $S = 3$.

30. When an anvil is dropped into a gorge, the speed with which it strikes the ground is directly proportional to the square root of the distance it falls. An anvil that falls 400 feet hits the ground at a speed of 160 feet per second. Find the height of a cliff over the gorge if a dropped anvil hits the ground at a speed of 128 feet per second.

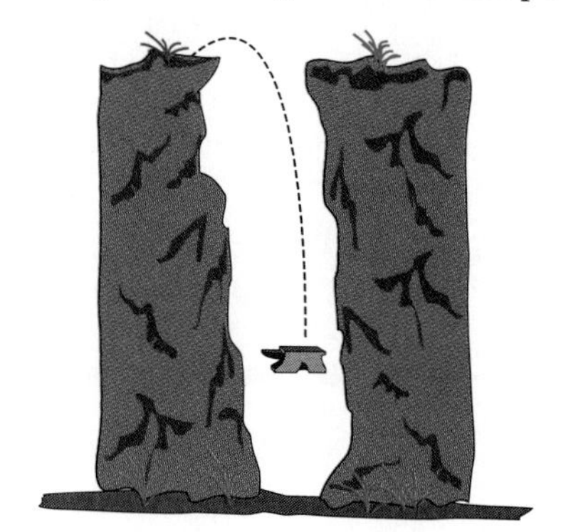

CHAPTER 7 CUMULATIVE REVIEW

1. Use the distributive property to remove parentheses.
  a. $5(x + 2)$ **b.** $-2(y + 0.3z - 1)$
  c. $-(x + y - 2z + 6)$

2. Solve for x: $\dfrac{x}{2} - 1 = \dfrac{2}{3}x - 3$.

3. Solve $3(x - 4) = 3x - 12$.

4. Solve $F = \dfrac{9}{5}C + 32$ for C.

5. Solve for x: $\dfrac{3}{x - 2} = \dfrac{5}{x + 1}$.

6. The circle graph given shows how much money homeowners spend annually on maintaining their homes. Use this graph to answer the questions below.

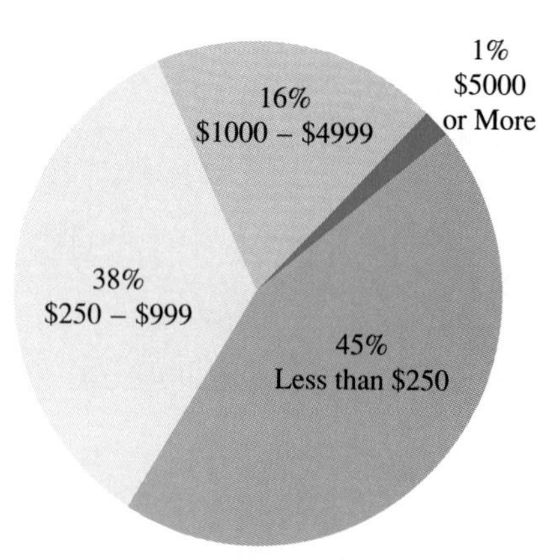

Source: *USA Today*, 1994.

a. What percent of homeowners spend less than \$250 on yearly home maintenance?
b. What percent of homeowners spend less than \$1000 per year on maintenance?
c. How many of the 22,000 homeowners in a town called Fairview might we expect to spend less than \$250 a year on home maintenance?
d. Find the number of degrees in the 16% sector.

7. Solve $|g| = -1$.

8. Solve for x: $-1 \le \frac{2x}{3} + 5 \le 2$.

9. Given $g(x) = x^2 - 3$, find the following.
a. $g(2)$ **b.** $g(-2)$ **c.** $g(0)$

10. Find the slope of the line through $(-1, 5)$ and $(2, -3)$. Graph the line.

11. Graph the line through $(-1, 5)$ with slope -2.

12. Evaluate each expression.
a. 2^3 **b.** 3^1 **c.** $(-4)^2$
d. -4^2 **e.** $\left(\frac{2}{3}\right)^3$ **f.** $2 \cdot 4^2$

13. Simplify each expression. Write answers using positive exponents only.
a. 3^{-2} **b.** $2x^{-3}$ **c.** $\frac{1}{2^{-5}}$

14. The CN Tower in Toronto, Ontario, is 1,821 feet tall and is the world's tallest self-supporting structure. An object is dropped from the top of this building. Neglecting air resistance, the height of the object at time t seconds is given by the polynomial function $P(t) = -16t^2 + 1821$. Find the height of the object when $t = 1$ second and when $t = 10$ seconds.

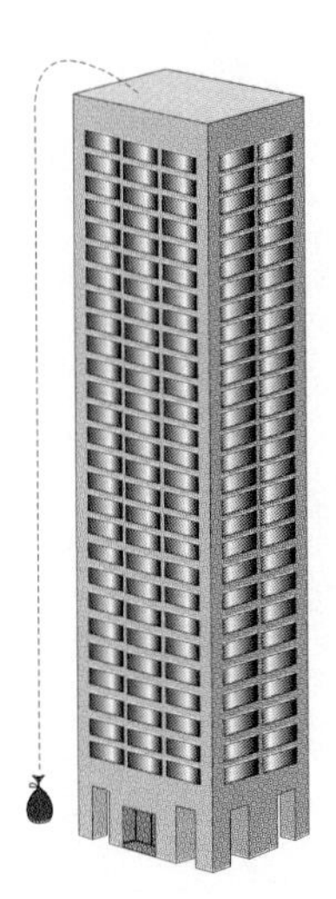

15. Divide $2x^2 - x - 10$ by $x + 2$.

16. Factor $3x^2 + 11x + 6$.

17. Solve $2x^2 = \frac{17}{3}x + 1$.

18. ICL Production Company uses the rational function $C(x) = \frac{2.6x + 10{,}000}{x}$ to describe the average cost per disc of pressing x compact discs. Find the average cost per disc for pressing:
a. 100 compact discs **b.** 1000 compact discs

19. Multiply.
a. $\frac{2x^2 + 3x - 2}{-4x - 8} \cdot \frac{16x^2}{4x^2 - 1}$
b. $(ac - ad + bc - bd) \cdot \frac{a + b}{d - c}$

20. Simplify each complex fraction.
a. $\frac{\frac{5x}{x + 2}}{\frac{10}{x - 2}}$ **b.** $\frac{x + \frac{1}{y}}{y + \frac{1}{x}}$

21. Solve $\frac{3}{x} - \frac{x + 21}{3x} = \frac{5}{3}$.

22. The intensity $I(x)$ in foot-candles of light x feet from its source is given by the rational

$$I(x) = \frac{320}{x^2}$$

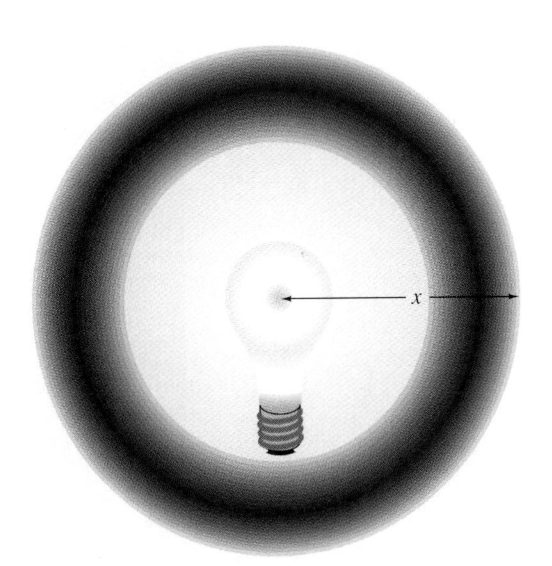

How far away is the source if the intensity of light is 5 foot-candles?

23. Suppose that y varies directly as x. If y is 5 when x is 30, find the constant of variation. Also, find y when $x = 90$.

CHAPTER 8

Roots and Radicals

Joe Gerken of Gerken's Nursery expands his business to include growing and selling Scotch pine Christmas trees. The efficient use of his land depends on an optimal pattern for planting the trees.

INTRODUCTION

In this chapter, we begin with a review of radical notation, and we then introduce **rational exponents.** As the name implies, rational exponents are exponents that are rational numbers. We present an interpretation of rational exponents that is consistent with the meaning and rules already established for integer exponents, and we present two forms of notation for rational exponents: the exponent notation and the radical notation. We conclude this chapter with **complex numbers,** a natural extension of the real number system that gives meaning to rational exponents applied to negative real numbers.

8.1 Introduction to Radicals

OBJECTIVES

Tape BA 26

1. Find square roots of perfect squares.
2. Identify rational and irrational numbers.
3. Graph square root functions.
4. Find cube roots of perfect cubes.
5. Find nth roots.

1 Recall that we can define finding the **root** of a number by its reverse operation, raising a number to a power. We begin with squares and square roots.

The square of 5 is $5^2 = 25$.
The square of -5 is $(-5)^2 = 25$.
The square of $\frac{1}{2}$ is $\left(\frac{1}{2}\right)^2 = \frac{1}{4}$.

The reverse operation of squaring a number is finding the **square root** of a number. For example:

A square root of 25 is 5 because $5^2 = 25$.
A square root of 25 is also -5 because $(-5)^2 = 25$.
A square root of $\frac{1}{4}$ is $\frac{1}{2}$ because $\left(\frac{1}{2}\right)^2 = \frac{1}{4}$.

Notice that both 5 and -5 are square roots of 25. We will use the symbol $\sqrt{}$ to denote the **positive** or **principal** square root of a number. For example,

$$\sqrt{25} = 5 \text{ since } 5^2 = 25$$

The symbol $-\sqrt{}$ will be used to denote the negative square root. For example,

$$-\sqrt{25} = -5$$

Square Root

The **positive** or **principal square root** of a positive number a is written as $\sqrt{a}$. The negative square root of a is written as $-\sqrt{a}$.

$$\sqrt{a} = b, \qquad \text{if } b^2 = a$$

Also, the square root of 0 is 0, written as $\sqrt{0} = 0$.

The symbol $\sqrt{}$ is called a **radical** or **radical sign.** The expression within or under a radical sign is called the **radicand.** An expression containing a radical is called a **radical expression.**

$\sqrt{a}$ ← radical sign; a ← radicand

EXAMPLE 1 Find each square root.

a. $\sqrt{36}$ **b.** $\sqrt{64}$ **c.** $-\sqrt{25}$ **d.** $\sqrt{\frac{9}{100}}$ **e.** $\sqrt{0}$

Solution:

a. $\sqrt{36} = 6$ because $6^2 = 36$ and 6 is positive.

b. $\sqrt{64} = 8$ because $8^2 = 64$ and 8 is positive.

c. $-\sqrt{25} = -5$. The negative sign in front of the radical indicates the negative square root of 25.

d. $\sqrt{\frac{9}{100}} = \frac{3}{10}$ because $\left(\frac{3}{10}\right)^2 = \frac{9}{100}$ and $\frac{3}{10}$ is positive.

e. $\sqrt{0} = 0$ because $0^2 = 0$. ■

Can we find the square root of a negative number, say $\sqrt{-4}$? That is, can we find a real number whose square is -4? No, there is no real number whose square is -4, and we say that $\sqrt{-4}$ is not a real number. In general:

The square root of a negative number is not a real number.

2 Recall that numbers such as 1, 4, 9, and 25 are called **perfect squares,** since $1^2 = 1$, $2^2 = 4$, $3^2 = 9$, and $5^2 = 25$. Square roots of perfect square radicands simplify to rational numbers. What happens when we try to simplify a root such as $\sqrt{3}$? Since 3 is not a perfect square, $\sqrt{3}$ is not a rational number. It is called an **irrational number,** and we can find a decimal **approximation** of it. To find decimal approximations, use the table in Appendix D or a calculator. For example, an approximation for $\sqrt{3}$ is

$$\sqrt{3} \approx 1.732$$

↑ approximation symbol

3 Recall that an equation in x and y describes a function if each x-value is paired with exactly one y-value. With this in mind, does the equation

$$y = \sqrt{x}$$

describe a function? Since $\sqrt{x}$ denotes the principal square root of x, then for every nonnegative number, there exists exactly one number, $\sqrt{x}$. Therefore, $y = \sqrt{x}$ describes a function, and we may write it as

$$f(x) = \sqrt{x}$$

We find function values as usual. For example,

$$f(0) = \sqrt{0} = 0$$
$$f(1) = \sqrt{1} = 1$$
$$f(4) = \sqrt{4} = 2$$
$$f(9) = \sqrt{9} = 3$$

Choosing perfect squares for x ensures us that $f(x)$ is a rational number, but it is important to stress that $f(x) = \sqrt{x}$ is defined for all nonnegative real numbers. For example,

$$f(3) = \sqrt{3} \approx 1.7$$

EXAMPLE 2 Graph the square root function $f(x) = \sqrt{x}$.

Solution: To graph, we discuss the domain, plot points, and connect the points with a smooth curve. The domain of this function is the set of all nonnegative numbers or $\{x \mid x \geq 0\}$. The table comes from our work above.

x	$f(x) = \sqrt{x}$
0	0
1	1
3	$\sqrt{3} \approx 1.7$
4	2
9	3

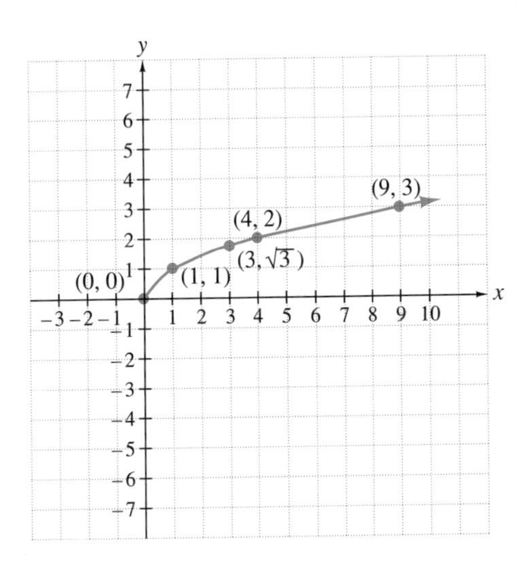

Notice that the graph of this function passes the vertical line test, as it should. ■

4 Finding roots can be extended to other roots such as cube roots. For example, since $2^3 = 8$, we call 2 the **cube root** of 8. In symbols, this is

$$\sqrt[3]{8} = 2$$

> **Cube Root**
>
> A **cube root** of a real number a is written as $\sqrt[3]{a}$, and
>
> $$\sqrt[3]{a} = b, \qquad \text{if } b^3 = a$$

From the above definition, we have

$$\sqrt[3]{27} = 3, \qquad \text{since } 3^3 = 27$$
$$\sqrt[3]{-64} = -4, \qquad \text{since } (-4)^3 = -64$$

Notice that, unlike with square roots, **it is possible to have a negative radicand when finding a cube root.** This is so because the cube of a negative number is a negative number. Therefore, the cube root of a negative number is a negative number.

EXAMPLE 3 Find the cube roots.

a. $\sqrt[3]{1}$ **b.** $\sqrt[3]{-27}$ **c.** $\sqrt[3]{\dfrac{1}{125}}$

Solution: **a.** $\sqrt[3]{1} = 1$ because $1^3 = 1$.

b. $\sqrt[3]{-27} = -3$ because $(-3)^3 = -27$.

c. $\sqrt[3]{\dfrac{1}{125}} = \dfrac{1}{5}$ because $\left(\dfrac{1}{5}\right)^3 = \dfrac{1}{125}$. ■

The equation $f(x) = \sqrt[3]{x}$ also describes a function. Here x may be any real number.

$$f(0) = \sqrt[3]{0} = 0$$
$$f(1) = \sqrt[3]{1} = 1$$
$$f(-1) = \sqrt[3]{-1} = -1$$
$$f(6) = \sqrt[3]{6} \approx 1.8$$
$$f(-6) = \sqrt[3]{-6} \approx -1.8$$
$$f(8) = \sqrt[3]{8} = 2$$
$$f(-8) = \sqrt[3]{-8} = -2$$

EXAMPLE 4 Graph the function $f(x) = \sqrt[3]{x}$.

Solution: To graph, we discuss the domain, plot points, and connect the points with a smooth curve. The domain of this function is the set of all real numbers. The table comes from our work above.

x	$f(x) = \sqrt[3]{x}$
0	0
1	1
-1	-1
6	$\sqrt[3]{6} \approx 1.8$
-6	$\sqrt[3]{-6} \approx -1.8$
8	2
-8	-2

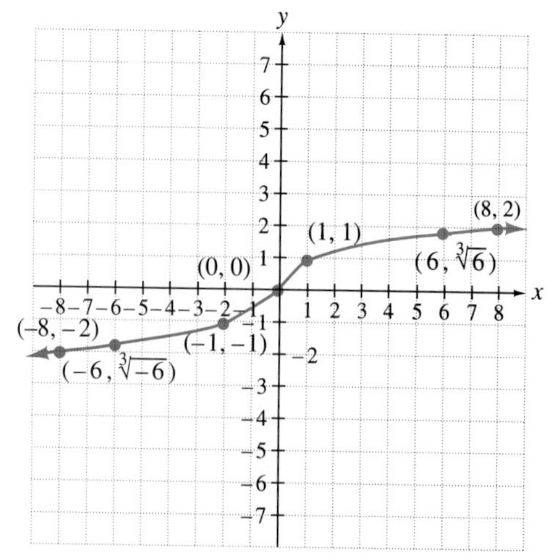

The graph of this function passes the vertical line test, as expected. ■

5 Just as we can raise a real number to powers other than 2 or 3, we can find roots other than square roots and cube roots. In fact, we can take the nth root of a number where n is any natural number. In symbols, the nth root of a is written as $\sqrt[n]{a}$, where n is called the **index.** The index 2 is usually omitted for square roots.

HELPFUL HINT

If the index is even, such as $\sqrt{\ }$, $\sqrt[4]{\ }$, $\sqrt[6]{\ }$, and so on, the radicand must be nonnegative for the root to be a real number. For example,

$\sqrt[4]{-16}$ is not a real number

$\sqrt[6]{-64}$ is not a real number

If the index is odd, such as $\sqrt[3]{\ }$, $\sqrt[5]{\ }$, and so on, the radicand may be any real number. For example,

$$\sqrt[3]{-64} = -3$$
$$\sqrt[5]{-32} = -2$$

EXAMPLE 5 Evaluate the following expressions.

a. $\sqrt[4]{16}$ **b.** $\sqrt[5]{-32}$ **c.** $-\sqrt[3]{8}$ **d.** $\sqrt[4]{-81}$

Solution:

a. $\sqrt[4]{16} = 2$ because $2^4 = 16$.
b. $\sqrt[5]{-32} = -2$ because $(-2)^5 = -32$.
c. $-\sqrt[3]{8} = -2$ since $\sqrt[3]{8} = 2$.
d. $\sqrt[4]{-81}$ is not a real number. ■

Recall that the notation $\sqrt{a^2}$ indicates the positive square root of a^2 only. For example,

$$\sqrt{(-3)^2} = \sqrt{9} = 3$$

In general, we write

$$\sqrt{a^2} = |a|$$

to ensure that the square root is positive. This is true when the index is any even positive integer.

If n is an even positive integer, then $\sqrt[n]{a^n} = |a|$.
If n is an odd positive integer, then $\sqrt[n]{a^n} = a$.

EXAMPLE 6 Simplify.

a. $\sqrt{(-3)^2}$ **b.** $\sqrt{x^2}$ **c.** $\sqrt[4]{(x-2)^4}$ **d.** $\sqrt[3]{(-5)^3}$

Solution:

a. $\sqrt{(-3)^2} = |-3| = 3$ The absolute value bars ensure us that our result is not negative.
b. $\sqrt{x^2} = |x|$
c. $\sqrt[4]{(x-2)^4} = |x-2|$
d. $\sqrt[3]{(-5)^3} = -5$ ■

EXERCISE SET 8.1

Evaluate each square root. See Example 1.

1. $\sqrt{16}$
2. $\sqrt{9}$
3. $\sqrt{81}$
4. $\sqrt{49}$
5. $\sqrt{\frac{1}{25}}$
6. $\sqrt{\frac{1}{64}}$
7. $-\sqrt{100}$
8. $-\sqrt{36}$

Graph each function. See Example 2.

9. $f(x) = \sqrt{x} + 2$
10. $f(x) = \sqrt{x} - 2$
11. $f(x) = \sqrt{x - 3}$; use the table:

x	$f(x)$
3	
4	
7	
12	

12. $f(x) = \sqrt{x + 1}$; use the table:

x	$f(x)$
-1	
0	
3	
8	

Find each cube root. See Example 3.

13. $\sqrt[3]{64}$
14. $\sqrt[3]{-1}$
15. $-\sqrt[3]{27}$
16. $-\sqrt[3]{8}$
17. $\sqrt[3]{\frac{1}{8}}$
18. $\sqrt[3]{\frac{1}{64}}$
19. $\sqrt[3]{-125}$
20. $\sqrt[3]{-27}$

Graph each function. See Example 4.

21. $f(x) = \sqrt[3]{x} + 1$
22. $f(x) = \sqrt[3]{x} - 2$

23. $g(x) = \sqrt[3]{x - 1}$; use the table:

x	$g(x)$
1	
2	
0	
9	
−7	

24. $g(x) = \sqrt[3]{x + 1}$; use the table:

x	$g(x)$
−1	
0	
−2	
7	
−9	

Find each root that is a real number. See Example 5.

25. $\sqrt[5]{32}$ **26.** $\sqrt[4]{-1}$ **27.** $\sqrt[4]{81}$ **28.** $\sqrt{121}$

29. $\sqrt{-4}$ **30.** $\sqrt[5]{\frac{1}{32}}$ **31.** $\sqrt[3]{\frac{1}{27}}$ **32.** $\sqrt[4]{256}$

33. $\sqrt{\frac{9}{25}}$ **34.** $\sqrt[3]{\frac{8}{27}}$ **35.** $-\sqrt{49}$ **36.** $-\sqrt[4]{625}$

Simplify each root. See Example 6.

37. $\sqrt{z^2}$ **38.** $\sqrt{y^{10}}$ **39.** $\sqrt[3]{x^3}$ **40.** $\sqrt[3]{z^6}$

41. $\sqrt{(x + 5)^2}$ **42.** $\sqrt{(y - 6)^2}$ **43.** $\sqrt[4]{(2z)^4}$ **44.** $\sqrt{(9x)^2}$

Find each root that is a real number.

45. $\sqrt{0}$ **46.** $\sqrt[3]{0}$ **47.** $-\sqrt[5]{\frac{1}{32}}$ **48.** $-\sqrt[3]{\frac{27}{125}}$

49. $\sqrt{-64}$ **50.** $-\sqrt[3]{-64}$ **51.** $-\sqrt{64}$ **52.** $\sqrt[6]{64}$

53. $-\sqrt{169}$ **54.** $\sqrt[4]{-16}$ **55.** $\sqrt{1}$ **56.** $\sqrt[3]{1}$

57. $\sqrt{\frac{25}{64}}$ **58.** $\sqrt{\frac{1}{100}}$ **59.** $-\sqrt[3]{-8}$ **60.** $-\sqrt[3]{-27}$

Graph each function.

61. $f(x) = \sqrt{x} - 3$ **62.** $f(x) = \sqrt{x} + 1$ **63.** $g(x) = \sqrt[3]{x} - 3$ **64.** $g(x) = \sqrt[3]{x} - 3$

65. $f(x) = \sqrt{x - 2}$ **66.** $f(x) = \sqrt{x + 5}$ **67.** $h(x) = \sqrt[3]{x + 4}$ **68.** $h(x) = \sqrt[3]{x + 5}$

Determine whether each square root is rational or irrational. If it is rational, find its **exact value.** *If it is irrational, use a calculator or the table in Appendix D to write a three-decimal-place* **approximation.**

69. $\sqrt{9}$ **70.** $\sqrt{8}$ **71.** $\sqrt{37}$ **72.** $\sqrt{36}$

73. $\sqrt{169}$ **74.** $\sqrt{160}$ **75.** $\sqrt{4}$ **76.** $\sqrt{27}$

Simplify each root.

77. $\sqrt[3]{x^{15}}$ **78.** $\sqrt[3]{y^{12}}$ **79.** $\sqrt{x^{12}}$ **80.** $\sqrt{z^{16}}$

81. $\sqrt{81x^2}$ **82.** $\sqrt{100z^4}$ **83.** $-\sqrt{144y^{14}}$ **84.** $-\sqrt{121z^{22}}$

85. $\sqrt{x^2 + 4x + 4}$ **86.** $\sqrt{x^2 - 6x + 9}$

Solve.

87. A fence is to be erected around a square garden with an area of 324 square feet. Each side of this garden has a length of $\sqrt{324}$ feet. Write a one-decimal-place approximation of this length.

88. A standard baseball diamond is a square with 90 foot sides connecting the bases. The distance from first base to third base is $90 \cdot \sqrt{2}$ feet. Approximate $\sqrt{2}$ accurate to two decimal places and use it to approximate the distance $90 \cdot \sqrt{2}$ feet.

89. The roof of the warehouse shown needs to be shingled. The total area of the roof is exactly $240\sqrt{41}$ square feet. Approximate this area to the nearest whole number.

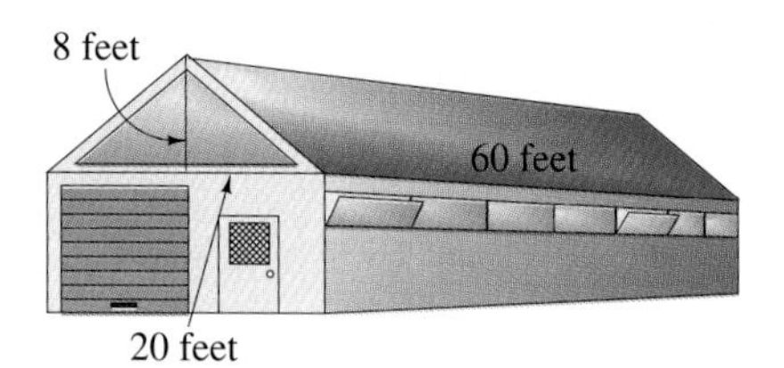

Skill Review

Write each integer as a product of two integers such that one of the factors is a perfect square. For example, in $18 = 9 \cdot 2$, 9 *is a perfect square.*

90. 50 **91.** 8 **92.** 32 **93.** 75

94. 28 **95.** 44 **96.** 27 **97.** 90

Write each integer as a product of two integers such that one of the factors is a perfect cube. For example, in $24 = 8 \cdot 3$, 8 *is a perfect cube.*

98. 16 **99.** 56 **100.** 54 **101.** 270

102. 80

Writing in Mathematics

103. Explain why the square root of a negative number is not a real number.

8.2 Simplifying Radicals

OBJECTIVES

1 Use the product rule to simplify radicals.

2 Use the quotient rule to simplify radicals.

Tape BA 26

1 Much of our work with expressions in this book has involved finding ways to write expressions in their simplest form. Writing radicals in simplest form requires recognizing several patterns, or rules, which we present here. Notice that

$$\sqrt{9 \cdot 16} = \sqrt{144} = 12$$

Also,

$$\sqrt{9} \cdot \sqrt{16} = 3 \cdot 4 = 12$$

Since both expressions simplify to 12, we can say that

$$\sqrt{9 \cdot 16} = \sqrt{9} \cdot \sqrt{16}$$

This suggests the following product rule for square roots.

Product Rule for Square Roots

If a and b are nonnegative numbers, then

$$\sqrt{a \cdot b} = \sqrt{a} \cdot \sqrt{b}$$

The product rule states that the square root of a product is equal to the product of the square roots. We use this rule to simplify radicals such as $\sqrt{20}$. A radical is **simplified** if the radicand has no perfect square factors other than 1. To simplify $\sqrt{20}$, factor 20 so that one of its factors is a perfect square factor.

$$\sqrt{20} = \sqrt{4 \cdot 5} \quad \text{Factor 20.}$$
$$= \sqrt{4} \cdot \sqrt{5} \quad \text{Use the product rule.}$$
$$= 2\sqrt{5} \quad \text{Write } \sqrt{4} \text{ as 2.}$$

The notation $2\sqrt{5}$ means $2 \cdot \sqrt{5}$. Since the radicand 5 has no perfect square factor other than 1, $2\sqrt{5}$ is in **simplified form.**

When factoring a radicand, look for at least one factor that is a perfect square. Review the table of perfect squares in Appendix D to help locate perfect square factors more quickly.

When simplifying a radical, realize that it does not mean getting a decimal approximation. The simplified form of a radical is an exact form and may still contain a radical.

HELPFUL HINT

When simplifying a radical, use **factors** of the radicand; **do not** write the radicand as a sum. For example, **do not** rewrite $\sqrt{20}$ as $\sqrt{4 + 16}$ because $\sqrt{4 + 16} \neq \sqrt{4} + \sqrt{16}$. Correctly simplified, $\sqrt{20} = \sqrt{4 \cdot 5} = \sqrt{4} \cdot \sqrt{5} = 2\sqrt{5}$.

EXAMPLE 1 Simplify each expression.

a. $\sqrt{54}$ **b.** $\sqrt{12}$ **c.** $\sqrt{200}$ **d.** $\sqrt{35}$

Solution: **a.** Try to factor 54 so that at least one of the factors is a perfect square. Since $54 = 9 \cdot 6$,

$$\sqrt{54} = \sqrt{9 \cdot 6} \quad \text{Factor.}$$
$$= \sqrt{9} \cdot \sqrt{6} \quad \text{Use the product rule.}$$
$$= 3\sqrt{6} \quad \text{Write } \sqrt{9} \text{ as 3.}$$

b.
$$\begin{aligned} \sqrt{12} &= \sqrt{4 \cdot 3} && \text{Factor 12.} \\ &= \sqrt{4} \cdot \sqrt{3} && \text{Use the product rule.} \\ &= 2\sqrt{3} && \text{Write } \sqrt{4} \text{ as 2.} \end{aligned}$$

c. The largest perfect square factor of 200 is 100.
$$\begin{aligned} \sqrt{200} &= \sqrt{100 \cdot 2} && \text{Factor 200.} \\ &= \sqrt{100} \cdot \sqrt{2} && \text{Use the product rule.} \\ &= 10\sqrt{2} && \text{Write } \sqrt{100} \text{ as 10.} \end{aligned}$$

d. The radicand 35 contains no perfect square factors other than 1. Thus, $\sqrt{35}$ is in simplified form. ■

2 Next let's examine the square root of a quotient.
$$\sqrt{\frac{16}{4}} = \sqrt{4} = 2$$
Also,
$$\frac{\sqrt{16}}{\sqrt{4}} = \frac{4}{2} = 2$$
Since both expressions equal 2, we have that
$$\sqrt{\frac{16}{4}} = \frac{\sqrt{16}}{\sqrt{4}}$$
This suggests the following quotient rule.

Quotient Rule for Square Roots

If a and b are nonnegative numbers and $b \neq 0$, then
$$\sqrt{\frac{a}{b}} = \frac{\sqrt{a}}{\sqrt{b}}$$

The quotient rule states that the square root of a quotient is equal to the quotient of the square roots.

EXAMPLE 2 Simplify the following.

a. $\sqrt{\frac{25}{36}}$ **b.** $\sqrt{\frac{3}{64}}$ **c.** $\sqrt{\frac{40}{81}}$

Solution: Use the quotient rule.

a. $\sqrt{\frac{25}{36}} = \frac{\sqrt{25}}{\sqrt{36}} = \frac{5}{6}$ **b.** $\sqrt{\frac{3}{64}} = \frac{\sqrt{3}}{\sqrt{64}} = \frac{\sqrt{3}}{8}$

c.
$$\begin{aligned} \sqrt{\frac{40}{81}} &= \frac{\sqrt{40}}{\sqrt{81}} && \text{Use the quotient rule.} \\ &= \frac{\sqrt{4} \cdot \sqrt{10}}{9} && \text{Write } \sqrt{81} \text{ as 9 and use the product rule.} \\ &= \frac{2\sqrt{10}}{9} && \text{Write } \sqrt{4} \text{ as 2.} \end{aligned}$$
■

For the remainder of this chapter, we will assume that variables represent positive real numbers. If this is so, we need not insert absolute value bars when simplifying even roots.

EXAMPLE 3 Simplify each expression. Assume that variables represent positive numbers only.

a. $\sqrt{x^5}$ **b.** $\sqrt{8y^2}$ **c.** $\sqrt{\frac{45}{x^6}}$

Solution: **a.** $\sqrt{x^5} = \sqrt{x^4 \cdot x} = \sqrt{x^4} \cdot \sqrt{x} = x^2\sqrt{x}$

b. $\sqrt{8y^2} = \sqrt{4 \cdot 2 \cdot y^2} = \sqrt{4y^2 \cdot 2} = \sqrt{4y^2} \cdot \sqrt{2} = 2y\sqrt{2}$

c. $\sqrt{\frac{45}{x^6}} = \frac{\sqrt{45}}{\sqrt{x^6}} = \frac{\sqrt{9 \cdot 5}}{x^3} = \frac{\sqrt{9} \cdot \sqrt{5}}{x^3} = \frac{3\sqrt{5}}{x^3}$ ■

The product and quotient rules also apply to roots other than square roots. In general, we have the following product and quotient rules for radicals:

Product Rule for Radicals

If $\sqrt[n]{a}$ and $\sqrt[n]{b}$ are real numbers, then

$$\sqrt[n]{a \cdot b} = \sqrt[n]{a} \cdot \sqrt[n]{b}$$

Quotient Rule for Radicals

If $\sqrt[n]{a}$ and $\sqrt[n]{b}$ are real numbers, then

$$\sqrt[n]{\frac{a}{b}} = \frac{\sqrt[n]{a}}{\sqrt[n]{b}}, \quad \text{providing } b \neq 0$$

For example, to simplify cube roots, look for perfect cube factors of the radicand. The number 8 is a perfect cube, since $2^3 = 8$.

To simplify $\sqrt[3]{48}$, factor 48 as $8 \cdot 6$.

$\sqrt[3]{48} = \sqrt[3]{8 \cdot 6}$	Factor 48.
$= \sqrt[3]{8} \cdot \sqrt[3]{6}$	Use the product rule.
$= 2\sqrt[3]{6}$	Write $\sqrt[3]{8}$ as 2.

$2\sqrt[3]{6}$ is in simplest form since the radicand 6 contains no perfect cube factors other than 1.

EXAMPLE 4 Simplify each expression.

a. $\sqrt[3]{54}$ **b.** $\sqrt[3]{18}$ **c.** $\sqrt[3]{\frac{7}{8}}$ **d.** $\sqrt[3]{\frac{40}{27}}$

Solution: **a.** $\sqrt[3]{54} = \sqrt[3]{27 \cdot 2} = \sqrt[3]{27} \cdot \sqrt[3]{2} = 3\sqrt[3]{2}$

b. The number 18 contains no perfect cube factors, so $\sqrt[3]{18}$ cannot be simplified further.

c. $\sqrt[3]{\frac{7}{8}} = \frac{\sqrt[3]{7}}{\sqrt[3]{8}} = \frac{\sqrt[3]{7}}{2}$

d. $\sqrt[3]{\frac{40}{27}} = \frac{\sqrt[3]{40}}{\sqrt[3]{27}} = \frac{\sqrt[3]{8 \cdot 5}}{3} = \frac{\sqrt[3]{8} \cdot \sqrt[3]{5}}{3} = \frac{2\sqrt[3]{5}}{3}$ ■

EXAMPLE 5 Simplify each expression. Assume that all variables represent positive numbers only.

a. $\sqrt[3]{x^5}$ **b.** $\sqrt[3]{40y^7}$ **c.** $\sqrt[3]{\frac{16x^3}{z^6}}$ **d.** $\sqrt[4]{16x^5}$

Solution:

a. $\sqrt[3]{x^5} = \sqrt[3]{x^3 \cdot x^2} = \sqrt[3]{x^3} \cdot \sqrt[3]{x^2} = x\sqrt[3]{x^2}$

b. $\sqrt[3]{40y^7} = \sqrt[3]{8 \cdot 5 \cdot y^6 \cdot y} = \sqrt[3]{8y^6 \cdot 5y} = \sqrt[3]{8y^6} \cdot \sqrt[3]{5y} = 2y^2\sqrt[3]{5y}$

c. $\sqrt[3]{\frac{16x^3}{z^6}} = \frac{\sqrt[3]{16x^3}}{\sqrt[3]{z^6}} = \frac{\sqrt[3]{8x^3 \cdot 2}}{z^2} = \frac{\sqrt[3]{8x^3} \cdot \sqrt[3]{2}}{z^2} = \frac{2x\sqrt[3]{2}}{z^2}$

d. $\sqrt[4]{16x^5} = \sqrt[4]{16x^4 \cdot x} = \sqrt[4]{16x^4} \cdot \sqrt[4]{x} = 2x\sqrt[4]{x}$ ■

MENTAL MATH

Simplify each expression. Assume that all variables represent nonnegative real numbers.

1. $\sqrt{9}$
2. $\sqrt{16}$
3. $\sqrt{x^2}$
4. $\sqrt{y^4}$
5. $\sqrt{0}$
6. $\sqrt{1}$
7. $\sqrt{25x^4}$
8. $\sqrt{49x^2}$

EXERCISE SET 8.2

Simplify each expression. See Example 1.

1. $\sqrt{20}$
2. $\sqrt{44}$
3. $\sqrt{18}$
4. $\sqrt{45}$
5. $\sqrt{50}$
6. $\sqrt{28}$
7. $\sqrt{33}$
8. $\sqrt{98}$

Simplify each expression. See Example 2.

9. $\sqrt{\frac{8}{25}}$
10. $\sqrt{\frac{63}{16}}$
11. $\sqrt{\frac{27}{121}}$
12. $\sqrt{\frac{24}{169}}$
13. $\sqrt{\frac{9}{4}}$
14. $\sqrt{\frac{100}{49}}$
15. $\sqrt{\frac{125}{9}}$
16. $\sqrt{\frac{27}{100}}$

Simplify each expression. Assume that all variables represent positive numbers only. See Example 3.

17. $\sqrt{x^7}$
18. $\sqrt{y^3}$
19. $\sqrt{\frac{88}{x^4}}$
20. $\sqrt{\frac{x^{11}}{81}}$

Simplify each expression. See Example 4.

21. $\sqrt[3]{24}$
22. $\sqrt[3]{81}$
23. $\sqrt[3]{\frac{5}{64}}$
24. $\sqrt[3]{\frac{32}{125}}$

Simplify each radical. Assume that all variables represent positive real numbers. See Example 5.

25. $\sqrt{\frac{3}{25}}$
26. $\sqrt{\frac{10}{9}}$
27. $\sqrt{\frac{49}{4x^2}}$
28. $\sqrt{\frac{81y^{12}}{z^4}}$
29. $\sqrt[3]{\frac{y^7}{8x^6}}$
30. $\sqrt[3]{\frac{-27}{x^9}}$

Simplify each radical. Assume that all variables represent positive real numbers.

31. $\sqrt{52}$ **32.** $\sqrt{75}$ **33.** $\sqrt{\frac{11}{36}}$ **34.** $\sqrt{\frac{30}{49}}$

35. $-\sqrt{\frac{27}{144}}$ **36.** $-\sqrt{\frac{84}{121}}$ **37.** $\sqrt[3]{\frac{15}{64}}$ **38.** $\sqrt[3]{\frac{4}{27}}$

39. $\sqrt[3]{80}$ **40.** $\sqrt[3]{108}$ **41.** $\sqrt[4]{48}$ **42.** $\sqrt[4]{162}$

43. $\sqrt{121}$ **44.** $\sqrt[3]{125}$ **45.** $\sqrt[3]{8x^3}$ **46.** $\sqrt{16x^8}$

47. $\sqrt{y^5}$ **48.** $\sqrt[3]{y^5}$ **49.** $\sqrt{20}$ **50.** $\sqrt{24}$

51. $\sqrt{25a^2b^3}$ **52.** $\sqrt{9x^5y^7}$ **53.** $\sqrt[3]{-27x^9}$ **54.** $\sqrt[3]{-8a^{21}b^6}$

55. $\sqrt[4]{a^{16}b^4}$ **56.** $\sqrt[4]{x^8y^{12}}$ **57.** $\sqrt[3]{50x^{14}}$ **58.** $\sqrt[3]{40y^{10}}$

59. $\sqrt[5]{-32x^{10}y}$ **60.** $\sqrt[5]{-243z^9}$ **61.** $-\sqrt{32a^8b^7}$ **62.** $-\sqrt{20ab^6}$

63. $\sqrt{\frac{6}{49}}$ **64.** $\sqrt{\frac{8}{81}}$ **65.** $\sqrt{\frac{5x^2}{4y^2}}$ **66.** $\sqrt{\frac{y^{10}}{9x^6}}$

67. $-\sqrt[3]{\frac{z^7}{27x^3}}$ **68.** $-\sqrt[3]{\frac{64a}{b^9}}$ **69.** $\sqrt[4]{\frac{x^7}{16}}$ **70.** $\sqrt[4]{\frac{y}{81x^4}}$

Solve.

71. If a cube is to have a volume of 80 cubic inches, then each side must be $\sqrt[3]{80}$ inches long. Simplify the radical representing the side length.

72. Jeannie Hull is swimming across a 40-foot-wide river, trying to head straight across to the opposite shore. The current is strong enough, however, to move her downstream 100 feet by the time she reaches land. (See the figure.) Because of the current, the actual distance she swam is $\sqrt{11{,}600}$ feet. Simplify this radical.

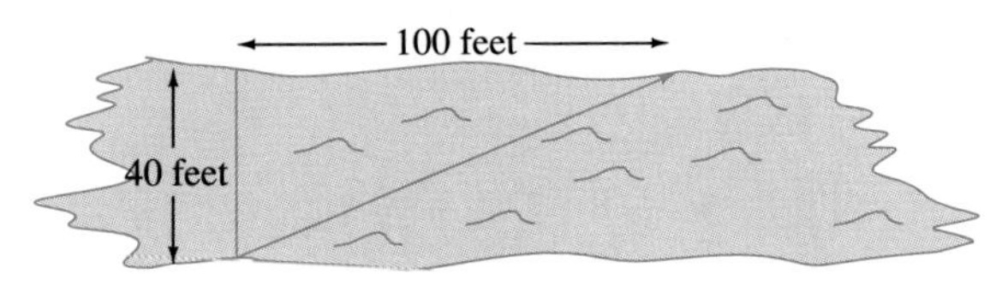

73. By using replacement values for a and b, show that $\sqrt{a^2 + b^2}$ does not equal $a + b$.

Skill Review

Perform the following operations. See Sections 1.5, 5.3, and 5.4.

74. $6x + 8x$ **75.** $(6x)(8x)$ **76.** $(2x + 3)(x - 5)$

77. $(2x + 3) + (x - 5)$ **78.** $9y^2 - 8y^2$ **79.** $(9y^2)(-8y^2)$

Factor. See Sections 6.2 and 6.3.

80. $2x^2 + 11x - 6$ **81.** $4x^2 - 7x + 4$

82. $4x^2 + 4x - 3$ **83.** $6x^2 + 5xy - 4y^2$

84. $6x^2 - 25xy + 4y^2$ **85.** $18x^2 - 60x + 50$

86. $2x^2 - 23xy - 39y^2$ **87.** $4x^2 - 28xy + 49y^2$

8.3
Adding and Subtracting Radical Expressions

OBJECTIVE 1 Add or subtract radical expressions.

Tape IA 21

1 We have learned that only like terms can be added or subtracted. To add or subtract like terms, we use the distributive property. For example,

$$2x + 3x = (2 + 3)x = 5x \quad \text{and} \quad 7x^2y - 4x^2y = (7 - 4)x^2y = 3x^2y$$

The distributive property can also be used to add **like radicals.**

Like Radicals

Radicals with the same index and the same radicand are like radicals.

For example, $2\sqrt{7} + 3\sqrt{7} = (2 + 3)\sqrt{7} = 5\sqrt{7}$. Also,

$$5\sqrt{3x} - 7\sqrt{3x} = (5 - 7)\sqrt{3x} = -2\sqrt{3x}$$

The expression $2\sqrt{7} + 2\sqrt[3]{7}$ cannot be simplifed since $2\sqrt{7}$ and $2\sqrt[3]{7}$ are not like radicals.

EXAMPLE 1 Add or subtract. Assume that all variables represent positive real numbers.

a. $\sqrt{20} + 2\sqrt{45}$ **b.** $\sqrt[3]{54} - 5\sqrt[3]{16} + \sqrt[3]{2}$ **c.** $\sqrt{27x} - 2\sqrt{9x} + \sqrt{72x}$
d. $\sqrt[3]{98} + \sqrt{98}$ **e.** $\sqrt[3]{48y^4} + \sqrt[3]{6y^4}$

Solution: First simplify each radical. Then add or subtract any like radicals.

a.
$$\begin{aligned}\sqrt{20} + 2\sqrt{45} &= \sqrt{4} \cdot \sqrt{5} + 2 \cdot \sqrt{9} \cdot \sqrt{5} \\ &= 2\sqrt{5} + 2 \cdot 3 \cdot \sqrt{5} \\ &= 2\sqrt{5} + 6\sqrt{5} = 8\sqrt{5}\end{aligned}$$

b.
$$\begin{aligned}\sqrt[3]{54} - 5\sqrt[3]{16} + \sqrt[3]{2} &= \sqrt[3]{27} \cdot \sqrt[3]{2} - 5 \cdot \sqrt[3]{8} \cdot \sqrt[3]{2} + \sqrt[3]{2} \\ &= 3\sqrt[3]{2} - 5 \cdot 2\sqrt[3]{2} + \sqrt[3]{2} \\ &= 3\sqrt[3]{2} - 10\sqrt[3]{2} + \sqrt[3]{2} \\ &= -6\sqrt[3]{2}\end{aligned}$$

c.
$$\begin{aligned}\sqrt{27x} - 2\sqrt{9x} + \sqrt{72x} &= \sqrt{9} \cdot \sqrt{3x} - 2 \cdot \sqrt{9} \cdot \sqrt{x} + \sqrt{36} \cdot \sqrt{2x} \\ &= 3\sqrt{3x} - 2 \cdot 3 \cdot \sqrt{x} + 6\sqrt{2x} \\ &= 3\sqrt{3x} - 6\sqrt{x} + 6\sqrt{2x}\end{aligned}$$

d. We can simplify $\sqrt{98}$, but since the indexes are different, these radicals cannot be added.

$$\sqrt[3]{98} + \sqrt{98} = \sqrt[3]{98} + \sqrt{49} \cdot \sqrt{2} = \sqrt[3]{98} + 7\sqrt{2}$$

e.
$$\begin{aligned}\sqrt[3]{48y^4} + \sqrt[3]{6y^4} &= \sqrt[3]{8y^3} \cdot \sqrt[3]{6y} + \sqrt[3]{y^3} \cdot \sqrt[3]{6y} \\ &= 2y\sqrt[3]{6y} + y\sqrt[3]{6y} \\ &= 3y\sqrt[3]{6y}\end{aligned}$$

■

The following summarizes how to simplify radical expressions.

To Simplify Radical Expressions

1. Simplify radical terms.
2. Add or subtract any like radicals.

EXAMPLE 2 Simplify.

a. $\frac{\sqrt{45}}{4} - \frac{\sqrt{5}}{3}$ **b.** $\sqrt[3]{\frac{7x}{8}} + 2\sqrt[3]{7x}$

Solution: **a.** $\frac{\sqrt{45}}{4} - \frac{\sqrt{5}}{3} = \frac{3\sqrt{5}}{4} - \frac{\sqrt{5}}{3} = \frac{3\sqrt{5} \cdot 3}{4 \cdot 3} - \frac{\sqrt{5} \cdot 4}{3 \cdot 4} = \frac{9\sqrt{5}}{12} - \frac{4\sqrt{5}}{12} = \frac{5\sqrt{5}}{12}$

b. $\sqrt[3]{\frac{7x}{8}} + 2\sqrt[3]{7x} = \frac{\sqrt[3]{7x}}{\sqrt[3]{8}} + 2\sqrt[3]{7x} = \frac{\sqrt[3]{7x}}{2} + 2\sqrt[3]{7x}$

$= \frac{\sqrt[3]{7x}}{2} + \frac{2\sqrt[3]{7x} \cdot 2}{2} = \frac{\sqrt[3]{7x}}{2} + \frac{4\sqrt[3]{7x}}{2} = \frac{5\sqrt[3]{7x}}{2}$ ■

MENTAL MATH

Simplify. Assume that all variables represent positive real numbers.

1. $2\sqrt{3} + 4\sqrt{3}$
2. $5\sqrt{7} + 3\sqrt{7}$
3. $8\sqrt{x} - 5\sqrt{x}$
4. $3\sqrt{y} + 10\sqrt{y}$
5. $7\sqrt[3]{x} + 5\sqrt[3]{x}$
6. $8\sqrt[3]{z} - 2\sqrt[3]{z}$

EXERCISE SET 8.3

Simplify. Assume that all variables represent positive real numbers. See Example 1.

1. $\sqrt{8} - \sqrt{32}$
2. $\sqrt{27} - \sqrt{75}$
3. $2\sqrt{2x^3} + 4x\sqrt{8x}$
4. $3\sqrt{45x^3} + x\sqrt{5x}$
5. $2\sqrt{50} - 3\sqrt{125} + \sqrt{98}$
6. $4\sqrt{32} - \sqrt{18} + 2\sqrt{128}$
7. $\sqrt[3]{16x} - \sqrt[3]{54x}$
8. $2\sqrt[3]{3a^4} - 3a\sqrt[3]{81a}$
9. $\sqrt{9b^3} - \sqrt{25b^3} + \sqrt{49b^3}$
10. $\sqrt{4x^7} + 9x^2\sqrt{x^3} - 5x\sqrt{x^5}$

Simplify. Assume that all variables represent positive real numbers. See Example 2.

11. $\frac{5\sqrt{2}}{3} + \frac{2\sqrt{2}}{5}$
12. $\frac{\sqrt{3}}{2} + \frac{4\sqrt{3}}{3}$
13. $\sqrt[3]{\frac{11}{8}} - \frac{\sqrt[3]{11}}{6}$
14. $\frac{2\sqrt[3]{4}}{7} - \frac{\sqrt[3]{4}}{14}$
15. $\frac{\sqrt{20x}}{9} + \sqrt{\frac{5x}{9}}$
16. $\frac{3x\sqrt{7}}{5} + \sqrt{\frac{7x^2}{100}}$

Simplify. Assume that all variables represent positive real numbers.

17. $7\sqrt{9} - 7 + \sqrt{3}$
18. $\sqrt{16} - 5\sqrt{10} + 7$
19. $2 + 3\sqrt{y^2} - 6\sqrt{y^2} + 5$
20. $3\sqrt{7} - \sqrt[3]{x} + 4\sqrt{7} - 3\sqrt[3]{x}$
21. $3\sqrt{108} - 2\sqrt{18} - 3\sqrt{48}$
22. $-\sqrt{75} + \sqrt{12} - 3\sqrt{3}$

23. $-5\sqrt[3]{625} + \sqrt[3]{40}$

24. $-2\sqrt[3]{108} - \sqrt[3]{32}$

25. $\sqrt{9b^3} - \sqrt{25b^3} + \sqrt{16b^3}$

26. $\sqrt{4x^7y^5} + 9x^2\sqrt{x^3y^5} - 5xy\sqrt{x^5y^3}$

27. $5y\sqrt{8y} + 2\sqrt{50y^3}$

28. $3\sqrt{8x^2y^3} - 2x\sqrt{32y^3}$

29. $\sqrt[3]{54xy^3} - 5\sqrt[3]{2xy^3} + y\sqrt[3]{128x}$

30. $2\sqrt[3]{24x^3y^4} + 4x\sqrt[3]{81y^4}$

31. $6\sqrt[3]{11} + 8\sqrt{11} - 12\sqrt{11}$

32. $3\sqrt[3]{5} + 4\sqrt{5}$

33. $-2\sqrt[4]{x^7} + 3\sqrt[4]{16x^7}$

34. $6\sqrt[3]{24x^3} - 2\sqrt[3]{81x^3} - x\sqrt[3]{3}$

35. $\dfrac{4\sqrt{3}}{3} - \dfrac{\sqrt{12}}{3}$

36. $\dfrac{\sqrt{45}}{10} + \dfrac{7\sqrt{5}}{10}$

37. $\dfrac{\sqrt[3]{8x^4}}{7} + \dfrac{3x\sqrt[3]{x}}{7}$

38. $\dfrac{\sqrt[4]{48}}{5x} - \dfrac{2\sqrt[4]{3}}{10x}$

39. $\sqrt{\dfrac{28}{x^2}} + \sqrt{\dfrac{7}{4x^2}}$

40. $\dfrac{\sqrt{99}}{5x} - \sqrt{\dfrac{44}{x^2}}$

41. $\sqrt[3]{\dfrac{16}{27}} - \dfrac{\sqrt[3]{54}}{6}$

42. $\dfrac{\sqrt[3]{3}}{10} + \sqrt[3]{\dfrac{24}{125}}$

43. $-\dfrac{\sqrt[3]{2x^4}}{9} + \sqrt[3]{\dfrac{250x^4}{27}}$

44. $\dfrac{\sqrt[3]{y^5}}{8} + \dfrac{5y\sqrt[3]{y^2}}{4}$

Solve.

45. Find the perimeter of the trapezoid.

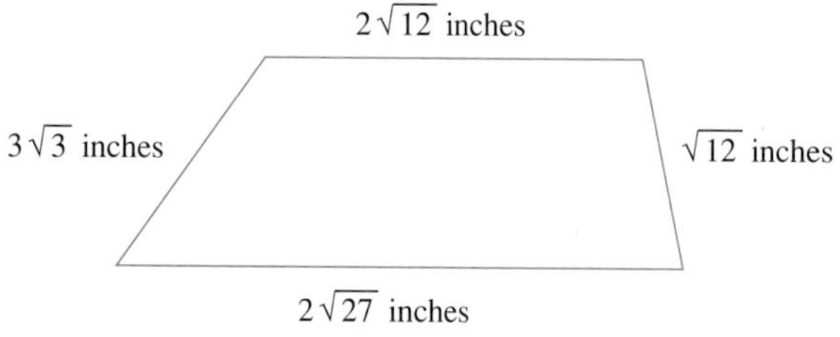

46. Find the perimeter of the triangle.

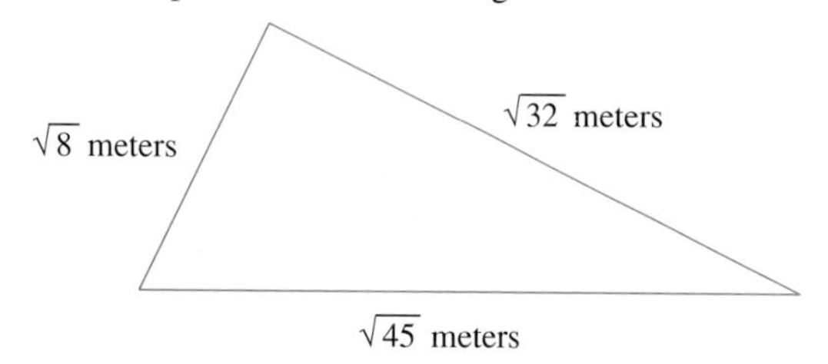

47. Baseboard needs to be installed around the perimeter of a rectangular room. Find how much baseboard should be ordered by finding the perimeter of the room.

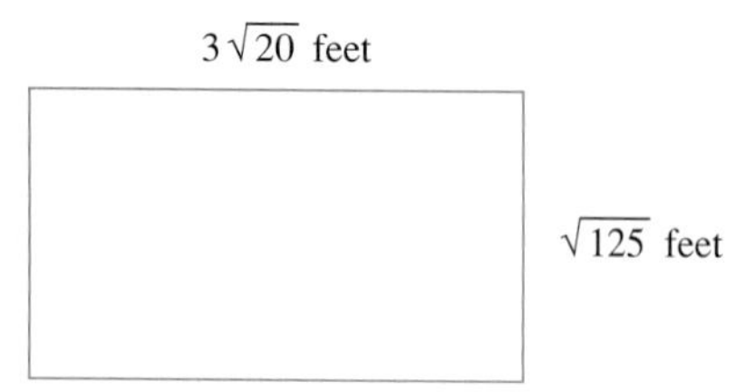

48. A border of wallpaper is to be used around the perimeter of the odd-shaped room shown. Find how much wallpaper border is needed by finding the perimeter of the room.

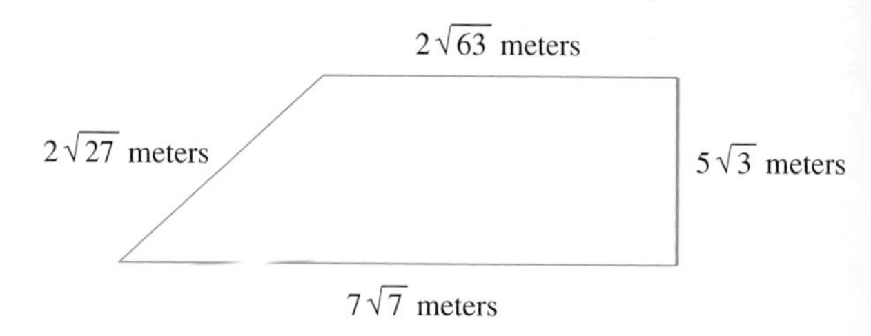

49. Find the perimeter of the rectangular picture frame.

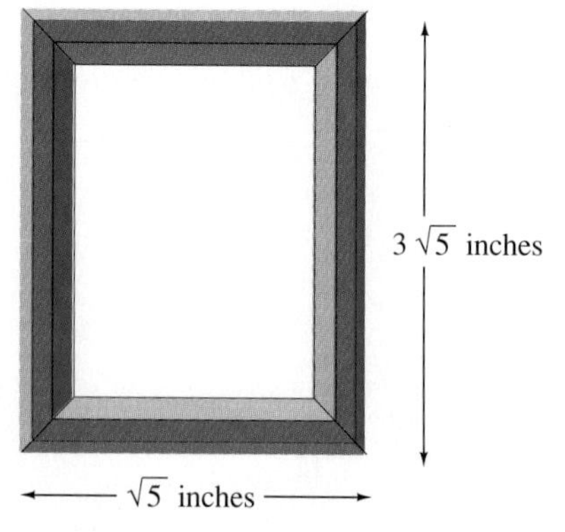

50. Find the perimeter of the plot of land.

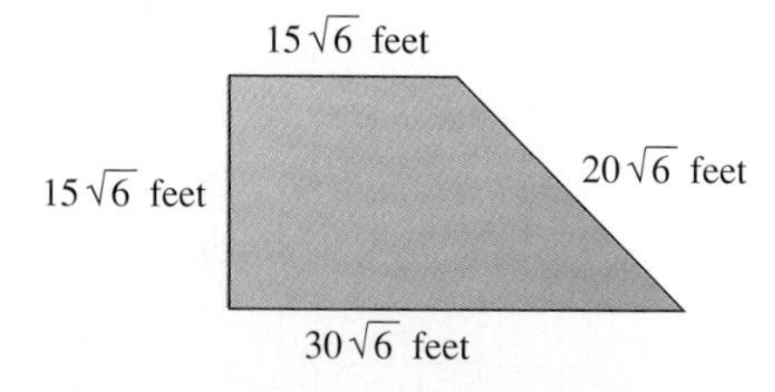

51. A water trough is to be made of wood. Each of the two triangular end pieces has an area of $\frac{3\sqrt{27}}{4}$ square feet. The two side panels are both rectangular. In simplest radical form, find the total area of the wood needed.

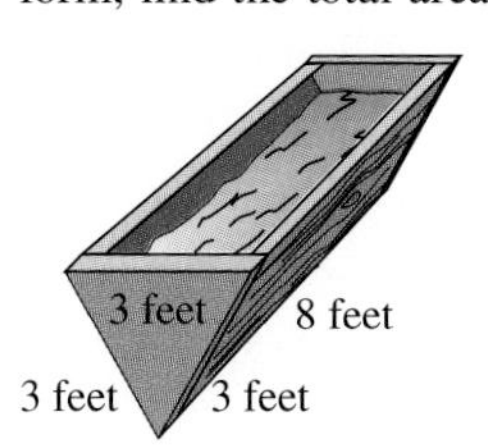

52. Eight wooden braces are to be attached along the diagonals of the vertical sides of a storage bin. Each of four of these diagonals has a length of $\sqrt{52}$ feet, whereas each of the other four has a length of $\sqrt{80}$ feet. In simplest radical form, find the total length of the wood needed for these braces.

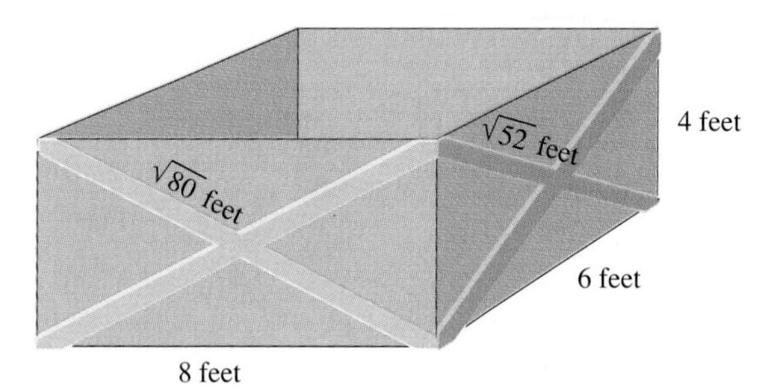

Skill Review

Multiply. See Section 5.4.

53. $(x + 4)(x - 4)$

54. $(3x + 2)(3x - 2)$

55. $(2x + 5)^2$

56. $(5xy - 4)(xy - 11)$

Find the domain of each function. See Sections 7.1 and 8.1.

57. $f(x) = \frac{1}{x}$

58. $g(x) = \sqrt{x}$

59. $h(x) = \sqrt{x - 1}$

60. $F(x) = \frac{x^2}{x^2 - 4}$

61. $f(x) = \sqrt[3]{x}$

62. $F(x) = \sqrt{x} + 10$

63. $h(x) = \frac{-2x}{x(x - 3)}$

64. $g(x) = x^2 + 2x + 1$

65. $f(x) = 2x - 6$

66. $g(x) = |x|$

Writing in Mathematics

67. In your own words, describe like radicals.

8.4 Multiplying and Dividing Radical Expressions

OBJECTIVES

1. Multiply radical expressions.
2. Rationalize denominators.
3. Rationalize numerators.

Tape IA 21

1 In this section, we multiply and divide radical expressions. Radical expressions are multiplied using many of the same properties used to multiply polynomial expressions. For instance, to multiply $\sqrt{2}(\sqrt{6} - 3\sqrt{2})$, we use the distributive property and multiply $\sqrt{2}$ by each term inside the parentheses.

$$\begin{aligned}\sqrt{2}(\sqrt{6} - 3\sqrt{2}) &= \sqrt{2}(\sqrt{6}) - \sqrt{2}(3\sqrt{2})\\ &= \sqrt{12} - 3\sqrt{2 \cdot 2}\\ &= \sqrt{4 \cdot 3} - 3 \cdot 2\\ &= 2\sqrt{3} - 6\end{aligned}$$

EXAMPLE 1 Multiply.

a. $\sqrt{3}(5 + \sqrt{30})$ **b.** $(\sqrt{5} - \sqrt{6})(\sqrt{7} + 1)$ **c.** $(7\sqrt{x} + 5)(3\sqrt{x} - \sqrt{5})$
d. $(4\sqrt{3} - 1)^2$ **e.** $(\sqrt{2x} - 5)(\sqrt{2x} + 5)$

Solution: **a.**
$$\begin{aligned}\sqrt{3}(5 + \sqrt{30}) &= \sqrt{3}(5) + \sqrt{3}(\sqrt{30})\\ &= 5\sqrt{3} + \sqrt{3 \cdot 30}\\ &= 5\sqrt{3} + \sqrt{3 \cdot 3 \cdot 10}\\ &= 5\sqrt{3} + 3\sqrt{10}\end{aligned}$$

b. To multiply, we can use the FOIL order.

$$\begin{aligned}(\sqrt{5} - \sqrt{6})(\sqrt{7} + 1) &= \overset{\text{First}}{\sqrt{5} \cdot \sqrt{7}} + \overset{\text{Outside}}{\sqrt{5} \cdot 1} - \overset{\text{Inside}}{\sqrt{6} \cdot \sqrt{7}} - \overset{\text{Last}}{\sqrt{6} \cdot 1}\\ &= \sqrt{35} + \sqrt{5} - \sqrt{42} - \sqrt{6}\end{aligned}$$

c.
$$\begin{aligned}(7\sqrt{x} + 5)(3\sqrt{x} - \sqrt{5}) &= 7\sqrt{x}(3\sqrt{x}) - 7\sqrt{x}(\sqrt{5}) + 5(3\sqrt{x}) - 5(\sqrt{5})\\ &= 21x - 7\sqrt{5x} + 15\sqrt{x} - 5\sqrt{5}\end{aligned}$$

d.
$$\begin{aligned}(4\sqrt{3} - 1)^2 &= (4\sqrt{3} - 1)(4\sqrt{3} - 1)\\ &= 4\sqrt{3}(4\sqrt{3}) - 4\sqrt{3}(1) - 1(4\sqrt{3}) - 1(-1)\\ &= 16 \cdot 3 - 4\sqrt{3} - 4\sqrt{3} + 1\\ &= 48 - 8\sqrt{3} + 1\\ &= 49 - 8\sqrt{3}\end{aligned}$$

e.
$$\begin{aligned}(\sqrt{2x} - 5)(\sqrt{2x} + 5) &= \sqrt{2x} \cdot \sqrt{2x} + 5\sqrt{2x} - 5\sqrt{2x} - 5 \cdot 5\\ &= 2x - 25\end{aligned}$$ ■

2 Often in mathematics, it is helpful to write a radical expression such as $\frac{\sqrt{3}}{\sqrt{2}}$ without a radical in the denominator. The process of writing this expression as an equivalent expression but without a radical in the denominator is called **rationalizing the denominator.** To rationalize the denominator of $\frac{\sqrt{3}}{\sqrt{2}}$, we use the fundamental principle of fractions and multiply the numerator and the denominator by $\sqrt{2}$.

$$\frac{\sqrt{3}}{\sqrt{2}} = \frac{\sqrt{3} \cdot \sqrt{2}}{\sqrt{2} \cdot \sqrt{2}} = \frac{\sqrt{6}}{\sqrt{4}} = \frac{\sqrt{6}}{2}$$

EXAMPLE 2 Rationalize the denominator of each expression.

a. $\frac{\sqrt{27}}{\sqrt{5}}$ **b.** $\frac{2\sqrt{16}}{\sqrt{9x}}$ **c.** $\sqrt[3]{\frac{1}{2}}$

Solution: **a.** First we simplify $\sqrt{27}$; then we rationalize the denominator.

$$\frac{\sqrt{27}}{\sqrt{5}} = \frac{\sqrt{9 \cdot 3}}{\sqrt{5}} = \frac{3\sqrt{3}}{\sqrt{5}}$$

To rationalize the denominator, multiply the numerator and denominator by $\sqrt{5}$.

$$\frac{3\sqrt{3}}{\sqrt{5}} = \frac{3\sqrt{3} \cdot \sqrt{5}}{\sqrt{5} \cdot \sqrt{5}} = \frac{3\sqrt{15}}{5}$$

b. First we simplify the radicals and then rationalize the denominator.

$$\frac{2\sqrt{16}}{\sqrt{9x}} = \frac{2(4)}{3\sqrt{x}} = \frac{8}{3\sqrt{x}}$$

To rationalize the denominator, multiply the numerator and denominator by $\sqrt{x}$. Then

$$\frac{8}{3\sqrt{x}} = \frac{8 \cdot \sqrt{x}}{3\sqrt{x} \cdot \sqrt{x}} = \frac{8\sqrt{x}}{3x}$$

c. $\sqrt[3]{\frac{1}{2}} = \frac{\sqrt[3]{1}}{\sqrt[3]{2}} = \frac{1}{\sqrt[3]{2}}$. Now we rationalize the denominator. Since this is a cube root, we want to multiply the denominator $\sqrt[3]{2}$ by a value that will make the radicand a perfect cube. If we multiply $\sqrt[3]{2}$ by $\sqrt[3]{2^2}$, we get $\sqrt[3]{2^3} = \sqrt[3]{8} = 2$.

$$= \frac{1 \cdot \sqrt[3]{2^2}}{\sqrt[3]{2} \cdot \sqrt[3]{2^2}} = \frac{\sqrt[3]{4}}{\sqrt[3]{8}} = \frac{\sqrt[3]{4}}{2}$$ ■

A different method is needed to rationalize a denominator that is a sum or difference of two terms. For example, let's rationalize the denominator of an expression like $\frac{5}{\sqrt{3} - 2}$. To eliminate the radical from this denominator, multiply both the numerator and denominator by $\sqrt{3} + 2$, the **conjugate** of $\sqrt{3} - 2$. In general, the conjugate of $a + b$ is $a - b$. Then

$$\frac{5}{\sqrt{3} - 2} = \frac{5\,(\sqrt{3} + 2)}{(\sqrt{3} - 2)\,(\sqrt{3} + 2)}$$

$$= \frac{5(\sqrt{3} + 2)}{\sqrt{3} \cdot \sqrt{3} + 2\sqrt{3} - 2\sqrt{3} - 4}$$

$$= \frac{5(\sqrt{3} + 2)}{3 - 4}$$

$$= \frac{5(\sqrt{3} + 2)}{-1}$$

$$= -5(\sqrt{3} + 2) \quad \text{or} \quad -5\sqrt{3} - 10$$

EXAMPLE 3 Rationalize each denominator.

a. $\frac{2}{3\sqrt{2} + 5}$ **b.** $\frac{\sqrt{6} + 2}{\sqrt{5} - \sqrt{3}}$ **c.** $\frac{7\sqrt{y}}{\sqrt{12x}}$ **d.** $\frac{2\sqrt{m}}{3\sqrt{x} + \sqrt{m}}$

Solution: **a.** Multiply the numerator and denominator by the conjugate of $3\sqrt{2} + 5$.

$$\frac{2}{3\sqrt{2}+5} = \frac{2\,(3\sqrt{2}-5)}{(3\sqrt{2}+5)\,(3\sqrt{2}-5)}$$

$$= \frac{2(3\sqrt{2}-5)}{(3\sqrt{2})^2 - 5^2}$$

Use the pattern discussed above to multiply $(3\sqrt{2} + 5)$ and $(3\sqrt{2} - 5)$.

$$= \frac{2(3\sqrt{2}-5)}{18-25}$$

$$= \frac{2(3\sqrt{2}-5)}{-7} \quad \text{or} \quad -\frac{2(3\sqrt{2}-5)}{7} \quad \text{or} \quad \frac{10-6\sqrt{2}}{7}$$

b. Multiply the numerator and denominator by the conjugate of $\sqrt{5} - \sqrt{3}$.

$$\frac{\sqrt{6}+2}{\sqrt{5}-\sqrt{3}} = \frac{(\sqrt{6}+2)\,(\sqrt{5}+\sqrt{3})}{(\sqrt{5}-\sqrt{3})\,(\sqrt{5}+\sqrt{3})}$$

$$= \frac{\sqrt{6}\sqrt{5}+\sqrt{6}\sqrt{3}+2\sqrt{5}+2\sqrt{3}}{(\sqrt{5})^2-(\sqrt{3})^2}$$

$$= \frac{\sqrt{30}+\sqrt{18}+2\sqrt{5}+2\sqrt{3}}{5-3}$$

$$= \frac{\sqrt{30}+3\sqrt{2}+2\sqrt{5}+2\sqrt{3}}{2}$$

c. Notice that the denominator of this example is **not the sum or difference of two terms.** For this reason, we simplify the radical expression and then multiply the numerator and denominator by a factor so that the resulting denominator is a rational expression.

$$\frac{7\sqrt{y}}{\sqrt{12x}} = \frac{7\sqrt{y}}{\sqrt{4}\sqrt{3x}} = \frac{7\sqrt{y}\cdot\sqrt{3x}}{2\sqrt{3x}\cdot\sqrt{3x}} = \frac{7\sqrt{3xy}}{2\cdot 3x} = \frac{7\sqrt{3xy}}{6x}$$

d. Multiply by the conjugate of $3\sqrt{x} + \sqrt{m}$ to eliminate the radicals from the denominator.

$$\frac{2\sqrt{m}}{3\sqrt{x}+\sqrt{m}} = \frac{2\sqrt{m}\,(3\sqrt{x}-\sqrt{m})}{(3\sqrt{x}+\sqrt{m})\,(3\sqrt{x}-\sqrt{m})} = \frac{6\sqrt{mx}-2m}{(3\sqrt{x})^2-(\sqrt{m})^2}$$

$$= \frac{6\sqrt{mx}-2m}{9x-m}$$

■

3 It is also often helpful to write an expression such as $\frac{\sqrt{3}}{\sqrt{2}}$ as an equivalent expression without a radical in the numerator. This process is called **rationalizing the numerator.** To rationalize the numerator of $\frac{\sqrt{3}}{\sqrt{2}}$, we multiply the numerator and denominator by $\sqrt{3}$.

$$\frac{\sqrt{3}}{\sqrt{2}} = \frac{\sqrt{3}\cdot\sqrt{3}}{\sqrt{2}\cdot\sqrt{3}} = \frac{\sqrt{9}}{\sqrt{6}} = \frac{3}{\sqrt{6}}$$

EXAMPLE 4 Rationalize the numerator of each expression.

a. $\dfrac{\sqrt{28}}{\sqrt{45}}$ **b.** $\dfrac{\sqrt[3]{2x^2}}{\sqrt[3]{5y}}$

Solution: **a.** First we simplify $\sqrt{28}$ and $\sqrt{45}$.

$$\frac{\sqrt{28}}{\sqrt{45}} = \frac{\sqrt{4 \cdot 7}}{\sqrt{9 \cdot 5}} = \frac{2\sqrt{7}}{3\sqrt{5}}$$

Next we rationalize the numerator by multiplying the numerator and denominator by $\sqrt{7}$.

$$\frac{2\sqrt{7}}{3\sqrt{5}} = \frac{2\sqrt{7} \cdot \sqrt{7}}{3\sqrt{5} \cdot \sqrt{7}} = \frac{2 \cdot 7}{3\sqrt{5 \cdot 7}} = \frac{14}{3\sqrt{35}}$$

b. The numerator and the denominator of this expression are already simplified. To rationalize the numerator, $\sqrt[3]{2x^2}$, we multiply the numerator and denominator by a factor that will make the radicand a perfect cube. If we multiply $\sqrt[3]{2x^2}$ by $\sqrt[3]{4x}$, we get $\sqrt[3]{8x^3} = 2x$.

$$\frac{\sqrt[3]{2x^2}}{\sqrt[3]{5y}} = \frac{\sqrt[3]{2x^2} \cdot \sqrt[3]{4x}}{\sqrt[3]{5y} \cdot \sqrt[3]{4x}} = \frac{\sqrt[3]{8x^3}}{\sqrt[3]{20xy}} = \frac{2x}{\sqrt[3]{20xy}}$$ ■

EXAMPLE 5 Rationalize the numerator of $\dfrac{\sqrt{x} + 2}{5}$.

Solution: Multiply the numerator and denominator by the conjugate of $\sqrt{x} + 2$.

$$\begin{aligned}\frac{\sqrt{x} + 2}{5} &= \frac{(\sqrt{x} + 2)(\sqrt{x} - 2)}{5(\sqrt{x} - 2)} \\ &= \frac{x - 2\sqrt{x} + 2\sqrt{x} - 4}{5(\sqrt{x} - 2)} \\ &= \frac{x - 4}{5(\sqrt{x} - 2)}\end{aligned}$$ ■

EXERCISE SET 8.4

Multiply, and then simplify if possible. See Example 1.

1. $\sqrt{7}(\sqrt{5} + \sqrt{3})$
2. $\sqrt{5}(\sqrt{15} - \sqrt{35})$
3. $(\sqrt{5} - \sqrt{2})^2$
4. $(3x - \sqrt{2})(3x - \sqrt{2})$
5. $\sqrt{3x}(\sqrt{3} - \sqrt{x})$
6. $\sqrt{5y}(\sqrt{y} + \sqrt{5})$
7. $(2\sqrt{x} - 5)(3\sqrt{x} + 1)$
8. $(8\sqrt{y} + z)(4\sqrt{y} - 1)$
9. $(\sqrt[3]{a} - 4)(\sqrt[3]{a} + 5)$
10. $(\sqrt[3]{a} + 2)(\sqrt[3]{a} + 7)$

Rationalize each denominator. See Example 2.

11. $\dfrac{1}{\sqrt{3}}$
12. $\dfrac{\sqrt{2}}{\sqrt{6}}$
13. $\sqrt{\dfrac{1}{5}}$
14. $\sqrt{\dfrac{1}{2}}$

15. $\frac{4}{\sqrt[3]{3}}$ **16.** $\frac{6}{\sqrt[3]{9}}$ **17.** $\frac{3}{\sqrt{8x}}$ **18.** $\frac{5}{\sqrt{27a}}$

19. $\frac{3}{\sqrt[3]{4x^2}}$ **20.** $\frac{5}{\sqrt[3]{3y}}$

Rationalize each denominator. See Example 3.

21. $\frac{5}{2 - \sqrt{7}}$ **22.** $\frac{3}{\sqrt{5} - 2}$

23. $\frac{-7}{\sqrt{x} - 3}$ **24.** $\frac{-8}{\sqrt{y} + 4}$

25. $\frac{\sqrt{2} - \sqrt{3}}{\sqrt{2} + \sqrt{3}}$ **26.** $\frac{\sqrt{3} + \sqrt{4}}{\sqrt{2} + \sqrt{3}}$

27. $\frac{\sqrt{a} + 1}{2\sqrt{a} - \sqrt{b}}$ **28.** $\frac{2\sqrt{a} - 3}{2\sqrt{a} - \sqrt{b}}$

Rationalize each numerator. See Example 4.

29. $\sqrt{\frac{5}{3}}$ **30.** $\sqrt{\frac{3}{2}}$ **31.** $\sqrt{\frac{18}{5}}$ **32.** $\sqrt{\frac{12}{7}}$

33. $\frac{\sqrt{4x}}{7}$ **34.** $\frac{\sqrt{3x^5}}{6}$ **35.** $\frac{\sqrt[3]{5y^2}}{\sqrt[3]{4x}}$ **36.** $\frac{\sqrt[3]{4x}}{\sqrt[3]{z^4}}$

Rationalize each numerator. See Example 5.

37. $\frac{2 - \sqrt{7}}{-5}$ **38.** $\frac{\sqrt{5} + 2}{\sqrt{2}}$ **39.** $\frac{\sqrt{x} + 3}{\sqrt{x}}$ **40.** $\frac{5 + \sqrt{2}}{\sqrt{2x}}$

Multiply, and then simplify if possible.

41. $6(\sqrt{2} - 2)$ **42.** $\sqrt{5}(6 - \sqrt{5})$

43. $\sqrt{2}(\sqrt{2} + x\sqrt{6})$ **44.** $\sqrt{3}(\sqrt{3} - 2\sqrt{5x})$

45. $(2\sqrt{7} + 3\sqrt{5})(\sqrt{7} - 2\sqrt{5})$ **46.** $(\sqrt{6} - 4\sqrt{2})(3\sqrt{6} + 1)$

47. $(\sqrt{x} - y)(\sqrt{x} + y)$ **48.** $(3\sqrt{x} + 2)(\sqrt{3x} - 2)$

49. $(\sqrt{3} + x)^2$ **50.** $(\sqrt{y} - 3x)^2$

51. $(\sqrt{5x} - 3\sqrt{2})(\sqrt{5x} - 3\sqrt{3})$ **52.** $(5\sqrt{3x} - \sqrt{y})(4\sqrt{x} + 1)$

53. $(\sqrt[3]{4} + 2)(\sqrt[3]{2} - 1)$ **54.** $(\sqrt[3]{3} + \sqrt[3]{2})(\sqrt[3]{9} - \sqrt[3]{4})$

55. $(\sqrt[3]{x} + 1)(\sqrt[3]{x} - 4\sqrt{x} + 7)$ **56.** $(\sqrt[3]{3x} + 3)(\sqrt[3]{2x} - 3x - 1)$

Rationalize each denominator.

57. $\sqrt{\frac{2}{5}}$ **58.** $\sqrt{\frac{3}{7}}$ **59.** $\frac{9}{\sqrt{3a}}$ **60.** $\frac{x}{\sqrt{5}}$

61. $\frac{3}{\sqrt[3]{2}}$ **62.** $\frac{5}{\sqrt[3]{9}}$ **63.** $\frac{8}{1 + \sqrt{10}}$ **64.** $\frac{-3}{\sqrt{6} - 2}$

65. $\frac{x}{\sqrt[3]{16x^2}}$ **66.** $\frac{3a}{\sqrt[3]{2a}}$ **67.** $\sqrt{\frac{18x^4y^6}{3z}}$ **68.** $\sqrt{\frac{8x^5y}{2z}}$

69. $\dfrac{\sqrt{2}-1}{\sqrt{2}+1}$
70. $\dfrac{\sqrt{8}-\sqrt{3}}{\sqrt{2}+\sqrt{3}}$
71. $\dfrac{\sqrt{x}+1}{\sqrt{x}-1}$
72. $\dfrac{\sqrt{x}+\sqrt{y}}{\sqrt{x}-\sqrt{y}}$

Rationalize each numerator.

73. $\sqrt{\dfrac{2}{5}}$
74. $\sqrt{\dfrac{3}{7}}$
75. $\dfrac{\sqrt{2x}}{11}$
76. $\dfrac{\sqrt{y}}{7}$
77. $\sqrt[3]{\dfrac{7}{8}}$
78. $\sqrt[3]{\dfrac{25}{2}}$
79. $\dfrac{2-\sqrt{11}}{6}$
80. $\dfrac{\sqrt{15}+1}{2}$
81. $\dfrac{\sqrt[3]{3x^5}}{10}$
82. $\sqrt[3]{\dfrac{9y}{7}}$
83. $\sqrt{\dfrac{18x^4y^6}{3z}}$
84. $\sqrt{\dfrac{8x^5y}{2z}}$
85. $\dfrac{\sqrt{2}-1}{\sqrt{2}+1}$
86. $\dfrac{\sqrt{8}-\sqrt{3}}{\sqrt{2}+\sqrt{3}}$
87. $\dfrac{\sqrt{x}+1}{\sqrt{x}-1}$
88. $\dfrac{\sqrt{x}+\sqrt{y}}{\sqrt{x}-\sqrt{y}}$

Solve.

89. Find the volume of the box.

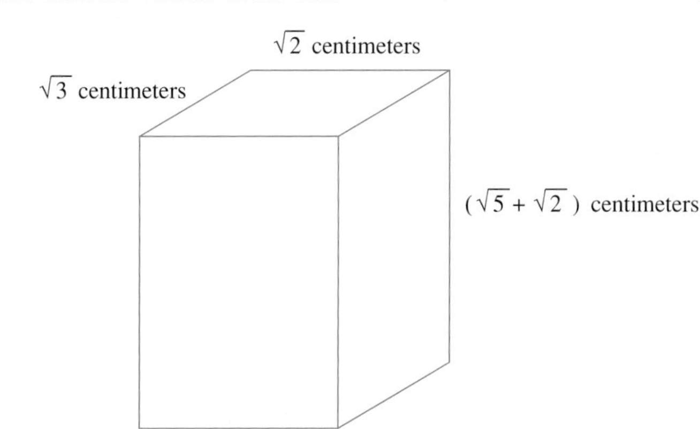

90. Find the area of the triangle.

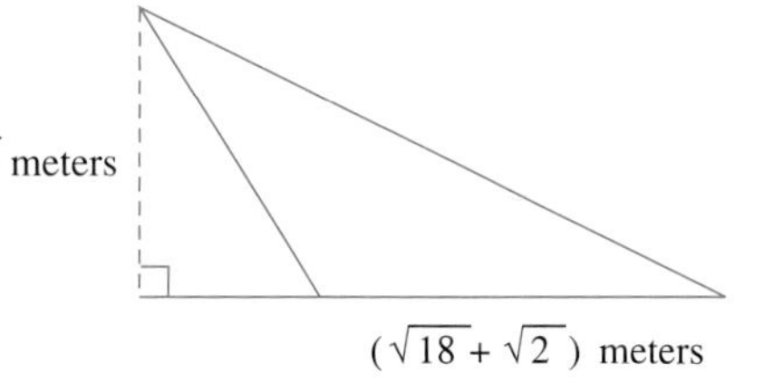

Skill Review

Evaluate. See Sections 5.1 and 5.2.

91. 5^{-1}
92. 7^{-2}
93. -2^{-3}
94. -3^{-4}

Simplify. Write answers with positive exponents. Assume that all variables represent nonzero real numbers.

95. $x^{-2}x^{-3}$
96. $\left(\dfrac{x}{3}\right)^{-3}$
97. $\dfrac{a^{-2}}{a^{-3}}$
98. $\dfrac{x^{-4}y^7}{x^3y}$
99. $\dfrac{(xy^3)^5}{(xy)^{-4}}$
100. $\dfrac{(rs)^{-3}}{(r^2s^3)^2}$
101. $\left(\dfrac{3x^2}{x^{-4}}\right)^{-2}$
102. $\left(\dfrac{2y^{-3}}{z}\right)^{-1}$
103. $\dfrac{(c^2)^4 \cdot c^{-5}}{c^{-10}}$
104. $\dfrac{(z^3)^5 \cdot z^{-7}}{(z^2)^{-2} \cdot z^6}$

Writing in Mathematics

105. Explain why rationalizing the denominator does not change the value of the original expression.

8.5 Rational Exponents

OBJECTIVES

Tape BA 29

1. Evaluate exponential expressions of the form $a^{1/n}$.
2. Evaluate exponential expressions of the form $a^{m/n}$.
3. Evaluate exponential expressions of the form $a^{-m/n}$.
4. Practice converting between radical notation and rational exponent notation.
5. Use rules for exponents to simplifying expressions containing fractional exponents.

1 Thus far, we have simplified exponential expressions containing integer exponents. In this section, we explore the meaning of expressions containing rational exponents such as

$$3^{1/2}, \quad 2^{-3/4}, \quad \text{and} \quad y^{5/6}$$

In simplifying fractional expressions, keep in mind that we want the rules for operating with rational exponents to be the same as the rules for operating with integer exponents. These rules are repeated here for review.

Summary of Exponent Rules

If m and n are integers and a, b, and c are real numbers, then

Product rule for exponents:	$a^m \cdot a^n = a^{m+n}$
Power rule for exponents:	$(a^m)^n = a^{m \cdot n}$
Power rules for products and quotients:	$(ab)^n = a^n b^n$ and $\left(\frac{a}{c}\right)^n = \frac{a^n}{c^n}, \quad c \neq 0$
Quotient rule for exponents:	$\frac{a^m}{a^n} = a^{m-n}, \quad a \neq 0$
Zero exponent:	$a^0 = 1, \quad a \neq 0$
Negative exponent:	$a^{-n} = \frac{1}{a^n}, \quad a \neq 0$

If the rule $(a^m)^n = a^{m \cdot n}$ is to hold for fractional exponents, it should be true that

$$(3^{1/2})^2 = 3^{1/2 \cdot 2} = 3^1 = 3$$

Also, we know that

$$(\sqrt{3})^2 = 3$$

Since both expressions simplify to 3, it would be reasonable to define

$$3^{1/2} \quad \text{as} \quad \sqrt{3}$$

In general, we have the following:

Definition of $a^{1/n}$

If n is a positive integer and $\sqrt[n]{a}$ is a real number, then

$$a^{1/n} = \sqrt[n]{a}$$

Notice that the denominator of the rational exponent corresponds to the index of the radical.

EXAMPLE 1 Simplify each expression.

a. $25^{1/2}$ **b.** $8^{1/3}$ **c.** $-16^{1/4}$ **d.** $(-27)^{1/3}$ **e.** $\left(\frac{1}{9}\right)^{1/2}$

Solution:

a. $25^{1/2} = \sqrt{25} = 5$

b. $8^{1/3} = \sqrt[3]{8} = 2$

c. In $-16^{1/4}$, the base of the exponent is 16. Thus, **the negative sign is not affected by the exponent;** so $-16^{1/4} = -\sqrt[4]{16} = -2$

d. $(-27)^{1/3} = \sqrt[3]{-27} = -3$

e. $\left(\frac{1}{9}\right)^{1/2} = \sqrt{\frac{1}{9}} = \frac{1}{3}$ ■

2 In Example 1, each rational exponent has a numerator of 1. What happens if the numerator is some other positive integer? Consider $8^{2/3}$. Since $\frac{2}{3}$ is the same $\frac{1}{3} \cdot 2$, we reason that

$$8^{2/3} = 8^{1/3 \cdot 2} = (8^{1/3})^2 = (\sqrt[3]{8})^2 = 2^2 = 4$$

The denominator 3 of the rational exponent corresponds to the index of the radical. The numerator 2 of the fractional exponent indicates that the base is to be squared.

Definition of $a^{m/n}$

If m and n are integers with $n > 0$ and a is positive, then

$$a^{m/n} = \sqrt[n]{a^m} = (\sqrt[n]{a})^m$$

EXAMPLE 2 Simplify each expression.

a. $4^{3/2}$ **b.** $(-27)^{2/3}$ **c.** $-16^{3/4}$

Solution:

a. $4^{3/2} = (4^{1/2})^3 = (\sqrt{4})^3 = 2^3 = 8$

b. $(-27)^{2/3} = (-27^{1/3})^2 = (\sqrt[3]{-27})^2 = (-3)^2 = 9$

c. The negative sign is **not** affected by the exponent since the base of the exponent is 16. Thus, $-16^{3/4} = -(16^{1/4})^3 = -(\sqrt[4]{16})^3 = -2^3 = -8$. ■

HELPFUL HINT

Recall that

$$-3^2 = -(3 \cdot 3) = -9$$

and

$$(-3)^2 = (-3)(-3) = 9$$

In other words, without parentheses the exponent 2 applies to the base of 3, **not** -3. The same is true of rational exponents. For example,

$$-16^{1/2} = -\sqrt{16} = -4$$

and

$$(-27)^{1/3} = \sqrt[3]{-27} = -3.$$

3 If the exponent is a negative rational number, use the following definition.

Definition of $a^{-m/n}$

If $a^{m/n}$ is a nonzero real number, then

$$a^{-m/n} = \frac{1}{a^{m/n}}$$

EXAMPLE 3 Simplify each expression.

a. $36^{-1/2}$ **b.** $16^{-3/4}$ **c.** $-9^{1/2}$ **d.** $32^{-4/5}$

Solution: **a.** $36^{-1/2} = \frac{1}{36^{1/2}} = \frac{1}{\sqrt{36}} = \frac{1}{6}$

b. $16^{-3/4} = \frac{1}{16^{3/4}} = \frac{1}{(\sqrt[4]{16})^3} = \frac{1}{2^3} = \frac{1}{8}$

c. $-9^{1/2} = -\sqrt{9} = -3$

d. $32^{-4/5} = \frac{1}{32^{4/5}} = \frac{1}{(\sqrt[5]{32})^4} = \frac{1}{2^4} = \frac{1}{16}$ ■

4 To emphasize that $a^{m/n}$ and $(\sqrt[n]{a})^m$ or $\sqrt[n]{a^m}$ are different notations that represent identical numbers, we practice converting between radical notation and rational exponent notation.

EXAMPLE 4 Write the following using radical notation.

a. $x^{1/5}$ **b.** $5x^{2/3}$ **c.** $3(p + q)^{1/2}$

Solution: **a.** $x^{1/5} = \sqrt[5]{x}$ **b.** $5x^{2/3} = 5\sqrt[3]{x^2}$ or $5x^{2/3} = 5(\sqrt[3]{x})^2$

c. $3(p + q)^{1/2} = 3\sqrt{p + q}$ ■

HELPFUL HINT

The **denominator** of a rational exponent is the index of the corresponding radical. For example, $x^{1/5} = \sqrt[5]{x}$ and $z^{2/3} = \sqrt[3]{z^2}$ or $z^{2/3} = (\sqrt[3]{z})^2$.

EXAMPLE 5 Write using rational exponents.

a. $\sqrt[5]{x}$ **b.** $\sqrt[3]{17x^2y^5}$ **c.** $\sqrt{x-5a}$ **d.** $3\sqrt{2p} - 5\sqrt[3]{p^2}$

Solution: **a.** $\sqrt[5]{x} = x^{1/5}$

b. $\sqrt[3]{17x^2y^5} = (17x^2y^5)^{1/3}$ We can further simplify this expression using a power rule for exponents.

$$(17x^2y^5)^{1/3} = 17^{1/3}x^{2/3}y^{5/3}$$

c. $\sqrt{x-5a} = (x-5a)^{1/2}$

d. $3\sqrt{2p} - 5\sqrt[3]{p^2} = 3(2p)^{1/2} - 5(p^2)^{1/3} = 3(2p)^{1/2} - 5p^{2/3}$ ■

We can use rational exponents to help us simplify radicals.

EXAMPLE 6 Use rational exponents to simplify.

a. $\sqrt[8]{x^4}$ **b.** $\sqrt[4]{r^2s^6}$

Solution: **a.** $\sqrt[8]{x^4} = x^{4/8} = x^{1/2} = \sqrt{x}$

b. $\sqrt[4]{r^2s^6} = (r^2s^6)^{1/4} = r^{2/4}s^{6/4} = r^{1/2}s^{3/2} = (rs^3)^{1/2} = \sqrt{rs^3}$ ■

5 It can be shown that the properties of integer exponents hold for rational exponents. By using these properties and definitions, we can now simplify expressions containing rational exponents.

EXAMPLE 7 Simplify the following expressions. Assume that all variables represent positive numbers. Write your answers using only positive exponents.

a. $x^{1/2}x^{1/3}$ **b.** $\dfrac{7^{1/3}}{7^{4/3}}$ **c.** $\dfrac{(2x^{2/5}y^{-1/3})^5}{x^2y}$

Solution: **a.** $x^{1/2}x^{1/3} = x^{(1/2+1/3)} = x^{3/6+2/6} = x^{5/6}$ **b.** $\dfrac{7^{1/3}}{7^{4/3}} = 7^{1/3-4/3} = 7^{-3/3} = 7^{-1} = \dfrac{1}{7}$

c. We begin by using the power rule $(ab)^m = a^mb^m$ to simplify the numerator.

$$\frac{(2x^{2/5}y^{-1/3})^5}{x^2y} = \frac{2^5(x^{2/5})^5(y^{-1/3})^5}{x^2y} = \frac{32x^2y^{-5/3}}{x^2y}$$

$= 32x^{2-2}y^{-5/3-3/3}$ Apply the quotient rule.

$= 32x^0y^{-8/3}$

$= \dfrac{32}{y^{8/3}}$ ■

EXAMPLE 8 Multiply. Assume that all variables represent positive numbers.

a. $z^{2/3}(z^{1/3} - z^5)$ **b.** $(x^{1/3} - 5)(x^{1/3} + 2)$

Solution: **a.** $z^{2/3}(z^{1/3} - z^5) = z^{2/3}z^{1/3} - z^{2/3}z^5$ Apply the distributive property.

$= z^{(2/3+1/3)} - z^{(2/3+5)}$ Use the product rule.

$= z^{3/3} - z^{(2/3+15/3)}$

$= z - z^{17/3}$

b. $(x^{1/3} - 5)(x^{1/3} + 2) = x^{2/3} + 2x^{1/3} - 5x^{1/3} - 10$

$= x^{2/3} - 3x^{1/3} - 10$ ■

EXERCISE SET 8.5

Simplify each expression. See Examples 1 and 2.

1. $8^{1/3}$ **2.** $16^{1/4}$ **3.** $9^{1/2}$ **4.** $16^{1/2}$
5. $16^{3/4}$ **6.** $27^{2/3}$ **7.** $32^{2/5}$ **8.** $64^{5/6}$

Simplify each expression. See Example 3.

9. $-16^{-1/4}$ **10.** $-8^{-1/3}$ **11.** $16^{-3/2}$ **12.** $27^{-4/3}$
13. $81^{-3/2}$ **14.** $32^{-2/5}$ **15.** $\left(\frac{4}{25}\right)^{-1/2}$ **16.** $\left(\frac{8}{27}\right)^{-1/3}$

Write using radical notation. See Example 4.

17. $a^{3/7}$ **18.** $b^{2/17}$ **19.** $2x^{5/3}$ **20.** $(2x)^{5/3}$
21. $(4t)^{-5/6}$ **22.** $4t^{-5/6}$ **23.** $(4x - 1)^{3/5}$ **24.** $(3x + 2y^2)^{9/5}$

Write using rational exponents. See Example 5.

25. $\sqrt{3}$ **26.** $\sqrt[3]{y}$ **27.** $\sqrt[3]{y^5}$ **28.** $\sqrt[4]{x^3}$
29. $\sqrt[5]{4y^7}$ **30.** $\sqrt[4]{11x^5}$ **31.** $\sqrt{(y + 1)^3}$ **32.** $\sqrt{(3 + y^2)^5}$
33. $2\sqrt{x} - 3\sqrt{y}$ **34.** $4\sqrt{2x} + \sqrt{xy}$

Use rational exponents to simplify each of the following. See Example 6.

35. $\sqrt[6]{x^3}$ **36.** $\sqrt[9]{a^3}$ **37.** $\sqrt[4]{16x^2}$ **38.** $\sqrt[8]{4y^2}$
39. $\sqrt[4]{(x + 3)^2}$ **40.** $\sqrt[6]{a^3b^6}$ **41.** $\sqrt[8]{x^4y^4}$ **42.** $\sqrt[9]{y^6z^3}$

Simplify each expression. Write using positive exponents. See Example 7.

43. $a^{2/3}a^{5/3}$ **44.** $b^{9/5}b^{8/5}$ **45.** $(4u^2v^{-6})^{3/2}$
46. $(32^{1/5}x^{2/3}y^{1/3})^3$ **47.** $\frac{b^{1/2}b^{3/4}}{-b^{1/4}}$ **48.** $\frac{a^{1/4}a^{-1/2}}{a^{2/3}}$

Multiply. See Example 8.

49. $y^{1/2}(y^{1/2} - y^{2/3})$ **50.** $x^{1/2}(x^{1/2} + x^{3/2})$ **51.** $x^{2/3}(2x - 2)$
52. $3x^{1/2}(x + y)$ **53.** $(2x^{1/3} + 3)(2x^{1/3} - 3)$ **54.** $(y^{1/2} + 5)(y^{1/2} + 5)$

Simplify each expression.

55. $81^{1/2}$ **56.** $(-27)^{1/3}$ **57.** $(-8)^{1/3}$ **58.** $36^{1/2}$
59. $-81^{1/4}$ **60.** $-64^{1/3}$ **61.** $\left(\frac{1}{81}\right)^{1/2}$ **62.** $\left(\frac{9}{16}\right)^{1/2}$
63. $\left(\frac{27}{64}\right)^{1/3}$ **64.** $\left(\frac{16}{81}\right)^{1/4}$ **65.** $9^{3/2}$ **66.** $16^{3/2}$
67. $64^{3/2}$ **68.** $64^{2/3}$ **69.** $-8^{2/3}$ **70.** $(-8)^{2/3}$
71. $4^{5/2}$ **72.** $9^{4/2}$ **73.** $\left(\frac{4}{9}\right)^{3/2}$ **74.** $\left(\frac{8}{27}\right)^{2/3}$
75. $\left(\frac{1}{81}\right)^{3/4}$ **76.** $\left(\frac{1}{32}\right)^{3/5}$ **77.** $4^{-1/2}$ **78.** $9^{-1/2}$
79. $125^{-1/3}$ **80.** $216^{-1/3}$ **81.** $625^{-3/4}$ **82.** $256^{-5/8}$

Simplify each expression. Write using positive exponents.

83. $x^{-5/3}x^{1/2}$ **84.** $x^{-1/3}x^{1/5}$ **85.** $\dfrac{w^{3/4}}{w^{1/4}}$ **86.** $\dfrac{y^{1/2}}{y^{2/3}}$

87. $(x^{-1/5}y^{1/12})^3$ **88.** $(a^{10}b^{15})^{-1/5}$ **89.** $(a^3b^9)^{-2/3}a^{2/3}$ **90.** $(x^4y^2z^6)^{-3/4}x^{1/2}$

91. $(16x^4y^2)^{1/2}$ **92.** $(9a^4b^8)^{1/4}$

Multiply.

93. $4x^{2/3}(6x^{5/2} - y^{2/3})$ **94.** $8x^{2/5}(6x^{3/2} - y^{1/2})$ **95.** $(x^{1/2} + 2)(x^{1/2} - 3)$

96. $(2a^{2/3} + b^{1/2})(3a^{2/3} + 2)$ **97.** $(x^{1/2} - 2x)^2$ **98.** $(3y^{1/4} + 2)^2$

Factor the common factor from the given expression.

99. $x^{3/4}$; $9x^{3/4} - x^{7/4}$ **100.** $y^{2/3}$; $7y^{5/3} - 3y^{2/3}$ **101.** $3x^{5/3}$; $9x^{5/3} + 15x^{7/3}$

102. $4x^{1/2}$; $8x^{1/2} - 16x^{3/2}$

Solve.

103. If a population grows at a rate of 8% annually, the formula $P = P_0(1.08)^N$ can be used to estimate the total population P after N years have passed, assuming that the original population is P_0. Find the population after $1\frac{1}{2}$ years if the original population of 10,000 people is growing at a rate of 8% annually.

Skill Review

Simplify. See Section 8.2.

104. $(\sqrt{x})^2$ **105.** $(\sqrt{y})^2$ **106.** $(\sqrt{y-1})^2$ **107.** $(\sqrt{z+2})^2$

Solve. See Sections 2.3 and 6.6.

108. $2x - 14 = 8^2$ **109.** $y - 2 = 5^2$ **110.** $x^2 - 4 = 3x$ **111.** $x^2 + 2x = 8$

112. $2x^2 - 5x - 3 = 0$ **113.** $3x^2 + x - 2 = 0$ **114.** $x + 5 = (x - 1)^2$ **115.** $x + 7 = (x + 5)^2$

Writing in Mathematics

116. Explain what each of the numbers 2, 3, and 4 signifies in the expression $4^{3/2}$.

117. Explain why $-4^{1/2}$ is a real number but $(-4)^{1/2}$ is not.

8.6 Solving Equations Containing Radicals and Problem Solving

OBJECTIVES

1. Solve equations containing radical expressions.
2. Find an unknown side of a right triangle using the Pythagorean theorem.

Tape IA 22

1 In this section, we present techniques to solve equations containing radical expressions such as

$$\sqrt{2x - 3} = 9$$

We use the power rule to help us solve these radical equations.

Power Rule

If both sides of an equation are raised to the same power, **all** solutions of the original equation are **among** the solutions of the new equation.

This rule **does not** say that raising both sides of an equation to a power yields an equivalent equation. A solution of the new equation **may or may not** be a solution of the original equation. Thus, **each solution of the new equation must be checked** to make sure it is a solution of the original equation.

EXAMPLE 1 Solve $\sqrt{2x - 3} = 9$ for x.

Solution: Use the power rule to square both sides of the equation to eliminate the radical.

$$\sqrt{2x - 3} = 9$$
$$(\sqrt{2x - 3})^2 = 9^2$$
$$2x - 3 = 81$$
$$2x = 84$$
$$x = 42$$

Now check the solution in the original equation.

$$\sqrt{2x - 3} = 9$$
$$\sqrt{2(42) - 3} = 9 \qquad \text{Let } x = 42.$$
$$\sqrt{84 - 3} = 9$$
$$\sqrt{81} = 9$$
$$9 = 9 \qquad \text{True.}$$

The solution checks, so we conclude that the solution set is $\{42\}$. ■

EXAMPLE 2 Solve $\sqrt{-10x - 1} + 3x = 0$ for x.

Solution: First isolate the radical on one side of the equation. To do this, we subtract $3x$ from both sides.

$$\sqrt{-10x - 1} + 3x = 0$$
$$\sqrt{-10x - 1} + 3x - 3x = 0 - 3x$$
$$\sqrt{-10x - 1} = -3x$$

Next use the power rule to eliminate the radical.

$$(\sqrt{-10x - 1})^2 = (-3x)^2$$
$$-10x - 1 = 9x^2$$

Since this is a quadratic equation, set the equation equal to 0 and try to solve by factoring.

$$9x^2 + 10x + 1 = 0$$

$$(9x + 1)(x + 1) = 0 \qquad \text{Factor.}$$

$$9x + 1 = 0 \quad \text{or} \quad x + 1 = 0 \qquad \text{Set each factor equal to 0.}$$

$$x = -\frac{1}{9} \quad \text{or} \quad x = -1$$

Possible solutions are $-\frac{1}{9}$ and -1. Now we check our work in the original equation.

Let $x = -\frac{1}{9}$

$$\sqrt{-10x - 1} + 3x = 0$$

$$\sqrt{-10\left(-\frac{1}{9}\right) - 1} + 3\left(-\frac{1}{9}\right) = 0$$

$$\sqrt{\frac{10}{9} - \frac{9}{9}} - \frac{3}{9} = 0$$

$$\sqrt{\frac{1}{9}} - \frac{1}{3} = 0$$

$$\frac{1}{3} - \frac{1}{3} = 0 \qquad \text{True.}$$

Let $x = -1$

$$\sqrt{-10x - 1} + 3x = 0$$

$$\sqrt{-10(-1) - 1} + 3(-1) = 0$$

$$\sqrt{10 - 1} - 3 = 0$$

$$\sqrt{9} - 3 = 0$$

$$3 - 3 = 0 \qquad \text{True.}$$

Both solutions check. The solution set is $\left\{-\frac{1}{9}, -1\right\}$. ■

> **To Solve a Radical Equation**
>
> *Step 1* Write the equation so that one radical with variables is by itself on one side of the equation.
> *Step 2* Raise each side of the equation to a power equal to the index of the radical.
> *Step 3* Add or subtract any like terms.
> *Step 4* If the equation still contains a radical term, repeat steps 1 through 3.
> *Step 5* Solve the equation.
> *Step 6* Check all proposed solutions in the original equation for extraneous solutions.

EXAMPLE 3 Solve for x: $\sqrt[3]{x + 1} + 5 = 3$.

Solution: First we isolate the radical by subtracting 5 from both sides of the equation.

$$\sqrt[3]{x + 1} + 5 = 3$$

$$\sqrt[3]{x + 1} = -2$$

Next raise both sides of the equation to the third power to eliminate the radical.

$$(\sqrt[3]{x+1})^3 = (-2)^3$$
$$x + 1 = -8$$
$$x = -9$$

The solution checks in the original equation, so the solution set is $\{-9\}$. ■

EXAMPLE 4 Solve $\sqrt{4 - x} = x - 2$ for x.

Solution:

$$\sqrt{4 - x} = x - 2$$
$$(\sqrt{4 - x})^2 = (x - 2)^2$$
$$4 - x = x^2 - 4x + 4$$
$$x^2 - 3x = 0 \quad \text{Write the quadratic equation in standard form.}$$
$$x(x - 3) = 0 \quad \text{Factor.}$$
$$x = 0 \quad \text{or} \quad x - 3 = 0$$
$$x = 3$$

Check the possible solutions.

$\sqrt{4 - x} = x - 2$		$\sqrt{4 - x} = x - 2$	
$\sqrt{4 - 0} = 0 - 2$	Let $x = 0$.	$\sqrt{4 - 3} = 3 - 2$	Let $x = 3$.
$2 = -2$	False.	$1 = 1$	True.

The proposed solution 3 checks, but 0 does not. When a proposed solution does not check, it is an **extraneous root or solution.** Since 0 is an extraneous solution, the solution set is $\{3\}$. ■

HELPFUL HINT

In Example 4, notice that $(x - 2)^2 = x^2 - 4x + 4$. Make sure that binomials are squared correctly.

EXAMPLE 5 Solve $\sqrt{2x + 5} + \sqrt{2x} = 3$.

Solution: Isolate a radical by subtracting $\sqrt{2x}$ from both sides.

$$\sqrt{2x + 5} + \sqrt{2x} = 3$$
$$\sqrt{2x + 5} = 3 - \sqrt{2x}$$

Use the power rule to begin eliminating the radicals. Square both sides.

$$(\sqrt{2x + 5})^2 = (3 - \sqrt{2x})^2$$
$$2x + 5 = 9 - 6\sqrt{2x} + 2x \quad \text{Multiply: } (3 - \sqrt{2x})(3 - \sqrt{2x}).$$

There is still a radical in the equation, so isolate the radical again. Then square both sides.

$$2x + 5 = 9 - 6\sqrt{2x} + 2x$$
$$6\sqrt{2x} = 4 \quad \text{Isolate the radical.}$$

$$(6\sqrt{2x})^2 = 4^2 \quad \text{Square both sides of the equation to eliminate the radical.}$$

$$36(2x) = 16$$

$$72x = 16$$

$$x = \frac{16}{72} \quad \text{Solve.}$$

$$x = \frac{2}{9}$$

The proposed solution, $\frac{2}{9}$, does check in the original equation, so the solution set is $\left\{\frac{2}{9}\right\}$. ■

HELPFUL HINT

Make sure that expressions are squared correctly. In Example 5, we squared $(3 - \sqrt{2x})$.

$$\begin{aligned}(3 - \sqrt{2x})^2 &= (3 - \sqrt{2x})(3 - \sqrt{2x}) \\ &= 3 \cdot 3 - 3\sqrt{2x} - 3\sqrt{2x} + \sqrt{2x} \cdot \sqrt{2x} \\ &= 9 - 6\sqrt{2x} + 2x\end{aligned}$$

2 Recall that the Pythagorean theorem states that in a right triangle, the length of the hypotenuse squared equals the sum of each of the lengths of the legs squared.

Pythagorean Theorem

If a and b are the lengths of the legs of a right triangle and c is the length of the hypotenuse, then $a^2 + b^2 = c^2$.

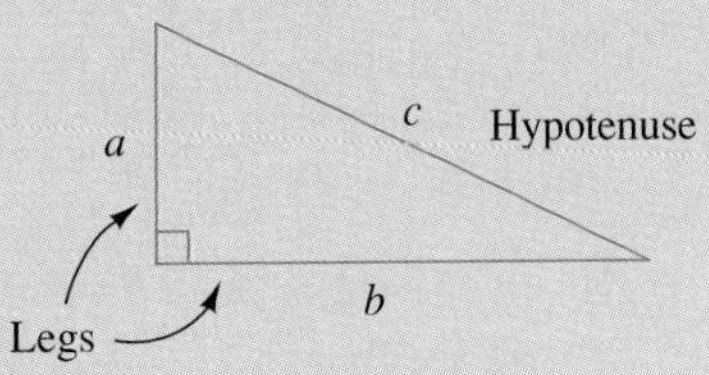

EXAMPLE 6 Find the length of the unknown leg of the following right triangle.

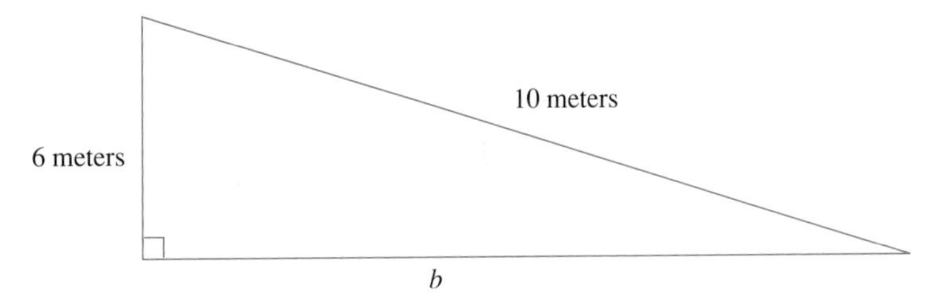

Solution: In the formula $a^2 + b^2 = c^2$, c is the hypotenuse. Let $c = 10$, the length of the hypotenuse, let $a = 6$, and solve for b. Then $a^2 + b^2 = c^2$ becomes

$$6^2 + b^2 = 10^2$$

$$36 + b^2 = 100$$

$$b^2 = 64 \qquad \text{Subtract 36 from both sides.}$$

$$b = 8 \quad \text{and} \quad b = -8 \qquad \text{Because } b^2 = 64.$$

Since we are solving for a length, we will list the positive solution only. The other leg of the triangle is 8 meters long. ■

EXAMPLE 7 A surveyor must determine the distance across a lake. He first inserts poles at points P and Q on opposite sides of the lake, as pictured. Perpendicular to line PQ, he finds another point R. If the length of $\overline{PR}$ is 320 feet and the length of $\overline{QR}$ is 240 feet, what is the distance across the lake?

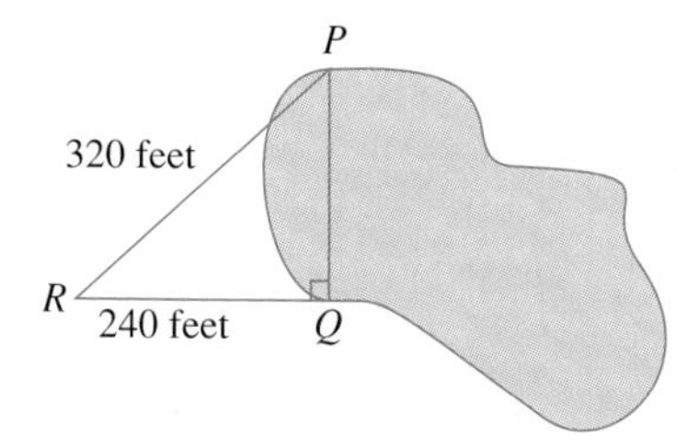

Solution: **1.** UNDERSTAND. Read and reread the problem. Guess a solution and check using the Pythagorean theorem.

2. ASSIGN. By creating a line perpendicular to line PQ, the surveyor deliberately constructed a right triangle. The hypotenuse, $\overline{PR}$, has a length of 320 feet, so we let $c = 320$ in the Pythagorean theorem. The side $\overline{QR}$ is one of the legs, so let $a = 240$, and $b =$ the unknown length.

3. ILLUSTRATE.

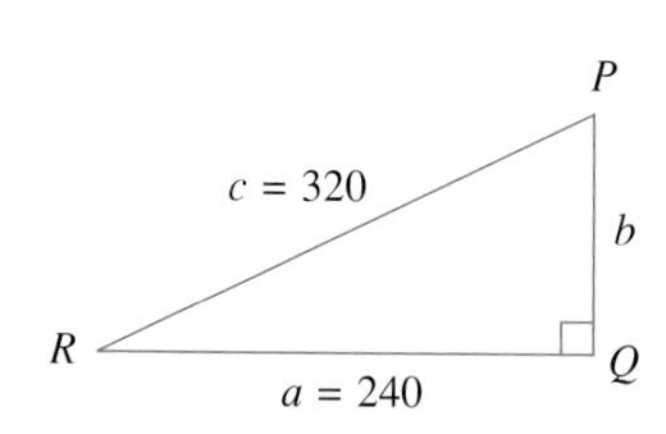

4. TRANSLATE. $a^2 + b^2 = c^2$ Use the Pythagorean theorem.

$240^2 + b^2 = 320^2$ Let $c = 320$ and $a = 240$.

5. COMPLETE. $57{,}600 + b^2 = 102{,}400$

$b^2 = 44{,}800$ Subtract 57,600 from both sides.

$b = \sqrt{44{,}800}$ Apply the square root property.

6. INTERPRET. *Check:* See that $240^2 + (\sqrt{44{,}800})^2 = 320^2$. *State:* The distance across the lake is **exactly** $\sqrt{44{,}800}$ feet. The surveyor can now use a calculator to find that $\sqrt{44{,}800}$ feet is **approximately** 211.6601 feet, so the distance across the lake is **roughly** 212 feet. ■

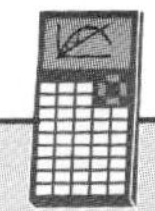

GRAPHING CALCULATOR BOX

We can use a grapher to solve equations such as radical equations. For example, to use a grapher to approximate the solutions of the equation solved in Example 4, we graph the following:

$$Y_1 = \sqrt{4 - x} \quad \text{and} \quad Y_2 = x - 2$$

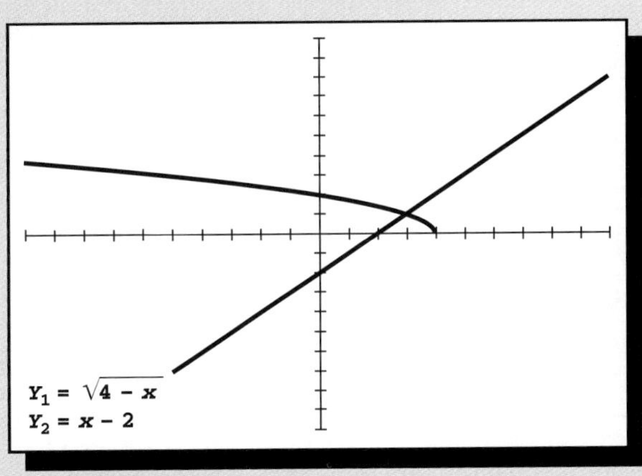

The x-value of the point of intersection is the solution. Use the Zoom and Trace features to see that the solution is 3.

Use a grapher to solve each radical equation. Round all solutions to the nearest hundredth.

1. $\sqrt{x + 7} = x$

2. $\sqrt{3x + 5} = 2x$

3. $\sqrt{2x + 1} = \sqrt{2x + 2}$

4. $\sqrt{10x - 1} = \sqrt{-10x + 10} - 1$

5. $1.2x = \sqrt{3.1x + 5}$

6. $\sqrt{1.9x^2 - 2.2} = -0.8x + 3$

EXERCISE SET 8.6

Solve. See Example 1.

1. $\sqrt{2x} = 4$

2. $\sqrt{3x} = 3$

3. $\sqrt{x - 3} = 2$

4. $\sqrt{x + 1} = 5$

5. $\sqrt{2x} = -4$

6. $\sqrt{5x} = -5$

Solve. See Example 2.

7. $\sqrt{4x - 3} - 5 = 0$

8. $\sqrt{x - 3} - 1 = 0$

9. $\sqrt{2x - 3} - 2 = 1$

10. $\sqrt{3x + 3} - 4 = 8$

Solve. See Example 3.

11. $\sqrt[3]{6x} = -3$

12. $\sqrt[3]{4x} = -2$

13. $\sqrt[3]{x - 2} - 3 = 0$

14. $\sqrt[3]{2x - 6} - 4 = 0$

Solve. See Example 4.

15. $\sqrt{13 - x} = x - 1$

16. $\sqrt{2x - 3} = 3 - x$

17. $x - \sqrt{4 - 3x} = -8$

18. $2x + \sqrt{x + 1} = 8$

Solve. See Example 5.

19. $\sqrt{y+5} = 2 - \sqrt{y-4}$

20. $\sqrt{x+3} + \sqrt{x-5} = 3$

21. $\sqrt{x-3} + \sqrt{x+2} = 5$

22. $\sqrt{2x-4} - \sqrt{3x+4} = -2$

Find the length of the unknown side in each triangle. See Examples 6 and 7.

23.

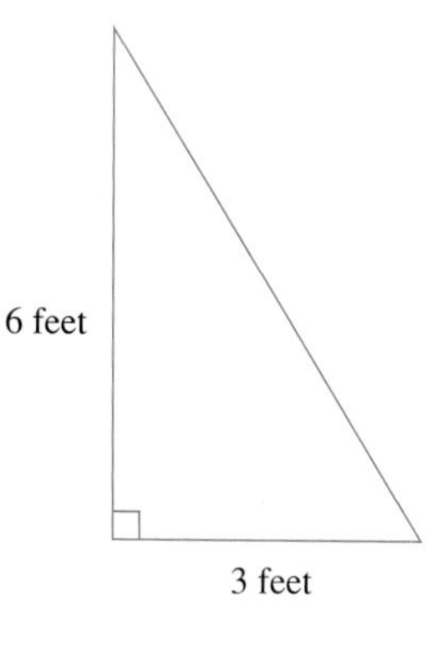

24.

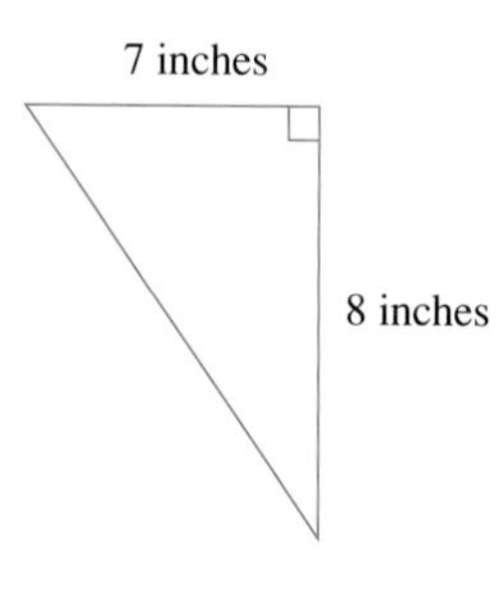

25.

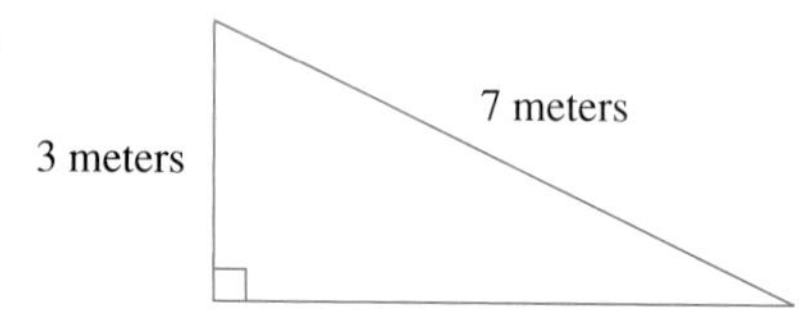

26.

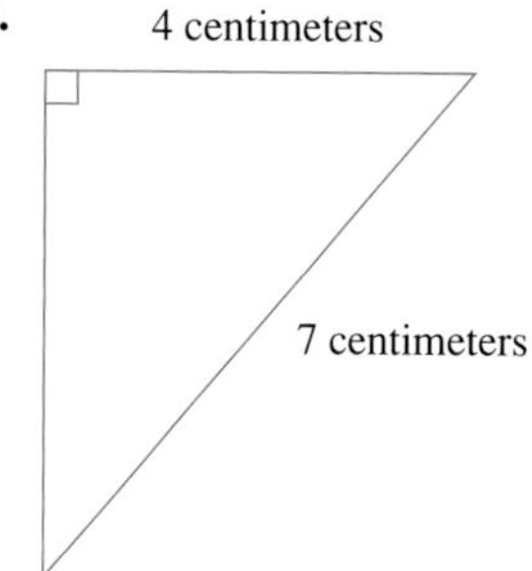

Solve.

27. $\sqrt{3x-2} - 5$

28. $\sqrt{5x-4} = 9$

29. $-\sqrt{2x} + 4 = -6$

30. $-\sqrt{3x+9} = -12$

31. $\sqrt{3x+1} + 2 = 0$

32. $\sqrt{3x+1} - 2 = 0$

33. $\sqrt[4]{4x+1} - 2 = 0$

34. $\sqrt[4]{2x-9} - 3 = 0$

35. $\sqrt{4x-3} = 5$

36. $\sqrt{3x+9} = 12$

37. $\sqrt[3]{6x-3} - 3 = 0$

38. $\sqrt[3]{3x} + 4 = 7$

39. $\sqrt[3]{2x-3} - 2 = -5$

40. $\sqrt[3]{x-4} - 5 = -7$

41. $\sqrt{x+4} = \sqrt{2x-5}$

42. $\sqrt{3y+6} = \sqrt{7y-6}$

43. $x - \sqrt{1-x} = -5$

44. $x - \sqrt{x-2} = 4$

45. $\sqrt[3]{-6x-1} = \sqrt[3]{-2x-5}$

46. $x + \sqrt{x+5} = 7$

47. $\sqrt{5x-1} - \sqrt{x} + 2 = 3$

48. $\sqrt{2x-1} - 4 = -\sqrt{x-4}$

49. $\sqrt{2x-1} = \sqrt{1-2x}$

50. $\sqrt{3x+4} - 1 = \sqrt{2x+1}$

Solve.

51. The cost $C(x)$ in dollars per day to operate a small delivery service is given by $C(x) = 80\sqrt[3]{x} + 500$, where x is the number of deliveries per day. In July, the manager decides that it is necessary to keep delivery costs below \$1620.00. Find the greatest number of deliveries this company can make per day and still keep overhead below \$1620.00.

52. The formula $v = \sqrt{2gh}$ relates the velocity v, in feet per second, of an object after it falls h feet accelerated by gravity g feet per second squared. If g is approximately 32 feet per second squared, find how far an object has fallen if its velocity is 80 feet per second.

53. Two tractors are pulling a tree stump from a field. If two forces A and B pull at right angles (90°) to each other, the size of the resulting force R is given by the formula

$$R = \sqrt{A^2 + B^2}$$

If tractor A is exerting 600 pounds of force and the resulting force is 850 pounds, find how much force tractor B is exerting.

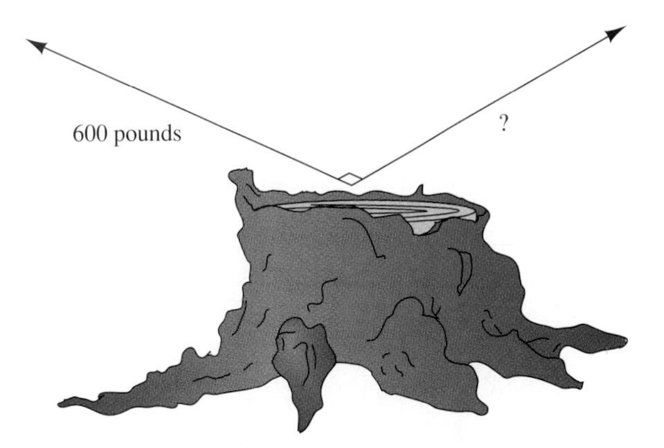

Find the length of the unknown side of each triangle.

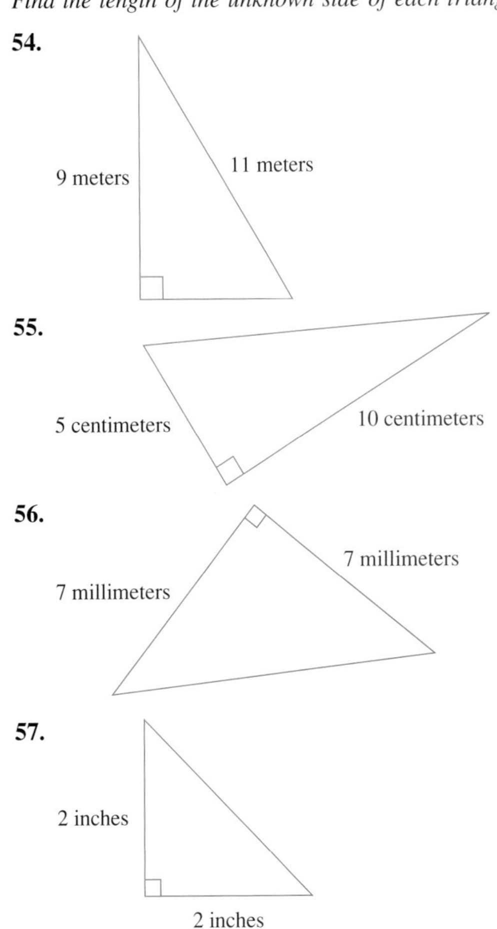

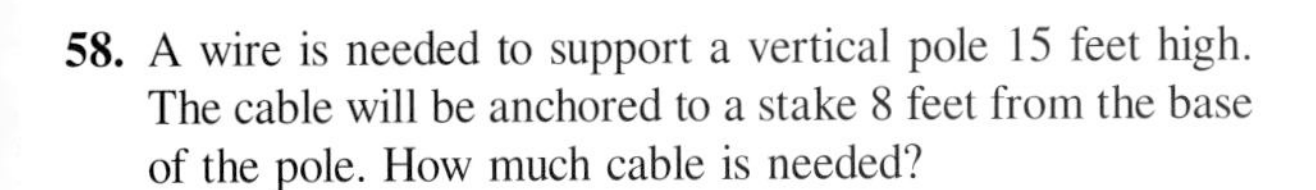

58. A wire is needed to support a vertical pole 15 feet high. The cable will be anchored to a stake 8 feet from the base of the pole. How much cable is needed?

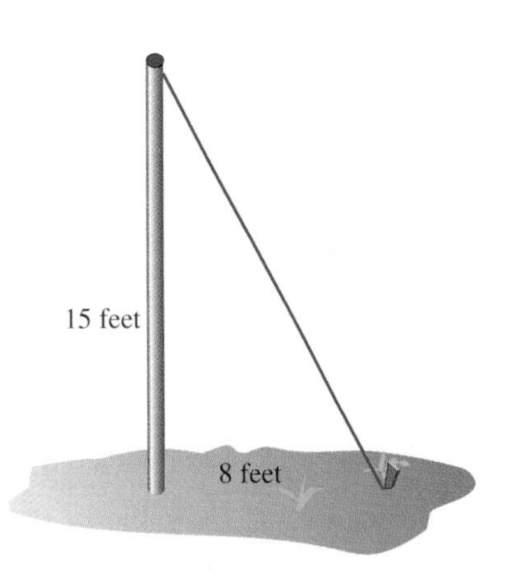

59. A spotlight is mounted on the eaves of a house 12 feet above the ground. A flower bed runs between the house and the sidewalk, so the closest the ladder can be placed to the house is 5 feet. How long a ladder is needed so that an electrician can reach the place where the light is to be mounted?

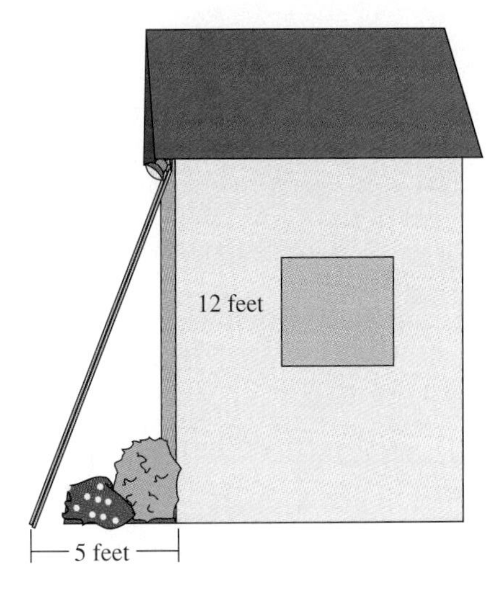

60. A furniture upholsterer wished to cut a cover from a piece of fabric that is 45 inches by 45 inches. The cover must be cut on the bias of the fabric. What is the longest strip that can be cut?

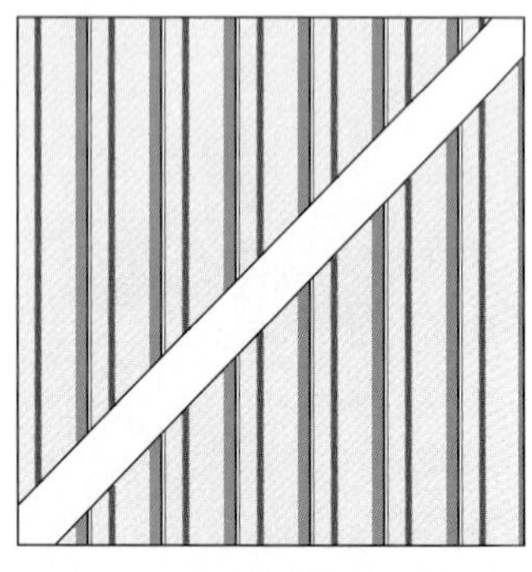

61. A wire is to be attached to support a telephone pole. Because of surrounding buildings, sidewalks, and roadway, the wire must be anchored exactly 15 feet from the base of the pole. Telephone company workers find they have only 30 feet of cable, and 2 feet of that must be used to attach the cable to the pole and to the stake on the ground. How high from the base of the pole can the wire be attached?

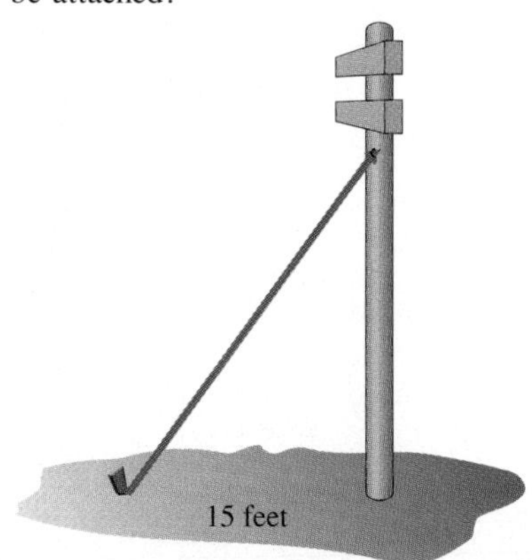

62. Railroad tracks are invariably made up of relatively short sections of rail connected by expansion joints. To see why this construction is necessary, consider a single rail 100 feet long (or 1200 inches). On an extremely hot day, suppose that it expands 1 inch in the hot sun to a new length of 1201 inches. Theoretically, the track would bow upward as pictured.

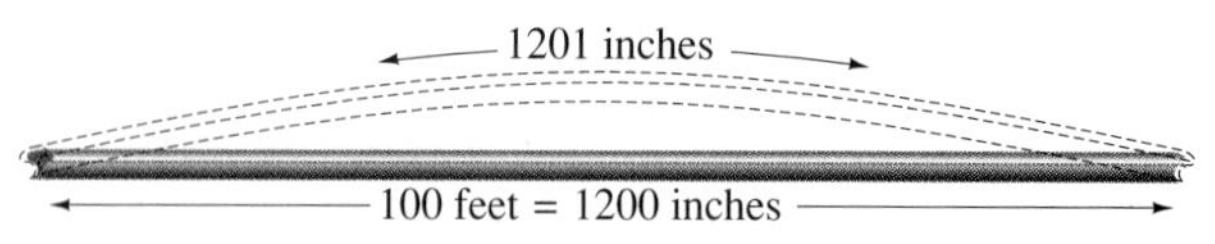

Let us approximate the bulge in the railroad this way.

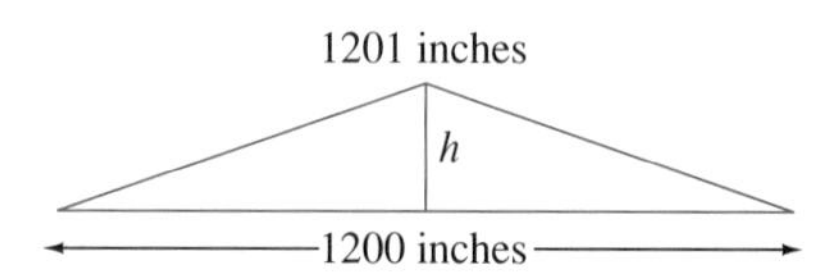

Calculate the height h of the bulge to the nearest tenth of an inch.

63. Police departments find it very useful to be able to approximate the speed of a car given the distance that the car skidded before coming to a stop. If the road surface is wet concrete, the function $S(x) = \sqrt{10.5x}$ is used where $S(x)$ is the speed of the car in miles per hour and x is the distance skidded in feet. Find how fast a car was moving to the nearest mile per hour if it skidded 280 feet on wet concrete.

64. The maximum distance $D(h)$ that a person can see from a height h kilometers above the ground is given by the function $D(h) = 111.7\sqrt{h}$. Find the height that would allow us to see 80 kilometers.

65. For a square-based pyramid, the formula $b = \sqrt{\frac{3V}{H}}$ describes the relationship between the length b of one side of the base, the volume V, and the height H. Find the volume if each edge of the base is 6 feet long and if the pyramid is 2 feet high.

66. The function $S(t) = 16t^2$ relates the distance $S(t)$, in feet, that an object falls in t seconds, assuming that air resistance does not slow down the object. Find how long, to the nearest hundredth of a second, it takes an object to reach the ground from the top of the Sears Tower in Chicago, a distance of 1454 feet.

Skill Review

Find the slope of the line containing each pair of points. See Section 4.4.

67. $(-1, 5), (0, 2)$ **68.** $(6, 9), (-3, 2)$ **69.** $(4, 7), (-1, 5)$ **70.** $(8, -3), (5, 0)$

Use the vertical line test to determine whether each graph represents the graph of a function. See Section 4.2.

71.

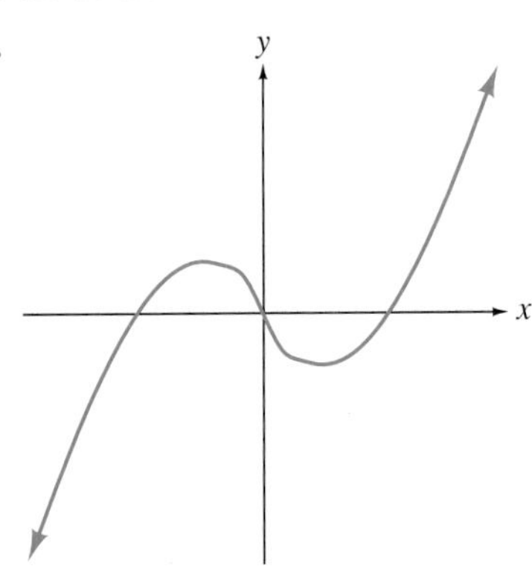

72.

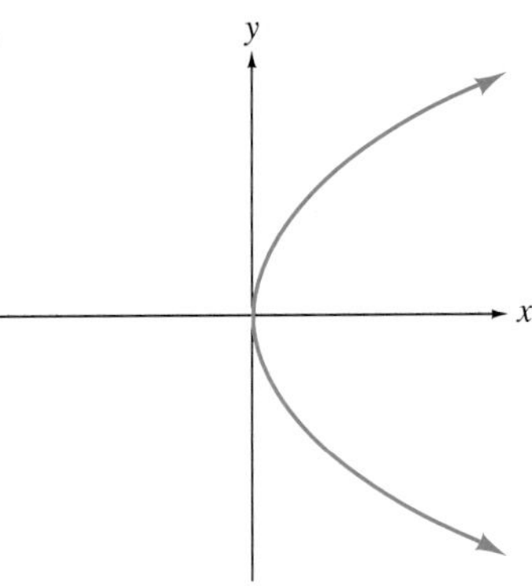

73.

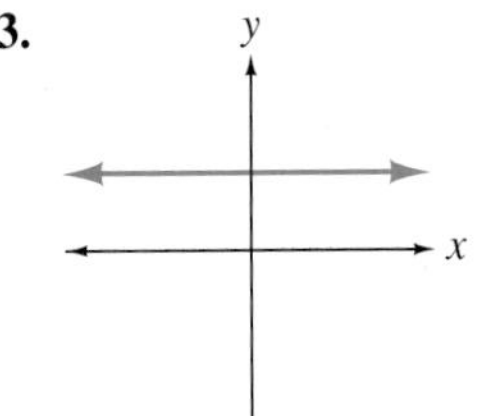

74.

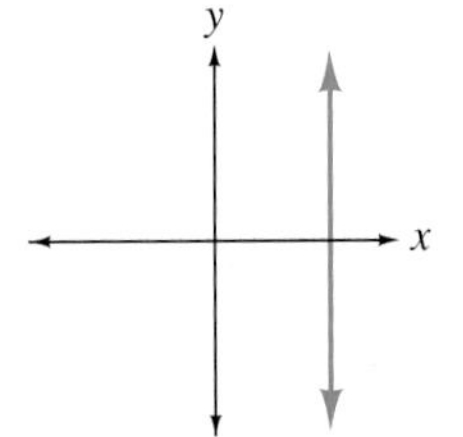

75.

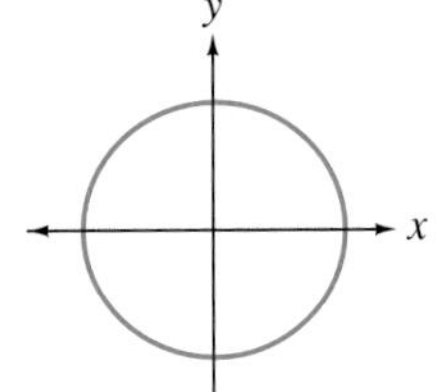

76.

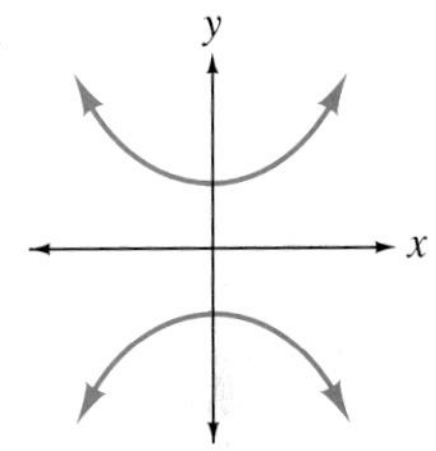

Simplify. See Section 7.4.

77. $\dfrac{\dfrac{x}{6}}{\dfrac{2x}{3} + \dfrac{1}{2}}$

78. $\dfrac{\dfrac{1}{y} + \dfrac{4}{5}}{\dfrac{-3}{20}}$

79. $\dfrac{\dfrac{z}{5} + \dfrac{1}{10}}{\dfrac{z}{20} - \dfrac{z}{5}}$

80. $\dfrac{\dfrac{1}{y} + \dfrac{1}{x}}{\dfrac{1}{y} - \dfrac{1}{x}}$

A Look Ahead

Solve. See the following example.

EXAMPLE Solve $(t^2 - 3t) - 2\sqrt{t^2 - 3t} = 0$.

Solution: Substitution can be used to make this problem somewhat simpler. Let $x = t^2 - 3t$.

$$(t^2 - 3t) - 2\sqrt{t^2 - 3t} = 0$$

$$x - 2\sqrt{x} = 0 \qquad \text{Let } x = t^2 - 3t.$$

$$x = 2\sqrt{x}$$

$$x^2 = (2\sqrt{x})^2$$

$$x^2 = 4x$$

$$x^2 - 4x = 0$$

$$x(x - 4) = 0$$

$$x = 0 \quad \text{or} \quad x - 4 = 0$$

$$x = 4$$

Now we "undo" the substitution.

$x = 0$ Replace x with $t^2 - 3t$.

$$t^2 - 3t = 0$$
$$t(t - 3) = 0$$
$$t = 0 \quad \text{or} \quad t - 3 = 0$$
$$t = 3$$

$x = 4$ Replace x with $t^2 - 3t$.

$$t^2 - 3t = 4$$
$$t^2 - 3t - 4 = 0$$
$$(t - 4)(t + 1) = 0$$
$$t - 4 = 0 \quad \text{or} \quad t + 1 = 0$$
$$t = 4 \quad \text{or} \quad t = -1$$

In this problem, we have four possible solutions: 0, 3, 4, and -1. All four solutions check in the original equation, so the solution set is $\{-1, 0, 3, 4\}$. ■

81. $3\sqrt{x^2 - 8x} = x^2 - 8x$

82. $\sqrt{(x^2 - x) + 7} = 2(x^2 - x) - 1$

83. $7 - (x^2 - 3x) = \sqrt{(x^2 - 3x) + 5}$

84. $x^2 + 6x = 4\sqrt{x^2 + 6x}$

8.7 Complex Numbers

OBJECTIVES

Tape IA 22

1. Define imaginary and complex numbers.
2. Add or subtract complex numbers.
3. Multiply complex numbers.
4. Divide complex numbers.
5. Raise i to powers.

1 Our work with radical expressions has excluded expressions such as $\sqrt{-16}$ because $\sqrt{-16}$ is not a real number: there is no real number whose square is -16. In this section, we discuss a number system that includes roots of negative numbers. This number system is the **complex number system,** and it includes real numbers. The complex number system allows us to solve equations such as $x^2 + 1 = 0$ that have no real number solutions. The set of complex numbers includes the **imaginary unit.**

Imaginary Unit

The imaginary unit, written i, is the number whose square is -1. That is,

$$i^2 = -1 \quad \text{and} \quad i = \sqrt{-1}$$

Using i, we can write $\sqrt{-16}$ as

$$\sqrt{-16} = \sqrt{16(-1)} = \sqrt{16}\,\sqrt{-1} = 4i$$

EXAMPLE 1 Write using i notation.

a. $\sqrt{-36}$

b. $\sqrt{-5}$

Solution: **a.** $\sqrt{-36} = \sqrt{36}\,\sqrt{-1} = 6i$

b. $\sqrt{-5} = \sqrt{-1(5)} = \sqrt{-1}\,\sqrt{5} = i\sqrt{5}$. Since $\sqrt{5}\,i$ can be easily confused with $\sqrt{5i}$, we write $\sqrt{5}\,i$ as $i\sqrt{5}$. ■

Now that we have practiced working with the imaginary unit, complex numbers will be defined.

> **Complex Numbers**
>
> A complex number is a number that can be written in the form $a + bi$, where a and b are real numbers.

The number a is the **real part** of $a + bi$, and the number b is the **imaginary part** of $a + bi$. Notice that the real numbers are a subset of the complex numbers since any real number can be written in the form of a complex number. For example,

$$16 = 16 + 0i$$

In general, a complex number $a + bi$ is a real number if $b = 0$. Also, a complex number is called a **pure imaginary number** if $a = 0$. For example,

$$3i = 0 + 3i \quad \text{and} \quad i\sqrt{7} = 0 + i\sqrt{7}$$

are imaginary numbers.

The following diagram shows the relationship between complex numbers and their subsets. The shaded region represents the set of real numbers.

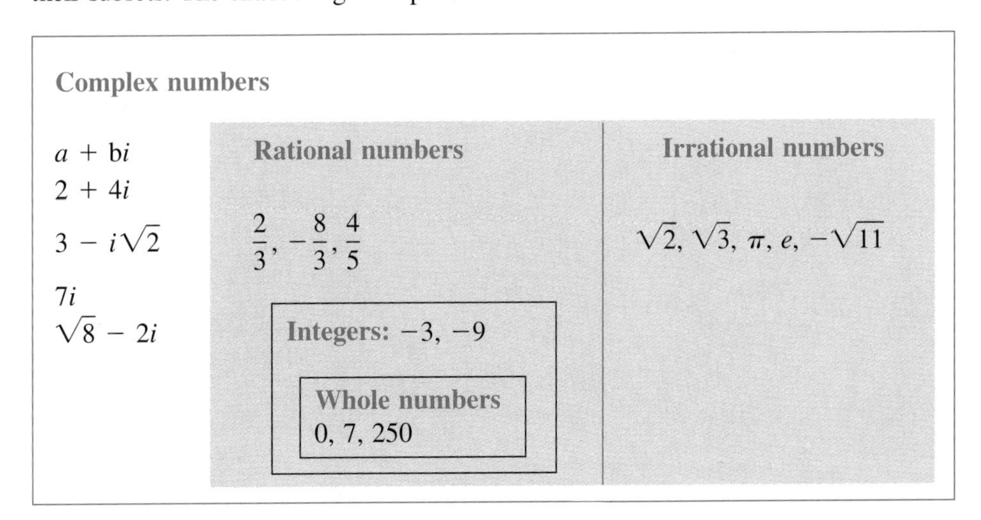

2 Two complex numbers $a + bi$ and $c + di$ are equal if and only if $a = c$ and $b = d$. Complex numbers can be added or subtracted by adding or subtracting their real parts and then adding or subtracting their imaginary parts.

Sum or Difference of Complex Numbers

If $a + bi$ and $c + di$ are complex numbers, then their sum is

$$(a + bi) + (c + di) = (a + c) + (b + d)i$$

Their difference is

$$(a + bi) - (c + di) = a + bi - c - di = (a - c) + (b - d)i$$

EXAMPLE 2 Add or subtract the complex numbers. Write the sum or difference in the form $a + bi$.

a. $(2 + 3i) + (-3 + 2i)$ **b.** $(5i) - (1 - i)$ **c.** $(-3 - 7i) - (-6)$

Solution: **a.** $(2 + 3i) + (-3 + 2i) = (2 - 3) + (3 + 2)i = -1 + 5i$

b. $5i - (1 - i) = 5i - 1 + i$

$$= -1 + (5 + 1)i$$

$$= -1 + 6i$$

c. $(-3 - 7i) - (-6) = -3 - 7i + 6$

$$= (-3 + 6) - 7i$$

$$= 3 - 7i$$ ■

3 We will use the relationship $i^2 = -1$ to simplify when multiplying two complex numbers.

EXAMPLE 3 Multiply the complex numbers. Write the product in the form $a + bi$.

a. $(2 - 5i)(4 + i)$ **b.** $(2 - i)^2$ **c.** $(7 + 3i)(7 - 3i)$

Solution: Multiply complex numbers as though they were binomials.

a. $(2 - 5i)(4 + i) = 2(4) + 2(i) - 5i(4) - 5i(i)$

$$= 8 + 2i - 20i - 5i^2$$

$$= 8 - 18i - 5(-1) \qquad \text{Let } i^2 = -1.$$

$$= 8 - 18i + 5$$

$$= 13 - 18i$$

b. $(2 - i)^2 = (2 - i)(2 - i)$

$$= 2(2) - 2(i) - 2(i) + i^2$$

$$= 4 - 4i + (-1) \qquad \text{Let } i^2 = -1.$$

$$= 3 - 4i$$

c. $(7 + 3i)(7 - 3i) = 7(7) - 7(3i) + 3i(7) - 3i(3i)$

$$= 49 - 21i + 21i - 9i^2$$

$$= 49 - 9(-1) \qquad \text{Let } i^2 = -1.$$

$$= 49 + 9$$

$$= 58$$ ■

From Example 3c, notice that the product of $7 + 3i$ and $7 - 3i$ is a real number. These two complex numbers are called **complex conjugates** of one another. In general, we have the following definition.

Complex Conjugates

The complex numbers $(a + bi)$ and $(a - bi)$ are called **complex conjugates** of each other, and $(a + bi)(a - bi) = a^2 + b^2$.

To see that the product of a complex number $a + bi$ and its conjugate $a - bi$ is the real number $a^2 + b^2$, we multiply.

$$\begin{aligned}(a + bi)(a - bi) &= a^2 - abi + abi - b^2i^2 \\ &= a^2 - b^2(-1) \\ &= a^2 + b^2\end{aligned}$$

4 We use complex conjugates to divide by a complex number.

EXAMPLE 4 Find each quotient. Write in the form $a + bi$.

a. $\dfrac{2 + i}{1 - i}$ **b.** $\dfrac{7}{3i}$

Solution: **a.** Multiply the numerator and denominator by the complex conjugate of $1 - i$ to eliminate the imaginary number in the denominator.

$$\begin{aligned}\frac{2 + i}{1 - i} &= \frac{(2 + i)(1 + i)}{(1 - i)(1 + i)} \\ &= \frac{2(1) + 2(i) + 1(i) + i^2}{1^2 - i^2} \\ &= \frac{2 + 3i - 1}{1 + 1} \\ &= \frac{1 + 3i}{2} \quad \text{or} \quad \frac{1}{2} + \frac{3}{2}i\end{aligned}$$

To check that $\dfrac{2 + i}{1 - i} = \dfrac{1}{2} + \dfrac{3}{2}i$, multiply $\left(\dfrac{1}{2} + \dfrac{3}{2}i\right)(1 - i)$ to verify that the product is $2 + i$.

b. Multiply the numerator and denominator by the conjugate of $3i$. Note that $3i = 0 + 3i$, so its conjugate is $0 - 3i$ or $-3i$.

$$\frac{7}{3i} = \frac{7(-3i)}{(3i)(-3i)} = \frac{-21i}{-9i^2} = \frac{-21i}{-9(-1)} = \frac{-21i}{9} = \frac{-7i}{3} \quad \text{or} \quad 0 - \frac{7}{3}i \quad ■$$

The product rule for radicals does not necessarily hold true for imaginary numbers. **To multiply imaginary numbers, each must be in complex form *bi*.** For example, to multiply $\sqrt{-4}$ and $\sqrt{-9}$, first write each number in the form bi.

$$\sqrt{-4}\,\sqrt{-9} = 2i(3i) = 6i^2 = -6$$

EXAMPLE 5 Multiply or divide the following as indicated.

a. $\sqrt{-3}\,\sqrt{-5}$ **b.** $\sqrt{-36}\,\sqrt{-1}$ **c.** $\sqrt{8}\,\sqrt{-2}$ **d.** $\dfrac{\sqrt{-125}}{\sqrt{5}}$

Solution: Write each imaginary number in the form bi first.

a. $\sqrt{-3}\sqrt{-5} = i\sqrt{3}\,(i\sqrt{5}) = i^2\sqrt{15} = -\sqrt{15}$

b. $\sqrt{-36}\sqrt{-1} = 6i(i) = 6i^2 = 6(-1) = -6$

c. $\sqrt{8}\sqrt{-2} = 2\sqrt{2}\,(i\sqrt{2}) = 2i(\sqrt{2}\sqrt{2}) = 2i(2) = 4i$

d. $\dfrac{\sqrt{-125}}{\sqrt{5}} = \dfrac{i\sqrt{125}}{\sqrt{5}} = i\sqrt{25} = 5i$ ■

5 We can use the fact that $i^2 = 1$ to find higher powers of i. To find i^3, we rewrite it as the product of i^2 and i.

$$i^3 = i^2 \cdot i = (-1)i = -i$$

Continue this process and find higher powers of i.

$$i^4 = i^2 \cdot i^2 = (-1)(-1) = 1$$

$$i^5 = i^4 \cdot i = 1 \cdot i = i$$

If we continue finding powers of i, we generate a pattern.

$i^1 = i$	$i^5 = i$	$i^9 = i$
$i^2 = -1$	$i^6 = -1$	$i^{10} = -1$
$i^3 = -i$	$i^7 = -i$	$i^{11} = -i$
$i^4 = 1$	$i^8 = 1$	$i^{12} = 1$

The values i, -1, $-i$, and 1 repeat as i is raised to higher and higher powers. This pattern allows us to find other powers of i.

EXAMPLE 6 Find the following powers of i.

a. i^7 **b.** i^{20} **c.** i^{46} **d.** i^{-12}

Solution: **a.** $i^7 = i^4 \cdot i^3 = 1(-i) = -i$ **b.** $i^{20} = (i^4)^5 = 1^5 = 1$

c. $i^{46} = i^{44} \cdot i^2 = (i^4)^{11} \cdot i^2 = 1^{11}(-1) = -1$

d. $i^{-12} = \dfrac{1}{i^{12}} = \dfrac{1}{(i^4)^3} = \dfrac{1}{(1)^3} = \dfrac{1}{1} = 1$ ■

MENTAL MATH

Simplify. See Example 1.

1. $\sqrt{-81}$ **2.** $\sqrt{-49}$ **3.** $\sqrt{-7}$ **4.** $\sqrt{-3}$

5. $-\sqrt{16}$ **6.** $-\sqrt{4}$ **7.** $\sqrt{-64}$ **8.** $\sqrt{-100}$

EXERCISE SET 8.7

Simplify. See Example 1.

1. $\sqrt{-24}$ **2.** $\sqrt{-32}$ **3.** $-\sqrt{-36}$ **4.** $-\sqrt{-121}$

5. $8\sqrt{-63}$ **6.** $4\sqrt{-20}$ **7.** $-\sqrt{54}$ **8.** $\sqrt{-63}$

Add or subtract. Write the sum or difference in the form $a + bi$. See Example 2.

9. $(4 - 7i) + (2 + 3i)$
10. $(2 - 4i) - (2 - i)$
11. $(6 + 5i) - (8 - i)$
12. $(8 - 3i) + (-8 + 3i)$
13. $6 - (8 + 4i)$
14. $(9 - 4i) - 9$

Multiply. Write the product in the form $a + bi$. See Example 3.

15. $6i(2 - 3i)$
16. $5i(4 - 7i)$
17. $(\sqrt{3} + 2i)(\sqrt{3} - 2i)$
18. $(\sqrt{5} - 5i)(\sqrt{5} + 5i)$
19. $(4 - 2i)^2$
20. $(6 - 3i)^2$

Write each quotient in the form $a + bi$. See Example 4.

21. $\dfrac{4}{i}$
22. $\dfrac{5}{6i}$
23. $\dfrac{7}{4 + 3i}$
24. $\dfrac{9}{1 - 2i}$
25. $\dfrac{3 + 5i}{1 + i}$
26. $\dfrac{6 + 2i}{4 - 3i}$

Multiply or divide. See Example 5.

27. $\sqrt{-2} \cdot \sqrt{-7}$
28. $\sqrt{-11} \cdot \sqrt{-3}$
29. $\sqrt{-5} \cdot \sqrt{-10}$
30. $\sqrt{-2} \cdot \sqrt{-6}$
31. $\sqrt{16} \cdot \sqrt{-1}$
32. $\sqrt{3} \cdot \sqrt{-27}$
33. $\dfrac{\sqrt{-9}}{\sqrt{3}}$
34. $\dfrac{\sqrt{49}}{\sqrt{-10}}$
35. $\dfrac{\sqrt{-80}}{\sqrt{-10}}$
36. $\dfrac{\sqrt{-40}}{\sqrt{-8}}$

Find each power of i. See Example 6.

37. i^8
38. i^{10}
39. i^{21}
40. i^{15}
41. i^{11}
42. i^{40}
43. i^{-6}
44. i^{-9}

Perform the indicated operation. Write the result in the form $a + bi$.

45. $(7i)(-9i)$
46. $(-6i)(-4i)$
47. $(6 - 3i) - (4 - 2i)$
48. $(-2 - 4i) - (6 - 8i)$
49. $(6 - 2i)(3 + i)$
50. $(2 - 4i)(2 - i)$
51. $(8 - \sqrt{-3}) - (2 + \sqrt{-12})$
52. $(8 - \sqrt{-4}) - (2 + \sqrt{-16})$
53. $(1 - i)(1 + i)$
54. $(6 + 2i)(6 - 2i)$
55. $\dfrac{16 + 15i}{-3i}$
56. $\dfrac{2 - 3i}{-7i}$
57. $(9 + 8i)^2$
58. $(4 - 7i)^2$
59. $\dfrac{2}{3 + i}$
60. $\dfrac{5}{3 - 2i}$
61. $(5 - 6i) - 4i$
62. $(6 - 2i) + 7i$
63. $\dfrac{2 - 3i}{2 + i}$
64. $\dfrac{6 + 5i}{6 - 5i}$
65. $(2 + 4i) + (6 - 5i)$
66. $(5 - 3i) + (7 - 8i)$

Skill Review

Recall that the sum of the measures of the angles of a triangle is 180°. Find the unknown angle in each triangle.

67.

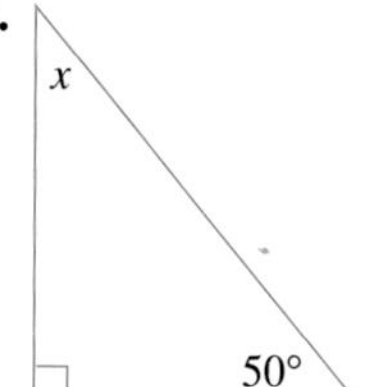

68.

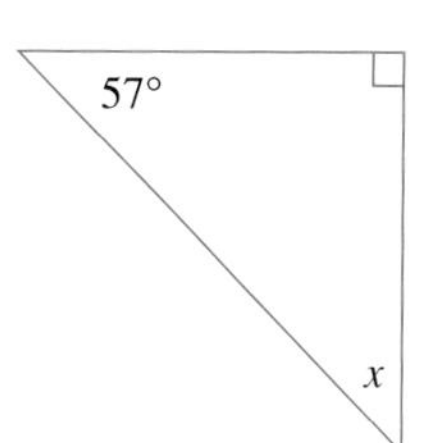

69.

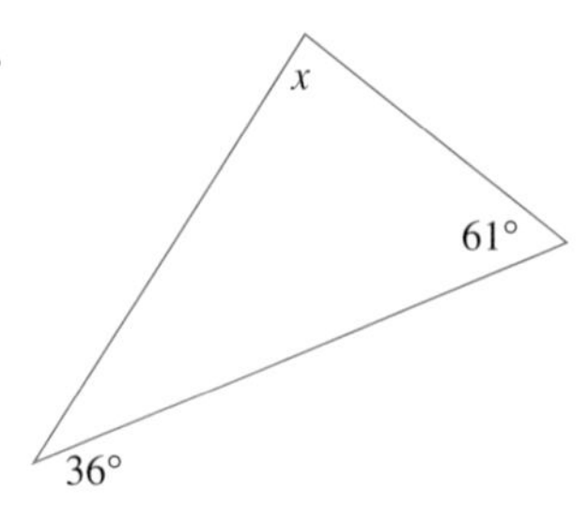

70.

15°
x
40°

Solve each compound inequality. See Section 3.6.

71. $8 < x - 7 < 12$

72. $0 < -3x + 5 < 8$

73. $-9 \le 2x + 3 \le 7$

74. $-2 < \frac{x}{2} - 1 < 1$

Perform the indicated operations. See Sections 7.2 and 7.3.

75. $\frac{15x}{x+8} \cdot \frac{2x+16}{3x}$

76. $\frac{9z+5}{15} \cdot \frac{5z}{81z^2 - 25}$

77. $\frac{5a+10}{18} \div \frac{a^2-4}{10a}$

78. $\frac{9}{x^2-1} \div \frac{12}{3x+3}$

79. $\frac{5}{x^2-3x+2} + \frac{1}{x-2}$

80. $\frac{4}{2x^2+5x-3} + \frac{2}{x+3}$

Writing in Mathematics

81. Describe how to find the conjugate of a complex number.

82. Explain why the product of a complex number and its complex conjugate is a real number.

CHAPTER 8 GLOSSARY

The complex numbers $(a + bi)$ and $(a - bi)$ are called **complex conjugates** of each other, and $(a + bi)(a - bi) = a^2 + b^2$.

A **complex number** is a number that can be written in the form $a + bi$, where a and b are real numbers.

The **conjugate** of $a + b$ is $a - b$.

A number b is the **cube root** of a if $b^3 = a$. That is, $\sqrt[3]{a} = b$ if $b^3 = a$.

The **imaginary unit,** i, is the number whose square is -1. That is, $i^2 = -1$ and $i = \sqrt{-1}$.

The positive integer n in $\sqrt[n]{\ }$ is called the **index.**

Radicals with the same index and the same radicand are **like radicals.**

The **positive** or **principal square root** of a positive real number a, written as $\sqrt{a}$, is the positive number whose square root equals a. That is, $\sqrt{a} = b$ if $b^2 = a$.

The expression $\sqrt[n]{a}$ is called a **radical expression.**

The symbol $\sqrt{\ }$ is called a **radical** or **radical sign.**

The expression within or under the radical sign is called the **radicand.**

Rationalizing the denominator is the process of writing an equivalent radical expression without a radical in the denominator.

Rationalizing the numerator is the process of writing an equivalent radical expression without a radical in the numerator.

CHAPTER 8 SUMMARY

PRODUCT RULE FOR RADICALS (8.2)

If $\sqrt[n]{a}$ and $\sqrt[n]{b}$ are real numbers, then $\sqrt[n]{a} \cdot \sqrt[n]{b} = \sqrt[n]{a \cdot b}$.

QUOTIENT RULE FOR RADICALS (8.2)

If $\sqrt[n]{a}$ and $\sqrt[n]{b}$ are real numbers, then $\dfrac{\sqrt[n]{a}}{\sqrt[n]{b}} = \sqrt[n]{\dfrac{a}{b}}$, providing $b \neq 0$.

DEFINITION OF $a^{1/n}$ (8.5)

If n is a positive integer and $\sqrt[n]{a}$ is a real number, then $a^{1/n} = \sqrt[n]{a}$.

DEFINITION OF $a^{m/n}$ (8.5)

If m and n are integers with $n > 0$ and a is positive, then $a^{m/n} = \sqrt[n]{a^m} = (\sqrt[n]{a})^m$.

DEFINITION OF $a^{-m/n}$ (8.5)

If $a^{m/n}$ is a nonzero real number, then $a^{-m/n} = \dfrac{1}{a^{m/n}}$.

THE PYTHAGOREAN THEOREM (8.6)

If a and b are the lengths of the legs of a right triangle and c is the length of the hypotenuse, then $a^2 + b^2 = c^2$.

CHAPTER 8 REVIEW

(8.1) *Find each root.*

1. $\sqrt{81}$ **2.** $-\sqrt{49}$ **3.** $\sqrt[3]{27}$ **4.** $\sqrt[4]{16}$

5. $-\sqrt{\dfrac{9}{64}}$ **6.** $\sqrt{\dfrac{36}{81}}$ **7.** $\sqrt[4]{-\dfrac{16}{81}}$ **8.** $\sqrt[3]{-\dfrac{27}{64}}$

Determine whether each of the following is rational or irrational. If rational, find the exact value. If irrational, use a calculator or Appendix D to find an approximation accurate to three decimal places.

9. $\sqrt{76}$ **10.** $\sqrt{576}$

Find the following roots. Assume that variables represent any real numbers.

11. $\sqrt{x^{12}}$ **12.** $\sqrt{x^8}$ **13.** $\sqrt{9x^6y^2}$

14. $\sqrt{25x^4y^{10}}$ **15.** $-\sqrt[3]{8x^6}$ **16.** $-\sqrt[4]{16x^8}$

Sketch the graph of each function.

17. $f(x) = \sqrt{x} + 3$ **18.** $g(x) = \sqrt{x + 2}$

19. $F(x) = \sqrt[3]{x} - 1$

20. $H(x) = \sqrt[3]{x + 1}$

(8.2) *Simplify each expression using the product rule. For the remainder of the review, assume that variables represent positive numbers only.*

21. $\sqrt{54}$
22. $\sqrt{88}$
23. $\sqrt{150x^3y^6}$
24. $\sqrt{92x^8y^5}$
25. $\sqrt[3]{54}$
26. $\sqrt[3]{88}$
27. $\sqrt[4]{48x^3y^6}$
28. $\sqrt[4]{162x^8y^5}$

Simplify each expression using the quotient rule.

29. $\sqrt{\dfrac{18}{25}}$
30. $\sqrt{\dfrac{75}{64}}$
31. $\sqrt{\dfrac{45x^2y^2}{4x^6}}$
32. $\sqrt{\dfrac{20x^5}{9x^2}}$
33. $\sqrt[4]{\dfrac{9}{16}}$
34. $\sqrt[3]{\dfrac{40}{27}}$
35. $\sqrt[3]{\dfrac{3y^6}{8x^3}}$
36. $\sqrt[4]{\dfrac{5x^6}{81x^8}}$

(8.3) *Add or subtract by combining like radicals.*

37. $3\sqrt[3]{2} + 2\sqrt[3]{3} - 4\sqrt[3]{2}$
38. $5\sqrt{2} + 2\sqrt[3]{2} - 8\sqrt{2}$
39. $\sqrt{6} + 2\sqrt[3]{6} - 4\sqrt[3]{6} + 5\sqrt{6}$
40. $3\sqrt{5} - \sqrt[3]{5} - 2\sqrt{5} + 3\sqrt[3]{5}$

Add or subtract by simplifying each radical and then combining like terms.

41. $\sqrt{28} + \sqrt{63} + \sqrt[3]{56}$
42. $\sqrt{75} + \sqrt{48} - \sqrt[4]{16}$
43. $\sqrt{\dfrac{5}{9}} - \sqrt{\dfrac{5}{36}}$
44. $\sqrt{\dfrac{11}{25}} + \sqrt{\dfrac{11}{16}}$
45. $2\sqrt[3]{125x^3} - 5x\sqrt[3]{8}$
46. $3\sqrt[3]{16x^4} - 2x\sqrt[3]{2x}$

(8.4) *Find the product and simplify if possible.*

47. $3\sqrt{10} \cdot 2\sqrt{5}$
48. $2\sqrt[3]{4} \cdot 5\sqrt[3]{6}$
49. $\sqrt{3}(2\sqrt{6} - 3\sqrt{12})$
50. $4\sqrt{5}(2\sqrt{10} - 5\sqrt{5})$
51. $(\sqrt{3} + 2)(\sqrt{6} - 5)$
52. $(2\sqrt{5} + 1)(4\sqrt{5} - 3)$

Find the quotient and simplify if possible.

53. $\dfrac{\sqrt{96}}{\sqrt{3}}$
54. $\dfrac{\sqrt{160}}{\sqrt{8}}$
55. $\dfrac{\sqrt{15x^6y}}{\sqrt{12x^3y^9}}$
56. $\dfrac{\sqrt{50xy^8}}{\sqrt{72x^7y^3}}$

Rationalize each denominator and simplify.

57. $\sqrt{\dfrac{5}{6}}$
58. $\sqrt{\dfrac{7}{10}}$
59. $\sqrt[3]{\dfrac{3}{2x}}$
60. $\sqrt{\dfrac{6x}{5y^2}}$
61. $\dfrac{3x}{\sqrt{5} - 2}$
62. $\dfrac{8}{\sqrt{10} - 3}$
63. $\dfrac{\sqrt{2}}{4 + \sqrt{x}}$
64. $\dfrac{3y}{5 - \sqrt{y}}$

Rationalize each numerator and simplify.

65. $\sqrt{\dfrac{7}{20y^2}}$
66. $\sqrt{\dfrac{5z}{12x^2}}$
67. $\sqrt[3]{\dfrac{7}{9}}$
68. $\sqrt[3]{\dfrac{2x^2}{3}}$
69. $\dfrac{\sqrt{6} + 2}{12}$
70. $\dfrac{\sqrt{15} - 3}{8x}$
71. $\dfrac{\sqrt{x} - 5}{2\sqrt{3}}$
72. $\dfrac{\sqrt{2} - y}{7y}$

(8.5) *Evaluate the following.*

73. $\left(\frac{1}{81}\right)^{1/4}$ **74.** $\left(-\frac{1}{27}\right)^{1/3}$ **75.** $(-27)^{-1/3}$ **76.** $(-64)^{-1/3}$

77. $-9^{3/2}$ **78.** $64^{-1/3}$ **79.** $(-25)^{5/2}$ **80.** $\left(\frac{25}{49}\right)^{-3/2}$

81. $\left(\frac{8}{27}\right)^{-2/3}$ **82.** $\left(-\frac{1}{36}\right)^{-1/4}$

Simplify each expression. Write using only positive exponents.

83. $a^{1/3}a^{4/3}a^{1/2}$ **84.** $\frac{b^{1/3}}{b^{4/3}}$ **85.** $(a^{-1/2})^{-2}$ **86.** $(x^{1/2}x^{3/4})^{-2}$

87. $(a^{1/2}a^{-2})^3$ **88.** $(a^8b^4c^4)^{3/4}$ **89.** $(x^{-3}y^6)^{1/3}$ **90.** $\left(\frac{b^{3/4}}{a^{-1/2}}\right)^8$

91. $\frac{x^{1/4}x^{-1/2}}{x^{2/3}}$ **92.** $x^{4/3}(x^{2/3} + x^{-1/3})$

Write using rational exponents.

93. $4x\sqrt[3]{(3x)^5}$ **94.** $\sqrt[5]{5x^2y^3}$

Write using radical notation.

95. $7x(3^{1/3}y^{2/3})$ **96.** $5(xy^2z^5)^{1/3}$

(8.6) *Solve each equation for the variable.*

97. $\sqrt[3]{4x} = -2$ **98.** $\sqrt{5x} = -5$

99. $\sqrt[3]{x - 2} = 3$ **100.** $\sqrt[3]{2x - 6} = 4$

101. $\sqrt{2x} + 1 = -4$ **102.** $\sqrt[4]{2x - 9} = 3$

103. $\sqrt[3]{3x - 9} - \sqrt[3]{2x + 12} = 0$ **104.** $\sqrt[3]{x - 12} - \sqrt[3]{5x + 16} = 0$

105. $\sqrt{x + 6} = \sqrt{x + 2}$ **106.** $2x - 5\sqrt{x} = 3$

Find each unknown length.

107.

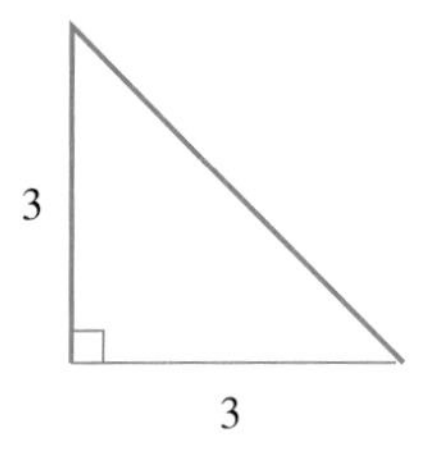

108.

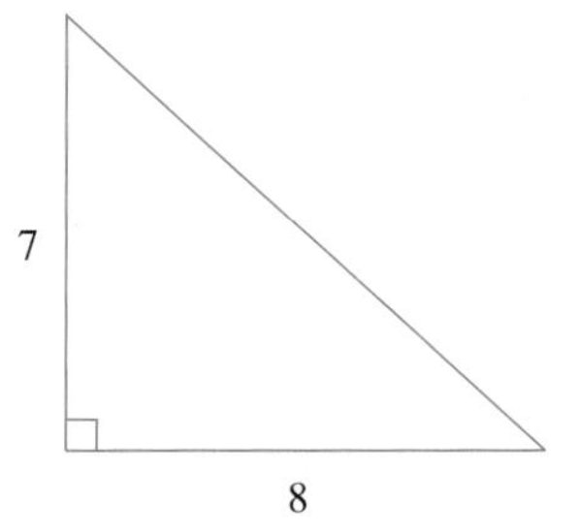

109. Beverly Hillis wants to determine the distance across a pond on her property. She is able to measure the distances shown on the following diagram. Find how wide the lake is to the nearest tenth of a foot.

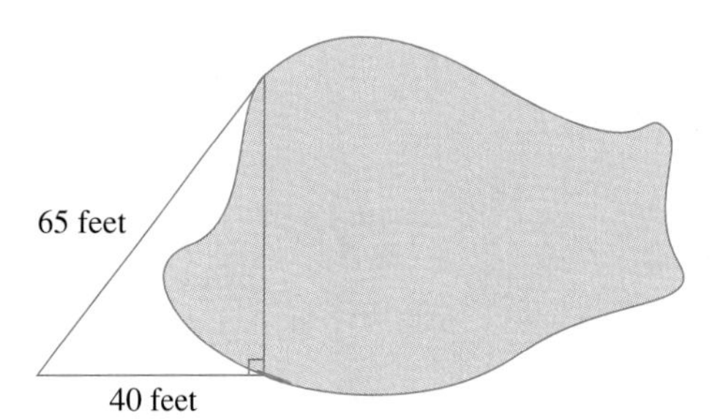

110. A pipefitter needs to connect two underground pipelines, which are offset by 3 feet, as pictured in the diagram. Neglecting the joints needed to join the pipes, find the length of the shortest possible connecting pipe rounded to the nearest hundredth of a foot.

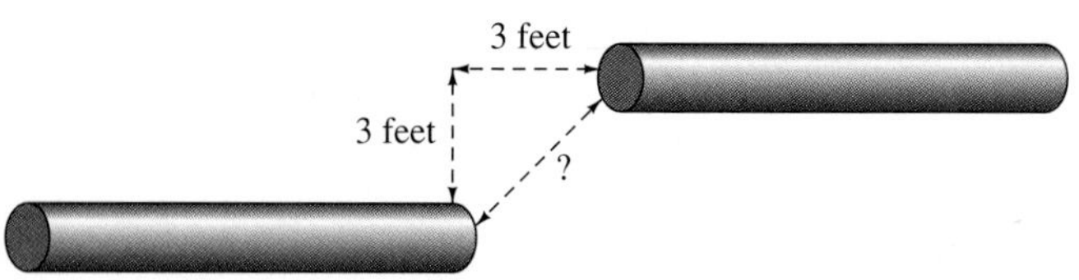

(8.7) *Perform the indicated operation and simplify. Write the result in the form a + bi.*

111. $\sqrt{-8}$
112. $-\sqrt{-6}$
113. $\sqrt{-4} + \sqrt{-16}$
114. $\sqrt{-25} - 2i$
115. $(12 - 6i) + (3 + 2i)$
116. $(-8 - 7i) - (5 - 4i)$
117. $(\sqrt{3} + \sqrt{2}) + (3\sqrt{2} - \sqrt{-8})$
118. $(3 - \sqrt{-72}) + (4 - \sqrt{-32})$
119. $2i(2 - 5i)$
120. $-3i(6 - 4i)$
121. $(3 + 2i)(1 + i)$
122. $(2 - 3i)(4 + \sqrt{-4})$
123. $\dfrac{2 + 3i}{2i}$
124. $\dfrac{1 + i}{-3i}$

CHAPTER 8 TEST

Simplify the following. Indicate if the expression is not a real number.

1. $\sqrt{16}$
2. $\sqrt[3]{125}$
3. $16^{3/4}$
4. $\left(\dfrac{9}{16}\right)^{1/2}$
5. $\sqrt[4]{-81}$
6. $27^{-2/3}$

Simplify each radical expression. Assume that all variables represent positive numbers only.

7. $\sqrt{54}$
8. $\sqrt{92}$
9. $\sqrt{3x^6}$
10. $\sqrt{8x^4y^7}$
11. $\sqrt{9x^9}$
12. $\sqrt[3]{40}$
13. $\sqrt[3]{8x^6y^{10}}$
14. $\sqrt{12} - 2\sqrt{75}$
15. $\sqrt{2x^2} + \sqrt[3]{54} - x\sqrt{18}$
16. $\sqrt{\dfrac{5}{16}}$
17. $\sqrt[3]{\dfrac{2x^3}{27}}$

18. Rationalize the denominator of $\sqrt[3]{\dfrac{3}{2x^2}}$ and simplify.

19. Rationalize the numerator of $\dfrac{\sqrt{6} + x}{8}$ and simplify.

Solve each of the following radical equations.

20. $\sqrt{x} + 8 = 11$
21. $\sqrt{3x - 6} = \sqrt{x + 4}$
22. $\sqrt{2x - 2} + 5 = x$

23. Express $\sqrt{-20}$ in terms of i and simplify.

Perform the indicated operations and simplify. Write each result in the form a + bi.

24. $(12 - 6i) - (12 - 3i)$
25. $(4 - 5i)^2$
26. $\dfrac{1 + 4i}{1 - i}$

Find each unknown length.

27.

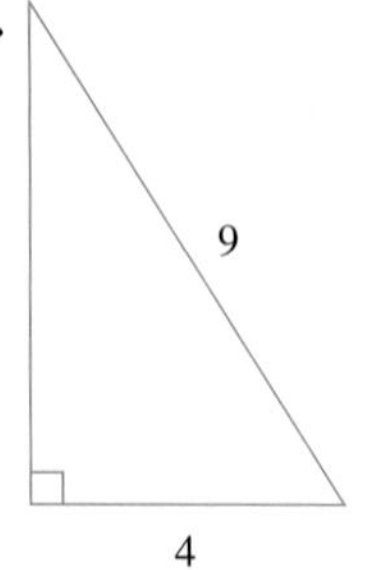

28.

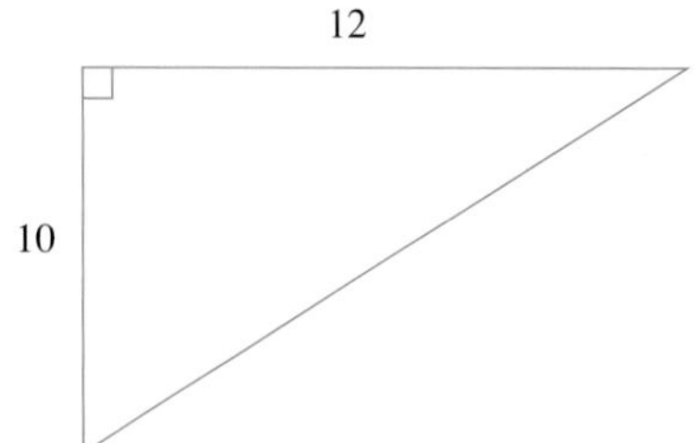

Solve.

29. The function $V(t) = \sqrt{2.5r}$ can be used to estimate the maximum safe velocity, V, in miles per hour, at which a car can travel if it is driven along a curved road with a **radius of curvature,** r, in feet. To the nearest whole number, find the maximum safe speed if a cloverleaf exit on an expressway has a radius of curvature of 300 feet.

30. Use the formula from Exercise 29 to find the radius of curvature if the safe velocity is 30 mph.

CHAPTER 8 CUMULATIVE REVIEW

1. Find each difference.
a. $-13 - (+4)$ **b.** $5 - (-6)$
c. $3 - 6$ **d.** $-1 - (-7)$

2. Simplify the following expressions.
a. $3(2x - 5) + 1$ **b.** $8 - (7x + 2) + 3x$
c. $-2(4x + 7) - (3x - 1)$

3. Solve $-2(x - 5) + 10 = -3(x + 2) + x$.

4. The following bar graph shows selected electricity rates per kilowatt hour.

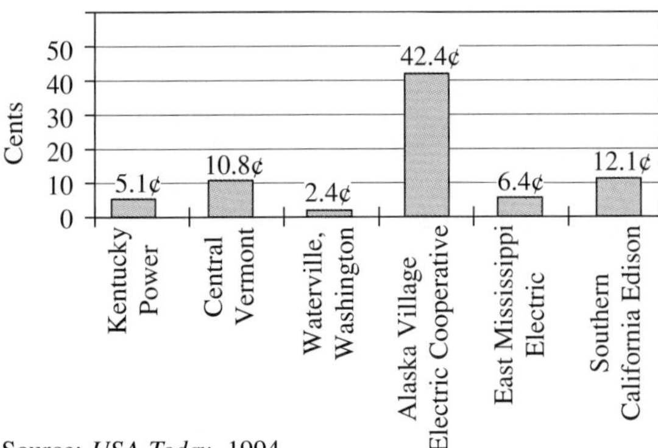

Source: *USA Today,* 1994.

a. Which company charges the highest rate?
b. Which company charges the lowest rate?
c. Find the difference in rates of the companies in parts a and b.
d. What is the electricity rate charged by Southern California Edison?

5. A chemist working on his doctorate degree at Massachusetts Institute of Technology needs 12 liters of a 50% acid solution for a lab experiment. The stockroom has only 40% and 70% solutions. How much of each solution should be mixed together to form 12 liters of a 50% solution?

6. Solve $|3x + 2| = |5x - 8|$.

7. Solve $5x - 3 \le 10$ or $x + 1 \ge 5$.

8. Graph the linear equation $y = \frac{1}{3}x$.

9. Graph $y = -3$.

10. Simplify each of the following expressions.
a. $(x^2)^5$ **b.** $(y^8)^2$ **c.** $[(-5)^3]^4$

11. Combine like terms.
a. $-12x^2 + 7x^2 - 6x$ **b.** $3xy - 2x + 5xy - x$

12. Multiply and simplify the product if possible.
a. $(x + 3)(2x + 5)$ **b.** $(2x^3 - 3)(5x^2 - 6x + 7)$

13. Use the remainder theorem and synthetic division to find $P(4)$ if
$$P(x) = 4x^6 - 25x^5 + 35x^4 + 17x^2$$

14. Factor $x^2 + 7x + 12$.

15. Solve $2x^2 + 9x - 5 = 0$.

16. Divide $\dfrac{8x^3 + 125}{x^4 + 5x^2 + 4} \div \dfrac{2x + 5}{2x^2 + 8}$.

17. Solve $\dfrac{x + 6}{x - 2} = \dfrac{2(x + 2)}{x - 2}$.

18. Evaluate the following expressions.
a. $\sqrt[4]{16}$ **b.** $\sqrt[5]{-32}$ **c.** $-\sqrt[3]{8}$ **d.** $\sqrt[4]{-81}$

19. Simplify each expression.
a. $\sqrt{54}$ **b.** $\sqrt{12}$ **c.** $\sqrt{200}$ **d.** $\sqrt{35}$

20. Rationalize each denominator.
a. $\dfrac{2}{3\sqrt{2} + 5}$ **b.** $\dfrac{\sqrt{6} + 2}{\sqrt{5} - \sqrt{3}}$
c. $\dfrac{7\sqrt{y}}{\sqrt{12x}}$ **d.** $\dfrac{2\sqrt{m}}{3\sqrt{x} + \sqrt{m}}$

21. Simplify each expression.
a. $25^{1/2}$ **b.** $8^{1/3}$ **c.** $-16^{1/4}$
d. $(-27)^{1/3}$ **e.** $\left(\dfrac{1}{9}\right)^{1/2}$

22. Solve $\sqrt{2x - 3} = 9$ for x.

23. A surveyor must determine the distance across a lake. He first inserts poles at points P and Q on opposite sides of the lake, as pictured. Perpendicular to line PQ, he finds another point R. If the length of $\overline{PR}$ is 320 feet and the length of $\overline{QR}$ is 240 feet, what is the distance across the lake?

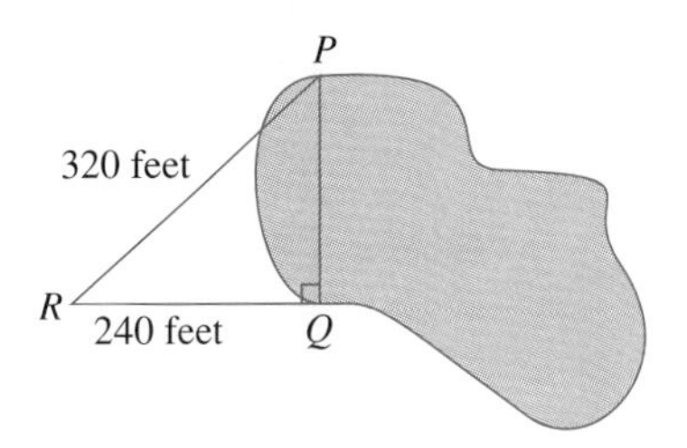

24. Find each quotient. Write in the form $a + bi$.
a. $\dfrac{2 + i}{1 - i}$ **b.** $\dfrac{7}{3i}$

CHAPTER 9

Quadratic Equations and Inequalities

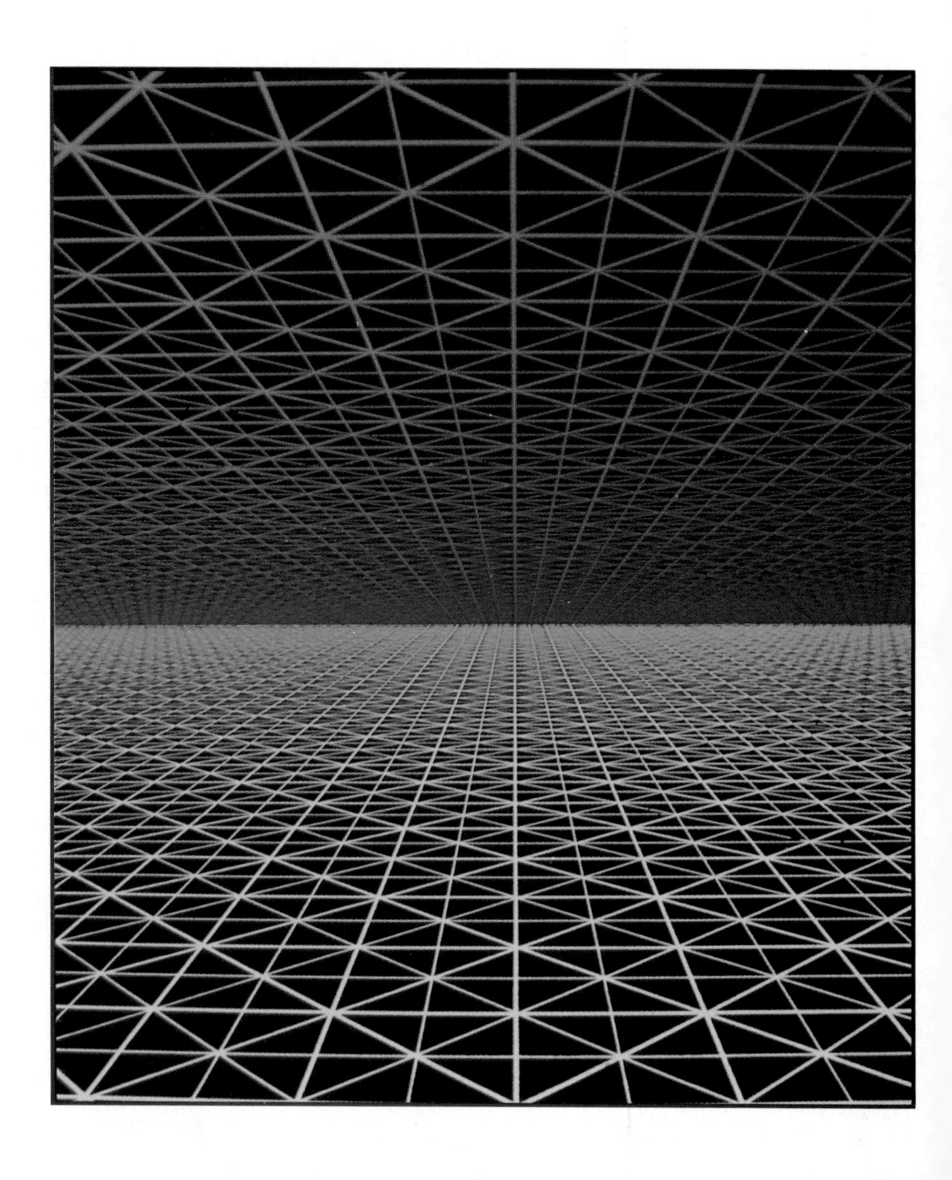

Amidst the disorder of these right triangles is a remarkable pattern, bridging geometry and the algebra of quadratic equations.

INTRODUCTION

An important part of the study of algebra is learning to model and solve problems. Often, the model of a problem is a quadratic equation or a function containing a second-degree polynomial. In this chapter, we continue the work begun in Section 6.6, when we solved quadratic equations in one variable by factoring. Two additional methods for solving quadratic equations are analyzed, as well as methods for solving nonlinear inequalities in one variable.

9.1 Solving Quadratic Equations by Completing the Square

OBJECTIVES

Tape IA 24

1. Use the square root property to solve quadratic equations.
2. Write perfect square trinomials.
3. Solve quadratic equations by completing the square.
4. Quadratic equations and problem solving.

1 In Chapter 6, we solved quadratic equations by factoring. Recall that a **quadratic** or **second-degree equation** is an equation that can be written in the form $ax^2 + bx + c = 0$, where a, b, and c are real numbers and a is not 0. To solve a quadratic equation such as $x^2 = 9$ by factoring, we use the zero-factor theorem. To use the zero-factor theorem, the equation must be written in standard form, $ax^2 + bx + c = 0$.

$$x^2 = 9$$

$$x^2 - 9 = 0 \qquad \text{Subtract 9 from both sides.}$$

$$(x + 3)(x - 3) = 0 \qquad \text{Factor.}$$

$$x + 3 = 0 \quad \text{or} \quad x - 3 = 0 \qquad \text{Set each factor equal to 0.}$$

$$x = -3 \qquad\quad x = 3 \qquad \text{Solve.}$$

The solution set is $\{-3, 3\}$, the positive and negative **square roots** of 9. Not all quadratic equations can be solved by factoring, so we need to explore other methods. Notice that the solutions of the equation $x^2 = 9$ are two numbers whose square is 9.

$$3^2 = 9 \quad \text{and} \quad (-3)^2 = 9$$

Thus we can solve the equation $x^2 = 9$ by taking the square root of both sides. Be sure to include both $\sqrt{9}$ and $-\sqrt{9}$ as solutions.

$$x^2 = 9$$

$$x = \pm\sqrt{9}$$

$$x = \pm 3$$

The notation ± 3 (read as plus or minus 3) indicates the pair of numbers $+3$ and -3.

This illustrates the square root property.

Square Root Property

If b is a real number and if $a^2 = b$, then $a = \pm\sqrt{b}$.

EXAMPLE 1 Use the square root property to solve $x^2 = 50$.

Solution:

$$x^2 = 50$$
$$x = \pm\sqrt{50}$$
$$x = \pm 5\sqrt{2}$$

The solution set is $\{5\sqrt{2}, -5\sqrt{2}\}$. ■

EXAMPLE 2 Use the square root property to solve $(x + 1)^2 = 12$ for x.

Solution: By the square root property, we have

$$(x + 1)^2 = 12$$

$x + 1 = \pm\sqrt{12}$ Apply the square root property.

$x + 1 = \pm 2\sqrt{3}$ Simplify the radical.

$x = -1 \pm 2\sqrt{3}$ Subtract 1 from both sides.

The solution set is $\{-1 + 2\sqrt{3}, -1 - 2\sqrt{3}\}$. ■

EXAMPLE 3 Solve $(2x - 5)^2 = -16$.

Solution:

$$(2x - 5)^2 = -16$$

$2x - 5 = \pm\sqrt{-16}$ Apply the square root property.

$2x - 5 = \pm 4i$ Simplify the radical.

$2x = 5 \pm 4i$ Add 5 to both sides.

$x = \dfrac{5 \pm 4i}{2}$ Divide both sides by 2.

The solution set is $\left\{\dfrac{5 + 4i}{2}, \dfrac{5 - 4i}{2}\right\}$. ■

2 Notice from Examples 2 and 3 that, if we write a quadratic equation so that one side is a binomial squared, we can solve using the square root property. To write the square of a binomial, we write perfect square trinomials. Recall that a perfect square trinomial is a trinomial that can be factored into two identical binomial factors.

Perfect Square Trinomials	*Factored Form*
$x^2 + 8x + 16$	$(x + 4)^2$
$x^2 - 6x + 9$	$(x - 3)^2$

Notice in each perfect square trinomial that **the constant term of the trinomial is the square of half the coefficient of the *x*-term.** For example,

$$x^2 + 8x + 16 \qquad\qquad x^2 - 6x + 9$$

$$\frac{1}{2}(8) = 4 \text{ and } 4^2 = 16 \qquad\qquad \frac{1}{2}(-6) = -3 \text{ and } (-3)^2 = 9$$

3 The process of writing a quadratic equation so that one side is a perfect square trinomial is called **completing the square.**

EXAMPLE 4 Solve $p^2 + 2p = 4$ by completing the square.

Solution: First add the square of half the coefficient of p to both sides so that the resulting trinomial will be a perfect square trinomial. The coefficient of p is 2.

$$\frac{1}{2}(2) = 1 \quad \text{and} \quad 1^2 = 1$$

Add 1 to both sides of the original equation.

$$p^2 + 2p = 4$$

$$p^2 + 2p + 1 = 4 + 1 \qquad \text{Add 1 to both sides.}$$

$$(p + 1)^2 = 5 \qquad \text{Factor.}$$

We may now apply the square root property and solve for p.

$$p + 1 = \pm\sqrt{5} \qquad \text{Use the square root property.}$$

$$p = -1 \pm \sqrt{5} \qquad \text{Subtract 1 from both sides.}$$

Notice that there are two solutions: $-1 + \sqrt{5}$ and $-1 - \sqrt{5}$. The solution set is $\{-1 + \sqrt{5}, -1 - \sqrt{5}\}$. ■

EXAMPLE 5 Solve $m^2 - 7m - 1 = 0$ for m by completing the square.

Solution: First add 1 to both sides of the equation so that the left side has no constant term.

$$m^2 - 7m - 1 = 0$$

$$m^2 - 7m = 1$$

Now find the proper constant term that makes the left side a perfect square trinomial by squaring half the coefficient of m. Add this result to both sides of the equation.

$$\frac{1}{2}(-7) = -\frac{7}{2} \quad \text{and} \quad \left(-\frac{7}{2}\right)^2 = \frac{49}{4}$$

$$m^2 - 7m + \frac{49}{4} = 1 + \frac{49}{4} \qquad \text{Add } \frac{49}{4} \text{ to both sides of the equation.}$$

$$\left(m - \frac{7}{2}\right)^2 = \frac{53}{4} \qquad \text{Factor the perfect square trinomial.}$$

$$m - \frac{7}{2} = \pm\sqrt{\frac{53}{4}} \qquad \text{Apply the square root property.}$$

$$m = \frac{7}{2} \pm \frac{\sqrt{53}}{2} \qquad \text{Add } \frac{7}{2} \text{ to both sides and simplify } \sqrt{\frac{53}{4}}.$$

$$m = \frac{7 \pm \sqrt{53}}{2} \qquad \text{Simplify.}$$

The solution set is $\left\{\frac{7 + \sqrt{53}}{2}, \frac{7 - \sqrt{53}}{2}\right\}$. ■

EXAMPLE 6 Solve $2x^2 - 8x + 3 = 0$.

Solution: Our procedure for finding the constant term to complete the square works only if the coefficient of the squared variable term is 1. Therefore, to solve this equation, the first step is to divide both sides by 2, the coefficient of x^2.

$$2x^2 - 8x + 3 = 0$$

$$x^2 - 4x + \frac{3}{2} = 0 \qquad \text{Divide both sides by 2.}$$

$$x^2 - 4x = -\frac{3}{2} \qquad \text{Subtract } \frac{3}{2} \text{ from both sides.}$$

Next find the square of half of -4.

$$\frac{1}{2}(-4) = -2 \quad \text{and} \quad (-2)^2 = 4$$

Add 4 to both sides of the equation to complete the square.

$$x^2 - 4x + 4 = -\frac{3}{2} + 4$$

$$(x - 2)^2 = \frac{5}{2} \qquad \text{Factor the perfect square.}$$

$$x - 2 = \pm\sqrt{\frac{5}{2}} \qquad \text{Apply the square root property.}$$

$$x - 2 = \pm\frac{\sqrt{10}}{2} \qquad \text{Rationalize the denominator.}$$

$$x = 2 \pm \frac{\sqrt{10}}{2} \qquad \text{Add 2 to both sides.}$$

$$= \frac{4}{2} \pm \frac{\sqrt{10}}{2} \qquad \text{Find the common denominator.}$$

$$= \frac{4 \pm \sqrt{10}}{2} \qquad \text{Simplify.}$$

The solution set is $\left\{\frac{4 + \sqrt{10}}{2}, \frac{4 - \sqrt{10}}{2}\right\}$. ■

The following steps may be used to solve a quadratic equation such as $ax^2 + bx + c = 0$ by completing the square.

To Solve a Quadratic Equation in x by Completing the Square

Step 1 If the coefficient of x^2 is 1, go to step 2. Otherwise, divide both sides of the equation by the coefficient of x^2.

Step 2 Isolate all variable terms on one side of the equation.

Step 3 Complete the square for the resulting binomial by adding the square of half of the coefficient of x to both sides of the equation.

Step 4 Factor the resulting perfect square trinomial.

Step 5 Apply the square root property to solve for x.

EXAMPLE 7 Solve $3x^2 - 9x + 8 = 0$ by completing the square.

Solution:

$$3x^2 - 9x + 8 = 0$$

$$x^2 - 3x + \frac{8}{3} = 0 \quad \text{Divide both sides of the equation by 3.}$$

$$x^2 - 3x = -\frac{8}{3} \quad \text{Subtract } \frac{8}{3} \text{ from both sides.}$$

Since $\frac{1}{2}(-3) = -\frac{3}{2}$ and $\left(-\frac{3}{2}\right)^2 = \frac{9}{4}$, we add $\frac{9}{4}$ to both sides of the equation.

$$x^2 - 3x + \frac{9}{4} = -\frac{8}{3} + \frac{9}{4}$$

$$\left(x - \frac{3}{2}\right)^2 = -\frac{5}{12} \quad \text{Factor the perfect square trinomial.}$$

$$x - \frac{3}{2} = \pm\sqrt{-\frac{5}{12}} \quad \text{Apply the square root property.}$$

$$x - \frac{3}{2} = \pm\frac{i\sqrt{5}}{2\sqrt{3}} \quad \text{Simplify the radical.}$$

$$x - \frac{3}{2} = \pm\frac{i\sqrt{15}}{6} \quad \text{Rationalize the denominator.}$$

$$x = \frac{3}{2} \pm \frac{i\sqrt{15}}{6} \quad \text{Add } \frac{3}{2} \text{ to both sides.}$$

$$= \frac{9}{6} \pm \frac{i\sqrt{15}}{6} \quad \text{Find a common denominator.}$$

$$= \frac{9 \pm i\sqrt{15}}{6} \quad \text{Simplify.}$$

The solution set is $\left\{\frac{9 + i\sqrt{15}}{6}, \frac{9 - i\sqrt{15}}{6}\right\}$. ■

4 Recall the **simple interest** formula $I = Prt$. If \$100 is invested at a simple interest rate of 5%, at the end of 3 years the interest I earned is:

$$I = P \cdot r \cdot t$$

or

$$I = 100 \cdot 0.05 \cdot 3 = \$15$$

and the new principal is

$$\$100 + \$15 = \$115$$

Most of the time, the interest computed on money borrowed or money deposited is **compound interest.** Compound interest, unlike simple interest, is computed on original principal **and** on interest already earned. To see the difference between simple interest and compound interest, suppose that \$100 is invested at a rate of 5% compounded annually. To find the amount of money at the end of 3 years, we calculate as follows:

$$\text{First year: Interest} = \$100 \cdot 0.05 \cdot 1 = \$5.00$$

$$\text{New principal} = \$100.00 + \$5.00 = \mathbf{\$105.00}$$

$$\text{Second Year: Interest} = \mathbf{\$105.00} \cdot 0.05 \cdot 1 = \$5.25$$

$$\text{New Principal} = \$105.00 + \$5.25 = \mathbf{\$110.25}$$

$$\text{Third Year: Interest} = \mathbf{\$110.25} \cdot 0.05 \cdot 1 \approx \$5.51$$

$$\text{New Principal} = \$110.25 + \$5.51 = \mathbf{\$115.76}$$

At the end of the third year, the compound interest earned is \$15.76, whereas the simple interest earned is \$15.

It would be tedious to calculate compound interest as we did above, so we use a compound interest formula. The formula for calculating compound interest for annually compounding is

$$A = P(1 + r)^t$$

For example, the amount of money A at the end of 3 years if the money is compounded annually is

$$A = \$100(1 + 0.05)^3 \approx \$100(1.1576) = \$115.76$$

as expected.

EXAMPLE 8 Find the interest rate r if \$2000 grows to \$2420 in 2 years compounded annually.

Solution: **1.** UNDERSTAND the problem. For this example, make sure that you understand the formula for compounding interest annually. Since a formula is known, we go to step 4.

4. TRANSLATE. Here we substitute given values into the formula:

$$A = P(1 + r)^t$$

$$2420 = 2000(1 + r)^2 \qquad \text{Let } A = 2420,\ P = 2000, \text{ and } t = 2.$$

5. COMPLETE. Solve the equation for r.

$$2420 = 2000(1 + r)^2$$

$$\frac{2420}{2000} = (1 + r)^2 \qquad \text{Divide both sides by 2000.}$$

$$\frac{121}{100} = (1 + r)^2 \qquad \text{Simplify the fraction.}$$

$$\pm\sqrt{\frac{121}{100}} = 1 + r \qquad \text{Apply the square root property.}$$

$$\pm\frac{11}{10} = 1 + r \qquad \text{Simplify.}$$

$$-1 \pm \frac{11}{10} = r$$

$$-\frac{10}{10} \pm \frac{11}{10} = r$$

$$\frac{1}{10} = r \quad \text{or} \quad -\frac{21}{10} = r$$

6. INTERPRET. The rate cannot be negative, so we reject $-\frac{21}{10}$ and *check* $\frac{1}{10} = 0.10 = 10\%$. If we invest \$2000 at 10% compounded annually, in 2 years the amount in the account would be $2000(1 + 0.10)^2 = 2420$ dollars, the desired amount.

State: The interest rate is 10%. ■

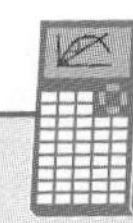

GRAPHING CALCULATOR BOX

In Section 6.6, we showed how we can use a grapher to approximate real number solutions of a quadratic equation written in standard form. We can also use a grapher to solve a quadratic equation when it is not written in standard form. For example, to solve $(x + 1)^2 = 12$, the quadratic equation in Example 2, we graph the following on the same set of axes. Use Xmin = −13, Xmax = 13, Ymin = −10, and Ymax = 10.

$$Y_1 = (x + 1)^2 \quad \text{and} \quad Y_2 = 12$$

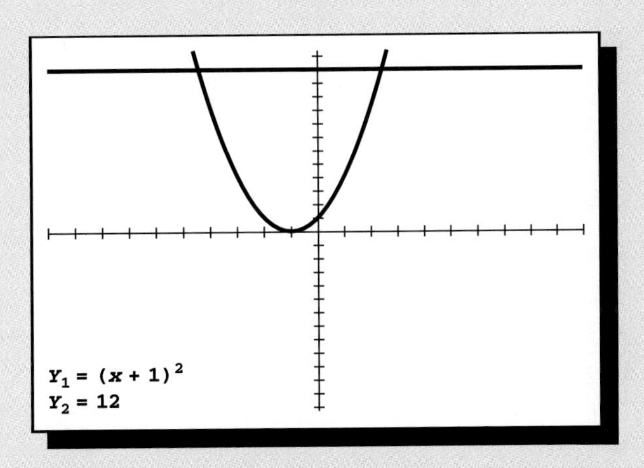

Use the Zoom and Trace features to locate the points of intersection of the graphs. The x-values of these points are the solutions of $(x + 1)^2 = 12$. The solutions rounded to two decimal points are 2.46 and −4.46.

Check to see that these numbers are approximations of the exact solutions $-1 \pm 2\sqrt{3}$.

Use a grapher to solve each quadratic equation. Round all solutions to the nearest hundredth.

1. $x(x - 5) = 8$

2. $x(x + 2) = 5$

3. $x^2 + 0.5x = 0.3x + 1$

4. $x^2 - 2.6x = -2.2x + 3$

5. Use a grapher and solve $(2x - 5)^2 = -16$, Example 3 in this section, using the window

Xmin = −20

Xmax = 20

Xscl = 1

Ymin = −20

Ymax = 20

Yscl = 1

Explain the results. Compare your results with the solution found in Example 3.

6. What are the advantages and disadvantages of using a grapher to solve quadratic equations?

EXERCISE SET 9.1

Use the square root property to solve each equation. These equations have real number solutions. See Example 1.

1. $x^2 = 16$
2. $x^2 = 49$
3. $x^2 - 7 = 0$
4. $x^2 - 11 = 0$
5. $x^2 = 18$
6. $y^2 = 20$
7. $3z^2 - 30 = 0$
8. $2x^2 = 4$

Use the square root property to solve each equation. See Example 1.

9. $x^2 + 9 = 0$
10. $x^2 + 4 = 0$
11. $x^2 - 6 = 0$
12. $y^2 - 10 = 0$
13. $2z^2 + 16 = 0$
14. $3p^2 + 36 = 0$

Use the square root property to solve each equation. These equations have real number solutions. See Example 2.

15. $(x + 5)^2 = 9$
16. $(y - 3)^2 = 4$
17. $(z - 6)^2 = 18$
18. $(y + 4)^2 = 27$
19. $(2x - 3)^2 = 8$
20. $(4x + 9)^2 = 6$

Use the square root property to solve each equation. See Example 3.

21. $(x - 1)^2 = -16$
22. $(y + 2)^2 = -25$
23. $(z + 7)^2 = 5$
24. $(x + 10)^2 = 11$
25. $(x + 3)^2 = -8$
26. $(y - 4)^2 = -18$

Add the proper constant to each binomial so that the resulting trinomial is a perfect square trinomial. Then factor the trinomial.

27. $x^2 + 16x$
28. $y^2 + 2y$
29. $z^2 - 12z$
30. $x^2 - 8x$
31. $p^2 + 9p$
32. $n^2 + 5n$
33. $x^2 + x$
34. $y^2 - y$

Solve each equation by completing the square. These equations have real number solutions. See Examples 4 and 5.

35. $x^2 + 8x = -15$
36. $y^2 + 6y = -8$
37. $x^2 + 6x + 2 = 0$
38. $x^2 - 2x - 2 = 0$
39. $x^2 + x - 1 = 0$
40. $x^2 + 3x - 2 = 0$

Solve each equation by completing the square. See Examples 4 and 5.

41. $y^2 + 2y + 2 = 0$

42. $x^2 + 4x + 6 = 0$

43. $x^2 - 6x + 3 = 0$

44. $x^2 - 7x - 1 = 0$

Solve each equation by completing the square. These equations have real number solutions. See Examples 6 and 7.

45. $x^2 + x - 1 = 0$

46. $y^2 + y - 7 = 0$

47. $3p^2 - 12p + 2 = 0$

48. $2x^2 + 14x - 1 = 0$

49. $4y^2 - 12y - 2 = 0$

50. $6x^2 - 3 = 6x$

51. $2x^2 + 7x = 4$

52. $3x^2 - 4x = 4$

Solve each equation by completing the square. See Examples 6 and 7.

53. $2a^2 + 8a = -12$

54. $3x^2 + 12x = -14$

55. $5x^2 + 15x - 1 = 0$

56. $16y^2 + 16y - 1 = 0$

57. $2x^2 - x + 6 = 0$

58. $4x^2 - 2x + 5 = 0$

Use the formula $A = P(1 + r)^t$ to solve Exercises 59 and 60. See Example 8.

59. Find the rate r to make \$3000 grow to \$4320 in 2 years.

60. Find the rate r to make \$800 grow to \$882 in 2 years.

Neglecting air resistance, the distance $s(t)$ traveled by a freely falling object is given by the function $s(t) = 16t^2$ where t is time in seconds. Use this formula to solve Exercises 61 to 64.

61. The height of the Columbia Seafirst Center in Seattle is 954 feet. How long would it take an object to fall from the top of the building?

62. The height of the First Interstate World Center in Los Angeles is 1017 feet. How long would it take an object to fall from the top of the building?

63. The height of the John Hancock Center in Chicago is 1127 feet. How long would it take an object to fall from the top of the building?

64. The height of the Carew Tower in Cincinnati is 568 feet. How long would it take an object to fall from the top of the building?

Solve each equation by completing the square. These equations have real number solutions.

65. $x^2 - 4x - 5 = 0$

66. $y^2 + 6y - 8 = 0$

67. $x^2 + 8x + 1 = 0$

68. $x^2 - 10x + 2 = 0$

69. $3y^2 + 6y - 4 = 0$

70. $2y^2 + 12y + 3 = 0$

71. $2x^2 - 3x - 5 = 0$

72. $5x^2 + 3x - 2 = 0$

Solve each equation by completing the square.

73. $x^2 + 10x + 28 = 0$

74. $y^2 + 8y + 18 = 0$

75. $z^2 + 3z - 4 = 0$

76. $y^2 + y - 2 = 0$

77. $2x^2 - 4x + 3 = 0$

78. $9x^2 - 36x = -40$

79. $3x^2 + 3x = 5$

80. $5y^2 - 15y = 1$

81. The area of a square room is 225 square feet. Find the dimensions of the room.

82. The area of a circle is 36π square inches. Find the radius of the circle.

83. An isosceles right triangle has legs of equal length. If the hypotenuse is 20 centimeters long, find the length of each leg.

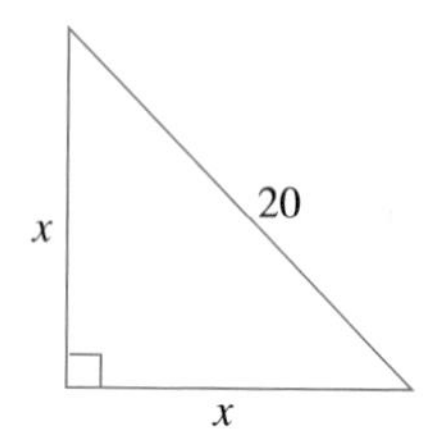

84. A 27-inch TV is advertised in the Daily Sentry newspaper. If 27 inches is the measure of the diagonal of the picture tube, find the measure of the side of the picture tube.

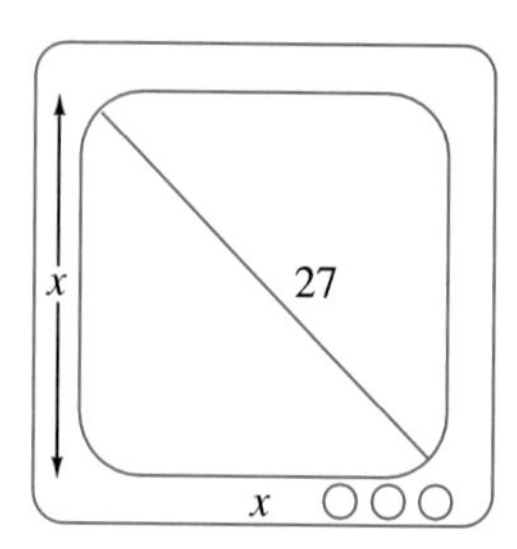

A common equation used in business is a demand equation. It expresses the relationship between the unit price of some commodity and the quantity demanded. For Exercises 85 and 86, p represents the unit price and x represents the quantity demanded in thousands.

85. A manufacturing company has found that the demand equation for a certain type of scissors is given by the equation $p = -x^2 + 47$. Find the demand for the scissors if the price is \$11 per pair.

86. Acme, Inc., sells desk lamps and has found that the demand equation for a certain style of desk lamp is given by the equation $p = -x^2 + 15$. Find the demand for the desk lamp if the price is \$7 per lamp.

Use the formula $A = P(1 + r)^t$ for Exercises 87 and 88.

87. Find the rate at which \$810 grows to \$1000 in 2 years.

88. Find the rate at which \$2000 grows to \$2880 in 2 years.

Skill Review

Simplify each expression. See Section 8.2.

89. $\frac{3}{4} - \sqrt{\frac{25}{16}}$

90. $\frac{3}{5} + \sqrt{\frac{16}{25}}$

91. $\frac{1}{2} - \sqrt{\frac{9}{4}}$

92. $\frac{9}{10} - \sqrt{\frac{49}{100}}$

Simplify each expression. See Section 8.4.

93. $\frac{6 + 4\sqrt{5}}{2}$

94. $\frac{10 - 20\sqrt{3}}{2}$

95. $\frac{3 - 9\sqrt{2}}{6}$

96. $\frac{12 - 8\sqrt{7}}{16}$

Evaluate $\sqrt{b^2 - 4ac}$ for each set of values. See Section 8.2.

97. $a = 2, b = 4, c = -1$

98. $a = 1, b = 6, c = 2$

99. $a = 3, b = -1, c = -2$

100. $a = 1, b = -3, c = -1$

Writing in Mathematics

101. In your own words, what is the difference between simple interest and compound interest?

102. If you are depositing money in an account paying 4%, would you prefer the interest to be simple or compound?

103. If you are borrowing money at a rate of 10%, would you prefer the interest to be simple or compound?

9.2 Solving Quadratic Equations by the Quadratic Formula

OBJECTIVES

Tape IA 23

1. Solve quadratic equations using the quadratic formula.
2. Determine the number and type of solutions of a quadratic equation using the discriminant.
3. Quadratic equations and problem solving.

1 Any quadratic equation can be solved by completing the square. Since the same sequence of steps is repeated each time we complete the square, let's complete the square for a general quadratic equation, $ax^2 + bx + c = 0$. By doing so, we find a pattern for the solutions of a quadratic equation known as the **quadratic formula.**

Recall that to complete the square for an equation such as $ax^2 + bx + c = 0$ we first divide both sides by the coefficient of x^2.

$$ax^2 + bx + c = 0$$

$$x^2 + \frac{b}{a}x + \frac{c}{a} = 0 \qquad \text{Divide both sides by } a\text{, the coefficient of } x^2.$$

$$x^2 + \frac{b}{a}x = -\frac{c}{a} \qquad \text{Subtract the constant } \frac{c}{a} \text{ from both sides.}$$

Find the square of half the coefficient of x, $\frac{b}{a}$.

$$\frac{1}{2}\left(\frac{b}{a}\right) = \frac{b}{2a} \quad \text{and} \quad \left(\frac{b}{2a}\right)^2 = \frac{b^2}{4a^2}$$

Add this result to both sides of the equation.

$$x^2 + \frac{b}{a}x + \frac{b^2}{4a^2} = -\frac{c}{a} + \frac{b^2}{4a^2} \qquad \text{Add } \frac{b^2}{4a^2} \text{ to both sides.}$$

$$x^2 + \frac{b}{a}x + \frac{b^2}{4a^2} = \frac{-c \cdot 4a}{a \cdot 4a} + \frac{b^2}{4a^2} \qquad \text{Find a common denominator on the right side.}$$

$$x^2 + \frac{b}{a}x + \frac{b^2}{4a^2} = \frac{b^2 - 4ac}{4a^2} \qquad \text{Simplify the right side.}$$

$$\left(x + \frac{b}{2a}\right)^2 = \frac{b^2 - 4ac}{4a^2} \qquad \text{Factor the perfect square trinomial on the left side.}$$

$$x + \frac{b}{2a} = \pm\sqrt{\frac{b^2 - 4ac}{4a^2}} \qquad \text{Apply the square root property.}$$

$$x + \frac{b}{2a} = \pm\frac{\sqrt{b^2 - 4ac}}{2a} \qquad \text{Simplify the radical.}$$

$$x = -\frac{b}{2a} \pm \frac{\sqrt{b^2 - 4ac}}{2a} \qquad \text{Subtract } \frac{b}{2a} \text{ from both sides.}$$

$$x = \frac{-b \pm \sqrt{b^2 - 4ac}}{2a} \qquad \text{Simplify.}$$

This equation identifies the solutions of the general quadratic equation in standard form and is called the quadratic formula. It can be used to solve any equation written in standard form $ax^2 + bx + c = 0$ as long as a is not 0.

Quadratic Formula

A quadratic equation written in the form $ax^2 + bx + c = 0$ has the solutions

$$x = \frac{-b \pm \sqrt{b^2 - 4ac}}{2a}$$

EXAMPLE 1 Solve $3x^2 + 16x + 5 = 0$ for x.

Solution: This equation is in standard form, so $a = 3$, $b = 16$, and $c = 5$. Substitute these values into the quadratic formula.

$$x = \frac{-b \pm \sqrt{b^2 - 4ac}}{2a} \qquad \text{Quadratic formula.}$$

$$= \frac{-16 \pm \sqrt{16^2 - 4(3)(5)}}{2(3)} \qquad \text{Let } a = 3,\ b = 16, \text{ and } c = 5.$$

$$= \frac{-16 \pm \sqrt{256 - 60}}{6}$$

$$= \frac{-16 \pm \sqrt{196}}{6} = \frac{-16 \pm 14}{6}$$

$$x = \frac{-16 + 14}{6} = -\frac{1}{3} \quad \text{or} \quad x = \frac{-16 - 14}{6} = -\frac{30}{6} = -5$$

The solution set is $\left\{-\frac{1}{3}, -5\right\}$. ■

HELPFUL HINT

A quadratic equation **should** be written in standard form $ax^2 + bx + c = 0$ to correctly identify a, b, and c in the quadratic formula.

EXAMPLE 2 Solve $2x^2 - 4x = 3$.

Solution: First write the equation in standard form by subtracting 3 from both sides.

$$2x^2 - 4x - 3 = 0$$

Now $a = 2$, $b = -4$, and $c = -3$. Substitute these values into the quadratic formula.

$$x = \frac{-b \pm \sqrt{b^2 - 4ac}}{2a}$$

$$= \frac{-(-4) \pm \sqrt{(-4)^2 - 4(2)(-3)}}{2(2)}$$

$$= \frac{4 \pm \sqrt{16 + 24}}{4}$$

$$= \frac{4 \pm \sqrt{40}}{4} = \frac{4 \pm 2\sqrt{10}}{4}$$

$$= \frac{2(2 \pm \sqrt{10})}{2 \cdot 2} = \frac{2 \pm \sqrt{10}}{2}$$

The solution set is $\left\{\frac{2 + \sqrt{10}}{2}, \frac{2 - \sqrt{10}}{2}\right\}$. ■

HELPFUL HINT

Consider the expression $\frac{4 \pm 2\sqrt{10}}{4}$ in the preceding example. Note that 2 **must** be factored out of both terms of the numerator **before** simplifying. Then

$$\frac{4 \pm 2\sqrt{10}}{4} = \frac{2(2 \pm \sqrt{10})}{2 \cdot 2} = \frac{2 \pm \sqrt{10}}{2}$$

EXAMPLE 3 Solve $\frac{1}{4}m^2 - m + \frac{1}{2} = 0$.

Solution: First we multiply both sides of the equation by 4 to clear fractions.

$$4\left(\frac{1}{4}m^2 - m + \frac{1}{2}\right) = 4 \cdot 0$$

$$m^2 - 4m + 2 = 0 \qquad \text{Simplify.}$$

Substitute $a = 1$, $b = -4$, and $c = 2$ into the quadratic formula and simplify.

$$m = \frac{-(-4) \pm \sqrt{(-4)^2 - 4(1)(2)}}{2(1)} = \frac{4 \pm \sqrt{16 - 8}}{2}$$

$$= \frac{4 \pm \sqrt{8}}{2} = \frac{4 \pm 2\sqrt{2}}{2} = \frac{2(2 \pm \sqrt{2})}{2} = 2 \pm \sqrt{2}$$

The solution set is $\{2 + \sqrt{2}, 2 - \sqrt{2}\}$. ■

EXAMPLE 4 Solve $p = -3p^2 - 3$.

Solution: The equation in standard form is $3p^2 + p + 3 = 0$. Thus, let $a = 3$, $b = 1$, and $c = 3$ in the quadratic formula.

$$p = \frac{-1 \pm \sqrt{1^2 - 4(3)(3)}}{2 \cdot 3} = \frac{-1 \pm \sqrt{1 - 36}}{6} = \frac{-1 \pm \sqrt{-35}}{6} = \frac{-1 \pm i\sqrt{35}}{6}$$

The solution set is $\left\{\frac{-1 + i\sqrt{35}}{6}, \frac{-1 - i\sqrt{35}}{6}\right\}$. ■

2 In the quadratic formula $x = \frac{-b \pm \sqrt{b^2 - 4ac}}{2a}$, the radicand $b^2 - 4ac$ is called the **discriminant** because, by knowing its value, we can **discriminate** among the possibilities and number of solutions of a quadratic equation. Possible values of the discriminant and their meanings are summarized next.

Discriminant

The following table corresponds the discriminant $b^2 - 4ac$ of a quadratic equation of the form $ax^2 + bx + c = 0$ with the number and type of solutions of the equation.

$b^2 - 4ac$	Number and Type of Solutions
Positive	Two real solutions
Zero	One real solution
Negative	Two complex but not real solutions

EXAMPLE 5 Use the discriminant to determine the number and type of solutions of each quadratic equation.

a. $x^2 + 2x + 1 = 0$ **b.** $3x^2 + 2 = 0$ **c.** $2x^2 - 7x - 4 = 0$

Solution: **a.** In $x^2 + 2x + 1 = 0$, $a = 1$, $b = 2$, and $c = 1$. Thus,

$$b^2 - 4ac = 4 - 4\,(1)(1) = 0$$

Since $b^2 - 4ac = 0$, this quadratic equation has one real solution.

b. In this equation, $a = 3$, $b = 0$, $c = 2$. Then $b^2 - 4ac = 0 - 4(3)(2) = -24$. Since $b^2 - 4ac$ is negative, there are two complex solutions.

c. In this equation, $a = 2$, $b = -7$, and $c = -4$. Then

$$b^2 - 4ac = 49 - 4(2)(-4) = 81$$

Since $b^2 - 4ac$ is positive, there are two real solutions. ■

3 The quadratic formula is useful in solving problems that are modeled by quadratic equations.

EXAMPLE 6 An experienced typist and an apprentice typist can process a document together in 6 hours. The experienced typist can process the document alone in 2 hours less time than the apprentice typist. Find the time that each person can process the document alone.

Solution: **1.** UNDERSTAND. Read and reread the problem. The key idea here is the relationship between the **time** (hours) it takes to complete the job and the **part of the job**

completed in one unit of time (hour). For example, because they can complete the job together in 6 hours, the **part of the job** they can complete in 1 hour is $\frac{1}{6}$.

2. ASSIGN. Let x represent the *time* in hours it takes the apprentice typist to complete the job alone. Then $x - 2$ represents the time in hours it takes the experienced typist to complete the job alone.

3. ILLUSTRATE. Here, we summarize the information discussed above in a chart.

	Hours to Complete	**Part of Job Completed in 1 Hour**
Apprentice Typist	x	$\frac{1}{x}$
Experienced Typist	$x - 2$	$\frac{1}{x-2}$
Together	6	$\frac{1}{6}$

4. TRANSLATE.

In words:	part of job apprentice typist completed in 1 hour	added to	part of job experienced typist completed in 1 hour	is equal to	part of job together in 1 hour
Translate:	$\frac{1}{x}$	$+$	$\frac{1}{x-2}$	$=$	$\frac{1}{6}$

5. COMPLETE.

$$\frac{1}{x} + \frac{1}{x-2} = \frac{1}{6}$$

$$6x(x-2)\left(\frac{1}{x} + \frac{1}{x-2}\right) = 6x(x-2)\left(\frac{1}{6}\right)$$ Multiply both sides by the LCD $6x(x - 2)$.

$$6x(x-2)\left(\frac{1}{x}\right) + 6x(x-2)\left(\frac{1}{x-2}\right) = 6x(x-2)\left(\frac{1}{6}\right)$$ Apply the distributive property.

$$6(x-2) + 6x = x(x-2)$$

$$6x - 12 + 6x = x^2 - 2x$$

$$0 = x^2 - 14x + 12$$

Substitute $a = 1$, $b = -14$, and $c = 12$ into the quadratic formula and simplify.

$$x = \frac{-(-14) \pm \sqrt{(-14)^2 - 4(1)(12)}}{2(1)} = \frac{14 \pm \sqrt{148}}{2}$$

Using a calculator or a square root table, we see that $\sqrt{148} \approx 12.2$ rounded to one decimal place. Thus,

$$x \approx \frac{14 \pm 12.2}{2}$$

$$x \approx \frac{14 + 12.2}{2} = 13.1 \quad \text{or} \quad x \approx \frac{14 - 12.2}{2} = 0.9$$

6. INTERPRET. *Check:* If the apprentice typist completes the job alone in 0.9 hours, the experienced typist completes the job alone in $x - 2 = 0.9 - 2 = -1.1$ hours. Since this is not possible, we reject the solution of 0.9. The approximate solution is thus 13.1.

State: The apprentice typist can complete the job alone in approximately 13.1 hours, and the experienced typist completes the job alone in approximately $x - 2 = 13.1 - 2 = 11.1$ hours. ■

EXERCISE SET 9.2

Use the quadratic formula to solve each equation. These equations have real number solutions. See Example 1.

1. $m^2 + 5m - 6 = 0$
2. $p^2 + 11p - 12 = 0$
3. $2y = 5y^2 - 3$
4. $5x^2 - 3 = 14x$
5. $x^2 - 6x + 9 = 0$
6. $y^2 + 10y + 25 = 0$

Use the quadratic formula to solve each equation. These equations have real number solutions. See Example 2.

7. $x^2 + 7x + 4 = 0$
8. $y^2 + 5y + 3 = 0$
9. $8m^2 - 2m = 7$
10. $11n^2 - 9n = 1$
11. $3m^2 - 7m = 3$
12. $x^2 - 13 = 5x$

Use the quadratic formula to solve each equation. These equations have real number solutions. See Example 3.

13. $\frac{1}{2}x^2 - x - 1 = 0$
14. $\frac{1}{6}x^2 + x + \frac{1}{3} = 0$
15. $\frac{2}{5}y^2 + \frac{1}{5}y = \frac{3}{5}$
16. $\frac{1}{8}x^2 + x = \frac{5}{2}$
17. $\frac{1}{3}y^2 - y - \frac{1}{6} = 0$
18. $\frac{1}{2}y^2 = y + \frac{1}{2}$

Use the quadratic formula to solve each equation. See Example 4.

19. $6 = -4x^2 + 3x$
20. $9x^2 + x + 2 = 0$
21. $(x + 5)(x - 1) = 2$
22. $x(x + 6) = 2$
23. $10y^2 + 10y + 3 = 0$
24. $3y^2 + 6y + 5 = 0$

Use the discriminant to determine the number and types of solutions of each equation. See Example 5.

25. $3x = -2x^2 + 7$
26. $3x^2 = 5 - 7x$
27. $6 = 4x - 5x^2$
28. $9x - 2x^2 + 5 = 0$
29. $5 - 4x + 12x^2 = 0$
30. $8x = 3 - 9x^2$
31. $4x^2 + 12x = -9$
32. $9x^2 + 1 = 6x$

Solve. Use a calculator or a table to approximate each solution to the nearest tenth. See Example 6.

33. Bill Shaughnessy and his son Billy can clean the house together in 4 hours. When he works alone, it takes the son an hour longer to clean than it takes his dad alone. Find how long to the nearest hundredth hour it takes the son to clean alone.

34. Scratchy and Freckles together eat a 50-pound bag of dog food in 30 days. Scratchy by himself eats a 50-pound bag in 2 weeks less time than Freckles by himself. How many days to the nearest whole day would a 50-pound bag of dog food last Freckles?

A rocket is launched from the top of an 80-foot cliff with an initial velocity of 120 feet per second. The height of the rocket h(t) after t seconds is given by the function

$$h(t) = -16t^2 + 120t + 80$$

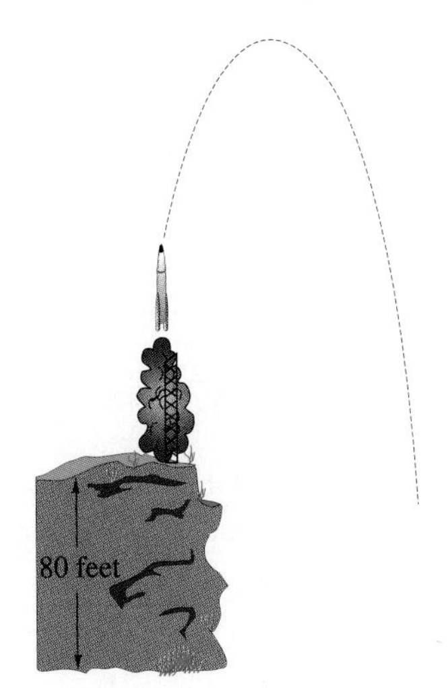

35. How long after the rocket is launched will it be 30 feet from the ground?

36. How long after the rocket is launched will it strike the ground? [Hint: The rocket will strike the ground when its height $h(t) = 0$.]

Use the quadratic formula to solve each equation. These equations have real number solutions.

37. $x^2 + 5x = -2$

38. $y^2 - 8 = 4y$

39. $2m^2 - 7m + 5 = 0$

40. $7p^2 - 12p + 5 = 0$

41. $\frac{x^2}{3} - x = \frac{5}{3}$

42. $\frac{x^2}{2} - 3 = -\frac{9}{2}x$

43. $6x^2 + 2x - 3 = 0$

44. $7x^2 + x = 2$

Use the quadratic formula to solve each equation.

45. $x^2 + 6x + 13 = 0$

46. $x^2 + 2x + 2 = 0$

47. $\frac{2}{5}y^2 + \frac{1}{5}y + \frac{3}{5} = 0$

48. $\frac{1}{8}x^2 + x + \frac{5}{2} = 0$

49. $\frac{1}{2}y^2 = y - \frac{1}{2}$

50. $\frac{2}{3}x^2 - \frac{20}{3}x = -\frac{100}{6}$

51. $(n - 2)^2 = 15n$

52. $\left(p - \frac{1}{2}\right)^2 = \frac{p}{2}$

Solve. If appropriate, use a calculator or a table to approximate each solution to the nearest tenth.

53. The product of a number and 4 less than the number is 96. Find the number.

54. A whole number increased by its square is two more than twice itself. Find the number.

55. An IBM computer and a Toshiba computer can complete a job together in 8 hours. The IBM computer alone can complete the job in 1 hour less time than the Toshiba computer alone. Find the time each computer can complete the job alone.

56. Two fax machines working together can fax a lengthy document in 75 minutes. Fax machine A alone can fax the document in 30 minutes less time than fax machine B alone. Find the time each machine can fax the document alone.

A ball is thrown downward from the top of a 180-foot building with an initial velocity of −20 feet per second. The height of the ball h(t) after t seconds is given by the function

$$h(t) = -16t^2 - 20t + 180$$

57. How long after the ball is thrown will it strike the ground?

58. How long after the ball is thrown will it be 50 feet from the ground?

The graph below shows the daily low temperatures for one week in New Orleans, Louisiana.

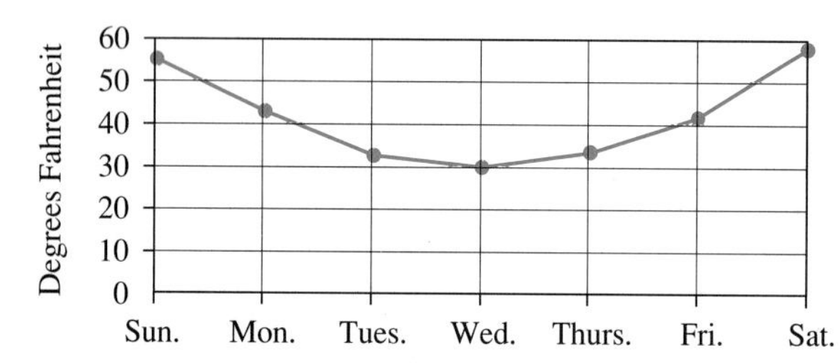

59. Which day of the week shows the greatest decrease in temperature low?

60. Which day of the week shows the greatest increase in temperature low?

61. Which day of the week had the lowest temperature?

62. Use the graph to estimate the low temperature on Thursday.

Notice that the shape of the temperature graph is similar to a parabola (see Section 6.7). In fact, this graph can be modeled by the quadratic function $f(x) = 3x^2 - 18x + 57$ where f(x) is the temperature in degrees Fahrenheit and x is the number of days from Sunday. Use this function to answer Exercises 63 and 64.

63. Use the quadratic function given to approximate the temperature on Thursday. Does your answer agree with the graph above?

64. Use the function given and the quadratic formula to find when the temperature was 35°F. [Hint: Let $f(x) = 35$ and solve for x.] Round your answer to one decimal place and interpret your result. Does your answer agree with the graph above?

Skill Review

Solve each equation. See Sections 7.5 and 8.6.

65. $\sqrt{5x - 2} = 3$

66. $\sqrt{y + 2} + 7 = 12$

67. $\frac{1}{x} + \frac{2}{5} = \frac{7}{x}$

68. $\frac{10}{z} = \frac{5}{z} - \frac{1}{3}$

Add or subtract as indicated. See Section 8.3.

69. $\sqrt{5} + 4\sqrt{5} - 7\sqrt{5}$

70. $2\sqrt{3x} - 3\sqrt{3x} + 4\sqrt{x}$

71. $\sqrt{12x} + \sqrt{27x}$

72. $\sqrt{45} - \sqrt{80}$

Factor. See Section 6.5.

73. $x^4 + x^2 - 20$

74. $2y^4 + 11y^2 - 6$

75. $z^4 - 13z^2 + 36$

76. $x^4 - 1$

A Look Ahead

Use the quadratic formula to solve each quadratic equation. See the following example.

EXAMPLE Solve $x^2 - 3\sqrt{2}x + 2 = 0$.

Solution: In this equation, $a = 1$, $b = -3\sqrt{2}$, and $c = 2$. By the quadratic formula, we have

$$x = \frac{-b \pm \sqrt{b^2 - 4ac}}{2a}$$

$$= \frac{3\sqrt{2} \pm \sqrt{(-3\sqrt{2})^2 - 4(1)(2)}}{2(1)} = \frac{3\sqrt{2} \pm \sqrt{18 - 8}}{2} = \frac{3\sqrt{2} \pm \sqrt{10}}{2}$$

The solution set is $\left\{\frac{3\sqrt{2} + \sqrt{10}}{2}, \frac{3\sqrt{2} - \sqrt{10}}{2}\right\}$. ■

77. $3x^2 - \sqrt{12}x + 1 = 0$

78. $5x^2 + \sqrt{20}x + 1 = 0$

79. $x^2 + \sqrt{2}x + 1 = 0$

80. $x^2 - \sqrt{2}x + 1 = 0$

81. $2x^2 - \sqrt{3}x - 1 = 0$

82. $7x^2 + \sqrt{7}x - 2 = 0$

9.3
Solving Equations in Quadratic Form

OBJECTIVES

1. Solve miscellaneous equations that are quadratic in form.
2. Solve geometric applications that lead to equations that are quadratic in form.

Tape IA 24

1 In this section, we discuss various types of equations that can be solved by the methods of this chapter.

The first example is an equation containing rational expressions that can be written in quadratic form.

EXAMPLE 1 Solve $\frac{3x}{x-2} - \frac{x+1}{x} = \frac{6}{x(x-2)}$.

Solution: In this equation, x cannot be either 2 or 0, because these values cause denominators to equal zero. To solve for x, first multiply both sides of the equation by $x(x - 2)$ to clear fractions. By the distributive property, this means that we multiply each term by $x(x - 2)$.

$$x(x-2)\left(\frac{3x}{x-2}\right) - x(x-2)\left(\frac{x+1}{x}\right) = x(x-2)\left[\frac{6}{x(x-2)}\right]$$

$$3x^2 - (x-2)(x+1) = 6 \quad \text{Simplify.}$$

$$3x^2 - (x^2 - x - 2) = 6 \quad \text{Multiply.}$$

$$3x^2 - x^2 + x + 2 = 6$$

$$2x^2 + x - 4 = 0 \quad \text{Simplify.}$$

$$x = \frac{-1 \pm \sqrt{1^2 - 4(2)(-4)}}{2 \cdot 2}$$ Let $a = 2$, $b = 1$, and $c = -4$ in the quadratic formula.

$$= \frac{-1 \pm \sqrt{1 + 32}}{4}$$ Simplify.

$$= \frac{-1 \pm \sqrt{33}}{4}$$

The solution set is $\left\{\frac{-1 + \sqrt{33}}{4}, \frac{-1 - \sqrt{33}}{4}\right\}$. ■

EXAMPLE 2 Solve $x^3 = 8$ for x.

Solution: Begin by subtracting 8 from both sides; then factor the resulting difference of cubes.

$$x^3 = 8$$

$$x^3 - 8 = 0$$

$$(x - 2)(x^2 + 2x + 4) = 0$$ Factor.

$$x - 2 = 0 \quad \text{or} \quad x^2 + 2x + 4 = 0$$ Set each factor equal to 0.

The solution of $x - 2 = 0$ is 2. Solve the second equation by the quadratic formula.

$$x = \frac{-2 \pm \sqrt{2^2 - 4(1)(4)}}{2 \cdot 1} = \frac{-2 \pm \sqrt{-12}}{2}$$ Let $a = 1$, $b = 2$, $c = 4$.

$$= \frac{-2 \pm 2i\sqrt{3}}{2} = \frac{2(-1 \pm i\sqrt{3})}{2}$$

$$= -1 \pm i\sqrt{3}$$

The solution set is $\{2, -1 + i\sqrt{3}, -1 - i\sqrt{3}\}$. ■

EXAMPLE 3 Solve $p^4 - 3p^2 - 4 = 0$.

Solution: First factor the trinomial.

$$p^4 - 3p^2 - 4 = 0$$

$$(p^2 - 4)(p^2 + 1) = 0$$

$$(p - 2)(p + 2)(p^2 + 1) = 0$$ Factor.

$$p - 2 = 0 \quad \text{or} \quad p + 2 = 0 \quad \text{or} \quad p^2 + 1 = 0$$ Set each factor equal to 0.

$$p = 2 \quad \text{or} \quad p = -2 \quad \text{or} \quad p^2 = -1$$

$$p = \pm\sqrt{-1} = \pm i$$

The solution set is $\{2, -2, i, -i\}$. ■

EXAMPLE 4 Solve $x^{2/3} - 5x^{1/3} + 6 = 0$.

Solution: The key to solving this equation is recognizing that $x^{2/3} = (x^{1/3})^2$. Replace $x^{1/3}$ with m so that

$$(x^{1/3})^2 - 5x^{1/3} + 6 = 0$$

becomes

$$m^2 - 5m + 6 = 0$$

Now solve by factoring.

$$m^2 - 5m + 6 = 0$$

$$(m - 3)(m - 2) = 0 \quad \text{Factor.}$$

$$m - 3 = 0 \quad \text{or} \quad m - 2 = 0 \quad \text{Set each factor equal to 0.}$$

$$m = 3 \quad \text{or} \quad m = 2$$

Since $m = x^{1/3}$, $x^{1/3} = 3$ or $x^{1/3} = 2$. To solve for x, cube both sides of each equation: $(x^{1/3})^3 = 3^3 = 27$ or $(x^{1/3})^3 = 2^3 = 8$. The solution set is $\{8, 27\}$. ■

2 Some applications lead to equations that are quadratic in form.

EXAMPLE 5 The hypotenuse of an isosceles right triangle is 2 centimeters longer than either of its legs. Find the perimeter of the triangle.

Solution: **1.** UNDERSTAND. Read and reread the problem. Recall that an isosceles right triangle has legs of equal length. Also recall the Pythagorean theorem for right triangles:

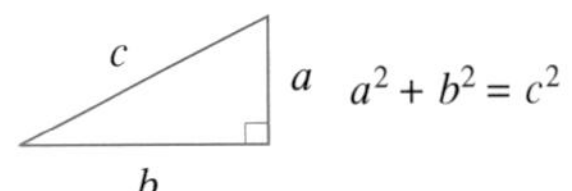

2. ASSIGN.

Let

$$x = \text{the length of each leg of the triangle}$$

so that

$$x + 2 = \text{the length of the hypotenuse of the triangle}$$

3. ILLUSTRATE. Draw a triangle and label it with the assigned variables.

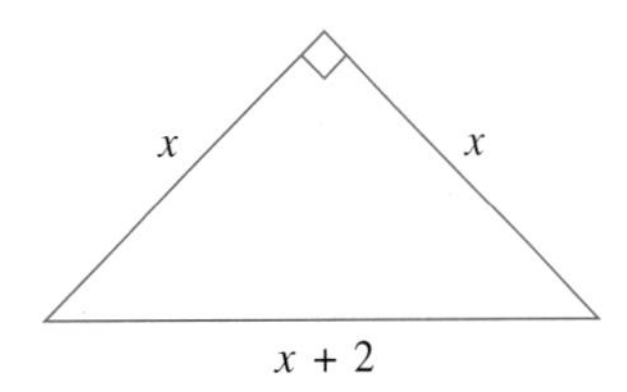

4. TRANSLATE. By the Pythagorean theorem,

In words: $(\text{leg})^2 + (\text{leg})^2 = (\text{hypotenuse})^2$

Translate: $x^2 + x^2 = (x + 2)^2$

5. COMPLETE. Solve the quadratic equation.

$$x^2 + x^2 = (x + 2)^2$$

$$x^2 + x^2 = x^2 + 4x + 4 \qquad \text{Square } (x + 2).$$

$$x^2 - 4x - 4 = 0 \qquad \text{Set the equation equal to 0.}$$

Next substitute $a = 1$, $b = -4$, and $c = -4$ in the quadratic formula.

$$x = \frac{4 \pm \sqrt{16 - 4(1)(-4)}}{2} = \frac{4 \pm \sqrt{32}}{2} = \frac{4 \pm 4\sqrt{2}}{2}$$

$$= \frac{2 \cdot (2 \pm 2\sqrt{2})}{2} = 2 \pm 2\sqrt{2}$$

Since the length of a side cannot be negative, we reject the solution $2 - 2\sqrt{2}$. The length of each leg is $(2 + 2\sqrt{2})$ centimeters. Since the hypotenuse is 2 centimeters longer than the legs, the hypotenuse is $(4 + 2\sqrt{2})$ centimeters.

6. INTERPRET. *Check* the lengths of the sides in the Pythagorean theorem. *State:* The perimeter of the triangle is $(2 + 2\sqrt{2})$ cm + $(2 + 2\sqrt{2})$ cm + $(4 + 2\sqrt{2})$ cm = $(8 + 6\sqrt{2})$ centimeters. ■

EXERCISE SET 9.3

Solve each equation. See Example 1.

1. $\frac{2}{x} + \frac{3}{x - 1} = 1$

2. $\frac{6}{x^2} = \frac{3}{x + 1}$

3. $\frac{3}{x} + \frac{4}{x + 2} = 2$

4. $\frac{5}{x - 2} + \frac{4}{x + 2} = 1$

5. $\frac{7}{x^2 - 5x + 6} = \frac{2x}{x - 3} - \frac{x}{x - 2}$

6. $\frac{11}{2x^2 + x - 15} = \frac{5}{2x - 5} - \frac{x}{x + 3}$

Solve each equation. See Example 2.

7. $y^3 - 1 = 0$

8. $x^3 + 8 = 0$

9. $x^4 + 27x = 0$

10. $y^5 + y^2 = 0$

11. $z^3 = 64$

12. $z^3 = -125$

Solve each equation. See Example 3.

13. $p^4 - 16 = 0$

14. $x^4 + 2x^2 - 3 = 0$

15. $4x^4 + 11x^2 = 3$

16. $z^4 = 81$

17. $z^4 - 13z^2 + 36 = 0$

18. $9x^4 + 5x^2 - 4 = 0$

Solve each equation. See Example 4.

19. $x^{2/3} - 3x^{1/3} - 10 = 0$

20. $x^{2/3} + 2x^{1/3} + 1 = 0$

21. $2x^{2/3} - 5x^{1/3} = 3$

22. $3x^{2/3} + 11x^{1/3} = 4$

23. $20x^{2/3} - 6x^{1/3} - 2 = 0$

24. $4x^{2/3} + 16x^{1/3} = -15$

Solve. See Example 5.

25. Uri Chechov's rectangular dog pen for his Irish setter must have an area of 400 square feet. Also, the length must be $2\frac{1}{2}$ times the width. Find the dimensions of the pen.

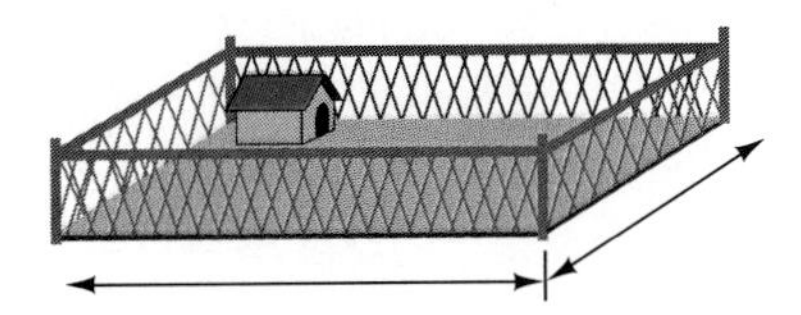

26. The hypotenuse of an isosceles right triangle is 5 inches longer than either of the legs. Find the length of the legs and the length of the hypotenuse.

Solve each equation.

27. $a^4 - 5a^2 + 6 = 0$

28. $x^4 - 12x^2 + 11 = 0$

29. $\frac{2x}{x-2} + \frac{x}{x+3} = \frac{-5}{x+3}$

30. $\frac{5}{x-3} + \frac{x}{x+3} = \frac{19}{x^2-9}$

31. $x^3 + 64 = 0$

32. $y^3 - 27 = 0$

33. $x^{2/3} - 8x^{1/3} + 15 = 0$

34. $x^{2/3} - 2x^{1/3} - 8 = 0$

35. $y^3 + 9y - y^2 - 9 = 0$

36. $x^3 + x - 3x^2 - 3 = 0$

37. $2x^{2/3} + 3x^{1/3} - 2 = 0$

38. $6x^{2/3} - 25x^{1/3} - 25 = 0$

39. $2x^3 - 250 = 0$

40. $8y^3 + 8 = 0$

41. $\frac{x}{x-1} + \frac{1}{x+1} = \frac{2}{x^2-1}$

42. $\frac{x}{x-5} + \frac{5}{x+5} = \frac{-1}{x^2-25}$

43. $p^4 - p^2 - 20 = 0$

44. $x^4 - 10x^2 + 9 = 0$

45. $2x^3 = -54$

46. $y^3 - 216 = 0$

47. $27y^4 + 15y^2 = 2$

48. $8z^4 + 14z^2 = -5$

Solve.

49. The base of a triangle is twice its height. If the area of the triangle is 42 square centimeters, find its base and height.

50. The width of a rectangle is $\frac{1}{3}$ its length. If its area is 12 square inches, find its length and width.

51. An entry in the Peach Festival Poster Contest must be rectangular and have an area of 1200 square inches. Furthermore, its length must be $1\frac{1}{2}$ times its width. Find the dimensions each entry must have.

52. A holding pen for cattle must be square and have a diagonal length of 100 meters.
a. Find the length of a side of the pen.
b. Find the area of the pen.

53. A rectangle is three times longer than it is wide. It has a diagonal of length 50 centimeters.
a. Find the dimensions of the rectangle.
b. Find the perimeter of the rectangle.

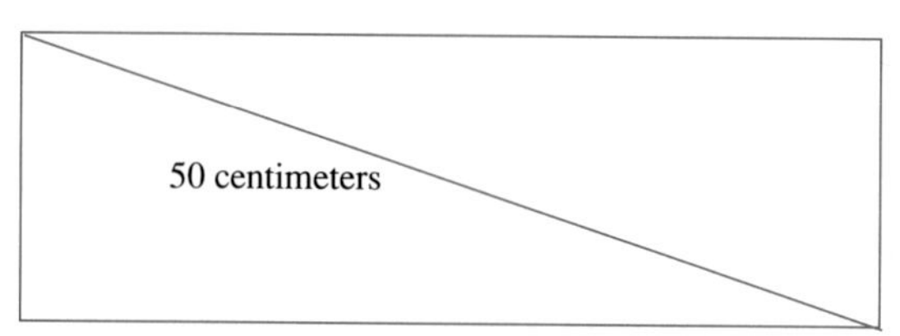

54. In Exercise Sets 2.5 and 3.1 we introduced the concept of the golden ratio. The exact value of the golden ratio is the positive solution to the quadratic equation $x = \frac{x + 1}{x}$. Find this value.

Skill Review

Solve each inequality. See Section 3.5.

55. $\frac{5x}{3} + 2 \leq 7$

56. $\frac{2x}{3} + \frac{1}{6} \geq 2$

57. $\frac{y - 1}{15} > \frac{-2}{5}$

58. $\frac{z - 2}{12} < \frac{1}{4}$

Find the domain and range of each relation graphed. Decide which relations are also functions. See Section 4.2.

59.

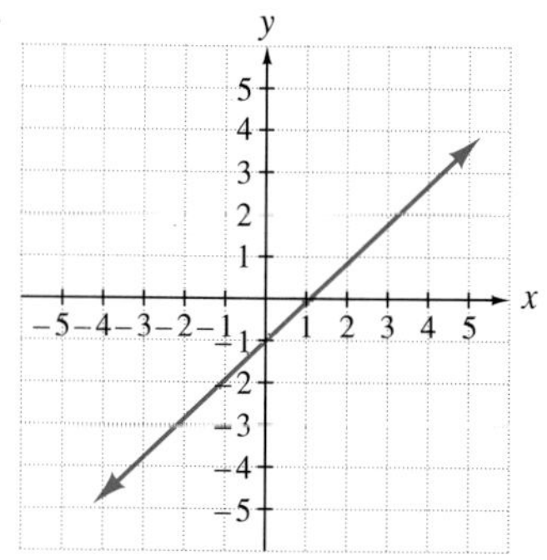

60.

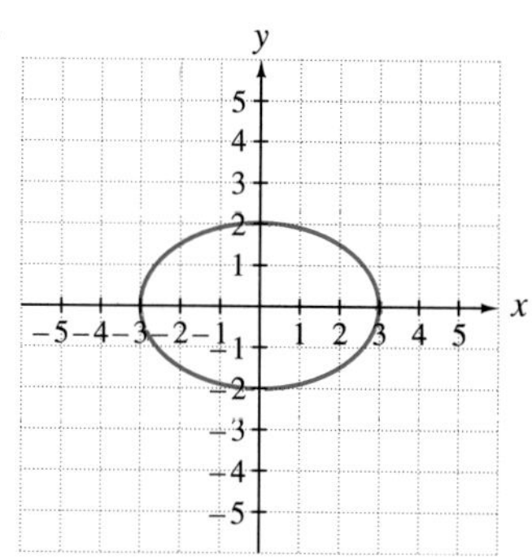

61.

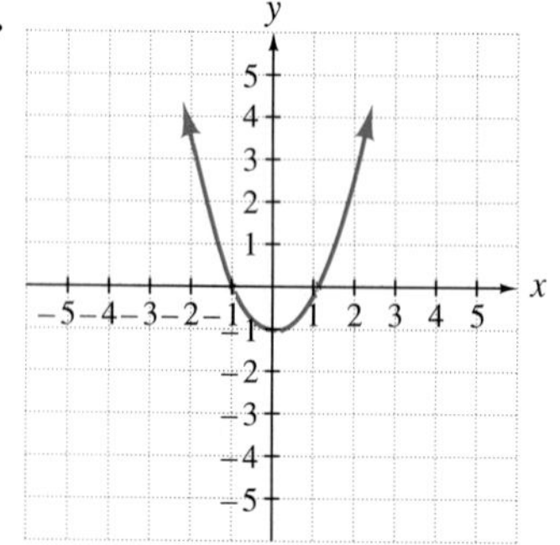

62.

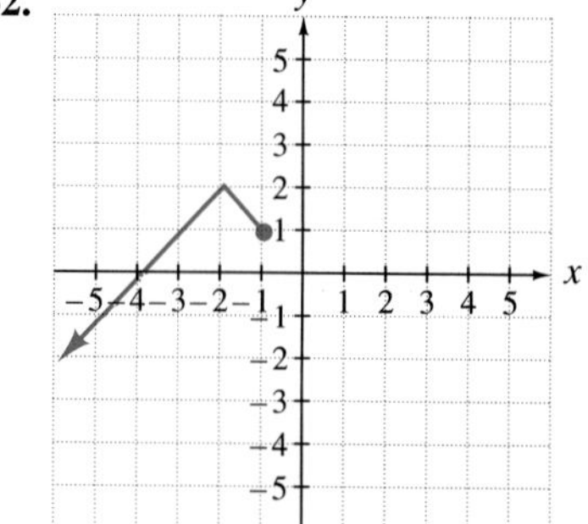

Sketch the graph of each quadratic function by plotting points. See Section 6.7.

63. $f(x) = x^2$

64. $f(x) = x^2 + 1$

9.4 Nonlinear Inequalities in One Variable

OBJECTIVES

1. Solve polynomial inequalities of degree 2 or greater.
2. Solve inequalities containing rational expressions with variables in the denominator.

Tape IA 23

1 Just as we can solve linear inequalities in one variable, so we can solve quadratic inequalities in one variable. A **quadratic inequality** is an inequality that can be written so that one side is a quadratic expression and the other side is 0. Here are examples of quadratic inequalities in one variable. Each is written in **standard form.**

$$x^2 - 10x + 7 \leq 0 \qquad 3x^2 + 2x - 6 > 0$$

$$2x^2 + 9x - 2 < 0 \qquad x^2 - 3x + 11 \geq 0$$

A solution of a quadratic inequality in one variable is a value of the variable that makes the inequality a true statement.

The value of an expression such as $x^2 - 3x - 10$ will sometimes be positive, sometimes negative, and sometimes 0, depending on the values substituted for x. To solve the inequality $x^2 - 3x - 10 < 0$, we are looking for all values of x that make the expression $x^2 - 3x - 10$ **less than 0,** or **negative.** To understand how we find these values, we'll study the graph of the quadratic function $y = x^2 - 3x - 10$.

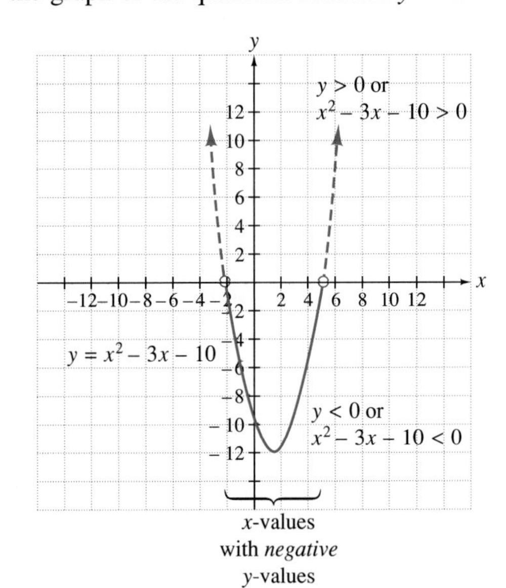

Notice that the x-values for which y or $x^2 - 3x - 10$ is positive are separated from the x-values for which y or $x^2 - 3x - 10$ is negative by the values for which y or $x^2 - 3x - 10$ is 0, the x-intercepts. Thus, the solution set of $x^2 - 3x - 10 < 0$ consists of all real numbers from -2 to 5, or $(-2, 5)$.

It is not necessary to graph $y = x^2 - 3x - 10$ to solve the related inequality $x^2 - 3x - 10 < 0$. We can simply draw a number line representing the x-axis and keep the following in mind: **Intervals on the number line where the value of $x^2 - 3x - 10$ is positive are separated from intervals on the number line where the value of $x^2 - 3x - 10$ is negative by values for which the expression is 0.** Find these values for which the expression is 0 by solving the related equation.

$$x^2 - 3x - 10 = 0$$

$$(x - 5)(x + 2) = 0 \qquad \text{Factor.}$$

$$x - 5 = 0 \quad \text{or} \quad x + 2 = 0 \qquad \text{Set each factor equal to 0.}$$

$$x = 5 \qquad\qquad x = -2 \qquad \text{Solve.}$$

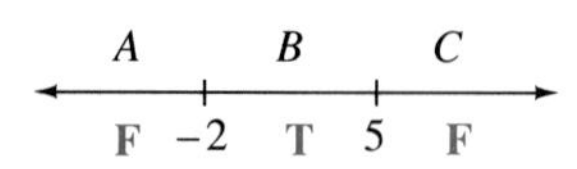

These two numbers divide the number line into three regions. We will call the regions A, B, and C. These regions are important because, if the value of $x^2 - 3x - 10$ is negative when one number from a region is substituted for x, then $x^2 - 3x - 10$ is negative for all numbers in the region. The same is true if the value of $x^2 - 3x - 10$ is positive.

To see whether the inequality $x^2 - 3x - 10 < 0$ is true or false in each region, choose a test point from each region and substitute its value for x in the inequality $x^2 - 3x - 10 < 0$. If the resulting inequality is true, the region containing the test point is a solution region.

	Test Point	$(x - 5)(x + 2) < 0$	
Region A	-3	$(-8)(-1) < 0$	False
Region B	0	$(-5)(2) < 0$	True
Region C	6	$(1)(8) < 0$	False

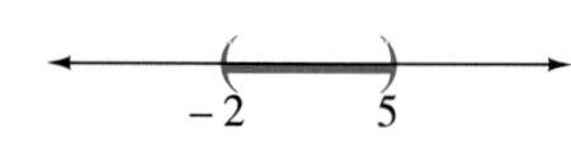

The points in region B satisfy the inequality. The numbers -2 and 5 are not included in the solution set since the inequality symbol is $<$. The solution set is $(-2, 5)$, and its graph is shown.

EXAMPLE 1 Solve $(x + 3)(x - 3) \geq 0$.

Solution: First solve the related equation $(x + 3)(x - 3) = 0$.

$$(x + 3)(x - 3) = 0$$

$$x + 3 = 0 \quad \text{or} \quad x - 3 = 0$$

$$x = -3 \qquad\qquad x = 3$$

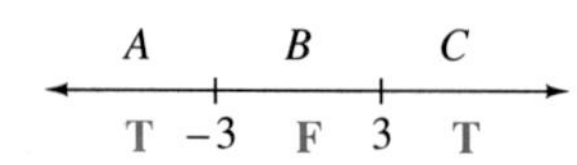

The two numbers -3 and 3 separate the number line into three regions.

Substitute the value of a test point from each region. If the value of the test point satisfies the inequality, the region containing the test point is a solution region.

	Test Point	$(x + 3)(x - 3) \geq 0$	
Region A	-4	$(-1)(-7) \geq 0$	True
Region B	0	$(3)(-3) \geq 0$	False
Region C	4	$(7)(1) \geq 0$	True

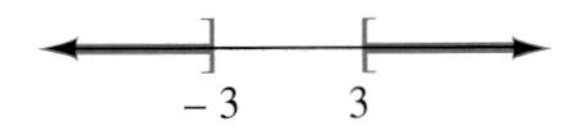

The points in regions A and C satisfy the inequality. The numbers -3 and 3 are included in the solution since the inequality symbol is $\geq$. The solution set is $(-\infty, -3] \cup [3, \infty)$, and its graph is shown. ■

The following steps may be used to solve a polynomial inequality.

> **To Solve a Polynomial Inequality**
>
> *Step 1* Write the inequality in standard form.
>
> *Step 2* Solve the related equation.
>
> *Step 3* Separate the number line into intervals using the solutions from step 2.
>
> *Step 4* For each interval, choose a test point and determine whether its value satisfies the **original inequality.**
>
> *Step 5* Write the solution set as the union of intervals whose test point is a solution.

EXAMPLE 2 Solve $x^2 - 4x \geq 0$.

Solution: First solve the related equation $x^2 - 4x = 0$.

$$x^2 - 4x = 0$$

$$x(x - 4) = 0$$

$$x = 0 \quad \text{or} \quad x = 4$$

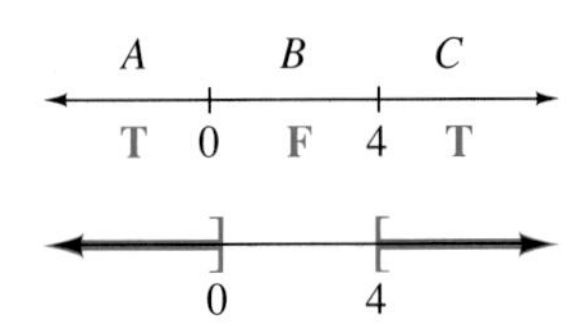

The numbers 0 and 4 separate the number line into three regions.

Check test points in each region in the original inequality. The points in regions A and C satisfy the inequality. The numbers 0 and 4 are included in the solution since the inequality symbol is $\geq$. The solution set is $(-\infty, 0] \cup [4, \infty)$, and its graph is shown. ■

EXAMPLE 3 Solve $(x + 2)(x - 1)(x - 5) \leq 0$.

Solution: First solve $(x + 2)(x - 1)(x - 5) = 0$. By inspection, the solutions are -2, 1, and 5. They separate the number line into four regions. Next check test points from each region.

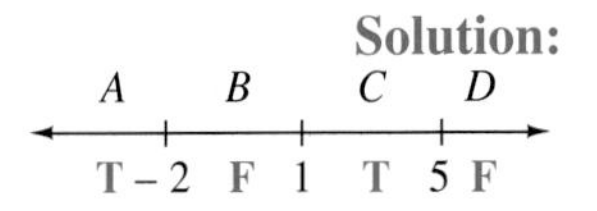

	Test Point	$(x + 2)(x - 1)(x - 5) \leq 0$	
Region A	-3	$(-1)(-4)(-8) \leq 0$	True
Region B	0	$(2)(-1)(-5) \leq 0$	False
Region C	2	$(4)(1)(-3) \leq 0$	True
Region D	6	$(8)(5)(1) \leq 0$	False

The solution set is $(-\infty, -2] \cup [1, 5]$, and its graph is shown. We include the numbers -2, 1, and 5 because the inequality symbol is $\leq$. ■

2 Inequalities containing rational expressions with variables in the denominator are solved using a similar procedure.

EXAMPLE 4 Solve $\frac{x+2}{x-3} \leq 0$.

Solution: First solve the related equation $\frac{x+2}{x-3} = 0$.

$$\frac{x+2}{x-3} = 0$$

$$x + 2 = 0 \qquad \text{Multiply both sides by the LCD, } x - 3.$$

$$x = -2$$

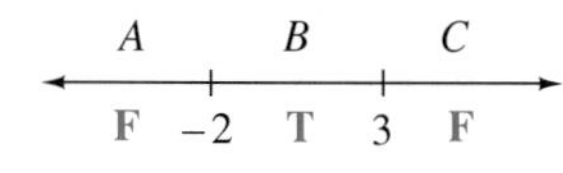

Also find all values that make the denominator equal to 0. To do this, solve $x - 3 = 0$ or $x = 3$. Place these numbers on a number line and proceed as before, checking test points in the original inequality.

Choose -3 from region A.

$$\frac{x+2}{x-3} \leq 0$$

$$\frac{-3+2}{-3-3} \leq 0$$

$$\frac{-1}{-6} \leq 0$$

$$\frac{1}{6} \leq 0 \qquad \text{False.}$$

Choose 0 from Region B.

$$\frac{x+2}{x-3} \leq 0$$

$$\frac{0+2}{0-3} \leq 0$$

$$-\frac{2}{3} \leq 0 \qquad \text{True.}$$

Choose 4 from region C.

$$\frac{x+2}{x-3} \leq 0$$

$$\frac{4+2}{4-3} \leq 0$$

$$6 \leq 0 \qquad \text{False.}$$

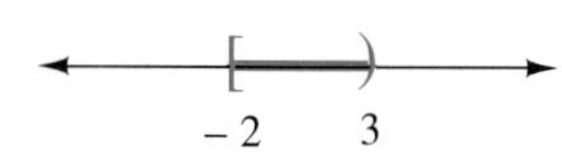

The solution set is $[-2, 3)$. This interval includes -2 because -2 satisfies the original inequality. This interval does not include 3, because 3 would make the denominator 0. ■

The following steps may be used to solve a rational inequality with variables in the denominator.

> **To Solve a Rational Inequality**
>
> *Step 1* Solve the related equation.
> *Step 2* Solve for values that make the denominator 0.
> *Step 3* Separate the number line into intervals using the solutions from steps 1 and 2.

Step 4 For each interval, choose a test point and determine whether its value satisfies the **original inequality.**

Step 5 Write the solution set as the union of intervals whose test point is a solution.

EXAMPLE 5 Solve $\frac{5}{x+1} < -2$.

Solution: First solve $\frac{5}{x+1} = -2$.

$$(x+1) \cdot \frac{5}{x+1} = (x+1) \cdot -2$$ Multiply both sides by the LCD, $x + 1$.

$$5 = -2x - 2$$ Simplify.

$$7 = -2x$$

$$-\frac{7}{2} = x$$

Next find values for x that make the denominator equal to 0.

$$x + 1 = 0$$

$$x = -1$$

A B C
F $-\frac{7}{2}$ T -1 F

Use these two solutions to divide a number line into three intervals, and choose test points.

Only a test point from region B satisfies the **original inequality.** The solution set is $\left(-\frac{7}{2}, -1\right)$, and its graph is shown. ■

$-\frac{7}{2}$ -1

EXERCISE SET 9.4

Solve each quadratic inequality. Write the solution set in interval notation and graph the solution. See Examples 1 and 2.

1. $(x + 1)(x + 5) > 0$ **2.** $(x + 1)(x + 5) \le 0$ **3.** $(x - 3)(x + 4) \le 0$ **4.** $(x + 4)(x - 1) > 0$

5. $x^2 - 7x + 10 \le 0$ **6.** $x^2 + 8x + 15 \ge 0$ **7.** $3x^2 + 16x < -5$ **8.** $2x^2 - 5x < 7$

Solve each inequality. Write the solution set in interval notation and graph the solution. See Example 3.

9. $(x - 6)(x - 4)(x - 2) > 0$ **10.** $(x - 6)(x - 4)(x - 2) \le 0$ **11.** $x(x - 1)(x + 4) \le 0$

12. $x(x - 6)(x + 2) > 0$

13. $(x^2 - 9)(x^2 - 4) > 0$

14. $(x^2 - 16)(x^2 - 1) \leq 0$

Solve each inequality. Write the solution set in interval notation and graph the solution. See Example 4.

15. $\frac{x + 7}{x - 2} < 0$

16. $\frac{x - 5}{x - 6} > 0$

17. $\frac{5}{x + 1} > 0$

18. $\frac{3}{y - 5} < 0$

19. $\frac{x + 1}{x - 4} \geq 0$

20. $\frac{x + 1}{x - 4} \leq 0$

Solve each inequality. Write the solution set in interval notation and graph the solution. See Example 5.

21. $\frac{3}{x - 2} < 4$

22. $\frac{-2}{y + 3} > 2$

23. $\frac{x^2 + 6}{5x} \geq 1$

24. $\frac{y^2 + 15}{8y} \leq 1$

Solve each inequality. Write the solution set in interval notation and graph the solution.

25. $(x - 8)(x + 7) > 0$

26. $(x - 5)(x + 1) < 0$

27. $(2x - 3)(4x + 5) \leq 0$

28. $(6x + 7)(7x - 12) > 0$

29. $x^2 > x$

30. $x^2 < 25$

31. $(2x - 8)(x + 4)(x - 6) \leq 0$

32. $(3x - 12)(x + 5)(2x - 3) \geq 0$

33. $6x^2 - 5x \geq 6$

34. $12x^2 + 11x \leq 15$

35. $x^4 - 26x^2 + 25 \geq 0$

36. $16x^4 - 40x^2 + 9 \leq 0$

37. $(2x - 7)(3x + 5) > 0$

38. $(4x - 9)(2x + 5) < 0$

39. $\frac{x}{x - 10} < 0$

40. $\dfrac{x + 10}{x - 10} > 0$

41. $\dfrac{x - 5}{x + 4} \geq 0$

42. $\dfrac{x - 3}{x + 2} \leq 0$

43. $\dfrac{x(x + 6)}{(x - 7)(x + 1)} \geq 0$

44. $\dfrac{(x - 2)(x + 2)}{(x + 1)(x - 4)} \leq 0$

45. $\dfrac{-1}{x - 1} > -1$

46. $\dfrac{4}{y + 2} < -2$

47. $\dfrac{x}{x + 4} \leq 2$

48. $\dfrac{4x}{x - 3} \geq 5$

49. $\dfrac{z}{z - 5} \geq 2z$

50. $\dfrac{p}{p + 4} \leq 3p$

51. $\dfrac{(x + 1)^2}{5x} > 0$

52. $\dfrac{(2x - 3)^2}{x} < 0$

Find all numbers that satisfy each of the following.

53. A number minus its reciprocal is less than zero. Find the numbers.

54. Twice a number added to its reciprocal is nonnegative. Find the numbers.

55. The total profit function $P(x)$ for a company producing x thousand units is given by

$$P(x) = -2x^2 + 26x - 44$$

Find the values of x for which the company makes a profit. [Hint: The company makes a profit when $P(x) > 0$.]

56. A projectile is fired straight up from the ground with an initial velocity of 80 feet per second. Its height $s(t)$ in feet at any time t is given by the function

$$s(t) = -16t^2 + 80t$$

Find the interval of time for which the height of the projectile is greater than 96 feet.

Skill Review

See Sections 4.1 and 4.2. Use the graph of $f(x) = 3x$ at right to sketch the graph of each function.

57. $f(x) = 3x + 1$

58. $g(x) = 3x - 2$

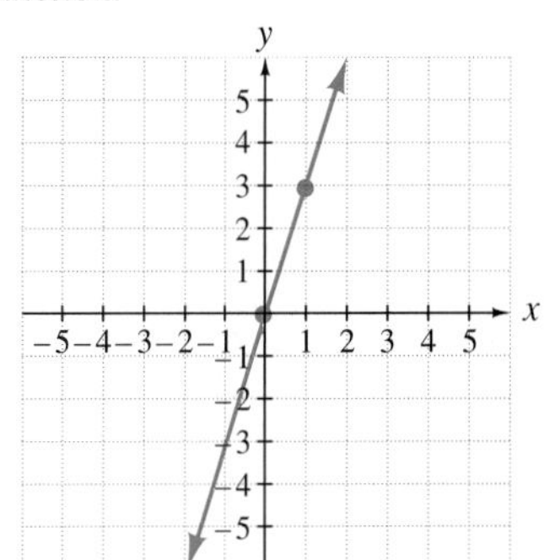

Use the graph of $f(x) = -2x$ at right to sketch the graph of each function.

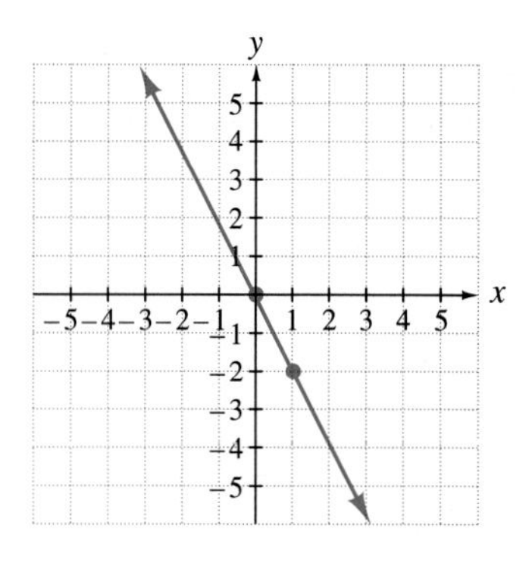

59. $H(x) = -2x + 3$

60. $F(x) = -2x - 1$

Use the graph of $f(x) = |x|$ at right to sketch the graph of each function.

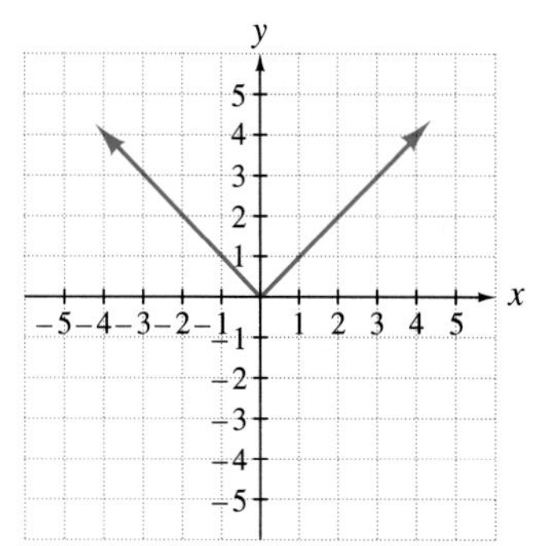

61. $g(x) = |x| + 2$

62. $H(x) = |x| - 2$

63. $F(x) = |x| - 1$

64. $h(x) = |x| + 5$

Use the graph of $f(x) = x^2$ at right to sketch the graph of each function.

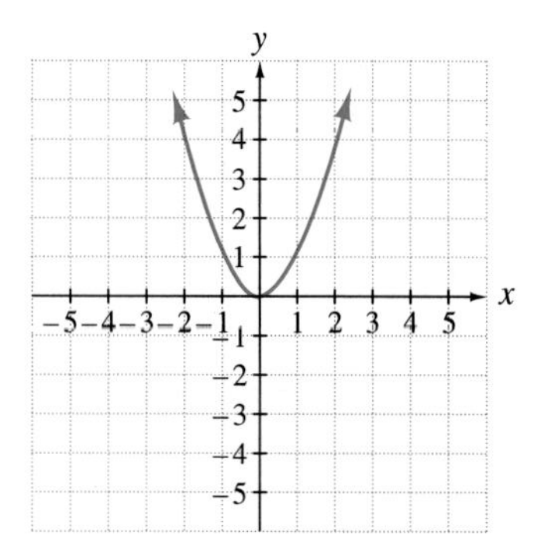

65. $F(x) = x^2 - 3$

66. $h(x) = x^2 - 4$

67. $H(x) = x^2 + 1$

68. $g(x) = x^2 + 3$

Writing in Mathematics

69. Explain why $\frac{x + 2}{x - 3} > 0$ and $(x + 2)(x - 3) > 0$ have the same solutions.

70. Explain why $\frac{x + 2}{x - 3} \geq 0$ and $(x + 2)(x - 3) \geq 0$ do not have the same solutions.

9.5 Quadratic Functions and Their Graphs

OBJECTIVES

Tape NA

1. Graph quadratic functions of the form $f(x) = x^2 + k$.
2. Graph quadratic functions of the form $f(x) = (x - h)^2$.
3. Graph quadratic functions of the form $f(x) = ax^2$.
4. Graph quadratic functions of the form $f(x) = a(x - h)^2 + k$.

1 We first graphed the quadratic equation $y = x^2$ in Section 4.1. In Section 4.2, we learned that this graph defined a function, and we wrote $y = x^2$ as $f(x) = x^2$. Quadratic functions and their graphs were studied further in Section 6.7. Throughout these sections, we discovered that the graph of a quadratic function is a parabola opening upward or downward. In this section, we further our study of quadratic functions and their graphs.

First let's recall the definition of a quadratic function.

Quadratic Function

A quadratic function is a function that can be written in the form $f(x) = ax^2 + bx + c$, where a, b, and c are real numbers and $a \neq 0$.

Notice that equations of the form $y = ax^2 + bx + c$, where $a \neq 0$, also define quadratic functions, since y is a function of x or $y = f(x)$.

Recall that if $a > 0$, the parabola opens upward, and if $a < 0$, the parabola opens downward. Also, the vertex of a parabola is the lowest point if the parabola opens upward and the highest point if the parabola opens downward. The axis of symmetry is the vertical line that passes through the vertex.

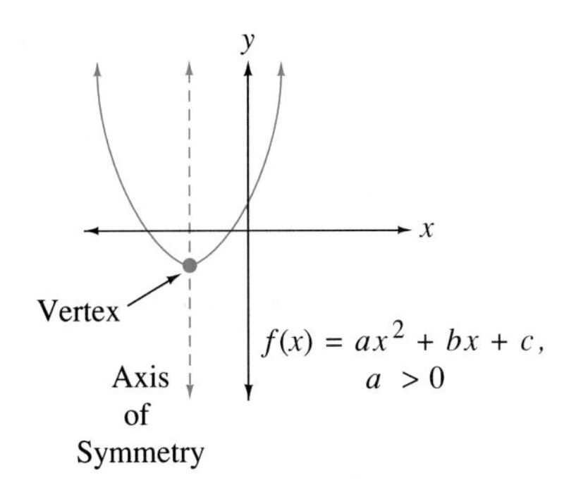

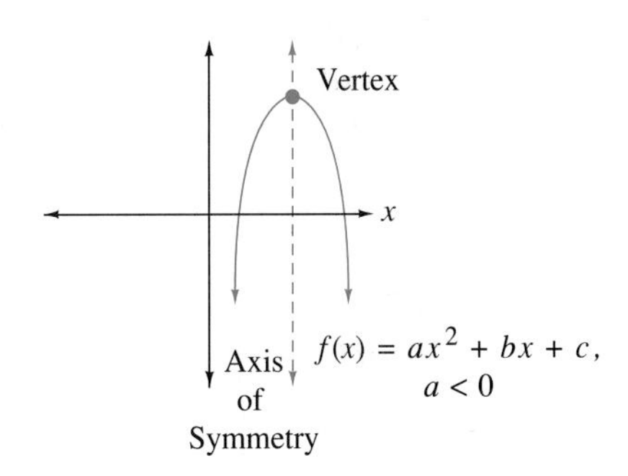

EXAMPLE 1 Sketch the graphs of $f(x) = x^2$ and $g(x) = x^2 + 1$ on the same set of axes.

Solution: By plotting points, we see that the graph of $g(x) = x^2 + 1$ is obtained by adding 1 to the $f(x)$ or y-value of each point of the graph of $f(x) = x^2$. In other words, the graph of $g(x) = x^2 + 1$ is the graph of $f(x) = x^2$ shifted upward 1 unit. The axis of symmetry for each graph is the y-axis.

x	$f(x) = x^2$
-2	4
-1	1
0	0
1	1
2	4

x	$g(x) = x^2 + 1$
-2	5
-1	2
0	1
1	2
2	5

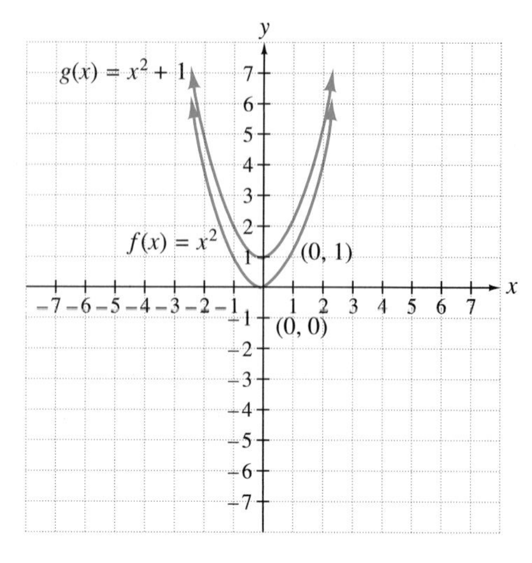

■

In general, we have the following properties.

> **Graphing the Parabola Defined by $f(x) = x^2 + k$**
>
> If k is positive, the graph of $f(x) = x^2 + k$ is the graph of $y = x^2$ shifted upward k units.
>
> If k is negative, the graph of $f(x) = x^2 + k$ is the graph of $y = x^2$ shifted downward $|k|$ units.
>
> The vertex is $(0, k)$ and the axis of symmetry is the y-axis.

EXAMPLE 2 Sketch the graph of each function.

a. $F(x) = x^2 + 2$ **b.** $g(x) = x^2 - 3$

Solution: **a.** The graph of $F(x) = x^2 + 2$ is obtained by shifting the graph of $y = x^2$ upward 2 units.

b. The graph of $g(x) = x^2 - 3$ is obtained by shifting the graph of $y = x^2$ downward 3 units.

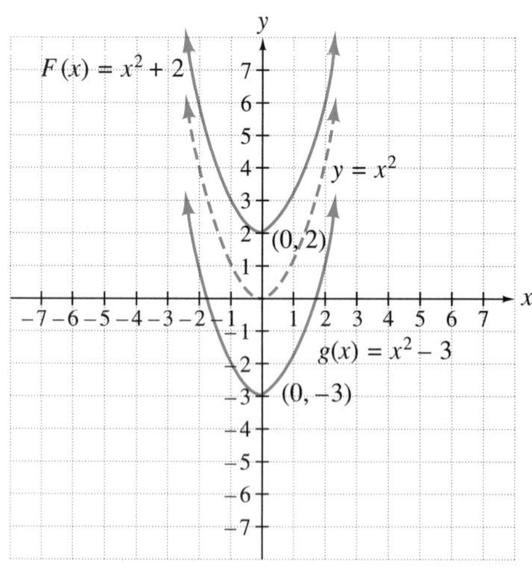

■

2

EXAMPLE 3 Sketch the graph of $f(x) = x^2$ and $g(x) = (x - 2)^2$ on the same set of axes.

Solution: By plotting points, we see that the graph of $g(x) = (x - 2)^2$ is obtained by adding 2 to the x-value of each point of the graph of $f(x) = x^2$. In other words, the graph of $g(x) = (x - 2)^2$ is the graph of $f(x) = x^2$ shifted to the right 2 units. The axis of symmetry for the graph of $g(x) = (x - 2)^2$ is also shifted 2 units to the right and is the line $x = 2$.

mx	$f(x) = x^2$
−2	4
−1	1
0	0
1	1
2	4

x	$g(x) = (x - 2)^2$
0	4
1	1
2	0
3	1
4	4

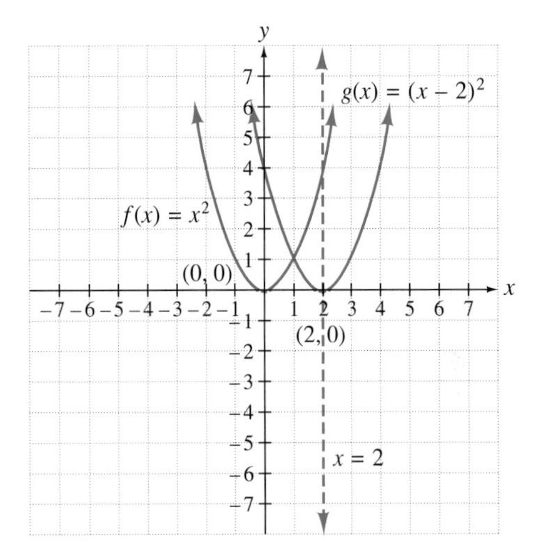

■

In general, we have the following properties.

> **Graphing the Parabola Defined by $f(x) = (x - h)^2$**
>
> If h is positive, the graph of $f(x) = (x - h)^2$ is the graph of $y = x^2$ shifted to the right h units.
> If h is negative, the graph of $f(x) = (x - h)^2$ is the graph of $y = x^2$ shifted to the left $|h|$ units.
> The vertex is $(h, 0)$ and the line of symmetry has the equation $x = h$.

EXAMPLE 4 Sketch the graph of each function.

a. $G(x) = (x - 3)^2$ **b.** $F(x) = (x + 1)^2$

Solution:

a. The graph of $G(x) = (x - 3)^2$ is obtained by shifting the graph of $y = x^2$ to the right 3 units.

b. The equation $F(x) = (x + 1)^2$ can be written as $F(x) = [x - (-1)]^2$. The graph of $F(x) = [x - (-1)]^2$ is obtained by shifting the graph of $y = x^2$ to the left 1 unit.

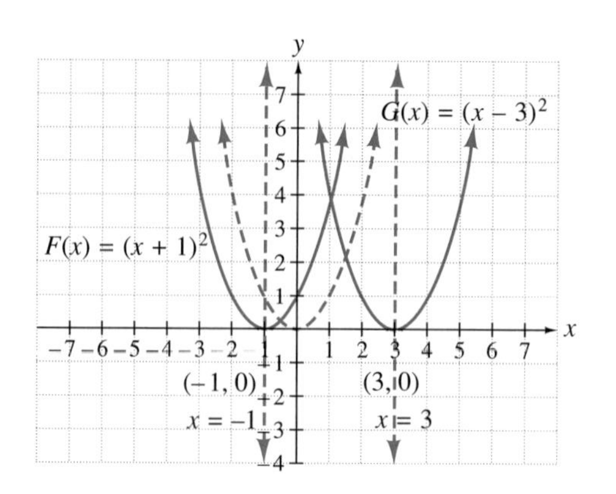

It is possible to combine vertical and horizontal shifts.

EXAMPLE 5 Sketch the graph of $F(x) = (x - 3)^2 + 1$.

Solution: The graph of $F(x) = (x - 3)^2 + 1$ is the graph of $y = x^2$ shifted 3 units to the right and 1 unit up. The vertex is then (3, 1), and the axis of symmetry is $x = 3$.

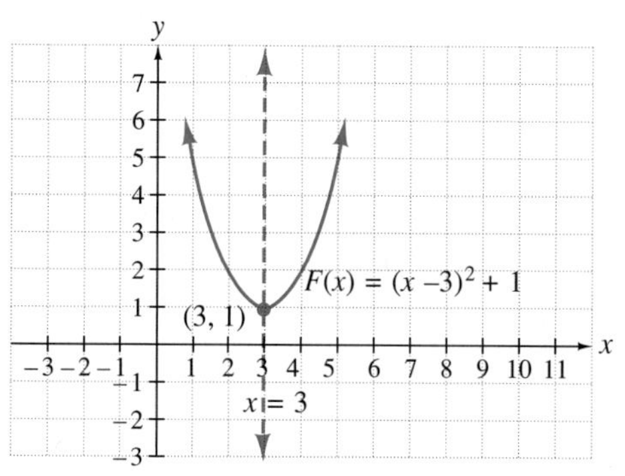

3 Next we discover the change in the shape of the graph when the coefficient of x^2 is not 1.

EXAMPLE 6 Sketch the graph of $f(x) = x^2$, $g(x) = 3x^2$, and $h(x) = \frac{1}{2}x^2$ on the same set of axes.

Solution: Compare the ordered pair solutions of each function with $f(x) = x^2$. We see that the y-value for corresponding x-values of each point of the graph of $g(x) = 3x^2$ is tripled

x	$f(x) = x^2$
−2	4
−1	1
0	0
1	1
2	4

x	$g(x) = 3x^2$
−2	12
−1	3
0	0
1	3
2	12

x	$h(x) = \frac{1}{2}x^2$
−2	2
−1	$\frac{1}{2}$
0	0
1	$\frac{1}{2}$
2	2

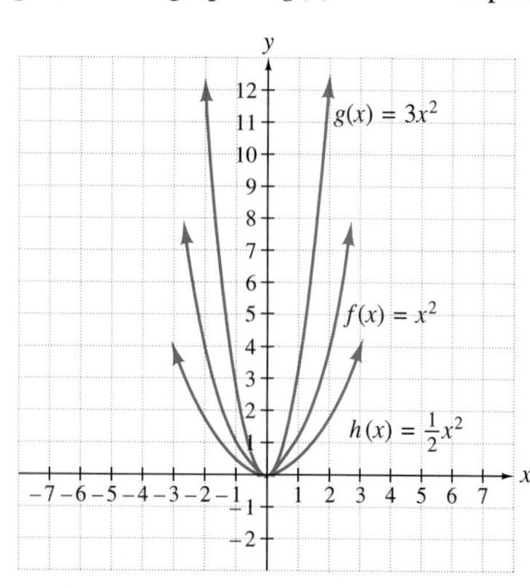

and that the y-value of each point of the graph of $h(x) = \frac{1}{2}x^2$ is halved. The vertex for each graph is (0, 0), and the axis of symmetry is the y-axis.

Notice that the graph of $g(x) = 3x^2$ is narrower than the graph of $f(x) = x^2$ and that the graph of $h(x) = \frac{1}{2}x^2$ is wider. ■

In general, we have the following properties.

> **Graphing the Parabola Defined by $f(x) = ax^2$**
>
> If a is positive, the parabola opens upward, and if a is negative, the parabola opens downward.
> If $|a| > 1$, the graph of the parabola is narrower than the graph of $y = x^2$.
> If $0 < |a| < 1$, the graph of the parabola is wider than the graph of $y = x^2$.

EXAMPLE 7 Sketch the graph of $f(x) = -2x^2$.

Solution: Because $a = -2$ is negative, this parabola opens downward. Also, $|-2| = 2$ and $2 > 1$, so the parabola appears narrower than the graph of $y = x^2$. The vertex is (0, 0), and the axis of symmetry is the y-axis. We verify this by plotting a few points.

x	$f(x) = -2x^2$
−2	−8
−1	−2
0	0
1	−2
2	−8

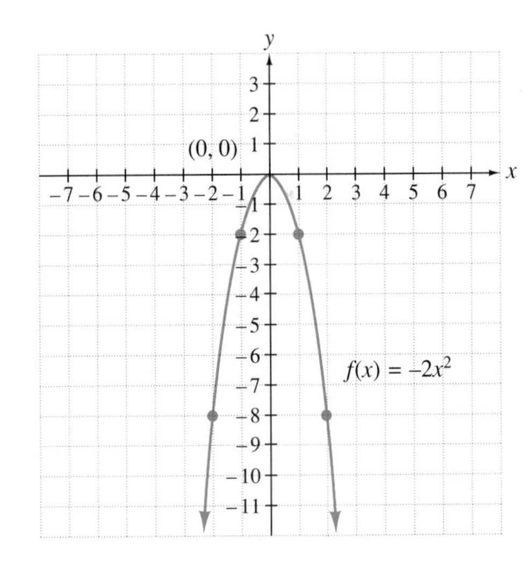

■

4

EXAMPLE 8 Sketch the graph of $g(x) = \frac{1}{2}(x + 2)^2 + 5$. Find the vertex and the axis of symmetry.

Solution: The function $g(x) = \frac{1}{2}(x + 2)^2 + 5$ may be written as $g(x) = \frac{1}{2}[x - (-2)]^2 + 5$. Thus, this graph is the same as the graph of $y = x^2$ shifted 2 units to the left and 5 units up, and it is wider because a is $\frac{1}{2}$. The vertex is $(-2, 5)$, and the axis of symmetry is $x = -2$. We plot a few points to verify.

x	$g(x) = \frac{1}{2}(x + 2)^2 + 5$
-4	7
-3	$5\frac{1}{2}$
-2	5
-1	$5\frac{1}{2}$
0	7

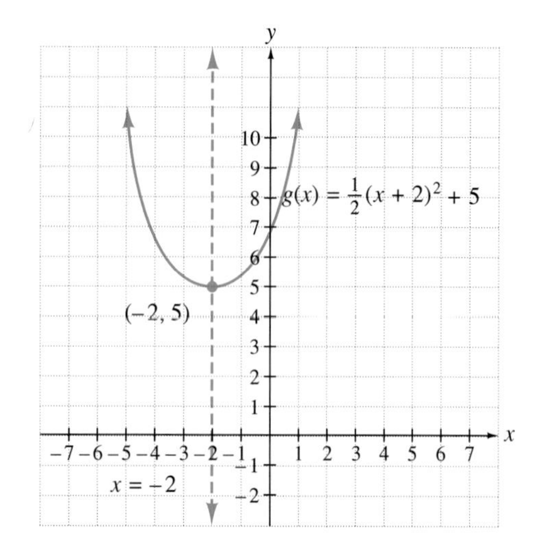

■

In general, the following holds.

Graph of a Quadratic Function

The graph of a quadratic function written in the form $f(x) = a(x - h)^2 + k$ is a parabola with vertex (h, k). If $a > 0$, the parabola opens upward, and if $a < 0$, the parabola opens downward. The axis of symmetry is the line whose equation is $x = h$.

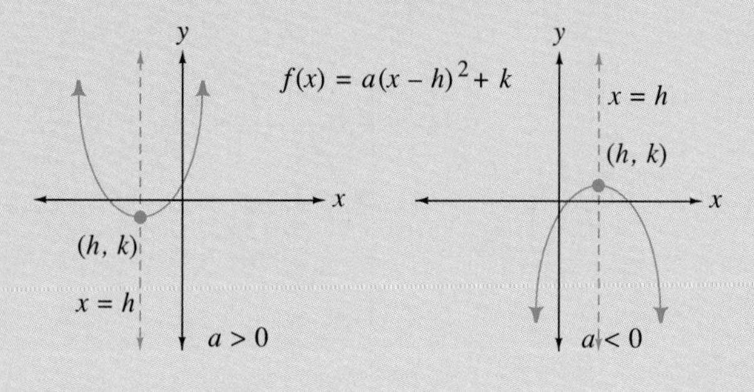

GRAPHING CALCULATOR BOX

Use a grapher to graph the first function of each pair below. Use its graph to predict the graph of the second function. Then check your prediction by graphing both on the same set of axes.

1. $F(x) = \sqrt{x}$; $G(x) = \sqrt{x} + 1$
2. $g(x) = x^3$; $H(x) = x^3 - 2$
3. $H(x) = |x|$; $f(x) = |x - 5|$
4. $h(x) = x^3 + 2$; $g(x) = (x - 3)^3 + 2$
5. $f(x) = |x + 4|$; $F(x) = |x + 4| + 3$
6. $G(x) = \sqrt{x} - 2$; $g(x) = \sqrt{x - 4} - 2$

MENTAL MATH

State the vertex of the graph of each quadratic function.

1. $f(x) = x^2$
2. $f(x) = -5x^2$
3. $g(x) = (x - 2)^2$
4. $g(x) = (x + 5)^2$
5. $f(x) = 2x^2 + 3$
6. $h(x) = x^2 - 1$
7. $g(x) = (x + 1)^2 + 5$
8. $h(x) = (x - 10)^2 - 7$

EXERCISE SET 9.5

Sketch the graph of each quadratic function. Label the vertex and sketch and label the axis of symmetry. See Examples 1 and 2.

1. $f(x) = x^2 - 1$
2. $g(x) = x^2 + 3$
3. $h(x) = x^2 + 5$
4. $h(x) = x^2 - 4$
5. $g(x) = x^2 + 7$
6. $f(x) = x^2 - 2$

See Examples 3 and 4.

7. $f(x) = (x - 5)^2$
8. $g(x) = (x + 5)^2$
9. $h(x) = (x + 2)^2$
10. $H(x) = (x - 1)^2$
11. $G(x) = (x + 3)^2$
12. $f(x) = (x - 6)^2$

See Example 5.

13. $f(x) = (x - 2)^2 + 5$
14. $g(x) = (x - 6)^2 + 1$
15. $h(x) = (x + 1)^2 + 4$

16. $G(x) = (x + 3)^2 + 3$

17. $g(x) = (x + 2)^2 - 5$

18. $h(x) = (x + 4)^2 - 6$

See Examples 6 and 7.

19. $g(x) = -x^2$

20. $f(x) = 5x^2$

21. $h(x) = \frac{1}{3}x^2$

22. $g(x) = -3x^2$

23. $H(x) = 2x^2$

24. $f(x) = -\frac{1}{4}x^2$

See Example 8.

25. $f(x) = 2(x - 1)^2 + 3$

26. $g(x) = 4(x - 4)^2 + 2$

27. $h(x) = -3(x + 3)^2 + 1$

28. $f(x) = -(x - 2)^2 - 6$

29. $H(x) = \frac{1}{2}(x - 6)^2 - 3$

30. $G(x) = \frac{1}{5}(x + 4)^2 + 3$

Sketch the graph of each quadratic function. Label the vertex and sketch and label the axis of symmetry.

31. $f(x) = -(x - 2)^2$

32. $g(x) = -(x + 6)^2$

33. $F(x) = -x^2 + 4$

34. $H(x) = -x^2 + 10$

35. $F(x) = 2x^2 - 5$

36. $g(x) = \frac{1}{2}x^2 - 2$

37. $h(x) = (x - 6)^2 + 4$

38. $f(x) = (x - 5)^2 + 2$

39. $F(x) = \left(x + \frac{1}{2}\right)^2 - 2$

40. $H(x) = \left(x + \frac{1}{4}\right)^2 - 3$

41. $F(x) = \frac{3}{2}(x + 7)^2 + 1$

42. $g(x) = -\frac{3}{2}(x - 1)^2 - 5$

43. $f(x) = \frac{1}{4}x^2 - 9$

44. $H(x) = \frac{3}{4}x^2 - 2$

45. $G(x) = 5\left(x + \frac{1}{2}\right)^2$

46. $F(x) = 3\left(x - \frac{3}{2}\right)^2$

47. $h(x) = -(x - 1)^2 - 1$

48. $f(x) = -3(x + 2)^2 + 2$

49. $g(x) = \sqrt{3}(x + 5)^2 + \frac{3}{4}$

50. $G(x) = \sqrt{5}(x - 7)^2 - \frac{1}{2}$

51. $h(x) = 10(x + 4)^2 - 6$

52. $h(x) = 8(x + 1)^2 + 9$ **53.** $f(x) = -2(x - 4)^2 + 5$ **54.** $G(x) = -4(x + 9)^2 - 1$

Write the equation of the parabola that has the same shape as $f(x) = 5x^2$ but with the following vertex.

55. (2, 3) **56.** (1, 6) **57.** (−3, 6) **58.** (4, −1)

The shifting properties covered in this section apply to the graphs of all functions. Given the graph of $y = f(x)$ below, sketch the graph of each of the following.

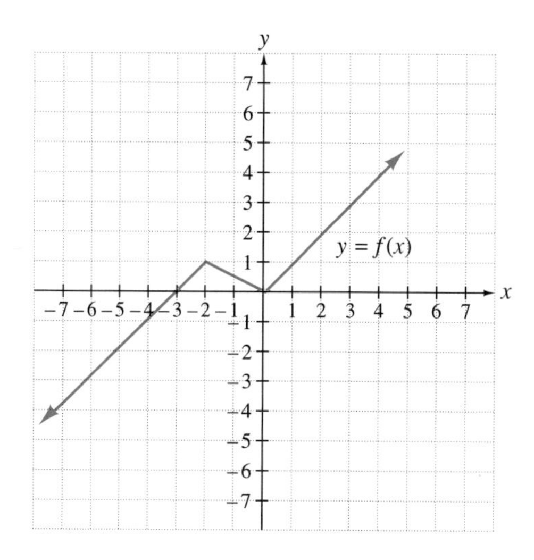

59. $y = f(x) + 1$ **60.** $y = f(x) - 2$ **61.** $y = f(x - 3)$ **62.** $y = f(x + 3)$

63. $y = f(x + 2) + 2$ **64.** $y = f(x - 1) + 1$

Skill Review

Add the proper constant to each binomial so that the resulting trinomial is a perfect square trinomial. See Section 9.1.

65. $x^2 + 8x$ **66.** $y^2 + 4y$ **67.** $z^2 - 16z$

68. $x^2 - 10x$ **69.** $y^2 + y$ **70.** $z^2 - 3z$

Solve by completing the square. *See Section 9.1.*

71. $x^2 + 4x = 12$

72. $y^2 + 6y = -5$

73. $z^2 + 10z - 1 = 0$

74. $x^2 + 14x + 20 = 0$

75. $z^2 - 8z = 2$

76. $y^2 - 10y = 3$

9.6 Further Graphing of Quadratic Functions

OBJECTIVES

Tape IA 25

1. Write quadratic functions in the form $y = a(x - h)^2 + k$.
2. Derive a formula for finding the vertex of a parabola.
3. Find the minimum or maximum value of a quadratic function.

1 We know that the graph of a quadratic function is a parabola. If a quadratic function is written in the form

$$f(x) = a(x - h)^2 + k$$

We can easily find the vertex (h, k) and graph the parabola. To write a quadratic function in this form, we complete the square.

EXAMPLE 1 Graph $f(x) = x^2 - 4x - 12$.

Solution: The graph of this quadratic function is a parabola. To find the vertex of the parabola, we complete the square on the binomial $x^2 - 4x$.

$$f(x) = (x^2 - 4x) - 12$$

Now add and subtract the square of half of -4.

$$\frac{1}{2}(-4) = -2 \quad \text{and} \quad (-2)^2 = 4$$

Add and subtract 4.

$$f(x) = (x^2 - 4x + 4) - 12 - 4$$

$$f(x) = (x - 2)^2 - 16 \qquad \text{Factor.}$$

From this equation, we can see that $h = 2$ and $k = -16$, so the vertex of the parabola is $(2, -16)$, and the axis of symmetry is the line $x = 2$.

Since $a > 0$, the parabola opens upward. This parabola opening upward with vertex $(2, -16)$ will have two x-intercepts. To find them, let $f(x)$ or $y = 0$.

$$0 = x^2 - 4x - 12$$

$$0 = (x - 6)(x + 2)$$

$$0 = x - 6 \quad \text{or} \quad 0 = x + 2$$

$$6 = x \quad \text{or} \quad -2 = x$$

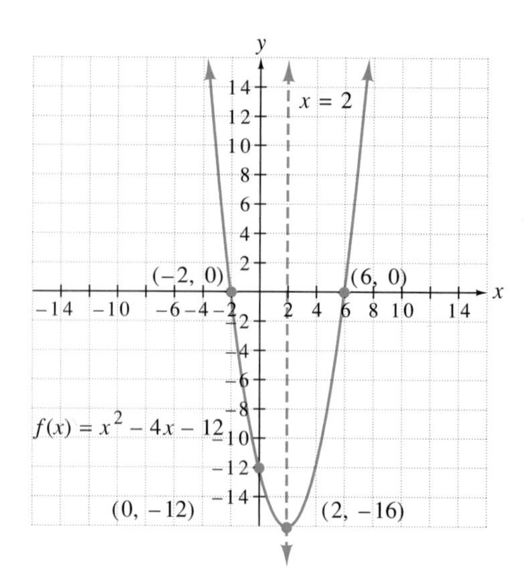

The two x-intercepts are 6 and -2. The sketch of $f(x) = x^2 - 4x - 12$ is shown. ■

EXAMPLE 2 Graph $f(x) = 3x^2 + 3x + 1$. Find the vertex and any intercepts.

Solution: Complete the square on x to write the equation in the form $f(x) = a(x - h)^2 + k$.

$$f(x) = 3x^2 + 3x + 1$$

Factor 3 from the terms $3x^2 + 3x$ so that the coefficient of x^2 is 1.

$$f(x) = 3(x^2 + x) + 1$$

Then $\frac{1}{2}(1) = \frac{1}{2}$ and $\left(\frac{1}{2}\right)^2 = \frac{1}{4}$.

$$f(x) = 3\left(x^2 + x + \frac{1}{4}\right) + 1 - 3\left(\frac{1}{4}\right)$$

$$= 3\left(x + \frac{1}{2}\right)^2 + \frac{4}{4} - \frac{3}{4}$$

$$= 3\left(x + \frac{1}{2}\right)^2 + \frac{1}{4}$$

Then $a = 3$, $h = -\frac{1}{2}$, and $k = \frac{1}{4}$. This means that the parabola opens upward with vertex $\left(-\frac{1}{2}, \frac{1}{4}\right)$ and the axis of symmetry is the line $x = -\frac{1}{2}$.

To find the y-intercept, let $x = 0$. Then

$$f(0) = 3(0)^2 + 3(0) + 1 = 1$$

This parabola has no x-intercepts since the vertex is in the second quadrant and opens upward. Use the vertex, axis of symmetry, and y-intercept to sketch the parabola.

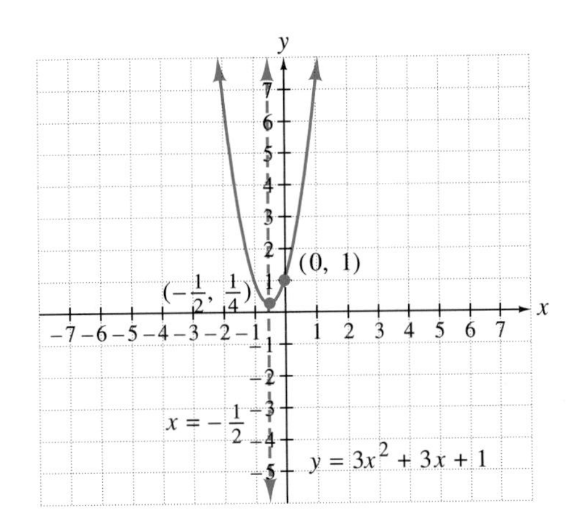

EXAMPLE 3 Graph $f(x) = -x^2 - 2x + 3$. Find the vertex and any intercepts.

Solution: To complete the square, first factor -1 from the terms $-x^2 - 2x$.

$$f(x) = -x^2 - 2x + 3$$
$$f(x) = -1(x^2 + 2x) + 3$$

Then $\frac{1}{2}(2) = 1$ and $1^2 = 1$.

$$f(x) = -1\,(x^2 + 2x \; + 1\,) + 3 - 1(-1)$$
$$= -1(x + 1)^2 + 4 \qquad \text{Factor.}$$

Since $a = -1$, the parabola opens downward with vertex $(-1, 4)$ and axis of symmetry $x = -1$.

To find the y-intercept, let $x = 0$ and solve for y. Then

$$f(0) = -0^2 - 2(0) + 3 = 3$$

Thus, 3 is the y-intercept.

To find the x-intercepts, let y or $f(x) = 0$ and solve for x.

$$f(x) = -x^2 - 2x + 3$$
$$0 = -x^2 - 2x + 3 \qquad \text{Let } f(x) = 0.$$

Divide both sides by -1 so that the coefficient of x^2 is 1.

$$\frac{0}{-1} = \frac{-x^2}{-1} - \frac{2x}{-1} + \frac{3}{-1} \qquad \text{Divide both sides by } -1.$$
$$0 = x^2 + 2x - 3 \qquad \text{Simplify.}$$
$$0 = (x + 3)(x - 1) \qquad \text{Factor.}$$
$$x + 3 = 0 \quad \text{or} \quad x - 1 = 0 \qquad \text{Set each factor equal to 0.}$$
$$x = -3 \qquad\qquad x = 1 \qquad \text{Solve.}$$

The x-intercepts are -3 and 1. Use these points to sketch the parabola.

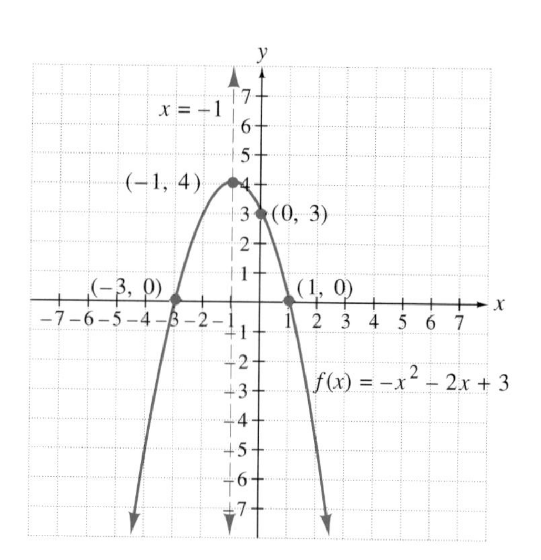

■

2 Recall in Section 6.7 that we introduced a formula for finding the vertex of a parabola. Now that we have practiced completing the square, we will now show that the x-coordinate of the vertex of the graph of $f(x) = ax^2 + bx + c$ can be found by the formula $x = \frac{-b}{2a}$. To do so, we complete the square on x and write the equation in the form $f(x) = a(x - h)^2 + k$.

First factor a from the terms $ax^2 + bx$.

$$f(x) = ax^2 + bx + c$$

$$f(x) = a\left(x^2 + \frac{b}{a}x\right) + c$$

Next complete the square by finding the square of half of $\frac{b}{a}$.

$$\frac{1}{2}\left(\frac{b}{a}\right) = \frac{b}{2a} \quad \text{and} \quad \left(\frac{b}{2a}\right)^2 = \frac{b^2}{4a^2}$$

$$f(x) = a\left(x^2 + \frac{b}{a}x + \frac{b^2}{4a^2}\right) + c - a\left(\frac{b^2}{4a^2}\right)$$

$$f(x) = a\left(x + \frac{b}{2a}\right)^2 + c - \frac{b^2}{4a}$$

Compare this form with $f(x) = a(x - h)^2 + k$ and see that h is $\frac{-b}{2a}$, which means the x-coordinate of the vertex of the graph of $f(x) = ax^2 + bx + c$ is $\frac{-b}{2a}$.

Vertex Formula

The graph of $f(x) = ax^2 + bx + c$, $a \neq 0$, is a parabola with vertex

$$\left(\frac{-b}{2a}, f\left(\frac{-b}{2a}\right)\right)$$

Let's use this formula to find the vertex of the parabola we graphed in Example 1. In the quadratic function $f(x) = x^2 - 4x - 12$, notice that $a = 1$, $b = -4$, and $c = -12$. Then

$$\frac{-b}{2a} = \frac{-(-4)}{2(1)} = 2$$

The x-value of the vertex is 2. To find the corresponding $f(x)$ or y-value, find $f(2)$. Then

$$f(2) = 2^2 - 4(2) - 12 = 4 - 8 - 12 = -16$$

The vertex is $(2, -16)$, and the axis of symmetry is the line $x = 2$. These results agree with our findings in Example 1.

3 The quadratic function whose graph is a parabola that opens upward has a minimum value, and the quadratic function whose graph is a parabola that opens downward has a maximum value. The $f(x)$ or y-value of the vertex is the minimum or maximum value, and the corresponding x-value tells where the minimum or maximum value occurs.

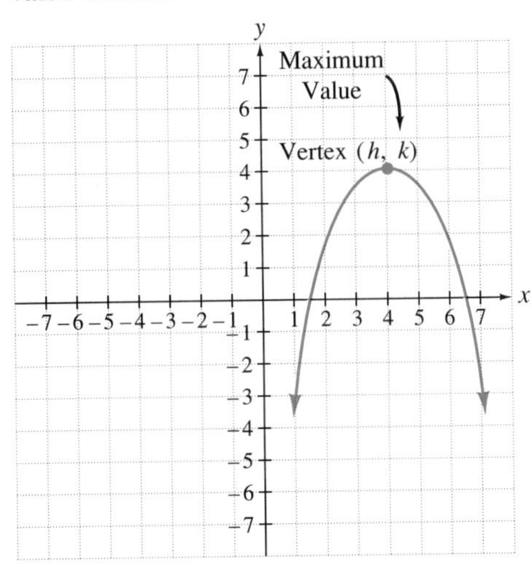

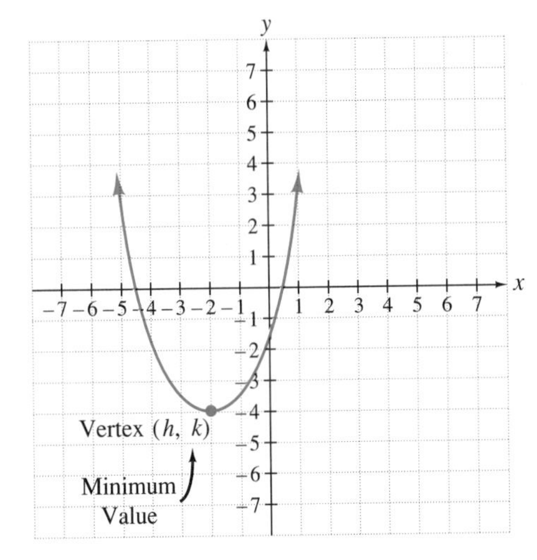

EXAMPLE 4 A rock is thrown upward from the ground. Its height in feet above ground after t seconds is given by the equation $f(t) = -16t^2 + 20t$. Find the maximum height of the rock.

Solution: **1.** UNDERSTAND. The maximum height of the rock is the largest value of $f(t)$. Since the equation $f(t) = -16t^2 + 20t$ is a quadratic function, its graph is a parabola. It opens downward since $-16 < 0$. Thus, the maximum value of $f(t)$ is the $f(t)$ or y-value of the vertex of its graph.

4. TRANSLATE. To find the vertex (h, k), notice that, for $f(t) = -16t^2 + 20t$, $a = -16$, $b = 20$, and $c = 0$. Use these values and the vertex formula

$$\left(\frac{-b}{2a}, f\left(\frac{-b}{2a}\right)\right)$$

5. COMPLETE. $$h = \frac{-b}{2a} = \frac{-20}{-32} = \frac{5}{8}$$

$$f\left(\frac{5}{8}\right) = -16\left(\frac{5}{8}\right)^2 + 20\left(\frac{5}{8}\right) = -16\left(\frac{25}{64}\right) + \frac{25}{2} = -\frac{25}{4} + \frac{50}{4} = \frac{25}{4}$$

6. INTERPRET. The graph of $f(t)$ is a parabola opening downward with vertex $\left(\frac{5}{8}, \frac{25}{4}\right)$. This means that the rock's maximum height is $\frac{25}{4}$ feet or $6\frac{1}{4}$ feet, which was reached in $\frac{5}{8}$ second. ■

EXERCISE SET 9.6

Find the vertex of the graph of each quadratic function. See Example 1.

1. $f(x) = x^2 + 8x + 7$
2. $f(x) = x^2 + 6x + 5$
3. $f(x) = -x^2 + 10x + 5$
4. $f(x) = -x^2 - 8x + 2$
5. $f(x) = 5x^2 - 10x + 3$
6. $f(x) = -3x^2 + 6x + 4$
7. $f(x) = -x^2 + x + 1$
8. $f(x) = x^2 - 9x + 8$

Find the vertex of the graph of each quadratic function. Determine whether the graph opens upward or downward, find any intercepts, and sketch the graph. See Examples 1 through 3.

9. $f(x) = x^2 + 4x - 5$
10. $f(x) = x^2 + 2x - 3$
11. $f(x) = -x^2 + 2x - 1$
12. $f(x) = -x^2 + 4x - 4$

13. $f(x) = x^2 - 4$
14. $f(x) = x^2 - 1$
15. $f(x) = 4x^2 + 4x - 3$
16. $f(x) = 2x^2 - x - 3$

Solve. See Example 4.

17. Find two positive numbers whose sum is 60 and whose product is as large as possible. [Hint: Let x and $60 - x$ be the two positive numbers. Their product can be described by the function $f(x) = x(60 - x)$.]

18. The length and width of a rectangle must have a sum of 40. Find the dimensions of the rectangle that will have maximum area. (Use the hint for Exercise 17.)

Find the vertex of the graph of each quadratic function. Determine whether the graph opens upward or downward, find any intercepts, and sketch the graph.

19. $f(x) = x^2 + 8x + 15$
20. $f(x) = x^2 + 10x + 9$
21. $f(x) = x^2 - 6x + 5$

22. $f(x) = x^2 - 4x + 3$

23. $f(x) = x^2 - 4x + 5$

24. $f(x) = x^2 - 6x + 11$

25. $f(x) = 2x^2 + 4x + 5$

26. $f(x) = 3x^2 + 12x + 16$

27. $f(x) = -2x^2 + 12x$

28. $f(x) = -4x^2 + 8x$

29. $f(x) = x^2 + 1$

30. $f(x) = x^2 + 4$

31. $f(x) = x^2 - 2x - 15$

32. $f(x) = x^2 - 4x + 3$

33. $f(x) = -5x^2 + 5x$

34. $f(x) = 3x^2 - 12x$

35. $f(x) = -x^2 + 2x - 12$

36. $f(x) = -x^2 + 8x - 17$

37. $f(x) = 3x^2 - 12x + 15$

38. $f(x) = 2x^2 - 8x + 11$

39. $f(x) = x^2 + x - 6$

40. $f(x) = x^2 + 3x - 18$

41. $f(x) = -2x^2 - 3x + 35$

42. $f(x) = 3x^2 - 13x - 10$

Solve.

43. If a projectile is fired straight upward from the ground with an initial speed of 96 feet per second, then its height h in feet after t seconds is given by the equation

$$h(t) = -16t^2 + 96t$$

Find the maximum height of the projectile.

44. If Rheam Gaspar throws a ball upward with an initial speed of 32 feet per second, then its height h in feet after t seconds is given by the equation

$$h(t) = -16t^2 + 32t$$

Find the maximum height of the ball.

45. The cost C in dollars of manufacturing x bicycles at Holladay's Production Plant is given by the function $C(x) = 2x^2 - 800x + 92{,}000$.

a. Find the number of bicycles that must be manufactured to minimize the cost.

b. Find the minimum cost.

46. The Utah Ski Club sells calendars to raise money. The profit P, in cents, from selling x calendars is given by the equation $P(x) = 360x - x^2$.

a. Find how many calendars must be sold to maximize profit.

b. Find the maximum profit.

Skill Review

Sketch the graph of each function. See Section 9.5.

47. $f(x) = x^2 + 2$

48. $f(x) = (x - 3)^2$

49. $g(x) = x + 2$

50. $h(x) = x - 3$

51. $f(x) = (x + 5)^2 + 2$

52. $f(x) = 2(x - 3)^2 + 2$

53. $f(x) = 3(x - 4)^2 + 1$

54. $f(x) = (x + 1)^2 + 4$

55. $f(x) = -(x - 4)^2 + \frac{3}{2}$

56. $f(x) = -2(x + 7)^2 + \frac{1}{2}$

Determine whether each graph is the graph of a function. See Section 4.2.

57.

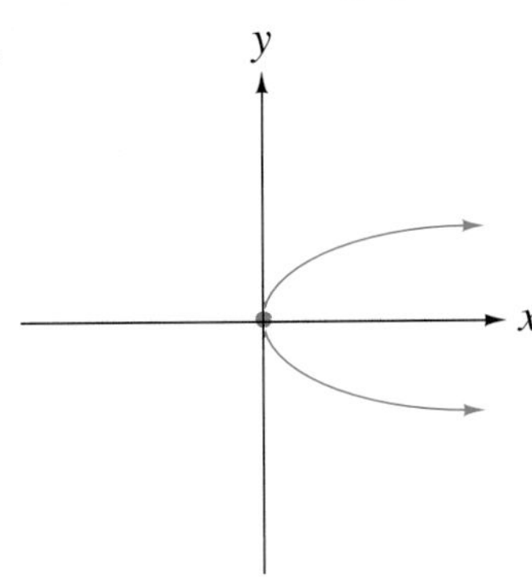

58.

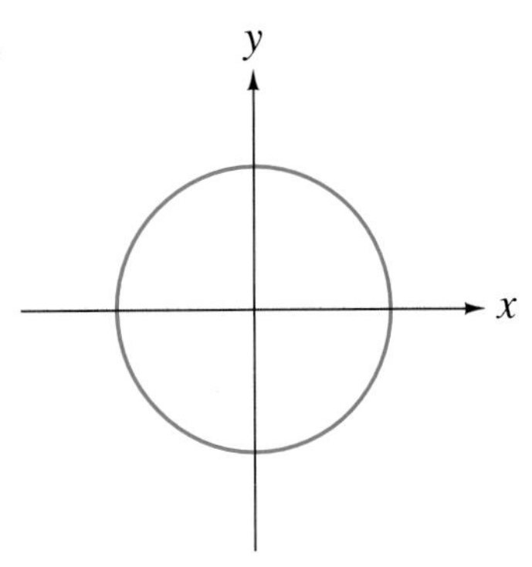

Writing in Mathematics

59. Consider a quadratic function whose graph is a parabola opening upward. Explain whether this function has a minimum value, a maximum value, or neither.

9.7 The Parabola and the Circle

OBJECTIVES

Tape IA 26

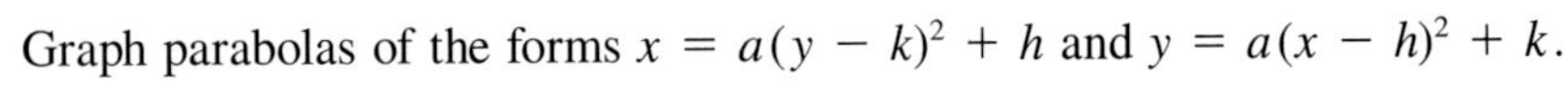

1. Graph parabolas of the forms $x = a(y - k)^2 + h$ and $y = a(x - h)^2 + k$.
2. Use the distance formula and the midpoint formula.
3. Graph circles of the form $(x - h)^2 + (y - k)^2 = r^2$.
4. Write the equation of a circle given its center and radius.
5. Find the center and the radius of a circle, given its equation.

We have analyzed some of the important connections between a parabola and its equation. Parabolas are interesting in their own right, but are more interesting still because they are part of a collection of curves known as **conic sections.** Conic sections derive their name because each conic section is the intersection of a right circular cone and a plane. The circle, parabola, ellipse, and hyperbola are the conic sections.

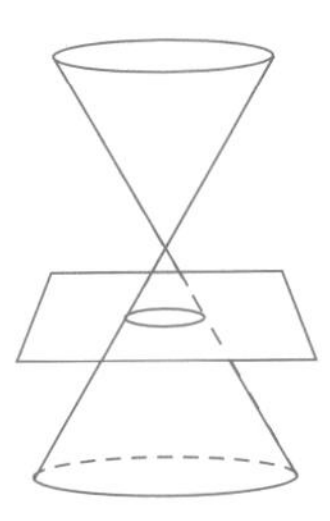

Circle

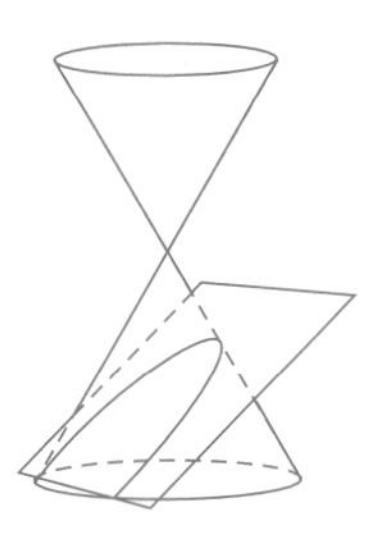

Parabola

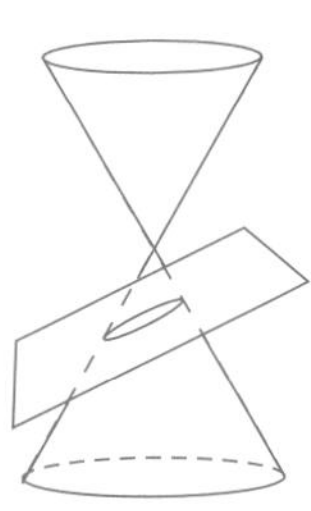

Ellipse

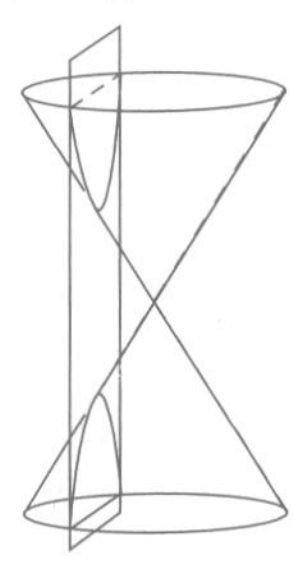

Hyperbola

1 Thus far, we have learned that $f(x)$ or $y = a(x - h)^2 + k$ is the equation of a parabola that opens upward if $a > 0$ or downward if $a < 0$. Parabolas can also open left or right, or even on a slant. Equations of these parabolas are not functions of x, of course, since a parabola opening any other way but upward or downward fails the vertical line test. In this section, we introduce parabolas opening to the left and to the right.

Just as $y = a(x - h)^2 + k$ is the equation of a parabola opening upward or downward, $x = a(y - k)^2 + h$ is the equation of a parabola opening to the right or to the left. The parabola opens to the right if $a > 0$ and to the left if $a < 0$. The parabola has vertex (h, k), and its axis of symmetry is the line $y = k$.

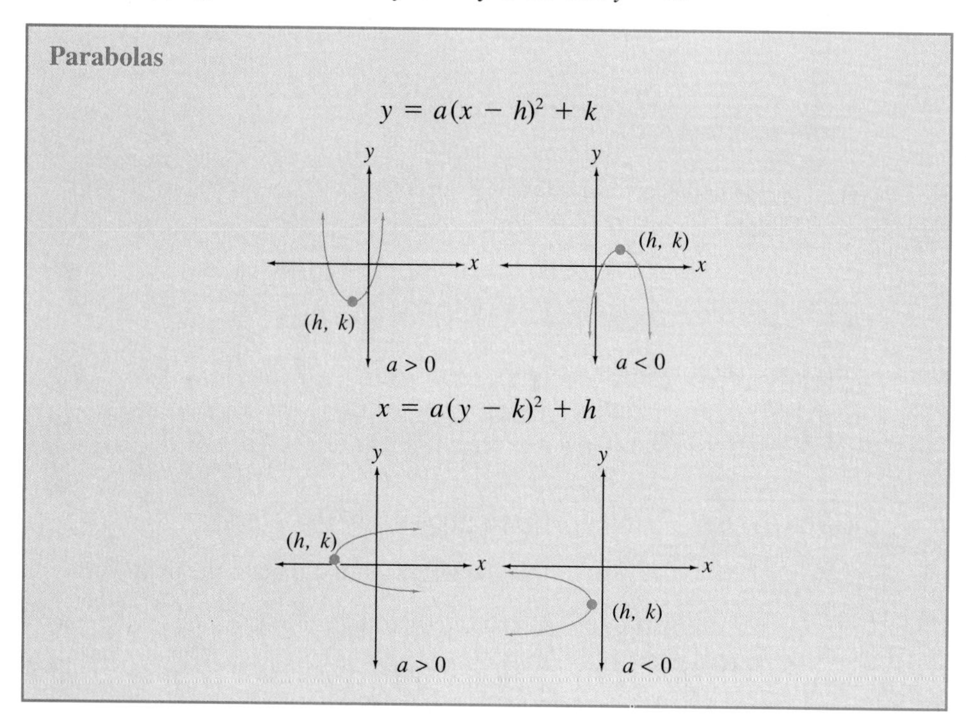

The forms $y = a(x - h)^2 + k$ and $x = a(y - k)^2 + h$ are both called **standard forms.**

EXAMPLE 1 Graph the parabola $x = 2y^2$.

Solution: The equation $x = 2y^2$ written in standard form is $x = 2(y - 0)^2 + 0$ with $a = 2$,

x	y
8	−2
2	−1
0	0
2	1
8	2

$h = 0$, and $k = 0$. Its graph is a parabola with vertex $(0, 0)$, and its axis of symmetry is the line $y = 0$. Since $a > 0$, this parabola opens to the right. The table shows a few more ordered pair solutions of $x = 2y^2$. Its graph is also shown. ■

EXAMPLE 2 Graph the parabola $x = -3(y - 1)^2 + 2$.

Solution: The equation $x = -3(y - 1)^2 + 2$ is in the form $x = a(y - k)^2 + h$ with $a = -3$, $k = 1$, and $h = 2$. Since $a < 0$, the parabola opens to the left. The vertex (h, k) is $(2, 1)$, and the axis of symmetry is the line $y = 1$. When $y = 0$, $x = -1$, the x-intercept. The parabola is sketched next.

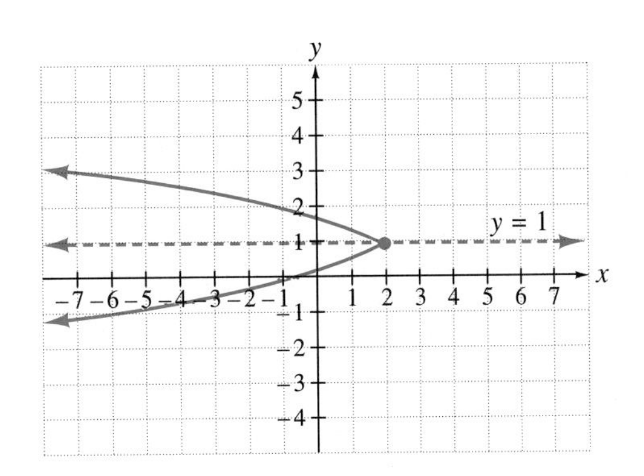

■

EXAMPLE 3 Graph $y = -x^2 - 2x + 15$.

Solution: Complete the square on x to write the equation in standard form.

$$y = -(x^2 + 2x) + 15$$

$$= -(x^2 + 2x + 1) + 15 + 1 \quad \text{Subtract and add 1.}$$

$$= -(x + 1)^2 + 16 \quad \text{Factor.}$$

The equation is now in the form $y = a(x - h)^2 + k$ with $a = -1$, $h = -1$, and $k = 16$. The parabola opens downward since $a < 0$ and has vertex $(-1, 16)$. Its axis of symmetry is the line $x = -1$. The y-intercept is 15.

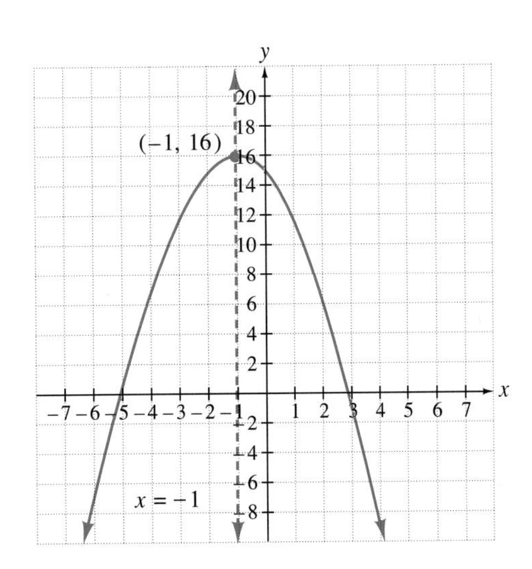

■

EXAMPLE 4 Graph $x = 2y^2 + 4y + 5$.

Solution: Complete the square on y to write the equation in standard form.

$$x = 2(y^2 + 2y) + 5$$

$$= 2(y^2 + 2y + 1) + 5 - 2(1) \quad \text{Add and subtract } 2(1) \text{ or } 2.$$

$$= 2(y + 1)^2 + 3 \quad \text{Factor.}$$

The equation is now in the form $x = a(y - k)^2 + h$ with $a = 2$, $k = -1$, and $h = 3$. The parabola opens to the right since $a > 0$ and has vertex $(3, -1)$. Its axis of symmetry is the line $y = -1$. The x-intercept is 5.

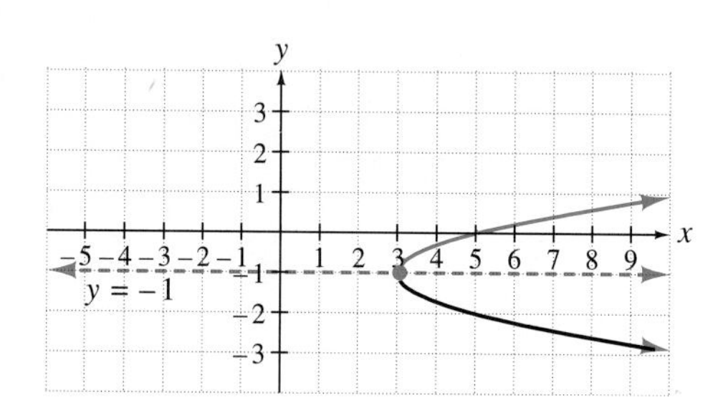

■

2 The Cartesian coordinate system helps us to visualize a distance between points. To find the distance between two points, we use the **distance formula,** which is derived from the Pythagorean theorem.

To find the distance d between two points (x_1, y_1) and (x_2, y_2) shown below, notice that the length of leg a is $x_2 - x_1$ and the length of leg b is $y_2 - y_1$.

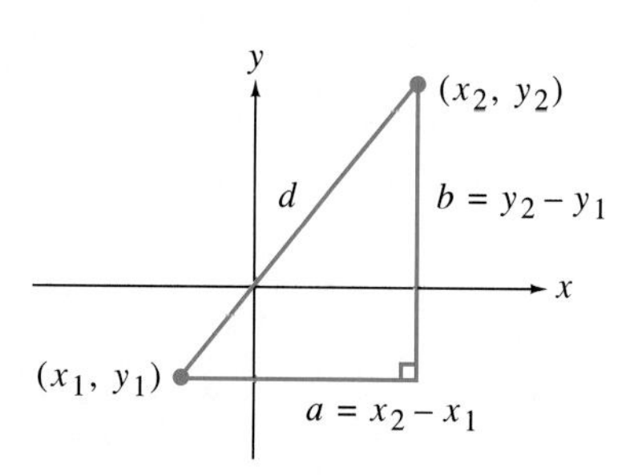

Thus, the Pythagorean theorem tells us that

$$d^2 = a^2 + b^2$$

or

$$d^2 = (x_2 - x_1)^2 + (y_2 - y_1)^2$$

or

$$d = \sqrt{(x_2 - x_1)^2 + (y_2 - y_1)^2}$$

This formula gives us the distance between any two points on the real plane.

Distance Formula

The distance between two points (x_1, y_1) and (x_2, y_2) is given by

$$d = \sqrt{(x_2 - x_1)^2 + (y_2 - y_1)^2}$$

EXAMPLE 5 Find the distance between the pair of points $(2, -5)$ and $(1, -4)$.

Solution: Use the distance formula. It makes no difference which point we call (x_1, y_1) and which point we call (x_2, y_2).

Let $(2, -5) = (x_1, y_1)$ and $(1, -4) = (x_2, y_2)$.

$$\begin{aligned} d &= \sqrt{(x_2 - x_1)^2 + (y_2 - y_1)^2} \\ &= \sqrt{(1 - 2)^2 + [-4 - (-5)]^2} \\ &= \sqrt{(-1)^2 + (1)^2} \\ &= \sqrt{1 + 1} \\ &= \sqrt{2} \end{aligned}$$

The distance between the two points is $\sqrt{2}$ units. ■

The **midpoint** of a line segment is the **point** located exactly halfway between the two end points of the line segment. On the following graph, the point M is the midpoint of line segment PQ. The distance between M and P equals the distance between M and Q.

The x-coordinate of M is at half the distance between the x-coordinates of P and Q, and the y-coordinate of M is at half the distance between the y-coordinates of P and Q.

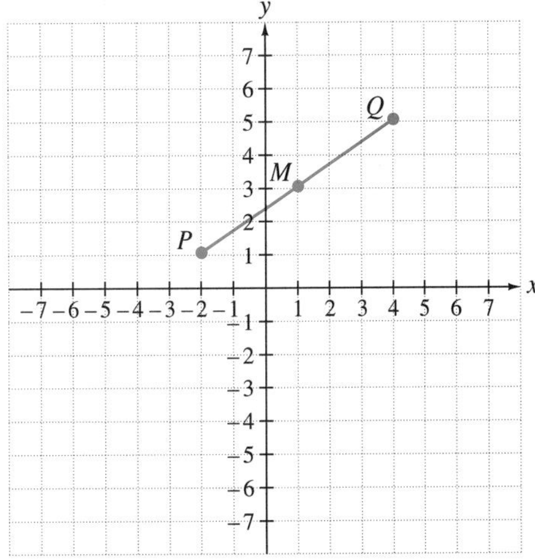

Midpoint Formula

The midpoint of the line segment whose end points are (x_1, y_1) and (x_2, y_2) is the point

$$\left(\frac{x_1 + x_2}{2}, \frac{y_1 + y_2}{2}\right)$$

EXAMPLE 6 Find the midpoint of the line segment joining points $P(-3, 3)$ and $Q(1, 0)$.

Solution: Use the midpoint formula. It makes no difference which point we call (x_1, y_1) or which point we call (x_2, y_2). Let $(-3, 3) = (x_1, y_1)$ and $(1, 0) = (x_2, y_2)$.

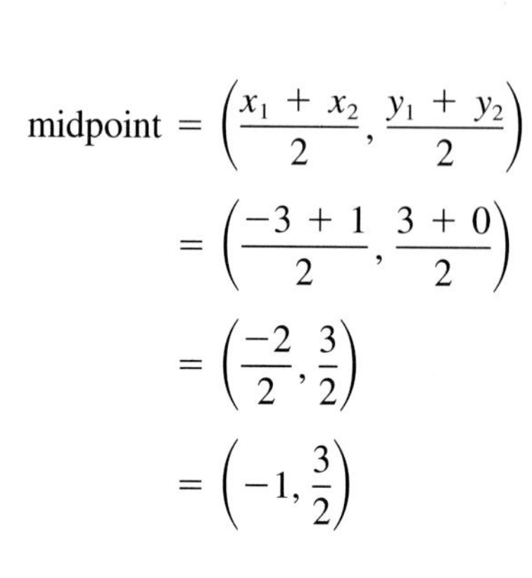

$$\text{midpoint} = \left(\frac{x_1 + x_2}{2}, \frac{y_1 + y_2}{2}\right)$$
$$= \left(\frac{-3 + 1}{2}, \frac{3 + 0}{2}\right)$$
$$= \left(\frac{-2}{2}, \frac{3}{2}\right)$$
$$= \left(-1, \frac{3}{2}\right)$$

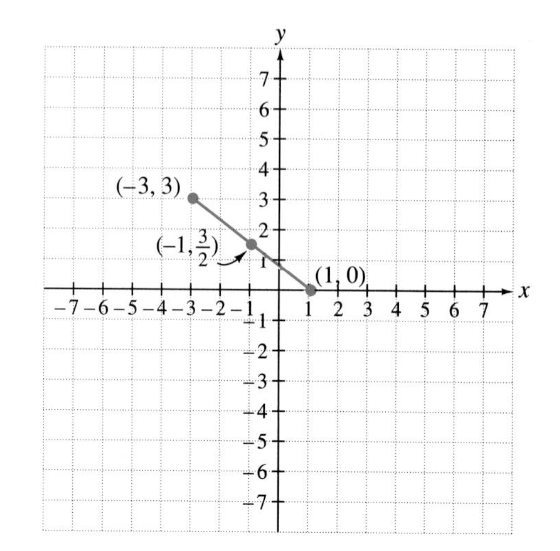

The midpoint of the segment is $\left(-1, \frac{3}{2}\right)$. ■

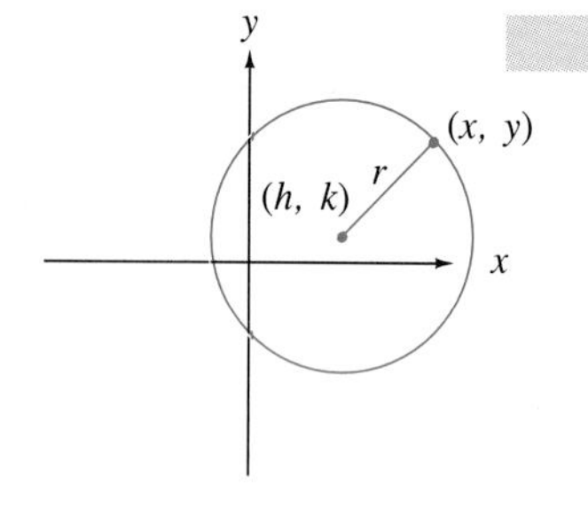

3 Another conic section is the **circle.** A circle is the set of all points in a plane that are the same distance from a fixed point called the **center.** The distance is called the **radius** of the circle. To find a standard equation for a circle, let (h, k) represent the center of the circle, and let (x, y) represent any point on the circle. The distance between (h, k) and (x, y) is defined to be the radius, r units. We can find this distance r by using the distance formula.

$$r = \sqrt{(x - h)^2 + (y - k)^2}$$

$$r^2 = (x - h)^2 + (y - k)^2 \qquad \text{Square both sides.}$$

Circle

The graph of $(x - h)^2 + (y - k)^2 = r^2$ is a circle with center (h, k) and radius r.

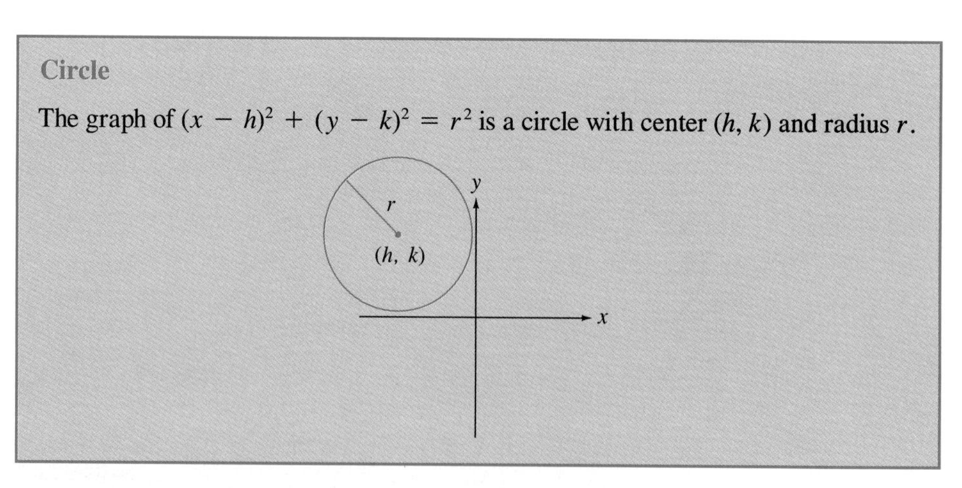

The form $(x - h)^2 + (y - k)^2 = r^2$ is called **standard form.**

EXAMPLE 7 Graph $x^2 + y^2 = 4$.

Solution: The equation can be written in standard form as

$$(x - 0)^2 + (y - 0)^2 = 2^2$$

The center of the circle is (0, 0), and the radius is 2. Its graph is shown. ■

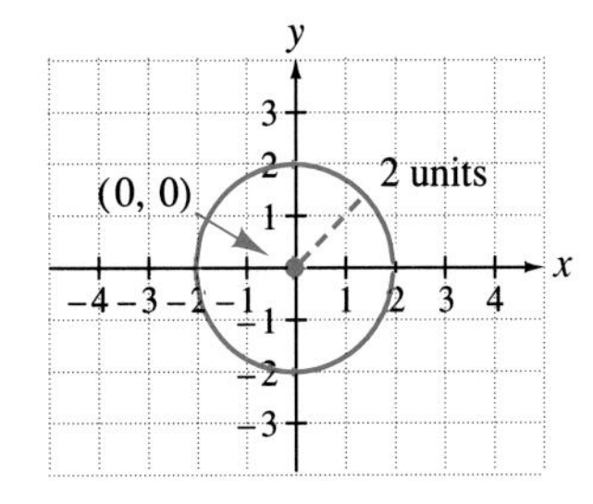

HELPFUL HINT

Notice the difference between the equation of a circle and the equation of a parabola. The equation of a circle contains both x^2 and y^2 terms with equal coefficients. The equation of a parabola has either an x^2 term or a y^2 term, but not both.

EXAMPLE 8 Graph $(x + 1)^2 + y^2 = 8$.

Solution: The equation can be written as $(x + 1)^2 + (y - 0)^2 = 8$ with $h = -1$, $k = 0$, and $r = \sqrt{8}$. The center is $(-1, 0)$, and the radius is $\sqrt{8} = 2\sqrt{2} \approx 2.8$.

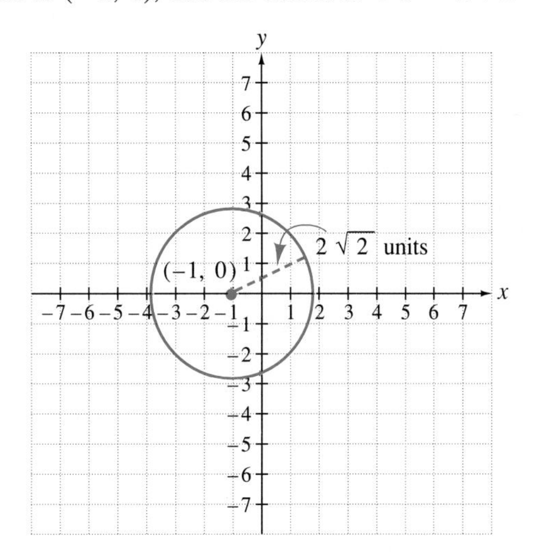

■

4 Next we practice writing the equation of a circle given its center and radius.

EXAMPLE 9 Find an equation of the circle with the center (−7, 3) and radius 10.

Solution: We are given that $h = -7$, $k = 3$, and $r = 10$. Using these values, we write the equation

$$(x - h)^2 + (y - k)^2 = r^2$$

or

$$[x - (-7)]^2 + (y - 3)^2 = 10^2$$

or

$$(x + 7)^2 + (y - 3)^2 = 100$$ ■

5 To write the equation of a circle in standard form, we complete the square on both x and y.

EXAMPLE 10 Graph $x^2 + y^2 + 4x - 8y = 16$.

Solution: Since this equation contains x^2 and y^2 terms whose coefficients are equal, its graph is a circle. To write the equation in standard form, group the terms involving x and the terms involving y, and then complete the square on each variable.

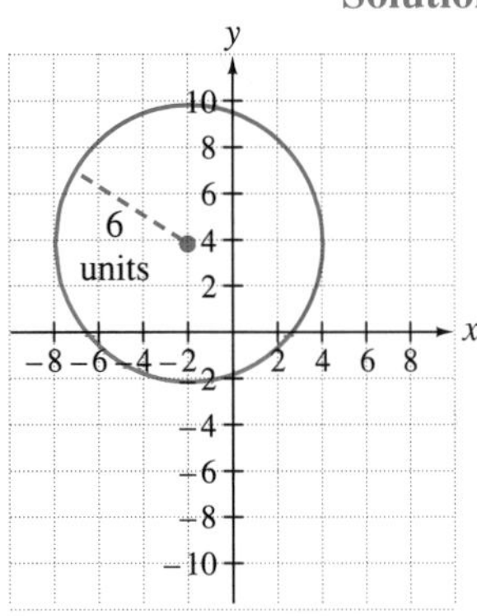

$$(x^2 + 4x) + (y^2 - 8y) = 16$$

Thus, $\frac{1}{2}(4) = 2$ and $2^2 = 4$. Also, $\frac{1}{2}(-8) = -4$ and $(-4)^2 = 16$. Add 4 and then 16 to both sides.

$$(x^2 + 4x + 4) + (y^2 - 8y + 16) = 16 + 4 + 16$$

$$(x + 2)^2 + (y - 4)^2 = 36 \qquad \text{Factor.}$$

This circle has the center $(-2, 4)$ and radius 6, as shown. ■

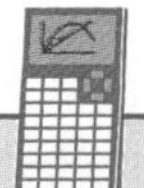

GRAPHING CALCULATOR BOX

To graph an equation such as $x^2 + y^2 = 7$ using a grapher, we first solve the equation for y.

$$x^2 + y^2 = 25$$

$$y^2 = 25 - x^2$$

$$y = \pm\sqrt{25 - x^2}$$

The graph of $y = \sqrt{25 - x^2}$ will be the top half of the circle, and the graph of $y = -\sqrt{25 - x^2}$ will be the bottom half of the circle.

To graph, press [Y=] and enter $Y_1 = \sqrt{25 - x^2}$ and $Y_2 = -\sqrt{25 - x^2}$.

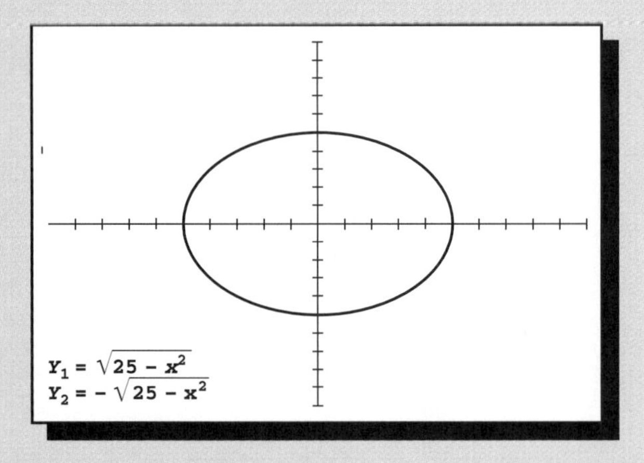

The graph does not appear to be a circle because we are currently using a standard window and the screen is rectangular. This causes the tick marks on the x-axis to be farther apart than the tick marks on the y-axis and thus creates the distorted circle. If we want the graph to appear circular, define a square window by using a feature of your grapher or redefine your window to show the x-axis from

-15 to 15 and the y-axis from -10 to 10. Using a square window, the graph appears as follows:

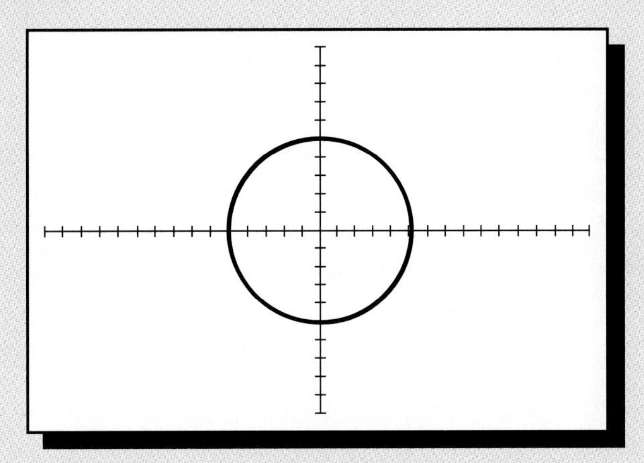

Use a grapher to graph each circle.

1. $x^2 + y^2 = 55$

2. $x^2 + y^2 = 20$

3. $7x^2 + 7y^2 - 89 = 0$

4. $3x^2 + 3y^2 - 35 = 0$

MENTAL MATH

The graph of each equation is a parabola. Determine whether the parabola opens upward, downward, to the left, or to the right.

1. $y = x^2 - 7x + 5$

2. $y = -x^2 + 16$

3. $x = -y^2 - y + 2$

4. $x = 3y^2 + 2y - 5$

5. $y = -x^2 + 2x + 1$

6. $x = -y^2 + 2y - 6$

EXERCISE SET 9.7

Sketch the graph of each equation. See Examples 1 and 2.

1. $x = 3y^2$

2. $x = -2y^2$

3. $x = (y - 2)^2 + 3$

4. $x = (y - 4)^2 - 1$

5. $y = 3(x - 1)^2 + 5$

6. $x = -4(y - 2)^2 + 2$

Find the vertex of the graph of each equation, and sketch its graph. See Examples 3 and 4.

7. $x = y^2 + 6y + 8$

8. $x = y^2 - 6y + 6$

9. $y = x^2 + 10x + 20$

10. $y = x^2 + 4x - 5$

11. $x = -2y^2 + 4y + 6$

12. $x = 3y^2 + 6y + 7$

Find the distance between each pair of points. See Example 5.

13. (5, 1) and (8, 5)

14. (2, 3) and (14, 8)

15. (−3, 2) and (1, −3)

16. (3, −2) and (−4, 1)

Find the midpoint of the line segment whose end points are given. See Example 6.

17. (6, −8), (2, 4)

18. (3, 9), (7, 11)

19. (−2, −1), (−8, 6)

20. (−3, −4), (6, −8)

Sketch the graph of each equation. See Examples 7 and 8.

21. $x^2 + y^2 = 9$

22. $x^2 + y^2 = 25$

23. $x^2 + (y - 2)^2 = 1$

24. $(x - 3)^2 + y^2 = 9$

25. $(x - 5)^2 + (y + 2)^2 = 1$

26. $(x + 3)^2 + (y + 3)^2 = 4$

Write an equation of the circle with the given center and radius. See Example 9.

27. (2, 3); 6
28. (−7, 6); 2
29. (0, 0); 2
30. (0, −6); 10
31. (−5, 4); $\sqrt{5}$
32. the origin; $\sqrt{7}$

The graph of each equation is a circle. Find the center and the radius, and then sketch. See Example 10.

33. $x^2 + y^2 + 6y = 0$
34. $x^2 + 10x + y^2 = 0$

35. $x^2 + y^2 + 2x - 4y = 4$
36. $x^2 + 6x - 4y + y^2 = 3$

Find the distance between each pair of points.

37. (6, 3) and (2, 4)
38. (8, 10) and (5, 8)
39. (−2, 1) and (−2, −9)
40. (0, −3) and (1, 5)
41. (4, −9) and (1, −8)
42. (3, −2) and (−6, −6)

Find the midpoint of the line segment whose end points are given.

43. (4, 9), (6, 5)
44. (10, 8), (4, 12)
45. (−3, 4), (−1, −2)
46. (7, 3), (−1, −3)
47. (−2, 5), (−1, 6)
48. (3, 0), (−3, 0)

Sketch the graph of each equation. If the graph is a parabola, find its vertex. If the graph is a circle, find its center and radius.

49. $x = y^2 + 2$
50. $x = y^2 - 3$
51. $y = (x + 3)^2 + 3$

52. $y = (x - 2)^2 - 2$
53. $x^2 + y^2 = 49$
54. $x^2 + y^2 = 1$

55. $x = (y - 1)^2 + 4$

56. $x = (y + 3)^2 - 1$

57. $(x + 3)^2 + (y - 1)^2 = 9$

58. $(x - 2)^2 + (y - 2)^2 = 16$

59. $x = -2(y + 5)^2$

60. $x = -(y - 1)^2$

61. $x^2 + (y + 5)^2 = 5$

62. $(x - 4)^2 + y^2 = 7$

63. $y = 3(x - 4)^2 + 2$

64. $y = 5(x + 5)^2 + 3$

65. $2x^2 + 2y^2 = \frac{1}{2}$

66. $\frac{x^2}{8} + \frac{y^2}{8} = 2$

67. $y = x^2 - 2x - 15$

68. $y = x^2 + 7x + 6$

69. $x^2 + y^2 + 6x + 10y - 2 = 0$

70. $x^2 + y^2 + 2x + 12y - 12 = 0$

71. $x = y^2 + 6y + 2$

72. $x = y^2 + 8y - 4$

73. $x^2 + y^2 - 8y + 5 = 0$

74. $x^2 - 10y + y^2 + 4 = 0$

75. $x = -2y^2 - 4y$

76. $x = -3y^2 + 30y$

77. $\frac{x^2}{3} + \frac{y^2}{3} = 2$

78. $5x^2 + 5y^2 = 25$

79. $y = 4x^2 - 40x + 105$

80. $y = 5x^2 - 20x + 16$

Solve.

81. Two surveyors need to find the distance across a lake. They placed a reference pole at point A in the diagram. Point B is 3 meters east and 1 meter north of the reference point A. Point C is 19 meters east and 13 meters north of point A. Find the distance across the lake, from B to C.

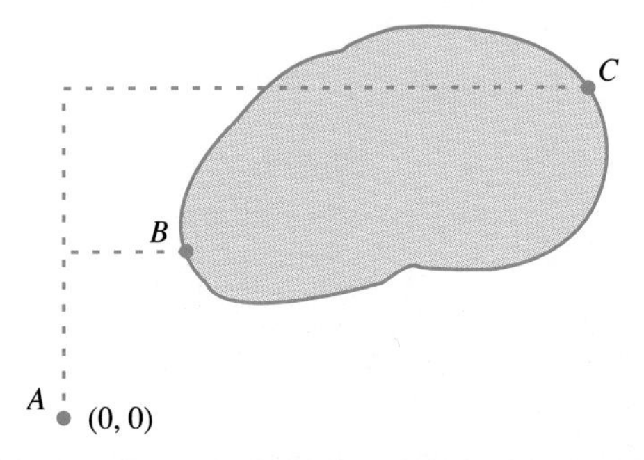

82. Determine whether the triangle with vertices (2, 6), (0, −2), and (5, 1) is an isosceles triangle.

83. Cindy Brown, an architect, is drawing plans on grid paper for a circular pool with a fountain in the middle. The paper is marked off in centimeters, where each centimeter represents a foot. On the paper, the diameter of the "pool" is 20 centimeters, and "fountain" is the point (0, 0).

a. Sketch the architect's drawing. Be sure to label the axes.

b. Write an equation describing the circular pool.

c. Cindy plans to place a circle of lights around the fountain so that each light is 5 feet from the fountain. Write an equation for the circle of lights, and sketch the circle on your drawing.

84. A bridge constructed over a bayou has a supporting arch in the shape of a parabola. Find an equation of the parabolic arch if the length of the road over the arch is 100 meters and the maximum height of the arch is 40 meters.

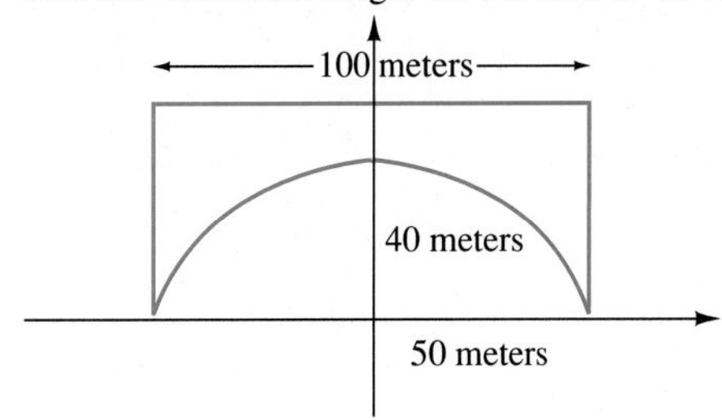

Skill Review

Graph each equation. See Section 4.3

85. $y = 2x + 5$

86. $y = -3x + 3$

87. $y = 3$

88. $x = -2$

Rationalize each denominator and simplify if possible. See Section 8.4.

89. $\dfrac{1}{\sqrt{3}}$

90. $\dfrac{\sqrt{5}}{\sqrt{8}}$

91. $\dfrac{4\sqrt{7}}{\sqrt{6}}$

92. $\dfrac{10}{\sqrt{5}}$

Writing in Mathematics

93. If you are given a list of equations of circles and parabolas and none are in standard form, explain how you would determine which is an equation of a circle and which is an equation of a parabola. Explain also how you would distinguish the upward or downward parabolas from the left or right parabolas.

9.8 The Ellipse and the Hyperbola

OBJECTIVES

Tape IA 26

1 Define the standard form equation for an ellipse.

2 Graph an ellipse.

3 Define the standard form equations for a hyperbola.

4 Graph a hyperbola.

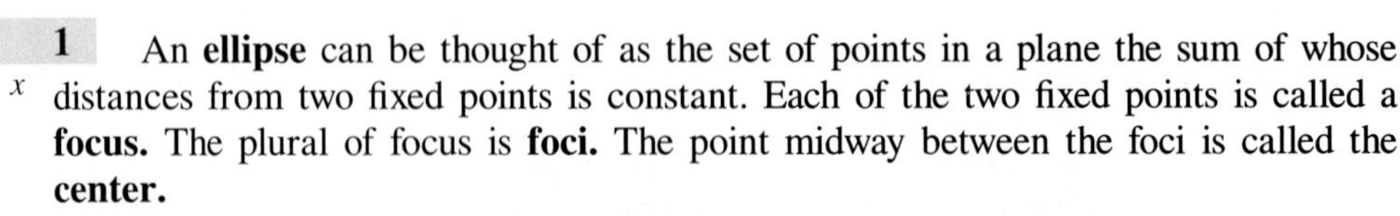

1 An **ellipse** can be thought of as the set of points in a plane the sum of whose distances from two fixed points is constant. Each of the two fixed points is called a **focus.** The plural of focus is **foci.** The point midway between the foci is called the **center.**

It can be shown that the **standard form** of an ellipse with center (0, 0) is $\dfrac{x^2}{a^2} + \dfrac{y^2}{b^2} = 1.$

Ellipse with Center (0, 0)

The graph of an equation of the form $\frac{x^2}{a^2} + \frac{y^2}{b^2} = 1$ is an ellipse with center (0, 0). The x-intercepts are a and $-a$, and the y-intercepts are b and $-b$.

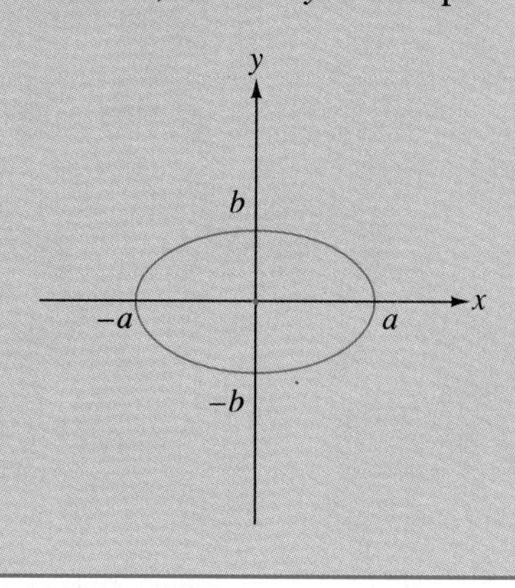

2 Next we practice graphing ellipses.

EXAMPLE 1 Graph $\frac{x^2}{9} + \frac{y^2}{16} = 1$.

Solution: The equation is of the form $\frac{x^2}{a^2} + \frac{y^2}{b^2} = 1$, with $a = 3$ and $b = 4$, so its graph is an ellipse with center (0, 0), x-intercepts 3 and -3, and y-intercepts 4 and -4 as graphed next.

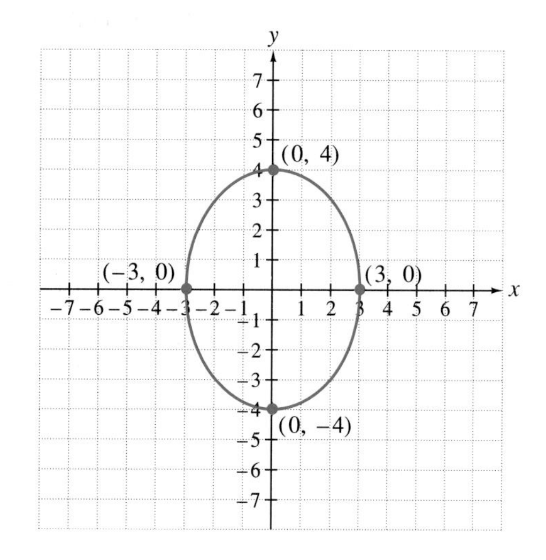

■

EXAMPLE 2 Graph the equation $4x^2 + 16y^2 = 64$.

Solution: Although this equation contains a sum of squared terms in x and y, this is not the equation of a circle since the coefficients of x^2 and y^2 are not the same. When this happens, the graph is an ellipse. Since the standard form of the equation of an ellipse has a 1 on one side, we first divide both sides of this equation by 64.

$$4x^2 + 16y^2 = 64$$

$$\frac{4x^2}{64} + \frac{16y^2}{64} = \frac{64}{64} \qquad \text{Divide both sides by 64.}$$

$$\frac{x^2}{16} + \frac{y^2}{4} = 1 \qquad \text{Simplify.}$$

We now recognize the equation of an ellipse with center (0, 0), x-intercepts 4 and -4, and y-intercepts 2 and -2.

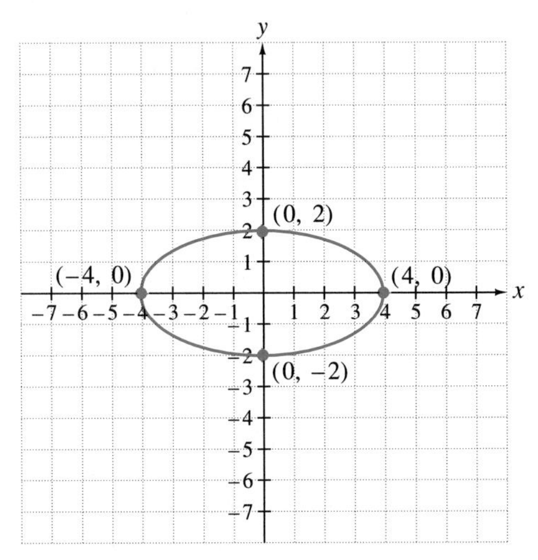

The center of an ellipse is not always (0, 0), as shown in the next example.

EXAMPLE 3 Graph $\dfrac{(x + 3)^2}{25} + \dfrac{(y - 2)^2}{36} = 1$.

Solution: This ellipse has center $(-3, 2)$. Also notice that $a = 5$ and $b = 6$. To find four points on the graph of the ellipse, first graph the center $(-3, 2)$. Since $a = 5$, count 5 units right and 5 units left of the point with coordinates $(-3, 2)$. Next, since $b = 6$, start at $(-3, 2)$ and count 6 units up and then 6 units down to find two more points on the ellipse.

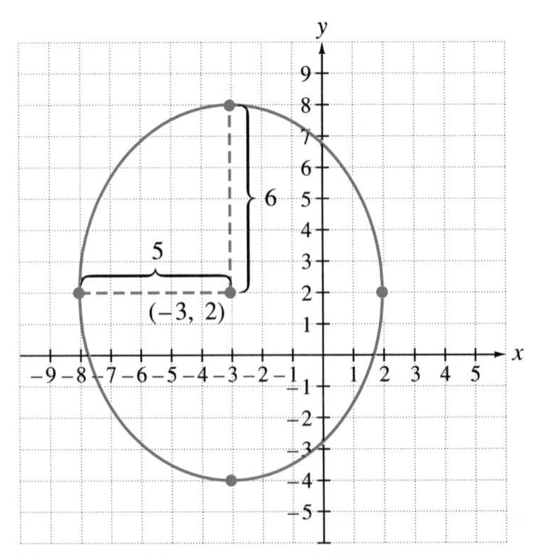

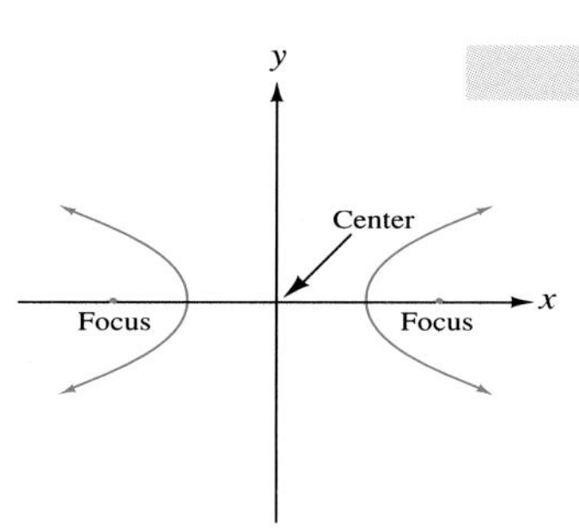

3 A final conic section is the **hyperbola.** A hyperbola is the set of points in a plane such that the absolute value of the difference of the distance from two fixed points is constant. Each of the two fixed points is called a **focus.** The point midway between the foci is called the **center.**

It can be shown that the graph of $\frac{x^2}{a^2} - \frac{y^2}{b^2} = 1$ is a hyperbola with x-intercepts a and $-a$. Also, the graph of $\frac{y^2}{b^2} - \frac{x^2}{a^2} = 1$ is a hyperbola with y-intercepts b and $-b$, as shown next. These equations are called **standard form** equations for a hyperbola.

Hyperbola with Center (0, 0)

The graph of an equation of the form $\frac{x^2}{a^2} - \frac{y^2}{b^2} = 1$ is a hyperbola with center $(0, 0)$ and x-intercepts a and $-a$.

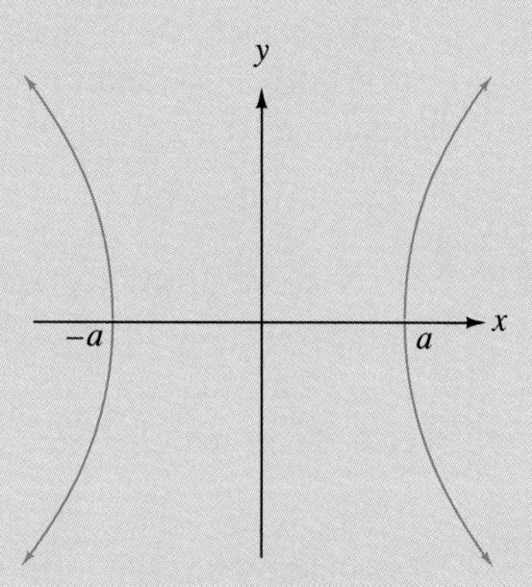

The graph of an equation of the form $\frac{y^2}{b^2} - \frac{x^2}{a^2} = 1$ is a hyperbola with center $(0, 0)$ and y-intercepts b and $-b$.

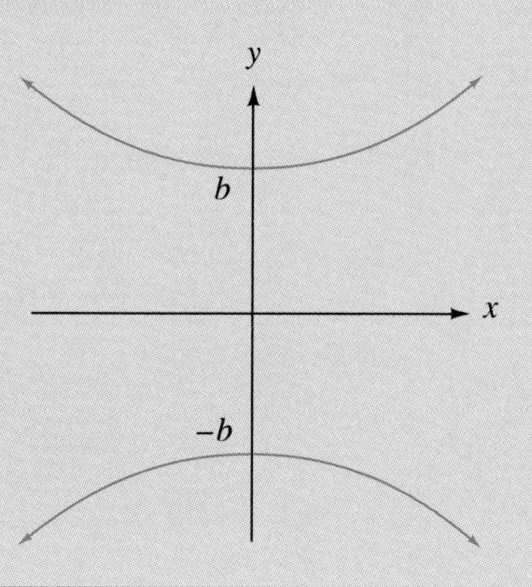

Graphing a hyperbola such as $\frac{y^2}{b^2} - \frac{x^2}{a^2} = 1$ is made easier by recognizing one of its important characteristics. Examining the following figure, notice how the sides of the branches of the hyperbola extend indefinitely and seem to more and more resemble the dashed lines in the figure. These dashed lines are called the **asymptotes** of the hyperbola.

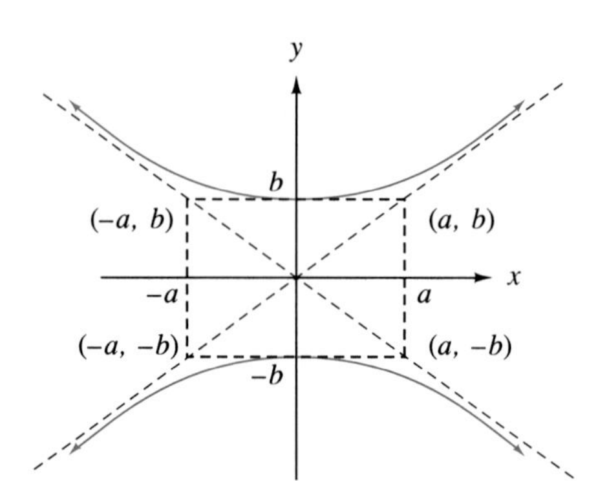

To sketch these lines, or asymptotes, draw a rectangle with vertices (a, b), $(-a, b)$, $(a, -b)$, and $(-a, -b)$. The asymptotes of this graph will be the extended diagonals of this rectangle.

4 Next we practice graphing hyperbolas.

EXAMPLE 4 Sketch the graph of $\frac{x^2}{16} - \frac{y^2}{25} = 1$.

Solution: This equation has the form $\frac{x^2}{a^2} - \frac{y^2}{b^2} = 1$, with $a = 4$ and $b = 5$. Thus, its graph is a hyperbola with center (0, 0) and x-intercepts 4 and -4. To aid in graphing the hyperbola, we first sketch its asymptotes. The diagonals of the rectangle with coordinates (4, 5), (4, −5), (−4, 5), and (−4, −5) can be used to sketch the asymptotes of the hyperbola. Then use the asymptotes to aid in sketching the hyperbola.

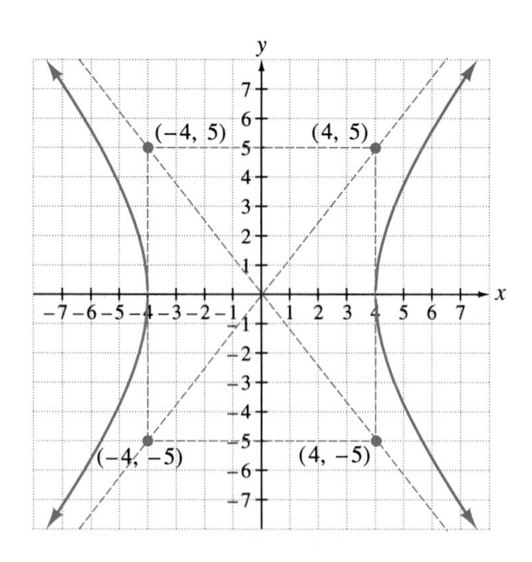

EXAMPLE 5 Sketch the graph of the equation $4y^2 - 9x^2 = 36$.

Solution: Since this is a difference of squared terms in x and y, its graph is a hyperbola, as opposed to an ellipse or a circle. The standard form of the equation of a hyperbola has a 1 on one side, so divide both sides of the equation by 36.

$$4y^2 - 9x^2 = 36$$

$$\frac{4y^2}{36} - \frac{9x^2}{36} = \frac{36}{36} \qquad \text{Divide both sides by 36.}$$

$$\frac{y^2}{9} - \frac{x^2}{4} = 1 \qquad \text{Simplify.}$$

The equation is of the form $\frac{y^2}{b^2} - \frac{x^2}{a^2} = 1$, with $a = 2$ and $b = 3$, so the hyperbola is centered at $(0, 0)$ with y-intercepts 3 and -3. The sketch of the hyperbola is shown.

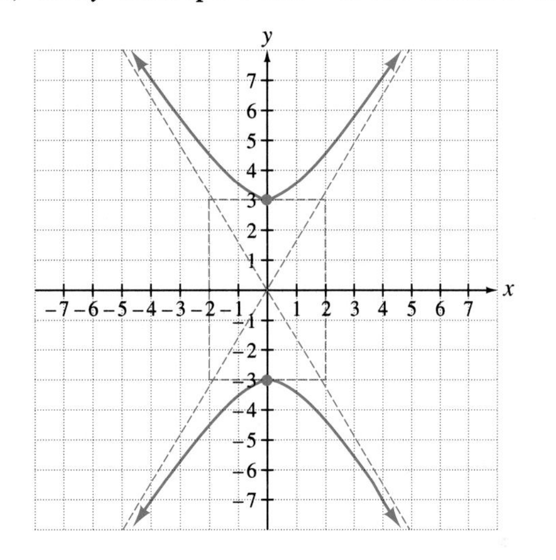

■

The following box provides a summary of conic sections.

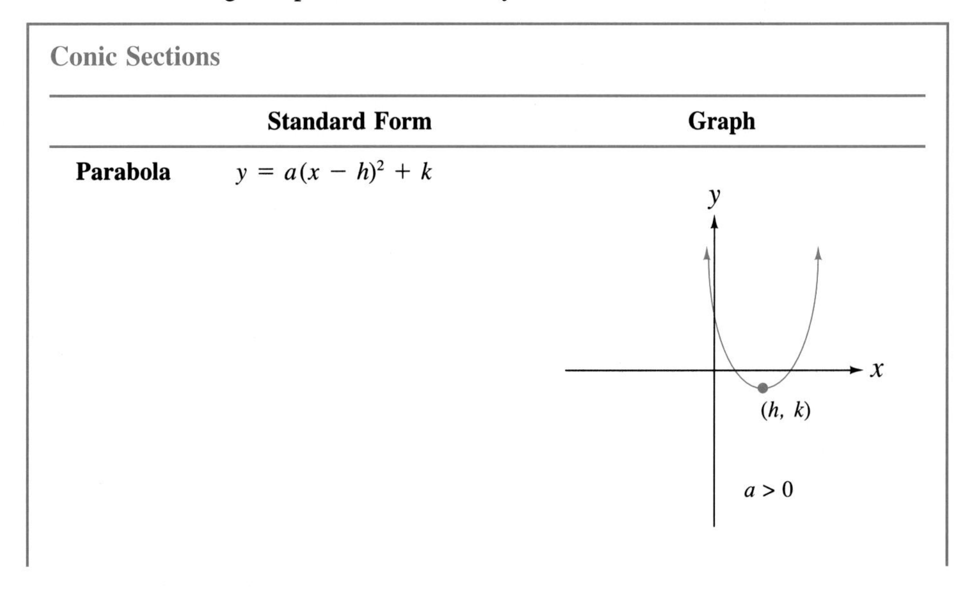

Conic Sections

	Standard Form	Graph
Parabola	$y = a(x - h)^2 + k$	(graph: y, x, (h, k), $a > 0$)

Conic Sections (cont.)

	Standard Form	Graph
Parabola	$x = a(y - k)^2 + h$	$a < 0$

y
x
(h, k)

Circle $(x - h)^2 + (y - k)^2 = r^2$

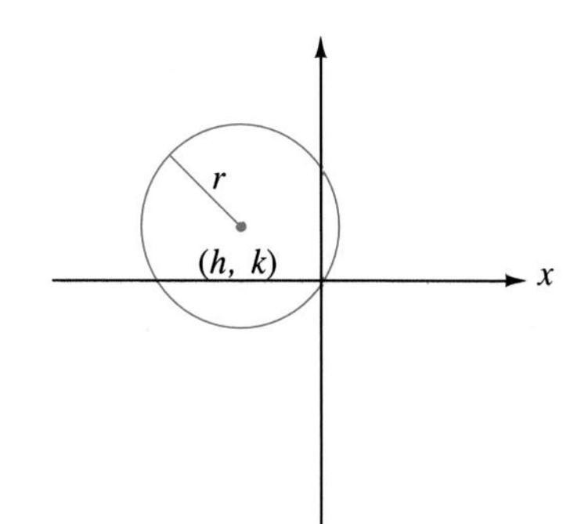

Ellipse $\frac{x^2}{a^2} + \frac{y^2}{b^2} = 1$

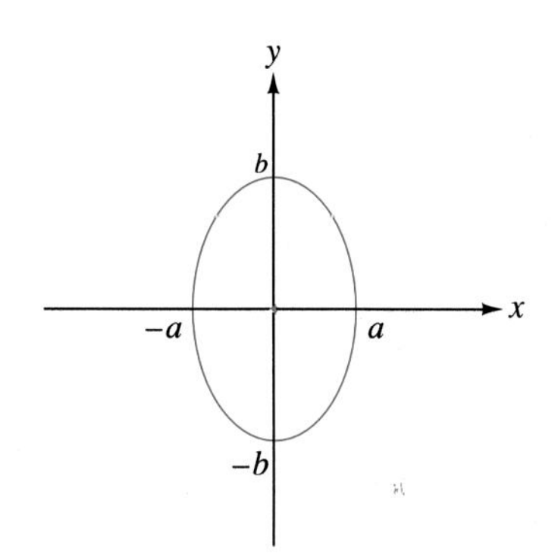

Hyperbola $\frac{x^2}{a^2} - \frac{y^2}{b^2} = 1$

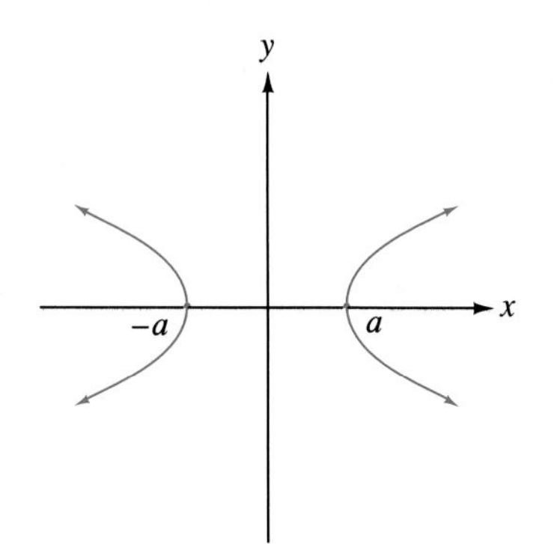

	Standard Form	Graph
Hyperbola	$\frac{y^2}{b^2} - \frac{x^2}{a^2} = 1$	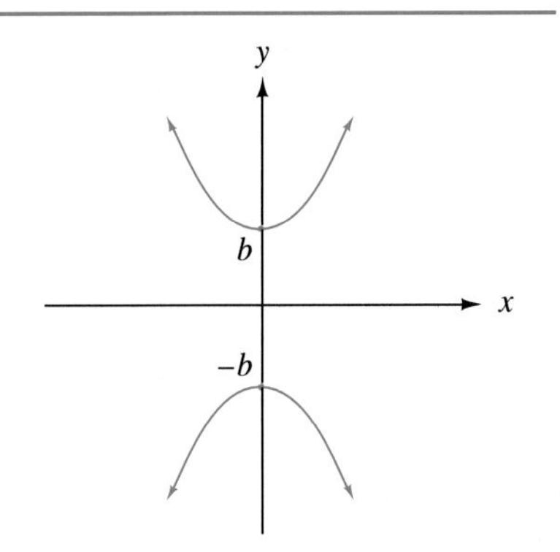

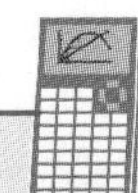

GRAPHING CALCULATOR BOX

To find the graph of an ellipse using a grapher, use the same procedure as for graphing a circle. For example, to graph $x^2 + 3y^2 = 22$, first solve for y.

$$3y^2 = 22 - x^2$$

$$y^2 = \frac{22 - x^2}{3}$$

$$y = \pm\sqrt{\frac{22 - x^2}{3}}$$

Next press the [Y=] key and enter $Y_1 = \sqrt{\frac{22 - x^2}{3}}$ and $Y_2 = -\sqrt{\frac{22 - x^2}{3}}$. The graph appears as follows:

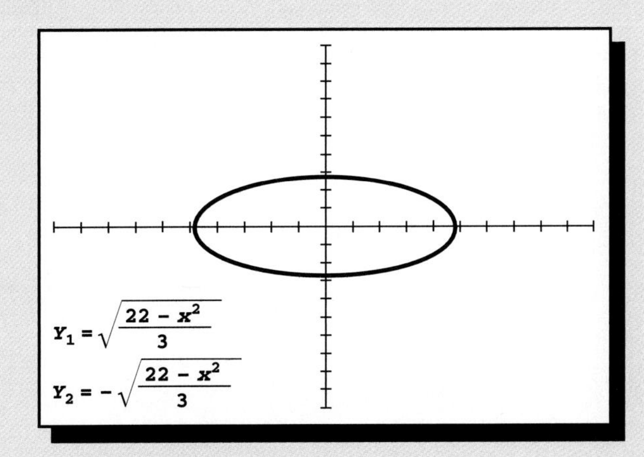

Use a grapher to graph each ellipse.

1. $10x^2 + y^2 = 32$

2. $20x^2 + 5y^2 = 100$

3. $7.3x^2 + 15.5y^2 = 95.2$

4. $18.8x^2 + 36.1y^2 = 205.8$

EXERCISE SET 9.8

Sketch the graph of each equation. See Example 1.

1. $\dfrac{x^2}{4} + \dfrac{y^2}{25} = 1$

2. $\dfrac{x^2}{9} + y^2 = 1$

3. $\dfrac{x^2}{16} + \dfrac{y^2}{9} = 1$

4. $x^2 + \dfrac{y^2}{4} = 1$

Sketch the graph of each equation. See Example 2.

5. $9x^2 + 4y^2 = 36$

6. $x^2 + 4y^2 = 16$

7. $4x^2 + 25y^2 = 100$

8. $36x^2 + y^2 = 36$

Sketch the graph of each equation. See Example 3.

9. $\dfrac{(x+1)^2}{36} + \dfrac{(y-2)^2}{49} = 1$

10. $\dfrac{(x-3)^2}{9} + \dfrac{(y+3)^2}{16} = 1$

11. $\dfrac{(x-1)^2}{4} + \dfrac{(y-1)^2}{25} = 1$

12. $\dfrac{(x+3)^2}{16} + \dfrac{(y+2)^2}{4} = 1$

Sketch the graph of each equation. See Example 4.

13. $\dfrac{x^2}{4} - \dfrac{y^2}{9} = 1$

14. $\dfrac{x^2}{36} - \dfrac{y^2}{36} = 1$

15. $\dfrac{y^2}{25} - \dfrac{x^2}{16} = 1$

16. $\dfrac{y^2}{25} - \dfrac{x^2}{49} = 1$

Sketch the graph of each equation. See Example 5.

17. $x^2 - 4y^2 = 16$

18. $4x^2 - y^2 = 36$

19. $16y^2 - x^2 = 16$

20. $4y^2 - 25x^2 = 100$

Identify whether each equation when graphed will be a parabola, circle, ellipse, or hyperbola. Sketch the graph of each equation.

21. $(x - 7)^2 + (y - 2)^2 = 4$

22. $y = x^2 + 4$

23. $y = x^2 + 12x + 36$

24. $\frac{x^2}{4} + \frac{y^2}{9} = 1$

25. $\frac{y^2}{9} - \frac{x^2}{9} = 1$

26. $\frac{x^2}{16} - \frac{y^2}{4} = 1$

27. $\frac{x^2}{16} + \frac{y^2}{4} = 1$

28. $x^2 + y^2 = 16$

29. $x = y^2 + 4y - 1$

30. $x = -y^2 + 6y$

31. $9x^2 - 4y^2 = 36$

32. $9x^2 + 4y^2 = 36$

33. $\frac{(x-1)^2}{49} + \frac{(y+2)^2}{25} = 1$

34. $y^2 = x^2 + 16$

35. $\left(x + \frac{1}{2}\right)^2 + \left(y - \frac{1}{2}\right)^2 = 1$

36. $y = -2x^2 + 4x - 3$

Solve.

37. A planet's orbit about the Sun can be described as an ellipse. Consider the Sun as the origin of a rectangular coordinate system. Suppose that the x-intercepts of the elliptical path of the planet are $\pm 130{,}000{,}000$ and that the y-intercepts are $\pm 125{,}000{,}000$. Write the equation of the elliptical path of the planet.

38. Comets orbit the Sun in elongated ellipses. Again consider the Sun as the origin of a rectangular coordinate system. Suppose that the equation of the path of a comet is

$$\frac{(x - 1{,}782{,}000{,}000)^2}{(3.42)(10^{23})} + \frac{(y - 356{,}400{,}000)^2}{(1.368)(10^{22})} = 1$$

Find the center of the path of the comet.

Skill Review

Solve each inequality. See Section 3.6.

39. $x < 5$ and $x < 1$

40. $x < 5$ or $x < 1$

41. $2x - 1 \geq 7$ or $-3x \leq -6$

42. $2x - 1 \geq 7$ and $-3x \leq -6$

Perform indicated operations. See Sections 5.1 and 5.3.

43. $(2x^3)(-4x^2)$

44. $2x^3 - 4x^3$

45. $-5x^2 + x^2$

46. $(-5x^2)(x^2)$

Show that the ordered pair given is a solution to both equations.

47. $(3, 7)$; $\begin{cases} x = y - 4 \\ y = 2x + 1 \end{cases}$

48. $(2, 4)$; $\begin{cases} 2x = y \\ y = 5x - 6 \end{cases}$

49. $(-1, 2)$; $\begin{cases} x + 2y = 3 \\ -x - y = -1 \end{cases}$

50. $(-5, -3)$; $\begin{cases} x + y = -8 \\ 3x - 2y = -9 \end{cases}$

A Look Ahead

Sketch the graph of each equation. See the following example.

EXAMPLE Sketch the graph of $\dfrac{(x-2)^2}{25} - \dfrac{(y-1)^2}{9} = 1$.

Solution: This hyperbola has center (2, 1). Notice that $a = 5$ and $b = 3$.

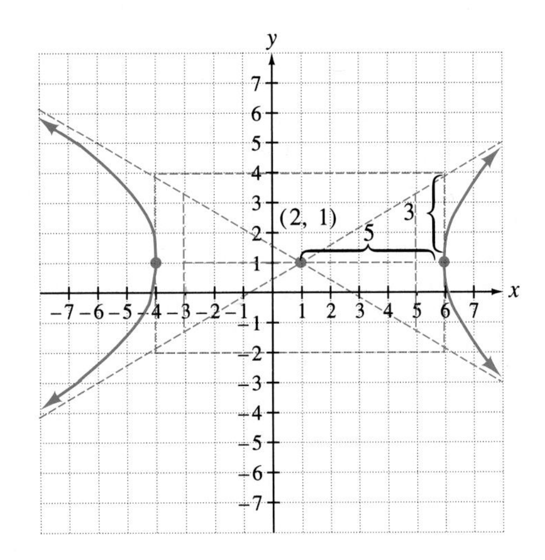

■

51. $\dfrac{(x-1)^2}{4} - \dfrac{(y+1)^2}{25} = 1$

52. $\dfrac{(x+2)^2}{9} - \dfrac{(y-1)^2}{4} = 1$

53. $\dfrac{y^2}{16} - \dfrac{(x+3)^2}{9} = 1$

54. $\dfrac{(y+4)^2}{4} - \dfrac{x^2}{25} = 1$

55. $\dfrac{(x+5)^2}{16} - \dfrac{(y+2)^2}{25} = 1$

56. $\dfrac{(x-3)^2}{9} - \dfrac{(y-2)^2}{4} = 1$

CHAPTER 9 GLOSSARY

A **circle** is the set of all points in a plane that are the same distance from a fixed point called the **center.** The distance is called the **radius** of the circle.

The process of writing a quadratic equation so that one side is a perfect square trinomial is called **completing the square.**

Conic sections derive their name because each conic section is the intersection of a right circular cone and a plane. The circle, parabola, ellipse, and hyperbola are the conic sections.

In the quadratic formula $x = \dfrac{-b \pm \sqrt{b^2 - 4ac}}{2a}$, the radicand $b^2 - 4ac$ is called the **discriminant** because, by knowing its value, we can **discriminate** among the possibilities and number of solutions of a quadratic equation.

An **ellipse** can be thought of as the set of points in a plane the sum of whose distances from two fixed points is constant. Each of the two fixed points is called a **focus.** The plural of focus is **foci.** The point midway between the foci is called the **center.**

A **hyperbola** is the set of points in a plane such that the absolute value of the difference of the distance from two fixed points is constant. Each of the two fixed points is called a **focus.** The point midway between the foci is called the **center.**

A **quadratic** or **second-degree equation** is an equation that can be written in the form $ax^2 + bx + c = 0$, where a, b, and c are real numbers and $a \neq 0$.

A **quadratic function** is a function that can be written in the form $f(x) = ax^2 + bx + c$, where a, b, and c are real numbers and $a \neq 0$.

A **quadratic inequality** is an inequality that can be written so that one side is a quadratic expression and the other side is 0.

CHAPTER 9 SUMMARY

SQUARE ROOT PROPERTY (9.1)

If b is a real number and if $a^2 = b$, then $a = \pm\sqrt{b}$.

QUADRATIC FORMULA (9.2)

A quadratic equation written in the form $ax^2 + bx + c = 0$ has the solutions

$$x = \frac{-b \pm \sqrt{b^2 - 4ac}}{2a}$$

GRAPH OF A QUADRATIC FUNCTION (9.5)

The graph of a quadratic function written in the form $f(x) = a(x - h)^2 + k$ is a parabola with vertex (h, k). If $a > 0$, the parabola opens upward, and if $a < 0$, the parabola opens downward. The axis of symmetry is the line whose equation is $x = h$.

VERTEX FORMULA (9.6)

The graph of $f(x) = ax^2 + bx + c$, $a \neq 0$, is a parabola with vertex

$$\left(\frac{-b}{2a}, f\left(\frac{-b}{2a}\right)\right)$$

CONIC SECTIONS (9.7) and (9.8)

	Standard Form	Graph
Parabola	$y = a(x - h)^2 + k$	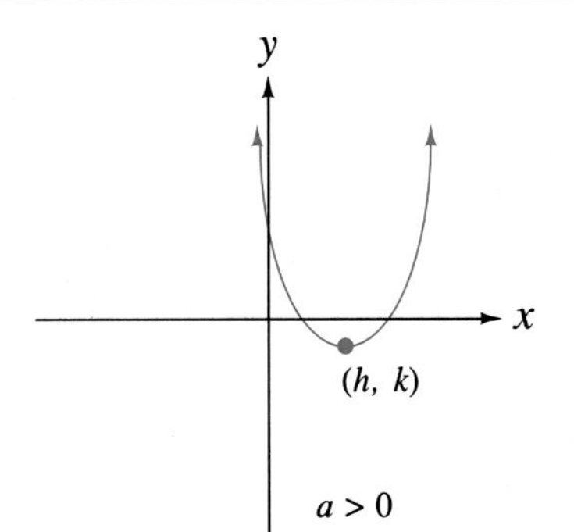
	$x = a(y - k)^2 + h$	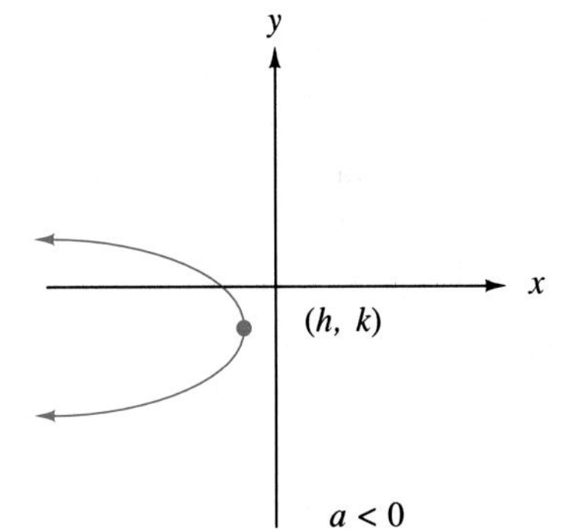
Circle	$(x - h)^2 + (y - k)^2 = r^2$	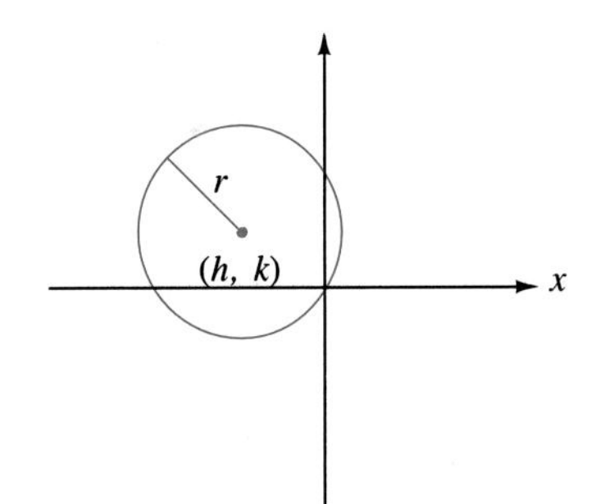
Ellipse	$\frac{x^2}{a^2} + \frac{y^2}{b^2} = 1$	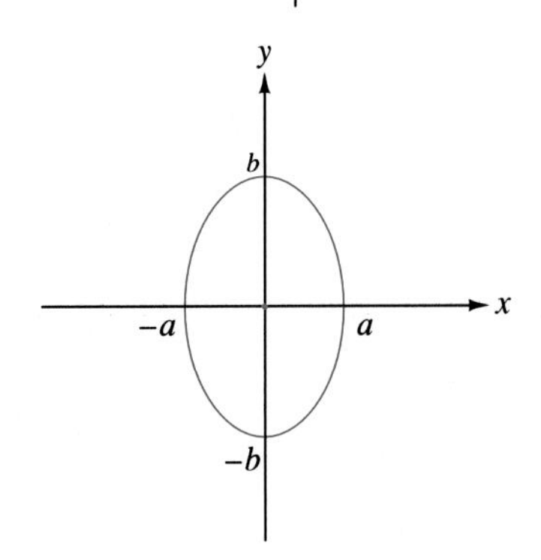

CONIC SECTIONS (9.7) and (9.8) (cont.)

	Standard Form	Graph
Hyperbola	$\frac{x^2}{a^2} - \frac{y^2}{b^2} = 1$	y, x, $-a$, a
	$\frac{y^2}{b^2} - \frac{x^2}{a^2} = 1$	y, x, b, $-b$

CHAPTER 9 REVIEW

(9.1) *Solve by factoring.*

1. $x^2 - 15x + 14 = 0$ **2.** $x^2 - x - 30 = 0$ **3.** $10x^2 = 3x + 4$ **4.** $7a^2 = 29a + 30$

Solve using the square root property.

5. $4m^2 = 196$ **6.** $9y^2 = 36$ **7.** $(9n + 1)^2 = 9$ **8.** $(5x - 2)^2 = 2$

Solve by completing the square.

9. $z^2 + 3z + 1 = 0$ **10.** $x^2 + x + 7 = 0$ **11.** $(2x + 1)^2 = x$ **12.** $(3x - 4)^2 = 10x$

(9.2) *If the discriminant of a quadratic equation has the given value, determine the number and type of solutions of the equation.*

13. -8 **14.** 48 **15.** 100 **16.** 0

Solve by using the quadratic formula.

17. $x^2 - 16x + 64 = 0$

18. $x^2 + 5x = 0$

19. $x^2 + 11 = 0$

20. $2x^2 + 3x = 5$

21. $6x^2 + 7 = 5x$

22. $9a^2 + 4 = 2a$

23. $(5a - 2)^2 - a = 0$

24. $(2x - 3)^2 = x$

(9.3) *Solve each equation for the variable.*

25. $x^3 = 27$

26. $y^3 = -64$

27. $\frac{5}{x} + \frac{6}{x - 2} = 3$

28. $\frac{7}{8} = \frac{8}{x^2}$

29. $x^4 - 21x^2 - 100 = 0$

30. $5(x + 3)^2 - 19(x + 3) = 4$

31. $x^{2/3} - 6x^{1/3} + 5 = 0$

32. $x^{2/3} - 6x^{1/3} = -8$

33. $a^6 - a^2 = a^4 - 1$

34. $(m - 4)^5 - 2(m - 4)^3 = (m - 4)^2 - 2$

Solve.

35. Find two consecutive even whole numbers such that the sum of their squares is 100.

36. A daycare center must have a play yard with an area of 1200 square feet. The rectangular shape of the property requires the length to be three times the width. Find the dimensions of the yard.

37. Cadets graduating from military school usually toss their hats high into the air at the end of the ceremony. One cadet threw her hat so that its distance $d(t)$ in feet above the ground t seconds after it was thrown was $d(t) = -16t^2 + 30t + 6$.

a. Find the distance above the ground of the hat 1 second after it was thrown.

b. Find the time it takes the hat to hit the ground.

38. The product of a number and 3 less than twice the number is 405. Find the number.

39. The middle number of three consecutive positive even integers is 26 less than $\frac{1}{10}$ the product of the other two. Find the integers.

40. The hypotenuse of an isosceles right triangle is 6 centimeters longer than either of the legs. Find the length of the legs.

41. A negative number decreased by its reciprocal is $-\frac{24}{5}$. Find the number.

42. One force of 30 pounds pulls an object to the left and another force of 40 pounds pulls the same object downward. Find the resulting force of the pulling. That is, find the length of the diagonal of the rectangle in the sketch.

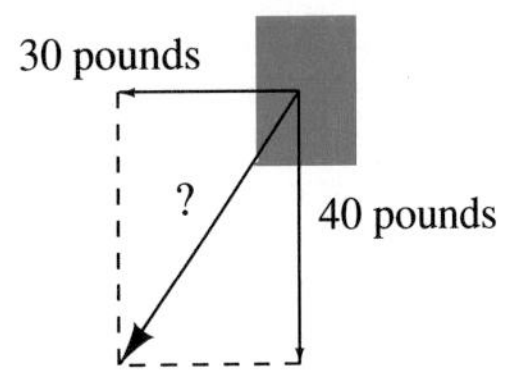

(9.4) *Solve each inequality for x. Write each solution set in interval notation and graph the solution set.*

43. $2x^2 - 50 \le 0$

44. $\frac{1}{4}x^2 < \frac{1}{16}$

45. $(2x - 3)(4x + 5) \ge 0$

46. $(x^2 - 16)(x^2 - 1) > 0$

47. $\dfrac{x - 5}{x - 6} < 0$

48. $\dfrac{x(x + 5)}{4x - 3} \geq 0$

49. $\dfrac{(4x + 3)(x - 5)}{x(x + 6)} > 0$

50. $(x + 5)(x - 6)(x + 2) \leq 0$

51. $\dfrac{x + 4}{x - \frac{1}{2}} \leq 0$

52. $\dfrac{(5x + 6)(x - 3)}{x(6x - 5)} < 0$

53. $\dfrac{x^2 + 10x + 25}{x^2 - 6x + 9} > 0$

54. $\dfrac{x^2 - 6x + 9}{x^2 + 10x + 25} < 0$

(9.5) *Sketch the graph of each function.*

55. $g(x) = x^2 + 5$

56. $H(x) = (x + 3)^2$

57. $f(x) = (x - 4)^2 - 2$

58. $f(x) = -3(x - 1)^2 + 1$

(9.6) *Sketch the graph of each function. Find the vertex and the intercepts.*

59. $f(x) = x^2 + 10x + 25$

60. $f(x) = -x^2 + 6x - 9$

61. $f(x) = 4x^2 - 1$

62. $f(x) = -5x^2 + 5$

63. Find two numbers whose product is as large as possible, given that their sum is 420.

64. Write an equation of a quadratic function whose graph is a parabola with vertex $(-3, 7)$ that passes through the origin.

(9.7) *Find the distance between each pair of points.*

65. $(-6, 3)$ and $(8, 4)$

66. $(3, 5)$ and $(8, 9)$

67. $(-4, -6)$ and $(-1, 5)$

68. $(-1, 5)$ and $(2, -3)$

Find the midpoint of the line segment whose end points are given.

69. $(2, 6)$ and $(-12, 4)$

70. $(-3, 8)$ and $(11, 24)$

71. $(-6, -5)$ and $(-9, 7)$

72. $(4, -6)$ and $(-15, 2)$

Write an equation of the circle with the given center and radius.

73. center $(-4, 4)$, radius 3

74. center $(5, 0)$, diameter 10

75. center $(-7, -9)$, radius $\sqrt{11}$

76. center $(0, 0)$, diameter 7

Sketch the graph of the equation. If the graph is a circle, find its center. If the graph is a parabola, find its vertex.

77. $x^2 + y^2 = 7$

78. $x = 2(y - 5)^2 + 4$

79. $x = -(y + 2)^2 + 3$

80. $(x - 1)^2 + (y - 2)^2 = 4$

81. $y = -x^2 + 4x + 10$

82. $x = -y^2 - 4y + 6$

83. $x^2 + y^2 + 2x + y = \frac{3}{4}$

84. $x^2 + y^2 + 3y = \frac{7}{4}$

85. $4x^2 + 4y^2 + 16x + 8y = 1$

86. $3x^2 + 6x + 3y^2 = 9$

87. $y = x^2 + 6x + 9$

88. $x = y^2 + 6y + 9$

Solve.

89. Write an equation of the circle centered at $(5.6, -2.4)$ with **diameter** 6.2.

(9.8) *Sketch the graph of each equation.*

90. $x^2 + \frac{y^2}{4} = 1$

91. $x^2 - \frac{y^2}{4} = 1$

92. $\frac{y^2}{4} - \frac{x^2}{16} = 1$

93. $\frac{y^2}{4} + \frac{x^2}{16} = 1$

94. $\frac{x^2}{5} + \frac{y^2}{5} = 1$

95. $\frac{x^2}{5} - \frac{y^2}{5} = 1$

96. $-5x^2 + 25y^2 = 125$

97. $4y^2 + 9x^2 = 36$

98. $\frac{(x-2)^2}{4} + (y-1)^2 = 1$

99. $\frac{(x+3)^2}{9} + \frac{(y-4)^2}{25} = 1$

100. $x^2 - y^2 = 1$

101. $36y^2 - 49x^2 = 1764$

102. $y^2 = x^2 + 9$

103. $x^2 = 4y^2 - 16$

104. $100 - 25x^2 = 4y^2$

Sketch the graph of each equation.

105. $y = x^2 + 4x + 6$

106. $y^2 = x^2 + 6$

107. $y^2 + x^2 = 4x + 6$

108. $y^2 + 2x^2 = 4x + 6$

109. $x^2 + y^2 - 8y = 0$

110. $x - 4y = y^2$

111. $x^2 - 4 = y^2$

112. $x^2 = 4 - y^2$

113. $6(x - 2)^2 + 9(y + 5)^2 = 36$

114. $36y^2 = 576 + 16x^2$

115. $\frac{x^2}{16} - \frac{y^2}{25} = 1$

116. $3(x - 7)^2 + 3(y + 4)^2 = 1$

CHAPTER 9 TEST

Solve each equation for the variable.

1. $5x^2 - 2x = 7$

2. $x^3 = -8$

3. $m^2 - m + 8 = 0$

4. $u^2 - 6u + 2 = 0$

5. $7x^2 + 8x + 1 = 0$

6. $a^2 - 3a = 5$

7. $\frac{4}{x+2} + \frac{2x}{x-2} = \frac{6}{x^2-4}$

8. $x^4 - 8x^2 - 9 = 0$

9. $x^6 + 1 = x^4 + x^2$

10. $(x + 1)^2 - 15(x + 1) + 56 = 0$

Solve the equation for the variable by completing the square.

11. $x^2 - 6x = -2$

12. $2a^2 + 5 = 4a$

Solve each inequality for x. Write the solution set in interval notation and then graph the solution set.

13. $(6x + 7)(7x - 12) \leq 0$

14. $(x^2 - 16)(x^2 - 25) > 0$

15. $\frac{2x - 11}{3x + 12} \geq 0$

16. $\frac{7x - 14}{x^2 - 9} \leq 0$

Solve.

17. Find the distance between the points $(-6, 3)$ and $(-8, -7)$.

18. Find the midpoint of the line segment whose end points are $(-2, -5)$ and $(-6, 12)$.

Sketch the graph of each equation. Label centers, intercepts, vertices, and asymptotes.

19. $y = 2(x + 3)^2 + 1$

20. $x^2 + y^2 = 36$

21. $x^2 - y^2 = 36$

22. $16x^2 + 9y^2 = 144$

23. $y = x^2 - 8x + 16$

24. $x^2 + y^2 + 6x = 16$

25. $x = y^2 + 8y - 3$

26. $\frac{(x - 4)^2}{16} + \frac{(y - 3)^2}{9} = 1$

Solve.

27. A stone is dropped from a bridge. The height in feet, $s(t)$, of the stone above the water t seconds after it is dropped is a function given by the equation $s(t) = -16t^2 + 256$.
a. Find the height of the bridge.
b. Find the time it takes the stone to hit the water.

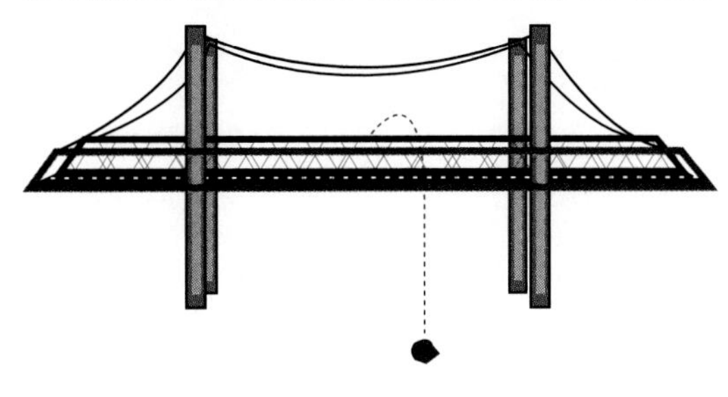

28. Find three consecutive odd whole numbers such that the product of the first and the third number is thirty-nine more than forty-two times the second number.

CHAPTER 9 CUMULATIVE REVIEW

1. Solve $y = mx + b$ for x.

2. To estimate the number of people in Jackson, population 50,000, who have no health insurance, 250 people were polled and 39 of those polled had no insurance. If this rate remains constant, how many people in the city might we expect to be uninsured?

3. Find 72% of 200.

4. Solve $|x - 3| = |5 - x|$.

5. Solve for x: $\left|2x - \dfrac{1}{10}\right| < -13$.

6. Graph the equation $y = |x|$.

7. Which of the following graphs are graphs of functions?

a.

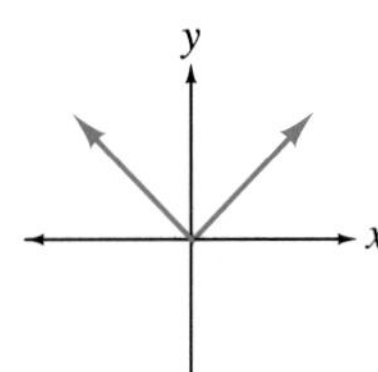

b.

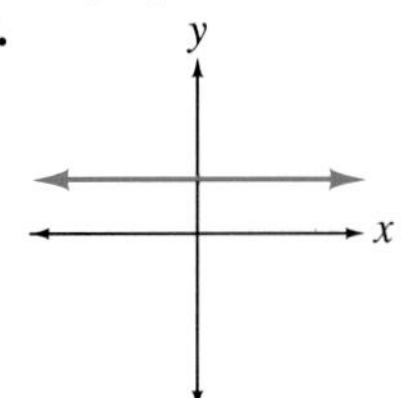

c.

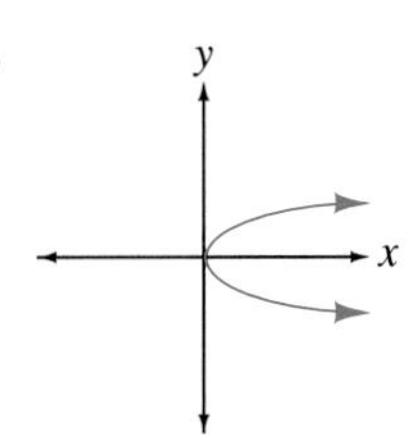

d.

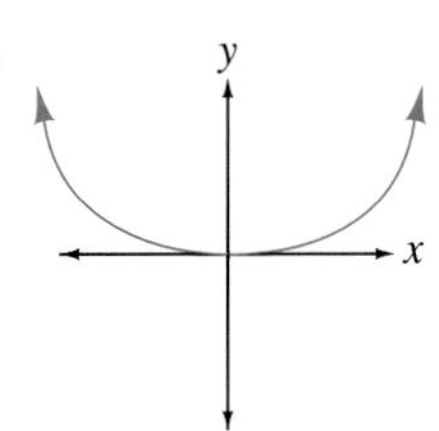

e.

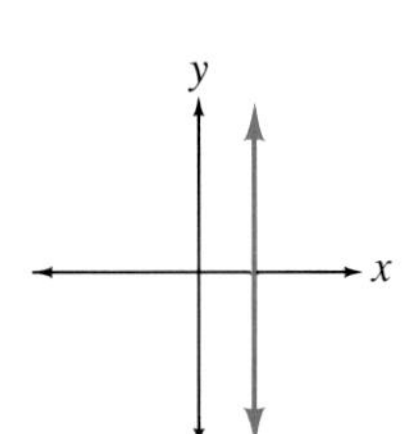

f.

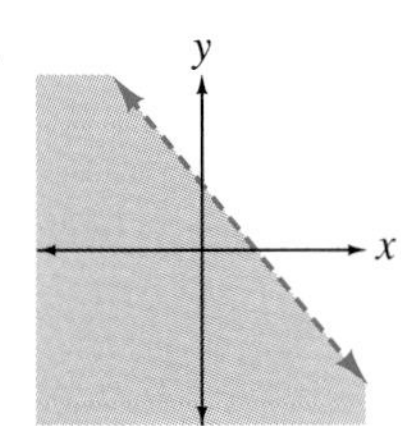

8. Find the slope and the y-intercept of the line whose equation is $3x - 4y = 4$.

9. Simplify each expression.
a. $(st)^4$ **b.** $\left(\dfrac{m}{n}\right)^7$ **c.** $(2a)^3$
d. $(-5x^2y^3z)^2$ **e.** $\left(\dfrac{2x^4}{3y^5}\right)^4$

10. Add.
a. $(7x^3y - xy^3 + 11) + (6x^3y - 4)$
b. $(3a^3 - b + 2a - 5) + (a + b + 5)$

11. Find the following products.
a. $(x - 3)(x + 3)$
b. $(4y + 1)(4y - 1)$
c. $(x^2 + 2y)(x^2 - 2y)$

12. Factor $ab - 6a + 2b - 12$.

13. Factor $y^3 - 27$.

14. Solve $x^3 = 4x$.

15. Find the domain of each rational function.
a. $f(x) = \dfrac{8x^3 + 7x^2 + 20}{2}$

b. $g(x) = \dfrac{5x^2 - 3}{x - 1}$

c. $f(x) = \dfrac{7x - 2}{2x^2 + 7x + 6}$

16. Find the number that, when subtracted from the numerator and added to the denominator of $\frac{9}{19}$, changes $\frac{9}{19}$ into a fraction equivalent to $\frac{1}{3}$.

17. Solve $\sqrt{-10x - 1} + 3x = 0$ for x.

18. Find the length of the unknown leg of the following right triangle.

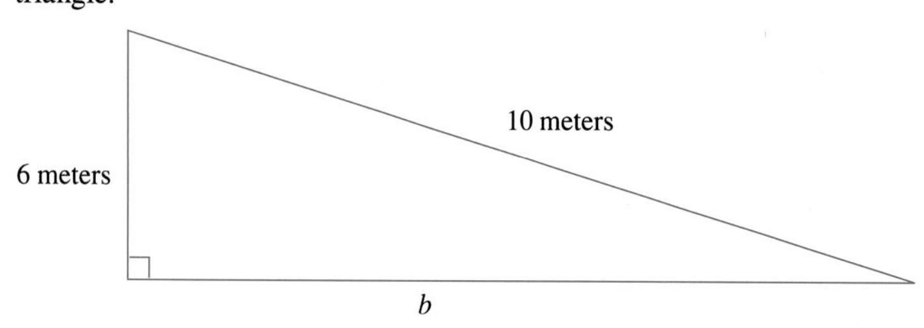

19. Use the square root property to solve $(x + 1)^2 = 12$ for x.

20. Solve $2x^2 - 4x = 3$.

21. Solve $(x + 2)(x - 1)(x - 5) \leq 0$.

22. Sketch the graph of each function.

a. $G(x) = (x - 3)^2$ **b.** $F(x) = (x + 1)^2$

23. Graph $f(x) = x^2 - 4x - 12$.

24. Find the distance between the pair of points $(2, -5)$ and $(1, -4)$.

25. Graph $x^2 + y^2 + 4x - 8y = 16$.

CHAPTER 10

Systems of Equations and Inequalities

When does an airplane in flight pass the point of no return?

INTRODUCTION

In this chapter, two or more equations in two or more variables are solved simultaneously. Such a collection of equations is called a **system of equations.** Systems of equations are good mathematical models for many real-world problems because these problems may involve several related patterns.

10.1
Solving Systems of Linear Equations in Two Variables

OBJECTIVES

1 Solve a system by graphing.

2 Solve a system by substitution.

3 Solve a system by elimination.

Tape IA 27

1 Two linear equations in the same two variables form a **system of linear equations.**

Systems of Linear Equations in Two Variables

$$\begin{cases} x - 2y = -7 \\ 3x + y = 0 \end{cases} \qquad \begin{cases} x = 5 \\ x + \dfrac{y}{2} = 9 \end{cases} \qquad \begin{cases} x - 3 = 2y + 6 \\ y = 1 \end{cases}$$

Recall that a solution of an equation in two variables is an ordered pair, (x, y), that makes the equation true. A **solution of a system** of two equations in two variables is an ordered pair, (x, y), that makes both equations true.

We can estimate the solutions of a system by graphing each equation on the same coordinate system and estimating the coordinates of any point of intersection. Since the graph of a linear equation in two variables is a line, graphing two such equations yields two lines in a plane.

EXAMPLE 1 Solve each system by graphing. If the system has just one solution, estimate the solution.

a. $\begin{cases} x + y = 2 \\ 3x - y = -2 \end{cases}$

b. $\begin{cases} x - 2y = 4 \\ x = 2y \end{cases}$

c. $\begin{cases} 2x + 4y = 10 \\ x + 2y = 5 \end{cases}$

Solution: **a.** $\begin{cases} x + y = 2 \\ 3x - y = -2 \end{cases}$

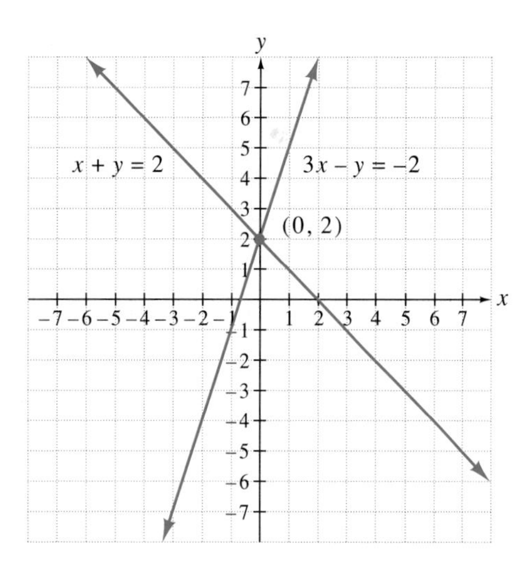

These lines intersect in one point as shown. The coordinates of the point of intersection appear to be $\{(0, 2)\}$. Check this estimated solution by replacing x with 0 and y with 2 in **both** equations.

$x + y = 2$	First equation.	$3x - y = -2$	Second equation.
$0 + 2 = 2$	Let $x = 0$ and $y = 2$.	$3 \cdot 0 - 2 = -2$	Let $x = 0$ and $y = 2$.
$2 = 2$	True.	$-2 = -2$	True.

The ordered pair (0, 2) does satisfy both equations. We therefore conclude that the solution set of the system is $\{(0, 2)\}$. A system that has at least one solution, such as this one, is said to be **consistent.**

b. $\begin{cases} x - 2y = 4 \\ x = 2y \end{cases}$

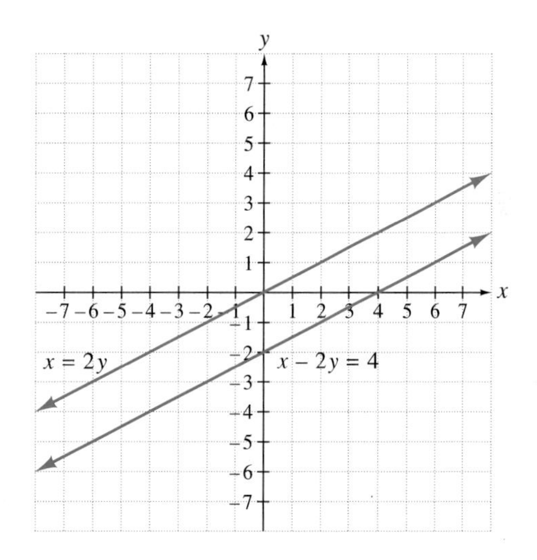

The lines appear to be parallel. To be sure, write each equation in point–slope form, $y = mx + b$.

$x - 2y = 4$	First equation.	$x = 2y$	Second equation.
$-2y = -x + 4$	Subtract x from both sides.	$\frac{1}{2}x = y$	Divide both sides by 2.

$$y = \frac{1}{2}x - 2 \qquad \text{Divide both sides by } -2. \qquad y = \frac{1}{2}x$$

The graphs of these equations have the same slope, $\frac{1}{2}$, but different y-intercepts, so these lines are parallel. Therefore, the system has no solution. Systems that have no solution are said to be **inconsistent.**

c. $\begin{cases} 2x + 4y = 10 \\ x + 2y = 5 \end{cases}$

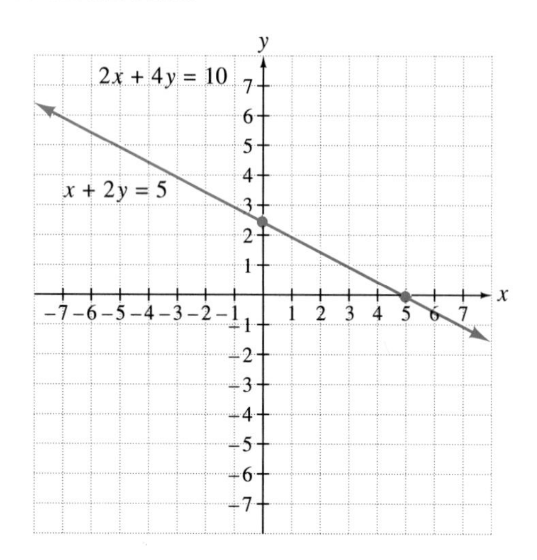

The graph of each equation is the same line. To see this, notice that if both sides of the second equation are multiplied by 2, the result is the first equation. This means that the equations have identical solutions. Any ordered pair solution of one equation will satisfy the other equation also. Therefore, there are an infinite number of solutions to this system. These equations are said to be **dependent equations.** The solution set of the system is $\{(x, y) \mid x + 2y = 5\}$ or, equivalently, $\{(x, y) \mid 2x + 4y = 10\}$, since the lines describe identical ordered pairs. Written this way, the solution set is read "the set of all ordered pairs (x, y), such that $2x + 4y = 10$." ■

We can summarize the information discovered in Example 1 as follows.

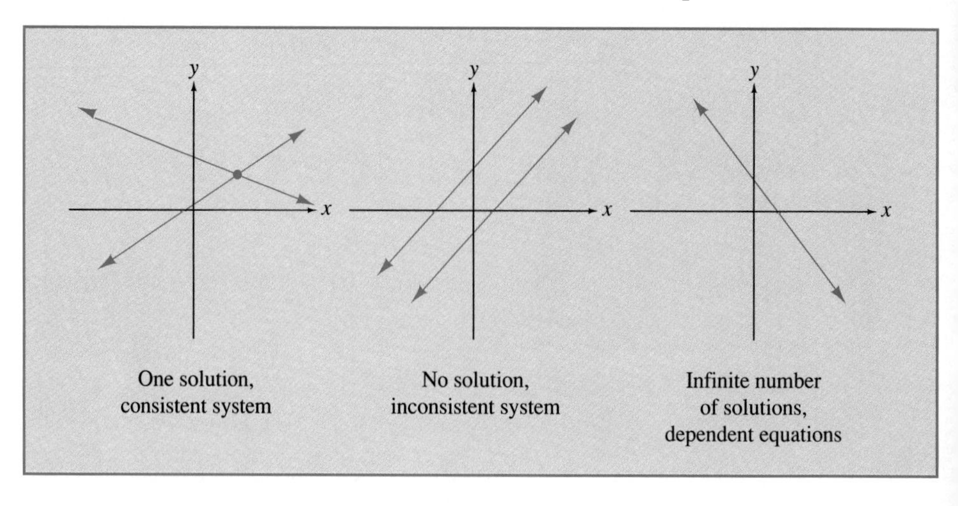

2 Graphing the equations of a system is not a reliable method for finding solutions of a system. We turn instead to two algebraic methods of solving systems. We use the first method, the **substitution method,** to solve the system

$$\begin{cases} y = x + 5 & \text{First equation.} \\ 3x = 2y - 9 & \text{Second equation.} \end{cases}$$

Remember that we are looking for an ordered pair, if there is one, that satisfies both equations. Satisfying the first equation, $y = x + 5$, means that y must be $x + 5$. **Substituting** the expression $x + 5$ for y in the second equation yields an equation in one variable, which we can solve for x.

$3x = 2y - 9$ Second equation.

$3x = 2(x + 5) - 9$ Replace y with $x + 5$ in the second equation.

$3x = 2x + 10 - 9$

$x = 1$

The x-coordinate of the solution of the system is 1. The y-coordinate is the y-value corresponding to the x-value, 1. Choose either equation of the system and solve for y when x is 1.

$y = x + 5$ First equation.

$= 1 + 5$ Let $x = 1$.

$= 6$

The y-coordinate is 6, so the solution of the system is (1, 6). This means that when both equations are graphed, the one point of intersection occurs at the point with coordinates (1, 6).

To Solve a System of Two Equations by the Substitution Method

Step 1 Solve one of the equations for one of its variables.

Step 2 Substitute the expression for the variable found in step 1 into the other equation.

Step 3 Find the value of one variable by solving the equation from step 2.

Step 4 Find the value of the other variable by substituting the value found in step 3 in any equation of the system.

Step 5 Check the ordered pair solution in **both** of the original equations.

EXAMPLE 2 Use the substitution method to solve the system

$$\begin{cases} 2x + 4y = -6 & \text{First equation.} \\ x - 2y = -5 & \text{Second equation.} \end{cases}$$

Solution: We begin by solving the second equation for x; the coefficient 1 of the x-term keeps us from introducing tedious fractions. The equation $x - 2y = -5$ solved for x is $x = 2y - 5$. Substitute $2y - 5$ for x in the first equation.

$$2x + 4y = -6 \qquad \text{First equation.}$$

$$2(2y - 5) + 4y = -6 \qquad \text{Substitute } 2y - 5 \text{ for } x.$$

$$4y - 10 + 4y = -6$$

$$8y = 4$$

$$y = \frac{4}{8} = \frac{1}{2} \qquad \text{Solve for } y.$$

The y-coordinate of the solution is $\frac{1}{2}$. To find the x-coordinate, replace y with $\frac{1}{2}$ in the equation $x = 2y - 5$.

$$x = 2y - 5$$

$$x = 2\left(\frac{1}{2}\right) - 5 = 1 - 5 = -4$$

The solution set is $\left\{\left(-4, \frac{1}{2}\right)\right\}$. To check, see that $\left(-4, \frac{1}{2}\right)$ satisfies both equations of the system. ■

EXAMPLE 3 Use substitution to solve the system

$$\begin{cases} -\dfrac{x}{6} + \dfrac{y}{2} = \dfrac{1}{2} \\ \dfrac{x}{3} - \dfrac{y}{6} = -\dfrac{3}{4} \end{cases}$$

Solution: First multiply each equation by its least common denominator to write this system as an equivalent system without fractions. We multiply the first equation by 6 and the second equation by 12. Then

$$\begin{cases} 6\left(-\dfrac{x}{6} + \dfrac{y}{2}\right) = 6\left(\dfrac{1}{2}\right) \\ 12\left(\dfrac{x}{3} - \dfrac{y}{6}\right) = 12\left(-\dfrac{3}{4}\right) \end{cases}$$

simplifies to

$$\begin{cases} -x + 3y = 3 & \text{First equation.} \\ 4x - 2y = -9 & \text{Second equation.} \end{cases}$$

We now solve the first equation for x.

$$-x + 3y = 3 \qquad \text{First equation.}$$

$$3y - 3 = x \qquad \text{Solve for } x.$$

Next replace x with $3y - 3$ in the second equation.

$$4x - 2y = -9 \qquad \text{Second equation.}$$

$$4(3y - 3) - 2y = -9$$

$$12y - 12 - 2y = -9$$

$$10y = 3$$

$$y = \frac{3}{10} \qquad \text{Solve for } y.$$

The y-coordinate is $\frac{3}{10}$. To find the x-coordinate, replace y with $\frac{3}{10}$ in the equation $x = 3y - 3$. Then

$$x = 3\left(\frac{3}{10}\right) - 3 = \frac{9}{10} - 3 = \frac{9}{10} - \frac{30}{10} = -\frac{21}{10}$$

The solution set is $\left\{\left(-\frac{21}{10}, \frac{3}{10}\right)\right\}$. ■

3 The **elimination** or **addition method** is a second algebraic technique for solving systems of equations. For this method, we rely on a version of the addition property of equality, which states that "equals added to equals are equal." In symbols,

$$\text{if } A = B \text{ and } C = D, \text{ then } A + C = B + D$$

EXAMPLE 4 Use the addition method to solve the system

$$\begin{cases} x - 5y = -12 & \text{First equation.} \\ -x + y = 4 & \text{Second equation.} \end{cases}$$

Solution: Since the left side of each equation is equal to the right side, we add equal quantities by adding the left sides of the equations and the right sides of the equations. This sum gives us an equation in one variable, y, which we can solve for y.

$$x - 5y = -12 \qquad \text{First equation.}$$

$$\underline{-x + y = 4} \qquad \text{Second equation.}$$

$$-4y = -8 \qquad \text{Add.}$$

$$y = 2 \qquad \text{Solve for } y.$$

The y-coordinate of the solution is 2. To find the corresponding x-coordinate, replace y with 2 in either original equation of the system.

$$-x + y = 4 \qquad \text{Second equation.}$$

$$-x + 2 = 4 \qquad \text{Let } y = 2.$$

$$-x = 2$$

$$x = -2$$

The solution set is $\{(-2, 2)\}$. Check to see that $(-2, 2)$ satisfies both equations of the system. ■

To Solve a System of Two Linear Equations by the Addition Method

Step 1 Rewrite each equation in standard form, $Ax + By = C$.

Step 2 If necessary, multiply one or both equations by some nonzero number so that the coefficient of one variable in one equation is the opposite of its coefficient in the other equation.

Step 3 Add the equations.

Step 4 Find the value of one variable by solving the equation from step 3.

Step 5 Find the value of the second variable by substituting the value found in step 4 into either of the original equations.

Step 6 Check the proposed solution in **both** of the original equations.

EXAMPLE 5 Use the addition method to solve the system

$$\begin{cases} 3x + \dfrac{y}{2} = 2 \\ 6x + y = 5 \end{cases}$$

Solution: If we add the two equations, the sum will still be an equation in two variables. Notice that if we multiply both sides of the first equation by -2, the coefficients of x will be opposites. Then

$$\begin{cases} -2\left(3x + \dfrac{y}{2}\right) = -2\,(2) \\ 6x + y = 5 \end{cases} \quad \text{simplifies to} \quad \begin{cases} -6x - y = -4 \\ 6x + y = 5 \end{cases}$$

Next add the left sides and add the right sides.

$$\begin{array}{r} -6x - y = -4 \\ 6x + y = 5 \\ \hline 0 = 1 \end{array} \qquad \text{False.}$$

The resulting equation, $0 = 1$, is false for all values of y or x. Thus, the system has no solution. The solution set is { } or Ø.

This system is inconsistent, and the graphs of the equations are parallel lines. ■

EXAMPLE 6 Use the addition method to solve the system

$$\begin{cases} 3x - 2y = 10 \\ 4x - 3y = 15 \end{cases}$$

Solution: To eliminate y when the equations are added, multiply both sides of the first equation by 3 and both sides of the second equation by -2. Then

$$\begin{cases} 3\,(3x - 2y) = 3\,(10) \\ -2\,(4x - 3y) = -2\,(15) \end{cases} \quad \text{simplifies to} \quad \begin{cases} 9x - 6y = 30 \\ -8x + 6y = -30 \end{cases}$$

$$x = 0 \quad \text{Add the equations.}$$

To find y, let $x = 0$ in either equation of the system.

$$3x - 2y = 10$$

$$3(0) - 2y = 10 \qquad \text{Let } x = 0.$$

$$-2y = 10$$

$$y = -5$$

The solution set is $\{(0, -5)\}$. Check to see that $(0, -5)$ satisfies both equations. ■

EXAMPLE 7 Use the addition method to solve

$$\begin{cases} -5x - 3y = 9 \\ 10x + 6y = -18 \end{cases}$$

Solution: To eliminate x when the equations are added, multiply both sides of the first equation by 2. Then

$$\begin{cases} 2(-5x - 3y) = 2(9) \\ 10x + 6y = -18 \end{cases} \text{ simplifies to } \begin{cases} -10x - 6y = 18 \\ 10x + 6y = -18 \end{cases}$$

$$0 = 0 \quad \text{Add the equations.}$$

The resulting equation, $0 = 0$, is true for all possible values of y or x. Notice in the original system that if both sides of the first equation are multiplied by 2, the result is the second equation. This means that the two equations are equivalent and that they have the same solution set. Thus, the equations of this system are dependent, and the solution set of the system is

$$\{(x, y) \mid -5x - 3y = 9\} \quad \text{or} \quad \{(x, y) \mid 10x + 6y = -18\}$$

The equations in the system are dependent, and the graphs are the same line. ■

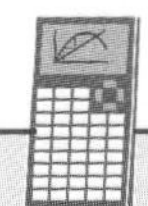

GRAPHING CALCULATOR BOX

We may use a grapher to approximate solutions of systems of equations by graphing each equation on the same set of axes and approximating any points of intersection.

Solve each system of equations. Approximate the solutions to two decimal places.

1. $\begin{cases} y = -2.68x + 1.21 \\ y = 5.22x - 1.68 \end{cases}$

2. $\begin{cases} y = 4.25x + 3.89 \\ y = -1.88x + 3.21 \end{cases}$

3. $\begin{cases} 4.3x - 2.9y = 5.6 \\ 8.1x + 7.6y = -14.1 \end{cases}$

4. $\begin{cases} -3.6x - 8.6y = 10 \\ -4.5x + 9.6y = -7.7 \end{cases}$

EXERCISE SET 10.1

Solve each system by graphing. See Example 1.

1. $\begin{cases} x + y = 1 \\ x - 2y = 4 \end{cases}$

2. $\begin{cases} 2x - y = 8 \\ x + 3y = 11 \end{cases}$

3. $\begin{cases} 2y - 4 = 0 \\ x + 2y = 5 \end{cases}$

4. $\begin{cases} 4x - y = 6 \\ x - y = 0 \end{cases}$

5. $\begin{cases} 3x - y = 4 \\ 6x - 2y = 4 \end{cases}$

6. $\begin{cases} -x + 3y = 6 \\ 3x - 9y = 9 \end{cases}$

Solve each system of equations by the substitution method. See Example 2.

7. $\begin{cases} x + y = 10 \\ y = 4x \end{cases}$

8. $\begin{cases} 5x + 2y = -17 \\ x = 3y \end{cases}$

9. $\begin{cases} 4x - y = 9 \\ 2x + 3y = -27 \end{cases}$

10. $\begin{cases} 3x - y = 6 \\ -4x + 2y = -8 \end{cases}$

Solve each system of equations by the substitution method. See Example 3.

11. $\begin{cases} \frac{1}{2}x + \frac{3}{4}y = -\frac{1}{4} \\ \frac{3}{4}x - \frac{1}{4}y = 1 \end{cases}$

12. $\begin{cases} \frac{2}{5}x + \frac{1}{5}y = -1 \\ x + \frac{2}{5}y = -\frac{8}{5} \end{cases}$

13. $\begin{cases} \frac{x}{3} + y = \frac{4}{3} \\ -x + 2y = 11 \end{cases}$

14. $\begin{cases} \frac{x}{8} - \frac{y}{2} = 1 \\ \frac{x}{3} - y = 2 \end{cases}$

Solve each system of equations by the elimination method. See Example 4.

15. $\begin{cases} 2x - 4y = 0 \\ x + 2y = 5 \end{cases}$

16. $\begin{cases} 2x - 3y = 0 \\ 2x + 6y = 3 \end{cases}$

17. $\begin{cases} 5x + 2y = 1 \\ x - 3y = 7 \end{cases}$

18. $\begin{cases} 6x - y = -5 \\ 4x - 2y = 6 \end{cases}$

Solve each system of equations by the elimination method. See Example 6.

19. $\begin{cases} 5x - 2y = 27 \\ -3x + 5y = 18 \end{cases}$

20. $\begin{cases} 3x + 4y = 2 \\ 2x + 5y = -1 \end{cases}$

21. $\begin{cases} 3x - 5y = 11 \\ 2x - 6y = 2 \end{cases}$

22. $\begin{cases} 6x - 3y = -3 \\ 4x + 5y = -9 \end{cases}$

Solve each system of equations. See Examples 5 and 7.

23. $\begin{cases} x - 2y = 4 \\ 2x - 4y = 4 \end{cases}$

24. $\begin{cases} -x + 3y = 6 \\ 3x - 9y = 9 \end{cases}$

25. $\begin{cases} 3x + y = 1 \\ 2y = 2 - 6x \end{cases}$

26. $\begin{cases} y = 2x - 5 \\ 8x - 4y = 20 \end{cases}$

Solve each system of equations.

27. $\begin{cases} 2x + 5y = 8 \\ 6x + y = 10 \end{cases}$

28. $\begin{cases} x - 4y = -5 \\ -3x - 8y = 0 \end{cases}$

29. $\begin{cases} x + y = 1 \\ x - 2y = 4 \end{cases}$

30. $\begin{cases} 2x - y = 8 \\ x + 3y = 11 \end{cases}$

31. $\begin{cases} \frac{1}{3}x + y = \frac{4}{3} \\ -\frac{1}{4}x - \frac{1}{2}y = -\frac{1}{4} \end{cases}$

32. $\begin{cases} \frac{3}{4}x - \frac{1}{2}y = -\frac{1}{2} \\ x + y = -\frac{3}{2} \end{cases}$

33. $\begin{cases} 2x + 6y = 8 \\ 3x + 9y = 12 \end{cases}$

34. $\begin{cases} x = 3y - 1 \\ 2x - 6y = -2 \end{cases}$

35. $\begin{cases} 4x + 2y = 5 \\ 2x + y = -1 \end{cases}$

36. $\begin{cases} 3x + 6y = 15 \\ 2x + 4y = 3 \end{cases}$

37. $\begin{cases} 10y - 2x = 1 \\ 5y = 4 - 6x \end{cases}$

38. $\begin{cases} 3x + 4y = 0 \\ 7x = 3y \end{cases}$

39. $\begin{cases} \frac{3}{4}x + \frac{5}{2}y = 11 \\ \frac{1}{16}x - \frac{3}{4}y = -1 \end{cases}$

40. $\begin{cases} \frac{2}{3}x + \frac{1}{4}y = -\frac{3}{2} \\ \frac{1}{2}x - \frac{1}{4}y = -2 \end{cases}$

41. $\begin{cases} x = 3y + 2 \\ 5x - 15y = 10 \end{cases}$

42. $\begin{cases} y = \frac{1}{7}x + 3 \\ x - 7y = -21 \end{cases}$

43. $\begin{cases} 2x - y = -1 \\ y = -2x \end{cases}$

44. $\begin{cases} x = \frac{1}{5}y \\ x - y = -4 \end{cases}$

45. $\begin{cases} 2x = 6 \\ y = 5 - x \end{cases}$

46. $\begin{cases} x = 3y + 4 \\ -y = 5 \end{cases}$

47. $\begin{cases} \frac{x + 5}{2} = \frac{6 - 4y}{3} \\ \frac{3x}{5} = \frac{21 - 7y}{10} \end{cases}$

48. $\begin{cases} \frac{y}{5} = \frac{8 - x}{2} \\ x = \frac{2y - 8}{3} \end{cases}$

49. $\begin{cases} 4x - 7y = 7 \\ 12x - 21y = 24 \end{cases}$

50. $\begin{cases} 2x - 5y = 12 \\ -4x + 10y = 20 \end{cases}$

51. $\begin{cases} \frac{2}{3}x - \frac{3}{4}y = -1 \\ -\frac{1}{6}x + \frac{3}{8}y = 1 \end{cases}$

52. $\begin{cases} \frac{1}{2}x - \frac{1}{3}y = -3 \\ \frac{1}{8}x + \frac{1}{6}y = 0 \end{cases}$

53. $\begin{cases} 0.7x - 0.2y = -1.6 \\ 0.2x - y = -1.4 \end{cases}$

54. $\begin{cases} -0.7x + 0.6y = 1.3 \\ 0.5x - 0.3y = -0.8 \end{cases}$

Solve.

55. The sum of two numbers is 45 and one number is twice the other. Find the numbers.

56. The difference between two numbers is 5. Twice the smaller number added to five times the larger number is 53. Find the numbers.

The double line graph below shows the number of pounds of fishery products from U.S. domestic catch and from imports. Use this graph to answer Exercises 57 and 58.

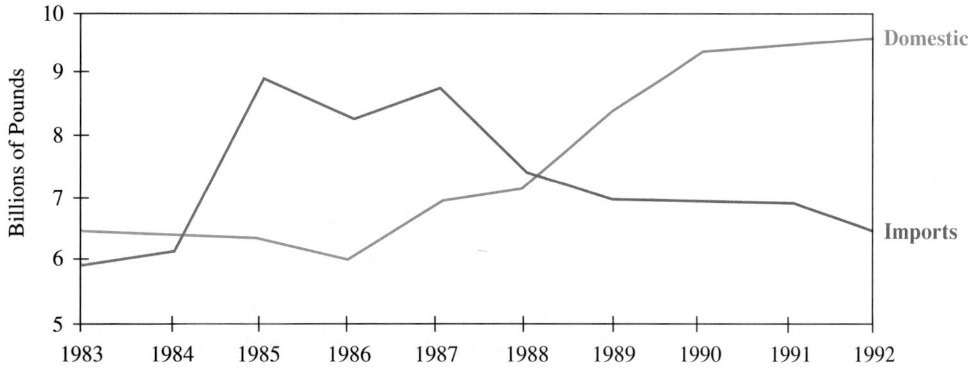

Source: U.S. Bureau of the Census, *Statistical Abstract of the United States: 1994,* 113th ed., Washington, DC, 1994.

57. In what year(s) is the number of pounds of fishery products imported equal to the number of pounds of domestic catch?

58. In what year(s) is the number of pounds of fishery products imported greater than the number of pounds of domestic catch?

Skill Review

Graph. See Sections 9.7 and 9.8.

59. $x^2 + y^2 = 9$

60. $\dfrac{x^2}{9} + \dfrac{y^2}{16} = 1$

61. $\dfrac{x^2}{25} + \dfrac{y^2}{4} = 1$

62. $(x - 2)^2 + (y + 4)^2 = 16$

Simplify. See Section 8.7.

63. $\sqrt{-4}$

64. $\sqrt{-25}$

65. $\sqrt{-20}$

66. $\sqrt{-18}$

Add each pair of equations.

67. $\begin{array}{r} 2x - y + 3z = 7 \\ \underline{5x + y - z = 10} \end{array}$

68. $\begin{array}{r} 5x + 2y - 3z = 20 \\ \underline{-5x + y - 2z = -3} \end{array}$

69. $\begin{array}{r} x + y - z = 2 \\ \underline{x + y + z = 2} \end{array}$

70. $\begin{array}{r} 3x + 3y + z = 3 \\ \underline{4x - 3y - z = -3} \end{array}$

A Look Ahead

Solve each system. See the following example.

EXAMPLE Solve the system $\begin{cases} -\dfrac{4}{x} - \dfrac{4}{y} = -11 \\ \dfrac{1}{x} + \dfrac{1}{y} = 1 \end{cases}$

Solution: First make the following substitution. Let $a = \dfrac{1}{x}$ and $b = \dfrac{1}{y}$ in both equations. Then

$$\begin{cases} -4\left(\dfrac{1}{x}\right) - 4\left(\dfrac{1}{y}\right) = -11 \\ \dfrac{1}{x} + \dfrac{1}{y} = 1 \end{cases} \quad \text{is equivalent to} \quad \begin{cases} -4a - 4b = -11 \\ a + b = 1 \end{cases}$$

We solve by the addition method, multiplying both sides of the second equation by 4 and adding the left sides and the right sides of the equations.

$$\begin{cases} -4a - 4b = -11 \\ 4\ (a + b) = 4\ (1) \end{cases} \quad \text{simplifies to} \quad \begin{cases} -4a - 4b = -11 \\ 4a + 4b = 4 \end{cases}$$

$$0 = -7 \quad \text{False.}$$

The equation $0 = -7$ is false for all values of a and hence for all values of $\dfrac{1}{x}$ and all values of x. This system has no solution. ■

71. $\begin{cases} \dfrac{1}{x} + y = 12 \\ \dfrac{3}{x} - y = 4 \end{cases}$

72. $\begin{cases} x + \dfrac{2}{y} = 7 \\ 3x + \dfrac{3}{y} = 6 \end{cases}$

73. $\begin{cases} \dfrac{1}{x} + \dfrac{1}{y} = 5 \\ \dfrac{1}{x} - \dfrac{1}{y} = 1 \end{cases}$

74. $\begin{cases} \dfrac{2}{x} + \dfrac{3}{y} = 5 \\ \dfrac{5}{x} - \dfrac{3}{y} = 2 \end{cases}$

75. $\begin{cases} \dfrac{2}{x} + \dfrac{3}{y} = -1 \\ \dfrac{3}{x} - \dfrac{2}{y} = 18 \end{cases}$

76. $\begin{cases} \dfrac{3}{x} - \dfrac{2}{y} = -18 \\ \dfrac{2}{x} + \dfrac{3}{y} = 1 \end{cases}$

77. $\begin{cases} \dfrac{2}{x} - \dfrac{4}{y} = 5 \\ \dfrac{1}{x} - \dfrac{2}{y} = \dfrac{3}{2} \end{cases}$

78. $\begin{cases} \dfrac{5}{x} + \dfrac{7}{y} = 1 \\ -\dfrac{10}{x} - \dfrac{14}{y} = 0 \end{cases}$

10.2 Solving Systems of Linear Equations in Three Variables

OBJECTIVES

1. Recognize a linear equation in three variables.
2. Solve a system of three linear equations in three variables.

Tape IA 27

1 In this section, the algebraic methods of solving systems of two linear equations in two variables are extended to systems of three linear equations in three variables. We call the equation $3x - y + z = -15$, for example, a **linear equation in three variables** since there are three variables and each variable is raised only to the power 1. A solution of this equation is an **ordered triple, (x, y, z),** that makes the equation a true statement. For example, the ordered triple $(2, 0, -21)$ is a solution of $3x - y + z = -15$ since replacing x with 2, y with 0, and z with -21 yields the true statement $3(2) - 0 + (-21) = -15$. The graph of this equation is a plane in three-dimensional space, just as the graph of a linear equation in two variables is a line in two-dimensional space.

Although we will not graph equations in three variables, visualizing the possible patterns of intersecting planes gives us insight into the possible patterns of solutions of a system of three three-variable linear equations. There are four possible patterns.

1. Three planes intersect at a single point. This point represents the single solution of the system. The system is **consistent.**

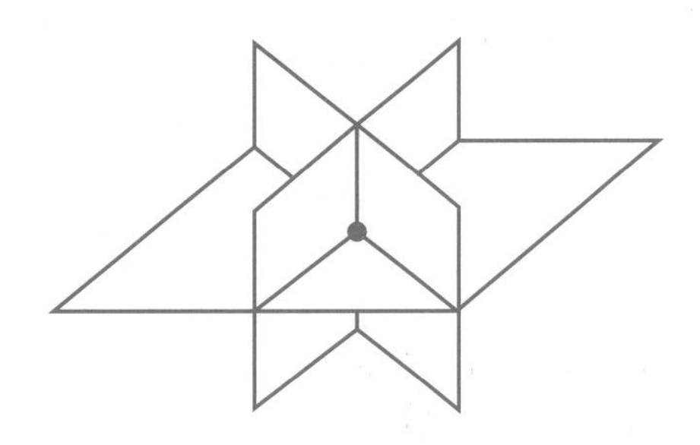

2. Three planes intersect at no points. This system has no solution. A few ways that this can occur are shown. This system is **inconsistent.**

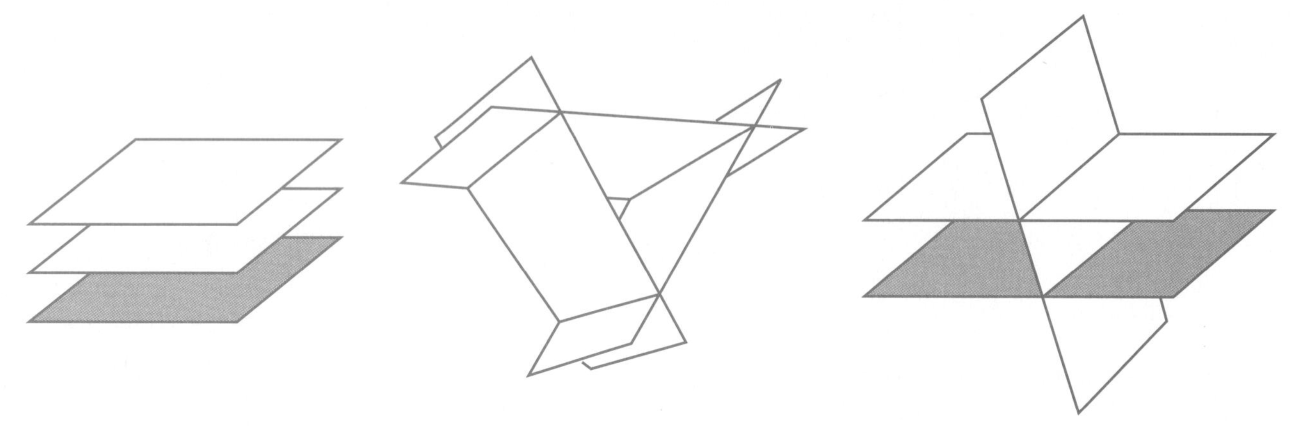

3. Three planes intersect at all the points of a single line. The system has infinitely many solutions. This system is **consistent.**

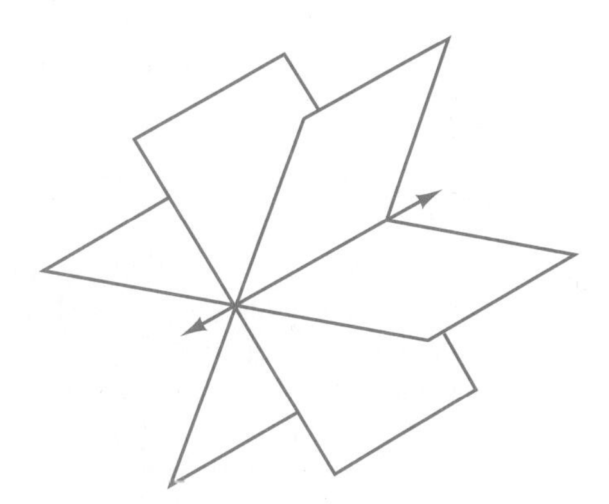

4. Three planes coincide at all points on the plane. The equations are **dependent.**

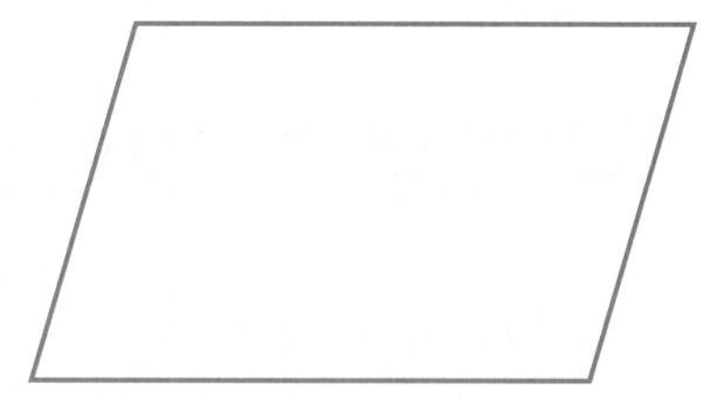

2 Using the addition method to solve a system in three variables, we reduce the system to a system in two variables.

EXAMPLE 1 Solve the system

$$\begin{cases} 3x - y + z = -15 & \text{Equation (1).} \\ x + 2y - z = 1 & \text{Equation (2).} \\ 2x + 3y - 2z = 0 & \text{Equation (3).} \end{cases}$$

Solution: Add equations (1) and (2) to eliminate z.

$$\begin{aligned} 3x - y + z &= -15 \\ x + 2y - z &= 1 \\ \hline 4x + y &= -14 \end{aligned}$$
Equation (4).

Next add two **other** equations and **eliminate** z **again.** To do so, multiply both sides of equation (1) by 2, and add this resulting equation to equation (3). Then

$$\begin{cases} 2(3x - y + z) = 2(-15) \\ 2x + 3y - 2z = 0 \end{cases} \text{ simplifies to } \begin{cases} 6x - 2y + 2z = -30 \\ 2x + 3y - 2z = 0 \end{cases}$$
$$8x + y = -30$$
Equation (5).

Now solve equations (4) and (5) for x and y. To solve by addition, multiply both sides of equation (4) by -1, and add this resulting equation to equation (5). Then

$$\begin{cases} -1(4x + y) = -1(-14) \\ 8x + y = -30 \end{cases} \text{ simplifies to } \begin{cases} -4x - y = 14 \\ 8x + y = -30 \end{cases}$$
$$4x = -16 \quad \text{Add the equations.}$$
$$x = -4 \quad \text{Solve for } x.$$

Replace x with -4 in equation (4) or (5).

$$\begin{aligned} 4x + y &= -14 && \text{Equation (4).} \\ 4(-4) + y &= -14 && \text{Let } x = -4. \\ y &= 2 && \text{Solve for } y. \end{aligned}$$

Finally, replace x with -4 and y with 2 in equation (1), (2), or (3).

$$\begin{aligned} x + 2y - z &= 1 && \text{Equation (2).} \\ -4 + 2(2) - z &= 1 && \text{Let } x = -4 \text{ and } y = 2. \\ -4 + 4 - z &= 1 \\ -z &= 1 \\ z &= -1 \end{aligned}$$

The solution is $(-4, 2, -1)$. To check, let $x = -4$, $y = 2$, and $z = -1$ in all three original equations of the system.

Equation (1)	*Equation (2)*	*Equation (3)*
$3x - y + z = -15$	$x + 2y - z = 1$	$2x + 3y - 2z = 0$
$3(-4) - 2 + (-1) = -15$	$-4 + 2(2) - (-1) = 1$	$2(-4) + 3(2) - 2(-1) = 0$
$-12 - 2 - 1 = -15$	$-4 + 4 + 1 = 1$	$-8 + 6 + 2 = 0$
$-15 = -15$	$1 = 1$	$0 = 0$
True.	True.	True.

All three statements are true, so the solution set is $\{(-4, 2, -1)\}$. ■

EXAMPLE 2 Solve the system

$$\begin{cases} 2x - 4y + 8z = 2 & (1) \\ -x - 3y + z = 11 & (2) \\ x - 2y + 4z = 0 & (3) \end{cases}$$

Solution: If equations (2) and (3) are added, x is eliminated, and the new equation is

$$-5y + 5z = 11 \qquad (4)$$

To eliminate x again, multiply both sides of equation (2) by 2, and add the resulting equation to equation (1). Then

$$\begin{cases} 2x - 4y + 8z = 2 \\ 2\,(-x - 3y + z) = 2\,(11) \end{cases} \text{ simplifies to } \begin{cases} 2x - 4y + 8z = 2 \\ -2x - 6y + 2z = 22 \end{cases}$$

$$-10y + 10z = 24 \qquad (5)$$

Next solve for y and z using equations (4) and (5). Multiply both sides of equation (4) by -2, and add the resulting equation to equation (5).

$$\begin{cases} -2\,(-5y + 5z) = -2\,(11) \\ -10y + 10z = 24 \end{cases} \text{ simplifies to } \begin{cases} 10y - 10z = -22 \\ -10y + 10z = 24 \end{cases}$$

$$0 = 2 \qquad \text{False.}$$

Since the statement is false, this system is inconsistent and has no solution. The solution set is the empty set { } or Ø. ■

The addition method is summarized next.

To Solve a System of Three Linear Equations by the Addition Method

Step 1 Eliminate any of the three variables from any pair of equations in the system.

Step 2 Choose any other pair of equations and eliminate the **same variable** as in step 1.

Step 3 Two equations in two variables should be obtained from step 1 and step 2. Use methods from Section 10.1 to solve this system for both variables.

Step 4 To solve for the third variable, substitute the values of the variables found in step 3 into any of the original equations.

EXAMPLE 3 Solve the system

$$\begin{cases} 2x + 4y = 1 & (1) \\ 4x - 4z = -1 & (2) \\ y - 4z = -3 & (3) \end{cases}$$

Solution: Notice that equation (2) has no term containing the variable y. Let's eliminate y using equations (1) and (3). Multiply both sides of equation (3) by -4, and add the resulting equation to equation (1). Then

$$\begin{cases} 2x + 4y = 1 \\ -4\,(y - 4z) = -4\,(-3) \end{cases} \text{ simplifies to } \begin{cases} 2x + 4y = 1 \\ -4y + 16z = 12 \end{cases}$$

$$2x + 16z = 13 \qquad (4)$$

Next solve for z using equations (4) and (2). Multiply both sides of equation (4) by -2, and add the resulting equation to equation (2).

$$\begin{cases} -2(2x + 16z) = -2(13) \\ 4x - 4z = -1 \end{cases} \quad \text{simplifies to} \quad \begin{cases} -4x - 32z = -26 \\ 4x - 4z = -1 \end{cases}$$

$$-36z = -27$$

$$z = \frac{3}{4}$$

Replace z with $\frac{3}{4}$ in equation (3), and solve for y.

$$y - 4\left(\frac{3}{4}\right) = -3 \quad \text{Let } z = \frac{3}{4} \text{ in equation (3).}$$

$$y - 3 = -3$$

$$y = 0$$

Replace y with 0 in equation (1), and solve for x.

$$2x + 4(0) = 1$$

$$2x = 1$$

$$x = \frac{1}{2}$$

The solution set is $\left\{\left(\frac{1}{2}, 0, \frac{3}{4}\right)\right\}$. Check to see that this solution satisfies all three equations of the system. ■

EXAMPLE 4 Solve the system

$$\begin{cases} x - 5y - 2z = 6 & (1) \\ -2x + 10y + 4z = -12 & (2) \\ \frac{1}{2}x - \frac{5}{2}y - z = 3 & (3) \end{cases}$$

Solution: Multiply both sides of equation (3) by 2 to eliminate fractions, and multiply both sides of equation (2) by $-\frac{1}{2}$ so that the coefficient of x is 1. The resulting system is then

$$\begin{cases} x - 5y - 2z = 6 & (1) \\ x - 5y - 2z = 6 & \text{Multiply (2) by } -\frac{1}{2}. \\ x - 5y - 2z = 6 & \text{Multiply (3) by 2.} \end{cases}$$

All three equations are identical, and therefore equations (1), (2), and (3) are all equivalent. There are infinitely many solutions of this system. The equations are dependent. The solution set can be written as $\{(x, y, z) \mid x - 5y - 2z = 6\}$. ■

EXERCISE SET 10.2

Solve each system. See Examples 1 and 3.

1. $\begin{cases} x + y = 3 \\ 2y = 10 \\ 3x + 2y - 3z = 1 \end{cases}$

2. $\begin{cases} 5x = 5 \\ 2x + y = 4 \\ 3x + y - 4z = -15 \end{cases}$

3. $\begin{cases} 2x + 2y + z = 1 \\ -x + y + 2z = 3 \\ x + 2y + 4z = 0 \end{cases}$

4. $\begin{cases} 2x - 3y + z = 5 \\ x + y + z = 0 \\ 4x + 2y + 4z = 4 \end{cases}$

Solve each system. See Examples 2 and 4.

5. $\begin{cases} x - 2y + z = -5 \\ -3x + 6y - 3z = 15 \\ 2x - 4y + 2z = -10 \end{cases}$

6. $\begin{cases} 3x + y - 2z = 2 \\ -6x - 2y + 4z = -2 \\ 9x + 3y - 6z = 6 \end{cases}$

7. $\begin{cases} 4x - y + 2z = 5 \\ 2y + z = 4 \\ 4x + y + 3z = 10 \end{cases}$

8. $\begin{cases} 5y - 7z = 14 \\ 2x + y + 4z = 10 \\ 2x + 6y - 3z = 30 \end{cases}$

Solve each system.

9. $\begin{cases} x + 5z = 0 \\ 5x + y = 0 \\ y - 3z = 0 \end{cases}$

10. $\begin{cases} x - 5y = 0 \\ x - z = 0 \\ -x + 5z = 0 \end{cases}$

11. $\begin{cases} 6x - 5z = 17 \\ 5x - y + 3z = -1 \\ 2x + y = -41 \end{cases}$

12. $\begin{cases} x + 2y = 6 \\ 7x + 3y + z = -33 \\ x - z = 16 \end{cases}$

13. $\begin{cases} x + y + z = 8 \\ 2x - y - z = 10 \\ x - 2y - 3z = 22 \end{cases}$

14. $\begin{cases} 5x + y + 3z = 1 \\ x - y + 3z = -7 \\ -x + y = 1 \end{cases}$

15. $\begin{cases} x + 2y - z = 5 \\ 6x + y + z = 7 \\ 2x + 4y - 2z = 5 \end{cases}$

16. $\begin{cases} 4x - y + 3z = 10 \\ x + y - z = 5 \\ 8x - 2y + 6z = 10 \end{cases}$

17. $\begin{cases} 2x - 3y + z = 2 \\ x - 5y + 5z = 3 \\ 3x - 14y + 14z = 9 \end{cases}$

18. $\begin{cases} 4x + y - z = 8 \\ x - y + 2z = 3 \\ 3x - y + z = 6 \end{cases}$

19. $\begin{cases} -2x - 4y + 6z = -8 \\ x + 2y - 3z = 4 \\ 4x + 8y - 12z = 16 \end{cases}$

20. $\begin{cases} -6x + 12y + 3z = -6 \\ 2x - 4y - z = 2 \\ -x + 2y + \dfrac{z}{2} = -1 \end{cases}$

21. $\begin{cases} 2x + 2y - 3z = 1 \\ y + 2z = -14 \\ 3x - 2y = -1 \end{cases}$

22. $\begin{cases} 7x + 4y = 10 \\ x - 4y + 2z = 6 \\ y - 2z = -1 \end{cases}$

23. $\begin{cases} \dfrac{3}{4}x - \dfrac{1}{3}y + \dfrac{1}{2}z = 9 \\ \dfrac{1}{6}x + \dfrac{1}{3}y - \dfrac{1}{2}z = 2 \\ \dfrac{1}{2}x - y + \dfrac{1}{2}z = 2 \end{cases}$

24. $\begin{cases} \dfrac{1}{3}x - \dfrac{1}{4}y + z = -9 \\ \dfrac{1}{2}x - \dfrac{1}{3}y - \dfrac{1}{4}z = -6 \\ x - \dfrac{1}{2}y - z = -8 \end{cases}$

Skill Review

Use the quadratic formula to solve each quadratic equation. See Section 9.2.

25. $x^2 + 4x + 1 = 0$ **26.** $2x^2 + x - 1 = 0$ **27.** $3x^2 - x + 2 = 0$ **28.** $x^2 - 2x + 3 = 0$

Simplify each complex fraction. See Section 7.4.

29. $\dfrac{\frac{1}{x}}{\frac{5}{x^2} - \frac{3}{x}}$ **30.** $\dfrac{\frac{2}{a} + \frac{1}{b}}{\frac{3}{b} - \frac{5}{a}}$

Simplify each radical. See Section 8.2.

31. $\sqrt{20}$ **32.** $\sqrt{18}$ **33.** $\sqrt[3]{54}$ **34.** $\sqrt[3]{24}$

35. $\sqrt[4]{32}$ **36.** $\sqrt[4]{162}$

A Look Ahead

Solve each system.

37. $\begin{cases} x + y - w = 0 \\ y + 2z + w = 3 \\ 2x + y + z = 1 \\ 2x - y - w = -1 \end{cases}$

38. $\begin{cases} 5x + 4y = 29 \\ y + z - w = -2 \\ 5x + z = 23 \\ y - z + w = 4 \end{cases}$

39. $\begin{cases} x + y + z + w = 5 \\ 2x + y + z + w = 6 \\ x + y + z = 2 \\ x + y = 0 \end{cases}$

40. $\begin{cases} 2x - z = -1 \\ y + z + w = 9 \\ y - 2w = -6 \\ x + y = 3 \end{cases}$

10.3 Applications of Linear Systems of Equations

OBJECTIVE

1 Solve problems that can be modeled by a system of linear equations.

Tape IA 27

1 We have solved problems by writing one-variable equations and solving for the variable. Some of these problems can be solved, perhaps more easily, by writing a system of equations, which we illustrate in this section.

EXAMPLE 1 A number is 4 less than a second number. Four times the first is 6 more than twice the second. Find the numbers.

Solution: 1. UNDERSTAND. Read and reread the problem and guess a solution. If one number is 10 and this is 4 less than a second number, the second number is 14. Four times the first number is 4(10), or 40. This is not equal to 6 more than twice the second number, which is $2(14) + 6$ or 34. Although we guessed incorrectly, we now have a better understanding of the problem.

2. ASSIGN. Since we are looking for two numbers, we will let

$$x = \text{first number}$$

$$y = \text{second number}$$

4. TRANSLATE. Since we have assigned two variables to this problem, we will translate our problem into two equations.

In words:	a number	is	4 less than a second number
Translate:	x	$=$	$y - 4$

Next we translate the second statement into an equation.

In words:	four times the first	is	6 more than twice the second
Translate:	$4x$	$=$	$2y + 6$

5. COMPLETE. Here we solve the system

$$\begin{cases} x = y - 4 \\ 4x = 2y + 6 \end{cases}$$

Since the first equation expresses x in terms of y, we will use substitution.
Substitute $y - 4$ for x in the second equation and solve for y.

$$4x = 2y + 6$$

$$4(y - 4) = 2y + 6 \qquad \text{Let } x = y - 4.$$

$$4y - 16 = 2y + 6$$

$$2y = 22$$

$$y = 11$$

Now replace y with 11 in the equation $x = y - 4$ and solve for x. Then $x = y - 4$ becomes $x = 11 - 4 = 7$. The solution of the system is (7, 11), corresponding to the numbers 7 and 11.

6. INTERPRET. *Checking,* notice that 7 **is** 4 less than 11, and 4 times 7 **is** 6 more than twice 11. The proposed numbers, 7 and 11, are correct.
State: The numbers are 7 and 11. ■

EXAMPLE 2 Two cars leave Indianapolis, one traveling east and the other west. After 3 hours they are 297 miles apart. If one car is traveling 5 mph faster than the other, what is the speed of each?

Solution: **1.** UNDERSTAND. Read and reread the problem. Let's guess an answer and use the formula $d = r \cdot t$ to check. Suppose that one car is traveling at a rate of 55 miles per hour. This means that the other car is traveling at a rate of 50 miles per hour since we are told that one car is traveling 5 mph faster than the other. To find the distance apart after 3 hours, we will first find the distance traveled by each car. One car's distance is rate · time = 55(3) = 165 miles. The other car's distance is rate · time = 50(3) = 150 miles. Since one car is traveling east and the other west, their distance apart will be the sum of their distances, or 165 miles + 150 miles = 315 miles, not the required distance of 297 miles.

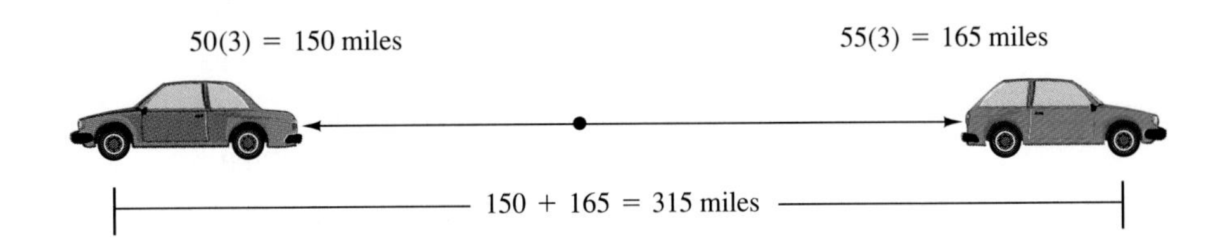

Now that we have a better understanding of the problem let's model it with a system of equations.

2. ASSIGN. Let

x = speed of one car

y = speed of the other car

3. ILLUSTRATE. We summarize the information on the following chart. The cars have each traveled 3 hours. Since distance = rate · time, we have that their distances are $3x$ and $3y$ miles, respectively.

	Rate ·	Time =	Distance
One Car	x	3	$3x$
Other Car	y	3	$3y$

4. TRANSLATE. We translate into two equations.

In words:	one car's distance	added to	the other car's distance	is	297
Translate:	$3x$	$+$	$3y$	$=$	297

In words:	one car	is	5 mph faster than the other
Translate:	x	$=$	$y + 5$

5. COMPLETE. Here we solve the system

$$\begin{cases} 3x + 3y = 297 \\ x = y + 5 \end{cases}$$

Again, the substitution method is appropriate. Replace x with $y + 5$ in the first equation and solve for y.

$$3x + 3y = 297$$

$$3(y + 5) + 3y = 297 \quad \text{Let } x = y + 5.$$

$$3y + 15 + 3y = 297$$

$$6y = 282$$

$$y = 47$$

To find x, replace y with 47 in the equation $x = y + 5$. Then $x = 47 + 5 = 52$. The cars are traveling at 52 mph and 47 mph.

6. INTERPRET. To *check,* notice that if one car travels 52 mph for 3 hours, the distance is $3(52) = 156$ miles. Also, the other car traveling for 3 hours at 47 mph travels a distance of $3(47) = 141$ miles. The sum of the distances $156 + 141$ is 297 miles, the required distance.

State: The cars are traveling at 52 mph and 47 mph. ■

EXAMPLE 3 Lynn Pike, a pharmacist, needs 70 liters of 50% alcohol solution. She has available a 30% alcohol solution and an 80% alcohol solution. How many liters of each solution should she mix to obtain 70 liters of a 50% alcohol solution?

Solution: **1.** UNDERSTAND. Read and reread the problem. Next let's guess the solution. Suppose that we need 20 liters of the 30% solution. Then we need $70 - 20 = 50$ liters of the 80% solution. To see if this gives us 70 liters of a 50% alcohol solution, let's find the amount of pure alcohol in each solution.

concentration rate	×	amount of solution	=	amount of pure alcohol
0.30	×	20 liters	=	6 liters
0.80	×	50 liters	=	40 liters
0.50	×	70 liters	=	35 liters

Since 6 liters + 40 liters = 46 liters and not 35 liters, our guess is incorrect, but we have gained some insight as to how we model and check this problem.

2. ASSIGN. Let

x = number of liters of 30% solution

y = number of liters of 80% solution

3. ILLUSTRATE. Use a table to organize the given data.

	Concentration Rate	Liters of Solution	Liters of Pure Alcohol
First Solution	30%	x	$0.30x$
Second Solution	80%	y	$0.80y$
Mixture Needed	50%	70	(0.50)(70)

4. TRANSLATE. We translate into two equations.

In words:	liters of 30% solution	+	liters of 80% solution	=	70
Translate:	x	+	y	=	70
In words:	alcohol in 30% solution	+	alcohol in 80% solution	=	alcohol in mixture
Translate:	$0.30x$	+	$0.80y$	=	$(0.50)(70)$

5. COMPLETE. Here we solve the system

$$\begin{cases} x + y = 70 \\ 0.30x + 0.80y = (0.50)(70) \end{cases}$$

To solve this system, use the addition method. Multiply both sides of the first equation by -3 and both sides of the second equation by 10. Then

$$\begin{cases} -3(x + y) = -3(70) \\ 10(0.30x + 0.80y) = 10(0.50)(70) \end{cases} \text{ simplifies to } \begin{cases} -3x - 3y = -210 \\ 3x + 8y = 350 \end{cases}$$

$$5y = 140$$
$$y = 28$$

Replace y with 28 in the equation $x + y = 70$ and find that $x + 28 = 70$ or $x = 42$.

6. INTERPRET. To *check,* recall how we checked our guess.

State: The pharmacist needs to mix 42 liters of 30% solution and 28 liters of 80% solution to obtain 70 liters of 50% solution. ■

EXAMPLE 4 A rectangular garden is to be completely fenced to keep animals out. The length of the garden is four times the width, and the garden requires 210 meters of fencing. Find the dimensions of the garden.

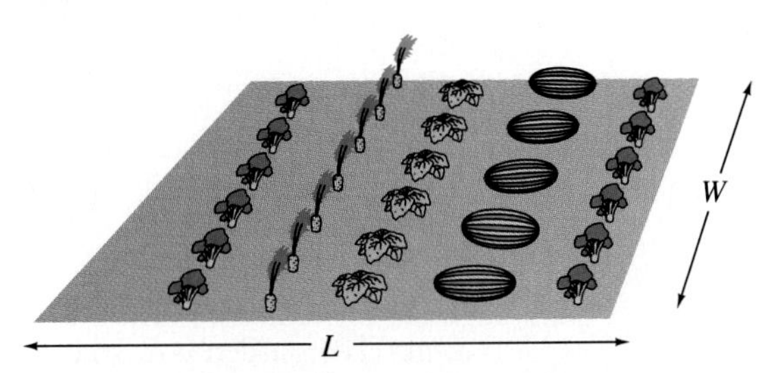

Solution: **1.** UNDERSTAND. Read and reread the problem. Fencing the garden has to do with perimeter. Recall that the formula for the perimeter of a rectangle is $P = 2W + 2L$. Draw a rectangle and guess a solution. If the width of the rectangle is 15 meters, its length is 4 times the width, or 4(15 meters) = 60 meters. The perimeter of this rectangle is $P = 2(15 \text{ meters}) + 2(60 \text{ meters}) = 150$ meters, not the required 210 meters. We now know that the width is greater than 15 meters.

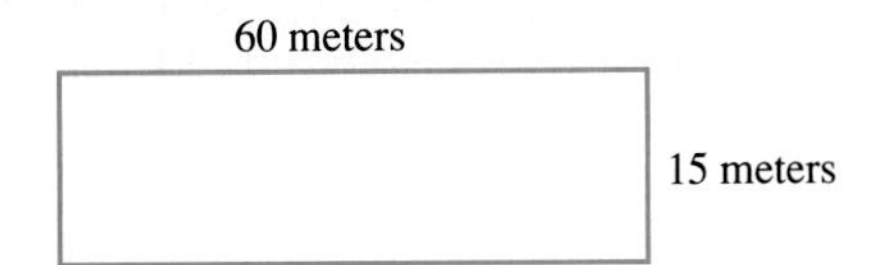

2. ASSIGN. Let

W = width of garden

L = length of garden

3. ILLUSTRATE. We draw a rectangle and label it.

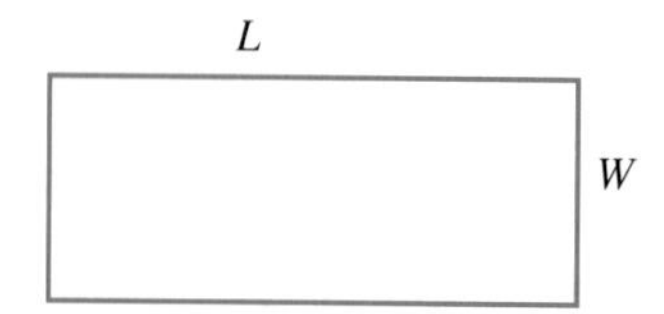

4. TRANSLATE.

In words:	length	=	4 times the width
Translate:	L	=	$4W$

In words:	the perimeter of the rectangle	=	210
Translate:	$2W + 2L$	=	210

5. COMPLETE. The system to solve is

$$\begin{cases} L = 4W \\ 2W + 2L = 210 \end{cases}$$

The substitution method is appropriate here. Replace L with $4W$ in the second equation.

$$2W + 2L = 210 \quad \text{Second equation.}$$
$$2W + 2(4W) = 210 \quad \text{Let } L = 4W.$$
$$10W = 210$$
$$W = 21$$

Replace W with 21 in $L = 4W$; thus, $L = 4(21) = 84$.

6. INTERPRET. To *check,* notice that the length is 4 times the width since $4(21) = 84$. Also, the perimeter of the resulting rectangle is $2(21) + 2(84) = 42 + 168 = 210$ meters, the required perimeter.

State: The garden is 21 meters by 84 meters. ■

The next problem is solved by using three variables.

EXAMPLE 5 The measure of the largest angle of a triangle is 80° more than the measure of the smallest angle, and the measure of the remaining angle is 10° more than the measure of the smallest angle. Find the measure of each angle.

Solution: **1.** UNDERSTAND. Read and reread the problem. Recall that the sum of the measures of the angles of a triangle is 180°. Then guess a solution. If the smallest angle measures **20°,** the measure of the largest angle is 80° more, or 20° + 80° = **100°.** The measure of the remaining angle is 10° more than the measure of the smallest angle, or

$20° + 10° =$ **30°.** The sum of these three angles is $20° + 100° + 30° = 150°$, not the required 180°. We now know that the measure of the smallest angle is greater than 20°.

2. ASSIGN. Let

$x =$ degree measure of the smallest angle

$y =$ degree measure of the largest angle.

$z =$ degree measure of the remaining angle

3. ILLUSTRATE.

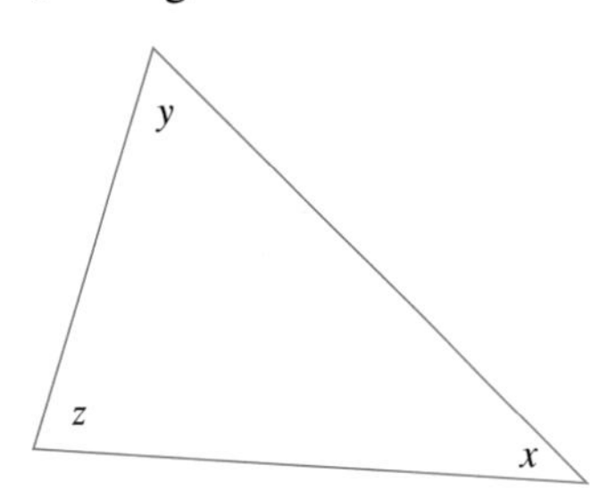

4. TRANSLATE. We translate into three equations.

In words:	the sum of the angles	=	180
Translate:	$x + y + z$	=	180

In words:	the largest angle	is	80 more than the smallest angle
Translate:	y	=	$x + 80$

In words:	the remaining angle	is	10 more than the smallest angle
Translate:	z	=	$x + 10$

5. COMPLETE. We solve the system

$$\begin{cases} x + y + z = 180 \\ y = x + 80 \\ z = x + 10 \end{cases}$$

Since y and z are both expressed in terms of x, we will solve using the substitution method.

Substitute $y = x + 80$ and $z = x + 10$ in the first equation. Then

$$x + y + z = 180$$

becomes

$$x + \overbrace{(x + 80)} + \overbrace{(x + 10)} = 180$$

$$3x + 90 = 180$$

$$3x = 90$$

$$x = 30$$

Then $y = x + 80 = 30 + 80 = 110$ and $z = x + 10 = 30 + 10 = 40$. The angles measure 30°, 40°, and 110°.

6. INTERPRET. To *check,* notice that $30° + 40° + 110° = 180°$. Also, the measure of the largest angle, 110°, is 80° more than the measure of the smallest angle, 30°. The measure of the remaining angle, 40°, is 10° more than the measure of the smallest angle, 30°.

State: The angles measure 30°, 40°, and 110°. ■

EXERCISE SET 10.3

Solve. See Example 1.

1. One number is two more than a second number. Twice the first is 4 less than 3 times the second. Find the numbers.

2. Three times one number minus a second is 8, and the sum of the numbers is 12. Find the numbers.

Solve. See Example 2.

3. A Delta 727 traveled 560 mph with the wind and 480 mph against the wind. Find the speed of the plane in still air and the speed of the wind.

4. Terry Watkins can row about 10.6 kilometers in 1 hour downstream and 6.8 kilometers upstream in 1 hour. Find how fast he can row in still water, and find the speed of the current.

Solve. See Example 3.

5. Find how many quarts of 4% butterfat milk and 1% butterfat milk should be mixed to yield 60 quarts of 2% butterfat milk.

6. A pharmacist needs 500 milliliters of a 20% phenobarbital solution, but has only 5% and 25% phenobarbital solutions available. Find how many milliliters of each he should mix to get the desired solution.

Solve. See Example 4.

7. Megan Sweet has 156 feet of fencing to make a rectangular pen for her German shepherds. If she wants the pen to be twice as long as it is wide, find the length and width of the pen.

8. A rectangular swimming pool is 22 feet longer than it is wide. The perimeter of the pool is 80 feet. Find the length and width of the pool.

Solve. See Example 5.

9. Chris Peckaitis bought some large frames for \$15 each and some small frames for \$8 each at a closeout sale. If she bought 22 frames for \$239, find how many of each type she bought.

10. Hilton University Drama Club sold 311 tickets for a play. Student tickets cost 50 cents each; nonstudent tickets cost \$1.50. If total receipts were \$385.50, find how many tickets of each type were sold.

Solve.

11. One number is two less than a second number. Twice the first is 4 more than 3 times the second. Find the numbers.

12. Twice one number plus a second number is 42, and the one number minus the second number is -6. Find the numbers.

13. Seven tablets and 4 pens cost \$6.40. Two tablets and 19 pens cost \$5.40. Find the price of each.

14. A candy shop manager mixes M&M's worth \$2.00 per pound with trail mix worth \$1.50 per pound. Find how many pounds of each she should use to get 50 pounds of a party mix worth \$1.80 per pound.

15. An airplane takes 3 hours to travel a distance of 2160 miles with the wind. The return trip takes 4 hours against the wind. Find the speed of the plane in still air and the speed of the wind.

16. Two cyclists start at the same point and travel in opposite directions. One travels 4 mph faster than the other. In 4 hours they are 112 miles apart. Find how fast each is traveling.

17. The perimeter of a quadrilateral (four-sided polygon) is 29 inches. The longest side is twice as long as the shortest side, and the other two sides are equally long and each are 2 inches longer than the shortest side. Find the length of all four sides.

18. The perimeter of a triangle is 93 centimeters. If two sides are equally long and the third side is 9 centimeters longer than the others, find the lengths of the three sides.

19. May Jones's change purse contains \$2.70 in nickels and dimes. If she has 39 coins in all, find how many of each type she has.

20. Gerhart Moore has a coin collection with a face value of \$5.52. If he has 48 coins in quarters and pennies, find how many of each type he has.

21. The sum of three numbers is 40. One number is five more than a second and twice the third. Find the numbers.

22. The sum of the digits of a three-digit number is 15. The tens-place digit is twice the hundreds-place digit, and the ones-place digit is 1 less than the hundreds-place digit. Find the three-digit number.

23. Jack Reinholt, a car salesman, has a choice of two pay arrangements: a weekly salary of \$200 plus 5% commission on sales, or a straight 15% commission. Find the amount of sales for which Jack's earnings are the same regardless of the pay arrangement.

24. Hertz car rental agency charges \$25 daily plus 10 cents per mile. Budget charges \$20 daily plus 25 cents per mile. Find the daily mileage for which the Budget charge for the day is twice that of the Hertz charge for the day.

25. Allan Little has 85 bills, some 10-dollar bills, and some 20-dollar bills, totaling \$1480. Find how many of each type there are.

26. A bank teller has 155 bills of \$1 and \$5 denominations with a total value of \$471. Find how many of each type of bill he has.

27. Mary Dooley has a collection of dimes, nickels, and pennies worth \$3.42. She has twice as many dimes as nickels and four more pennies than nickels. Find how many of each she has.

28. Carroll Blakemore, a drafting student, bought 3 templates and a pencil one day for \$6.45. Another day he bought 2 pads of paper and 4 pencils for \$7.50. If the price of a pad of paper is three times the price of a pencil, find the price of each type of item.

29. Rabbits in a lab are to be kept on a strict daily diet to include 30 grams of protein, 16 grams of fat, and 24 grams of carbohydrates. The scientist has only three food mixes available with the following grams of nutrients per unit.

	Protein	Fat	Carbohydrate
Mix A	4	6	3
Mix B	6	1	2
Mix C	4	1	12

Find how many units of each mix are needed daily to meet each rabbit's dietary needs.

30. Gerry Gundersen mixes different solutions with concentrations of 25%, 40%, and 50% to get 200 liters of a 32% solution. If he uses twice as much of the 25% solution as of the 40% solution, find how many liters of each kind he uses.

Skill Review

Factor completely. See Section 6.1.

31. $x^2 + xy + 3x + 3y$

32. $ab + 4a - 2b - 8$

Multiply. See Section 5.4.

33. $(x + 5)^2$

34. $(y - 3)^2$

35. $(2x - y)^2$

36. $(3x + 4)^2$

Evaluate. See Section 1.7.

37. $(-1)(-5) - (6)(3)$

38. $(2)(-8) - (-4)(1)$

39. $(4)(-10) - (2)(-2)$

40. $(-7)(3) - (-2)(-6)$

41. $(-3)(-3) - (-1)(-9)$

42. $(5)(6) - (10)(10)$

10.4 Solving Systems of Equations by Determinants

OBJECTIVES

Tape IA 28

1. Define and evaluate a 2×2 determinant.
2. Use Cramer's rule to solve a system of two linear equations in two variables.
3. Define and evaluate a 3×3 determinant.
4. Use Cramer's rule to solve a linear system of three equations in three variables.

1 Three methods for solving systems of two linear equations in two variables have been shown: graphically, by substitution, and by elimination. Now we will analyze another method called **Cramer's rule.** First we introduce determinants. A **determinant** is a real number associated with a square array of numbers written between two vertical bars.

$$\begin{vmatrix} 1 & 6 \\ 5 & 2 \end{vmatrix} \qquad \begin{vmatrix} 2 & 4 & 1 \\ 0 & 5 & 2 \\ 3 & 6 & 9 \end{vmatrix}$$

Each number in the array is called an **element.** The numbers in a horizontal line form a **row;** those in a vertical line form a **column.** A second-order determinant or a 2×2 (read as 2 by 2) determinant has 2 rows and 2 columns. The value of a 2×2 determinant is defined next.

Value of a 2×2 Determinant

$$\begin{vmatrix} a & b \\ c & d \end{vmatrix} = ad - bc$$

EXAMPLE 1 Find the value of each determinant.

a. $\begin{vmatrix} -1 & 2 \\ 3 & -4 \end{vmatrix}$ **b.** $\begin{vmatrix} 2 & 0 \\ 7 & -5 \end{vmatrix}$

Solution: First identify the values of a, b, c, and d.

a. Here $a = -1$, $b = 2$, $c = 3$, and $d = -4$.

$$\begin{vmatrix} -1 & 2 \\ 3 & -4 \end{vmatrix} = ad - bc = (-1)(-4) - (2)(3) = -2$$

b. In this example, $a = 2$, $b = 0$, $c = 7$, and $d = -5$.

$$\begin{vmatrix} 2 & 0 \\ 7 & -5 \end{vmatrix} = ad - bc = 2(-5) - (0)(7) = -10 \quad ■$$

2 To develop Cramer's rule, we solve by elimination the system $\begin{cases} ax + by = h \\ cx + dy = k \end{cases}$.
First, eliminate y by multiplying both sides of the first equation by d and both sides of

the second equation by $-b$ so that the coefficients of y are opposites. The result is the following system. We then add the two equations and solve for x.

$$\begin{cases} d\,(ax + by) = d \cdot h \\ -b\,(cx + dy) = -b \cdot k \end{cases} \quad \text{simplifies to} \quad \begin{cases} adx + bdy = hd \\ -bcx - bdy = -kb \end{cases}$$

$$adx - bcx = hd - kb$$

Add the equations.

$$(ad - bc)\,x = hd - kb$$

$$x = \frac{hd - kb}{ad - bc}$$

Solve for x.

When we replace x with $\frac{hd - kb}{ad - bc}$ in the equation $ax + by = h$ and solve for y, we find that $y = \frac{ak - ch}{ad - bc}$.

Notice that the numerator of the value of x can be written as a determinant.

$$hd - kb = \begin{vmatrix} h & b \\ k & d \end{vmatrix}$$

The numerator of the value of y can be written as the determinant

$$ak - ch = \begin{vmatrix} a & h \\ c & k \end{vmatrix}$$

The denominator of the values of x and y is the same and can be written as

$$ad - bc = \begin{vmatrix} a & b \\ c & d \end{vmatrix}$$

This means that the values of x and y can be written as

$$x = \frac{\begin{vmatrix} h & b \\ k & d \end{vmatrix}}{\begin{vmatrix} a & b \\ c & d \end{vmatrix}} \quad \text{and} \quad y = \frac{\begin{vmatrix} a & h \\ c & k \end{vmatrix}}{\begin{vmatrix} a & b \\ c & d \end{vmatrix}}$$

For convenience, we will call the determinants D, D_x, and D_y.

x-coefficients
↓

$$\begin{vmatrix} a & b \\ c & d \end{vmatrix} = D \qquad \begin{vmatrix} h & b \\ k & d \end{vmatrix} = D_x \qquad \begin{vmatrix} a & h \\ c & k \end{vmatrix} = D_y$$

↑ y-coefficients

↑ x-column replaced by constants

↑ y-column replaced by constants

These determinant formulas for the coordinates of the solution of a system are known as **Cramer's rule.**

Cramer's Rule for Two Linear Equations in Two Variables

The solution of the system $\begin{cases} ax + by = h \\ cx + dy = k \end{cases}$ is given by

$$x = \frac{\begin{vmatrix} h & b \\ k & d \end{vmatrix}}{\begin{vmatrix} a & b \\ c & d \end{vmatrix}} = \frac{D_x}{D}; \qquad y = \frac{\begin{vmatrix} a & h \\ c & k \end{vmatrix}}{\begin{vmatrix} a & b \\ c & d \end{vmatrix}} = \frac{D_y}{D}$$

as long as $D = ad - bc$ is not 0.

When $D = 0$, the system is either inconsistent or the equations are dependent. When this happens, use another method to see which is true.

EXAMPLE 2 Use Cramer's rule to solve each system.

a. $\begin{cases} 3x + 4y = -7 \\ x - 2y = -9 \end{cases}$

b. $\begin{cases} 5x + y = 5 \\ -7x - 2y = -7 \end{cases}$

Solution: **a.** Find D, D_x, and D_y.

$$\begin{matrix} a & b & h \\ \downarrow & \downarrow & \downarrow \end{matrix}$$
$$\begin{cases} 3x + 4y = -7 \\ x - 2y = -9 \end{cases}$$
$$\begin{matrix} \uparrow & \uparrow & \uparrow \\ c & d & k \end{matrix}$$

$$D = \begin{vmatrix} 3 & 4 \\ 1 & -2 \end{vmatrix} = 3(-2) - 4(1) = -10$$

$$D_x = \begin{vmatrix} -7 & 4 \\ -9 & -2 \end{vmatrix} = (-7)(-2) - 4(-9) = 50$$

$$D_y = \begin{vmatrix} 3 & -7 \\ 1 & -9 \end{vmatrix} = 3(-9) - (-7)(1) = -20$$

Then $x = \frac{D_x}{D} = \frac{50}{-10} = -5$ and $y = \frac{D_y}{D} = \frac{-20}{-10} = 2$. The solution set is $\{(-5, 2)\}$. As always, check the solution in both original equations.

b. $\begin{cases} 5x + y = 5 \\ -7x - 2y = -7 \end{cases}$ Find D, D_x, and D_y.

$$D = \begin{vmatrix} 5 & 1 \\ -7 & -2 \end{vmatrix} = 5(-2) - (-7)(1) = -3$$

$$D_x = \begin{vmatrix} 5 & 1 \\ -7 & -2 \end{vmatrix} = 5(-2) - (-7)(1) = -3$$

$$D_y = \begin{vmatrix} 5 & 5 \\ -7 & -7 \end{vmatrix} = 5(-7) - 5(-7) = 0$$

$$x = \frac{D_x}{D} = \frac{-3}{-3} = 1, \qquad y = \frac{D_y}{D} = \frac{0}{-3} = 0$$

The solution set is $\{(1, 0)\}$. ■

3 Three-by-three determinants can be used to solve systems of three equations in three variables. Finding the value of a 3×3 determinant, however, is considerably more complex than the 2×2 case.

The Value of a 3 × 3 Determinant

$$\begin{vmatrix} a_1 & b_1 & c_1 \\ a_2 & b_2 & c_2 \\ a_3 & b_3 & c_3 \end{vmatrix} = a_1 \cdot \begin{vmatrix} b_2 & c_2 \\ b_3 & c_3 \end{vmatrix} - a_2 \cdot \begin{vmatrix} b_1 & c_1 \\ b_3 & c_3 \end{vmatrix} + a_3 \cdot \begin{vmatrix} b_1 & c_1 \\ b_2 & c_2 \end{vmatrix}$$

The value of a 3×3 determinant, then, is based on three 2×2 determinants. Each of these 2×2 determinants is called a **minor,** and every element of a 3×3 determinant has a minor associated with it. For example, the minor of c_2 has no row or column containing c_2.

$$\begin{matrix} a_1 & b_1 & c_1 \\ a_2 & b_2 & c_2 \\ a_3 & b_3 & c_3 \end{matrix}$$

The minor of c_2 is

$$\begin{vmatrix} a_1 & b_1 \\ a_3 & b_3 \end{vmatrix}$$

Also, the minor of element a_1 is the 2×2 determinant that has no row or column containing a_1.

$$\begin{matrix} a_1 & b_1 & c_1 \\ a_2 & b_2 & c_2 \\ a_3 & b_3 & c_3 \end{matrix}$$

The minor of a_1 is

$$\begin{vmatrix} b_2 & c_2 \\ b_3 & c_3 \end{vmatrix}$$

So the value of a 3×3 determinant can be written as

$$a_1 \cdot (\text{minor of } a_1) - a_2(\text{minor of } a_2) + a_3(\text{minor of } a_3)$$

We call this finding the value of the determinant by **expanding** by the minors of the first column. The value of a determinant can be found by expanding by the minors of any row or column. The following array of signs is helpful in determining whether to add or subtract the product of an element and its minor.

$$\begin{vmatrix} + & - & + \\ - & + & - \\ + & - & + \end{vmatrix}$$

If an element is in a position marked $+$, we add. If marked $-$, we subtract.

EXAMPLE 3 Find the value of the determinant by expanding by the minors of the given row or column.

$$\begin{vmatrix} 0 & 5 & 1 \\ 1 & 3 & -1 \\ -2 & 2 & 4 \end{vmatrix}$$

a. First column **b.** Second row

Solution: **a.** The elements of the first column are 0, 1, and −2. The first column of the array of signs is +, −, +.

$$0 \cdot \begin{vmatrix} 3 & -1 \\ 2 & 4 \end{vmatrix} - 1 \cdot \begin{vmatrix} 5 & 1 \\ 2 & 4 \end{vmatrix} + (-2) \cdot \begin{vmatrix} 5 & 1 \\ 3 & -1 \end{vmatrix}$$
$$= 0(12 + 2) - 1(20 - 2) + (-2)(-5 - 3)$$
$$= 0 - 18 + 16 = -2$$

b. The elements of the second row are 1, 3, and −1. This time, the signs begin with − and again alternate.

$$-1 \cdot \begin{vmatrix} 5 & 1 \\ 2 & 4 \end{vmatrix} + 3 \cdot \begin{vmatrix} 0 & 1 \\ -2 & 4 \end{vmatrix} - (-1) \cdot \begin{vmatrix} 0 & 5 \\ -2 & 2 \end{vmatrix}$$
$$= -1(20 - 2) + 3(0 - (-2)) - (-1)(0 - (-10))$$
$$= -18 + 6 + 10 = -2$$

Notice that the determinant has the same value regardless of the row or column you select to expand by. ■

4 A system of three equations in three variables may also be solved with Cramer's rule. Using the elimination process to solve a system with unknown constants as coefficients leads to the following relationships.

Cramer's Rule for Three Equations in Three Variables

The solution of the system $\begin{cases} a_1x + b_1y + c_1z = k_1 \\ a_2x + b_2y + c_2z = k_2 \\ a_3x + b_3y + c_3z = k_3 \end{cases}$ is given by $x = \dfrac{D_x}{D}$, $y = \dfrac{D_y}{D}$, and $z = \dfrac{D_z}{D}$, where

$$D = \begin{vmatrix} a_1 & b_1 & c_1 \\ a_2 & b_2 & c_2 \\ a_3 & b_3 & c_3 \end{vmatrix} \qquad D_x = \begin{vmatrix} k_1 & b_1 & c_1 \\ k_2 & b_2 & c_2 \\ k_3 & b_3 & c_3 \end{vmatrix}$$

$$D_y = \begin{vmatrix} a_1 & k_1 & c_1 \\ a_2 & k_2 & c_2 \\ a_3 & k_3 & c_3 \end{vmatrix} \qquad D_z = \begin{vmatrix} a_1 & b_1 & k_1 \\ a_2 & b_2 & k_2 \\ a_3 & b_3 & k_3 \end{vmatrix}$$

as long as D is not 0.

EXAMPLE 4 Use Cramer's rule to solve the system

$$\begin{cases} x - 2y + z = 4 \\ 3x + y - 2z = 3 \\ 5x + 5y + 3z = -8 \end{cases}$$

Solution: First evaluate D, D_x, D_y, and D_z. We will expand D by the minors of the first column.

$$D = \begin{vmatrix} 1 & -2 & 1 \\ 3 & 1 & -2 \\ 5 & 5 & 3 \end{vmatrix}$$

$$= 1 \cdot \begin{vmatrix} 1 & -2 \\ 5 & 3 \end{vmatrix} - 3 \cdot \begin{vmatrix} -2 & 1 \\ 5 & 3 \end{vmatrix} + 5 \cdot \begin{vmatrix} -2 & 1 \\ 1 & -2 \end{vmatrix}$$

$$= 1(3 - (-10)) - 3(-6 - 5) + 5(4 - 1)$$

$$= 13 + 33 + 15 = 61$$

$$D_x = \begin{vmatrix} 4 & -2 & 1 \\ 3 & 1 & -2 \\ -8 & 5 & 3 \end{vmatrix}$$

$$= 4 \cdot \begin{vmatrix} 1 & -2 \\ 5 & 3 \end{vmatrix} - 3 \cdot \begin{vmatrix} -2 & 1 \\ 5 & 3 \end{vmatrix} + (-8) \cdot \begin{vmatrix} -2 & 1 \\ 1 & -2 \end{vmatrix}$$

$$= 4(3 - (-10)) - 3(-6 - 5) + (-8)(4 - 1)$$

$$= 52 + 33 - 24 = 61$$

$$D_y = \begin{vmatrix} 1 & 4 & 1 \\ 3 & 3 & -2 \\ 5 & -8 & 3 \end{vmatrix}$$

$$= 1 \cdot \begin{vmatrix} 3 & -2 \\ -8 & 3 \end{vmatrix} - 3 \cdot \begin{vmatrix} 4 & 1 \\ -8 & 3 \end{vmatrix} + 5 \cdot \begin{vmatrix} 4 & 1 \\ 3 & -2 \end{vmatrix}$$

$$= 1(9 - 16) - 3(12 + 8) + 5(-8 - 3)$$

$$= -7 - 60 - 55 = -122$$

$$D_z = \begin{vmatrix} 1 & -2 & 4 \\ 3 & 1 & 3 \\ 5 & 5 & -8 \end{vmatrix}$$

$$= 1 \cdot \begin{vmatrix} 1 & 3 \\ 5 & -8 \end{vmatrix} - 3 \cdot \begin{vmatrix} -2 & 4 \\ 5 & -8 \end{vmatrix} + 5 \cdot \begin{vmatrix} -2 & 4 \\ 1 & 3 \end{vmatrix}$$

$$= 1(-8 - 15) - 3(16 - 20) + 5(-6 - 4)$$

$$= -23 + 12 - 50 = -61$$

From these determinants, we calculate the solution:

$$x = \frac{D_x}{D} = \frac{61}{61} = 1, \qquad y = \frac{D_y}{D} = \frac{-122}{61} = -2, \qquad z = \frac{D_z}{D} = \frac{-61}{61} = -1$$

The solution set of the system is $\{(1, -2, -1)\}$. Check this solution by verifying that it satisfies each equation of the system. ■

EXERCISE SET 10.4

Find the value of each determinant. See Example 1.

1. $\begin{vmatrix} 3 & 5 \\ -1 & 7 \end{vmatrix}$ **2.** $\begin{vmatrix} -5 & 1 \\ 0 & -4 \end{vmatrix}$ **3.** $\begin{vmatrix} 9 & -2 \\ 4 & -3 \end{vmatrix}$

4. $\begin{vmatrix} 4 & 0 \\ 9 & 8 \end{vmatrix}$ **5.** $\begin{vmatrix} -2 & 9 \\ 4 & -18 \end{vmatrix}$ **6.** $\begin{vmatrix} -40 & 8 \\ 70 & -14 \end{vmatrix}$

Use Cramer's rule, if possible, to solve each system of linear equations. See Example 2.

7. $\begin{cases} 2y - 4 = 0 \\ x + 2y = 5 \end{cases}$ **8.** $\begin{cases} 4x - y = 5 \\ 3x - 3 = 0 \end{cases}$ **9.** $\begin{cases} 3x + y = 1 \\ 2y = 2 - 6x \end{cases}$

10. $\begin{cases} y = 2x - 5 \\ 8x - 4y = 20 \end{cases}$ **11.** $\begin{cases} 5x - 2y = 27 \\ -3x + 5y = 18 \end{cases}$ **12.** $\begin{cases} 4x - y = 9 \\ 2x + 3y = -27 \end{cases}$

Find the value of each determinant. See Example 3.

13. $\begin{vmatrix} 2 & 1 & 0 \\ 0 & 5 & -3 \\ 4 & 0 & 2 \end{vmatrix}$ **14.** $\begin{vmatrix} -6 & 4 & 2 \\ 1 & 0 & 5 \\ 0 & 3 & 1 \end{vmatrix}$ **15.** $\begin{vmatrix} 4 & -6 & 0 \\ -2 & 3 & 0 \\ 4 & -6 & 1 \end{vmatrix}$

16. $\begin{vmatrix} 5 & 2 & 1 \\ 3 & -6 & 0 \\ -2 & 8 & 0 \end{vmatrix}$ **17.** $\begin{vmatrix} 3 & 6 & -3 \\ -1 & -2 & 3 \\ 4 & -1 & 6 \end{vmatrix}$ **18.** $\begin{vmatrix} 2 & -2 & 1 \\ 4 & 1 & 3 \\ 3 & 1 & 2 \end{vmatrix}$

Use Cramer's rule, if possible, to solve each system of linear equations. See Example 4.

19. $\begin{cases} 3x + z = -1 \\ -x - 3y + z = 7 \\ 3y + z = 5 \end{cases}$ **20.** $\begin{cases} 4y - 3z = -2 \\ 8x - 4y = 4 \\ -8x + 4y + z = -2 \end{cases}$

21. $\begin{cases} x + y + z = 8 \\ 2x - y - z = 10 \\ x - 2y + 3z = 22 \end{cases}$ **22.** $\begin{cases} 5x + y + 3z = 1 \\ x - y - 3z = -7 \\ -x + y = 1 \end{cases}$

Find the value of each determinant.

23. $\begin{vmatrix} 10 & -1 \\ -4 & 2 \end{vmatrix}$ **24.** $\begin{vmatrix} -6 & 2 \\ 5 & -1 \end{vmatrix}$ **25.** $\begin{vmatrix} 1 & 0 & 4 \\ 1 & -1 & 2 \\ 3 & 2 & 1 \end{vmatrix}$ **26.** $\begin{vmatrix} 0 & 1 & 2 \\ 3 & -1 & 2 \\ 3 & 2 & -2 \end{vmatrix}$

27. $\begin{vmatrix} \frac{3}{4} & \frac{5}{2} \\ -\frac{1}{6} & \frac{7}{3} \end{vmatrix}$ **28.** $\begin{vmatrix} \frac{5}{7} & \frac{1}{3} \\ \frac{6}{7} & \frac{2}{3} \end{vmatrix}$ **29.** $\begin{vmatrix} 4 & -2 & 2 \\ 6 & -1 & 3 \\ 2 & 1 & 1 \end{vmatrix}$ **30.** $\begin{vmatrix} 1 & 5 & 0 \\ 7 & 9 & -4 \\ 3 & 2 & -2 \end{vmatrix}$

31. $\begin{vmatrix} -2 & 5 & 4 \\ 5 & -1 & 3 \\ 4 & 1 & 2 \end{vmatrix}$ **32.** $\begin{vmatrix} 5 & -2 & 4 \\ -1 & 5 & 3 \\ 1 & 4 & 2 \end{vmatrix}$

Use Cramer's rule, if possible, to solve each system of linear equations.

33. $\begin{cases} 2x - 5y = 4 \\ x + 2y = -7 \end{cases}$

34. $\begin{cases} 3x - y = 2 \\ -5x + 2y = 0 \end{cases}$

35. $\begin{cases} 4x + 2y = 5 \\ 2x + y = -1 \end{cases}$

36. $\begin{cases} 3x + 6y = 15 \\ 2x + 4y = 3 \end{cases}$

37. $\begin{cases} 2x + 2y + z = 1 \\ -x + y + 2z = 3 \\ x + 2y + 4z = 0 \end{cases}$

38. $\begin{cases} 2x - 3y + z = 5 \\ x + y + z = 0 \\ 4x + 2y + 4z = 4 \end{cases}$

39. $\begin{cases} \frac{2}{3}x - \frac{3}{4}y = -1 \\ -\frac{1}{6}x + \frac{3}{4}y = \frac{5}{2} \end{cases}$

40. $\begin{cases} \frac{1}{2}x - \frac{1}{3}y = -3 \\ \frac{1}{8}x + \frac{1}{6}y = 0 \end{cases}$

41. $\begin{cases} 0.7x - 0.2y = -1.6 \\ 0.2x - y = -1.4 \end{cases}$

42. $\begin{cases} -0.7x + 0.6y = 1.3 \\ 0.5x - 0.3y = -0.8 \end{cases}$

43. $\begin{cases} -2x + 4y - 2z = 6 \\ x - 2y + z = -3 \\ 3x - 6y + 3z = -9 \end{cases}$

44. $\begin{cases} -x - y + 3z = 2 \\ 4x + 4y - 12z = -8 \\ -3x - 3y + 9z = 6 \end{cases}$

45. $\begin{cases} x - 2y + z = -5 \\ 3y + 2z = 4 \\ 3x - y = -2 \end{cases}$

46. $\begin{cases} 4x + 5y = 10 \\ 3y + 2z = -6 \\ x + y + z = 3 \end{cases}$

Skill Review

Multiply both sides of equation (1) by 2, and add the resulting equation to equation (2). See Section 10.2.

47. $\begin{array}{ll} 3x - y + z = 2 & (1) \\ -x + 2y + 3z = 6 & (2) \end{array}$

48. $\begin{array}{ll} 2x + y + 3z = 7 & (1) \\ -4x + y + 2z = 4 & (2) \end{array}$

Multiply both sides of equation (1) by −3, and add the resulting equation to equation (2). See Section 10.2.

49. $\begin{array}{ll} x + 2y - z = 0 & (1) \\ 3x + y - z = 2 & (2) \end{array}$

50. $\begin{array}{ll} 2x - 3y + 2z = 5 & (1) \\ x - 9y + z = -1 & (2) \end{array}$

Multiply. See Section 8.7.

51. $(5 + i)(2 - 3i)$

52. $(7 - 2i)(1 + i)$

53. $(3 - 4i)(3 + 4i)$

54. $(2 + 9i)(2 + 9i)$

A Look Ahead

Find the value of each determinant. See the following example.

EXAMPLE Evaluate the determinant

$$\begin{vmatrix} 2 & 0 & -1 & 3 \\ 0 & 5 & -2 & -1 \\ 3 & 1 & 0 & 1 \\ 4 & 2 & -2 & 0 \end{vmatrix}$$

Solution: To evaluate a 4 × 4 determinant, select any row or column and expand by the minors. The array of signs for a 4 × 4 determinant is the same as for a 3 × 3 determinant except expanded. We expand using the fourth row.

$$\begin{vmatrix} 2 & 0 & -1 & 3 \\ 0 & 5 & -2 & -1 \\ 3 & 1 & 0 & 1 \\ 4 & 2 & -2 & 0 \end{vmatrix} \leftarrow = -4 \cdot \begin{vmatrix} 0 & -1 & 3 \\ 5 & -2 & -1 \\ 1 & 0 & 1 \end{vmatrix} + 2 \cdot \begin{vmatrix} 2 & -1 & 3 \\ 0 & -2 & -1 \\ 3 & 0 & 1 \end{vmatrix} - (-2) \cdot \begin{vmatrix} 2 & 0 & 3 \\ 0 & 5 & -1 \\ 3 & 1 & 1 \end{vmatrix} + 0 \cdot \begin{vmatrix} 2 & 0 & -1 \\ 0 & 5 & -2 \\ 3 & 1 & 0 \end{vmatrix}$$

Now find the value of each 3 × 3 determinant. The value of the 4 × 4 determinant is

$$-4(12) + 2(17) + 2(-33) + 0 = -80$$

55. $\begin{vmatrix} 5 & 0 & 0 & 0 \\ 0 & 4 & 2 & -1 \\ 1 & 3 & -2 & 0 \\ 0 & -3 & 1 & 2 \end{vmatrix}$

56. $\begin{vmatrix} 1 & 7 & 0 & -1 \\ 1 & 3 & -2 & 0 \\ 1 & 0 & -1 & 2 \\ 0 & -6 & 2 & 4 \end{vmatrix}$

57. $\begin{vmatrix} 4 & 0 & 2 & 5 \\ 0 & 3 & -1 & 1 \\ 0 & 0 & 2 & 0 \\ 0 & 0 & 0 & 1 \end{vmatrix}$

58. $\begin{vmatrix} 2 & 0 & -1 & 4 \\ 6 & 0 & 4 & 1 \\ 2 & 4 & 3 & -1 \\ 4 & 0 & 5 & -4 \end{vmatrix}$

Writing in Mathematics

59. If the elements in a single row of a determinant are all zero, what is the value of the determinant? Explain your answer. Is the answer the same if "row" is replaced by "column"?

10.5 Solving Systems of Equations by Matrices

OBJECTIVES

Tape IA 28

1 Use matrices to solve a system of two equations.

2 Use matrices to solve a system of three equations.

By now, you have seen that the solution of a system of equations depends on the coefficients of the equations in the system. Cramer's rule gives formulas for the coordinates of the solution in terms of these coefficients. In this section, we introduce solving a system of equations by a **matrix** (plural **matrices**).

1 A matrix is a rectangular array of numbers. The following are examples of matrices.

$$\begin{bmatrix} 1 & 0 \\ 0 & 1 \end{bmatrix} \qquad \begin{bmatrix} 2 & 1 & 3 & -1 \\ 0 & -1 & 4 & 5 \\ -6 & 2 & 1 & 0 \end{bmatrix} \qquad \begin{bmatrix} a & b & c \\ d & e & f \end{bmatrix}$$

2 × 2 matrix
2 rows, 2 columns

3 × 4 matrix
3 rows, 4 columns

2 × 3 matrix
2 rows, 3 columns

To see the relationship between systems of equations and matrices, consider this system of equations, written in standard form.

$$\begin{cases} 2x - 3y = 6 \\ x + y = 0 \end{cases}$$

A corresponding matrix associated with this system is

$$\left[\begin{array}{cc:c} 2 & -3 & 6 \\ 1 & 1 & 0 \end{array}\right]$$

The coefficients of each variable are placed to the left of a vertical dashed line. The constants are placed to the right. This 2 × 3 matrix is called the **augmented matrix of the system.** Observe that the rows of this augmented matrix correspond to the equations in the system. The first equation corresponds to the first row; the second equation corresponds to the second row.

The method of solving systems by matrices is to write the augmented matrix as an equivalent matrix from which we can easily find the solution. Two matrices are equivalent if they represent systems that have the same solution set. The following **row operations** can be performed on matrices, and the result is an equivalent matrix.

> **Elementary Row Operations**
>
> **1.** Any two rows in a matrix may be interchanged.
>
> **2.** The elements of any row may be multiplied (or divided) by the same nonzero number.
>
> **3.** The elements of any row may be multiplied (or divided) by a nonzero number and added to its corresponding elements in any other row.

EXAMPLE 1 Solve the system using matrices.

$$\begin{cases} x + 3y = 5 \\ 2x - y = -4 \end{cases}$$

Solution: The augmented matrix is $\left[\begin{array}{cc|c} 1 & 3 & 5 \\ 2 & -1 & -4 \end{array}\right]$. Use elementary row operations to write an equivalent matrix that has 1's along the main diagonal and 0's below each 1 in the main diagonal. The main diagonal of a matrix is the left-to-right diagonal starting with row 1, column 1. For the matrix given, the element in the first row, first column is already 1, as desired. Next we write an equivalent matrix with a 0 below the 1. To do this, multiply row 1 by -2 and add to row 2. **We will only change row 2.**

$$\left[\begin{array}{cc|c} 1 & 3 & 5 \\ \underset{\substack{\uparrow \\ \text{row 1} \\ \text{element}}}{-2(1)} + \underset{\substack{\uparrow \\ \text{row 2} \\ \text{element}}}{2} & \underset{\substack{\uparrow \\ \text{row 1} \\ \text{element}}}{-2(3)} + \underset{\substack{\uparrow \\ \text{row 2} \\ \text{element}}}{(-1)} & \underset{\substack{\uparrow \\ \text{row 1} \\ \text{element}}}{-2(5)} + \underset{\substack{\uparrow \\ \text{row 2} \\ \text{element}}}{(-4)} \end{array}\right] \quad \text{simplifies to} \quad \left[\begin{array}{cc|c} 1 & 3 & 5 \\ 0 & -7 & -14 \end{array}\right]$$

Now continue down the main diagonal and change the -7 to a 1 by use of an elementary row operation. Divide row 2 by -7. Then

$$\left[\begin{array}{cc|c} 1 & 3 & 5 \\ \frac{0}{-7} & \frac{-7}{-7} & \frac{-14}{-7} \end{array}\right] \quad \text{simplifies to} \quad \left[\begin{array}{cc|c} 1 & 3 & 5 \\ 0 & 1 & 2 \end{array}\right]$$

This last matrix corresponds to the system

$$\begin{cases} 1x + 3y = 5 \\ 0x + 1y = 2 \end{cases} \quad \text{or} \quad \begin{cases} x + 3y = 5 \\ y = 2 \end{cases}$$

To find x, let $y = 2$ in the first equation, $x + 3y = 5$.

$$x + 3y = 5 \qquad \text{First equation.}$$

$$x + 3(2) = 5 \qquad \text{Let } y = 2.$$

$$x = -1$$

The solution set is $\{(-1, 2)\}$. Check to see that this ordered pair satisfies both equations. ■

2 Solving a system of three equations in three variables using matrices means writing the corresponding matrix and finding an equivalent matrix that has 1's along the main diagonal and 0's below the 1's.

EXAMPLE 2 Solve the system using matrices.

$$\begin{cases} x + 2y + z = 2 \\ -2x - y + 2z = 5 \\ x + 3y - 2z = -8 \end{cases}$$

Solution: The corresponding matrix is $\left[\begin{array}{ccc|c} 1 & 2 & 1 & 2 \\ -2 & -1 & 2 & 5 \\ 1 & 3 & -2 & -8 \end{array}\right]$. Our goal is to write an equivalent matrix with 1's on the main diagonal and 0's below the 1's. The element in row 1, column 1 is already 1. Next get 0's for each element in the rest of column 1. To do this, first multiply the elements of row 1 by 2, and add the new elements to row 2. Also, we multiply the elements of row 1 by -1 and add the new elements to the elements of row 3. We **do not change row 1.** Then

$$\left[\begin{array}{ccc|c} 1 & 2 & 1 & 2 \\ 2(1) - 2 & 2(2) - 1 & 2(1) + 2 & 2(2) + 5 \\ -1(1) + 1 & -1(2) + 3 & -1(1) - 2 & -1(2) - 8 \end{array}\right] \text{ simplifies to } \left[\begin{array}{ccc|c} 1 & 2 & 1 & 2 \\ 0 & 3 & 4 & 9 \\ 0 & 1 & -3 & -10 \end{array}\right]$$

Now continue down the diagonal and use elementary row operations to get 1 where the element 3 is now. To do this, interchange rows 2 and 3.

$$\left[\begin{array}{ccc|c} 1 & 2 & 1 & 2 \\ 0 & 3 & 4 & 9 \\ 0 & 1 & -3 & -10 \end{array}\right] \text{ is equivalent to } \left[\begin{array}{ccc|c} 1 & 2 & 1 & 2 \\ 0 & 1 & -3 & -10 \\ 0 & 3 & 4 & 9 \end{array}\right]$$

Next we want the row 3, column 2 element to be 0. Multiply the elements of row 2 by -3, and add the new elements to the elements of row 3.

$$\left[\begin{array}{ccc|c} 1 & 2 & 1 & 2 \\ 0 & 1 & -3 & -10 \\ -3(0) + 0 & -3(1) + 3 & -3(-3) + 4 & -3(-10) + 9 \end{array}\right] \text{ simplifies to } \left[\begin{array}{ccc|c} 1 & 2 & 1 & 2 \\ 0 & 1 & -3 & -10 \\ 0 & 0 & 13 & 39 \end{array}\right]$$

Finally, we divide the elements of row 3 by 13, so that the final main diagonal element is 1.

$$\left[\begin{array}{ccc|c} 1 & 2 & 1 & 2 \\ 0 & 1 & -3 & -10 \\ \frac{0}{13} & \frac{0}{13} & \frac{13}{13} & \frac{39}{13} \end{array}\right] \text{ simplifies to } \left[\begin{array}{ccc|c} 1 & 2 & 1 & 2 \\ 0 & 1 & -3 & -10 \\ 0 & 0 & 1 & 3 \end{array}\right]$$

This matrix corresponds to the system

$$\begin{cases} x + 2y + z = 2 \\ y - 3z = -10 \\ z = 3 \end{cases}$$

We identify the z-coordinate of the solution as 3. Replace z with 3 in the second equation and solve for y.

$$y - 3z = -10 \qquad \text{Second equation.}$$
$$y - 3(3) = -10 \qquad \text{Let } z = 3.$$
$$y = -1$$

To find x, let $z = 3$ and $y = -1$ in the first equation.

$$x + 2y + z = 2 \qquad \text{First equation.}$$
$$x + 2(-1) + 3 = 2 \qquad \text{Let } z = 3 \text{ and } y = -1.$$
$$x = 1$$

The solution set is $\{(1, -1, 3)\}$. Check to see that it satisfies the original system. ∎

EXAMPLE 3 Solve the system using matrices.

$$\begin{cases} 2x - y = 3 \\ 4x - 2y = 5 \end{cases}$$

Solution: The corresponding augmented matrix is $\left[\begin{array}{cc|c} 2 & -1 & 3 \\ 4 & -2 & 5 \end{array}\right]$. To get 1 in the row 1, column 1 position, divide the elements of row 1 by 2.

$$\left[\begin{array}{cc|c} \frac{2}{2} & -\frac{1}{2} & \frac{3}{2} \\ 4 & -2 & 5 \end{array}\right] \text{ simplifies to } \left[\begin{array}{cc|c} 1 & -\frac{1}{2} & \frac{3}{2} \\ 4 & -2 & 5 \end{array}\right]$$

To get 0 under the 1, multiply the elements of row 1 by -4 and add the new elements to the elements of row 2.

$$\left[\begin{array}{cc|c} 1 & -\frac{1}{2} & \frac{3}{2} \\ -4(1) + 4 & -4\left(-\frac{1}{2}\right) - 2 & -4\left(\frac{3}{2}\right) + 5 \end{array}\right] \text{ simplifies to } \left[\begin{array}{cc|c} 1 & -\frac{1}{2} & \frac{3}{2} \\ 0 & 0 & -1 \end{array}\right]$$

The corresponding system is $\begin{cases} x - \frac{1}{2}y = \frac{3}{2} \\ 0 = -1 \end{cases}$. The equation $0 = -1$ is false for all values of y or x, and hence the system is inconsistent and has no solution. ∎

EXERCISE SET 10.5

Solve each system of linear equations using matrices. See Example 1.

1. $\begin{cases} x + y = 1 \\ x - 2y = 4 \end{cases}$

2. $\begin{cases} 2x - y = 8 \\ x + 3y = 11 \end{cases}$

3. $\begin{cases} 2y - 4 = 0 \\ x + 2y = 0 \end{cases}$

4. $\begin{cases} 4x - y = 5 \\ 3x - 3 = 0 \end{cases}$

Solve each system of linear equations using matrices. See Example 2.

5. $\begin{cases} x + y = 3 \\ 2y = 10 \\ 3x + 2y - 4z = 12 \end{cases}$

6. $\begin{cases} 5x = 5 \\ 2x + y = 4 \\ 3x + y - 5z = -15 \end{cases}$

7. $\begin{cases} 2y - z = -7 \\ x + 4y + z = -4 \\ 5x - y + 2z = 13 \end{cases}$

8. $\begin{cases} 4y + 3z = -2 \\ 5x - 4y = 1 \\ -5x + 4y + z = -3 \end{cases}$

Solve each system of linear equations using matrices. See Example 3.

9. $\begin{cases} x - 2y = 4 \\ 2x - 4y = 4 \end{cases}$

10. $\begin{cases} -x + 3y = 6 \\ 3x - 9y = 9 \end{cases}$

11. $\begin{cases} 3x - 3y = 9 \\ 2x - 2y = 6 \end{cases}$

12. $\begin{cases} 9x - 3y = 6 \\ -18x + 6y = -12 \end{cases}$

Solve each system of linear equations using matrices.

13. $\begin{cases} x - 4 = 0 \\ x + y = 1 \end{cases}$

14. $\begin{cases} 3y = 6 \\ x + y = 7 \end{cases}$

15. $\begin{cases} x + y + z = 2 \\ 2x - z = 5 \\ 3y + z = 2 \end{cases}$

16. $\begin{cases} x + 2y + z = 5 \\ x - y - z = 3 \\ y + z = 2 \end{cases}$

17. $\begin{cases} 5x - 2y = 27 \\ -3x + 5y = 18 \end{cases}$

18. $\begin{cases} 4x - y = 9 \\ 2x + 3y = -27 \end{cases}$

19. $\begin{cases} 4x - 7y = 7 \\ 12x - 21y = 24 \end{cases}$

20. $\begin{cases} 2x - 5y = 12 \\ -4x + 10y = 20 \end{cases}$

21. $\begin{cases} 4x - y + 2z = 5 \\ 2y + z = 4 \\ 4x + y + 3z = 10 \end{cases}$

22. $\begin{cases} 5y - 7z = 14 \\ 2x + y + 4z = 10 \\ 2x + 6y - 3z = 30 \end{cases}$

23. $\begin{cases} 4x + y + z = 3 \\ -x + y - 2z = -11 \\ x + 2y + 2z = -1 \end{cases}$

24. $\begin{cases} x + y + z = 9 \\ 3x - y + z = -1 \\ -2x + 2y - 3z = -2 \end{cases}$

Skill Review

Solve. See Section 2.5.

25. Twice a number subtracted from 25 is 13. Find the number.

26. Three-fourths of a number is 21. Find the number.

27. Five less than four times a number is three more than twice the same number. Find the number.

28. Eight more than a number is 2 less than 3 times the number. Find the number.

29. A rectangle is 4 kilometers longer than it is wide. The perimeter is 28 kilometers. Find its dimensions.

30. The perimeter of a painting shaped like an equilateral triangle (three sides equal) is 81 inches. Find the length of a side.

Sketch the graph of each equation. See Sections 9.7 and 9.8.

31. $(x - 5)^2 + (y + 1)^2 = 9$ **32.** $\frac{x^2}{9} + \frac{y^2}{4} = 1$ **33.** $x^2 - 3y = 1$ **34.** $x^2 + y^2 = 5$

10.6 Solving Nonlinear Systems of Equations

OBJECTIVES

1. Solve a nonlinear system by substitution.
2. Solve a nonlinear system by elimination.

Tape IA 29

In Section 10.1, we used graphing, substitution, and elimination methods to find solutions of systems of linear equations. We now apply these same methods to nonlinear equations. A **nonlinear system** of equations is a system of equations at least one of which is not linear. Since we will be graphing the equations in each system, we are interested in real number solutions only.

1

EXAMPLE 1 Solve the system

$$\begin{cases} x^2 - 3y = 1 \\ x - y = 1 \end{cases}$$

Solution: We can solve this system by substitution if we solve one equation for one of the variables. Solving the first equation for x would be a poor choice since we would then need to take a square root. We solve the second equation for y.

$x - y = 1$ Second equation.

$x - 1 = y$ Solve for y.

Replace y with $x - 1$ in the first equation, and then solve for x.

$x^2 - 3y = 1$ First equation.

$x^2 - 3(x - 1) = 1$ Replace y with $x - 1$.

$x^2 - 3x + 3 = 1$

$x^2 - 3x + 2 = 0$

$(x - 2)(x - 1) = 0$

$x = 2$ or $x = 1$

Let $x = 2$ and then $x = 1$ in the equation $y = x - 1$ to find corresponding y-values.

Let $x = 2$.	Let $x = 1$.
$y = x - 1$	$y = x - 1$
$y = 2 - 1 = 1$	$y = 1 - 1 = 0$

The solution set is $\{(2, 1), (1, 0)\}$. Check both solutions in both equations. Both solutions will satisfy both equations, so both are solutions of the system. The graph of each equation in the system is next.

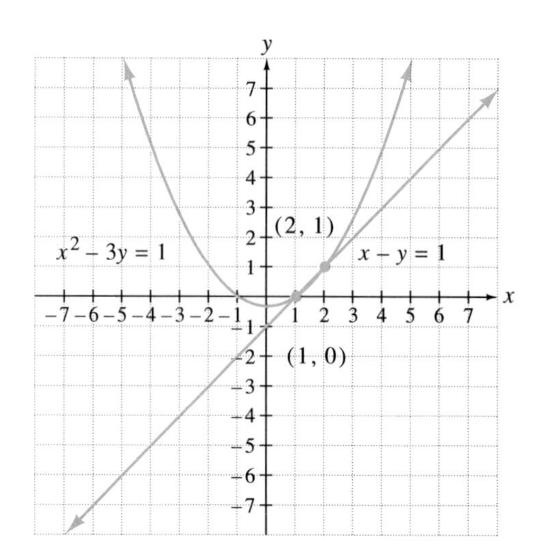

■

EXAMPLE 2 Solve the system

$$\begin{cases} y = \sqrt{x} \\ x^2 + y^2 = 6 \end{cases}$$

Solution: This system is ideal for substitution since y is expressed in terms of x in the first equation. Substitute $\sqrt{x}$ for y in the second equation, and solve for x.

$$x^2 + y^2 = 6$$

$$x^2 + (\sqrt{x})^2 = 6 \qquad \text{Let } y = \sqrt{x}.$$

$$x^2 + x = 6$$

$$x^2 + x - 6 = 0$$

$$(x + 3)(x - 2) = 0$$

$$x = -3 \quad \text{or} \quad x = 2$$

Let $x = -3$, and then $x = 2$ in the first equation to find corresponding y-values.

Let $x = -3$.	Let $x = 2$.
$y = \sqrt{x}$	$y = \sqrt{x}$
$y = \sqrt{-3}$	$y = \sqrt{2}$
Not a real number.	

Since we are interested in real number solutions, the only solution is $(2, \sqrt{2})$. The solution set is $\{(2, \sqrt{2})\}$. Check to see that this solution satisfies both equations. The graph of each equation in this system is shown next.

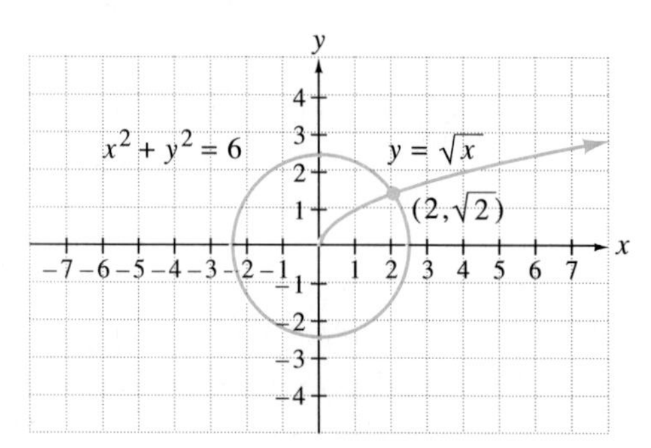

■

2

EXAMPLE 3 Solve the system

$$\begin{cases} x^2 + 2y^2 = 10 \\ x^2 - \quad y^2 = \quad 1 \end{cases}$$

Solution: Use addition or elimination to solve this system. To eliminate x^2 when we add the two equations, multiply both sides of the second equation by -1. Then

$$\begin{cases} x^2 + 2y^2 = 10 \\ -1\,(x^2 - y^2) = -1 \cdot 1 \end{cases} \quad \text{simplifies to} \quad \begin{cases} x^2 + 2y^2 = 10 \\ -x^2 + y^2 = -1 \end{cases}$$

$$3y^2 = 9$$

$$y^2 = 3$$

$$y = \pm\sqrt{3}$$

To find corresponding x-values, let $y = \sqrt{3}$ and $y = -\sqrt{3}$ in either original equation. We choose the second equation.

Let $y = \sqrt{3}$.	Let $y = -\sqrt{3}$.
$x^2 - y^2 = 1$	$x^2 - y^2 = 1$
$x^2 - (\sqrt{3})^2 = 1$	$x^2 - (-\sqrt{3})^2 = 1$
$x^2 - 3 = 1$	$x^2 - 3 = 1$
$x^2 = 4$	$x^2 = 4$
$x = \pm\sqrt{4} = \pm 2$	$x = \pm\sqrt{4} = \pm 2$

The solution set is $\{(2, \sqrt{3}), (-2, \sqrt{3}), (2, -\sqrt{3}), (-2, -\sqrt{3})\}$. Check all four ordered pairs in both equations of the system. The graph of each equation in this system is shown next.

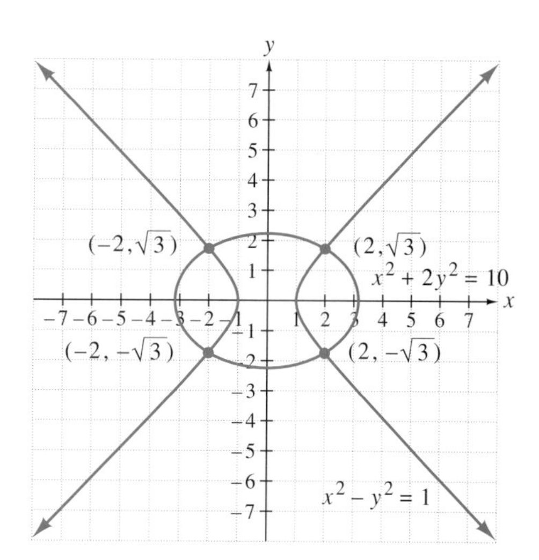

■

EXAMPLE 4 Solve the system

$$\begin{cases} x^2 + y^2 = 4 \\ x + y = 3 \end{cases}$$

Solution: The addition method is not a good choice here, since x and y are squared in the first equation but not in the second equation. Use the substitution method and solve the second equation for x.

$$x + y = 3 \qquad \text{Second equation.}$$
$$x = 3 - y$$

Let $x = 3 - y$ in the first equation.

$$x^2 + y^2 = 4 \qquad \text{First equation.}$$
$$(3 - y)^2 + y^2 = 4$$
$$9 - 6y + y^2 + y^2 = 4$$
$$2y^2 - 6y + 5 = 0$$

By the quadratic formula, where $a = 2$, $b = -6$, and $c = 5$, we have

$$y = \frac{6 \pm \sqrt{(-6)^2 - 4 \cdot 2 \cdot 5}}{2 \cdot 2} = \frac{6 \pm \sqrt{-4}}{4}$$

Since $\sqrt{-4}$ is not a real number, there is no solution. Graphically, the circle and the line do not intersect, as shown.

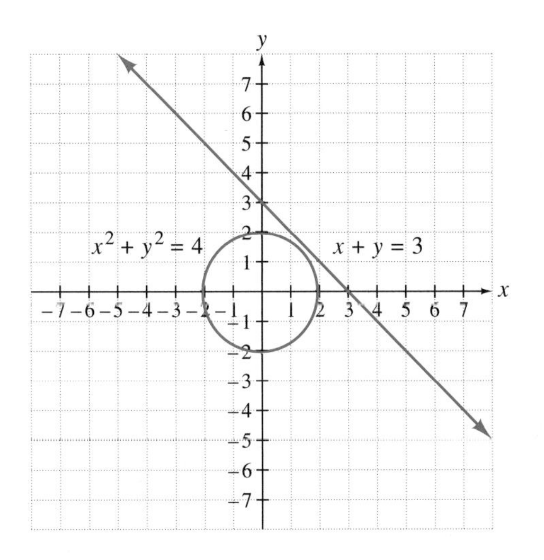

■

EXERCISE SET 10.6

Solve each nonlinear system of equations. See Examples 1 and 2.

1. $\begin{cases} x^2 + y^2 = 25 \\ 4x + 3y = 0 \end{cases}$

2. $\begin{cases} x^2 + y^2 = 25 \\ 3x + 4y = 0 \end{cases}$

3. $\begin{cases} x^2 + 4y^2 = 10 \\ y = x \end{cases}$

4. $\begin{cases} 4x^2 + y^2 = 10 \\ y = x \end{cases}$

5. $\begin{cases} y^2 = 4 - x \\ x - 2y = 4 \end{cases}$

6. $\begin{cases} x^2 + y^2 = 4 \\ x + y = -2 \end{cases}$

Solve each system of equations. See Example 3.

7. $\begin{cases} x^2 + y^2 = 9 \\ 16x^2 - 4y^2 = 64 \end{cases}$

8. $\begin{cases} 4x^2 + 3y^2 = 35 \\ 5x^2 + 2y^2 = 42 \end{cases}$

Solve each system of equations. See Example 4.

9. $\begin{cases} x^2 + 2y^2 = 2 \\ x - y = 2 \end{cases}$

10. $\begin{cases} x^2 + 2y^2 = 2 \\ x^2 - 2y^2 = 6 \end{cases}$

Solve each system of equations.

11. $\begin{cases} y = x^2 - 3 \\ 4x - y = 6 \end{cases}$

12. $\begin{cases} y = x + 1 \\ x^2 - y^2 = 1 \end{cases}$

13. $\begin{cases} y = x^2 \\ 3x + y = 10 \end{cases}$

14. $\begin{cases} 6x - y = 5 \\ xy = 1 \end{cases}$

15. $\begin{cases} y = 2x^2 + 1 \\ x + y = -1 \end{cases}$

16. $\begin{cases} x^2 + y^2 = 9 \\ x + y = 5 \end{cases}$

17. $\begin{cases} y = x^2 - 4 \\ y = x^2 - 4x \end{cases}$

18. $\begin{cases} x = y^2 - 3 \\ x = y^2 - 3y \end{cases}$

19. $\begin{cases} 2x^2 + 3y^2 = 14 \\ -x^2 + y^2 = 3 \end{cases}$

20. $\begin{cases} 4x^2 - 2y^2 = 2 \\ -x^2 + y^2 = 2 \end{cases}$

21. $\begin{cases} x^2 + y^2 = 1 \\ x^2 + (y + 3)^2 = 4 \end{cases}$

22. $\begin{cases} x^2 + 2y^2 = 4 \\ x^2 - y^2 = 4 \end{cases}$

23. $\begin{cases} y = x^2 + 2 \\ y = -x^2 + 4 \end{cases}$

24. $\begin{cases} x = -y^2 - 3 \\ x = y^2 - 5 \end{cases}$

25. $\begin{cases} 3x^2 + y^2 = 9 \\ 3x^2 - y^2 = 9 \end{cases}$

26. $\begin{cases} x^2 + y^2 = 25 \\ x = y^2 - 5 \end{cases}$

27. $\begin{cases} x^2 + 3y^2 = 6 \\ x^2 - 3y^2 = 10 \end{cases}$

28. $\begin{cases} x^2 + y^2 = 1 \\ y = x^2 - 9 \end{cases}$

29. $\begin{cases} x^2 + y^2 = 36 \\ y = \frac{1}{6}x^2 - 6 \end{cases}$

30. $\begin{cases} x^2 + y^2 = 16 \\ y = -\frac{1}{4}x^2 + 4 \end{cases}$

Skill Review

Graph each inequality in two variables. See Section 4.6.

31. $x > -3$

32. $y \le 1$

33. $y < 2x - 1$

34. $3x - y \le 4$

Find the perimeter of each geometric figure. See Section 5.3.

35.

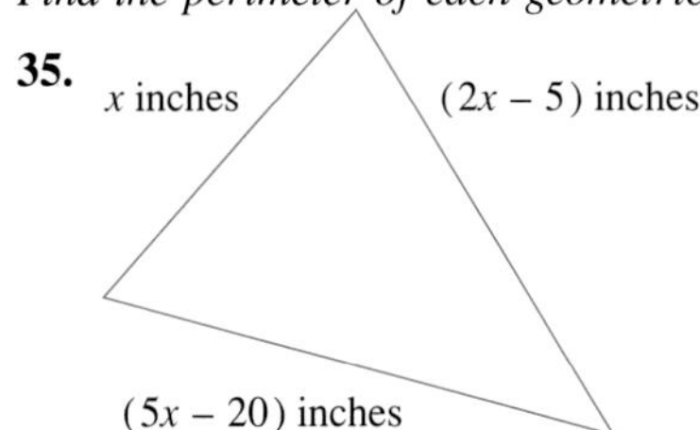

36.

37.

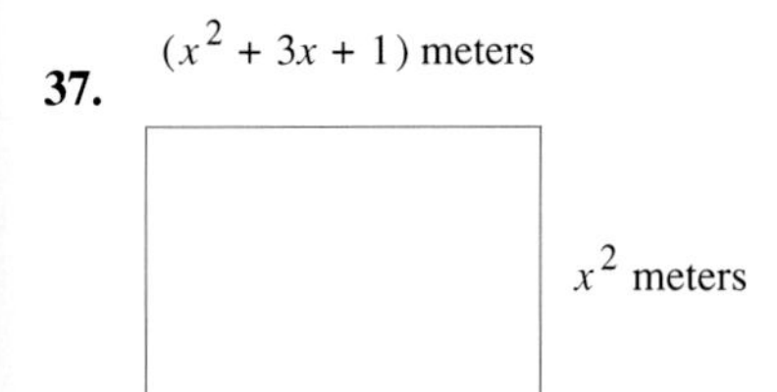

38.

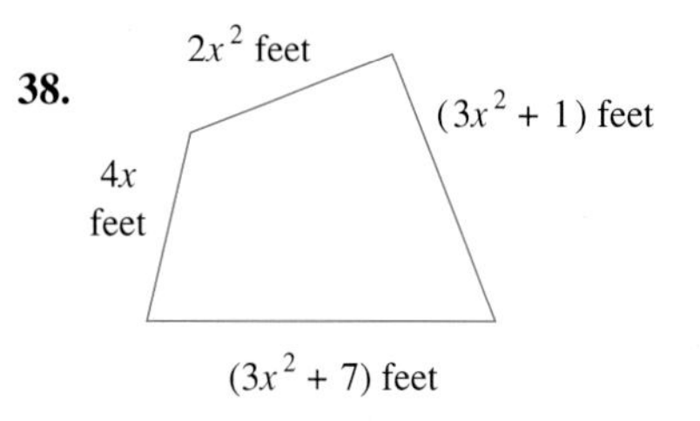

10.7 Nonlinear Inequalities and Systems of Inequalities

OBJECTIVES

1. Sketch the graph of a nonlinear inequality.
2. Sketch the solution set of a system of nonlinear inequalities.

Tape IA 29

1 We can graph a nonlinear inequality in two variables such as $\frac{x^2}{9} + \frac{y^2}{16} \le 1$ in a way similar to the way we graphed a linear inequality in two variables in Section 4.6. First we graph the related equation $\frac{x^2}{9} + \frac{y^2}{16} = 1$. The graph of the equation is our boundary. Then, using test points, we shade the region whose points satisfy the inequality.

EXAMPLE 1 Graph $\frac{x^2}{9} + \frac{y^2}{16} \le 1$.

Solution: First graph the equation $\frac{x^2}{9} + \frac{y^2}{16} = 1$. We graph using a solid line since the graph of $\frac{x^2}{9} + \frac{y^2}{16} \le 1$ includes the graph of $\frac{x^2}{9} + \frac{y^2}{16} = 1$. The graph is an ellipse, and it divides the plane into two regions, the "inside" and the "outside" of the ellipse. To determine which region contains the solutions, select a test point in one of the regions and determine whether the coordinates of the point satisfy the inequality. We choose (0, 0) as the test point.

$$\frac{x^2}{9} + \frac{y^2}{16} \le 1$$

$$\frac{0^2}{9} + \frac{0^2}{16} \le 1 \qquad \text{Let } x = 0 \text{ and } y = 0.$$

$$0 \le 1 \qquad \text{True.}$$

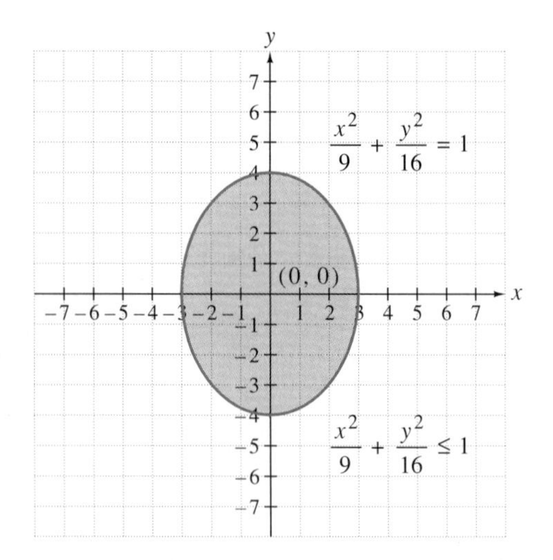

Since this statement is true, we shade the region containing (0, 0). The graph of the solution set is the set of points on and within the ellipse, as shaded in the figure. ■

EXAMPLE 2 Graph $4y^2 > x^2 + 16$.

Solution: The related equation is $4y^2 = x^2 + 16$ or $\frac{y^2}{4} - \frac{x^2}{16} = 1$, which is a hyperbola. Graph the hyperbola using a dashed line since the graph of $4y^2 > x^2 + 16$ does **not** include the graph of $4y^2 = x^2 + 16$. The hyperbola divides the plane into three regions. Select a test point in each region to determine whether that region contains solutions of the inequality.

Test region A with (0, 4)	Test region B with (0, 0)	Test region C with (0, −4)
$4y^2 > x^2 + 16$	$4y^2 > x^2 + 16$	$4y^2 > x^2 + 16$
$4(4)^2 > 0^2 + 16$	$4(0)^2 > 0^2 + 16$	$4(-4)^2 > 0^2 + 16$
$64 > 16$ True.	$0 > 16$ False.	$64 > 16$ True.

The graph of the solution set includes the shaded regions only, not the boundary.

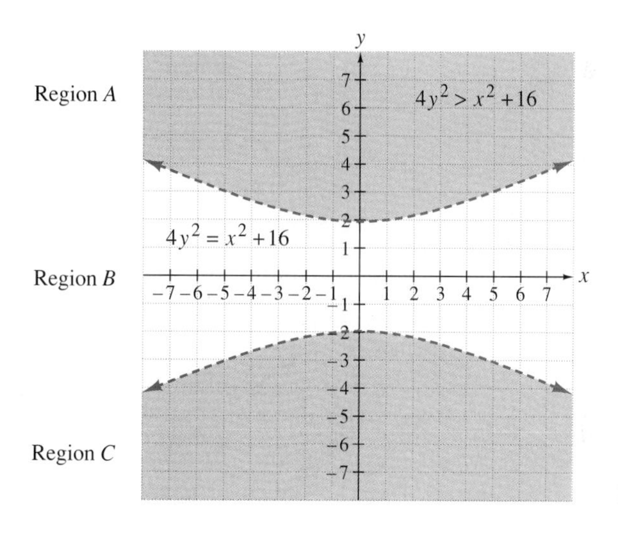

■

2 In Section 4.6, we looked at the intersection of graphs of inequalities in two variables. Although we did not identify them as such, we now can recognize these sets of inequalities as systems of inequalities. The graph of the solution set of a system of inequalities is the intersection of the graphs of the inequalities.

EXAMPLE 3 Graph the system

$$\begin{cases} x \leq 1 - 2y \\ y \leq x^2 \end{cases}$$

Solution: Graph each inequality on the same set of axes. The intersection is the darkest shaded region along with its boundary lines. The coordinates of the points of intersection can be found by solving the related system

$$\begin{cases} x = 1 - 2y \\ y = x^2 \end{cases}$$

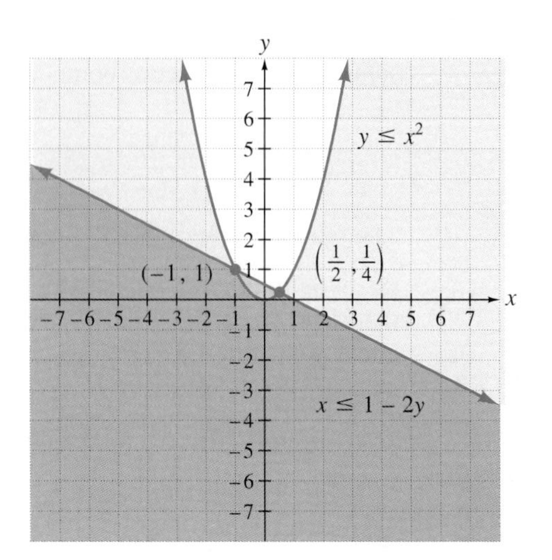

■

EXAMPLE 4 Graph the system

$$\begin{cases} x^2 + y^2 < 25 \\ \dfrac{x^2}{9} - \dfrac{y^2}{25} < 1 \\ y < x + 3 \end{cases}$$

Solution: Graph each inequality. The graph of $x^2 + y^2 < 25$ contains points "inside" the circle that has center (0, 0) and radius 5. The graph of $\frac{x^2}{9} - \frac{y^2}{25} < 1$ is the region between the two branches of the hyperbola with x-intercepts -3 and 3 and center (0, 0). The graph of $y < x + 3$ is the region "below" the line with slope 1 and y-intercept 3. The graph of the solution set of the system is the intersection of all the graphs, the darkest shaded region shown. The boundary of this region is not part of the solution.

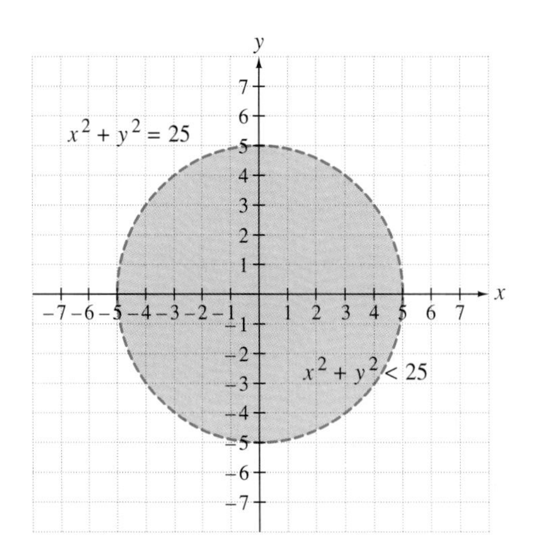

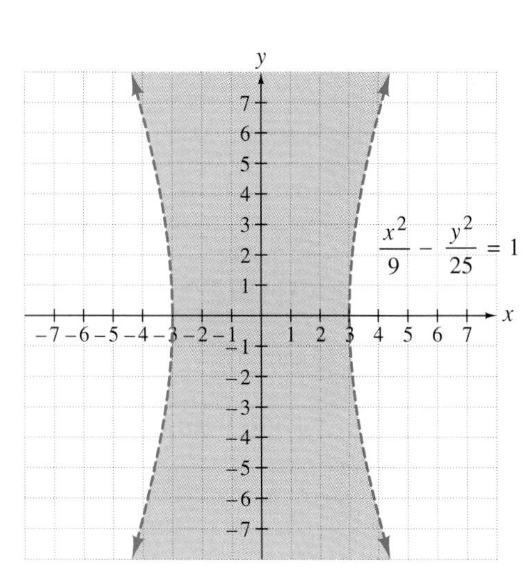

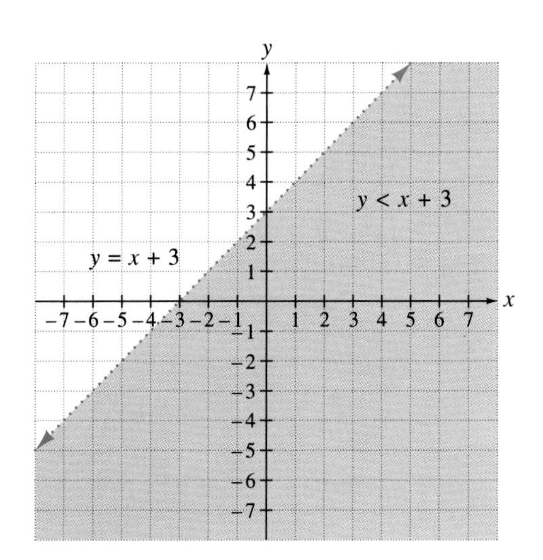

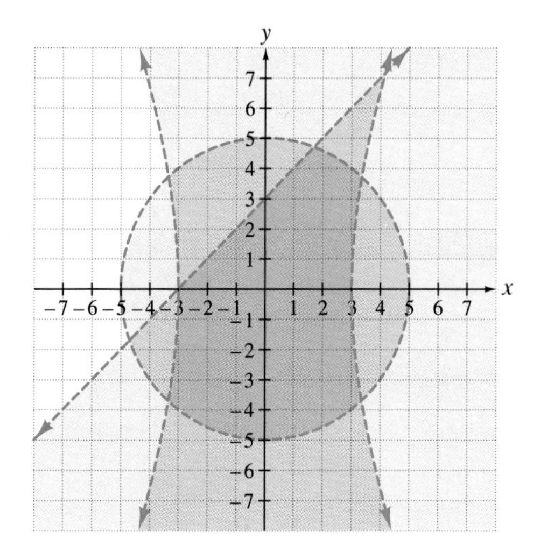

EXERCISE SET 10.7

Graph each inequality. See Examples 1 and 2.

1. $y < x^2$

2. $y < -x^2$

3. $x^2 + y^2 \geq 16$

4. $x^2 + y^2 < 36$

5. $\frac{x^2}{4} - y^2 < 1$

6. $x^2 - \frac{y^2}{9} \geq 1$

Graph the solution of each system. See Examples 3 and 4.

7. $\begin{cases} 2x - y < 2 \\ y \leq -x \end{cases}$

8. $\begin{cases} x - 2y > 4 \\ y > -x^2 \end{cases}$

9. $\begin{cases} 4x + 3y \geq 12 \\ x^2 + y^2 < 16 \end{cases}$

10. $\begin{cases} 3x - 4y \le 12 \\ x^2 + y^2 < 16 \end{cases}$

11. $\begin{cases} x^2 + y^2 \le 9 \\ x^2 + y^2 \ge 1 \end{cases}$

12. $\begin{cases} x^2 + y^2 \ge 9 \\ x^2 + y^2 \ge 16 \end{cases}$

Graph each inequality.

13. $y > (x - 1)^2 - 3$

14. $y > (x + 3)^2 + 2$

15. $x^2 + y^2 \le 9$

16. $x^2 + y^2 > 4$

17. $y > -x^2 + 5$

18. $y < -x^2 + 5$

19. $\frac{x^2}{4} + \frac{y^2}{9} \le 1$

20. $\frac{x^2}{25} + \frac{y^2}{4} \ge 1$

21. $\frac{y^2}{4} - x^2 \le 1$

22. $\frac{y^2}{16} - \frac{x^2}{9} > 1$

23. $y < (x - 2)^2 + 1$

24. $y > (x - 2)^2 + 1$

25. $y \le x^2 + x - 2$

26. $y > x^2 + x - 2$

Graph the solution of each system.

27. $\begin{cases} y > x^2 \\ y \geq 2x + 1 \end{cases}$

28. $\begin{cases} y \leq -x^2 + 3 \\ y \leq 2x - 1 \end{cases}$

29. $\begin{cases} x > y^2 \\ y > 0 \end{cases}$

30. $\begin{cases} x < (y + 1)^2 + 2 \\ x + y \geq 3 \end{cases}$

31. $\begin{cases} x^2 + y^2 > 9 \\ y > x^2 \end{cases}$

32. $\begin{cases} x^2 + y^2 \leq 9 \\ y < x^2 \end{cases}$

33. $\begin{cases} \dfrac{x^2}{4} + \dfrac{y^2}{9} \geq 1 \\ x^2 + y^2 \geq 4 \end{cases}$

34. $\begin{cases} x^2 + (y - 2)^2 \geq 9 \\ \dfrac{x^2}{4} + \dfrac{y^2}{25} < 1 \end{cases}$

35. $\begin{cases} x^2 - y^2 \geq 1 \\ y \geq 0 \end{cases}$

36. $\begin{cases} x^2 - y^2 \geq 1 \\ x \geq 0 \end{cases}$

37. $\begin{cases} x + y \geq 1 \\ 2x + 3y < 1 \\ x > -3 \end{cases}$

38. $\begin{cases} x - y < -1 \\ 4x - 3y > 0 \\ y > 0 \end{cases}$

39. $\begin{cases} x^2 - y^2 < 1 \\ \dfrac{x^2}{16} + y^2 \leq 1 \\ x \geq -2 \end{cases}$

40. $\begin{cases} x^2 - y^2 \geq 1 \\ \dfrac{x^2}{16} + \dfrac{y^2}{4} \leq 1 \\ y \geq 1 \end{cases}$

Skill Review

Solve for y. See Section 2.4.

41. $x + 3y = 5x$

42. $2y + 7x = 3x$

43. $a - 2y = c$

44. $ax + by = c$

Solve each inequality. See Section 9.4.

45. $(x - 1)(x + 2) > 0$

46. $(x + 5)(x - 3) \leq 0$

47. $\dfrac{x}{x - 3} < 0$

48. $\dfrac{5}{x + 2} > 0$

CHAPTER 10 GLOSSARY

When a system has at least one solution, the system is said to be **consistent.** Systems that have no solution are said to be **inconsistent.**

Equations whose graphs are the same are **dependent equations.**

A **determinant** is a real number associated with a square array of numbers written between two vertical bars.

Each number in the array of a determinant is called an **element.**

A **matrix** (plural **matrices**) is a rectangular array of numbers.

Systems in which at least one equation is not linear are **nonlinear systems.**

A **solution of a system** of two equations in two variables is an ordered pair, (x, y), that makes both equations true.

Two or more equations in two or more variables considered simultaneously form a **system of equations.**

CHAPTER 10 SUMMARY

(10.1)

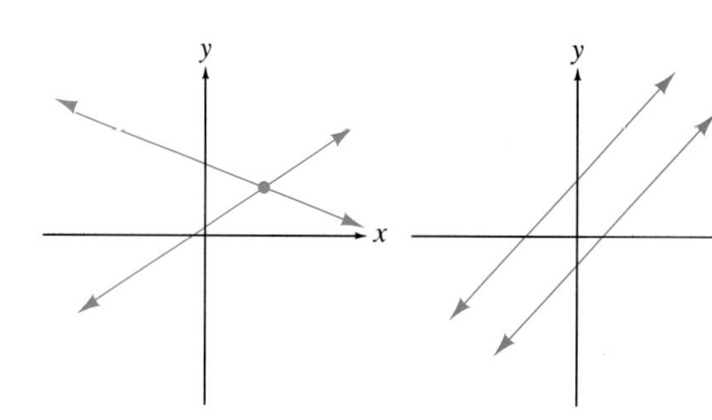

One solution, consistent system

No solution, inconsistent system

Infinite number of solutions, dependent equations

THE VALUE OF A 2 × 2 DETERMINANT (10.4)

$$\begin{vmatrix} a & b \\ c & d \end{vmatrix} = ad - bc$$

CRAMER'S RULE FOR TWO LINEAR EQUATIONS IN TWO VARIABLES (10.4)

The solution of the system $\begin{cases} ax + by = h \\ cx + dy = k \end{cases}$ is given by

$$x = \frac{\begin{vmatrix} h & b \\ k & d \end{vmatrix}}{\begin{vmatrix} a & b \\ c & d \end{vmatrix}} = \frac{D_x}{D}; \qquad y = \frac{\begin{vmatrix} a & h \\ c & k \end{vmatrix}}{\begin{vmatrix} a & b \\ c & d \end{vmatrix}} = \frac{D_y}{D}$$

as long as $D = ad - bc$ is not 0.

VALUE OF A 3 × 3 DETERMINANT (10.4)

$$\begin{vmatrix} a_1 & b_1 & c_1 \\ a_2 & b_2 & c_2 \\ a_3 & b_3 & c_3 \end{vmatrix} = a_1 \cdot \begin{vmatrix} b_2 & c_2 \\ b_3 & c_3 \end{vmatrix} - a_2 \cdot \begin{vmatrix} b_1 & c_1 \\ b_3 & c_3 \end{vmatrix} + a_3 \cdot \begin{vmatrix} b_1 & c_1 \\ b_2 & c_2 \end{vmatrix}$$

CRAMER'S RULE FOR THREE EQUATIONS IN THREE VARIABLES (10.4)

The solution of the system $\begin{cases} a_1x + b_1y + c_1z = k_1 \\ a_2x + b_2y + c_2z = k_2 \\ a_3x + b_3y + c_3z = k_3 \end{cases}$ is given by $x = \dfrac{D_x}{D}$, $y = \dfrac{D_y}{D}$, and $z = \dfrac{D_z}{D}$, where

$$D = \begin{vmatrix} a_1 & b_1 & c_1 \\ a_2 & b_2 & c_2 \\ a_3 & b_3 & c_3 \end{vmatrix} \qquad D_x = \begin{vmatrix} k_1 & b_1 & c_1 \\ k_2 & b_2 & c_2 \\ k_3 & b_3 & c_3 \end{vmatrix}$$

$$D_y = \begin{vmatrix} a_1 & k_1 & c_1 \\ a_2 & k_2 & c_2 \\ a_3 & k_3 & c_3 \end{vmatrix} \qquad D_z = \begin{vmatrix} a_1 & b_1 & k_1 \\ a_2 & b_2 & k_2 \\ a_3 & b_3 & k_3 \end{vmatrix}$$

as long as D is not 0.

ELEMENTARY ROW OPERATIONS (10.5)

1. Any two rows in a matrix may be interchanged.
2. The elements of any row may be multiplied (or divided) by the same nonzero number.
3. The elements of any row may be multiplied (or divided) by a nonzero number and added to its corresponding elements in any other row.

CHAPTER 10 REVIEW

(10.1) *Solve each system of equations in two variables by each of three methods: (1) graphing, (2) substitution, and (3) elimination.*

1. $\begin{cases} 3x + 10y = 1 \\ x + 2y = -1 \end{cases}$

2. $\begin{cases} y = \dfrac{1}{2}x + \dfrac{2}{3} \\ 4x + 6y = 4 \end{cases}$

3. $\begin{cases} 2x - 4y = 22 \\ 5x - 10y = 16 \end{cases}$

4. $\begin{cases} 3x - 6y = 12 \\ 2y = x - 4 \end{cases}$

5. $\begin{cases} \frac{1}{2}x - \frac{3}{4}y = -\frac{1}{2} \\ \frac{1}{8}x + \frac{3}{4}y = \frac{19}{8} \end{cases}$

(10.2) *Solve each system of equations in three variables.*

6. $\begin{cases} x + z = 4 \\ 2x - y = 4 \\ x + y - z = 0 \end{cases}$

7. $\begin{cases} 2x + 5y = 4 \\ x - 5y + z = -1 \\ 4x - z = 11 \end{cases}$

8. $\begin{cases} 4y + 2z = 5 \\ 2x + 8y = 5 \\ 6x + 4z = 1 \end{cases}$

9. $\begin{cases} 5x + 7y = 9 \\ 14y - z = 28 \\ 4x + 2z = -4 \end{cases}$

10. $\begin{cases} 3x - 2y + 2z = 5 \\ -x + 6y + z = 4 \\ 3x + 14y + 7z = 20 \end{cases}$

11. $\begin{cases} x + 2y + 3z = 11 \\ y + 2z = 3 \\ 2x - 2z = 10 \end{cases}$

12. $\begin{cases} 7x - 3y + 2z = 0 \\ 4x - 4y - z = 2 \\ 5x + 2y + 3z = 1 \end{cases}$

13. $\begin{cases} x - 3y - 5z = -5 \\ 4x - 2y + 3z = 13 \\ 5x + 3y + 4z = 22 \end{cases}$

(10.3) *Solve the following applications using systems of equations.*

14. The sum of three numbers is 98. The sum of the first and second is two more than the third number, and the second is four times the first. Find the numbers.

15. Alice Dreyfus's coin purse has 95 coins in it, all dimes, nickels, and pennies, worth \$4.03 total. There are twice as many pennies as dimes and one fewer nickel than dimes. Find how many of each type of coin the purse contains.

16. One number is 3 times a second number, and twice the sum of the numbers is 168. Find the numbers.

17. Sue Maller is 16 years older than Pat O'Brien. In 15 years Sue will be twice as old as Pat is then. How old is each now?

18. Two cars leave Chicago, one traveling east and the other west. After 4 hours they are 492 miles apart. If one car is traveling 7 mph faster than the other, find the speed of each.

19. The foundation for a rectangular Hardware Warehouse has a length three times the width and is 296 feet around. Find the dimensions of the building.

20. James Wilson has available a 10% alcohol solution and a 60% alcohol solution. Find how many liters of each solution he should mix to make 50 liters of 40% alcohol solution.

21. An employee at See's Candy Store needs a special mixture of candy. She has creme-filled chocolates that sell for \$3.00 per pound, chocolate-covered nuts that sell for \$2.70 per pound, and chocolate-covered raisins that sell for \$2.25 per pound. She wants to have twice as many raisins as nuts in the mixture. Find how many pounds of each she should use to make 45 pounds worth \$2.80 per pound.

22. Chris Oliver has \$2.77 in his coin jar, all in pennies, nickels, and dimes. If he has 53 coins in all and four more nickels than dimes, find how many of each type of coin he has.

23. If \$10,000 and \$4000 are invested so that \$1250 is earned in one year, and if the rate of interest on the larger investment is 2% more than that of the smaller investment, find the rates of interest.

24. The perimeter of a triangle is 73 centimeters. If two sides are of equal length and the third side is 7 centimeters longer than the others, find the lengths of the three sides.

25. The sum of three numbers is 295. One number is five more than a second and twice the third. Find the numbers.

(10.4) *Evaluate.*

26. $\begin{vmatrix} -1 & 3 \\ 5 & 2 \end{vmatrix}$

27. $\begin{vmatrix} 3 & -1 \\ 2 & 5 \end{vmatrix}$

28. $\begin{vmatrix} 2 & -1 & -3 \\ 1 & 2 & 0 \\ 3 & -2 & 2 \end{vmatrix}$

29. $\begin{vmatrix} -2 & 3 & 1 \\ 4 & 4 & 0 \\ 1 & -2 & 3 \end{vmatrix}$

Use Cramer's rule to solve each system of equations.

30. $\begin{cases} 3x - 2y = -8 \\ 6x + 5y = 11 \end{cases}$

31. $\begin{cases} 6x - 6y = -5 \\ 10x - 2y = 1 \end{cases}$

32. $\begin{cases} 3x + 10y = 1 \\ x + 2y = -1 \end{cases}$

33. $\begin{cases} y = \dfrac{1}{2}x + \dfrac{2}{3} \\ 4x + 6y = 4 \end{cases}$

34. $\begin{cases} 2x - 4y = 22 \\ 5x - 10y = 16 \end{cases}$

35. $\begin{cases} 3x - 6y = 12 \\ 2y = x - 4 \end{cases}$

36. $\begin{cases} x + z = 4 \\ 2x - y = 0 \\ x \quad y - z = 0 \end{cases}$

37. $\begin{cases} 2x + 5y = 4 \\ x - 5y + z = -1 \\ 4x - z = 11 \end{cases}$

38. $\begin{cases} x + 3y - z = 5 \\ 2x - y - 2z = 3 \\ x + 2y + 3z = 4 \end{cases}$

39. $\begin{cases} 2x - z = 1 \\ 3x - y + 2z = 3 \\ x + y + 3z = -2 \end{cases}$

40. $\begin{cases} x + 2y + 3z = 14 \\ y + 2z = 3 \\ 2x - 2z = 10 \end{cases}$

41. $\begin{cases} 5x + 7y = 9 \\ 14y - z = 28 \\ 4x + 2z = -4 \end{cases}$

(10.5) *Solve each system using matrices.*

42. $\begin{cases} 3x + 10y = 1 \\ x + 2y = -1 \end{cases}$

43. $\begin{cases} 3x - 6y = 12 \\ 2y = x - 4 \end{cases}$

44. $\begin{cases} 3x - 2y = -8 \\ 6x + 5y = 11 \end{cases}$

45. $\begin{cases} 6x - 6y = -5 \\ 10x - 2y = 1 \end{cases}$

46. $\begin{cases} 3x - 6y = 0 \\ 2x + 4y = 5 \end{cases}$

47. $\begin{cases} 5x - 3y = 10 \\ -2x + y = -1 \end{cases}$

48. $\begin{cases} 0.2x - 0.3y = -0.7 \\ 0.5x + 0.3y = 1.4 \end{cases}$

49. $\begin{cases} 3x + 2y = 8 \\ 3x - y = 5 \end{cases}$

50. $\begin{cases} x + z = 4 \\ 2x - y = 0 \\ x + y - z = 0 \end{cases}$

51. $\begin{cases} 2x + 5y = 4 \\ x - 5y + z = -1 \\ 4x - z = 11 \end{cases}$

52. $\begin{cases} 3x - y = 11 \\ x + 2z = 13 \\ y - z = -7 \end{cases}$

53. $\begin{cases} 5x + 7y + 3z = 9 \\ 14y - z = 28 \\ 4x + 2z = -4 \end{cases}$

54. $\begin{cases} 7x - 3y + 2z = 0 \\ 4x - 4y - z = 2 \\ 5x + 2y + 3z = 1 \end{cases}$

55. $\begin{cases} x + 2y + 3z = 14 \\ y + 2z = 3 \\ 2x - 2z = 10 \end{cases}$

(10.6) *Solve each system of equations.*

56. $\begin{cases} y = 2x - 4 \\ y^2 = 4x \end{cases}$

57. $\begin{cases} x^2 + y^2 = 4 \\ x - y = 4 \end{cases}$

58. $\begin{cases} y = x + 2 \\ y = x^2 \end{cases}$

59. $\begin{cases} y = x^2 - 5x + 1 \\ y = -x + 6 \end{cases}$

60. $\begin{cases} 4x - y^2 = 0 \\ 2x^2 + y^2 = 16 \end{cases}$

61. $\begin{cases} x^2 + 4y^2 = 16 \\ x^2 + y^2 = 4 \end{cases}$

62. $\begin{cases} x^2 + y^2 = 10 \\ 9x^2 + y^2 = 18 \end{cases}$

63. $\begin{cases} x^2 + 2y = 9 \\ 5x - 2y = 5 \end{cases}$

64. $\begin{cases} y = 3x^2 + 5x - 4 \\ y = 3x^2 - x + 2 \end{cases}$

65. $\begin{cases} x^2 - 3y^2 = 1 \\ 4x^2 + 5y^2 = 21 \end{cases}$

(10.7) *Graph the inequality or system of inequalities.*

66. $y \le -x^2 + 3$

67. $x^2 + y^2 < 9$

68. $x^2 - y^2 < 1$

69. $\dfrac{x^2}{4} + \dfrac{y^2}{9} \ge 1$

70. $\begin{cases} 2x \le 4 \\ x + y \ge 1 \end{cases}$

71. $\begin{cases} 3x + 4y \le 12 \\ x - 2y > 6 \end{cases}$

72. $\begin{cases} y > x^2 \\ x + y \ge 3 \end{cases}$

73. $\begin{cases} x^2 + y^2 \le 16 \\ x^2 + y^2 \ge 4 \end{cases}$

74. $\begin{cases} x^2 + y^2 < 4 \\ x^2 - y^2 \le 1 \end{cases}$

75. $\begin{cases} x^2 + y^2 < 4 \\ y \ge x^2 - 1 \\ x \ge 0 \end{cases}$

CHAPTER 10 TEST

Evaluate each determinant.

1. $\begin{vmatrix} 4 & -7 \\ 2 & 5 \end{vmatrix}$

2. $\begin{vmatrix} 4 & 0 & 2 \\ 1 & -3 & 5 \\ 0 & -1 & 2 \end{vmatrix}$

Solve each system of equations graphically and then solve by the addition method or the substitution method.

3. $\begin{cases} 2x - y = -1 \\ 5x + 4y = 17 \end{cases}$

4. $\begin{cases} 7x - 14y = 5 \\ x = 2y \end{cases}$

Solve each system.

5. $\begin{cases} 4x - 7y = 29 \\ 2x + 5y = -11 \end{cases}$

6. $\begin{cases} 15x + 6y = 15 \\ 10x + 4y = 10 \end{cases}$

7. $\begin{cases} 2x - 3y = 4 \\ 3y + 2z = 2 \\ x - z = -5 \end{cases}$

8. $\begin{cases} 3x - 2y - z = -1 \\ 2x - 2y = 4 \\ 2x - 2z = -12 \end{cases}$

Solve each system using Cramer's rule.

9. $\begin{cases} 3x - y = 7 \\ 2x + 5y = -1 \end{cases}$

10. $\begin{cases} 4x - 3y = -6 \\ -2x + y = 0 \end{cases}$

11. $\begin{cases} x + y + z = 4 \\ 2x + 5y = 1 \\ x - y - 2z = 0 \end{cases}$

12. $\begin{cases} 3x + 2y + 3z = 3 \\ x - z = 9 \\ 4y + z = -4 \end{cases}$

Solve each system using matrices.

13. $\begin{cases} x - y = -2 \\ 3x - 3y = -6 \end{cases}$

14. $\begin{cases} x + 2y = -1 \\ 2x + 5y = -5 \end{cases}$

15. $\begin{cases} x - y - z = 0 \\ 3x - y - 5z = -2 \\ 2x + 3y = -5 \end{cases}$

16. $\begin{cases} 2x - y + 3z = 4 \\ 3x - 3z = -2 \\ -5x + y = 0 \end{cases}$

Solve each system.

17. $\begin{cases} x^2 + y^2 = 169 \\ 5x + 12y = 0 \end{cases}$

18. $\begin{cases} x^2 + y^2 = 26 \\ x^2 - y^2 = 24 \end{cases}$

19. $\begin{cases} y = x^2 - 5x + 6 \\ y = 2x \end{cases}$

20. $\begin{cases} x^2 + 4y^2 = 5 \\ y = x \end{cases}$

21. $\begin{cases} 2x + 5y \geq 10 \\ y \geq x^2 + 1 \end{cases}$

22. $\begin{cases} \dfrac{x^2}{4} + y^2 \leq 1 \\ x + y > 1 \end{cases}$

23. $\begin{cases} x^2 + y^2 > 1 \\ \dfrac{x^2}{4} - y^2 \geq 1 \end{cases}$

24. $\begin{cases} x^2 + y^2 \geq 4 \\ x^2 + y^2 < 16 \\ y \geq 0 \end{cases}$

Solve.

25. Dean Swift jogs clockwise around a 4-mile track at the same time Tom Brackett bicycles counterclockwise, going 10 mph faster than Dean. They meet after 10 minutes. Find how fast each person travels and find how far Dean jogs.

26. A student whose average score is between 70 and 79 inclusive in a mathematics class receives a C. If Jamie Calvo has grades of 70, 64, 85, and 73 on the first four tests, find the lowest score possible on the fifth test so that Jamie gets a C grade. (Assume no rounding off in averaging.)

CHAPTER 10 CUMULATIVE REVIEW

1. Solve $2x + 7 \leq x - 11$, and graph the solution.
2. Solve for y: $|y - 3| > 7$.
3. Find the equation of the line containing the point (4, 4) and parallel to the line $2x + 3y = -6$.
4. Simplify each product.
   a. $4^2 \cdot 4^5$ **b.** $x^2 \cdot x^5$ **c.** $y^3 \cdot y$
   d. $y^3 \cdot y^2 \cdot y^7$ **e.** $(-5)^7 \cdot (-5)^8$
5. Subtract $4x^3y^2 - 3x^2y^2 + 2y^2$ from $10x^3y^2 - 7x^2y^2$.
6. Find the quotient: $\dfrac{3x^5y^2 - 15x^3y - 6x}{6x^2y^3}$.
7. If $f(x) = x^2$ and $g(x) = x + 1$, find the following.
   a. $(f \circ g)(x)$ **b.** $(g \circ f)(2)$.
8. Factor $8x^2 - 22x + 5$.
9. Factor $64x^3 + 1$.
10. Solve $x^3 + 5x^2 = x + 5$.
11. Write each rational expression in lowest terms.
    a. $\dfrac{2 + x}{x + 2}$ **b.** $\dfrac{2 - x}{x - 2}$ **c.** $\dfrac{18 - 2x^2}{x^2 - 2x - 3}$
12. Simplify $\dfrac{x^{-1} + 2xy^{-1}}{x^{-2} - x^{-2}y^{-1}}$.
13. Rationalize the numerator of $\dfrac{\sqrt{x} + 2}{5}$.
14. Simplify each expression.
    a. $36^{-1/2}$ **b.** $16^{-3/4}$ **c.** $-9^{1/2}$ **d.** $32^{-4/5}$
15. Solve $3x^2 - 9x + 8 = 0$ by completing the square.
16. An experienced typist and an apprentice typist can process a document in 6 hours. The experienced typist can process the document alone in 2 hours less time than the apprentice typist. Find the time that each typist can process the document alone.
17. Solve $\dfrac{x + 2}{x - 3} \leq 0$.
18. A rock is thrown upward from the ground. Its height in feet above ground after t seconds is given by the equation $f(t) = -16t^2 + 20t$. Find the maxiumum height of the rock.
19. Graph $\dfrac{(x + 3)^2}{25} + \dfrac{(y - 2)^2}{36} = 1$.
20. Use the substitution method to solve the system

$$\begin{cases} 2x + 4y = -6 & \text{First equation.} \\ x - 2y = -5 & \text{Second equation.} \end{cases}$$

21. Two cars leave Indianapolis, one traveling east and the other west. After 3 hours they are 297 miles apart. If one car is traveling 5 mph faster than the other, what is the speed of each?
22. Solve the system using matrices.

$$\begin{cases} x + 2y + z = 2 \\ -2x - y + 2z = 5 \\ x + 3y - 2z = -8 \end{cases}$$

23. Graph $4y^2 > x^2 + 16$.

CHAPTER 11

Exponential and Logarithmic Functions

Like many "third-world" countries, Kenya faces surging population growth. Populations typically grow "exponentially," so that the growth can become unmanageable.

INTRODUCTION

In this chapter, we discuss two closely related functions: exponential and logarithmic functions. These functions are vital in applications in economics, finance, engineering, the sciences, education, and other fields. Models of tumor growth and learning curves are two examples of the uses of exponential and logarithmic functions.

11.1 Inverse Functions

OBJECTIVES

Tape IA 30

1. Determine whether a function is a one-to-one function.
2. Use the horizontal line test to test whether a function is a one-to-one function.
3. Define the inverse of a function.
4. Find the equation of the inverse of a function.

1 The set $f = \{(0, 1), (2, 2), (-3, 5), (7, 6)\}$ is a function since each x-value corresponds to a unique y-value. For this particular function f, each y-value also corresponds to a unique x-value. When this happens, we call the function a **one-to-one function.**

> **One-to-One Function**
>
> If f is a function, then f is a **one-to-one-function** if each y-value corresponds to a unique x-value.

EXAMPLE 1 Determine whether each function is one-to-one.

a. $f = \{(6, 2), (5, 4), (-1, 0), (7, 3)\}$

b. $g = \{(3, 9), (-4, 2), (-3, 9), (0, 0)\}$

c. $h = \{(1, 1), (2, 2), (10, 10), (-5, -5)\}$

Solution: **a.** f is one-to-one since each y-value corresponds to only one x-value.

b. g is not one-to-one because in $(3, 9)$ and $(-3, 9)$ the y-value 9 corresponds to two different x-values.

c. h is a one-to-one function since each y-value corresponds to only one x-value. ■

2 Recall that we recognize the graph of a function when it passes the vertical line test. Since every x-value of the function corresponds to exactly one y-value, each vertical line, for which all the x-values are the same, intersects the function's graph at most once. Is the function shown below a one-to-one function? The answer is no. To see why, notice that the y-value of the ordered pair $(-3, 3)$, for example, is the same as the y-value of the ordered pair $(3, 3)$. This function is therefore not one-to-one.

To test whether a graph is the graph of a one-to-one function, apply the vertical line test to see if it is a function, and then apply a similar **horizontal line test** to see if it is a one-to-one function.

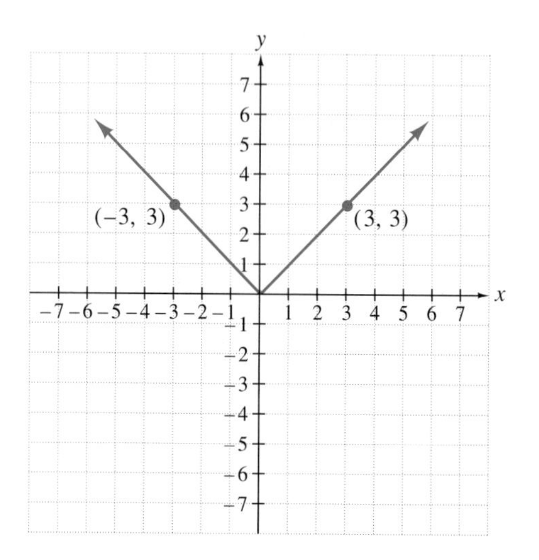

Horizontal Line Test

If every horizontal line intersects the graph of a function at most once, then the function is a one-to-one function.

EXAMPLE 2 Determine whether each graph is the graph of a one-to-one function.

a.

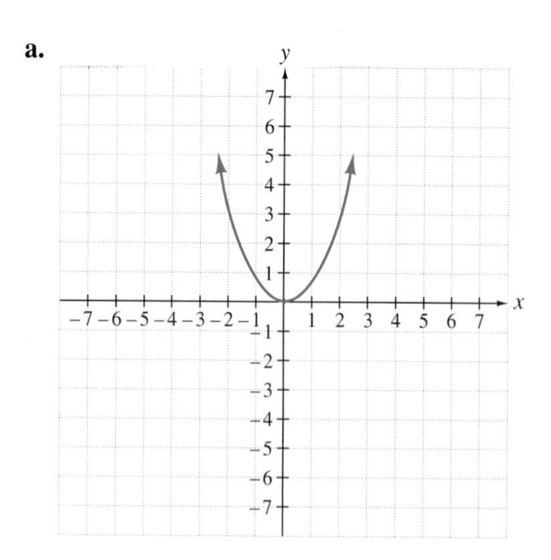

b.

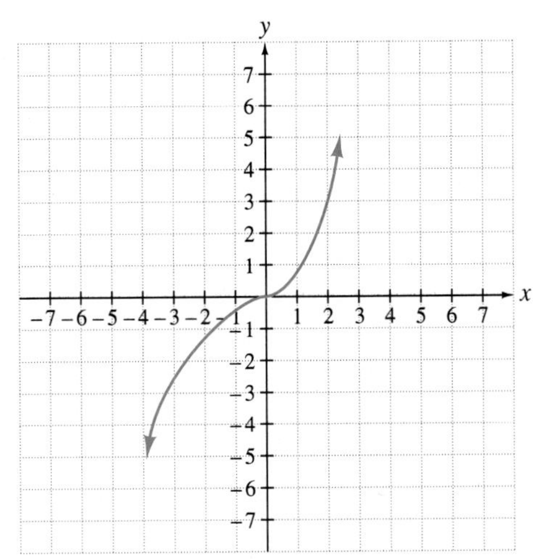
y
x
7 6 5 4 3 2 1 −1 −2 −3 −4 −5 −6 −7
−7 −6 −5 −4 −3 −2 −1 1 2 3 4 5 6 7

c.

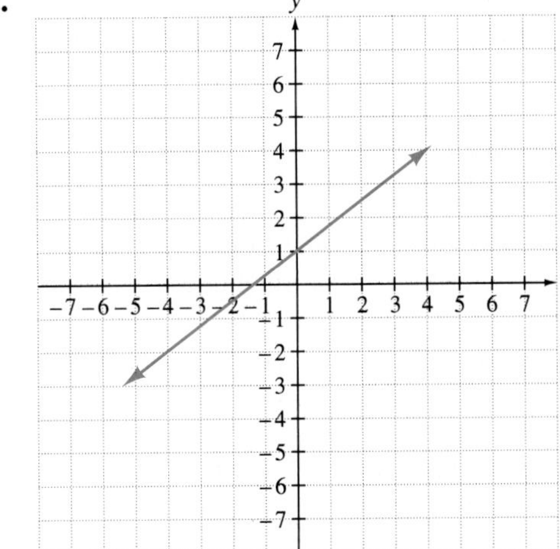
y
x
7 6 5 4 3 2 1 −1 −2 −3 −4 −5 −6 −7
−7 −6 −5 −4 −3 −2 −1 1 2 3 4 5 6 7

d.

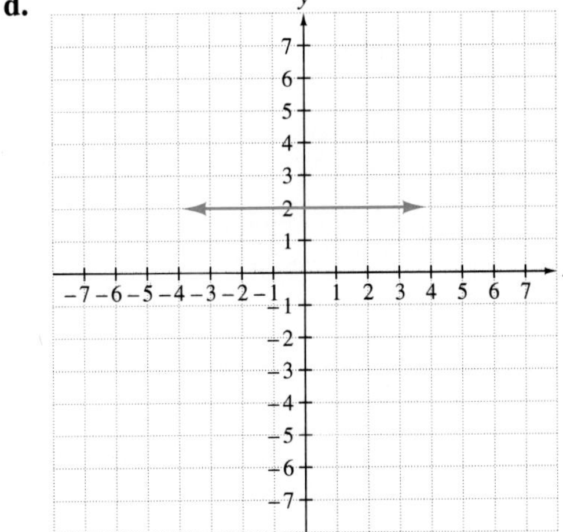
y
x
7 6 5 4 3 2 1 −1 −2 −3 −4 −5 −6 −7
−7 −6 −5 −4 −3 −2 −1 1 2 3 4 5 6 7

e.

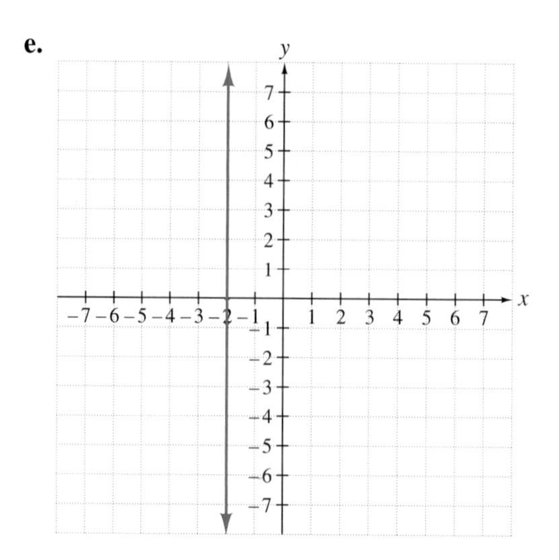

Solution: Graphs **a, b, c, d** all pass the vertical line test so only these graphs are graphs of functions. But, of these, only **b** and **c** pass the horizontal line test, so only **b** and **c** are graphs of one-to-one-functions. ■

HELPFUL HINT

All linear equations are one-to-one functions except those whose graphs are horizontal or vertical lines. A vertical line does not pass the vertical line test and hence is not the graph of a function. A horizontal line is the graph of a function, but does not pass the horizontal line test and hence is not the graph of a one-to-one function.

3 One-to-one functions are special in that their graphs pass the vertical and horizontal line tests. They are special, too, in that for each one-to-one function we can find its **inverse function** by switching the order of the coordinates of the ordered pairs of the function. If the function f is $\{(2, -3), (5, 10), (9, 1)\}$, for example, then its inverse function is $\{(-3, 2), (10, 5), (1, 9)\}$. For a function f, we use the notation f^{-1}, read "f inverse," to denote its inverse function. Notice that since the coordinates of each ordered pair have been switched the domain of f is the range of f^{-1}, and the range of f is the domain of f^{-1}.

Inverse Function

The inverse of a one-to-one function f is the one-to-one function f^{-1} that is the set of all ordered pairs (y, x) where (x, y) belongs to f.

HELPFUL HINT

The symbol f^{-1} is the single symbol used to denote the inverse of the function f. It is read as "f inverse". This symbol **does not mean** $\frac{1}{f}$.

4 If a one-to-one function f is defined as a set of ordered pairs, we can find f^{-1} by interchanging the x and y coordinates of the ordered pairs. If a one-to-one function f is given in the form of an equation, we can find f^{-1} using a similar procedure.

To Find the Inverse of a One-to-One Function $f(x)$

Step 1 Replace $f(x)$ by y.
Step 2 Interchange x and y.
Step 3 Solve the equation for y.
Step 4 Replace y with the notation $f^{-1}(x)$.

EXAMPLE 3 Find the equation of the inverse of $f(x) = 3x - 5$. Graph f and f^{-1} on the same set of axes.

Solution: First, let $y = f(x)$.

$$f(x) = 3x - 5$$
$$y = 3x - 5$$

Next, interchange x and y and solve for y.

$$x = 3y - 5 \quad \text{Interchange } x \text{ and } y.$$
$$3y = x + 5$$
$$y = \frac{x+5}{3} \quad \text{Solve for } y.$$

Let $y = f^{-1}(x)$.

$$f^{-1}(x) = \frac{x+5}{3}$$

Now graph $f(x)$ and $f^{-1}(x)$ on the same set of axes. Both $f(x) = 3x - 5$ and $f^{-1}(x) = \frac{x+5}{3}$ are linear functions, so each graph is a line.

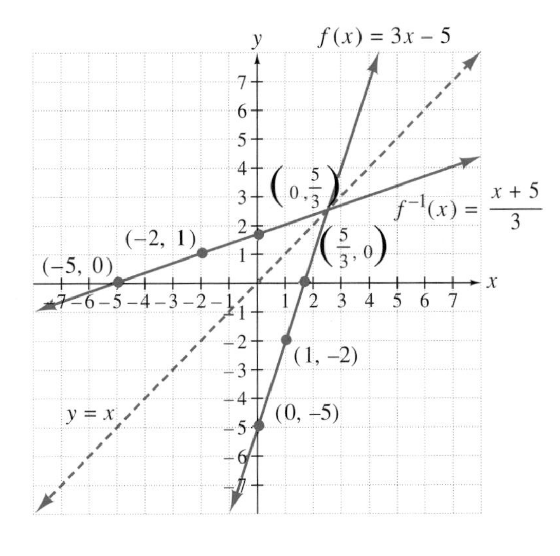

$f(x) = 3x - 5$

x	$y = f(x)$
1	-2
0	-5
$\frac{5}{3}$	0

$f^{-1}(x) = \frac{x+5}{3}$

x	$y = f^{-1}(x)$
0	$\frac{5}{3}$
-5	0
-2	1

■

Notice that the graphs of f and f^{-1} are mirror images of each other, and the "mirror" is the line $y = x$. This is true for every function and its inverse. For this reason, we say the graphs of f and f^{-1} are **symmetric about the line** $y = x$.

EXAMPLE 4 Find the equation of the inverse of $g(x) = \dfrac{-2x + 1}{3}$. Graph g and g^{-1} on the same set of axes.

Solution: The function g is a one-to-one function since its graph is a line that is neither vertical nor horizontal. Let $y = g(x)$ and obtain

$$y = \frac{-2x + 1}{3}$$

Next, switch variables and solve for y.

$$x = \frac{-2y + 1}{3} \qquad \text{Interchange } x \text{ and } y.$$

$$3x = -2y + 1 \qquad \text{Multiply both sides by 3.}$$

$$2y = -3x + 1$$

$$y = \frac{-3x + 1}{2} \qquad \text{Solve for } y.$$

Let $y = g^{-1}(x)$.

$$g^{-1}(x) = \frac{-3x + 1}{2}$$

The graphs of g and g^{-1} are two lines symmetric about the line $y = x$.

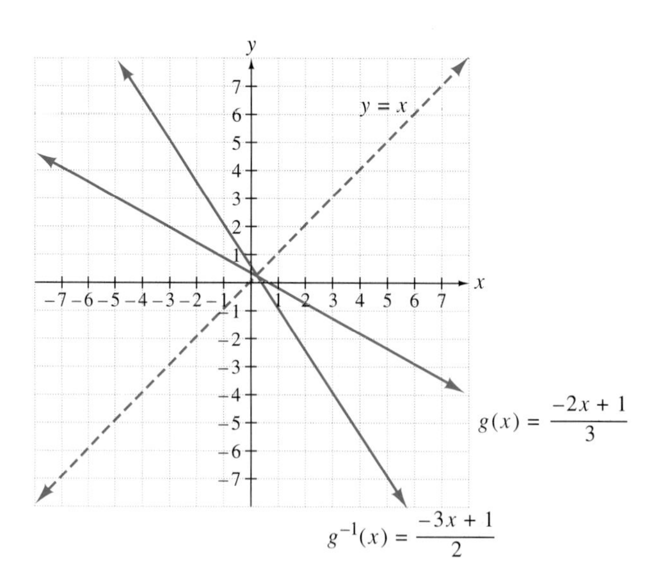

■

EXERCISE SET 11.1

Determine whether each function is a one-to-one function. See Example 1.

1. $f = \{(-1, -1), (1, 1), (0, 2), (2, 0)\}$

2. $g = \{(8, 6), (9, 6), (3, 4), (-4, 4)\}$

3. $h = \{(10, 10)\}$

4. $r = \{(1, 2), (3, 4), (5, 6), (6, 7)\}$

5. $f = \{(11, 12), (4, 3), (3, 4), (6, 6)\}$

6. $g = \{(0, 3), (3, 7), (6, 7), (-2, -2)\}$

Determine whether each graph is the graph of a one-to-one function. See Example 2.

7.

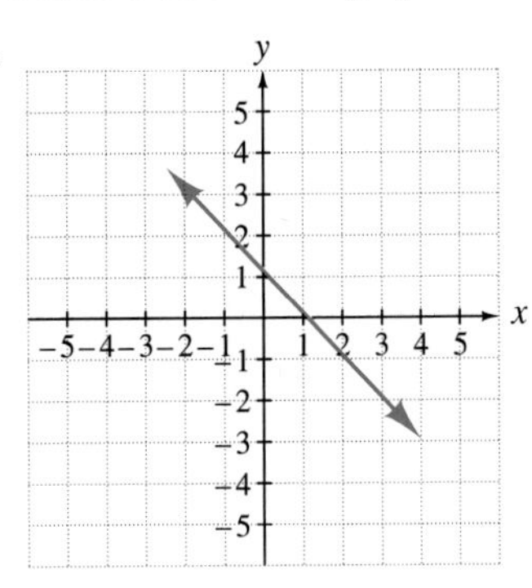

8.

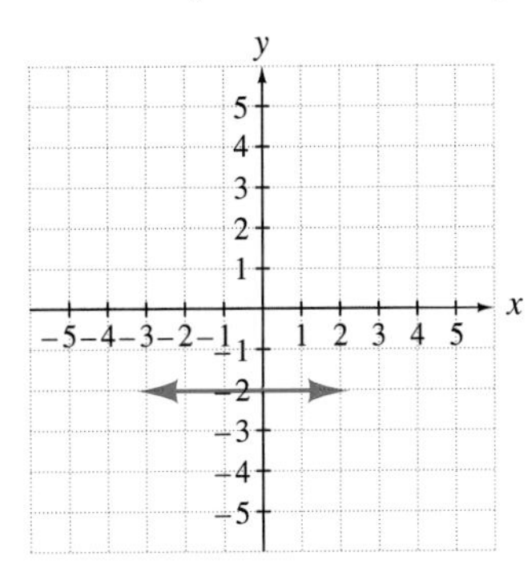

9.

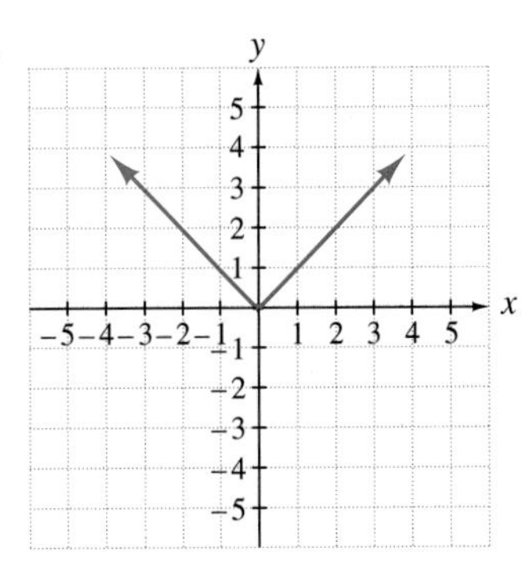

10.

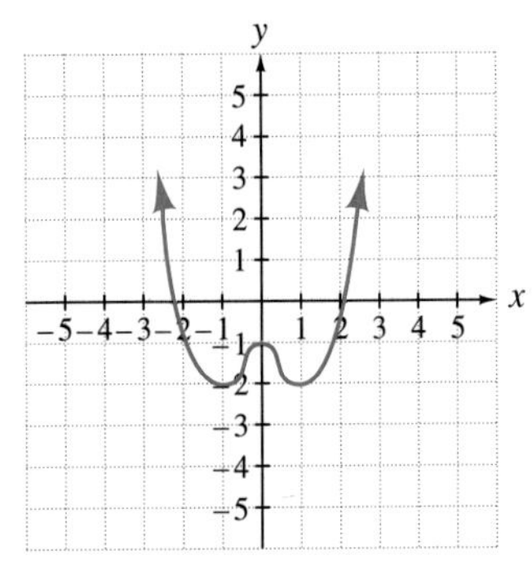

11.

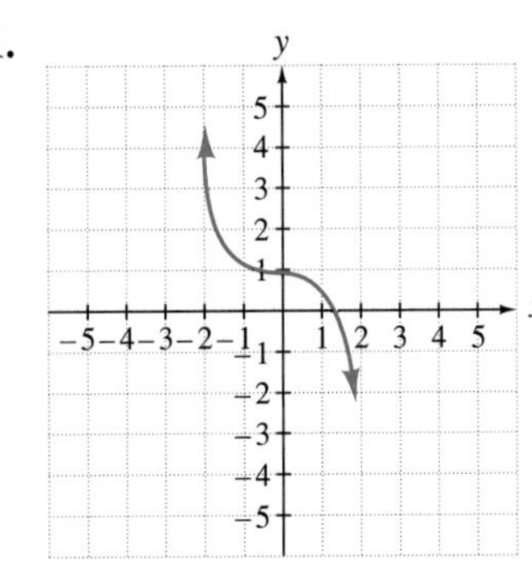

12.

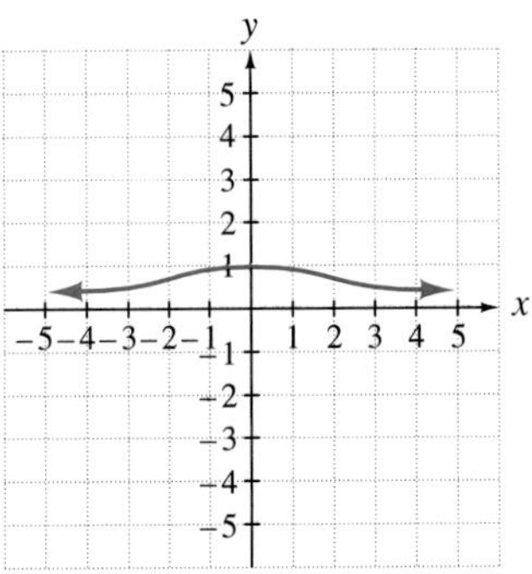

13.

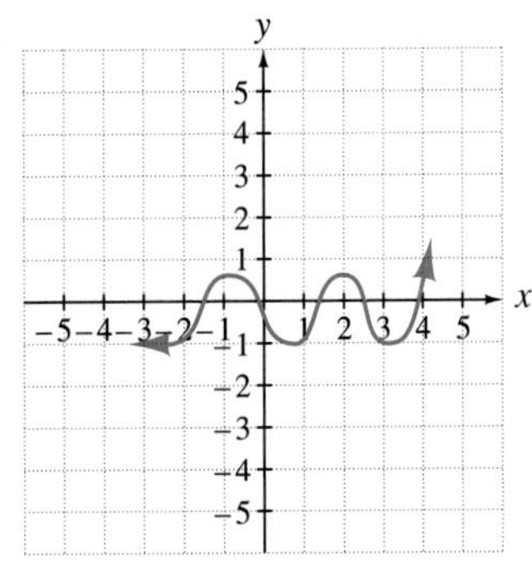

14.

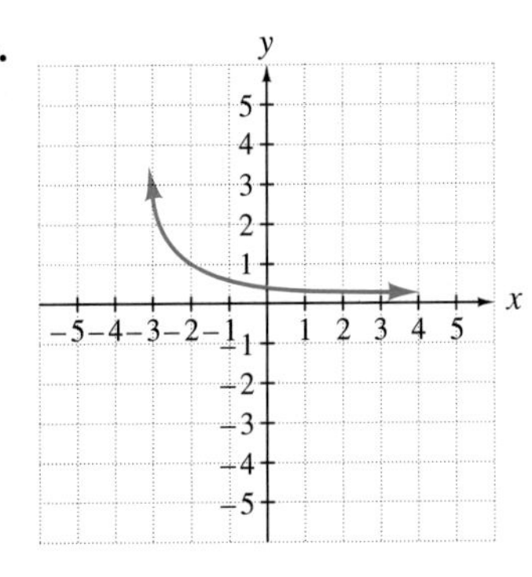

Each function is one-to-one. Find the inverse of each function and graph the function and its inverse on the same set of axes. See Examples 3 and 4.

15. $f(x) = x + 4$

16. $f(x) = x - 5$

17. $f(x) = 2x - 3$

18. $f(x) = 4x + 9$

19. $f(x) = \frac{12x - 4}{3}$

20. $f(x) = \frac{7x + 5}{11}$

Determine whether each function is one-to-one. If a function is one-to-one, list the elements of its inverse function.

21. $h = \{(-3, 9), (-2, 4), (-1, 1), (0, 0), (1, 1)\}$

22. $g = \{(-5, 2), (0, 7), (2, 9), (1, 8)\}$

23. $f = \{(1, 1), (2, 1), (3, 2), (4, 2)\}$

24. $g = \{(7, 3), (4, 0), (3, -1), (-3, -7)\}$

25. $f = \{(-4, -8), (-6, -12), (-8, -16), (-9, -18)\}$

26. $h = \{(-4, -3), (-3, -2), (-2, -1), (-1, -1)\}$

Sketch the graph of each equation and decide if it is a one-to-one function.

27. $x = y^2$

28. $y = x^2$

29. $y = 5x$

30. $y = -4x + 1$

31. $x = 7$

32. $y = -3$

Each function is one-to-one. Find the inverse of each function and graph the function and its inverse on the same set of axes.

33. $f(x) = 3x + 1$

34. $f(x) = -2x + 5$

35. $f(x) = \frac{x - 2}{5}$

36. $f(x) = \frac{4x - 3}{2}$

37. $g(x) = \frac{1}{2}x - 4$

38. $g(x) = \frac{3}{2}x - \frac{5}{4}$

Skill Review

Find the unknown side of each right triangle. See Section 6.6.

39.

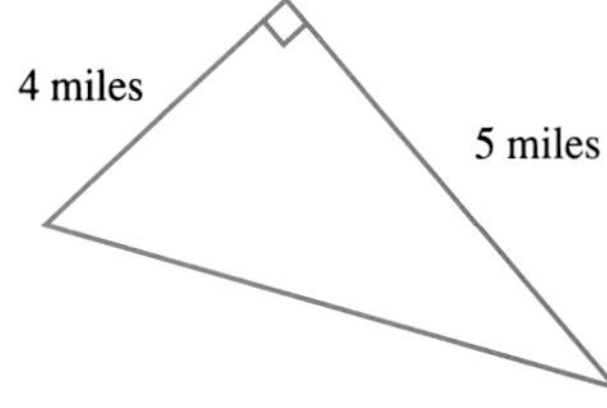

40.

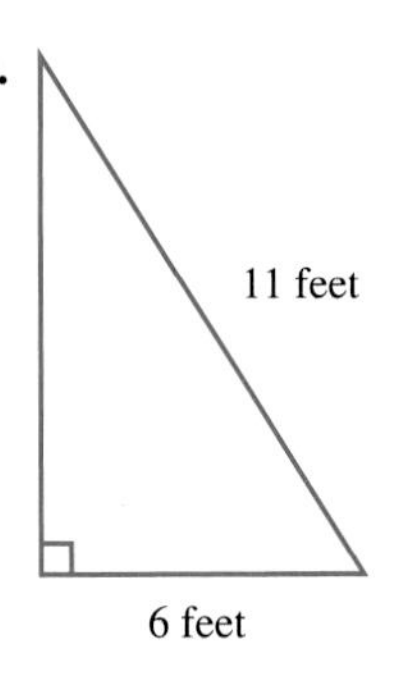

Evaluate. See Sections 5.1 and 5.2.

41. 5^0

42. -2^0

43. $3x^0$

44. $(6x)^0 + 6x^0$

45. 2^{-2}

46. 2^{-1}

47. 2^3

48. 2^4

Writing in Mathematics

49. Discuss the purpose of the vertical line test and the horizontal line test.

50. Explain whether this statement is true: Every straight line is the graph of a one-to-one function.

11.2 Exponential Functions

OBJECTIVES

Tape IA 30

1. Identify exponential functions.
2. Graph exponential functions.
3. Solve equations of the form $b^x = b^y$.

1 In earlier chapters, we gave meaning to exponential expressions such as 2^x, where x is a rational number. For example,

$$2^3 = 2 \cdot 2 \cdot 2 = 8$$

It is beyond the scope of this book to give meaning to 2^x if x is irrational. As long as the base b is positive, we will assume that b^x is a real number for all real numbers x. We also assume that rules of exponents apply whether x is rational or irrational as long as b is positive. The equation $y = b^x$ is called an **exponential function.**

Exponential Function

A function of the form

$$f(x) = b^x$$

is an exponential function where $b > 0$, b is not 1, and x is a real number.

2

EXAMPLE 1 Graph the exponential functions defined by $f(x) = 2^x$ and $g(x) = 3^x$ on the same set of axes.

Solution: Graph each function by plotting points. Set up a table of values for each of the two functions:

$f(x) = 2^x$

x	$f(x)$
0	1
1	2
2	4
3	8
-1	$\frac{1}{2}$
-2	$\frac{1}{4}$

$g(x) = 3^x$

x	$g(x)$
0	1
1	3
2	9
3	27
-1	$\frac{1}{3}$
-2	$\frac{1}{9}$

If each set of points is plotted and connected with a smooth curve, the following graphs result:

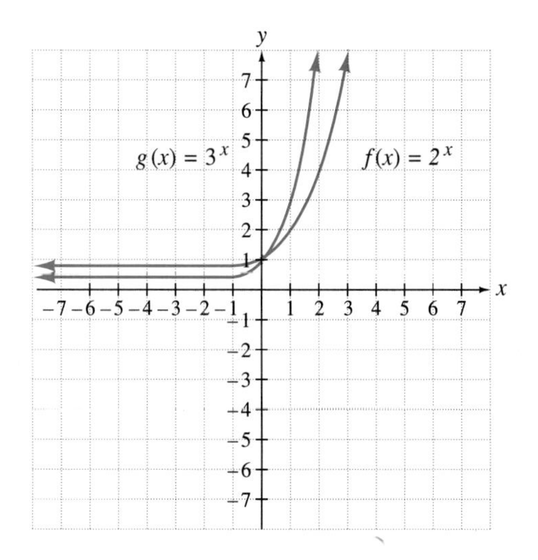

■

A number of things should be noted about these two graphs of exponential functions. First, the graphs confirm that $f(x) = 2^x$ and $g(x) = 3^x$ define one-to-one functions since each graph passes the vertical and horizontal line tests. The y-intercept of each graph is 1, but neither graph has an x-intercept. From the graph, we can also see that the domain of each function is all real numbers and the range is $\{y \mid y > 0\}$. We can also see that as x increases, y increases also.

EXAMPLE 2 Graph the exponential functions $y = \left(\frac{1}{2}\right)^x$ and $y = \left(\frac{1}{3}\right)^x$ on the same set of axes.

Solution: As before, plot points and connect them with a smooth curve.

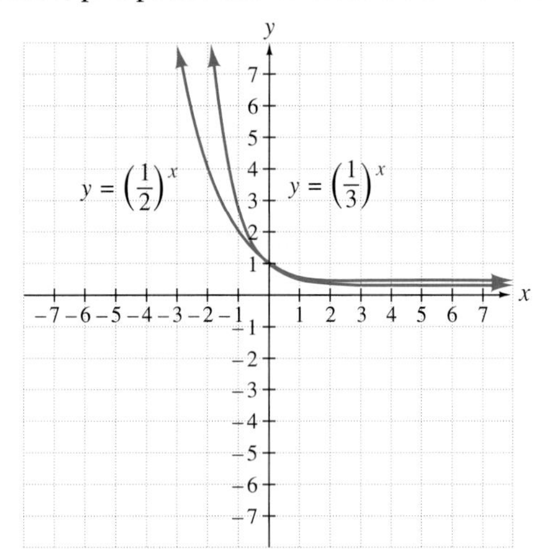

$y = \left(\frac{1}{2}\right)^x$

x	y
0	1
1	$\frac{1}{2}$
2	$\frac{1}{4}$
3	$\frac{1}{8}$
−1	2
−2	4

$y = \left(\frac{1}{3}\right)^x$

x	y
0	1
1	$\frac{1}{3}$
2	$\frac{1}{9}$
3	$\frac{1}{27}$
−1	3
−2	9

Each function again is a one-to-one function. The y-intercept of both is 1. The domain is the set of all real numbers and the range is $\{y \mid y > 0\}$. ■

Notice the difference between the graphs of Example 1 and the graphs of Example 2. An exponential function is always increasing if the base is larger than 1. When the base is between 0 and 1, the graph is decreasing. The following figures summarize these characteristics of exponential functions.

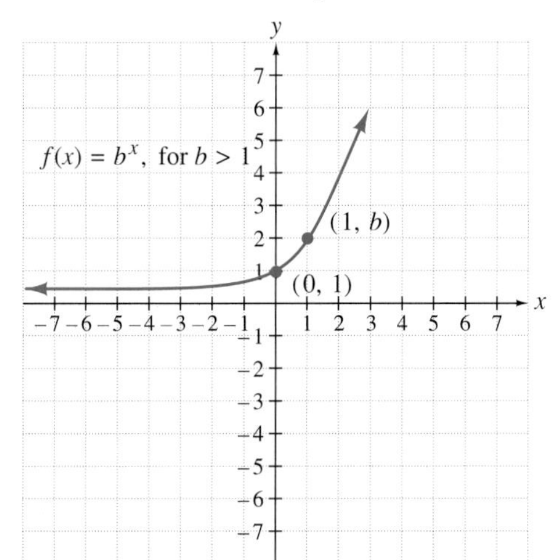

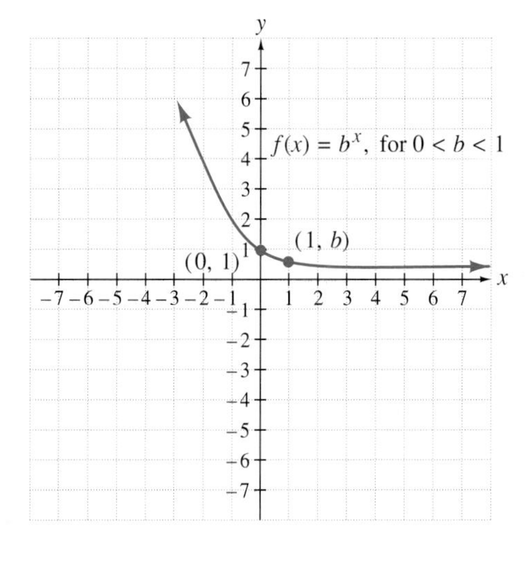

3 Because an exponential function $y = b^x$ is a one-to-one function, we have the following property.

> **Uniqueness of b^x**
>
> Let $b > 0$ and $b \neq 1$. If
>
> $$b^x = b^y, \quad \text{then} \quad x = y$$

We can use this property to solve exponential equations.

EXAMPLE 3 Solve each equation for x.

a. $2^x = 16$ **b.** $9^x = 27$ **c.** $4^{x+3} = 8^x$

Solution: **a.** We write 16 as a power of 2 and then use the uniqueness of b^x to solve.

$$2^x = 16$$
$$2^x = 2^4$$

Since the bases are the same, by the uniqueness of b^x, we have then that the exponents are equal. Thus,

$$x = 4$$

b. Notice that 9 and 27 are both powers of 3.

$$9^x = 27$$
$$(3^2)^x = 3^3 \quad \text{Write 9 and 27 as powers of 3.}$$
$$3^{2x} = 3^3$$
$$2x = 3 \quad \text{Apply the uniqueness of } b^x.$$
$$x = \frac{3}{2} \quad \text{Divide by 2.}$$

To check, replace x with $\frac{3}{2}$ in the original expression $9^x = 27$.

c. Write both 4 and 8 as powers of 2.

$$4^{x+3} = 8^x$$
$$(2^2)^{x+3} = (2^3)^x$$
$$2^{2x+6} = 2^{3x}$$
$$2x + 6 = 3x \quad \text{Apply the uniqueness of } b^x.$$
$$6 = x \quad \text{Subtract } 2x \text{ from both sides.}$$

■

There is one major problem with the preceding technique. Often the two expressions in the equation cannot easily be written as powers of a common base. We explore how to solve an equation like $4 = 3^x$ with the help of **logarithms** later.

The natural world abounds with patterns that can be modeled by exponential functions. To make these applications realistic, numbers are used that warrant a calculator. The first application has to do with exponential decay.

EXAMPLE 4 As a result of the Chernobyl nuclear accident, radioactive debris was carried through the atmosphere. One immediate concern was the impact it had on the milk supply. The percent y of radioactive material in raw milk after t days is estimated by $y = 100\,(2.7)^{-0.1t}$. Estimate the expected percent of radioactive material in the milk after 30 days.

Solution: Replace t with 30 in the given equation.

$$y = 100(2.7)^{-0.1t}$$
$$= 100(2.7)^{-3}$$

To **approximate** the percent y, press the following keys on your calculator.

2.7 | y^x | 3 | +/− | = | × | 100 | =

The display should read

5.0805263

Thus nearly 5% of the radioactive material still contaminates the milk supply after 30 days. ■

The exponential function defined by $A = P\left(1 + \frac{r}{n}\right)^{nt}$ models the pattern relating the dollars A accrued (or owed) after P dollars are invested (or loaned) at a rate of interest r compounded n times each year for t years.

EXAMPLE 5 Find the amount owed at the end of 5 years if $1600 is loaned at a rate of 9% compounded monthly.

Solution: Use the formula $A = P\left(1 + \frac{r}{n}\right)^{nt}$, with the following values:

$P = \$1600$ (the amount of the loan)

$r = 9\% = 0.09$ (the rate of interest)

$n = 12$ (the number of times interest is compounded each year)

$t = 5$ (the duration of the loan)

$$A = P\left(1 + \frac{r}{n}\right)^{nt} \qquad \text{Compound interest formula.}$$

$$= 1600\left(1 + \frac{0.09}{12}\right)^{12(5)} \qquad \text{Substitute known values.}$$

$$= 1600(1.0075)^{60}$$

To **approximate** A, press the following keys on your calculator.

1.0075 | y^x | 60 | = | × | 1600 | =

The display should read

2505.0896

Thus the amount A owed is approximately $2505.09. ■

EXERCISE SET 11.2

Graph each exponential function. See Example 1.

1. $y = 4^x$

2. $y = 5^x$

3. $y = 1 + 2^x$

4. $y = 3^x - 1$

Graph each exponential function. See Example 2.

5. $y = \left(\frac{1}{4}\right)^x$

6. $y = \left(\frac{1}{5}\right)^x$

7. $y = \left(\frac{1}{2}\right)^x - 2$

8. $y = \left(\frac{1}{3}\right)^x + 2$

Solve each equation for x. See Example 3.

9. $3^x = 27$

10. $6^x = 36$

11. $16^x = 8$

12. $64^x = 16$

13. $32^{2x-3} = 2$

14. $9^{2x+1} = 81$

15. $\frac{1}{4} = 2^{3x}$

16. $\frac{1}{27} = 3^{2x}$

Solve. See Example 4.

17. One type of uranium has a daily radioactive decay rate of 0.4%. If 30 pounds of this uranium is available today, find how much will still remain after 50 days. Use $y = 30(2.7)^{-0.004t}$ and let t be 50.

18. The nuclear waste from an atomic energy plant decays at a rate of 3% each century. If 150 pounds of nuclear waste is disposed of, find how much of it will still remain after 10 centuries. Use $y = 150(2.7)^{-0.03t}$ and let t be 10.

Solve. Use $A = P\left(1 + \frac{r}{n}\right)^{nt}$. *See Example 5.*

19. Find the amount Erica owes at the end of 3 years if $6000 is loaned to her at a rate of 8% compounded monthly.

20. Find the amount owed at the end of 5 years if $3000 is loaned at a rate of 10% compounded quarterly.

Graph each exponential function.

21. $y = -2^x$

22. $y = -3^x$

23. $y = 3^x - 2$

24. $y = 2^x - 3$

25. $y = -\left(\frac{1}{4}\right)^x$

26. $y = -\left(\frac{1}{5}\right)^x$

27. $y = \left(\frac{1}{3}\right)^x + 1$

28. $y = \left(\frac{1}{2}\right)^x - 2$

Solve each equation for x.

29. $5^x = 625$

30. $2^x = 64$

31. $4^x = 8$

32. $32^x = 4$

33. $27^{x+1} = 9$

34. $125^{x-2} = 25$

35. $81^{x-1} = 27^{2x}$

36. $4^{3x-7} = 32^{2x}$

Write an exponential equation defining the function whose graph is given.

37.

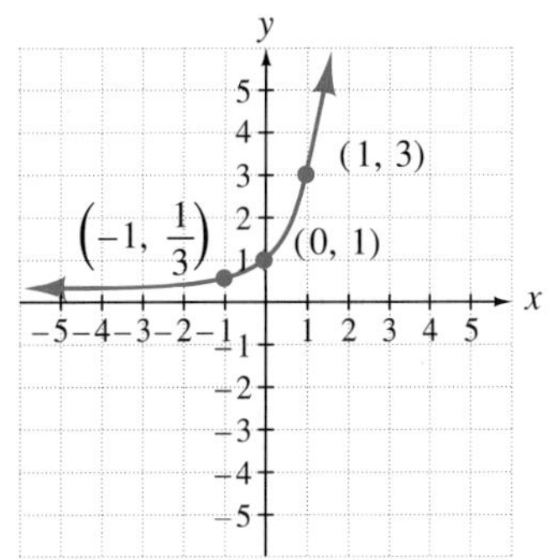

38.

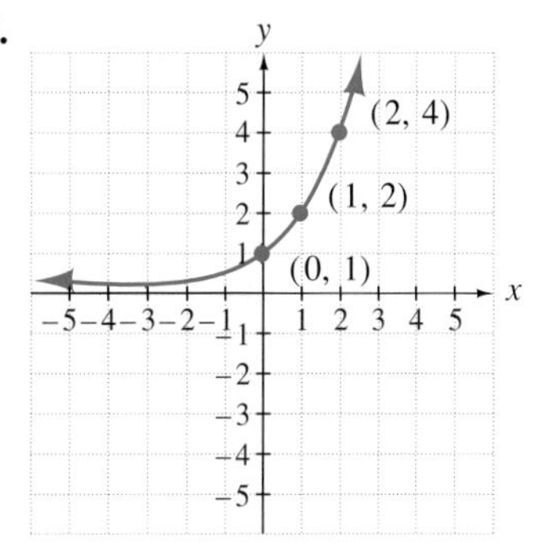

39.

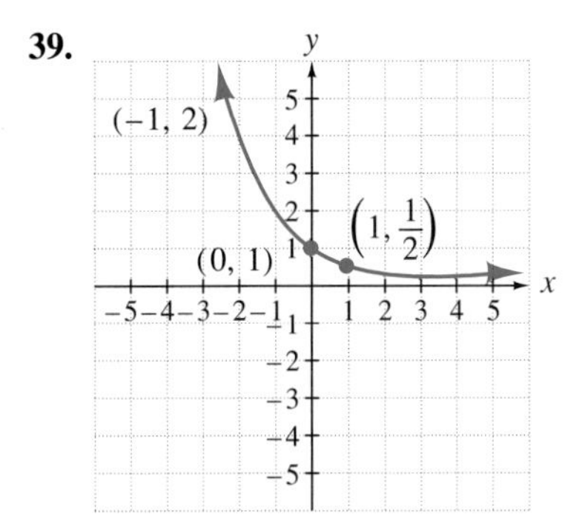

40.

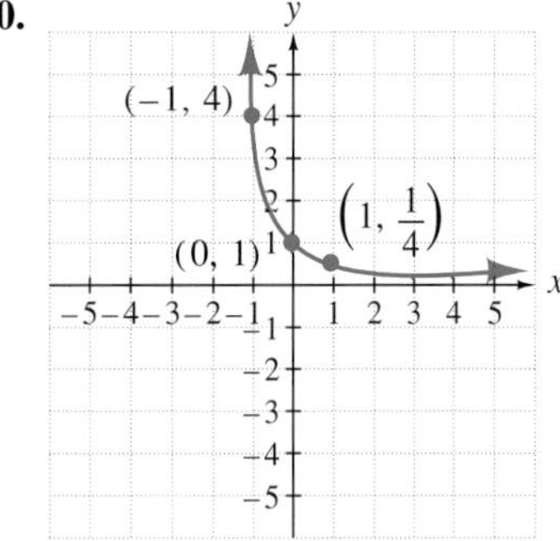

Solve.

41. The size of the rat population of a wharf area grows at a rate of 8% monthly. If there are 200 rats in January, find how many rats should be expected by next January. Use $y = 200(2.7)^{0.08t}$.

42. National Park Service personnel are trying to increase the size of the bison population of Theodore Roosevelt National Park. If 260 bison currently live in the park, and if the population's rate of growth is 2.5% annually, find how many bison there should be in 10 years. Use $y = 260(2.7)^{0.025t}$.

43. A rare isotope of a nuclear material is very unstable, decaying at a rate of 15% each second. Find the percent of this isotope remaining 10 seconds after 5 grams of the isotope is created. Use $y = 5(2.7)^{-0.15t}$.

44. An accidental spill of 75 grams of radioactive material in a local stream has led to the presence of radioactive debris decaying at a rate of 4% each day. Find the percent of this debris still remaining after 14 days. Use $y = 75(2.7)^{-0.04t}$.

45. Due to economic conditions, Mexico City is growing at a rate of 5.2% annually. If there were 7,000,000 residents of Mexico City in 1990, find how many are living in the city in 1995 (to the nearest ten thousand). Use $y = 7{,}000{,}000(2.7)^{0.052t}$.

46. An unusually wet spring has caused the size of the Cape Cod mosquito population to increase by 8% each day. If an estimated 200,000 mosquitoes are on Cape Cod on May 12, find how many thousands of mosquitoes will inhabit the Cape on May 25. Use $y = 200{,}000(2.7)^{0.08t}$.

47. Find the total amount Janina has in a college savings account if \$2000 was invested and earned 6% compounded semiannually for 12 years.

48. Find the amount accrued if \$500 is invested and earns 7% compounded monthly for 4 years.

Skill Review

Solve each equation. See Sections 2.3 and 6.6.

49. $5x - 2 = 18$

50. $3x - 7 = 11$

51. $3x - 4 = 3(x + 1)$

52. $2 - 6x = 6(1 - x)$

53. $x^2 + 6 = 5x$

54. $18 = 11x - x^2$

By inspection, find the value for x that makes each statement true.

55. $2^x = 8$

56. $3^x = 9$

57. $5^x = \frac{1}{5}$

58. $4^x = 1$

Writing in Mathematics

59. Explain why the graph of an exponential function $y = b^x$ contains the point $(1, b)$.

60. Explain why an exponential function $y = b^x$ has a y-intercept of 1.

11.3 Logarithmic Functions

OBJECTIVES

Tape IA 31

1. Write exponential equations using logarithmic notation, and write logarithmic equations using exponential notation.
2. Solve logarithmic equations by using exponential notation.
3. Identify and graph logarithmic functions.

1 Recall from the last section that $y = 2^x$ is a one-to-one function. Let's begin this section by finding the inverse of this one-to-one function. To find its inverse, interchange variables and then solve for y.

$$y = 2^x$$
$$x = 2^y$$

At this point, we have no method for solving for y until we develop in this section a new notation. We use the symbol $\mathbf{log}_b\, \boldsymbol{x}$ to mean the **exponent that raises $\boldsymbol{b}$ to $\boldsymbol{x}$.**

> **Logarithmic Definition**
>
> If $b > 0$ and $b \neq 1$, then
>
> $$x = b^y \quad \text{is equivalent to} \quad y = \log_b x$$

Before returning to the function $x = 2^y$ and solving it for y in terms of x, let's practice using the new notation $\log_b x$. The expression $\log_b x = y$ is read "the logarithm of x to the base b is y" or "the logarithm base b of x is y."

It is important to be able to write exponential equations using logarithmic notation, and vice versa. The following table shows examples of both forms.

Logarithmic Form		*Exponential Form*
$\log_3 9 = 2$	is equivalent to	$3^2 = 9$
$\log_6 1 = 0$	is equivalent to	$6^0 = 1$
$\log_2 8 = 3$	is equivalent to	$2^3 = 8$

Logarithmic Form		*Exponential Form*
$\log_4 \frac{1}{16} = -2$	is equivalent to	$4^{-2} = \frac{1}{16}$
$\log_8 2 = \frac{1}{3}$	is equivalent to	$8^{1/3} = 2$

HELPFUL HINT

Notice that the base of the logarithmic expression is the base of the exponential expression.

EXAMPLE 1 Find the value of each logarithmic expression.

a. $\log_4 16$ **b.** $\log_{10} \frac{1}{10}$ **c.** $\log_9 3$

Solution: **a.** $\log_4 16 = 2$ because $4^2 = 16$. **b.** $\log_{10} \frac{1}{10} = -1$ because $10^{-1} = \frac{1}{10}$.

c. $\log_9 3 = \frac{1}{2}$ because $9^{1/2} = \sqrt{9} = 3$. ■

2 The ability to interchange the logarithmic and exponential forms of an expression can be very helpful when solving logarithmic equations.

EXAMPLE 2 Solve each equation for x.

a. $\log_4 \frac{1}{4} = x$ **b.** $\log_5 x = 3$ **c.** $\log_x 25 = 2$

d. $\log_3 1 = x$ **e.** $\log_b 1 = x$

Solution: **a.** $\log_4 \frac{1}{4} = x$ is equivalent to $4^x = \frac{1}{4}$. Solve $4^x = \frac{1}{4}$ for x.

$$4^x = \frac{1}{4}$$

$$4^x = 4^{-1}$$

Since the bases are the same, by the uniqueness of b^x, we have that

$$x = -1$$

The solution set is $\{-1\}$. To check, see that $\log_4 \frac{1}{4} = -1$, since $4^{-1} = \frac{1}{4}$.

b. $\log_5 x = 3$ is equivalent to $5^3 = x$ or

$$x = 125$$

The solution set is $\{125\}$.

c. $\log_x 25 = 2$ is equivalent to $x^2 = 25$.

$$x^2 = 25$$

$$x = \pm\sqrt{25} \quad \text{or} \quad \pm 5$$

Since the base b of a logarithm must be positive, the solution set contains 5 only. The solution set is $\{5\}$.

d. $\log_3 1 = x$ is equivalent to $3^x = 1$. Either solve this equation by inspection or solve by writing 1 as 3^0 as shown.

$$3^x = 3^0 \qquad \text{Since } 3^0 \text{ equals } 1.$$
$$x = 0 \qquad \text{Apply the uniqueness of } b^x.$$

The solution set is $\{0\}$.

e. $\log_b 1 = x$ is equivalent to $b^x = 1$.

$$b^x = b^0 \qquad \text{Write 1 as } b^0.$$
$$x = 0 \qquad \text{Apply the uniqueness of } b^x.$$

The solution set is $\{0\}$. ■

In Example 3e we proved an important property of logarithms. That is, $\log_b 1$ is always 0. This property as well as two important others are summarized next.

Properties of Logarithms

If b is a real number, $b > 0$ and $b \neq 1$, then

1. $\log_b 1 = 0$
2. $\log_b b^x = x$
3. $b^{\log_b x} = x$

To see that $\log_b b^x = x$, change the logarithmic form to exponential form. Then, $\log_b b^x = x$ only when $b^x = b^x$. In exponential form, the expressions are identical, so in logarithmic form, the expressions are identical.

EXAMPLE 3 Simplify.

a. $\log_3 3^2$ **b.** $\log_7 7^{-1}$ **c.** $5^{\log_5 3}$ **d.** $2^{\log_2 6}$

Solution: **a.** From property 2, $\log_3 3^2 = 2$. **b.** From property 2, $\log_7 7^{-1} = -1$.
c. From property 3, $5^{\log_5 3} = 3$. **d.** From property 3, $2^{\log_2 6} = 6$. ■

3 Having gained proficiency with the notation $\log_b a$, we return to the function $x = 2^y$ and others like it. We know the function $x = 2^y$ is the inverse of the function $y = 2^x$, and we would like to solve $x = 2^y$ for y in terms of x. We use logarithmic notation to do so.

$$2^y = x$$

is equivalent to

$$\log_2 x = y$$

because $\log_2 x = y$ only when $2^y = x$. We now have an expression of y in terms of x. This equation, $y = \log_2 x$, defines a function that is the inverse function of the function $y = 2^x$. The function $y = \log_2 x$ is called a **logarithmic function.**

> **Logarithmic Function**
>
> If x is a positive real number, b is a constant positive real number, and b is not 1, then a **logarithmic function** is a function that can be defined by
>
> $$f(x) = \log_b x$$
>
> The domain of f is the positive real numbers, and the range of f is the real numbers.

We can explore logarithmic functions by graphing them.

EXAMPLE 4 Graph the logarithmic function $y = \log_2 x$.

Solution: Write the equation using exponential notation as $2^y = x$. Find some points that satisfy this equation. Plot the points and connect them with a smooth curve. The domain of this function is $\{x \mid x > 0\}$ and the range of the function is all real numbers, **R.**

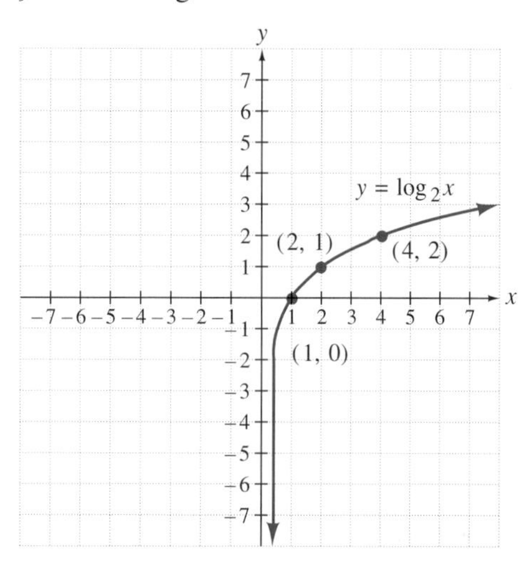

■

Recall that the graph of a function and its inverse are symmetric about the line $y = x$. Next, we graph the exponential function $y = 2^x$ and its inverse function $y = \log_2 x$ or $x = 2^y$ on the same set of axes to illustrate this.

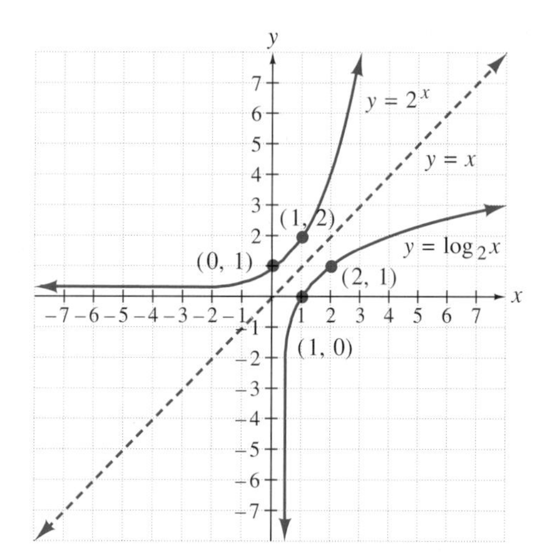

EXAMPLE 5 Graph the logarithmic function $f(x) = \log_{1/3} x$.

Solution: Replace $f(x)$ with y, and write the result using exponential notation.

$$f(x) = \log_{1/3} x$$

$$y = \log_{1/3} x \qquad \text{Replace } f(x) \text{ with } y.$$

$$\left(\frac{1}{3}\right)^y = x \qquad \text{Write in exponential form.}$$

Find points that satisfy $\left(\frac{1}{3}\right)^y = x$, plot these points, and connect them with a smooth curve as shown in the figure.

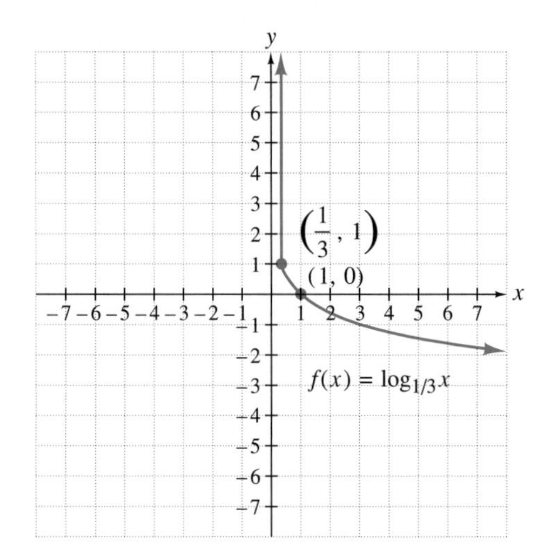

The domain of this function is $\{x \mid x > 0\}$ and the range is the set of all real numbers. ■

The figure shown summarizes characteristics of logarithmic functions.

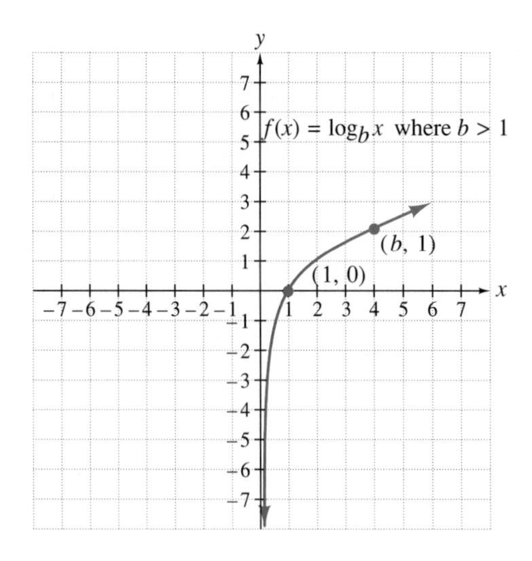

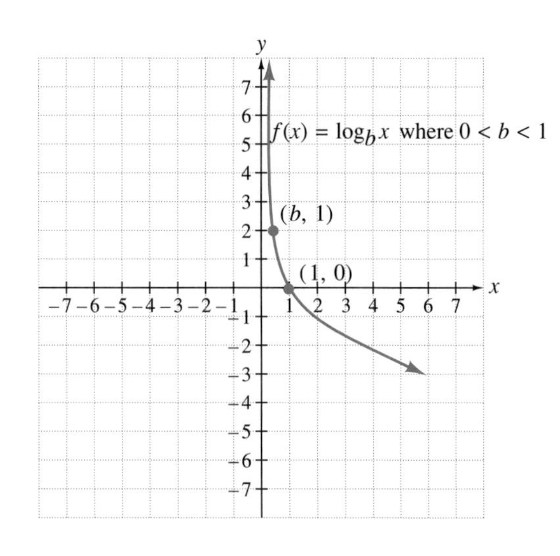

EXERCISE SET 11.3

Find the value of each logarithmic expression. See Example 1.

1. $\log_2 8$

2. $\log_4 64$

3. $\log_3 \frac{1}{9}$

4. $\log_2 \frac{1}{32}$

5. $\log_{25} 5$

6. $\log_8 \frac{1}{2}$

7. $\log_{1/2} 2$

8. $\log_{2/3} \frac{4}{9}$

Solve each equation for x. See Example 2.

9. $\log_3 9 = x$

10. $\log_2 8 = x$

11. $\log_3 x = 4$

12. $\log_2 x = 3$

13. $\log_x 49 = 2$

14. $\log_x 8 = 3$

15. $\log_2 \frac{1}{8} = x$

16. $\log_3 \frac{1}{81} = x$

Simplify. See Example 3.

17. $\log_5 5^3$

18. $\log_6 6^2$

19. $2^{\log_2 3}$

20. $7^{\log_7 4}$

21. $\log_9 9$

22. $\log_8 (8)^{-1}$

Graph each logarithmic function. Label any intercepts. See Example 4.

23. $y = \log_3 x$

24. $y = \log_2 x$

25. $f(x) = \log_{1/4} x$

26. $f(x) = \log_{1/2} x$

Graph each function and its inverse function on the same set of axes. Label any intercepts. See Example 5.

27. $y = 4^x; y = \log_4 x$

28. $y = 3^x; y = \log_3 x$

29. $y = \left(\frac{1}{3}\right)^x; y = \log_{1/3} x$

30. $y = \left(\frac{1}{2}\right)^x; y = \log_{1/2} x$

Find the value of each expression.

31. $\log_7 1$

32. $\log_9 9$

33. $\log_2 2^4$

34. $\log_6 6^{-2}$

35. $\log_{10} 100$

36. $\log_{10} \frac{1}{10}$

37. $3^{\log_3 5}$

38. $5^{\log_5 7}$

39. $\log_3 81$

40. $\log_2 16$

41. $\log_4 \frac{1}{64}$

42. $\log_3 \frac{1}{9}$

Solve each equation for x.

43. $\log_3 \frac{1}{27} = x$

44. $\log_5 \frac{1}{125} = x$

45. $\log_8 x = \frac{1}{3}$

46. $\log_9 x = \frac{1}{2}$

47. $\log_4 16 = x$

48. $\log_2 16 = x$

49. $\log_{3/4} x = 3$

50. $\log_{2/3} x = 2$

51. $\log_x 100 = 2$

52. $\log_x 27 = 3$

Graph each logarithmic function. Label any intercepts.

53. $f(x) = \log_5 x$

54. $f(x) = \log_6 x$

55. $f(x) = \log_{1/6} x$

56. $f(x) = \log_{1/5} x$

57. The formula $\log_{10}(1 - k) = \frac{-0.3}{H}$ models the relationship between the half-life H of a radioactive material and its rate of decay k. Find the rate of decay of the iodine isotope I-131 if its half-life is 8 days.

Skill Review

Simplify each rational expression. See Section 7.1.

58. $\frac{x + 3}{3 + x}$

59. $\frac{x - 5}{5 - x}$

60. $\frac{x^2 - 8x + 16}{2x - 8}$

61. $\frac{x^2 - 3x - 10}{2 + x}$

Add or subtract as indicated. See Section 7.3.

62. $\frac{2}{x} + \frac{3}{x^2}$

63. $\frac{3x}{x + 3} + \frac{9}{x + 3}$

64. $\frac{m^2}{m + 1} - \frac{1}{m + 1}$

65. $\frac{5}{y + 1} - \frac{4}{y - 1}$

Writing in Mathematics

66. Explain why the graph of the function $y = \log_b x$ contains the point $(1, 0)$ no matter what b is.

67. Explain why $\log_b a$ is not defined if a is negative.

11.4 Properties of Logarithms

OBJECTIVES

Tape IA 31

1. Apply the product property of logoarithms.
2. Apply the quotient property of logarithms.
3. Apply the power property of logarithms.

In the previous section we explored some basic properties of logarithms. We now introduce and explore additional properties.

1 The first of these properties is called the **product property of logarithms,** because it deals with the logarithm of a product.

> **Product Property of Logarithm**
>
> If x, y, and b are positive real numbers, $b \neq 1$, then
>
> $$\log_b xy = \log_b x + \log_b y$$

To prove this, let $\log_b x = M$ and $\log_b y = N$. Now write each logarithm using exponential notation.

$$\log_b x = M \quad \text{is equivalent to} \quad b^M = x$$
$$\log_b y = N \quad \text{is equivalent to} \quad b^N = y$$

Multiply the left sides and the right sides of the exponential equations and we have that

$$xy = (b^M)(b^N) = b^{M+N}$$

Now, write the equation $xy = b^{M+N}$ in equivalent logarithmic form.

$$xy = b^{M+N} \quad \text{is equivalent to} \quad \log_b xy = M + N$$

But since $M = \log_b x$ and $N = \log_b y$, we can write

$$\log_b xy = M + N$$

as

$$\log_b xy = \log_b x + \log_b y \qquad \text{Let } M = \log_b x \text{ and } N = \log_b y.$$

The logarithm of a product is the sum of the logarithms of the factors. This property is sometimes used to simplify logarithmic expressions.

EXAMPLE 1 Use the product rule to simplify. Assume that variables represent positive numbers.

a. $\log_{11} 10 + \log_{11} 3$ **b.** $\log_3 \frac{1}{2} + \log_3 12$ **c.** $\log_2 (x + 2) + \log_2 x$

Solution: All terms have a common logarithmic base.

a. $\log_{11} 10 + \log_{11} 3 = \log_{11} (10 \cdot 3)$ Apply the product property.

$= \log_{11} 30$

b. $\log_3 \frac{1}{2} + \log_3 12 = \log_3 \left(\frac{1}{2} \cdot 12\right) = \log_3 6$

c. $\log_2 (x + 2) + \log_2 x = \log_2 [(x + 2) \cdot x] = \log_2 (x^2 + 2x)$ ■

2 The second property is the **quotient property of logarithms.**

Quotient Property of Logarithms

If x, y, and b are positive real numbers, $b \neq 1$, then

$$\log_b \frac{x}{y} = \log_b x - \log_b y$$

The proof of the quotient property of logarithms is similar to the proof of the product rule. Notice that the quotient property says that the logarithm of a quotient is the difference of the logarithms of the dividend and divisor.

EXAMPLE 2 Use the quotient property to simplify. Assume that x represents positive numbers only.

a. $\log_{10} 27 - \log_{10} 3$ **b.** $\log_5 8 - \log_5 x$ **c.** $\log_4 25 + \log_4 3 - \log_4 5$
d. $\log_3 (x^2 + 5) - \log_3 (x^2 + 1)$

Solution: All terms have a common logarithmic base.

a. $\log_{10} 27 - \log_{10} 3 = \log_{10} \frac{27}{3} = \log_{10} 9$

b. $\log_5 8 - \log_5 x = \log_5 \frac{8}{x}$

c. Use both the product and quotient properties.

$\log_4 25 + \log_4 3 - \log_4 5 = \log_4 (25 \cdot 3) - \log_4 5$ Apply the product property.

$= \log_4 75 - \log_4 5$ Simplify.

$= \log_4 \frac{75}{5}$ Apply the quotient property.

$= \log_4 15$ Simplify.

d. $\log_3 (x^2 + 5) - \log_3 (x^2 + 1) = \log_3 \frac{x^2 + 5}{x^2 + 1}$ Apply the quotient rule.

3 The third and final property we introduce is called the **power property of logarithms.**

Power Property of Logarithms

If x and b are positive real numbers, $b \neq 1$, and r is a real number, then

$$\log_b x^r = r \log_b x$$

For example,

$$\log_3 2^4 = 4 \log_3 2, \qquad \log_5 x^3 = 3 \log_5 x, \qquad \log_4 \sqrt{x} = \log_4 (x)^{1/2} = \frac{1}{2} \log_4 x$$

EXAMPLE 3 Write each as the logarithm of a single expression. Assume that x is a positive number.

a. $2 \log_5 3 + 3 \log_5 2$ **b.** $2 \log_9 x + \log_9 (x + 1)$

Solution: Notice that the terms have a common logarithmic base.

a. $2 \log_5 3 + 3 \log_5 2 = \log_5 3^2 + \log_5 2^3$ Apply the power property.

$= \log_5 9 + \log_5 8$

$= \log_5 (9 \cdot 8)$ Apply the product property.

$= \log_5 72$

b. $2 \log_9 x + \log_9 (x + 1) = \log_9 x^2 + \log_9 (x + 1)$ Apply the power property.

$= \log_9 [x^2 (x + 1)]$ Apply the product property.

■

EXAMPLE 4 Use properties of logarithms to write each expression as a sum or difference of multiples of logarithms.

a. $\log_3 \frac{5 \cdot 7}{4}$ **b.** $\log_2 \frac{x^5}{y^2}$

Solution: **a.** $\log_3 \frac{5 \cdot 7}{4} = \log_3 (5 \cdot 7) - \log_3 4$

$= \log_3 5 + \log_3 7 - \log_3 4$

b. $\log_2 \frac{x^5}{y^2} = \log_2 (x^5) - \log_2 (y^2)$ Apply the quotient property.

$= 5 \log_2 x - 2 \log_2 y$ Apply the power property. ■

> **HELPFUL HINT**
>
> Notice that we are not able to simplify further a logarithmic expression such as $\log_5 (2x - 1)$. None of the basic properties gives a way to write the logarithm of a difference in some equivalent form.

EXAMPLE 5 If $\log_b 2 = 0.43$ and $\log_b 3 = 0.68$, use the properties of logarithms to evaluate.

a. $\log_b 6$ **b.** $\log_b 9$ **c.** $\log_b \sqrt{2}$

Solution: **a.** $\log_b 6 = \log_b (2 \cdot 3)$ Write 6 as $2 \cdot 3$.

$= \log_b 2 + \log_b 3$ Apply the product property.

$= 0.43 + 0.68$ Substitute given values.

$= 1.11$

b. $\log_b 9 = \log_b 3^2$ Write 9 as 3^2.

$= 2 \log_b 3$

$= 2(0.68)$ Substitute 0.68 for $\log_b 3$.

$= 1.36$

c. First, recall that $\sqrt{2} = 2^{1/2}$. Then

$$\log_b \sqrt{2} = \log_b 2^{1/2} = \frac{1}{2} \log_b 2$$

$$= \frac{1}{2}(0.43)$$

$$= 0.215$$ ■

Next we summarize the basic properties of logarithms that we developed so far.

> **Properties of Logarithms**
>
> If x, y, and b are positive real numbers, $b \neq 1$, and r is a real number, then:
>
> **1.** $\log_b 1 = 0$
> **2.** $\log_b b^x = x$
> **3.** $b^{\log_b x} = x$
> **4.** $\log_b xy = \log_b x + \log_b y$ Product property.
> **5.** $\log_b \frac{x}{y} = \log_b x - \log_b y$ Quotient property.
> **6.** $\log_b x^r = r \log_b x$ Power property.

EXERCISE SET 11.4

Write each as the logarithm of a single expression. Assume that variables represent positive numbers. See Example 1.

1. $\log_5 2 + \log_5 7$
2. $\log_3 8 + \log_3 4$
3. $\log_4 9 + \log_4 x$
4. $\log_2 x + \log_2 y$
5. $\log_{10} 5 + \log_{10} 2 + \log_{10} (x^2 + 2)$
6. $\log_6 3 + \log_6 (x + 4) + \log_6 5$

Write each as the logarithm of a single expression. Assume that variables represent positive numbers. See Example 2.

7. $\log_5 12 - \log_5 4$
8. $\log_7 20 - \log_7 4$
9. $\log_2 x - \log_2 y$
10. $\log_3 12 - \log_3 z$
11. $\log_4 2 + \log_4 10 - \log_4 5$
12. $\log_6 18 + \log_6 2 - \log_6 9$

Write each as the logarithm of a single expression. Assume that variables represent positive numbers. See Example 3.

13. $2 \log_2 5$
14. $3 \log_5 2$
15. $3 \log_5 x + 6 \log_5 z$
16. $2 \log_7 y + 6 \log_7 z$
17. $\log_{10} x - \log_{10} (x + 1) + \log_{10} (x^2 - 2)$
18. $\log_9 (4x) - \log_9 (x - 3) + \log_9 (x^3 + 1)$

Write each expression as a sum or difference of multiples of logarithms. Assume that variables represent positive numbers. See Example 4.

19. $\log_3 \frac{4y}{5}$
20. $\log_4 \frac{2}{9z}$
21. $\log_2 \left(\frac{x^3}{y}\right)$
22. $\log_5 \left(\frac{x}{y^4}\right)$
23. $\log_b \sqrt{7x}$
24. $\log_b \sqrt{\frac{3}{y}}$

If $\log_b 3 \approx 0.5$ and $\log_b 5 \approx 0.7$, approximate the following. See Example 5.

25. $\log_b \frac{5}{3}$
26. $\log_b 25$
27. $\log_b 15$
28. $\log_b \frac{3}{5}$
29. $\log_b \sqrt[3]{5}$
30. $\log_b \sqrt[4]{3}$

Write each as the logarithm of a single expression. Assume that variables represent positive numbers.

31. $\log_4 5 + \log_4 7$
32. $\log_3 2 + \log_3 5$
33. $\log_3 8 - \log_3 2$
34. $\log_5 12 - \log_5 3$
35. $\log_7 6 + \log_7 3 - \log_7 4$
36. $\log_8 5 + \log_8 15 - \log_8 20$
37. $3 \log_4 2 + \log_4 6$
38. $2 \log_3 5 + \log_3 2$
39. $3 \log_2 x + \frac{1}{2} \log_2 x - 2 \log_2 (x + 1)$
40. $2 \log_5 x + \frac{1}{3} \log_5 x - 3 \log_5 (x + 5)$
41. $2 \log_8 x - \frac{2}{3} \log_8 x + 4 \log_8 x$
42. $5 \log_6 x - \frac{3}{4} \log_6 x + 3 \log_6 x$

Write each expression as a sum or difference of multiples of logarithms. Assume that variables represent positive numbers.

43. $\log_7 \frac{5x}{4}$
44. $\log_9 \frac{7}{y}$
45. $\log_5 x^3 (x + 1)$
46. $\log_2 y^3 z$
47. $\log_6 \frac{x^2}{x + 3}$
48. $\log_3 \frac{(x + 5)^2}{x}$

If $\log_b 2 = 0.43$ and $\log_b 3 = 0.68$, evaluate the following.

49. $\log_b 8$

50. $\log_b 81$

51. $\log_b \frac{3}{9}$

52. $\log_b \frac{4}{32}$

53. $\log_b \sqrt{\frac{2}{3}}$

54. $\log_b \sqrt{\frac{3}{2}}$

Skill Review

55. Graph the functions $y = 10^x$ and $y = \log_{10} x$ on the same set of axes. See Section 11.3.

Solve each equation. See Section 8.6.

56. $\sqrt{3x + 1} = 2$

57. $\sqrt{2x - 3} = 5$

58. $\sqrt{x^2 + 9} = x + 3$

59. $\sqrt{x^2 - 2} = x - 2$

Evaluate each expression. See Section 11.3.

60. $\log_{10} 100$

61. $\log_{10} \frac{1}{10}$

62. $\log_e e^2$

63. $\log_e \sqrt{e}$

Writing in Mathematics

64. Explain whether the quotient property of logarithms can be applied to $\frac{\log_4 18}{\log_4 6}$.

65. Explain whether the product property of logarithms can be applied to $(\log_3 6) \cdot (\log_3 4)$.

11.5 Common Logarithms, Natural Logarithms, and Change of Base

OBJECTIVES

Tape IA 32

1. Identify common logarithms and approximate them by calculator.
2. Evaluate common logarithms of powers of 10.
3. Identify natural logarithms and approximate them by calculator.
4. Evaluate natural logarithms of powers of e.
5. Apply the change of base formula.

In this section we look closely at two particular logarithmic bases. These two logarithmic bases are used so frequently that logarithms to their bases are given special names. **Common logarithms** are logarithms to base 10. **Natural logarithms** are logarithms to base e, which we introduce in this section. Because of the wide availability and low cost of calculators today, the work in this section is based on using calculators, which typically have both the common "log" and the natural "log" keys.

1 Logarithms to base 10, common logarithms, are used frequently because our number system is a base 10 decimal system. The notation $\log x$ means the same as $\log_{10} x$.

Common Logarithms

$$\log x \text{ means } \log_{10} x$$

EXAMPLE 1 Use a calculator to approximate log 7 to 4 decimal places.

Solution: Press the following sequence of keys:

[7] [log]

(Some calculators require [7] [log] [=].) The number [0.845098] should appear in the display. To four decimal places,

$$\log 7 \approx 0.8451$$

Some scientific calculators do not have a [log] key, but do have a [10^x] key. If this is the case, then log 7 is approximated by pressing

[7] [INV] [10^x]

This sequence is based on the fact that the functions $y = \log_{10} x$ or $y = \log x$ and $y = 10^x$ are inverses of each other. ■

2 To evaluate the common log of a power of 10, a calculator is not needed. According to the property of logarithms

$$\log_b b^x = x$$

it follows that

$$\log 10^x = x$$

because the base of this logarithm is understood to be 10.

EXAMPLE 2 Find the exact value of each logarithm.

a. $\log 10$ **b.** $\log 1000$ **c.** $\log \frac{1}{10}$ **d.** $\log \sqrt{10}$

Solution: **a.** $\log 10 = \log 10^1 = 1$ **b.** $\log 1000 = \log 10^3 = 3$

c. $\log \frac{1}{10} = \log 10^{-1} = -1$ **d.** $\log \sqrt{10} = \log 10^{1/2} = \frac{1}{2}$ ■

As we will soon see, equations containing common logs are useful models of many natural phenomena.

EXAMPLE 3 Solve $\log x = 1.2$ for x. Give an exact solution, and then approximate the solution to four decimal places.

Solution: Write the logarithmic equation using exponential notation. Keep in mind that the base of a common log is understood to be 10.

$$\log x = 1.2$$

$$10^{1.2} = x \qquad \text{Write using exponential notation.}$$

$$15.848932 \approx x \qquad \text{Press } \boxed{10}\ \boxed{y^x}\ \boxed{1.2}\ \boxed{=}.$$

The exact solution is $10^{1.2}$. To four decimal places, $x \approx 15.8489$. ■

3 **Natural logarithms** are also frequently used, especially to describe natural events; hence the label "natural logarithm." Natural logarithms are logarithms to the base e, which is a constant approximately equal to 2.7183. The number e is an irrational number like π. The notation $\log_e x$ is usually abbreviated to $\ln x$.

Natural Logarithms

$$\ln x \text{ means } \log_e x$$

EXAMPLE 4 Use a calculator to approximate ln 8 to four decimal places.

Solution: Press the following sequence of keys:

$$\boxed{8}\ \boxed{\ln}$$

The display should show $\boxed{2.0794415}$. To four decimal places,

$$\ln 8 \approx 2.0794$$

Some scientific calculators do not have a $\boxed{\ln}$ key, but they have an $\boxed{e^x}$ key instead. If this is the case, then ln 8 is approximated by pressing

$$\boxed{8}\ \boxed{\text{INV}}\ \boxed{e^x}$$

This sequence is based on the fact that the functions $y = \log_e x$ or $y = \ln x$ and $y = e^x$ are inverses of each other. ■

4 As a result of the property $\log_b b^x = x$, we know that $\log_e e^x = x$ or $\ln e^x = x$.

EXAMPLE 5 Find the exact value of each natural logarithm.

a. $\ln e^3$ **b.** $\ln \sqrt[5]{e}$

Solution: **a.** $\ln e^3 = 3$ **b.** $\ln \sqrt[5]{e} = \ln e^{1/5} = \frac{1}{5}$ ■

EXAMPLE 6 Solve the equation $\ln 3x = 5$ for x. Give an exact solution and then approximate the solution to four decimal places.

Solution: Write the equation using exponential notation. Keep in mind that the base of a natural logarithm is understood to be e.

$$\ln 3x = 5$$

$$e^5 = 3x \qquad \text{Write using exponential notation.}$$

$$\frac{e^5}{3} = x$$

The exact solution is $\frac{e^5}{3}$. To four decimal places, $x \approx 49.4711$. ■

The Richter scale measures the intensity or magnitude of an earthquake. The formula for the magnitude R of an earthquake is $R = \log\left(\frac{a}{T}\right) + B$, where a is the amplitude in micrometers of the vertical motion of the ground at the recording station, T is the number of seconds between successive seismic waves, and B is an adjustment factor that takes into account the weakening of the seismic wave as the distance increases from the epicenter of the earthquake.

EXAMPLE 7 Find the magnitude of an earthquake on the Richter scale if a recording station measures an amplitude of 300 micrometers and 2.5 seconds between waves. Assume that B is 4.2. Approximate the solution to the nearest tenth.

Solution: Substitute the known values into the formula for earthquake intensity.

$$R = \log\left(\frac{a}{T}\right) + B \qquad \text{Richter scale formula.}$$

$$= \log\left(\frac{300}{2.5}\right) + 4.2 \qquad \text{Let } a = 300,\ T = 2.5, \text{ and } B = 4.2.$$

$$= \log(120) + 4.2$$

$$\approx 2.1 + 4.2 \qquad \text{Approximate log 120 by 2.1.}$$

$$= 6.3$$

This earthquake had a magnitude of 6.3 on the Richter scale. ■

Recall from Section 11.2 the formula $A = P\left(1 + \frac{r}{n}\right)^{nt}$ for compound interest, where n represents the number of compoundings per year. When interest is compounded continuously, the formula $A = Pe^{rt}$ is used.

EXAMPLE 8 Find the amount owed at the end of 5 years if \$1600 is loaned at a rate of 9% compounded continuously.

Solution: Use the formula $A = Pe^{rt}$, where

$P = \$1600$ (the size of the loan)

$r = 9\% = 0.09$ (the rate of interest)

$t = 5$ (the 5-year duration of the loan)

$$A = Pe^{rt}$$

$$= 1600e^{0.09(5)} \qquad \text{Substitute in known values.}$$

$$= 1600e^{0.45}$$

Now use a calculator to approximate the solution. Press the keys

[0.45] [e^x] [×] [1600] [=]

Thus

$$A \approx 2509.30$$

The total amount of money owed is approximately \$2509.30. ■

5 Calculators are handy tools for approximating natural and common logarithms. Unfortunately, some calculators cannot be used to approximate logarithms to bases other than e or 10, at least not directly. In such cases, we use the change of base formula.

Change of Base

If a, b, and c are positive real numbers and neither b nor c is 1, then

$$\log_b a = \frac{\log_c a}{\log_c b}$$

EXAMPLE 9 Approximate $\log_5 3$ to four decimal places.

Solution: Use the change of base property to write $\log_5 3$ as a quotient of logarithms to base 10.

$$\log_5 3 = \frac{\log 3}{\log 5} \quad \text{Use the change of base property.}$$

$$\approx \frac{0.4771213}{0.69897} \quad \text{Approximate logarithms by calculator.}$$

$$\approx 0.6826063 \quad \text{Simplify by calculator.}$$

To four decimal places, $\log_5 3 \approx 0.6826$. ■

EXERCISE SET 11.5

Use a calculator to approximate each logarithm to four decimal places. See Example 1.

1. $\log 8$ **2.** $\log 6$ **3.** $\log 2.31$ **4.** $\log 4.86$

Find the exact value. See Example 2.

5. $\log 100$ **6.** $\log 10{,}000$ **7.** $\log\left(\frac{1}{1000}\right)$ **8.** $\log\left(\frac{1}{100}\right)$

Solve each equation for x. Give an exact solution and a four-decimal-place approximation. See Example 3.

9. $\log x = 1.3$ **10.** $\log x = 2.1$ **11.** $\log 2x = 1.1$ **12.** $\log 3x = 1.3$

Use a calculator to approximate each logarithm to four decimal places. See Example 4.

13. $\ln 2$ **14.** $\ln 3$ **15.** $\ln 0.0716$ **16.** $\ln 0.0032$

Find the exact value. See Example 5.

17. $\ln e^2$ **18.** $\ln e^4$ **19.** $\ln \sqrt[4]{e}$ **20.** $\ln \sqrt[5]{e}$

Solve each equation for x. Give an exact solution and a four-decimal-place approximation. See Example 6.

21. $\ln x = 1.4$ **22.** $\ln x = 2.1$ **23.** $\ln (3x - 4) = 2.3$ **24.** $\ln (2x + 5) = 3.4$

Use the formula $R = \log\left(\frac{a}{T}\right) + B$ to find the intensity R on the Richter scale of the earthquakes fitting the descriptions given. See Example 7.

25. Amplitude a is 200 micrometers, time T between waves is 1.6 seconds, and B is 2.1.

26. Amplitude a is 150 micrometers, time T between waves is 3.6 seconds, and B is 1.9.

Solve. See Example 8.

27. Find the amount of money Paul Banks owes at the end of 4 years if 6% interest is compounded continuously on his $2000 debt.

28. Find the amount of money a $2500 certificate of deposit is redeemable for if it has been paying 10% interest compounded continuously for 3 years.

Approximate each logarithm to four decimal places. See Example 9.

29. $\log_2 3$ **30.** $\log_3 2$ **31.** $\log_{1/2} 5$ **32.** $\log_{1/3} 2$

Use a calculator to approximate each logarithm to four decimal places.

33. $\log 12.6$ **34.** $\log 25.9$ **35.** $\ln 5$ **36.** $\ln 7$

37. $\log 41.5$ **38.** $\ln 41.5$

Find the exact value.

39. $\log 10^3$ **40.** $\ln e^5$ **41.** $\ln e^2$ **42.** $\log 10^7$

43. $\log 0.0001$ **44.** $\log 0.001$ **45.** $\ln \sqrt{e}$ **46.** $\log \sqrt{10}$

Solve each equation for x. Give an exact solution and a four-decimal-place approximation.

47. $\log x = 2.3$ **48.** $\log x = 3.1$

49. $\ln x = -2.3$ **50.** $\ln x = -3.7$

51. $\log (2x + 1) = -0.5$ **52.** $\log (3x - 2) = -0.8$

53. $\ln 4x = 0.18$ **54.** $\ln 3x = 0.76$

Approximate each logarithm to four decimal places.

55. $\log_4 9$ **56.** $\log_9 4$

57. $\log_3 \frac{1}{6}$ **58.** $\log_6 \frac{2}{3}$

59. $\log_8 6$ **60.** $\log_6 8$

Use the formula $R = \log\left(\frac{a}{T}\right) + B$ to find the intensity R on the Richter scale of the earthquakes fitting the descriptions given.

61. Amplitude a is 400 micrometers, time T between waves is 2.6 seconds, and B is 3.1.

62. Amplitude a is 450 micrometers, time T between waves is 4.2 seconds, and B is 2.7.

Solve.

63. Find how much money Dana Jones has after 12 years if $1400 is invested at 8% interest compounded continuously.

64. Determine the size of an account, where $3500 earns 6% interest compounded continuously for 1 year.

Psychologists call the graph of the formula

$$t = \frac{1}{c}\ln\left(\frac{A}{A - N}\right)$$

the learning curve, since the formula relates time t passed, in weeks, to a measure N of learning achieved, to a measure A of maximum learning possible, to a measure c of an individual's learning style.

65. Norman is learning to type. If he wants to type at a rate of 50 words per minute (N is 50), and his expected maximum rate is 75 words per minute (A is 75), find how many weeks it takes him to achieve his goal. Assume that c is 0.09.

66. Janine is working on her dictation skills. She wants to take dictation at a rate of 150 words per minute and believes that the maximum rate she could ever hope for is 210 words per minute. Find how many weeks it takes her to achieve the 150-word level if c is 0.07.

67. An experiment with teaching chimpanzees sign language shows that a typical chimp can master a maximum of 65 signs. Find how many weeks it should take a chimpanzee to achieve mastery of 30 signs if c is 0.03.

68. A psychologist is measuring human capability to memorize nonsense syllables. Find how many minutes it should take a subject to learn 15 nonsense syllables if the maximum possible to learn is 24 syllables and c is 0.17.

Skill Review

Solve each equation for x. See Sections 2.3 and 6.6.

69. $6x - 3(2 - 5x) = 6$

70. $2x + 3 = 5 - 2(3x - 1)$

71. $2x + 3y = 6x$

72. $4x - 8y = 10x$

73. $x^2 + 7x = -6$

74. $x^2 + 4x = 12$

Solve each system of equations. See Section 10.1.

75. $\begin{cases} x + 2y = -4 \\ 3x - y = 9 \end{cases}$

76. $\begin{cases} 5x + y = 5 \\ -3x - 2y = -10 \end{cases}$

Writing in Mathematics

77. On a calculator, press $\boxed{-3}$ $\boxed{\log}$. Describe what happens and explain why it does.

78. Without using a calculator, explain which of log 50 and ln 50 must be larger.

11.6 Exponential and Logarithmic Functions and Problem Solving

OBJECTIVES

1. Solve exponential equations.
2. Solve logarithmic equations.
3. Solve problems that can be modeled by exponential and logarithmic equations.

Tape IA 32

1 In Section 11.2, we solved exponential equations like $2^x = 16$ by writing 16 as a power of 2 and applying the uniqueness of b^x.

$$2^x = 16$$

$$2^x = 2^4 \quad \text{Write 16 as } 2^4.$$

$$x = 4 \quad \text{Apply the uniqueness of } b^x.$$

To solve an equation like $3^x = 7$, we use the following.

Uniqueness of $\log_b a$ as a Logarighm to Base b

If a, b, and c are real numbers such that $\log_b a$ and $\log_b c$ are real numbers and b is not 1, then

$$\log_b a = \log_b c \text{ only when } a = c$$

EXAMPLE 1 Solve $3^x = 7$.

Solution: To solve, use the uniqueness of logarithms and take the logarithm of both sides. For this example, we use the common logarithm.

$$3^x = 7$$

$$\log 3^x = \log 7 \quad \text{Take the common log of both sides.}$$

$$x \log 3 = \log 7 \quad \text{Apply the power property of logarithms.}$$

$$x = \frac{\log 7}{\log 3} \quad \text{Divide both sides by log 3.}$$

The exact solution is $\frac{\log 7}{\log 3}$. If a decimal approximation is preferred, $\frac{\log 7}{\log 3} \approx \frac{0.845098}{0.4771213} \approx 1.7712$ to four decimal places. The solution set is $\left\{\frac{\log 7}{\log 3}\right\}$ or **approximately** $\{1.7712\}$. ■

2 By applying the appropriate properties of logarithms, a broad variety of logarithmic equations can be solved.

EXAMPLE 2 Solve $\log_4 (x - 2) = 2$ for x.

Solution: First, write the equation using exponential notation.

$$\log_4 (x - 2) = 2$$

$$4^2 = x - 2$$

$$16 = x - 2$$

$$18 = x \quad \text{Add 2 to both sides.}$$

To check, replace x with 18 in the **original equation.**

$$\log_4 (x - 2) = 2$$

$$\log_4 (18 - 2) = 2 \quad \text{Let } x = 18.$$

$$\log_4 16 = 2$$

$$4^2 = 16 \quad \text{True.}$$

The solution set is $\{18\}$. ■

EXAMPLE 3 Solve $\log_2 x + \log_2 (x - 1) = 1$ for x.

Solution: Apply the product rule to the left side of the equation.

$$\log_2 x + \log_2 (x - 1) = 1$$

$$\log_2 x(x - 1) = 1 \quad \text{Use the product rule.}$$

$$\log_2 (x^2 - x) = 1$$

Next, write the equation using exponential notation and solve for x.

$$2^1 = x^2 - x$$

$$0 = x^2 - x - 2 \quad \text{Subtract 2 from both sides.}$$

$$0 = (x - 2)(x + 1) \quad \text{Factor.}$$

$$0 = x - 2 \quad \text{or} \quad 0 = x + 1 \quad \text{Set each factor equal to 0.}$$

$$2 = x \qquad -1 = x$$

Verify that 2 satisfies the original equation. Now check -1 by replacing x with -1 in the original equation.

$$\log_2 x + \log_2 (x - 1) = 1$$

$$\log_2 (-1) + \log_2 (-1 - 1) = 1 \quad \text{Let } x = -1.$$

Because the logarithm of a negative number is undefined, -1 is an extraneous solution. The solution set is $\{2\}$. ■

EXAMPLE 4 Solve $\log(x + 2) - \log x = 2$ for x.

Solution: Apply the quotient property of logarithms to the left side of the equation.

$$\log(x + 2) - \log x = 2$$

$$\log \frac{x + 2}{x} = 2 \quad \text{Apply the quotient property.}$$

$$10^2 = \frac{x + 2}{x} \quad \text{Write using exponential notation.}$$

$$100 = \frac{x + 2}{x}$$

$$100x = x + 2 \quad \text{Multiply both sides by } x.$$

$$99x = 2 \quad \text{Subtract } x \text{ from both sides.}$$

$$x = \frac{2}{99} \quad \text{Divide both sides by 99.}$$

Verify that the solution set is $\left\{\frac{2}{99}\right\}$. ■

3 Throughout this chapter we have emphasized that logarithmic and exponential functions are used in a variety of scientific, technical, and business settings. A few examples are shown.

EXAMPLE 5 The population of lemmings varies according to the relationship $y = y_0e^{0.15t}$. In this formula, t is time in months, and y_0 is some initial population at time 0. Estimate the population in 6 months if there are originally 5000 lemmings.

Solution: Substitute 5000 for y_0 and 6 for t.

$$y = y_0e^{0.15t}$$

$$= 5000e^{0.15(6)} \qquad \text{Let } t = 6 \text{ and } y_0 = 5000.$$

$$= 5000e^{0.9} \qquad \text{Multiply.}$$

To find an approximation for $5000e^{0.9}$, press these keys on your calculator:

$$\boxed{0.9}\ \boxed{e^x}\ \boxed{\times}\ \boxed{5000}\ \boxed{=}$$

Then $y \approx 12{,}298.016$. In 6 months the population should be approximately 12,300 lemmings. ■

EXAMPLE 6 How long does it take an investment of \$2000 to double if it is invested at 5% interest compounded quarterly? The necessary formula to use is $A = P\left(1 + \frac{r}{n}\right)^{nt}$, where A is the accrued (or owed) amount, P is the principal invested, r is the rate of interest, n is the number of compounding periods per year, and t is the number of years.

Solution: We are given that $P = \$2000$ and $r = 5\% = 0.05$. Compounding quarterly means 4 times a year so that $n = 4$. The investment is to double, so A must be \$4000. Substitute these values and solve for t.

$$A = P\left(1 + \frac{r}{n}\right)^{nt}$$

$$4000 = 2000\left(1 + \frac{0.05}{4}\right)^{4t} \qquad \text{Substitute in known values.}$$

$$4000 = 2000(1.0125)^{4t} \qquad \text{Simplify } 1 + \frac{0.05}{4}.$$

$$2 = (1.0125)^{4t} \qquad \text{Divide both sides by 2000.}$$

$$\log 2 = \log 1.0125^{4t} \qquad \text{Apply the uniqueness of logarithms.}$$

$$\log 2 = 4t(\log 1.0125) \qquad \text{Apply the power property.}$$

$$\frac{\log 2}{(4 \log 1.0125)} = t \qquad \text{Divide both sides by 4 log 1.0125.}$$

$$13.949408 \approx t \qquad \text{Approximate by calculator.}$$

Thus it takes nearly 14 years for the money to double in value. ■

EXERCISE SET 11.6

Solve each equation. Give an exact solution, and also approximate the solution to four decimal places. See Example 1.

1. $3^x = 6$

2. $4^x = 7$

3. $3^{2x} = 3.8$

4. $5^{3x} = 5.6$

5. $2^{x-3} = 5$

6. $8^{x-2} = 12$

Solve each equation. See Example 2.

7. $\log_2 (x + 5) = 4$

8. $\log_6 (x^2 - x) = 1$

9. $\log_3 x^2 = 4$

10. $\log_2 x^2 = 6$

Solve each equation. See Examples 3 and 4.

11. $\log_4 2 + \log_4 x = 0$

12. $\log_3 5 + \log_3 x = 1$

13. $\log_2 6 - \log_2 x = 3$

14. $\log_4 10 - \log_4 x = 2$

15. $\log_4 x + \log_4 (x + 6) = 2$

16. $\log_3 x + \log_3 (x + 6) = 3$

17. $\log_5 (x + 3) - \log_5 x = 2$

18. $\log_6 (x + 2) - \log_6 x = 2$

Solve. See Example 5.

19. The size of the wolf population at Isle Royale National Park increases at a rate of 4.3% per year. If the size of the current population is 83 wolves, find how many there should be in 5 years. Use $y = y_0\, e^{0.043t}$.

20. The number of victims of a flu epidemic is increasing at a rate of 7.5% per week. If 20,000 persons are currently infected, find in how many days we can expect 45,000 to have the flu. Use $y = y_0\, e^{0.075t}$.

Solve. Use the formula $A = P\left(1 + \frac{r}{n}\right)^{nt}$ *to solve the compound interest problem. See Example 6.*

21. Find how long it takes $1000 to double if it is invested at 8% interest compounded semiannually.

22. Find how long it takes $1000 to double if it is invested at 8% interest compounded monthly.

Solve each equation. Give an exact solution, and also approximate the solution to four decimal places.

23. $9^x = 5$

24. $3^x = 11$

25. $4^{x+7} = 3$

26. $6^{x+3} = 2$

27. $7^{3x-4} = 11$

28. $5^{2x-6} = 12$

29. $e^{6x} = 5$

30. $e^{2x} = 8$

Solve each equation.

31. $\log_3 (x - 2) = 2$

32. $\log_2 (x - 5) = 3$

33. $\log_4 (x^2 - 3x) = 1$

34. $\log_8 (x^2 - 2x) = 1$

35. $\log_3 5 + \log_3 x = 2$

36. $\log_5 2 + \log_5 x = 3$

37. $3 \log_8 x - \log_8 x^2 = 2$

38. $2 \log_6 x - \log_6 x = 3$

39. $\log_2 x + \log_2 (x + 5) = 1$

40. $\log_4 x + \log_4 (x + 7) = 1$

41. $\log_4 x - \log_4 (2x - 3) = 3$

42. $\log_2 x - \log_2 (3x + 5) = 4$

43. $\log_2 x + \log_2 (3x + 1) = 1$

44. $\log_3 x + \log_3 (x - 8) = 2$

Solve.

45. The size of the population of Senegal is increasing at a rate of 2.6% per year. If 7,000,000 people lived in Senegal in 1986, find how many inhabitants there will be by 2000, rounded to the nearest ten thousand. Use $y = y_0 e^{0.026t}$.

46. In 1986, 784 million people were citizens of India. Find how long it will take India's population to reach a size of 1000 million (that is, 1 billion) if the population size is growing at a rate of 2.1% per year. Round to the nearest tenth. Use $y = y_0 e^{0.021t}$.

Use the formula $A = P\left(1 + \frac{r}{n}\right)^{nt}$ *to solve these compound interest problems. See Example 6.*

47. Find how long it takes $600 to double if it is invested at 7% interest compounded monthly.

48. Find how long it takes $600 to double if it is invested at 12% interest compounded monthly.

49. Find how long it takes a $1200 investment to earn $200 interest if it is invested at 9% interest compounded quarterly.

50. Find how long it takes a $1500 investment to earn $200 interest if it is invested at 10% compounded semiannually.

The formula $w = 0.00185h^{2.67}$ is used to estimate the normal weight w of a boy h inches tall. Use this formula to solve the height–weight problem.

51. Find the expected height of a boy who normally should weigh 85 pounds.

52. Find the expected height of a boy who normally should weigh 140 pounds.

The formula $P = 14.7e^{-0.21x}$ gives the average atmospheric pressure P, in pounds per square inch, at an altitude x, in miles above sea level. Use this formula to solve these pressure problems.

53. Find the average atmospheric pressure of Denver, which is 1 mile above sea level.

54. Find the average atmospheric pressure of Pikes Peak, which is 2.7 miles above sea level.

55. Find the elevation of a Delta jet if the atmospheric pressure outside the jet is 7.5 lb/in.^{2}.

56. Find the elevation of a remote Himalayan peak if the atmospheric pressure atop the peak is 6.5 lb/in.^{2}.

Skill Review

If $x = -2$, $y = 0$, and $z = 3$, find the value of each expression. See Sections 1.4 and 1.7.

57. $\dfrac{x^2 - y + 2z}{3x}$

58. $\dfrac{x^3 - 2y + z}{2z}$

59. $\dfrac{3z - 4x + y}{x + 2z}$

60. $\dfrac{4y - 3x + z}{2x + y}$

Find the inverse function of each one-to-one function. See Section 11.1.

61. $f(x) = 5x + 2$

62. $f(x) = \dfrac{x - 3}{4}$

CHAPTER 11 GLOSSARY

Common logarithms are logarithms to base 10.

A function of the form $f(x) = b^x$ is an **exponential function** where $b > 0$, b is not 1, and x is a real number.

If x is a positive real number, b is a constant positive real number, and b is not 1, then a **logarithmic function** is a function that can be defined by $f(x) = \log_b x$.

The **inverse of a one-to-one function** f is the one-to-one function f^{-1} that is the set of all ordered pairs (y, x) where (x, y) belongs to f.

Natural logarithms are logarithms to base e.

If f is a function, then f is a **one-to-one-function** if each y-value corresponds to a unique x-value.

CHAPTER 11 SUMMARY

HORIZONTAL LINE TEST (11.1)

If every horizontal line intersects the graph of a function at most once, then the function is a one-to-one function.

TO FIND THE INVERSE OF A ONE-TO-ONE FUNCTION $f(x)$ (11.1)

1. Replace $f(x)$ by y.
2. Interchange x and y.
3. Solve the equation for y.
4. Replace y with the notation $f^{-1}(x)$.

LOGARITHMIC DEFINITION (11.3)

If $b > 0$ and $b \neq 1$, then

$$x = b^y \quad \text{is equivalent to} \quad y = \log_b x$$

PROPERTIES OF LOGARITHMS (11.4)

If x, y, and b are positive real numbers, $b \neq 1$, and r is a real number, then:

1. $\log_b 1 = 0$
2. $\log_b b^x = x$
3. $b^{\log_b x} = x$
4. $\log_b xy = \log_b x + \log_b y$ Product property.
5. $\log_b \frac{x}{y} = \log_b x - \log_b y$ Quotient property.
6. $\log_b x^r = r \log_b x$ Power property.

CHANGE OF BASE (11.5)

If a, b, and c are positive real numbers and neither b nor c is 1, then

$$\log_b a = \frac{\log_c a}{\log_c b}$$

CHAPTER 11 REVIEW

(11.1) *Determine whether each function is a one-to-one function. If so, list the elements of its inverse.*

1. $h = \{(-9, 14), (6, 8), (-11, 12), (15, 15)\}$

2. $f = \{(-5, 5), (0, 4), (13, 5), (11, -6)\}$

Determine whether each function is a one-to-one function.

3.

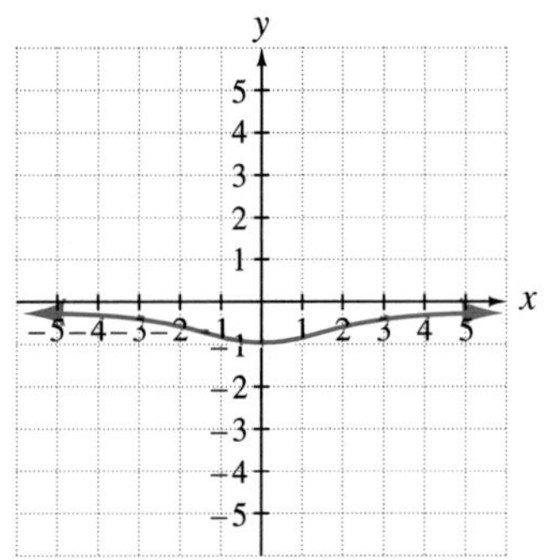

4.

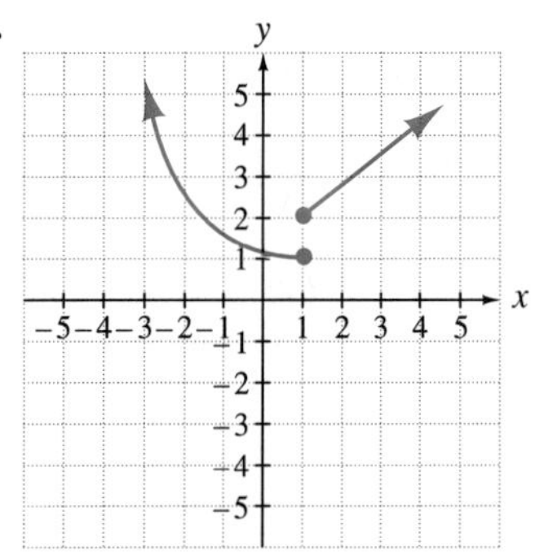

5.

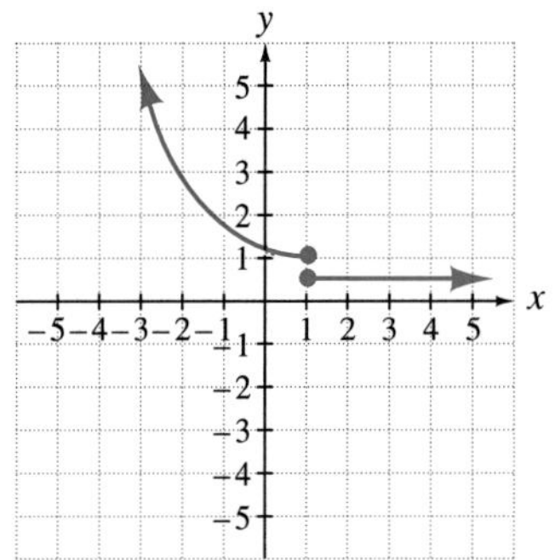

6.

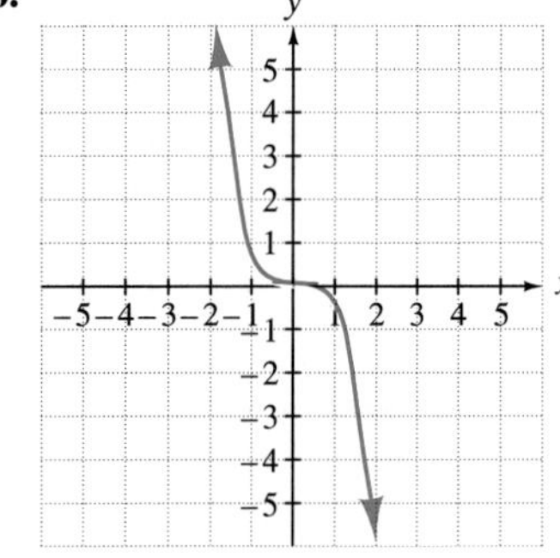

Find an equation defining the inverse function of the given function.

7. $f(x) = 6x + 11$

8. $f(x) = 12x$

9. $q(x) = mx + b$

10. $g(x) = \dfrac{12x - 7}{6}$

11. $r(x) = \dfrac{13}{2}x - 4$

On the same set of axes, graph the given one-to-one function and its inverse.

12. $g(x) = \sqrt{x}$

13. $h(x) = 5x - 5$

(11.2) *Solve each equation for x.*

14. $4^x = 64$

15. $3^x = \dfrac{1}{9}$

16. $2^{3x} = \dfrac{1}{16}$

17. $5^{2x} = 125$

18. $9^{x+1} = 243$

19. $8^{3x-2} = 4$

Graph each exponential function.

20. $y = 3^x$

21. $y = \left(\dfrac{1}{3}\right)^x$

22. $y = 4 \cdot 2^x$

23. $y = 2^x + 4$

Use the formula $A = P\left(1 + \dfrac{r}{n}\right)^{nt}$ *to solve the interest problems. In this formula:*

A = amount accrued (or owed)

P = principal invested (or loaned)

r = rate of interest

n = number of compounding periods per year

t = time in years

24. Find the amount if \$1600 is invested at 9% interest compounded semiannually for 7 years.

25. \$800 is invested in a 7% certificate of deposit for which interest is compounded quarterly. Find the value of this certificate at the end of 5 years.

(11.3) *Write each equation using logarithmic notation.*

26. $49 = 7^2$

27. $2^{-4} = \dfrac{1}{16}$

Write each logarithmic equation using exponential notation.

28. $\log_{\frac{1}{2}} 16 = -4$

29. $\log_{0.4} 0.064 = 3$

Solve for x.

30. $\log_4 x = -3$

31. $\log_3 x = 2$

32. $\log_3 1 = x$

33. $\log_4 64 = x$

34. $\log_x 64 = 2$

35. $\log_x 81 = 4$

36. $\log_4 4^5 = x$

37. $\log_7 7^{-2} = x$

38. $5^{\log_5 4} = x$

39. $2^{\log_2 9} = x$

40. $\log_2 (3x - 1) = 4$

41. $\log_3 (2x + 5) = 2$

42. $\log_4 (x^2 - 3x) = 1$

43. $\log_8 (x^2 + 7x) = 1$

Graph each pair of equations on the same coordinate system.

44. $y = 2^x$ and $y = \log_2 x$

45. $y = \left(\frac{1}{2}\right)^x$ and $y = \log_{\frac{1}{2}} x$

(11.4) *Write each of the following as single logarithms.*

46. $\log_3 8 + \log_3 4$

47. $\log_2 6 + \log_2 3$

48. $\log_7 15 - \log_7 20$

49. $\log 18 - \log 12$

50. $\log_{11} 8 + \log_{11} 3 - \log_{11} 6$

51. $\log_5 14 + \log_5 3 - \log_5 21$

52. $2 \log_5 x - 2 \log_5 (x + 1) + \log_5 x$

53. $4 \log_3 x - \log_3 x + \log_3 (x + 2)$

Use properties of logarithms to write each expression as a sum or difference of multiples of logarithms.

54. $\log_3 \frac{x^3}{x + 2}$

55. $\log_4 \frac{x + 5}{x^2}$

56. $\log_2 \frac{3x^2y}{z}$

57. $\log_7 \frac{yz^3}{x}$

If $\log_b 2 = 0.36$ and $\log_b 5 = 0.83$, find the following.

58. $\log_b 50$

59. $\log_b \frac{4}{5}$

(11.5) *Use a calculator to approximate the logarithm to four decimal places.*

60. $\log 3.6$

61. $\log 0.15$

62. $\ln 1.25$

63. $\ln 4.63$

Find the exact value.

64. $\log 1000$

65. $\log \frac{1}{10}$

66. $\ln \left(\frac{1}{e}\right)$

67. $\ln (e^4)$

Solve each equation for x.

68. $\ln (2x) = 2$

69. $\ln (3x) = 1.6$

70. $\ln (2x - 3) = -1$

71. $\ln (3x + 1) = 2$

Use the formula $\ln \frac{I}{I_0} = -kx$ *to solve radiation problems. In this formula:*

x = depth in millimeters

I = intensity of radiation

I_0 = initial intensity

k = a constant measure dependent on the material

72. Find the depth at which the intensity of the radiation passing through a lead shield is reduced to 3% of the original intensity if the value of k is 2.1.

73. If k is 3.2, find the depth at which 2% of the original radioactivity will penetrate.

Approximate the logarithm to four decimal places.

74. $\log_5 1.6$

75. $\log_3 4$

Use the formula $A = Pe^{rt}$ *to solve the interest problems in which interest is compounded continuously. In this formula:*

A = amount accrued (or owed)

P = principal invested (or loaned)

r = rate of interest

t = time in years

76. Chase Manhattan Bank offers a 5-year 6% continuously compounded investment option. Find the amount accrued if $1450 is invested.

77. Find the amount that a $940 investment grows to if it is invested at 11% compounded continuously for 3 years.

(11.6) *Solve each exponential equation for x. Give an exact solution and also approximate the solution to four decimal places.*

78. $3^{2x} = 7$

79. $6^{3x} = 5$

80. $3^{2x+1} = 6$

81. $4^{3x+2} = 9$

82. $5^{3x-5} = 4$

83. $8^{4x-2} = 3$

84. $2 \cdot 5^{x-1} = 1$

85. $3 \cdot 4^{x+5} = 2$

Solve the equation for x.

86. $\log_5 2 + \log_5 x = 2$

87. $\log_3 x + \log_3 10 = 2$

88. $\log (5x) - \log (x + 1) = 4$

89. $\ln (3x) - \ln (x - 3) = 2$

90. $\log_2 x + \log_2 2x - 3 = 1$

91. $-\log_6 (4x + 7) + \log_6 x = 1$

Use the formula $y = y_0 e^{kt}$ *to solve the population growth problems. In this formula:*

y = size of population

y_0 = initial count of population

k = rate of growth

t = time

92. The population of mallard ducks in Nova Scotia is expected to grow at a rate of 6% per week during the spring migration. If 155,000 ducks are already in Nova Scotia, find how many are expected by the end of 4 weeks.

93. The population of Indonesia is growing at a rate of 1.7% per year. If the population in 1986 was 176,800,000, find the expected population by the year 2000.

94. Anaheim, California, is experiencing an annual growth rate of 3.16%. If 230,000 people now live in Anaheim, find how long it will take for the size of the population to be 500,000.

95. Memphis, Tennessee, is growing at a rate of 0.36% per year. Find how long it will take the population of Memphis to increase from 650,000 to 700,000.

96. Egypt's population is increasing at a rate of 2.1% per year. Find how long it will take for its 50,500,000-person population to double in size.

97. The greater Mexico City area had a population of 16.9 million in 1985. How long will it take the city to triple in population if its growth rate is 3.4% annually?

Use the compound interest equation $A = P\left(1 + \frac{r}{n}\right)^{nt}$ *to solve the following. (See directions for Exercises 24 and 25 for an explanation of this formula.)*

98. Find how long it will take a $5000 investment to grow to $10,000 if it is invested at 8% interest compounded quarterly.

99. An investment of $6000 has grown to $10,000 while the money was invested at 6% interest compounded monthly. Find how long it was invested.

CHAPTER 11 TEST

On the same set of axes, graph the given one-to-one function and its inverse.

1. $7x - 14 = f(x)$

Determine whether the given graph is the graph of a one-to-one function.

2.

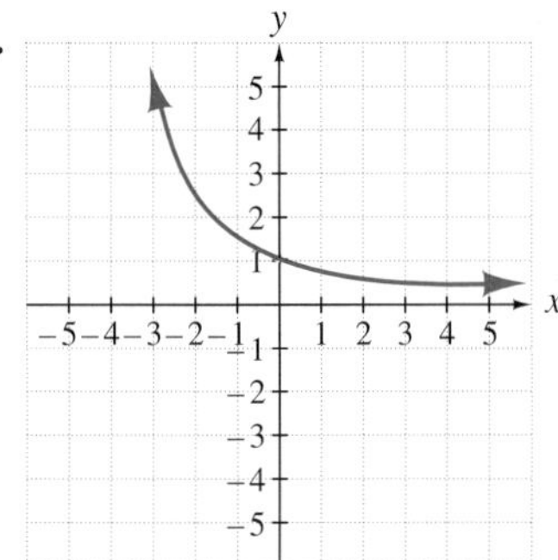

3.

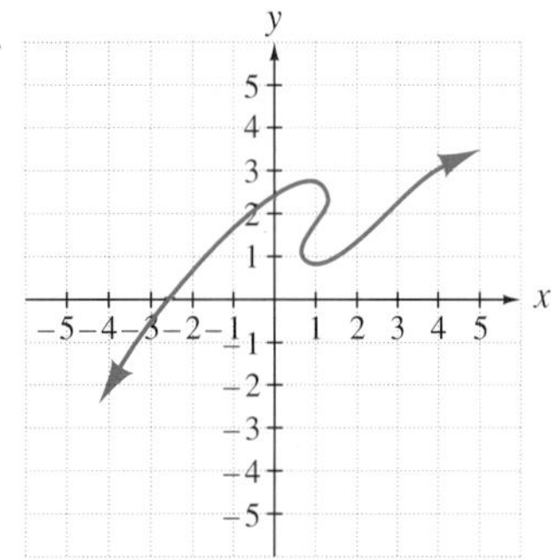

Find a set of ordered pairs or an equation that defines the inverse function of the given function.

4. $y = 6 - 2x$

5. $f = \{(0, 0), (2, 3), (-1, 5)\}$

Use the properties of logarithms to write each expression as a single logarithm.

6. $\log_3 6 + \log_3 4$

7. $\log_5 x + 3 \log_5 x - \log_5 (x + 1)$

Solve.

8. Write the expression $\log_6 \frac{2x}{y^3}$ as the sum or difference of multiples of logarithms.

9. If $\log_b 3 = 0.79$ and $\log_b 5 = 1.16$, find the value of $\log_b \left(\frac{3}{25}\right)$.

10. Approximate $\log_7 8$ to four decimal places.

11. Solve $8^{x-1} = \frac{1}{64}$ for x. Give an exact solution.

12. Solve $3^{2x+5} = 4$ for x. Give an exact solution, and also approximate the solution to four decimal places.

Solve each logarithmic equation for x. Give an exact solution.

13. $\log_3 x = -2$

14. $\ln \sqrt{e} = x$

15. $\log_8 (3x - 2) = 2$

16. $\log_5 x + \log_5 3 = 2$

17. $\log_4 (x + 1) - \log_4 (x - 2) = 3$

18. Solve $\ln (3x + 7) = 1.31$ accurate to four decimal places.

19. Graph $y = \left(\frac{1}{2}\right)^x + 1$.

20. Graph the functions $y = 3^x$ and $y = \log_3 x$ on the same coordinate system.

Use the formula $A = P\left(1 + \frac{r}{n}\right)^{nt}$ *to solve Exercises 21 and 22.*

21. Find the amount in the account if \$4000 is invested for 3 years at 9% interest compounded monthly.

22. Find how long it will take \$2000 to grow to \$3000 if the money is invested at 7% interest compounded semiannually.

Use the population growth formula $y = y_0e^{kt}$ *to solve Exercises 23 and 24.*

23. The prairie dog population of the Grand Rapids area now stands at 57,000 animals. If the population is growing at a rate of 2.6% annually, find how many prairie dogs there will be 5 years from now.

24. In an attempt to save an endangered species of wood duck, naturalists would like to increase their population from 400 to 1000 ducks. If the annual population growth rate is 6.2%, find how long it will take the naturalists to reach their goal.

25. The formula $\log(1 + k) = \frac{0.3}{D}$ relates the doubling time D, in days, and growth rate k for a population of mice. Find the rate at which the population is increasing if the doubling time is 56 days.

CHAPTER 11 CUMULATIVE REVIEW

1. Solve $|x| \leq 3$.

2. Graph $f(x) = 3x + 6$.

3. Find the slope of the line $x = 5$.

4. Simplify each expression. Write answers with positive exponents.

a. $\left(\frac{2}{3}\right)^{-4}$ b. $2^{-1} + 4^{-1}$ c. $(-2)^{-4}$

5. Use synthetic division to divide $2x^3 - x^2 - 13x + 1$ by $x - 3$.

6. Factor $x^2 - 8x + 15$.

7. Write each rational expression in lowest terms.

a. $\frac{24x^6y^5}{8x^7y}$ b. $\frac{2x^2}{10x^3 - 2x^2}$

8. Solve $\frac{8x}{5} + \frac{3}{2} = \frac{3x}{5}$.

9. Rationalize the denominator of each expression.

a. $\frac{\sqrt{27}}{\sqrt{5}}$ b. $\frac{2\sqrt{16}}{\sqrt{9x}}$ c. $\sqrt[3]{\frac{1}{2}}$

10. Solve $\sqrt{4 - x} = x - 2$ for x.

11. Solve $m^2 - 7m - 1 = 0$ for m by completing the square.

12. Solve $x^3 = 8$ for x.

13. Sketch the graph of $f(x) = -2x^2$.

14. Sketch the graph of $\frac{x^2}{16} - \frac{y^2}{25} = 1$.

15. Solve the system:

$$\begin{cases} 3x - y + z = -15 \\ x + 2y - z = 1 \\ 2x + 3y - 2z = 0 \end{cases}$$

16. Find the value of each determinant.

a. $\begin{vmatrix} -1 & 2 \\ 3 & -4 \end{vmatrix}$ b. $\begin{vmatrix} 2 & 0 \\ 7 & -5 \end{vmatrix}$

17. Solve the system:

$$\begin{cases} y = \sqrt{x} \\ x^2 + y^2 = 6 \end{cases}$$

18. Find the equation of the inverse of $f(x) = 3x - 5$. Graph f and f^{-1} on the same set of axes.

19. Find the amount owed at the end of 5 years if \$1600 is loaned at a rate of 9% compounded monthly.

20. Solve $\log_4(x - 2) = 2$ for x.

CHAPTER 12

Sequences, Series, and the Binomial Theorem

Drone bees originate from single-parent families, in the literal sense of the word, creating family trees quite distinct from peoples'. Both kinds of family trees, though, are described by sequences.

INTRODUCTION

Having explored in some depth the concept of function, we turn now in this final chapter to sequences. In one sense, a sequence is simply an ordered list of numbers. In another sense, a sequence is itself a function. Phenomena modeled by sequences are everywhere around us in the mathematical world. The starting place for all mathematics is, after all, the sequence of natural numbers: 1, 2, 3, 4, and so on.

Sequences lead us to **series,** which are a sum of ordered numbers. Through series we gain new insight, for example, about the expansion of a binomial $(a + b)^n$, the concluding topic for this book.

12.1
Sequences

OBJECTIVES

Tape IA 33

1. Write the terms of a sequence given its general term.
2. Find the general term of a sequence.
3. Solve applications involving sequences.

A town has a present population of 100,000 and is growing by 5% each year. After the first year, the town's population is

$$100{,}000 + 0.05(100{,}000) = 105{,}000$$

After the second year, the town's population is

$$105{,}000 + 0.05(105{,}000) = 110{,}250$$

After the third year, the town's population is

$$110{,}250 + 0.05(110{,}250) = 115{,}762.5$$

The town's population can be written as the **sequence** of numbers

$$105{,}000,\ 110{,}250,\ 115{,}762.5, \ldots$$

Another sequence of numbers is 2, 4, 8, 16, ...

Infinite Sequence

An infinite sequence is a function whose domain is the set of natural numbers 1, 2, 3, 4, and so on.

1 The **general term** of the sequence 2, 4, 8, 16, . . . is 2^n, where n is a natural number. Since 2^n is a function, we can write it as $f(n) = 2^n$, where $n = 1, 2, 3, \ldots$. Instead, we use the notation

$$a_n = 2^n$$

The **domain** of this function or sequence is the set of natural numbers. The **range** of this sequence is the set of function values $a_1, a_2, a_3, \ldots$. These values are called the **terms** of the sequence.

EXAMPLE 1 Write the first five terms of the sequence whose general term is

$$a_n = n^2 - 1$$

Solution: Evaluate a_n, when n is 1, 2, 3, 4, and 5.

$$a_n = n^2 - 1$$

$a_1 = 1^2 - 1 = 0$ Replace n with 1.

$a_2 = 2^2 - 1 = 3$ Replace n with 2.

$a_3 = 3^2 - 1 = 8$ Replace n with 3.

$a_4 = 4^2 - 1 = 15$ Replace n with 4.

$a_5 = 5^2 - 1 = 24$ Replace n with 5.

Thus the first five terms of the sequence $a_n = n^2 - 1$ are 0, 3, 8, 15, and 24. ■

EXAMPLE 2 If the general term of a sequence is $a_n = \frac{(-1)^n}{3n}$, find:

a. the first term of the sequence **b.** a_8
c. the one-hundredth term of the sequence **d.** a_{15}

Solution: **a.** $a_1 = \frac{(-1)^1}{3(1)} = -\frac{1}{3}$ Replace n with 1.

b. $a_8 = \frac{(-1)^8}{3(8)} = \frac{1}{24}$ Replace n with 8.

c. $a_{100} = \frac{(-1)^{100}}{3(100)} = \frac{1}{300}$ Replace n with 100.

d. $a_{15} = \frac{(-1)^{15}}{3(15)} = -\frac{1}{45}$ Replace n with 15. ■

2 Suppose we know the first few terms of a sequence and want to find the general term.

EXAMPLE 3 Find the general term a_n for the sequence whose first four terms are $\frac{1}{2}, \frac{1}{4}, \frac{1}{8}, \frac{1}{16}$.

Solution: Notice that the denominators double each time.

$$\frac{1}{2}, \frac{1}{2 \cdot 2}, \frac{1}{2(2 \cdot 2)}, \frac{1}{2(2 \cdot 2 \cdot 2)}$$

$$= \frac{1}{2}, \frac{1}{2^2}, \frac{1}{2^3}, \frac{1}{2^4}$$

We might suppose then that the general term is $a_n = \frac{1}{2^n}$. ■

3 Sequences model many phenomena of the physical world, as illustrated by the following example.

EXAMPLE 4 The amount of weight, in pounds, a puppy gains in each month of its first year is modeled by a sequence whose general term is $a_n = n + 4$, where n is the number of the month. Write the first five terms of the sequence, and find how much weight the puppy should gain in its fifth month.

Solution: Evaluate $a_n = n + 4$ when n is 1, 2, 3, 4, and 5.

$$a_1 = 1 + 4 = 5$$
$$a_2 = 2 + 4 = 6$$
$$a_3 = 3 + 4 = 7$$
$$a_4 = 4 + 4 = 8$$
$$a_5 = 5 + 4 = 9$$

The puppy should gain 9 pounds in its fifth month. ■

EXERCISE SET 12.1

Write the first five terms of each sequence whose general term is given. See Example 1.

1. $a_n = n + 4$

2. $a_n = 5 - n$

3. $a_n = (-1)^n$

4. $a_n = (-2)^n$

5. $a_n = \dfrac{1}{n + 3}$

6. $a_n = \dfrac{1}{7 - n}$

Find the indicated term for each sequence whose general term is given. See Example 2.

7. $a_n = 3n^2$; a_5

8. $a_n = -n^2$; a_{15}

9. $a_n = 6n - 2$; a_{20}

10. $a_n = 100 - 7n$; a_{50}

11. $a_n = \dfrac{n + 3}{n}$; a_{15}

12. $a_n = \dfrac{n}{n + 4}$; a_{24}

Find a general term a_n for each sequence whose first four terms are given. See Example 3.

13. 3, 7, 11, 15

14. 2, 7, 12, 17

15. $-2, -4, -8, -16$

16. $-4, 16, -64, 256$

17. $\dfrac{1}{3}, \dfrac{1}{9}, \dfrac{1}{27}, \dfrac{1}{81}$

18. $\dfrac{2}{5}, \dfrac{2}{25}, \dfrac{2}{125}, \dfrac{2}{625}$

Solve. See Example 4.

19. The distance, in feet, that a thermos dropped from a cliff falls in each consecutive second is modeled by a sequence whose general term is $a_n = 32n - 16$, when n is the number of seconds. Find the distance the thermos falls in the second, third, and fourth seconds.

20. A culture of bacteria triples every hour so that its size is modeled by the sequence $a_n = 50(3)^{n-1}$, where n is the number of the hour just beginning. Find the size of the culture at the beginning of the fourth hour. Find the size of the culture originally.

21. Mrs. Laser agrees to give her son Mark an allowance of \$0.10 on the first day of his 14-day vacation, \$0.20 on the second day, \$0.40 on the third day, and so on. Write an equation of a sequence whose terms correspond to Mark's allowance. Find the allowance Mark receives on the last day of his vacation.

22. A small theater has 10 rows with 12 seats in the first row, 15 seats in the second row, 18 seats in the third row, and so on. Write an equation of a sequence whose terms correspond to the seats in each row. Find the number of seats in the eighth row.

Write the first five terms of each sequence whose general term is given.

23. $a_n = 2n$

24. $a_n = -6n$

25. $a_n = -n^2$

26. $a_n = n^2 + 2$

27. $a_n = 2^n$

28. $a_n = 3^{n-2}$

29. $a_n = 2n + 5$

30. $a_n = 1 - 3n$

31. $a^n = (-1)^n n^2$

32. $a_n = (-1)^{n+1}(n - 1)$

Find the indicated term for each sequence whose general term is given.

33. $a_n = (-3)^n$; a_6

34. $a_n = 5^{n+1}$; a_3

35. $a_n = \dfrac{n-2}{n+1}$; a_6

36. $a_n = \dfrac{n+3}{n+4}$; a_8

37. $a_n = \dfrac{(-1)^n}{n}$; a_8

38. $a_n = \dfrac{(-1)^n}{2n}$; a_{100}

39. $a_n = -n^2 + 5$; a_{10}

40. $a_n = 8 - n^2$; a_{20}

41. $a_n = \dfrac{(-1)^n}{n+6}$; a_{19}

42. $a_n = \dfrac{n-4}{(-2)^n}$; a_6

43. The number of cases of a new infectious disease is doubling every year so that the number of cases is modeled by a sequence whose general term is $a_n = 75(2)^{n-1}$, where n is the number of the year just beginning. Find how many cases there are at the beginning of the sixth year. Find how many cases there were when the disease was first discovered.

44. A new college had an initial enrollment of 2700 students in 1995, and each year the enrollment increases by 150 students. Find the enrollment for each of 5 years, beginning with 1995.

45. An endangered species of sparrow had an estimated population numbering 800 in 1994, and scientists predict that its population will decrease by half each year. Estimate the population in 1998. Estimate the year the sparrow will be extinct.

46. A **Fibonacci sequence** is a special type of sequence in which the first two terms are 1 and each term thereafter is the sum of the two previous terms: 1, 1, 2, 3, 5, 8. Many plants and animals seem to grow according to a Fibonacci sequence, including pine cones, pineapple scales, nautilus shells, and certain flowers. Write the first 15 terms of the Fibonacci sequence.

Skill Review

Sketch the graph of each quadratic function. See Section 9.5.

47. $f(x) = (x-1)^2 + 3$

48. $f(x) = (x-2)^2 + 1$

49. $f(x) = 2(x+4)^2 + 2$

50. $f(x) = 3(x-3)^2 + 4$

Find the distance between each pair of points. See Section 9.7.

51. $(-4, -1)$ and $(-7, -3)$

52. $(-2, -1)$ and $(-1, 5)$

53. $(2, -7)$ and $(-3, -3)$

54. $(10, -14)$ and $(5, -11)$

12.2 Arithmetic and Geometric Sequences

OBJECTIVES

1. Identify arithmetic sequences and their common differences.
2. Identify geometric sequences and their common ratios.

Tape IA 33

1 Find the first four terms of the sequence whose general term is

$$a_n = 5 + (n-1)3.$$

$a_1 = 5 + (1-1)3 = 5$ Replace n with 1.

$a_2 = 5 + (2-1)3 = 8$ Replace n with 2.

$a_3 = 5 + (3-1)3 = 11$ Replace n with 3.

$a_4 = 5 + (4-1)3 = 14$ Replace n with 4.

The first four terms are 5, 8, 11, and 14. Notice that each term after the first is the sum of 3 and the previous term; that is, $a_n = 3 + \underbrace{a_{n-1}}_{\text{previous term}}$. When this happens, we call the sequence an **arithmetic sequence,** or an **arithmetic progression.** The constant difference d in successive terms is called the **common difference.** In this example, d is 3.

Arithmetic Sequence and Common Difference

An **arithmetic sequence** is a sequence in which each term after the first differs from the preceding term by a constant amount d. The constant d is called the **common difference.**

The sequence 2, 6, 10, 14, 18, . . . is an arithmetic sequence. Its common difference is 4. Given the first term, a_1, and the common difference d of an arithmetic sequence, we can find any term of the sequence.

EXAMPLE 1 Write the first five terms of the arithmetic sequence whose first term is 7 and whose common difference is 2.

Solution:

$$a_1 = 7$$

$$a_2 = 7 + 2 = 9 \qquad a_2 = a_1 + d$$

$$a_3 = 9 + 2 = 11 \qquad a_3 = a_1 + 2d$$

$$a_4 = 11 + 2 = 13 \qquad a_4 = a_1 + 3d$$

$$a_5 = 13 + 2 = 15 \qquad a_5 = a_1 + 4d$$

The first five terms are 7, 9, 11, 13, 15. ■

The pattern on the right suggests that the general term a_n of an arithmetic sequence is given by

$$a_n = a_1 + (n - 1)d$$

General Term of an Arithmetic Sequence

The general term a_n of an arithmetic sequence is given by

$$a_n = a_1 + (n - 1)d$$

where a_1 is the first term and d is the common difference.

EXAMPLE 2 Consider the arithmetic sequence whose first term is 3 and common difference is -5.

a. Write an expression for the general term a_n.

b. Find the twentieth term of this sequence.

Solution: **a.** Since this is an arithmetic sequence, the general term a_n is given by $a_n = a_1 + (n - 1)d$. Here, $a_1 = 3$ and $d = -5$ so that

$$a_n = 3 + (n - 1)(-5)$$

b. $a_{20} = 3 + (20 - 1)(-5) = 3 + 19(-5) = -92$
The twentieth term is -92. ■

EXAMPLE 3 Find the eleventh term of the arithmetic sequence whose first three terms are 2, 9, 16.

Solution: Since the sequence is arithmetic, the eleventh term is

$$a_{11} = a_1 + (11 - 1)d = a_1 + 10d$$

We know a_1 is the first term of the sequence, so $a_1 = 2$. Also, d is the constant difference of terms, so $d = a_2 - a_1 = 9 - 2 = 7$. Thus

$$\begin{aligned} a_{11} &= a_1 + 10d \\ &= 2 + 10 \cdot 7 \\ &= 72 \end{aligned}$$

■

EXAMPLE 4 If the third term of an arithmetic progression is 12 and the eighth term is 27, find the fifth term.

Solution: We need to find a_1 and d to write the general term, which then enables us to find a_5, the fifth term. The two given terms a_3 and a_8 lead to a system of linear equations.

$$\begin{cases} a_3 = a_1 + (3 - 1)d \\ a_8 = a_1 + (8 - 1)d \end{cases} \quad \text{or} \quad \begin{cases} 12 = a_1 + 2d \\ 27 = a_1 + 7d \end{cases}$$

Next, we solve the system $\begin{cases} 12 = a_1 + 2d \\ 27 = a_1 + 7d \end{cases}$ by addition. Multiply both sides of the second equation by -1 so that

$$\begin{cases} 12 = a_1 + 2d \\ -1\,(27) = -1\,(a_1 + 7d) \end{cases} \quad \text{simplifies to} \quad \begin{cases} 12 = a_1 + 2d \\ -27 = -a_1 - 7d \end{cases}$$

$$-15 = -5d$$

$$3 = d \qquad \text{Divide both sides by } -5.$$

To find a_1, let $d = 3$ in $12 = a_1 + 2d$. Then

$$\begin{aligned} 12 &= a_1 + 2(3) \\ 12 &= a_1 + 6 \\ 6 &= a_1 \end{aligned}$$

Thus $a_1 = 6$ and $d = 3$, so

$$\begin{aligned} a_n &= 6 + (n - 1)(3) \\ a_5 &= 6 + (5 - 1)(3) = 18 \end{aligned}$$

■

EXAMPLE 5 Donna has an offer for a job starting at \$20,000 per year and guaranteeing her a raise of \$800 per year for the next 5 years. Write the general term for the arithmetic sequence modeling Donna's potential annual salaries and find her salary for the fourth year.

Solution: The first term $a_1 = 20{,}000$ and $d = 800$. So

$$a_n = 20{,}000 + (n - 1)(800) \quad \text{and} \quad a_4 = 20{,}000 + (3)(800) = 22{,}400$$

Her salary for the fourth year will be \$22,400. ■

2 We now investigate a **geometric sequence,** also called a **geometric progression.** In the sequence 5, 15, 45, 135, . . . , each term after the first is the **product** of 3 and the preceding term. This pattern of multiplying by a constant to get the next term defines a geometric sequence.

Geometric Sequence and Common Ratio

A **geometric sequence** is a sequence in which each term after the first is obtained by multiplying the preceding term by a constant r. The constant r is called the **common ratio.**

The sequence $12, 6, 3, \frac{3}{2}, \ldots$ is geometric since each term after the first is the product of the previous term and $\frac{1}{2}$.

EXAMPLE 6 Write the first five terms of a geometric sequence whose first term is 7 and whose common ratio is 2.

Solution:

$$\begin{aligned} a_1 &= 7 & & \\ a_2 &= 7(2) = 14 & a_2 &= a_1(r) \\ a_3 &= 14(2) = 28 & a_3 &= a_1(r^2) \\ a_4 &= 28(2) = 56 & a_4 &= a_1(r^3) \\ a_5 &= 56(2) = 112 & a_5 &= a_1(r^4) \end{aligned}$$

The first five terms are 7, 14, 28, 56, and 112. ■

Notice that the pattern on the right suggests that the general term of a geometric sequence is given by $a_n = a_1 r^{n-1}$.

General Term of a Geometric Sequence

The general term a_n of a geometric sequence is given by

$$a_n = a_1 r^{n-1}$$

where a_1 is the first term and r is the common ratio.

EXAMPLE 7 Find the eighth term of the geometric sequence whose first term is 12 and whose common ratio is $\frac{1}{2}$.

Solution: Since this is a geometric sequence, the general term a_n is given by

$$a_n = a_1 r^{n-1}$$

Here $a_1 = 12$ and $r = \frac{1}{2}$, so $a_n = 12\left(\frac{1}{2}\right)^{n-1}$. Evaluate a_n when $n = 8$.

$$a_8 = 12\left(\frac{1}{2}\right)^{8-1} = 12\left(\frac{1}{2}\right)^7 = 12\left(\frac{1}{128}\right) = \frac{3}{32} \quad ■$$

EXAMPLE 8 Find the fifth term of the geometric sequence whose first three terms are 2, -6, and 18.

Solution: Since the sequence is geometric, the fifth term must be a_1r^{5-1}, or $2r^4$. We know that r is the common ratio of terms, so r must be $\frac{-6}{2}$, or -3. Thus

$$a_5 = 2r^4$$
$$a_5 = 2(-3)^4 = 162 \quad \blacksquare$$

EXAMPLE 9 If the second term of a geometric sequence is $\frac{5}{4}$ and the third term is $\frac{5}{16}$, find the first term and the common ratio.

Solution: Notice that $\frac{5}{16} \div \frac{5}{4} = \frac{1}{4}$, so $r = \frac{1}{4}$. Then

$$a_2 = a_1\left(\frac{1}{4}\right)^1$$
$$\frac{5}{4} = a_1\left(\frac{1}{4}\right) \quad \text{or} \quad a_1 = 5 \qquad \text{Replace } a_2 \text{ with } \frac{5}{4}.$$

The first term is 5. ■

EXAMPLE 10 A bacteria culture growing under controlled conditions doubles each day. Find how large the culture is at the beginning of day 7 if it measured 10 units at the beginning of day 1.

Solution: Since the culture doubles its size each day, the sizes are modeled by a geometric sequence. Here $a_1 = 10$ and $r = 2$. Thus

$$a_n = a_1r^{n-1} = 10(2)^{n-1} \quad \text{and} \quad a_7 = 10(2)^{7-1} = 640$$

The bacteria measures 640 units at the beginning of day 7. ■

EXERCISE SET 12.2

Write the first five terms of the arithmetic or geometric sequence whose first term a_1 and common difference d or common ratio r are given. See Examples 1 and 6.

1. $a_1 = 4; d = 2$

2. $a_1 = 3; d = 10$

3. $a_1 = 6; d = -2$

4. $a_1 = -20; d = 3$

5. $a_1 = 1; r = 3$

6. $a_1 = -2; r = 2$

7. $a_1 = 48; r = \frac{1}{2}$

8. $a_1 = 1; r = \frac{1}{3}$

Find the indicated term of each sequence. See Examples 2 and 7.

9. The eighth term of the arithmetic sequence whose first term is 12 and whose common difference is 3.

10. The twelfth term of the arithmetic sequence whose first term is 32 and whose common difference is -4.

11. The fourth term of the geometric sequence whose first term is 7 and whose common ratio is -5.

12. The fifth term of the geometric sequence whose first term is 3 and whose common ratio is 3.

13. The fifteenth term of the arithmetic sequence whose first term is -4 and whose common difference is -4.

14. The sixth term of the geometric sequence whose first term is 5 and whose common ratio is -4.

Find the indicated term of each sequence. See Examples 3 and 8.

15. The ninth term of the arithmetic sequence 0, 12, 24, ...

16. The thirteenth term of the arithmetic sequence -3, 0, 3, ...

17. The twenty-fifth term of the arithmetic sequence 20, 18, 16, ...

18. The ninth term of the geometric sequence 5, 10, 20, ...

19. The fifth term of the geometric sequence 2, -10, 50, ...

20. The sixth term of the geometric sequence $\frac{1}{2}, \frac{3}{2}, \frac{9}{2}, \ldots$.

Find the indicated term of each sequence. See Examples 4 and 9.

21. The eighth term of the arithmetic sequence whose fourth term is 19 and whose fifteenth term is 52.

22. If the second term of an arithmetic sequence is 6 and the tenth term is 30, find the twenty-fifth term.

23. If the second term of an arithmetic progression is -1 and the fourth term is 5, find the ninth term.

24. If the second term of a geometric progression is 15 and the third term is 3, find a_1 and r.

25. If the second term of a geometric progression is $-\frac{4}{3}$ and the third term is $\frac{8}{3}$, find a_1 and r.

26. If the third term of a geometric sequence is 4 and the fourth term is -12, find a_1 and r.

Solve. See Examples 5 and 10.

27. An auditorium has 54 seats in the first row, 58 seats in the second row, 62 seats in the third row, and so on. Find the general term of this arithmetic sequence and the number of seats in the twentieth row.

28. A triangular display of cans in a grocery store has 20 cans in the first row, 17 cans in the next row, and so on, in an arithmetic sequence. Find the general term and the number of cans in the fifth row. Find how many rows there are in the display and how many cans are in the top row.

29. The initial size of a virus culture is 6 units and it triples its size every day. Find the general term of the geometric sequence modeling the culture's size.

30. A real estate investment broker predicts that a certain property will increase in value 15% each year. Thus the yearly property values can be modeled by a geometric sequence whose common ratio r is 1.15. If the initial property value was \$500,000, write the first four terms of the sequence and predict the value at the end of the third year.

Given are the first three terms of a sequence that is either arithmetic or geometric. Based on these terms, if a sequence is arithmetic, find a_1 and d. If a sequence is geometric, find a_1 and r.

31. 2, 4, 6

32. 8, 16, 24

33. 5, 10, 20

34. 2, 6, 18

35. $\frac{1}{2}, \frac{1}{10}, \frac{1}{50}$

36. $\frac{2}{3}, \frac{4}{3}, 2$

37. $x, 5x, 25x$

38. $y, -3y, 9y$

39. $p, p + 4, p + 8$

40. $t, t - 1, t - 2$

Find the indicated term of each sequence.

41. The twenty-first term of the arithmetic sequence whose first term is 14 and whose common difference is $\frac{1}{4}$.

42. The fifth term of the geometric sequence whose first term is 8 and whose common ratio is -3.

43. The fourth term of the geometric sequence whose first term is 3 and whose common ratio is $-\frac{2}{3}$.

44. The fourth term of the arithmetic sequence whose first term is 9 and whose common difference is 5.

45. The fifteenth term of the arithmetic sequence $\frac{3}{2}, 2, \frac{5}{2}, \ldots$.

46. The eleventh term of the arithmetic sequence $2, \frac{5}{3}, \frac{4}{3}, \ldots$.

47. The sixth term of the geometric sequence $24, 8, \frac{8}{3}, \ldots$.

48. The eighteenth term of the arithmetic sequence 5, 2, -1, ...

49. If the third term of an arithmetic progression is 2 and the seventeenth term is -40, find the tenth term.

50. If the third term of a geometric sequence is -28 and the fourth term is -56, find a_1 and r.

51. A rubber ball is dropped from a height of 486 feet, and each time it bounces back one-third the height from which it last fell. Write out the first five terms of this geometric sequence and find the general term. Find how many bounces it takes for the ball to rebound less than 1 foot.

52. On the first swing, the length of the arc through which a pendulum swings is 50 inches. The length of each successive swing is 80% of the preceding swing. Determine whether this sequence is arithmetic or geometric. Find the length of the fourth swing.

53. Jose takes a job with a monthly starting salary of \$1000 and guaranteeing him a monthly raise of \$125 during his first year of training. Find the general term of this arithmetic sequence and his salary at the end of his training.

54. At the beginning of Claudia's exercise program, she rides 15 minutes on the Lifecycle. Each week she increases her riding time by 5 minutes. Write the general term of this arithmetic sequence, and find her riding time after 7 weeks. Find how many weeks it takes her to reach a riding time of 1 hour.

55. If an element has a half-life of 3 hours, then x grams of the element dwindles to $\frac{x}{2}$ grams after 3 hours. If a nuclear reactor has 400 grams of a radioactive material with a half-life of 3 hours, find the amount of radioactive material after 12 hours.

Skill Review

Evaluate. See Section 1.7.

56. $5(1) + 5(2) + 5(3) + 5(4)$

57. $\frac{1}{3(1)} + \frac{1}{3(2)} + \frac{1}{3(3)}$

58. $2(2 - 4) + 3(3 - 4) + 4(4 - 4)$

59. $3^0 + 3^1 + 3^2 + 3^3$

60. $\frac{1}{4(1)} + \frac{1}{4(2)} + \frac{1}{4(3)}$

61. $\frac{8-1}{8+1} + \frac{8-2}{8+2} + \frac{8-3}{8+3}$

Writing in Mathematics

62. Explain why 14, 10, 6 may be the first three terms of an arithmetic sequence when it appears we are subtracting instead of adding to get the next term.

63. Explain why 80, 20, 5 may be the first three terms of a geometric sequence when it appears we are dividing instead of multiplying to get the next term.

64. Describe a situation in your life that can be modeled by a geometric sequence. Write an equation for the sequence.

12.3 Series

OBJECTIVES

1 Identify series.

2 Use summation notation.

3 Find partial sums.

Tape IA 34

1 A person who conscientiously saves money by saving first \$100, and then saving \$10 more each month than he saved the preceding month, is saving money according to the arithmetic sequence

$$a_n = 100 + 10(n - 1)$$

Following this sequence, he can predict how much money he should save for any particular month. But if he also wants to know how much money **in total** he has saved, say on the fifth month, he must find the **sum** of the first five terms of the sequence

$$\underbrace{100}_{a_1} + \underbrace{100 + 10}_{a_2} + \underbrace{100 + 20}_{a_3} + \underbrace{100 + 30}_{a_4} + \underbrace{100 + 40}_{a_5}$$

A sum of the terms of a sequence is called a **series** (plural is also series). As our example here suggests, series are frequently used to model financial and natural phenomena.

A series is a **finite series** if it is the sum of only the first k terms, for some natural number k. A series is an **infinite series** if it is the sum of all the terms. For example,

Sequence	*Series*	
5, 9, 13	$5 + 9 + 13$	Finite, k is 3
5, 9, 13, . . .	$5 + 9 + 13 + \cdots$	Infinite
$4, -2, 1, -\frac{1}{2}, \frac{1}{4}$	$4 + (-2) + 1 + \left(-\frac{1}{2}\right) + \left(\frac{1}{4}\right)$	Finite, k is 5
4, −2, 1, . . .	$4 + (-2) + 1 + \cdots$	Infinite
3, 6, . . . , 99	$3 + 6 + \cdots + 99$	Finite, k is 33

2 A shorthand notation for denoting a series when the general term of the sequence is known is called **summation notation.** The Greek uppercase letter **sigma** Σ is used to mean "sum." The expression $\sum_{n=1}^{5} (3n + 1)$ is read "the sum of $3n + 1$ as n goes from 1 to 5" and tells us to find the sum of the first five terms of the sequence whose general term is $a_n = 3n + 1$. Often, the variable i is used instead of n when we use summation notation: $\sum_{i=1}^{5} (3i + 1)$. Whether we use n, i, k, or some other variable, the variable is called the **index of summation.** The equation $i = 1$ below Σ indicates the beginning value for i, and the number 5 above the Σ indicates the ending value for i. Thus the terms of the sequence are found by successively replacing i with the natural numbers 1, 2, 3, 4, 5. To find the sum, we write out the terms and then add:

$$\begin{aligned}\sum_{i=1}^{5} (3i + 1) &= (3 \cdot 1 + 1) + (3 \cdot 2 + 1) + (3 \cdot 3 + 1) \\ &\quad + (3 \cdot 4 + 1) + (3 \cdot 5 + 1) \\ &= 4 + 7 + 10 + 13 + 16 = 50\end{aligned}$$

EXAMPLE 1 Evaluate.

a. $\sum_{i=0}^{6} \frac{i - 2}{2}$ **b.** $\sum_{i=3}^{5} 2^i$

Solution: **a.**
$$\begin{aligned}\sum_{i=0}^{6} \frac{i - 2}{2} &= \frac{0 - 2}{2} + \frac{1 - 2}{2} + \frac{2 - 2}{2} + \frac{3 - 2}{2} + \frac{4 - 2}{2} + \frac{5 - 2}{2} + \frac{6 - 2}{2} \\ &= (-1) + \left(-\frac{1}{2}\right) + 0 + \frac{1}{2} + 1 + \frac{3}{2} + 2 \\ &= \frac{7}{2} \text{ or } 3\frac{1}{2}\end{aligned}$$

b. $\sum_{i=3}^{5} 2^i = 2^3 + 2^4 + 2^5$

$= 8 + 16 + 32$

$= 56$ ■

EXAMPLE 2 Write each series using summation notation.

a. $3 + 6 + 9 + 12 + 15$ **b.** $\frac{1}{2} + \frac{1}{4} + \frac{1}{8} + \frac{1}{16}$

Solution: **a.** Since each term is the **sum** of the preceding term and 3, the terms correspond to the first five terms of an arithmetic sequence with $a_1 = 3$, $d = 3$, and $a_n = 3 + (n - 1)3$. Thus

$$3 + 6 + 9 + 12 + 15 = \sum_{i=1}^{5} 3 + (i - 1)3$$

b. Since each term is the **product** of the preceding term and $\frac{1}{2}$, these terms correspond to the first four terms of a geometric sequence. Here $a_1 = \frac{1}{2}$, $r = \frac{1}{2}$, and $a_n = \left(\frac{1}{2}\right)\left(\frac{1}{2}\right)^{n-1} = \left(\frac{1}{2}\right)^n$. So

$$\frac{1}{2} + \frac{1}{4} + \frac{1}{8} + \frac{1}{16} = \sum_{i=1}^{4} \left(\frac{1}{2}\right)^i$$ ■

3 If we want the sum of the first n terms of a sequence, we find a **partial sum,** S_n. Thus

$$S_1 = a_1$$

$$S_2 = a_1 + a_2$$

$$S_3 = a_1 + a_2 + a_3$$

and so on. In general, S_n is the sum of the first n terms of a sequence.

EXAMPLE 3 Find the sum of the first three terms of the sequence whose general term is $a_n = \frac{n + 3}{2n}$.

Solution: $S_3 = \sum_{i=1}^{3} \frac{i + 3}{2i} = 2 + \frac{5}{4} + 1 = 4\frac{1}{4}$. ■

The next example illustrates how these sums model real-life phenomena.

EXAMPLE 4 The number of babies born per year in the San Diego Zoo's gorilla house is a sequence defined by $a_n = n(n - 1)$, where n is the number of the year. Find the **total** number of baby gorillas born in the **first 4 years.**

Solution: To solve, find the sum,

$$\sum_{i=1}^{4} i(i - 1)$$

$$\sum_{i=1}^{4} i(i - 1) = 0 + 2 + 6 + 12 = 20$$

There are 20 gorillas born in the first 4 years. ■

EXERCISE SET 12.3

Evaluate. See Example 1.

1. $\sum_{i=1}^{4} (i - 3)$
2. $\sum_{i=1}^{5} (i + 6)$
3. $\sum_{i=4}^{7} (2i + 4)$
4. $\sum_{i=2}^{3} (5i - 1)$
5. $\sum_{i=2}^{4} (i^2 - 3)$
6. $\sum_{i=3}^{5} i^3$
7. $\sum_{i=1}^{3} \left(\frac{1}{i + 5}\right)$
8. $\sum_{i=2}^{4} \left(\frac{2}{i + 3}\right)$

Write each series using summation notation. See Example 2.

9. $1 + 3 + 5 + 7 + 9$
10. $4 + 7 + 10 + 13$
11. $4 + 12 + 36 + 108$
12. $5 + 10 + 20 + 40 + 80 + 160$
13. $12 + 9 + 6 + 3 + 0 + (-3)$
14. $5 + 1 + (-3) + (-7)$

Find each partial sum. See Example 3.

15. Find the sum of the first four terms of the sequence whose general term is $a_n = -2n$.
16. Find the sum of the first three terms of the sequence whose general term is $a_n = -\frac{n}{3}$.
17. Find the sum of the first five terms of the sequence whose general term is $a_n = (n - 1)^2$.
18. Find the sum of the first three terms of the sequence whose general term is $a_n = (n + 4)^2$.

Solve. See Example 4.

19. A gardener is making a triangular planting with 1 tree in the first row, 2 trees in the second row, 3 trees in the third row, and so on for 10 rows. Write the sequence describing the trees in each row. Find the total number of trees planted.
20. Some surfers at the beach form a human pyramid with 2 surfers in the top row, 3 surfers in the second row, 4 surfers in third row, and so on. If there are 6 rows in the pyramid, write the sequence describing the number of surfers in each row in the pyramid. Find the total number of surfers.
21. A fungus culture starts with 6 units and doubles every day. Write the sequence describing the growth of this fungus. Find the number of fungus units at the beginning of the fifth day.
22. A bacteria colony begins with 100 bacteria and doubles every 6 hours. Write the sequence describing the growth of the bacteria. Find the number of bacteria after 24 hours.

Evaluate.

23. $\sum_{i=1}^{3} \frac{1}{6i}$
24. $\sum_{i=1}^{3} \frac{1}{3i}$
25. $\sum_{i=2}^{6} 3i$
26. $\sum_{i=3}^{6} -4i$
27. $\sum_{i=3}^{5} i(i + 2)$
28. $\sum_{i=2}^{4} i(i - 3)$
29. $\sum_{i=1}^{5} 2^i$
30. $\sum_{i=1}^{4} 3^{i-1}$
31. $\sum_{i=1}^{4} \frac{4i}{i + 3}$
32. $\sum_{i=2}^{5} \frac{6 - i}{6 + i}$

Write each series using summation notation.

33. $12 + 4 + \frac{4}{3} + \frac{4}{9}$
34. $80 + 20 + 5 + \frac{5}{4} + \frac{5}{16}$
35. $1 + 4 + 9 + 16 + 25 + 36 + 49$
36. $1 + (-4) + 9 + (-16)$

Find each partial sum.

37. Find the sum of the first two terms of the sequence whose general term is $a_n = (n + 2)(n - 5)$.
38. Find the sum of the first two terms of the sequence whose general term is $a_n = n(n - 6)$.
39. Find the sum of the first six terms of the sequence whose general term is $a_n = (-1)^n$.
40. Find the sum of the first seven terms of the sequence whose general term is $a_n = (-1)^{n-1}$.

41. Find the sum of the first four terms of the sequence whose general term is $a_n = (n + 3)(n + 1)$.

42. Find the sum of the first five terms of the sequence whose general term is $a_n = \frac{(-1)^n}{2n}$.

Solve.

43. A bacteria colony begins with 50 bacteria and doubles every 12 hours. Write the sequence describing the growth of the bacteria. Find the number of bacteria after 48 hours.

44. The number of otters born each year in a new aquarium forms a sequence whose general term is $a_n = (n - 1)(n + 3)$. Find the number of otters born in the third year, and find the total number of otters born in the first three years.

45. The number of opossums killed each month on a new highway describes the sequence whose general term is $a_n = (n + 1)(n + 2)$, where n is the number of the months. Find the number of opossums killed in the fourth month, and find the total number killed in the first four months.

46. In 1988 the population of an endangered fish was estimated by environmentalists to be decreasing each year. The size of the population in a given year is $24 - 4n$ thousand fish fewer than the previous year. Find the decrease in population in 1990, if year 1 is 1988. Find how many total fish died from 1988 through 1990.

47. The amount of decay in pounds of a radioactive isotope each year is given by the sequence whose general term is $a_n = 100(0.5)^n$, where n is the number of the year. Find the amount of decay in the fourth year, and find the total amount of decay in the first four years.

48. Susan has a choice between two job offers. Job A has an annual starting salary of \$20,000 with guaranteed annual raises of \$1200 for the next four years, while job B has an annual starting salary of \$18,000 with guaranteed annual raises of \$2500 for the next four years. Compare the fifth partial sums for each sequence to determine which job would pay Susan more money over the next 5 years.

49. A pendulum swings a length of 40 inches on its first swing. Each successive swing is $\frac{4}{5}$ of the preceding swing. Find the length of the fifth swing, and find the total length swung during the first five swings. (Round to the nearest tenth of an inch.)

Skill Review

Evaluate. See Section 1.7.

50. $\dfrac{5}{1 - \frac{1}{2}}$

51. $\dfrac{-3}{1 - \frac{1}{7}}$

52. $\dfrac{\frac{1}{3}}{1 - \frac{1}{10}}$

53. $\dfrac{\frac{6}{11}}{1 - \frac{1}{10}}$

54. $\dfrac{3(1 - 2^4)}{1 - 2}$

55. $\dfrac{2(1 - 5^3)}{1 - 5}$

56. $\dfrac{10}{2}(3 + 15)$

57. $\dfrac{12}{2}(2 + 19)$

Writing in Mathematics

58. Explain the difference between a sequence and a series.

12.4 Partial Sums of Arithmetic and Geometric Sequences

OBJECTIVES

1. Find the partial sum of an arithmetic sequence.
2. Find the partial sum of a geometric sequence.
3. Find the infinite series of a geometric sequence.

Tape IA 34

1 Partial sums S_n are relatively easy to find when n is small, that is, when the number of terms to add is small. But when n is large, finding S_n can be tedious. For large n, S_n is still relatively easy to find if the addends are terms of an arithmetic sequence or a geometric sequence.

For an arithmetic sequence, $a_n = a_1 + (n - 1)d$ for some first term a_1 and some common difference d. So S_n, the sum of the first n terms, is

$$S_n = a_1 + (a_1 + d) + (a_1 + 2d) + \cdots + a_n$$

We might also find S_n by "working backward" from the nth term a_n, finding the preceding term a_{n-1}, by subtracting d each time.

$$S_n = a_n + (a_n - d) + (a_n - 2d) + \cdots + a_1$$

Now add left sides of these two equations, and add right sides:

$$2S_n = (a_1 + a_n) + (a_1 + a_n) + (a_1 + a_n) + \cdots + (a_1 + a_n)$$

The d terms subtract out, leaving n sums of the first term a_1 and last term a_n. Thus we write

$$2S_n = n(a_1 + a_n)$$

or

$$S_n = \frac{n}{2}(a_1 + a_n)$$

Partial Sum S_n of an Arithmetic Sequence

The partial sum S_n of the first n terms of an arithmetic sequence is given by

$$S_n = \frac{n}{2}(a_1 + a_n)$$

where a_1 is the first term of the sequence and a_n is the nth term.

EXAMPLE 1 Use the partial sum formula to find the sum of the first six terms of the arithmetic sequence 2, 5, 8, 11, 14, 17, ...

Solution: Use the formula for S_n of an arithmetic sequence, replacing n with 6, a_1 with 2, and a_n with 17.

$$S_n = \frac{n}{2}(a_1 + a_n) = \frac{6}{2}(2 + 17) = 3(19) = 57 \quad ■$$

EXAMPLE 2 Find the sum of the first 30 positive integers.

Solution: Because 1, 2, 3, . . . , 30 is an arithmetic sequence, use the formula for S_n with $n = 30$, $a_1 = 1$, and $a_n = 30$. Thus

$$S_n = \frac{n}{2}(a_1 + a_n) = \frac{30}{2}(1 + 30) = 15(31) = 465 \quad ■$$

EXAMPLE 3 Rolls of carpet are stacked in 20 rows with 3 rolls in the top row, 4 rolls in the next row, and so on, forming an arithmetic sequence. Find the total number of carpet rolls if there are 22 rolls in the last row.

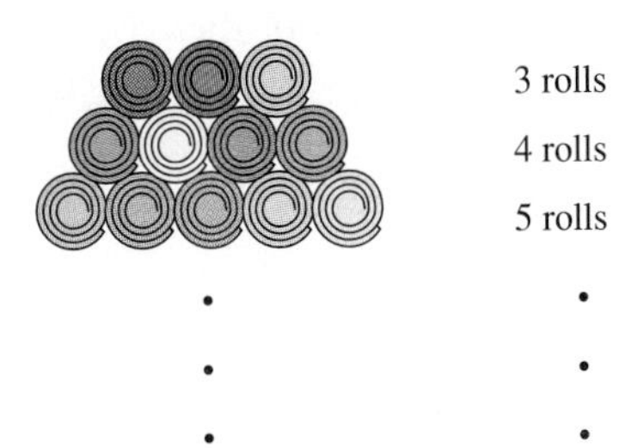

Solution: The list 3, 4, 5, . . . , 22 is the first 20 terms of an arithmetic sequence. Use the formula for S_n with $a_1 = 3$, $a_n = 22$, and $n = 20$ terms. Thus,

$$S_{20} = \frac{20}{2}(3 + 22) = 10(25) = 250$$

There are a total of 250 rolls of carpet in the display. ■

2 It is also useful to have a formula for the partial sum of the first n terms of a geometric series. To derive the formula, we write

$$S_n = \underset{\substack{\uparrow \\ \text{1st} \\ \text{term}}}{a_1} + \underset{\substack{\uparrow \\ \text{2nd} \\ \text{term}}}{a_1 r} + \underset{\substack{\uparrow \\ \text{3rd} \\ \text{term}}}{a_1 r^2} + \cdots + \underset{\substack{\uparrow \\ n\text{th} \\ \text{term}}}{a_1 r^{n-1}}$$

Multiply each side of the equation by $-r$:

$$-rS_n = -a_1 r - a_1 r^2 - a_1 r^3 - \cdots - a_1 r^n$$

Add the two equations.

$$S_n - rS_n = a_1 + (a_1 r - a_1 r) + (a_1 r^2 - a_1 r^2) + (a_1 r^3 - a_1 r^3) + \cdots - a_1 r^n$$

$$S_n - rS_n = a_1 - a_1 r^n$$

Now factor each side.

$$(1 - r)S_n = a_1 (1 - r^n)$$

Solve for S_n by dividing both sides by $1 - r$.

$$S_n = \frac{a_1(1 - r^n)}{1 - r}$$

as long as r is not 1.

Partial Sum S_n of a Geometric Sequence

The partial sum S_n of the first n terms of a geometric sequence, whose first term is a_1, is given by

$$S_n = \frac{a_1(1 - r^n)}{1 - r}$$

as long as the common ratio of the sequence r is not 1.

EXAMPLE 4 Find the sum of the first six terms of the geometric sequence 5, 10, 20, 40, 80, 160.

Solution: Use the S_n formula for the sum of the terms of a geometric sequence. Here, $n = 6$, the first term $a_1 = 5$, and the common ratio $r = 2$.

$$S_n = \frac{a_1(1 - r^n)}{1 - r}$$

$$S_6 = \frac{5(1 - 2^6)}{1 - 2} = \frac{5(-63)}{-1} = 315 \quad ■$$

EXAMPLE 5 A grant from an alumnus to a university specified that the university will receive \$800,000 during the first year and 75% of the preceding year's donation during each of the next five years. Find the total amount donated during the 6 years.

Solution: The donations model the first six terms of a geometric sequence. Evaluate S_n when $n = 6$, $a_1 = 800{,}000$, and $r = 0.75$.

$$S_6 = \frac{800{,}000[1 - (0.75)^6]}{1 - 0.75}$$

$$= \$2{,}630{,}468.75$$

The total amount donated during the 6 years is \$2,630,468.75. ■

3 Is it possible to find the sum of all the terms of an infinite sequence? Examine the partial sums of the geometric sequence $\frac{1}{2}, \frac{1}{4}, \frac{1}{8}, \ldots$.

$$S_1 = \frac{1}{2}$$

$$S_2 = \frac{1}{2} + \frac{1}{4} = \frac{3}{4}$$

$$S_3 = \frac{1}{2} + \frac{1}{4} + \frac{1}{8} = \frac{7}{8}$$

$$S_4 = \frac{1}{2} + \frac{1}{4} + \frac{1}{8} + \frac{1}{16} = \frac{15}{16}$$

$$S_5 = \frac{1}{2} + \frac{1}{4} + \frac{1}{8} + \frac{1}{16} + \frac{1}{32} = \frac{31}{32}$$

$$\vdots$$

$$S_{10} = \frac{1}{2} + \frac{1}{4} + \frac{1}{8} + \cdots + \frac{1}{2^{10}} = \frac{1023}{1024}$$

Even though each partial sum is larger than the preceding partial sum, we see that each partial sum is closer to 1 than the preceding partial sum. If n gets larger and larger, then S_n gets closer and closer to 1. In general, if $|r| < 1$, the following formula will give the sum of an infinite geometric series.

Sum of the Terms of an Infinite Geometric Sequence

The sum of the terms of an infinite geometric sequence with first term a_1 and common ratio r with $|r| < 1$ is given by S_∞, where

$$S_\infty = \frac{a_1}{1 - r}$$

If $|r| \geq 1$, S_∞ does not exist.

What happens for other values of r? For example, in the following geometric sequence $r = 3$.

$$6, 18, 54, 162, \ldots$$

Here, as n increases, the sum S_n increases also. This time, though, S_n does not approach a number but will increase without bound.

EXAMPLE 6 Find the sum of the terms of the geometric sequence $2, \frac{2}{3}, \frac{2}{9}, \frac{2}{27}, \ldots$.

Solution: For this geometric sequence, $r = \frac{1}{3}$. Since $|r| < 1$, we may use the formula S_∞ of a geometric sequence with $a_1 = 2$ and $r = \frac{1}{3}$.

$$S_\infty = \frac{a_1}{1 - r} = \frac{2}{1 - \frac{1}{3}} = \frac{2}{\frac{2}{3}} = 3 \quad ■$$

The formula for the sum of the terms of an infinite geometric sequence can be used to write a repeating decimal as a fraction. For example,

$$0.33\overline{3} = \frac{3}{10} + \frac{3}{100} + \frac{3}{1000} + \cdots$$

This sum is the sum of the terms of an infinite geometric sequence whose first term $a_1 = \frac{3}{10}$ and whose common ratio $r = \frac{1}{10}$. Using the formula for S_∞,

$$S_\infty = \frac{a_1}{1 - r} = \frac{\frac{3}{10}}{1 - \frac{1}{10}} = \frac{1}{3}$$

So $0.33\overline{3} = \frac{1}{3}$.

EXAMPLE 7 A pendulum swings through an arc of 24 inches on its first pass. On each pass thereafter, the length of the arc is 75% of the length of the arc on the preceding pass. Find the total distance the pendulum travels before it comes to rest.

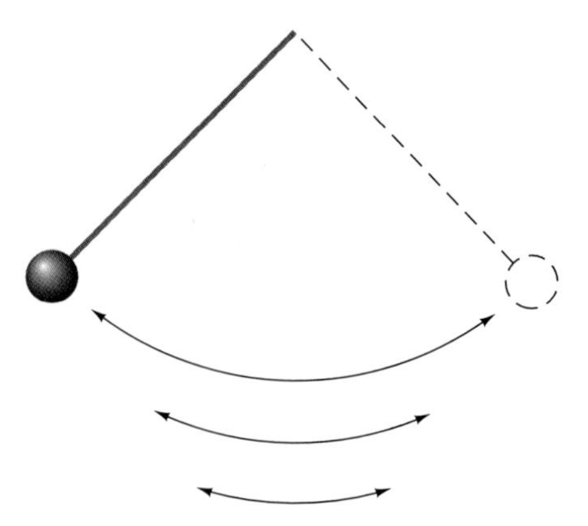

Solution: We must find the sum of the terms of an infinite geometric sequence whose first term a_1 is 24 and whose common ratio r is 0.75. Since $|r| < 1$, we may use the formula for S_∞.

$$S_\infty = \frac{a_1}{1-r} = \frac{24}{1-0.75} = \frac{24}{0.25} = 96$$

The pendulum travels a total distance of 96 inches before it comes to rest. ■

EXERCISE SET 12.4

Use the partial sum formula to find the partial sum of the given arithmetic sequence. See Example 1.

1. Find the sum of the first six terms of the arithmetic sequence 1, 3, 5, 7, ...
2. Find the sum of the first seven terms of the arithmetic sequence −7, −11, −15, ...
3. Find the sum of the first six terms of the arithmetic sequence 3, 6, 9, ...
4. Find the sum of the first four terms of the arithmetic sequence −4, −8, −12, ...

Solve. See Example 2.

5. Find the sum of the first ten positive integers.
6. Find the sum of the first eight negative integers.
7. Find the sum of the first four positive odd integers.
8. Find the sum of the first five negative odd integers.

Solve. See Example 3.

9. Modern Car Company has come out with a new car model. Market analysts predict 4000 cars will be sold in the first month and that sales will drop by 50 cars per month after that during the first year. Write out the first five terms of the sequence, and find the number of sold cars predicted for the twelfth month. Find the total number of sold cars predicted for the first year.
10. A fax company charges \$3 for the first page sent and \$0.10 less than the preceding for each additional page sent. The cost per page forms an arithmetic sequence. Write the first five terms of this sequence, and use a partial sum to find the cost of sending a nine-page document.
11. Sal has two job offers: Firm *A* starts at \$22,000 per year and guarantees raises of \$1000 per year, while Firm *B* starts at \$20,000 and guarantees raises of \$1200 per year. Over a 10-year period, determine the more profitable offer.
12. The game of pool uses 15 balls numbered 1 to 15. In the game of rotation, when a player sinks a ball, the player receives as many points as the number on the ball. Use an arithmetic series to find the score of a player who sinks all 15 balls.

Use the partial sum formula to find the partial sum of the given geometric sequence. See Example 4.

13. Find the sum of the first five terms of the geometric sequence 4, 12, 36, ...

14. Find the sum of the first eight terms of the geometric sequence $-1, 2, -4, \ldots$.

15. Find the sum of the first four terms of the geometric sequence $2, \frac{2}{5}, \frac{2}{25}, \ldots$.

16. Find the sum of the first five terms of the geometric sequence $\frac{1}{3}, -\frac{2}{3}, \frac{4}{3}, \ldots$.

Solve. See Example 5.

17. A woman made $30,000 during the first year she owned her business and made an additional 10% over the previous year in each subsequent year. Find how much she made during her fourth year of business. Find her total earnings during the first four years.

18. A parachutist in free fall falls 16 feet during the first second, 48 feet during the second second, 80 feet during the third second, and so on. Find how far she falls during the eighth second. Find the total distance she falls during the first 8 seconds.

19. A trainee in a computer company takes 0.9 times as long to assemble each computer as he took to assemble the preceding computer. If it took him 30 minutes to assemble the first computer, find how long it takes him to assemble the fifth computer. Find the total time he takes to assemble the first five computers (round to the nearest minute).

20. On a gambling trip to Reno, Carol doubled her bet each time she lost. If her first losing bet was $5 and she lost six consecutive bets, find how much she lost on the sixth bet. Find the total amount lost on these six bets.

Find the sum of the terms of each infinite geometric sequence. See Example 6.

21. 12, 6, 3, . . .

22. 45, 15, 5, . . .

23. $\frac{1}{10}, \frac{1}{100}, \frac{1}{1000}, \ldots$

24. $\frac{3}{5}, \frac{3}{20}, \frac{3}{80}, \ldots$

Solve. See Example 7.

25. A ball is dropped from a height of 20 feet and repeatedly rebounds to a height that is $\frac{4}{5}$ of its previous height. Find the total distance the ball covers before it comes to rest.

26. A rotating flywheel coming to rest makes 300 revolutions in the first minute, and in each minute thereafter makes $\frac{2}{5}$ as many revolutions as in the preceding minute. Find how many revolutions the wheel makes before coming to rest.

Solve.

27. Find the sum of the first ten terms of the sequence $-4, 1, 6, \ldots$.

28. Find the sum of the first twelve terms of the sequence $-3, -13, -23, \ldots$.

29. Find the sum of the first seven terms of the sequence $3, \frac{3}{2}, \frac{3}{4}, \ldots$.

30. Find the sum of the first five terms of the sequence $-2, -6, -18, \ldots$.

31. Find the sum of the first five terms of the sequence $-12, 6, -3, \ldots$.

32. Find the sum of the first four terms of the sequence $-\frac{1}{4}, -\frac{3}{4}, -\frac{9}{4}, \ldots$.

33. Find the sum of the first twenty terms of the sequence $\frac{1}{2}, \frac{1}{4}, 0, \ldots$.

34. Find the sum of the first fifteen terms of the sequence $-5, -9, -13, \ldots$.

35. If a_1 is 8 and r is $-\frac{2}{3}$, find S_3.

36. If a_1 is 10 and d is $-\frac{1}{2}$, find S_{18}.

Find the sum of the terms of each infinite geometric sequence.

37. $-10, -5, -\frac{5}{2}, \ldots$

38. $-16, -4, -1, \ldots$

39. $2, -\frac{1}{4}, \frac{1}{32}, \ldots$

40. $-3, \frac{3}{5}, -\frac{3}{25}, \ldots$

41. $\frac{2}{3}, -\frac{1}{3}, \frac{1}{6}, \ldots$

42. $6, -4, \frac{8}{3}, \ldots$

Solve.

43. In the pool game of rotation, player A sinks balls numbered 1 to 9, and player B sinks the rest of the balls. Use arithmetic series to find each player's score (see Exercise 12).

44. A godfather deposited \$250 in a savings account on the day his godchild was born. On each subsequent birthday he deposited \$50 more than he deposited the previous year. Find how much money he deposited on his godchild's twenty-first birthday. Find the total amount deposited over the 21 years.

45. During the holiday rush a business can rent a computer system for \$200 the first day, with the rent decreasing \$5 for each additional day. Find how much rent is paid for 20 days during the holiday rush.

46. Spraying a field with insecticide killed 6400 weevils the first day, 1600 the second day, 400 the third day, and so on. Find the total number of weevils killed during the first 5 days.

47. A college student humorously asks his parents to charge him room and board according to this geometric sequence: \$0.01 for the first day of the month, \$0.02 for the second day, \$0.04 for the third day, and so on. Find the total room and board he would pay for 30 days.

48. A bank attracted 80 new customers the first day following its television advertising campaign, 120 the second day, 160 the third day, and so on, in an arithmetic sequence. Find how many new customers were attracted during the first 5 days following its television campaign.

49. Write $0.88\overline{8}$ as an infinite geometric series and use the formula for S_∞ to write it as a rational number.

50. Write $0.54\overline{54}$ as an infinite geometric series and use the formula S_∞ to write it as a rational number.

Skill Review

Evaluate.

51. $6 \cdot 5 \cdot 4 \cdot 3 \cdot 2 \cdot 1$

52. $8 \cdot 7 \cdot 6 \cdot 5 \cdot 4 \cdot 3 \cdot 2 \cdot 1$

53. $\frac{3 \cdot 2 \cdot 1}{2 \cdot 1}$

54. $\frac{5 \cdot 4 \cdot 3 \cdot 2 \cdot 1}{3 \cdot 2 \cdot 1}$

Multiply. See Section 5.4.

55. $(x + 5)^2$

56. $(x - 2)^2$

57. $(2x - 1)^3$

58. $(3x + 2)^3$

Writing in Mathematics

59. Explain whether the sequence 5, 5, 5, . . . is arithmetic, geometric, neither, or both.

60. Describe a situation in everyday life that can be modeled by an infinite geometric series.

12.5 The Binomial Theorem

OBJECTIVES

Tape IA 35

1. Use Pascal's triangle to expand binomials.
2. Evaluate factorials.
3. Use the binomial theorem to expand binomials.
4. Find the nth term in the expansion of a binomial raised to a positive power.

In this section, we learn how to easily expand binomials of the form $(a + b)^n$. First, we review the patterns in the expansions of $(a + b)^n$.

$(a+b)^0 = 1$	1 term
$(a+b)^1 = a + b$	2 terms
$(a+b)^2 = a^2 + 2ab + b^2$	3 terms
$(a+b)^3 = a^3 + 3a^2b + 3ab^2 + b^3$	4 terms
$(a+b)^4 = a^4 + 4a^3b + 6a^2b^2 + 4ab^3 + b^4$	5 terms
$(a+b)^5 = a^5 + 5a^4b + 10a^3b^2 + 10a^2b^3 + 5ab^4 + b^5$	6 terms

Notice the following patterns:

1. The expansion of $(a + b)^n$ contains $n + 1$ terms. For example, for $(a + b)^3$, $n = 3$, and the expansion contains $3 + 1$ or 4 terms.
2. The first term of the expansion of $(a + b)^n$ is a^n and the last term is b^n.
3. The powers of a decrease by 1 for each term, while the powers of b increase by 1 for each term.
4. The sum of the variable exponents for each term is n.

1 There are more patterns in the coefficients of the terms as well. Written in a triangular array, the coefficients are called **Pascal's triangle.**

$(a+b)^0$:						1					
$(a+b)^1$:					1		1				
$(a+b)^2$:				1		2		1			
$(a+b)^3$:			1		3		3		1		
$(a+b)^4$:		1		4		6		4		1	
$(a+b)^5$:	1		5		10		10		5		1

Each row in Pascal's triangle begins and ends with 1. Any other number is the sum of the two closest numbers above it. For example, we can write the next row, the seventh row, by beginning with 1, adding the two numbers closest above in row 6, and ending with 1.

	1		5		10		10		5		1	
1		6		15		20		15		6		1

We can use Pascal's triangle and the patterns noted to expand $(a + b)^n$ without actually multiplying.

EXAMPLE 1 Expand $(a + b)^6$.

Solution: Using the seventh row of Pascal's triangle as the coefficients and following the patterns noted, $(a + b)^6$ can be expanded as

$$a^6 + 6a^5b + 15a^4b^2 + 20a^3b^3 + 15a^2b^4 + 6ab^5 + b^6$$ ■

2 For large n, using Pascal's triangle to find coefficients for $(a + b)^n$ can be tedious. An alternative method for determining these coefficients is based on the concept of a **factorial.**

A **factorial of n,** written $n!$ (read "n factorial") is the product of the first n consecutive natural numbers.

Factorial of n: $n!$

If n is a natural number, then $n! = n(n-1)(n-2)(n-3) \cdot \cdots \cdot 3 \cdot 2 \cdot 1$. The factorial of 0, written $0!$, is defined to be 1.

For example, $3! = 3 \cdot 2 \cdot 1 = 6$, $5! = 5 \cdot 4 \cdot 3 \cdot 2 \cdot 1 = 120$, and $0! = 1$.

EXAMPLE 2 Evaluate each expression.

a. $\frac{5!}{6!}$ **b.** $\frac{10!}{7!3!}$

c. $\frac{3!}{2!1!}$ **d.** $\frac{7!}{7!0!}$

Solution: **a.** $\frac{5!}{6!} = \frac{5 \cdot 4 \cdot 3 \cdot 2 \cdot 1}{6 \cdot 5 \cdot 4 \cdot 3 \cdot 2 \cdot 1} = \frac{1}{6}$

b. $\frac{10!}{7!3!} = \frac{10 \cdot 9 \cdot 8 \cdot 7!}{7! \cdot 3 \cdot 2 \cdot 1} = \frac{10 \cdot 9 \cdot 8}{3 \cdot 2 \cdot 1}$

$= 10 \cdot 3 \cdot 4 = 120$

c. $\frac{3!}{2!1!} = \frac{3 \cdot 2 \cdot 1}{2 \cdot 1 \cdot 1} = 3$

d. $\frac{7!}{7!0!} = \frac{7!}{7! \cdot 1} = 1$ ■

HELPFUL HINT

We can use a calculator with a factorial key to evaluate the factorial. A calculator uses scientific notation for large results.

3 It can be proved, though we won't do so here, that the coefficients of terms in the expansion of $(a+b)^n$ can be expressed in terms of factorials. Following patterns 1 through 4 given earlier and using the factorial expressions of the coefficients, we have the **binomial theorem.**

Binomial Theorem

If n is a positive integer, then

$$(a+b)^n = a^n + \frac{n}{1!}a^{n-1}b^1 + \frac{n(n-1)}{2!}a^{n-2}b^2 + \frac{n(n-1)(n-2)}{3!}a^{n-3}b^3 + \cdots + b^n$$

We call the formula for $(a+b)^n$ given by the binomial theorem the **binomial formula.**

EXAMPLE 3 Use the binomial theorem to expand $(x + y)^{10}$.

Solution: Let $a = x$, $b = y$, and $n = 10$ in the binomial formula.

$$(x + y)^{10} = x^{10} + \frac{10}{1!}x^9y + \frac{10 \cdot 9}{2!}x^8y^2 + \frac{10 \cdot 9 \cdot 8}{3!}x^7y^3 + \frac{10 \cdot 9 \cdot 8 \cdot 7}{4!}x^6y^4$$

$$+ \frac{10 \cdot 9 \cdot 8 \cdot 7 \cdot 6}{5!}x^5y^5 + \frac{10 \cdot 9 \cdot 8 \cdot 7 \cdot 6 \cdot 5}{6!}x^4y^6$$

$$+ \frac{10 \cdot 9 \cdot 8 \cdot 7 \cdot 6 \cdot 5 \cdot 4}{7!}x^3y^7$$

$$+ \frac{10 \cdot 9 \cdot 8 \cdot 7 \cdot 6 \cdot 5 \cdot 4 \cdot 3}{8!}x^2y^8$$

$$+ \frac{10 \cdot 9 \cdot 8 \cdot 7 \cdot 6 \cdot 5 \cdot 4 \cdot 3 \cdot 2}{9!}xy^9 + y^{10}$$

$$= x^{10} + 10x^9y + 45x^8y^2 + 120x^7y^3 + 210x^6y^4 + 252x^5y^5 + 210x^4y^6$$

$$+ 120x^3y^7 + 45x^2y^8 + 10xy^9 + y^{10} \quad \blacksquare$$

EXAMPLE 4 Use the binomial theorem to expand $(x + 2y)^5$.

Solution: Let $a = x$ and $b = 2y$ in the binomial formula.

$$(x + 2y)^5 = x^5 + \frac{5}{1!}x^4(2y) + \frac{5 \cdot 4}{2!}x^3(2y)^2 + \frac{5 \cdot 4 \cdot 3}{3!}x^2(2y)^3$$

$$+ \frac{5 \cdot 4 \cdot 3 \cdot 2}{4!}x(2y)^4 + (2y)^5$$

$$= x^5 + 10x^4y + 40x^3y^2 + 80x^2y^3 + 80xy^4 + 32y^5 \quad \blacksquare$$

EXAMPLE 5 Use the binomial theorem to expand $(3m - n)^4$.

Solution: Let $a = 3m$ and $b = -n$ in the binomial formula.

$$(3m - n)^4 = (3m)^4 + \frac{4}{1!}(3m)^3(-n) + \frac{4 \cdot 3}{2!}(3m)^2(-n)^2$$

$$+ \frac{4 \cdot 3 \cdot 2}{3!}(3m)(-n)^3 + (-n)^4$$

$$= 81m^4 - 108m^3n + 54m^2n^2 - 12mn^3 + n^4 \quad \blacksquare$$

4 Sometimes it is convenient to find a specific term of a binomial expansion without writing out the entire expansion. By studying the expansion of binomials, a pattern forms for each term. This pattern is most easily stated for the $(r + 1)$st term.

> **$(r + 1)$st Term in a Binomial Expansion**
>
> The $(r + 1)$st term of the expansion of $(a + b)^n$ is $\dfrac{n!}{r!(n - r)!}a^{n-r}b^r$.

EXAMPLE 6 Find the eighth term in the expansion of $(2x - y)^{10}$.

Solution: Use the formula, with $n = 10$, $a = 2x$, $b = -y$, and $r + 1 = 8$. Notice that, since $r + 1 = 8$, $r = 7$.

$$\begin{aligned}\frac{n!}{r!(n-r)!}a^{n-r}b^r &= \frac{10!}{7!3!}(2x)^3(-y)^7 \\ &= 120(8x^3)(-y)^7 \\ &= -960x^3y^7\end{aligned}$$ ■

EXERCISE SET 12.5

Use Pascal's triangle to expand the binomial. See Example 1.

1. $(m + n)^3$
2. $(x + y)^4$
3. $(c + d)^5$
4. $(a + b)^6$
5. $(y - x)^5$
6. $(q - r)^7$

Evaluate each expression. See Example 2.

7. $\frac{8!}{7!}$
8. $\frac{6!}{0!}$
9. $\frac{7!}{5!}$
10. $\frac{8!}{5!}$
11. $\frac{10!}{7!2!}$
12. $\frac{9!}{5!3!}$
13. $\frac{8!}{6!0!}$
14. $\frac{10!}{4!6!}$

Use the binomial formula to expand each binomial. See Examples 3 and 4.

15. $(a + b)^7$
16. $(x + y)^8$
17. $(a + 2b)^5$
18. $(x + 3y)^6$

Use the binomial formula to expand each binomial. See Example 5.

19. $(2a - b)^5$
20. $(5x - y)^4$
21. $(c - 2d)^6$
22. $(m - 3n)^4$

Find the indicated term. See Example 6.

23. The third term of the expansion of $(x + y)^4$
24. The fourth term of the expansion of $(a + b)^8$
25. The second term of the expansion of $(a + 3b)^{10}$
26. The third term of the expansion of $(m + 5n)^7$

Expand each binomial.

27. $(q + r)^9$
28. $(b + c)^6$
29. $(4a + b)^5$
30. $(3m + n)^4$
31. $(5a - 2b)^4$
32. $(m - 4)^6$
33. $(2a + 3b)^3$
34. $(4 - 3x)^5$
35. $(x + 2)^5$
36. $(3 + 2a)^4$

Find the indicated term.

37. The fifth term of the expansion of $(c - d)^5$
38. The fourth term of the expansion of $(x - y)^6$
39. The eighth term of the expansion of $(2c + d)^7$
40. The tenth term of the expansion of $(5x - y)^9$
41. The fourth term of the expansion of $(2r - s)^5$
42. The first term of the expansion of $(3q - 7r)^6$

Skill Review

Sketch the graph of each function. Decide whether each function is one-to-one. See Sections 9.5 and 11.1.

43. $f(x) = |x|$

44. $g(x) = 3(x - 1)^2$

45. $H(x) = 2x + 3$

46. $F(x) = -2$

47. $f(x) = x^2 + 3$

48. $h(x) = -(x + 1)^2 - 4$

Writing in Mathematics

49. Explain how to generate a row of Pascal's triangle.

CHAPTER 12 GLOSSARY

An **arithmetic sequence** is a sequence in which each term after the first differs from the preceding term by a constant amount d. The constant d is called the **common difference.**

The **domain** of a sequence is the set of natural numbers.

A series is a **finite series** if it is the sum of only the first k terms, for some natural number k.

A **geometric sequence** is a sequence for which each term after the first is obtained by multiplying the preceding term by a constant r. The constant r is called the **common ratio.**

A series is an **infinite series** if it is the sum of all the terms of a sequence.

An **infinite sequence** or **sequence** is a function whose domain is the set of natural numbers.

Written in a triangular array, the coefficients of the expansions of the binomial $(a + b)^n$ for $n = 1, 2, 3$, and so on, are called **Pascal's triangle.**

A shorthand notation for denoting a series when the general term of the sequence is known is called **summation notation.** The Greek uppercase letter **sigma** Σ is used to mean "sum."

The **range** of a sequence is the set of function values $a_1, a_2, a_3, \ldots$. These values are called the **terms** of the sequence.

A sum of the terms of a sequence is called a **series** (plural is also series).

CHAPTER 12 SUMMARY

GENERAL TERM OF AN ARITHMETIC SEQUENCE (12.2)

The general term a_n of an arithmetic sequence is given by

$$a_n = a_1 + (n - 1)d$$

where a_1 is the first term and d is the common difference.

GENERAL TERM OF A GEOMETRIC SEQUENCE (12.2)

The general term a_n of a geometric sequence is given by

$$a_n = a_1 r^{n-1}$$

where a_1 is the first term and r is the common ratio.

PARTIAL SUM S_n OF AN ARITHMETIC SEQUENCE (12.4)

The partial sum S_n of the first n terms of an arithmetic sequence is given by

$$S_n = \frac{n}{2}(a_1 + a_n)$$

PARTIAL SUM S_n OF A GEOMETRIC SEQUENCE (12.4)

The partial sum S_n of the first n terms of a geometric series, whose first term is a_1 is given by

$$S_n = \frac{a_1(1 - r^n)}{1 - r}$$

as long as the common ratio of the sequence r is not 1.

SUM OF THE TERMS OF AN INFINITE GEOMETRIC SEQUENCE (12.4)

The sum of the terms of an infinite geometric sequence with first term a_1 and common ratio r with $|r| < 1$ is given by S_∞, where

$$S_\infty = \frac{a_1}{1 - r}$$

If $|r| \geq 1$, S_∞ is not a real number.

FACTORIAL OF n: n! (12.5)

If n is a natural number, then $n! = n(n - 1)(n - 2)(n - 3) \cdot \cdots \cdot 3 \cdot 2 \cdot 1$. The factorial of 0, written $0!$, is defined to be 1.

BINOMIAL THEOREM (12.5)

If n is a positive integer, then

$$(a + b)^n = a^n + \frac{n}{1!}a^{n-1}b^1 + \frac{n(n - 1)}{2!}a^{n-2}b^2 + \frac{n(n - 1)(n - 2)}{3!}a^{n-3}b^3 + \cdots + b^n$$

CHAPTER 12 REVIEW

(12.1) *Find the indicated term(s) of the given sequence.*

1. The first five terms of the sequence $a_n = -3n^2$.
2. The first five terms of the sequence $a_n = n^2 + 2n$.
3. The one-hundredth term of the sequence $a_n = \frac{(-1)^n}{100}$.
4. The fiftieth term of the sequence $a_n = \frac{2n}{(-1)^2}$.
5. The general term a_n of the sequence $\frac{1}{6}, \frac{1}{12}, \frac{1}{18}, \ldots$.
6. The general term a_n of the sequence $-1, 4, -9, 16, \ldots$.

Solve the following applications.

7. The distance in feet that an olive falling from rest in a vacuum will travel during each second is given by an arithmetic sequence whose general term is $a_n = 32n - 16$, where n is the number of the second. Find the distance the olive will fall during the fifth, sixth, and seventh seconds.

8. A culture of yeast doubles every day in a geometric progression whose general term is $a_n = 100(2)^{n-1}$, where n is the number of the day just beginning. Find how many days it takes the yeast culture to measure at least 10,000. Find the original measure of the yeast culture.

9. A Center for Disease Control (CDC) reported that a new type of immune system virus infected approximately 450 people during 1994, the year it was first discovered. The CDC predicts that during the next decade the virus will infect three times as many people each year as the year before. Write out the first five terms of this geometric sequence, and predict the number of infected people in 1998.

10. The first row of an amphitheater contains 50 seats, and each row thereafter contains 8 additional seats. Write the first ten terms of this arithmetic progression, and find the number of seats in the tenth row.

(12.2)

11. Find the first five terms for the geometric sequence whose first term is -2 and whose common ratio is $\frac{2}{3}$.

12. Find the first five terms for the arithmetic sequence whose first term is 12 and whose common difference is -1.5.

13. Find the thirtieth term of the arithmetic sequence whose first term is -5 and whose common difference is 4.

14. Find the eleventh term of the arithmetic sequence whose first term is 2 and whose common difference is $\frac{3}{4}$.

15. Find the twentieth term of the arithmetic sequence whose first three terms are 12, 7, and 2.

16. Find the sixth term of the geometric sequence whose first three terms are 4, 6, and 9.

17. If the fourth term of an arithmetic sequence is 18 and the twentieth term is 98, find the first term and the common difference.

18. If the third term of a geometric sequence is -48 and the fourth term is 192, find the first term and the common ratio.

19. Find the general term of the sequence $\frac{3}{10}, \frac{3}{100}, \frac{3}{1000}, \ldots$.

20. Find a general term that satisfies the terms shown for the sequence 50, 58, 66, ...

Determine which of the following sequences are arithmetic or geometric. If a sequence is arithmetic, find a_1 and d. If a sequence is geometric, find a_1 and r. If neither, write neither.

21. $\frac{8}{3}, 4, 6, \ldots$

22. $-10.5, -6.1, -1.7$

23. $7x, -14x, 28x$

24. $3x^2, 9x^4, 81x^8, \ldots$

Solve the following applications.

25. To test the bounce of a racquetball, the ball is dropped from a height of 8 feet. The ball is judged "good" if it rebounds at least 75% of its previous height with each bounce. Write out the first six terms of this geometric sequence (round to the nearest tenth). Determine if a ball is "good" that rebounds to a height of 2.5 feet after the fifth bounce.

26. A display of oil cans in an auto parts store has 25 cans in the bottom row, 21 cans in the next row, and so on, in an arithmetic progression. Find the general term, and the number of cans in the top row.

27. Suppose that you save \$1 the first day of a month, \$2 the second day, \$4 the third day, continuing to double your savings each day. Write the general term of this geometric sequence and find the amount you will save on the tenth day. Estimate the amount you will save on the thirtieth day of the month, and check your estimate with a calculator.

28. On the first swing, the length of an arc through which a pendulum swings is 30 inches. The length of the arc for each successive swing is 70% of the preceding swing. Find the length of the arc for the fifth swing.

29. Rosa takes a job that has a monthly starting salary of \$900 and guarantees her a monthly raise of \$150 during her 6-month training period. Find the general term of this sequence and her salary at the end of her training.

30. A sheet of paper is $\frac{1}{512}$-inch thick. By folding the sheet in half, the total thickness will be $\frac{1}{256}$ inch: a second fold produces a total thickness of $\frac{1}{128}$ inch. Estimate the thickness of the stack after 15 folds, and then check your estimate with a calculator.

(12.3) *Write out the terms and find the sum for each of the following.*

31. $\sum_{i=1}^{5} 2i - 1$

32. $\sum_{i=1}^{5} i(i + 2)$

33. $\sum_{i=2}^{4} \frac{(-1)^i}{2i}$

34. $\sum_{i=3}^{5} 5(-1)^{i-1}$

Find the partial sum of the given sequence.

35. S_4 of the sequence $a_n = (n - 3)(n + 2)$

36. S_6 of the sequence $a_n = n^2$

37. S_5 of the sequence $a_n = -8 + (n - 1)3$

38. S_3 of the sequence $a_n = 5(4)^{n-1}$

Write the sum using Σ notation.

39. $1 + 3 + 9 + 27 + 81 + 243$

40. $6 + 2 + (-2) + (-6) + (-10) + (-14) + (-18)$

41. $\frac{1}{4} + \frac{1}{16} + \frac{1}{64} + \frac{1}{256}$

42. $1 + \left(-\frac{3}{2}\right) + \frac{9}{4}$

Solve.

43. A yeast colony begins with 20 yeast and doubles every 8 hours. Write the sequence that describes the growth of the yeast, and find the total yeast after 48 hours.

44. The number of cranes born each year in a new aviary forms a sequence whose general term is $a_n = n^2 + 2n - 1$. Find the number of cranes born in the fourth year, and the total number of cranes born in the first four years.

45. Harold has a choice between two job offers. Job *A* has an annual starting salary of \$19,500 with guaranteed annual raises of \$1100 for the next four years, while job *B* has an annual starting salary of \$21,000 with guaranteed annual raises of \$700 for the next four years. Compare the salaries for the fifth year under each job offer.

46. A sample of radioactive waste is decaying so that the amount decaying in kilograms during year n is $a_n = 200(0.5)^n$. Find the amount of decay in the third year, and the total amount of decay in the first three years.

(12.4) *Find the partial sum of the given sequence.*

47. The sixth partial sum of the sequence 15, 19, 23, ...

48. The ninth partial sum of the sequence 5, −10, 20, ...

49. The sum of the first 30 odd positive integers.

50. The sum of the first 20 positive multiples of 7.

51. The sum of the first 20 terms of the sequence 8, 5, 2, ...

52. The sum of the first eight terms of the sequence $\frac{3}{4}, \frac{9}{4}, \frac{27}{4}, \ldots$.

53. S_4 if $a_1 = 6$ and $r = 5$.

54. S_{100} if $a_1 = -3$ and $d = -6$.

Find the sum of each infinite geometric sequence.

55. $5, \frac{5}{2}, \frac{5}{4}, \ldots$

56. $18, -2, \frac{2}{9}, \ldots$

57. $-20, -4, -\frac{4}{5}, \ldots$

58. 0.2, 0.02, 0.002, . . .

Solve.

59. A frozen yogurt store owner cleared \$20,000 the first year he owned his business and made an additional 15% over the previous year in each subsequent year. Find how much he made during his fourth year of business. Find his total earnings during the first 4 years (round to the nearest dollar).

60. On his first morning in a television assembly factory, a trainee takes 0.8 times as long to assemble each television as he took to assemble the one before. If it took him 40 minutes to assemble the first television, find how long it takes him to assemble the fourth television. Find the total time he takes to assemble the first four televisions (round to the nearest minute).

61. During the harvest season a farmer can rent a combine machine for \$100 the first day, with the rent decreasing \$7 for each additional day. Find how much rent the farmer pays for the seventh day. Find how much total rent the farmer pays for 7 days.

62. A rubber ball is dropped from a height of 15 feet and rebounds 80% of its previous height after each bounce. Find the total distance the ball travels before it comes to rest.

63. Spraying a pond once with insecticide killed 1800 mosquitoes the first day, 600 the second day, 200 the third day, and so on. Find the total number of mosquitoes killed during the first 6 days after the spraying (round to the nearest unit).

64. See Exercise 63. Find the day on which the insecticide is no longer effective, and find the total number of mosquitoes killed (round to the nearest mosquito).

65. Use the formula S_∞ to write $0.55\overline{5}$ as a fraction.

66. A movie theater has 27 seats in the first row, 30 seats in the second row, 33 seats in the third row, and so on. Find the total number of seats in the theater if there are 20 rows.

(12.5) *Use Pascal's triangle to expand the binomial.*

67. $(x + z)^5$

68. $(y - r)^6$

69. $(2x + y)^4$

70. $(3y - z)^4$

Use the binomial formula to expand the following.

71. $(b + c)^8$

72. $(x - w)^7$

73. $(4m - n)^4$

74. $(p - 2r)^5$

Find the indicated term.

75. The fourth term of the expansion of $(a + b)^7$.

76. The eleventh term of the expansion of $(y + 2z)^{10}$.

CHAPTER 12 TEST

Find the indicated term(s) of the given sequence.

1. The first five terms of the sequence $a_n = \dfrac{(-1)^n}{n + 4}$.

2. The first five terms of the sequence $a_n = \dfrac{3}{(-1)^n}$.

3. The eightieth term of the sequence $a_n = 10 + 3(n - 1)$.

4. The two-hundredth term of the sequence $a_n = (n + 1)(n - 1)(-1)^n$.

5. The general term of the sequence $\dfrac{2}{5}, \dfrac{2}{25}, \dfrac{2}{125}, \ldots$.

6. The general term of the sequence $-9, 18, -27, 36, \ldots$.

Find the partial sum of the given sequence.

7. S_5 of the sequence $a_n = 5(2)^{n-1}$

8. S_{30} of the sequence $a_n = 18 + (n - 1)(-2)$

9. S_∞ of the sequence $a_1 = 24$ and $r = \dfrac{1}{6}$

10. S_∞ of the sequence $\dfrac{3}{2}, -\dfrac{3}{4}, \dfrac{3}{8}, \ldots$

11. $\sum_{i=1}^{4} i(i - 2)$

12. $\sum_{i=2}^{4} 5(2)^i (-1)^{i-1}$

Expand the binomial using Pascal's triangle.

13. $(a - b)^6$

14. $(2x + y)^5$

Expand the binomial using the binomial formula.

15. $(y + z)^8$

16. $(2p + r)^7$

Solve the following applications.

17. The population of a small town is growing yearly according to the sequence defined by $a_n = 250 + 75(n - 1)$, where n is the number of the year just beginning. Predict the population at the beginning of the tenth year. Find the town's initial population.

18. A gardener is making a triangular planting with one shrub in the first row, three shrubs in the second row, five shrubs in the third row, and so on, for eight rows. Write the finite series of this sequence and find the total number of shrubs planted.

19. A pendulum swings through an arc of length 80 centimeters on its first swing. On each successive swing, the length of the arc is $\frac{3}{4}$ the length of the arc on the preceding swing. Find the length of the arc on the fourth swing, and find the total arc lengths for the first four swings.

20. See Exercise 19. Find the total arc lengths before the pendulum comes to rest.

21. A parachutist in free fall falls 16 feet during the first second, 48 feet during the second second, 80 feet during the third second, and so on. Find how far he falls during the tenth second. Find the total distance he falls during the first 10 seconds.

22. Use the formula S_∞ to write $0.42\overline{42}$ as a fraction.

CHAPTER 12 CUMULATIVE REVIEW

1. Solve for x: $2(x + 3) > 2x + 1$.

2. Graph the linear equation $y = 3x + 2$.

3. Find the following products.
a. $2x(5x - 4)$ **b.** $-3x^2(4x^2 - 6x + 1)$
c. $-xy(7x^2y + 3xy - 11)$

4. Factor $9x^2 - 36$.

5. Find the lengths of the sides of a right triangle if the lengths can be expressed by three consecutive even integers.

6. Perform the indicated operation.
a. $\frac{2}{x^2y} + \frac{5}{3x^3y}$ **b.** $\frac{3z}{z + 2} + \frac{2z}{z - 2}$
c. $\frac{5k}{k^2 - 4} - \frac{2}{k^2 + k - 2}$

7. Simplify.
a. $\frac{\sqrt{45}}{4} - \frac{\sqrt{5}}{3}$ **b.** $\sqrt[3]{\frac{7x}{8}} + 2\sqrt[3]{7x}$

8. Multiply the complex numbers. Write the product in the form $a + bi$.
a. $(2 - 5i)(4 + i)$ **b.** $(2 - i)^2$
c. $(7 + 3i)(7 - 3i)$

9. Solve $p^4 - 3p^2 - 4 = 0$.

10. Sketch the graph of $g(x) = \frac{1}{2}(x + 2)^2 + 5$. Find the vertex and the axis of symmetry.

11. Graph $(x + 1)^2 + y^2 = 8$.

12. Use the addition method to solve the system

$$\begin{cases} 3x - 2y = 10 \\ 4x - 3y = 15 \end{cases}$$

13. Solve each equation for x.
a. $2^x = 16$ **b.** $9^x = 27$ **c.** $4^{x+3} = 8^x$

14. Find the value of each logarithmic expression.
a. $\log_4 16$ **b.** $\log_{10} \frac{1}{10}$ **c.** $\log_9 3$

15. Solve $\log(x + 2) - \log x = 2$ for x.

16. The population of lemmings varies according to the relationship $y = y_0e^{0.15t}$. In this formula, t is time in months and y_0 is some initial population at time 0. Estimate the population in 6 months if there are originally 5000 lemmings.

17. If the general term of a sequence is $a_n = \frac{(-1)^n}{3n}$, find:
a. The first term of the sequence **b.** a_8
c. The one-hundredth term of the sequence **d.** a_{15}

18. Find the eighth term of the geometric sequence whose first term is 12 and whose common ratio is $\frac{1}{2}$.

19. Evaluate.
a. $\sum_{i=0}^{6} \frac{i - 2}{2}$ **b.** $\sum_{i=3}^{5} 2^i$

20. Find the sum of the terms of the geometric sequence $2, \frac{2}{3}, \frac{2}{9}, \frac{2}{27}, \ldots$.

21. Use the binomial theorem to expand $(x + 2y)^5$.

APPENDIX A

Operations on Decimals

To **add** or **subtract** decimals, write the numbers vertically with decimal points lined up. Add or subtract as with whole numbers and place the decimal point in the answer directly below the decimal points in the problem.

EXAMPLE 1 Add: 5.87 + 23.279 + 0.003.

Solution:

$$\begin{array}{r} 5.87 \\ 23.279 \\ +\ \ 0.003 \\ \hline 29.152 \end{array} \quad \blacksquare$$

EXAMPLE 2 Subtract: 32.15 − 11.237.

Solution:

$$\begin{array}{rrcrrr} & & {\scriptstyle 1} & {\scriptstyle 11} & {\scriptstyle 4} & {\scriptstyle 10} \\ & 3 & \not{2}\,. & \not{1} & \not{5} & \not{0} \\ - & 1 & 1\,. & 2 & 3 & 7 \\ \hline & 2 & 0\,. & 9 & 1 & 3 \end{array} \quad \blacksquare$$

To **multiply** decimals, multiply the numbers as if they were whole numbers. The decimal point in the product is placed so that the number of decimal places in the product is the same as the sum of the number of decimal places in the factors.

EXAMPLE 3 Multiply: 0.072 × 3.5.

Solution:

$$\begin{array}{rl} 0.072 & \quad \text{3 decimal places} \\ \times\ \ 3.5 & \quad \text{1 decimal place} \\ \hline 360 & \\ 216 & \\ \hline 0.2520 & \quad \text{4 decimal places} \end{array} \quad \blacksquare$$

To **divide** decimals, move the decimal point in the divisor to the right of the last digit. Move the decimal point in the dividend the same number of places that the decimal point in the divisor was moved. The decimal point in the quotient lies directly above the decimal point in the dividend.

EXAMPLE 4 Divide 9.46 ÷ 0.04.

Solution:

```
     236.5
04.)946.0
   -8
    14
   -12
     26
    -24
      20
     -20
```

APPENDIX A EXERCISE SET

Perform the indicated operations.

1. 9.076 + 8.004

2. $\begin{array}{r} 6.3 \\ \times 0.05 \\ \hline \end{array}$

3. $\begin{array}{r} 27.006 \\ -14.2 \\ \hline \end{array}$

4. $\begin{array}{r} 0.0036 \\ 7.12 \\ 32.502 \\ +\ 0.05 \\ \hline \end{array}$

5. $\begin{array}{r} 107.92 \\ +\ 3.04 \\ \hline \end{array}$

6. 7.2 ÷ 4

7. 10 − 7.6

8. 40 ÷ 0.25

9. 126.32 − 97.89

10. $\begin{array}{r} 3.62 \\ 7.11 \\ 12.36 \\ 4.15 \\ +\ 2.29 \\ \hline \end{array}$

11. $\begin{array}{r} 3.25 \\ \times .70 \\ \hline \end{array}$

12. $\begin{array}{r} 26.014 \\ -\ 7.8 \\ \hline \end{array}$

13. 8.1 ÷ 3

14. $\begin{array}{r} 1.2366 \\ 0.005 \\ 15.17 \\ +\ 0.97 \\ \hline \end{array}$

15. 55.405 − 6.1711

16. 8.09 + 0.22

17. 60 ÷ 0.75

18. 20 − 12.29

19. 7.612 ÷ 100

20. $\begin{array}{r} 8.72 \\ 1.12 \\ 14.86 \\ 3.98 \\ +\ 1.99 \\ \hline \end{array}$

21. 12.312 ÷ 2.7

22. 0.443 ÷ 100

23. $\begin{array}{r} 569.2 \\ 71.25 \\ +\ 8.01 \\ \hline \end{array}$

24. 3.706 − 2.91

25. $768 - 0.17$

26. $63 \div 0.28$

27. $12 + 0.062$

28. $0.42 + 18$

29. $76 - 14.52$

30. $1.1092 \div 0.47$

31. $3.311 \div 0.43$

32. $7.61 + 0.0004$

33.
$$\begin{array}{r} 762.12 \\ 89.7 \\ +\ 11.55 \\ \hline \end{array}$$

34. $444 \div 0.6$

35. $23.4 - 0.821$

36. $3.7 + 5.6$

37. $476.12 - 112.97$

38. $19.872 \div 0.54$

39. $0.007 + 7$

40.
$$\begin{array}{r} 51.77 \\ +\ 3.6 \\ \hline \end{array}$$

APPENDIX **B**

Review of Angles, Lines, and Special Triangles

The word **geometry** is formed from the Greek words, **geo,** meaning earth, and **metron,** meaning measure. Geometry literally means to measure the earth.

This section contains a review of some basic geometric ideas. It will be assumed that fundamental ideas of geometry such as point, line, ray, and angle are known. In this appendix, the notation $\angle 1$ is read "angle 1" and the notation $m\angle 1$ is read "the measure of angle 1."

We first review types of angles.

Angles

A **right angle** is an angle whose measure is 90°. A right angle can be indicated by a square drawn at the vertex of the angle, as shown below. An angle whose measure is more than 0° but less than 90° is called an **acute angle.**

An angle whose measure is greater than 90° but less than 180° is called an **obtuse angle.**

An angle whose measure is 180° is called a **straight angle.**

Two angles are said to be **complementary** if the sum of their measures is 90°. Each angle is called the **complement** of the other.

Two angles are said to be **supplementary** if the sum of their measures is 180°. Each angle is called the **supplement** of the other.

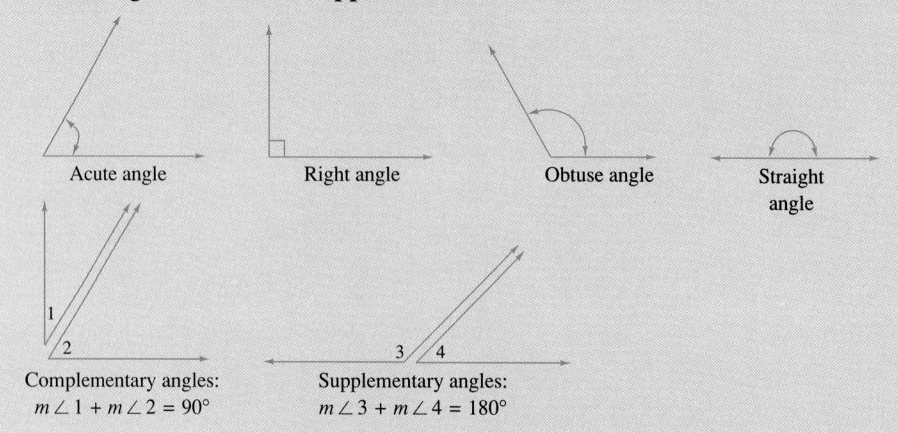

Complementary angles: $m\angle 1 + m\angle 2 = 90°$

Supplementary angles: $m\angle 3 + m\angle 4 = 180°$

EXAMPLE 1 If an angle measures 28°, find its complement.

Solution: Two angles are complementary if the sum of their measures is 90°. The complement of a 28° angle is an angle whose measure is $90° - 28° = 62°$. To check, notice that $28° + 62° = 90°$. ■

Plane is an undefined term that we will describe. A plane can be thought of as a flat surface with infinite length and width, but no thickness. A plane is two dimensional. The arrows in the following diagram indicate that a plane extends indefinitely and has no boundaries.

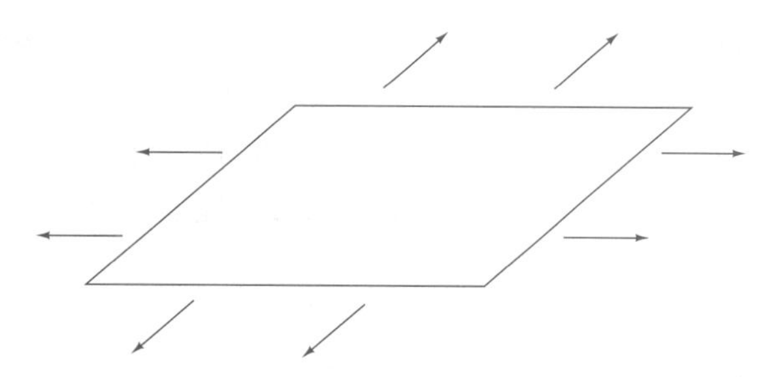

Figures that lie on a plane are called **plane figures.** (See the description of common plane figures in Appendix C.) Lines that lie in the same plane are called **coplanar.**

Lines

Two lines are **parallel** if they lie in the same plane but never meet.

Intersecting lines meet or cross in one point.

Two lines that form right angles when they intersect are said to be **perpendicular.**

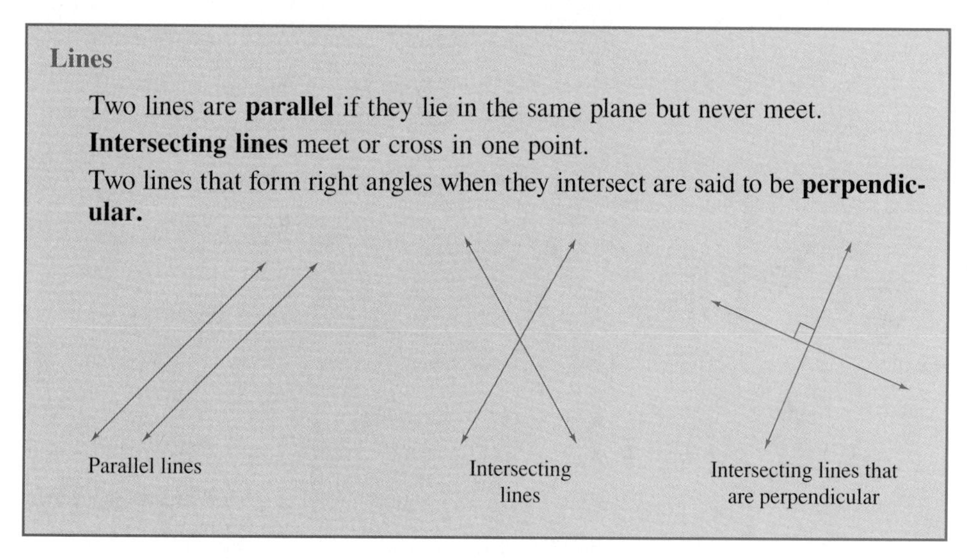

Two intersecting lines form **vertical angles.** Angles 1 and 3 are vertical angles. Also angles 2 and 4 are vertical angles. It can be shown that **vertical angles have equal measures.**

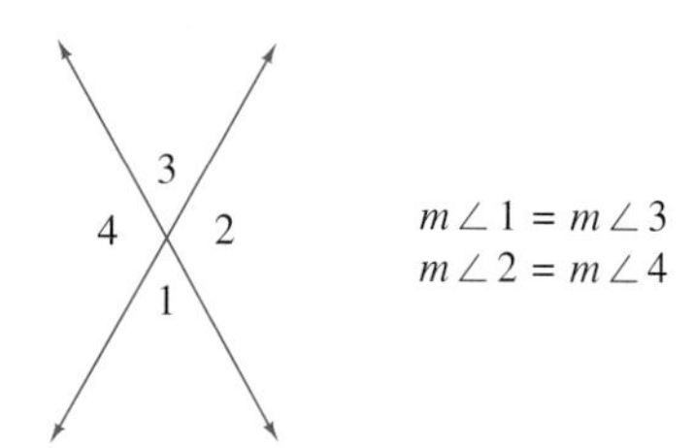

$m\angle 1 = m\angle 3$
$m\angle 2 = m\angle 4$

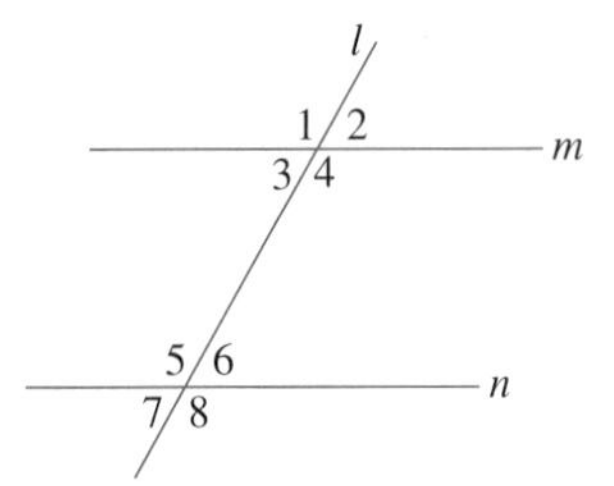

Adjacent angles have the same vertex and share a side. Angles 1 and 2 are adjacent angles. Other pairs of adjacent angles are angles 2 and 4, angles 3 and 4, and angles 3 and 1.

A **transversal** is a line that intersects two or more lines in the same plane. Line l is a transversal that intersects lines m and n. The eight angles formed are numbered and certain pairs of these angles are given special names.

Corresponding angles: $\angle 1$ and $\angle 5$, $\angle 3$ and $\angle 7$, $\angle 2$ and $\angle 6$, and $\angle 4$ and $\angle 8$.
Exterior angles: $\angle 1$, $\angle 2$, $\angle 7$, and $\angle 8$.
Interior angles: $\angle 3$, $\angle 4$, $\angle 5$, and $\angle 6$.
Alternate interior angles: $\angle 3$ and $\angle 6$, $\angle 4$ and $\angle 5$.

These angles and parallel lines are related in the following manner.

Parallel Lines Cut by a Transversal

1. If two parallel lines are cut by a transversal, then
   a. **corresponding angles are equal** and
   b. **alternate interior angles are equal.**
2. If corresponding angles formed by two lines and a transversal are equal, then the lines are parallel.
3. If alternate interior angles formed by two lines and a transversal are equal, then the lines are parallel.

EXAMPLE 2 Given that lines m and n are parallel and that the measure of angle 1 is 100°, find the measures of angles 2, 3, and 4.

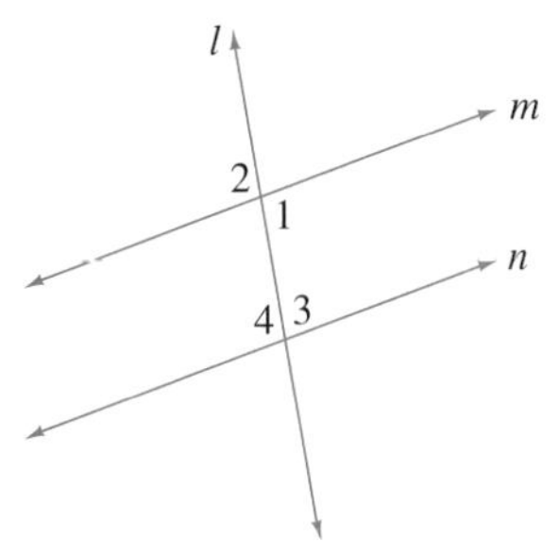

Solution: $m\angle 2 = 100°$, since angles 1 and 2 are vertical angles.

$m\angle 4 = 100°$, since angles 1 and 4 are alternate interior angles.

$m\angle 3 = 180° - 100° = 80°$, since angles 4 and 3 are supplementary angles. ■

A **polygon** is the union of three or more coplanar line segments that intersect each other only at each end point, with each end point shared by exactly two segments.

A **triangle** is a polygon with three sides. The sum of the measures of the three angles of a triangle is 180°. In the following figure, $m\angle 1 + m\angle 2 + m\angle 3 = 180°$.

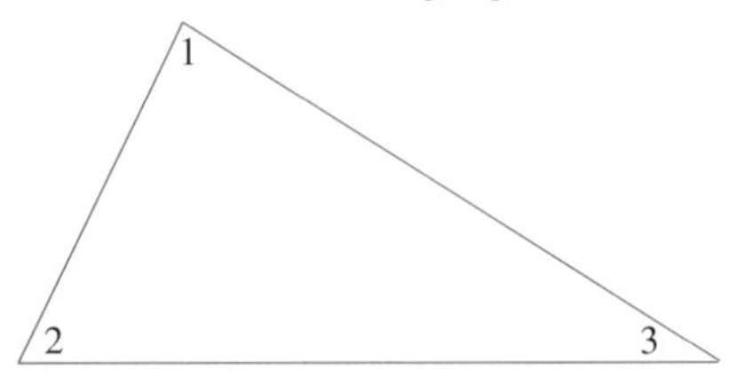

EXAMPLE 3 Find the measure of the third angle of the triangle shown.

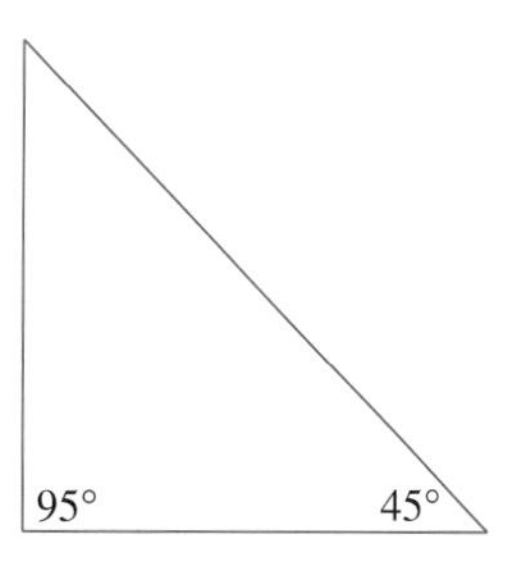

Solution: The sum of the measures of the angles of a triangle is 180°. Since one angle measures 45° and the other angle measures 95°, the third angle measures $180° - 45° - 95° = 40°$. ■

Two triangles are **congruent** if they have the same size and the same shape. In congruent triangles, the measures of corresponding angles are equal and the lengths of corresponding sides are equal. The following triangles are congruent.

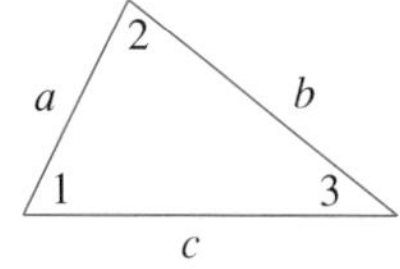

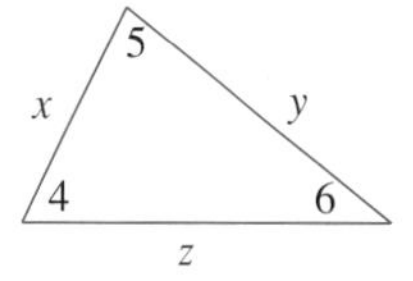

Corresponding angles are equal: $m\angle 1 = m\angle 4$, $m\angle 2 = m\angle 5$, and $m\angle 3 = m\angle 6$. Also, lengths of corresponding sides are equal: $a = x$, $b = y$, and $c = z$.

Any one of the following may be used to determine whether two triangles are congruent.

Congruent Triangles

1. If the measures of two angles of a triangle equal the measures of two angles of another triangle and the lengths of the sides between each pair of angles are equal, the triangles are congruent.

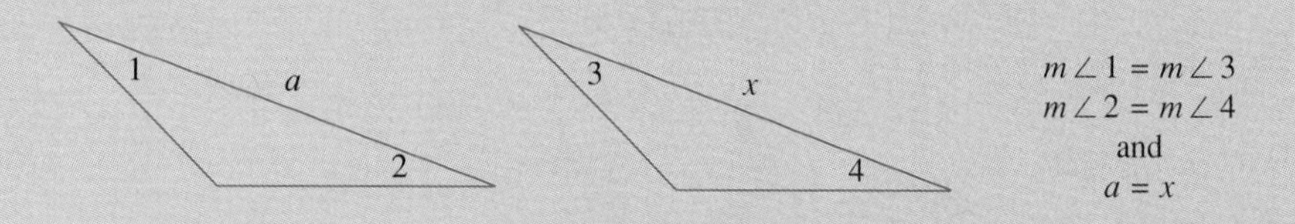

2. If the lengths of the three sides of a triangle equal the lengths of corresponding sides of another triangle, the triangles are congruent.

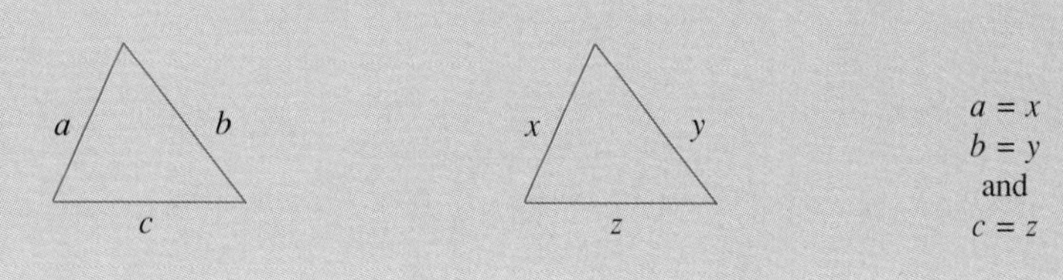

3. If the lengths of two sides of a triangle equal the lengths of corresponding sides of another triangle, and the measures of the angles between each pair of sides are equal, the triangles are congruent.

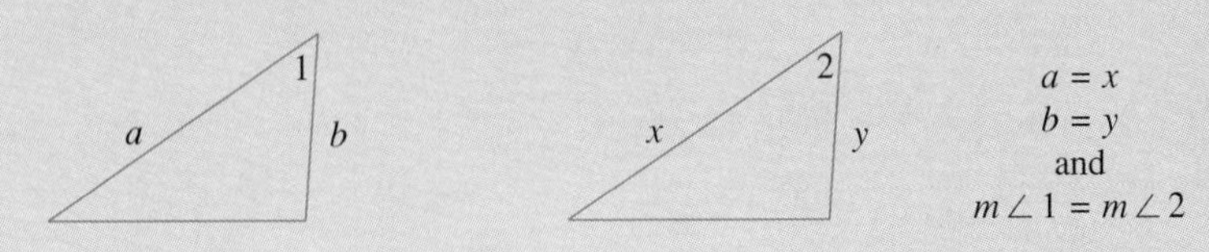

Two triangles are similar if they have the same shape. In similar triangles, the measures of corresponding angles are equal and corresponding sides are in proportion. The following triangles are similar. (All similar triangles drawn in this appendix will be oriented the same.)

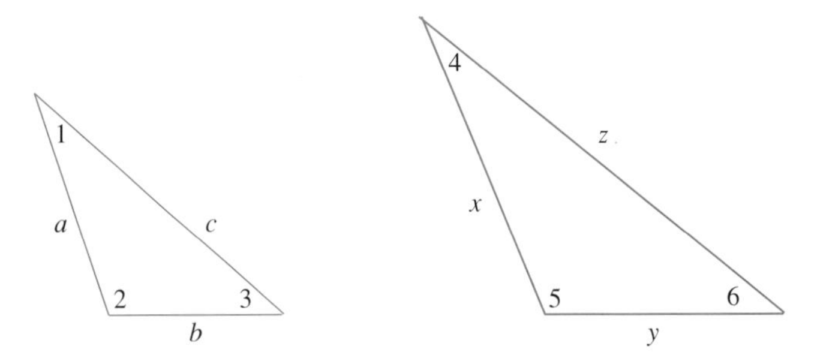

Corresponding angles are equal: $m\angle 1 = m\angle 4$, $m\angle 2 = m\angle 5$, and $m\angle 3 = m\angle 6$. Also, corresponding sides are proportional: $\frac{a}{x} = \frac{b}{y} = \frac{c}{z}$.

Any one of the following may be used to determine whether two triangles are similar.

Similar Triangles

1. If the measures of two angles of a triangle equal the measures of two angles of another triangle, the triangles are similar.

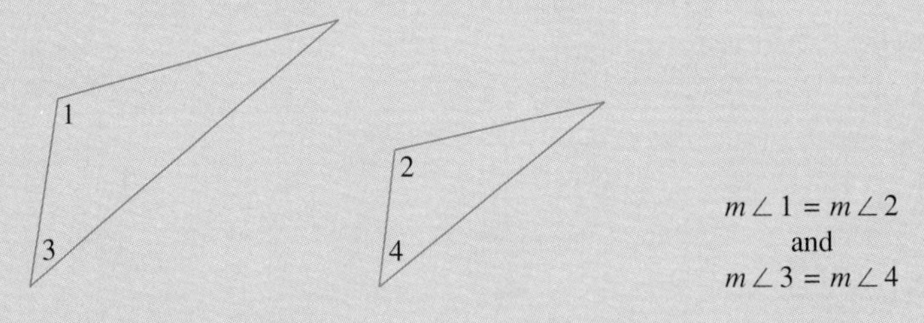

2. If three sides of one triangle are proportional to three sides of another triangle, the triangles are similar.

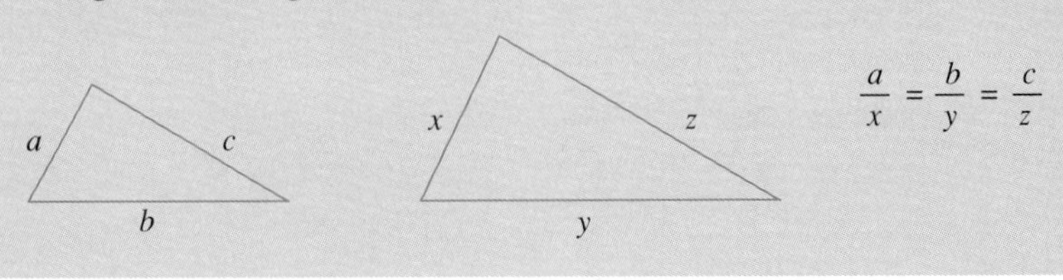

3. If two sides of a triangle are proportional to two sides of another triangle and the measures of the included angles are equal, the triangles are similar.

$m\angle 1 = m\angle 2$
and
$\frac{a}{x} = \frac{b}{y}$

EXAMPLE 4 Given that the following triangles are similar, find the missing length x.

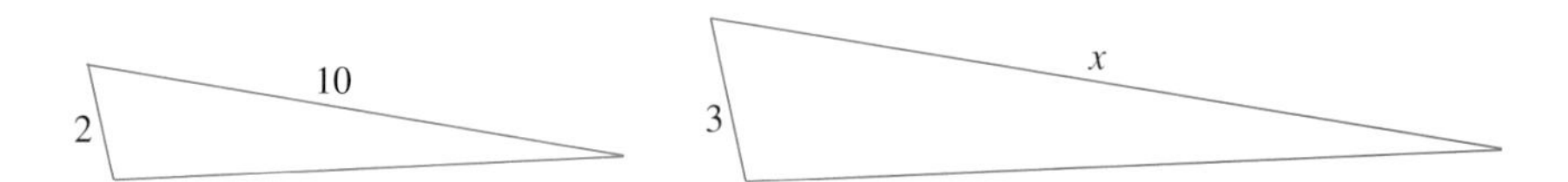

Solution: Since the triangles are similar, corresponding sides are in proportion. Thus, $\frac{2}{3} = \frac{10}{x}$. To solve this equation for x, we multiply both sides by the LCD $3x$.

$$3x\left(\frac{2}{3}\right) = 3x\left(\frac{10}{x}\right)$$

$$2x = 30$$

$$x = 15$$

The missing length is 15 units. ■

A **right triangle** contains a right angle. The side opposite the right angle is called the **hypotenuse,** and the other two sides are called the **legs.** The **Pythagorean theorem** gives a formula that relates the lengths of the three sides of a right triangle.

The Pythagorean Theorem

If a and b are the lengths of the legs of a right triangle, and c is the length of the hypotenuse, then $a^2 + b^2 = c^2$.

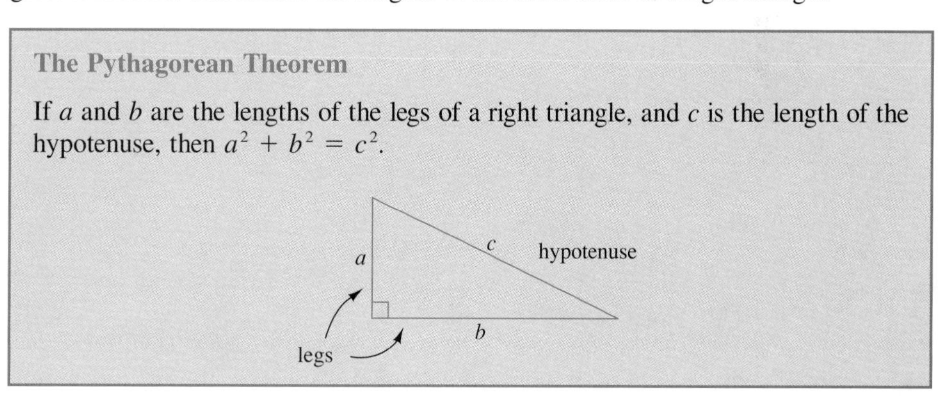

EXAMPLE 5 Find the length of the hypotenuse of a right triangle whose legs have lengths of 3 centimeters and 4 centimeters.

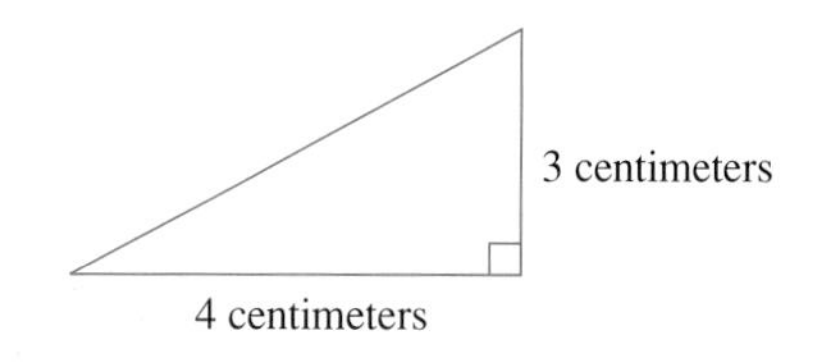

Solution: Because we have a right triangle, we use the Pythagorean theorem. The legs are 3 centimeters and 4 centimeters, so let $a = 3$ and $b = 4$ in the formula.

$$a^2 + b^2 = c^2$$
$$3^2 + 4^2 = c^2$$
$$9 + 16 = c^2$$
$$25 = c^2$$

Since c represents a length, we assume that c is positive. Thus, if c^2 is 25, c must be 5. The hypotenuse has a length of 5 centimeters. ■

APPENDIX B EXERCISE SET

Find the complement of each angle. See Example 1.

1. 19° **2.** 65° **3.** 70.8°

4. $45\frac{2}{3}°$ **5.** $11\frac{1}{4}°$ **6.** 19.6°

Find the supplement of each angle.

7. 150° **8.** 90° **9.** 30.2°

10. 81.9° **11.** $79\frac{1}{2}°$ **12.** $165\frac{8}{9}°$

13. If lines m and n are parallel, find the measures of angles 1 through 7. See Example 2.

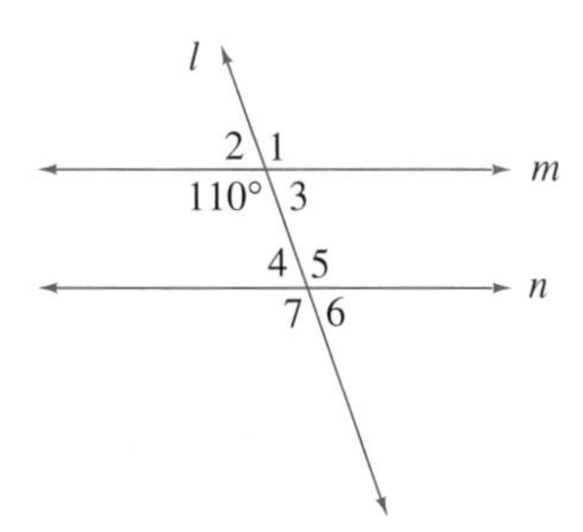

14. If lines m and n are parallel, find the measures of angles 1 through 5. See Example 2.

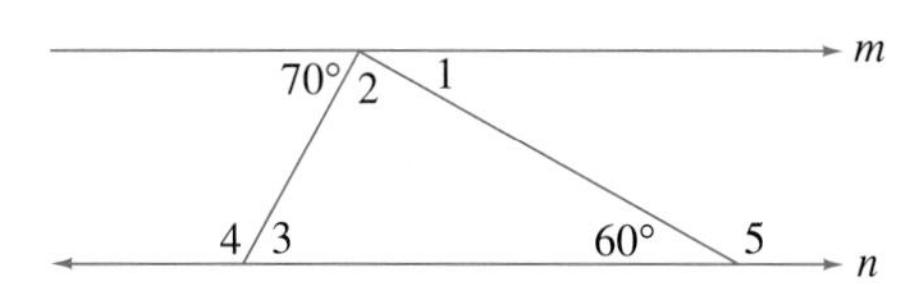

In each of the following, the measures of two angles of a triangle are given. Find the measure of the third angle. See Example 3.

15. 11°, 79° **16.** 8°, 102° **17.** 25°, 65°

18. 44°, 19° **19.** 30°, 60° **20.** 67°, 23°

In each of the following, the measure of one angle of a right triangle is given. Find the measures of the other two angles.

21. 45° **22.** 60° **23.** 17°

24. 30° **25.** $39\frac{3}{4}°$ **26.** 72.6°

Given that each of the following pairs of triangles is similar, find the missing lengths. See Example 4.

27.

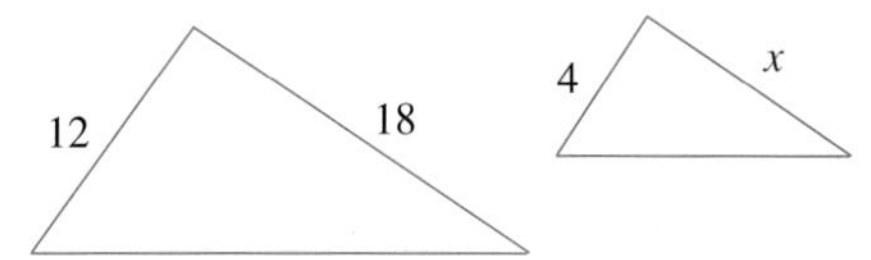

28.

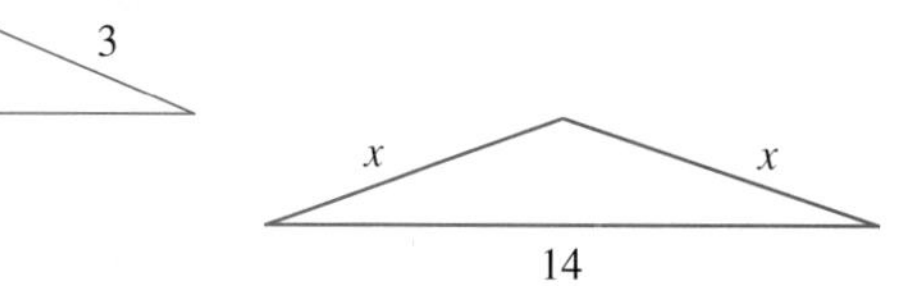

29.

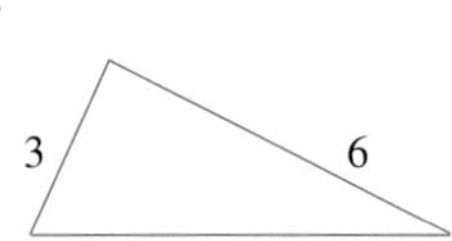

30.

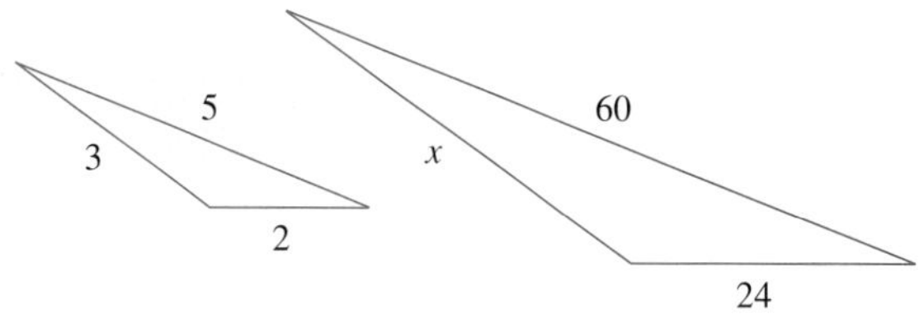

Use the Pythagorean theorem to find the missing lengths in the right triangles. See Example 5.

31.

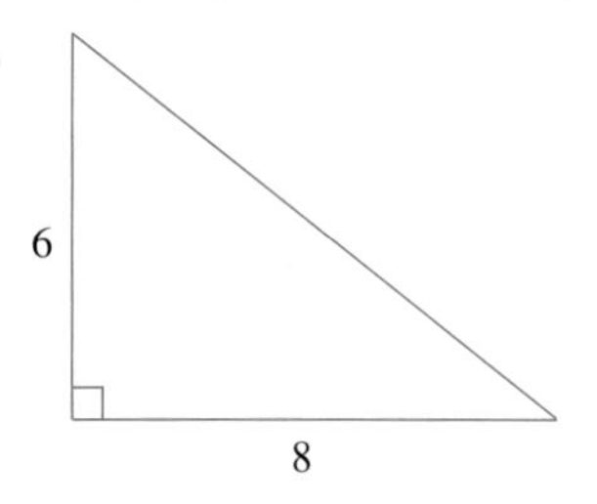

32.

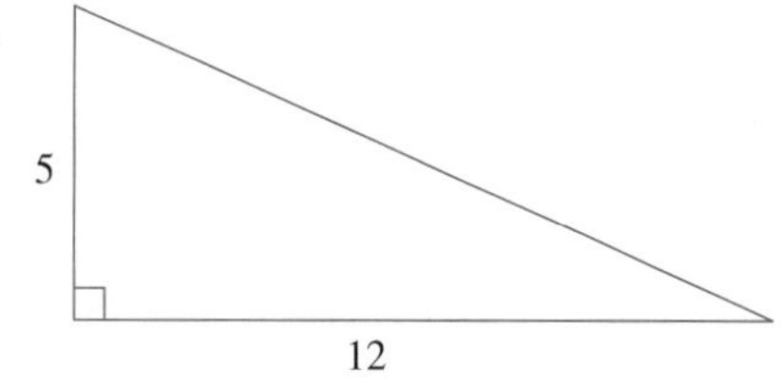

33.

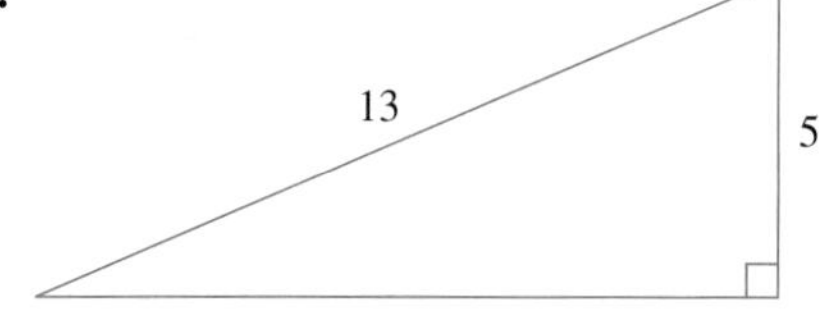

34.

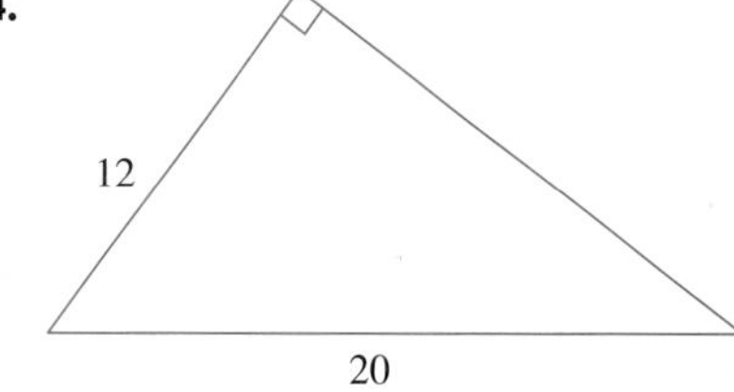

APPENDIX C

Review of Geometric Figures

Plane figures have length and width but no thickness or depth.		
Name	**Description**	**Figure**
Polygon	Union of three or more coplanar line segments that intersect each other only at each end point, with each end point shared by two segments.	
Triangle	Polygon with three sides (sum of measures of three angles is 180°).	
Scalene triangle	Triangle with no sides of equal length.	
Isosceles triangle	Triangle with two sides of equal length.	
Equilateral triangle	Triangle with all sides of equal length.	
Right triangle	Triangle that contains a right angle.	leg, hypotenuse, leg
Quadrilateral	Polygon with four sides (sum of measures of four angles is 360°).	

Name	Description	Figure
Plane figures have length and width but no thickness or depth.		
Trapezoid	Quadrilateral with exactly one pair of opposite sides parallel.	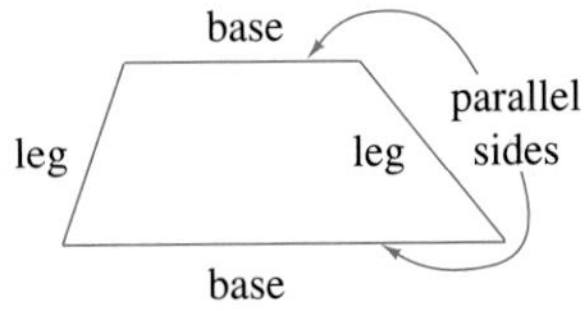 base; leg; leg; parallel sides; base
Isosceles trapezoid	Trapezoid with legs of equal length.	
Parallelogram	Quadrilateral with both pair of opposite sides parallel and equal in length.	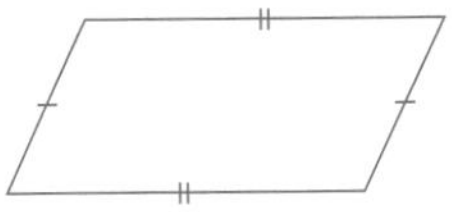
Rhombus	Parallelogram with all sides of equal length.	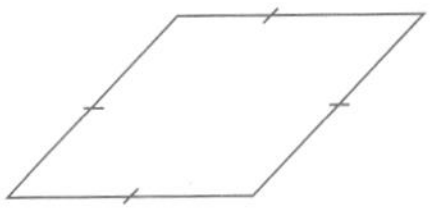
Rectangle	Parallelogram with four right angles.	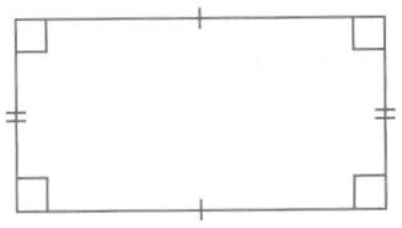
Square	Rectangle with all sides of equal length.	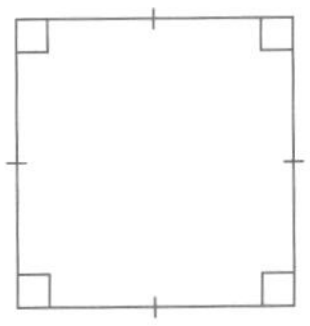
Circle	All points in a plane the same distance from a fixed point called the **center**.	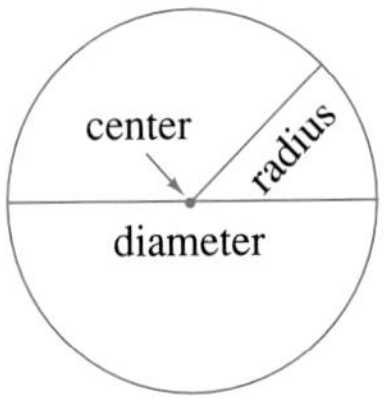 center; radius; diameter

Solid figures have length, width, and height or depth.

Name	Description	Figure
Rectangular solid	A solid with six sides, all of which are rectangles.	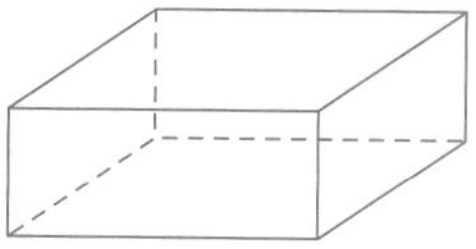
Cube	A rectangular solid whose six sides are squares.	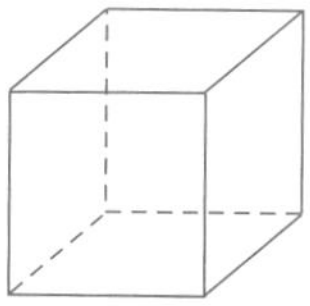
Sphere	All points the same distance from a fixed point, called the **center**.	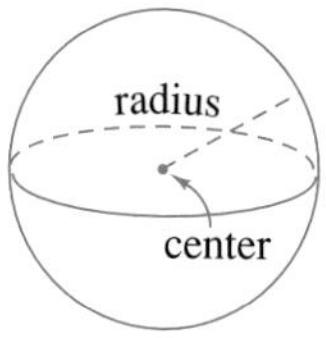 radius center
Right circular cylinder	A cylinder consisting of two circular bases that are perpendicular to its altitude.	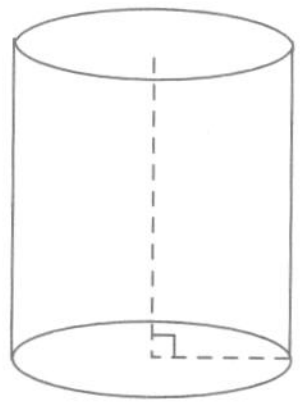
Right circular cone	A cone with a circular base that is perpendicular to its altitude.	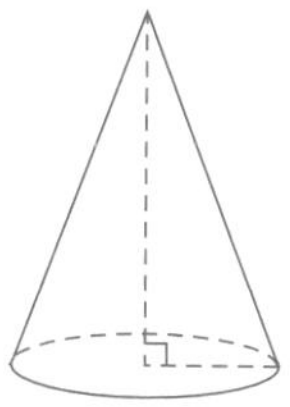

APPENDIX D

Table of Squares and Square Roots

n	n^2	$\sqrt{n}$	n	n^2	$\sqrt{n}$
1	1	1.000	51	2,601	7.141
2	4	1.414	52	2,704	7.211
3	9	1.732	53	2,809	7.280
4	16	2.000	54	2,916	7.348
5	25	2.236	55	3,025	7.416
6	36	2.449	56	3,136	7.483
7	49	2.646	57	3,249	7.550
8	64	2.828	58	3,364	7.616
9	81	3.000	59	3,481	7.681
10	100	3.162	60	3,600	7.746
11	121	3.317	61	3,721	7.810
12	144	3.464	62	3,844	7.874
13	169	3.606	63	3,969	7.937
14	196	3.742	64	4,096	8.000
15	225	3.873	65	4,225	8.062
16	256	4.000	66	4,356	8.124
17	289	4.123	67	4,489	8.185
18	324	4.243	68	4,624	8.246
19	361	4.359	69	4,761	8.307
20	400	4.472	70	4,900	8.367
21	441	4.583	71	5,041	8.426
22	484	4.690	72	5,184	8.485
23	529	4.796	73	5,329	8.544
24	576	4.899	74	5,476	8.602
25	625	5.000	75	5,625	8.660
26	676	5.099	76	5,776	8.718
27	729	5.196	77	5,929	8.775
28	784	5.292	78	6,084	8.832
29	841	5.385	79	6,241	8.888
30	900	5.477	80	6,400	8.944
31	961	5.568	81	6,561	9.000
32	1,024	5.657	82	6,724	9.055
33	1,089	5.745	83	6,889	9.110
34	1,156	5.831	84	7,056	9.165
35	1,225	5.916	85	7,225	9.220
36	1,296	6.000	86	7,396	9.274
37	1,369	6.083	87	7,569	9.327
38	1,444	6.164	88	7,744	9.381
39	1,521	6.245	89	7,921	9.434
40	1,600	6.325	90	8,100	9.487
41	1,681	6.403	91	8,281	9.539
42	1,764	6.481	92	8,464	9.592
43	1,849	6.557	93	8,649	9.644
44	1,936	6.633	94	8,836	9.695
45	2,025	6.785	95	9,025	9.747
46	2,116	6.782	96	9,216	9.798
47	2,209	6.856	97	9,409	9.849
48	2,304	6.928	98	9,604	9.899
49	2,401	7.000	99	9,801	9.950
50	2,500	7.071	100	10,000	10.000

APPENDIX E

Answers to Selected Exercises

CHAPTER 1
Real Numbers and Their Properties

Exercise Set 1.1

1. $<$ **3.** $>$ **5.** $=$ **7.** $<$ **9.** true **11.** false **13.** false **15.** true **17.** $8 < 12$ **19.** $5 \geq 4$ **21.** $2 + 3 < 6$ **23.** $\frac{10}{2} = 5$ **25.** $4 \leq (3)(5)$ **27.** $<$ **29.** $>$ **31.** $<$ **33.** $=$ **35.** $>$ **37.** true **39.** false **41.** true **43.** $25 \leq 25$ **45.** $6 > 0$ **47.** $b \geq a$ **49.** $y < x$ **51.** $4 > 2.5$ **53.** $8 \leq 12$ **55.** $5 \leq 6$ **57.** $5 + 6 > 10$ **59.** $(3)(5) > 12$ **61.** $\frac{12}{6} > 1$ **63.** $3 > 2$ **65.** $4 = 4 + 0$ **67.** $a \leq 5$ **69.** $c < d$ **71.** $150 > 57$ **73.** $32 < 212$ **75.** $30 \leq 45$ **77.** 90 **79.** Grades are increasing.

Exercise Set 1.2

1. whole, integers, rational, real **3.** integers, rational, real **5.** natural, whole, integers, rational, real **7.** rational, real **9.** $<$ **11.** $>$ **13.** $=$ **15.** $<$ **17.** $=$ **19.** $>$ **21.** $<$ **23.** integers, rational, real **25.** irrational, real **27.** natural, whole, integers, rational, real **29.** true **31.** false **33.** true **35.** true **37.** false **39.** true **41.** false **43.** $>$ **45.** $>$ **47.** $<$ **49.** $<$ **51.** $>$ **53.** $=$ **55.** false **57.** true **59.** false **61.** false **63.** $0 > -26.7$ **65.** Sun **67.** Sun **69.** Tuesday **71.** $7°$

Exercise Set 1.3

1. $\frac{3}{8}$ **3.** $\frac{5}{7}$ **5.** $2 \cdot 2 \cdot 5$ **7.** $3 \cdot 5 \cdot 5$ **9.** $3 \cdot 3 \cdot 5$ **11.** $\frac{1}{2}$ **13.** $\frac{2}{3}$ **15.** $\frac{3}{7}$ **17.** $\frac{3}{5}$ **19.** $\frac{3}{8}$ **21.** $\frac{1}{2}$ **23.** $\frac{6}{7}$ **25.** 15 **27.** $\frac{1}{6}$ **29.** $\frac{25}{27}$ **31.** $\frac{3}{5}$ **33.** 1 **35.** $\frac{1}{3}$ **37.** $\frac{9}{35}$ **39.** $\frac{21}{30}$ **41.** $\frac{4}{18}$ **43.** $\frac{16}{20}$ **45.** $\frac{23}{21}$ **47.** $\frac{2}{3}$ **49.** $\frac{5}{66}$ **51.** $\frac{7}{5}$ **53.** $\frac{5}{7}$ **55.** $\frac{65}{21}$ **57.** $\frac{2}{5}$ **59.** $\frac{9}{7}$ **61.** $\frac{3}{4}$ **63.** $\frac{17}{3}$ **65.** $\frac{7}{26}$ **67.** 1 **69.** $\frac{1}{5}$ **71.** $\frac{31}{6}$ **73.** $\frac{17}{18}$ **75.** $\frac{1}{5}$ **77.** $\frac{3}{8}$ **79.** $8\frac{1}{2}$ pounds **81.** 60% **83.** $4\frac{1}{4}$ in.

Exercise Set 1.4

1. 243 **3.** 27 **5.** 1 **7.** 5 **9.** $\frac{1}{125}$ **11.** $\frac{16}{81}$ **13.** 7 **15.** $\frac{1}{3}$ **17.** 4 **19.** 3 **21.** 17 **23.** 20 **25.** 10 **27.** 21 **29.** 50 **31.** $\frac{25}{24}$ **33.** $\frac{7}{18}$ **35.** $\frac{21}{8}$ **37.** $\frac{7}{5}$ **39.** 88 **41.** 11 **43.** 50 **45.** 42

47. 9 **49.** 21 **51.** 16 **53.** $\frac{3}{4}$ **55.** 30 **57.** 225 **59.** 12 **61.** 3 **63.** 9 **65.** $\frac{17}{24}$ **67.** $\frac{7}{6}$

69. 6 **71.** $\frac{7}{5}$ **73.** $\frac{27}{55}$ **75.** 2.86 **77.** $\frac{4}{3}$ **79.** 11 **81.** 330 **83.** 17 **85.** 3.752 **87.** 5

Exercise Set 1.5

1. 1 **3.** 11 **5.** 8 **7.** 10 **9.** 15 **11.** $\frac{1}{2}x$ **13.** $\frac{x}{9}$ **15.** $8x$ **17.** no **19.** yes **21.** yes

23. $2x = 17$ **25.** $x - 7 = 0$ **27.** $8 + 2x = 42$ **29.** 0 **31.** $\frac{9}{5}$ **33.** 5 **35.** 42 **37.** 15 **39.** 100

41. 120 **43.** 36 **45.** $\frac{17}{14}$ **47.** $\frac{19}{42}$ **49.** yes **51.** yes **53.** no **55.** $7(x + 19)$ **57.** $4x - x = 75.6$

59. $12 - x$ **61.** $\frac{3}{4}(x + 1) = 9$ **63.** $3x + 1\frac{11}{12} = x + 2$ **65.** $\frac{11}{14}$ yd **67.** 55 square in. **69.** 51 mph

Exercise Set 1.6

1. 9 **3.** −14 **5.** −15 **7.** −16 **9.** 11 **11.** $2\frac{5}{8}$ **13.** −6 **15.** 2 **17.** 0 **19.** −6 **21.** −2

23. 0 **25.** $-\frac{2}{3}$ **27.** −10 **29.** −5 **31.** 19 **33.** −15 **35.** −34 **37.** −5 **39.** 3 **41.** −45

43. −4 **45.** 9 **47.** −3 **49.** −16 **51.** 2 **53.** −10 **55.** $\frac{11}{6}$ **57.** −100° **59.** −9 **61.** −11

63. 11 **65.** 5 **67.** $-\frac{43}{30}$ **69.** −0.8 **71.** −13 **73.** −17 **75.** 14 **77.** 48 **79.** 15 **81.** $\frac{7}{12}$

83. −2.9 **85.** −19 **87.** −93 **89.** −48 **91.** −573 **93.** −1 **95.** −23 **97.** −9 **99.** −7 **101.** $\frac{7}{5}$

103. 11 **105.** Monday 7°; Tuesday 4°; Wednesday −9°; Thursday 3°; Friday −6°; Saturday 6° **107.** Wednesday

109. yes **111.** no **113.** no **115.** −308 ft **117.** 146 ft **119.** $-4\frac{1}{8}$

Exercise Set 1.7

1. −12 **3.** 42 **5.** 0 **7.** −18 **9.** $-\frac{2}{3}$ **11.** 2 **13.** −30 **15.** 90 **17.** 16 **19.** −9 **21.** 3

23. 5 **25.** 0 **27.** undefined **29.** 16 **31.** −3 **33.** $-\frac{16}{7}$ **35.** −21 **37.** 41 **39.** −134 **41.** 3

43. −1 **45.** 12 **47.** −14 **49.** −6 **51.** undefined **53.** 5 **55.** 0 **57.** −36 **59.** −125

61. 16 **63.** −16 **65.** 2 **67.** $\frac{6}{5}$ **69.** −5 **71.** 18 **73.** −30 **75.** −24 **77.** $-\frac{1}{2}$ **79.** $-4\frac{1}{5}$

81. −8.372 **83.** 14 **85.** 22 **87.** 2 **89.** 3 **91.** −3 **93.** $\frac{2}{3}$ **95.** $-\frac{15}{3} + (-2), -7$

97. $2[-5 + (-3)], -16$ **99.** yes **101.** yes **103.** yes **105.** Wednesday, October 6

Exercise Set 1.8

1. commutative property of multiplication **3.** associative property of addition **5.** distributive property **7.** $18 + 3x$

9. $-2y + 2z$ **11.** $-21y + 35$ **13.** −16 **15.** 8 **17.** −9 **19.** $\frac{3}{2}$ **21.** $-\frac{6}{5}$ **23.** $\frac{1}{6}$ **25.** $-\frac{1}{2}$

27. $\frac{2}{3} \cdot \frac{3}{2} = 1$ **29.** $(-3)(-4)$ **31.** $(3 + 8) + 9$ **33.** associative property of multiplication

35. identity property of addition **37.** distributive property **39.** associative property of multiplication

41. $5x + 20m + 10$ **43.** $-4 + 8m - 4n$ **45.** $-5x - 2$ **47.** $-r + 3 + 7p$ **49.** 0 **51.** -2 **53.** -8
55. -3 **57.** 2 **59.** 5 **61.** 3 **63.** -1 **65.** Undefined, there is none. **67.** $-\frac{5}{3}$ **69.** $\frac{6}{23}$ **71.** $y + 0 = y$
73. $xa + xb$ **75.** $(b + c)a$ **77.** yes **79.** no

Chapter 1 Review

1. $<$ **3.** $=$ **5.** $4 \geq 3$ **7.** $8 + 4 \leq 12$ **9.** $7 = 3 + 4$ **11.** $50 > 40$
13. a. $\{2, 5\}$ **b.** $\{2, 5\}$ **c.** $\{-3, 2, 5\}$ **d.** $\left\{-3, -1.6, 2, 5, \frac{11}{2}, 15.1\right\}$ **e.** $\{\sqrt{5}, 2\pi\}$ **f.** $\left\{-3, -1.6, 2, 5, \frac{11}{2}, 15.1, \sqrt{5}, 2\pi\right\}$
15. $>$ **17.** $<$ **19.** Wednesday **21.** $2 \cdot 2 \cdot 2 \cdot 3 \cdot 5$ **23.** $\frac{4}{3}$ **25.** $\frac{3}{5}$ **27.** $4\frac{1}{2}$ **29.** 70 **31.** 37 **33.** $\frac{5}{2}$
35. 18 **37.** 5 **39.** 63° **41.** 3 **43.** -17 **45.** -5 **47.** 3.9 **49.** -11 **51.** -11 **53.** 4 **55.** 1
57. -48 **59.** 3 **61.** -36 **63.** undefined **65.** 6 **67.** $-\frac{1}{6}$ **69.** commutative property of addition
71. distributive property **73.** associative property of addition **75.** distributive property **77.** multiplicative inverse
79. commutative property of addition

Chapter 1 Test

1. $|-7| > 5$ **2.** $(9 + 5) \geq 4$ **3.** -5 **4.** -11 **5.** -14 **6.** -39 **7.** 12 **8.** -2
9. undefined **10.** 4 **11.** $-\frac{1}{3}$ **12.** $4\frac{5}{8}$ **13.** $\frac{51}{40}$ **14.** -32 **15.** -48 **16.** 3 **17.** 0 **18.** $>$ **19.** $>$
20. $>$ **21.** $=$ **22. a.** $\{1, 7\}$ **b.** $\{0, 1, 7\}$ **c.** $\{-5, -1, 0, 1, 7\}$ **d.** $\left\{-5, -1, 0, \frac{1}{4}, 1, 7, 11.6\right\}$ **e.** $\{\sqrt{7}, 3\pi\}$
f. $\left\{-5, -1, \frac{1}{4}, 0, 1, 7, 11.6, \sqrt{7}, 3\pi\right\}$ **23.** 40 **24.** 12 **25.** 22 **26.** -1
27. associative property of addition **28.** commutative property of multiplication **29.** distributive property
30. multiplicative inverse **31.** 9 **32.** -3 **33.** second down **34.** yes **35.** 17° **36.** $-\$420$

CHAPTER 2
Equations and Problem Solving

Mental Math, Sec. 2.1

1. -7 **3.** 1 **5.** 17 **7.** like **9.** unlike **11.** like

Exercise Set 2.1

1. $15y$ **3.** $13w$ **5.** $-7b - 9$ **7.** $-m - 6$ **9.** $5y - 20$ **11.** $7d - 11$ **13.** $-3x + 2y - 1$
15. $2x + 14$ **17.** $10x - 3$ **19.** $-4x - 9$ **21.** $2x - 4$ **23.** $\frac{3}{4}x + 12$ **25.** $12x - 2$ **27.** $5x^2$
29. $4x - 3$ **31.** $8x - 53$ **33.** -8 **35.** $7.2x - 5.2$ **37.** $k - 6$ **39.** $0.9m + 1$ **41.** $-12y + 16$
43. $x + 5$ **45.** -11 **47.** $1.3x + 3.5$ **49.** $x + 2$ **51.** $-15x + 18$ **53.** $2k + 10$ **55.** $-3x + 5$
57. $-4m - 3$ **59.** $8x + 48$ **61.** $x - 10$ **63.** $(18x - 2)$ ft **65.** $\frac{7x}{6}$ **67.** 2 **69.** -25 **71.** -32
73. 5 hours **75.** 2 hours **77.** between hours 6 and 7 **79.** $5m^4p^2 - 5m^2p^4$ **81.** $14y^2 - 14xy^2$
83. $-11c^3d + 3c$

Mental Math, Sec. 2.2

1. $\{2\}$ **3.** $\{12\}$ **5.** $\{17\}$

Exercise Set 2.2

1. $\{-13\}$ **3.** $\{-14\}$ **5.** $\{-8\}$ **7.** $\{11\}$ **9.** $\{0\}$ **11.** $\{-3\}$ **13.** $\{16\}$ **15.** $\{-4\}$ **17.** $\{0\}$ **19.** $\{12\}$ **21.** $\{10\}$ **23.** $\{-12\}$ **25.** $\{3\}$ **27.** $\{-2\}$ **29.** $\{0\}$ **31.** $\{11\}$ **33.** $\{-30\}$ **35.** $\{-7\}$ **37.** $2x + 7 = x + 6; x = -1$ **39.** $\frac{1}{2}x = -5; x = -10$ **41.** $\{-2\}$ **43.** $\{-10\}$ **45.** $\{8.9\}$ **47.** $\{-6\}$ **49.** $\{4\}$ **51.** $\left\{\frac{3}{4}\right\}$ **53.** $\{-0.7\}$ **55.** $\{-1\}$ **57.** $\{12\}$ **59.** $\{2\}$ **61.** $\{0.2\}$ **63.** $\{-12\}$ **65.** $\{-1\}$ **67.** $\{0\}$ **69.** $\{0\}$ **71.** $\{-5\}$ **73.** $\{0\}$ **75.** 6 meters, 13 meters, 16 meters **77.** $\frac{21}{29}$ **79.** $\frac{27}{40}$ **81.** $\frac{4}{5}$ **83.** $>$ **85.** $=$ **87.** $=$ **89.** $<$

Exercise Set 2.3

1. $\{2\}$ **3.** $\{-5\}$ **5.** $\{10\}$ **7.** $\{2\}$ **9.** $\{-9\}$ **11.** $\left\{-\frac{10}{7}\right\}$ **13.** $\left\{\frac{1}{6}\right\}$ **15.** $\{1\}$ **17.** $\{x \mid x$ is a real number$\}$ **19.** $\{\ \}$ **21.** $\{\ \}$ **23.** $2(x + 6) = 3(x + 4); x = 0$ **25.** $(180 - x)°$ **27.** $12 - z$ **29.** $\{4\}$ **31.** $\{4\}$ **33.** $\{-4\}$ **35.** $\{3\}$ **37.** $\{-2\}$ **39.** $\{4\}$ **41.** $\left\{\frac{7}{3}\right\}$ **43.** $\{\ \}$ **45.** $\left\{\frac{9}{5}\right\}$ **47.** $\left\{\frac{4}{19}\right\}$ **49.** $\{1\}$ **51.** $\{\ \}$ **53.** $\left\{\frac{7}{2}\right\}$ **55.** $\{-17\}$ **57.** $\left\{\frac{19}{6}\right\}$ **59.** $\{3\}$ **61.** $\{13\}$ **63.** $\{-8\}$ **65.** $\left\{-\frac{3}{47}\right\}$ **67.** $3x - 1 = 2x; \{1\}$ **69.** $30 - x$ **71.** $8 - x$ **73.** $x + 55 = 2x - 90; \{145\}$; yes **75.** midway **77.** -1 **79.** $\frac{1}{5}$ **81.** $(6x - 8)$ meters

Exercise Set 2.4

1. $h = 3$ **3.** $I = 800$ **5.** $r = 15$ **7.** 500 seconds **9.** $d = 2585$; 3000 miles **11.** 25,120 miles **13.** $-109.3°$F **15.** $-10°$C **17.** \$500,000 **19.** $h = \frac{f}{5g}$ **21.** $W = \frac{V}{LH}$ **23.** $y = 7 - 3x$ **25.** $R = \frac{A - p}{PT}$ **27.** $A = \frac{3V}{h}$ **29.** $2x + 2$ **31.** $360 - 9x$ **33.** $b = 5$ **35.** $h = 4$ **37.** $h = 15$ **39.** $r = 2000$ mph **41.** $w = 30$ ft **43.** $P = \$500{,}000$ **45.** $V = 800$ cubic ft **47.** 96 piranhas **49.** $t = 2.25$ hours **51.** $-40°$F **53.** 10.8 **55.** $V = 33{,}493{,}333{,}333$ cubic miles **57.** $V = 0.325$ cubic ft **59.** $C = 25.6°$ **61.** $t = 6.25$ hours **63.** $t = \frac{D}{r}$ **65.** $R = \frac{I}{PT}$ **67.** $y = \frac{9x - 16}{4}$ **69.** $a = P - b - c$ **71.** $B = \frac{T - 2C}{AC}$ **73.** $r = \frac{C}{2\pi}$ **75.** $w = \frac{uv}{v - u}$ **77.** $n + 284$ **79.** $x + 2$ **81.** $x + 2$ **83.** $3x + 6$ **85.** $805x - 25$ **87.** $\frac{9}{x + 5}$ **89.** $3(x + 4)$ **91.** $2(10 + 4x)$ **93.** $3(x - 12)$

Exercise Set 2.5

1. length $= 78$ ft; width $= 52$ ft **3.** shortest side $= 6$ ft **5.** U.S. team, 37 gold medals; Unified team, 45 gold medals **7.** 12 hours **9.** 65°; 115° **11.** 13 in. **13.** 98 points; 99 points **15.** no **17.** 15 ft by 24 ft **19.** 70°; 20° **21.** height $= 34$ in.; diameter $= 49$ in. **23.** 18, 20, and 22 **25.** 18 ft, 36 ft, 48 ft **27.** 27, 28, and 29 **29.** 38°, 38°, and 104° **31.** 5 ft; 16 ft **33.** son \$5000; husband \$10,000 **35.** -4 **37.** -6 **39.** -4 **41.** $\frac{1}{2}(x - 1) = 37$ **43.** $\frac{3(x + 2)}{5} = 0$

Exercise Set 2.6

1. 5.1¢ per Kilowatt hour **3.** Alaska Village Electric Cooperative, Southern California Edison **5.** \$110 million **7.** *Snow White and the Seven Dwarfs* **9.** 58 beats per minute **11.** 27 beats per minute

13. quadrant I

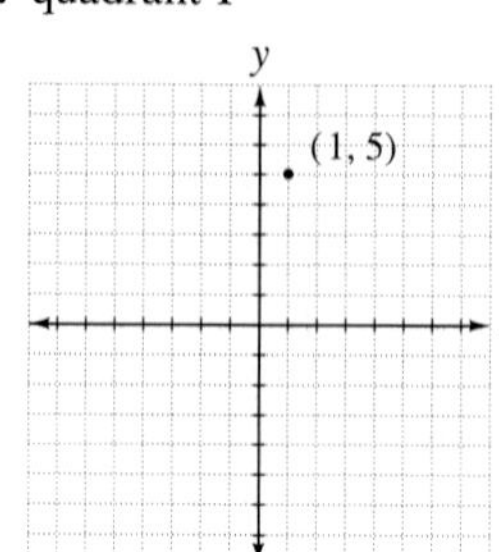

15. 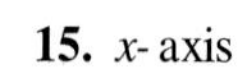 x-axis

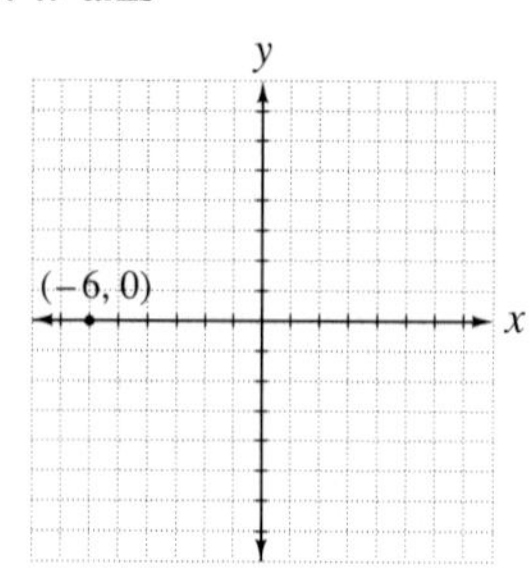

17. quadrant IV

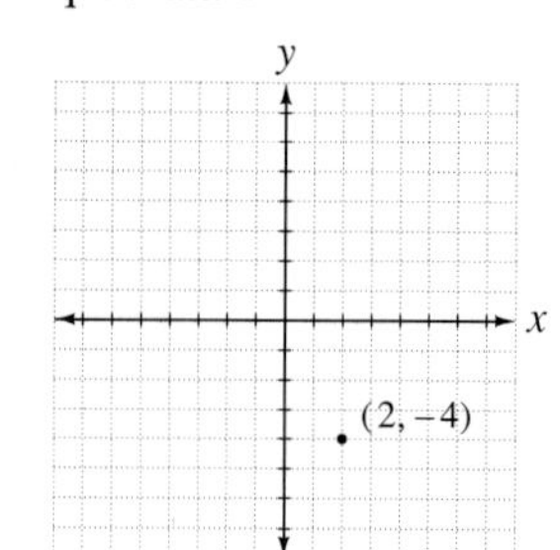

19. x-axis

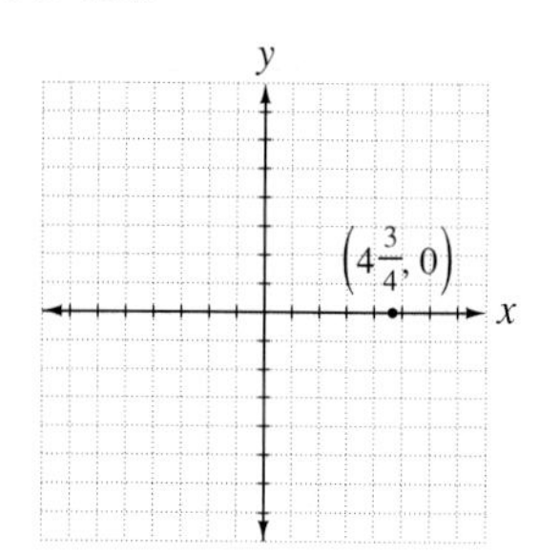

21. Cleveland Indians **23.** Both teams have gone 11 years without being in a playoff. **25.** 3 years

27. $A(0, 0)$; $B\left(3\frac{1}{2}, 0\right)$; $C(3, 2)$; $D(-1, 3)$; $E(-2, -2)$; $F(0, -1)$; $G(2, -1)$ **29.** 1975 **31.** 37 pounds

33. 51 pounds **35.** From 1975 to 1980 **37.** less beef consumed; more chicken consumed

39. The point $(2, -1)$ lies in the fourth quadrant.

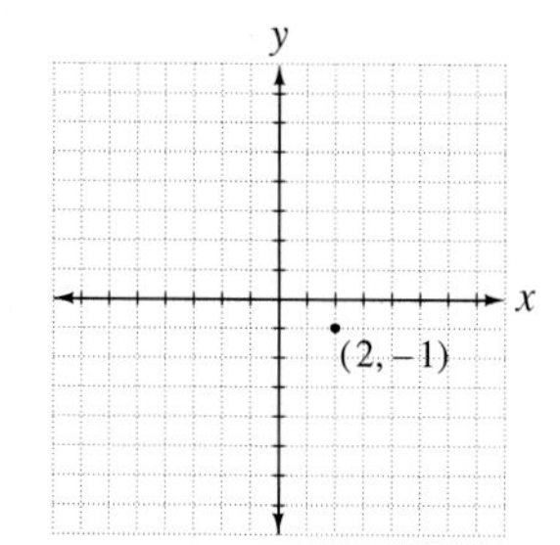

41. The point $(-3, -1)$ lies in the third quadrant.

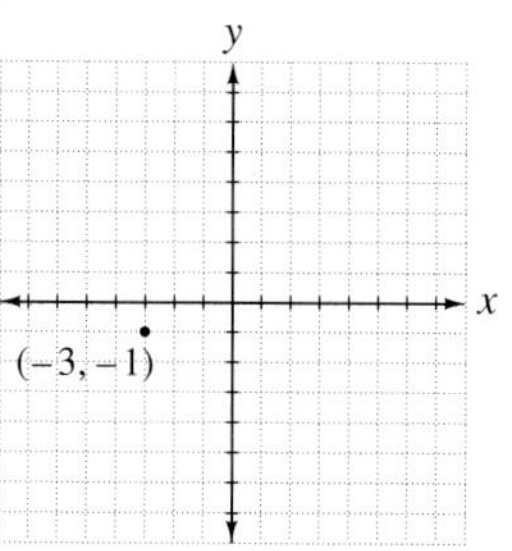

43. The point $(2, 3)$ lies in the first quadrant.

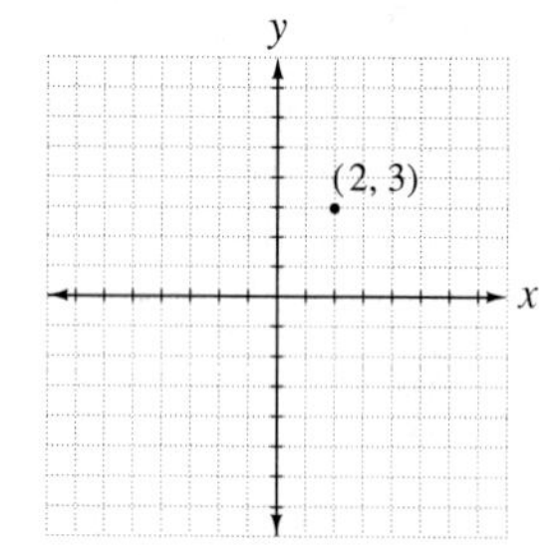

45. The point $(-2, 4)$ lies in the second quadrant.

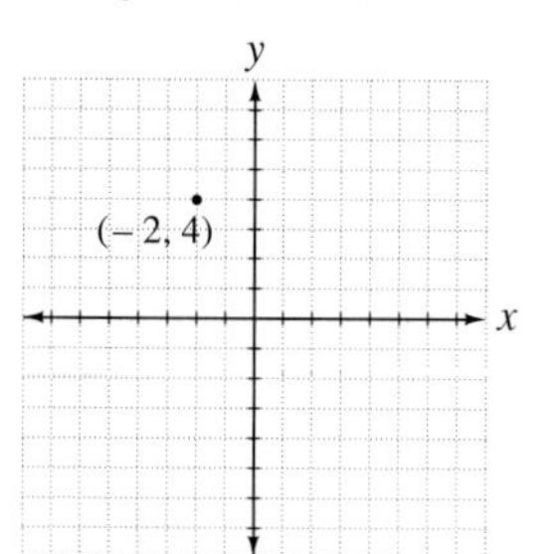

47. The point $(5, 0)$ lies on the x-axis.

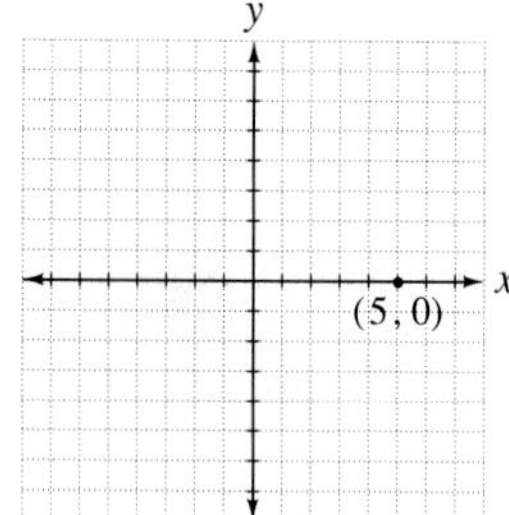

49. quadrant IV

51. quadrants I and IV **53.** $\{-5\}$ **55.** $\left\{-\frac{1}{10}\right\}$ **57.** $\frac{1}{5}$ **59.** $-\frac{3}{10}$

Chapter 2 Review

1. $6x$ **3.** $4x - 2$ **5.** $3n - 18$ **7.** $-6x + 7$ **9.** $3x - 7$ **11.** $10 - x$ **13.** $\{4\}$ **15.** $\{-6\}$ **17.** $\{-9\}$

19. $\{-12\}$ **21.** $\left\{-\frac{19}{2}\right\}$ **23.** $\{14\}$ **25.** $2x - 5 = 3x - 6$; $x = 1$ **27.** 3 **29.** -4 **31.** $\{-4\}$ **33.** $\{-3\}$

35. $\{\ \}$

37. $\left\{-\frac{8}{9}\right\}$ **39.** $\{-1\}$ **41.** $\{0\}$ **43.** $\left\{-\frac{6}{23}\right\}$ **45.** $\left\{-\frac{2}{5}\right\}$ **47.** $\{0.7\}$ **49.** $\frac{5}{7}$ **51.** 5, 7, and 9
53. $s = \frac{r+5}{vt}$ **55.** $y = \frac{2+3x}{6}$ **57.** $\pi = \frac{C}{2r}$ **59.** $32\frac{2}{9}°C$ **61.** 30 meters **63.** −40, −38, and −36
65. $A(5, 2)$ **67.** $C(3, -1)$ **69.** $E(-5, -2)$ **71.** $G(-1, 0)$
73. quadrant II

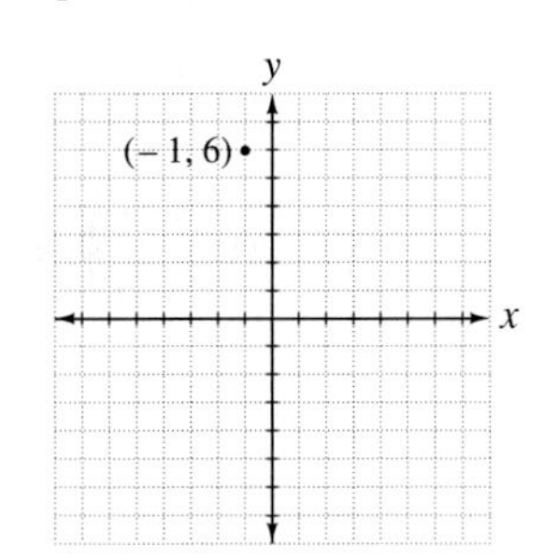

75.

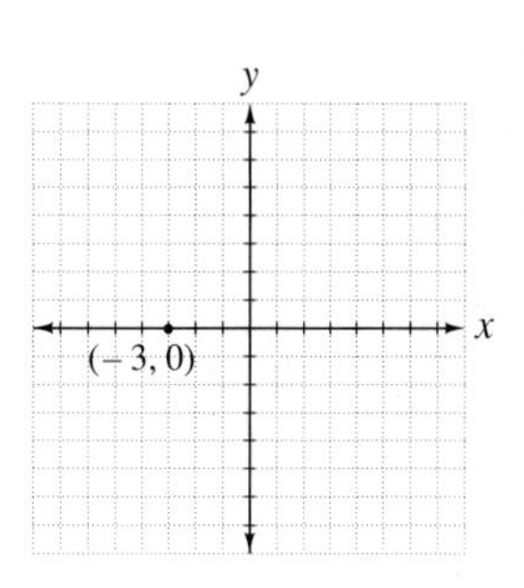

77.

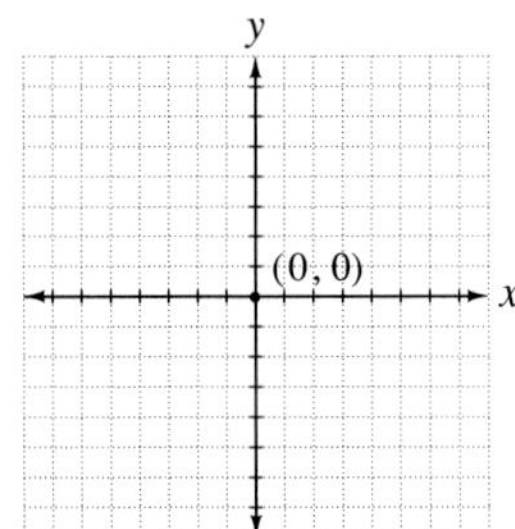

79. quadrant I

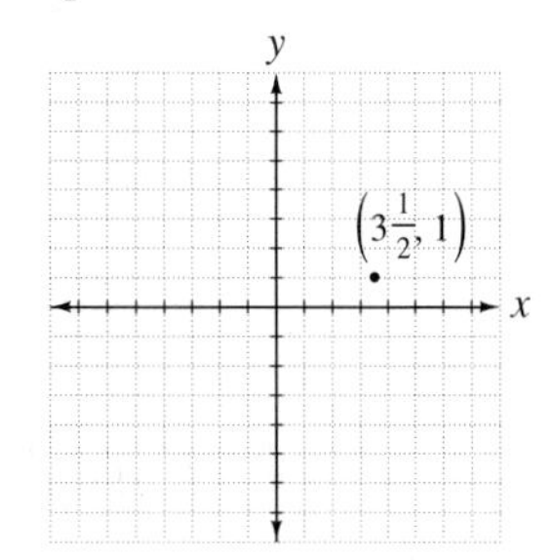

81. 20 mpg **83.** 3.5 mpg **85.** 1986 to 1987

Chapter 2 Test

1. $y - 10$ **2.** $2x - 9$ **3.** $-2x + 10$ **4.** $-15y + 1$ **5.** $\{-5\}$ **6.** $\{8\}$ **7.** $\left\{\frac{7}{10}\right\}$ **8.** $\{0\}$ **9.** $\{27\}$
10. $\left\{-\frac{19}{6}\right\}$ **11.** $\{3\}$ **12.** $\{\ \}$ **13.** $\left\{\frac{3}{11}\right\}$ **14.** $\left\{\frac{2}{17}\right\}$ **15.** $\{1\}$ **16.** $\left\{\frac{25}{7}\right\}$ **17.** 21 **18.** 7 gallons
19. −11 and −9 **20.** Winter 78 in., Sotomayor 92 in. **21.** $h = \frac{V}{\pi r^2}$ **22.** $t = \frac{W}{6b}$ **23.** $h = \frac{5g - p}{2}$
24. $y = \frac{3x - 10}{4}$

25. quadrant III

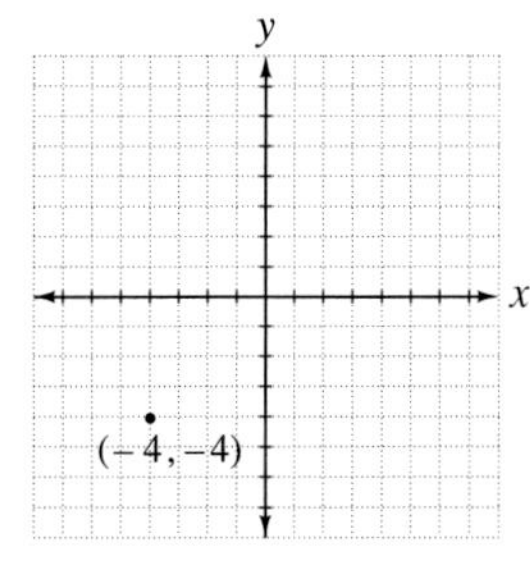

26. quadrant IV

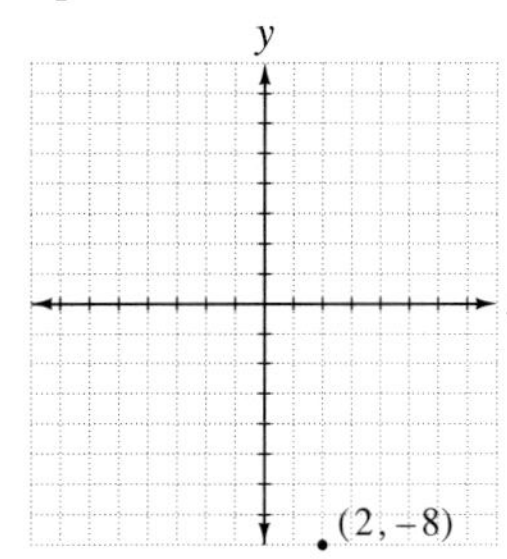

27. y-axis

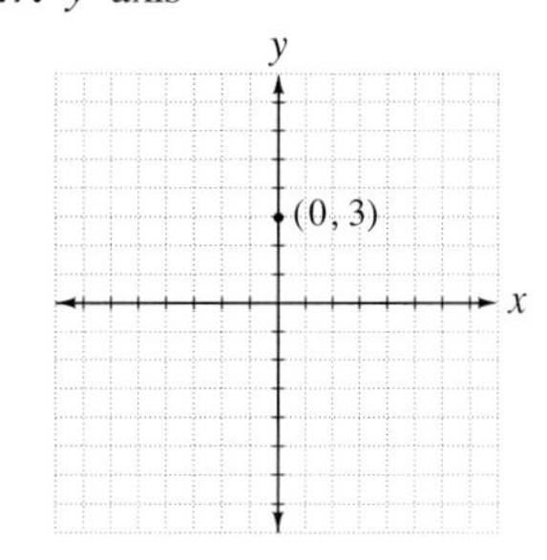

28. x-axis

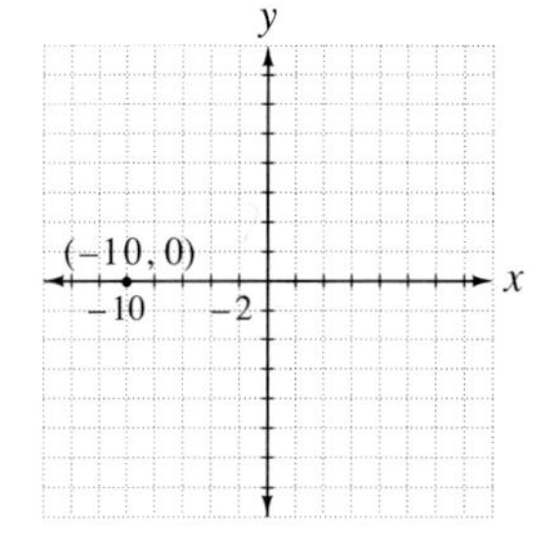

29. \$8 billion **30.** \$3 billion **31.** \$2.5 billion **32.** 1993 to 1994

Chapter 2 Cumulative Review

1. a. $9 \leq 11$ **b.** $8 > 1$ **c.** $3 \neq 4$; *Sec. 1.1, Ex. 3* **2. a.** $(2)(3) = 6$ **b.** $8 - 4 \leq 4$ **c.** $\frac{10}{2} \neq 3$; *Sec. 1.1, Ex. 4*

3. a. $-1 < 0$ **b.** $7 = \frac{14}{2}$ **c.** $-5 > -6$; *Sec. 1.2, Ex. 2* **4. a.** 4 **b.** 5 **c.** 0; *Sec. 1.2, Ex. 3*

5. a. $\frac{6}{7}$ **b.** $\frac{11}{27}$ **c.** $\frac{22}{5}$; *Sec. 1.3, Ex. 2* **6. a.** $\frac{6}{7}$ **b.** $\frac{1}{2}$ **c.** 1 **d.** $\frac{4}{3}$; *Sec. 1.3, Ex. 5*

7. a. 3 **b.** 5 **c.** $\frac{1}{2}$; *Sec 1.4, Ex. 2* **8. a.** 27 **b.** 2 **c.** 29 **d.** 48 **e.** $\frac{1}{4}$; *Sec. 1.4, Ex. 4*

9. a. 4 **b.** $\frac{9}{4}$ **c.** $\frac{5}{2}$ **d.** 5; *Sec. 1.5, Ex. 1* **10. a.** -12 **b.** -3; *Sec. 1.6, Ex. 8*

11. a. $4x$ **b.** $11y^2$ **c.** $8x^2 - x$; *Sec. 2.1, Ex. 3* **12.** $-2x - 1$; *Sec. 2.1, Ex. 7* **13.** $\{-3\}$; *Sec. 2.2, Ex. 2*

14. $\{-11\}$; *Sec. 2.2, Ex. 5* **15.** $\{1\}$; *Sec. 2.3, Ex. 2* **16.** $\left\{\frac{21}{11}\right\}$; *Sec. 2.3, Ex. 4* **17. a.** 5 **b.** $8-x$; *Sec. 2.3, Ex. 8*

18. 5; *Sec. 2.4, Ex. 2* **19.** $W = \frac{P - 2l}{2}$; *Sec. 2.4, Ex. 7* **20.** 44 Republican, 56 Democrat; *Sec. 2.5, Ex.2*

21. a. 80 beats per minute **b.** 5 minutes before lighting a cigarette **c.** between 0 and 5 minutes; *Sec. 2.6, Ex.3*

22. quadrant IV

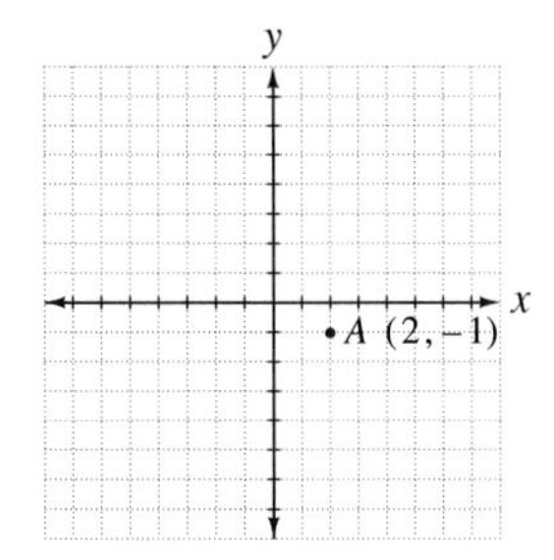

y- axis

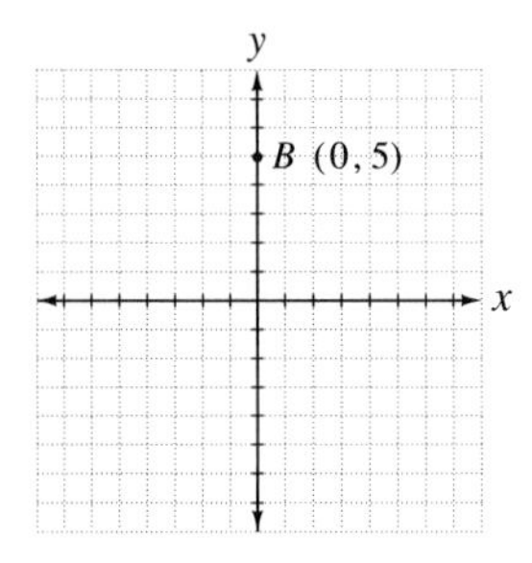

quadrant III

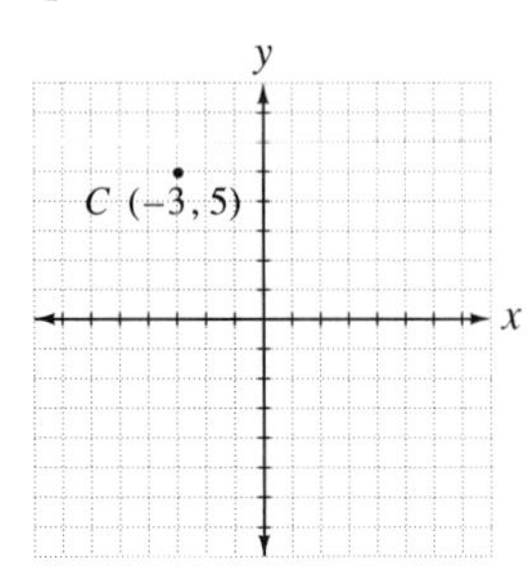

x- axis

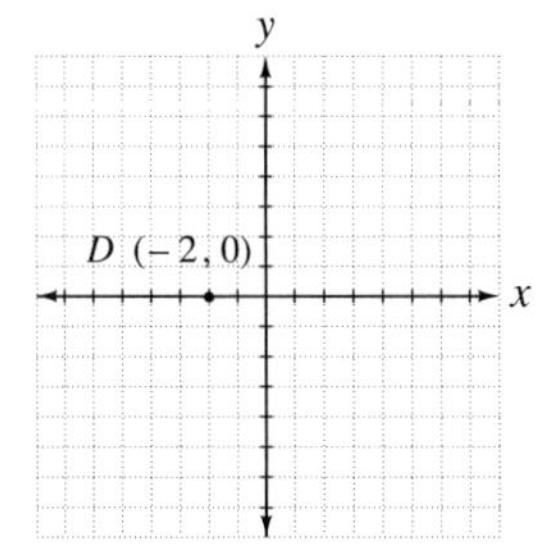

quadrant III

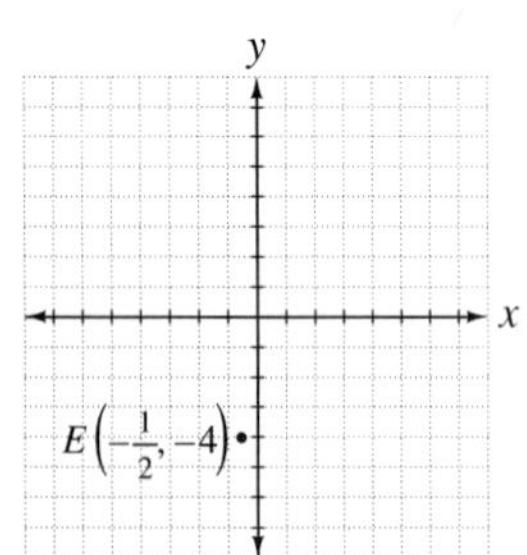

Sec 2.6, Ex. 4

CHAPTER 3
Inequalities, Absolute Value, and Problem Solving

Exercise Set 3.1

1. $\frac{2}{15}$ **3.** $\frac{\$15}{1}$ **5.** $\frac{50}{1}$ **7.** $\{4\}$ **9.** $\left\{\frac{50}{9}\right\}$ **11.** $\left\{\frac{14}{9}\right\}$ **13.** $\{5\}$ **15.** $\left\{-\frac{2}{3}\right\}$ **17.** 30 ft **19.** $\frac{15}{2}$ cm **21.** 123 pounds **23.** approximately 3441 pounds **25.** $\left\{\frac{21}{4}\right\}$ **27.** $\{30\}$ **29.** $\{7\}$ **31.** $\left\{-\frac{1}{3}\right\}$ **33.** $\{-3\}$ **35.** 165 calories **37.** 20 tablespoons **39.** 31 mpg **41.** 3833 women **43.** 26 students **45.** 245 miles **47.** 9 gallons **49.** \$19.13 per hour **51.** \$1348.58 per week **53.** $8\frac{1}{4}$ km **55.** 50 ft **57.** 3 hours 45 minutes **59.** 1471 megawatts **61.** approximately 4400 megawatts **63.** $\frac{x}{\text{population of town}} = \frac{1000}{560{,}000}$ **65.** 0 **67.** -3 **69.** 18 **71.** -32 **73.** 16

Exercise Set 3.2

1. 1.20 **3.** 0.225 **5.** 0.0012 **7.** 75% **9.** 200% **11.** 12.5% **13.** 38% **15.** 54% **17.** 136.8° **19.** answers vary **21.** 11.2 **23.** 55% **25.** 180 **27.** \$39 decrease, \$117 sale price **29.** 15.2% increase **31.** 4.6 **33.** 50 **35.** 30% **37.** 647.5 ft **39.** 55.40% **41.** 54 people **43.** No, many people use several medications. **45.** 75% increase **47.** 9.6% **49.** 26.9%; yes **51.** 17.1% **53.** \$280 **55.** 24%; no **57.** 166,567% **59.** 88% **61.** 18 **63.** More supermarkets were offering services in 1992 than in 1990. **65.** -14 **67.** -54 **69.** 155 **71.** \$1000 **73.** 10.7 hours

Exercise Set 3.3

1. first 3 hours, 55 mph; last hour, 35 mph **3.** 240 miles **5.** 4 ounces of 20%; 2 ounces of 50% **7.** 0.8 liter **9.** mutual fund \$14,000; CD \$10,000 **11.** \$13,200 **13.** \$13,500 @ 9%; \$27,000 @ 10%; \$40,500 @ 11% **15.** \$3500 **17.** 105 miles **19.** 25 pounds of \$6/pound coffee; 75 pounds of \$5.85/pound coffee **21.** \$7000 @ 11% profit; \$3000 @ 4% loss **23.** \$8500 @ 10%; \$17,000 @ 12% **25.** 3 hours **27.** 55 mph **29.** 60 cc **31.** 7 pounds **33.** \$5000 @ 12%; \$15,000 @ 4% **35.** \$4500 @ 9%; \$9000 @ 10%; \$13,500 @ 11% **37.** 2 hours $37\frac{1}{2}$ minutes **39.** 5 gallons **41.** 2.2 mph and 3.3 mph **43.** 27.5 miles **45.** 3 hours **47.** 200 pounds ground sirloin; 300 pounds hamburger **49.** 200 bikes **51.** 2 gallons **53.** 25 skateboards **55.** 800 books **57.** -4 **59.** -12 **61.** -4 **63.** $\{-2\}$ **65.** $\{0\}$ **67.** $\{18\}$ **69.** $A(2, 2)$; $B(0, 0)$; $C(-3, 2)$; $D(-6, -1)$; $E(4, -4)$; $F(0, -3)$

Mental Math Sec. 3.4

1. 7 **3.** -5 **5.** -6 **7.** 12

Exercise Set 3.4

1. $\{7, -7\}$ **3.** $\{4, -4\}$ **5.** $\{7, -2\}$ **7.** $\{8, 4\}$ **9.** $\{5, -5\}$ **11.** $\{3, -3\}$ **13.** $\{0\}$ **15.** $\{\ \}$ **17.** $\left\{\frac{1}{5}\right\}$ **19.** $\left\{9, -\frac{1}{2}\right\}$ **21.** $\left\{-\frac{5}{2}\right\}$ **23.** $\{4, -4\}$ **25.** $\{0\}$ **27.** $\{\ \}$ **29.** $\left\{0, \frac{14}{3}\right\}$ **31.** $\{2, -2\}$ **33.** $\{\ \}$ **35.** $\{7, -1\}$ **37.** $\{\ \}$ **39.** $\{\ \}$ **41.** $\left\{-\frac{1}{8}\right\}$ **43.** $\left\{\frac{1}{2}, -\frac{5}{6}\right\}$ **45.** $\left\{2, -\frac{12}{5}\right\}$ **47.** $\{3, -2\}$ **49.** $\left\{-8, \frac{2}{3}\right\}$ **51.** $\{\ \}$ **53.** $\{4\}$ **55.** $\{13, -8\}$ **57.** $\{3, -3\}$ **59.** $\{8, -7\}$ **61.** $\{2, 3\}$ **63.** $\left\{2, -\frac{10}{3}\right\}$ **65.** $\left\{\frac{3}{2}\right\}$ **67.** $\{\ \}$ **69.** -5 **71.** 8 **73.** $-\frac{1}{3}$ **75.** $\frac{9}{5}$ **77.** $\{-6\}$ **79.** $\{5\}$

Exercise Set 3.5

1. $(-\infty, -3)$ **3.** $[0.3, \infty)$ **5.** $(5, \infty)$ **7.** $(-2, 5)$ **9.** $(-1, 5)$ **11.** $(-\infty, -3)$ **13.** $[-5, \infty)$ **15.** $[-2, \infty)$ **17.** $(-\infty, 11]$ **19.** $\left[\frac{8}{3}, \infty\right)$ **21.** $(-13, \infty)$ **23.** $(-\infty, 7]$ **25.** $(-\infty, \infty)$ **27.** $\{\ \}$

29. 1040 pounds luggage and cargo **31.** 18 ounces

33. $(-1, \infty)$ **35.** $(-\infty, 2]$ **37.** $(2, \infty)$ **39.** $[-8, \infty)$ **41.** $(-\infty, -1]$ **43.** $(0, \infty)$ **45.** $(-2, \infty)$ **47.** $(-\infty, -2]$ **49.** $(-\infty, -8]$ **51.** $\left[-\frac{3}{5}, \infty\right)$ **53.** $[-9, \infty)$ **55.** $(38, \infty)$ **57.** $[0, \infty)$ **59.** $(-\infty, -5]$ **61.** $\left(-\infty, \frac{1}{4}\right)$ **63.** $(-\infty, -1]$ **65.** $\left[-\frac{79}{3}, \infty\right)$ **67.** $(-\infty, -15)$

69. Minimum score is 30. **71.** 4.57 minutes **73.** 193 **75.** Maximum length is 35 cm. **77.** 5 **79.** 2 **81.** 2 **83.** -2 **85.** $[0, 5]$ **87.** $[3, 10)$

89. 32 million **91.** 1992

Exercise Set 3.6

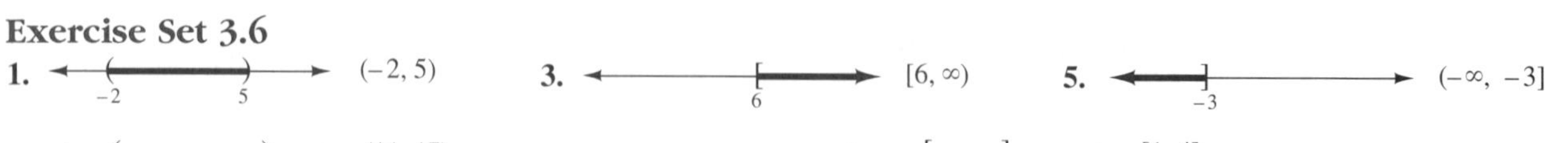

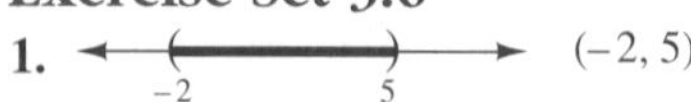

1. $(-2, 5)$ **3.** $[6, \infty)$ **5.** $(-\infty, -3]$ **7.** $(11, 17)$ **9.** $[1, 4]$ **11.** $\left[-3, \frac{3}{2}\right]$ **13.** $[-21, -9]$

15. $(-\infty, -1) \cup (0, \infty)$ **17.** $[2, \infty)$

19. $(-\infty, \infty)$ **21.** $(-1, 2)$

23. $(-\infty, \infty)$ **25.** $[-1, \infty)$ **27.** $[-5, \infty)$

29. $\left[\frac{3}{2}, 6\right]$ **31.** $\left(\frac{5}{4}, \frac{11}{4}\right)$ **33.** { }

35. $(-7, 0)$ **37.** $\left(-5, \frac{5}{2}\right)$

39. $\left(0, \frac{14}{3}\right)$ **41.** $(-\infty, -3]$

43. $(-\infty, -1] \cup \left(\frac{29}{7}, \infty\right)$ **45.** { }

47. $\left[-\frac{1}{2}, \frac{3}{2}\right)$ **49.** $\left(-\frac{4}{3}, \frac{7}{3}\right)$

51. $(6, 12)$

53. $-38.2° \le F \le 113°$ **55.** $89.3 \le x \le 100$ **57.** $0.924 \le d \le 0.987$ **59.** 3 **61.** 7 **63.** 8 **65.** 2

67. $\{21, -17\}$ **69.** { } **71.** $\{-4\}$ **73.** $(6, \infty)$ **75.** $[3, 7]$

77. $(-\infty, -1)$

Exercise Set 3.7

1. $[-4, 4]$ **3.** $(1, 5)$ **5.** $(-5, -1)$

7. $[-10, 3]$ **9.** $[-5, 5]$ **11.** { }

13. $[0, 12]$ **15.** $(-\infty, -3) \cup (3, \infty)$

17. $(-\infty, -24) \cup (4, \infty)$ **19.** $(-\infty, -4) \cup (4, \infty)$

21. $(-\infty, \infty)$ **23.** $\left(-\infty, \frac{2}{3}\right) \cup (2, \infty)$

25. $\{0\}$ **27.** $\left(-\infty, -\frac{3}{8}\right) \cup \left(-\frac{3}{8}, \infty\right)$

29. $[-2, 2]$ **31.** $(-\infty, -1) \cup (1, \infty)$

33. $(-5, 11)$ **35.** $\left(-\infty, \frac{2}{3}\right) \cup (2, \infty)$

37. { } **39.** $(-\infty, \infty)$ **41.** $[-2, 9]$

43. $(-\infty, -11] \cup [1, \infty)$ **45.** $(-\infty, 0) \cup (0, \infty)$

47. $(-\infty, \infty)$ **49.** $\left[-\frac{1}{2}, 1\right]$

51.  $(-\infty, -3) \cup (0, \infty)$ **53.** { }

55.  $(-\infty, \infty)$ **57.** 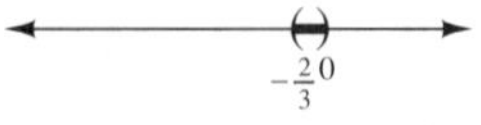 $\left(-\frac{2}{3}, 0\right)$ **59.**  $(-\infty, \infty)$

61.  $(-\infty, -1) \cup (1, \infty)$ **63.**  $(-\infty, -12) \cup (0, \infty)$

65. 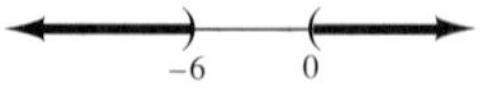 $(-\infty, -6) \cup (0, \infty)$ **67.**  $\left(-\frac{31}{5}, \frac{11}{5}\right)$

69.  $[-1, 8]$ **71.**  $\left[-\frac{23}{8}, \frac{17}{8}\right]$

73. $(-2, 5)$ **75.** $\{5, -2\}$ **77.** $(-\infty, -7] \cup [17, \infty)$ **79.** $\left\{-\frac{9}{4}\right\}$ **81.** $(-2, 1)$ **83.** $\left\{2, \frac{4}{3}\right\}$ **85.** { } **87.** $\left\{\frac{19}{2}, -\frac{17}{2}\right\}$ **89.** $\left(-\infty, -\frac{25}{3}\right) \cup \left(\frac{35}{3}, \infty\right)$ **91.** {1, 3, 5, 7, 9} **93.** { } **95.** 34 **97.** 1 **99.** 35%
101. 4.39 billion **103.** Each x increases by 1. Each y increases by 3. **103.**

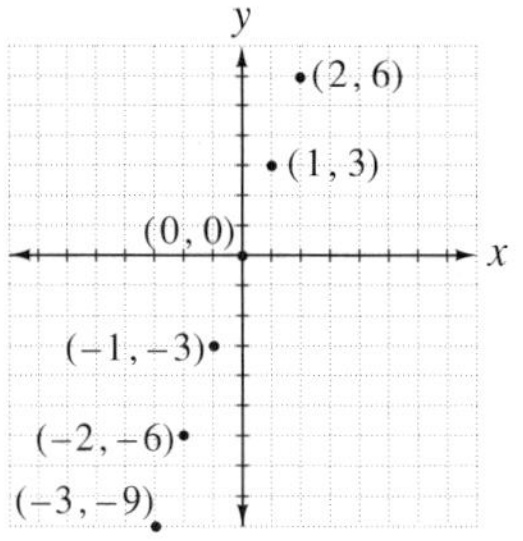

Chapter 3 Review

1. $\frac{20}{1}$ **3.** {6} **5.** {312.5} **7.** {9} **9.** {3} **11.** 675 parts **13.** 157 letters **15.** 21 cm **17.** 93.5
19. 70% **21.** 1280 **23.** 6% **25.** 120 travelers **27.** 14.3% **29.** 2 hours **31.** \$7000 @ 11%; \$3000 @ 4%
33. 66.7 gallons **35.** 13.75 miles **37.** 50 cL of 15% solution; 100 cL of 45% solution **39.** {5, 11} **41.** $\left\{-1, \frac{11}{3}\right\}$
43. $\left\{-\frac{1}{6}\right\}$ **45.** { } **47.** {1, 5} **49.** { } **51.** $\left\{-10, -\frac{4}{3}\right\}$ **53.** $(-\infty, -4]$ **55.** $(-17, \infty)$ **57.** $(-\infty, 4]$
59. $(-\infty, 1)$ **61.** $(2, \infty)$ **63.** $260° \le C \le 538°$ **65.** $\$1750 \le x \le \$3750$ **67.** $[-4, 0]$ **69.** $\left[-2, -\frac{9}{5}\right)$
71. $\left(-\frac{3}{5}, 0\right)$ **73.** $\left[-\frac{4}{3}, \frac{7}{6}\right]$ **75.** $(-\infty, \infty)$ **77.** $(5, \infty)$ **79.**  $(-\infty, -4] \cup [1, \infty)$

81.  $(-3, 3)$ **83.**  $(-\infty, \infty)$

85. 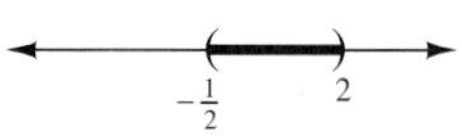 $\left(-\frac{1}{2}, 2\right)$ **87.** { }

Chapter 3 Test

1. {−6} **2.** $\left\{\frac{2}{3}, 1\right\}$ **3.** { } **4.** {−17, 11} **5.** $(5, \infty)$ **6.** $[2, \infty)$ **7.** $\left(\frac{3}{2}, 5\right]$ **8.** $(-\infty, -2) \cup \left(\frac{4}{3}, \infty\right)$
9. $[5, \infty)$ **10.** $[4, \infty)$ **11.** $[-3, -1)$ **12.** $(-\infty, \infty)$ **13.** 170 **14.** 25% **15.** 90 **16.** 12.9%
17. 81.3% **18.** \$5.93 billion **19.** 24.12° **20.** more than 850 **21.** \$8500 @ 10%; \$17,000 @ 12%
22. 2 hours 5 minutes **23.** 20 cars **24.** 6.25 liters

Chapter 3 Cumulative Review

1. a. {11, 112} **b.** {0, 11, 112} **c.** {−3, −2, 0, 11, 112} **d.** $\left\{-3, -2, 0, \frac{1}{4}, 11, 112\right\}$ **e.** $\{\sqrt{2}\}$

f. $\left\{-3, -2, 0, \frac{1}{4}, \sqrt{2}, 11, 112\right\}$; *Sec. 1.2, Ex. 1* **2. a.** $40 = 2 \cdot 2 \cdot 2 \cdot 5$ **b.** $63 = 3 \cdot 3 \cdot 7$; *Sec. 1.3, Ex. 1*
3. a. $\frac{64}{25}$ **b.** $\frac{1}{20}$ **c.** $\frac{5}{4}$; *Sec. 1.3, Ex. 4* **4. a.** 9 **b.** 125 **c.** 16; *Sec. 1.4, Ex. 1* **5. a.** yes; *Sec. 1.5, Ex. 3*
6. a. -4 **b.** 8 **c.** -3; *Sec. 1.6, Ex. 2* **7. a.** 0 **b.** -24 **c.** 45 **d.** -8 **e.** 54; *Sec. 1.7, Ex. 2*
8. a. -6 **b.** -24 **c.** $\frac{3}{4}$; *Sec. 1.7, Ex. 7* **9. a.** $2x + 2y$ **b.** $15 - 5z$ **c.** $5x + 5y - 5z$
d. $-2 + y$ **e.** $-3 - x + w$; *Sec. 1.8, Ex. 3* **10. a.** unlike **b.** like **c.** like **d.** like; *Sec. 2.1, Ex. 2*
11. a. $2x + 6$ **b.** $\frac{x-4}{7}$ **c.** $8 + 3x$; *Sec. 2.1, Ex. 8* **12.** {17}; *Sec. 2.2, Ex. 1*
13. $2x + 7 = x - 3$; $x = -10$; *Sec. 2.3, Ex. 7* **14.** {−2}; *Sec. 2.3, Ex. 3* **15.** 15°C; *Sec. 2.4, Ex. 3*
16. {63}; *Sec. 3.1, Ex. 2* **17.** $62.45; *Sec. 3.1, Ex. 5* **18.** 87.5%; *Sec. 3.2, Ex. 5*
19. $\left\{-2, \frac{4}{5}\right\}$; *Sec. 3.4, Ex. 2* **20.** 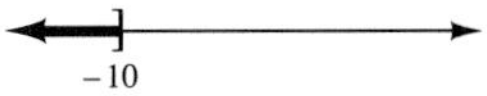

$(-\infty, -10]$ *Sec. 3.5, Ex. 4*
21. at most 64 people; *Sec. 3.5, Ex. 11* **22.** $x < 4$, $(-\infty, 4)$; *Sec. 3.6, Ex. 1* **23.** (4, 8); *Sec. 3.7, Ex. 2*

CHAPTER 4
Graphing Equations

Mental Math, Sec. 4.1

1. answers vary, Ex. (5, 5), (2, 8) **3.** answers vary, Ex. (9, 2), (−1, −8) **5.** answers vary, Ex. (2, 10), (0, 0)

Exercise Set 4.1

1. yes, no, yes **3.** no, yes, yes **5.** no, yes, yes
7. (0, 2), (6, 0), (3, 1)

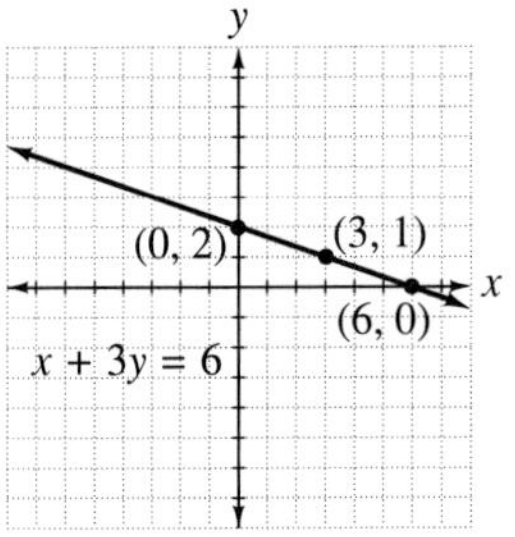

9. (0, −12), (5, −2), (−3, −18)

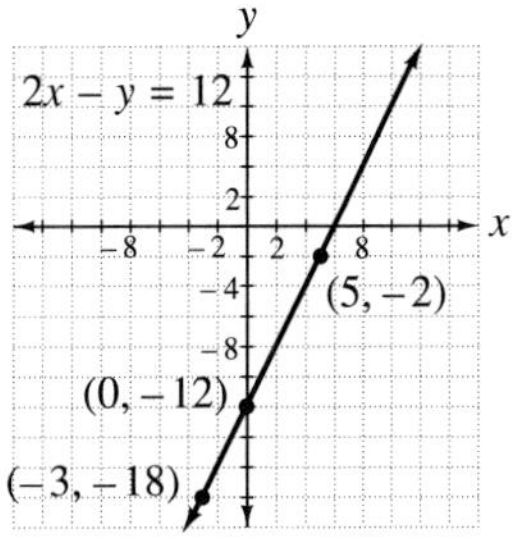

11. $\left(0, \frac{5}{7}\right)$, $\left(\frac{5}{2}, 0\right)$, (−1, 1)

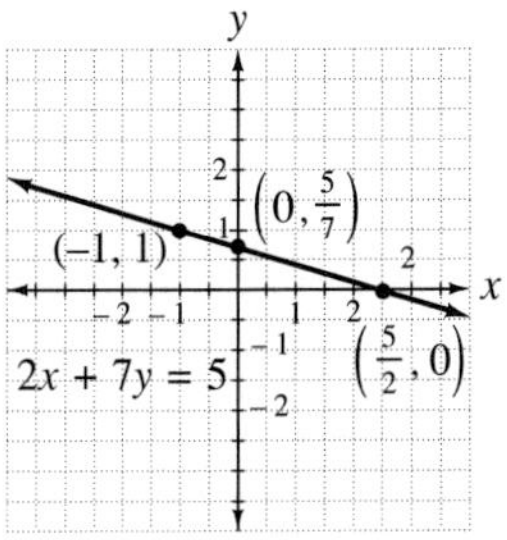

13.

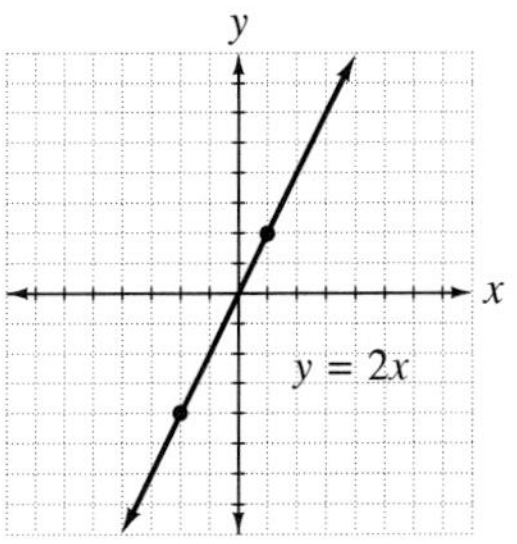

15.

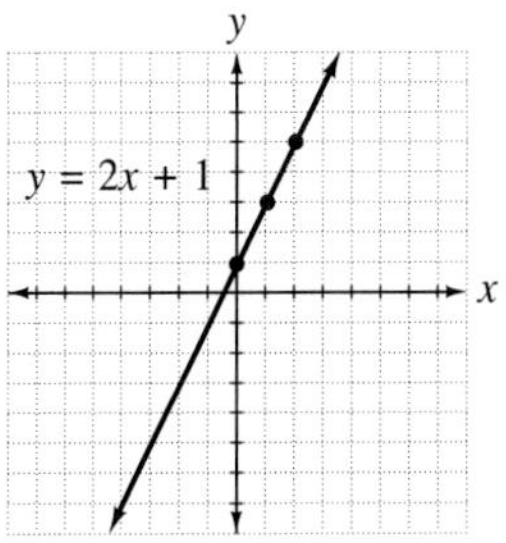

17.

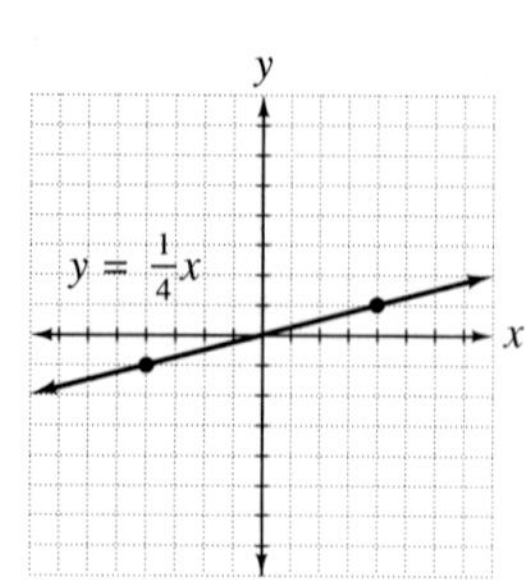

19.

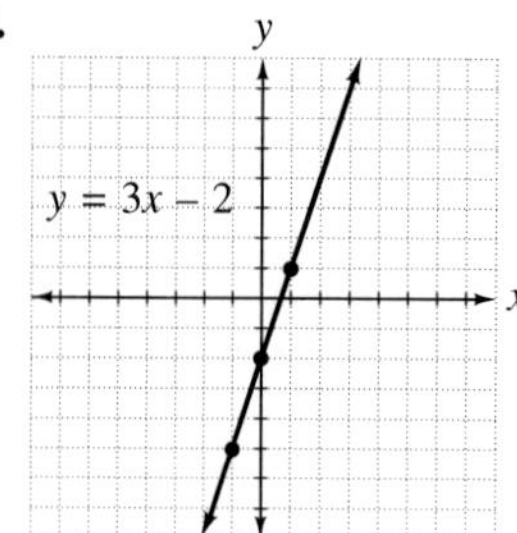

21.

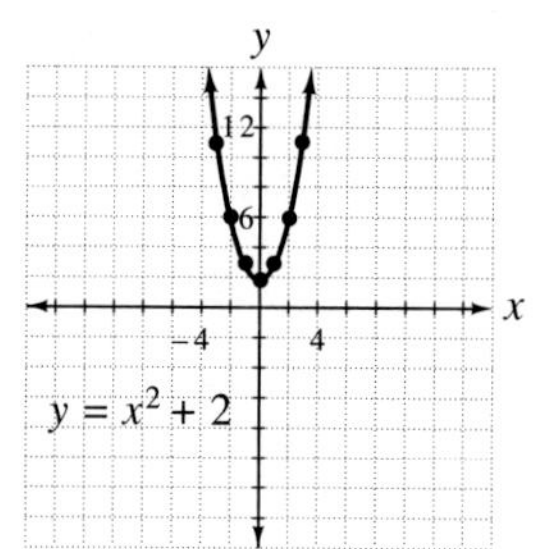

23.

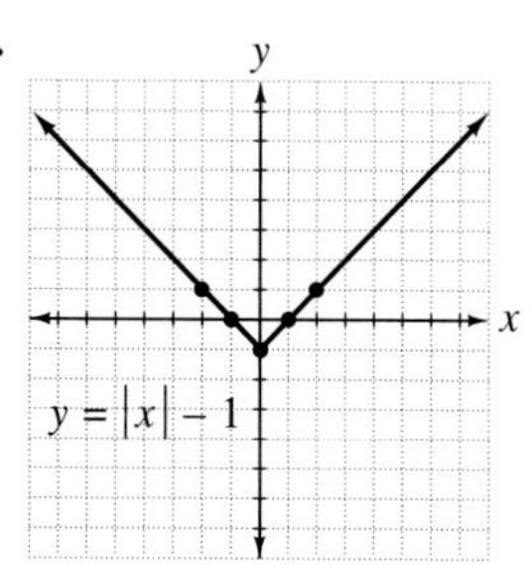

25.

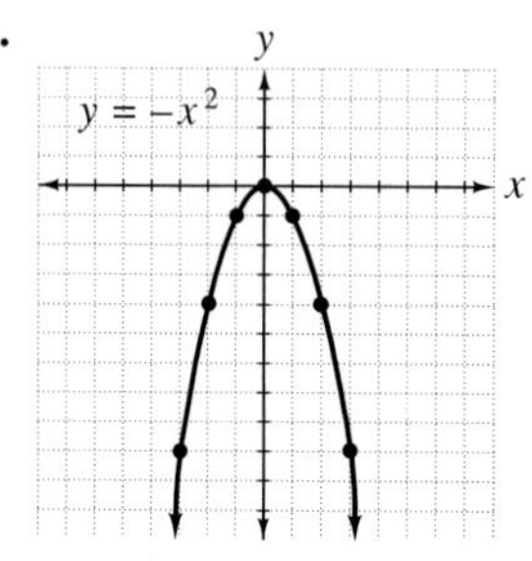

27.

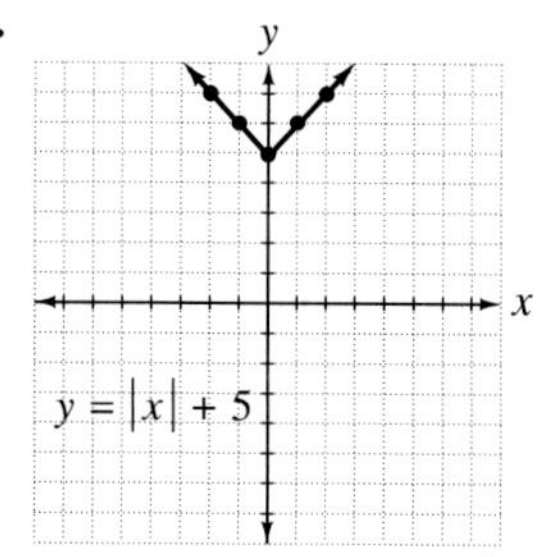

29. yes, no **31.** no, no **33.** yes, yes **35.** yes, yes **37.** yes, no

39. linear

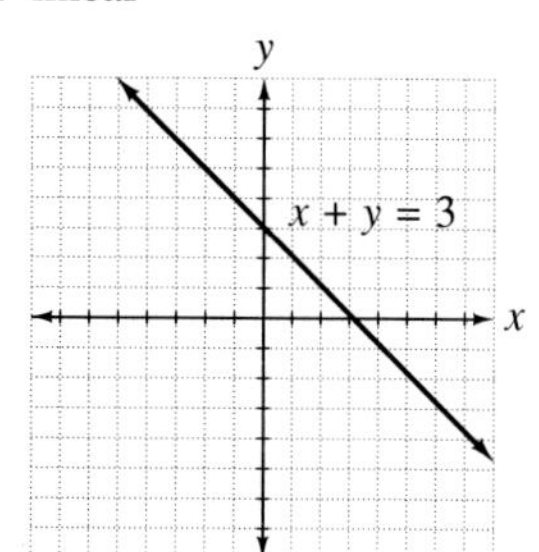

41. linear

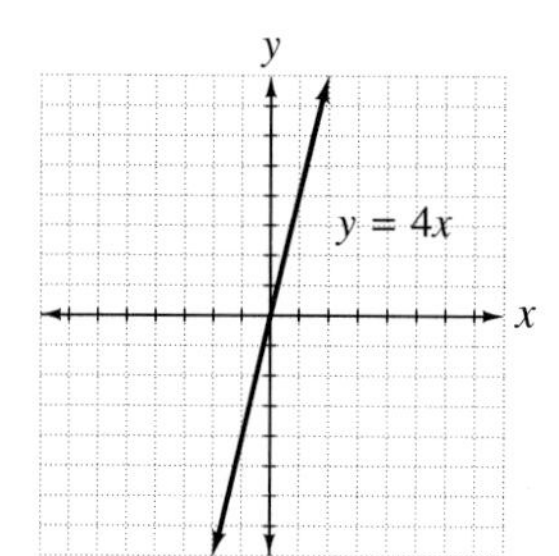

43. linear

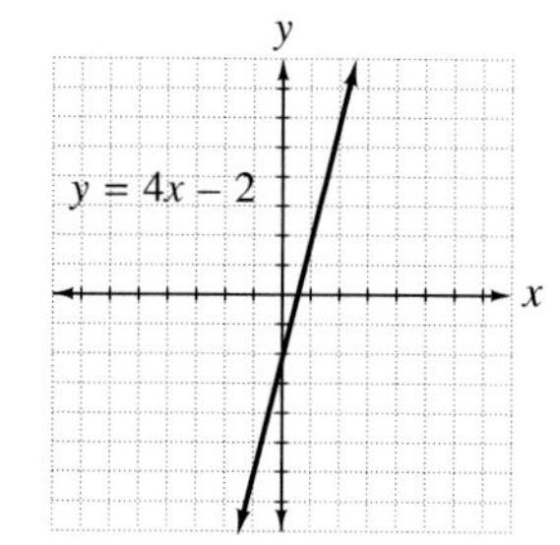

45. not linear

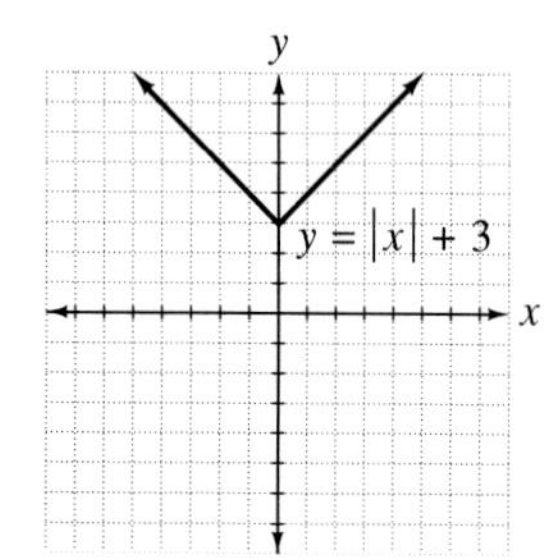

47. linear

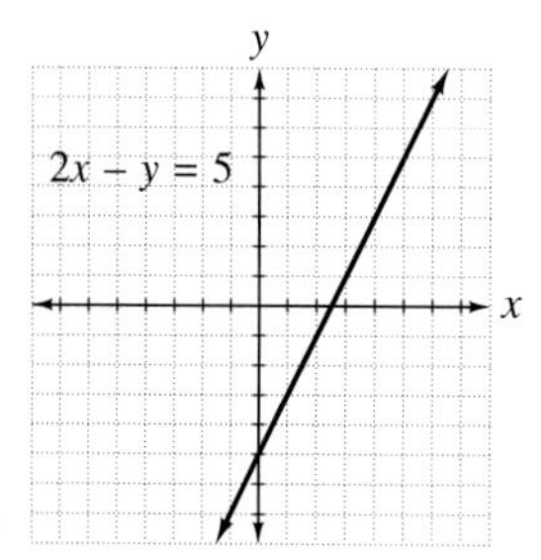

49. not linear

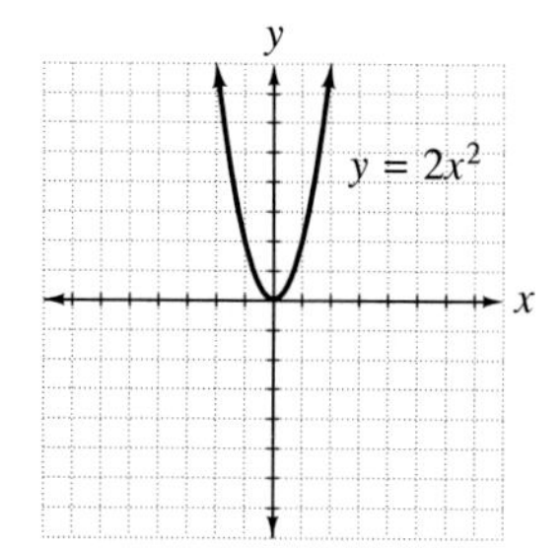

51. not linear

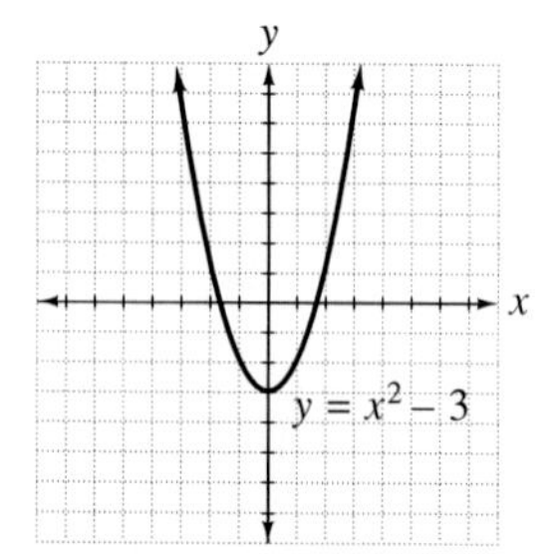

53. linear

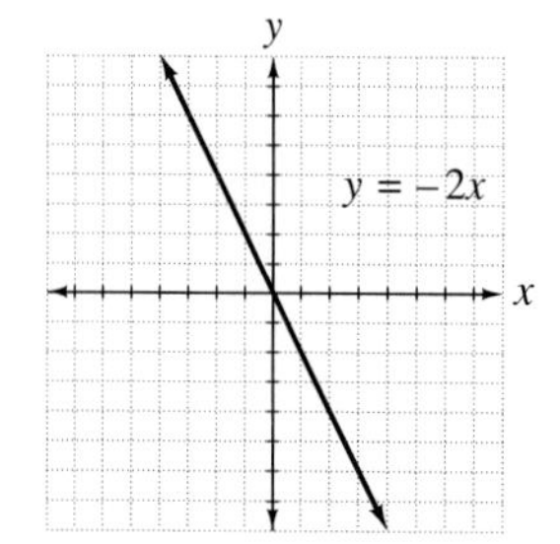

55. linear

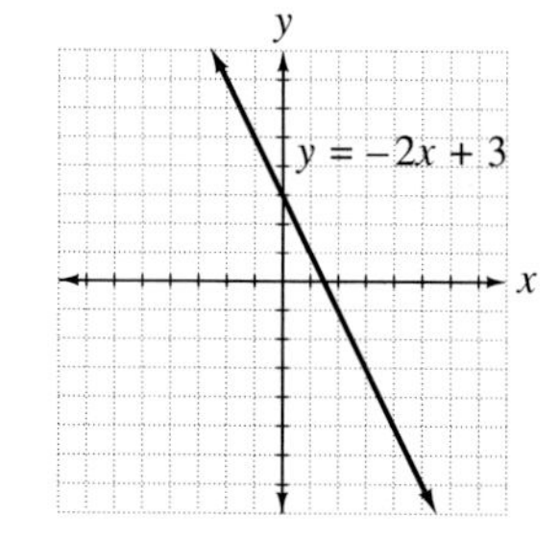

57. \$7000 **59.** \$500 **61.** Depreciation is the same from year to year.

63.

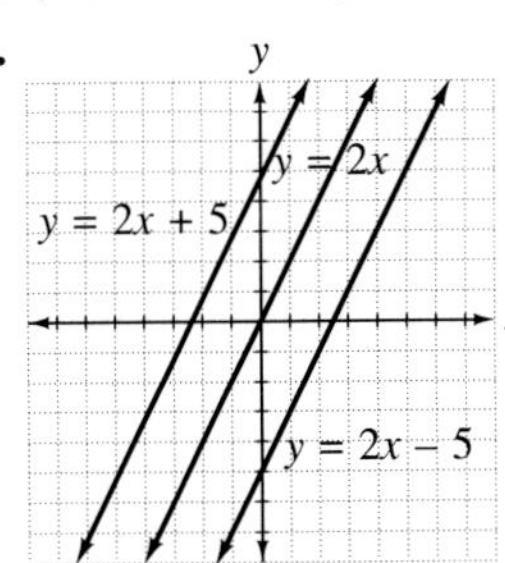

65.

67.

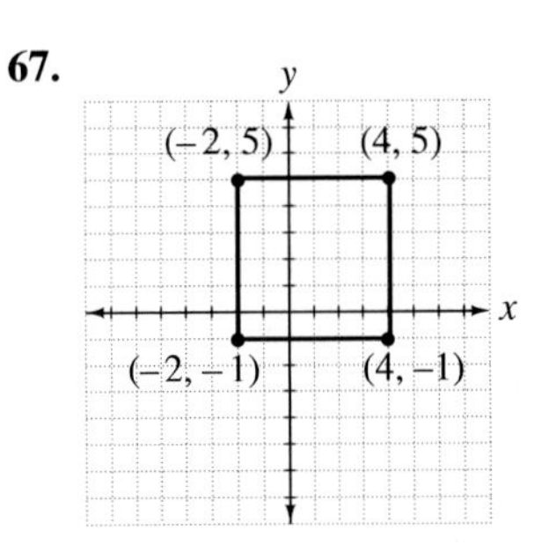

69. $\{-5\}$ **71.** $\left\{-\frac{1}{10}\right\}$ **73.** $\{x \mid x \leq -5\}$ **75.** $\{x \mid x < -4\}$

Exercise Set 4.2

1. domain: $\{-7, 0, 2, 10\}$; range: $\{-7, 0, 4, 10\}$ **3.** domain: $\{0, 1, 5\}$; range: $\{-2\}$ **5.** domain: $\{-1, 1, 2, 3\}$; range: $\{1, 2\}$ **7.** domain: $\{32°, 104°, 212°, 50°\}$; range: $\{0°, 40°, 100°, 10°\}$ **9.** yes **11.** no **13.** yes **15.** no **17.** yes **19.** yes **21.** yes **23.** no **25.** yes **27.** domain: $[2, 7]$; range: $[1, 6]$; no **29.** domain: $\{-2\}$; range: $(-\infty, \infty)$; no **31.** domain: $(-\infty, \infty)$; range: $(-\infty, 3]$; yes **33. a.** -9 **b.** -5 **c.** 1 **35. a.** 11 **b.** $2\frac{1}{16}$ **c.** 11 **37. a.** -8 **b.** -216 **c.** 0 **39. a.** 7 **b.** 7 **c.** 0 **41.** 6.3 billion **43.** domain: $\{-1, 0, -2, 5\}$; range: $\{7, 6, 2\}$; yes **45.** domain: $\{-2, 6, -7\}$; range: $\{4, -3, -8\}$; no **47.** domain: $\{1\}$; range: $\{1, 2, 3, 4\}$; no **49.** domain: $\left\{\frac{3}{2}, 0\right\}$; range: $\left\{\frac{1}{2}, -7, \frac{4}{5}\right\}$; no **51.** domain: $\{-3, 0, 3\}$; range: $\{-3, 0, 3\}$; yes **53.** domain: $\{2, -1, 5, 100\}$; range: $\{0\}$; yes **55.** It is a function. **57.** It is not a function

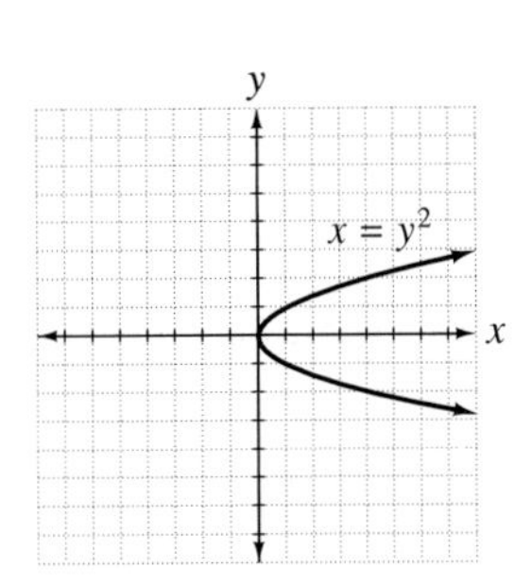

59. domain: $[0, \infty)$; range: $(-\infty, \infty)$; no **61.** domain: $[-1, 1]$; range: $(-\infty, \infty)$; no **63.** domain: $(-\infty, \infty)$; range: $(-\infty, -3] \cup [3, \infty)$; no **65. a.** 0 **b.** 10 **c.** -10 **67. a.** 245 **b.** 5 **c.** $3\frac{1}{2}$ **69. a.** 0 **b.** 6 **71. a.** 6 **b.** 6 **c.** 6 **73.** 25π square cm **75.** 2744 cubic in. **77.** 166.38 cm **79.** 163.2 mg **81.** $(-1, 4)$ **83.** $(-\infty, 5)$ **85.** 45°, 135° **87. a.** 11 **b.** $2a + 7$ **89. a.** 16 **b.** $a^2 + 7$

Exercise Set 4.3

1.

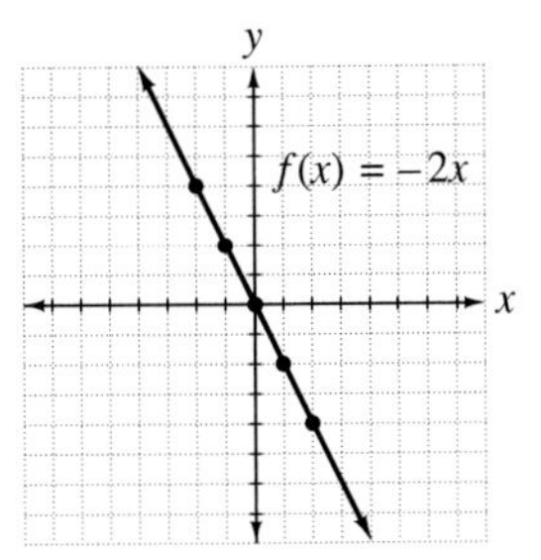

3.

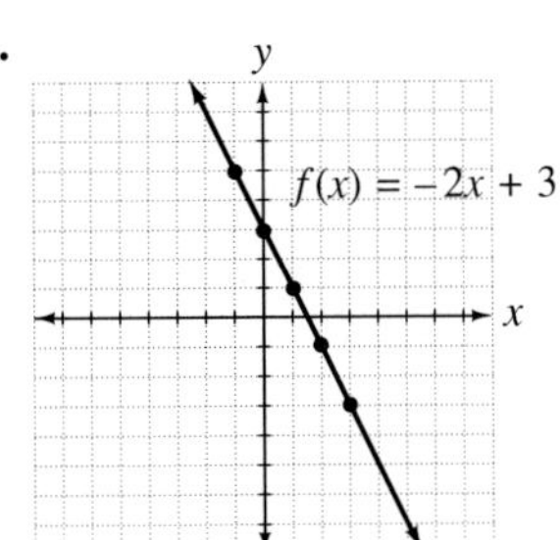

5.

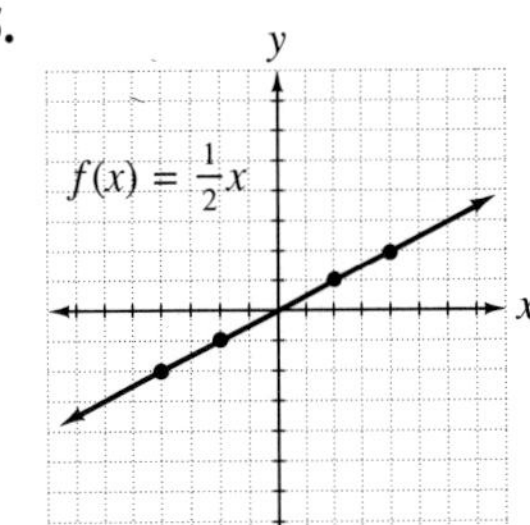

7.

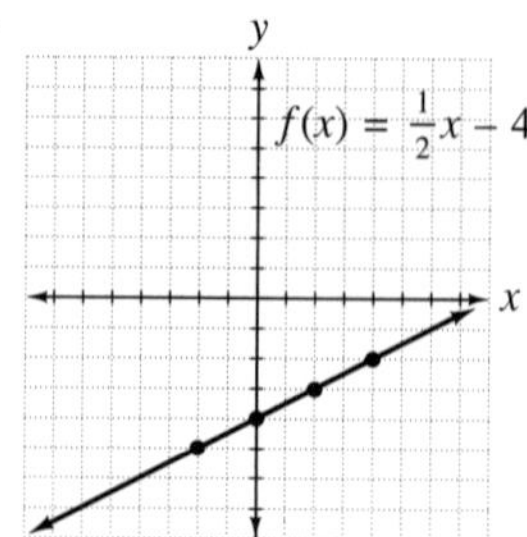

9.

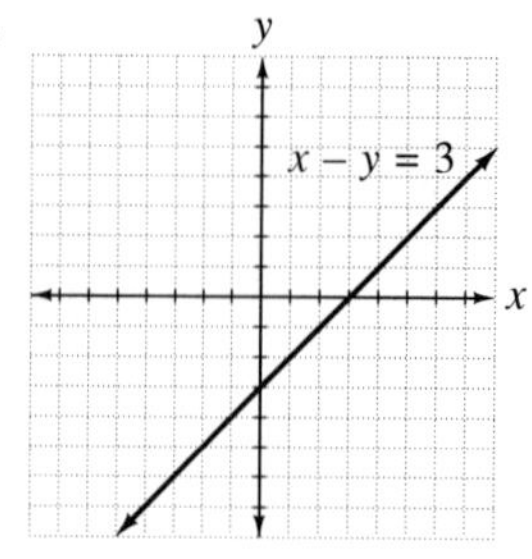

11.

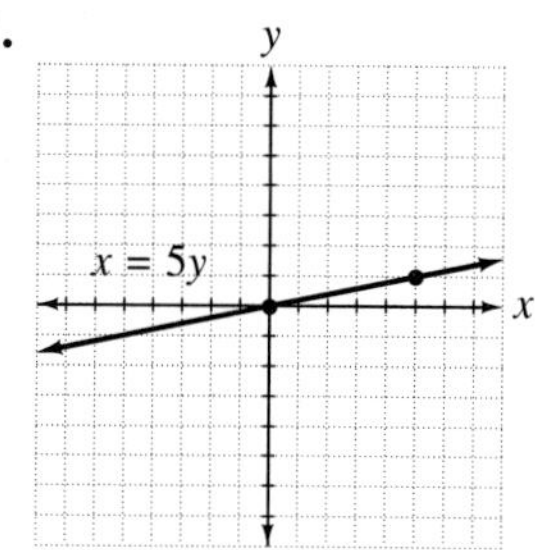

13.

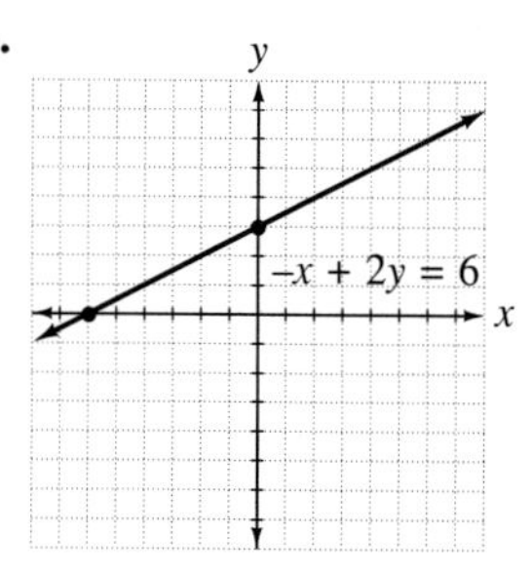

15.

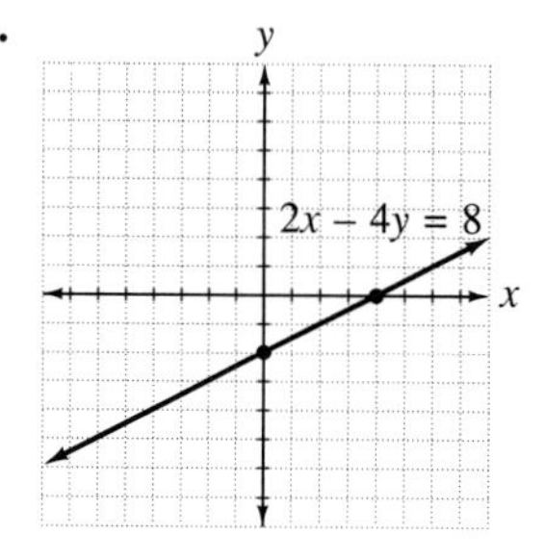

17.

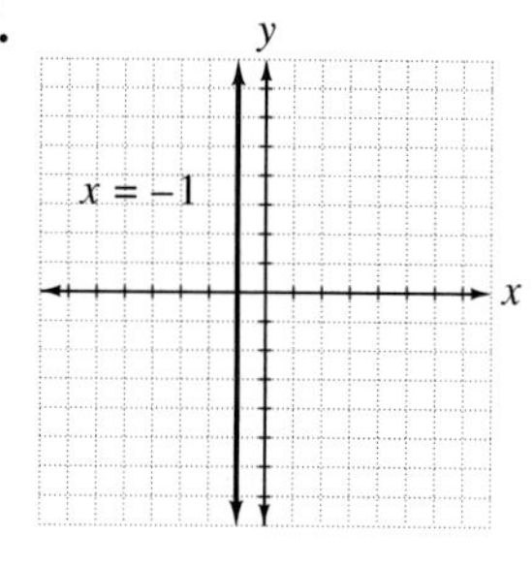

19.

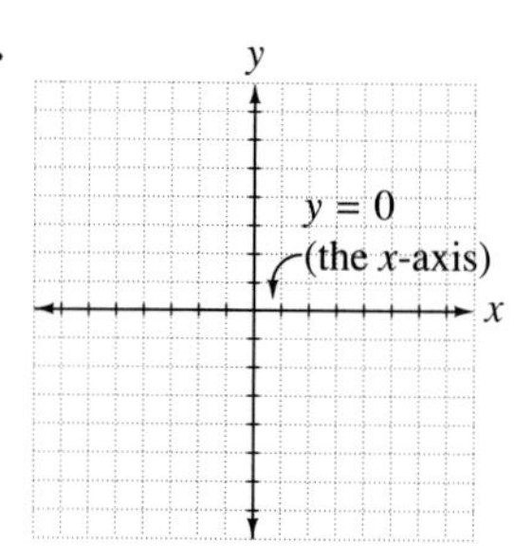

21.

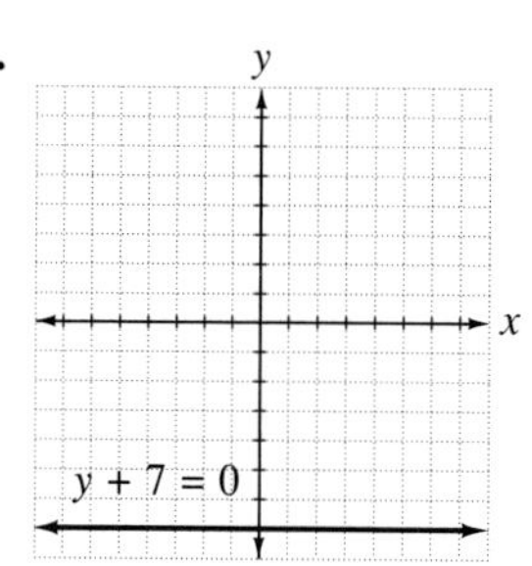

23.

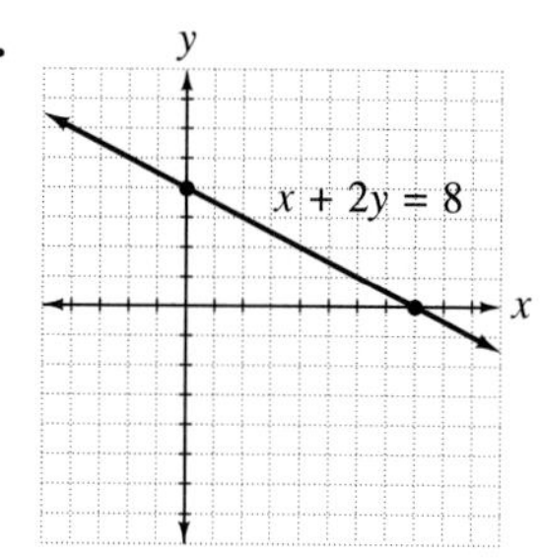

25.

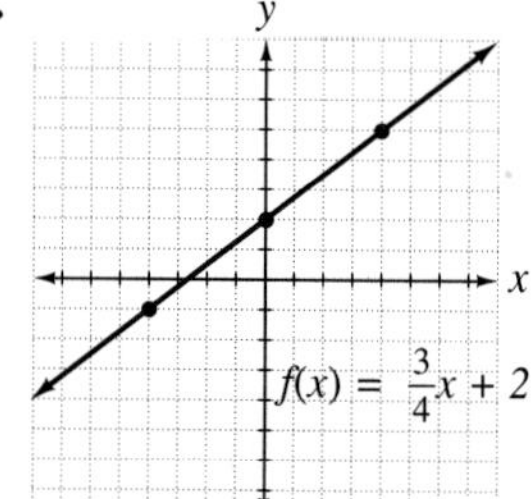

27.

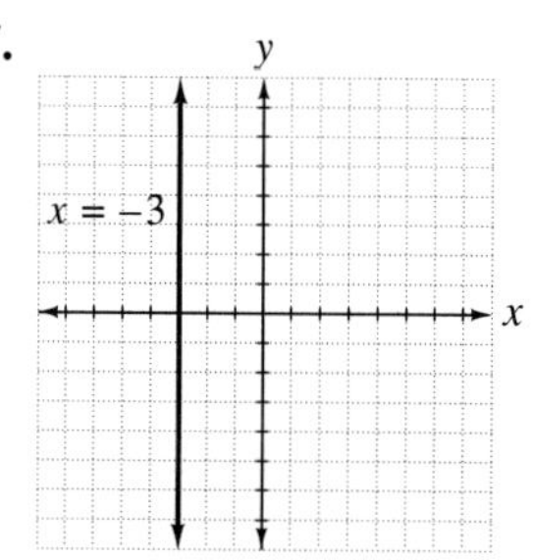

29.

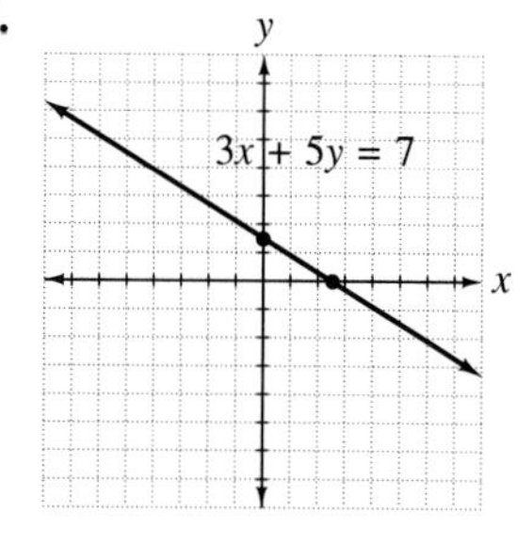

31.

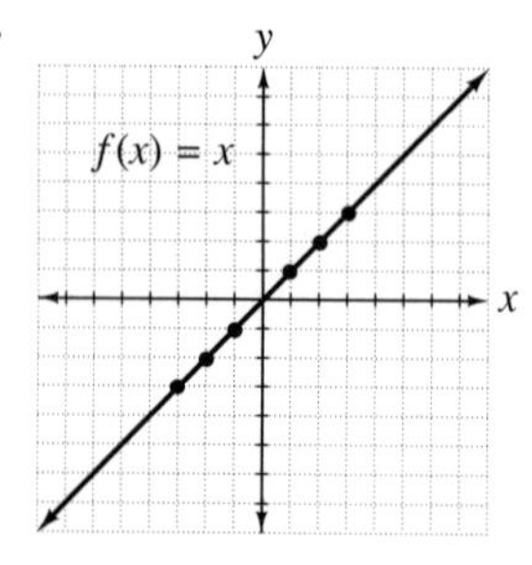

33.

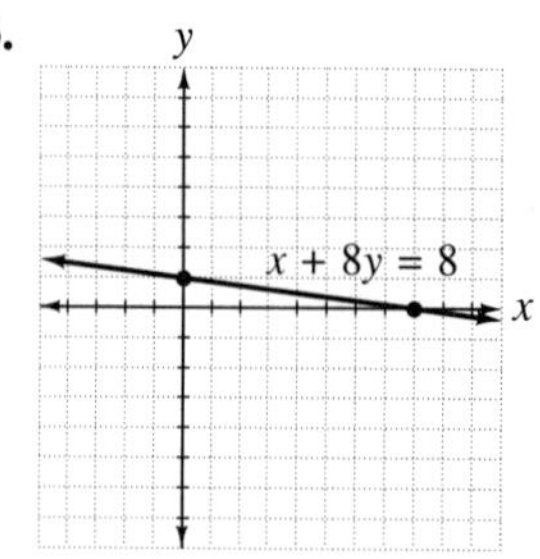

35.

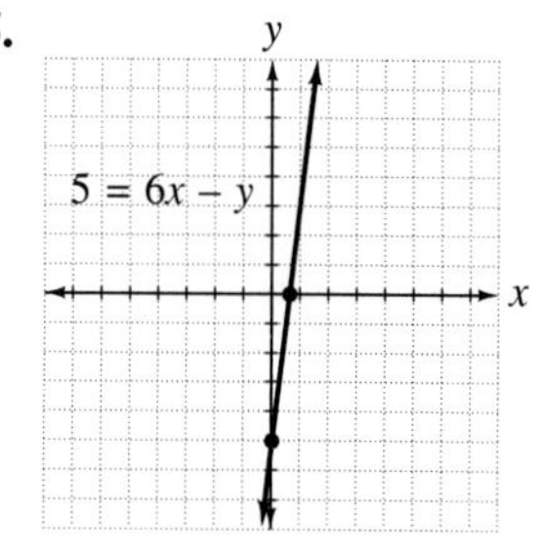

37.

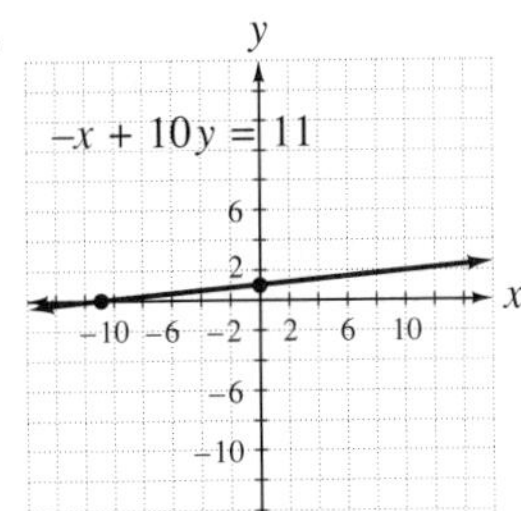

39.

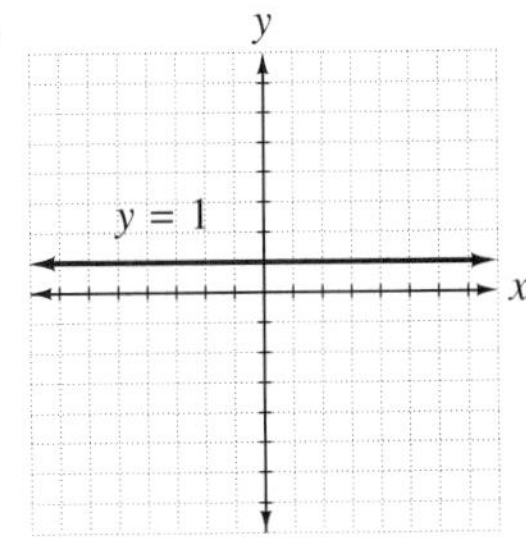

41.

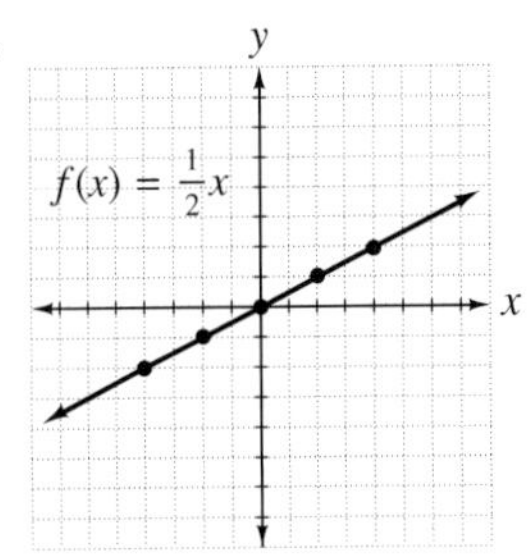

43.

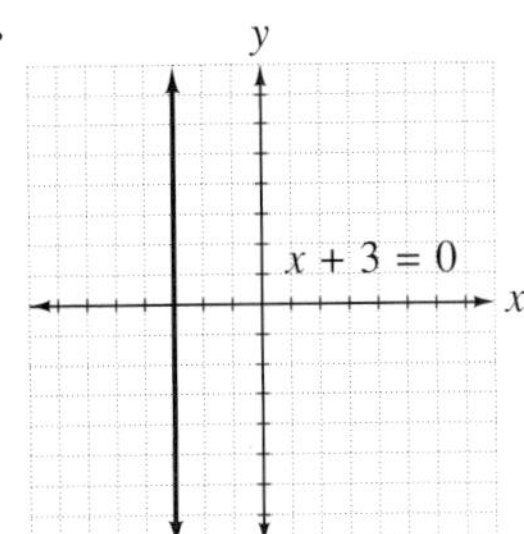

45.

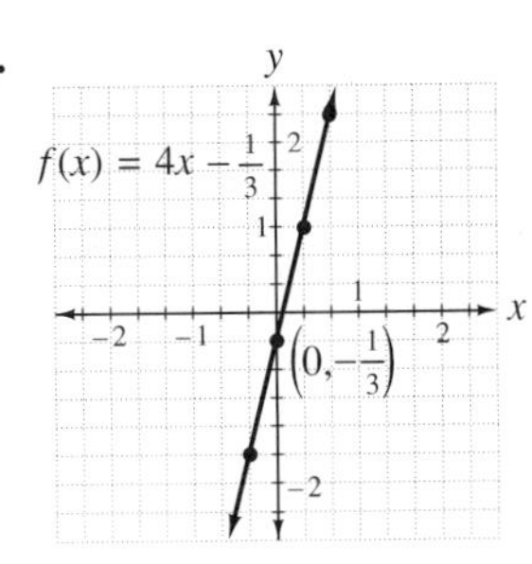

47.

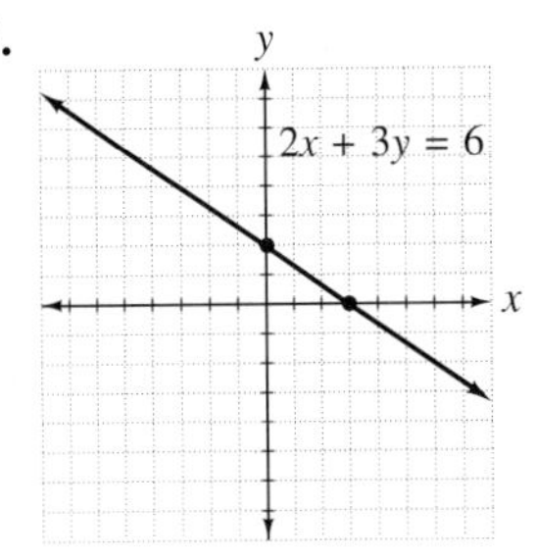

49. a.

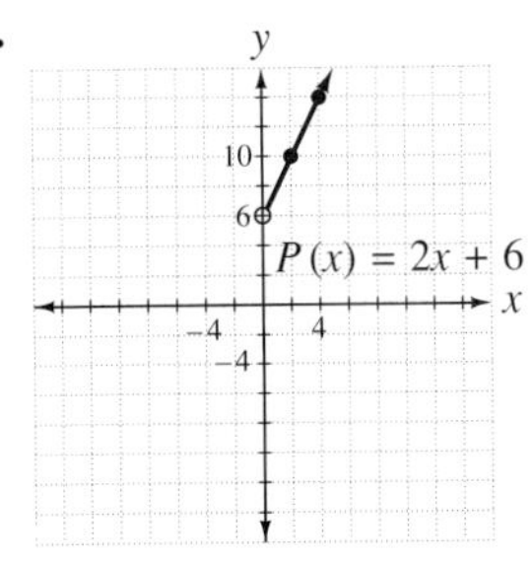

b. $P(4) = 14$ in.

51. a. (0, 500) **b.** (750, 0) **c.** 466 **53.** $\{-3, 9\}$ **55.** $(-\infty, -4) \cup (-1, \infty)$ **57.** $\left[\frac{2}{3}, 2\right]$ **59.** $\frac{3}{2}$ **61.** 6 **63.** $-\frac{6}{5}$

Mental Math Sec. 4.4

1. upward **3.** horizontal

Exercise Set 4.4

1. $\frac{8}{7}$ **3.** -1 **5.** $-\frac{1}{4}$ **7.** undefined **9.** $m = 5, b = -2$ **11.** $m = -2, b = 7$ **13.** $m = \frac{2}{3}, b = -\frac{10}{3}$ **15.** $m = \frac{1}{2}, b = 0$ **17.** neither **19.** parallel **21.** perpendicular **23.** undefined **25.** 0 **27.** undefined **29.** $-\frac{2}{3}$ **31.** undefined **33.** 0 **35.** 2 **37.** -1 **39.** $m = -1, b = 5$ **41.** $m = \frac{6}{5}, b = 6$ **43.** $m = 3, b = 9$ **45.** $m = 0, b = 4$ **47.** $m = 7, b = 0$ **49.** $m = 0, b = -6$ **51.** undefined, no y-intercept **53.** $\frac{2}{3}$ **55.** $\frac{4}{33}$ **57.** $-\frac{7}{2}$ **59.** $\frac{2}{7}$ **61.** $-\frac{1}{5}$ **63.** $\frac{1}{3}$ **65.** 6 **67.** -1 **69.** \$70 **71.** \$90 **73.** from 1990 to 1991 **75.** $y = 5x + 32$ **77.** $y = -3x - 30$

Mental Math Sec. 4.5

1. $m = -4, b = 12$ **3.** $m = 5, b = 0$ **5.** $m = \frac{1}{2}, b = 6$ **7.** parallel **9.** neither

Exercise Set 4.5

1. $y = -x + 1$ **3.** $y = 2x + \frac{3}{4}$ **5.** $y = \frac{2}{7}x$ **7.**

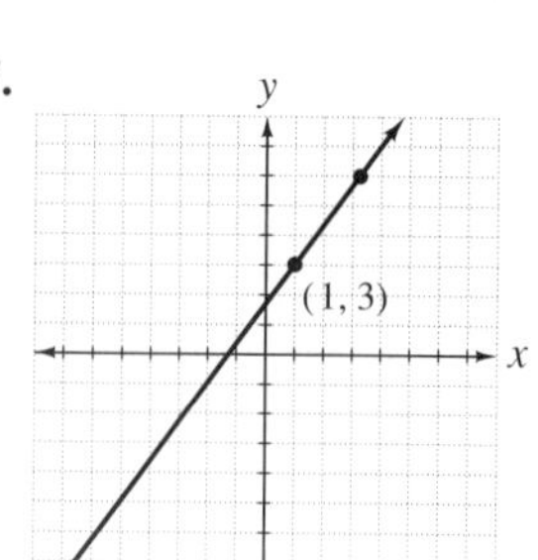

9.

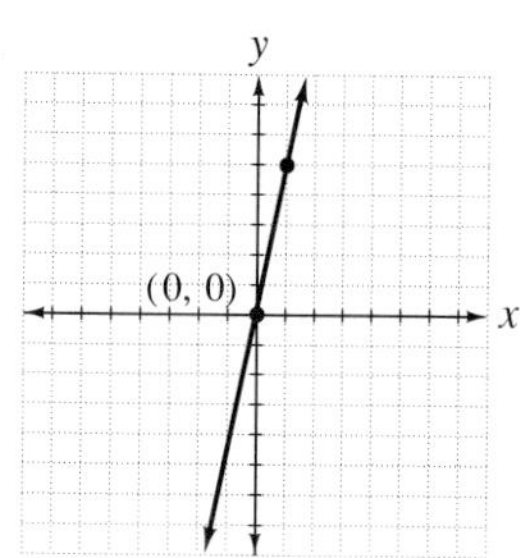

11.

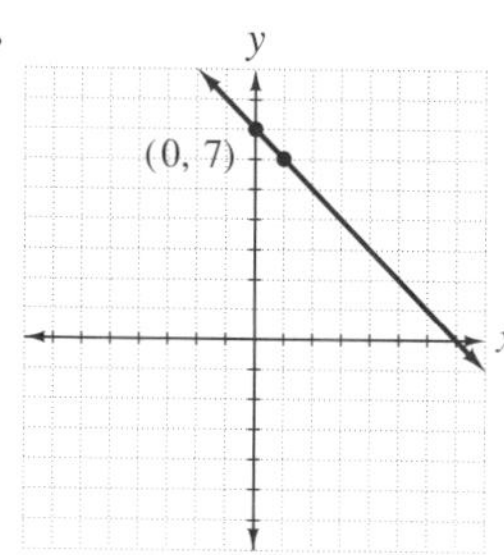

13. $6x - y = 10$ **15.** $8x + y = -13$ **17.** $x - 2y = 17$ **19.** $f(x) = 2x - 4$

21. $f(x) = 8x + 11$ **23.** $f(x) = \frac{4}{3}x + \frac{1}{3}$ **25.** $x = 0$ **27.** $y = 3$ **29.** $x = -7$ **31.** $f(x) = 4x - 4$

33. $f(x) = -3x + 1$ **35.** $f(x) = -\frac{3}{2}x - 6$ **37.** $3x + 6y = 10$ **39.** $x - y = -16$ **41.** $x + y = 17$

43. $y = 7$ **45.** $4x + 7y = -18$ **47.** $2x - y = -7$ **49.** $x + y = 7$ **51.** $x + 2y = 22$ **53.** $2x + 7y = -42$

55. $4x + 3y = -20$ **57.** $x = -2$ **59.** $x + 2y = 2$ **61.** $y = 12$ **63. a.** $R(x) = 32x$ **b.** 128 ft per second

65. a. $P(x) = 12{,}000x + 18{,}000$ **b.** \$102,000 **c.** 9 years **67.** $K = C + 273$ **69.** $-\frac{1}{8}$ **71.** $-\frac{1}{8}$

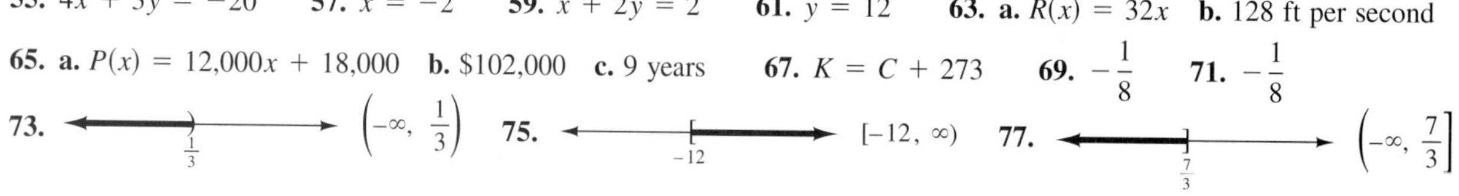

Exercise Set 4.6

1.

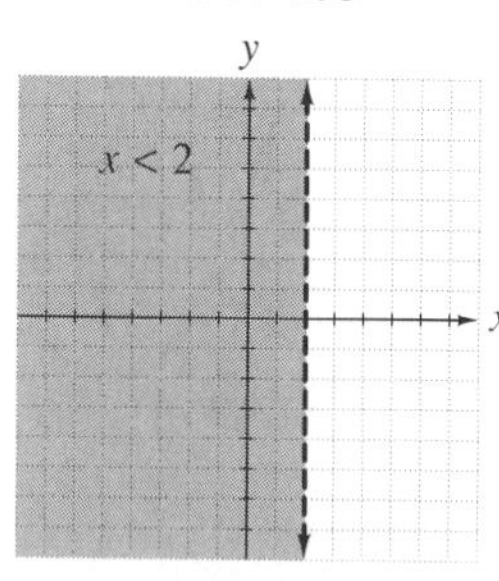

3.

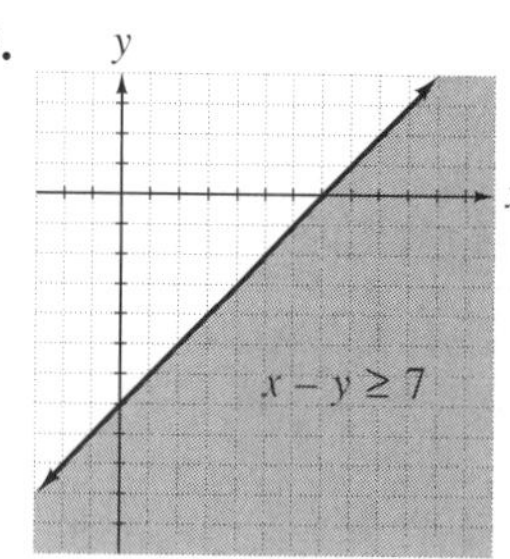

5.

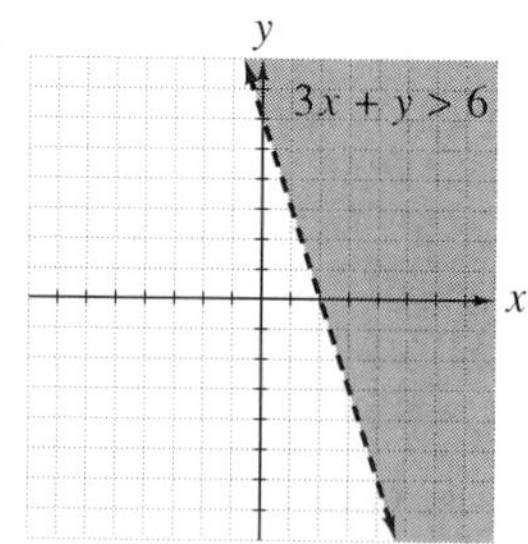

7.

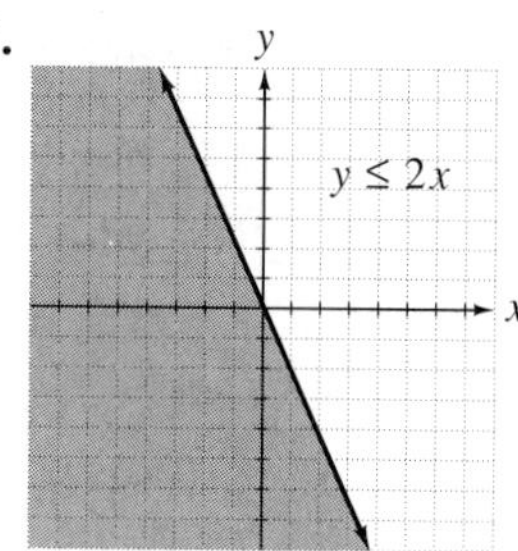

9.

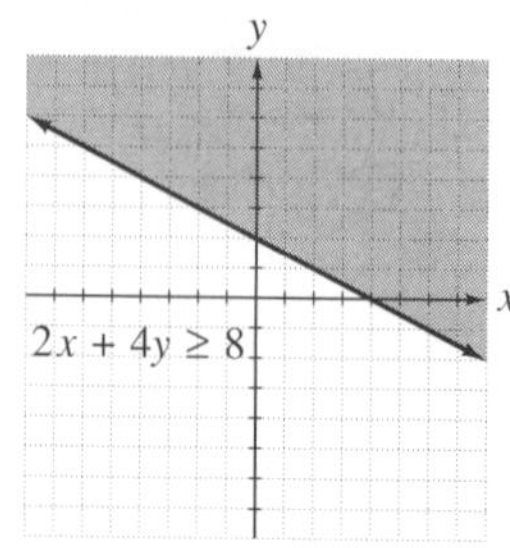

11.

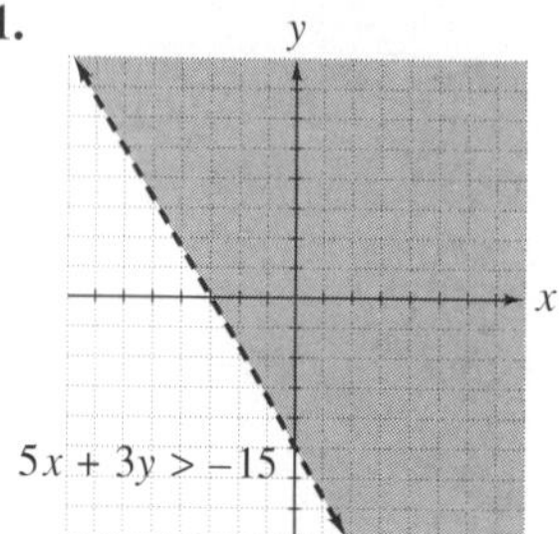

13.

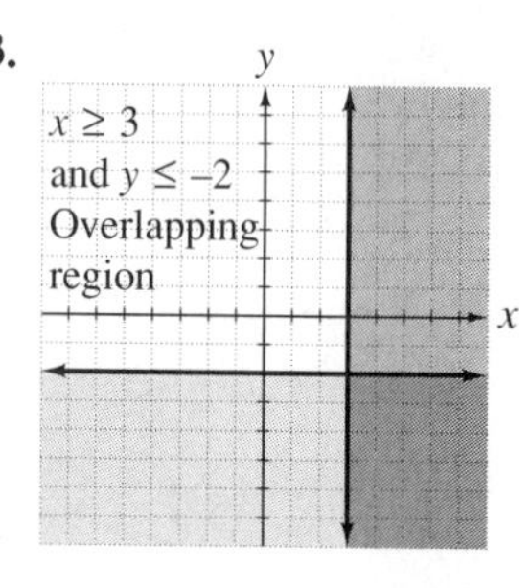

15.

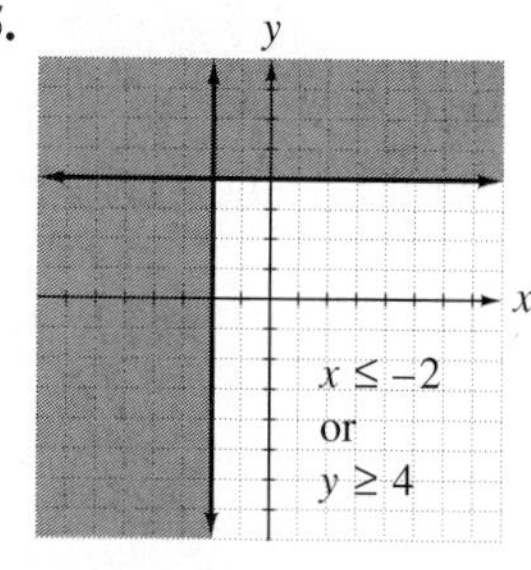

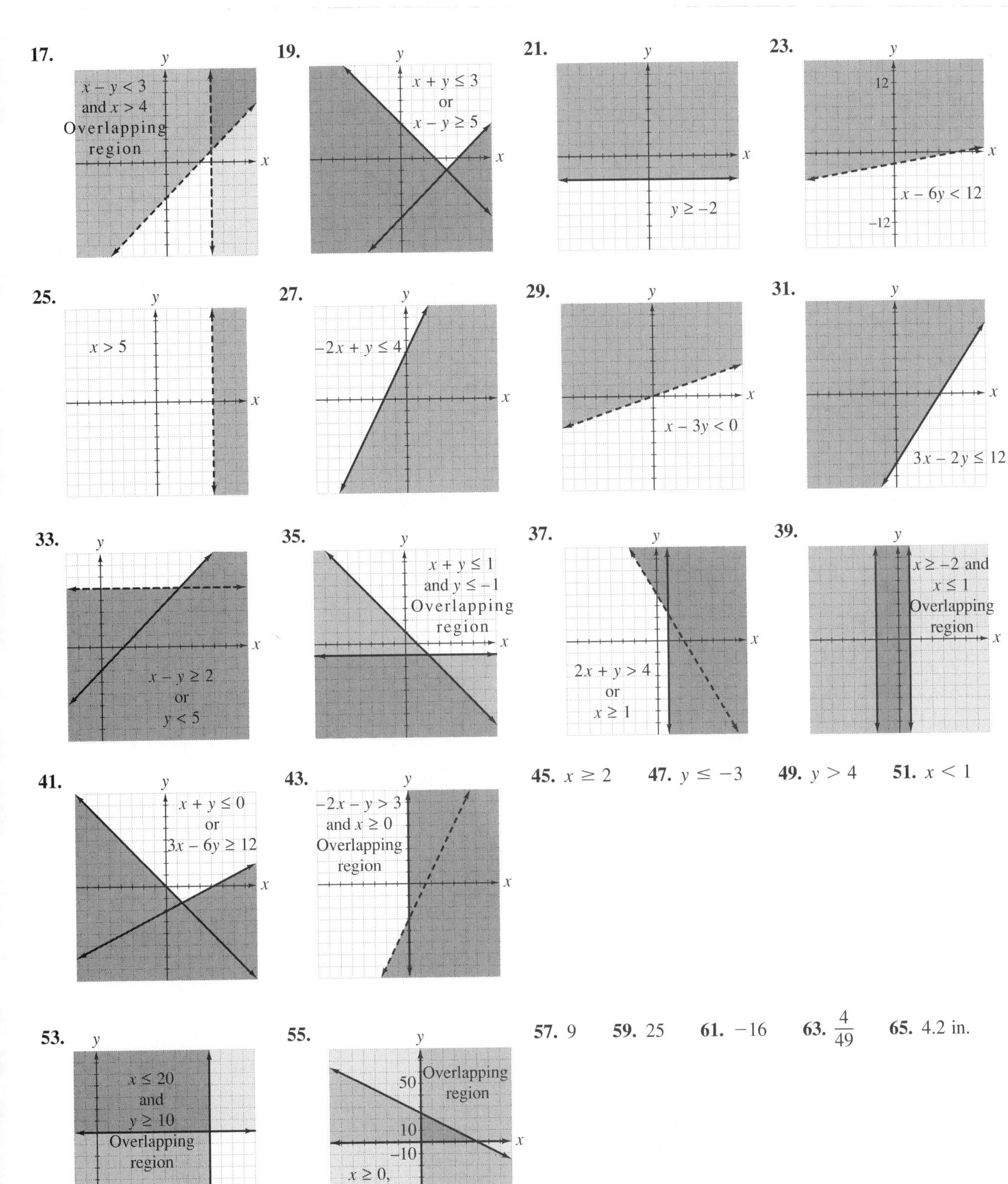

45. $x \ge 2$ **47.** $y \le -3$ **49.** $y > 4$ **51.** $x < 1$

57. 9 **59.** 25 **61.** -16 **63.** $\frac{4}{49}$ **65.** 4.2 in.

67. domain: $(-\infty, 2] \cup [2, \infty)$; range: $(-\infty, \infty)$; no

Chapter 4 Review

1. no, yes **3.** yes, yes **5.** (7, 44) **7. a.** (−3, 0) **b.** (1, 3) **c.** (9, 9) **9. a.** (0, 0) **b.** (10, 5) **c.** (−10, −5)

11. linear

$x - 3y = 2$

13. not linear

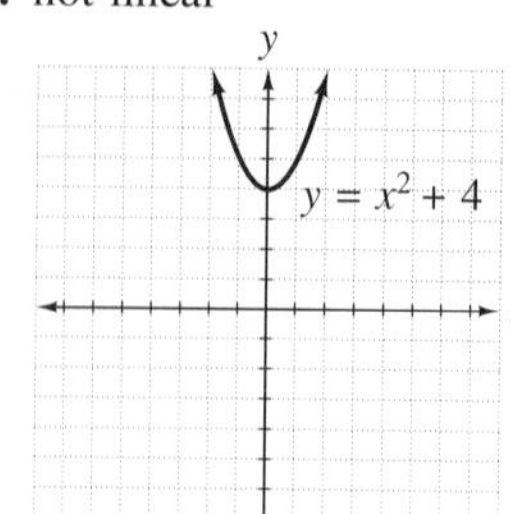

15. linear

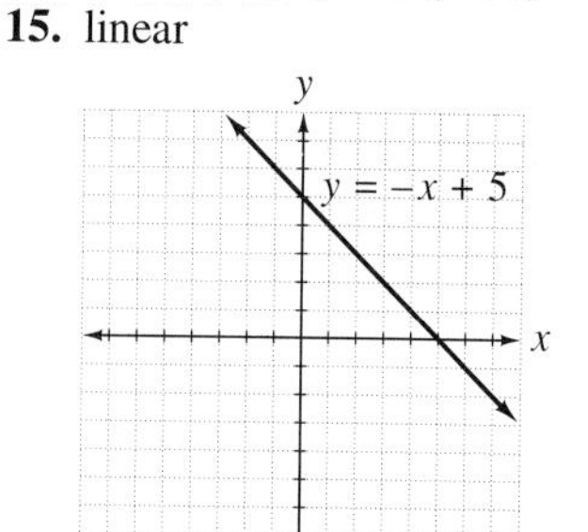

17. domain: $\left\{\frac{3}{4}, -12, 25\right\}$; range: $\left\{-\frac{1}{2}, 6, 0, 25\right\}$; no

19. domain: {triangle, square, rectangle, parallelogram}; range: {3, 4}; yes **21.** domain: {−3}; range: $(-\infty, \infty)$; no

23. domain: $[-1, 1]$; range: $[-1, 1]$; no **25. a.** −11 **b.** 4 **c.** −5 **27. a.** 5 **b.** 5 **c.** −10 **29.** 5080 pounds

31.

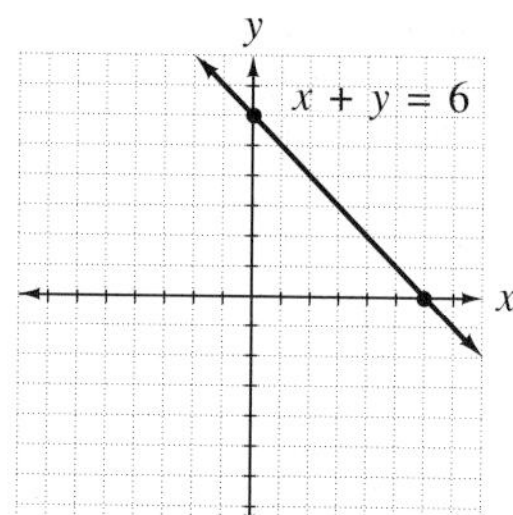

33.

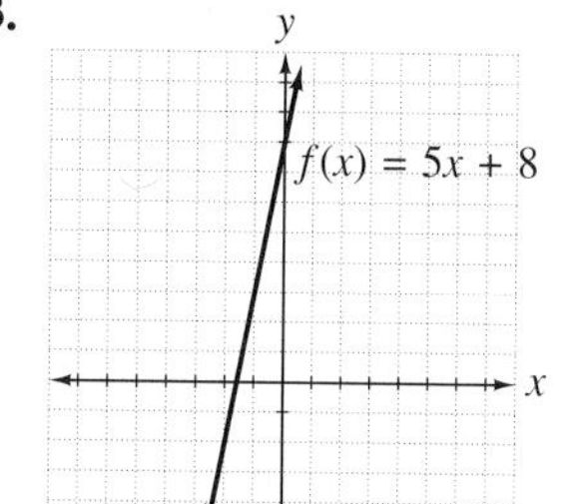

35.

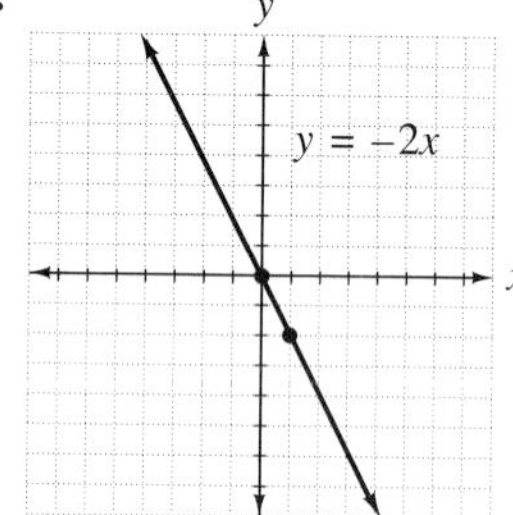

37.

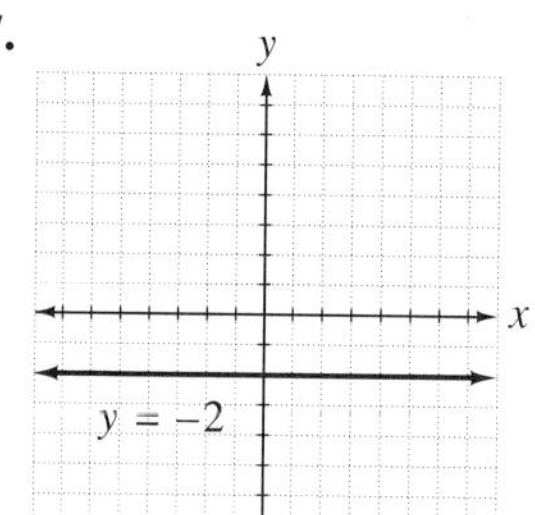

39.

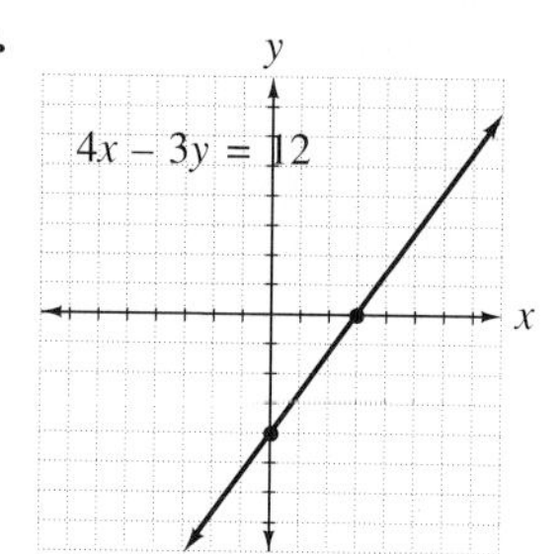

41. $\frac{5}{3}$ **43.** −1 **45.** $m = \frac{1}{2}, b = -2$ **47.** $m = \frac{4}{5}, b = -\frac{9}{5}$ **49.** undefined, no y-intercept **51.** parallel

53. $x = -2$ **55.** $y = 5$ **57.** $f(x) = 2x - 12$ **59.** $f(x) = -x - 2$ **61.** $f(x) = -\frac{3}{2}x - 8$

63. $f(x) = -\frac{3}{2}x - 1$ **65.** $f(x) = -11x - 52$ **67.** $f(x) = -5$

69.

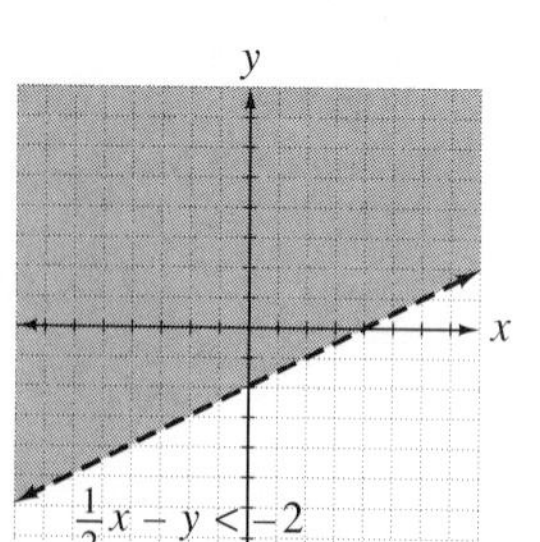

71.

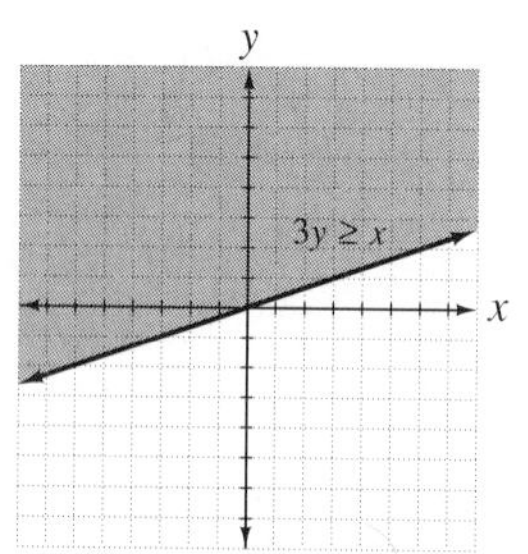

73.

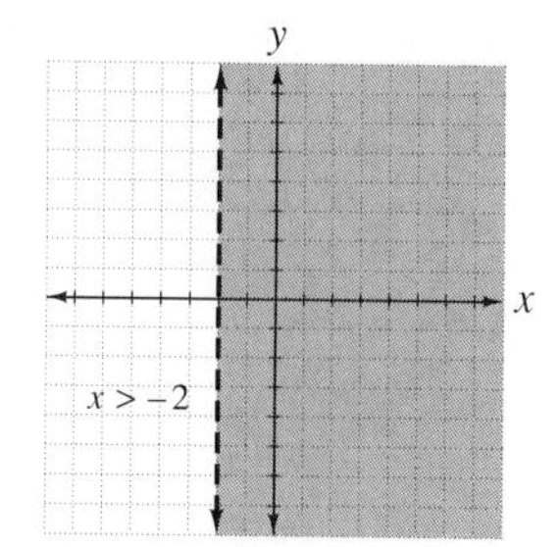

75.

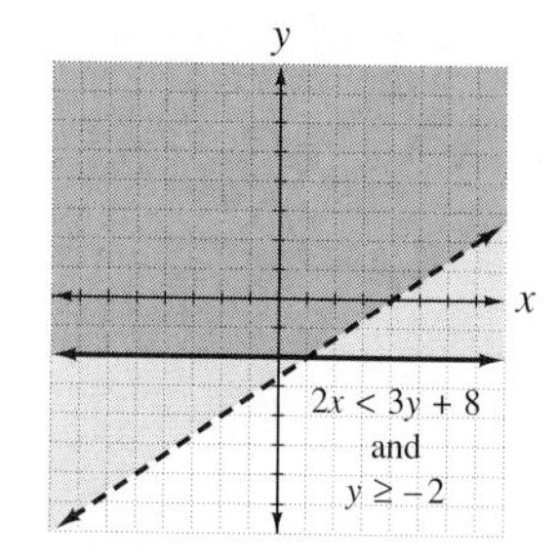

Chapter 4 Test

1. (1, 1) **2.** (−4, 17) **3.** no **4.** no **5.** −3 **6.** −1

7.

$f(x) = -2x + 8$

8.

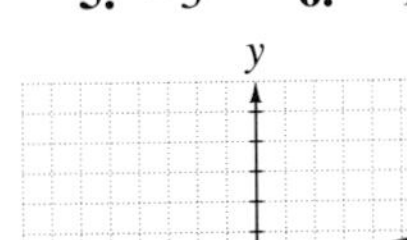
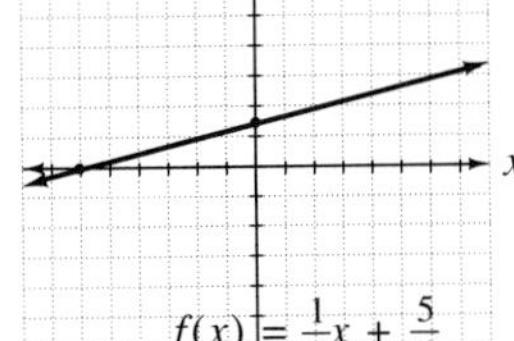

9.

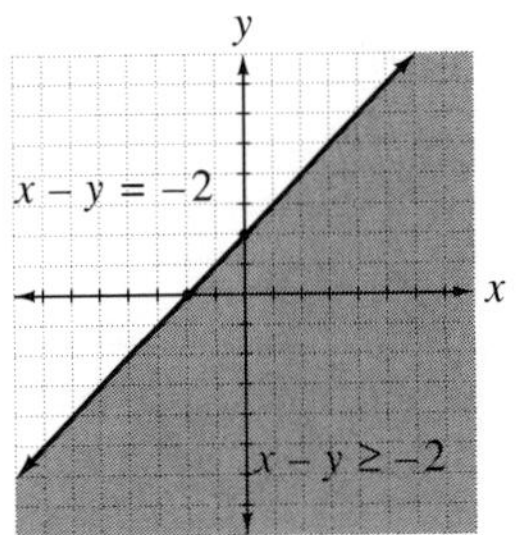

10.

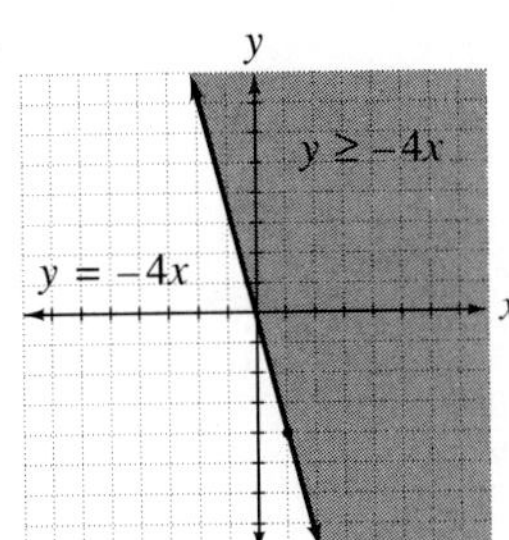

11.

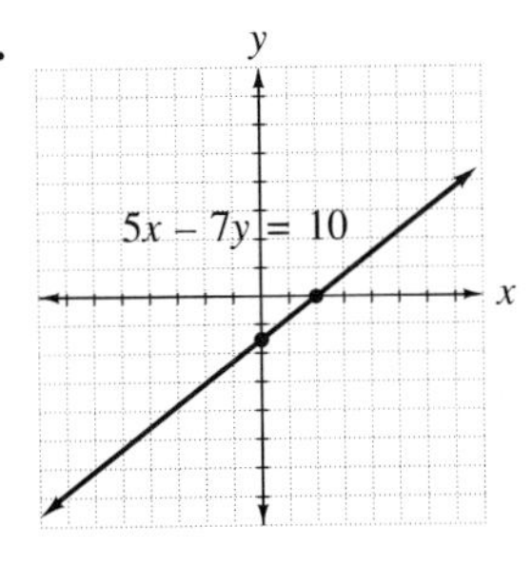

12.

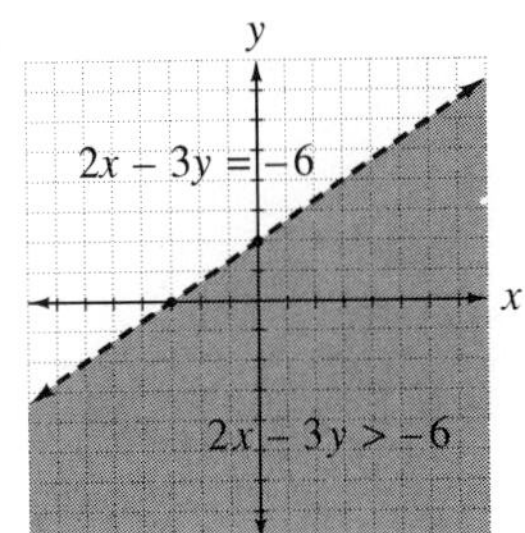

13.

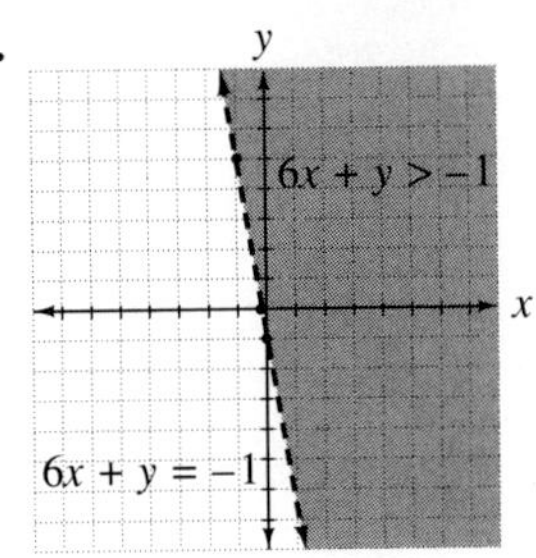

14.

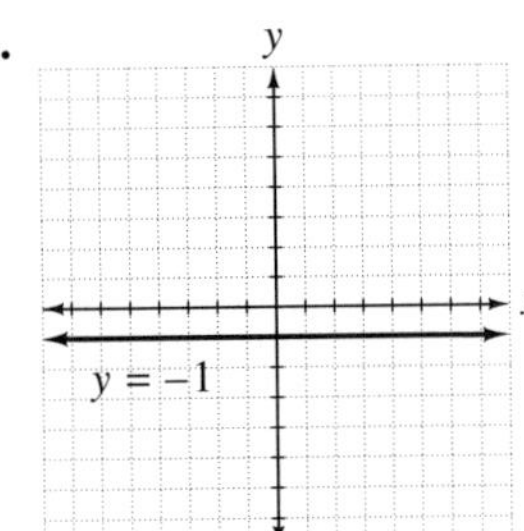

15.

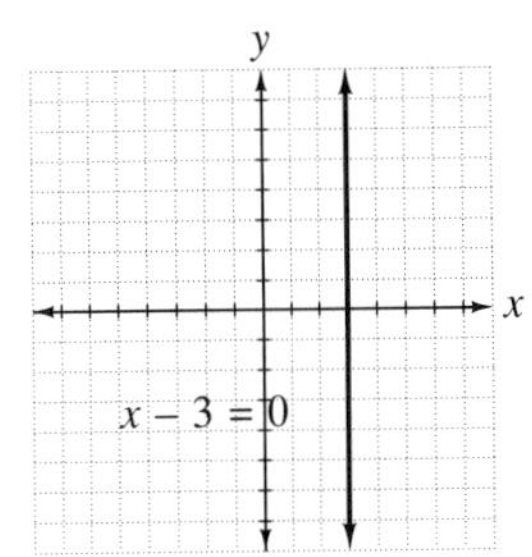

16.

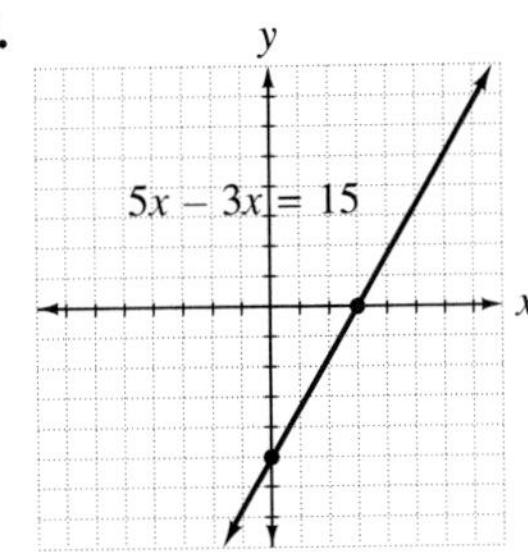

17.

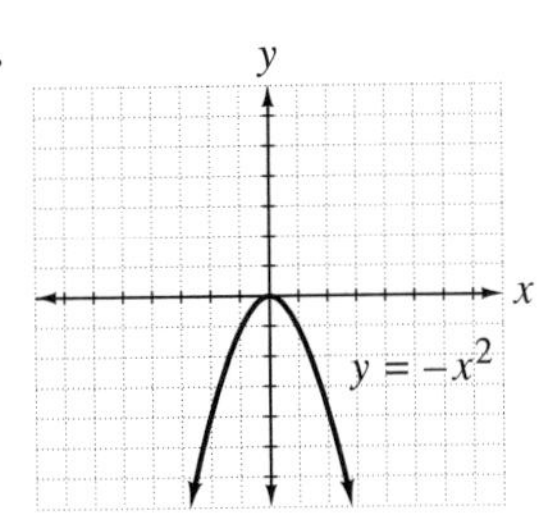

18.

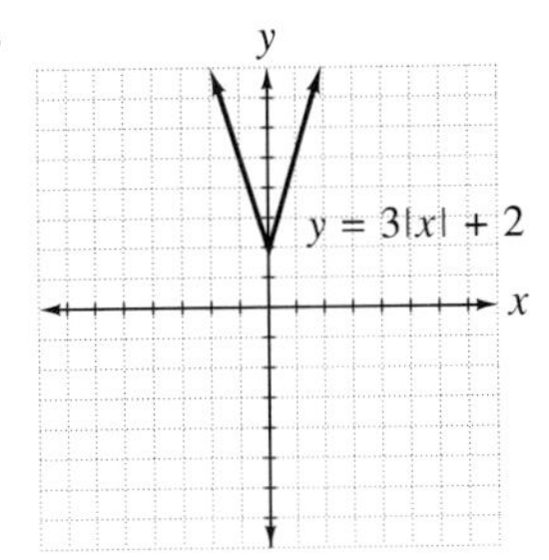

19. domain: $(-\infty, \infty)$; range: $(-\infty, \infty)$; yes **20.** domain: $[0, \infty)$; range: $(-\infty, \infty)$; no **21.** domain: $\{-2\}$; range: $(-\infty, \infty)$; no **22. a.** −8 **b.** −3.6 **c.** −4 **23. a.** 0 **b.** 0 **c.** 60 **24. a.** 6 **b.** 6 **c.** 6 **25.** $f(x) = -\frac{7}{6}x$ **26.** $f(x) = -8x + 11$ **27.** $f(x) = -1$ **28.** $f(x) = \frac{1}{8}x + 12$ **29.** $f(x) = -\frac{1}{3}x + \frac{5}{3}$ **30.** $f(x) = -\frac{1}{2}x - \frac{1}{2}$ **31.** 15.7 cm **32.** 50.3 meters

Chapter 4 Cumulative Review

1. $\frac{2}{39}$; *Sec. 1.3, Ex. 3* **2. a.** 3 **b.** 1 **c.** 2; *Sec. 1.4, Ex. 3* **3. a.** $x + 3$ **b.** $3x$ **c.** $2x$ **d.** $10 - x$ **e.** $5x + 7$; *Sec. 1.5, Ex. 2* **4. a.** −10 **b.** 17 **c.** −21 **d.** −12; *Sec. 1.6, Ex. 6* **5. a.** −8 **b.** 2; *Sec. 1.7, Ex. 6* **6. a.** −3 **b.** 22 **c.** 2 **d.** −1; *Sec. 2.1, Ex. 1* **7.** {6}; *Sec. 2.2, Ex. 4* **8.** 79.2 years; *Sec. 2.4, Ex. 1*

9. length 10 ft, width 4 ft; *Sec. 2.5, Ex. 3* **10. a.** Aladdin **b.** \$72 million; *Sec. 2.6, Ex. 2* **11.** 24 cm; *Sec. 3.1, Ex. 4* **12. a.** 0.35 **b.** 0.895 **c.** 1.5; *Sec. 3.2, Ex. 1* **13.** 7 hours; *Sec. 3.3, Ex. 1* **14.** $\{1, -1\}$; *Sec. 3.4, Ex. 4* **15.** $\left(\frac{5}{2}, \infty\right)$; *Sec. 3.5, Ex. 8* **16.**

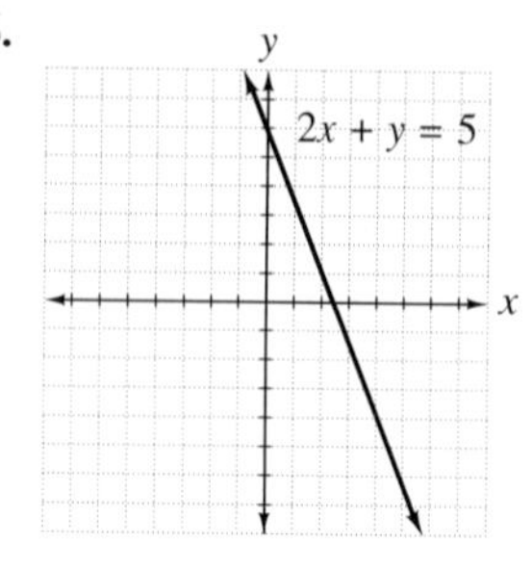

17. a. yes **b.** no; *Sec. 4.2, Ex. 2*

18. *Sec. 4.1, Ex. 2*

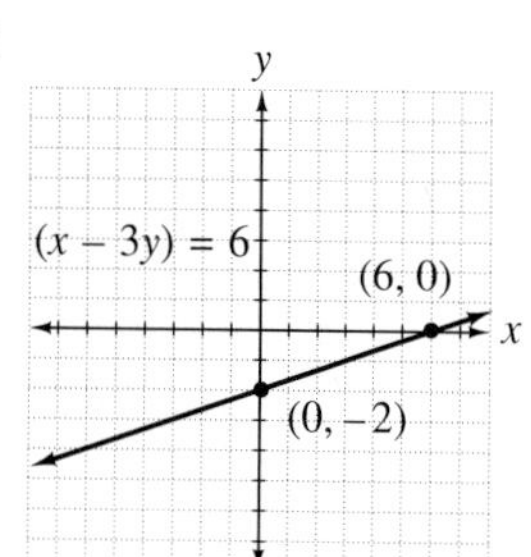

19. $m = 2$; *Sec. 4.4, Ex. 2*

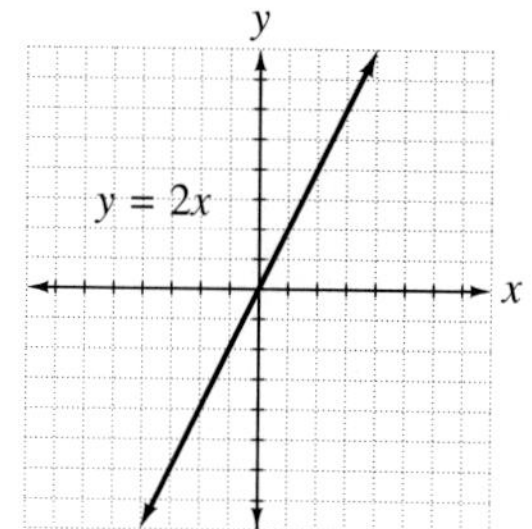

20. $f(x) = \frac{1}{5}x + \frac{23}{5}$; *Sec. 4.5, Ex. 4* **21.** *Sec. 4.6, Ex. 1*

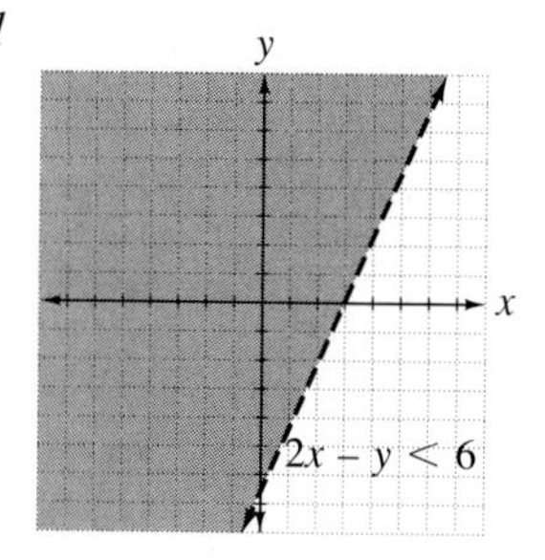

CHAPTER 5
Exponents

Mental Math, Sec. 5.1

1. base = 3; exponent = 2 **3.** base = −3; exponent = 6 **5.** base = 4; exponent = 2 **7.** base = 5; exponent = 3

Exercise Set 5.1

1. 49 **3.** −5 **5.** −16 **7.** 16 **9.** x^7 **11.** $15y^5$ **13.** $-24z^{20}$ **15.** p^7q^7 **17.** $\frac{m^9}{n^9}$ **19.** $x^{10}y^{15}$ **21.** $\frac{4x^2z^2}{y^{10}}$ **23.** x^2 **25.** 4 **27.** p^6q^5 **29.** $\frac{y^3}{2}$ **31.** 1 **33.** −2 **35.** 2 **37.** $5p^2q$ **39.** $81x^2yz^2$ **41.** $6^5m^4n^3$ **43.** ab^4 **45.** −5 **47.** −64 **49.** 1 **51.** 40 **53.** b^6 **55.** a^9 **57.** $-16x^7$ **59.** cannot be combined **61.** $64a^3$ **63.** $36x^2y^2z^6$ **65.** $\frac{y^{15}}{8x^{12}}$ **67.** x **69.** $2x^2y$ **71.** $\frac{2a^5y^5}{7}$ **73.** $-\frac{3x^3y^4}{2}$ **75.** $\frac{a^{15}}{q^{10}}$ **77.** $\frac{y^{18}}{x^3}$ **79.** $\frac{y^2}{25x^2}$ **81.** $243x^7$ **83.** $\frac{xy}{7}$ **85.** $-\frac{25}{q}$ **87.** $20x^5$ square ft **89.** $27y^{12}$ cubic ft **91.** −2 **93.** $\frac{3}{5}$ **95.** yes **97.** 19,470 **99.** x^{9a} **101.** a^{5b} **103.** x^{5a} **105.** $x^{5a^2}y^{5ab}z^{5ac}$

Mental Math, Sec. 5.2

1. $\frac{5}{x^2}$ **3.** y^6 **5.** $4y^3$

Exercise Set 5.2

1. $\frac{1}{4^3} = \frac{1}{64}$ **3.** $\frac{7}{x^3}$ **5.** $4^3 = 64$ **7.** $\frac{5}{6}$ **9.** p^3 **11.** $\frac{q^4}{p^5}$ **13.** $\frac{1}{x^3}$ **15.** z^3 **17.** a^{30} **19.** $\frac{1}{x^{10}y^6}$ **21.** $\frac{z^2}{4}$ **23.** 7.8×10^4 **25.** 1.67×10^{-6} **27.** 9.3×10^7 **29.** 786,000,000 **31.** 0.0000000008673 **33.** 6,250,000,000,000,000,000 **35.** 80,000,000,000 **37.** 15,000 **39.** $\frac{1}{9}$ **41.** $-p^4$ **43.** -2 **45.** r^6 **47.** $\frac{1}{x^{15}y^9}$ **49.** $\frac{4}{3}$ **51.** $\frac{1}{2^5x^5} = \frac{1}{32x^5}$ **53.** $\frac{49a^4}{b^6}$ **55.** $a^{24}b^8$ **57.** x^9y^{19} **59.** $-\frac{y^8}{8x^2}$ **61.** 6.35×10^{-3} **63.** 1.16×10^6 **65.** 2.0×10^7 **67.** 0.033 **69.** 20,320 **71.** 9,460,000,000,000 km **73.** 0.000036 **75.** 0.00000000000000000028 **77.** 0.0000005 **79.** 200,000 **81.** 10 **83.** 0.00008 **85.** 11,000,000 **87.** 2×10^{-3} seconds **89.** 1.512×10^{10} cubic ft **91.** $\frac{8}{x^6y^3}$ cubic meters **93.** $-2x - 5$ **95.** $-y + 26$ **97.** $P = 12x$ **99.** $\{3\}$ **101.** $\left\{\frac{23}{2}\right\}$ **103.** a^m **105.** $27y^{6z}$ **107.** y^{5a} **109.** $z^{(-6a-4)}$

Exercise Set 5.3

1. 0 **3.** 2 **5.** 3 **7.** 1; binomial **9.** 2; trinomial **11.** 3; monomial **13.** 3; none **15.** 57 **17.** 499 **19.** 1 **21.** 1757 ft **23.** 1245 ft **25.** $6y$ **27.** $11x - 3$ **29.** $xy + 2x - 1$ **31.** $18y^2 - 17$ **33.** $3x^2 - 3xy + 6y^2$ **35.** $x^2 - 4x + 8$ **37.** $y^2 + 3$ **39.** $-2x^2 + 5x$ **41.** $-2x^2 - 4x + 15$ **43.** $4x - 13$ **45.** $x^2 + 2$ **47.** $12x^3 + 8x + 8$ **49.** $7x^3 + 4x^2 + 8x - 10$ **51.** $-18y^2 + 11y + 14$ **53.** $-x^3 + 8a - 12$ **55.** $5x^2 - 9x - 3$ **57.** $-3x^2 + 3$ **59.** $2x^3 + 3x^2 + 8xy - 3$ **61.** $7y^2 - 3$ **63.** $5x^2 + 22x + 16$ **65.** $-q^4 + q^2 - 3q + 5$ **67.** $15x^2 + 8x - 6$ **69.** $x^4 - 7x^2 + 5$ **71.** 15 **73.** 38 **75.** 7 **77.** 3 **79.** $(x^2 + 7x + 4)$ ft **81.** $(3y^2 + 4y + 11)$ meters **83.** $(4x^2 - 2x + 1)$ cm **85. a.** 284 ft **b.** 536 ft **c.** 756 ft **d.** 944 ft **87.** 19 seconds **89.** \$26,000 **91.** $6x^2$ **93.** $-10y^3$ **95.** $-12x^8$ **97.** $\left\{\frac{9}{5}\right\}$ **99.** $\{2\}$ **101.** 240 **103. a.** $2a - 3$ **b.** $-2x - 3$ **c.** $2x + 2h - 3$ **105. a.** $4a$ **b.** $-4x$ **c.** $4x + 4h$ **107. a.** $4a - 1$ **b.** $-4x - 1$ **c.** $4x + 4h - 1$

Exercise Set 5.4

1. $-12x^5$ **3.** $9x^2y^3z^3$ **5.** $12x^2 + 21x$ **7.** $-24x^2y - 6xy^2$ **9.** $-4a^3bx - 4a^3by + 12ab$ **11.** $2x^2 - 2x - 12$ **13.** $2x^4 + 3x^3 - 2x^2 + x + 6$ **15.** $x^4 + 4x^3 + 2x^2 - 4x + 1$ **17.** $3x^2 + 14x + 8$ **19.** $2x^3 - 14x^2 + 22x - 10$ **21.** $a^4 - 6a^3 + 26a + 15$ **23.** $6a^4 + a^3b + 4a^2b^2 + b^4$ **25.** $x^2 + x - 12$ **27.** $4x^2 - 24x + 32$ **29.** $3x^2 + 8x - 3$ **31.** $9x^2 - \frac{1}{4}$ **33.** $x^2 + 8x + 16$ **35.** $36y^2 - 1$ **37.** $9x^2 - 6xy + y^2$ **39.** $9b^2 - 36y^2$ **41.** $16b^2 + 32b + 16$ **43.** $4s^2 - 12s + 8$ **45.** $x^2y^2 - 4xy + 4$ **47.** $a^2 - 3a$ **49.** $a^2 + 7a + 10$ **51.** $-12ab^2$ **53.** $4x^2y + 4y^2 + 4yz$ **55.** $9x^2 + 18x + 5$ **57.** $10x^5 + 8x^4 + 2x^3 + 25x^2 + 20x + 5$ **59.** $49x^2 - 9$ **61.** $9x^3 + 30x^2 + 12x - 24$ **63.** $16x^2 - \frac{2}{3}x - \frac{1}{6}$ **65.** $36x^2 + 12x + 1$ **67.** $x^4 - 4y^2$ **69.** $-30a^4b^4 + 36a^3b^2 + 36a^2b^3$ **71.** $2a^2 - 12a + 16$ **73.** $49a^2b^2 - 9c^2$ **75.** $-9x^2 - 6x + 15$ **77.** $m^2 - 8m + 16$ **79.** $9 + 12y - 6c + 4y^2 - 4yc + c^2$ **81.** $25x^2 - 20xy + 4y^2 - 16$ **83.** $9x^2 + 6x + 1$ **85.** $y^2 - 7y + 12$ **87.** $-4x^2 - 4xy - y^2 + 16$ **89.** $2x^3 + 2x^2y + x^2 + xy - x - y$ **91.** $9x^4 + 12x^3 - 2x^2 - 4x + 1$ **93.** $12x^3 - 2x^2 + 13x + 5$ **95. a.** $2x + 10$ **b.** $x^2 + 10x + 25$ **97. a.** $3a^2 + 1$ **b.** $3a^2 - 6a + 4$ **99.** $(4x^2 - 25)$ square yd **101.** $(6x^2 - 4x)$ square in. **103.** $(6x^3 + 36x^2 - 6x)$ cubic ft **105.** $\frac{x^2y^2}{4}$ **107.** $-9x^3y^4$ **109.** $\{-4, 14\}$ **111.** $\{0, -28\}$ **113.** $(-13, 7)$ **115.** $(-\infty, 1] \cup [3, \infty)$ **117.** $5x^{-3} + 15x^{-2} + 10x^{-1}$ **119.** $x^{-4} - 4y^2$ **121.** $3a^2 + 3ay^{-2} + ay^{-4} + y^{-6}$ **123.** $18a^{n+5} - 36a^4$ **125.** $9x^{2y} + 42x^y + 49$

Exercise Set 5.5

1. $2b^5$ **3.** $\frac{2y^3z^2}{x^4b}$ **5.** $-\frac{1}{3y^2}$ **7.** $2a + 4$ **9.** $3ab + 4$ **11.** $2x^2 + 3x - 2$ **13.** $x^2 + 2x + 1$ **15.** $x + 1 + \frac{1}{x+2}$ **17.** $2x - 8$ **19.** $x - \frac{1}{2}$ **21.** $2x^2 - \frac{1}{2}x + 5$ **23.** $\frac{5b^5}{2a^3}$ **25.** $x^3y^3 - 1$ **27.** $a + 3$ **29.** $2x + 5$ **31.** $4y - 6y^2$ **33.** $2x + 23 + \frac{130}{x-5}$ **35.** $10x + 3y - 6x^2y^2$ **37.** $2x + 4$ **39.** $y + 5$ **41.** $2x + 3$ **43.** $2x^2 - 8x + 38 - \frac{156}{x+4}$ **45.** $3x + 3 - \frac{1}{x-1}$ **47.** $-2x^3 + 3x^2 - x + 4$ **49.** $3x^3 + 5x + 4 - \frac{2x}{x^2-2}$ **51.** $x - \frac{5}{3x^2}$ **53.** $(3x^3 + x - 4)$ ft **55.** $(7x - 10)$ in. **57.** $(6x^3 - 2x + 1)$ ft **59.** $(x - 5)$ cm **61.** $P(-2) = -13$; $P(-2) =$ the remainder **63.** 18 to 24 years old **65.** answers vary **67.** Vermont 95 ft; Louisiana −8 ft **69.**

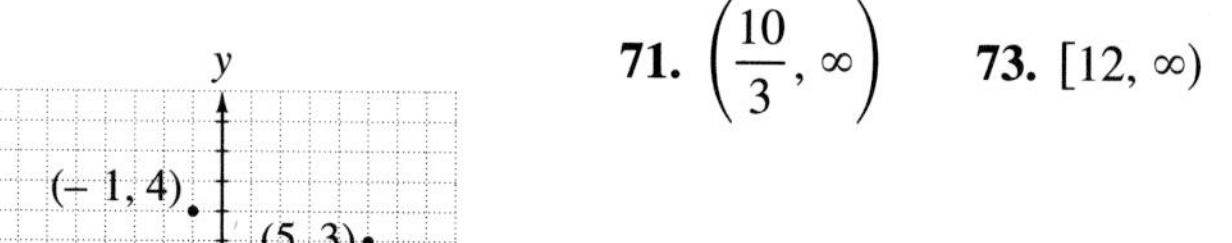

71. $\left(\frac{10}{3}, \infty\right)$ **73.** $[12, \infty)$ **75.** $2x^2 + \frac{1}{2}x - 5$ **77.** $2x^3 + \frac{9}{2}x^2 + 10x + 21 + \frac{42}{x-2}$ **79.** $3x^4 - 2x$

Exercise Set 5.6

1. $x + 8$ **3.** $x - 1$ **5.** $x^2 - 5x - 23 - \frac{41}{x-2}$ **7.** $4x + 8 + \frac{7}{x-2}$ **9.** 3 **11.** 73 **13.** −8 **15.** $x^2 + \frac{2}{x-3}$ **17.** $6x + 7 + \frac{1}{x+1}$ **19.** $2x^3 - 3x^2 + x - 4$ **21.** $3x - 9 + \frac{12}{x+3}$ **23.** $3x^2 - \frac{9}{x}x + \frac{7}{4} + \frac{47}{8x-4}$ **25.** $3x^2 + 3x - 3$ **27.** $3x^2 + 4x - 8 + \frac{20}{x+1}$ **29.** $x^2 + x + 1$ **31.** $x - 6$ **33.** 1 **35.** −133 **37.** 3 **39.** $\frac{-187}{81}$ **41.** $\frac{95}{32}$ **43.** $(x + 3)(x^2 + 4) = x^3 + 3x^2 + 4x + 12$ **45.** $P(c) = 0$ **47.** $x^3 + 2x^2 + 7x + 28$ **49.** $(x - 1)$ meters **51.** $2x^2 + 2$ **53.** $6x^2 + 14$ **55.** $x^4 + 6x^2 + 9$ **57.** $\frac{x-5}{3x^2+5}$ **59.** = **61.** =

Exercise Set 5.7

1. 15 **3.** 3 **5.** 4 **7.** $\frac{1}{3}$ **9.**

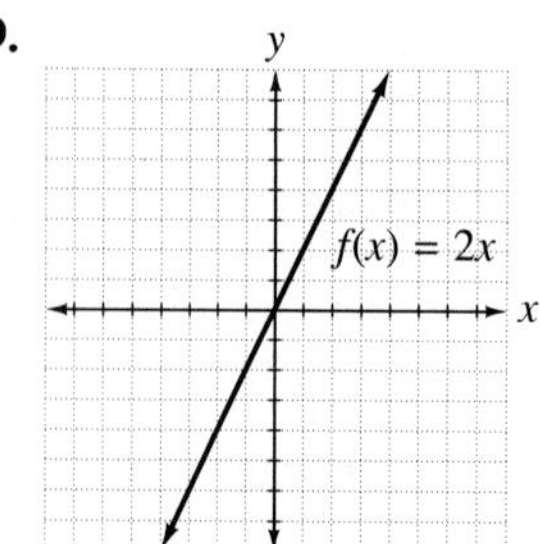

11.

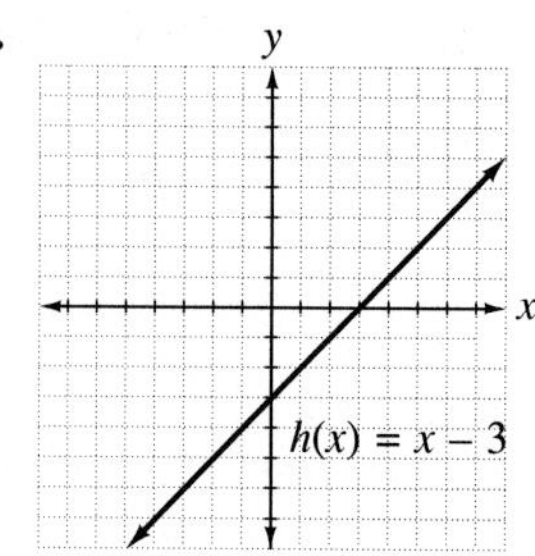

13.

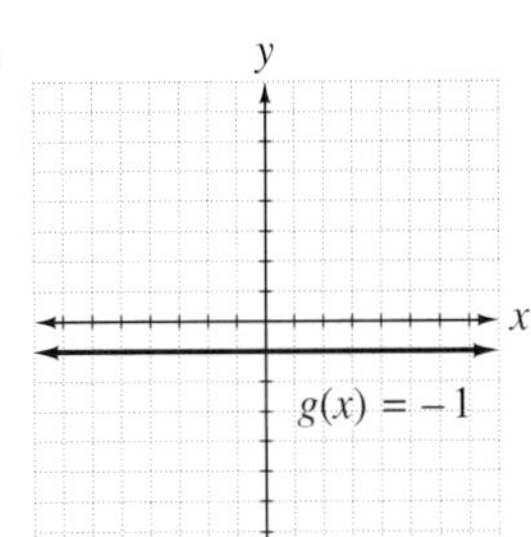

15. $x^2 + 5x + 1$ **17.** $x^2 - 5x + 1$ **19.** $\frac{5x}{x^2 + 1}$ **21.** $(f \circ g)(x) = 25x^2 + 1$

23. 5 **25.** 51 **27.** 0 **29.** $(f + g)(x) = x^2 - 2x + 2$ **31.** $\left(\frac{f}{g}\right)(x) = -\frac{2x}{x^2 + 2}$
33. $(g - h)(x) = x^2 - 4x - 1$ **35.** 19 **37.** $(f \circ g)(x) = -2x^2 - 4$ **39.** 27 **41.** $(h + f)(x) = 2x + 3$
43. $(f \circ f)(x) = 4x$ **45.** $f(a + b) = -2a - 2b$ **47.** $\left(\frac{f}{h}\right)(x) = -\frac{2x}{4x + 3}$ **49.** $(g \circ h)(x) = 16x^2 + 24x + 11$

51.

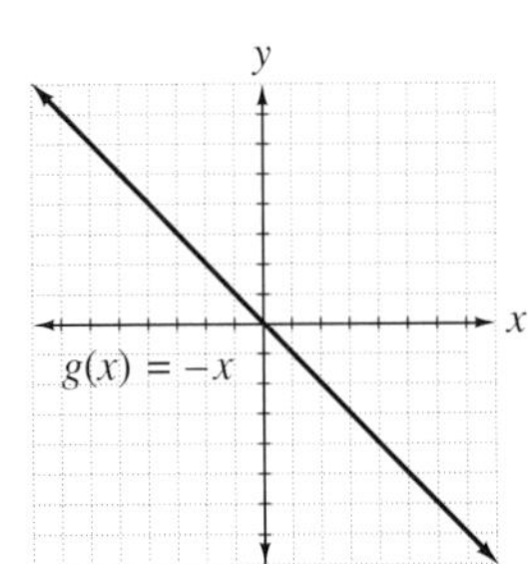

53.

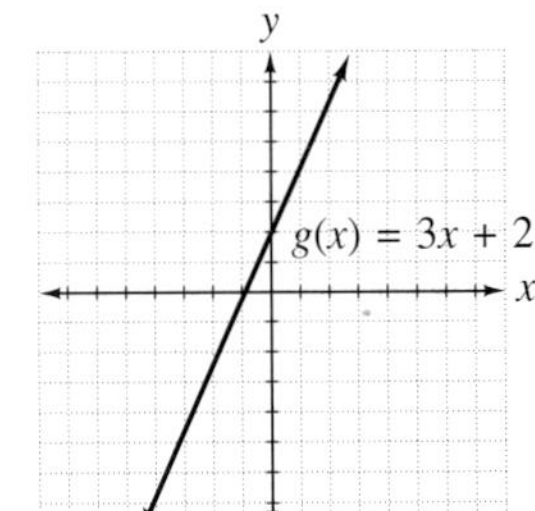

55.

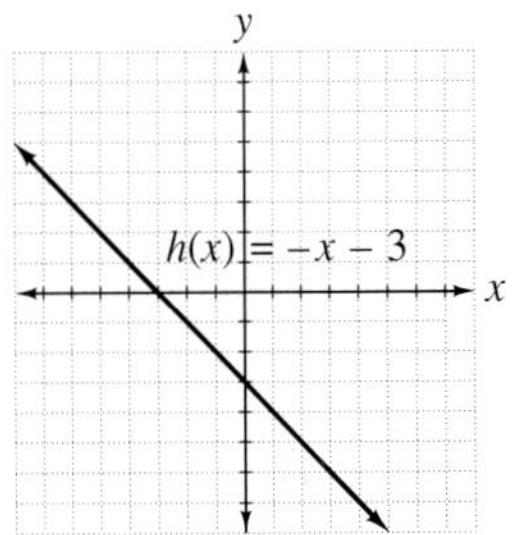

57.

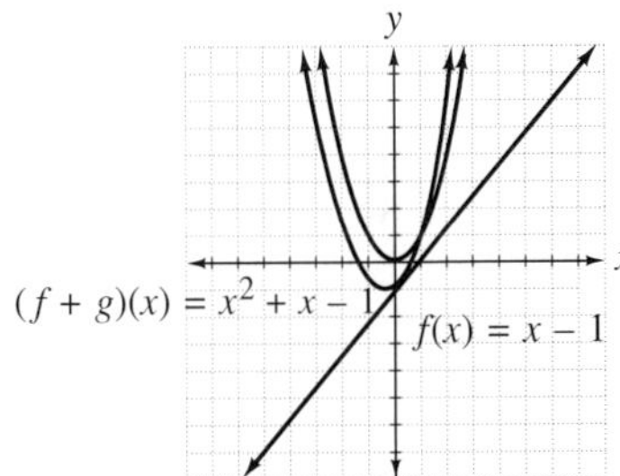

59.

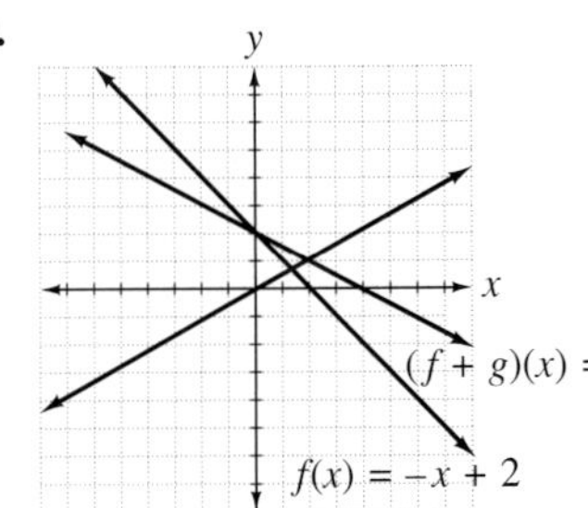

61. 51 pounds

63. beef consumption decreasing; chicken consumption increasing **65.** $(f \circ g)(x) = x$; $(g \circ f)(x) = x$
67. $P(x) = R(x) - C(x)$ **69.** 100π cm^2 **71. a.** $W(x) = 2x$ **b.** 130 pounds **73.** $63x^2 - 7x$
75. $81x^2y - 36xy$ **77.** $\{0\}$ **79.** $\left\{\frac{17}{6}\right\}$ **81.** -6 **83.** $y = x + 6$

Chapter 5 Review

1. base = 3; exponent = 2 **3.** base = 5; exponent = 4 **5.** 36 **7.** -65 **9.** 1 **11.** 8 **13.** $-10x^5$ **15.** $\frac{b^4}{16}$
17. $\frac{x^6y^6}{4}$ **19.** $40a^{19}$ **21.** $\frac{3}{64}$ **23.** $\frac{1}{x}$ **25.** 5 **27.** 1 **29.** $6a^6b^9$ **31.** $\frac{1}{49}$ **33.** $\frac{2}{x^4}$ **35.** 125
37. $1\frac{1}{16} = \frac{17}{16}$ **39.** $8q^3$ **41.** $\frac{s^4}{r^3}$ **43.** $-\frac{3x^3}{4r^4}$ **45.** $\frac{x^{15}}{8}$ **47.** $\frac{a^2b^2c^4}{9}$ **49.** $\frac{5}{x^3}$ **51.** c^4 **53.** $\frac{1}{x^6y^{13}}$ **55.** $-\frac{15}{16}$
57. a^{11m} **59.** $27x^3y^{6z}$ **61.** 2.7×10^{-4} **63.** 8.08×10^7 **65.** 2.58×10^8 **67.** 867,000 **69.** 0.00086
71. 100,000,000,000,000,000,000 **73.** 0.016 **75.** 4 **77.** 1 **79.** 1 **81.** $-4xy^3 - 3x^3y$ **83.** $-4x^2 + 10y^2$
85. $-4x^3 + 4x^2 + 16xy - 9x + 18$ **87.** $x^2 - 6x + 3$ **89.** 290 **91.** 110 **93.** 601
95. $-24x^2y^4 + 36x^2y^3 - 6xy^3$ **97.** $2x^2 + x - 36$ **99.** $9x^2a^2 - 24xab + 16b^2$ **101.** $15x^2 + 18x - 81$
103. $9x^2 - 6xy + y^2$ **105.** $x^2 - 9y^2$ **107.** $-9a^2 + 6ab - b^2 + 16$ **109.** $y^4 - 18y^3 + 87y^2 - 54y + 9$

111. $16y^6 + 24x^2y^3 + 9x^4$ **113.** $16x^2y^{2z} - 8xy^zb + b^2$ **115.** **a.** $5a^2 + 2a$ **b.** $5a^2 + 10ah + 5h^2 + 2a + 2h$ **c.** $25x^4 + 20x^3 + 4x^2$ **117.** $\frac{9b^2z^3}{4a}$ **119.** $\frac{3}{b} + 4b$ **121.** $2x^3 + 6x^2 + 17x + 56 + \frac{156}{x - 3}$

123. $x^2 + \frac{7}{2}x - \frac{1}{4} + \frac{15}{8x - 4}$ **125.** $3x^2 + 6$ **127.** $3x^2 - \frac{5}{2}x - \frac{1}{4} - \frac{5}{8x + 12}$ **129.** $x^2 + 3x + 9 - \frac{54}{x - 3}$

131. $3x^3 - 6x^2 + 10x - 20 + \frac{50}{x + 2}$ **133.** -9323 **135.** $\frac{365}{32}$ **137.** $x^2 + x - 1$ **139.** $x^2 - 2x + 2 - \frac{2}{x + 1}$

141. $x^2 + 2x - 1$ **143.** 18 **145.** -2 **147.**

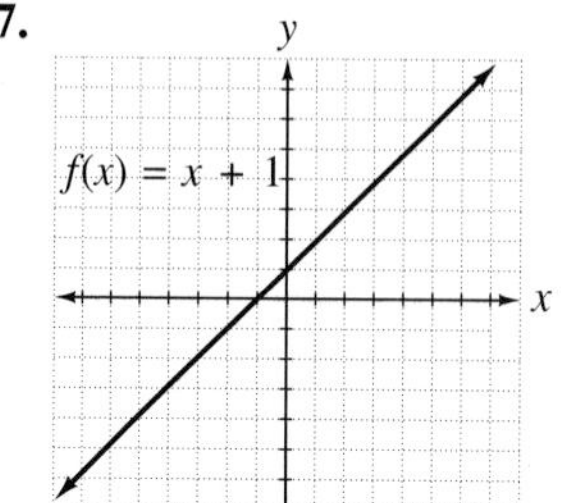

149.

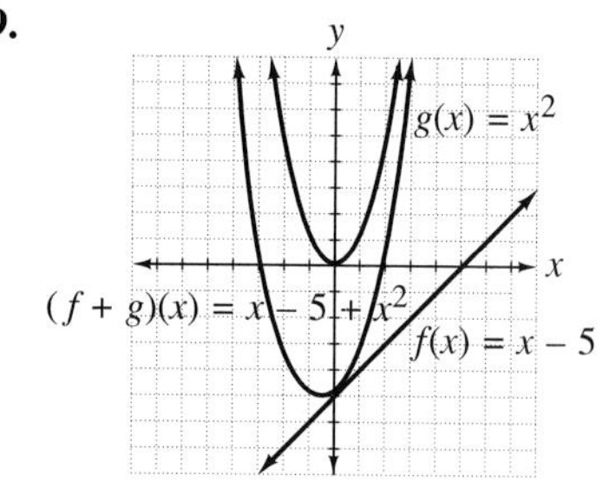

Chapter 5 Test

1. -8 **2.** $\frac{1}{36}$ **3.** $-12x^2z$ **4.** $\frac{1}{81x^2}$ **5.** $-\frac{y^{40}}{z^5}$ **6.** $\frac{12x^5z^3}{y^7}$ **7.** 6.3×10^8 **8.** 1.2×10^{-2} **9.** 0.000005 **10.** 0.0009 **11.** $-5x^3 - 11x - 9$ **12.** $14x^6 - 3xy^2 + 6y^2 + 15$ **13.** 95 **14.** $-12x^2y - 3xy^2$ **15.** $12x^2 - 5x - 28$ **16.** $81x^4 + 72x^2y + 36x^2 + 16y^2 + 16y + 4$ **17.** $16x^2 - 8xy + y^2 - 24x + 6y + 9$

18. $\frac{4xy}{3z} + \frac{3}{z} + \frac{1}{3x}$ **19.** $x^5 + 5x^4 + 8x^3 + 16x^2 + 33x + 63 + \frac{128}{x - 2}$ **20.** $4x^3 - \frac{1}{3}x^2 + \frac{16}{9}x + \frac{5}{27} - \frac{71}{81\left(x - \frac{2}{3}\right)}$

21. 91 **22.** $(h - g)(x) = x^2 - 7x + 12$ **23.** $(h \cdot f)(x) = x^3 - 6x^2 + 5x$ **24.** $(g \circ f)(x) = x - 7$ **25.** $(g \circ h)(x) = x^2 - 6x - 2$ **26.**

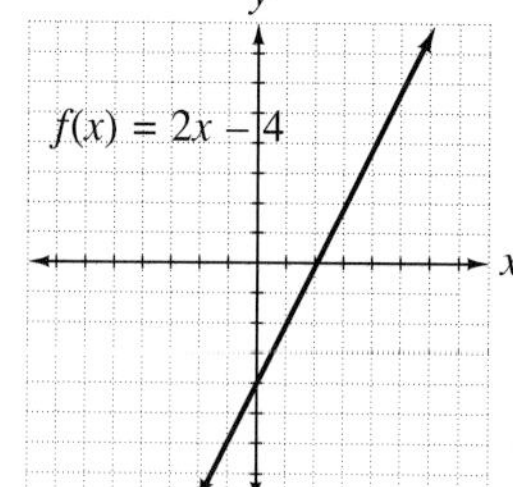

27. **a.** $2a^2 - 4$ **b.** $2a^2 + 8a + 4$ **c.** $4x^4 - 16x^2 + 16$

28. $x^2 - 8x + 16$ **29.** $P(x) = 1.2x - 10{,}000$; 14,000

Chapter 5 Cumulative Review

1. **a.** $\frac{13}{20}$ **b.** $\frac{12}{11}$ **c.** $\frac{5}{4}$; *Sec. 1.3, Ex. 7* **2.** **a.** $\frac{15}{x} = 4$ **b.** $12 - 3 = x$ **c.** $4x + 17 = 21$; *Sec. 1.5, Ex. 4*

3. -12; *Sec. 1.6, Ex. 7* **4.** $x = -12$; *Sec. 2.2, Ex. 7* **5.** no; *Sec. 2.4, Ex. 4* **6.** 138 minutes; *Sec. 2.5, Ex. 4*

7. $\{-2, 2\}$; *Sec. 3.4, Ex. 1* **8.**

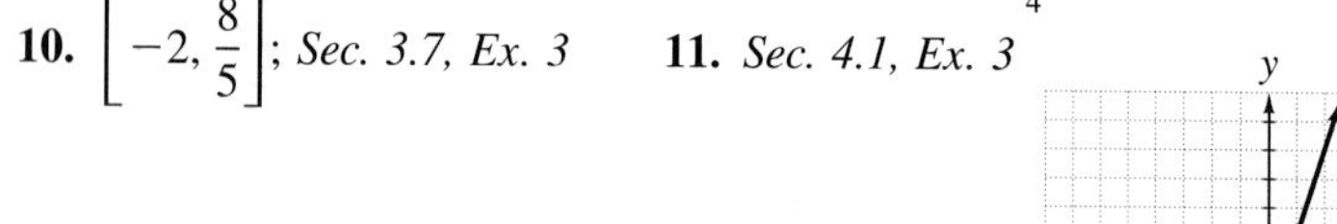

$(-\infty, 4]$; *Sec. 3.5, Ex. 6* **9.** $(-3, 2)$; *Sec. 3.6, Ex. 2*

10. $\left[-2, \frac{8}{5}\right]$; *Sec. 3.7, Ex. 3* **11.** *Sec. 4.1, Ex. 3* **12.** *Sec. 4.3, Ex. 5*

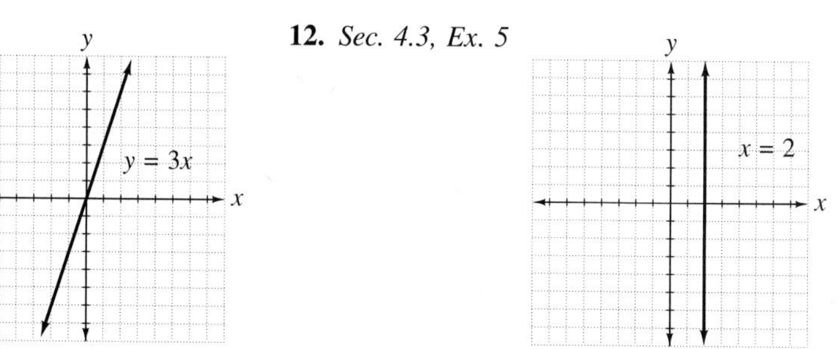

13. $2x + y = 3$; *Sec. 4.5, Ex. 2* **14.** $-6x^7$; *Sec. 5.1, Ex. 3* **15. a.** x^3 **b.** 4^4 **c.** -27 **d.** $2x^4y$; *Sec. 5.1, Ex. 6*
16. a. y^3 **b.** $\frac{q^9}{p^4}$ **c.** $\frac{1}{x^{12}}$; *Sec. 5.2, Ex. 3* **17.** $12x^3 - 12x^2 - 9x + 2$; *Sec. 5.3, Ex. 7*
18. $6x^2 - 29x + 28$; *Sec. 5.4, Ex. 7* **19.** $\frac{7a}{2b} - 1$; *Sec. 5.5, Ex. 3* **20.** $3x^2 + 2x + 3 + \frac{-6x + 9}{x^2 - 1}$; *Sec. 5.5, Ex. 7*
21. $x^3 - 4x^2 - 3x + 11 + \frac{12}{x + 2}$; *Sec. 5.6, Ex. 2* **22. a.** -3 **b.** $2a + 5$ **c.** $2x + 2h + 5$; *Sec. 5.7, Ex. 1*

CHAPTER 6
Factoring Polynomials

Mental Math, Sec. 6.1

1. 6 **3.** 5 **5.** x **7.** $7x$

Exercise Set 6.1

1. 6 **3.** 28 **5.** 4 **7.** 6 **9.** a^3 **11.** y^2z^2 **13.** $3x^2y$ **15.** $5xz^3$ **17.** $6(3x - 2)$ **19.** $4y^2(1 - 4xy)$
21. $2x^3(3x^2 - 4x + 1)$ **23.** $4ab(2a^2b^2 - ab + 1 + 4b)$ **25.** $(x + 3)(6 + 5a)$ **27.** $(z + 7)(2x + 1)$
29. $(x^2 + 5)(3x - 2)$ **31.** $(a + 2)(b + 3)$ **33.** $(a - 2)(c + 4)$ **35.** $(x - 2)(2y - 3)$ **37.** $(4x - 1)(3y - 2)$
39. $3(2x^3 + 3)$ **41.** $x^2(x + 3)$ **43.** $4a(2a^2 - 1)$ **45.** $-4xy(5x - 4y^2)$ **47.** $5ab^2(2ab + 1 - 3b)$
49. $3b(3ac^2 + 2a^2c - 2a + c)$ **51.** $(y - 2)(4x - 3)$ **53.** $(n - 8)(2m - 1)$ **55.** $3x^2y^2(5x - 6)$
57. $(2x + 3)(3y + 5)$ **59.** $(x + 3)(y - 5)$ **61.** $(2a - 3)(3b - 1)$ **63.** $(6x + 1)(2y + 3)$ **65.** $(2x + 3y)(x + 2)$
67. $(5x - 3)(x + y)$ **69.** $(x^2 + 4)(x + 3)$ **71.** $(x^2 - 2)(x - 1)$ **73.** $2\pi r(r + h)$ **75.** $A = P(1 + RT)$
77. $h(t) = -16t(t - 4)$ **79.** $55x^7$ **81.** $125x^6$ **83.** $x^2 - 3x - 10$ **85.** $x^2 + 5x + 6$ **87.** $y^2 - 4y + 3$
89. $8x^2 + 2x - 7$ **91.** $x^n(x^{2n} - 2x^n + 5)$ **93.** $2x^{3a}(3x^{5a} - x^{2a} - 2)$ **95.** $x^{-2}(3 + 8x)$ **97.** $xy^{-3}(3x + 2y^2)$
99. $2x^{-2}y^{-4}(3y^3 - 1 + 4x^3y^2)$

Mental Math, Sec. 6.2

1. $(x + 5)$ **3.** $(x - 3)$ **5.** $(x + 2)$

Exercise Set 6.2

1. $(x + 6)(x + 1)$ **3.** $(x + 5)(x + 4)$ **5.** $(x - 5)(x - 3)$ **7.** $(x - 9)(x - 1)$ **9.** not factorable
11. $(x - 6)(x + 3)$ **13.** not factorable **15.** $(x + 5y)(x + 3y)$ **17.** $(x - y)(x - y)$ **19.** $(x - 4y)(x + y)$
21. $2(z + 8)(z + 2)$ **23.** $2x(x - 5)(x - 4)$ **25.** $7(x + 3y)(x - y)$ **27.** $(x + 12)(x + 3)$ **29.** $(x - 2)(x + 1)$
31. $(r - 12)(r - 4)$ **33.** $(x - 7)(x + 3)$ **35.** $(x + 5y)(x + 2y)$ **37.** not factorable **39.** $2(t + 8)(t + 4)$
41. $x(x - 6)(x + 4)$ **43.** $(x - 9)(x - 7)$ **45.** $(x + 2y)(x - y)$ **47.** $3(x + 5)(x - 2)$ **49.** $3(x - 18)(x - 2)$
51. $(x - 24)(x + 6)$ **53.** $6x(x + 4)(x + 5)$ **55.** $2t^3(t - 4)(t - 3)$ **57.** $5xy(x - 8y)(x + 3y)$ **59.** $4(x^2 + x - 3)$
61. $(x - 3)(x - 7)$ **63.** $2y(x + 5)(x + 10)$ **65.** $-12y^3(x^2 + 2x + 3)$ **67.** $(x + 1)(y - 5)(y + 3)$ **69.** $\frac{1}{25}$
71. $\frac{1}{x^{15}}$ **73.** 1.248×10^6 **75.** $3x^2 + 13x - 10$ **77.** $y^2 - 2y + 1$ **79.** $20x^2 - 11x - 3$ **81.** $25x^2 + 10x + 1$

Mental Math, Sec. 6.3

1. Yes **3.** No **5.** Yes

Exercise Set 6.3

1. $(2x + 3)(x + 5)$ **3.** $(2x + 1)(x - 5)$ **5.** $(2y + 3)(y - 2)$ **7.** $(4a - 3)^2$ **9.** $(9r - 8)(4r + 3)$
11. $(5x + 1)(2x + 3)$ **13.** $3(7x + 5)(x - 3)$ **15.** $2(2x - 3)(3x + 1)$ **17.** $x(4x + 3)(x - 3)$ **19.** $(x + 11)^2$
21. $(x - 8)^2$ **23.** $(4y - 5)^2$ **25.** $(xy - 5)^2$ **27.** $(2x + 11)(x - 9)$ **29.** $(2x - 7)(2x + 3)$
31. $(6x - 7)(5x - 3)$ **33.** $(4x - 9)(6x - 1)$ **35.** $(3x - 4y)^2$ **37.** $(x - 7y)^2$ **39.** $(2x + 5)(x + 1)$

41. not factorable **43.** $(5 - 2y)(2 + y)$ **45.** $(4x + 3y)^2$ **47.** $2y(4x - 7)(x + 6)$ **49.** $(3x - 2)(x + 1)$ **51.** $(xy + 2)^2$ **53.** $(7y + 3x)^2$ **55.** $3(x^2 - 14x + 21)$ **57.** $(7a - 6)(6a - 1)$ **59.** $(6x - 7)(3x + 2)$ **61.** $(5p - 7q)^2$ **63.** $(5x + 3)(3x - 5)$ **65.** $(7t + 1)(t - 4)$ **67.** $-3xy^2(4x - 5)(x + 1)$ **69.** $-2pq(p - 3q)(15p + q)$ **71.** $(y - 1)^2(4x^2 + 10x + 25)$ **73.** $y(3x - 8)(x - 1)$ **75.** $2(x + 3)(x - 2)$ **77.** $(x + 2)(x - 7)$ **79.** $(2x^3 - 3)(x^3 + 3)$ **81.** $2x(6y^2 - z)^2$ **83.** $h(3h + 4)(h - 2)$ **85.** $x^3 - 8$ **87.** $x^2 - 9$ **89.** $25a^2 - b^2$ **91.** -9 **93.** -8 **95.**

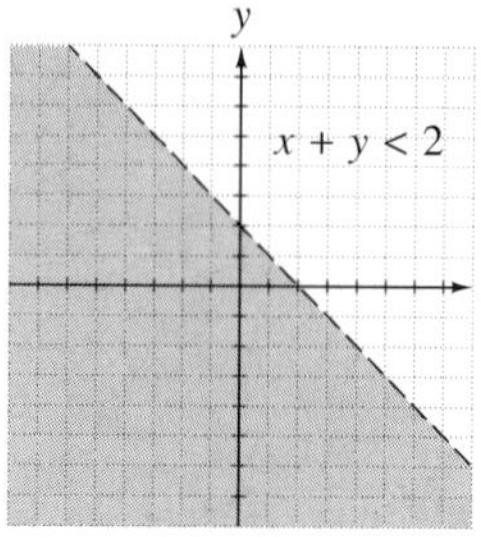

97.

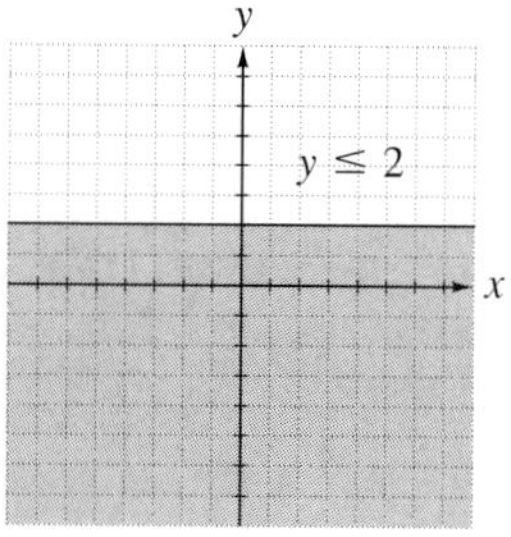

99. $(x^n + 2)(x^n + 8)$ **101.** $(x^n - 6)(x^n + 3)$ **103.** $(2x^n + 1)(x^n + 5)$ **105.** $(2x^n - 3)^2$

Mental Math, Sec. 6.4

1. 1^2 **3.** 9^2 **5.** 3^2 **7.** 1^3 **9.** 2^3

Exercise Set 6.4

1. $(5y - 3)(5y + 3)$ **3.** $(11 - 10x)(11 + 10x)$ **5.** $3(2x - 3)(2x + 3)$ **7.** $(13a - 7b)(13a + 7b)$ **9.** $(xy - 1)(xy + 1)$ **11.** $(a + 3)(a^2 - 3a + 9)$ **13.** $(2a + 1)(4a^2 - 2a + 1)$ **15.** $5(k + 2)(k^2 - 2k + 4)$ **17.** $(xy - 4)(x^2y^2 + 4xy + 16)$ **19.** $(x + 5)(x^2 - 5x + 25)$ **21.** $3x(2x - 3y)(4x^2 + 6xy + 9y^2)$ **23.** $(x - 2)(x + 2)$ **25.** $(9 - p)(9 + p)$ **27.** $(2r - 1)(2r + 1)$ **29.** $(3x - 4)(3x + 4)$ **31.** not factorable **33.** $(3 - t)(9 + 3t + t^2)$ **35.** $8(r - 2)(r^2 + 2r + 4)$ **37.** $(t - 7)(t^2 + 7t + 49)$ **39.** $(x - 13y)(x + 13y)$ **41.** $(xy - z)(xy + z)$ **43.** $(xy + 1)(x^2y^2 - xy + 1)$ **45.** $(s - 4t)(s^2 + 4st + 16t^2)$ **47.** $2(3r - 2)(3r + 2)$ **49.** $x(3y - 2)(3y + 2)$ **51.** $25y^2(y - 2)(y + 2)$ **53.** $xy(x - 2y)(x + 2y)$ **55.** $4s^3t^3(2s^3 + 25t^3)$ **57.** $xy^2(27xy - 1)$ **59.** $(x - 2)(x + 2)(x^2 + 4)$ **61.** $(a - 2 - b)(a + 2 + b)$ **63.** $(x - 2)^2(x + 1)(x + 3)$ **65.** $(x + 6)$ **67.** $(2x + y)$ **69.** $\pi(R - r)(R + r)$ **71.** $(x + y)(x - y)$ **73.** $4x^3 + 2x^2 - 1 + \frac{3}{x}$ **75.** $2x + 1$ **77.** $3x + 4 - \frac{2}{x + 3}$ **79.**

$(7, 8)$ **81.**

$[-2, 28]$ **83.**

$[-5, \infty)$ **85.**

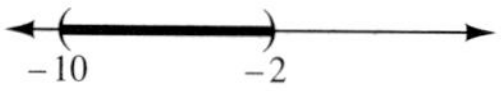

$(-10, -2)$ **87.** $(x^n - 5)(x^n + 5)$ **89.** $(x^n - 3)(x^n + 3)$ **91.** $(6x^n - 7)(6x^n + 7)$ **93.** $(x^n - 2)(x^n + 2)(x^{2n} + 4)$

Exercise Set 6.5

1. $(x + 3)(x - 3)$ **3.** $(x - 9)(x + 1)$ **5.** prime polynomial **7.** $(x - 4 + y)(x - 4 - y)$ **9.** $x(x - 1)(x^2 + x + 1)$ **11.** $2xy(7x - 1)$ **13.** $2ab(4a - 3b)$ **15.** $x^3(x - 1)$ **17.** $x^2(x + 2)(x - 2)$ **19.** $4(x + 2)(x - 2)$ **21.** $x(1 + 3x)(1 - 3x)$ **23.** $(3x - 11)(x + 1)$ **25.** $4(x + 3)(x - 1)$ **27.** $(4x + 3)(x + 5)$ **29.** $(3x^2 - 4)(2 + y)$ **31.** $2(x + 3)(y + 4)$ **33.** $(x + 2)(x - 2)(x + 3)$ **35.** $(2x + 9)^2$ **37.** $(3x - 5)^2$ **39.** prime polynomial **41.** $3(4x - 1)^2$ **43.** prime polynomial **45.** $2(x^2 + 1)(x + 1)(x - 1)$ **47.** $(a - 2b)(a^2 + 2ab + 4b^2)$ **49.** $(5 - x)(25 + 5x + x^2)$ **51.** $(2x + 3y)(4x^2 - 6xy + 9y^2)$ **53.** $2(a + 4b)(a^2 - 4ab + 16b^2)$ **55.** $6a^2(b^2 + 2a)(b^2 - 2a)$ **57.** $2y^3(2xy + 1)(2xy - 1)$ **59.** $(x + 3 + y)(x + 3 - y)$ **61.** $2(x - 5 + y)(x - 5 - y)$ **63.** $3(a - 1 + b)(a - 1 - b)$ **65.** $x(3x - y)(9x^2 + 3xy + y^2)$ **67.** $2y(2xy + 3)(4x^2y^2 - 6xy + 9)$ **69.** $8x^2(2y - 1)(4y^2 + 2y + 1)$ **71.** $3(x + 1)(3x + 2)$ **73.** $(x^2 + 7)(x + 1)(x - 1)$ **75.** $(x^3 + 3)(x + 1)(x^2 - x + 1)$ **77.** $(x + y + 3)(x + y - 3)$ **79.** $(a - 3)(a^2 - 9a + 21)$ **81.** $(x + 5 + y)(x^2 + 10x - xy - 5y + y^2 + 25)$ **83.** $2(x + 2)(4x^2 - 2x + 7)$ **85.** $-x(x^2 + 9x + 27)$ **87.** $12(x + 3)$ sq in. **89.** -1 **91.** -17 **93.** $\{-7\}$ **95.** $\{3\}$ **97.** $\{0\}$ **99.** $\{-4\}$ **101.** $h = 5$ ft **103.** $(x^n + 5)(x^n - 5)$ **105.** $(x^{2n} + 4)(x^n + 2)(x^n - 2)$ **107.** $(y^n - 3)(y^{2n} + 3y^n + 9)$

Mental Math, Sec. 6.6

1. $\{3, -5\}$ **3.** $\{3, -7\}$ **5.** $\{0, 9\}$

Exercise Set 6.6

1. $\left\{-3, \frac{4}{3}\right\}$ **3.** $\left\{\frac{5}{2}, -\frac{3}{4}\right\}$ **5.** $\{-3, -8\}$ **7.** $\left\{\frac{1}{4}, -\frac{2}{3}\right\}$ **9.** $\{1, 9\}$ **11.** $\left\{\frac{3}{5}, -1\right\}$ **13.** $\{0\}$ **15.** $\{6, -3\}$ **17.** $\left\{\frac{2}{5}, -\frac{1}{2}\right\}$ **19.** $\left\{\frac{3}{4}, -\frac{1}{2}\right\}$ **21.** $\left\{-2, 7, \frac{8}{3}\right\}$ **23.** $\{0, 3, -3\}$ **25.** $\{2, 1, -1\}$ **27.** 5 seconds **29.** 5 sides **31.** 15 cm, 20 cm, 25 cm **33.** 36 ft **35.** $\left\{-\frac{7}{2}, 10\right\}$ **37.** $\{0, 5\}$ **39.** $\{-3, 5\}$ **41.** $\left\{-\frac{1}{2}, \frac{1}{3}\right\}$ **43.** $\{-4, 9\}$ **45.** $\left\{\frac{4}{5}\right\}$ **47.** $\{-5, 0, 2\}$ **49.** $\left\{-3, 0, \frac{4}{5}\right\}$ **51.** $\{\ \}$ **53.** $\{-7, 4\}$ **55.** $\{4, 6\}$ **57.** $\left\{-\frac{1}{2}\right\}$ **59.** $\{-4, -3, 3\}$ **61.** $\{-5, 0, 5\}$ **63.** $\{-6, 5\}$ **65.** $\left\{-\frac{1}{3}, 0, 1\right\}$ **67.** $\left\{-\frac{1}{3}, 0\right\}$ **69.** $\left\{-\frac{7}{8}\right\}$ **71.** $\left\{\frac{31}{4}\right\}$ **73.** $\{1\}$ **75.** -10 and -11 **77.** 10 km **79.** 105 units **81.** 2 in. **83.** ascending, 2 seconds; descending, 3 seconds **85.** 12 mm, 16 mm, 20 mm **87.** $\frac{4}{7}$ **89.** $\frac{3}{2}$ **91.** $\frac{2y}{x}$ **93.** $\frac{5a^2}{b^2}$ **95.** no **97.** yes

Mental Math, Sec. 6.7

1. upward **3.** downward

Exercise Set 6.7

1.

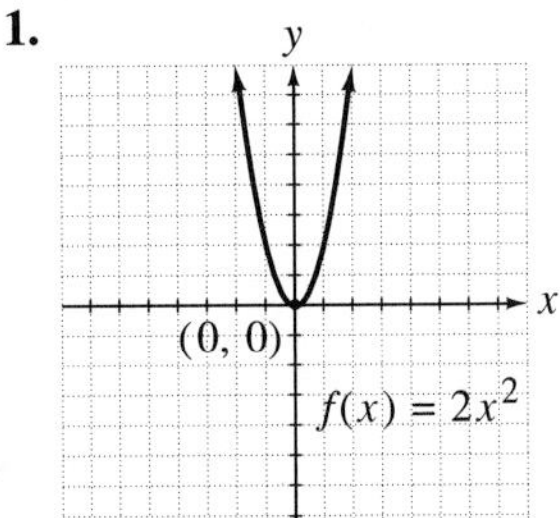

3.

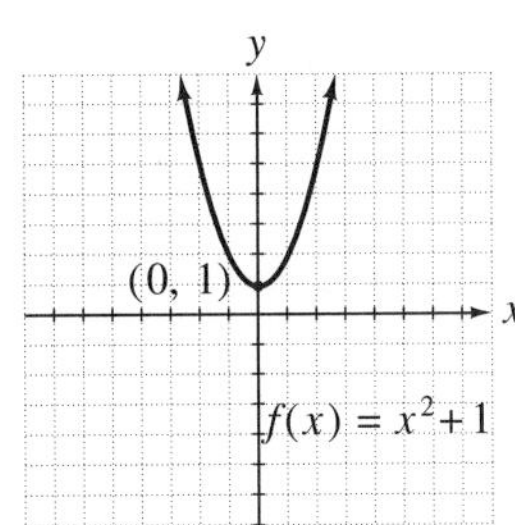

5.

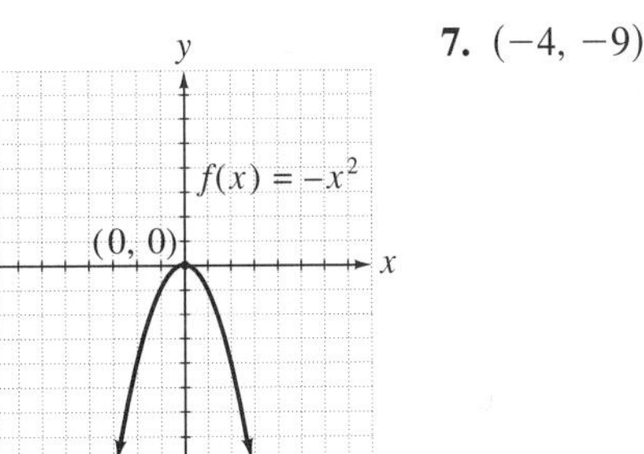

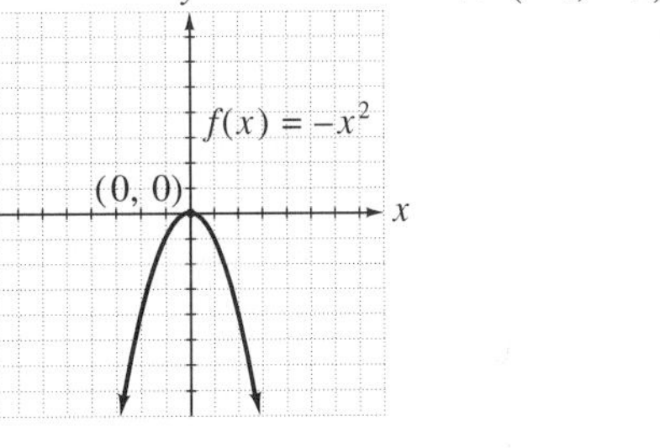

7. $(-4, -9)$

9. $(-1, 1)$ **11.** $(5, 30)$

13.

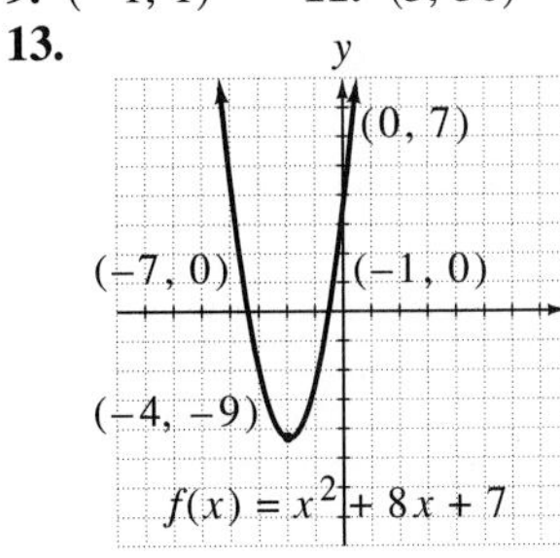

15.

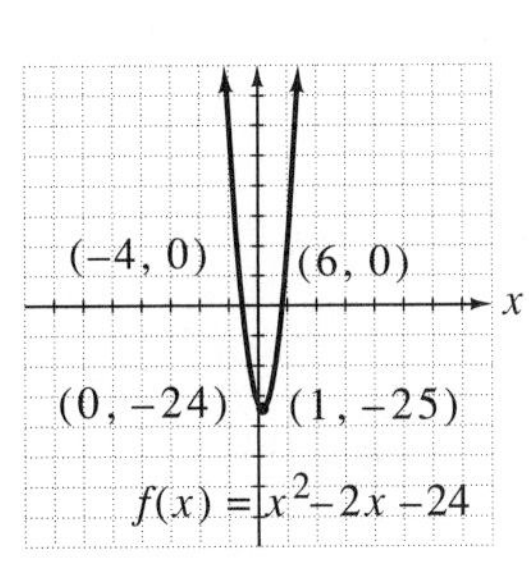

17.

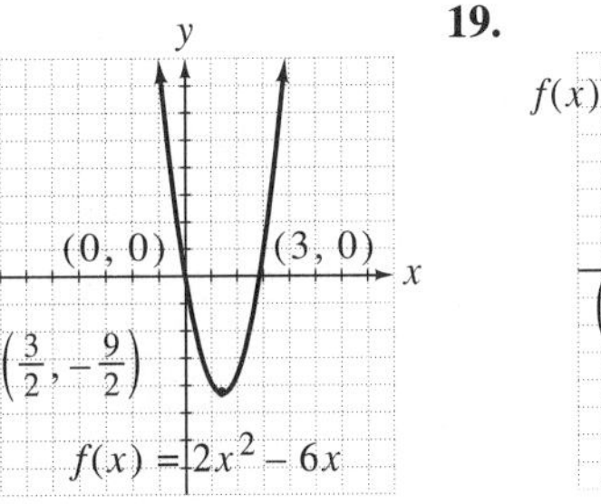

19.

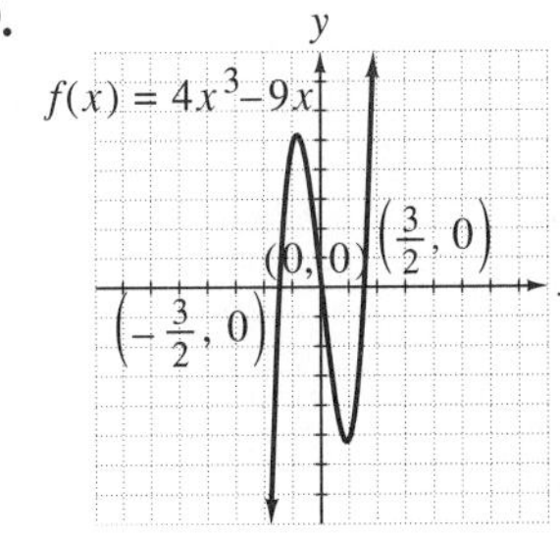

21.

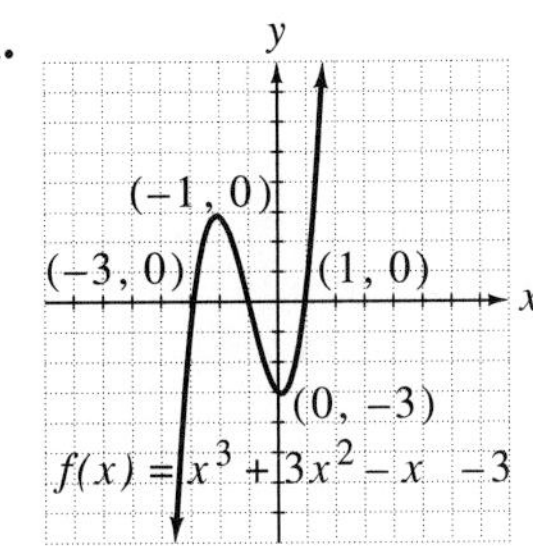

23.

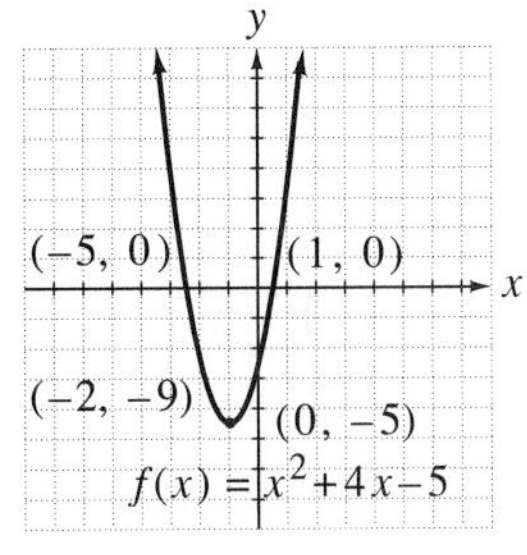

25.

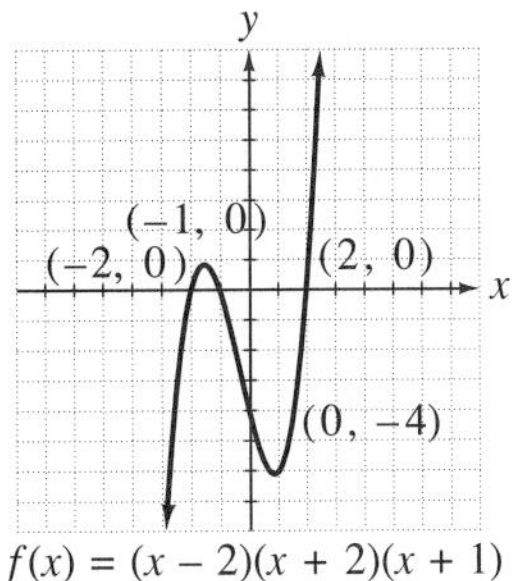

27.

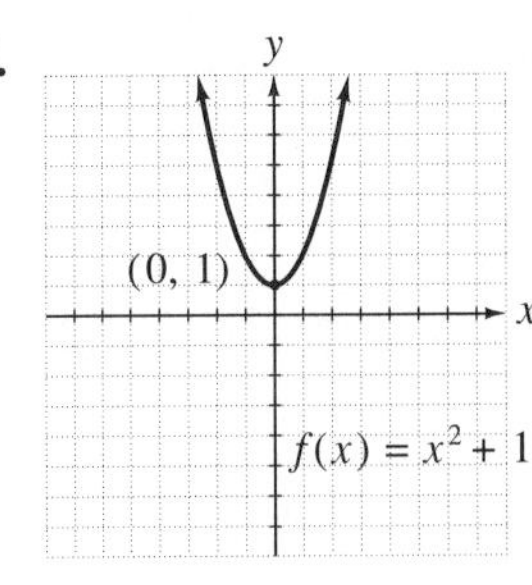

29.

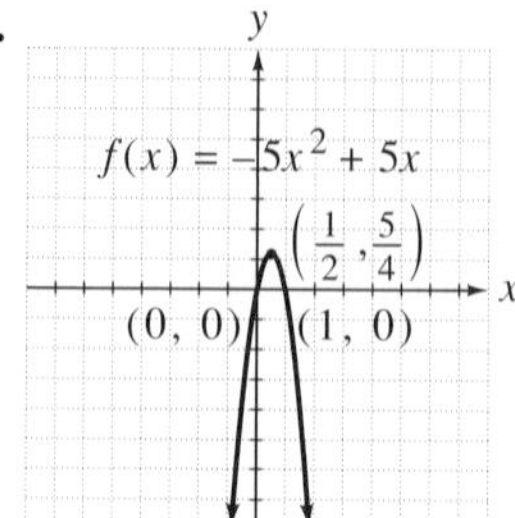

31.

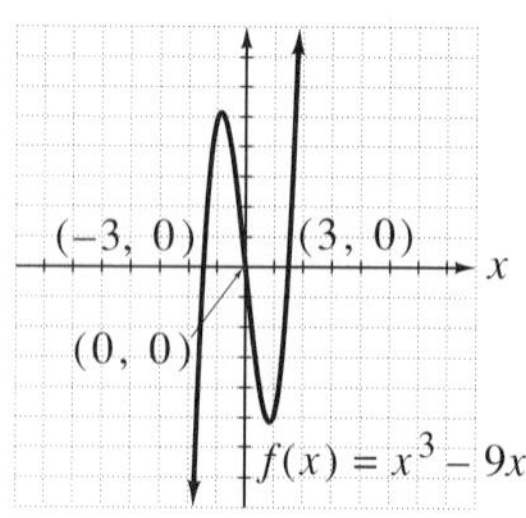

33.

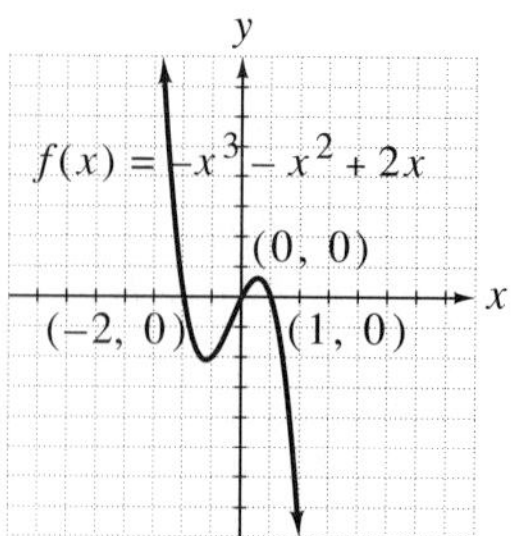

35.

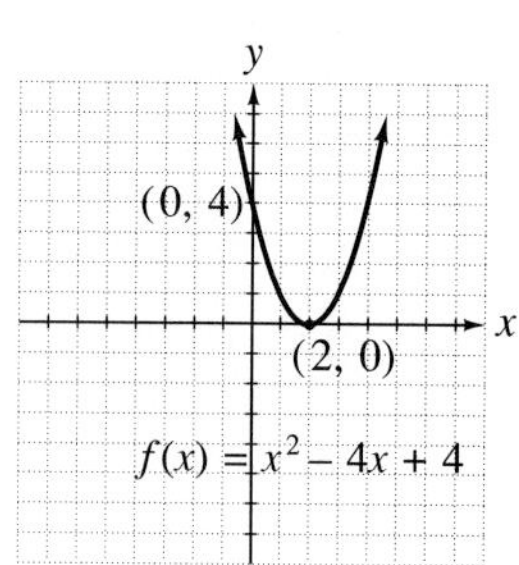

37.

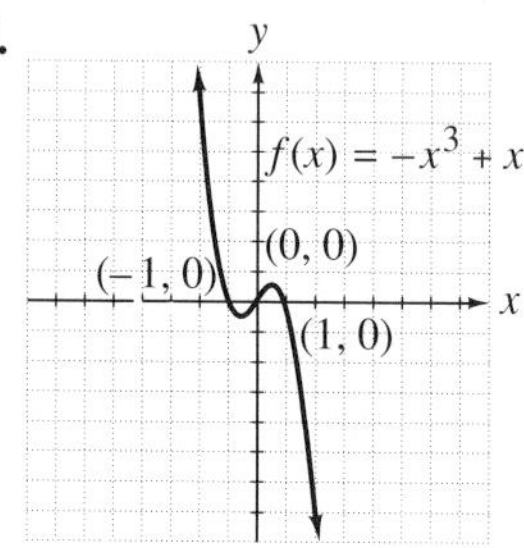

39.

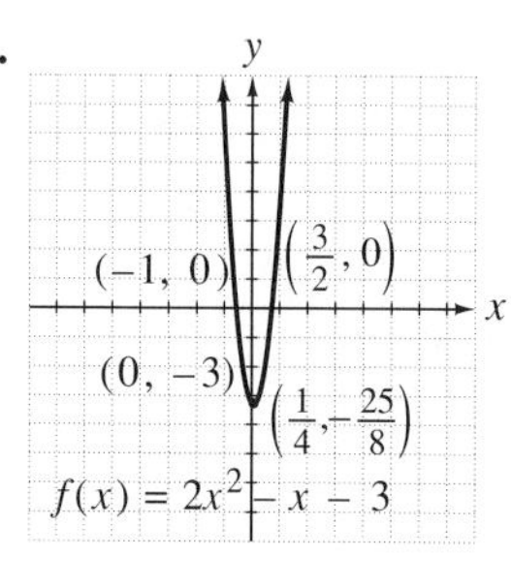

41.

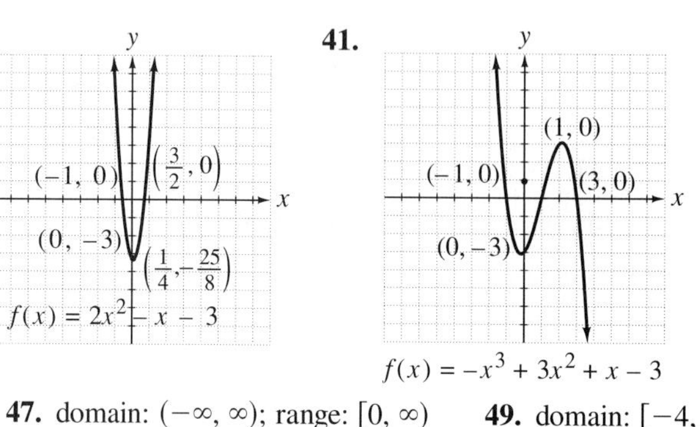

43.

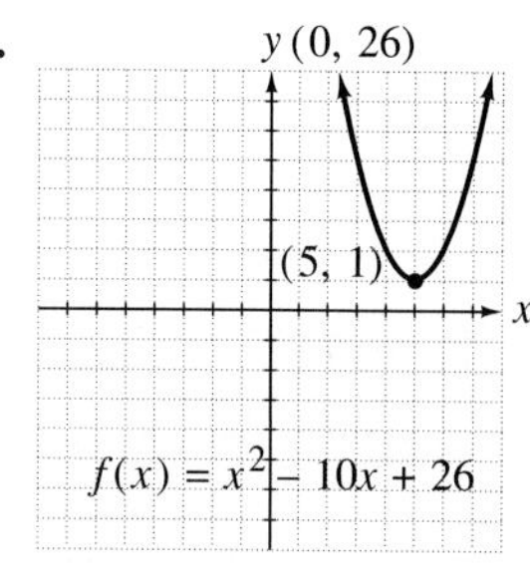

45.

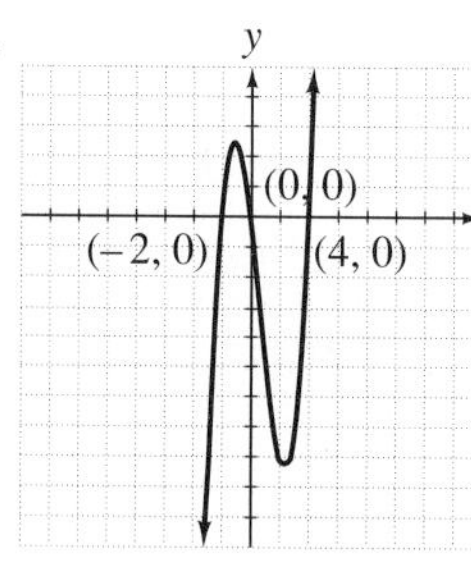

$f(x) = x(x - 4)(x + 2)$

47. domain: $(-\infty, \infty)$; range: $[0, \infty)$ **49.** domain: $[-4, 3]$; range: $[-4, 4]$ **51.** $3x^3y$ **53.** $-16x^3y^3$ **55.** $x^2 + 12x + 36$ **57.** $4x^2 - 4x + 1$

Chapter 6 Review

1. 12 **3.** x^2 **5.** $2x$ **7.** $8x^2(2x - 3)$ **9.** $3xy^2z(5x^2y^2 - z + 2x)$ **11.** $2ab(3b + 4 - 2ab)$ **13.** $(a + 3b)(6a - 5)$ **15.** $(x - 6)(y + 3)$ **17.** $(p - 5)(q - 3)$ **19.** $(x^2 - 2)(x - 1)$ **21.** $y(2x + y)$ **23.** $(x - 8)(x - 3)$ **25.** $(x - 6)(x + 1)$ **27.** $(x + 6y)(x - 2y)$ **29.** $3y(x + y)^2$ **31.** $4(8 + 3x - x^2)$ **33.** prime **35.** $(3x + 4y)(2x - y)$ **37.** $2(3x - 5)^2$ **39.** $(2x - 7y)^2$ **41.** $3xy(2x + 3y)(6x - 5y)$ **43.** $(3t - 5s)(3t + 5s)$ **45.** $(x - 2y)(x^2 + 2xy + 4y^2)$ **47.** $2x(x^2 + 4)$ **49.** $(3x - 2y)(3x + 2y)$ **51.** prime **53.** $3xy(2y - 1)$ **55.** $25(x + 2)(x - 2)$ **57.** $(3x - 4)(x - 2)$ **59.** $(x + 2)(x - 2)(y + 3)$ **61.** prime **63.** $(2x + 5)(2x - 5)$ **65.** $(2x - y)(4x^2 + 2xy + y^2)$ **67.** $2x^3y(x + 3y)(x^2 - 3xy + 9y^2)$ **69.** $4(a - 3 - b)(a - 3 + b)$ **71.** $2x^3y^5(y - 4)(y^2 + 4y + 16)$ **73.** $\left\{-7, \frac{1}{3}\right\}$ **75.** $\left\{0, \frac{9}{2}, 4\right\}$ **77.** $\{0, 6\}$ **79.** $\left\{-\frac{1}{3}, 2\right\}$ **81.** $\{-4, 1\}$ **83.** $\{0, -3, 6\}$ **85.** $\{-2, 0, 1\}$ **87.** 36 ft **89.** 19 and 20 **91.** width: 2 meters, length: 8 meters

93.

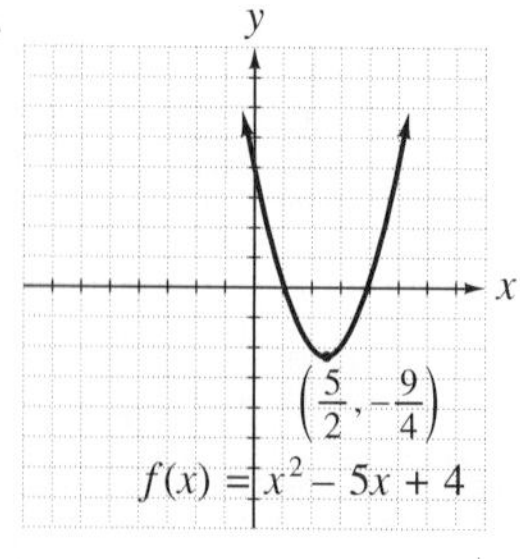

95.

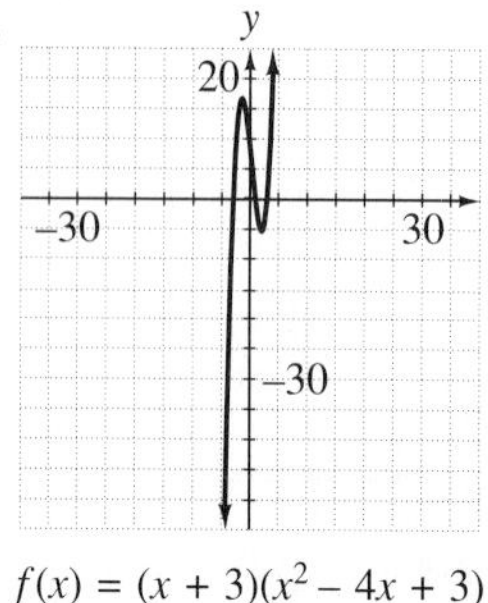

$f(x) = (x + 3)(x^2 - 4x + 3)$

97.

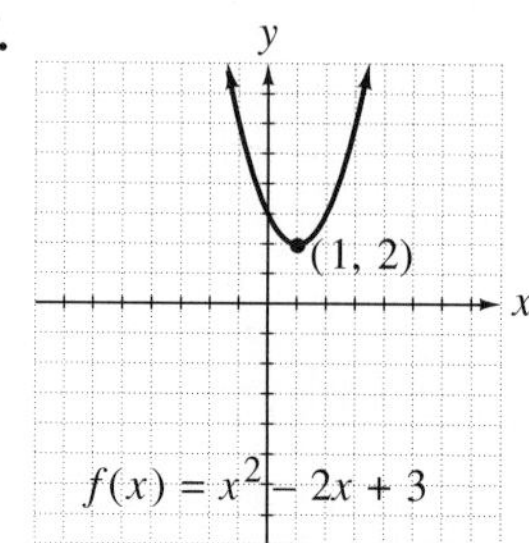

99.

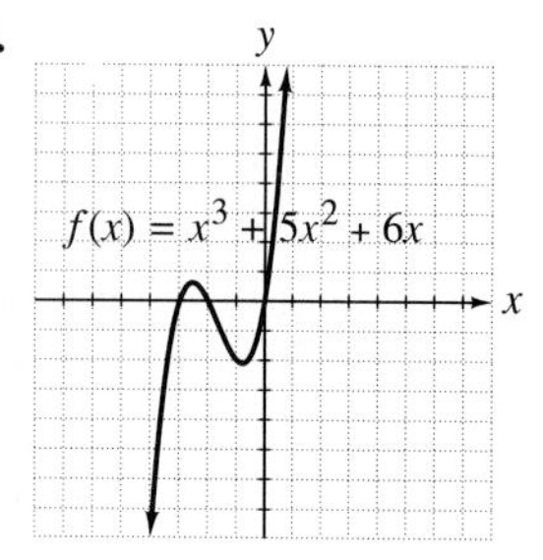

Chapter 6 Test

1. $4x^2y(4x - 3y^3)$ **2.** $5ab^2(4a^2 - 7b)$ **3.** $(x + 3)(y - 5)$ **4.** $(x^2 + 3)(x + 2)$ **5.** $(x - 3)(x - 8)$ **6.** $(x - 15)(x + 2)$ **7.** $(2x - 1)(x + 9)$ **8.** $(2y + 5)^2$ **9.** $3(2x + 1)(x - 3)$ **10.** $x(x^2 + 1)(x^2 + 2)$ **11.** $(2x + 5)(2x - 5)$ **12.** $3(x + 1)(3x - 1)$ **13.** $(x + 4)(x^2 - 4x + 16)$ **14.** $(2x - 5)(4x^2 + 10x + 25)$ **15.** $3y(x + 3y)(x - 3y)$ **16.** $(4x + 3)(2x - 3)$ **17.** $6(x^2 + 4)$ **18.** $4y^2(x^2 - 2y)(x^4 + 2x^2y + 4y^2)$ **19.** $(x + y - 1)(x - y - 1)$ **20.** $4(2x - 1)(x - 5)$ **21.** $\left\{4, -\frac{8}{7}\right\}$ **22.** $\{2, 3\}$ **23.** $\{0, -2, 2\}$ **24.** $\left\{-\frac{5}{2}, -2, 2\right\}$ **25.** $\left\{\frac{5}{3}, -1\right\}$ **26.** width: 6 ft, length: 11 ft **27.** $(x + 2y)(x - 2y)$ **28.** 11 seconds

29.

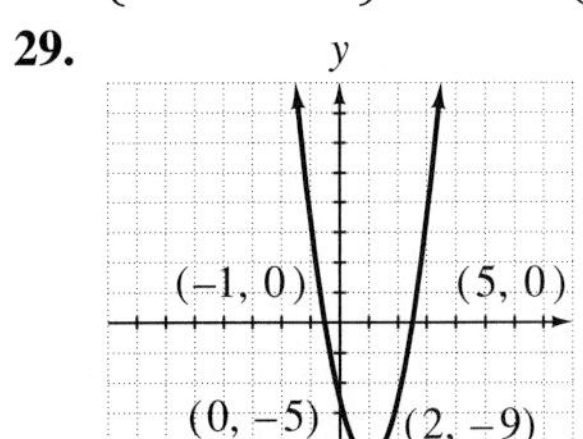

30.

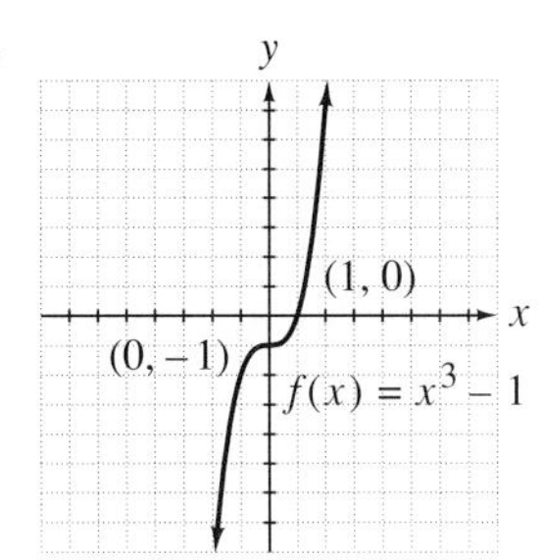

Chapter 6 Cumulative Review

1. $\frac{8}{20}$; *Sec. 1.3, Ex. 6* **2. a.** -12 **b.** -9; *Sec. 1.6, Ex. 3*

3. a. $(-3 + 2) + 4 = -3 + (2 + 4)$ **b.** $(-3 \cdot 2) \cdot 4 = -3 \cdot (2 \cdot 4)$; *Sec. 1.8, Ex. 2*
$-1 + 4 = -3 + 6$ $\quad -6 \cdot 4 = -3 \cdot 8$
$3 = 3$ $\quad -24 = -24$

4. a. $5x + 7$ **b.** $-4a - 1$ **c.** $4y - 3y^2$ **d.** $7.3x - 6$; *Sec. 2.1, Ex. 4* **5.** $\{140\}$; *Sec. 2.2, Ex. 6* **6. a.** $\frac{2}{5}$ **b.** $\frac{3}{4}$; *Sec. 3.1, Ex.1* **7.** 21.2%; *Sec. 3.2, Ex. 7* **8.** $\{-20, 24\}$; *Sec. 3.4, Ex. 3* **9. a.** yes **b.** no **c.** yes; *Sec. 4.1, Ex. 1* **10. a.** \$11.5 billion **b.** \$13.4 billion; *Sec. 4.2, Ex. 8* **11.** $y = \frac{1}{4}x - 3$

12. *Sec. 4.6, Ex. 2*

13. a. $\frac{2x}{25}$ **b.** $\frac{b^3}{27a^6}$ **c.** $\frac{16y^7}{x^5}$ **d.** $\frac{y^{18}}{z^{36}}$ **e.** $-8x^6y^6$; *Sec. 5.2, Ex. 4* **14. a.** 5 **b.** 35; *Sec. 5.3, Ex. 3* **15.** $2x - 3 + \frac{-3}{3x - 5}$; *Sec. 5.5, Ex. 6* **16. a.** 5 **b.** 5; *Sec. 5.6, Ex. 3* **17.** $5x^2$; *Sec. 6.1, Ex. 2* **18.** $(r - 7)(r + 6)$; *Sec. 6.2, Ex. 4* **19. a.** $4ab(2a - 1)$ **b.** $9(2x + 1)(2x - 1)$ **c.** $(2x - 7)(x + 1)$; *Sec. 6.5, Ex. 1* **20.** $\left\{-\frac{1}{2}, 4\right\}$; *Sec. 6.6, Ex. 3* **21.** 4 seconds; *Sec. 6.6, Ex. 8*

CHAPTER 7 Simplifying Rational Expressions

Mental Math, Sec. 7.1

1. $x \neq 0$ **3.** $x \neq 0, x \neq 1$

Exercise Set 7.1

1. $f(2) = \frac{10}{3}, f(0) = -8, f(-1) = -\frac{7}{3}$ **3.** $g(3) = -\frac{17}{48}, g(-2) = \frac{2}{7}, g(1) = -\frac{3}{8}$ **5.** $\{x \mid x \text{ is a real number}\}$ **7.** $\{t \mid t \text{ is a real number and } t \neq 0\}$ **9.** $\{x \mid x \text{ is a real number and } x \neq 2, x \neq -2\}$ **11.** $\frac{5x^2}{9}$ **13.** $\frac{x^4}{2y^2}$

15. $\frac{1}{2(q-1)}$ **17.** 1 **19.** $\frac{-1}{1+x}$ **21.** $\frac{-1}{x-7}$ **23.** $\frac{1}{x^2-3x+9}$ **25.** $\frac{2(x^2+2x+4)}{3}$ **27.** $\frac{y+2}{x-3}$ **29.** $\frac{10y^2z}{4y^3z}$ **31.** $\frac{3x^2+15x}{2x^2+9x-5}$ **33.** $\frac{x^2-4}{x+2}$ **35.** $\{x \mid x$ is a real number and $x \neq 7\}$ **37.** $\{x \mid x$ is a real number and $x \neq 0, x \neq -3\}$ **39.** $\{x \mid x$ is a real number and $x \neq -2, x \neq 0, x \neq 1\}$ **41.** **a.** \$33.33 million **b.** \$88.89 million **c.** \$128.57 million **d.** 55.56 million **43.** $\frac{2}{9}$ **45.** $2b$ **47.** -1 **49.** $\frac{4}{3}$ **51.** -2 **53.** $\frac{x+1}{x-3}$ **55.** $\frac{2x+6}{x-3}$ **57.** $\frac{3}{x}$ **59.** $\frac{1}{x-2}$ **61.** $\frac{x+1}{x^2+1}$ **63.** x^2-4 **65.** $x-4$ **67.** $\frac{2x+1}{x-1}$ **69.** $-x^2-5x-25$ **71.** $\frac{4x^2+6x+9}{2}$ **73.** $\frac{x+5}{x^2+5}$ **75.** $\frac{30m^2}{6m^3}$ **77.** $\frac{35}{5m-10}$ **79.** $\frac{y^2+8y+16}{y^2-16}$ **81.** $\frac{12x^2+24x}{x^2+4x+4}$ **83.** $\frac{x^2-2x+4}{x^3+8}$ **85.** $\frac{ab-3a}{ab-3a+2b-6}$ **87.** $\frac{8}{15}$ **89.** $\frac{2}{7}$ **91.** $\frac{1}{3}$ **93.**

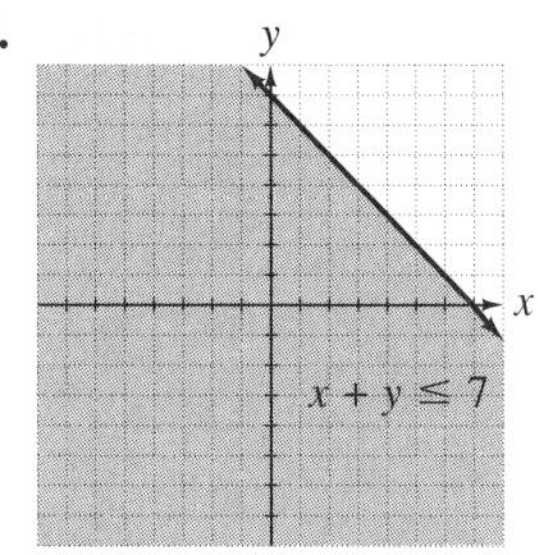

95. 28.1% **97.** 49.8% **99.** 1 **101.** x^k-3 **103.** irreducible

Exercise Set 7.2

1. $\frac{x}{2}$ **3.** $\frac{c}{4}$ **5.** $\frac{x}{3}$ **7.** $-\frac{4}{5}$ **9.** $-\frac{6a}{2a+1}$ **11.** $\frac{1}{6}$ **13.** $-\frac{8}{3}$ **15.** $\frac{x}{2}$ **17.** $\frac{4}{c}$ **19.** $\frac{x}{3}$ **21.** $\frac{5}{3}$ **23.** $\frac{4}{(x+2)(x+3)}$ **25.** $\frac{1}{2}$ **27.** $\frac{49y^7}{2x^2}$ **29.** $-\frac{2}{3x^3y^2}$ **31.** $\frac{4}{ab^6}$ **33.** $\frac{1}{4a(a-b)}$ **35.** $\frac{x^2+5x+6}{4}$ **37.** $\frac{3}{2(x-1)}$ **39.** $\frac{4a^2}{a-b}$ **41.** $\frac{2x^2-18}{5(x^2-8x-15)}$ **43.** $\frac{x+2}{x+3}$ **45.** $\frac{3b}{a-b}$ **47.** $\frac{3a}{a-b}$ **49.** $\frac{1}{4}$ **51.** -1 **53.** $\frac{8}{3}$ **55.** $\frac{8a-16}{3(a+2)}$ **57.** $\frac{8}{x^2y}$ **59.** $\frac{(y+5)(2x-1)}{(y+2)(5x+1)}$ **61.** $\frac{15a+10}{a}$ **63.** $\frac{5x^2-2}{(x-1)^2}$ **65.** $\frac{5}{x-2}$ square meters **67.** $\frac{x^3-3x+2}{x^5}$ ft **69.** $\frac{7}{5}$ **71.** $\frac{1}{12}$ **73.** $\frac{11}{16}$ **75.** $x^2-5x-2+\frac{-6}{x-1}$ **77.** 5 people **79.** 14 people **81.** 16.7% **83.** $2x^2(x^n+2)$ **85.** $\frac{1}{10y(y^n+3)}$ **87.** $\frac{y^n+1}{2(y^n-1)}$

Exercise Set 7.3

1. $-\frac{3}{x}$ **3.** $\frac{x+2}{x-2}$ **5.** $x-2$ **7.** $\frac{1}{2-x}$ **9.** $35x$ **11.** $x(x+1)$ **13.** $(x+7)(x-7)$ **15.** $6(x+2)(x-2)$ **17.** $(a+b)(a-b)^2$ **19.** $\frac{17}{6x}$ **21.** $\frac{35-4y}{14y^2}$ **23.** $\frac{-13x+4}{x^2-16}$ **25.** $\frac{2x+4}{(x-5)(x+4)}$ **27.** 0 **29.** $\frac{x^2}{x-1}$ **31.** $\frac{-}{x}$ **33.** $\frac{y^2+2y+10}{(y+4)(y-4)(y-2)}$ **35.** $\frac{5x^2+5x-20}{(3x+2)(x+3)(2x-5)}$ **37.** $\frac{x^2+5x+21}{(x-2)(x+1)(x+3)}$ **39.** $\frac{-2x+5}{2(x+1)}$ **41.** $\frac{2x^2+2x-42}{(x+3)^2(x-3)}$ **43.** $\frac{3}{x^2y^3}$ **45.** $-\frac{5}{x}$ **47.** $\frac{25}{6(x+5)}$ **49.** $\frac{-2x-1}{x^2(x-3)}$ **51.** $\frac{2ab-b^2}{(a+b)(a-b)}$ **53.** $\frac{2x+16}{(x+2)^2(x-}$ **55.** $\frac{5a+1}{(a+1)^2(a-1)}$ **57.** $\frac{2x^2+9x-18}{6x^2}$ **59.** $\frac{4}{3}$ **61.** $\frac{4a^2}{9(a-1)}$ **63.** 4 **65.** $\frac{6x}{(x+3)(x-3)^2}$ **67.** $-\frac{4}{x-1}$ **69.** $-\frac{32}{x(x+2)(x-2)}$ **71.** $\frac{4x}{x+5}$ ft; $\frac{x^2}{x^2+10x+25}$ square ft **73.** $\frac{3x^2+6x-12}{x(x+2)(x-2)}$

75. $\frac{x}{2(x-2)}$ **77.** 10 **79.** $4 + x^2$ **81.** 10 **83.** 2 **85.** 3 **87.** 5 meters **89.** $\frac{3}{2x}$ **91.** $\frac{4-3x}{x^2}$
93. $\frac{1-3x}{x^3}$

Exercise Set 7.4

1. $\frac{5}{6}$ **3.** $\frac{8}{5}$ **5.** 4 **7.** $\frac{7}{13}$ **9.** $\frac{4}{x}$ **11.** $\frac{9x-18}{9x^2-4}$ **13.** $\frac{1-x}{1+x}$ **15.** $\frac{xy^2}{x^2+y^2}$ **17.** $\frac{2b^2+3a}{b^2-ab}$
19. $\frac{x}{x^2-1}$ **21.** $\frac{x+1}{x+2}$ **23.** $\frac{10}{69}$ **25.** $\frac{2x+2}{2x-1}$ **27.** $\frac{x^2+x}{6}$ **29.** $\frac{x}{2-3x}$ **31.** $-\frac{y}{x+y}$ **33.** $-\frac{2x^3}{xy-y^2}$
35. $\frac{2x+1}{y}$ **37.** $\frac{x-3}{9}$ **39.** $\frac{1}{x+2}$ **41.** $\frac{x}{5x-10}$ **43.** $\frac{x-2}{2x-1}$ **45.** $-\frac{x^2+4}{4x}$ **47.** $\frac{x-3y}{x+3y}$
49. $\frac{1+a}{1-a}$ **51.** $\frac{x^2+6xy}{2y}$ **53.** $\frac{5a}{2a+4}$ **55.** $5xy^2 + 2x^2y$ **57.** $\frac{xy}{2x+5y}$ **59.** $\frac{xy}{x+y}$ **61.** $x^2 + x$
63. a. $\frac{1}{a+h}$ **b.** $\frac{1}{a}$ **c.** $\frac{\frac{1}{a+h}-\frac{1}{a}}{h}$ **d.** $\frac{-1}{a(a+h)}$ **65. a.** $\frac{3}{a+h+1}$ **b.** $\frac{3}{a+1}$ **c.** $\frac{\frac{3}{a+h+1}-\frac{3}{a+1}}{h}$
d. $\frac{-3}{(a+h+1)(a+1)}$ **67.** $\frac{R_1R_2}{R_2+R_1}$ **69.** $\left\{\frac{13}{3}\right\}$ **71.** $\{-3, 1\}$ **73.** $\{-1\}$ **75.** $(2y+1)(4y^2-2y+1)$
77. $(a-3)(a^2+3a+9)$ **79.** $(x-1)(x+y)$ **81.** $2x(x+4)(x-4)$ **83.** $\frac{1+x}{2+x}$ **85.** 5 **87.** $\frac{4x-9y}{4x}$

Exercise Set 7.5

1. {72} **3.** {2} **5.** {6} **7.** {2} **9.** {3} **11.** { } **13.** {15} **15.** {4} **17.** { } **19.** {1} **21.** {1}
23. {−3} **25.** $\left\{\frac{5}{3}\right\}$ **27.** {10, 2} **29.** {2} **31.** {3} **33.** { } **35.** { } **37.** {−1} **39.** {9} **41.** {1, 7}
43. $\left\{\frac{1}{10}\right\}$ **45.** $\left\{-\frac{2}{3}\right\}$ **47.** $\frac{5}{2x}$ **49.** $-\frac{y}{x}$ **51.** $\frac{-a^2+31a+10}{5(a-6)(a+1)}$ **53.** $\left\{-\frac{3}{13}\right\}$ **55.** $\frac{-a-8}{4a(a-2)}$
57. $\frac{x^2-3x+10}{2(x+3)(x-3)}$ **59.** $\{x \mid x \neq 2 \text{ and } x \neq -1\}$ **61.** $\frac{22z-45}{z(3z-9)}$ **63.** $\left\{\frac{1}{9}, -\frac{1}{4}\right\}$ **65.** {3, 2} **67.** 73 and 74
69. $\frac{1}{2}$, 2 **71.** −9 **73.** −9 **75.** 1% **77.** 67% **79.** {−1, 0} **81.** {−2}

Exercise Set 7.6

1. $C = \frac{5}{9}(F - 32)$ **3.** $R = \frac{R_1R_2}{R_1+R_2}$ **5.** $n = \frac{2S}{a+L}$ **7.** 1 and 5 **9.** 5 **11.** 6 ohms
13. $\frac{1}{r} = \frac{1}{r_1} + \frac{1}{r_2} + \frac{1}{r_3}$; $r = \frac{15}{13}$ ohms **15.** 15.6 hours **17.** 10 minutes **19.** 200 mph **21.** 15 mph
23. $h = \frac{2A}{a+b}$ **25.** $T_2 = T_1 - \frac{LH}{kA}$ **27.** $r = \frac{E}{I} - R$ **29.** $a_1 = S(1-r) + a_nr$ **31.** $M = -\frac{Fr^2}{Gm}$
33. −8 and −7 **35.** 36 minutes **37.** 45 mph and 60 mph **39.** 10 mph, 8 mph **41.** 2 hours **43.** 135 hours
45. 12 miles **47.** $2\frac{2}{9}$ hours **49.** by jet, 3 hours; by car, 4 hours **51.** $1\frac{1}{2}$ minutes **53.** 63 mph **55.** {−5}
57. {2} **59.** $x(2x+3)(x-6)$ **61.** $(x+2)(x^2+2x+4)$ **63.** 25% **65.** 84%

Exercise Set 7.7

1. $A = kB$ **3.** $X = \frac{k}{Z}$ **5.** $N = kP^2$ **7.** $T = \frac{k}{R}$ **9.** $P = kR$ **11.** $A = 45$ **13.** $V = 24$ cubic meters
15. $H = 10$ **17.** 72 amps **19.** $x = kyz$ **21.** $r = kst^3$ **23.** $Q = 7$ **25.** $M = \frac{48}{5}$ **27.** $B = 2$
29. $W = 2.7$ **31.** 4.05 pounds **33.** $0.80 **35.** 108 cars **37.** 16 pounds **39.** 7.91 tons **41.** multiplied by 9
43. divided by 4 **45.** $C = 8\pi$ in.; $A = 16\pi$ in.^{2} **47.** $C = 18\pi$ cm; $A = 18\pi$ cm^2 **49.** $\frac{9}{5}$ **51.** undefined

53.

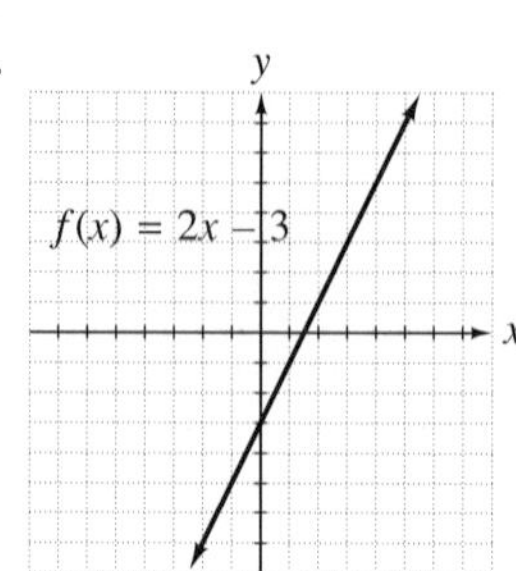

55.

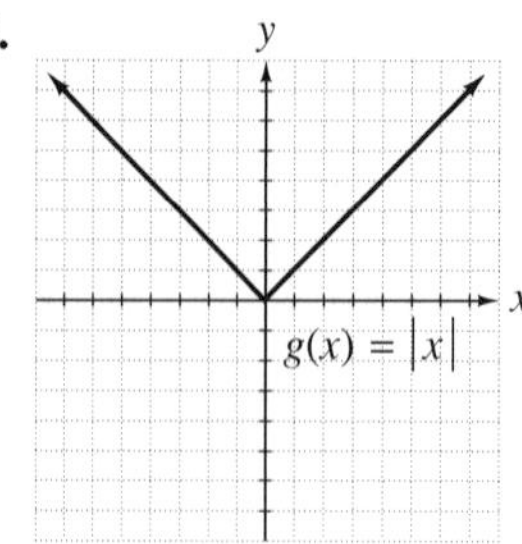

57.

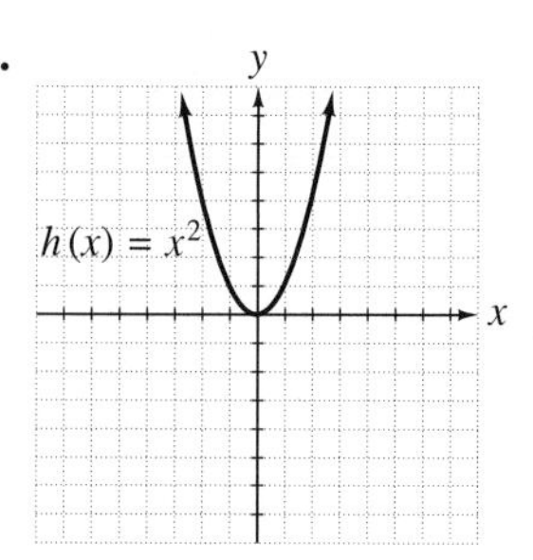

Chapter 7 Review

1. $\{x \mid x \text{ is a real number}\}$ **3.** $\{x \mid x \text{ is a real number and } x \neq 5\}$ **5.** $\{x \mid x \text{ is a real number and } x \neq 0, x \neq -8\}$ **7.** $\frac{x^2}{3}$ **9.** $\frac{9m^2p}{5}$ **11.** $\frac{1}{5}$ **13.** $\frac{1}{x-1}$ **15.** $\frac{2(x-3)}{x-4}$ **17. a.** $C(50) = 119$; **b.** $C(1000) = 784$ **19.** $\frac{2x^3}{z^3}$ **21.** $\frac{2}{5}$ **23.** $\frac{1}{6}$ **25.** $\frac{3x}{16}$ **27.** $\frac{3c^2}{14a^2b}$ **29.** $\frac{x^2+9x+20}{3}$ **31.** $\frac{7x-28}{2(x-2)}$ **33.** $-\frac{1}{x}$ **35.** $\frac{8}{9a^2}$ **37.** $\frac{6}{a}$ **39.** $60x^2y^5$ **41.** $5x(x-5)$ **43.** $\frac{2}{5}$ **45.** $\frac{2}{x^2}$ **47.** $\frac{1}{x-2}$ **49.** $\frac{5x^2-3y^2}{15x^4y^3}$ **51.** $\frac{-x+5}{x^2-1}$ **53.** $\frac{2x^2-5x-4}{x-3}$ **55.** $\frac{3x^2-7x-4}{27x^3-64}$ **57.** $\frac{-12}{x(x+1)(x-3)}$ **59.** $\frac{60+4x-x^2}{15x^2}$ **61.** $\frac{-10x-25}{(x+5)^2}$ **63.** $\frac{2}{3}$ **65.** $\frac{2}{15-2x}$ **67.** $\frac{y}{2}$ **69.** $\frac{20x-15}{2(5x^2-2)}$ **71.** $\frac{5x^2y+x^2}{3y}$ **73.** $\frac{1+x}{1-x}$ **75.** $\frac{x-1}{3x-1}$ **77.** $-\frac{x^2+9}{6x}$ **79. a.** $\frac{3}{a+h}$ **b.** $\frac{3}{a}$ **c.** $\frac{\frac{3}{a+h}-\frac{3}{a}}{h}$ **d.** $\frac{-3}{a(a+h)}$ **81.** $\{6\}$ **83.** $\{3\}$ **85.** $\{-7, 7\}$ **87.** $\frac{2x+5}{x(x-7)}$ **89.** $\frac{-5x-30}{2x(x-3)}$ **91.** $R_2 = \frac{RR_1}{R_1 - R}$ **93.** $\frac{A-P}{Pt} = r$ **95.** 1 or 2 **97.** Mary, 18 years; Mark, 24 years **99.** 14 and 15 **101.** $1\frac{23}{37}$ hours **103.** 10 hours **105.** 6 ohms **107.** 45 mph **109.** $63\frac{2}{3}$ mph, 45 mph **111.** $C = 4$ **113.** 64 square in.

Chapter 7 Test

1. $\{x \mid x \text{ is a real number and } x \neq 1\}$ **2.** $\{x \mid x \text{ is a real number and } x \neq -3, x \neq -1\}$ **3.** $\frac{5x^3}{3}$ **4.** $-\frac{7}{8}$ **5.** $\frac{2}{x-3}$ **6.** $\frac{x}{x+9}$ **7.** $\frac{x+2}{5}$ **8.** $\frac{3x+3}{10}$ **9.** $\frac{5}{3x}$ **10.** $\frac{4a^3b^4}{c^6}$ **11.** $\frac{x+2}{2(x+3)}$ **12.** $\frac{-8x-36}{5}$ **13.** $75x^3$ **14.** $\frac{3}{x^3}$ **15.** -1 **16.** $\frac{5x-2}{(x-3)(x+2)(x-2)}$ **17.** $\frac{x-1}{4x^2-2x+1}$ **18.** $\frac{-x+30}{6(x-7)}$ **19.** $\frac{3}{2}$ **20.** $\frac{5x+3y}{xy}$ **21.** $\frac{3-y}{x+y}$ **22.** $\frac{1}{5}$ **23.** $\frac{64}{3}$ **24.** $\frac{7y^2+4y}{6}$ **25.** $\frac{x^2-6x+9}{x-2}$ **26.** 5 **27.** $\frac{6}{7}$ hour **28.** $W = 16$ **29.** $Q = 9$ **30.** 256 ft

Chapter 7 Cumulative Review

1. a. $5x + 10$ **b.** $-2y - 0.6z + 2$ **c.** $-x - y + 2z - 6$; *Sec. 2.1, Ex. 5* **2.** $\{12\}$; *Sec. 2.3, Ex. 1* **3.** $\{x \mid x \text{ is a real number}\}$; *Sec. 2.3, Ex. 6* **4.** $C = \frac{5}{9}(F - 32)$; *Sec. 2.4, Ex. 8* **5.** $\left\{\frac{13}{2}\right\}$; *Sec. 3.1, Ex. 3* **6. a.** 45% **b.** 83% **c.** 9900 **d.** 57.6°; *Sec. 3.2, Ex. 4* **7.** { }; *Sec. 3.4, Ex. 6* **8.** $\left[-9, -\frac{9}{2}\right]$; *Sec. 3.6, Ex. 3*

9. a. 1 **b.** 1 **c.** −3; *Sec. 4.2, Ex. 7* **10.** $-\frac{8}{3}$; *Sec. 4.4, Ex. 1* **11.** *Sec. 4.5, Ex. 2*

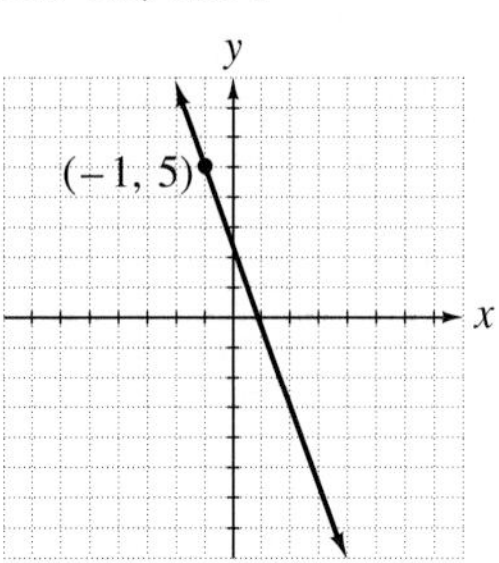

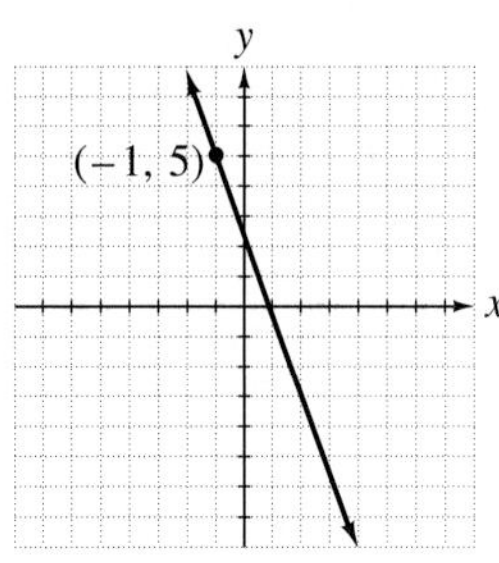

12. a. 8 **b.** 3 **c.** 16 **d.** −16 **e.** $\frac{8}{27}$ **f.** 32; *Sec. 5.1, Ex. 1* **13. a.** $\frac{1}{9}$ **b.** $\frac{2}{x^3}$ **c.** 2^5; *Sec. 5.2, Ex. 1*
14. a. 1805 ft **b.** 221 ft; *Sec. 5.3, Ex. 4* **15.** $2x - 5$; *Sec. 5.5, Ex. 5* **16.** $(3x + 2)(x + 3)$; *Sec. 6.3, Ex. 1*
17. $\left\{-\frac{1}{6}, 3\right\}$; *Sec. 6.6, Ex. 5* **18. a.** $C(100) = 102.6$ **b.** $C(1000) = 12.6$; *Sec. 7.1, Ex. 1*
19. a. $\frac{-4x^2}{2x + 1}$ **b.** $-(a + b)^2$; *Sec. 7.2, Ex. 2* **20. a.** $\frac{x(x - 2)}{2(x + 2)}$ **b.** $\frac{x}{y}$; *Sec. 7.4, Ex. 2* **21.** $\{-2\}$; *Sec. 7.5, Ex. 1*
22. 8 ft; *Sec. 7.6, Ex. 3* **23.** $k = \frac{1}{6}, y = 15$; *Sec. 7.7, Ex. 1*

CHAPTER 8
Roots and Radicals

Exercise Set 8.1

1. 4 **3.** 9 **5.** $\frac{1}{5}$ **7.** −10 **9.**

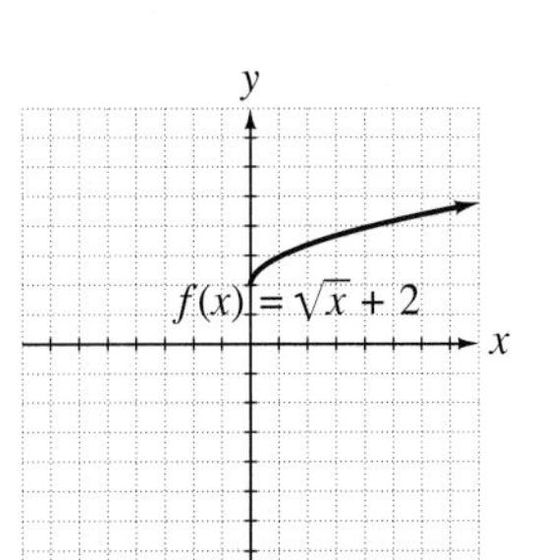

11.

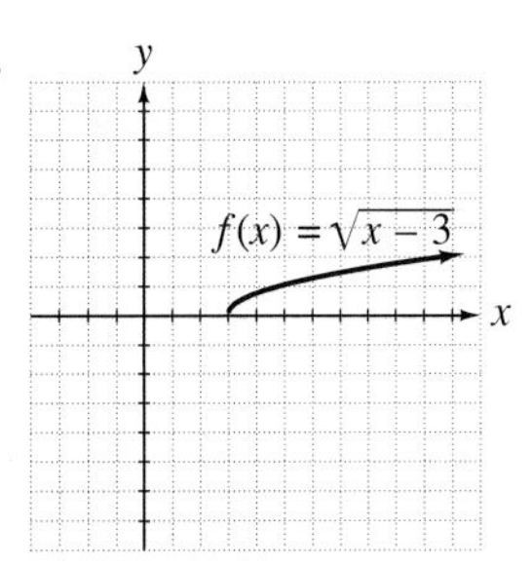

13. 4 **15.** −3

17. $\frac{1}{2}$ **19.** −5 **21.**

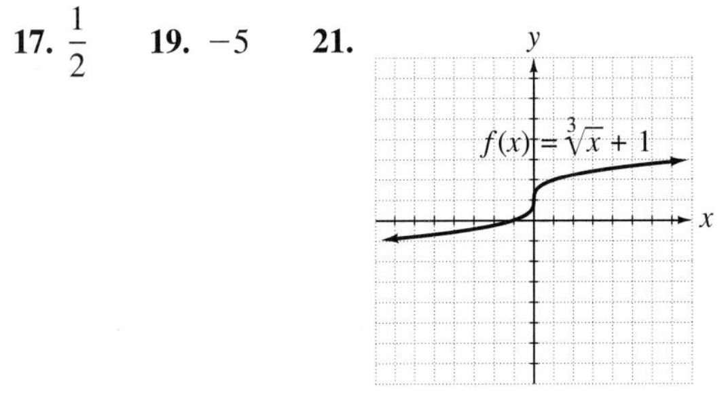

23.

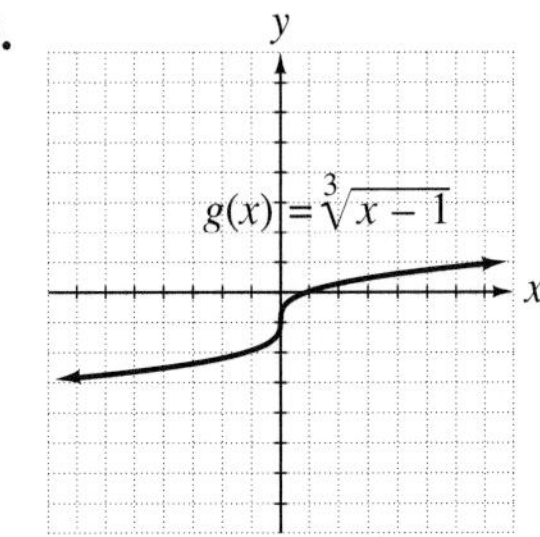

25. 2 **27.** 3

29. not a real number **31.** $\frac{1}{3}$ **33.** $\frac{3}{5}$ **35.** -7 **37.** $|z|$ **39.** x **41.** $|x+5|$ **43.** $2|z|$ **45.** 0 **47.** $-\frac{1}{2}$
49. not a real number **51.** -8 **53.** -13 **55.** 1 **57.** $\frac{5}{8}$ **59.** 2 **61.**

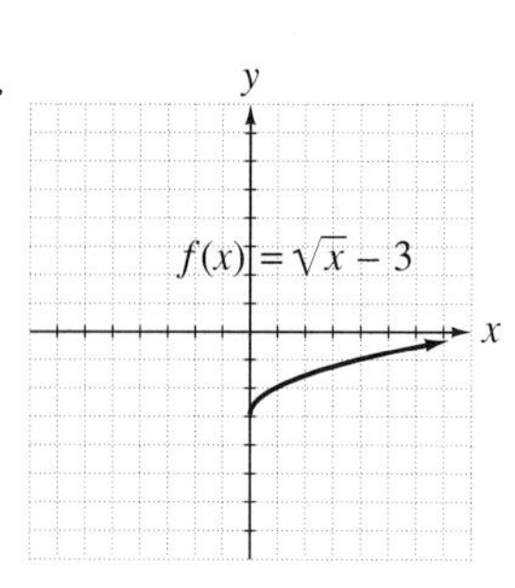

63.

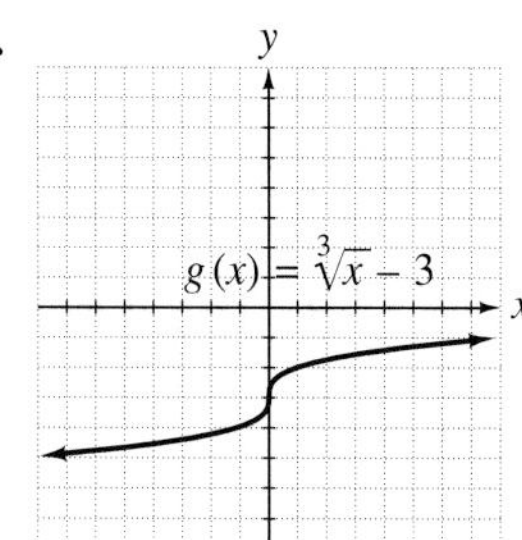

65.

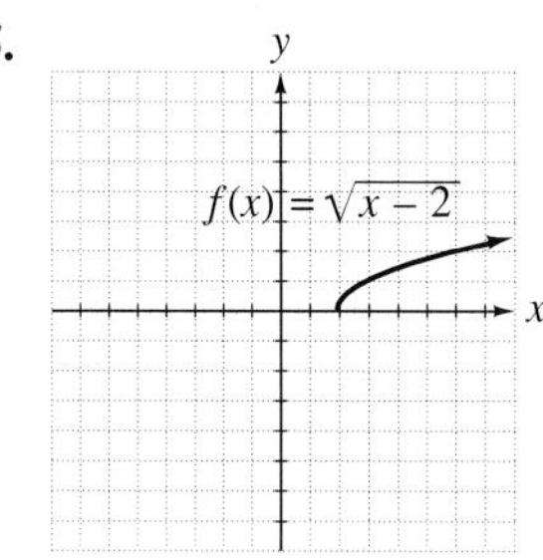

67.

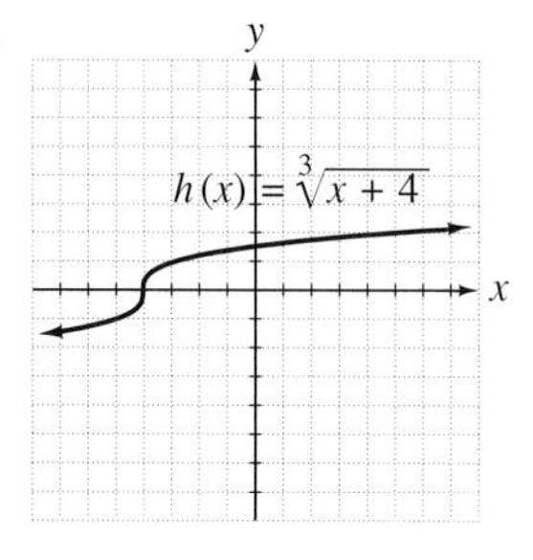

69. rational, 3 **71.** irrational, 6.083 **73.** rational, 13 **75.** rational, 2 **77.** x^5 **79.** x^6 **81.** $9|x|$
83. $-12|y^7|$ **85.** $|x+2|$ **87.** 18.0 **89.** 1537 square ft **91.** $8 = 4 \cdot 2$ **93.** $75 = 25 \cdot 3$
95. $44 = 4 \cdot 11$ **97.** $90 = 9 \cdot 10$ **99.** $56 = 8 \cdot 7$ **101.** $270 = 27 \cdot 10$

Mental Math, Sec. 8.2

1. 3 **3.** x **5.** 0 **7.** $5x^2$

Exercise Set 8.2

1. $2\sqrt{5}$ **3.** $3\sqrt{2}$ **5.** $5\sqrt{2}$ **7.** $\sqrt{33}$ **9.** $\frac{2\sqrt{2}}{5}$ **11.** $\frac{3\sqrt{3}}{11}$ **13.** $\frac{3}{2}$ **15.** $\frac{5\sqrt{5}}{3}$ **17.** $x^3\sqrt{x}$ **19.** $\frac{2\sqrt{22}}{x^2}$
21. $2\sqrt[3]{3}$ **23.** $\frac{\sqrt[3]{5}}{4}$ **25.** $\frac{\sqrt{3}}{5}$ **27.** $\frac{7}{2x}$ **29.** $\frac{y^2\sqrt[3]{y}}{2x^2}$ **31.** $2\sqrt{13}$ **33.** $\frac{\sqrt{11}}{6}$ **35.** $-\frac{\sqrt{3}}{4}$ **37.** $\frac{\sqrt[3]{15}}{4}$
39. $2\sqrt[3]{10}$ **41.** $2\sqrt[4]{3}$ **43.** 11 **45.** $2x$ **47.** $y^2\sqrt{y}$ **49.** $2\sqrt{5}$ **51.** $5ab\sqrt{b}$ **53.** $-3x^3$ **55.** a^4b
57. $x^4\sqrt[3]{50x^2}$ **59.** $-2x^2\sqrt[5]{y}$ **61.** $-4a^4b^3\sqrt{2b}$ **63.** $\frac{\sqrt{6}}{7}$ **65.** $\frac{x\sqrt{5}}{2y}$ **67.** $\frac{-z^2\sqrt[3]{z}}{3x}$ **69.** $\frac{x\sqrt[4]{x^3}}{2}$ **71.** $2\sqrt[3]{10}$
73. Let $a = 2$, $b = 3$ **75.** $48x^2$ **77.** $3x - 2$ **79.** $-72y^4$ **81.** not factorable **83.** $(3x + 4y)(2x - y)$
$\sqrt{2^2 + 3^2} \neq 2 + 3$ **85.** $2(3x - 5)^2$ **87.** $(2x - 7y)^2$
$\sqrt{4 + 9} \neq 5$
$\sqrt{13} \neq 5$

Mental Math, Sec. 8.3

1. $6\sqrt{3}$ **3.** $3\sqrt{x}$ **5.** $12\sqrt[3]{x}$

Exercise Set 8.3

1. $-2\sqrt{2}$ **3.** $10x\sqrt{2x}$ **5.** $17\sqrt{2} - 15\sqrt{5}$ **7.** $-\sqrt[3]{2x}$ **9.** $5b\sqrt{b}$ **11.** $\frac{31\sqrt{2}}{15}$ **13.** $\frac{\sqrt[3]{11}}{3}$ **15.** $\frac{5\sqrt{5x}}{9}$
17. $14 + \sqrt{3}$ **19.** $7 - 3y$ **21.** $6\sqrt{3} - 6\sqrt{2}$ **23.** $-23\sqrt[3]{5}$ **25.** $2b\sqrt{b}$ **27.** $20y\sqrt{2y}$ **29.** $2y\sqrt[3]{2x}$
31. $6\sqrt[3]{11} - 4\sqrt{11}$ **33.** $4x\sqrt[4]{x^3}$ **35.** $\frac{2\sqrt{3}}{3}$ **37.** $\frac{5x\sqrt[3]{x}}{7}$ **39.** $\frac{5\sqrt{7}}{2x}$ **41.** $\frac{\sqrt[3]{2}}{6}$ **43.** $\frac{14x\sqrt[3]{2x}}{9}$ **45.** $15\sqrt{3}$ in.

47. $22\sqrt{5}$ ft **49.** $8\sqrt{5}$ in. **51.** $\left(48 + \frac{9\sqrt{3}}{2}\right)$ square ft. **53.** $x^2 - 16$ **55.** $4x^2 + 20x + 25$
57. $\{x \mid x \text{ is a real number and } x \neq 0\}$ **59.** $\{x \mid x \geq 1\}$ **61.** $\{x \mid x \text{ is a real number}\}$
63. $\{x \mid x \text{ is a real number and } x \neq 0, x \neq 3\}$ **65.** $\{x \mid x \text{ is a real number}\}$

Exercise Set 8.4

1. $\sqrt{35} + \sqrt{21}$ **3.** $7 - 2\sqrt{10}$ **5.** $3\sqrt{x} - x\sqrt{3}$ **7.** $6x - 13\sqrt{x} - 5$ **9.** $\sqrt[3]{a^2} + \sqrt[3]{a} - 20$ **11.** $\frac{\sqrt{3}}{3}$
13. $\frac{\sqrt{5}}{5}$ **15.** $\frac{4\sqrt[3]{9}}{3}$ **17.** $\frac{3\sqrt{2x}}{4x}$ **19.** $\frac{3\sqrt[3]{2x}}{2x}$ **21.** $-\frac{5(2 + \sqrt{7})}{3}$ **23.** $\frac{7(3 + \sqrt{x})}{9 - x}$ **25.** $-5 + 2\sqrt{6}$
27. $\frac{2a + 2\sqrt{a} + \sqrt{ab} + \sqrt{b}}{4a - b}$ **29.** $\frac{5}{\sqrt{15}}$ **31.** $\frac{6}{\sqrt{10}}$ **33.** $\frac{4x}{7\sqrt{4x}}$ **35.** $\frac{5y}{\sqrt[3]{100xy}}$ **37.** $\frac{3}{10 + 5\sqrt{7}}$
39. $\frac{x - 9}{x - 3\sqrt{x}}$ **41.** $6\sqrt{2} - 12$ **43.** $2 + 2x\sqrt{3}$ **45.** $-16 - \sqrt{35}$ **47.** $x - y^2$ **49.** $3 + 2\sqrt{3x} + x^2$
51. $5x - 3\sqrt{15x} - 3\sqrt{10x} + 9\sqrt{6}$ **53.** $-\sqrt[3]{4} + 2\sqrt[3]{2}$ **55.** $\sqrt[3]{x^2} - 4\sqrt[6]{x^5} + 8\sqrt[3]{x} - 4\sqrt{x} + 7$ **57.** $\frac{\sqrt{10}}{5}$
59. $\frac{3\sqrt{3a}}{a}$ **61.** $\frac{3\sqrt[3]{4}}{2}$ **63.** $\frac{-8(1 - \sqrt{10})}{9}$ **65.** $\frac{\sqrt[3]{4x}}{4}$ **67.** $\frac{x^2y^3\sqrt{6z}}{z}$ **69.** $3 - 2\sqrt{2}$ **71.** $\frac{x + 2\sqrt{x} + 1}{x - 1}$
73. $\frac{2}{\sqrt{10}}$ **75.** $\frac{2x}{11\sqrt{2x}}$ **77.** $\frac{7}{2\sqrt[3]{49}}$ **79.** $\frac{-7}{12 + 6\sqrt{11}}$ **81.** $\frac{3x^2}{10\sqrt[3]{9x}}$ **83.** $\frac{6x^2y^3}{\sqrt{6z}}$ **85.** $\frac{1}{3 - 2\sqrt{2}}$
87. $\frac{x - 1}{x - 2\sqrt{x} + 1}$ **89.** $(\sqrt{30} + 2\sqrt{3})$ cm³ **91.** $\frac{1}{5}$ **93.** $-\frac{1}{8}$ **95.** $\frac{1}{x^5}$ **97.** a **99.** x^9y^{19}
101. $\frac{1}{9x^{12}}$ **103.** c^3

Exercise Set 8.5

1. 2 **3.** 3 **5.** 8 **7.** 4 **9.** $-\frac{1}{2}$ **11.** $\frac{1}{64}$ **13.** $\frac{1}{729}$ **15.** $\frac{5}{2}$ **17.** $\sqrt[7]{a^3}$ **19.** $2\sqrt[3]{x^5}$ **21.** $\frac{1}{\sqrt[6]{(4t)^5}}$
23. $\sqrt[5]{(4x - 1)^3}$ **25.** $3^{1/2}$ **27.** $y^{5/3}$ **29.** $4^{1/5}y^{7/5}$ **31.** $(y + 1)^{3/2}$ **33.** $2x^{1/2} - 3y^{1/2}$ **35.** $x^{1/2}$ **37.** $2x^{1/2}$
39. $(x + 3)^{1/2}$ **41.** $x^{1/2}y^{1/2}$ **43.** $a^{7/3}$ **45.** $\frac{8u^3}{v^9}$ **47.** $-b$ **49.** $y - y^{7/6}$ **51.** $2x^{5/3} - 2x^{2/3}$ **53.** $4x^{2/3} - 9$
55. 9 **57.** −2 **59.** −3 **61.** $\frac{1}{9}$ **63.** $\frac{3}{4}$ **65.** 27 **67.** 512 **69.** −4 **71.** 32 **73.** $\frac{8}{27}$ **75.** $\frac{1}{27}$
77. $\frac{1}{2}$ **79.** $\frac{1}{5}$ **81.** $\frac{1}{125}$ **83.** $\frac{1}{x^{7/6}}$ **85.** $w^{1/2}$ **87.** $\frac{y^{1/4}}{x^{3/5}}$ **89.** $\frac{1}{a^{4/3}b^6}$ **91.** $4x^2y$ **93.** $24x^{19/6} - 4x^{2/3}y^{2/3}$
95. $x - x^{1/2} - 6$ **97.** $x - 4x^{3/2} + 4x^2$ **99.** $x^{3/4}(9 - x)$ **101.** $3x^{5/3}(3 + 5x^{2/3})$ **103.** 11,224 **105.** y
107. $z + 2$ **109.** {27} **111.** {−4, 2} **113.** $\left\{-1, \frac{2}{3}\right\}$ **115.** {−6, −3}

Exercise Set 8.6

1. {8} **3.** {7} **5.** { } **7.** {7} **9.** {6} **11.** $\left\{-\frac{9}{2}\right\}$ **13.** {29} **15.** {4} **17.** {−4} **19.** { }
21. {7} **23.** $3\sqrt{5}$ ft **25.** $2\sqrt{10}$ meters **27.** {9} **29.** {50} **31.** { } **33.** $\left\{\frac{15}{4}\right\}$ **35.** {7} **37.** {5}
39. {−12} **41.** {9} **43.** {−3} **45.** {1} **47.** {1} **49.** $\left\{\frac{1}{2}\right\}$ **51.** fewer than 2744 deliveries
53. $50\sqrt{145}$ pounds **55.** $5\sqrt{5}$ cm **57.** $2\sqrt{2}$ in. **59.** 13 ft **61.** 23.6 ft **63.** 54.22 mph **65.** 24 cubic ft.
67. −3 **69.** $\frac{2}{5}$ **71.** yes **73.** yes **75.** no **77.** $\frac{x}{4x + 3}$ **79.** $-\frac{4z + 2}{3z}$ **81.** {−1, 0, 8, 9}
83. {−1, 4}

Mental Math, Sec. 8.7

1. $9i$ **3.** $i\sqrt{7}$ **5.** -4 **7.** $8i$

Exercise Set 8.7

1. $2i\sqrt{6}$ **3.** $-6i$ **5.** $24i\sqrt{7}$ **7.** $-3\sqrt{6}$ **9.** $6 - 4i$ **11.** $-2 + 6i$ **13.** $-2 - 4i$ **15.** $18 + 12i$ **17.** 7 **19.** $12 - 16i$ **21.** $-4i$ **23.** $\frac{28}{25} - \frac{21}{25}i$ **25.** $4 + i$ **27.** $-\sqrt{14}$ **29.** $-5\sqrt{2}$ **31.** $4i$ **33.** $i\sqrt{3}$ **35.** $2\sqrt{2}$ **37.** 1 **39.** i **41.** $-i$ **43.** -1 **45.** 63 **47.** $2 - i$ **49.** 20 **51.** $6 - 3i\sqrt{3}$ **53.** 2 **55.** $-5 + \frac{16}{3}i$ **57.** $17 + 144i$ **59.** $\frac{3}{5} - \frac{1}{5}i$ **61.** $5 - 10i$ **63.** $\frac{1}{5} - \frac{8}{5}i$ **65.** $8 - i$ **67.** $40°$ **69.** $83°$ **71.** $(15, 19)$ **73.** $[-6, 2]$ **75.** 10 **77.** $\frac{25a}{9(a - 2)}$ **79.** $\frac{4 + x}{(x - 2)(x - 1)}$

Chapter 8 Review

1. 9 **3.** 3 **5.** $-\frac{3}{8}$ **7.** not a real number **9.** irrational, 8.718 **11.** x^6 **13.** $3|x^3y|$ **15.** $-2x^2$

17. **19.**

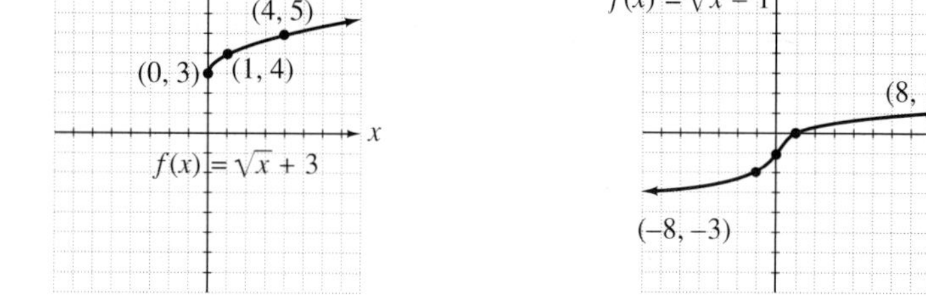

21. $3\sqrt{6}$ **23.** $5xy^3\sqrt{6x}$ **25.** $3\sqrt[3]{2}$ **27.** $2y\sqrt[4]{3x^3y^2}$ **29.** $\frac{3\sqrt{2}}{5}$ **31.** $\frac{3y\sqrt{5}}{2x^2}$ **33.** $\frac{\sqrt[4]{9}}{2}$ **35.** $\frac{y^2\sqrt[3]{3}}{2x}$ **37.** $-\sqrt[3]{2} + 2\sqrt[3]{3}$ **39.** $6\sqrt{6} - 2\sqrt[3]{6}$ **41.** $5\sqrt{7} + 2\sqrt[3]{7}$ **43.** $\frac{\sqrt{5}}{6}$ **45.** 0 **47.** $30\sqrt{2}$ **49.** $6\sqrt{2} - 18$ **51.** $3\sqrt{2} - 5\sqrt{3} + 2\sqrt{6} - 10$ **53.** $4\sqrt{2}$ **55.** $\frac{x\sqrt{5x}}{2y^4}$ **57.** $\frac{\sqrt{30}}{6}$ **59.** $\frac{\sqrt[3]{12x^2}}{2x}$ **61.** $3x(\sqrt{5} + 2)$ **63.** $\frac{4\sqrt{2} - \sqrt{2x}}{16 - x}$ **65.** $\frac{7}{2y\sqrt{35}}$ **67.** $\frac{7}{\sqrt[3]{441}}$ **69.** $\frac{1}{6(\sqrt{6} - 2)}$ **71.** $\frac{x - 25}{2\sqrt{3x} + 10\sqrt{3}}$ **73.** $\frac{1}{3}$ **75.** $-\frac{1}{3}$ **77.** -27 **79.** not a real number **81.** $\frac{9}{4}$ **83.** $a^{13/6}$ **85.** a **87.** $\frac{1}{a^{9/2}}$ **89.** $\frac{y^2}{x}$ **91.** $\frac{1}{a^{11/12}}$ **93.** $4(3^{5/3}x^{8/3})$ **95.** $7x\sqrt[3]{3y^2}$ **97.** $\{-2\}$ **99.** $\{29\}$ **101.** $\{\ \}$ **103.** $\{21\}$ **105.** $\{\ \}$ **107.** $3\sqrt{2}$ **109.** 51.2 ft **111.** $2\sqrt{2}i$ **113.** $6i$ **115.** $15 - 4i$ **117.** $(4\sqrt{2} + \sqrt{3}) - 2\sqrt{2}i$ **119.** $10 + 4i$ **121.** $1 + 5i$ **123.** $\frac{3}{2} - i$

Chapter 8 Test

1. 4 **2.** 5 **3.** 8 **4.** $\frac{3}{4}$ **5.** not a real number **6.** $\frac{1}{9}$ **7.** $3\sqrt{6}$ **8.** $2\sqrt{23}$ **9.** $x^3\sqrt{3}$ **10.** $2x^2y^3\sqrt{2y}$ **11.** $3x^4\sqrt{x}$ **12.** $2\sqrt[3]{5}$ **13.** $2x^2y^3\sqrt[3]{y}$ **14.** $-8\sqrt{3}$ **15.** $3\sqrt[3]{2} - 2x\sqrt{2}$ **16.** $\frac{\sqrt{5}}{4}$ **17.** $\frac{x\sqrt[3]{2}}{3}$ **18.** $\frac{\sqrt[3]{12x}}{2x}$ **19.** $\frac{6 - x^2}{8(\sqrt{6} - x)}$ **20.** $\{9\}$ **21.** $\{5\}$ **22.** $\{9\}$ **23.** $2i\sqrt{5}$ **24.** $-3i$ **25.** $-9 - 40i$ **26.** $-\frac{3}{2} + \frac{5}{2}i$ **27.** $\sqrt{65}$ **28.** $2\sqrt{61}$ **29.** 27 mph **30.** 360 ft

Chapter 8 Cumulative Review

1. a. -17 **b.** 11 **c.** -3 **d.** 6; *Sec. 1.6, Ex. 6*
2. a. $6x - 14$ **b.** $6 - 4x$ **c.** $-11x - 13$; *Sec. 2.1, Ex. 6* **3.** $\{\ \}$; *Sec. 2.3, Ex. 5*
4. a. Alaska Village Electric Coop **b.** Waterville, Wash. **c.** 40¢ **d.** 12.1¢ per kilowatt hour; *Sec. 2.6, Ex. 1*

5. 8 liters of 40% solution, 4 liters of 70% solution; *Sec. 3.3, Ex. 2* **6.** $\left\{\frac{3}{4}, 5\right\}$; *Sec. 3.4, Ex. 8*

7. $\left(-\infty, \frac{13}{5}\right] \cup [4, \infty)$; *Sec. 3.6, Ex. 4* **8.** *Sec. 4.1, Ex. 4* **9.** *Sec. 4.3, Ex. 6*

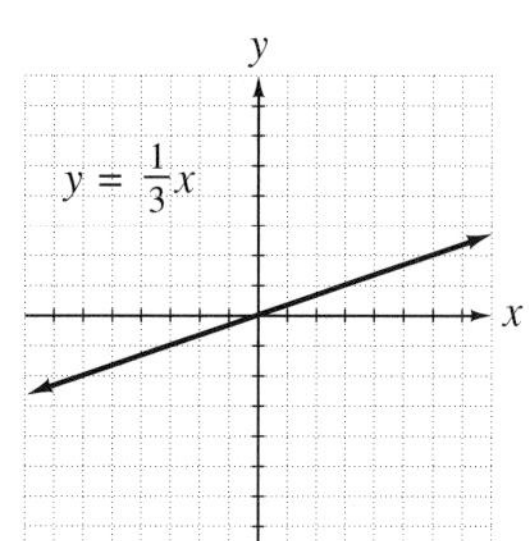

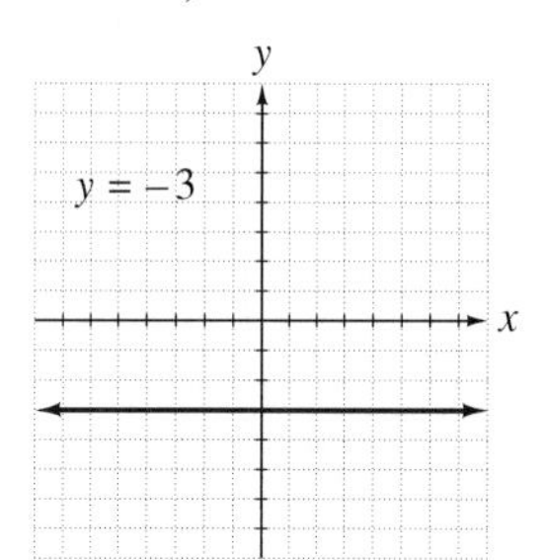

10. a. x^{10} **b.** y^{16} **c.** $(-5)^{12}$; *Sec. 5.1, Ex. 4* **11. a.** $-5x^2 - 6x$ **b.** $8xy - 3x$; *Sec. 5.3, Ex. 5*
12. a. $2x^2 + 11x + 15$ **b.** $10x^5 - 12x^4 + 14x^3 - 15x^2 + 18x - 21$; *Sec. 5.4, Ex. 3* **13.** 16; *Sec. 5.6, Ex. 4*
14. $(x + 3)(x + 4)$; *Sec. 6.2, Ex. 1* **15.** $\left\{-5, \frac{1}{2}\right\}$; *Sec. 6.6, Ex. 2* **16.** $\frac{2(4x^2 - 10x + 25)}{(x^2 + 1)}$; *Sec. 7.2, Ex. 4*
17. { }; *Sec. 7.5, Ex. 3* **18. a.** 2 **b.** −2 **c.** −2 **d.** not a real number; *Sec. 8.1, Ex. 5*
19. a. $3\sqrt{6}$ **b.** $2\sqrt{3}$ **c.** $10\sqrt{2}$ **d.** $\sqrt{35}$; *Sec. 8.2, Ex. 1*
20. a. $\frac{10 - 6\sqrt{2}}{7}$ **b.** $\frac{\sqrt{30}}{2} + \frac{3\sqrt{2}}{2} + \sqrt{5} + \sqrt{3}$ **c.** $\frac{7\sqrt{3xy}}{6x}$ **d.** $\frac{6\sqrt{mx} - 2m}{9x - m}$; *Sec. 8.4, Ex. 3*
21. a. 5 **b.** 2 **c.** −2 **d.** −3 **e.** $\frac{1}{3}$; *Sec. 8.5, Ex. 1* **22.** {42}; *Sec. 8.6, Ex. 1* **23.** $80\sqrt{7}$ ft; *Sec. 8.6, Ex. 7*
24. a. $\frac{1}{2} + \frac{3}{2}i$ **b.** $-\frac{7}{3}i$; *Sec. 8.7, Ex. 4*

CHAPTER 9
Quadratic Equations and Inequalities

Exercise Set 9.1

1. {4, −4} **3.** $\{\sqrt{7}, -\sqrt{7}\}$ **5.** $\{3\sqrt{2}, -3\sqrt{2}\}$ **7.** $\{\sqrt{10}, -\sqrt{10}\}$ **9.** $\{3i, -3i\}$ **11.** $\{\sqrt{6}, -\sqrt{6}\}$
13. $\{2i\sqrt{2}, -2i\sqrt{2}\}$ **15.** {−8, −2} **17.** $\{6 - 3\sqrt{2}, 6 + 3\sqrt{2}\}$ **19.** $\left\{\frac{3 - 2\sqrt{2}}{2}, \frac{3 + 2\sqrt{2}}{2}\right\}$
21. $\{1 - 4i, 1 + 4i\}$ **23.** $\{-7 - \sqrt{5}, -7 + \sqrt{5}\}$ **25.** $\{-3 - 2i\sqrt{2}, -3 + 2i\sqrt{2}\}$
27. $x^2 + 16x + 65 = (x + 8)^2$ **29.** $z^2 - 12z + 36 = (z - 6)^2$ **31.** $p^2 + 9p + \frac{81}{4} = \left(p + \frac{9}{2}\right)^2$
33. $x^2 + x + \frac{1}{4} = \left(x + \frac{1}{2}\right)^2$ **35.** {−5, −3} **37.** $\{-3 - \sqrt{7}, -3 + \sqrt{7}\}$ **39.** $\left\{\frac{-1 - \sqrt{5}}{2}, \frac{-1 + \sqrt{5}}{2}\right\}$
41. $\{-1 - i, -1 + i\}$ **43.** $\{3 + \sqrt{6}, 3 - \sqrt{6}\}$ **45.** $\left\{\frac{-1 - \sqrt{5}}{2}, \frac{-1 + \sqrt{5}}{2}\right\}$ **47.** $\left\{\frac{6 - \sqrt{30}}{3}, \frac{6 + \sqrt{30}}{3}\right\}$
49. $\left\{\frac{3 - \sqrt{11}}{2}, \frac{3 + \sqrt{11}}{2}\right\}$ **51.** $\left\{\frac{1}{2}, -4\right\}$ **53.** $\{-2 - i\sqrt{2}, -2 + i\sqrt{2}\}$ **55.** $\left\{\frac{-15 - 7\sqrt{5}}{10}, \frac{-15 + 7\sqrt{5}}{10}\right\}$
57. $\left\{\frac{1 - i\sqrt{47}}{4}, \frac{1 + i\sqrt{47}}{4}\right\}$ **59.** 20% **61.** 7.72 seconds **63.** 8.39 seconds **65.** {−1, 5}
67. $\{-4 - \sqrt{15}, -4 + \sqrt{15}\}$ **69.** $\left\{\frac{-3 - \sqrt{21}}{3}, \frac{-3 + \sqrt{21}}{3}\right\}$ **71.** $\left\{-1, \frac{5}{2}\right\}$ **73.** $\{-5 - i\sqrt{3}, -5 + i\sqrt{3}\}$

75. $\{-4, 1\}$ **77.** $\left\{1 - \frac{\sqrt{2}}{2}i, 1 + \frac{\sqrt{2}}{2}i\right\}$ **79.** $\left\{-\frac{1}{2} - \frac{\sqrt{69}}{6}, -\frac{1}{2} + \frac{\sqrt{69}}{6}\right\}$ **81.** 15 ft by 15 ft **83.** $10\sqrt{2}$ cm

85. 6000 scissors **87.** 11% **89.** $-\frac{1}{2}$ **91.** -1 **93.** $3 + 2\sqrt{5}$ **95.** $\frac{1 - 3\sqrt{2}}{2}$ **97.** $2\sqrt{6}$ **99.** 5

Exercise Set 9.2

1. $\{-6, 1\}$ **3.** $\left\{-\frac{3}{5}, 1\right\}$ **5.** $\{3\}$ **7.** $\left\{\frac{-7 - \sqrt{33}}{2}, \frac{-7 + \sqrt{33}}{2}\right\}$ **9.** $\left\{\frac{1 - \sqrt{57}}{8}, \frac{1 + \sqrt{57}}{8}\right\}$

11. $\left\{\frac{7 - \sqrt{85}}{6}, \frac{7 + \sqrt{85}}{6}\right\}$ **13.** $\{1 - \sqrt{3}, 1 + \sqrt{3}\}$ **15.** $\left\{-\frac{3}{2}, 1\right\}$ **17.** $\left\{\frac{3 - \sqrt{11}}{2}, \frac{3 + \sqrt{11}}{2}\right\}$

19. $\left\{\frac{3 - i\sqrt{87}}{8}, \frac{3 + i\sqrt{87}}{8}\right\}$ **21.** $\{-2 - \sqrt{11}, -2 + \sqrt{11}\}$ **23.** $\left\{\frac{-5 - i\sqrt{5}}{10}, \frac{-5 + i\sqrt{5}}{10}\right\}$

25. two real solutions **27.** two complex solutions **29.** two complex solutions **31.** one real solution

33. 8.53 hours **35.** 7.89 seconds **37.** $\left\{\frac{-5 - \sqrt{17}}{2}, \frac{-5 + \sqrt{17}}{2}\right\}$ **39.** $\left\{\frac{5}{2}, 1\right\}$ **41.** $\left\{\frac{3 - \sqrt{29}}{2}, \frac{3 + \sqrt{29}}{2}\right\}$

43. $\left\{\frac{-1 - \sqrt{19}}{6}, \frac{-1 + \sqrt{19}}{6}\right\}$ **45.** $\{-3 - 2i, -3 + 2i\}$ **47.** $\left\{\frac{-1 - i\sqrt{23}}{4}, \frac{-1 + i\sqrt{23}}{4}\right\}$ **49.** $\{1\}$

51. $\left\{\frac{19 - \sqrt{345}}{2}, \frac{19 + \sqrt{345}}{2}\right\}$ **53.** 12 or -8 **55.** Toshiba, 16.5 hours; IBM, 15.5 hours **57.** 2.8 seconds

59. Sunday to Monday **61.** Wednesday **63.** $f(4) = 33$; yes **65.** $\left\{\frac{11}{5}\right\}$ **67.** $\{15\}$ **69.** $-2\sqrt{5}$ **71.** $5\sqrt{3x}$

73. $(x^2 + 5)(x + 2)(x - 2)$ **75.** $(z + 3)(z - 3)(z + 2)(z - 2)$ **77.** $\left\{\frac{\sqrt{3}}{3}\right\}$

79. $\left\{\frac{-\sqrt{2} - i\sqrt{2}}{2}, \frac{-\sqrt{2} + i\sqrt{2}}{2}\right\}$ **81.** $\left\{\frac{\sqrt{3} - \sqrt{11}}{4}, \frac{\sqrt{3} + \sqrt{11}}{4}\right\}$

Exercise Set 9.3

1. $\{3 - \sqrt{7}, 3 + \sqrt{7}\}$ **3.** $\left\{\frac{3 - \sqrt{57}}{4}, \frac{3 + \sqrt{57}}{4}\right\}$ **5.** $\left\{\frac{1 - \sqrt{29}}{2}, \frac{1 + \sqrt{29}}{2}\right\}$ **7.** $\left\{1, \frac{-1 - i\sqrt{3}}{2}, \frac{-1 + i\sqrt{3}}{2}\right\}$

9. $\left\{0, -3, \frac{3 - 3i\sqrt{3}}{2}, \frac{3 + 3i\sqrt{3}}{2}\right\}$ **11.** $\{4, -2 - 2i\sqrt{3}, -2 + 2i\sqrt{3}\}$ **13.** $\{-2, 2, -2i, 2i\}$

15. $\left\{-\frac{1}{2}, \frac{1}{2}, -i\sqrt{3}, i\sqrt{3}\right\}$ **17.** $\{-3, 3, -2, 2\}$ **19.** $\{125, -8\}$ **21.** $\left\{-\frac{1}{8}, 27\right\}$ **23.** $\left\{-\frac{1}{125}, \frac{1}{8}\right\}$

25. width, $4\sqrt{10}$ ft; length, $10\sqrt{10}$ ft **27.** $\{-\sqrt{2}, \sqrt{2}, -\sqrt{3}, \sqrt{3}\}$ **29.** $\left\{\frac{-9 - \sqrt{201}}{6}, \frac{-9 + \sqrt{201}}{6}\right\}$

31. $\{-4, 2 - 2i\sqrt{3}, 2 + 2i\sqrt{3}\}$ **33.** $\{27, 125\}$ **35.** $\{1, -3i, 3i\}$ **37.** $\left\{\frac{1}{8}, -8\right\}$

39. $\left\{5, \frac{-5 + 5i\sqrt{3}}{2}, \frac{-5 - 5i\sqrt{3}}{2}\right\}$ **41.** $\{-3\}$ **43.** $\{-\sqrt{5}, \sqrt{5}, -2i, 2i\}$ **45.** $\left\{-3, \frac{3 - 3i\sqrt{3}}{2}, \frac{3 + 3i\sqrt{3}}{2}\right\}$

47. $\left\{-\frac{1}{3}, \frac{1}{3}, -\frac{i\sqrt{6}}{3}, \frac{i\sqrt{6}}{3}\right\}$ **49.** base, $2\sqrt{42}$ cm; height, $\sqrt{42}$ cm **51.** width, $20\sqrt{2}$ in.; height, $30\sqrt{2}$ in.

53. a. width, $5\sqrt{10}$ cm; length, $15\sqrt{10}$ cm; **b.** perimeter, $40\sqrt{10}$ cm **55.** $(-\infty, 3]$

57. $(-5, \infty)$ **59.** domain: $\{x \mid x \text{ is a real number}\}$
range: $\{y \mid y \text{ is a real number}\}$; yes **61.** domain: $\{x \mid x \text{ is a real number}\}$
range: $\{y \mid y \geq -1$; yes$\}$ **63.**

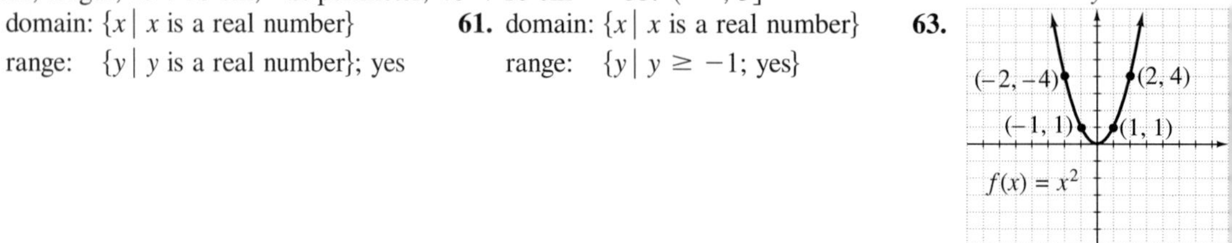

Exercise set 9.4

1. $(-\infty, -5) \cup (-1, \infty)$

3. $[-4, 3]$

5. $[2, 5]$

7. $\left(-5, -\frac{1}{3}\right)$

9. $(2, 4) \cup (6, \infty)$

11. $(-\infty, -4] \cup [0, 1]$

13. $(-\infty, -3) \cup (-2, 2) \cup (3, \infty)$

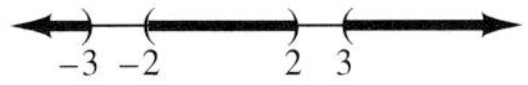

15. $(-7, 2)$

17. $(-1, \infty)$

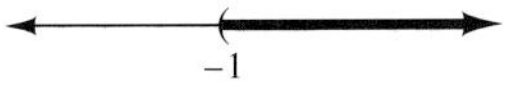

19. $(-\infty, -1] \cup (4, \infty)$

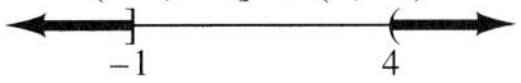

21. $(-\infty, 2) \cup \left(\frac{11}{4}, \infty\right)$

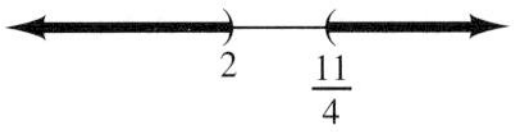

23. $(0, 2] \cup [3, \infty)$

25. $(-\infty, -7) \cup (8, \infty)$

27. $\left[-\frac{5}{4}, \frac{3}{2}\right]$

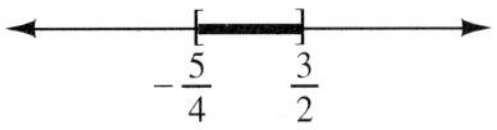

29. $(-\infty, 0) \cup (1, \infty)$

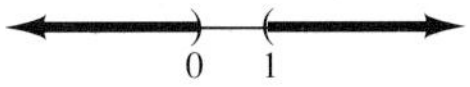

31. $(-\infty, -4] \cup [4, 6]$

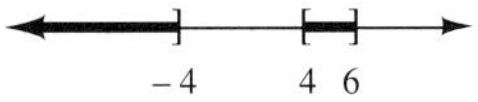

33. $\left(-\infty, -\frac{2}{3}\right] \cup \left[\frac{3}{2}, \infty\right)$

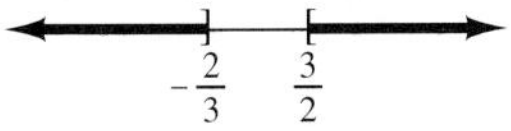

35. $(-\infty, -5] \cup [-1, 1] \cup [5, \infty)$

37. $\left(-\infty, -\frac{5}{3}\right) \cup \left(\frac{7}{2}, \infty\right)$

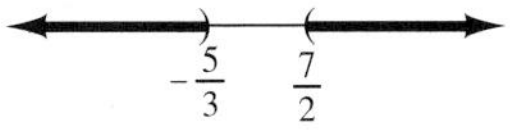

39. $(0, 10)$

41. $(-\infty, -4) \cup [5, \infty)$

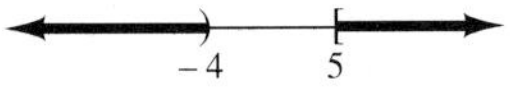

43. $(-\infty, -6] \cup (-1, 0] \cup (7, \infty)$

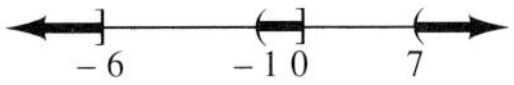

45. $(-\infty, 1) \cup (2, \infty)$

47. $(-\infty, -8] \cup (-4, \infty)$

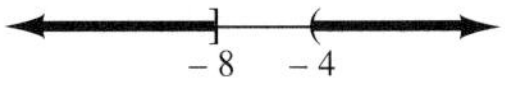

49. $(-\infty, 0) \cup \left(5, \frac{11}{2}\right]$

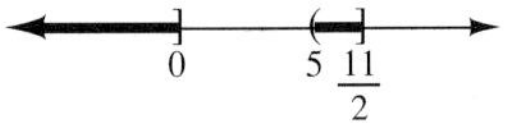

51. $(0, \infty)$

53. the number is any number less than -1 or between 0 and 1.

55. $(2, 11)$

57.

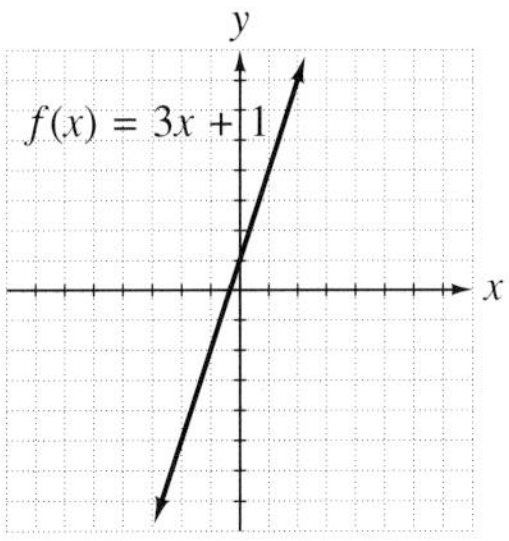

59.

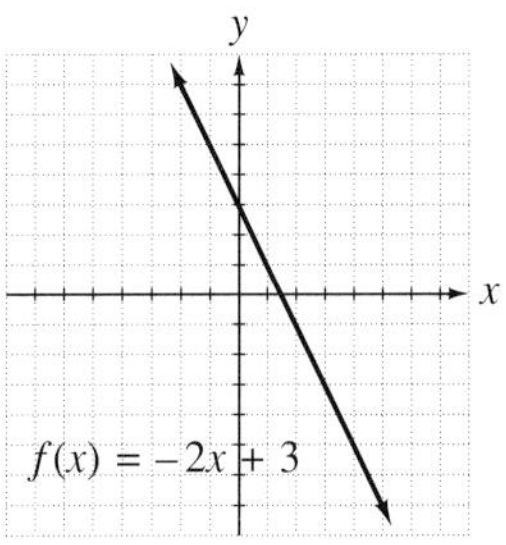

61.

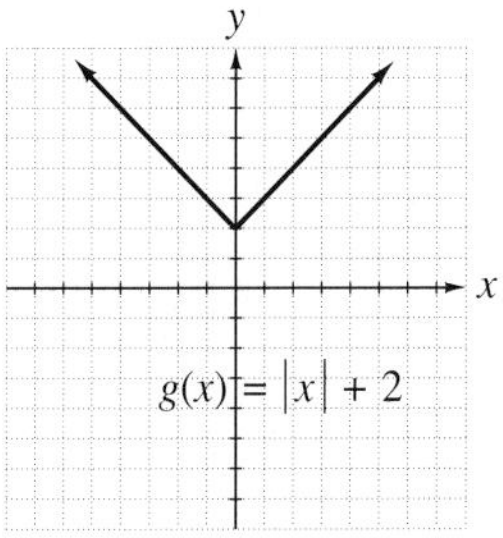

63.

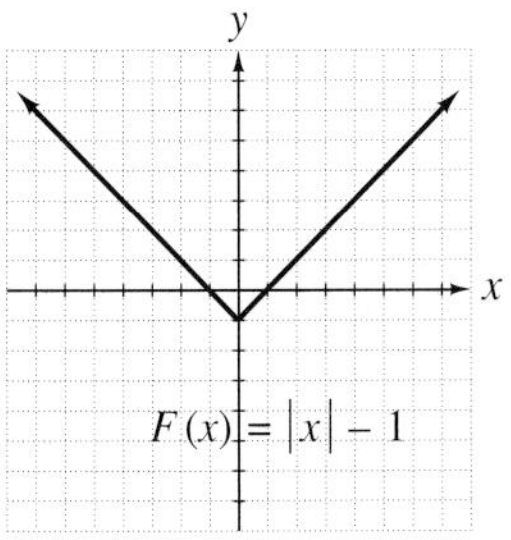

65.

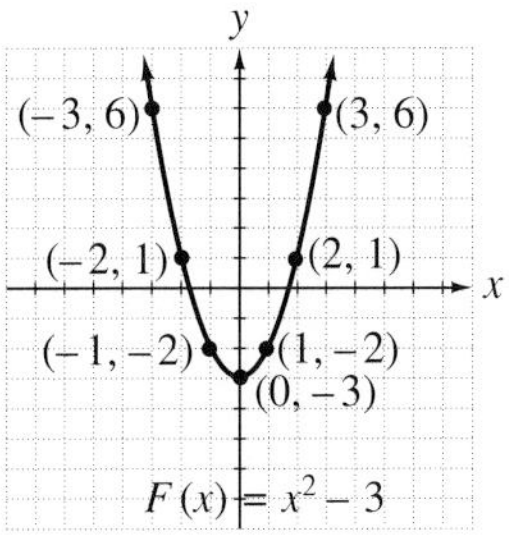

67.

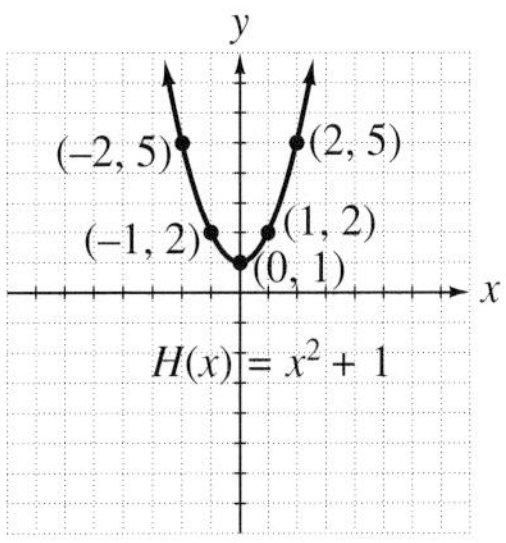

Mental Math, Sec. 9.5

1. (0, 0) **3.** (2, 0) **5.** (0, 3) **7.** (−1, 5)

Exercise Set 9.5

1.

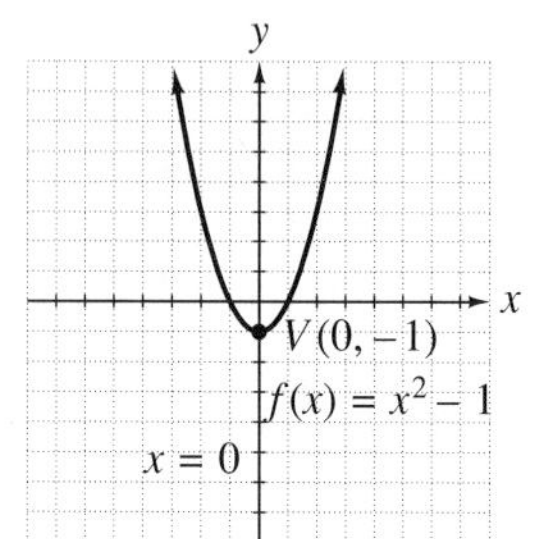

3.

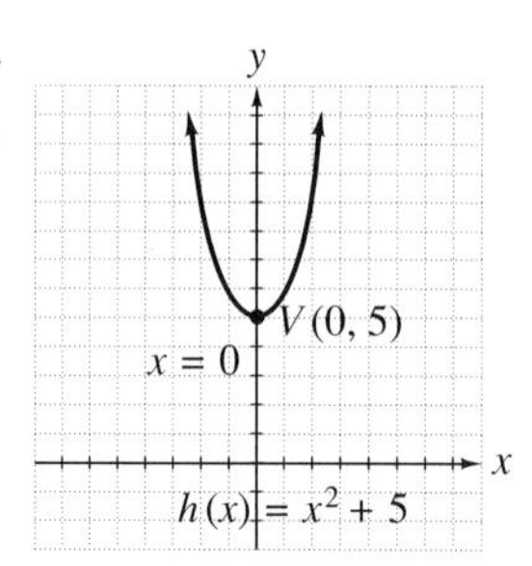

5.

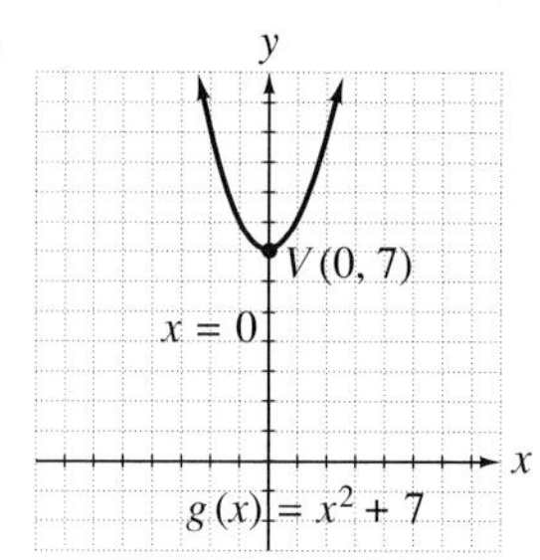

7.

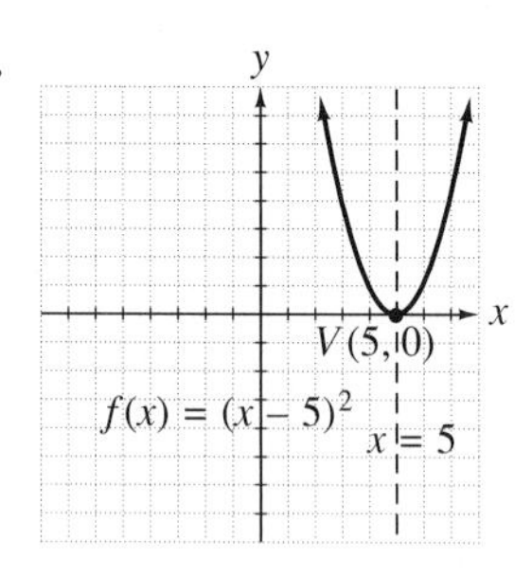

9.

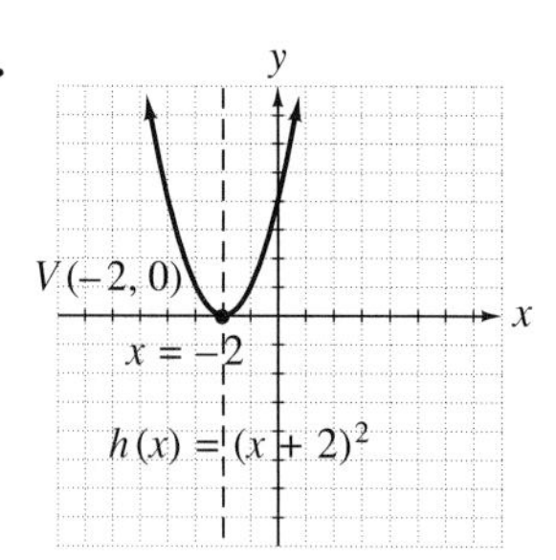

11.

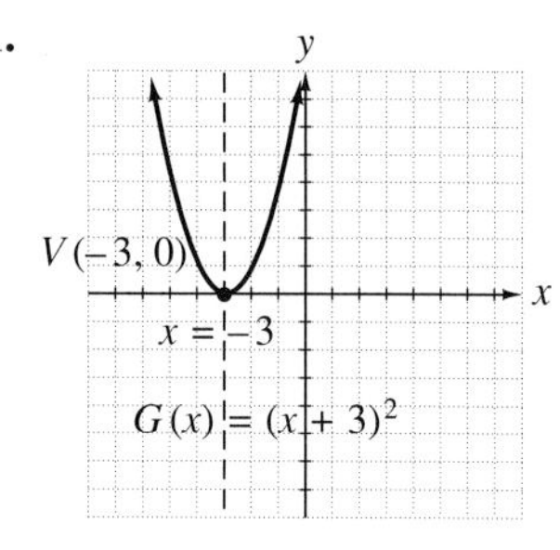

13.

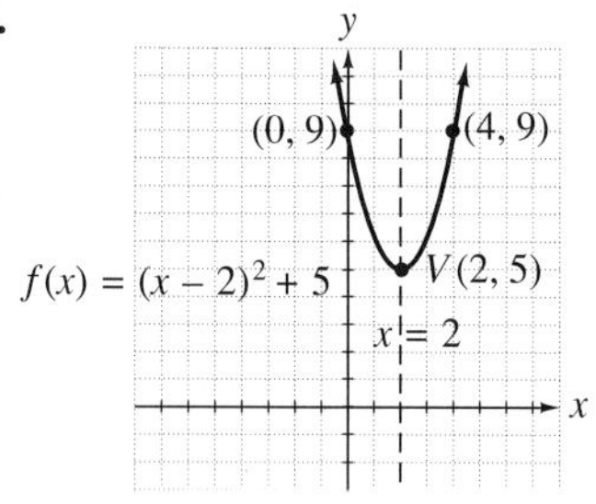

15.

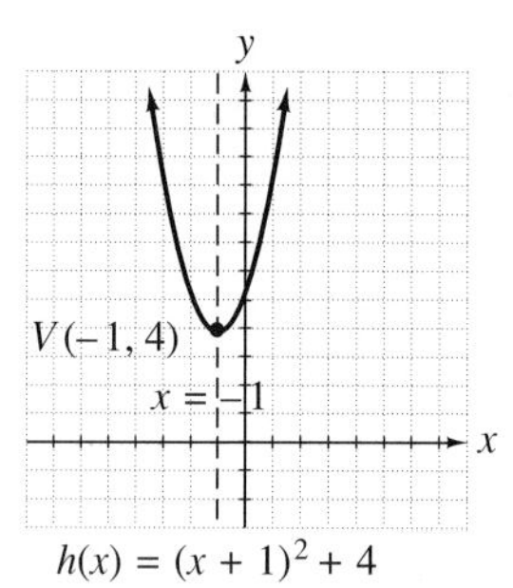

17.

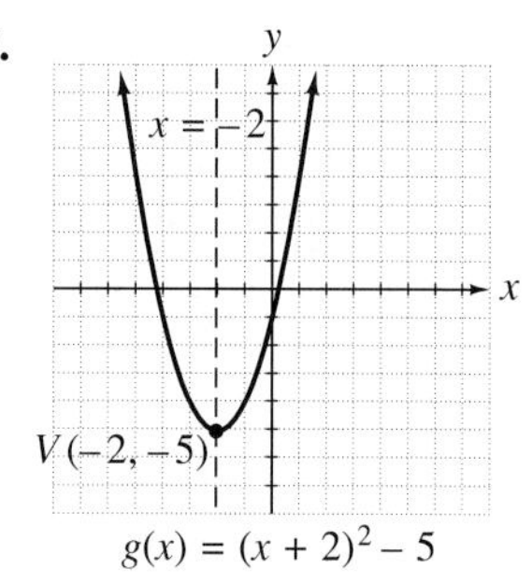

19.

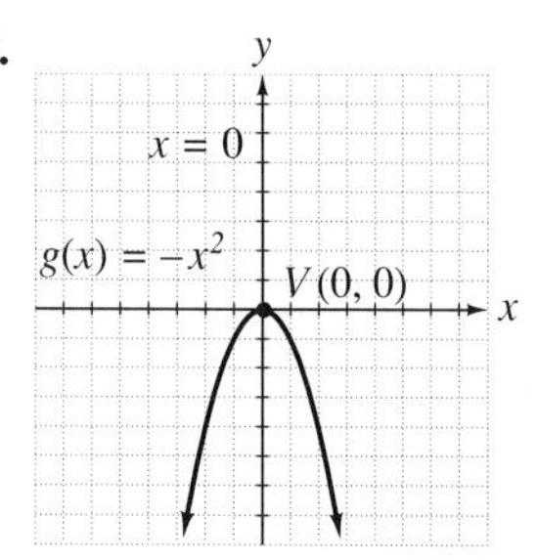

21.

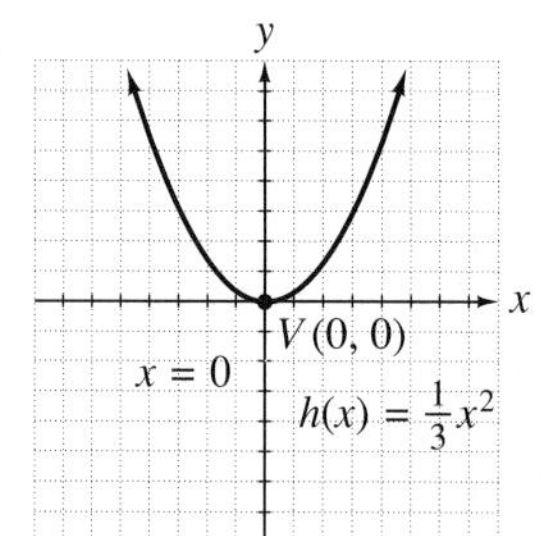

23.

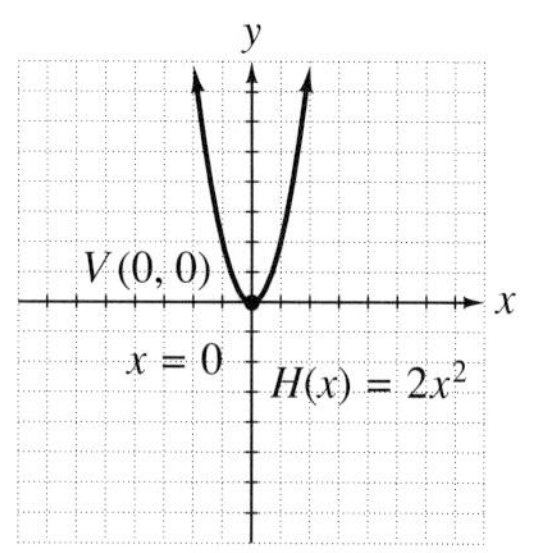

25.

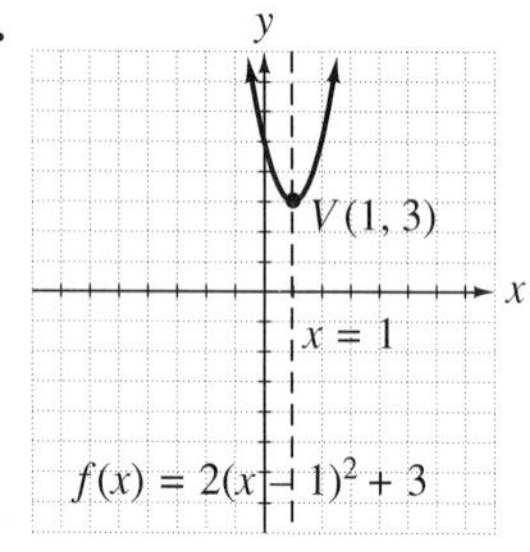

27.

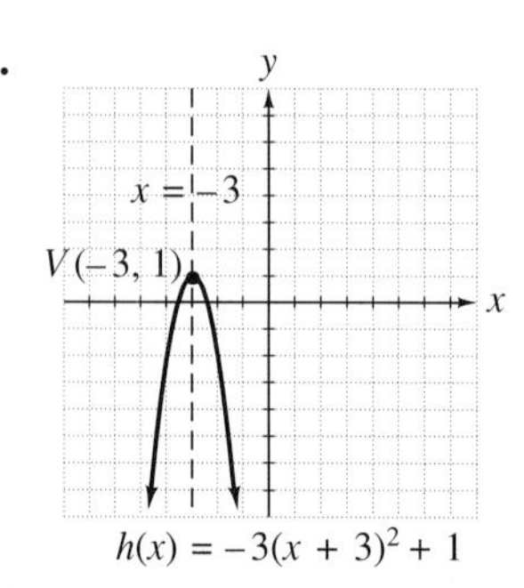

29.

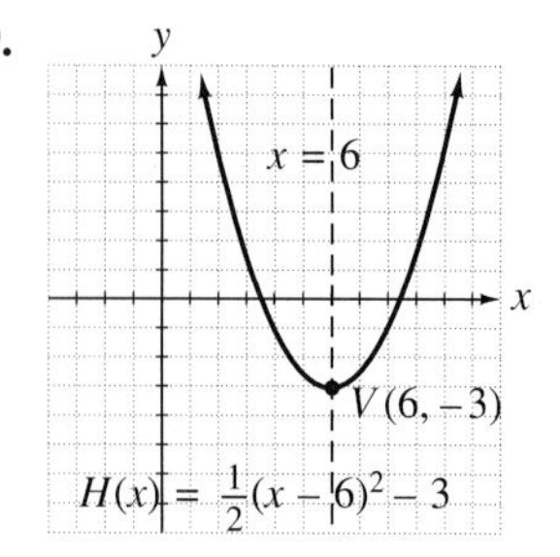

31.

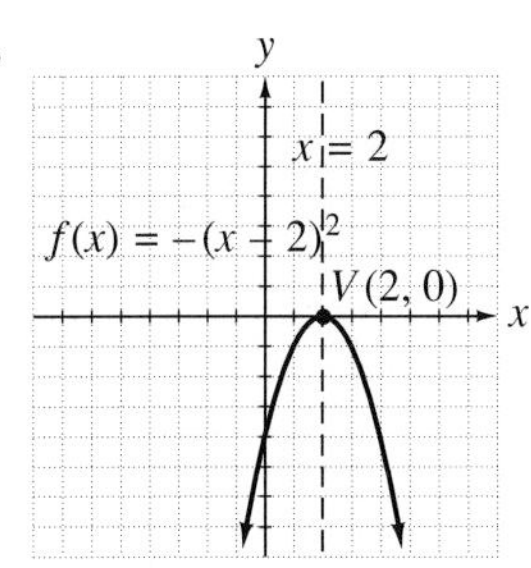

33.

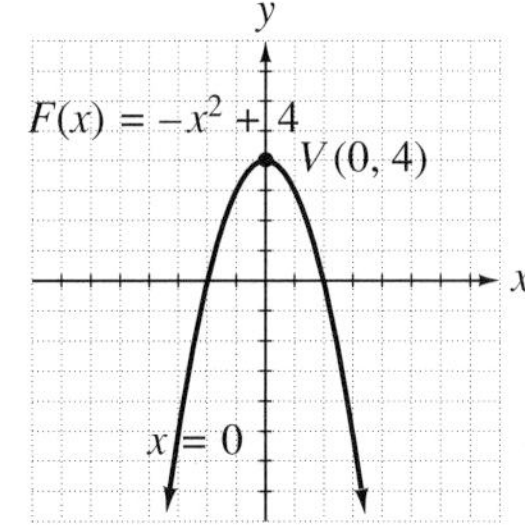

35.

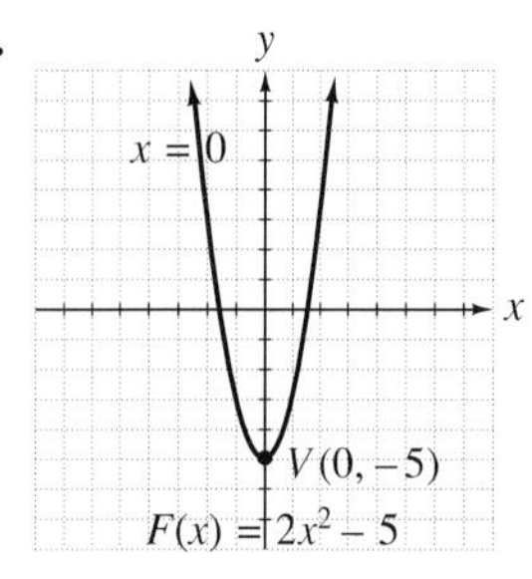

37.

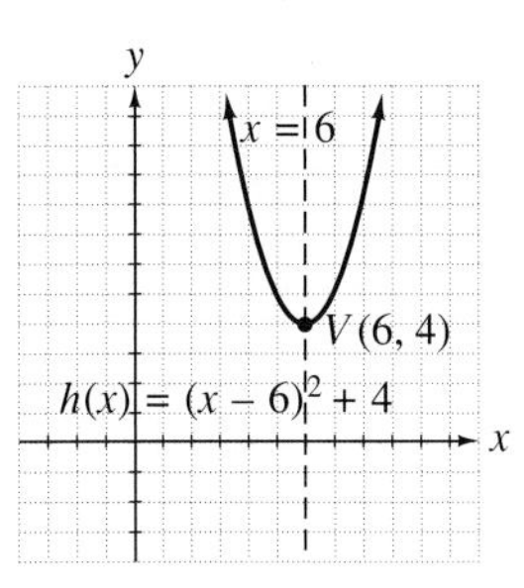

39.

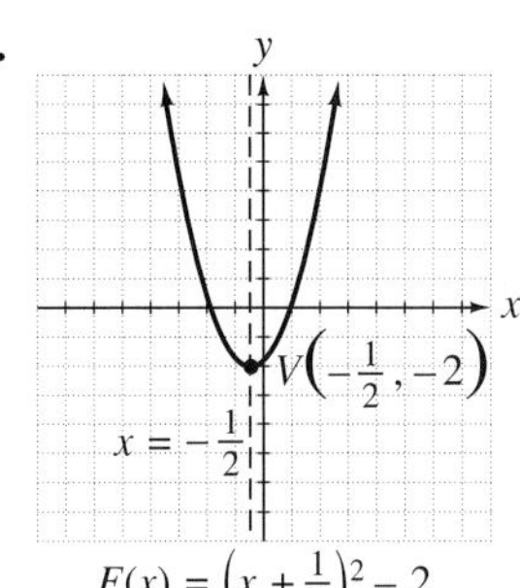

$F(x) = \left(x + \frac{1}{2}\right)^2 - 2$

41.

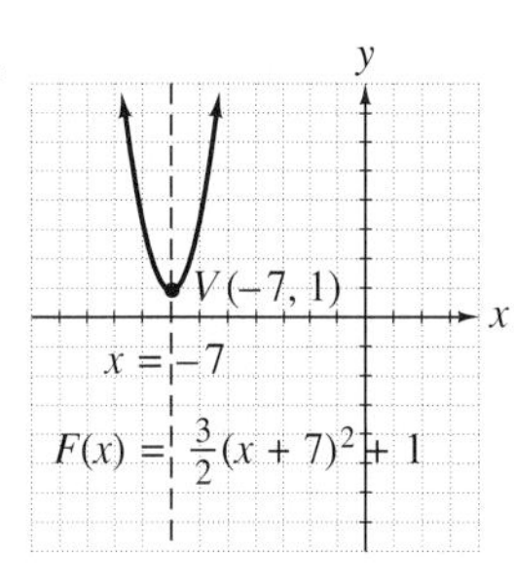

43.

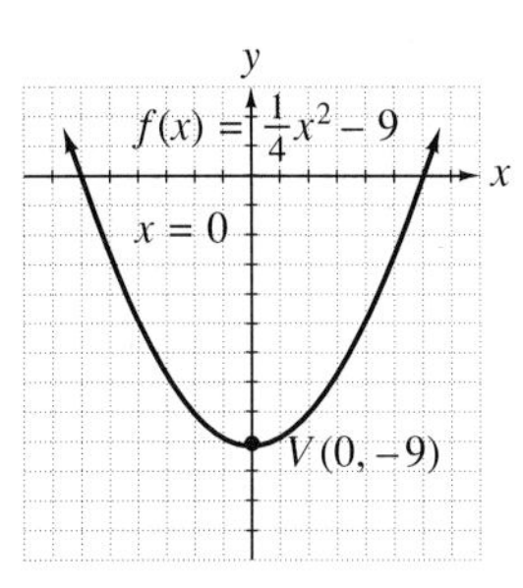

45.

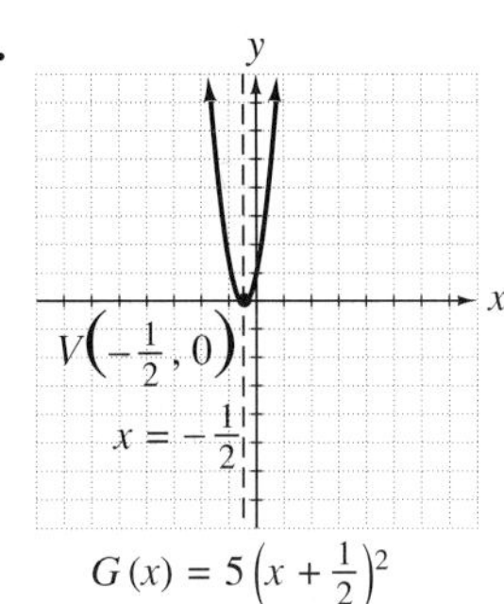

$G(x) = 5\left(x + \frac{1}{2}\right)^2$

47.

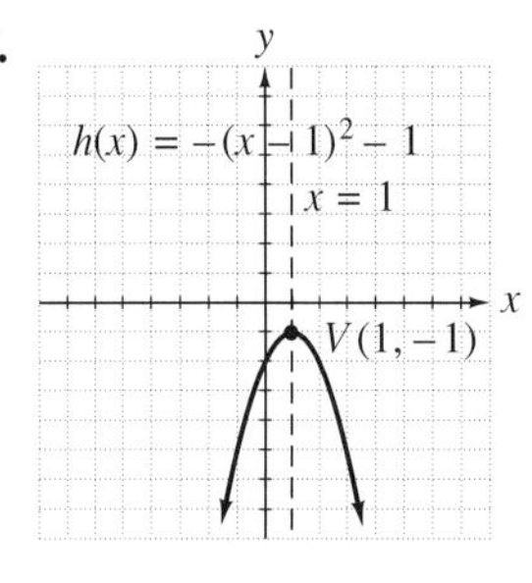

49.

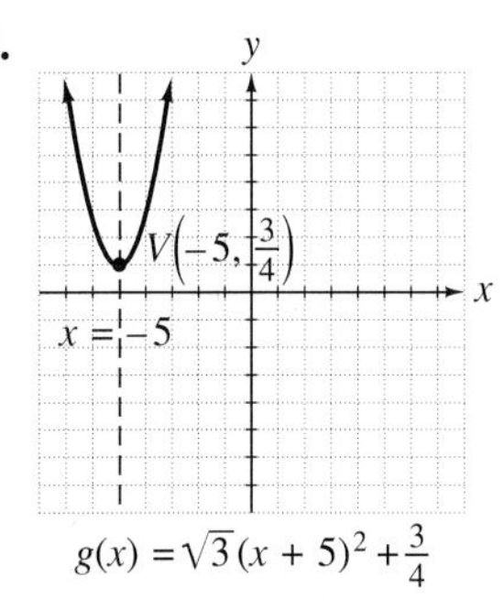

$g(x) = \sqrt{3}(x + 5)^2 + \frac{3}{4}$

51.

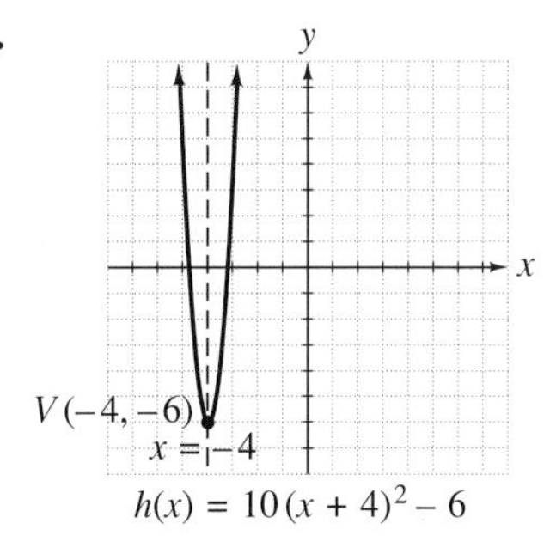

$h(x) = 10(x + 4)^2 - 6$

53.

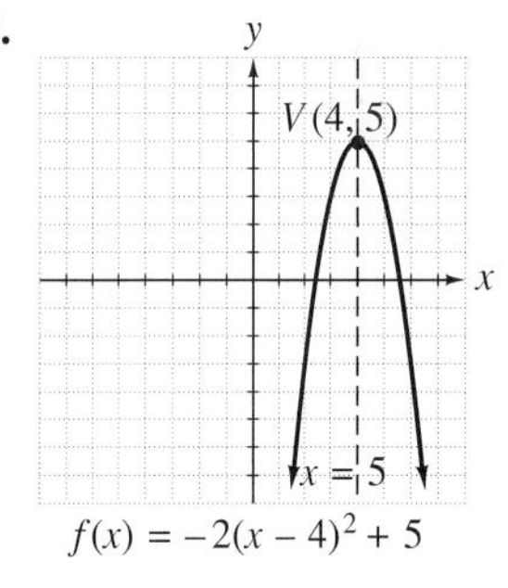

$f(x) = -2(x - 4)^2 + 5$

55. $f(x) = 5(x - 2)^2 + 3$ **57.** $f(x) = 5(x + 3)^2 + 6$

59.

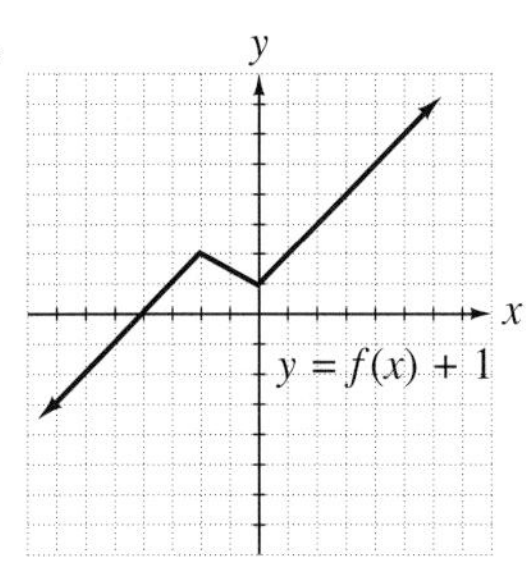

61.

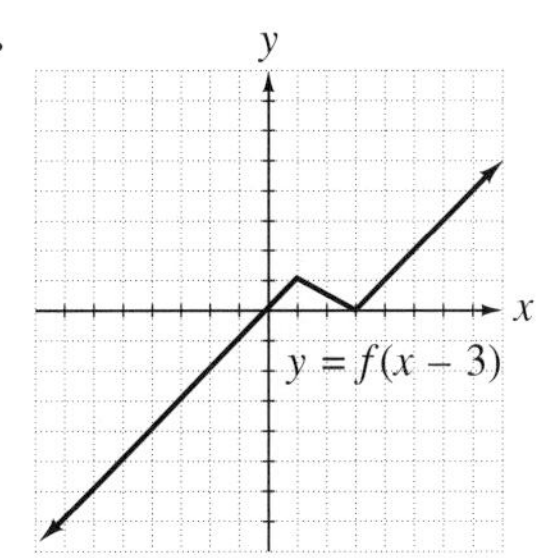

63.

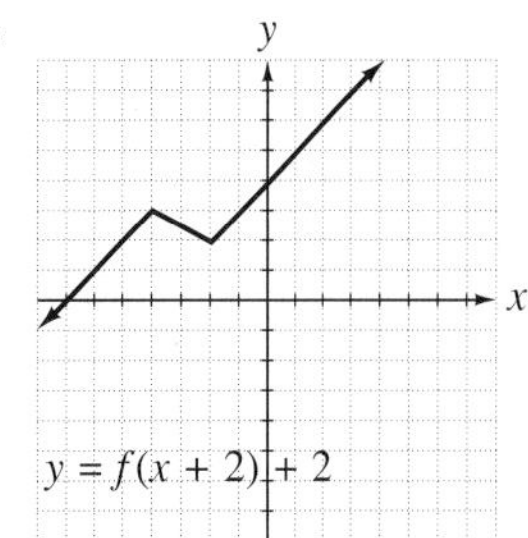

65. $x^2 + 8x + 16$ **67.** $z^2 - 16z + 64$ **69.** $y^2 + y + \frac{1}{4}$ **71.** $\{-6, 2\}$ **73.** $\{-5 - \sqrt{26}, -5 + \sqrt{26}\}$
75. $\{4 - 3\sqrt{2}, 4 + 3\sqrt{2}\}$

Exercise Set 9.6

1. $(-4, -9)$ **3.** $(5, 30)$ **5.** $(1, -2)$ **7.** $\left(\frac{1}{2}, \frac{5}{4}\right)$

9.

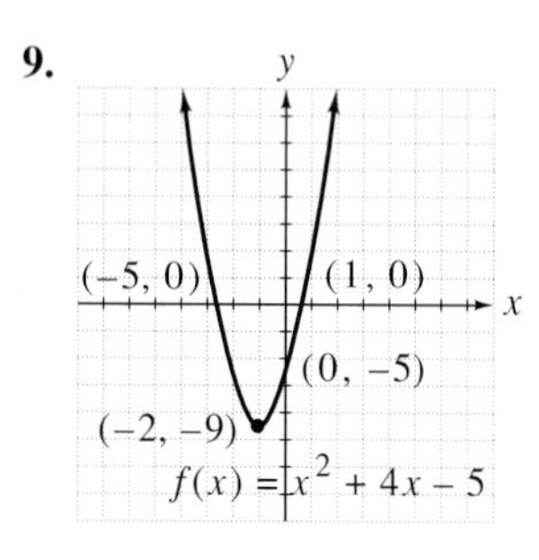

11.

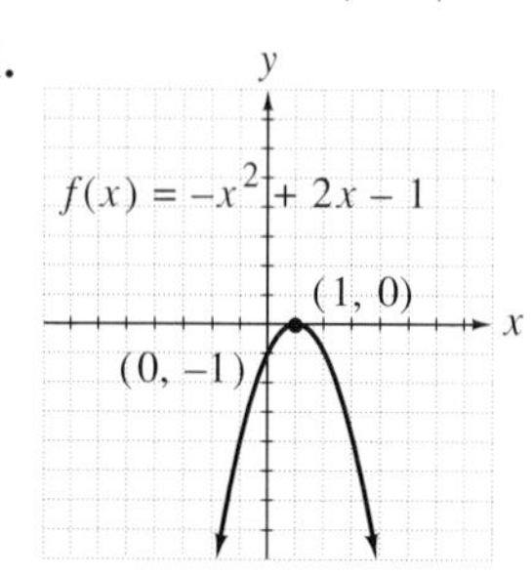

13.

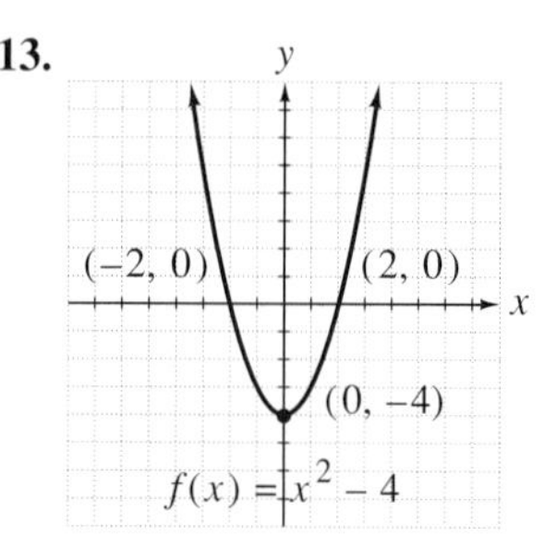

15.

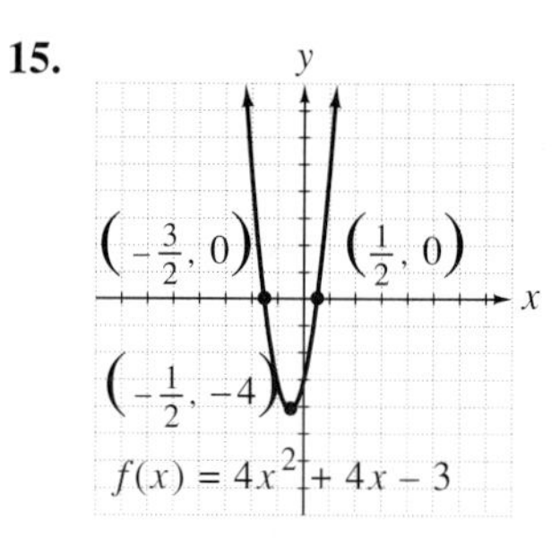

17. 30 and 30

19.

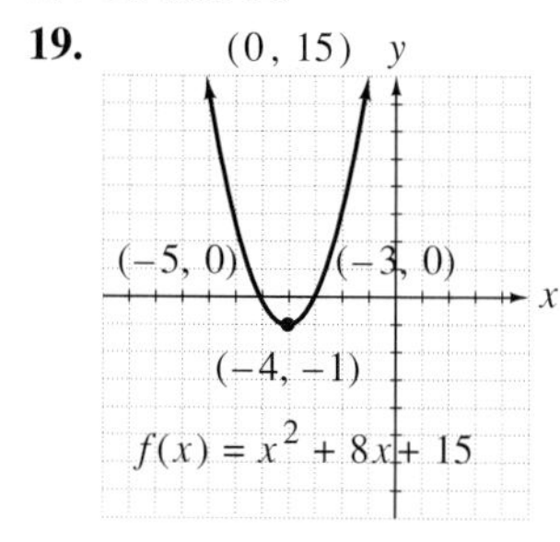

21.

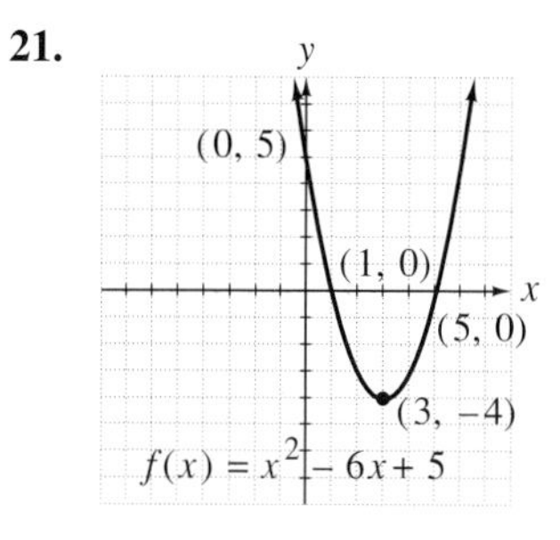

23.

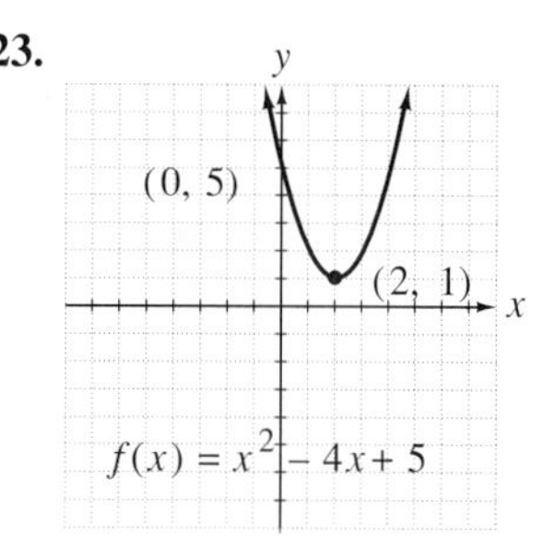

25.

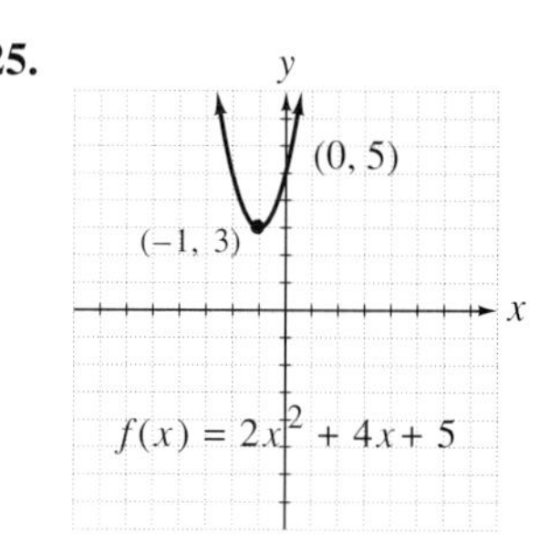

27.

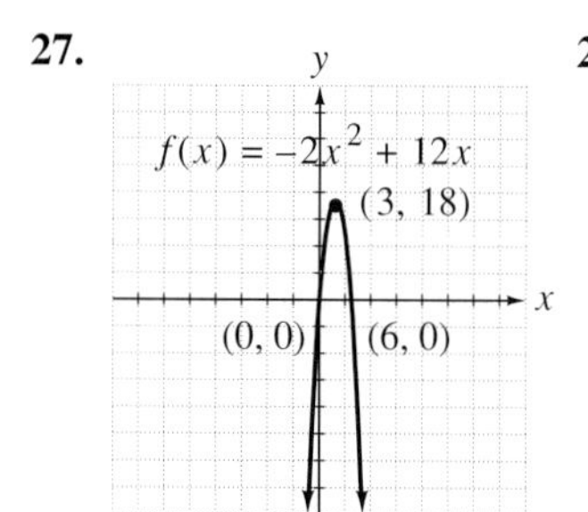

29.

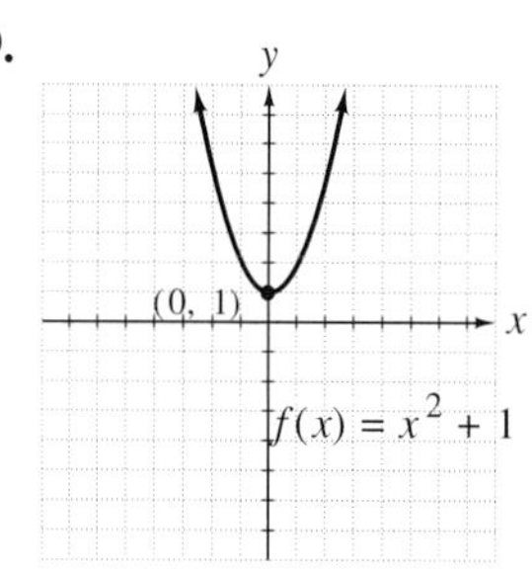

31.

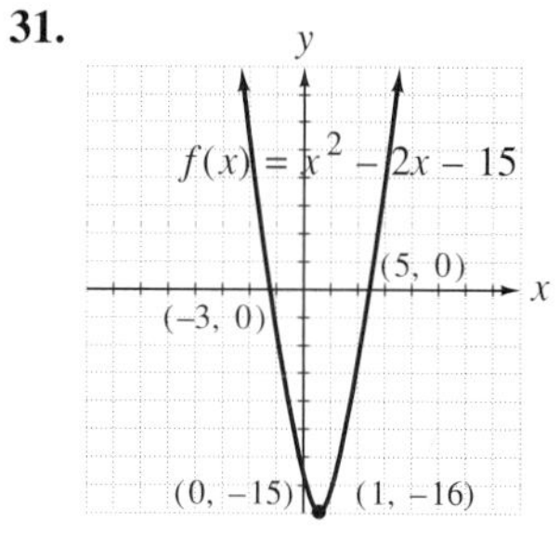

33.

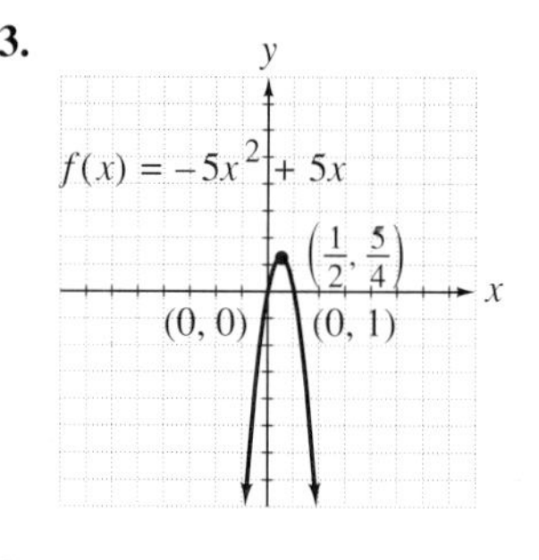

35.

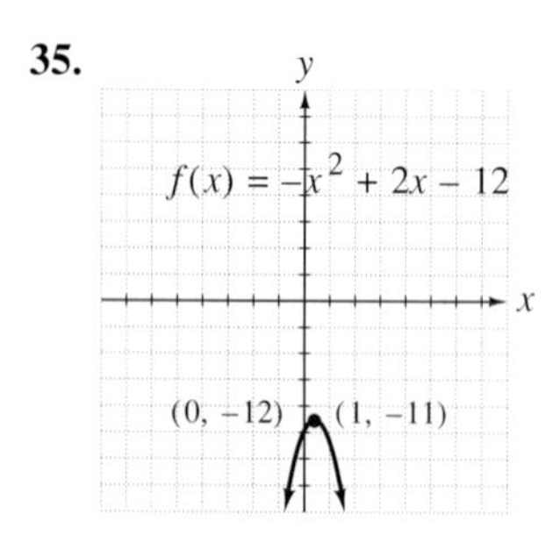

37.

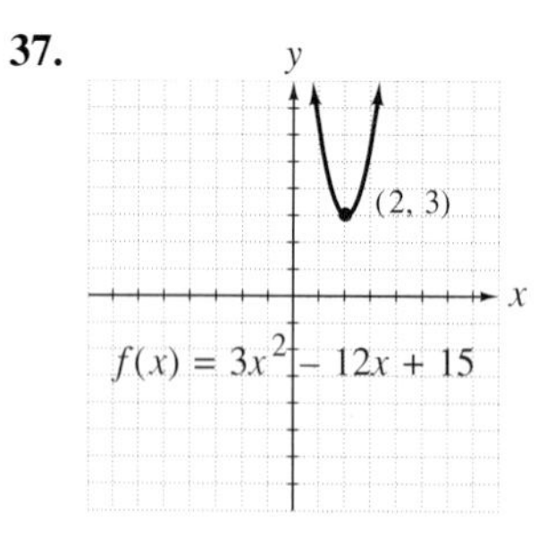

39.

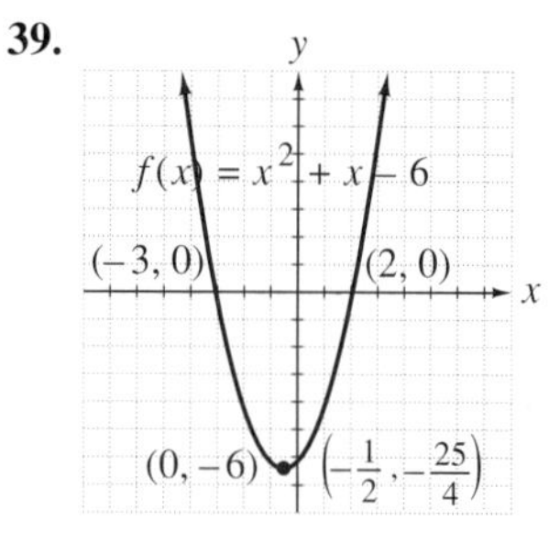

41.

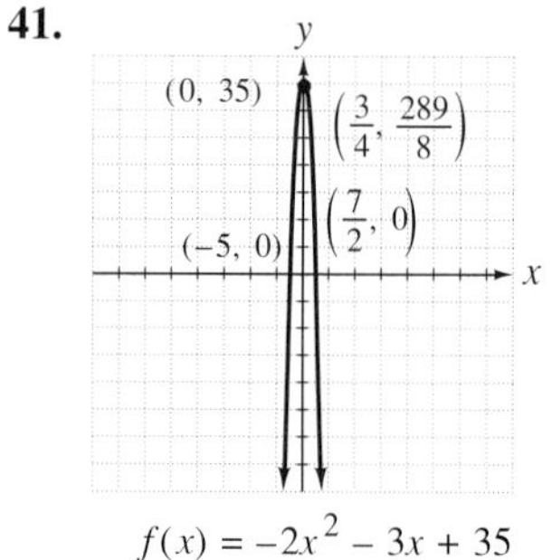

$f(x) = -2x^2 - 3x + 35$

43. 144 ft **45.** **a.** 200 bicycles **b.** $12,000

47.

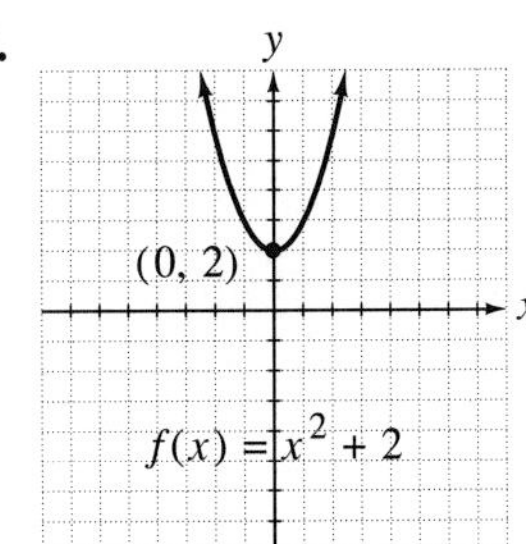

49.

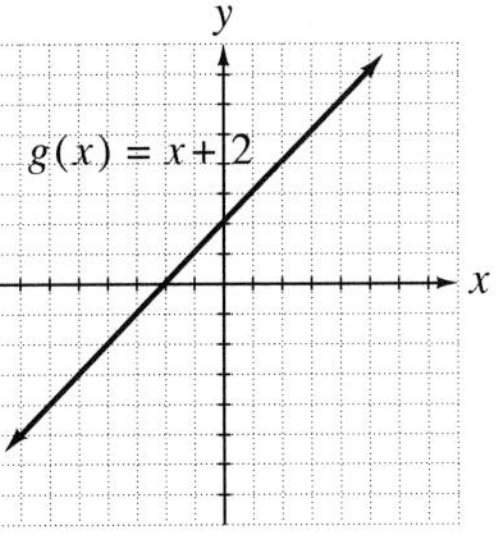
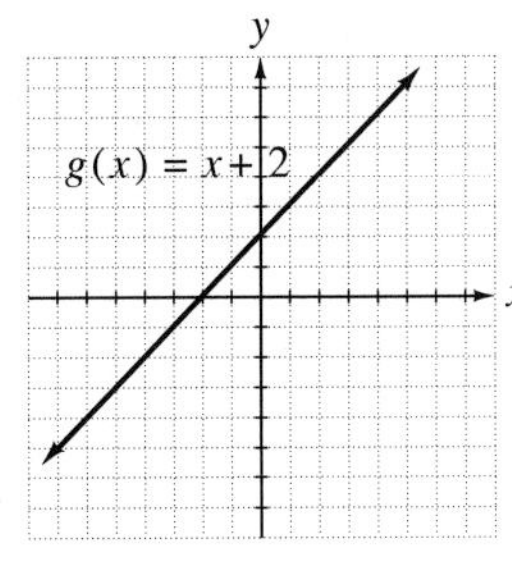

51. $y = (x + 5)^2 + 2$
vertex $(-5, 2)$
opens upward
goes through $(0, 27)$

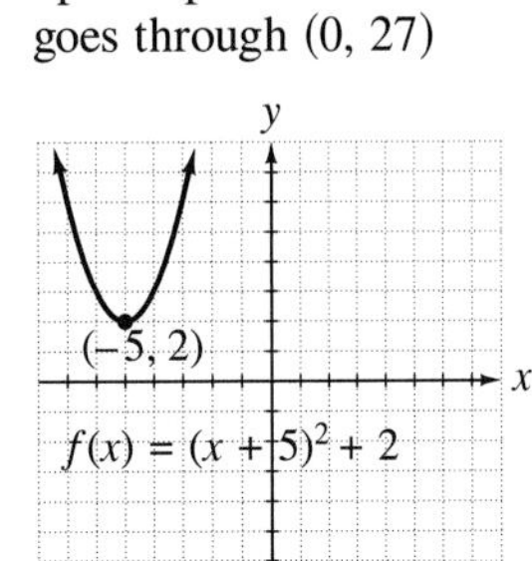

53. $y = 3(x - 4)^2 + 1$
vertex $(4, 1)$
opens upward
goes through $(0, 49)$

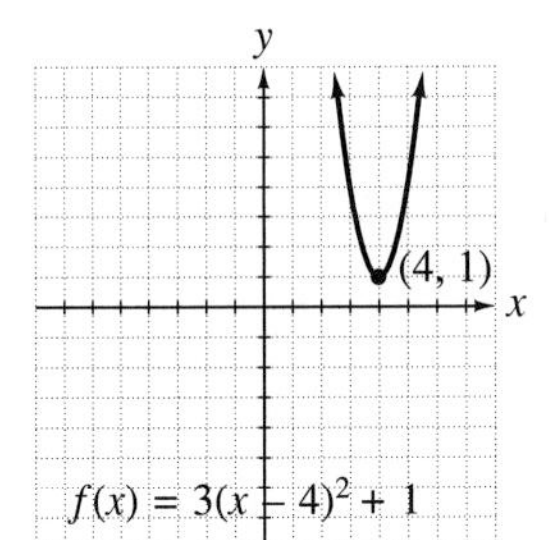

55. y, x, $f(x) = -(x - 4)^2 + \frac{3}{2}$, $\left(4, \frac{3}{2}\right)$

57. no

Mental Math, Sec. 9.7

1. upward **3.** to the left **5.** downward

Exercise Set 9.7

1.

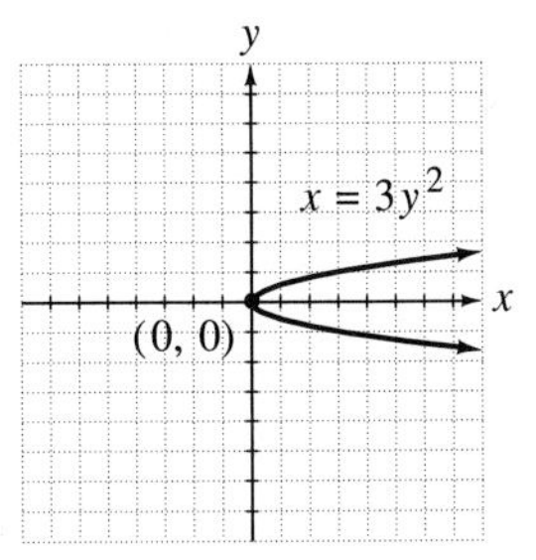

3.

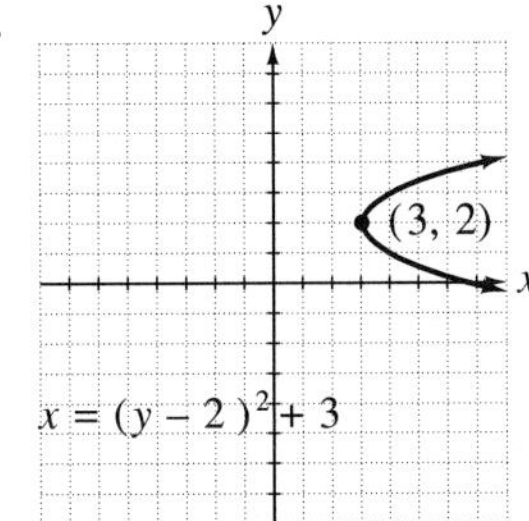

5.

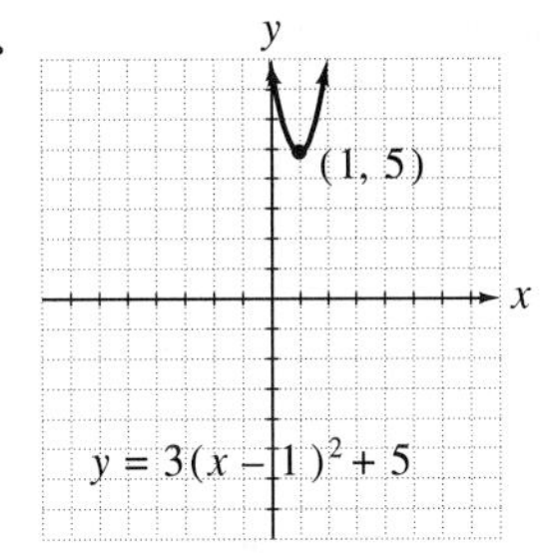

7.

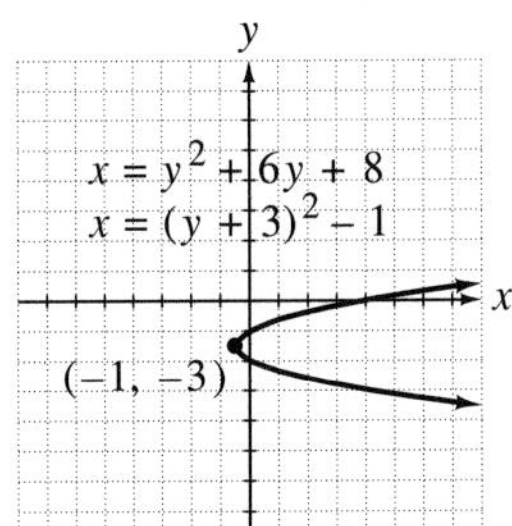

9.

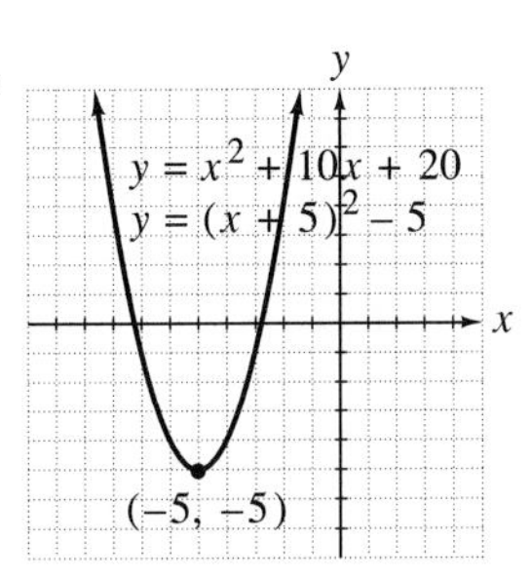

11.

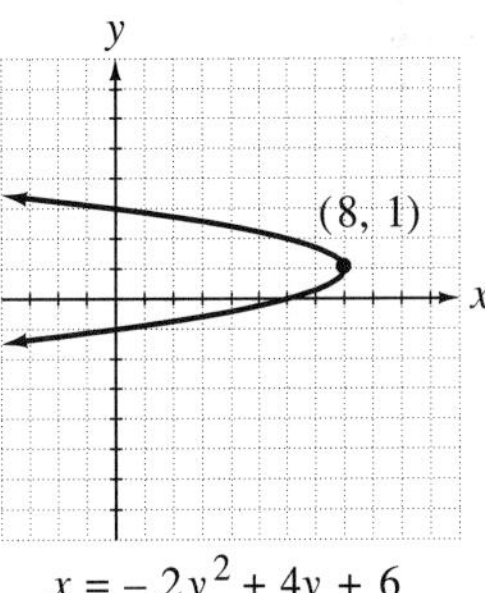

$x = -2y^2 + 4y + 6$
$x = -2(y - 1)^2 + 8$

13. 5 **15.** $\sqrt{41}$ **17.** $(4, -2)$ **19.** $\left(-5, \frac{5}{2}\right)$

21.

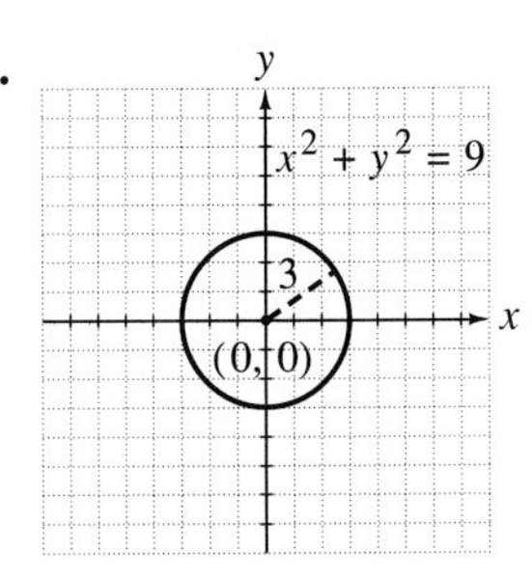

23.

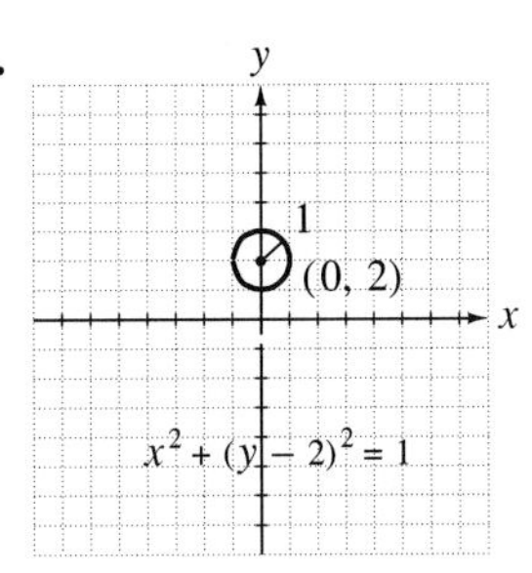

25.

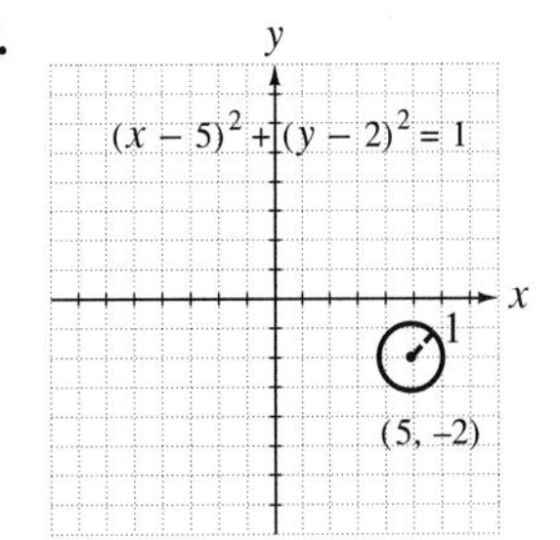

27. $(x - 2)^2 + (y - 3)^2 = 36$ **29.** $x^2 + y^2 = 4$ **31.** $(x + 5)^2 + (y - 4)^2 = 5$

33.

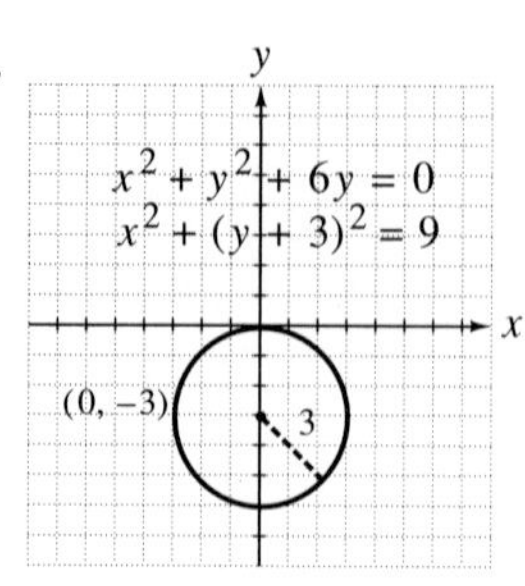

35.

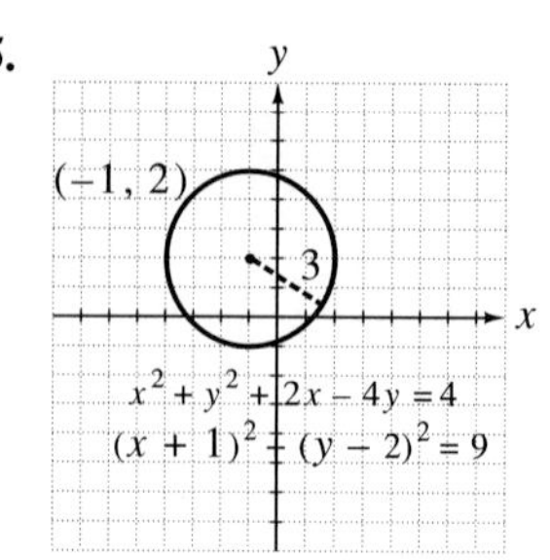

37. $\sqrt{17}$ **39.** 10 **41.** $\sqrt{10}$ **43.** (5, 7) **45.** (−2, 1) **47.** $\left(-\frac{3}{2}, \frac{11}{2}\right)$

49.

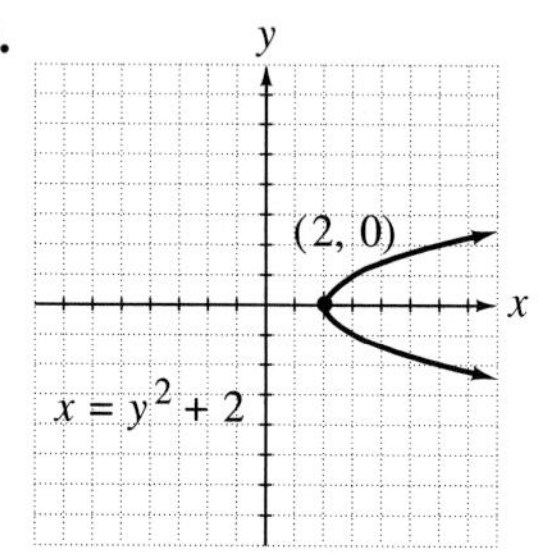

51.

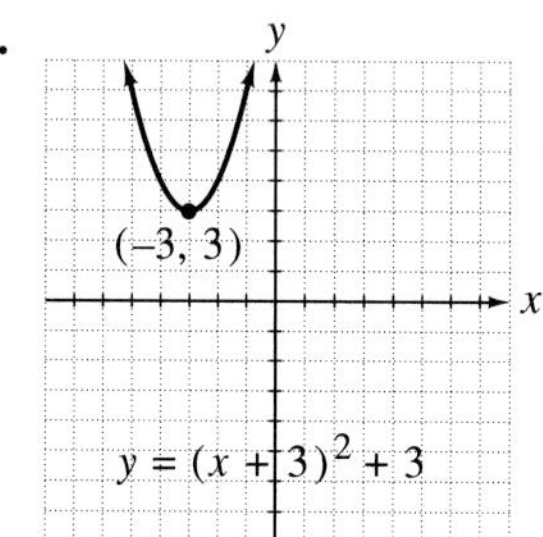

53.

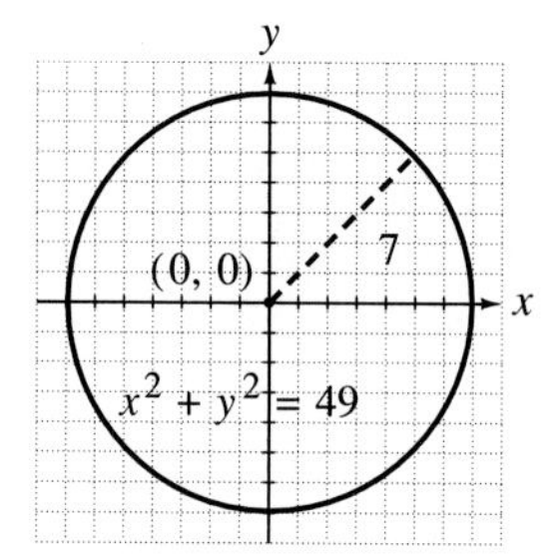

55.

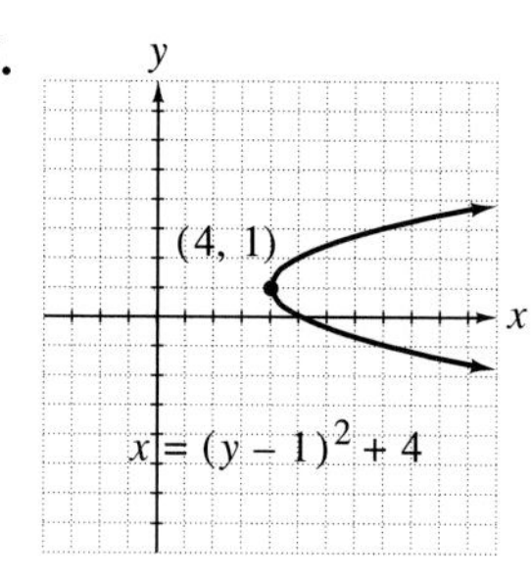

57.

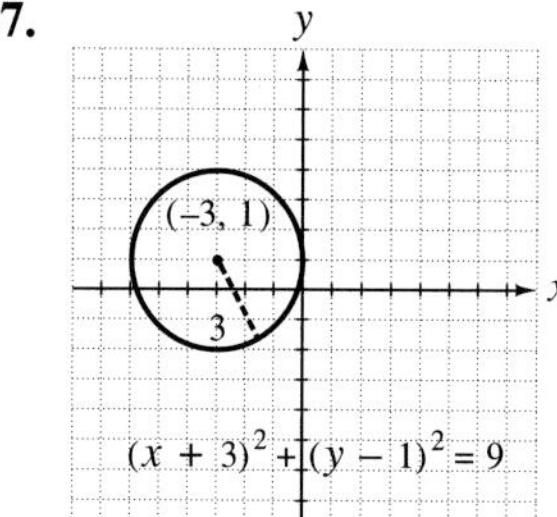

59.

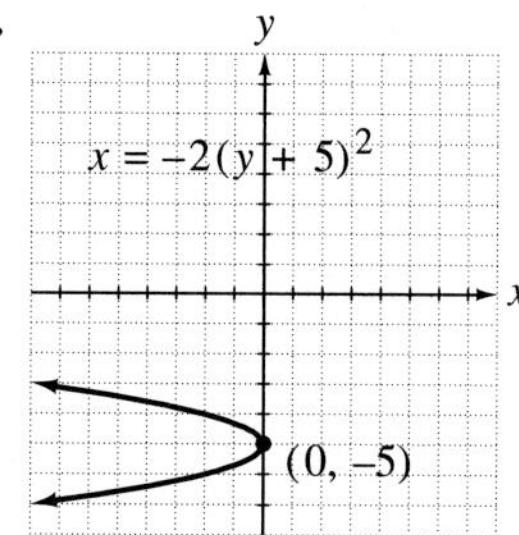

61.

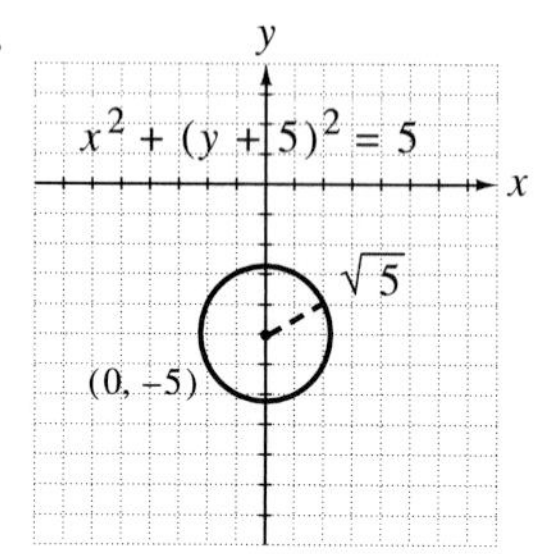

63.

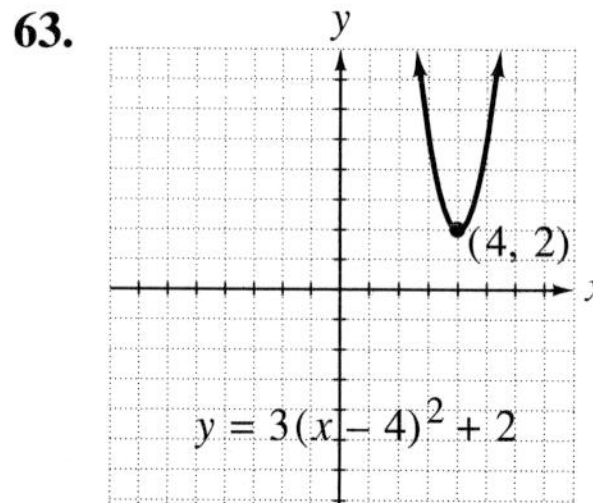

65.

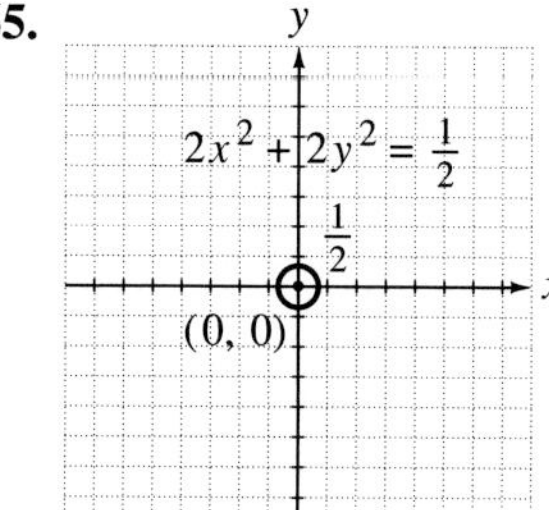

67.

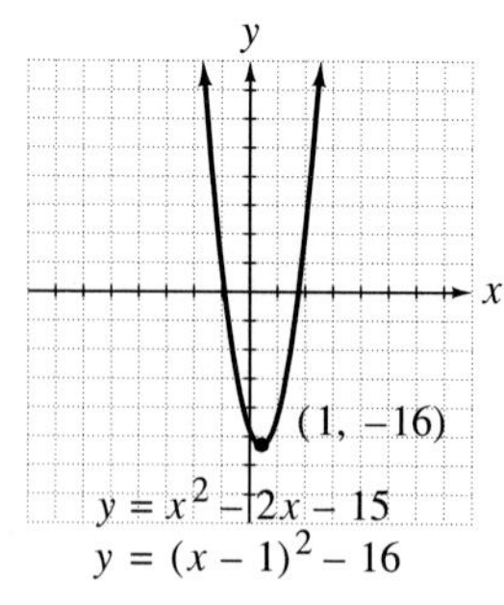

69.

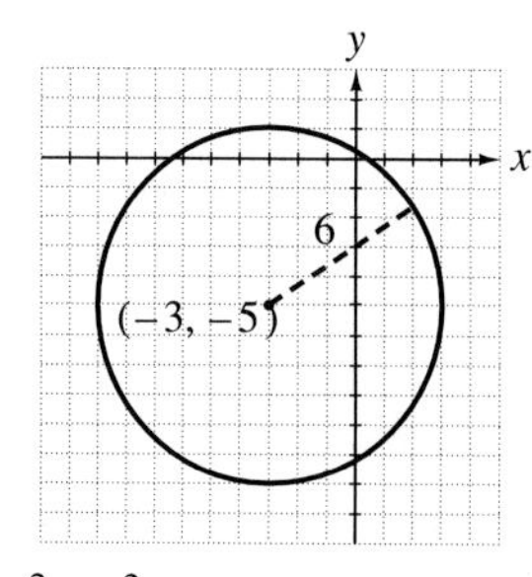

71.

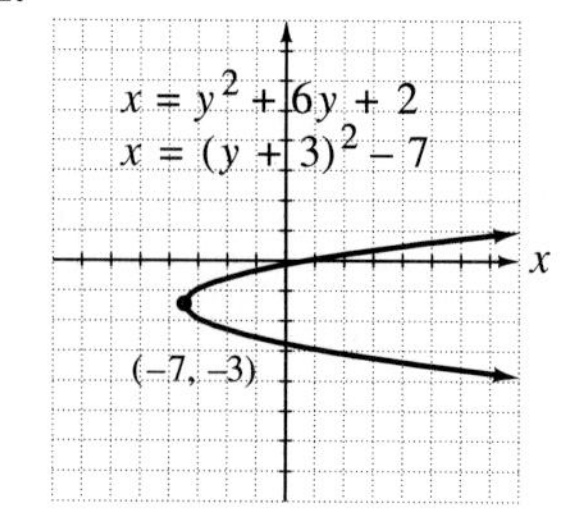

73.

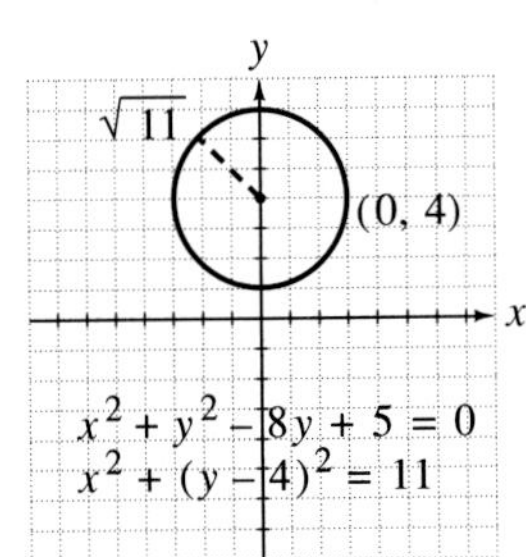

75.

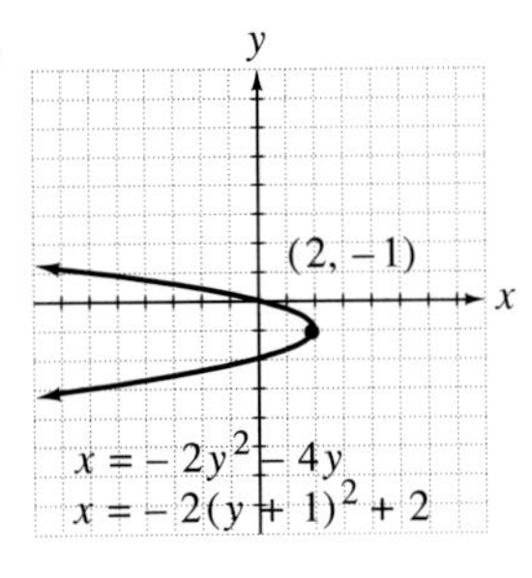

77.

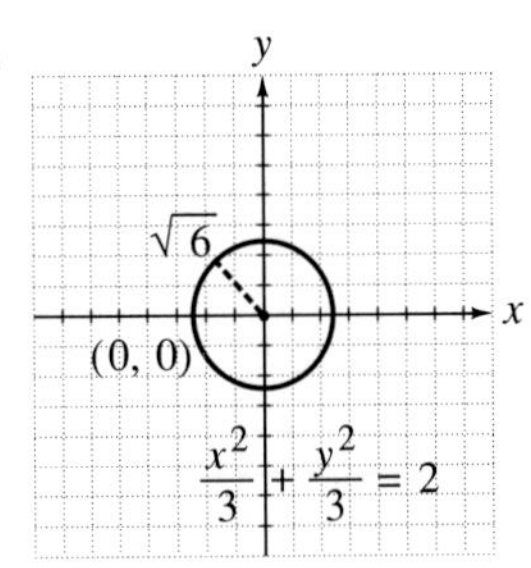

79.

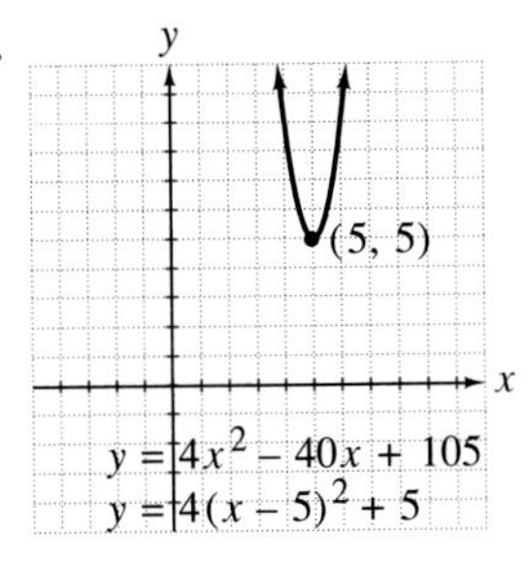

81. 20 meters

83.

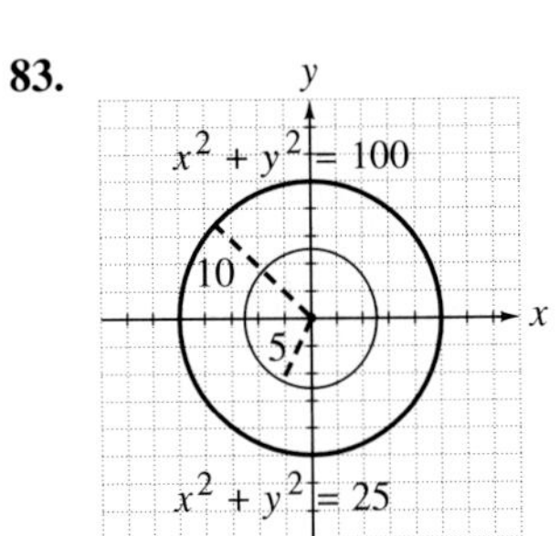

85.

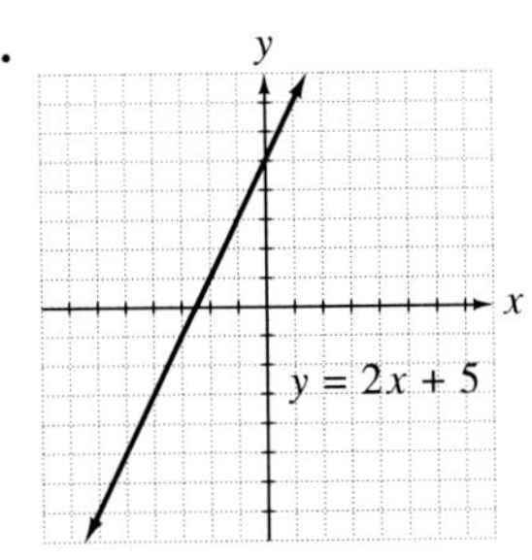

87.

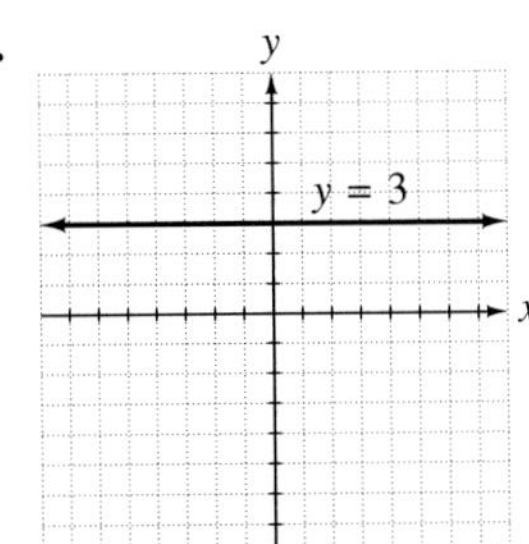

89. $\frac{\sqrt{3}}{3}$ **91.** $\frac{2\sqrt{42}}{3}$

Exercise Set 9.8

1.

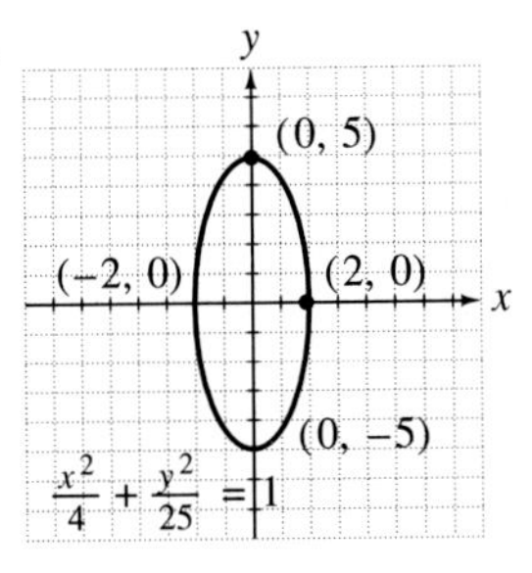

3.

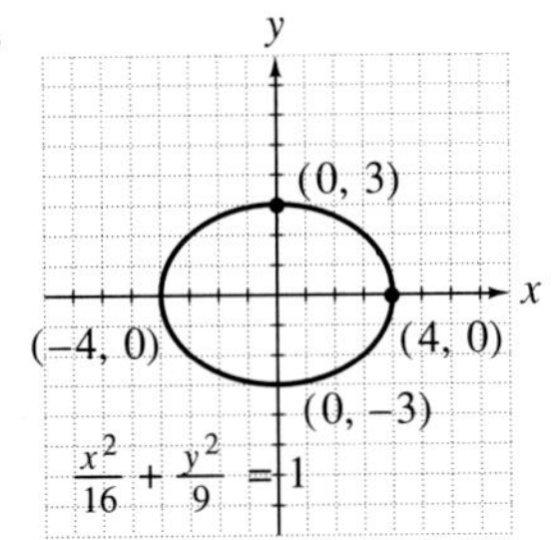

5.

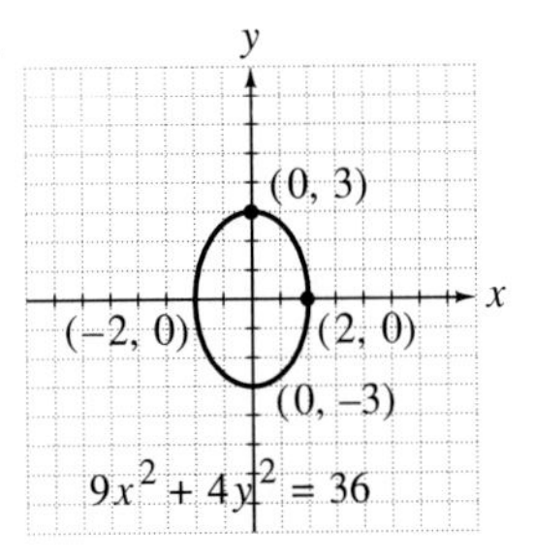

7.

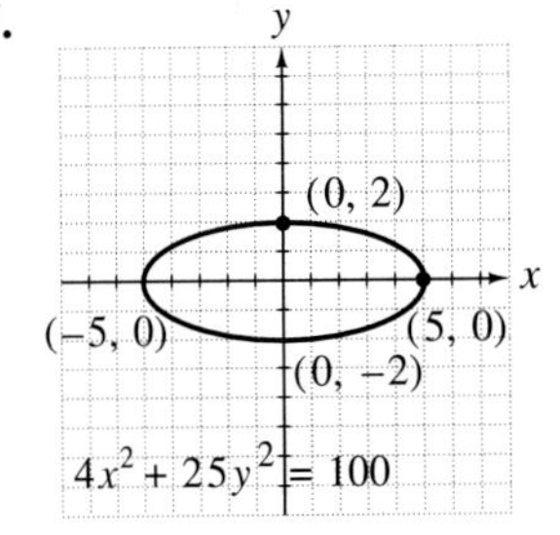

9.

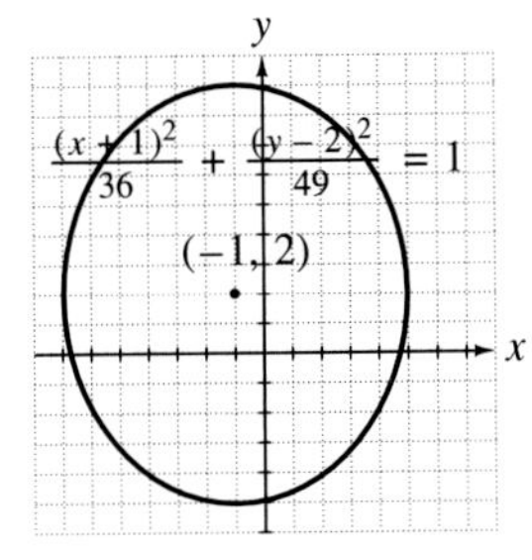

11.

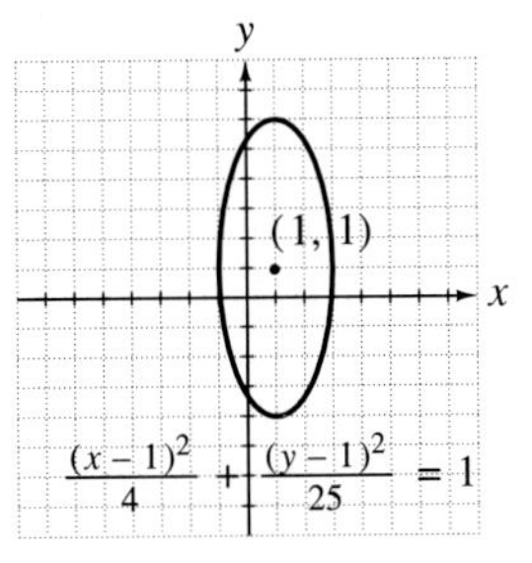

13.

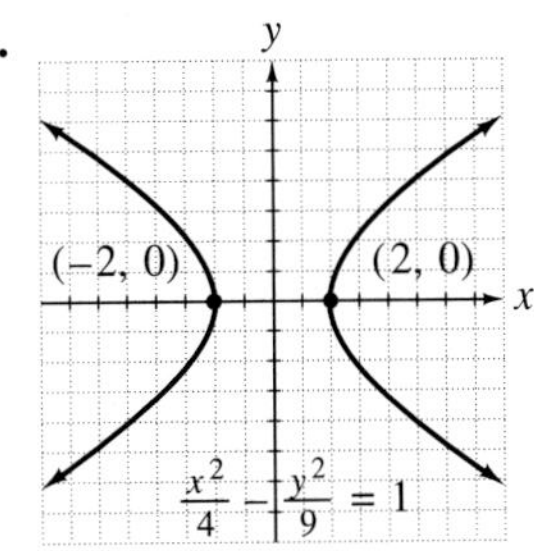

15.

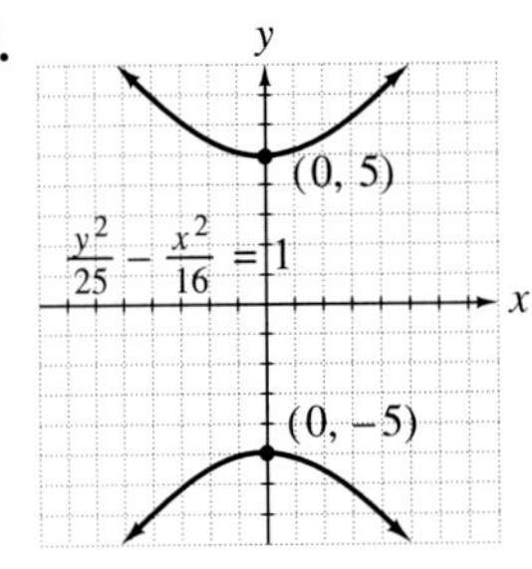

17.

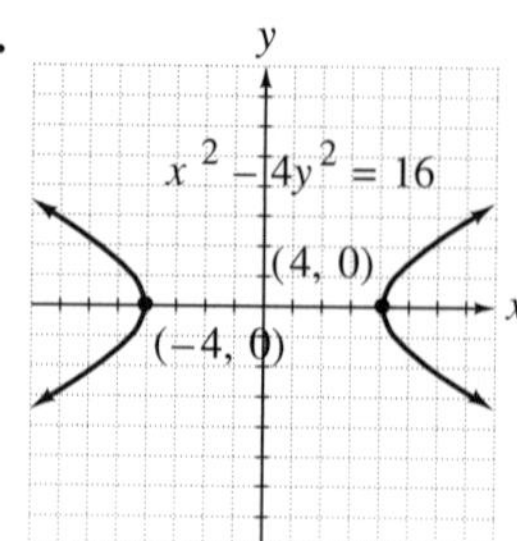

19.

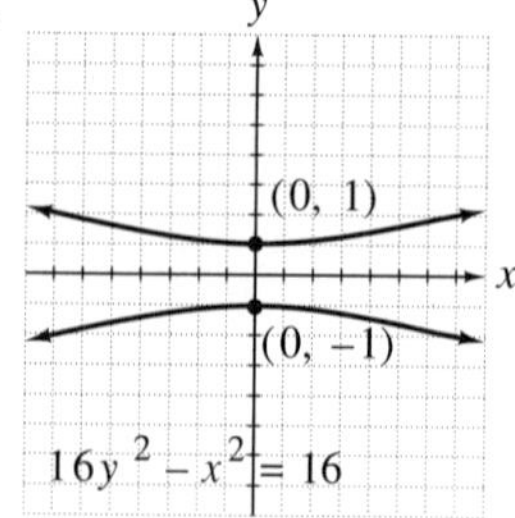

21. circle;

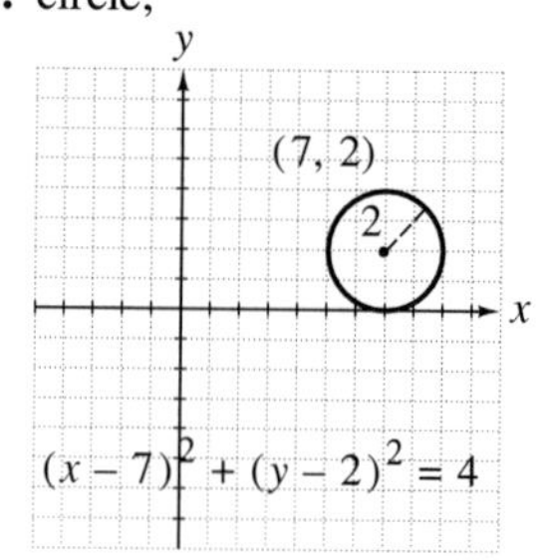

23. parabola;

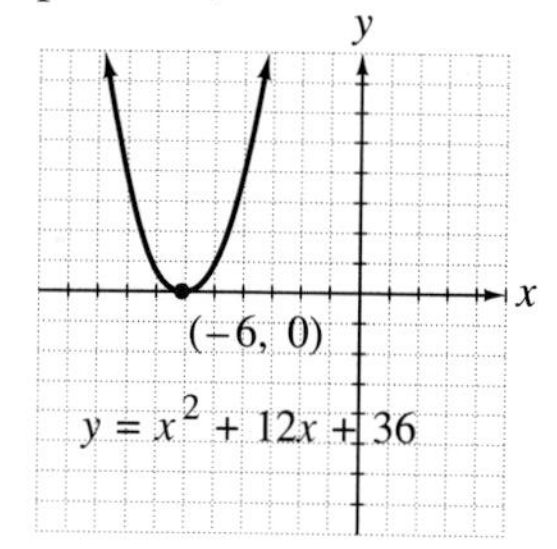

25. hyperbola;

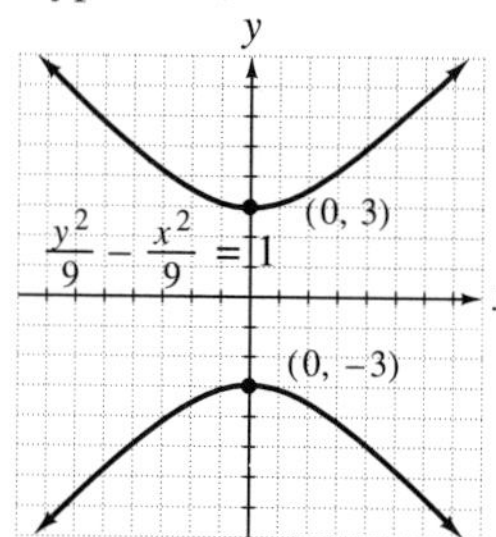

27. ellipse;

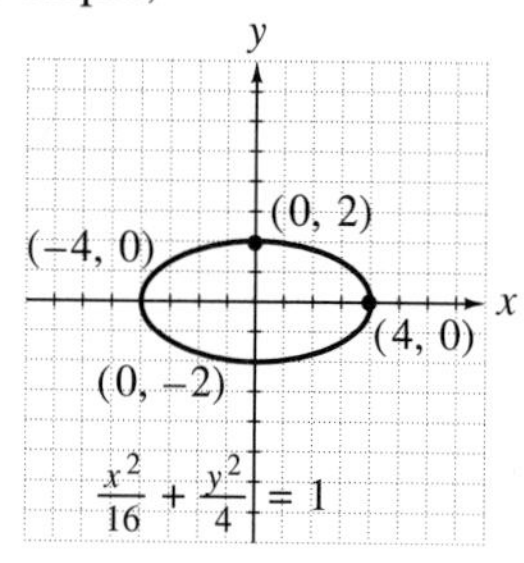

29. parabola;

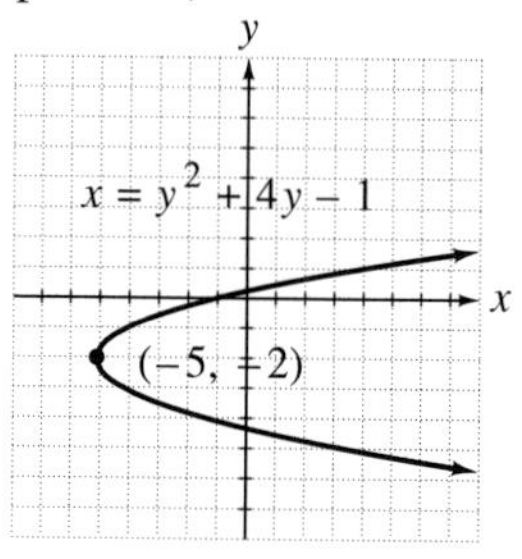

31. hyperbola;

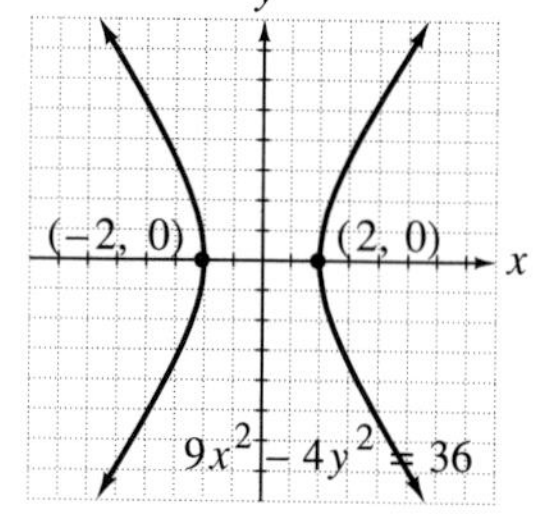

33. ellipse;

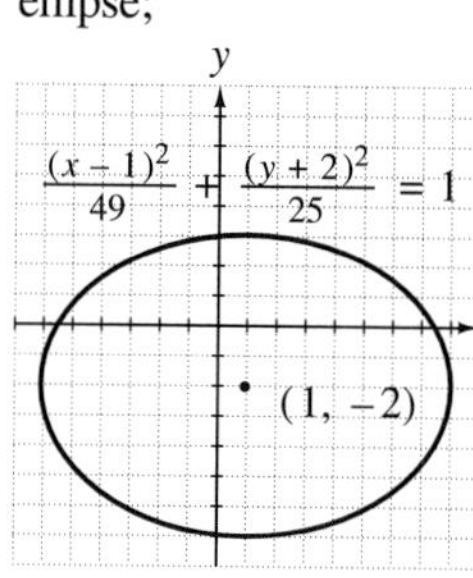

35. circle;

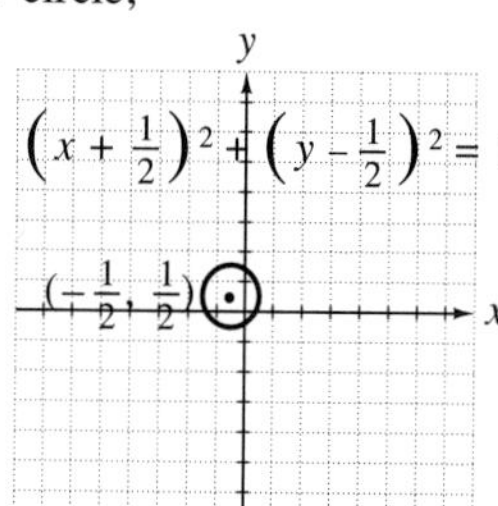

37. $\dfrac{x^2}{1.69 \times 10^{16}} + \dfrac{y^2}{1.5625 \times 10^{16}} = 1$

39. $(-\infty, 1)$ **41.** $[2, \infty)$ **43.** $-8x^5$ **45.** $-4x^2$

47. $(3, 7)$; $x = y - 4$; $3 = 7 - 4$; $3 = 3$; $y = 2x + 1$; $7 = 2(3) + 1$; $7 = 7$

49. $(-1, 2)$; $x + 2y = 3$; $-1 + 2(2) = 3$; $-1 + 4 = 3$; $3 = 3$; $-x - y = -1$; $-(-1) - 2 = -1$; $1 - 2 = -1$; $-1 = -1$

51.

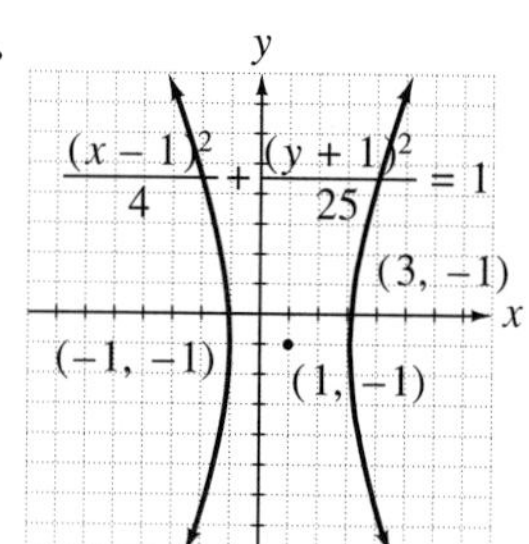

53.

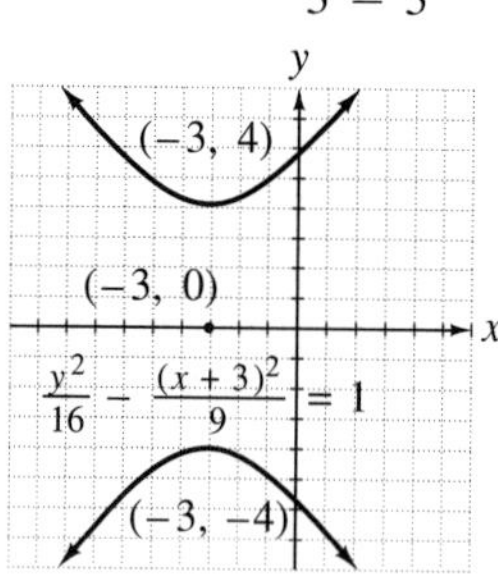

55.

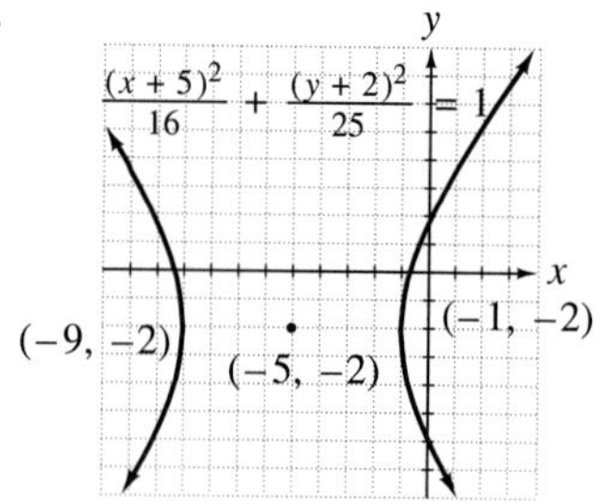

Chapter 9 Review

1. $\{14, 1\}$ **3.** $\left\{\frac{4}{5}, -\frac{1}{2}\right\}$ **5.** $\{-7, 7\}$ **7.** $\left\{-\frac{4}{9}, \frac{2}{9}\right\}$ **9.** $\left\{\frac{-3-\sqrt{5}}{2}, \frac{-3+\sqrt{5}}{2}\right\}$

11. $\left\{\frac{-3-i\sqrt{7}}{8}, \frac{-3+i\sqrt{7}}{8}\right\}$ **13.** two complex solutions **15.** two real solutions **17.** $\{8\}$

19. $\{-i\sqrt{11}, i\sqrt{11}\}$ **21.** $\left\{\frac{5-i\sqrt{143}}{12}, \frac{5+i\sqrt{143}}{12}\right\}$ **23.** $\left\{\frac{21-\sqrt{41}}{50}, \frac{21+\sqrt{41}}{50}\right\}$

25. $\left\{3, \frac{-3 + 3i\sqrt{3}}{2}, \frac{-3 - 3i\sqrt{3}}{2}\right\}$ **27.** $\left\{\frac{2}{3}, 5\right\}$ **29.** $\{-5, 5, -2i, 2i\}$ **31.** $\{1, 125\}$ **33.** $\{-1, 1, -i, i\}$

35. 6 and 8 **37. a.** 20 ft; **b.** $\frac{15 + \sqrt{321}}{16}$ seconds **39.** The integers are 20, 22, and 24.

41. The number is -5. **43.** $[-5, 5]$ **45.** $\left(-\infty, -\frac{5}{4}\right] \cup \left[\frac{3}{2}, \infty\right)$ **47.** $(5, 6)$

49. $(-\infty, -6) \cup \left(-\frac{3}{4}, 0\right) \cup (5, \infty)$ **51.** $\left[-4, \frac{1}{2}\right)$ **53.** $\{x \mid x \neq -5 \text{ and } x \neq 3\}$

55.

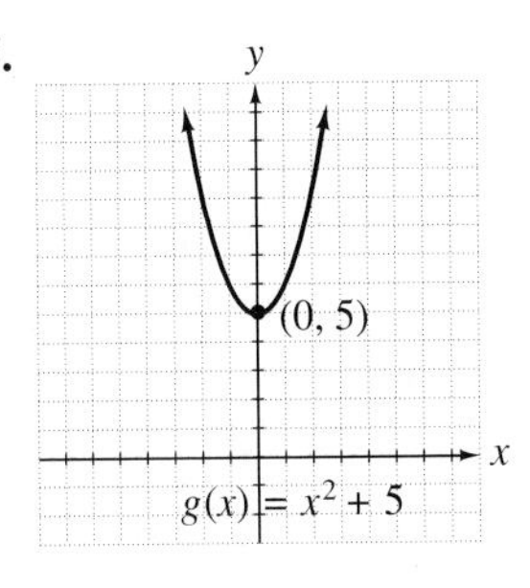

57.

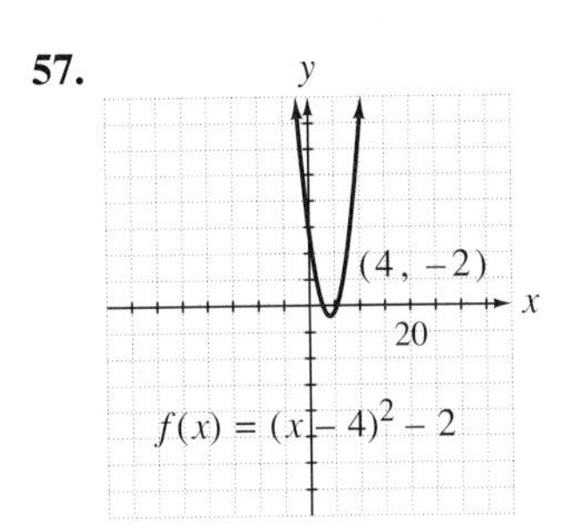

59.

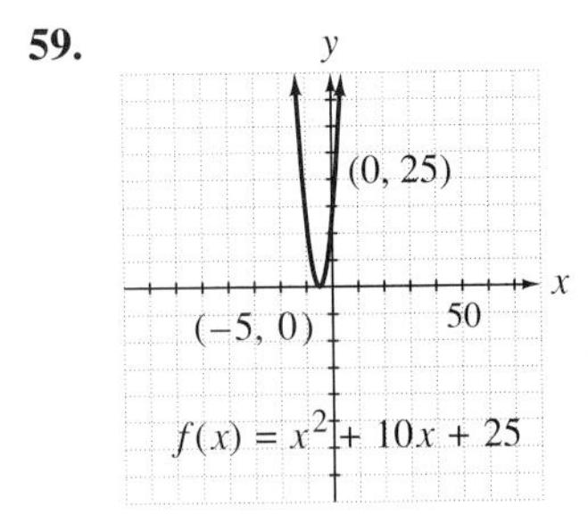

61.

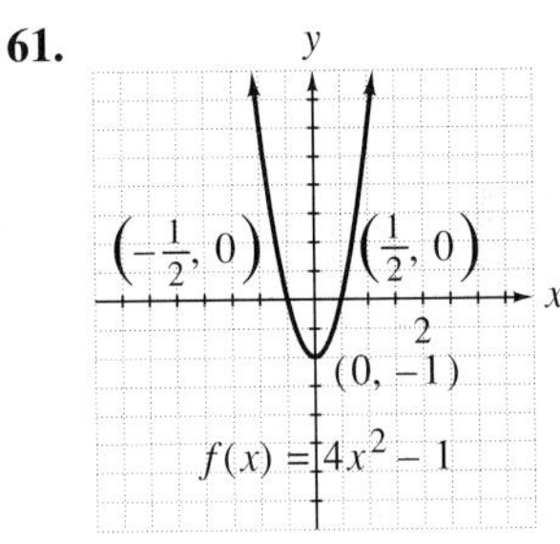

63. The numbers are both 210. **65.** $\sqrt{197}$ **67.** $\sqrt{130}$ **69.** $(-5, 5)$ **71.** $\left(-\frac{15}{2}, 1\right)$

73. $(x + 4)^2 + (y - 4)^2 = 9$ **75.** $(x + 7)^2 + (y + 9)^2 = 11$

77.

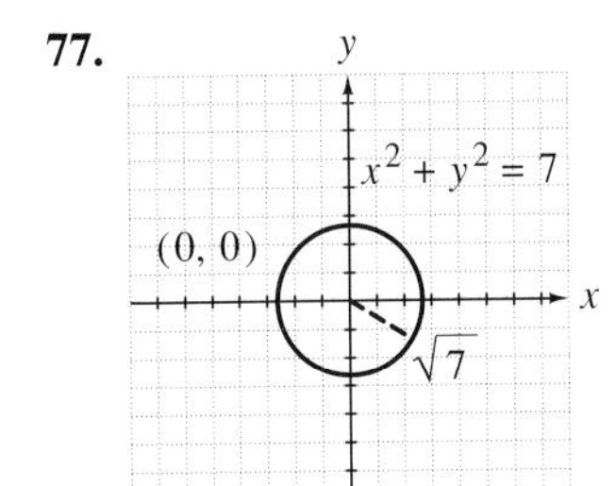

79.

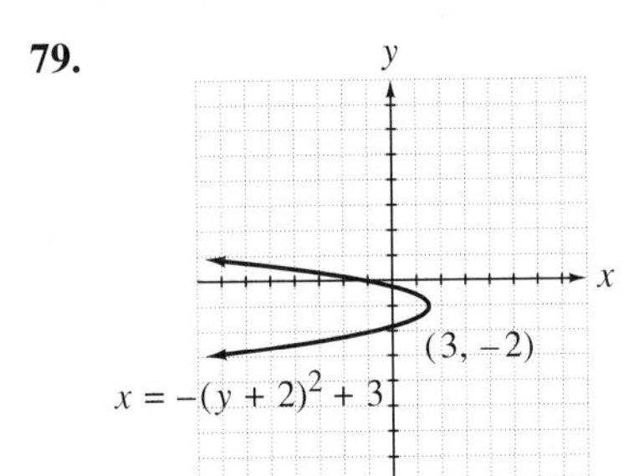

81.

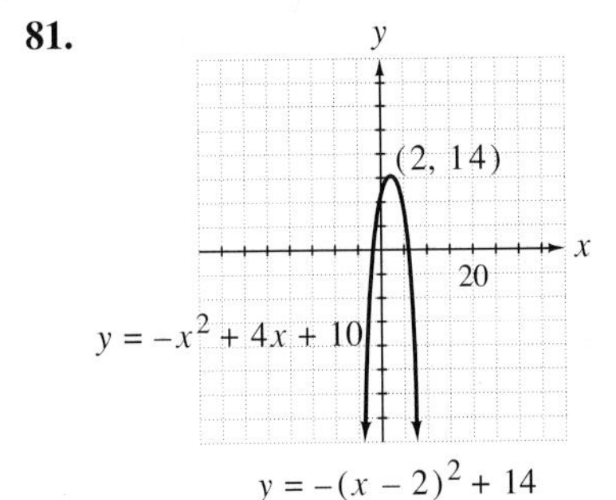

83.

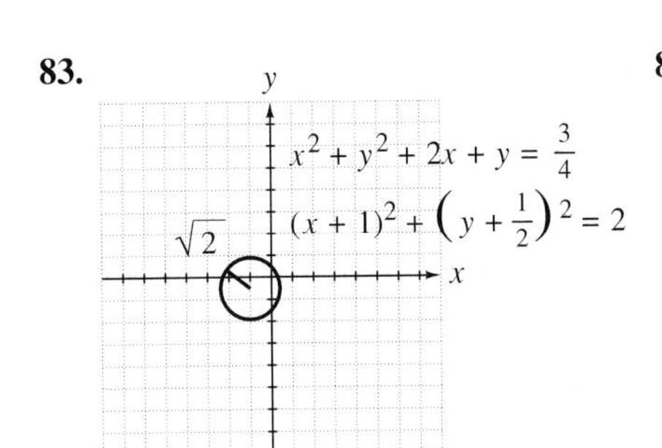

85.

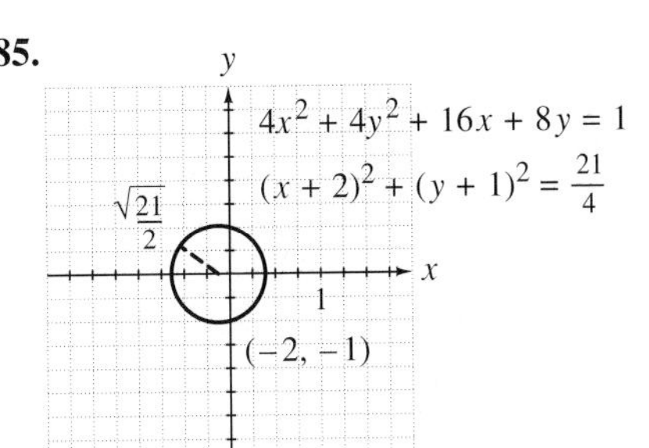

87.

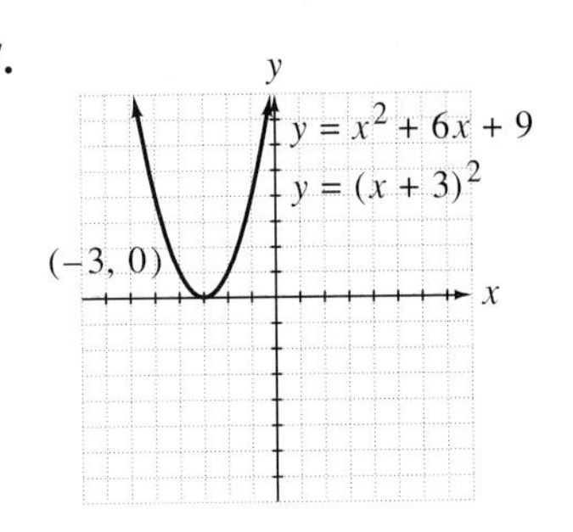

89. $(x - 5.6)^2 + (y + 2.4)^2 = 9.61$

91.

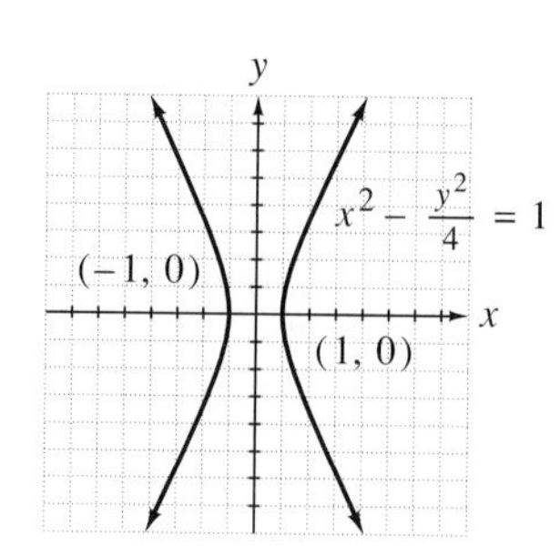

93.

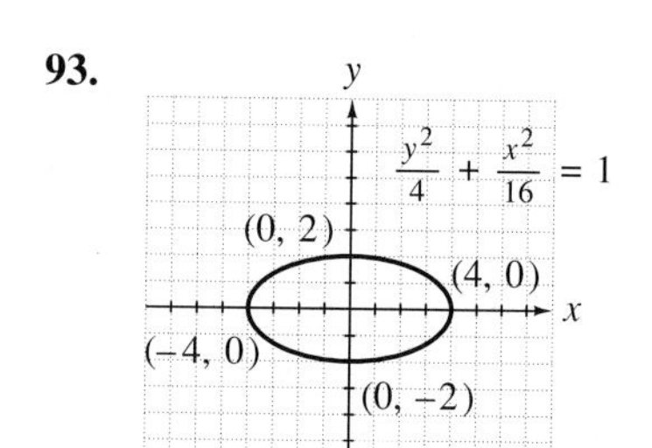

95.

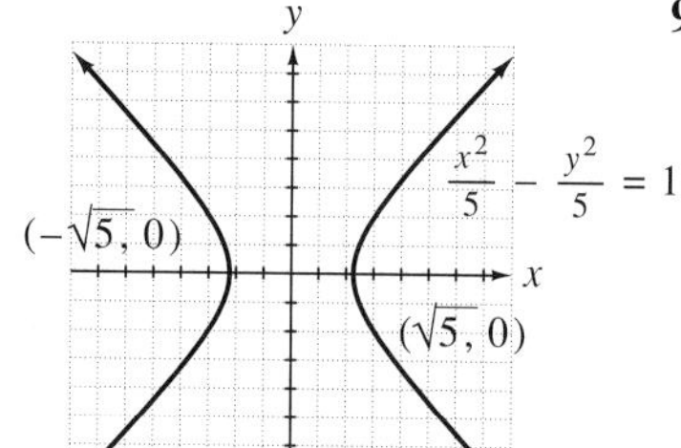

97.

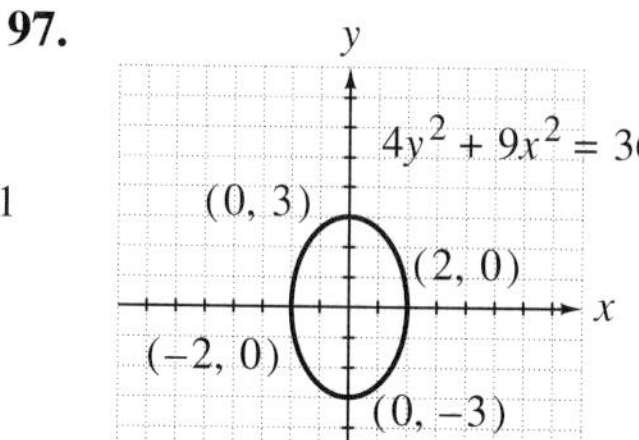

99.

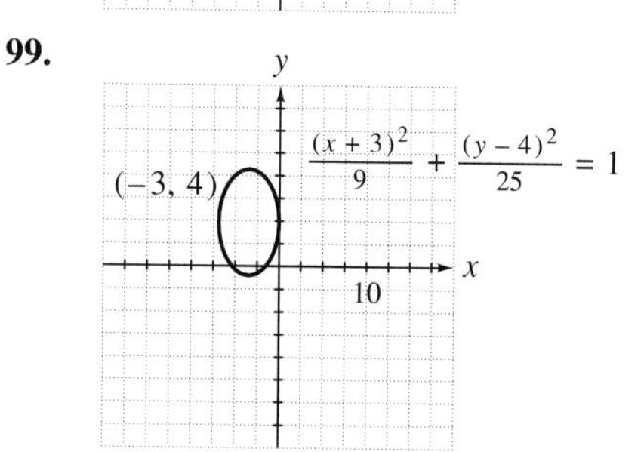

101.

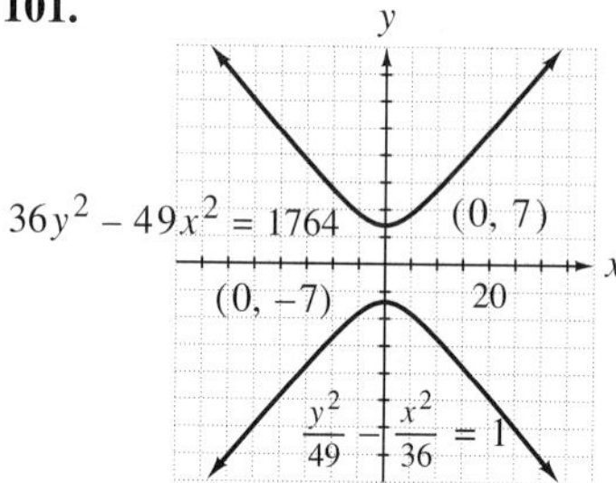

103.

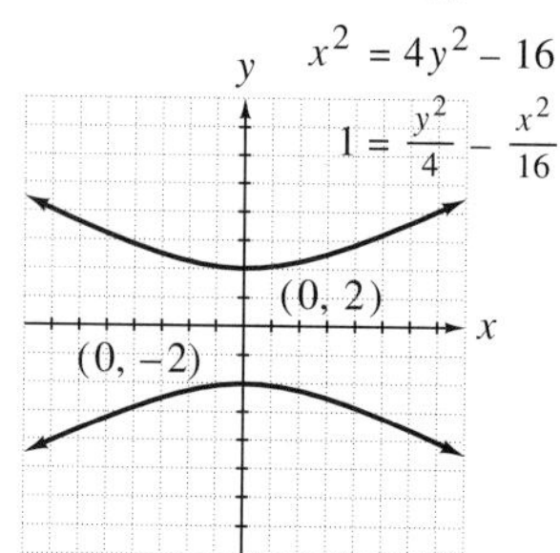

105.

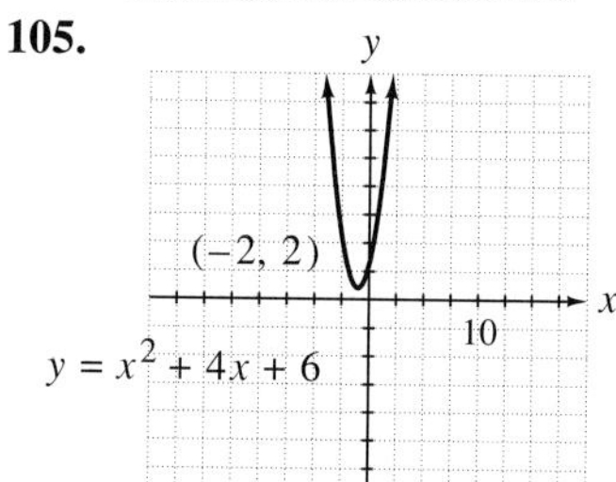

107.

$y^2 + x^2 = 4x + 6$

$(x - 2)^2 + y^2 = 10$

$\sqrt{10}$

(2, 0)

109.

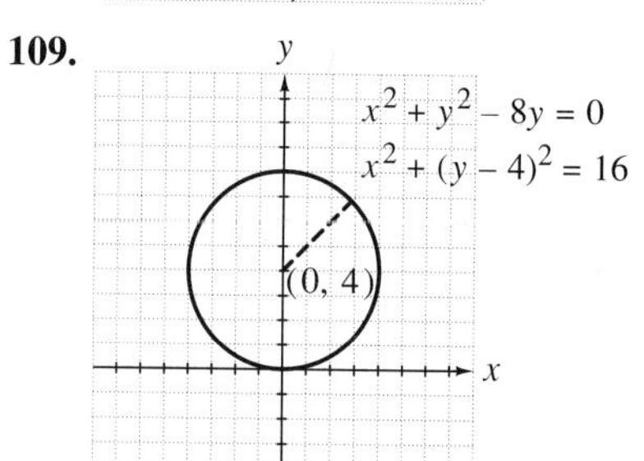

111.

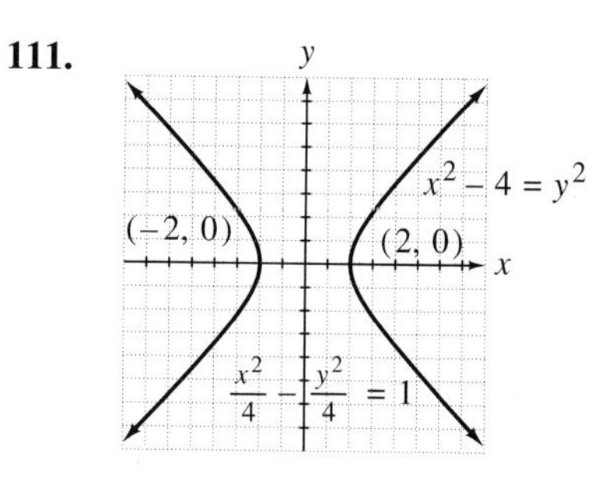

113.

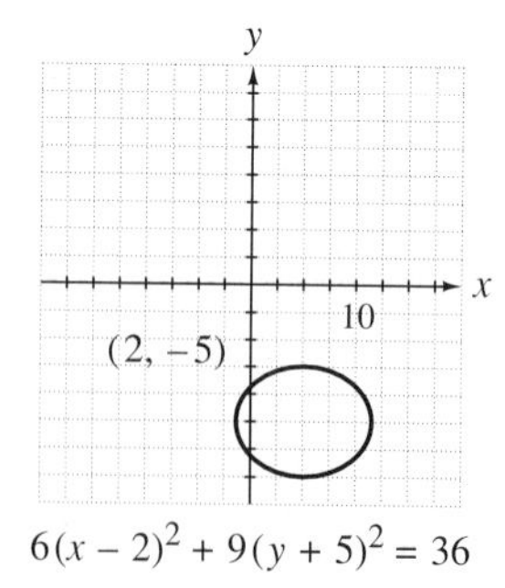

115.

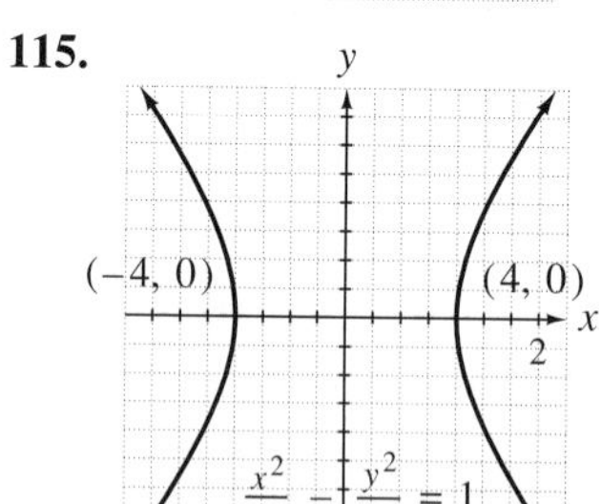

Chapter 9 Test

1. $\left\{\frac{7}{5}, -1\right\}$ **2.** $\left\{-2, 1 - i\sqrt{3}, 1 + i\sqrt{3}\right\}$ **3.** $\left\{\frac{1 + i\sqrt{31}}{2}, \frac{1 - i\sqrt{31}}{2}\right\}$ **4.** $\{3 - \sqrt{7}, 3 + \sqrt{7}\}$

5. $\left\{-\frac{1}{7}, -1\right\}$ **6.** $\left\{\frac{3 + \sqrt{29}}{2}, \frac{3 - \sqrt{29}}{2}\right\}$ **7.** $\{-2 - \sqrt{11}, -2 + \sqrt{11}\}$ **8.** $\{-3, 3, -i, i\}$ **9.** $\{-1, 1, -i, i\}$

10. $\{6, 7\}$ **11.** $\{3 - \sqrt{7}, 3 + \sqrt{7}\}$ **12.** $\left\{\frac{2 - i\sqrt{6}}{2}, \frac{2 + i\sqrt{6}}{2}\right\}$ **13.** $\left[-\frac{7}{6}, -\frac{12}{7}\right]$

14. $(-\infty, -5) \cup (-4, 4) \cup (5, \infty)$ **15.** $(-\infty, -4) \cup \left[\frac{11}{2}, \infty\right)$ **16.** $(-\infty, -3) \cup [2, 3)$ **17.** $2\sqrt{26}$ **18.** $\left(-4, \frac{7}{2}\right)$

19.

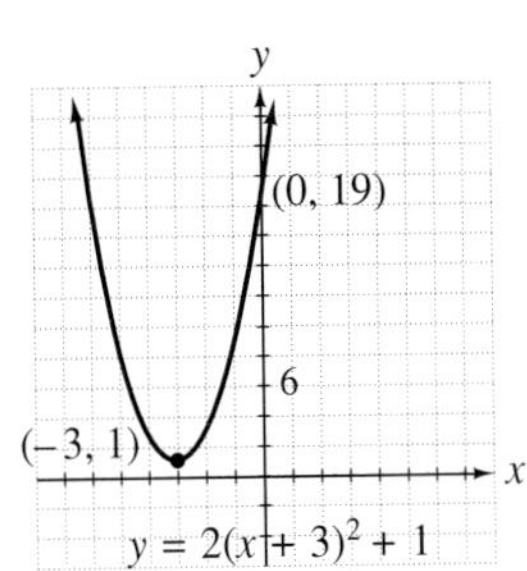

20.

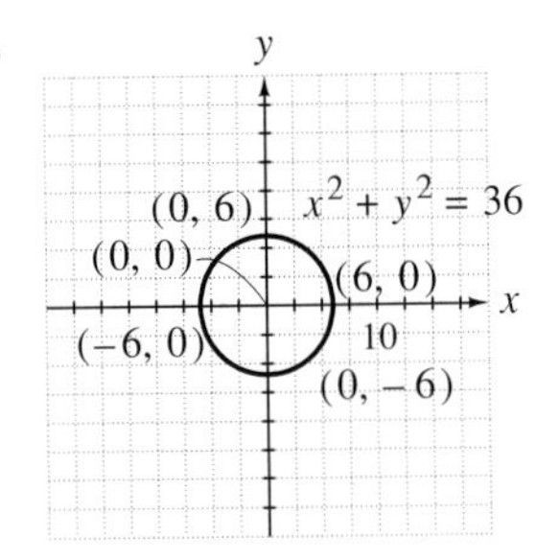

21.

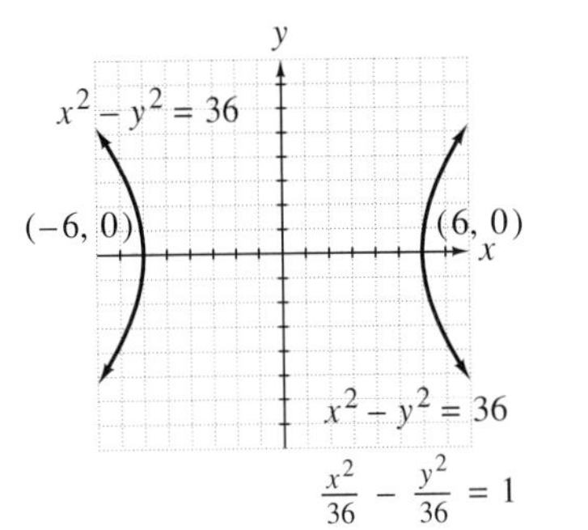

22.

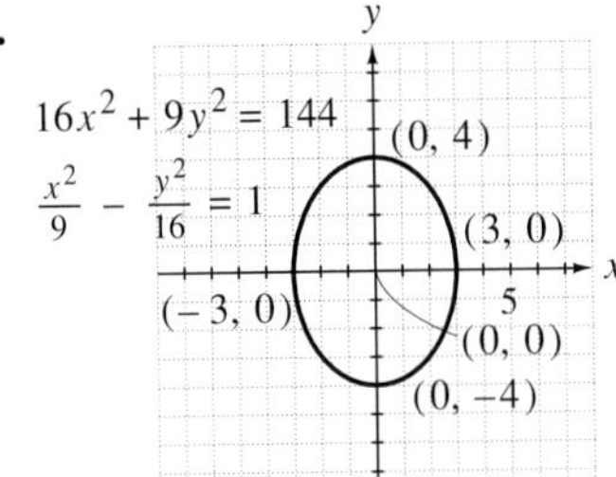

23.

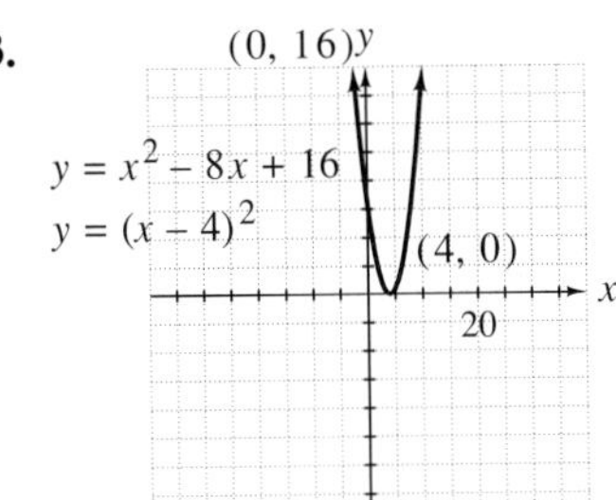

24.

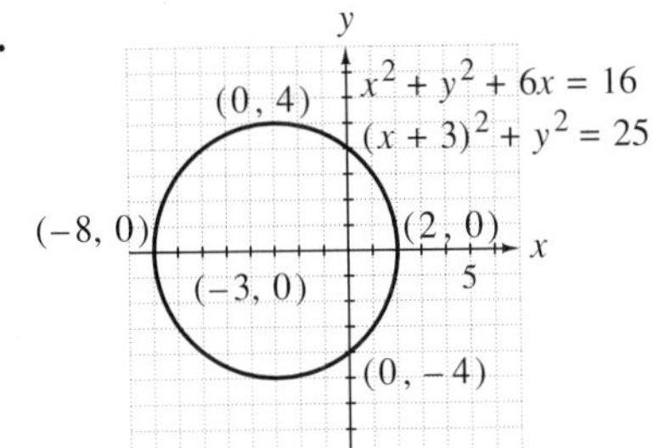

25.

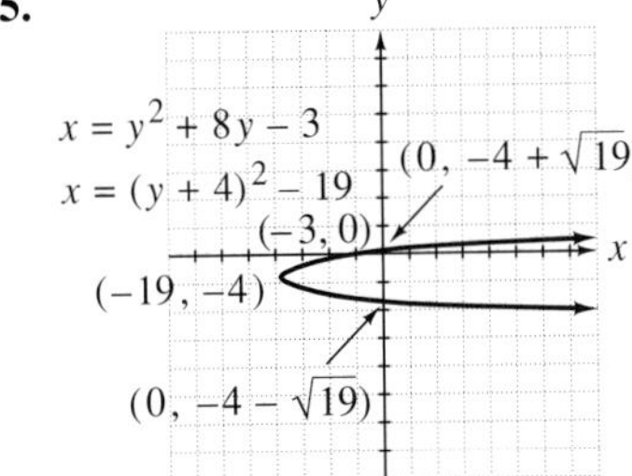

26.

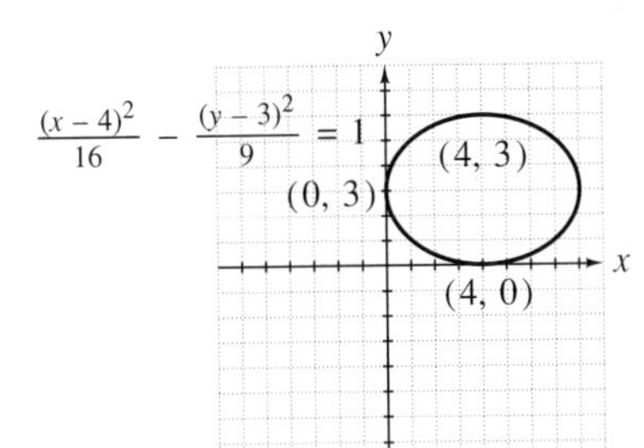

27. a. 256 ft; **b.** 4 seconds

28. 41, 43, and 45

Chapter 9 Cumulative Review

1. $x = \frac{y - b}{m}$; *Sec. 2.4, Ex. 6* **2.** 7800; *Sec. 3.1, Ex. 6* **3.** 144; *Sec. 3.2, Ex. 3* **4.** $\{4\}$; *Sec. 3.4, Ex. 9* **5.** $\{\ \}$; *Sec. 3.7, Ex. 4* **6.** *Sec. 4.1, Ex. 6*

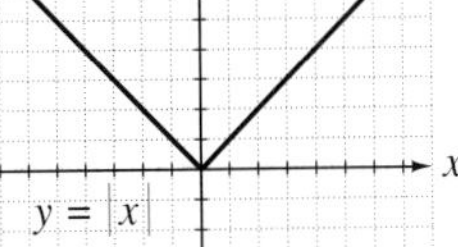

7. a. yes **b.** yes **c.** no **d.** yes **e.** no **f.** no; *Sec. 4.2, Ex. 5* **8.** $m = \frac{3}{4}$; $b = -1$; *Sec. 4.4, Ex. 4* **9. a.** s^4t^4 **b.** $\frac{m^7}{n^7}$ **c.** $8a^3$ **d.** $25x^4y^6z^2$ **e.** $\frac{16x^{16}}{81y^{20}}$; *Sec. 5.1, Ex. 5* **10. a.** $13x^3y - xy^3 + 7$ **b.** $3a^3 + 3a$; *Sec. 5.3, Ex. 6*

11. a. $x^2 - 9$ **b.** $16y^2 - 1$ **c.** $x^4 - 4y^2$; *Sec. 5.4, Ex. 9* **12.** $(b - 6)(a + 2)$; *Sec. 6.1, Ex. 8*
13. $(y - 3)(y^2 + 3y + 9)$; *Sec. 6.4, Ex. 6* **14.** $\{-2, 0, 2\}$; *Sec. 6.6, Ex. 6*
15. a. domain: $\{x \mid x \text{ is a real number}\}$ **b.** domain: $\{x \mid x \text{ is a real number and } x \neq 1\}$
c. domain: $\{x \mid x \text{ is a real number and } x \neq -\frac{3}{2}, x \neq -2\}$; *Sec. 7.1, Ex. 2*
16. 2; *Sec. 7.6, Ex. 2* **17.** $\left\{-\frac{1}{9}, -1\right\}$; *Sec. 8.6, Ex. 2* **18.** 8 meters; *Sec. 8.6, Ex. 6*
19. $\{-1 - 2\sqrt{3}, -1 + 2\sqrt{3}\}$; *Sec. 9.1, Ex. 2* **20.** $\left\{\frac{2 - \sqrt{10}}{2}, \frac{2 + \sqrt{10}}{2}\right\}$; *Sec. 9.2, Ex. 2*
21. $(-\infty, -2] \cup [1, 5]$; *Sec. 9.4, Ex. 3* **22. a.**

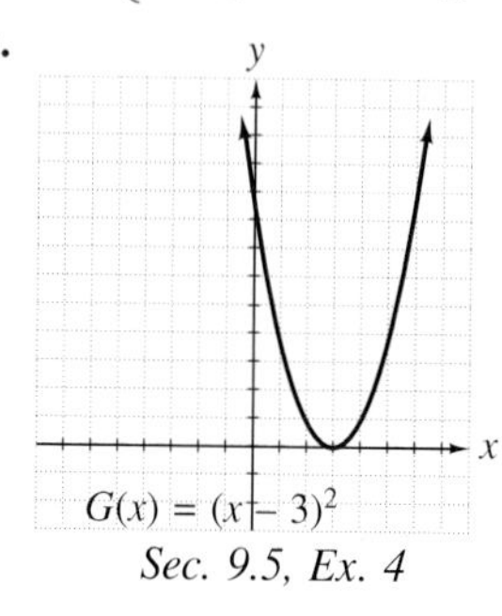

b.

y
x
$F(x) = (x + 1)^2$

Sec. 9.5, Ex. 4

23. $f(x) = x^2 - 4x - 12$
$f(x) = x^2 - 4x + 4 - 4 - 12$
$f(x) = (x - 2)^2 - 16$
$V(2, -16)$
Sec. 9.6, Ex. 1

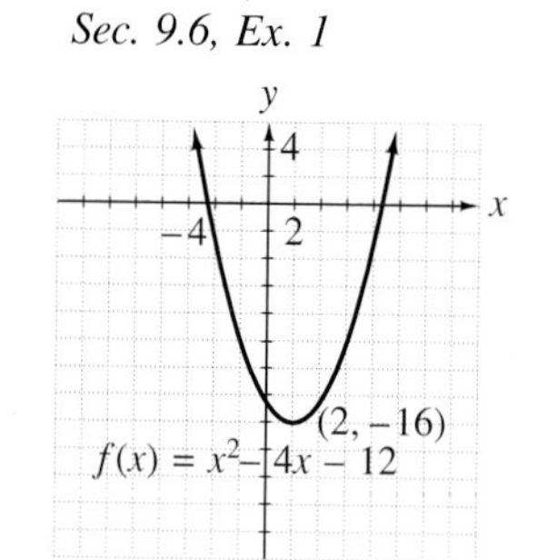

24. $\sqrt{2}$; *Sec. 9.7, Ex. 5*

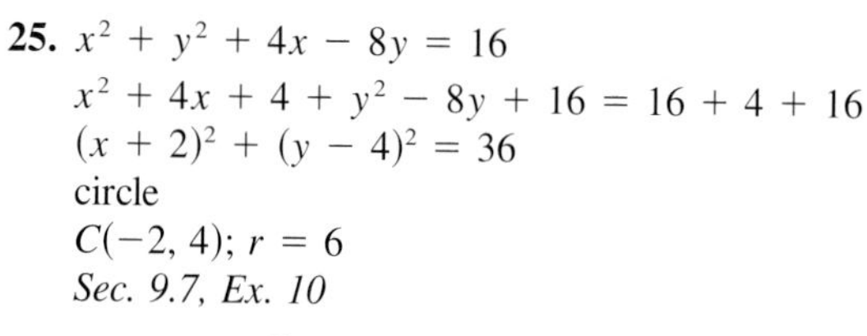
25. $x^2 + y^2 + 4x - 8y = 16$
$x^2 + 4x + 4 + y^2 - 8y + 16 = 16 + 4 + 16$
$(x + 2)^2 + (y - 4)^2 = 36$
circle
$C(-2, 4)$; $r = 6$
Sec. 9.7, Ex. 10

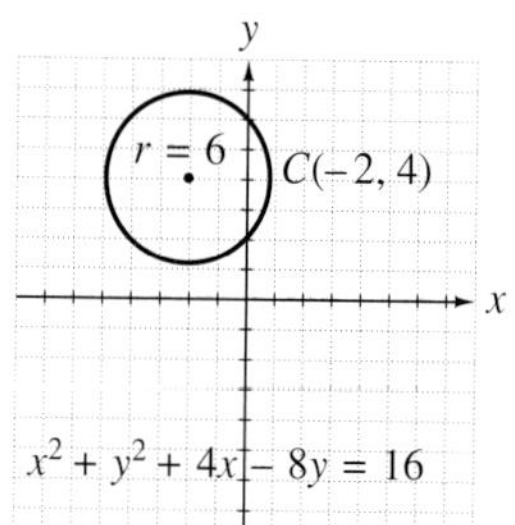

CHAPTER 10
Systems of Equations and Inequalities

Exercise Set 10.1

1.

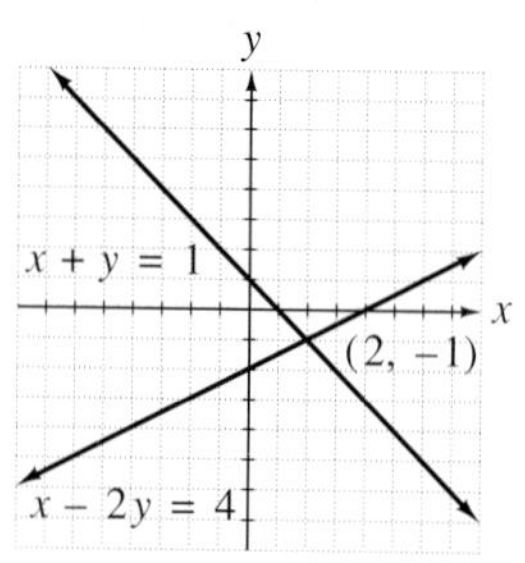

3.

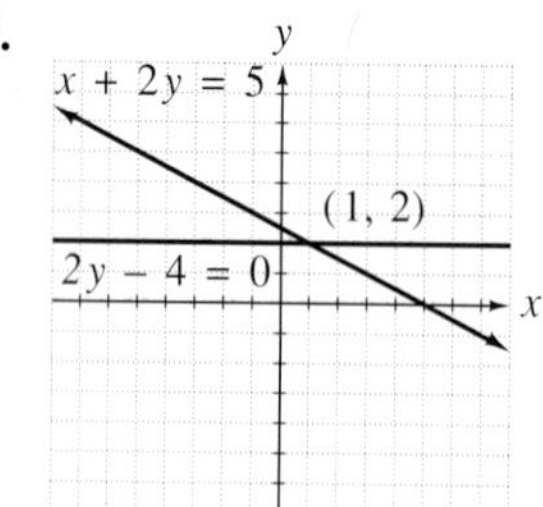

5.

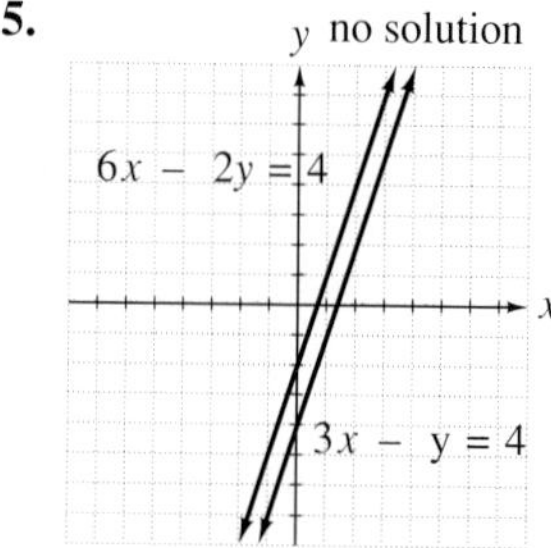

7. $\{(2, 8)\}$ **9.** $\{(0, -9)\}$ **11.** $\{(1, -1)\}$ **13.** $\{(-5, 3)\}$ **15.** $\left\{\left(\frac{5}{2}, \frac{5}{4}\right)\right\}$ **17.** $\{(1, -2)\}$ **19.** $\{(9, 9)\}$ **21.** $\{(7, 2)\}$ **23.** $\{\ \}$ **25.** $\{(x, y) \mid 3x + y = 1\}$ **27.** $\left\{\left(\frac{3}{2}, 1\right)\right\}$ **29.** $\{(2, -1)\}$ **31.** $\{(-5, 3)\}$ **33.** $\{(x, y) \mid 3x + 9y = 12\}$ **35.** $\{\ \}$ **37.** $\left\{\left(\frac{1}{2}, \frac{1}{5}\right)\right\}$ **39.** $\{(8, 2)\}$ **41.** $\{(x, y) \mid x = 3y + 2\}$ **43.** $\left\{\left(-\frac{1}{4}, \frac{1}{2}\right)\right\}$ **45.** $\{(3, 2)\}$ **47.** $\{(7, -3)\}$ **49.** $\{\ \}$ **51.** $\{(3, 4)\}$ **53.** $\{(-2, 1)\}$ **55.** 15 and 30 **57.** 1984 and 1988 **59.** **61.**

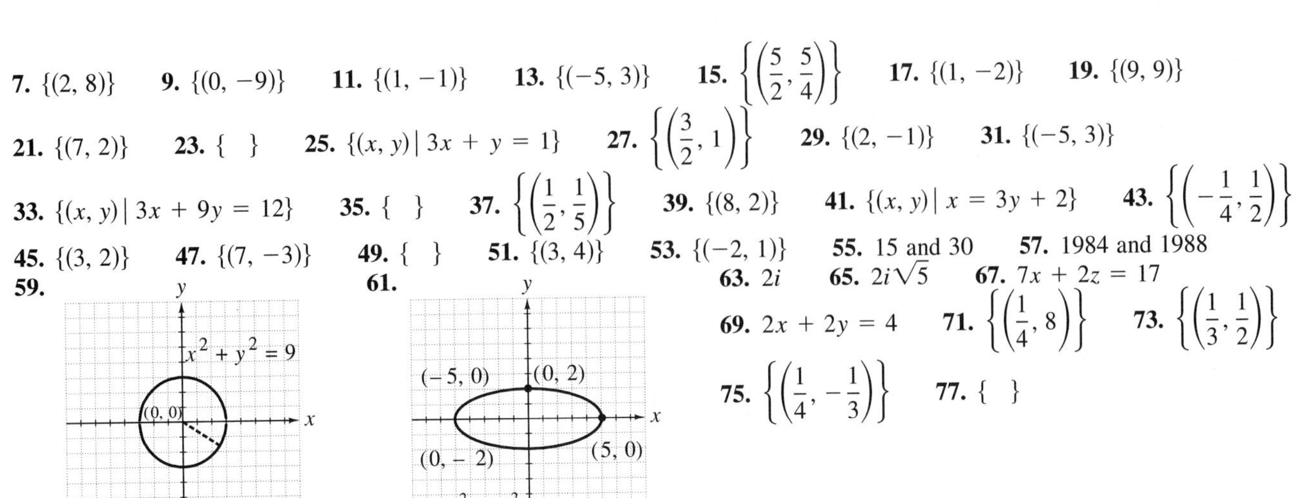

63. $2i$ **65.** $2i\sqrt{5}$ **67.** $7x + 2z = 17$ **69.** $2x + 2y = 4$ **71.** $\left\{\left(\frac{1}{4}, 8\right)\right\}$ **73.** $\left\{\left(\frac{1}{3}, \frac{1}{2}\right)\right\}$ **75.** $\left\{\left(\frac{1}{4}, -\frac{1}{3}\right)\right\}$ **77.** $\{\ \}$

Exercise Set 10.2

1. $\{(-2, 5, 1)\}$ **3.** $\{(-2, 3, -1)\}$ **5.** $\{(x, y, z) \mid x - 2y + z = -5\}$ **7.** $\{\ \}$ **9.** $\{(0, 0, 0)\}$ **11.** $\{(-3, -35, -7)\}$ **13.** $\{(6, 22, -20)\}$ **15.** $\{\ \}$ **17.** $\{(3, 2, 2)\}$ **19.** $\{(x, y, z) \mid x + 2y - 3z = 4\}$ **21.** $\{(-3, -4, -5)\}$ **23.** $\{(12, 6, 4)\}$ **25.** $\{-2 - \sqrt{3}, -2 + \sqrt{3}\}$ **27.** $\left\{\frac{1 - i\sqrt{23}}{6}, \frac{1 + i\sqrt{23}}{6}\right\}$ **29.** $\frac{x}{5 - 3x}$ **31.** $2\sqrt{5}$ **33.** $3\sqrt[3]{2}$ **35.** $2\sqrt[4]{2}$ **37.** $\{(1, 1, 0, 2)\}$ **39.** $\{(1, -1, 2, 3)\}$

Exercise Set 10.3

1. 10 and 8 **3.** plane, 520 mph; wind, 40 mph **5.** 20 quarts of 4%; 40 quarts of 1% **7.** length, 52 ft; width, 26 ft **9.** 9 large frames; 13 small frames **11.** −10 and −8 **13.** tablets, \$0.80; pens, \$0.20 **15.** plane, 630 mph; wind, 90 mph **17.** 5 in., 7 in., 7 in., and 10 in. **19.** 24 nickels; 15 dimes **21.** 18, 13, and 9 **23.** \$2000 in sales **25.** 22 \$10 bills; 63 \$20 bills **27.** 26 dimes, 13 nickels, and 17 pennies **29.** Two units of mix A, 3 units of mix B, and 1 unit of mix C **31.** $(x + 3)(x + y)$ **33.** $x^2 + 10x + 25$ **35.** $4x^2 - 4xy + y^2$ **37.** −13 **39.** −36 **41.** 0

Exercise Set 10.4

1. 26 **3.** −19 **5.** 0 **7.** $\{(1, 2)\}$ **9.** $\{(x, y) \mid 3x + y = 1\}$ **11.** $\{(9, 9)\}$ **13.** 8 **15.** 0 **17.** 54 **19.** $\{(-2, 0, 5)\}$ **21.** $\{(6, -2, 4)\}$ **23.** 16 **25.** 15 **27.** $\frac{13}{6}$ **29.** 0 **31.** 56 **33.** $\{(-3, -2)\}$ **35.** $\{\ \}$ **37.** $\{(-2, 3, -1)\}$ **39.** $\{(3, 4)\}$ **41.** $\{(-2, 1)\}$ **43.** $\{(x, y, z) \mid x - 2y + z = -3\}$ **45.** $\{(0, 2, -1)\}$ **47.** $5x + 5z = 10$ **49.** $-5y + 2z = 2$ **51.** $13 - 13i$ **53.** 25 **55.** −125 **57.** 24

Exercise Set 10.5

1. $\{(2, -1)\}$ **3.** $\{(-4, 2)\}$ **5.** $\{(-2, 5, -2)\}$ **7.** $\{(1, -2, 3)\}$ **9.** $\{\ \}$ **11.** $\{(x, y) \mid x - y = 3\}$ **13.** $\{(4, -3)\}$ **15.** $\{(2, 1, -1)\}$ **17.** $\{(9, 9)\}$ **19.** $\{\ \}$ **21.** $\{\ \}$ **23.** $\{(1, -4, 3)\}$ **25.** 6 **27.** 4 **29.** width, 5 km; length, 9 km

31. $(x - 5)^2 + (y + 1)^2 = 9$
$C(5, -1);\ r = 3$

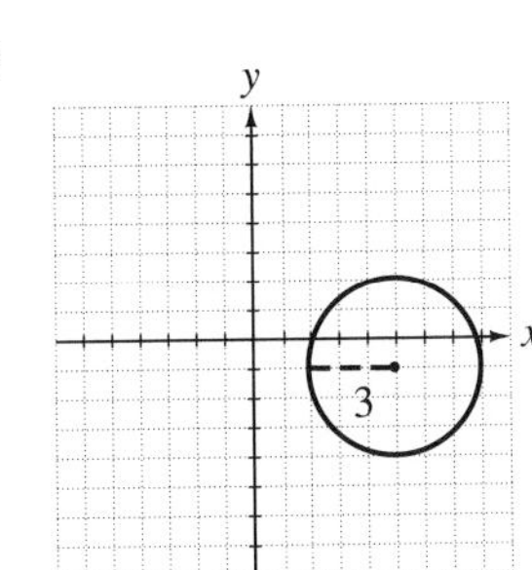

33. $x^2 - 3y = 1;\ V\left(0, -\frac{1}{3}\right)$

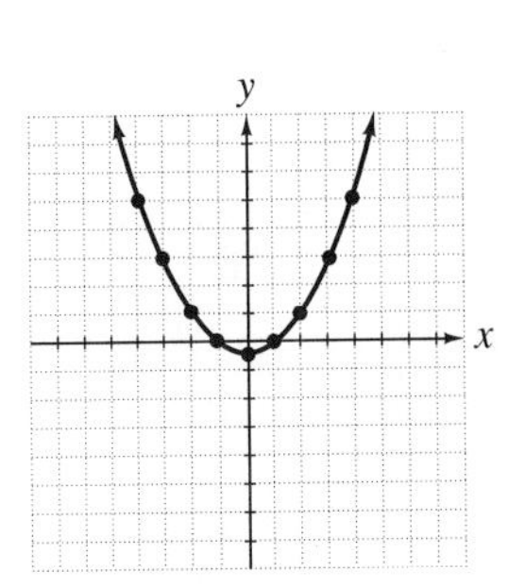

Exercise Set 10.6

1. $\{(3, -4), (-3, 4)\}$ **3.** $\{(\sqrt{2}, \sqrt{2}), (-\sqrt{2}, -\sqrt{2})\}$ **5.** $\{(4, 0), (0, -2)\}$
7. $\{(-\sqrt{5}, -2), (-\sqrt{5}, 2), (\sqrt{5}, -2), (\sqrt{5}, 2)\}$ **9.** $\{\ \}$ **11.** $\{(1, -2), (3, 6)\}$ **13.** $\{(2, 4), (-5, 25)\}$
15. $\{\ \}$ **17.** $\{(1, -3)\}$ **19.** $\{(-1, -2), (-1, 2), (1, -2), (1, 2)\}$ **21.** $\{(0, -1)\}$ **23.** $\{(-1, 3), (1, 3)\}$
25. $\{(-\sqrt{3}, 0), (\sqrt{3}, 0)\}$ **27.** $\{\ \}$ **29.** $\{(-6, 0), (6, 0), (0, -6)\}$

31.

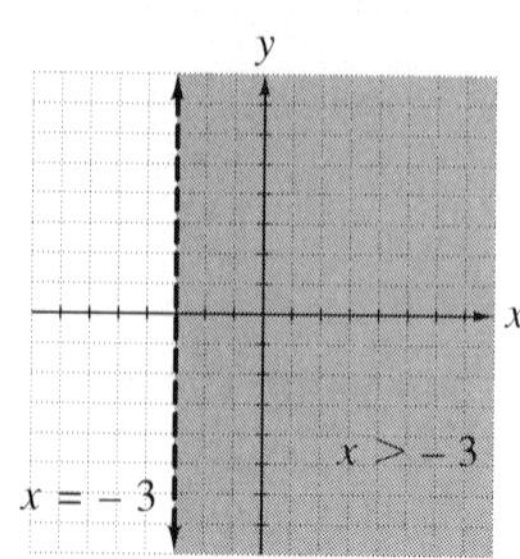

33.

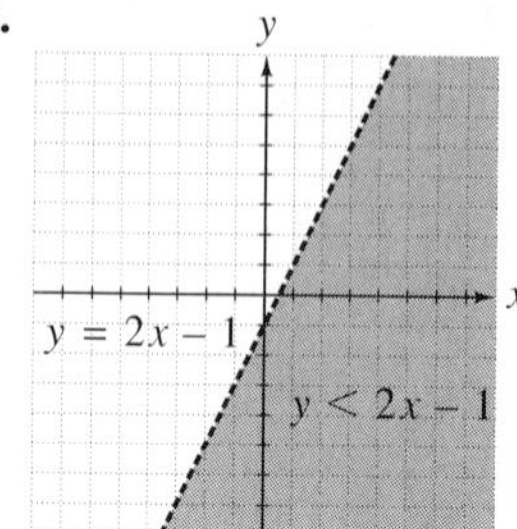

35. $(8x - 25)$ in. **37.** $(4x^2 + 6x + 2)$ meters

Exercise Set 10.7

1.

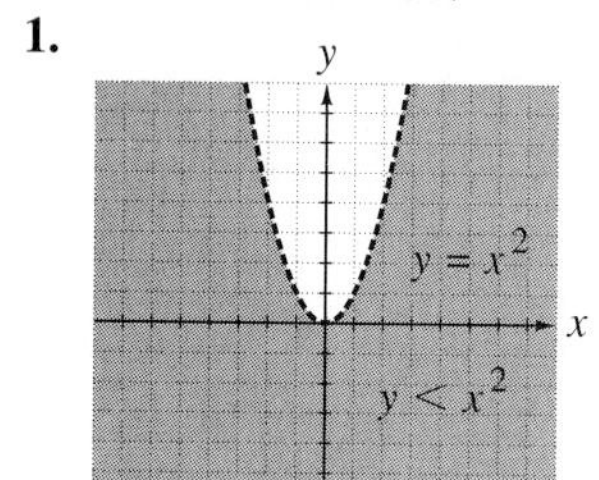

3.

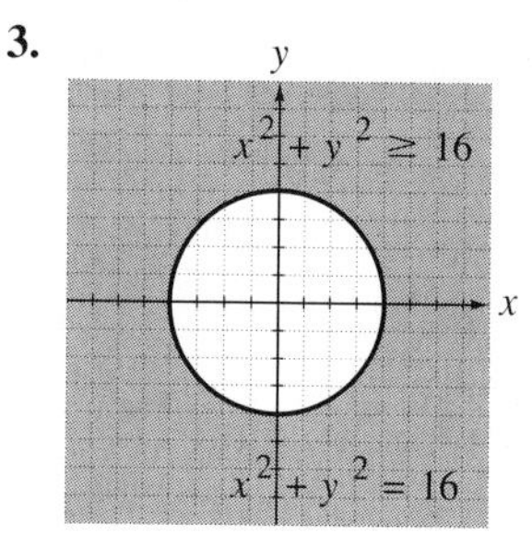

5.

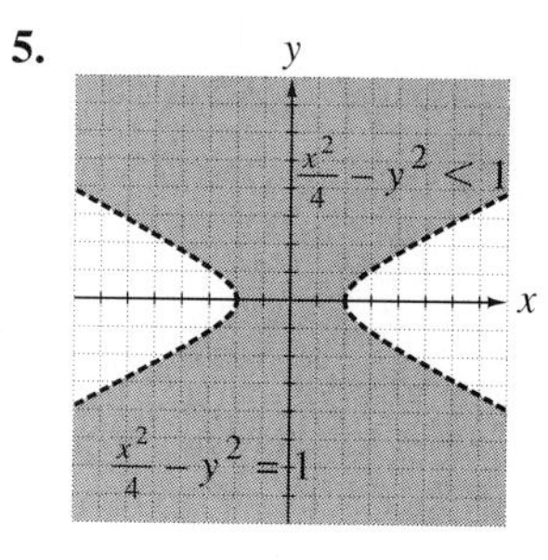

7.

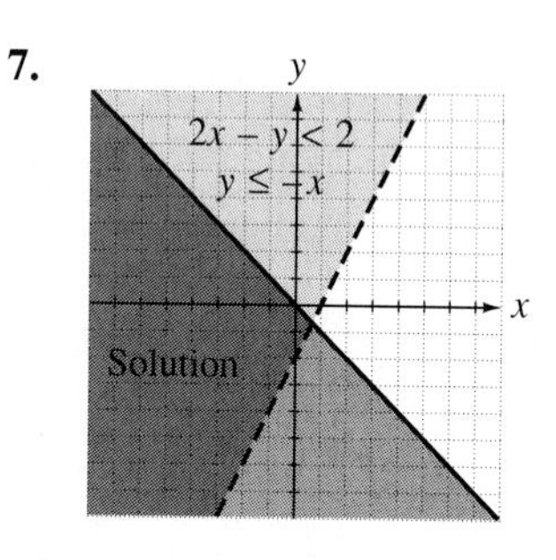

9.

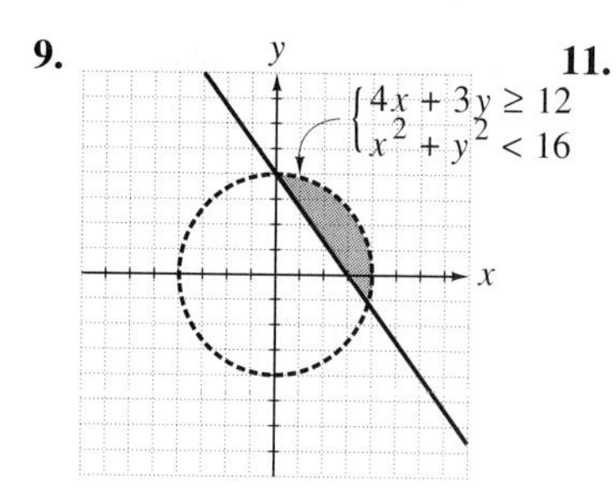

11.

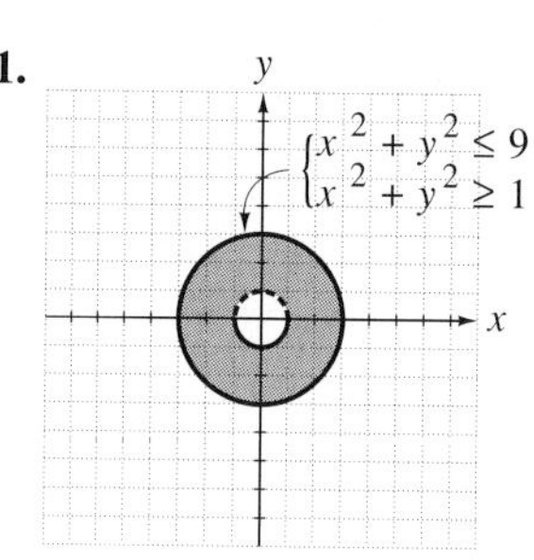

13.

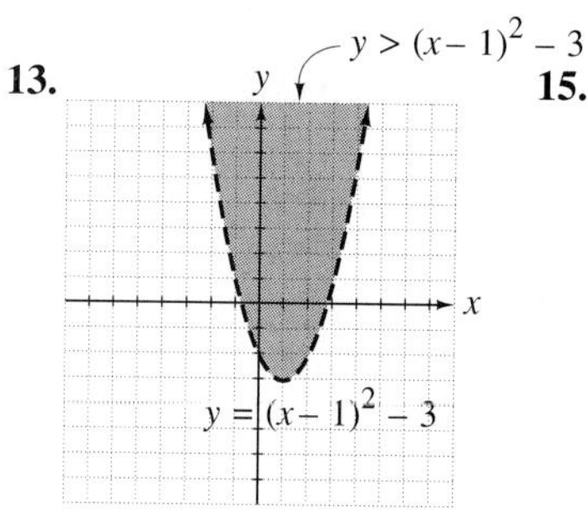

15.

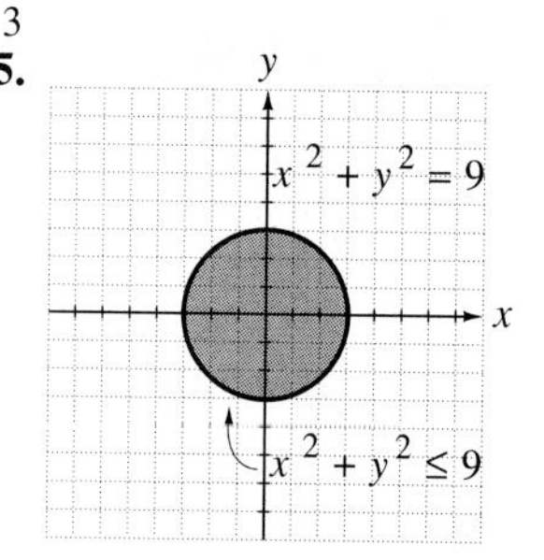

17.

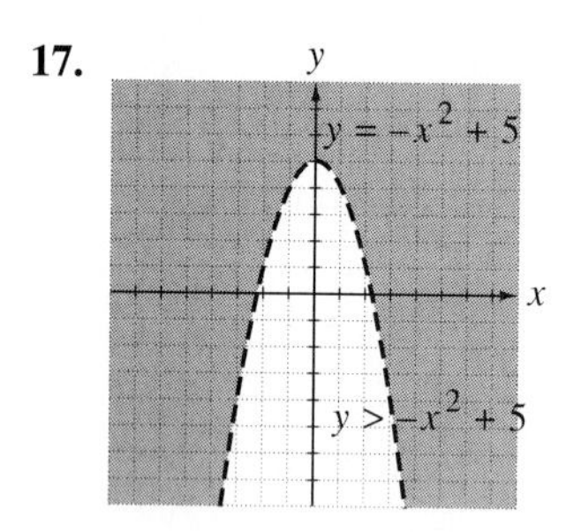

19.

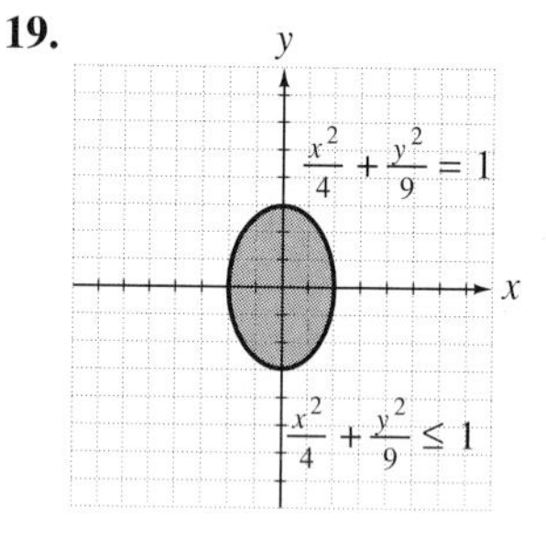

21.

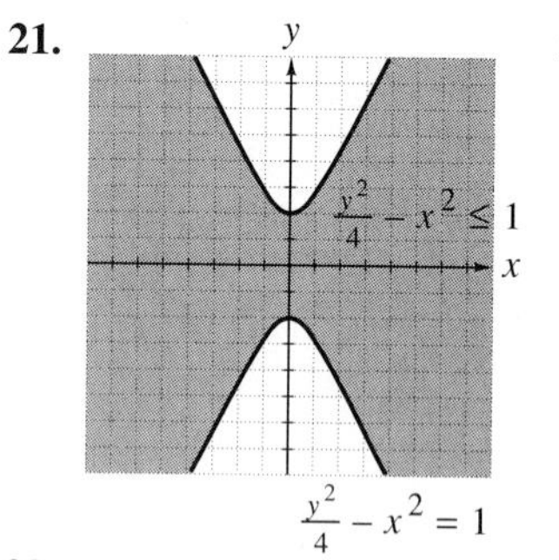

23.

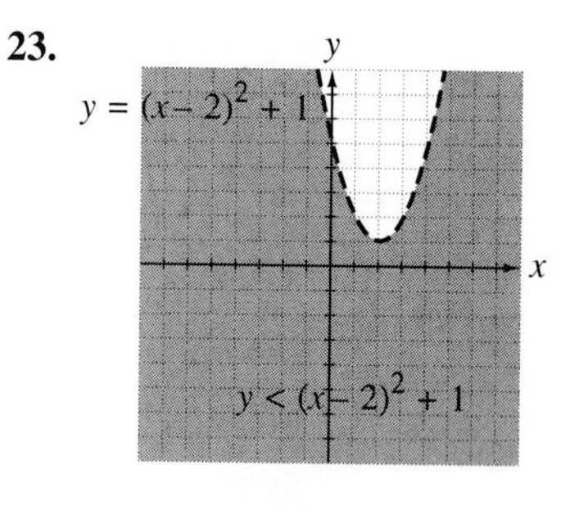

25.

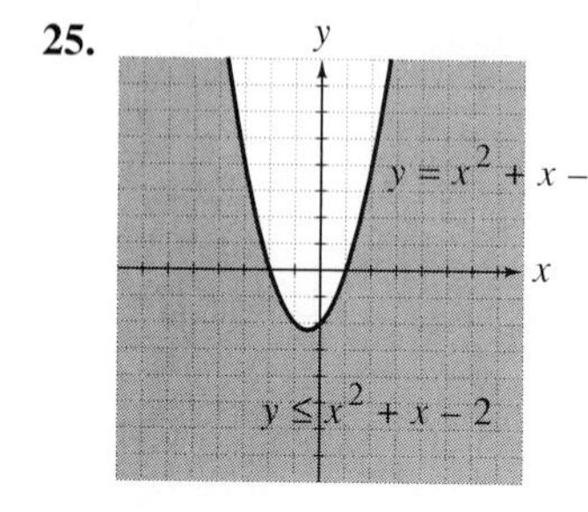

27.

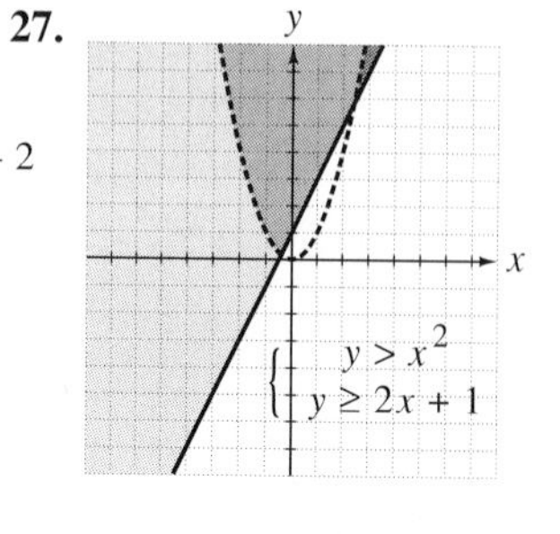

29.

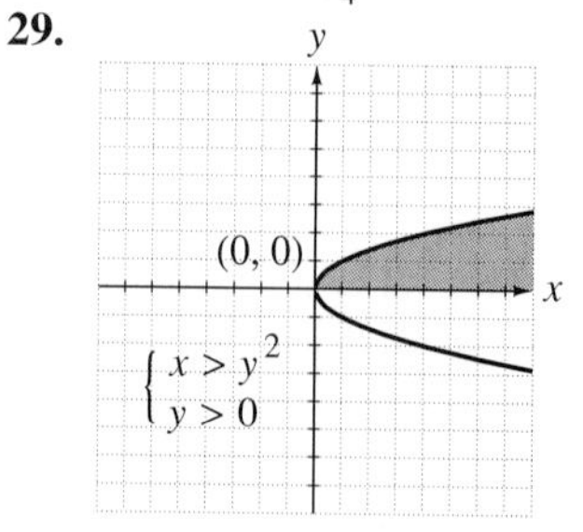

31.

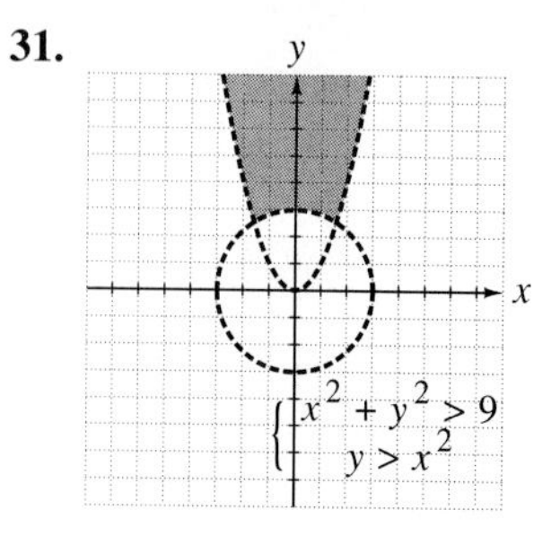

33.

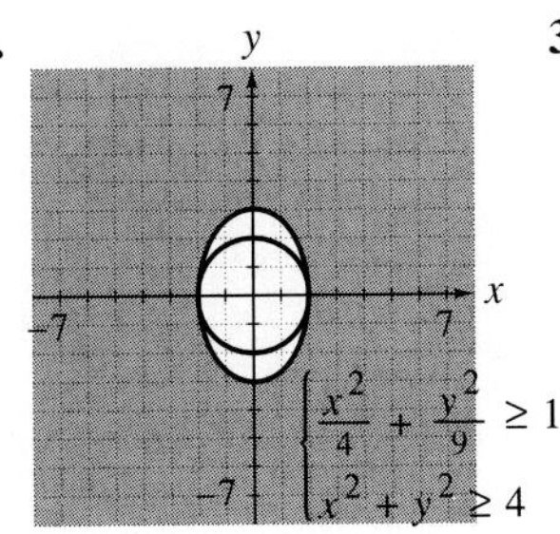

35.

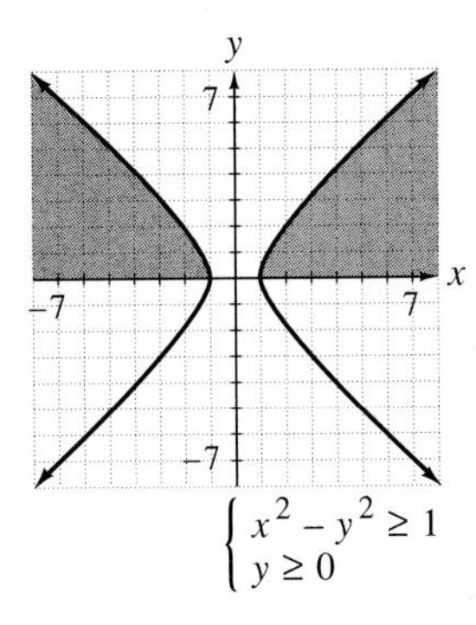

37.

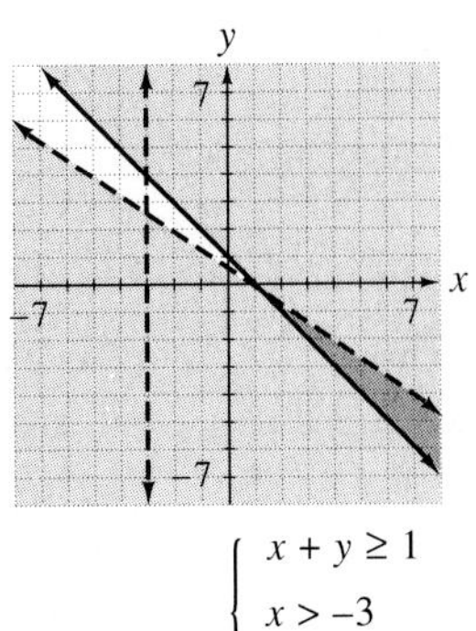

39.

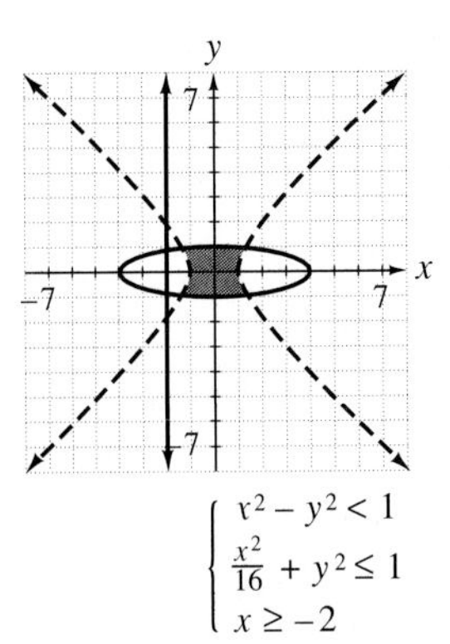

41. $y = \frac{4x}{3}$ **43.** $y = \frac{a - c}{2}$ **45.** $(-\infty, -2) \cup (1, \infty)$ **47.** $(0, 3)$

Chapter 10 Review

1. $\{(-3, 1)\}$;

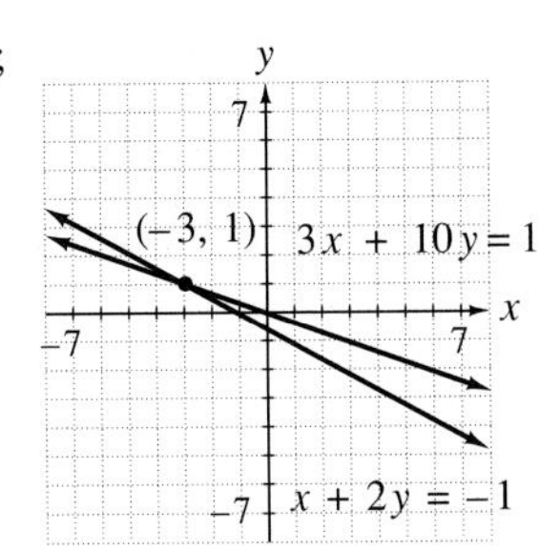

3. { };

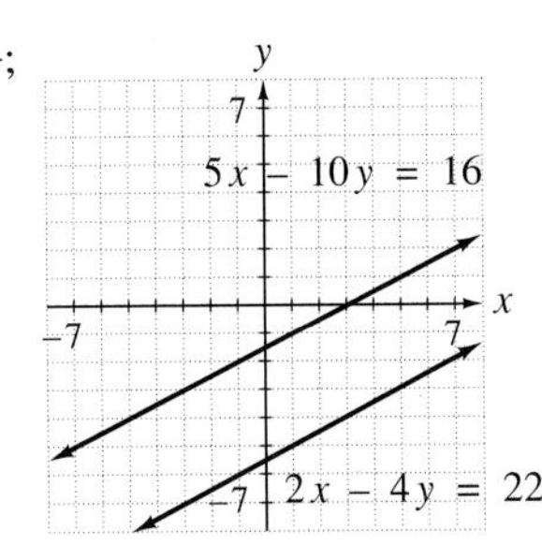

5. $\left\{\left(3, \frac{8}{3}\right)\right\}$;

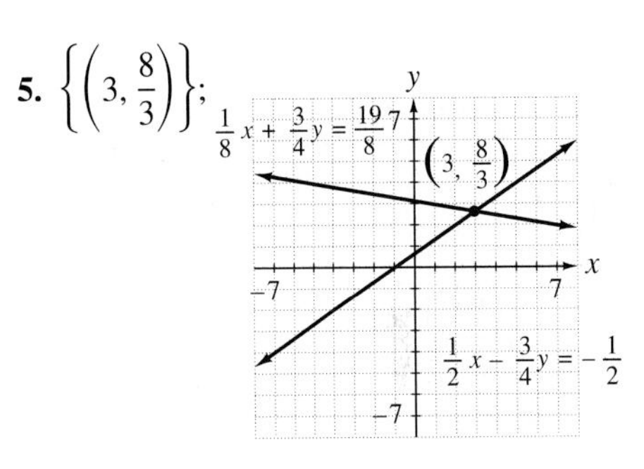

7. $\{(2, 0, -3)\}$ **9.** $\{(-1, 2, 0)\}$ **11.** $\{(x, y, z) \mid x + 2y + 3z = 11\}$ **13.** $\{(3, 1, 1)\}$ **15.** 24 dimes, 23 nickels, and 48 pennies **17.** Sue is 17 years old and Pat is 1 year old **19.** width, 37 ft; length, 111 ft **21.** 30 pounds of creme-filled; 5 pounds of chocolate-covered nuts; 10 pounds of chocolate-covered raisins **23.** larger investment, 9.5%; smaller investment, 7.5% **25.** 120, 115, and 60 **27.** 17 **29.** −72 **31.** $\left\{\left(\frac{1}{3}, \frac{7}{6}\right)\right\}$ **33.** $\left\{\left(0, \frac{2}{3}\right)\right\}$ **35.** $\{(x, y) \mid x - 2y = 4\}$ **37.** $\{(2, 0, -3)\}$ **39.** $\left\{\left(\frac{3}{7}, -2, -\frac{1}{7}\right)\right\}$ **41.** $\{(-1, 2, 0)\}$ **43.** $\{(x, y) \mid x - 2y = 4\}$ **45.** $\left\{\left(\frac{1}{3}, \frac{7}{6}\right)\right\}$ **47.** $\{(-7, -15)\}$ **49.** $\{(2, 1)\}$ **51.** $\{(2, 0, -3)\}$ **53.** $\{(-1, 2, 0)\}$ **55.** { } **57.** { } **59.** $\{(5, 1), (-1, 7)\}$ **61.** $\{(0, 2), (0, -2)\}$ **63.** $\left\{\left(2, \frac{5}{2}\right), (-7, -20)\right\}$ **65.** $\{(-2, -1), (-2, 1), (2, -1), (2, 1)\}$

67.

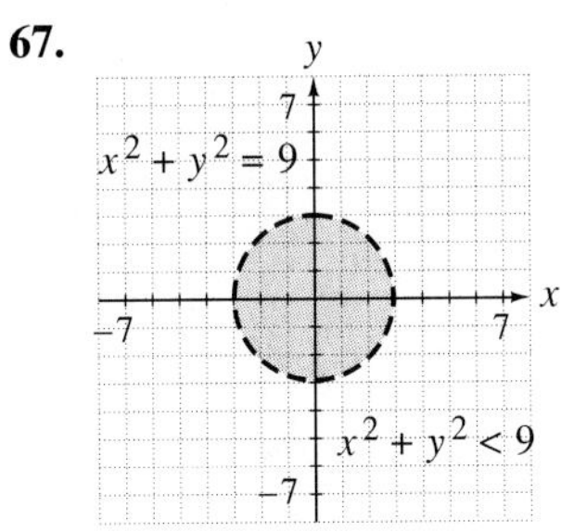

69.

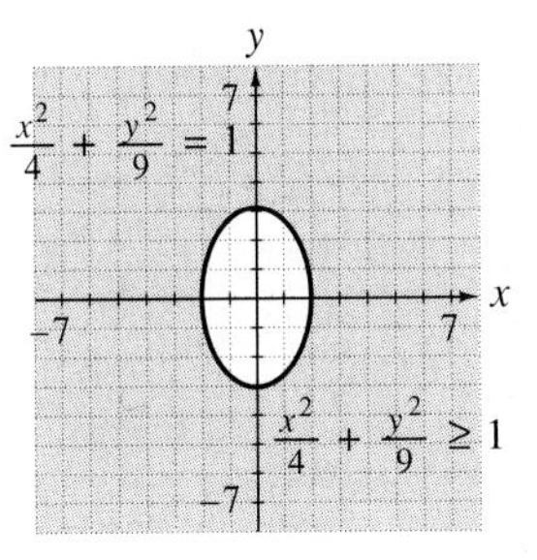

71.

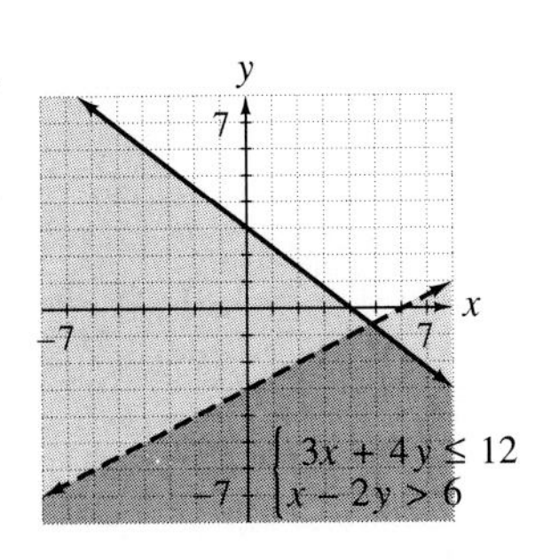

73.

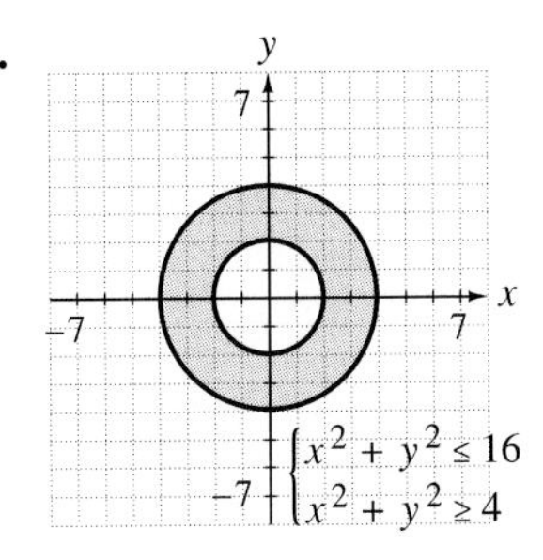

75.

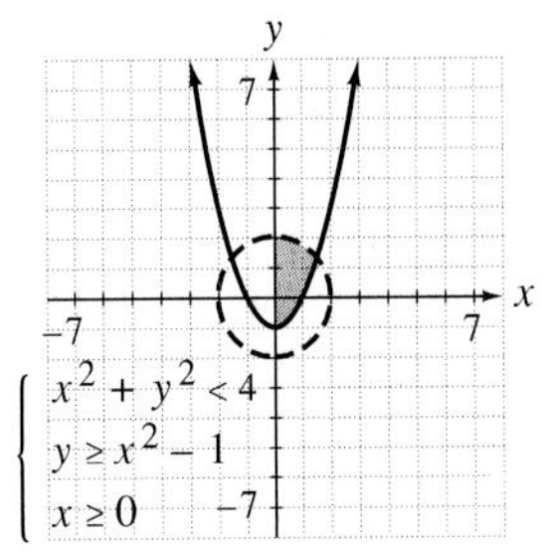

Chapter 10 Test

1. 34 **2.** -6 **3.** $\{(1, 3)\}$

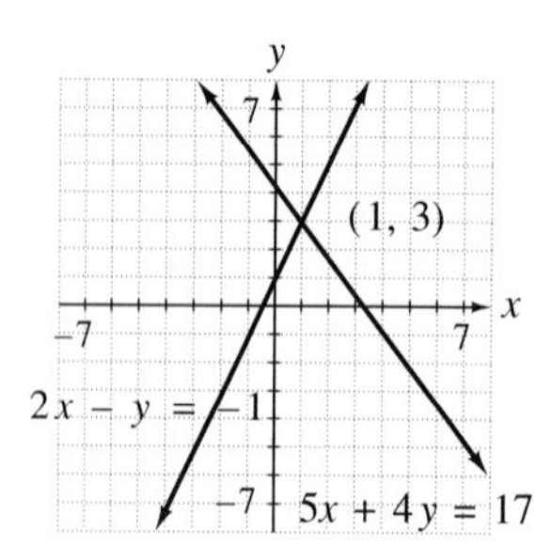

4. $\{\ \}$

$7x - 14y = 5$
$x = 2y$

5. $\{(2, -3)\}$ **6.** $\{(x, y) \mid 10x + 4y = 10\}$ **7.** $\{(-1, -2, 4)\}$ **8.** $\{\ \}$ **9.** $\{(2, -1)\}$ **10.** $\{3, 6\}$ **11.** $\{(3, -1, 2)\}$ **12.** $\{(5, 0, -4)\}$ **13.** $\{(x, y) \mid x - y = -2\}$ **14.** $\{(5, -3)\}$ **15.** $\{(-1, -1, 0)\}$ **16.** $\{\ \}$ **17.** $\{(-12, 5), (12, -5)\}$ **18.** $\{(-5, -1), (-5, 1), (5, -1), (5, 1)\}$ **19.** $\{(6, 12), (1, 2)\}$ **20.** $\{(1, 1), (-1, -1)\}$

21. $\begin{cases} 2x + 5y \geq 10 \\ y \geq x^2 + 1 \end{cases}$

22. $\begin{cases} \frac{x^2}{4} + y^2 \leq 1 \\ x + y > 1 \end{cases}$

23. $\begin{cases} x^2 + y^2 > 1 \\ \frac{x^2}{4} - y^2 \geq 1 \end{cases}$

24.

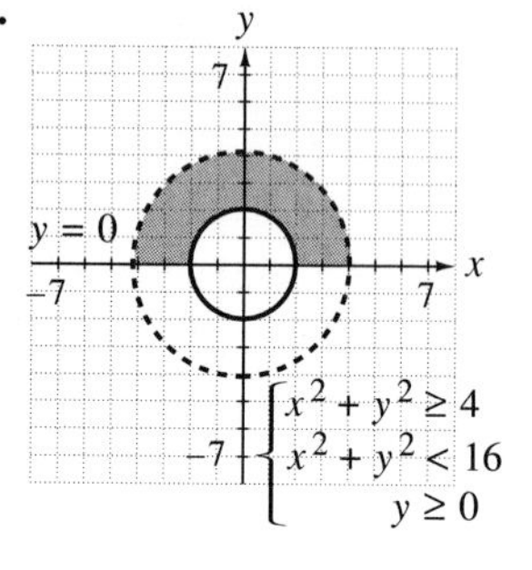

25. Dean jogs 7 mph, Tom bicycles 17 mph, and Dean jogged 7/6 miles. **26.** 58

Chapter 10 Cumulative Review

1. $(-\infty, -18]$; -18; *Sec. 3.5, Ex. 7* **2.** $(-\infty, -4) \cup (10, \infty)$; *Sec. 3.7, Ex. 5* **3.** $2x + 3y = 20$; *Sec. 4.5, Ex. 6*

4. a. 4^7 **b.** x^7 **c.** y^4 **d.** y^{12} **e.** $(-5)^{15}$; *Sec. 5.1, Ex. 2* **5.** $6x^3y^2 - 4x^2y^2 - 2y^2$; *Sec. 5.3, Ex. 10*

6. $\frac{x^3}{2y} - \frac{5x}{2y^2} - \frac{1}{xy^3}$; *Sec. 5.5, Ex. 4* **7. a.** $x^2 + 2x + 1$ **b.** 5; *Sec. 5.7, Ex. 4* **8.** $(4x - 1)(2x - 5)$; *Sec. 6.3, Ex. 2*

9. $(4x + 1)(16x^2 - 4x + 1)$; *Sec. 6.4, Ex. 7* **10.** $\{-5, -1, 1\}$; *Sec. 6.6, Ex. 7*

11. a. 1 **b.** -1 **c.** $\frac{-2x - 6}{x + 1}$; *Sec. 7.1, Ex. 4* **12.** $\frac{xy + 2x^3}{y - 1}$; *Sec. 7.4, Ex. 3* **13.** $\frac{x - 4}{5\sqrt{x} - 10}$; *Sec. 8.4, Ex. 5*

14. a. $\frac{1}{6}$ **b.** $\frac{1}{8}$ **c.** -3 **d.** $\frac{1}{16}$; *Sec. 8.5, Ex. 3* **15.** $\left\{\frac{9 - i\sqrt{15}}{6}, \frac{9 + i\sqrt{15}}{6}\right\}$; *Sec. 9.1, Ex. 7*

16. 11.1 hours and 13.1 hours; *Sec. 9.2, Ex. 6* **17.** $[-2, 3)$; *Sec. 9.4, Ex. 4* **18.** $6\frac{1}{4}$ ft; *Sec. 9.6, Ex. 4*

19. $\frac{(x + 3)^2}{25} + \frac{(y-2)^2}{36} = 1$; *Sec. 9.8, Ex. 3*

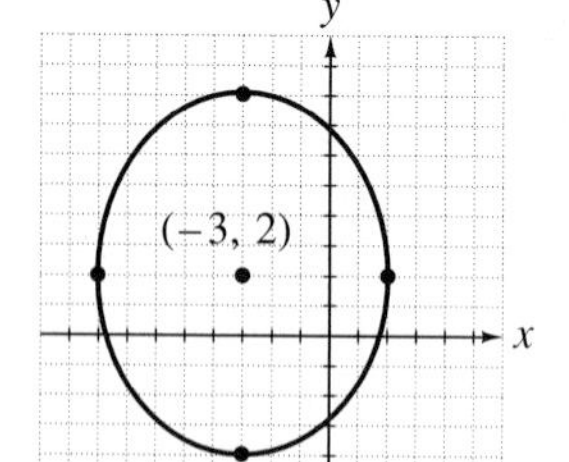

20. $\left\{\left(-4, \frac{1}{2}\right)\right\}$; *Sec. 10.1, Ex. 2*

21. 52 mph and 47 mph; *Sec. 10.3, Ex.2*

22. $\{(1, -1, 3)\}$; *Sec. 10.5, Ex. 2*

23. $4y^2 > x^2 + 16$; *Sec. 10.7, Ex. 2*

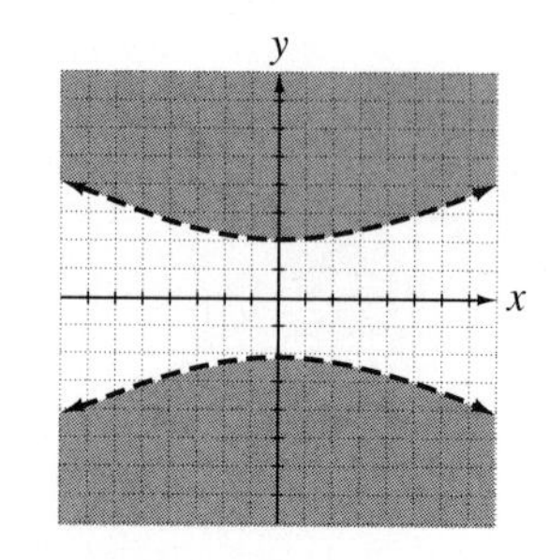

CHAPTER 11
Exponential and Logarithmic Functions

Exercise Set 11.1

1. one-to-one **3.** one-to-one **5.** one-to-one **7.** one-to-one **9.** not one-to-one **11.** one-to-one **13.** not one-to-one

15. $f^{-1}(x) = x - 4$

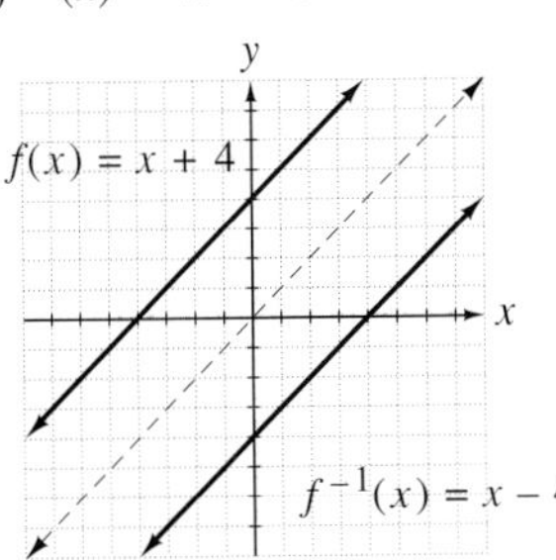

17. $f^{-1}(x) = \frac{x + 3}{2}$

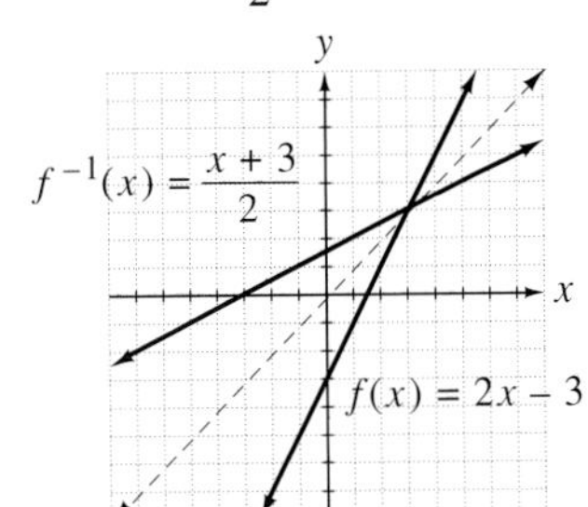

19. $f^{-1}(x) = \frac{3x + 4}{12}$

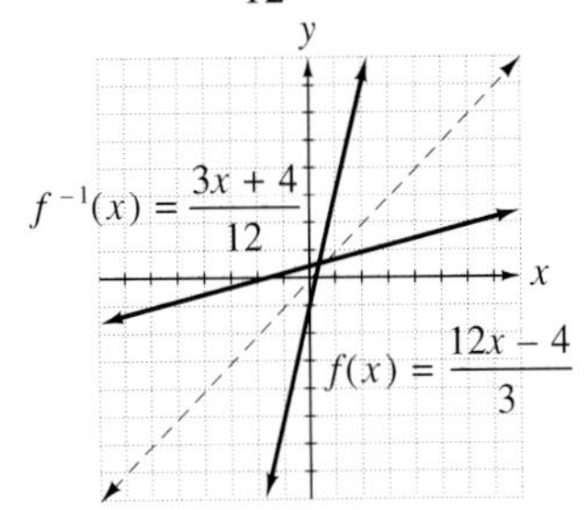

21. not one-to-one **23.** not one-to-one **25.** one-to-one; $f^{-1} = \{(-8, -4), (-12, -6), (-16, -8), (-18, -9)\}$

27. not a function

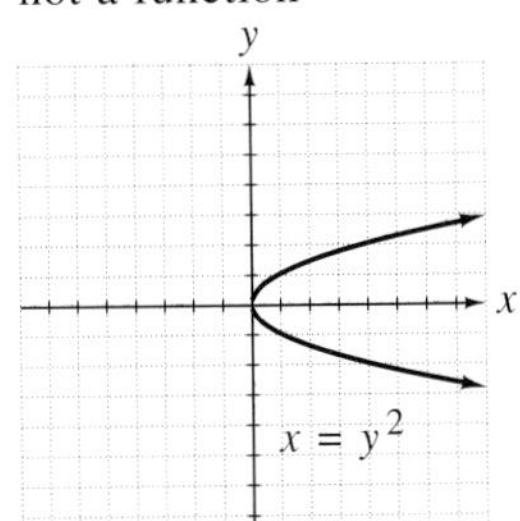

29. one-to-one

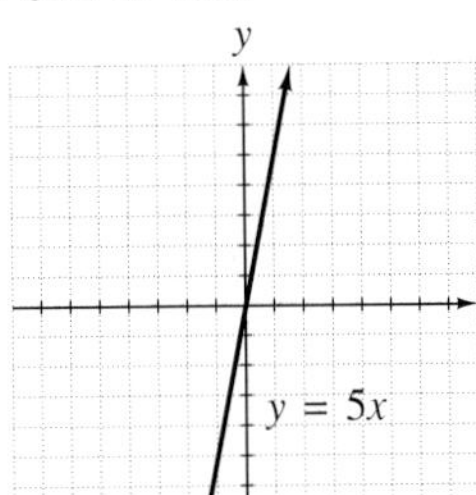

31. not a function

33. $f^{-1}(x) = \frac{x - 1}{3}$

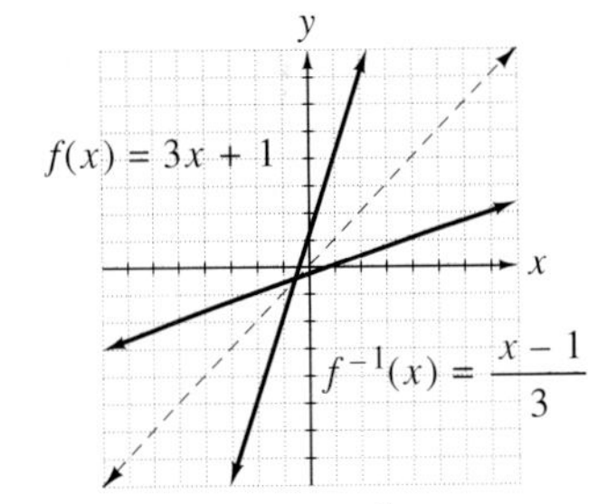

35. $f^{-1}(x) = 5x + 2$

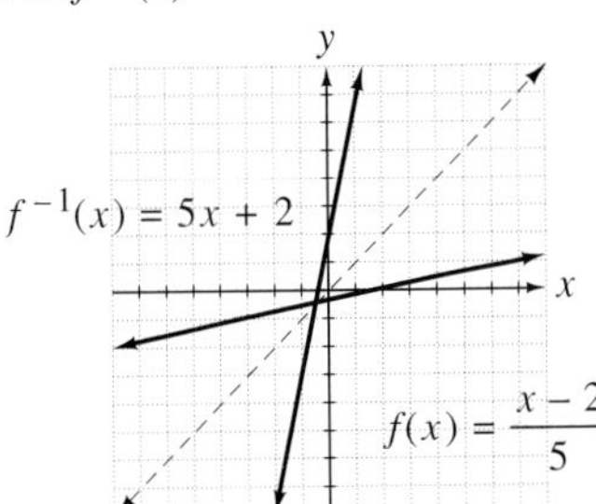

37. $g^{-1}(x) = 2x + 8$

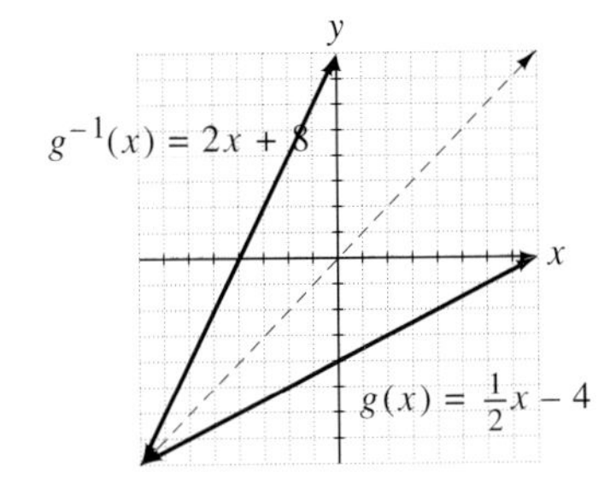

39. $\sqrt{41}$ miles **41.** 1 **43.** 3 **45.** $\frac{1}{4}$ **47.** 8

Exercise Set 11.2

1.

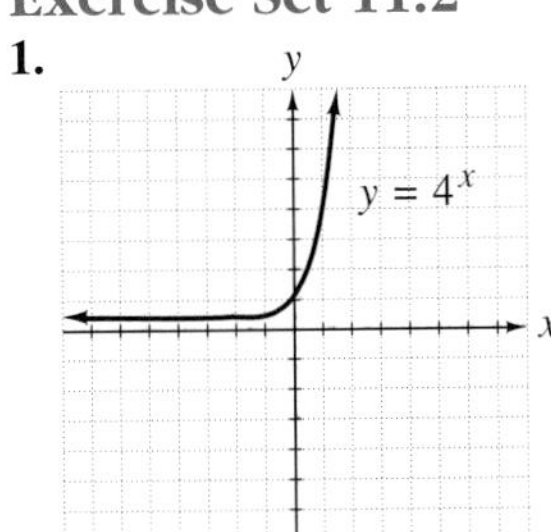

3.

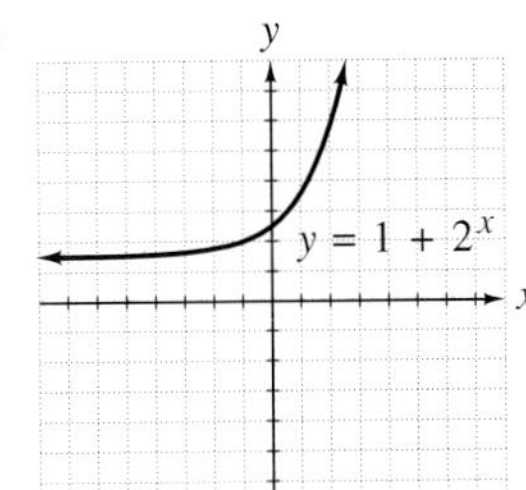

5.

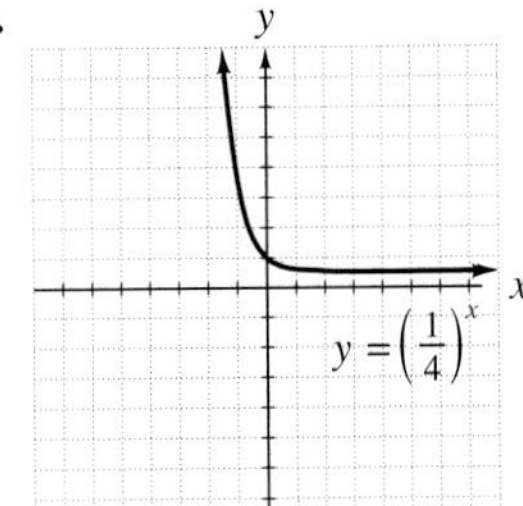

7.

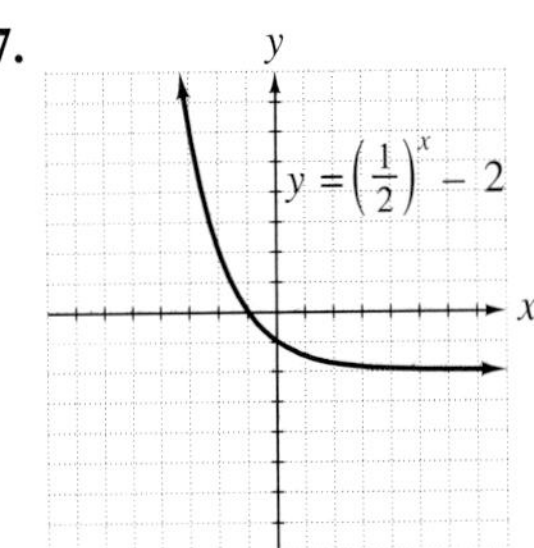

9. $\{3\}$ **11.** $\left\{\frac{3}{4}\right\}$ **13.** $\left\{\frac{8}{5}\right\}$ **15.** $\left\{-\frac{2}{3}\right\}$ **17.** ~24.6 pounds **19.** ~$7621.42

21.

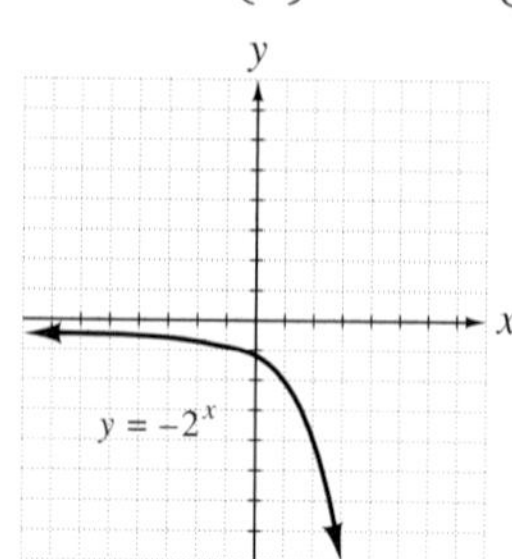

23.

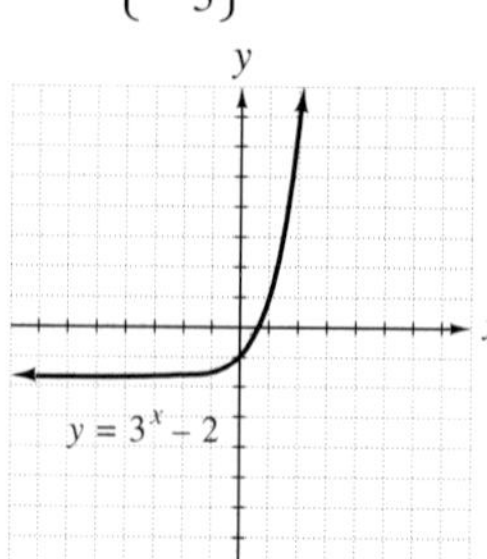

25.

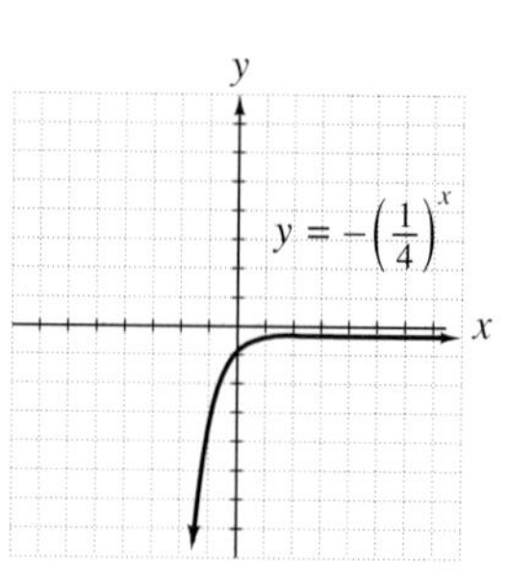

27.

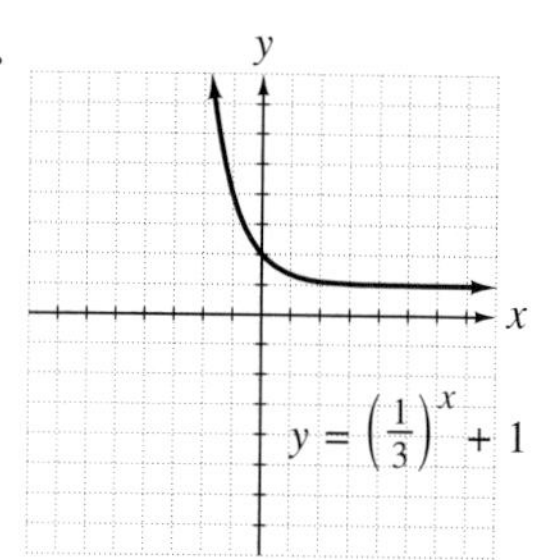

29. $\{4\}$ **31.** $\left\{\frac{3}{2}\right\}$ **33.** $\left\{-\frac{1}{3}\right\}$ **35.** $\{-2\}$ **37.** $y = 3^x$ **39.** $y = \left(\frac{1}{2}\right)^x$ **41.** ~519 rats **43.** ~22.5%
45. ~9,060,000 **47.** ~$4065.59 **49.** $\{4\}$ **51.** $\{\ \}$ **53.** $\{2, 3\}$ **55.** $\{3\}$ **57.** $\{-1\}$

Exercise Set 11.3

1. 3 **3.** −2 **5.** $\frac{1}{2}$ **7.** −1 **9.** $\{2\}$ **11.** $\{81\}$ **13.** $\{7\}$ **15.** $\{-3\}$ **17.** 3 **19.** 3 **21.** 1

23.

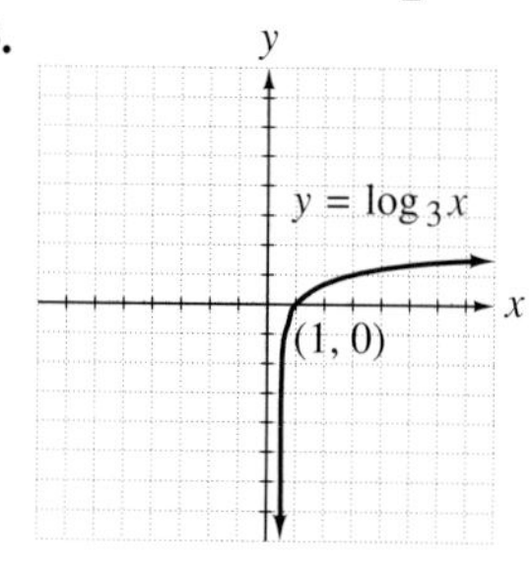

25.

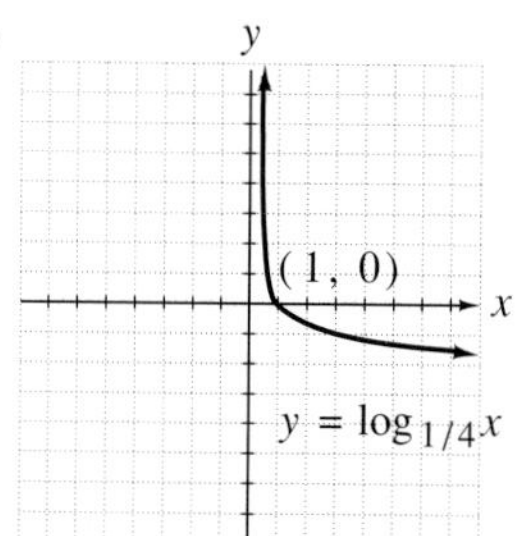

27.

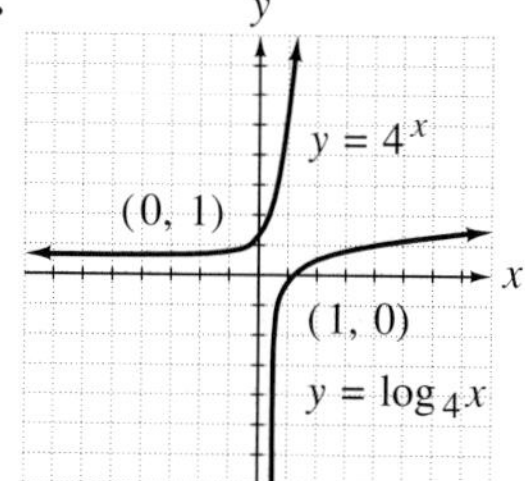

29.

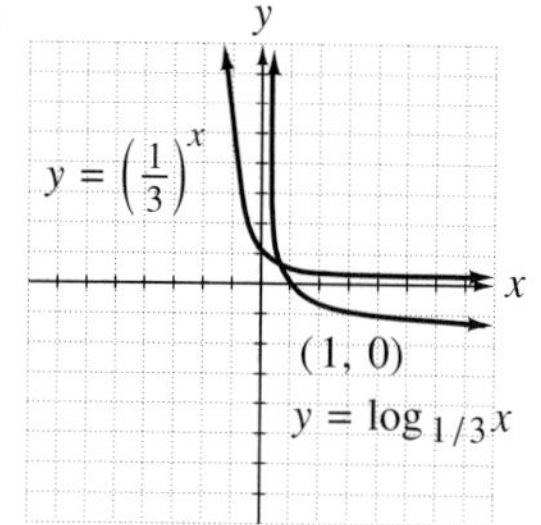

31. 0 **33.** 4 **35.** 2 **37.** 5 **39.** 4 **41.** −3 **43.** $\{-3\}$ **45.** $\{2\}$ **47.** $\{2\}$ **49.** $\left\{\frac{27}{64}\right\}$ **51.** $\{10\}$

53.

y
(1, 0)
x
$f(x) = \log_5(x)$

55.

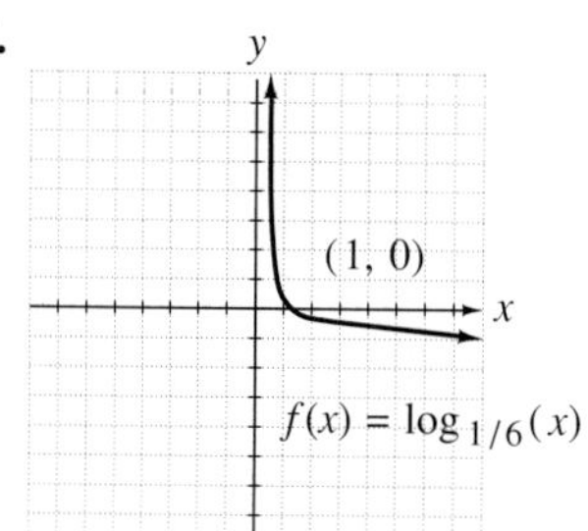

57. 0.0827 **59.** −1 **61.** $x - 5$ **63.** 3 **65.** $\frac{y-9}{y^2-1}$

Exercise Set 11.4

1. $\log_5 14$ **3.** $\log_4 (9x)$ **5.** $\log_{10} (10x^2 + 20)$ **7.** $\log_5 3$ **9.** $\log_2 \left(\frac{x}{y}\right)$ **11.** 1 **13.** $\log_2 25$

15. $\log_5 (x^3z^6)$ **17.** $\log_{10}\left(\frac{x^3 - 2x}{x + 1}\right)$ **19.** $\log_3 4 + \log_3 y - \log_3 5$ **21.** $3 \log_2 x - \log_2 y$

23. $\frac{1}{2}\log_b 7 + \frac{1}{2}\log_b x$ **25.** 0.2 **27.** 1.2 **29.** 0.23 **31.** $\log_4 35$ **33.** $\log_3 4$ **35.** $\log_7 \left(\frac{9}{2}\right)$ **37.** $\log_4 48$

39. $\log_2 \left[\frac{x^{7/2}}{(x + 1)^2}\right]$ **41.** $\log_8 (x^{16/3})$ **43.** $\log_7 5 + \log_7 x - \log_7 4$ **45.** $3 \log_5 x + \log_5 (x + 1)$

47. $2 \log_6 x - \log_6 (x + 3)$ **49.** 1.29 **51.** −0.68 **53.** −0.125

55.

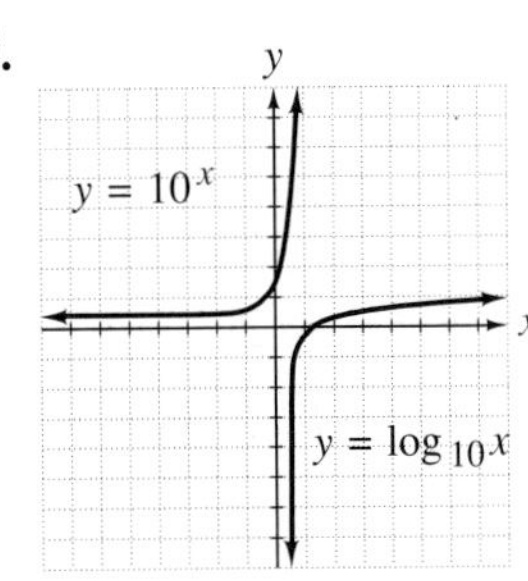

57. {14} **59.** { } **61.** -1 **63.** $\frac{1}{2}$

Exercise Set 11.5

1. 0.9031 **3.** 0.3636 **5.** 2 **7.** -3 **9.** $10^{1.3} \approx 19.9526$ **11.** $\frac{10^{1.1}}{2} \approx 6.2946$ **13.** 0.6931 **15.** -2.6367 **17.** 2 **19.** $\frac{1}{4}$ **21.** $e^{1.4} \approx 4.0552$ **23.** $\frac{4 + e^{2.3}}{3} \approx 4.6581$ **25.** 4.2 **27.** \$2542.50 **29.** 1.5850 **31.** -2.3219 **33.** 1.1004 **35.** 1.6094 **37.** 1.6180 **39.** 3 **41.** 2 **43.** -4 **45.** $\frac{1}{2}$ **47.** $10^{2.3} \approx 199.5262$ **49.** $e^{-2.3} \approx 0.1003$ **51.** $\frac{10^{-0.5} - 1}{2} \approx -0.3419$ **53.** $\frac{e^{0.18}}{4} \approx 0.2993$ **55.** 1.5850 **57.** -1.6309 **59.** 0.8617 **61.** 5.3 **63.** \$3656.38 **65.** 13 weeks **67.** 21 weeks **69.** $\left\{\frac{4}{7}\right\}$ **71.** $x = \frac{3y}{4}$ **73.** $\{-6, -1\}$ **75.** $\{(2, -3)\}$

Exercise Set 11.6

1. $\left\{\frac{\log 6}{\log 3}\right\}$; {1.6309} **3.** $\left\{\frac{\log 3.8}{2 \log 3}\right\}$; {0.6076} **5.** $\left\{3 + \frac{\log 5}{\log 2}\right\}$; {5.3219} **7.** {11} **9.** $\{9, -9\}$ **11.** $\left\{\frac{1}{2}\right\}$ **13.** $\left\{\frac{3}{4}\right\}$ **15.** {2} **17.** $\left\{\frac{1}{8}\right\}$ **19.** 103 wolves **21.** 8.8 years **23.** $\left\{\frac{\log 5}{\log 9}\right\}$; {0.7325} **25.** $\left\{\frac{\log 3}{\log 4} - 7\right\}$; $\{-6.2075\}$ **27.** $\left\{\frac{1}{3}\left(4 + \frac{\log 11}{\log 7}\right)\right\}$; {1.7441} **29.** $\left\{\frac{\ln 5}{6}\right\}$; {0.2682} **31.** {11} **33.** $\{4, -1\}$ **35.** $\left\{\frac{9}{5}\right\}$ **37.** {64} **39.** $\left\{\frac{-5 + \sqrt{33}}{2}\right\}$ **41.** $\left\{\frac{192}{127}\right\}$ **43.** $\left\{\frac{2}{3}\right\}$ **45.** 10,073,520 inhabitants **47.** 10 years **49.** $1\frac{3}{4}$ years **51.** 56 in. **53.** 11.9 pounds/in.^{2} **55.** 3.2 miles **57.** $-\frac{5}{3}$ **59.** $\frac{17}{4}$ **61.** $f^{-1}(x) = \frac{x - 2}{5}$

Chapter 11 Review

1. one-to-one; $h^{-1} = \{(14, -9), (8, 6), (12, -11), (15, 15)\}$ **3.** not one-to-one **5.** not one-to-one **7.** $f^{-1}(x) = \frac{x - 11}{6}$ **9.** $q^{-1}(x) = \frac{x - b}{m}$ **11.** $r^{-1}(x) = \frac{2(x + 4)}{13}$

13.

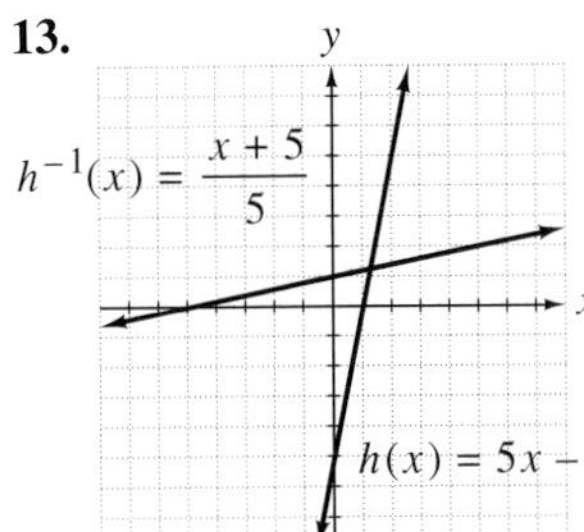

15. $\{-2\}$ **17.** $\left\{\frac{3}{2}\right\}$ **19.** $\left\{\frac{8}{9}\right\}$

21.

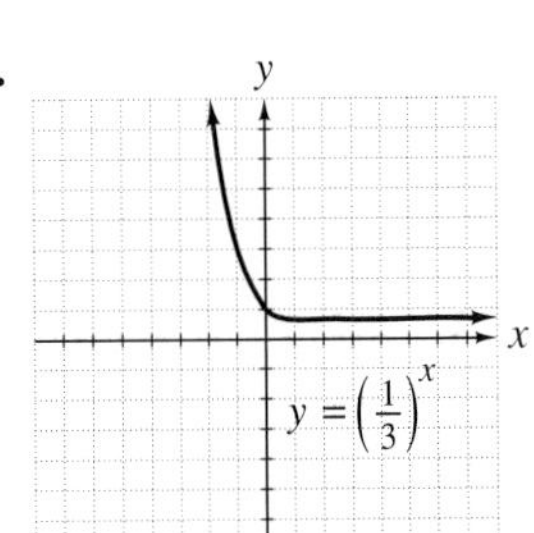

23.

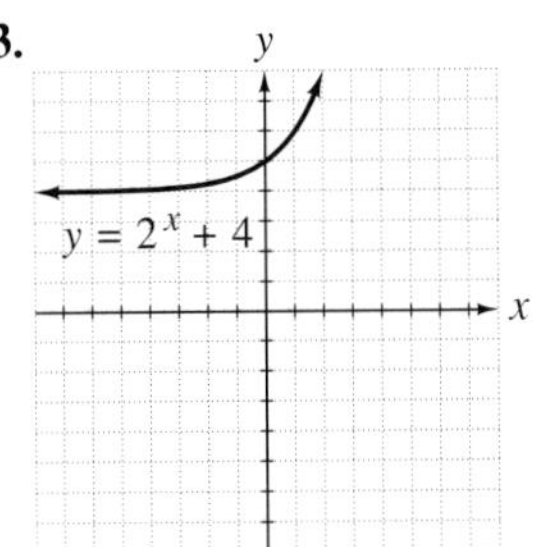

25. \$1131.82 **27.** $\log_2\left(\frac{1}{16}\right) = -4$ **29.** $0.4^3 = 0.064$ **31.** {9} **33.** {3} **35.** {3} **37.** {−2} **39.** {9}

41. {2} **43.** {−8, 1} **45.**

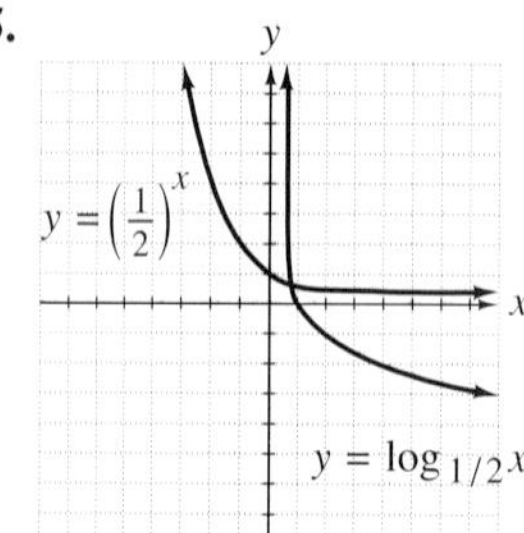

47. $\log_2 18$ **49.** $\log\left(\frac{3}{2}\right)$ **51.** $\log_5 2$ **53.** $\log_3(x^4 + 2x^3)$ **55.** $\log_4(x + 5) - 2\log_4 x$ **57.** $\log_7 y + 3\log_7 z - \log_7 x$ **59.** −0.11 **61.** −0.8239 **63.** 1.5326 **65.** −1 **67.** 4 **69.** $\left\{\frac{e^{1.6}}{3}\right\}$ **71.** $\left\{\frac{e^2 - 1}{3}\right\}$ **73.** 1.22 mm **75.** 1.2619 **77.** \$1307.51 **79.** $\left\{\frac{\log 5}{3\log 6}\right\}$; {0.2994}

81. $\left\{\frac{1}{3}\left(\frac{\log 9}{\log 4} - 2\right)\right\}$; {−0.1383} **83.** $\left\{\frac{1}{4}\left(\frac{\log 3}{\log 8} + 2\right)\right\}$; {0.6321} **85.** $\left\{\frac{\log \frac{2}{3}}{\log 4} - 5\right\}$; {−5.2925} **87.** $\left\{\frac{9}{10}\right\}$ **89.** $\left\{\frac{3e^2}{e^2 - 3}\right\}$ **91.** { } **93.** 224,310,000 **95.** 21 years **97.** 33 years **99.** 8.6 years

Chapter 11 Test

1.

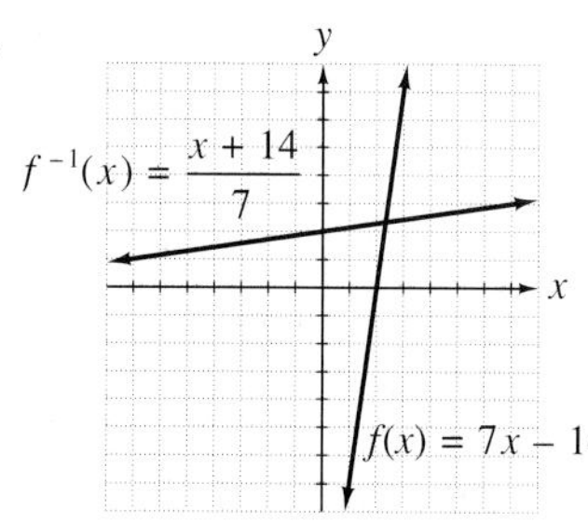

2. one-to-one **3.** not one-to-one **4.** $f^{-1}(x) = \frac{-x + 6}{2}$ **5.** $f^{-1} = \{(0, 0), (3, 2), (5, -1)\}$ **6.** $\log_3 24$ **7.** $\log_5\left(\frac{x^4}{x + 1}\right)$ **8.** $\log_6 2 + \log_6 x - 3\log_6 y$ **9.** −1.53 **10.** 1.0686 **11.** {−1} **12.** $\left\{\frac{1}{2}\left(\frac{\log 4}{\log 3} - 5\right)\right\}$; {−1.8691} **13.** $\left\{\frac{1}{9}\right\}$ **14.** $\left\{\frac{1}{2}\right\}$ **15.** {22} **16.** $\left\{\frac{25}{3}\right\}$

17. $\left\{\frac{43}{21}\right\}$ **18.** {−1.0979} **19.**

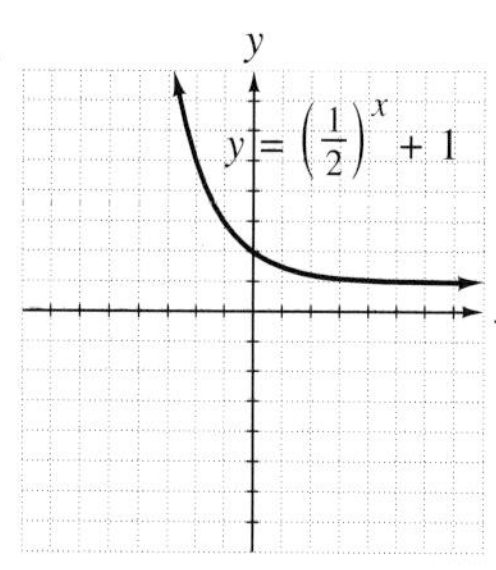

20.

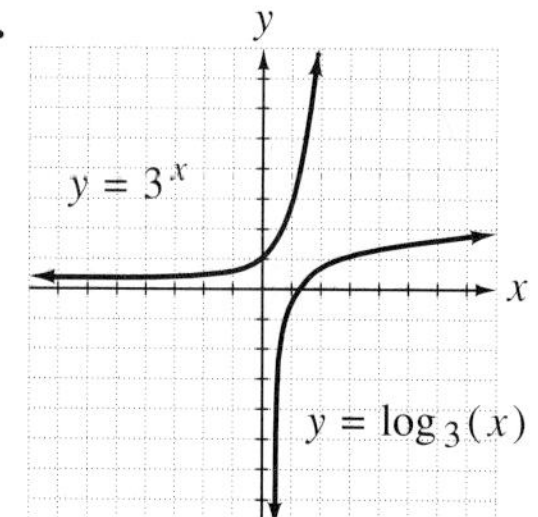

21. \$5234.58 **22.** 6 years **23.** 64,913 prairie dogs **24.** 15 years **25.** 1.2%

Chapter 11 Cumulative Review

1. [−3, 3]; *Sec. 3.7, Ex. 1*
2. *Sec. 4.3, Ex. 1*

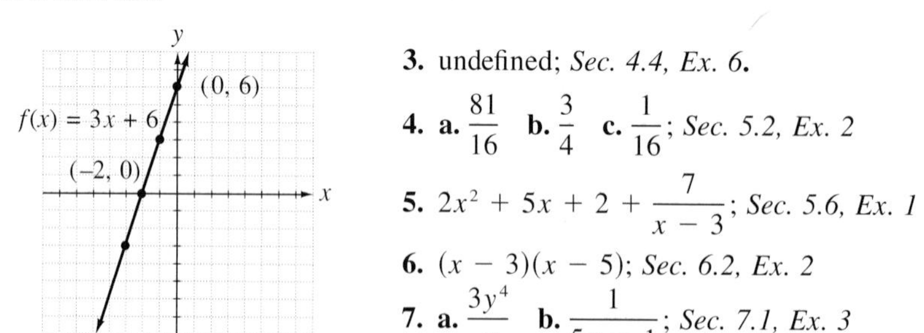

3. undefined; *Sec. 4.4, Ex. 6.*
4. a. $\frac{81}{16}$ **b.** $\frac{3}{4}$ **c.** $\frac{1}{16}$; *Sec. 5.2, Ex. 2*
5. $2x^2 + 5x + 2 + \frac{7}{x - 3}$; *Sec. 5.6, Ex. 1*
6. $(x - 3)(x - 5)$; *Sec. 6.2, Ex. 2*
7. a. $\frac{3y^4}{x}$ **b.** $\frac{1}{5x - 1}$; *Sec. 7.1, Ex. 3*
8. $\left\{-\frac{3}{2}\right\}$; *Sec. 7.5, Ex. 1*

9. a. $\frac{3\sqrt{15}}{5}$ **b.** $\frac{8\sqrt{x}}{3x}$ **c.** $\frac{\sqrt[3]{4}}{2}$; *Sec. 8.4, Ex. 2* **10.** {3}; *Sec. 8.6, Ex. 4*

11. $\left\{\frac{7-\sqrt{53}}{2}, \frac{7+\sqrt{53}}{2}\right\}$; *Sec. 9.1, Ex. 5* **12.** $\{2, -1 - i\sqrt{3}, -1 + i\sqrt{3}\}$; *Sec. 9.3, Ex. 2*

13. $f(x) = -2x^2$
$V(0, 0)$; *Sec. 9.5, Ex. 7*

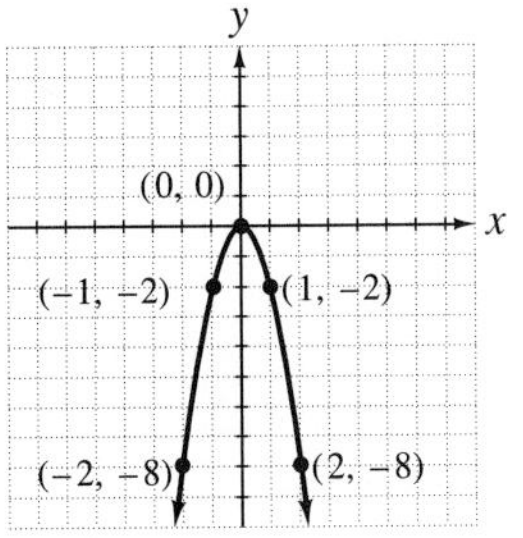

14. $\frac{x^2}{16} - \frac{y^2}{25} = 1$;
Sec. 9.8, Ex. 4

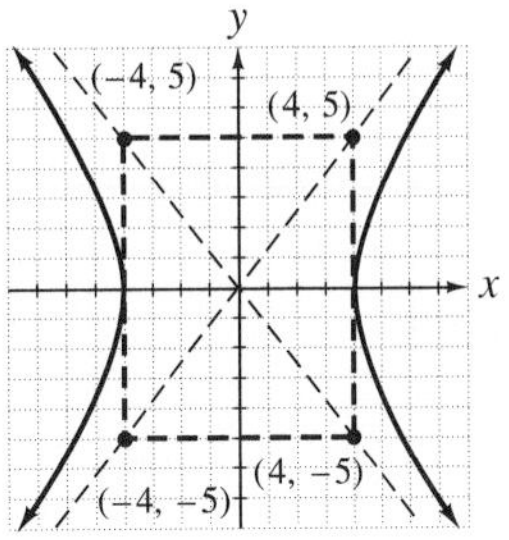

15. {(−4, 2, −1)}; *Sec. 10.2, Ex. 1* **16. a.** −2 **b.** −10; *Sec. 10.4, Ex. 1* **17.** $\{(2, \sqrt{2})\}$; *Sec. 10.6, Ex. 2*

18. $f^{-1}(x) = \frac{x+5}{3}$;
Sec. 11.1, Ex. 13

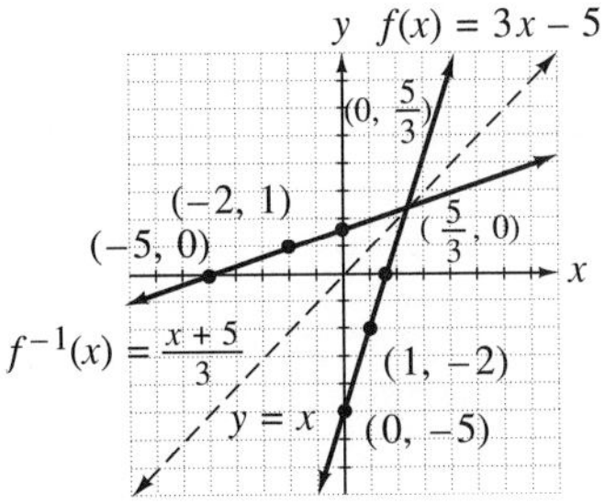

19. $2505.09; *Sec. 11.2, Ex. 5* **20.** {18}; *Sec. 11.6, Ex. 2*

CHAPTER 12
Sequences, Series, and the Binomial Theorem

Exercise Set 12.1

1. 5, 6, 7, 8, 9 **3.** −1, 1, −1, 1, −1 **5.** $\frac{1}{4}, \frac{1}{5}, \frac{1}{6}, \frac{1}{7}, \frac{1}{8}$ **7.** 75 **9.** 118 **11.** $\frac{6}{5}$ **13.** $a_n = 4n - 1$ **15.** $a_n = -2^n$ **17.** $a_n = \frac{1}{3^n}$ **19.** 48 ft, 80 ft, and 112 ft **21.** $a_n = 0.10(2)^{n-1}$, $819.20 **23.** 2, 4, 6, 8, 10 **25.** −1, −4, −9, −16, −25 **27.** 2, 4, 8, 16, 32 **29.** 7, 9, 11, 13, 15 **31.** −1, 4, −9, 16, −25 **33.** 729 **35.** $\frac{4}{7}$ **37.** $\frac{1}{8}$ **39.** −95 **41.** $-\frac{1}{25}$ **43.** 2400 cases; 75 cases originally **45.** 50; extinct in 2004

47.

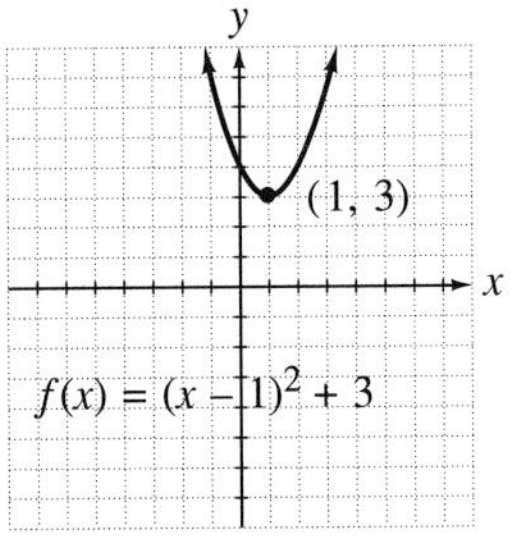

49.

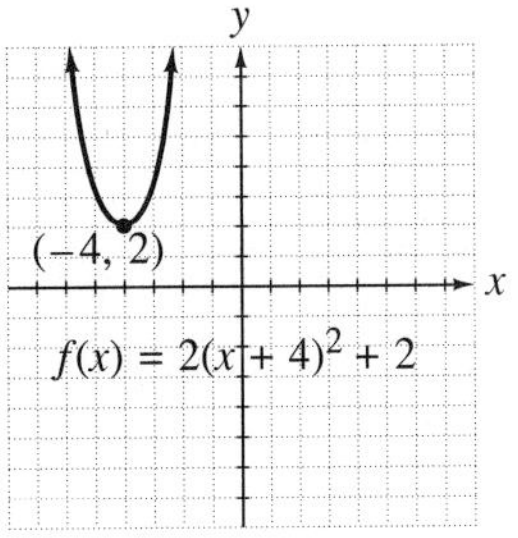

51. $\sqrt{13}$ **53.** $\sqrt{41}$

Exercise Set 12.2

1. 4, 6, 8, 10, 12 **3.** 6, 4, 2, 0, −2 **5.** 1, 3, 9, 27, 81 **7.** 48, 24, 12, 6, 3 **9.** 33 **11.** −875 **13.** −60 **15.** 96 **17.** −28 **19.** 1250 **21.** 31 **23.** 20 **25.** $a_1 = \frac{2}{3}$ and $r = -2$ **27.** $a_n = 4n + 50$; 130 seats

29. $a_n = 6(3)^{n-1} = 2(3)^n$ **31.** $a_1 = 2; d = 2$ **33.** $a_1 = 5; r = 2$ **35.** $a_1 = \frac{1}{2}; r = \frac{1}{5}$ **37.** $a_1 = x; r = 5$ **39.** $a_1 = p; d = 4$ **41.** 19 **43.** $-\frac{8}{9}$ **45.** $\frac{17}{2}$ **47.** $\frac{8}{81}$ **49.** -19 **51.** 486, 162, 54, 18, 6, 2; $a_n = \frac{486}{3^{n-1}}$; seven bounces **53.** $1000 + (n - 1)125$; \$2375 **55.** 25 grams **57.** $\frac{11}{18}$ **59.** 40 **61.** $\frac{907}{495}$

Exercise Set 12.3

1. -2 **3.** 60 **5.** 20 **7.** $\frac{73}{168}$ **9.** $\sum_{i=1}^{5} (2i - 1)$ **11.** $\sum_{i=1}^{4} 4(3)^{i-1}$ **13.** $\sum_{i=1}^{6} (-3i + 15)$ **15.** -20 **17.** 30 **19.** 1, 2, 3, · · ·, 10; 55 trees **21.** $a_n = 6(2)^{n-1}$; 96 units **23.** $\frac{11}{36}$ **25.** 60 **27.** 74 **29.** 62 **31.** $\frac{241}{35}$ **33.** $\sum_{i=1}^{4} \frac{4}{3^{i-2}}$ **35.** $\sum_{i=1}^{7} i^2$ **37.** -24 **39.** 0 **41.** 82 **43.** $a_n = 50(2)^n$; n represents the number of 12-hour periods; 800 bacteria **45.** 30 opossums; 68 opossums **47.** 6.25 pounds; 93.75 pounds **49.** 16.4 in.; 134.5 in. **51.** $-\frac{7}{2}$ **53.** $\frac{20}{33}$ **55.** 62 **57.** 126

Exercise Set 12.4

1. 36 **3.** 63 **5.** 55 **7.** 16 **9.** 4000, 3950, 3900, 3850, 3800; 3450 cars; 44,700 cars **11.** Firm A (Firm A, \$265,000; Firm B, \$254,000) **13.** 484 **15.** 2.496 **17.** \$39,930; \$139,230 **19.** 20 min; 123 min **21.** 24 **23.** $\frac{1}{9}$ **25.** 180 ft **27.** 185 **29.** $\frac{381}{64}$ or 5.95 **31.** $-\frac{33}{4}$ or -8.25 **33.** $-\frac{75}{2}$ **35.** $\frac{56}{9}$ **37.** -20 **39.** $\frac{16}{9}$ **41.** $\frac{4}{9}$ **43.** player A, 45 points; player B, 75 points **45.** \$3050 **47.** \$10,737,418.23 **49.** $\frac{8}{10} + \frac{8}{100} + \frac{8}{1000} + \cdots; \frac{8}{9}$ **51.** 720 **53.** 3 **55.** $x^2 + 10x + 25$ **57.** $8x^3 + 12x^2 + 6x - 1$

Exercise Set 12.5

1. $m^3 + 3m^2n + 3mn^2 + n^3$ **3.** $c^5 + 5c^4d + 10c^3d^2 + 10c^2d^3 + 5cd^4 + d^5$ **5.** $y^5 - 5y^4x + 10y^3x^2 - 10y^2x^3 + 5yx^4 - x^5$ **7.** 8 **9.** 42 **11.** 360 **13.** 56 **15.** $a^7 + 7a^6b + 21a^5b^2 + 35a^4b^3 + 35a^3b^4 + 21a^2b^5 + 7ab^6 + b^7$ **17.** $a^5 + 10a^4b + 40a^3b^2 + 80a^2b^3 + 80ab^4 + 32b^5$ **19.** $32a^5 - 80a^4b + 80a^3b^2 - 40a^2b^3 + 10ab^4 - b^5$ **21.** $c^6 - 12c^5d + 60c^4d^2 - 160c^3d^3 + 240c^2d^4 - 192cd^5 + 64d^6$ **23.** $6x^2y^2$ **25.** $30a^9b$ **27.** $q^9 + 9q^8r + 36q^7r^2 + 84q^6r^3 + 126q^5r^4 + 126q^4r^5 + 84q^3r^6 + 36q^2r^7 + 9qr^8 + r^9$ **29.** $1024a^5 + 1280a^4b + 640a^3b^2 + 160a^2b^3 + 20ab^4 + b^5$ **31.** $625a^4 - 1000a^3b + 600a^2b^2 - 160ab^3 + 16b^4$ **33.** $8a^3 + 36a^2b + 54ab^2 + 27b^3$ **35.** $x^5 + 10x^4 + 40x^3 + 80x^2 + 80x + 32$ **37.** $5cd^4$ **39.** d^7 **41.** $-40r^2s^3$

43. $f(x) = |x|$; not one-to-one

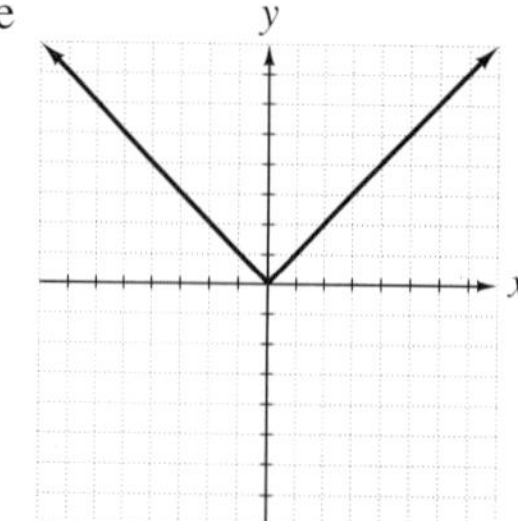

45. $H(x) = 2x + 3$; one-to-one

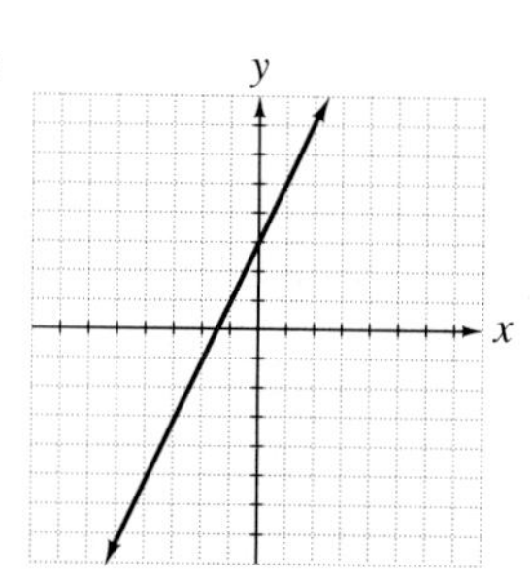

47. $f(x) = x^2 + 3$; not one-to-one

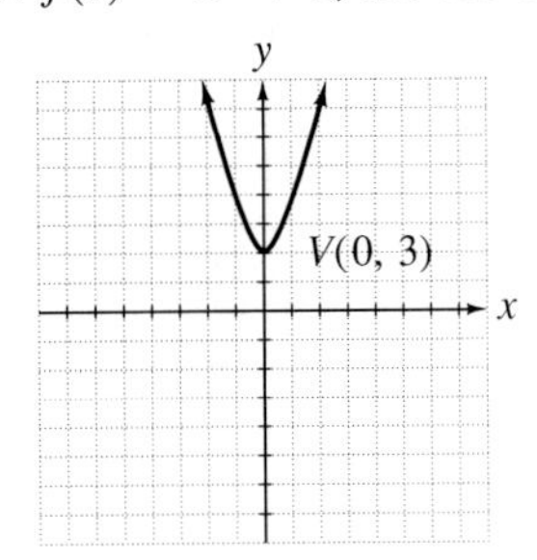

Chapter 12 Review

1. $-3, -12, -27, -48, -75$ **3.** $\frac{1}{100}$ **5.** $a_n = \frac{1}{6n}$ **7.** 144 ft, 176 ft, 208 ft
9. 450, 1350, 4050, 12,150, 36,450; 36,450 infected people in 1998 **11.** $-2, -\frac{4}{3}, -\frac{8}{9}, -\frac{16}{27}, -\frac{32}{81}$ **13.** 111
15. -83 **17.** $a_1 = 3$; $d = 5$ **19.** $a_n = \frac{3}{10^n}$ **21.** $a_1 = \frac{8}{3}, r = \frac{3}{2}$ **23.** $a_1 = 7x, r = -2$
25. 8, 6, 4.5, 3.4, 2.5, 1.9; good **27.** $a_n = 2^{n-1}$, \$512, \$536,870,912 **29.** $a_n = 150n + 750$; \$1650/month
31. $1 + 3 + 5 + 7 + 9 = 25$ **33.** $\frac{1}{4} - \frac{1}{6} + \frac{1}{8} = \frac{5}{24}$ **35.** -4 **37.** -10 **39.** $\sum_{i=1}^{6} 3^{i-1}$ **41.** $\sum_{i=1}^{4} \frac{1}{4^i}$
43. $a_n = 20(2)^n$; n represents the number of 8-hour periods; 1280 yeast **45.** Job A, \$23,900; Job B, \$23,800 **47.** 150
49. 900 **51.** -410 **53.** 936 **55.** 10 **57.** -25 **59.** \$30,418; \$99,868 **61.** \$58; \$553
63. 2696 mosquitoes **65.** $\frac{5}{9}$ **67.** $x^5 + 5x^4z + 10x^3z^2 + 10x^2z^3 + 5xz^4 + z^5$
69. $16x^4 + 32x^3y + 24x^2y^2 + 8xy^3 + y^4$
71. $b^8 + 8b^7c + 28b^6c^2 + 56b^5c^3 + 70b^4c^4 + 56b^3c^5 + 28b^2c^6 + 8bc^7 + c^8$
73. $256m^4 - 256m^3n + 96m^2n^2 - 16mn^3 + n^4$ **75.** $35a^4b^3$

Chapter 12 Test

1. $-\frac{1}{5}, \frac{1}{6}, -\frac{1}{7}, \frac{1}{8}, -\frac{1}{9}$ **2.** $-3, 3, -3, 3, -3$ **3.** 247 **4.** 39,999 **5.** $a_n = \frac{2}{5}\left(\frac{1}{5}\right)^{n-1}$
6. $a_n = (-1)^n\, 9n$ **7.** 155 **8.** -330 **9.** $\frac{144}{5}$ **10.** 1 **11.** 10 **12.** -60
13. $a^6 - 6a^5b + 15a^4b^2 - 20a^3b^3 + 15a^2b^4 - 6ab^5 + b^6$ **14.** $32x^5 + 80x^4y + 80x^3y^2 + 40x^2y^3 + 10xy^4 + y^5$
15. $y^8 + 8y^7z + 28y^6z^2 + 56y^5z^3 + 70y^4z^4 + 56y^3z^5 + 28y^2z^6 + 8yz^7 + z^8$
16. $128p^7 + 448p^6r + 672p^5r^2 + 560p^4r^3 + 280p^3r^4 + 84p^2r^5 + 14pr^6 + r^7$ **17.** 925 people; 250 people initially
18. $1 + 3 + 5 + 7 + 9 + 11 + 13 + 15$; 64 shrubs **19.** 33.75 cm; 218.75 cm **20.** 320 cm **21.** 304 ft; 1600 ft
22. $\frac{14}{33}$

Chapter 12 Cumulative Review

1. $(-\infty, \infty)$; *Sec. 3.5, Ex. 10*
2. *Sec. 4.1, Ex. 5*

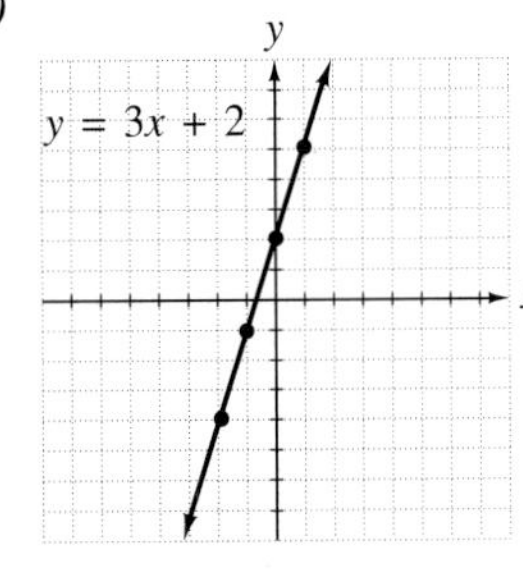

3. a. $10x^2 - 8x$ **b.** $-12x^4 + 18x^3 - 3x^2$
c. $-7x^3y^2 - 3x^2y^2 + 11xy$; *Sec. 5.4, Ex. 2*
4. $9(x + 2)(x - 2)$; *Sec. 6.4, Ex. 3*
5. 6, 8, and 10; *Sec. 6.6, Ex. 9*
6. a. $\frac{6x + 5}{3x^3y}$ **b.** $\frac{5z^2 - 2z}{(z + 2)(z - 2)}$
c. $\frac{5k^2 - 7k + 4}{(k + 2)(k - 2)(k - 1)}$; *Sec. 7.3, Ex. 3*
7. a. $\frac{5\sqrt{5}}{12}$ **b.** $\frac{5\sqrt[3]{7x}}{2}$; *Sec. 8.3, Ex. 2*

8. a. $13 - 18i$ **b.** $3 - 4i$ **c.** 58; *Sec. 8.7, Ex. 3* **9.** $\{-2, 2, -i, i\}$; *Sec. 9.3, Ex. 3*
10. *Sec. 9.5, Ex. 8* **11.** $(x + 1)^2 + y^2 = 8$; *Sec. 9.7, Ex. 8*

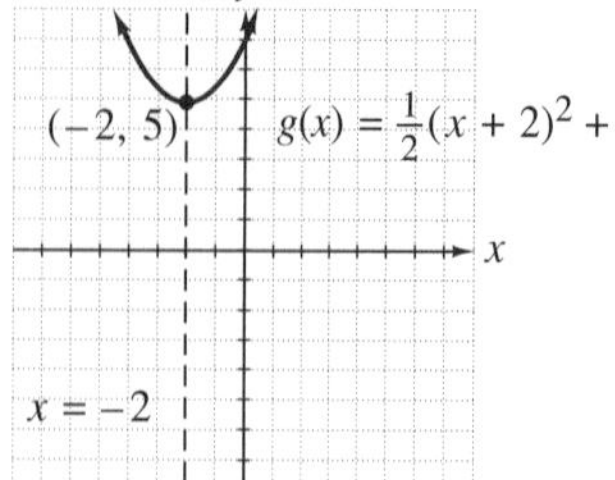

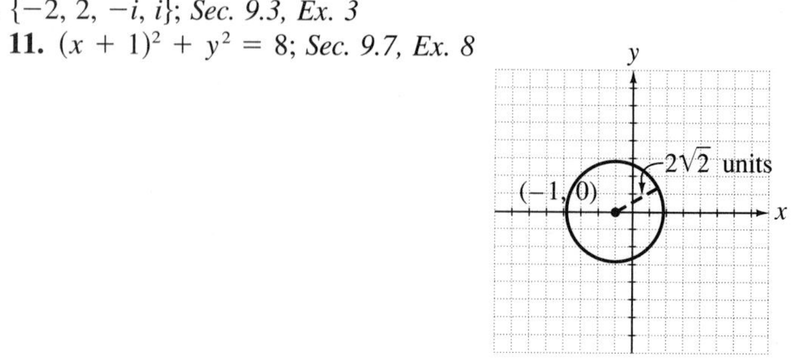

12. $\{(0, -5)\}$; *Sec. 10.1, Ex. 6*
13. a. $\{4\}$ **b.** $\left\{\frac{3}{2}\right\}$ **c.** $\{6\}$; *Sec. 11.2, Ex. 3* **14. a.** 2 **b.** -1 **c.** $\frac{1}{2}$; *Sec. 11.3, Ex. 1*
15. $\left\{\frac{2}{99}\right\}$; *Sec. 11.6, Ex. 4* **16.** approximately 12,300 lemmings; *Sec. 11.6, Ex. 5*
17. a. $-\frac{1}{3}$ **b.** $\frac{1}{24}$ **c.** $\frac{1}{300}$ **d.** $-\frac{1}{45}$; *Sec. 12.1, Ex. 2* **18.** $\frac{3}{32}$; *Sec. 12.2, Ex. 7*
19. a. $\frac{7}{2}$ **b.** 56; *Sec. 12.3, Ex. 1* **20.** 3; *Sec. 12.4, Ex. 6*
21. $x^5 + 10x^4y + 40x^3y^2 + 80x^2y^3 + 80xy^4 + 32y^5$; *Sec. 12.5, Ex. 4*

Appendix A Exercise Set

1. 17.08 **2.** 0.315 **3.** 12.806 **4.** 39.6756 **5.** 110.96 **6.** 1.8 **7.** 2.4 **8.** 160 **9.** 28.43 **10.** 29.53 **11.** 227.5 **12.** 18.214 **13.** 2.7 **14.** 17.3816 **15.** 49.2339 **16.** 8.31 **17.** 80 **18.** 7.71 **19.** 0.07612 **20.** 30.67 **21.** 4.56 **22.** 0.00443 **23.** 648.46 **24.** 0.796 **25.** 767.983 **26.** 225 **27.** 12.062 **28.** 18.42 **29.** 61.48 **30.** 2.36 **31.** 7.7 **32.** 7.6104 **33.** 863.37 **34.** 740 **35.** 22.579 **36.** 9.3 **37.** 363.15 **38.** 36.8 **39.** 7.007 **40.** 55.37

Appendix B Exercise Set

1. 71° **2.** 25° **3.** 19.2° **4.** $44\frac{1}{3}°$ **5.** $78\frac{3}{4}°$ **6.** 70.4° **7.** 30° **8.** 90° **9.** 149.8° **10.** 98.1° **11.** $100\frac{1}{2}°$ **12.** $14\frac{1}{9}°$ **13.** $\angle 1 = 110°$; $\angle 2 = 70°$; $\angle 3 = 70°$; $\angle 4 = 70°$; $\angle 5 = 110°$; $\angle 6 = 70°$; $\angle 7 = 110°$ **14.** $\angle 1 = 60°$; $\angle 2 = 50°$; $\angle 3 = 70°$; $\angle 4 = 110°$; $\angle 5 = 120°$ **15.** 90° **16.** 70° **17.** 90° **18.** 117° **19.** 90° **20.** 90° **21.** 45°; 90° **22.** 30°; 90° **23.** 73°; 90° **24.** 60°; 90° **25.** $50\frac{1}{4}°$; 90° **26.** 17.4°; 90° **27.** $x = 6$ **28.** $x = 6$ **29.** $x = 4\frac{1}{2}$ **30.** $x = 36$ **31.** 10 **32.** 13 **33.** 12 **34.** 16

Index